P9-DNA-665

Psychological Science

SECOND CANADIAN EDITION

Michael S. Gazzaniga
UNIVERSITY OF CALIFORNIA, SANTA BARBARA

Todd F. Heatherton
DARTMOUTH COLLEGE

Steven J. Heine
UNIVERSITY OF BRITISH COLUMBIA

Daniel C. McIntyre
CARLETON UNIVERSITY

Psychological Science

SECOND CANADIAN EDITION

W. W. Norton & Company
New York • London

W. W. Norton & Company has been independent since its founding in 1923, when William Warder Norton and Mary D. Herter Norton first published lectures delivered at the People's Institute, the adult education division of New York City's Cooper Union. The Nortons soon expanded their program beyond the Institute, publishing books by celebrated academics from America and abroad. By mid-century, the two major pillars of Norton's publishing program—trade books and college texts—were firmly established. In the 1950s, the Norton family transferred control of the company to its employees, and today—with a staff of four hundred and a comparable number of trade, college, and professional titles published each year—W. W. Norton & Company stands as the largest and oldest publishing house owned wholly by its employees.

Editor: Jon Durbin
Managing Editor, College: Marian Johnson
Project Editors: Kim Yi and Carla L. Talmadge
Electronic Media Editor: Denise Shanks
Editorial Assistants: Kelly Rolf and Robert Haber
Copy Editor: Barbara Gerr
Director of Manufacturing, College: Roy Tedoff
Book Designer: Rubina Yeh
Photo Researchers: Neil Hoos, Kelly Rolf, Kelly Mitchell, and Robert Haber
Illustrations: Frank Forney
Composition by GGS Book Services
Manufacturing by R. R. Donnelley

Printed in China.
Second Canadian Edition.

The text of this book is composed in Benton Modern Two with the display type in Polymer.

Library of Congress Cataloging-in-Publication Data
Psychological science/Michael S. Gazzaniga . . . [et al.].—2nd Canadian ed.
p. cm.
Includes bibliographical references and indexes.
ISBN 13: 978-0-393-92962-1
ISBN 10: 0-393-92962-0
1. Psychology—Textbooks. I. Gazzaniga, Michael S.

BF121.G393 2007
150—dc22

2006053059

W. W. Norton & Company, Inc., 500 Fifth Avenue, New York, NY 10110
www.wwnorton.com
W. W. Norton & Company Ltd., Castle House, 75/76 Wells Street, London W1T 3QT

1 2 3 4 5 6 7 8 9 0

We dedicate this book to Francesca and Zachary Gazzaniga, Sarah and James Heatherton, Nariko Takayanagi and Seiji Heine, and Jamie, Jenifer, and Tobi McIntyre.

Michael S. Gazzaniga (Ph.D., California Institute of Technology) is Distinguished Professor and Director of the Sage Center for the Study of the Mind at the University of California, Santa Barbara. He founded and presides over the Cognitive Neuroscience Institute and is founding editor-in-chief of the *Journal of Cognitive Neuroscience.* He is President of the American Psychological Society and a member of the American Academy of Arts and Sciences. His research focuses on split-brain patients. He has held positions at the University of California, Santa Barbara; New York University; the State University of New York, Stony Brook; Cornell University Medical College; and the University of California, Davis. He has written many notable books including most recently, *The Ethical Brain.*

Todd F. Heatherton (Ph.D., University of Toronto) is the Champion International Professor of Psychological and Brain Sciences at Dartmouth College. His research examines processes related to self, in particular self-regulation, self-esteem, and self-referential processing. This research is grounded in the traditions of personality and social psychology although the guiding theories, as well as the techniques and methodologies that he uses, are strongly influenced by research in cognitive neuroscience. He has been on the executive committees of the Society of Personality and Social Psychology (SPSP), the Association of Researchers in Personality, and the International Society of Self & Identity. He is Associate Editor of the Journal of Cognitive Neuroscience, and on several editorial boards. He has served as Head Tutor and chair of the undergraduate program in Psychology at Harvard University and department chair at Dartmouth College. He received the Petra Shattuck Award for Teaching Excellence from the Harvard Extension School in 1994. In 2005, he received the award for Distinguished Service on Behalf of Social and Personality Psychology from SPSP for his role in creating the annual SPSP meeting.

Steven J. Heine (Ph.D., University of British Columbia) is Associate Professor of Psychology and Distinguished University Scholar at the University of British Columbia. His research largely focuses on cultural psychology, in particular, he has explored how cultural experiences shape the ways that people understand themselves and the world around them. He has also been on the faculty at the University of Pennsylvania, and has had visiting positions at Kyoto University and Tokyo University. He has received the Distinguished Scientist Early Career Award for Social Psychology from the American Psychological Association and the Early Career Award from the International Society of Self and Identity. He is also the author of the textbook *Cultural Psychology,* published by Norton.

Daniel C. McIntyre (Ph.D., University of Waterloo) is a Chancellor's Professor in Psychology in the Institute of Neuroscience at Carleton University, Ottawa, Ontario, Canada. His work on the functional anatomy of epilepsy and many related themes is principally funded by research grants from the Canadian Institutes of Health Research and the Natural Sciences and Engineering Research Council of Canada. He serves on several editorial boards, and in 2003 received the coveted American Epilepsy Society research award in basic science. Having received several teaching awards over the years, he was also voted in 2005 by TVO as one of Ontario's best lecturers.

CONTENTS IN BRIEF

CONTENTS

3 GENETIC AND BIOLOGICAL FOUNDATIONS 73

4 THE BRAIN AND CONSCIOUSNESS 117

5 SENSATION, PERCEPTION, AND ATTENTION 159

9 MOTIVATION 343

10 EMOTIONS AND HEALTH 385

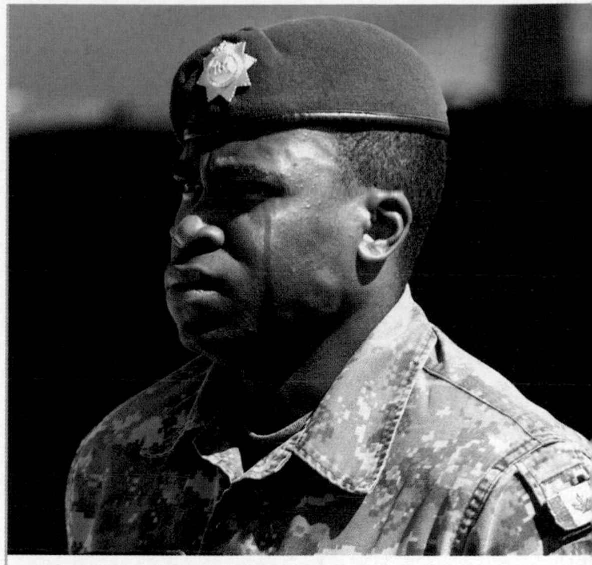

11 HUMAN DEVELOPMENT 433

12 PERSONALITY 483

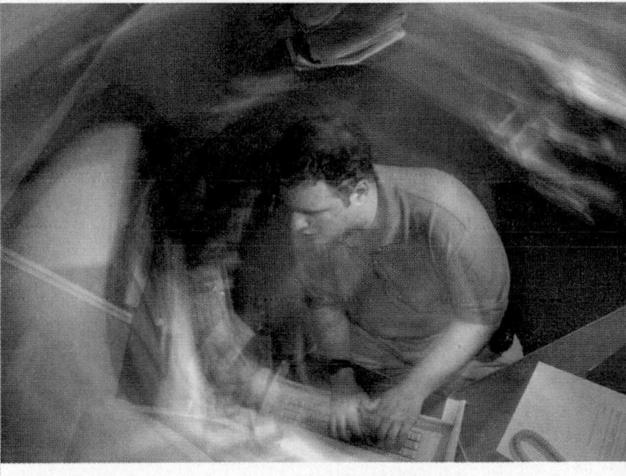

15 SOCIAL PSYCHOLOGY 611

16 CULTURAL PSYCHOLOGY 665

Psychological Science *in Action*

Thinking Critically

Profiles in Psychological Science

PREFACE

Just over seven years ago we joined forces with W. W. Norton to create an introductory psychology textbook that would capture the excitement of twenty-first-century psychological science. Our vision for the book was bold, as we sought to break the mold of the homogenized and encyclopedic textbooks that focus on classic themes and standard topics. Thanks to an editorial team that shared our vision, the First Edition of *Psychological Science* achieved our principal goals, the most important of which was to produce a readable book that focused on cutting-edge psychological and brain sciences. This approach resonated with faculty and students alike, as it captured the excitement of contemporary research being conducted in labs around the world. Instructors were especially enthusiastic about the levels-of-analysis approach to studying human behaviour, which has become the cutting-edge way to investigate behaviour in the fields of psychology and neuroscience. It is clear that there is an eager audience for a rigorous and scientific treatment of psychology that is accessible for the majority of students.

We frankly were thrilled with the response our First Edition received. We tried a lot of new approaches and most of them worked. Many of our colleagues told us that they especially liked the declarative writing style that focused on answers to current scientific questions driving psychological research. The "ask and answer" approach that serves as the pedagogical foundation of the book captured students' interest and kept the material engaging. It helped students take home the idea that psychological scientists are addressing big questions and making progress in solving them.

Even though the book received considerable praise, as authors we felt that some features and material didn't quite satisfy what we were trying to achieve. For instance, we sometimes went into too much technical detail and occasionally our coverage was a bit unbalanced. As we noted in the preface to the First Edition, our greatest challenge was in deciding what not to include rather than what to include. In retrospect, we sometimes included material that was unnecessary and other topics needed to be included or covered in more depth. Our core goal for *Psychological Science* in its current and future editions is to provide students with what they really need to know to be informed about psychological science. As we prepared this revision we spent considerable time addressing that very question. We sought advice from many colleagues, even repeatedly asking some of you, "Do students really need to know this at the introductory level?" If the answer was no, we eliminated that material so that we could more thoroughly treat essential topics that required greater attention. We thank all our colleagues and reviewers who helped us sort out these issues and provided good advice on how to improve the text.

Everything we tried to do in the First Edition was first and foremost for the benefit of students, and they continue to be our primary audience in the Second Edition. In response to their comments, we now include even more vivid examples and case histories, especially in the *Profiles in Psychological Science* units in each chapter, as well as descriptions of how psychological scientists are having an impact on their daily lives. Each chapter's *Psychological Science in Action* unit introduces students to how knowledge from psychological research is used in various professions (e.g., marketing, forensics, sports, robotics) and how it has helped answer important policy questions, such as how young children should be taught to read. In developing these units, we place psychological science in a human context by emphasizing the extent to which it applies to students' everyday lives. We have expanded the number of citations to the research literature, but we have avoided overloading the text with unnecessary citations—the focus remains on the words and ideas. We believe students will respond favourably to the changes we have made in this edition. We look forward to their responses to our efforts.

This Second Edition also includes changes that will appeal to instructors. Building on the "ask and answer" approach, we sought to have one or two broad themes drive each chapter. The themes reflect overarching questions that have played prominent roles throughout the history of psychological science, such as Chapter 3's discussion of the role of environment in gene expression, which provides new insights to longstanding questions about nature and nurture, and Chapter 4's discussion of whether psychological functions are localized in specific brain regions and which regions are involved in conscious experience. This strategy demonstrates the cumulative nature of psychological science as well as the multiple perspectives that researchers and teachers bring to the subject matter. Another good example of this occurs in Chapter 15, where we highlight the idea that people consistently underestimate the power of situations to affect human behaviour, including their own, and discuss how much social cognition occurs automatically and without conscious awareness or intent, as people quickly evaluate and categorize others. Each chapter is now more tightly conceived and unified; each has its own unique emphasis. The chapter conclusions then take up these themes to provide students with a take-home message. When students finish a chapter, they should have a clear sense of the big questions and issues driving contemporary research in that area, as well as an understanding of what is known and what remains to be discovered.

Given the popularity of our levels-of-analysis approach, we worked hard to integrate more social and cultural examples throughout the first half of the book. We have streamlined our discussion to the three essential levels (biological, individual, social), but also expanded coverage within each level. We believe that the Second Edition balances social and cultural psychology with the biological. This increased emphasis on culture throughout the book begins in Chapter 1 with a foundational discussion on how culture is adaptive and concludes with the first ever full chapter on the burgeoning field of cultural psychology in any leading introductory text. We also increased the prominence of critical thinking throughout the book. Based on reviewer and instructor suggestions, we have made it a centrepiece of the new edition. To each chapter we added a *Thinking Critically* unit that takes on a controversial topic and demonstrates the value of critical analysis. We have also added short-answer critical thinking questions in the margins of each chapter. Finally, at the end of each chapter, we introduce a study feature called Applying the Principles that asks students to answer a series of questions designed to see if they can apply the chapter concepts to the real world. Recognizing that introductory psychology is the only exposure to the subject that many students will have, we believe it to be vitally important that students understand what it means to be consumers of psychological research, including recognizing the value of good, solid science.

In response to comments by reviewers and instructors and our own sense of the strengths and weaknesses of the First Edition, we made both major and minor changes throughout the book. We scrutinized each chapter and in a few cases decided that they needed to be reconsidered altogether. The chapters on cognition (Chapter 8) and sensation and perception (Chapter 5) have been completely reconceptualized and rewritten. The material in these chapters is both more student-friendly, in terms of the level of technical detail, and more applicable to students' lives. In Chapter 4, we introduced an innovative and appealing discussion of functional neuroanatomy and how it gives rise to consciousness. This is a fresh approach that really brings the anatomy home to the students and helps them see how understanding the brain helps with understanding the mind, both when it is awake and when it is in an altered state of consciousness, such as sleep. In the previous edition there were two developmental chapters and two social chapters. In this edition we consider these as unified areas, no longer separating, for example, cognitive and social development. In each case we took the best material from the two chapters, reframed the major themes and issues, and built a more concise and integrated chapter. We now have one unified development chapter (Chapter 11), with a greater lifespan emphasis, and one unified social chapter (Chapter 15), which focuses more on how context shapes behaviour. In keeping with our levels-of-analysis approach, we moved the social chapter to the end of the book, followed by the new chapter on cultural psychology where we use them to summarize many of the accomplishments of psychological science in understanding what it means to be humans.

The new Canadian Edition builds on and expands on the strengths of the First Edition. Canadian research psychology comes from a long and rich tradition. It has made many significant contributions to the larger field, particularly in the areas of neuroscience and social psychology. The First Edition of this text seamlessly incorporated a significant amount of research and historical material from Canadian researchers in these areas and across the discipline. Canadian researchers are also at the forefront of the burgeoning new field of cultural psychology, which is why we have included the first ever chapter on that subject in the new Canadian Edition. In addition to this path breaking new chapter, we have also incorporated more Canadian research across the domains and have included many more Canadian popular culture examples and statistics to make the material even more interesting and relevant for Canadian students.

We noted in the preface to the First Edition that writing this book had changed us as scientists and inspired us to more fully cross levels of analysis in our own research, including collaborating on using neuroscience methods to understand the self. The cross-fertilization of our interests has continued. Mike's most recent book, *The Ethical Brain*, considers the fundamental social nature of human existence. Todd has become even more interested in understanding the neural basis of social and personality phenomena. Steve continues his exploration of the relationship between biology, cognition, and culture. Two summers ago, Steve cosponsored a summer conference at UBC, which focused on ways to integrate evolutionary and cultural approaches to the study of the human mind and behaviour. Dan continues his research on understanding the mechanisms underlying neuroplasticity as reflected in both epilepsy and learning and memory. He was also pleased to receive a 3M teaching award this past year. While continuing to keep busy on the research and administrative fronts, we truly enjoy teaching both graduate and undergraduate students. They keep us energized and teach us new ideas as we move forward with our own lives and careers. We hope that our own enthusiasm for the science of psychology inspires students to think about the big issues and questions that fascinate those of us who have dedicated our lives to understanding the mind and behaviour.

Tour of *Psychological Science*

CHAPTER 1 (*Introduction to Psychological Science*) sets the stage for the book. We introduce students to psychological science by providing engaging examples of psychological questions and answers. We continue to emphasize the four guiding themes that characterize psychological science: (1) It is a cumulative science, with principles established through incremental advances in knowledge obtained through research. (2) A biological revolution has energized psychological research. (3) The human mind is adaptive, which means that the brain has evolved to solve everyday problems and also that culture provides adaptive solutions. (4) Psychological science crosses different levels of analysis and perspectives, from the biological to the social, from genes to culture. In the new edition, we have simplified our levels-of-analysis approach so that it now focuses on three broad levels of analysis: biological, individual, and social. Our first discussion from this perspective explores the psychological aspects of music. There is also a much greater emphasis on culture throughout the book. In this chapter, we present a foundational discussion that considers differences between how people in eastern and western cultures view and reason about the physical world. We then briefly trace the intellectual origins and historical background of psychological science and go on to describe how psychological knowledge is applied to solving real-world problems, such as treating mental disorders. We end the chapter by emphasizing the importance of critical thinking for consumers of psychological research.

In **CHAPTER** 2 (*Research Methodology*), we lay the foundations for understanding the methods of psychological science. We cover the major methods in sufficient detail so that students understand the techniques and strategies used to examine psychological questions. In the new edition we place a stronger emphasis on the role of critical thinking in testing theories about the mind, brain, and behaviour. We also expand the description of how research with animals, including manipulating their genes, is an important tool for addressing fundamental psychological questions. As part of the emphasis on critical thinking, we highlight the need for researchers to pay attention to potential cultural differences when drawing conclusions about human nature. We also offer a unified presentation of the methods for assessing the working brain (i.e., electrophysiology, imaging, neuropsychology). Pulling the descriptions of these methods together will help allow students understand the full range of tools available to psychological scientists.

In **CHAPTER** 3 (*Genetic and Biological Foundations*) we have substantially reconsidered the role of genes in psychological processes. For years, psychologists focused on whether people possessed certain types of genes, such as genes for psychological disorders or intelligence, with an emphasis on the extent to which human traits were hardwired. Recent cutting-edge research is emphasizing gene expression, what turns genes "on" and "off." The major lesson is that the environment, including social factors, plays an important role in gene expression, and therefore in how the environment affects the mind, brain, and behaviour. It is as much gene expression as gene possession that makes us who we are as human beings. We provide a compelling example of how antisocial behaviour results from the interaction of genes and environment. We also describe cutting-edge research on how gene expression can be modified, such as through small RNA activity. Throughout the chapter we have reduced the level of technical difficulty by focusing on what students really need to know in order to understand basic human physiology. For example, we eliminated the in-depth description of metabotropic receptors as well as distinctions between peptide and steroid hormones. We continue to enliven the discussion of the nervous system by focusing on the influence of neurotransmission on everyday mental activity and behaviour.

CHAPTER 4 (*The Brain and Consciousness*) has been transformed in ways that we believe will excite teachers and students alike. Rather than focusing solely on the brain, we now explore how brain activity gives rise to the unique human capacity of consciousness. After briefly discussing evolving views of the localization of function, we emphasize that mental activity results from the integrated actions of specific and localized structures that are distributed throughout the brain. We also discuss the idea that we are not conscious of much of our mental activity, with social psychologists such as Dan Wegner and John Bargh demonstrating that much of human behaviour occurs automatically and without intention. This raises interesting philosophical questions for students, such as the degree to which we are able to consciously will our actions. In keeping with the idea that the brain gives rise to conscious experience, we then discuss sleep in the context of brain activity. After all, many regions of the brain are more active when we sleep than when we are awake. Our goal with this fresh approach is to emphasize the importance of brain function to human experience.

CHAPTER 5 (*Sensation, Perception, and Attention*) provides a foundation for understanding how the brain senses and perceives the world. This chapter has been revised substantially to make the material more accessible to students. The section on psychophysics and signal detection now emphasizes basic concepts and methodological approaches rather than mathematical equations of difference thresholds. We reorganized the discussion of the five senses, beginning with taste and building up to vision, which helps students appreciate the commonalities in the various perceptual systems. A theme that runs throughout the chapter is that perception is an active constructive process that occurs in the brain. Our goal for this chapter is to foster an appreciation for how the brain takes in ambiguous sensory information and constructs rich and meaningful experiences that allow us to navigate the world around us. Because perception is important to social interaction, we discuss face perception in this chapter. Finally, we consider how attentional processes influence how people perceive the world around them, emphasizing that the ability to selectively attend to important sensory stimuli is crucial for survival.

CHAPTER 6 (*Learning and Reward*) covers the foundations of principles of learning and the role of learning in solving adaptive challenges. We emphasize that the study of learning has been a powerful force in psychological science for more than a century, and that the principles of classical and operant conditioning are important throughout psychology and allied sciences. This chapter provides a good demonstration of our theme that psychological science is cumulative, with researchers responding to and expanding on their predecessors to understand the complexities of learning. We have expanded the discussion of how economic principles can inform the understanding of reinforcement as well as how learning can be passed on through cultural transmission. In keeping with our goal of limiting technical jargon, we removed the discussion of the various subcomponents of the mesolimbic dopamine pathway, instead focusing more directly on the importance of dopamine to reward. We also describe new studies that use genetic manipulation to explore how long-term potentiation works.

CHAPTER 7 (*Memory*) continues to provide a cutting-edge treatment of how people encode, store, and retrieve different types of information. We have included new findings on emotional memory, on reconsolidation, and on H.M.'s recently discovered acquisition of new semantic knowledge. A central focus of the chapter is the neurological basis of memory, with classic and contemporary examples of how brain injury interferes with explicit recall. We also continue to emphasize the practical aspects of human memory, such as the ability to serve as eyewitnesses and the role of motivation and social context in shaping what we remember.

CHAPTER 8 (*Thinking and Intelligence*) has been completely reconceptualized and expanded. It now incorporates a much greater emphasis on the intelligent use of information and the idea that for the most part thinking is adaptive. We present a more streamlined discussion of representational knowledge and emphasize how knowledge structures, such as scripts and schemas, guide our behaviour, often without our conscious awareness. The idea that unconscious processes play prominent roles in cognition is an important theme throughout the chapter. We have significantly modified the discussion of problem solving and decision making to emphasize more recent research on heuristic processes. We have expanded the discussion of deductive and inductive reasoning and provided more detail on how people solve problems in everyday life. We also consider at some length the idea that more thought does not necessarily make for better decisions. The revised chapter has a much more thorough discussion of what it means to be intelligent, with an increased focus on fluid intelligence and the prominent role it plays in dealing with modern society. We describe recent research on the neural and cognitive bases of intelligence and discuss the role that genes and environment play in shaping intelligence.

CHAPTER 9 (*Motivation*) presents an overview of the factors that motivate behaviour. This chapter demonstrates the need to consider biological, psychological, and social factors in order to develop a satisfactory understanding of why people "choose" to engage in specific behaviours. We consider motivation in its adaptive context, with an interpretation based on evolutionary principles. The section on sleep, which was moved to Chapter 4, has been replaced by expanded discussions of addiction, sexual behaviour, and the human need to belong. Each of these is considered across the multiple levels of analysis. The chapter invites students to reflect on what motivates them from day to day, how they deal with competing motives, and how setting and achieving goals can help them lead meaningful lives.

CHAPTER 10 (*Emotions and Health*) explores how emotions influence the human experience, including where they come from and how they are experienced, as well as how people cope with stress and handle negative events. In this revised chapter, we consider behaviours that significantly influence physical health, particularly those that lead to healthy living. We also include more discussion of how stress affects the heart and the immune system. In a new section, we look at how daily behaviours, such as eating, exercising, and smoking, play an important role in the most common causes of death in our society, and we consider evidence that emotional states often encourage people to develop bad habits.

CHAPTER 11 (*Human Development*) provides a thorough discussion of physical, cognitive, and social development. We have merged the two development chapters from the previous edition into one unified treatment of developmental issues across the life span. The driving idea behind this merger is that in looking at the factors that shape us, it is impossible to separate the cognitive from the social. We emphasize the multiple forces, from genes to culture, that interact to produce unique humans. Taking a life span approach, we consider how humans change as they grow, from infancy to old age. We now offer greater coverage of middle age and old age, examining the important roles that family and career have in our lives. The message of the final section is that, although physical and cognitive declines are an inevitable part of the aging process, contemporary research finds that older people are surprisingly satisfied with their lives.

CHAPTER 12 (*Personality*) remains a somewhat radical departure from historical treatments of personality. Most textbooks describe the area of personality as it existed more than 40 years ago, with an emphasis on unconscious Freudian processes and social learning experiences. Yet recent research has provided compelling evidence that human personality is determined to a large extent by genetic and physiological mechanisms. This chapter focuses on research by contemporary psychologists who have made considerable strides in understanding the development and structure of human personality. We cover traditional topics, such as Freudian and trait theory, but we emphasize more current research on the cognitive and biological factors involved in personality.

CHAPTER 13 (*Disorders of Mind and Body*) introduces students to the best-known forms of psychological disorders, especially those that we know to be of interest to students. In this revised chapter we have streamlined the discussion of legal issues and expanded the discussion of assessment techniques, such as interviews and behavioural assessments. Following a brief review of the issues surrounding determining whether a condition reflects a psychological disorder, we describe the essential features of each major disorder and discuss various etiological theories. We provide a thorough discussion of the biological, psychological, and social factors that have been implicated in the development of schizophrenia. We describe recent evidence that some indications of schizophrenia and other forms of adult psychopathology can be observed in childhood, such as in home movies. Vivid case studies are used throughout to illustrate symptoms and assessment. We include expanded coverage of childhood disorders, such as ADHD and autism, which are typically ignored in introductory textbooks; it is our experience that students find these disorders especially interesting. The overall theme of the chapter is that to really understand psychopathology we need to recognize the social, psychological, and biological factors that contribute to the expression of mental illness in those who are susceptible.

CHAPTER 14 (*Treating Disorders of Mind and Body*) examines the theoretical basis of psychotherapy, as well as typical treatments and outcomes. Following a discussion of treatment goals, we introduce students to the most common types of therapy. Although we mention classic psychological treatments, we emphasize the types of treatments most widely used by contemporary therapists. We consider treatments for anxiety disorders, depression, and schizophrenia in detail, with special attention to the empirical evidence for successful outcomes. One theme running through the chapter is that treatments are tailored specifically for a client with specific psychological symptoms. That is, there is no overall grand theory that guides treatment, but rather treatment is based on evidence of its effectiveness, which is established through rigorous research. Another important point is the preference for psychological treatments over pharmacological treatments when both are effective, as psychological treatments have fewer side effects and are longer lasting. We also discuss evidence that both psychological and pharmacological treatments lead to changes in brain functioning. As in the previous chapter, we include discussion of issues specifically related to children, such as how behaviour therapy can be used to treat autism. The take-home message of the chapter is that although we have a ways to go, psychological scientists have discovered a great deal about how to improve the quality of life for people with serious disorders of mind and body.

CHAPTER 15 (*Social Psychology*) is a new chapter that merges the best sections from the two social chapters in the previous edition. An important theme that guides the chapter is that people have evolved as social animals and that much of human behaviour and experience is shaped by social context; cultural rules and norms determine a great deal of our behaviour. Another important theme of the chapter is that people reliably underestimate the power of situations in affecting human behaviour. Numerous vivid examples, such as the prisoner abuse at Abu Ghraib, help students appreciate the effect of social context on behaviour. We also emphasize that a great deal of social cognition occurs automatically and without conscious awareness or intent, especially in how we think about others and ourselves. In this chapter we have narrowed our focus to the social self, with an increased emphasis on self-relevant motives and the extent to which these motives are true across cultures. We discuss research that crosses levels of analysis from the social to the biological, as we explore recent neuroimaging studies on the self, attitudes, and stereotyping. And we expand the section on relationships, with a greater emphasis on romantic relationships. This unified chapter describes many of the important studies that have been conducted by social psychologists and demonstrates how the principles that developed out of these studies are relevant to students' everyday lives.

CHAPTER 16 (*Cultural Psychology*) is an exciting new chapter created for the Canadian Edition. As far as we know, it's the first ever chapter fully dedicated to cultural psychology to appear in a mainstream introductory psychology text. It serves two important purposes in the text. First, it serves as the capstone chapter for the text reinforcing the author's belief that a levels-of-analysis approach is the best way to understand human behaviour. Second, the elevation of cultural psychology to a chapter level subject reinforces our goal to present students with the most cutting-edge material possible, while showing that psychology is a dynamic research-driven field. The larger intellectual goal for the chapter is to help students understand why culture matters when we study human behaviour. As part of explaining why culture matters, we ask five basic questions: What is culture? How does culture affect the mind? What happens to people's sense of self when they move to a different culture? How does culture affect how we think and behave? How does culture influence mental health? An important theme that guides the chapter is how a person's sense of self relates to others. As we discussed in the social chapter, every person is unique with their own ideas, beliefs and goals, and yet each of us is a social creature fundamentally connected with others. Research has shown that the type of culture we belong to, either independent or interdependent, has dramatic effects on our sense of self and our thoughts and behaviours. Moreover, we are now beginning to see that this difference in self-concept underlies many of the differences in psychological processes across cultures. A second theme throughout this chapter is that both universal and culturally specific psychologies exist. People the world over share the same biology, and at some level of analysis, the psychological experiences of people around the world are universally similar. But at other levels of analysis, we see pronounced cultural variation in psychological processes. This chapter focuses on the interesting ways that people with different cultural experiences differ in their psychological processes.

Although it may not appear as obvious to a student how the fields of cultural psychology and neuroscience can be relevant when studying the same subject, we believe that it is only through the use of integrative approaches like levels of analysis that we will truly be able to understand complex human behaviours and thoughts.

A Pedagogical Program That Reinforces *Psychological Science*'s Core Principles

I. Overview

Psychological Science's chapters are built around four to six major principles that are addressed through the "ask and answer" approach. Each is first raised in the form of a question ("ask"). Each major section then discusses one of these questions ("answer").

OUTLINING THE PRINCIPLES: At one level, this feature serves as a simple outline or road map for the chapter. At another level, it clarifies the major principles that will be covered in the chapter. Major heads are questions. Minor heads are declarative statements that reveal the current state of knowledge about the larger principles and concepts.

CHAPTER TIMELINES: Psychological science is built on cumulative knowledge and experience. This is one of the major themes of the text. Basic principles, both new and old, inspire and guide thinking and research in the field. The timelines highlight major developments within the various domains of psychology.

CRITICAL THINKING QUESTIONS: These questions are designed to encourage students to think integratively and beyond the text. Appearing in the margins throughout each chapter, they selectively highlight core principles and are ideal as short-answer questions on quizzes or for discussion in small groups.

"ASK AND ANSWER" RUNNING HEADS: *Psychological Science*'s left-hand running heads emphasize the larger topics, with the right-hand running heads repeating the questions that are explored in each section. The heads help students stay focused on the larger issue as they try to see the forest for the trees.

REVIEWING THE PRINCIPLES: These boxes are a critical component of the "ask and answer" approach and appear at the end of each major section. They repeat the question that governed the section and provide a basic answer that highlights key points for students to remember.

DEFINING THE PRINCIPLES: *Psychological Science* has a marginal glossary running throughout each chapter as well as a glossary at the end of the book and on the companion Web site. Many books highlight too many key terms for students to memorize. In keeping with this book's focus on core principles and concepts, *Psychological Science* highlights approximately 30 key terms per chapter.

CONCLUSION: *Psychological Science*'s brief chapter conclusions sum up the big ideas and concepts and remind students how the book's four key themes wove their way through the chapter.

SUMMARIZING THE PRINCIPLES: Designed to follow the outline for each chapter, this feature provides a detailed review. Students are reminded again of the major questions that governed each section and the key pieces of research-based knowledge that provide insight into those questions.

APPLYING THE PRINCIPLES: *Psychological Science* encourages students to apply core principles to real-world settings. Multiple-choice Applying the Principles questions at the end of each chapter highlight core principles students need to learn. Answers appear at the end of the section.

II. *Psychological Science*'s Art and Citations

A DYNAMIC ART PROGRAM: The visual materials in *Psychological Science* add substantially to the students' experience. The book contains a variety of visual materials, from photographs to tables and charts to drawn art. The emphasis in *Psychological Science*, however, is clearly on the drawn art. Having used many general psychology books ourselves, we wanted to take our text in a new direction. By featuring drawn art, *Psychological Science* is able to convey precisely, accurately, and meaningfully what the students need to gain from every image. This high level of precision can't be achieved with the stock photographs common in many texts.

SELECTIVE USE OF CITATIONS: *Psychological Science* embraces the notion that students should be introduced to material in a narrative style that focuses on ideas, concepts, and empirical findings rather than on specific researchers. At the same time, the text includes enough citations that students can pursue topics of interest and appreciate that psychological science is based on published empirical research. We have selected essential citations that we hope most teachers would agree are central to a first-year student's exploration of psychological science, and we have cited them in a way that should not distract from the narrative voice.

III. *Psychological Science*'s Special Topic Units

Special Topic Units amplify the text's basic strengths by challenging students to think critically and to understand the impact that the field of psychology has on their lives and world. Because each unit is seamlessly integrated into the narrative rather than set off like traditional feature boxes, it remains an important part of the learning experience.

PSYCHOLOGICAL SCIENCE IN ACTION units highlight major areas of applied research and possible careers in fields such as organizational psychology, sports psychology, consumer psychology, forensic psychology, and clinical psychology. Among these units are "Sports Psychologists Help Athletes Find the Zone," "Motivation on the Job," and "Forensic Assessment and Profiling."

PROFILES IN PSYCHOLOGICAL SCIENCE units focus on the human perspective, thereby broadening the book's rich presentation of science and research. Each of these units offers an individual case study to show how people's lives can intersect with science. Among these units are "One Boy's Journey out of Autism's Grasp," "Making Marriage Last," and "What Makes Killers Kill?"

THINKING CRITICALLY units contain stories from newspapers and magazines on controversial topics that enable students to formulate their own conclusions based on the information available. Among these units are "Should Drugs Be Used to Treat Adolescent Depression?" "Should Memory Be Altered by Drugs?" and "Are Self-Serving Biases Universal?"

Psychological Science's Ancillaries Focus on Research in Action

ZAPS: The Norton Psychology Labs wwnorton.com/zaps

ZAPS: THE NORTON PSYCHOLOGY LABS give students a firm understanding of fundamental concepts by allowing them to experience psychological phenomena firsthand, taking the role of either subject or researcher in engaging, diverse experiments and demonstrations. Ideal for introductory psychology courses, these groundbreaking online labs allow students to see the significance of psychological research.

A **class results** feature allows instructors to collect aggregate ZAPS data for their class in real time.

Visit the free ZAPS demo site at wwnorton.com/zaps.

The Norton Psychology Reader edited by Gary Marcus (New York University)

0–393–92712–X / paper / 375 pages

The Norton Psychology Reader offers a diverse collection of popular readings (8–10 pages in length) sure to enliven classroom discussion and inform student research, most of them written in the last decade. Available alone or as an ideal complement to *Psychological Science*, Second Canadian Edition, this engaging reader highlights the most exciting ideas in the field today. Authors range from Steven Pinker to Natalie Angier, from Malcolm Gladwell to Robert Cialdini.

Norton/Discovery Education Psychology DVD created by Patrick Carroll (University of Texas, Austin)

With over 90 minutes of video footage, this useful classroom resource illustrates the exciting possibilities of psychological research through a diverse video-clip collection of research being performed. Free to qualified adopters.

Student Web site to accompany *Psychological Science*, Second Canadian Edition wwnorton.com/psychsci

This innovative online learning tool offers a wealth of resources for study and review. In addition to timelines, vocabulary flashcards, and detailed chapter reviews, the site includes:

- **Diagnostic Quizzes** with smart feedback that allow students to pinpoint strengths and weaknesses
- **Activities**
- The **Norton Gradebook** feature, which allows students and instructors to effectively track quiz results
- **Animations** that clarify difficult psychological concepts
- **Streaming Video Exercises** that are derived from the Norton Discovery Channel Psychology DVD
- **Psychology in the News**, a unique section, updated weekly, that gathers relevant news coverage and analysis

Study Guide to accompany *Psychological Science*, Second Canadian Edition

by Brett Beck and Eileen Astor-Stetson (both of Bloomsburg University), and Heather Schellinck (Dalhousie University)

Created by highly successful instructors of large lecture classes, this carefully crafted study aid offers a guide to each chapter of the textbook, with helpful study advice, completion questions, key-figure exercises, multiple-choice self-tests, and thought questions.

Instructor's Resource Manual to accompany *Psychological Science*, Second Canadian Edition

by Margaret Forgie (Lethbridge University), Heather Schellinck (Dalhousie University), and George Spilich (Washington College)

Prepared by three master teachers who cultivate active learning environments in their own classrooms, the Instructor's Resource Manual provides teaching suggestions from Professors Heatherton and Gazzaniga; Chapter Overviews by Professor Heatherton; Chapter Objectives; Ticket-In, Ticket-Out Assignments; Five-Minute Lecture Launchers; Classroom Demonstrations and Activities; Discussion Topics; and a substantial Link Library. Available to all adopters.

Test Bank

by Robert Kleck (Dartmouth College), Margaret Lynch (San Francisco State University), and Heather Schellinck (Dalhousie University)

These Test-Item Files provide over 2,000 test questions, classified by chapter topic and question type (factual, vocabulary, applied, and conceptual) in the user-friendly Exam*View* program. Available to all adopters.

Norton Media Library

The Norton Media Library features Web-ready teaching materials for WebCT, Blackboard, or personal-course page formats. Ideal for both classroom learning and distance education, this comprehensive online resource offers tools for instructors such as test-item files, problem sets, lecture outlines, PowerPoint slides of textbook art and figures, and glossaries.

Norton Gradebook wwnorton.com/college/nrl/gradebook

The Norton Gradebook is an online resource that allows instructors and students to store and track their online quiz results effectively. Student results from each quiz are uploaded to the password-protected Gradebook, where instructors can access and sort them by section, chapter, book, student name, and date. Students can access the Gradebook to review their personal results. Registration for the Gradebook is instant and no setup is required.

PowerPoint Lecture Outlines with "clicker" questions

These lecture PowerPoints, complete with artwork from the text, follow the book's "ask and answer" approach. Detailed information in the notes section provides instructors with an invaluable pedagogical tool. The Applying the Principles questions from the text are included as "clicker" questions at the end of each chapter.

WebCT ePacks and Blackboard Course Cartridges

Instructors can use as is or tailor to suit their own needs.

Studying the Mind DVD

Filmed at Dartmouth College's Summer Institute for Cognitive Neuroscience, and featuring original footage exclusive to Norton, this DVD was created to bring examples of current brain-science research into the introductory psychology lecture. These five-to-seven-minute segments feature such well-known neuroscientific researchers as Marcus Raichle, Robert Knight, Mark D'Esposito, Michael Gazzaniga, John Gabrieli, Elizabeth Phelps, Marcia Johnson, Morris Moscovitch, Helen Neville, Denise Parks, and Patricia Reuter-Lorenz. Free to qualified adopters.

Norton Resource Library wwnorton.com/nrl

A Web site designed to house all electronic *Psychological Science* resources, from the test bank to PowerPoint slides to video clips. All of these resources are readily uploadable into WebCT and Blackboard environments. Available to all adopters.

Norton Video Library

A collection of first-rate documentary films focusing on psychological science drawn from the Films for Sciences and Humanities catalog and other fine video collections. Available to qualifying adopters.

ACKNOWLEDGMENTS

We begin once again by thanking our families, who have put up with us spending long hours away from them for the past three years. We are both very fortunate to have unwavering family support. We also are grateful to the many colleagues who gave us constructive feedback and advice. Special acknowledgment needs to be made to six individuals: Elizabeth Phelps, for helping us develop a clear vision for the cognition chapter, Peter Tse and Howard Hughes for their suggestions for making the sensation and perception chapter clear and accessible, Jay Hull for assisting us in making hard decisions regarding how best to merge the social chapters, and Gary Marcus for pushing us to think about gene expression in new ways We also benefited from the astute guidance of John Caccioppo, Jamie Pennebaker, Fernanda Ferreira, Tara Callaghan, Margaret Lynch, Jim Enns, Wendi Gardner, Lauretta Reeves, and many others who were willing to discuss their teaching goals for introductory psychology and their beliefs about what works and doesn't work in introductory textbooks.

Producing a psychology textbook is a team enterprise and relies on the efforts of many individuals, especially those who help with the ancillaries. We are grateful to Margaret Lynch for her work on the Applying the Principles questions and on the Test-Item File. Beginning with the First Edition, Margaret has pushed us to make sure that the book speaks clearly to the students while never taking them for granted. Robert Kleck also deserves special mention for overseeing the development of the test-item file. Our colleagues from Bloomsburg University, Brett Beck, Eileen Astor-Stetson, and Connie Schick, provided key advice about the First Edition and also developed the Study Guide. Patrick Carroll from the University of Texas, Austin, was masterful in identifying video material for the DVD that accompanies the book. Caton Roberts, from the University of Wisconsin–Madison, put together a first-rate set of Powerpoint Lecture Outlines. Heather Schellinck from Dalhousie University skillfully updated the ancillaries for the Canadian Edition, including authoring all the content for the new cultural psychology chapter We thank Lauretta Reeves from University of Texas, Austin, for her key role in helping to develop ZAPS: The Norton Psychology Labs, Margaret Forgie from the University of Lethbridge for bringing many fresh new teaching insights to the Instructor's Manual for the Second Edition, and, finally, Gary Marcus of NYU for creating *The Norton Psychology Reader*, a wonderful new companion text for introductory psychology courses. All of these individuals have lent their talent and time to create a strong support package for the Second Edition.

Once again, many people at Norton served critical roles in supporting and publishing the Second Edition. Chief among them is Jon Durbin, editor extraordinaire. Jon brings unflagging energy and incredible focus to his books. He is building an exciting psychology list at Norton that captures trends in the field like neuroscience and culture, while pushing the envelope hard to build a strong list across the subfields of psychology. Through his persistence and insight, Jon has become a true insider in the field. He is one in a million: the Lance Armstrong of editors.

Kim Yi and Carla Talmadge, the project editors, did a superb job of keeping the entire project on track, once again with good humour and amazing attention to detail. They both have a knack for identifying troublesome sentences, and best of all for fixing them. They sure know how to keep the authors happy and the project moving along efficiently. Barbara Gerr did a great job in copy editing the manuscript, including making sure the meaning was clear and the level of writing was consistent across chapters. Marian Johnson provided excellent advice on writing style throughout. Kelly Rolf and Robert Haber were a godsend in keeping everything organized and on schedule and also are to be recognized for their ability to find interesting photos and cartoons. Neil Hoos and Kelly Mitchell from the photo research department added a great deal in rounding out the photo program. Rubina Yeh developed a striking new design for the book that holds to our original vision for an uncluttered text in which all of the elements have strong pedagogical value. Her clean design helps students focus on what is important. It is also extremely attractive. Roy Tedoff shepherded the book very effectively through the production process. We thank Denise Shanks for taking charge of the ancillaries to make sure we have the strongest possible support package. Many of the Norton sales representatives have also been especially helpful, including Peter Ruscitti, Peter McCullough, Chris Curcio, Scott Berzon, Kym Silvasy-Neale, Jason Flippen, Andrea Haver, Gordon Lee, John Darger, John Kelly, and Annie Stewart. We also offer special thanks to Steve Dunn (Director of Sales and Marketing) and Nicole Albas (National Sales Manager) for their sound insights and unflagging support of the text through its first two editions. Finally, we continue to be grateful to Roby Harrington and Drake McFeely for their faith in us.

Consultants and Reviewers for the Second Edition

Consultants

Tara Callaghan, St. Francis Xavier University
Fernanda Ferreira, Michigan State University
James Enns, University of British Columbia
Steven Heine, University of British Columbia
Howard Hughes, Dartmouth College
Jay Hull, Dartmouth College
Margaret Lynch, San Francisco State University
Gary Marcus, New York University
James Pennebaker, University of Texas, Austin
Elizabeth Phelps, New York University
Lauretta Reeves, University of Texas, Austin
Peter Tse, Dartmouth College

Reviewers

Rahan Ali, Pennsylvania State University
Caroyln Barry, Loyola College
Scott Bates, Utah State University
Joe Bilotta, Western Kentucky University
Tom Capo, University of Maryland
Graham Cousens, Macalester College
Dale Dagenbach, Wake Forest University
Suzanne Delaney, University of Arizona
Wendy Domjan, University of Texas, Austin
Valerie Farmer-Dougan, Illinois State University
Greg Feist, University of California, Davis
Holly Filcheck, Louisiana State University
Joseph Fitzgerald, Wayne State University
Trisha Folds-Bennett, College of Charleston
Margaret Forgie, University of Lethbridge
David Funder, University of California, Riverside
Margaret Gatz, University of Southern California
Katherine Gibbs, University of California, Davis
Bryan Gibson, Central Michigan University
Laura Gonnerman, Lehigh University
Tom Guilmette, Providence College
Erin Hardin, Texas Tech University
John Henderson, Michigan State University
Sarah Hodges, University of Oregon
Cynthia Hoffman, Indiana University
Don Hoffman, University of California, Irvine
Steve Joordans, University of Toronto, Scarborough
Mark Lamaukis, San Diego State University
Lori Lange, University of North Florida
Ting Lei, Borough of Manhattan Community College
Carol Lemley, Elizabethtown College
Liang Lou, Grand Valley State
Margaret Lynch, University of California, San Francisco
Karl Maier, Salisbury University
Leonard Mark, Miami University (Ohio)
Paul Merritt, George Washington University
Paul McCormack, St. Francis Xavier University
David McDonald, University of Missouri Columbia
Patricia McMullen, Dalhousie University
Hal Miller, Brigham Young University
Joe Morrissey, SUNY-Binghamton
Todd Nelson, Cal State-Stanislaus
Jackie Pope-Tarrance, Western Kentucky University

Gabriel Radvansky, Notre Dame University
Patty Randolph, Western Kentucky University
William Rogers, Grand Valley State University
Sharleen Sakai, Michigan State University
Heather Schellinck, Dalhousie University
Richard Schiffman, Rutgers University
Jeniffer Siciliani-Pride, University of Missouri, St. Louis
Sheldon Solomon, Skidmore College
Sue Spaulding, University of North Carolina, Charlotte
Faye Steur, College of Charleston
Dawn Strongin, Cal State-Stanislaus
George Taylor, University of Missouri, St. Louis
Lee Thompson, Case Western Reserve University
Rob Tigner, Truman State College
Shaun Vecera, University of Iowa
Gordon Whitman, Tidewater Community College
Nicole Wilson, University of California, Santa Cruz
Al Witkofsky, Salisbury University
Dahlia Zaidel, University of California, Los Angeles

Consultants and Reviewers for the First Edition

Consultants

Alan Baddelay, Bristol University
Lori Badura, State University of New York, Buffalo
Mahzarin Banaji, Harvard University
Colin Blakemore, Oxford University
Randy Buckner, Washington University
Tara Callaghan, St. Francis Xavier University
Jennifer Campbell, University of British Columbia
Jonathan Cheek, Wellesley College
Dennis Cogan, Texas Tech University
Martin Conway, Bristol University
Michael Corballis, University of Auckland
James Enns, University of British Columbia
Raymond Fancher, York University
Fernanda Ferreira, Michigan State University
Vic Ferreira, University of California, San Diego
Christine Gancarz, Southern Methodist University
Wendi Gardner, Northwestern University
Rick O. Gilmore, Penn State University
James Gross, Stanford University
John Hallonquist, University of the Caribou
Mikki Hebl, Rice University
Mark Henn, University of New Hampshire
Don Hoffman, University of California, Irvine
James Hoffman, University of Delaware
Jake Jacobs, University of Arizona
Thomas Joiner, Florida State University
Dacher Keltner, University of California, Berkeley
Marke Leary, Wake Forest University
Monica Luciana, University of Minnesota
Neil Macrae, Bristol University/Dartmouth College
Julie Norem, Wellesley College
Lauretta Reeves, Rowan University
Jennifer Richeson, Dartmouth College
Paul Rozin, University of Pennsylvania
Constantine Sedikedes, University of Southhampton
Allison Sekuler, McMaster University
Andrew Shatte, University of Pennsylvania
Diane Tice, Case Western Reserve University
David Uttal, Northwestern University
Elaine Walker, Emory University

Reviewers

Gordon A. Allen, Miami University of Ohio
Ron Apland, Malaspina College
John P. Broida, University of Southern Maine
Katherine Cameron, Washington College
Timoth Cannon, University of Scranton
Charles Carver, University of Miami
Stephen Clark, Keene State College
Joseph Dien, Tulane University
Jack Dovidio, Colgate University
Margaret Forgie, University of Lethbridge
David Funder, University of California, Riverside
Peter Gerhardstein, Binghamton University
Peter Graf, University of British Columbia
Leonard Green, Washington University
Norman Henderson, Oberlin College
Terence Hines, Pace University
Sara Hodges, University of Oregon
Mike Kerchner, Washington College
Rondall Khoo, Western Connecticut State University
Charles Leith, Northern Michigan University
Margaret Lynch, San Fransisco State University
James Gross, Stanford University
Dan McAdams, Northwestern University
Doug McCann, York University
Bill McKeachie, University of Michigan
Judy Miller, Oberlin College
Zehra Peynircioglu, The American University
Brady Phelps, South Dakota State University
Alex Rothman, University of Minnesota
Juan Salinas, University of Texas, Austin

Margaret Sereno, University of Oregon
Arthur Shimamura, University of California, Berkeley
Rebecca Shiner, Colgate University
Reid Skeel, Central Michigan University
Dennison Smith, Oberlin College
Mark Snyder, University of Minnesota
George Taylor, University of Missouri

Robin R. Vallacher, Florida Atlantic University
Benjamin Walker, Georgetown University
Brian Wandell, Stanford University
Kevin Weinfurt, Duke University
Doug Whitman, Wayne State University
Maxine Gallander Wintre, York University
Claire Wiseman, Trinity College

Dear Student:

Our most important overarching goal for this textbook was to write it first and foremost for you, the student reader. We know that many of you are drawn to psychology to find out more about what makes you and those you know tick. We also know from our own recent teaching experiences that many of you are highly interested in learning more about how the human mind works and what that means for you in everyday life. Thus, as you search for insights into the human experience, we have made every effort to focus on core psychological principles and ideas to provide a starting point (and sometimes an end point) for your quest. Our focus on principles is reinforced by the "ask and answer" approach that serves as the pedagogical foundation for the book. Each chapter consists of a series of "big questions" that focus on major psychological principles and concepts. Each section of a chapter answers one of these questions. Our use of declarative headings reflects our belief that psychological scientists have made headway in providing answers to these questions—perhaps not the final answers, as new research helps shape our thinking, but answers that summarize what psychological scientists have discovered about mind, brain, and behaviour.

Psychological Science is intended both for those of you who wish to pursue careers in psychology and for those of you for whom this course will be your only exposure to psychology as a science. While using this text, you will gain an integrated grounding in traditional psychology as well as an introduction to new approaches within psychological science. The material is by nature intellectually challenging, but we have tried to make it accessible and enjoyable for you as well as directly applicable to your life. We hope that *Psychological Science* spurs on your curiosity about psychological phenomena and that you will learn to think critically about issues and themes in psychological science. In the end (or the beginning!), we hope that you will also develop greater self-understanding and understanding of others.

Before you begin, please take a few minutes to study the following pages so that you can gain a full understanding of how to get the most out of reading *Psychological Science*.

Overview—A Focus on the Principles

Psychological Science's chapters are built around core principles. The "ask and answer" approach serves as the structure and foundation for every chapter. Each chapter focuses on approximately four to six major principles, which are first raised in the form of questions ("ask"). Each major section in a chapter then looks to "answer" one of these questions. Here is how it works in action.

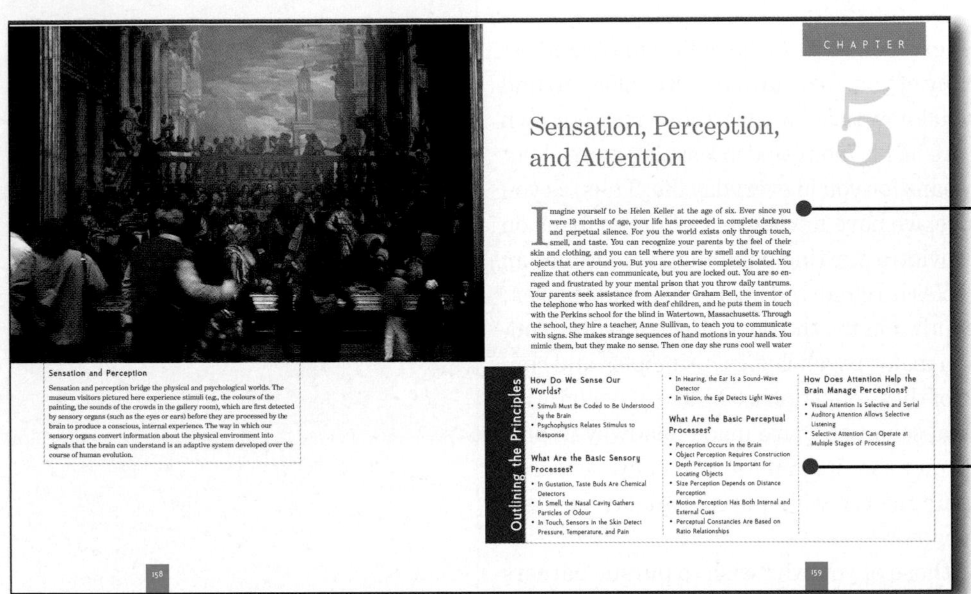

CHAPTER 5

Sensation, Perception, and Attention

Imagine yourself to be Helen Keller at the age of six. Ever since you were 19 months of age, your life has proceeded in complete darkness and perpetual silence. For you the world exists only through touch, smell, and taste. You can recognize your parents by the feel of their skin and clothing, and you can tell where you are by smell and by touching objects that are around you. But you are otherwise completely isolated. You realize that others can communicate, but you are locked out. You are so enraged and frustrated by your mental prison that you throw daily tantrums. Your parents seek assistance from Alexander Graham Bell, the inventor of the telephone who has worked with deaf children, and he puts them in touch with the Perkins school for the blind in Watertown, Massachusetts. Through the school, they hire a teacher, Anne Sullivan, to teach you to communicate with signs. She makes strange sequences of hand motions in your hands. You mimic them, but they make no sense. Then one day she runs cool well water

Sensation and Perception Sensation and perception bridge the physical and psychological worlds. The museum visitors pictured here experience stimuli (e.g., the colours of the painting, the sounds of the crowds in the gallery room), which are first detected by sensory organs (such as the eyes or ears) before they are processed by the brain to produce a conscious, internal experience. The way in which our sensory organs convert information about the physical environment into signals that the brain can understand is an adaptive system developed over the course of human evolution.

Outlining the Principles

How Do We Sense Our Worlds?
- Stimuli Must Be Coded to Be Understood by the Brain
- Psychophysics Relates Stimulus to Response

What Are the Basic Sensory Processes?
- In Gustation, Taste Buds Are Chemical Detectors
- In Smell, the Nasal Cavity Gathers Particles of Odour
- In Touch, Sensors in the Skin Detect Pressure, Temperature, and Pain
- In Hearing, the Ear Is a Sound-Wave Detector
- In Vision, the Eye Detects Light Waves

What Are the Basic Perceptual Processes?
- Perception Occurs in the Brain
- Object Perception Requires Construction
- Depth Perception Is Important for Locating Objects
- Size Perception Depends on Distance Perception
- Motion Perception Has Both Internal and External Cues
- Perceptual Constancies Are Based on Ratio Relationships

How Does Attention Help the Brain Manage Perceptions?
- Visual Attention Is Selective and Serial
- Auditory Attention Allows Selective Listening
- Selective Attention Can Operate at Multiple Stages of Processing

158 159

A Chapter Opening Vignette begins each chapter. The vignettes are drawn from a variety of sources, including news media, research journals, and history. They highlight a major theme, issue, or tension point that will be discussed throughout the chapter.

An Outlining the Principles box appears on the second page of every chapter opener. At one level, this feature serves as a simple outline or road map for the chapter. At another level, it reveals the major principles that will be discussed in the chapter. By studying the major headings, you can see which questions ("ask") will drive the chapter. The subheadings appear in the form of declarative statements ("answer") that reveal our current state of knowledge about the question.

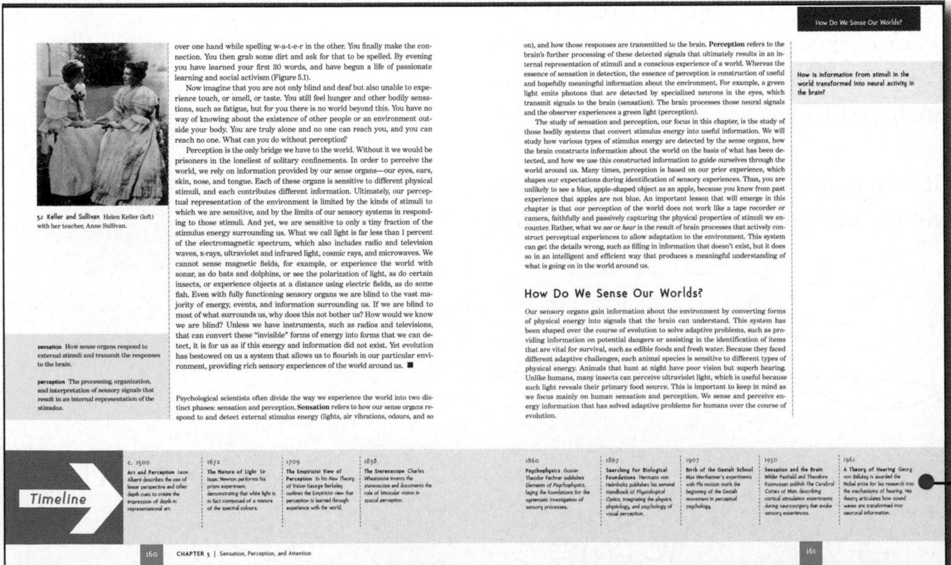

How Do We Sense Our Worlds?

over one hand while spelling w-a-t-e-r in the other. You finally make the connection. You then grab some dirt and ask for that to be spelled. By evening you have learned your first 30 words, and have begun a life of passionate learning and social activism (Figure 5.1).

Now imagine that you are not only blind and deaf but also unable to experience touch, or smell, or taste. You still feel hunger and other bodily sensations, such as fatigue, but for you there is no world beyond this. You have no way of knowing about the existence of other people or an environment outside your body. You are truly alone and no one can reach you, and you can reach no one. What can you do without perception?

Perception is the only bridge we have to the world. Without it we would be prisoners in the loneliest of solitary confinements. In order to perceive the world, we rely on information provided by our sense organs—our eyes, ears, skin, nose, and tongue. Each of these organs is sensitive to different physical stimuli, and each contributes different information. Ultimately, our perceptual representation of the environment is limited by the kinds of stimuli to which we are sensitive, and by the limits of our sensory systems in responding to those stimuli. And yet, we are sensitive to only a tiny fraction of the stimulus energy surrounding us. What we call light is far less than 1 percent of the electromagnetic spectrum, which also includes radio and television waves, x-rays, ultraviolet and infrared light, cosmic rays, and microwaves. We cannot sense magnetic fields, for example, or experience the world with sonar, as do bats and dolphins, or see the polarization of light, as do certain insects, or experience objects at a distance using electric fields, as do some fish. Even with fully functioning sensory organs we are blind to the vast majority of energy, events, and information surrounding us. If we are blind to most of what surrounds us, why does this not bother us? How would we know we are blind? Unless we have instruments, such as radios and televisions, that can convert these "invisible" forms of energy into forms that we can detect, it is for us as if this energy and information did not exist. Yet evolution has bestowed on us a system that allows us to flourish in our particular environment, providing rich sensory experiences of the world around us. ■

5.1 **Keller and Sullivan** Helen Keller (left) with her teacher, Anne Sullivan.

sensation How sense organs respond to external stimuli and transmit the responses to the brain.

perception The processing, organization, and interpretation of sensory signals that result in an internal representation of the stimulus.

Psychological scientists often divide the way we experience the world into two distinct phases: sensation and perception. **Sensation** refers to how our sense organs respond to and detect external stimulus energy (lights, air vibrations, odours, and so on), and how those responses are transmitted to the brain. **Perception** refers to the brain's further processing of these detected signals that ultimately results in an internal representation of stimuli and a conscious experience of a world. Whereas the essence of sensation is detection, the essence of perception is construction of useful and hopefully meaningful information about the environment. For example, a green light emits photons that are detected by specialized neurons in the eyes, which transmit signals to the brain (sensation). The brain processes those neural signals and the observer experiences a green light (perception).

The study of sensation and perception, our focus in this chapter, is the study of those bodily systems that convert stimulus energy into useful information. We will study how various types of stimulus energy are detected by the sense organs, how the brain constructs information about the world on the basis of what has been detected, and how we use this constructed information to guide ourselves through the world around us. Many times, perception is based on our prior experience, which shapes our expectations during identification of sensory experiences. Thus, you are unlikely to see a blue, apple-shaped object as an apple, because you know from past experience that apples are not blue. An important lesson that will emerge in this chapter is that our perception of the world does not work like a tape recorder or camera, faithfully and passively capturing the physical properties of stimuli we encounter. Rather, what we *see* or *hear* is the result of brain processes that actively construct perceptual experiences to allow adaptation to the environment. This system can get the details wrong, such as filling in information that doesn't exist, but it does so in an intelligent and efficient way that produces a meaningful understanding of what is going on in the world around us.

How is information from stimuli in the world transformed into neural activity in the brain?

How Do We Sense Our Worlds?

Our sensory organs gain information about the environment by converting forms of physical energy into signals that the brain can understand. This system has been shaped over the course of evolution to solve adaptive problems, such as providing information on potential dangers or assisting in the identification of items that are vital for survival, such as edible foods and fresh water. Because they faced different adaptive challenges, each animal species is sensitive to different types of physical energy. Animals that hunt at night have poor vision but superb hearing. Unlike humans, many insects can perceive ultraviolet light, which is useful because such light reveals their primary food source. This is important to keep in mind as we focus mainly on human sensation and perception. We sense and perceive energy information that has solved adaptive problems for humans over the course of evolution.

Timeline

c. 1500 **Art and Perception** Leon Alberti describes the use of linear perspective and other depth cues to create the impression of depth in representational art.

1672 **The Nature of Light** Sir Isaac Newton performs his prism experiment, demonstrating that white light is in fact composed of a mixture of the spectral colours.

1709 **The Empiricist View of Perception** In his *New Theory of Vision* George Berkeley outlines the Empiricist view that perception is learned through experience with the world.

1838 **The Stereoscope** Charles Wheatstone invents the stereoscope and documents the role of binocular vision in spacial perception.

1860 **Psychophysics** Gustav Theodor Fechner publishes *Elements of Psychophysics*, laying the foundations for the systematic investigation of sensory processes.

1867 **Searching for Biological Foundations** Hermann von Helmholtz publishes his seminal *Handbook of Physiological Optics*, integrating the physics, physiology, and psychology of visual perception.

1907 **Birth of the Gestalt School** Max Wertheimer's experiments with Phi motion mark the beginning of the Gestalt movement in perceptual psychology.

1950 **Sensation and the Brain** Wilder Penfield and Theodore Rasmussen publish *The Cerebral Cortex of Man*, describing cortical stimulation experiments during neurosurgery that evoke sensory experiences.

1961 **A Theory of Hearing** Georg von Békésy is awarded the Nobel prize for his research into the mechanisms of hearing. His theory articulates how sound waves are transformed into neuronal information.

160 CHAPTER 5 | Sensation, Perception, and Attention 161

A Chapter Timeline appears on the bottom of the page in the opening section of every chapter. Psychological science is built on cumulative knowledge and experience. This is one of the major themes of the text. Basic principles, both new and old, inspire and guide thinking and research in the field. The timelines highlight major developments within the various domains of psychology. By studying them, you will see more clearly how various principles have been established, challenged, and modified.

What Are the Basic Sensory Processes?

the hair. Other receptors are capsules in the skin that respond to continued vibration, sudden movements, and steady pressure. In terms of temperature, there appear to be separate receptors for hot and cold, although both can be triggered at the same time by intense stimuli. The simultaneous activation of hot and cold receptors can produce strange sensory experiences, such as a feeling of wetness. The integration of these various signals produces haptic experiences. For instance, stroking multiple pressure points can produce a tickling sensation, which can be pleasant or unpleasant depending on the mental state of the person being tickled. This raises the age-old question of why you can't tickle yourself. Indeed, recent imaging research finds that the brain areas involved in touch sensation respond less to self-produced tactile stimulation than to external tactile stimulation (Blakemore, Wolpert, & Frith, 1998).

TWO TYPES OF PAIN Pain is part of a warning system that stops you from continuing activities that may harm you. Whether the message is to remove your hand from a hot burner or to stop running when you have damaged a tendon, pain signals that you should quit doing whatever you are doing. Children born with a rare genetic disorder that leaves them insensitive to pain usually die young, no matter how carefully they are supervised. They simply do not know how to avoid activities that harm them or to report when they are not feeling well (Melzack & Wall, 1982). As with other sensory systems, the actual experience of pain is created by the brain. For instance, as you recall from Chapter 4, people with amputated limbs sometimes feel phantom pain in the nonexistent limb. The person really feels pain, but the pain occurs because of a misinterpretation of neural activity (Melzack, 1992).

Can we override pain?

Most experiences of pain result from damaging the skin, which activates haptic receptors. The nerve fibres that convey pain information are thinner than those for temperature and touch and are found in all body tissues that sense pain: skin, muscles, membranes around bones and joints, organs, and so on. Two kinds of nerve fibres have been identified for pain: fast fibres for sharp, immediate pain, and slow fibres for chronic, dull, steady pain. An important distinction between these fibres is the myelination of their axons that travel from the pain receptors to the spinal cord. Recall from Chapter 3 that myelination speeds up neural signals; fast-acting axons are myelinated, whereas the axons of slow-acting receptors are not. Think of when you touch a hot pan. You feel two kinds of pain: a sharp, fast, localized pain at the point where your skin touches the pan, followed by a slow, dull, more diffuse burning pain. The fast-action receptors are activated by strong physical pressure and temperature extremes, whereas the slow-acting receptors are activated by chemical changes in tissue when skin is damaged. From an adaptive perspective, fast pain leads to recoil from harmful objects and therefore is protective, whereas slow pain keeps us from using the affected body part and therefore helps in recuperation.

GATE-CONTROL THEORY The brain regulates the experience of pain, sometimes producing it, sometimes suppressing it. The *gate-control theory of pain,* formulated by Ronald Melzack (Figure 5.5) and Patrick Wall (1982) at McGill University in Montreal, states that in order for us to experience pain, pain receptors must be activated and a neural "gate" in the spinal cord must allow these signals through to the brain. One way to close the gate is to stimulate other haptic receptors, which overwhelms the signals from pain receptors. This is why rubbing an aching muscle helps to reduce pain or why scratching an itch is so satisfying. Experienced nurses often vigorously rub the skin surrounding the injection site, which overwhelms the haptic signals and reduces the sting of the needle. A number of cognitive states can also close the gate, such as when people are distracted by more pressing concerns. Athletes sometimes

5.5 Richard Melzack A professor at McGill University in Montreal, Quebec, Melzack has produced much of the world's most groundbreaking work on pain. In 1965, he produced one of the most influential papers ever written on the subject, revolutionizing the way clinicians think about the topic of pain.

169

"Ask and Answer" Running Heads help you stay focused on the larger issue as you try to see the forest for the trees. *Psychological Science*'s left-hand running heads emphasize the larger topics, with the right-hand running heads repeating the questions that are explored in each section.

Critical Thinking Questions are designed to encourage you to think integratively and beyond the text. Appearing in the margins throughout each chapter, they selectively highlight core principles and are ideal for discussion in small groups.

Reviewing the Principles boxes are a critical component of the "ask and answer" approach and appear at the end of each major section. They repeat the question that governed the section and provide a basic answer that highlights key points for you to remember.

Defining the Principles refers to the marginal glossary that runs throughout each chapter. Many books highlight an overabundance of key terms for students to memorize. In keeping with this book's focus on core principles and concepts, *Psychological Science* highlights approximately 30 key terms per chapter. The full glossary appears at the end of the book and on the companion Web site.

REVIEWING
the Principles | How Do We Sense Our Worlds?

The study of sensation focuses on how our sense organs respond to and detect external stimulus energy. Stimuli need to be transduced in order for the brain to use that information. Sensory coding for quantitative factors, such as intensity and loudness, depends on the number of neurons firing and how frequently they fire. Qualitative aspects, such as colour or bitterness, are coded by the integration of activation across specific receptors. The development of psychophysical methods allowed psychological scientists to study psychological reaction to physical events. Psychophysical methods can be used to determine thresholds for detecting events and for noticing change. These thresholds can be influenced by situational factors and by biases in human judgment.

What Are the Basic Sensory Processes?

How does information about the world get into the brain? Helen Keller's perceptual experience was restricted because her sensory systems were damaged. The neurons in her brain that construct visual and auditory perceptions did not receive any sensory signals. As you learned in Chapter 3, no neuron beyond the sensory organs responds directly to events in the world. All that a neuron in the brain receives is signals from other neurons. In other words, a neuron in the brain does not respond to events in the world, it responds to its input from other neurons. As a result, you could say that your experience is a construction of your brain and resides inside your skull, despite the illusion that your experience is of objects or events out in the space around you. Neurons are talking to neurons in total darkness, and yet from this emerges our conscious experience of the outside world. How is this information transmitted to the brain? In this section, we will review the first step in this process, how stimuli are detected and sent to the brain for each of the five primary senses: taste, smell, touch, hearing, and vision.

In Gustation, Taste Buds Are Chemical Detectors

The job of our sense of taste, or *gustation,* is to keep poisons out of our digestive systems while allowing good food in. The stimuli for taste are chemical substances from food that dissolve in saliva, though how these stimuli work is still largely a mystery. The taste receptors are part of the **taste buds** of the tongue and mouth. People differ greatly in how many taste buds they have, ranging from 500 to 10,000, mostly located on the tongue, but also spread throughout the mouth and even the throat. Each taste bud has about 50 receptor cells. *Microvilli*—short, hairlike structures at the tip of each taste bud—come into direct contact with saliva. When stimulated, they send electrical signals to a brainstem region called the medulla and from there to the thalamus and cortex, which ultimately produces the experience of taste.

taste buds Sensory receptors that transduce taste information.

We have several different kinds of taste receptors. These taste buds each respond most strongly to one of four "primary" taste sensations: sweet, sour, salty, and bitter. Every taste experience is composed of a mixture of these four qualities. This experience reflects a general principle of sensory experiences, in that a near-infinite variety of perceptual experiences arises from the activation of unique combinations

166 CHAPTER 5 | Sensation, Perception, and Attention

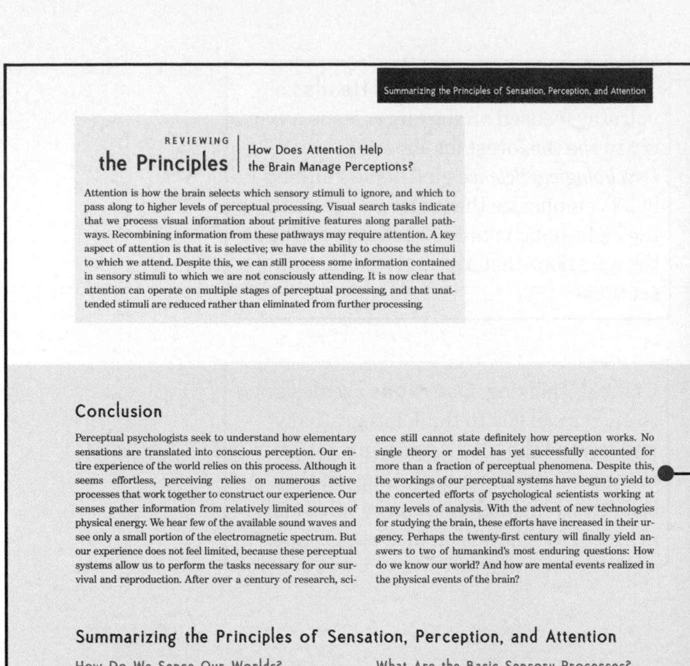

A Brief Conclusion sums up the big ideas and concepts and reminds you how the book's four key themes wove their way through the chapter.

The Summarizing the Principles section provides a more detailed review. Designed to follow the outline for each chapter, you are reminded again of the major questions that governed each section and the key pieces of research-based knowledge that provide insight into those questions.

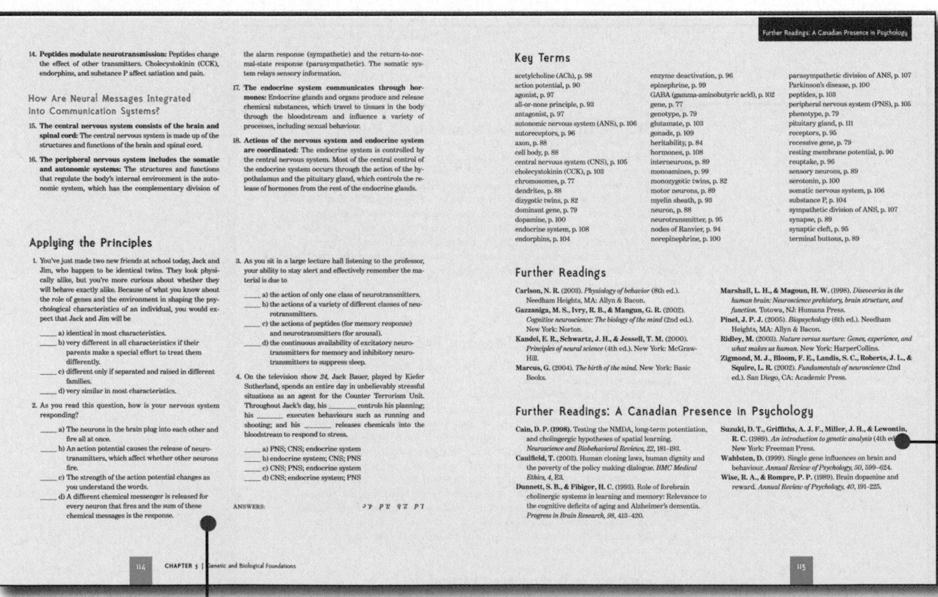

Further Readings represent psychological writing at its best. If any of the topics in a chapter interested you, we encourage you to follow up with one of the suggested reading titles. There are many wonderful popular psychology books that provide keen insights and pleasurable excursions into human behaviour.

Applying the Principles questions highlight the core principles that you need to learn. These multiple-choice questions at the end of each chapter encourage you to apply these principles to real-world settings. The answers appear at the end of the section.

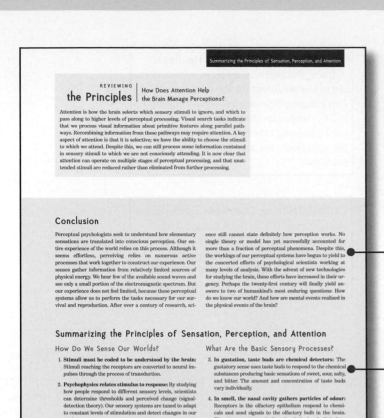

A Focus on Critical Thinking and Applications

SPECIAL TOPIC UNITS apply psychological science to your life. These units amplify the text's basic strengths by challenging you to think critically and to understand the impact that the field of psychology has on your life and world.

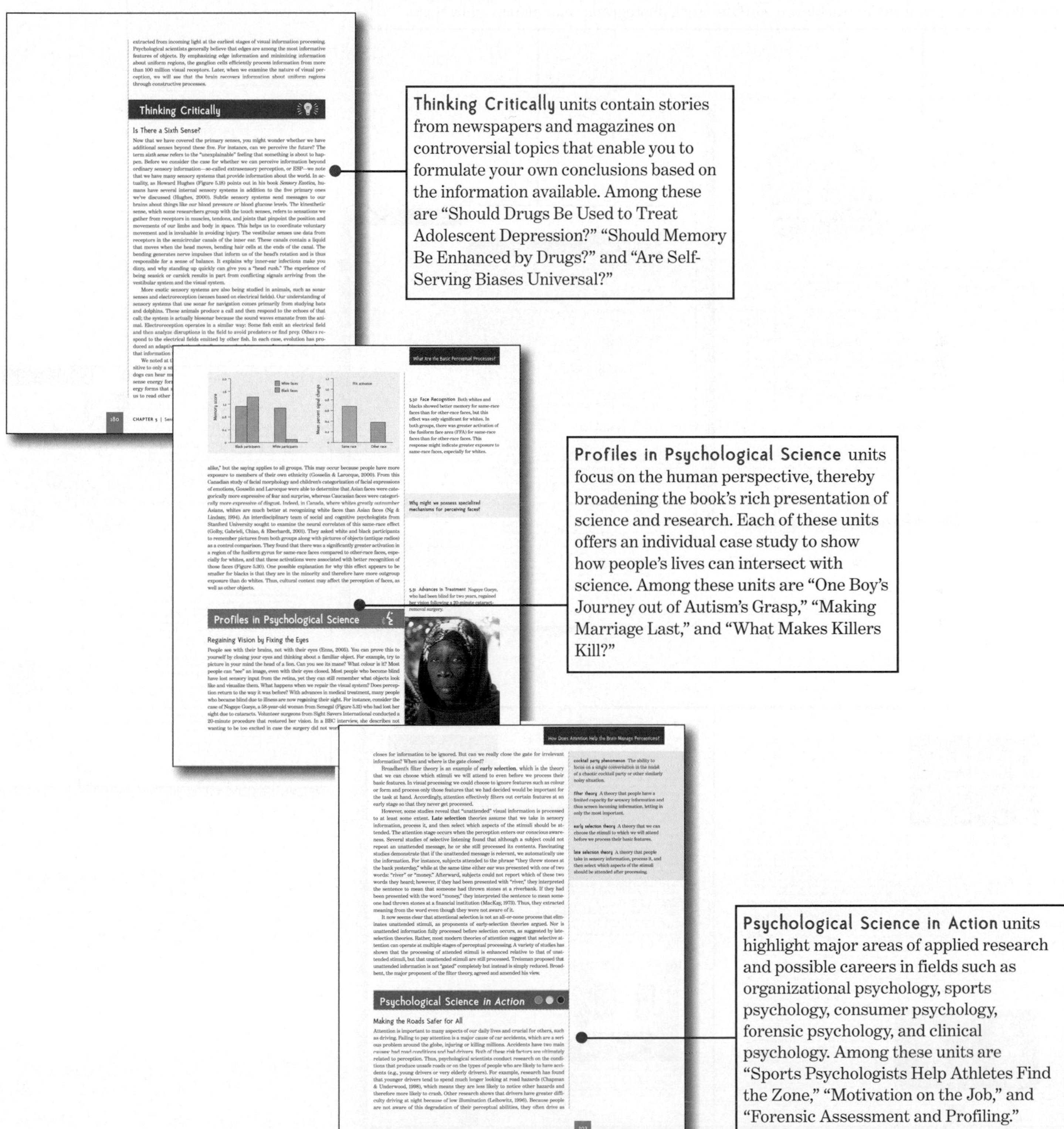

Thinking Critically units contain stories from newspapers and magazines on controversial topics that enable you to formulate your own conclusions based on the information available. Among these are "Should Drugs Be Used to Treat Adolescent Depression?" "Should Memory Be Enhanced by Drugs?" and "Are Self-Serving Biases Universal?"

Profiles in Psychological Science units focus on the human perspective, thereby broadening the book's rich presentation of science and research. Each of these units offers an individual case study to show how people's lives can intersect with science. Among these units are "One Boy's Journey out of Autism's Grasp," "Making Marriage Last," and "What Makes Killers Kill?"

Psychological Science in Action units highlight major areas of applied research and possible careers in fields such as organizational psychology, sports psychology, consumer psychology, forensic psychology, and clinical psychology. Among these units are "Sports Psychologists Help Athletes Find the Zone," "Motivation on the Job," and "Forensic Assessment and Profiling."

A Dynamic Art Program

The Art in *Psychological Science* enhances your reading experience. The text contains a variety of visual materials, from photographs to tables and charts to drawn art. The emphasis, however, is clearly on the drawn art. By featuring drawn art, *Psychological Science* is able to convey precisely, accurately, and meaningfully what you need to gain from every image. This high level of precision can't be achieved with the stock photographs common in many texts.

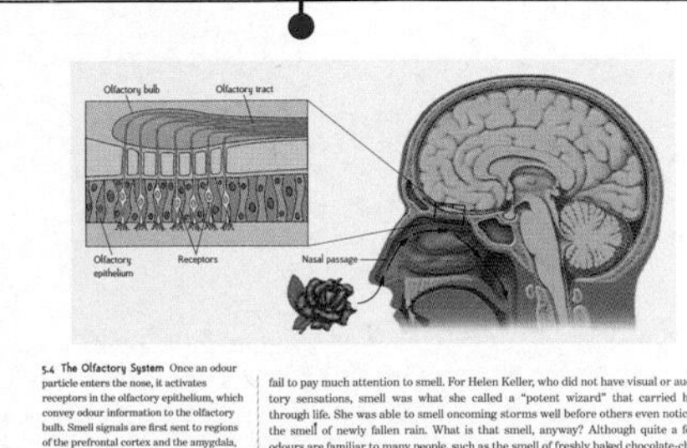

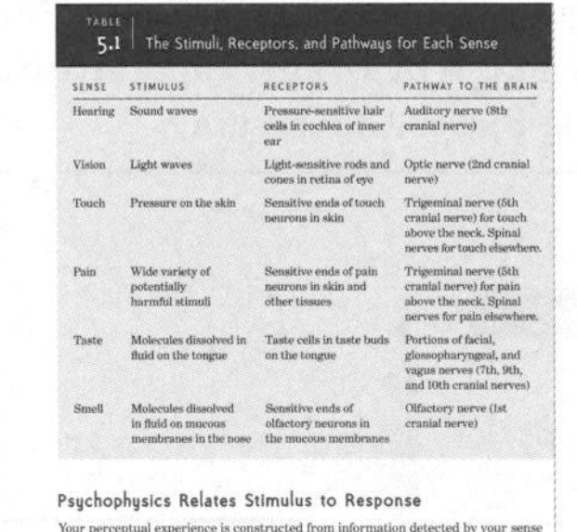

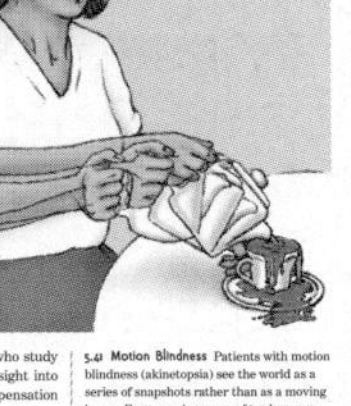

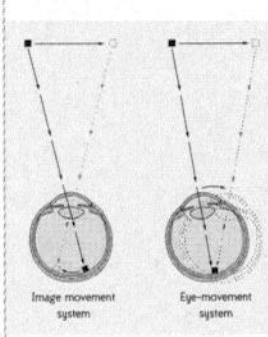

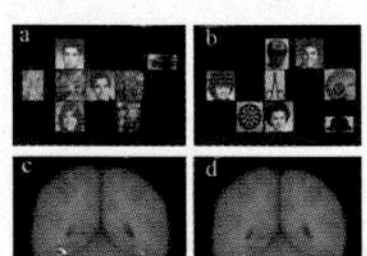

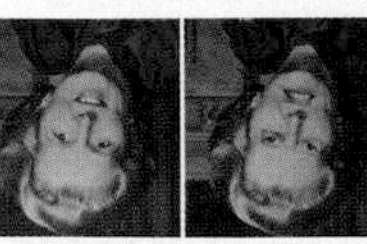

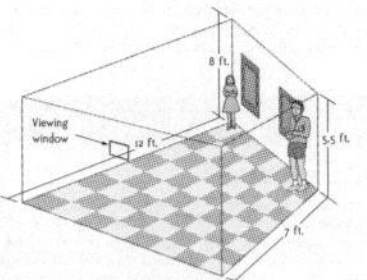

Psychological Science's Student Supplements Will Help You Succeed

Study Guide to accompany *Psychological Science* 

by Brett Beck and Eileen Astor-Stetson (both of Bloomsburg University), and Heather Schellinck (Dalhousie University)

Created by highly successful instructors of large lectures classes, this carefully crafted study aid offers a guide to each chapter of the textbook, with helpful study advice, completion questions, key-figure exercises, multiple-choice self-tests, and thought questions.

Student Web site to accompany *Psychological Science* 

wwnorton.com/psychsci

This innovative online learning tool offers a wealth of resources for study and review. In addition to timelines, vocabulary flashcards, and detailed chapter reviews, the site includes:

- **Diagnostic Quizzes** with smart feedback that allow students to pinpoint strengths and weaknesses
- **Activities**
- The **Norton Gradebook** feature, which allows students and instructors to effectively track quiz results
- **Animations** that clarify difficult psychological concepts
- **Streaming Video Exercises** that are derived from the Norton Discovery Channel Psychology DVD
- **Psychology in the News**, a unique section, updated weekly, that gathers relevant news coverage and analysis

Psychological Science

SECOND CANADIAN EDITION

Exploring the Mind and Brain

Throughout the history of psychology, there has been debate about whether psychological characteristics are due more to *nature* or *nurture*. It is now clear that culture, social environment, and biology all play important roles in the development of one's behaviour and mental life. Individuals who share genes, such as these quadruplets, are particularly interesting to psychological scientists studying the impact of biology and environment.

CHAPTER 1

Introduction to Psychological Science

Around midnight on February 4, 1999, four members of a special police Street Crime Unit drove down Wheeler Avenue in the Bronx (New York City), looking for a rape suspect. They saw 22-year-old Amadou Diallo (Figure 1.1), a West African immigrant, standing in the doorway to his apartment building. The officers say they told Diallo to "freeze" but then saw him reach into his pants pocket. The officers, believing that Diallo was reaching for a weapon, opened fire. Within approximately 5 seconds, the four officers fired a total of 41 shots at the unarmed Diallo, 19 of which struck him. Diallo died at the scene. Neighbors say that Diallo probably didn't understand the word "freeze," since English was not his first language. Others suggest that Diallo might have been reaching for his wallet to prove his identity.

Outlining the Principles

What Are the Themes of Psychological Science?

- The Principles of Psychological Science Are Cumulative
- A New Biological Revolution Is Energizing Research
- The Mind Is Adaptive
- Psychological Science Crosses Levels of Analysis

What Are the Intellectual Origins of Psychology?

- The Nature-Nurture Debate Considers the Impact of Biology and Environment
- The Mind-Body Problem Has Challenged Philosophers and Psychologists
- Evolutionary Theory Introduces Natural Selection

How Did the Scientific Foundations of Psychology Develop?

- Experimental Psychology Begins with Structuralism
- Functionalism Addresses the Purpose of Behaviour
- Gestalt Psychology Emphasizes Patterns and Context in Learning
- The Unconscious Influences Everyday Mental Life
- Most Behaviour Can Be Modified by Reward and Punishment
- How People Think Affects Behaviour
- Social Situations Shape Behaviour
- Psychological Therapy Is Based on Science

How Can We Apply Psychological Science?

- Subdisciplines Focus on Different Levels of Analysis
- Psychological Knowledge Is Used in Many Professions
- People Are Intuitive Psychological Scientists
- Psychological Sciences Require Critical Thinking

1.1 **Amadou Diallo** Diallo, a native of West Africa, was shot by police officers who were searching for a serial rapist. The innocent Diallo died at the scene. Did racism play a part in this tragic death?

No one will ever know. The four officers were tried for murder, and all were acquitted.

Psychologists are interested in understanding how people perceive, think, and act in a wide range of situations. A situation such as the killing of Amadou Diallo is especially interesting to psychologists because it allows them to consider thought and behaviour in the context of a real-life event. The Diallo case involves psychological phenomena such as emotion, memory, visual perception, decision making, social interaction, cultural differences, prejudice, group behaviour, and mental trauma. Within this menu of topics, psychologists would be interested in knowing, for example, how the emotional state of the officers affected their decision making at the scene. They would also want to study the accuracy of the defendants' and eyewitnesses' accounts of the shooting. In addition, they would want to examine group behaviour. Did the fact that many officers arrived at the scene simultaneously affect individual behaviour? Did the high-profile demonstrations outside the courthouse influence the trial? Finally, psychologists would also want to know if prejudice played a role. Did prejudice affect the way the officers identified and approached the suspect? Did being the target of prejudice affect Diallo's reaction to the police? Did prejudice affect the jury's decision? Many Bronx residents charged that the Diallo shooting was motivated by racism. The defense lawyers asserted that race had little to do with the shooting. How can we know? Is this a problem unique to the United States? Clearly not, as a private member's bill (C-296) was introduced to the House of Commons in November 2004 in an attempt to eliminate racial profiling in Canada.

Studying the factors underlying racial attitudes is difficult, since most people deny holding beliefs that are not socially acceptable. How, then, can we "peek inside" the mind to discover what people are thinking and feeling? ■

Over the decades, psychology has offered many powerful insights into prejudice. Employing a range of strategies—from self-reports to observational studies to experiments—researchers have discovered a number of important findings about prejudice. For instance, prejudice influences people's beliefs and expectations about others, which leads to racist behaviour. Racism has strong negative psychological effects on the targets of prejudice, whose behaviour is shaped in part by others' beliefs about them. Perhaps most important, however, psychologists have also discovered methods for reducing stereotyping and prejudice. These findings have given us

1637
Cartesian Dualism René Descartes proposes that mind and body are interrelated but separate entities, with each affecting the other. This idea challenges the long-held belief that the mind, or soul, is master of the body.

1859
The Mind Evolved Charles Darwin's groundbreaking theory of natural selection lays the foundation for a biological basis of behaviour and the notion that the human mind has evolved along with physical characteristics and behaviours.

1879
Psychology Adopts the Tools of Science Wilhelm Wundt sets up the first psychological laboratory and starts to measure behaviour. His first methods involve asking people to reflect on their mental experiences.

1890
Principles of Psychology Influenced by Darwin's theory, William James argues in his groundbreaking *Principles of Psychology* that there is a need to understand the adaptive functions of behaviour.

glimpses inside the human mind, but they have not provided a complete picture of how the mind works and its relevance to our own behaviour.

A number of important developments over the past few years have opened new doors for studying mental life that should really deepen our understanding of the human mind and behaviour. We now have methods for observing the working brain in action. A collection of techniques known as *brain imaging* involves assessing changes in metabolic activity in the brain, such as noting where blood flows as people process information. These changes in blood flow represent changes in brain activity that indicate which parts of the brain are involved in certain behaviours or mental activities. For instance, the cognitive neuroscientist Elizabeth Phelps and the social psychologist Mahzarin Banaji and their colleagues used brain imaging to study racial attitudes. They showed white college students pictures of familiar and unfamiliar black and white faces while they used a technique known as *functional magnetic resonance imaging (fMRI)* to scan the brain (Phelps et al., 2000). For some of the research participants, the unfamiliar black faces activated a structure in the brain called the *amygdala*, which is involved in detecting threat; this activation indicates a fear response. This does not mean that the white students who responded in this way are necessarily afraid of unfamiliar black people, but it does indicate that those faces activate the brain region that detects threat.

It is important to point out that this response did not occur in all student participants. Rather, it occurred only in those who also showed signs of holding negative attitudes about blacks, as measured by a computer program called the *implicit attitudes test (IAT)*. This test indirectly assesses how people associate positive and negative words with certain groups of people. Those who were found to hold negative attitudes on the computer test were the most likely to show activation of the amygdala when observing unfamiliar black faces (Figure 1.2). In a second study, the researchers showed a new group of white students pictures of familiar black and white faces; this time activation of the amygdala did not occur. This is encouraging news. It suggests that increasing familiarity reduces the fear response, which may indicate a reduction in the likelihood of prejudice and discrimination.

The research just described reveals a great deal about the current state of psychology and touches on a number of issues that are central to this book. Psychology has formally been around as a discipline of study for just over 100 years, and in that time we've learned a great deal about such basic mental processes as learning, memory, emotion, and perception. Psychologists have documented changes in individuals from birth to old age and have developed elaborate theories about how humans do things such as acquire language and

1.2 Composite Maps of the Brain Composite maps displaying brain regions in which greater activation in response to black versus white faces is associated with an indirect measure of racial attitudes. The areas indicate activation of the amygdala, especially on the left half of the brain, as well as another area, the anterior cingulate, which is also associated with emotional responses.

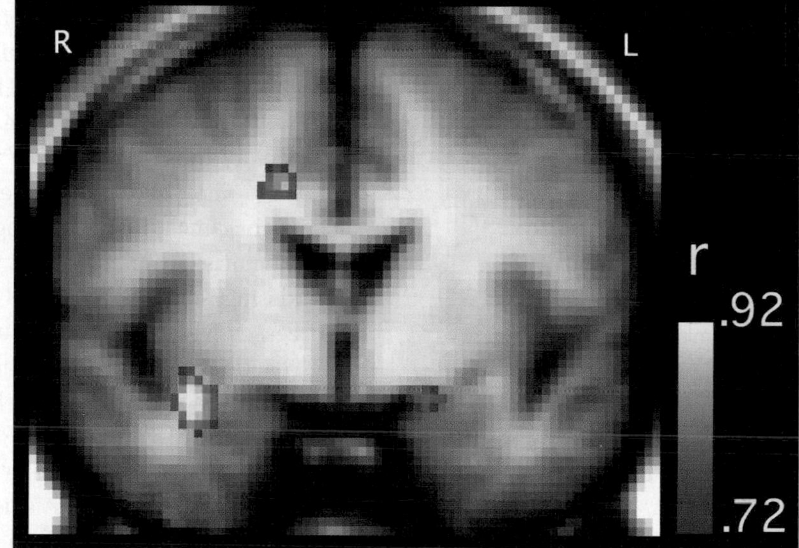

1891
Psychology in Canada James Mark Baldwin founded the psychology laboratory at the University of Toronto, the first of its kind in the British Commonwealth.

1900
The Role of the Unconscious Sigmund Freud introduces the idea of the unconscious and the role it plays in our everyday mental life. His theory of psychoanalysis captures clinicians and scientists alike for almost half a century.

1912
Gestalt Psychology German psychologist Max Wertheimer proposes that perception is a subjective experience that cannot be understood by looking at its elementary components. That is, the whole is different from the sum of its parts.

1925
Behaviourism John B. Watson and later B. F. Skinner argue that all behaviours can be understood as a result of learning, and that to predict behaviour one has to look at environmental forces.

1940s
Physiological Psychology Karl Lashley and Roger Sperry conduct systematic brain research using animals as a way to gain insight into how the human mind works.

solve difficult problems. They have explored the importance of the social world in which we live, such as how people are influenced by the presence of others and the circumstances that lead them to prefer some types of people to others. Psychologists have also examined how the environment helps shape each person in a unique way that produces what we call "personality." The goal for psychology today is to understand people by considering both individual factors, such as how a person's brain processes information about others, and contextual factors, such as how societal beliefs shape how we behave toward those other people. You will learn what has been discovered about each of these topics in this book and about the future directions of research. You will learn too about how genes and biology influence human mental life, an area of research that has deepened psychology over the last few years.

The ultimate ambition of psychologists is to explain effectively the most important human behaviours in real-life contexts. To do this, we need to study behaviour from its most fundamental aspects: biological, individual, and social. It's precisely for this reason that it is so exciting on a daily basis to learn more about how the brain works. New technologies and methods have allowed our understanding to move forward in ways that were unimaginable even 20 years ago. These technologies have allowed psychological scientists to begin examining which parts of the brain are involved when we perform certain tasks and interact with our social worlds. For instance, did you know that your brain chemistry is being altered with each concept or story that you remember while reading this chapter? Every time we remember a story or a joke or feel an intense emotion, that experience is processed and stored in our brains. One major challenge for psychologists is to figure out what brain mechanisms are involved when we interact with our environment, as well as how the environment influences our brain mechanisms. Of course, understanding people in their real lives requires more than just studying the activity of their brains. People live in societies in which they interact regularly with others in multiple and complex ways. Another major challenge for psychologists is to understand how people are shaped by the cultures in which they live.

psychological science The study of mind, brain, and behaviour.

Psychological science is the study of mind, brain, and behaviour. Let's explore each of these terms. *Mind* refers to mental activity, such as your thoughts and feelings. The perceptual experiences that you have while interacting with the world (sight, smell, taste, hearing, and touch) are examples of the mind in action, as are memories, thinking about what you want to have for lunch, and how you feel about kissing someone you find attractive. Mental activity results from biological processes within the *brain,* such as the action of nerve cells and associated chemical reactions. You will read later about the long-standing controversy among scholars about the nature of the relationship between the brain and the mind. For now it is sufficient to note that the "mind is what the brain does" (Kosslyn & Koenig, 1995, p. 4). In other words, it is the physical brain that enables the mind.

What is mind?

The term *behaviour* is used to describe a wide variety of actions, from the subtle to the complex, that occur in organisms from ants to humans. For many years psychol-

1944
Social Dynamics Kurt Lewin introduces field theory, in which he argues that situational dynamics play an important role in predicting human behaviour.

1957
The Cognitive Revolution At Harvard University, George A. Miller launches the field of cognitive psychology, which will later be formalized by Ulric Neisser in his integrative 1967 book, *Cognitive Psychology*.

1960s
Biology of Mental Disorders Advances in drug-treatment trials support theories of a biological basis for many types of mental disorders. Disorders such as depression and schizophrenia are linked to neurochemical abnormalities.

1990s
The Brain Enables Mind A number of fields including neuroscience, cognitive psychology, computer science, and neurology forge a new interdisciplinary field, cognitive neuroscience.

1990s
The Rise of Culture Scholars such as Hazel Markus, Shinobu Kitayama, Steven Heine, and Richard Nisbett are among those whose work helps increase recognition that culture shapes mental processes.

ogists tended to focus mainly on behaviour rather than on mental states. They did this in large part because they had few objective techniques for assessing the mind. The advent of technology to observe the working brain in action has allowed psychological scientists to study mental states such as consciousness, which in turn allows us to develop a fuller understanding of human behaviour.

What Are the Themes of Psychological Science?

For as long as people have been thinking, they've been thinking about other people. We have a strong desire to figure out other people, to understand their motives, thoughts, desires, intentions, moods, actions, and so on. We want to know why they remember some details and conveniently forget others, or why they engage in behaviours that are self-destructive. We want to know whether others are friends or foes, leaders or followers, likely to spurn us or fall in love with us. Our social interactions require us to use our impressions of others to categorize them, to make predictions about their intentions and actions. Essentially, people are constantly trying to figure out what makes other people tick. People who do this for a living are known as *psychological scientists.* They use the methods of science to understand how people think, feel, and act. This crucial point will be emphasized throughout this book. As you will read in Chapter 2, the scientific method refers to the use of objective, systematic procedures that lead to an accurate understanding of what is being studied. It involves careful observations of the natural world to examine how things work. Psychological scientists typically conduct experiments to understand human behaviour. For instance, they might study prejudice by looking at how interracial interactions affect people's thoughts, feelings, or behaviours compared with interactions between those of the same race. Through the application of the scientific method, psychological scientists have learned a great deal about the mind, brain, and behaviour.

Throughout this book we will highlight four major themes that have defined this field in the past and present. These themes guide and direct the way psychological scientists study the mind, brain, and behaviour.

The Principles of Psychological Science Are Cumulative

The first theme is that research on mind, brain, and behaviour has accumulated over time to produce the *principles* of psychological science. Throughout history there have been occasional leaps in scientific knowledge, such as the recognition of gravitational forces, the discovery of penicillin, and the recent mapping of the human genome. But more typically, science progresses in smaller, incremental steps, as knowledge accumulates based on systematic study of questions raised by what is already known. In this way, science builds on the foundation of shared knowledge. Understanding the human mind and behavior is no different.

So that you can appreciate the questions that drive contemporary research, we will describe the basic principles that make up the foundation of psychological knowledge. In doing so, we focus on what is *known* in psychological science. For example, the behavioural properties of memory are quite well known, and today no psychologist has to demonstrate that it is easier to *recognize* old information than it is to *recall* old information. We've known that for more than half a century; it is one of the many principles we will use in this book. As scientists expand the outer edge of knowledge, their search for the unknown is still rooted in the basic principles of psychological science. To demonstrate this, we will describe cutting-edge research that is building on these principles, focusing on the research that we believe will establish the future foundations of psychological science. In Chapter 7 you will see that our modern understanding of memory builds on findings that have accumulated over the past century.

One consequence of focusing on principles is that it can at times make things seem simpler than they really are. Most psychological phenomena involve complexities that we do not have the space to discuss. Complexity is inherent in science as ideas and theories are modified by new information that describes the conditions under which phenomena exist. Consider gravity, a basic force that has been recognized for hundreds of years. Gravity does not work in a uniform fashion; its force depends on the properties of the earth itself. On earth if you drop an apple, it falls to the ground. That is the principle. Likewise, when we say that recognition is easier than recall, there are certainly conditions under which this is not true. But these complexities are of greater interest to the scientific researcher than to the student learning the material for the first time. So if you enjoy your introductory psychology course, and this book, then by all means consider taking advanced psychology courses in which you will explore the complexities of the field in greater depth. For now, it is time to focus on the principles.

A New Biological Revolution Is Energizing Research

The second theme is this: A new biological revolution of profound significance is in progress at the dawn of the twenty-first century, bringing with it a deeper understanding of the human mind and behaviour. Since the time of Aristotle, philosophers and other scholars have asked questions about basic psychological phenomena, but they lacked the methods to examine scientifically many of these fundamental questions, such as *What is consciousness? Where does emotion come from, and how does it affect cognitive processes? How are memories stored in the brain?* In the end, they were left with philosophical speculation. This began to change around the beginning of the twentieth century, and in the last 20 years or so there has been a tremendous growth in understanding of the biological bases of these mental activities. This interest in biology permeates all areas of psychological science, from locating the neural, or brain, correlates of how we identify friends to discovering the neurochemical problems that produce various psychological disorders. Three developments have set the stage for a new biological revolution contributing to our understanding psychological phenomena.

BRAIN CHEMISTRY The first major development in the biological revolution is an understanding of brain chemistry. The brain works through the actions of chemicals known as *neurotransmitters,* which communicate messages between nerve cells. Over the last 30 years psychological scientists have made tremendous progress in identifying these chemicals and their functions. Although it was long believed that no more than a handful of neurotransmitters were involved in brain activity, it is now known that hundreds of different substances play critical roles in mental activity and behaviour. For instance, we now know that people have better memories for events that happen when they are aroused than when they are calm, because chemicals involved in responding to stimuli influence the neural mechanisms involved in memory. Understanding the chemical processes of the brain has provided many new insights into mental activity and behaviour and has also been useful for developing treatments to help people with various psychological disorders.

THE HUMAN GENOME The second major development in the biological revolution is the enormous progress in understanding the influence of genetic processes. Not only have scientists been able to map out the *human genome,* the basic genetic code or blueprint for the human body, but they have also developed various techniques that allow them to discover the link between genes and behaviour. For instance, to study the effects of a gene on memory, researchers have been able to breed mice that either lack a specific gene or have new genes inserted. These mice subsequently show either impaired memory or improved memory, respectively. By identifying the genes that are involved in memory, researchers may soon be able to develop therapies based on genetic manipulation that will assist people who have memory problems, such as those who have Alzheimer's disease.

What role does genetics play in mind and behaviour?

Of course, the idea that a single gene causes a specific behaviour is overly simplistic. Almost all psychological and biological activity is affected by the actions of multiple genes. No one specific gene is solely responsible for memory, or racist attitudes, or shyness. Nonetheless, evidence is accumulating that genes are involved in many of these various processes. You will see in Chapter 3 that many physical and mental characteristics are to some degree inherited. You will also see that scientists are beginning to understand how situational contexts, such as the presence of other people, influence how genes are expressed and therefore how they affect behaviour. It is the expression of genes that helps give rise to mind and behaviour. Mapping the human genome has provided scientists with the foundational knowledge to study how specific genes affect thoughts, actions, feelings, and various disorders. Although many of the fantastic possibilities for correcting genetic defects are decades away, the methods used by scientists to study the influence of genetic processes have provided fresh insights into mental activity.

WATCHING THE WORKING BRAIN The development of methods for assessing the brain in action has provided the third major impetus to the biological revolution in psychology. The principles of how cells operate in the brain to influence behaviour have been studied with increased effectiveness for more than a century, but only since the late 1980s have researchers been able to study the working brain as it performs its vital psychological functions. Using the methods of brain science, or *neuroscience,* psychological scientists have been able to address some of the most central questions of human experience, such as how different brain regions interact to produce perceptual experience, how various types of memory are similar or different, and how conscious experience involves changes in brain activity.

Knowing where in the brain something happens doesn't by itself tell you very much, but knowing that there are consistent patterns of brain activation associated with specific mental tasks provides evidence that the two are connected. Indeed, for more than 100 years scientists have disagreed about whether psychological processes are located in specific parts of the brain or distributed throughout the brain. We now know that there is some *localization* of function, but that many different brain regions participate to produce behaviour and mental activity. The use of brain imaging has allowed psychological scientists to make tremendous strides in understanding mental states such as those involved in paying full attention to one thing while ignoring

Great moments in evolution

other things, the study of which has been central to psychology for more than a century (Posner & DiGirolamo, 2000). The progress in understanding the neural basis of mental life has been rapid and dramatic. This new knowledge is being used throughout psychology; for example, as demonstrated in the opening paragraphs of this chapter, social psychologists have been able to identify and better understand the neural correlates of racism. The 1990s were labeled the decade of the brain, for good reason.

The Mind Is Adaptive

evolutionary theory In psychological science, a theory that emphasizes the inherited, adaptive value of behaviour and mental activity throughout the entire history of a species.

natural selection Darwin's theory that those who inherit characteristics that help them adapt to their particular environment have a selective advantage over those who do not.

adaptations In evolutionary theory, the physical characteristics, skills, or abilities that increase the chances of reproduction or survival and are therefore likely to be passed along to future generations.

The third theme of psychological science is that the mind has been shaped by evolution. Humans are products of both biological and cultural evolution, each of which exerts an influence over how people think and behave. From an **evolutionary theory** perspective, the brain is an organ that has evolved over millions of years to solve problems related to survival and reproduction. During the course of human evolution, those ancestors who were able to solve survival problems and adapt to their environments were most likely to reproduce and pass along their genes. That is, those who inherited characteristics that helped them survive in their particular environments had a selective advantage over those who did not, which is the basis of the process of **natural selection**. Random gene mutations endowed some of our ancestors with physical characteristics, skills, and abilities, known as **adaptations**, that increased their chances of survival and reproduction, which means that their genes were passed along to future generations. Of course, if the environment changes, what once was adaptive might become maladaptive. The ability to store fat in the body may have been adaptive when the food supply was scarce, but it may be maladaptive when food is abundant. Further complexities in the process of natural selection are discussed in Chapter 3.

Why is evolutionary theory important to understanding mental activity?

Modern evolutionary theory has driven the field of biology for years. Only recently, however, have psychologists taken it up. Rather than being a specific area of scientific inquiry, evolutionary theory represents a way of thinking that can be used to understand many different aspects of mind and behaviour (Buss, 1999). Three aspects of evolutionary theory are particularly helpful in this regard.

SOLVING ADAPTIVE PROBLEMS Over the last five million years that humans have been evolving, adaptive behaviours have been built into our bodies and brains. A corollary to this notion is the idea that the body contains specialized mechanisms that have evolved to solve problems that required adaptation. For instance, a mechanism that produces calluses has evolved to protect the skin from the abuses of hard work, and these calluses are useful when humans need to engage in physical labour to survive. Likewise, the brain has evolved specialized circuits or structures that solve adaptive problems (Cosmides & Tooby, 2001).

Evolutionary theory is especially useful for thinking about adaptive problems that occur regularly and have the potential to affect whether one survives and reproduces, such as mechanisms for eating, sex, language and communication, emotions, and aggression. Accordingly, the evolutionary approach is particularly relevant to social behaviour. For decades we have known that situational and cultural contexts are influential in the development of social behaviour(s) and attitudes. Now evidence is accumulating that many of these behaviours can also be considered adaptive solutions to recurring human problems. For example, humans have a fundamental need to belong to their group, and therefore behaviours that lead to possible social exclusion are discouraged in all societies (Baumeister & Leary, 1995). People who lie, cheat, or steal drain group resources and thereby possibly decrease survival and

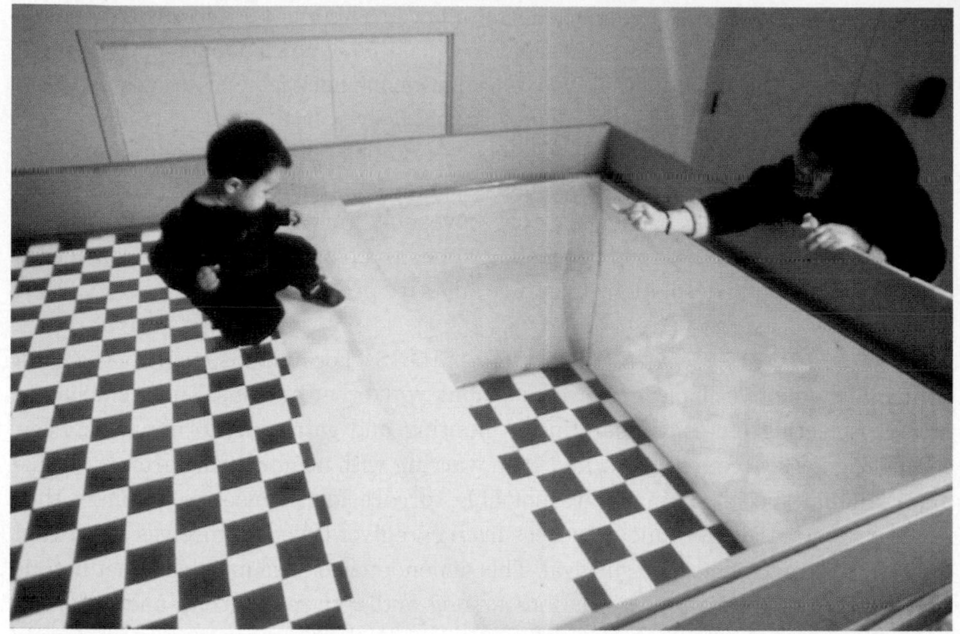

1.3 Infants and Height Although there is a glass covering the visual cliff, infants will not crawl onto the glass, even if encouraged by their mothers. Infants become wary of heights at about the same age they learn to crawl.

reproduction for other group members. Some evolutionary psychologists believe that humans have "cheater detectors" that are on the lookout for this sort of behaviour in others (Cosmides & Tooby, 2000).

Evolutionary theory can be applied to many topics in psychology. The capacity to see well, remember where food was abundant, recognize dangerous objects, understand the basic laws of physics (such as the effects of gravity if one walks off a cliff), and so on were also critical to survival and therefore can be considered from an evolutionary perspective. According to evolutionary theory, solutions to these adaptive problems are built into the brain and therefore require no special training. Young infants develop a fear of heights at about the same time they learn to crawl, even though they have little personal experience with heights or gravity (Figure 1.3). Such built-in mechanisms assist in solving recurring problems that faced our ancestors over the course of human evolution.

MODERN MINDS IN STONE AGE SKULLS According to evolutionary theory, we must seek to understand the challenges that faced our early ancestors to understand much of our current behaviour, whether adaptive or maladaptive. To provide perspective on this, consider the following: The human brain evolved slowly over millions of years, and many of the adaptive problems faced by early humans no longer exist, or at least no longer present the threat they once did. Humans began evolving about five million years ago, but modern humans (*Homo sapiens*) can be traced back only about 100,000 years, to the Pleistocene era. If we compare the history of the earth to a 24-hour timescale, the arrival of humans has occurred in the last 30 seconds. The fact that the human brain adapted to accommodate the needs of Pleistocene hunter-gatherers means we should be looking at what life was like then, and look to understand how the brain works within the context of the environmental pressures that the brains of Pleistocene-era humans faced. For instance, people like sweet foods, especially those that are high in fat. These foods are highly caloric and eating them would have provided substantial survival value in prehistoric times. In other words, a preference for fatty-sweet foods would have been adaptive. Today, many societies have an abundance of foods that are high in fat and sugar. That we still enjoy them and eat them, sometimes to excess, may now be maladaptive in that it can produce obesity. Nonetheless, our evolutionary heritage encourages us to eat foods that had survival value when food was relatively scarce.

Are all behaviours adaptive?

Many of our current behaviours, of course, do not reflect our evolutionary heritage. Reading books, driving cars, using computers, talking on telephones, and watching television are behaviours that have only very recently become part of human experience. Rather than being adaptations, such behaviours can be considered *by-products* of adaptive solutions to earlier adaptive problems.

CULTURE PROVIDES ADAPTIVE SOLUTIONS For humans, many of the most demanding adaptive challenges involve dealing with other humans. These challenges include selecting mates, cooperating in hunting and gathering, forming alliances, competing for scarce resources, and even warring with neighbouring groups. Unlike many animal species, humans are not able to care for themselves at birth; they require substantial effort and resources from caregivers, who themselves are reliant on other group members for survival. This dependency on group living is not unique to humans, but the nature of relations among and between ingroup and outgroup members is especially complex in human societies. The complexity of living in groups gives rise to **culture**, which refers to the beliefs, values, rules, norms, and customs that exist within a group of people who share a common language and environment, the assumption being that the various aspects of culture are transmitted from one generation to the next through learning. For instance, our musical and food preferences, our ways of expressing emotion, our tolerance of body odours, and so on are strongly affected by the culture in which we are raised. Many of these cultural rules reflect adaptive solutions that have been worked out by previous generations. Successful methods of hunting and gathering, beliefs about supernatural forces that shape behaviour, and the capacity to communicate with other group members reflect knowledge that is handed down through the generations, helping future generations solve recurring adaptive problems.

culture The beliefs, values, rules, and customs that exist within a group of people who share a common language and environment and that are transmitted through learning from one generation to the next.

In contrast to biological evolution in humans, which has taken place over several million years, cultural evolution has occurred over a much shorter period, with the most dramatic changes coming in the last few thousand years. Although humans have changed only modestly in physical terms in the last two or three thousand years, they have changed profoundly in terms of how they live together; even within the last century there have been dramatic changes in how human societies interact. The flows of peoples, commodities, and financial instruments, often referred to as globalization, have increased in velocity and scale in the past century in ways that were unimaginable nearly 500 years ago, when explorers from Europe and China first set out on their global journeys. Some commentators have suggested that we may indeed be witnessing the rise of a global culture as a result of this newest form of globalization. Most recently, the Internet has created a worldwide network of humans, in essence creating a new form of culture with its own rules, values, and customs.

Over the past decade there has been a growing recognition that culture plays a foundational role in shaping how people view and reason about the world around them and that different cultures possess strikingly different minds. Research by the social psychologist Richard Nisbett has demonstrated that people from most Asian countries have a view of the world that is quite different from the worldview of people from most European and North American countries. The work of Nisbett has suggested that Westerners tend to miss the forest for the trees, focusing on single elements in the forefront, whereas those from Eastern cultures tend to overlook single trees, focusing on the entire forest in the background. Nisbett and his colleagues (2001) have documented numerous ways in which the thinking styles of these groups differ. The general pattern is that Westerners are much more analytic, breaking complex ideas into simpler components, categorizing information, and using logic and rules to explain behaviour, whereas Easterners tend to be more holistic in their thinking, seeing everything in front of them as an inherently complicated whole, with all elements affecting all other elements.

In his book *The Geography of Thought,* Nisbett (2003) argues that these essential cultural differences date back to ancient Greek and Chinese societies, from roughly the eighth through the third centuries B.C.E., with the Greeks focusing on personal freedom, logic, and debate and the Chinese focusing on harmonious relationships with family and other villagers. To this day, the general findings from cultural psychologists are that Westerners tend to be "independent" and autonomous, stressing their individuality, whereas Easterners tend to be more "interdependent" on each other, stressing their sense of being part of a collective. Moreover, Canadian psychologist Steven Heine (2003) has found that Westerners are more likely to emphasize their personal strengths whereas Easterners are more likely to emphasize their need for self-improvement.

How does culture shape the way we interact with the world?

The culture in which we live shapes many aspects of daily life. Pause for a moment and think about the following questions: How do we decide what is most important in our lives? How do we want to relate to members of our families? Our friends? Our colleagues at work? How should we spend our leisure time? How do we define ourselves in relationship to our own cultures—or across cultures? For instance, the increased participation of women in the work force has changed the nature of contemporary Western culture in numerous ways, from a fundamental change in how women are viewed to more practical changes such as people marrying and having children later in life, a greater number of children in day care, and a greater reliance on convenient, fast foods. Culture shapes beliefs and values, such as the extent to which people should emphasize their own interests versus the interests of their families or the group. As we've already seen, this is magnified even more when we compare phenomena across cultures. Cultural rules are learned as *norms,* which specify how people ought to behave in different contexts, such as not laughing at funerals and keeping quiet in libraries. There are also material aspects of culture, such as media, technology, health care, and transportation. Many of us find it hard to imagine life without cars, televisions, and telephones, yet we recognize that each of these inventions has changed the fundamental ways in which people interact. Psychological scientists have played a significant role in providing a better understanding of the relationship between culture and behaviour, showing that it's important to consider behavioural phenomena in their cultural context.

The accumulating evidence indicates that the human mind is adaptive in both biological and cultural terms, providing solutions to survival and reproductive challenges as well as a strong framework for a shared social understanding of how the world works. A true understanding of mind, brain, and behaviour requires careful consideration of the evolution of biology and culture. We will focus on how culture is studied in psychological science in Chapter 16.

Psychological Science Crosses Levels of Analysis

The fourth theme of psychological science is that the mind and behaviour can be studied on many levels of analysis. As we'll see throughout this book, psychological researchers working on the same problem often ask different questions and work at

CATEGORY	LEVELS	WHAT IS STUDIED?
Social	Cultural	Norms, beliefs, values, symbols, ethnicity
	Interpersonal	Groups, relationships, persuasion, influence, workplace
Individual	Individual Differences	Personality, gender, developmental age groups, self-concept
	Perception and Cognition	Thinking, decision making, language, memory, seeing, hearing
	Behaviour	Observable actions, responses, physical movements
Biological	Brain Systems	Neuroanatomy, animal research, brain imaging
	Neurochemical	Neurotransmitters and hormones, animal studies, drug studies
	Genetic	Gene mechanisms, heritability, twin and adoption studies

1.4 Categories of Analysis Psychological scientists study three basic categories of behaviour and mental life: social, individual, and biological.

different levels. Indeed, many times psychologists collaborate with researchers from other sciences, such as biology, computer science, physics, anthropology, and sociology. This *interdisciplinary* effort shares the common goal of understanding the mind and behaviour. Using the principles of the scientific method, psychological scientists must break down behavioural phenomena into their component parts in order to understand them. Only then can we begin to develop a full understanding of how the biological, individual, and social components influence our specific behaviours.

How can the same problem be studied at multiple levels?

Three broadly defined categories of analysis will be evident throughout this book, reflecting research methods that emphasize the various aspects of studying mind and behaviour. These categories can be further broken down in terms of levels of analysis; the most common levels are shown in Figure 1.4. *Social* aspects involve an examination of how cultural and social contexts affect the ways people interact and influence each other. *Individual* aspects concern individual differences in personality and mental processes that concern how we perceive and know our worlds. The *biological* aspects deal with how the physical body contributes to mind and behaviour, such as the neurochemical and genetic processes that occur with the body and brain. Within these broader categories, psychological scientists typically focus on a certain level of analysis, such as at the cultural or neurochemical level (see Figure 1.4).

To understand how different types of psychological scientists work at each of these levels, let's look at the study of music. What type of music do you like? The enjoyment of music is a fascinating aspect of human life. Music is present in many aspects of daily life and is important to most people (Renfrow & Gosling, 2003). There are many questions to be asked about the musical experience, such as how preferences vary among individuals and across cultures, how music affects emotional states and thought processes, and even how the brain perceives sound as music rather than noise. For instance, suppose you wanted to know how often people listen to music. One survey of 2,465 English adolescents found that they reported listening to 2.45 hours of music each day, and they said they did so because it allowed them to project a desired "image" to the world and helped them to satisfy emotional needs (North, Hargreaves, & O'Neill, 2000). In a study conducted at the *behavioural* level of analysis, researchers used a computerized tape recorder to sample students' activities throughout the day and found that they listened to music during 14 percent of their waking hours (Mehl & Pennebaker, 2003)

What effect does listening to music have on people? At the *cognitive* level of analysis, researchers have used laboratory experiments to study the effects of music on mood, memory, decision making, and a variety of other mental states and processes

(Krumhansl, 2002). For instance, "Russia under the Mongolian Yoke" from Prokofiev's *Field of the Dead,* played at half speed, reliably puts people into negative moods. Mood may be affected not only by the tempo of the music, but also by whether it is in major or minor mode. At least for Western music, major mode is typically associated with positive moods and minor mode is associated with sad moods. This emotional response appears to be learned rather than innate, since very young children do not discriminate between modes, but by age seven or eight, children can reliably distinguish the mood effects of major and minor modes (Gregory, Worrall, & Sarge, 1996). Why does music affect mood? One way to study this is to go to the *neurochemical* level of analysis and look at the effect of music on brain chemistry. Indeed, it appears that pleasant music may be associated with increased activation of one brain chemical, serotonin, that is known to be relevant to mood (Evers & Suhr, 2000). These mood effects of music can even change how people behave. In one field study, shoppers were more likely to purchase expensive wine while listening to classical music than while listening to Top 40 music (Areni & Kim, 1993).

Researchers can also examine the effects of music at a *brain-systems* level of analysis (Peretz & Zatorre, 2005). Does perceiving music use the same brain circuits as, say, perceiving the sound of cars or spoken language? It turns out that the processing of musical information is similar to general auditory processing, but it likely also uses different brain mechanims. Case reports of patients with certain types of brain injury indicate that some lost the ability to hear tones and melody but not speech or environmental sounds. One 35-year-old woman who had brain damage lost the ability to recognize even familiar tunes, a condition known as *amusia,* even though other aspects of her memory and language systems were intact (Peretz, 1996). There are also case studies of so-called musical savants, who can play piano concertos by ear or sing elaborate songs but who otherwise show gross intellectual impairments, such as being unable to communicate with others or even respond to their own names. Such findings suggest that there are specific brain regions responsible for the processing of music. Researchers at McGill University are using methods such as brain imaging to understand which parts of the brain are involved in perceiving and enjoying music, and how the perception of music might differ from the perception of other sounds (Blood & Zatorre, 2001).

Finally, some researchers have examined aspects of music at the *genetic* level of analyis. One study of attitudes in 800 British twins (Martin et al., 1986) revealed that about half the variability in the liking of jazz is determined by genetic influence. Moreover, there are anecdotal reports of musical expertise running in families, such as the Bach family, which might be due to an inborn ability to develop absolute pitch, also called perfect pitch. The evidence suggests that the ability to identify or produce the pitch of a sound without any reference point appears to depend both on genes and on musical experience during childhood (Levitin & Rogers, 2005; Zatorre, 2003). Only some people develop absolute pitch, and people who do not receive formal music training before age 9 or 10 are never able to develop it.

As these examples show, researchers are crossing different levels of analysis to gain a greater understanding of music. Throughout the text, we will include more examples of this kind of effort. Of course, each of the levels can be studied independent of the others. Throughout the history of psychology, this has been the favoured approach. Only recently have researchers started to explain behaviour in terms of several levels of analysis. It is this crossing of the levels of analysis that many modern-day psychological scientists find so captivating because it helps provide an increasingly complete picture of important behavioural and mental processes.

REVIEWING the Principles | What Are the Themes of Psychological Science?

Psychological science is the study of mind, brain, and behaviour. Four themes characterize psychological science. (1) It is cumulative, in that principles are established on the basis of incremental advances in knowledge obtained through research. (2) A biological revolution has been energizing psychological research. Increasing knowledge of the neurochemistry of mental disorders, the mapping of the human genome, and the invention of imaging technologies that allow researchers to observe the working brain in action have provided psychological scientists with the methods to examine how the brain enables the mind. (3) Psychological science has also been heavily influenced in recent years by evolutionary psychology, which argues that the brain has evolved to solve adaptive problems. (4) Finally, although psychological scientists share the common goal of understanding mind, brain, and behaviour, they do so by focusing on the same problems at different levels of analysis.

What Are the Intellectual Origins of Psychology?

Psychology is a young science that addresses questions that have challenged great minds for millennia. Many of these issues reflect long-standing philosophical questions about the nature of human experience. Psychology's roots are in philosophy and medicine, and many of the earliest psychologists were trained in one of these disciplines. In this section we will consider some of the "grand" questions and issues that have shaped psychological debate over centuries.

The Nature–Nurture Debate Considers the Impact of Biology and Environment

nature-nurture debate The arguments concerning whether psychological characteristics are biologically innate or acquired through education, experience, and culture.

How do we disentangle nature and nurture?

From the time of the ancient Greeks, there has been a debate about whether psychological characteristics are more due to *nature* or *nurture,* that is, whether they are biologically innate or acquired through education, experience, or culture. The **nature-nurture debate** has taken one form or another throughout psychology's history and will likely continue as researchers explore how thoughts, feelings, and behaviours are influenced by genes and by the culture or society in which one lives. It is now widely recognized that both nature and nurture are important, but more significant, recent advances in scientific knowledge have allowed researchers to specify *when* either nature or nurture is important, as well as how they interact. It is the relative importance of nature and nurture in determining mind and behaviour that captivates the interest of psychological scientists.

As an example of the changing influences of nature and nurture, consider just two mental disorders—schizophrenia and bipolar disorder (you will read much more about them in Chapter 13). *Schizophrenia* is a disorder in which people have unusual thoughts, such as believing they are God, or experience unusual sensations, such as hearing voices. In *bipolar disorder* a person has dramatic mood swings, from feeling extremely sad (depressed) to feeling euphoric (manic). Prior to the 1950s, it was generally believed that these two mental disorders, among others, resulted from bad parenting or other environmental circumstances—that is, the causes were all nurture. But in the late 1950s and 1960s a variety of drugs were discovered that could alleviate the symptoms of these disorders; more recent research showed that these conditions are also heritable. Psychological scientists now believe that many mental disorders result as much from the way the brain is "wired" (nature) as from the way people are treated (nurture). Rapid advancements in understanding the biological basis of mental disorders have led to effective treatments that allow people to live normal lives. So is it all nature rather than nurture? Of course not. Both schizophrenia and bipolar disorder are more likely in certain environments, suggesting that they can be triggered by the situation. Many mental disorders result from events that happen in people's lives, such as combat soldiers who develop *posttraumatic stress disorder* (*PTSD*), in which people have intrusive and unwanted memories of their traumatic experiences. But recent research also indicates that some people inherit a genetic predisposition to develop PTSD, so nurture activates nature. Thus, nature and nurture are tightly interwoven and inseparable. Moreover, the social environment also plays an important role in whether treatment is successful; negative comments from family members predict poor outcomes. Psychological science depends on understanding both the genetic basis of human nature and the environment that shapes it.

The Mind-Body Problem Has Challenged Philosophers and Psychologists

mind-body problem A fundamental psychological issue that considers whether mind and body are separate and distinct or whether the mind is simply the subjective experience of the physical brain.

Close your eyes and think about yourself for a second. Where do your thoughts reside? If you are like most people, you have a subjective sense that your mind is floating somewhere around your head, perhaps a few inches inside your skull, or perhaps even a few inches above or in front of your forehead. But why do you feel like your mind is in your head? The mind has been viewed throughout history as residing in many organs of the body, especially the liver or the heart. What is the relationship between the mental activity of your mind and the physical workings of your body? The **mind-body problem** is perhaps the quintessential psychological issue: whether the mind and body are separate and distinct, or whether the mind is simply the subjective experience of the physical brain.

For most of human history, scholars have believed that the mind and body are separate entities, with the mind very much in control of the body. This belief was held, in part, because of strong theological beliefs that the existence of a *divine and immortal soul* is what separates humans from animals. Even early theorists who challenged church doctrine were careful to avoid being too controversial. Leonardo da Vinci conducted experiments around 1500 to make his anatomical drawings more accurate, which offended the church because they violated the presumed sanctity of the human body. His dissections led him to many conclusions about the workings of the brain, including the idea that all sensory messages, such as vision, touch, and smell, arrived at one location in the brain (Figure 1.5), the *sensus communis,* which he believed to be the home of thought and judgment. (This is possibly why we call using good judgment *common sense* [Blakemore, 1983].)

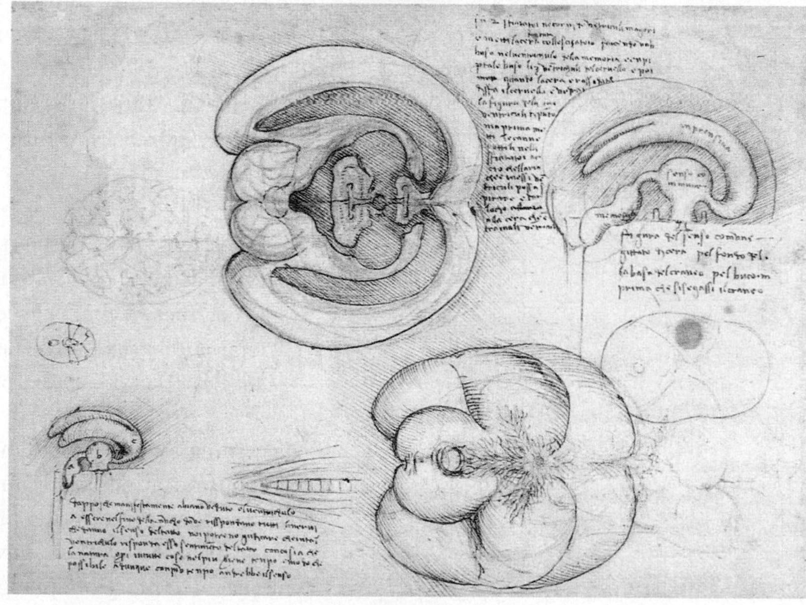

1.5 Leonardo da Vinci and the Brain This drawing by Leonardo da Vinci dates from around 1506. He used a wax cast to study the brain. He found that the various sensory images arrived in the middle region of the brain, which he referred to as the *sensus communis,* or common sense.

dualism The philosophical idea that the mind exists separately from the physical body.

It was René Descartes, the great French philosopher, who promoted the first influential theory that mind and body were separate yet intertwined (the theory known as **dualism**). The notion that the mind and body were separate was not novel, of course, but how Descartes connected them was at the time quite radical. The body, he argued, was nothing more than an organic machine, governed by "reflex," which Descartes defined as a "unit of mechanical, predictable, deterministic action" (Figure 1.6). For Descartes, many mental functions, such as memory and imagination, were the result of bodily functions. Linking some mental states with the body was a fundamental departure from the earlier views of dualism, in which all mental states were separate from bodily functions. In keeping with prevailing religious beliefs, however, Descartes concluded that the rational mind, which controlled volitional action, was divine and separate from the body. Thus, his view of dualism kept the distinction between mind and body, but he assigned to the body many of the mental functions previously considered the sovereign domain of the mind.

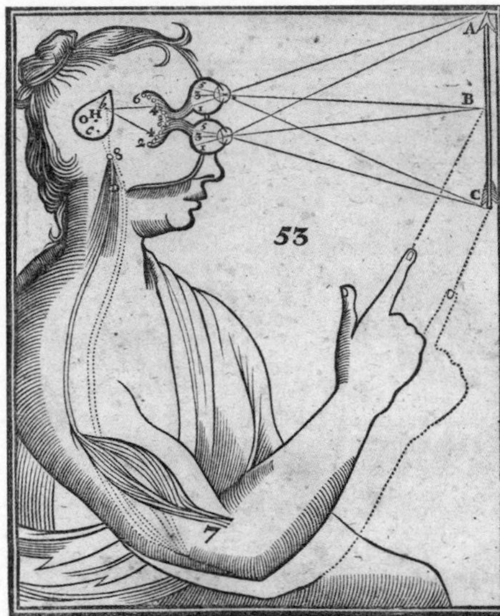

1.6 Descartes's "Reflex" Theory A woodcut illustrating Descartes's "reflex" theory of biological function.

Descartes's most radical idea was to suggest that although the mind could affect the body, the body could also affect the mind. For instance, he believed that passions such as love, hate, and sadness arose from the body and influenced mental states, although the body acted on these passions through its own mechanisms. In this way, Descartes brought mind and body closer together by focusing on their interactions.

Evolutionary Theory Introduces Natural Selection

One of the major intellectual events that shaped the future of psychological science was the publication of Charles Darwin's *On the Origin of Species* in 1859. In the book he outlined a theory of evolution that relied on the process of natural selection (as described earlier). Earlier naturalists and philosophers, including his own grandfather Erasmus Darwin, had all discussed the possibility that species might evolve, but it wasn't until Charles Darwin came along that the mechanism of evolution became clear. How did Darwin develop his theory of natural selection?

Charles Darwin (Figure 1.7) attended Cambridge and traveled as a naturalist aboard the *Beagle* from 1832 to 1837, collecting information about finches on the Galápagos Islands. When he later analyzed this information, he discovered that each island had slightly different species of finches (Figure 1.8). It seemed clear to him that these different finches must have descended from the same species, but he wondered what could account for the small variations. His famous notebooks from this era trace his thinking about how this must have occurred. He found a suitable explanation in Thomas Malthus's essay on population:

1.7 Charles Darwin Darwin's theory of natural selection has had a huge impact on how psychologists think about the mind. This portrait is reported to be his favorite picture of himself.

> I happened to read for amusement "Malthus on population," and being well prepared to appreciate the struggle for existence which everywhere goes on from long-continued observation of the habits of animals and plants, it at once struck me that under these circumstances favourable variations would tend to be preserved and unfavourable ones to be destroyed. The result of this would be the formation of a new species. (Darwin, 1887, p. 68)

Darwin called this mechanism for evolution *natural selection,* the process by which random mutations in organisms that are adaptive are passed along and mutations that hinder reproduction are not. Thus as species struggle to survive, those that are better adapted to their environment will leave more offspring, and those offspring will produce more offspring, and so on (i.e., the notion of *survival of the fittest*).

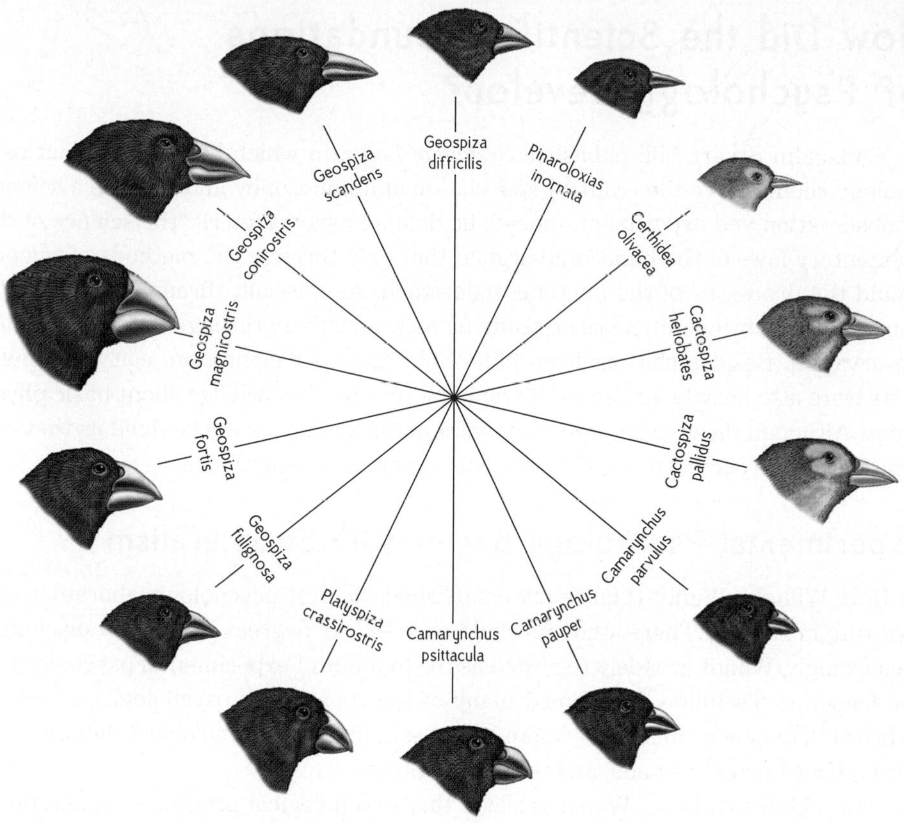

1.8 Finches on the Galápagos Islands The variation in these different species of finches on the Galápagos Islands was powerful evidence for some kind of selection process working on a common ancestor.

A further implication of Darwin's theory was that inheritable *individual differences* provide the basis of evolutionary development. This idea was seized on by his cousin Francis Galton, who proposed that some differences were psychological in nature (e.g., intelligence), and that they could be measured and tested. The *mental testing movement* followed in Galton's wake. Ultimately, the idea of natural selection has had a profound impact on science, philosophy, and society.

REVIEWING the Principles | What Are the Intellectual Origins of Psychology?

The origins of psychological science can be found in the major philosophical questions that have endured for centuries. For instance, the nature-nurture debate involves determining the extent to which mind and behaviour are predetermined by biology or are developed and shaped by environment. The mind-body problem tackles the question of how mental activity is related to brain functioning. Darwin's theory of evolution set the stage for a new understanding of the origins of mind and behaviour.

How Did the Scientific Foundations of Psychology Develop?

How can we study mental activity?

In 1843, John Stuart Mill published *System of Logic,* in which he declared that psychology should leave the realm of speculation and philosophy and become a science of observation and experiment. Indeed, he defined psychology as "the science of the elementary laws of the mind" and argued that only through the methods of science would the processes of the mind be understood. As a result, throughout the 1800s there was a growing emphasis on studying mental activity through careful scientific observation. Besides shifting from philosophy to experimentation, early psychologists were also heavily influenced by rapid increases in knowledge about basic physiology. Although these two approaches were not originally as intertwined as they are now, both were central to the development of psychological science.

introspection A systematic examination of subjective mental experiences that requires people to inspect and report on the content of their thoughts.

structuralism An approach to psychology based on the idea that conscious experience can be broken down into its basic underlying components or elements.

Experimental Psychology Begins with Structuralism

In 1879, Wilhelm Wundt (Figure 1.9) established the first psychology laboratory and institute in Leipzig. There students could earn higher degrees in the new discipline. Accordingly, Wundt is widely regarded as the founder of experimental psychology as an academic discipline. He trained many of the great early psychologists, such as Edward Titchener, many of whom then established psychological laboratories throughout Europe, Canada, and the United States.

What are the problems with the method of introspection?

Through his training, Wundt realized that psychological processes, as the products of physiological actions in the brain, take time to occur. Thus, to study the mind he would present a subject with two psychological tasks: one that was simple and one that was more complex. He would then measure the speed at which subjects completed the tasks. By subtracting the easier task from the complex task, Wundt could infer how much time a particular mental event took to occur. Wundt was not satisfied with simply studying mental reaction times: he wanted to measure conscious experiences. To do this, he developed the method of **introspection**, a systematic examination of subjective mental experiences that required people to inspect and report on the content of their thoughts, such as describing the "blueness" of the sky. Using introspection, Wundt asked people to compare their subjective experiences as they contemplated a series of objects, such as by stating which one they found more pleasant.

1.9 Wilhelm Wundt Wundt (far right), the founder of modern experimental psychology, at work with his collaborators in his later years.

Edward Titchener used methods such as introspection to develop a new school of thought that became known as *structuralism.* The basic idea of **structuralism** is that conscious experience can be studied when it is broken down into its underlying components or elements. Just as when you know the ingredients and recipe you can bake a cake, Titchener believed that understanding the basic elements of consciousness would provide the scientific basis for understanding the mind. Titchener argued that one could take a stimulus such as a musical tone and by introspection analyze its "quality," "intensity," "duration," and "clarity." Wundt ultimately rejected this particular use of introspection, but Titchener relied on the method throughout his career. The problem with this approach is that experience is subjective. Each person brings to introspection a unique perceptual system, and it is difficult to determine whether subjects are using the criteria in a similar way. Accordingly, over the course of time introspection was largely abandoned in psychology. Nonetheless, Wundt, Titchener, and other structuralists were important because of their goal to develop a pure science of psychology with its own self-contained vocabulary and set of rules.

Functionalism Addresses the Purpose of Behaviour

One of the main critics of structuralism was William James (Figure 1.10), a brilliant scholar whose work has had an enormous, enduring impact on psychology. James abandoned a career in medicine to teach physiology at Harvard University in 1873. He was an excellent lecturer and was among the first professors at Harvard to openly welcome questions from students. His personal interests were more philosophical than physiological; he was captivated by the nature of conscious experience. He gave his first lecture on psychology in 1875, which he later quipped was also the first lecture on psychology he had ever heard. James's charm and brilliance were best expressed in his classic *Principles of Psychology,* published in 1890. Good genes perhaps contributed to his writing skills; his brother was the famous novelist Henry James. *Principles of Psychology* was an immediate hit with students and became the most influential book in the early history of psychology. To this day psychologists find rich delight in reading James's penetrating analysis of the human mind; it is amazing how many of his central ideas have held up over time.

1.10 William James James was highly influenced by Darwin and is credited with "naturalizing" the mind. His book *The Principles of Psychology* remains a classic.

James criticized structuralism's failure to capture the most important aspects of mental experience. He believed that the mind could not be broken down into its elements because the mind was much more complex than its elements. For instance, he noted that the mind consisted of a continuous series of thoughts that are ever-changing. This **stream of consciousness** could not be frozen in time, according to James, and so the strategies used by the structuralists were sterile and artificial. He likened psychologists who used the structural approach to someone who tried to understand a house by studying each of its bricks individually. More important to James was that the bricks worked together to form a house, and that a house has a particular function. James stressed that the important point was not the elements that made up the mind, but rather the mind's usefulness to people.

stream of consciousness A phrase coined by William James to describe one's continuous series of ever-changing thoughts.

functionalism An approach to psychology concerned with the adaptive purpose, or function, of mind and behaviour.

James was also heavily influenced by Darwinian thinking, and he argued that psychologists ought to examine the *functions* served by the mind. His approach to psychology, which became known as **functionalism**, was more concerned with how the mind operates than with what the mind contains. According to functionalism, the mind came into existence over the course of human evolution, and it works the way it does because it is useful for preserving life and passing along genes to future generations. In other words, it helps organisms adapt to environmental demands.

Many of the functionalists were concerned with applying psychological research to the real world. After all, if behaviour served a purpose, the purpose ought to be reflected in daily human life. Thus, for example, John Dewey, the great educator, tested functionalist theories in his classrooms, where he stressed teaching students according to how the mind processed information rather than simply through repetitive drill learning. This *progressive* approach to education emphasized divergent thinking and creativity rather than the rote learning of conventional knowledge that might be incorrect anyway (Hothersall, 1995). William James was also interested in applying the functional approach to the study of real-world phenomena, such as the nature of religious experience. Yet the broad-ranging subjects to which functionalism was applied led to criticism that it was not sufficiently rigorous, and therefore functionalism slowly lost steam as a movement within psychology. However, the functional approach has returned to psychological science within the past few decades, as more and more researchers consider the adaptiveness of the behaviours and mental processes they study.

Profiles in Psychological Science

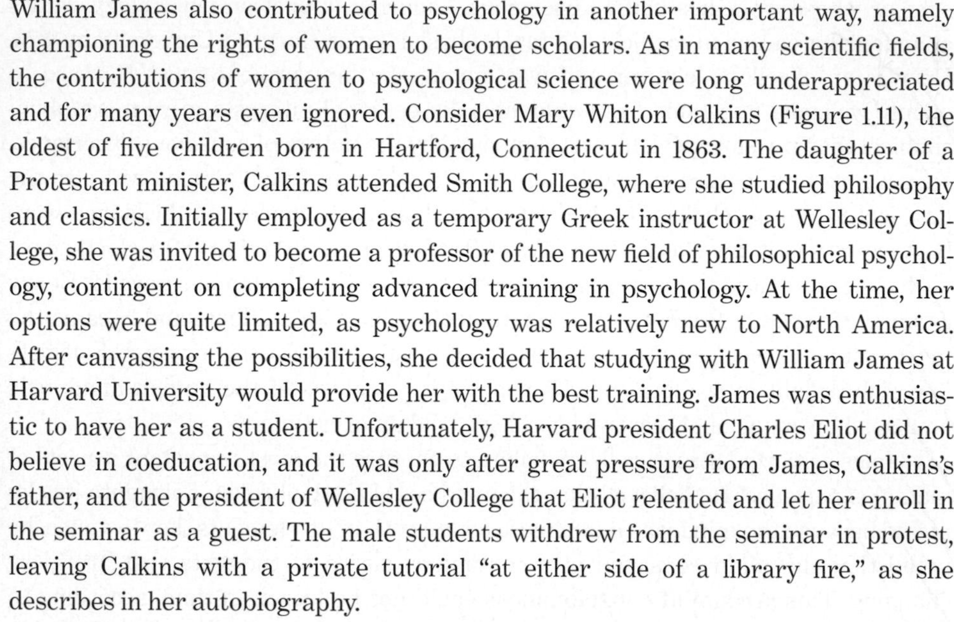

Early Women Pioneers

1.11 Mary Whiton Calkins Calkins was a pioneer of psychological science.

William James also contributed to psychology in another important way, namely championing the rights of women to become scholars. As in many scientific fields, the contributions of women to psychological science were long underappreciated and for many years even ignored. Consider Mary Whiton Calkins (Figure 1.11), the oldest of five children born in Hartford, Connecticut in 1863. The daughter of a Protestant minister, Calkins attended Smith College, where she studied philosophy and classics. Initially employed as a temporary Greek instructor at Wellesley College, she was invited to become a professor of the new field of philosophical psychology, contingent on completing advanced training in psychology. At the time, her options were quite limited, as psychology was relatively new to North America. After canvassing the possibilities, she decided that studying with William James at Harvard University would provide her with the best training. James was enthusiastic to have her as a student. Unfortunately, Harvard president Charles Eliot did not believe in coeducation, and it was only after great pressure from James, Calkins's father, and the president of Wellesley College that Eliot relented and let her enroll in the seminar as a guest. The male students withdrew from the seminar in protest, leaving Calkins with a private tutorial "at either side of a library fire," as she describes in her autobiography.

Calkins continued her psychological studies with a number of other mentors at Harvard, including the famous German psychophysicist Hugo Munsterberg. With him, she completed all of the requirements for a Ph.D., including scoring higher than her male classmates on the qualifying exam, but Harvard denied her the degree, offering instead a Ph.D. from Radcliffe, the women's school affiliated with Harvard. Calkins refused the degree, bristling at the unequal treatment she received, describing differential education of men and women as artificial and illogical. Efforts to have Harvard overturn its earlier decision continue to this day, as does Harvard's refusal to grant her the degree that she earned (Harvard did not grant a Ph.D. to a woman until 1963).

Mary Whiton Calkins had an enormously productive career as a professor at Wellesley College, including writing an introductory psychology textbook in 1901, being the first woman to set up a psychology laboratory, publishing more than 100 articles, and being elected in 1905 the first woman president of the American Psychological Association. Her major research interest focused on the "self," which she believed could be studied using the methods of science. She made a number of other important contributions to the early science of psychology, although in her later years she became somewhat disenchanted by the rise of behaviourism and its dismissal of concepts such as "self." Professor Calkins retired from Wellesley in 1927 and died in 1930.

Margaret Flay Washburn was the first woman to be officially granted a Ph.D. in psychology. She was awarded her doctorate in 1921 at Cornell University, where she studied with Edward Titchener. She spent most of her career at Vassar College, which she had attended as an undergraduate, and also set up a scholarship fund for women. The likely first Canadian woman to be awarded a Ph.D. for psychological research was Emma S. Baker at the University of Toronto. The philosophy department, however, granted her the degree; the psychology department was not yet established. Baker, who completed her work in the early 1900s, went on to become principal at a number of colleges, including Mt. Allison in New Brunswick (Green, 2002).

Gestalt theory A theory based on the idea that the whole of personal experience is different from simply the sum of its constituent elements.

Of course, many other women contributed to the history of psychology, and thankfully their contributions are now readily acknowledged. Today, women make up approximately three-quarters of psychology majors and nearly half of all psychology doctorates in the work force (Frincke & Pate, 2004). ■

1.12 What Do You See? If you look carefully, you will see that these fragments make up a picture of a dog.

Gestalt Psychology Emphasizes Patterns and Context in Learning

Another of the schools of thought that arose in opposition to structuralism was the *Gestalt* school, founded by Max Wertheimer in 1912 and further expanded by Wolfgang Köhler, among others. According to **Gestalt theory**, the whole of personal experience is different from simply the sum of its constituent elements, or *the whole is different from the sum of the parts.* If you show people a triangle, they see a triangle, not three lines on a piece of paper, as would be the case for the trained observers in one of Titchener's structuralist experiments. So, for instance, look at Figure 1.12. What do you see? The elements of the picture are organized by the mind, automatically and with little effort, to produce the percept of a dog sniffing the ground. The picture is processed and experienced as a unified whole. Experimentally, the Gestalt psychologists relied not on trained observers but on the observations of ordinary people in investigating subjective experience. This unstructured reporting of experience was called the *phenomenological* approach, referring to the totality of subjective conscious experience.

The Gestalt movement reflected an important idea that was at the heart of criticisms of structuralism: The perception of objects is subjective and dependent on context. Two people can look at an object and see different things. Indeed, one person can look at an object and see it in completely different ways, as in Figure 1.13. Note that you can alternate between seeing the face, the profiles, or the candlestick, but that it is difficult to perceive the image in all three ways at the same time. Thus, your mind organizes the scene into a perceptual whole, so that you see the picture in a specific way. The important lesson of Gestalt psychology is that the mind perceives the world in an organized fashion that cannot be broken down into its constituent elements. The Gestalt perspective has had an impact on many areas of psychology, such as the study of vision and our understanding of human personality.

1.13 Face(s)? This drawing by Stanford psychologist Roger Shepard can be viewed as either a face behind a candlestick or two separate profiles.

The Unconscious Influences Everyday Mental Life

Twentieth-century psychology was profoundly influenced by one of its most famous thinkers, Sigmund Freud (Figure 1.14). Freud was trained in medicine and began his career working with people who had neurological disorders, such as paralysis of various body parts. He found that many of his patients had few medical reasons for their paralysis, and he soon came to believe that their conditions were caused by psychological factors.

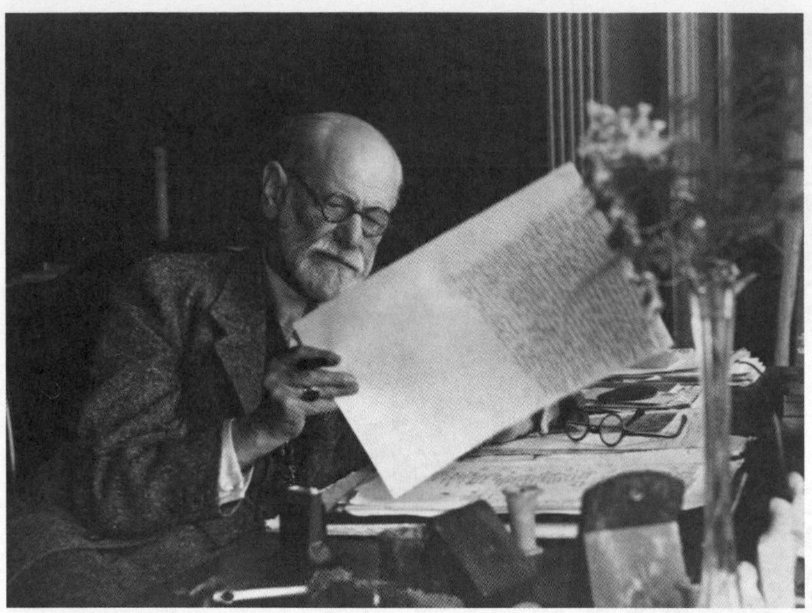

1.14 Sigmund Freud The father of psychoanalytic theory had a huge influence in the early days of psychology. Today his theories have largely been abandoned.

At the end of the nineteenth century, the structuralists and functionalists were focusing on conscious experience, but Freud made the deduction that much of human behaviour is determined by mental processes that operate below the level of conscious awareness, at the level of the **unconscious**. Freud believed that these unconscious mental forces were often in conflict, which produced psychological discomfort and in some cases even apparent psychological disorders.

From his theories, Freud pioneered the clinical case study approach and developed the therapeutic method of **psychoanalysis**, which involves trying to bring the contents of the patients' unconscious into conscious awareness so that their conflicts could be dealt with in a constructive manner. For example, he analyzed the apparent symbolic content in his patients' dreams in search of hidden conflicts. He also used a technique called *free association,* in which people would simply talk about whatever they wanted to for as long as they wanted to. Freud believed that through free association people would eventually reveal the unconscious conflicts that were causing them problems. He eventually extended his theories to account for general psychological functioning.

Freud's influence was considerable, not only on the psychologists who followed him but also on the public's view of psychology. Many people unfamiliar with psychology imagine that most psychologists have patients lie on couches and probe their innermost thoughts. In fact, today relatively few psychologists follow Freudian thinking. The basic problem with many of Freud's original ideas, such as the meaning of dreams, is that they are extremely difficult to test using the methods of science (as you will see in the next chapter). And although Freud's idea that mental processes occur below the level of conscious awareness is now widely accepted in psychological science, the unconscious processes studied by contemporary scientists share only a passing resemblance to the unconscious sexual conflicts that permeated Freudian theorizing.

Most Behaviour Can Be Modified by Reward and Punishment

Psychology's focus on mental processes, both conscious and unconscious, was soon challenged as inherently unscientific by John B. Watson (Figure 1.15), an American psychologist who developed the approach known as **behaviourism**, which emphasizes the role that environmental forces have on producing behaviour. Watson believed that if psychology was to be a science, it had to stop trying to study mental events that could not be directly observed, and he therefore was scornful of methods such as introspection and free association.

The intellectual question that was most central to Watson and his followers was the nature-nurture question. For Watson and other behaviourists, it was all nurture. Heavily influenced by the work of Russian physiologist Ivan Pavlov, Watson believed that all behaviour was caused by environmental factors: Understanding the environmental *stimuli,* or triggers, was all one needed to predict a behavioural *response.* Watson's behaviourism was mainly concerned with how animals acquired new behaviours, which is known as learning (discussed in Chapter 6). Psychologists greeted Watson's approach with great enthusiasm. Many had grown dissatisfied with the ambiguous

unconscious A term that identifies mental processes that operate below the level of conscious awareness.

psychoanalysis A method developed by Sigmund Freud that attempts to bring the contents of the unconscious into conscious awareness so that conflicts can be revealed.

behaviourism A psychological approach that emphasizes the role of environmental forces in producing behaviour.

methods used by those studying mental processes; they longed to be taken more seriously as scientists, which they believed would happen if they focused on studying only observable behaviours.

It was B. F. Skinner who became famous for taking up the mantle of behaviourism. Like Watson, Skinner denied the existence of mental states, writing in his provocative *Beyond Freedom and Dignity* (1971) that concepts referring to mental processes were of no scientific value for explaining behaviour. Rather, Skinner believed that mental states were nothing more than an illusion. Skinner was interested in how repeated behaviours were shaped or influenced by the events or consequences that followed them. For instance, an animal would learn to perform a behaviour if doing so in the past had led to a positive outcome, such as receiving food.

Behaviourism dominated psychological research well into the early 1960s. In many ways, these were extremely productive times for psychologists. Many of the basic principles established by behaviourists continue to be viewed as critical to understanding the mind, brain, and behaviour. At the same time, sufficient evidence has accumulated to show that thought processes do influence outcomes, and few psychologists today describe themselves as strict behaviourists.

1.15 **John B. Watson** Watson, who spent most of his adult life in advertising, was a proponent of behaviourism. His views were amplified by thousands of psychologists, including B. F. Skinner.

How People Think Affects Behaviour

While much of psychology was focused on studying observable behaviour, during the first half of the twentieth century evidence slowly emerged showing that how humans perceived situations could influence behaviour and that learning was not as simple as the behaviourists believed. In the late 1920s, the Gestalt theorist Wolfgang Köhler found that chimpanzees could solve a problem in their efforts to reach a banana. The chimpanzees had to figure out how to connect two sticks, which would then allow them to reach the banana and draw it close. The animals tried various things until suddenly they seemed to have insight, as evidenced by their reaching the banana and using the strategy perfectly on subsequent tasks. At around the same time, learning theorists such as Edward Tolman were showing that animals could learn by observation, which made little sense (according to behaviourist theory) given that the observing animals were not being rewarded—it was all going on in their heads. Other psychologists conducting research on memory, language, and child development showed that the simple laws of behaviourism could not explain such things as why cultural influences alter how people remember a story, why grammar develops in a systematic fashion, and why children go through stages of development in which they interpret the world in different ways. All of this suggested that mental functions were important for understanding behaviour.

In 1957, George A. Miller (Figure 1.16) and colleagues established the Center for Cognitive Studies at Harvard University and launched the *cognitive revolution* in psychology. Ulric Neisser integrated a wide range of cognitive phenomena in his classic 1967 book, which named and defined the field. *Cognitive psychology* is concerned with higher-order mental functions, such as intelligence, thinking, language, memory, and decision making (you will read more about cognitive psychology in Chapter 8). Cognitive research has shown that the way people think about things influences their behaviour. A number of events occurring in the 1950s set the stage for the rise of cognitive science, with perhaps the growing use of computers leading the way. Computers operate according to software programs, which dictate rules for how information is processed. Cognitive psychologists such as Alan Newell and the Nobel laureate Herbert Simon applied this process to their explanation of how the mind works. These *information-processing* theories of cognition viewed the brain as the hardware that ran the mind as software; the brain takes in information as a

How might a computer analogy be useful for thinking about the mind?

1.16 George A. Miller Miller launched the cognitive revolution by establishing the Center for Cognitive Science at Harvard University in 1957.

code, processes it, stores relevant sections, and retrieves stored information as required. There was an early recognition that the brain was important to cognition, but many cognitive psychologists preferred to focus exclusively on the software, with little interest in the specific brain mechanisms involved. In the early 1980s, cognitive psychologists joined forces with neuroscientists, computer scientists, and philosophers to develop an integrated view of mind and brain. The field of *cognitive neuroscience* emerged during the 1990s as one of the most exciting fields of science.

Social Situations Shape Behaviour

While tracing the origins of the field of psychology, we have focused a great deal on the relationship between the mind and the brain, and the role of evolution and learning on human behaviour. After World War II, psychologists broadened their efforts to better understand human behaviour in the real world. In particular, the atrocities committed in World War II led a number of psychologists and thinkers to begin researching topics such as authority, obedience, and group behaviour. These topics are the province of *social psychology,* which focuses on the power of situation and how people are shaped through their interactions with others.

In 1962 Adolf Eichmann, one of Hitler's chief lieutenants, was hanged for "causing the killing of millions of Jews." Shortly before his death, Eichmann claimed, "I am not the monster I am made out to be. I am the victim of a fallacy." The atrocities committed in Nazi Germany compelled psychologists to consider whether evil is an integral part of human nature. Why had apparently normal Germans willingly participated in the murder of innocent men, women, and children? Researchers, many influenced by Freudian ideas, initially sought to understand what types of people would commit evil acts. They concluded that certain types, especially those raised by unusually strict parents, did display a slightly greater willingness to follow orders. But social psychology shows that almost all people are strongly influenced by social situations. Indeed, as you will read in Chapter 15, people will send painful electric shocks to innocent others if directed to do so by someone in apparent authority. Although it does not excuse anything, Eichmann was correct that people were overlooking the power of the situation in explaining his heinous actions.

Kurt Lewin (Figure 1.17), who was trained as a Gestalt psychologist, can be credited with bringing a scientific, experimental approach to the field of social psychology. His *field theory* emphasized the interplay between people—their biology, habits, and beliefs—and their environment, such as social situations and group dynamics. This perspective allowed psychologists to begin examining some of the most complex forms of human mental activity, such as how people's attitudes shape behaviour, why they are prejudiced against other groups, how they are influenced by other people, and why they are attracted to some people and repelled by others. As we saw so vividly in the case of Amadou Diallo, human beings navigate through the social world, and psychological science recognizes the importance of fully considering the situation in order to predict and understand behaviour.

Psychological Therapy Is Based on Science

Earlier we mentioned that the physician Sigmund Freud developed psychoanalysis to treat people with psychological disorders. Although credit must be given to Freud for recognizing the impact of psychological factors on human functioning, his method of therapy is no longer widely used. In the middle of the twentieth century, a humanistic approach to the treatment of psychological disorders, led by Carl Rogers and Abraham Maslow, stressed how people can come to know and accept themselves in order to reach their unique potentials. Some of the techniques developed by Rogers, such as specific ways of questioning and listening during therapy, are staples of modern treatment. Only in the last four decades, however, has a scientific approach to the study of psychological disorders emerged.

Throughout psychology's history, the methods developed to treat psychological disorders mirrored advances in psychological science. For instance, the rise of behaviourism led to a group of therapies designed to modify behaviour rather than deal with underlying mental conflicts. Behavioural modification methods continue to be highly effective in a number of situations, from training those with intellectual impairments to treating patients who are especially anxious and fearful. The cognitive revolution in scientific thinking led therapists to recognize the important role of thought processes in mental disorders. Pioneers such as Aaron T. Beck developed therapies to correct faulty cognitions. These cognitive therapies are especially effective for treating conditions such as depression and eating disorders. For some mental disorders, such as schizophrenia, the only truly effective treatments are drugs that alter brain chemistry. For those with these disorders, the biological revolution in psychological science has been a godsend. But drug therapy is less preferable than cognitive and behavioural therapies for other disorders because of the potential for side effects and for patients to develop a reliance on the drugs. In many situations, a combination of drug and cognitive-behavioural therapy may be the best treatment plan. Today, most therapeutic approaches to dealing with psychological disorders consist of two key factors that reflect the origins of the field: adopting a widely recognized treatment of choice that scientific research has demonstrated to be clinically effective and recognizing that each person is a unique individual with specific issues and needs. Probably the greatest change in the field of clinical psychology over the course of the twentieth century comes from our realization, via scientific research, that there is no universal treatment or approach for all psychological disorders—contrary to the thinking of the early giants such as Freud, Rogers, and Skinner. Instead, as you will see in Chapter 14, different treatments are most effective for different disorders.

1.17 Kurt Lewin Lewin is the founder of modern social psychology. He pioneered the use of theory, using experimentation to test hypotheses.

REVIEWING the Principles | How Did the Scientific Foundations of Psychology Develop?

The foundations of psychological science emerged from philosophy and physiology. Early experimental psychologists included some who believed it necessary to reduce mental processes into their constituent, "structural" parts, and others who argued that it was more important to understand how the mind functions than what it contains. During this period, most research was aimed at understanding the subjective mind, such as Freud's emphasis on the unconscious and the Gestalt movement's focus on perception. It was the behaviourists who claimed that the study of mind was too subjective and therefore unscientific. Accordingly, for the first half of the twentieth century most psychologists studied only observable behaviours. The cognitive revolution in the 1960s returned the mind to centre stage, and research on mental processes such as memory, language, and decision making blossomed. Throughout the last century, some psychologists have emphasized the social context of behaviour and mental activity. Advances in psychological science have informed the treatment of psychological disorders.

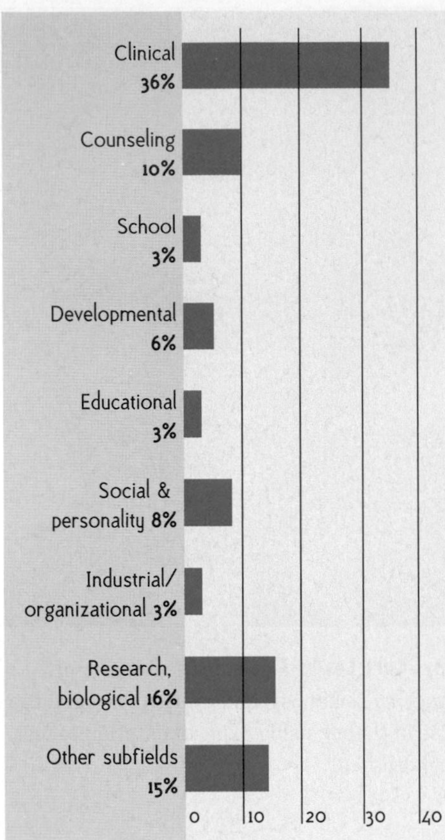

1.18 How Psychology Is Applied Psychologists pursue their interests by studying a variety of topics within different settings.

How Can We Apply Psychological Science?

As you will discover, psychological science covers broad ground. Many different types of researchers study the mind, brain, and behaviour at different conceptual levels. This diversity means that psychological scientists pursue research on topics that touch all aspects of human life.

The history of psychology in Canada closely parallels that in the United States, in both its research diversity and its practitioner forms. The list of readings at the end of the chapter contains a sampling of articles and books on psychology in Canada.

Subdisciplines Focus on Different Levels of Analysis

The term *psychologist* is used broadly to describe people whose careers involve predicting behaviour or understanding mental life (Figure 1.18). We use the term **psychological scientist** to refer to those who use the methods of science to study the brain, the mind, and behaviour. Psychological scientists work in many settings, such as clinics, schools, businesses, universities, and colleges. There are also **psychological practitioners**, who apply the findings of psychological science in order to do such things as help people in need of psychological treatment, design safe and pleasant work environments, counsel people on career paths, or help teachers design better classroom curricula. The distinction between science and practice can be fuzzy, since many psychological scientists are also practitioners. For example, many *clinical psychologists* both study and treat people with psychological disorders.

As we discussed earlier in this chapter, psychological scientists pursue their research interests at different levels of analysis; not surprisingly, the different levels of analysis tend to be associated with different subdisciplines. Let's apply these levels of analysis to the example of music, which we looked at earlier in this chapter, and examine how different types of psychological scientists might study it. *Social psychologists* focus on the influences that other situations and people have on how we act, think, and feel. Thus, researchers in this area might look at how cultural values or peer groups shape musical preferences. *Personality psychologists* are interested in individual preferences, such as how certain types of people might prefer certain types of music, whereas *developmental psychologists* address changes in mind and behaviour over the life span and so they might study how children learn the basic structure of music or how musical preferences change as one grows older.

Cognitive psychology is concerned with mental processes such as thinking, perceiving, remembering, and decision making, and so researchers in this area might be interested in how people perceive various aspects of music as well as how music changes the way people think. Those in *cognitive neuroscience* would seek to understand the brain systems involved in the perception of music and how they might differ from the parts of the brain involved in ordinary auditory processing. *Behavioural neuroscientists* study the biological mechanisms responsible for behaviour, and therefore might be interested in how music affects the body and brain. Music changes our physical state: classical music may soothe us, rock music might rev us up. As mentioned earlier, music also changes brain chemistry in a way that affects mood. Finally, *experimental psychopathologists,* who study abnormal or disordered behaviour, might use music to induce a sad mood in people in order to examine the mental processes associated with feeling depressed.

Many of the chapters of this book reflect the work of the different subdisciplines of psychological science and consequently show the use of different levels of analysis. The choice of which level to study is based on the scientists' particular interests, as well as their general theoretical approaches and how they were trained. But more

psychological scientist One who uses the methods of science to study the interplay between brain, mind, and behaviour and how the social environment affects these processes.

psychological practitioners Those who apply findings from psychological science in order to assist people in their daily lives.

Why does crossing levels of analysis provide better insights about the mind?

and more, psychological science emphasizes examining behaviour across multiple levels in an integrated fashion. Thus, psychologists interested in understanding the hormonal basis of obesity interact with geneticists who explore the heritability of obesity, as well as social psychologists who study human beliefs about eating. Crossing the levels of analysis provides more insights than working within one level. The Gestalt psychologists were right in asserting that the whole is different from the sum of its parts. Throughout this book, you will see how this multilevel approach has led to breakthroughs in understanding psychological activity.

Psychological Science *in Action*

Sports Psychologists Help Athletes Find the Zone

As we have seen, psychological practitioners apply what is known from psychological science to help people perform at their best. A good example of a popular applied area is sports psychology. Elite athletes work for years to perfect physical skills necessary to excel at their sports. Their daily practices hone their skills so that they can perform at peak competitive levels. Yet many professionals go through slumps in which they can look like blundering amateurs. Consider David Duval, a golfer who within recent years was on the leaderboard at every tournament, yet has fallen out of contention for the last few seasons, finishing nearly last in the 2005 British Open at St. Andrews. Golf analysts have suggested that he has not lost his skills but rather his mental edge. Elite performances require not only physical prowess, but also a mind-set that one is a champion. Helping elite athletes think of themselves as winners is part of the job of sports psychologists, professionals who study the behavioural factors that influence and are influenced by participation in sports and exercise. Sports psychologists help athletes do a number of things, such as maintaining focus during competition, communicating with teammates, building feelings of confidence, motivating themselves to achieve their goals, and dealing with the intensity of their emotions. Specific exercises help athletes to overcome doubts, use relaxation to calm their minds, and develop strategies for focusing solely on the task at hand. In other words, they help athletes get into "the zone."

1.19 Mike Weir For the world's top athletes, psychology can be the difference between taking home the trophy and leaving empty-handed. Weir, winner of the 2003 Masters and many other prestigious golf tournaments, has worked with sports psychologist Rich Gordin since 1997 to help improve his game.

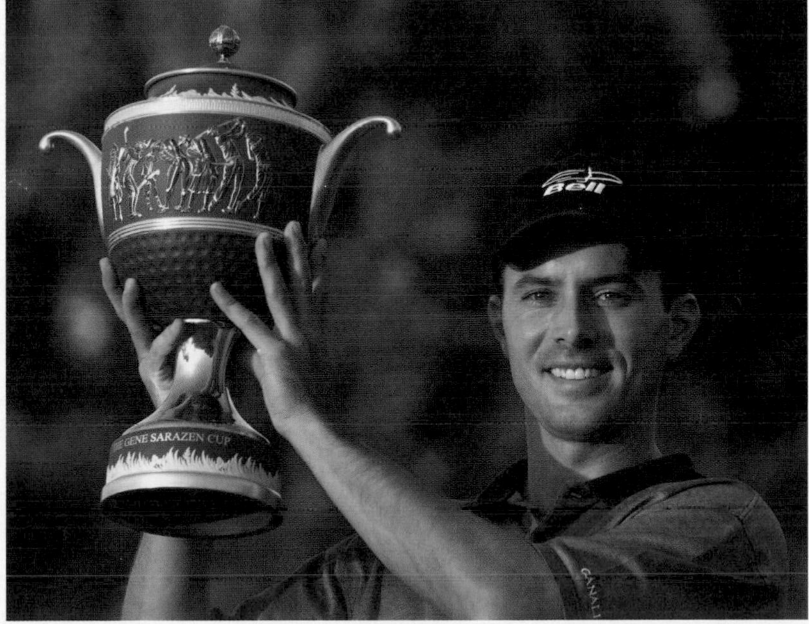

Many athletes work with a sports psychologist as part of their regular training. But sometimes athletes going through slumps require stronger assistance. Golfers like David Duval and Mike Weir (Figure 1.19) often seek out the assistance of the sports psychologist Bob Rotella, known affectionately as "Doc" on the PGA Tour. Rotella, a native of Vermont who earned his doctorate in psychology from the University of Connecticut, has worked with many PGA golfers as well as professional athletes in baseball, hockey, and basketball, and with members of the U.S. Olympic ski and equestrian teams. He is considered one of the pioneers in sports psychology, conducting research on exercise and sports at the University of Virginia. He now is a best-selling author and frequent television commentator. Rotella helps athletes train their minds to focus on their goals and teaches them to deal with their doubts, worries, and frustrations. Like many sports psychologists, he emphasizes relaxation as a key to preparing for competition. One of his favorite psychologists is William James, who often espoused the view that people by and large become what they think about themselves. For Rotella, this means that how athletes view themselves, their beliefs, and their performance expectations shapes how they actually perform. His approach to working with athletes, therefore, is to help them learn to think of themselves as champions.

Sports psychologists receive extensive training to be able to help their clients. To be certified, they need to demonstrate expert knowledge in the research base of psychological principles that apply to sports, which include understanding motor behaviour, human physiology, principles of learning, motivation, and counseling skills. More than 50 programs around the world specialize in training practitioners of sports psychology, one of the fastest-growing areas of applied psychology. ■

Psychological Knowledge Is Used in Many Professions

Psychology is one of the most popular majors in many colleges. Not only is the subject material fascinating and personally relevant, but psychological science serves as excellent training for many professions. For instance, physicians need to know a lot more than anatomy and chemistry. They need to know how to relate to their patients, how the patients' behaviours are linked to health, and what motivates or discourages patients from seeking medical care or following treatment protocols. Understanding the aging brain and how it affects visual perception, memory, and motor movement is vital for those who treat elderly patients. Psychological scientists make major contributions to research on human physical and mental health. Much of the psychological research you will read about in this book is being applied today to make people's lives better.

How do we use knowledge acquired from psychological science?

Psychological science is equally useful for anyone whose career involves needing to understand people. In order to persuade jurors, lawyers need to know how groups make decisions. Advertisers need to know how attitudes are formed or changed, and to what extent people's attitudes predict their behaviour. Politicians use psychological techniques of impression management to make themselves attractive to voters. The general usefulness of understanding mental activity may also explain psychology's popularity on campuses. It can help you understand your motives, your personality, even why you remember some things and forget others.

Of course, some of you will be so fascinated by psychological science that you will devote your lives to studying the mind and behaviour. We, the authors, understand how you feel. There is tremendous excitement in psychological science as we unravel the very nature of what it means to be a human being. Although the foundations of

"For God's sake, think! Why is he being so nice to you?"

psychological science are growing and established scientific principles of mind, brain, and behaviour exist, there is still a tremendous amount to learn about issues such as how nature and nurture interact and how the brain enables the mind. Indeed, contemporary psychological science is providing new insights into problems and issues that the great scholars of the past struggled to solve.

People Are Intuitive Psychological Scientists

By our very nature, we humans are intuitive psychological scientists who develop hypotheses about and try to predict the behaviour of others. People choose marriage partners they expect will best meet their emotional, sexual, and support needs. Defensive driving relies on an intuitive understanding of when other people are likely to make mistakes while driving. People are also pretty good at predicting whether others are kind, would make good teachers, or are trustworthy. But people can't intuitively know if taking certain herbs will increase memory, or whether playing music to newborns makes them more intelligent, or whether mental illness results from too much or too little of a certain brain chemical.

One of the most important goals of our textbook is to provide a basic education about the methods of psychological science for those students whose only exposure to psychology is through this introductory course. Although psychologists make important contributions to understanding and treating mental illness, most of psychological science has little to do with couches or dreams. Instead, it has everything to do with understanding mental activity, social interactions, and how people acquire behaviours. To understand what makes you tick and what makes other people tick, you will need to be introduced to the basic operating manual for the human mind and behaviour. This is found in the psychological sciences.

Psychological Sciences Require Critical Thinking

One of the hallmarks of science is skepticism. You should be skeptical of dramatic new findings reported in science until you have convincing evidence to support the claims. Science progresses carefully and often slowly, and good science takes time. The ability to think skeptically is often referred to as **critical thinking**, a systematic way of evaluating information in order to reach reasonable conclusions. Being a critical thinker involves considering alternative explanations, looking for holes in evidence, and using logic and reasoning to see whether the information makes sense. It also involves considering whether the information might be biased, such as by personal or political agendas. Mostly, critical thinking involves healthy questioning and keeping an open mind. This doesn't mean just questioning information that doesn't fit in with your worldviews, but also thinking critically about all information, and perhaps especially that which fits your preconceptions. After all, this is where you may be least motivated to think critically.

critical thinking A systematic way of evaluating information in order to reach reasonable conclusions.

Suppose, for instance, you read about a study that purports to demonstrate that people who share your political views are much more intelligent than those with a different political orientation. As a critical thinker, you would want to know whether the political affiliation of the researchers influenced the study as well as how they measured intelligence, such as whether they measured knowledge about events that matter more to one political group than to others. Before you take solace in the findings, it is important to make sure that those findings are real. The only way to do so is to evaluate the findings using critical thinking. Developing critical thinking skills will be useful to you not only in your other classes but also in any career that you choose. Being a critical thinker means making good decisions based on reasonable and logical conclusions about evidence.

Thinking Critically

Does Listening to Mozart Make You Smarter?

As you read this book you will learn how to separate the believable from the incredible. You will learn to spot badly designed experiments, and you will develop the skills necessary to critically evaluate claims made in the popular media. The information in this book will also provide a state-of-the-art background in psychological science so that you will know whether or not a claim is consistent with what we know about psychology. The media love a good story, and findings from psychological research are often very provocative. Unfortunately, media reports can be distorted or even flat-out wrong. You will be consumers of psychological science throughout your lives, and you will need to be skeptical of overblown media of "brand-new" findings obtained by "groundbreaking" research. For instance, a study conducted by a cognitive psychologist found that playing Mozart to research participants led them to score slightly higher on a test that is loosely related to intelligence. The media jumped onto the so-called Mozart effect with abandon; the result has been that many parents are determinedly playing Mozart to young infants, and even to fetuses (Figure 1.20). Indeed, governors of two U.S. states instituted programs that provided free Mozart CDs to every newborn. Web sites make bold claims about the power of Mozart, including fantastical assertions that listening to Mozart can cure neurological illness and other maladies. What is the power of Mozart for the developing mind?

What is critical thinking?

Let's step back and critically evaluate the research underlying the Mozart effect. In 1993, researchers played the first 10 minutes of the Mozart Sonata for Two Pianos in D Major (K. 448) to a group of college students. Compared with students who listened to relaxation instructions or who sat in silence, those who heard Mozart performed slightly better on a task that involved folding and cutting paper, a task that is part of a larger overall measure of intelligence. This modest increase lasted for about 10 to 15 minutes (Rauscher, Shaw, & Ky, 1993).

Unfortunately, subsequent research largely failed to replicate the findings. Following a careful review of the studies testing the "Mozart effect," Christopher Chabris and his colleagues (1999) have concluded that the effect of listening to Mozart is quite trivial and certainly unlikely to increase intelligence among listeners. According to Chabris and his team, listening to Mozart appears to enhance only certain types of motor skills but does not influence measures more commonly associated with intelligence. Thus, it is unclear whether the effects have anything to do with intelligence. Another important question from a critical thinking perspective is whether it was the music or some other aspect of the situation that led to better performance on the folding and cutting task. A team of Canadian researchers has shown that the effect may be due simply to the fact that the music is more uplifting than sitting in silence or relaxing, and thus it is the increase in positive mood that is largely responsible for better performance (Thompson, Schellenberg, & Husain, 2001). Recall from earlier in this chapter that music has many effects on mood and cognition.

1.20 The Mozart Effect A much publicized research study done in the early 1990s concluded that infants benefited from the "Mozart effect," but subsequent findings have been extremely critical of this theory.

It is also critical to note that all of the studies to date have been conducted on college students, yet all of the publicity concerns whether playing Mozart to infants increases their intelligence. Most likely it does not. John Bruer, in his important book

The Myth of the First Three Years (1999), notes that people have taken leaps past the data in suggesting that experiences during the first three years shape or alter the developing brain in any meaningful or permanent way. Now that you are a bit more informed, what do you think? Does listening to Mozart increase your intelligence? Does it have any positive impact on your physical or mental health? ■

REVIEWING **the Principles** | How Can We Apply Psychological Science?

Psychological scientists study the mind, brain, and behaviour across different levels of analysis and often identify themselves within various subdisciplines of psychology. However, there is a growing tendency for researchers to cross traditional areas to collaborate with researchers from different subdisciplines. The content of psychological science is of interest and value to many professions, which may explain why psychology is one of the most popular majors on most college campuses. Using critical thinking skills and understanding the methods of psychological science are important for evaluating research that is reported in the popular media.

Conclusion

The rich history of psychology is clear. Its founders grappled with the most important and interesting issues there are. Who are we? How does the brain get the job done? How much of us is predetermined by our genes? How does our personality affect our lives? What are the social forces that influence our personal decisions?

As Darwin predicted over a century ago, the concept of natural selection is becoming an integral part of psychological science. In general terms, psychologists have come to appreciate that to understand behaviour, they first must understand what it is designed for. Did this behaviour solve a problem for our human ancestors? Once we have a sense of what something is for, then we can ask, "How does it work?" These questions remain at the forefront of research today as psychological scientists seek to unravel the mysteries of mind, brain, and behaviour. These basic questions apply to all levels of analysis: biological, individual, and social. Furthermore, research across the multiple levels of analysis is invigorating psychological science and providing new insights into age-old questions. In the chapters ahead, you will learn not only how psychological science operates but also many remarkable things about the mind, brain, and behaviour. It might just be one of the most fascinating explorations of your life.

Summarizing the Principles of Psychological Science

What Are the Themes of Psychological Science?

1. **The principles of psychological science are cumulative:** Psychological science relies on empirical evidence and builds on previous discoveries in a dynamic way.
2. **The new biological revolution is energizing research:** Discoveries in brain activity have been enhanced by the discoveries of neurotransmitters. Mapping of the human genome has furthered the role of genetics in analyzing disease and behaviour. Tremendous advances in brain imaging have revealed the working brain.
3. **The mind is adaptive:** The brain has evolved to solve survival problems and adapt to environments. Many modern behaviours are by-products of adaptation.
4. **Psychological science crosses levels of analysis:** Psychological scientists examine behaviour from a variety of levels

of analysis: social (culture and interpersonal behaviour), individual (personality, and perception and cognition), and biological (brain systems, neurochemistry, and genetics).

What Are the Intellectual Origins of Psychology?

5. **The nature-nurture debate considers the impact of biology and environment:** Historically, philosophy and science have been concerned with the relative influence of biology and environment on behaviour.

6. **The mind-body problem has challenged philosophers and psychologists:** A persistent challenge exists in reconciling the relationship of the mind to the physical activity of the brain.

7. **Evolutionary theory introduces natural selection:** Darwin's theory of evolution assumes that behaviours that promote reproductive success and survival are passed on and, therefore, that individual differences are the basis of evolutionary change.

How Did the Scientific Foundations of Psychology Develop?

8. **Experimental psychology begins with structuralism:** Using the technique of introspection, scientists attempted to understand conscious experience by reducing it to its basic elements, its structure. This brings about a strong scientific approach.

9. **Functionalism addresses the purpose of behaviour:** Functionalists argued that the mind is best understood by examining the functions or purposes it serves, not its structure.

10. **Gestalt psychology emphasizes patterns and context in learning:** The assertion that the whole experience (the Gestalt) is different from the sum of its parts led to an approach emphasizing the subjective experience of perception.

11. **The unconscious influences everyday mental life:** The psychoanalytic assumption that unconscious processes not readily available to our awareness influence behaviour had an enormous impact on psychology.

12. **Most behaviour can be modified by reward and punishment:** Discoveries that behaviour is changed by its consequences caused behaviourism to dominate psychology until the 1960s.

13. **How people think affects behaviour:** The computer analogy of the brain and the cognitive revolution led to the information processing perspective.

14. **Social situations shape behaviour:** Work in social psychology has highlighted how other people and situations are powerful forces in shaping behaviour.

15. **Psychological therapy is based on science:** Scientific research over the course of the twentieth century taught psychological scientists that there is no universal treatment for psychological disorders. Instead, different treatments are effective for different disorders.

How Can We Apply Psychological Science?

16. **Subdisciplines focus on different levels of analysis:** Different approaches to understanding behaviour are the focus of the various areas of psychology and are often combined.

17. **Psychological knowledge is used in many professions:** Because psychology focuses on human behaviour, it is of interest to many students and professionals.

18. **People are intuitive psychological scientists:** Humans naturally explain and predict the behaviour of others.

19. **Psychological sciences require critical thinking:** Skepticism is an important element of science and requires critical thinking and the evaluation of evidence and conclusions.

Applying the Principles

1. You are watching a baseball game and comment that the pitching expertise accounts for the score; someone else says it's the hitting; and a third person asserts that it is the payroll. This is similar to what theme of psychological science?

   _____ a) science is cumulative
   _____ b) the biological revolution
   _____ c) the brain is adaptive
   _____ d) there are different levels of analysis

2. It is freezing outside, so you slip on a heavy coat before heading out. Your cousin visiting from Hawaii has never even owned a heavy coat and decides to stay in. He complains that he is cold even in your apartment, which is

heated to 68 degrees. Your different behaviours and the ability to think about and solve the problem of how to deal with cold temperatures demonstrate which of the following basic principles of psychological science?

_____ a) the role of both the brain and the environment in shaping behaviours for survival
_____ b) the influence of environmental experience alone on survival
_____ c) the ability of the brain alone to determine survival behaviours
_____ d) the belief that nature has a greater influence on behaviour than nurture

3. If you attempt to understand a behaviour—for example, bowling a perfect strike or solving a puzzle—by breaking it down into empirical evidence defined by the stimulus (the pins; the puzzle pieces), the response (perfect release of the ball; fitting together pieces), and the consequence (the strike score; the completed puzzle); you would be using the scientific approach of _____, which has its roots in _____.

_____ a) Gestalt; psychoanalysis
_____ b) behaviourism; functionalism
_____ c) psychoanalysis; structuralism
_____ d) behaviourism; structuralism

4. Having been a student for some time, you have an intuitive sense for what a course will be like from watching the professor during the first couple of sessions. The accuracy of your judgment could be the focus of all but which of the following psychologists?

_____ a) cognitive psychologist
_____ b) personality psychologist
_____ c) social psychologist
_____ d) psychopathologist (clinician)

ANSWERS: 1.d 2.b 3.b 4.d

Key Terms

adaptations, p. 10
behaviourism, p. 24
critical thinking, p. 31
culture, p. 12
dualism, p. 18
evolutionary theory, p. 10
functionalism, p. 21
Gestalt theory, p. 23
introspection, p. 20
mind-body problem, p. 17
natural selection, p. 10
nature-nurture debate, p. 16
psychoanalysis, p. 24
psychological practitioners, p. 28
psychological science, p. 6
psychological scientist, p. 28
stream of consciousness, p. 21
structuralism, p. 20
unconscious, p. 24

Further Readings

Aronson, E. (2003). *The social animal* (9th ed.). New York: Freeman.

Bruer, J. (1999). *The myth of the first three years: A new understanding of early brain development and lifelong learning.* New York: Free Press.

Darwin, C. (1964). *On the origin of species.* Cambridge, MA: Harvard University Press. (Original work published 1859)

Gazzaniga, M. S. (1998). *The mind's past.* Berkeley, CA: University of California Press.

Hothersall, D. (2003). *History of psychology* (4th ed.). Boston, MA: McGraw-Hill.

James, W. (1983). *Principles of psychology.* Cambridge, MA: Harvard University Press. (Original work published 1890)

Further Readings: A Canadian Presence in Psychology

Adair, J. G., Paivio, A. & Ritchie, P. (1966). Psychology in Canada. *Annual Review of Psychology, 47,* 341–370.

Canadian Psychological Association. (1992). The history of psychology in Canada. *Canadian Psychology, 33,* 2.

Cherry, F. (1994). Archivist's corner. *History and Philosophy of Psychology Bulletin, 6(1),* 12.

Ferguson, G. A. (1982). Psychology at McGill. In M. J. Wright & C. R. Myers, (Eds.), *History of academic psychology in Canada* (pp. 33–67). Toronto: Hogrefe.

Green, C. D. (2002). Introduction to "Experimental psychology and the laboratory in Toronto" by Albert H. Abbott (1900). Retrieved from http://psychclassics.yorku.ca/Abbott/intro.htm

Tolman, C. (1994). Archivist's corner: Notes on some sources for the history of psychology in English Canada. *History and Philosophy of Psychology Bulletin, 6(2),* 10–11.

Wright, M. J. & Myers, C. R. (1982). *History of academic psychology in Canada.* Toronto: Hogrefe.

Creative Approaches to Psychological Research

A good theory is one that produces a wide variety of testable hypotheses. In the early twentieth century, psychologist Jean Piaget theorized that cognitive development occurs in a fixed series of "stages" from birth to adolescence. He designed experiments to test his theory. In one such test, similar to the one shown here, children of different ages were asked to compare the volume of equal amounts of liquid poured into different-sized glasses. Piaget found that the responses were consistent within an age group, and only children at a certain age, or "stage," were able to grasp that the volume of liquid in each glass was equal. Studies in psychological science have subsequently shown that some aspects of this theory were wrong.

CHAPTER

2

Research Methodology

In the late 1800s, the Italian physiologist Angelo Mosso began studying the relationship between mind and brain by examining how mental activity affects the flow of blood to the brain. He was working on the assumption that because the brain is the organ responsible for generating thoughts, feelings, and emotions, mental activity should affect how the brain functions. In one interesting experiment, Mosso had his participant lie on a table that was carefully balanced on a fulcrum. Remarkably, as soon as the subject engaged in emotional or intellectual activity, such as being afraid or doing mental arithmetic, the table began to tilt down toward the head! Mosso concluded that because mental activity increases the physiological activity of the brain, the tilt in the table was due to an increase in the amount of blood that was flowing into the subject's brain.

Mosso's experiment was a harbinger of the modern experimental techniques that allow us today to visually image mental activity. For example, by injecting research participants with small doses of radioactively labeled

Outlining the Principles

What Is Scientific Inquiry?

- Scientific Questions Can Be Objectively Answered
- The Empirical Process Depends on Theories, Hypotheses, and Research

What Are the Types of Studies in Psychological Research?

- An Experiment Involves Manipulating Conditions
- Correlational Designs Examine How Variables Are Related
- Descriptive Studies Observe and Classify Behaviour

What Are the Data-Collection Methods of Psychological Science?

- Observing Is an Unobtrusive Strategy
- Asking Takes a More Active Approach
- Case Studies Examine Individual Lives
- Response Performance Measures Stimulus Processing
- Body and Brain Activity Can Be Directly Measured
- Research with Animals Provides Important Data
- There Are Ethical Issues to Consider

How Are Data Analyzed and Evaluated?

- Good Research Requires Valid, Reliable, and Accurate Data
- Descriptive Statistics Provide a Summary
- Correlations Describe the Relationships between Variables
- Inferential Statistics Predict Chance Results

water, scientists can now track changes in the flow of blood in the brain during mental tasks, such as trying to remember what you ate for breakfast or looking at a picture of someone who is afraid.

Although dramatic improvements have been made in the research technology since Mosso's table experiment, his study does share many of the goals of modern psychological research. Research always begins with a question that intrigues the researcher, who then designs and conducts a study to answer the question. Whether the study provides a useful answer depends on how well it is designed. In this chapter we will examine the different ways researchers study the questions of psychological science. You will learn the qualities that separate good studies from bad, as well as the types of questions that can be addressed through psychological research. ■

During the last century psychologists have made a number of important discoveries about the mind, brain, and behaviour. These discoveries have emerged through careful scientific research. Throughout your life you have observed people and developed your own explanations for why they think and act the way they do. Unfortunately, many commonsense explanations for behaviour are either inadequate or flat-out wrong. For example, most people believe that brainstorming leads to creative solutions, that opposites attract, that venting anger is a good idea, that children with low self-esteem are especially aggressive, and that there are shooting streaks in basketball. None of these intuitions is supported by scientific research. In the previous chapter, we noted the importance of using critical thinking, which is a systematic way of scrutinizing evidence to see if it makes sense. A key aspect of critical thinking is maintaining skepticism, which means being wary of subjective beliefs, such as intuition, that can be biased and are often based on limited information. Intuition and hunches help guide scientific questions rather than answers. To answer the questions, scientists use objective, systematic procedures so that they can develop an accurate understanding of the phenomena they study.

So how do psychologists use the methods of science to study behaviour? Let's suppose you are concerned about drunk driving. Thus, you decide you want to conduct a study that examines the effects of alcohol on driving skills. You begin with an idea that alcohol probably impairs motor skills and coordination. Thus, you are testing the hypothesis that alcohol affects behaviour in a specific way. You then set up a study in which you give people alcohol and measure some behavioural response that you believe is related to driving skill, such as how long they can balance on one leg or how quickly they can respond to a command to press a brake pedal. Basically, you have designed an *experiment* in which you altered, or *manipulated,* the situation in order to see how the change affects mental state or behaviour. Of course, you must also mea-

1861
Neuropsychological Assessment Paul Broca's autopsy of his patient Tan reveals frontal lobe damage, which supports the idea that mental functions are localized in the brain.

1879
Reaction Times Wilhelm Wundt and others at the University of Leipzig use reaction-time measures to study mental processes. These measures assume the time it takes to respond to a stimulus corresponds to the time it takes for mental activity.

1880
Case Studies Josef Breuer, a Viennese physician, treats a young patient who is suffering from a psychological affliction. Sigmund Freud develops many of his ideas about psychoanalysis by analyzing the case of Anna O.

1896
Correlation Coefficient Karl Pearson, an English statistician, builds on Galton's ideas and develops the correlation coefficient, a statistic that measures the degree of association between two variables; it is still widely used in psychological science.

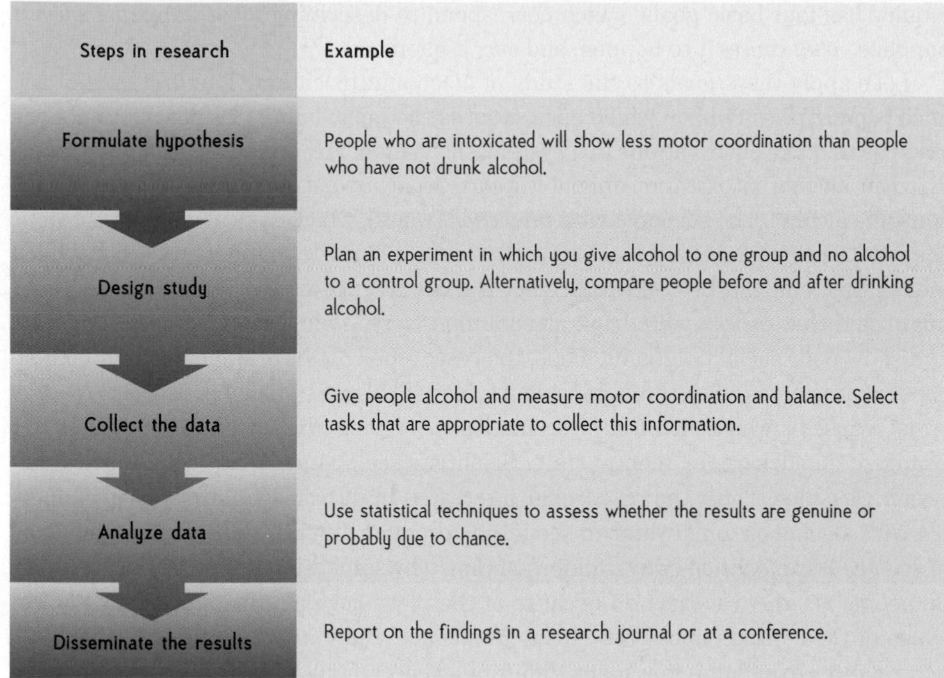

Steps in research
Example
Formulate hypothesis
People who are intoxicated will show less motor coordination than people who have not drunk alcohol.
Design study
Plan an experiment in which you give alcohol to one group and no alcohol to a control group. Alternatively, compare people before and after drinking alcohol.
Collect the data
Give people alcohol and measure motor coordination and balance. Select tasks that are appropriate to collect this information.
Analyze data
Use statistical techniques to assess whether the results are genuine or probably due to chance.
Disseminate the results
Report on the findings in a research journal or at a conference.

2.1 The Methods of Psychological Science A flowchart showing the normal progression of psychological science.

sure the same behaviours when your research participants have not been drinking to see if the manipulation changed behaviour. Alternatively, you could compare your group's behaviour after drinking to that of another group of people who did not consume alcohol, known as a *control group.*

Once you have collected the performance information, or *data,* you use *statistical procedures* to assess whether the results you obtained were due to drinking alcohol or simply due to chance. The results of your statistical analysis will provide an answer to your research question, which you can then share with the world. Although this example provides a general overview of scientific research (Figure 2.1), you need to address many important details to make sure your study adequately tests your question.

What Is Scientific Inquiry?

Science assumes that the world works in an orderly fashion based on physical laws: Gravity causes apples to fall to the ground, and the Bernoulli principle explains how aircraft can fly. Although the various fields of science differ somewhat in the particular techniques and strategies they use, all share a general approach known as the *scientific method,* which is an objective examination of the natural world. Scientific

1897
First Social Psychology Experiment The first observational study of social behaviour is undertaken by Norman Triplett, who discovers that bicycle racers pedal more quickly in the presence of other cyclists than when on their own.

1924
The Electrical Brain Hans Berger records the first electrical signals of the brain from the intact skull of a human. The electroencephalogram is used to measure the activity of the brain during mental processes.

1925
The Foundation of Experimental Design Sir Ronald Fisher publishes *Statistical Methods for Research Workers,* which establishes the foundation for contemporary experimental design.

1933
Testing Hypotheses Jerzy Neyman and Egon Pearson, the son of Karl Pearson, detail the central concepts behind statistical hypothesis testing in a paper presented to the Cambridge Philosophical Society.

1950s
Recording Individual Neurons Experimenting on cats, David Hubel and Torsten Wiesel make the first recordings from individual neurons in the cerebral cortex.

inquiry has four basic goals, which correspond to describing *what* happens, *when* it happens, *what causes* it to happen, and *why* it happens.

Let's apply these goals to the study of alcohol intoxication. The first goal of science is *description* (what), which asks, what is the behaviour? The description stage leads us to make observations in a systematic fashion. For instance, a scientist interested in alcohol intoxication might watch restaurant patrons to see whether they consume alcohol. The second goal is *prediction* (when), which asks when the behaviour occurs. Observation can reveal that two events tend to happen together, which means that one can be used to predict the occurrence of the other. A researcher might find that people who drink alcohol tend to stumble, have impaired language, and show poor social judgment. Thus, the researcher might predict that drinking alcohol is associated with those behaviours. The third goal of science is *causal control* (what causes), which involves systematically varying the situation to produce a change in behaviours or mental state. Scientists seek cause-effect relationships between variables. Thus, the researcher interested in intoxication might provide people with alcohol or tonic water to see if they produce different effects. The final goal of science is *explanation* (why), understanding why something happens. For instance, do people act the way they do because of the physiological effects of alcohol or because of their beliefs about the effects of alcohol? Fully understanding a phenomenon requires examining it at many different levels of analysis, which is the essence of scientific inquiry.

Scientific Questions Can Be Objectively Answered

What defines an empirical question?

Scientific inquiry is the study of empirical questions—questions that can be answered by observing and measuring the world around us. Can some psychics communicate with the dead? Well, perhaps, but this is impossible to study because it would require the ability to assess what dead people were saying, assuming that dead people can speak, which itself is impossible to investigate using the methods of science. Does alcohol impair the ability to drive? This is a question that can be answered by measurement. And the answer is yes: A review of 112 studies found that the skills necessary to drive a car can become impaired after people consume even small amounts of alcohol (Moskowitz & Fiorentino, 2000). The key point is that scientific questions are those that can be tested and either confirmed or shown to be false. The answer to an empirical question does not depend on what the respondent or scientist *thinks* about the question. Whether a researcher believes that consuming alcohol is a good or bad thing is irrelevant.

replication Repetition of an experiment to confirm the results.

Because empirical questions can be answered by objective means, the same answer should be attained by any and all individuals willing to take the trouble, at least in theory. That is, studies that are properly designed and conducted can be repeated, by the same researcher or by others, and produce the same findings. **Replication**, in

1960s
Experimenter Expectancy Robert Rosenthal shows that the outcome of an experiment can be influenced by the researcher's beliefs or expectancies. Others develop techniques to minimize bias in research.

1969
Identifying Cognitive Operations Saul Sternberg updates how to use reaction-time measures to infer the sequence of cognitive processing stages employed during a task, laying the groundwork for human brain-mapping techniques.

1970s
Understanding Relationships Elaine Hatfield and Ellen Berscheid pioneer the scientific study of love. Their research provides important insights into the study of interpersonal relationships.

1986
Brain Imaging Researchers lay out the theoretical foundation for tracking changes in blood flow in the brain. This leads to the development of brain-imaging technologies that associate brain regions with specific mental processes.

1990s
Gene Manipulation Methods are rapidly developed that allow researchers to delete (knock out) or substitute (knock in) genes with great specificity. These techniques provide experimental methods for determining how genes influence psychological processes such as learning and memory.

which other researchers repeat and confirm the results of earlier studies, helps build the principles of psychological science. Studies that are either poorly designed or that obtain positive results due to chance will tend not to be replicated, and they will ultimately be viewed with skepticism.

The Empirical Process Depends on Theories, Hypotheses, and Research

Formally speaking, the empirical process reflects a dynamic interaction between three essential elements. First, there is a **theory**, which is an idea or a model of how something in the world works. It consists of interconnected ideas and concepts, explains what is observed, and makes predictions about future events. Second, there is a **hypothesis**, which is a specific prediction of what should be observed if the theory under consideration is correct. In this manner, the hypothesis serves as a direct test of the theory. If the theory is reasonably accurate, the prediction framed in the hypothesis should be correct. Third, there must be **research**, which involves the systematic and careful collection of **data**, or objective information, to examine or test whether the given hypothesis—and ultimately, the corresponding theory—is indeed tenable. Once the findings of the study are in, the researcher returns to the original theory to evaluate the implications of the data obtained. Either the findings support the theory or they require that the theory be modified to take them into account. As shown in Figure 2.2, good research reflects a dynamic, cyclical interaction among these three elements.

theory A model of interconnected ideas and concepts that explains what is observed and makes predictions about future events.

hypothesis A specific prediction of what should be observed in the world if a theory is correct.

research Scientific process that involves the systematic and careful collection of data.

data Objective observations or measurements.

THEORIES SHOULD GENERATE HYPOTHESES How can we decide whether a theory is any good? First, a good theory is one that produces a wide variety of *testable* hypotheses. For instance, in the early twentieth century the great developmental psychologist Jean Piaget proposed a theory of infant-child development that suggested that cognitive development occurs in a fixed series of "stages," from birth to adolescence. A good scientific theory, Piaget's model led to a number of hypotheses regarding the specific kinds of behaviours that should be observed within each stage of development. In the decades since its proposal, the theory has generated thousands of scientific papers (many of which uncovered flaws in Piaget's theory, as we discuss in Chapter 11). On the other hand, in his legendary treatise *The Interpretation of Dreams,* Sigmund Freud outlined his theory that all dreams represent the fulfillment of a wish. From a scientific perspective, Freud's theory wasn't a good one, because it generated few if any testable hypotheses regarding the actual function of dreams. As a result, researchers were left with no way to evaluate properly whether the wish-fulfillment theory was either reasonable or accurate. Indeed, on being presented with a patient's dream that clearly contained no hint of wish fulfillment, Freud went so far as to claim that the dream's actual wish was to prove his theory wrong!

THE VALUE OF UNEXPECTED FINDINGS It is important to note that research does not always proceed in a neat and orderly fashion. On the contrary, many significant findings are the result of *serendipity*—the unexpected stumbling upon something important. Typically, chance findings occur when some event happens that was not part of the original plan of the study. For example, in the late 1950s, when the Harvard physiologists

2.2 The Empirical Process The empirical process reflects a cyclical relationship between a theory, testable hypotheses derived from the theory, research conducted to test the hypotheses, and adjustments to the theory as findings prompt reevaluation. A good theory will evolve over time and the result is an increasingly accurate model of the world.

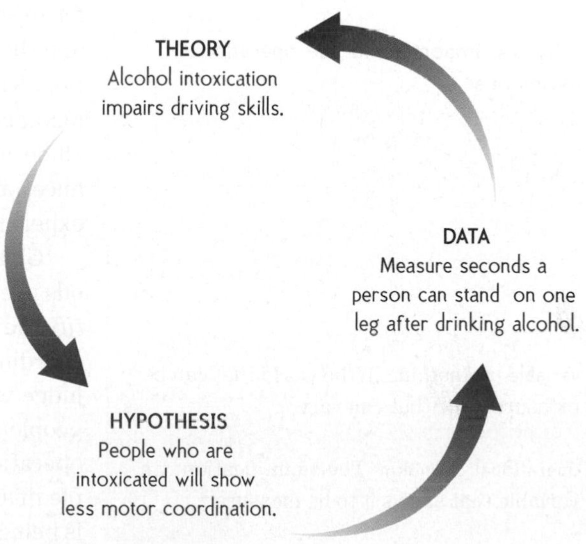
THEORY
Alcohol intoxication impairs driving skills.
DATA
Measure seconds a person can stand on one leg after drinking alcohol.
HYPOTHESIS
People who are intoxicated will show less motor coordination.

Torsten Wiesel and David Hubel were recording the activity of cats' nerve cells in brain areas associated with vision, they were hoping to find that these cells would respond when the cats were observing slides of dots. After much disappointing work that produced no significant activity in these cells, the projector suddenly jammed between slides, and the cells unexpectedly began to fire at an astonishing rate! The jammed slide had produced a visual "edge" on the screen, leading Wiesel and Hubel to discover that instead of responding to simple dots, cells in the brain area they were studying respond to lines and edges. A Nobel prize ultimately recognized the importance of this serendipitous finding.

REVIEWING the Principles | What Is Scientific Inquiry?

Our subjective beliefs, such as intuitions, can be biased or based on limited information. To explain behaviour, scientists must use objective, systematic procedures to measure it. The four goals of psychological science are to describe (what), predict (when), control (what causes), and explain (why) behaviour and mental activity. The empirical process is based on the use of theories to generate hypotheses that can be tested by collecting objective data. Theories in turn must be adjusted and refined as new findings confirm or disprove different hypotheses. Generally, effective theories will generate a number of testable hypotheses.

What Are the Types of Studies in Psychological Research?

Once the researcher has defined the hypothesis, the next issue to be addressed is the type of study design that ought to be used. There are essentially three main types of designs to choose from: *experimental, correlational,* and *descriptive.* These designs differ in the extent to which the researcher has control over the variables in the study and therefore in the extent to which the research can make conclusions about causation. **Variables** are things in the world that can be measured and that can vary. For instance, some of the variables in the alcohol and driving study example would include amount of alcohol consumed, level of intoxication, motor coordination, balance, and so on. Notice that the term *variable* can refer to something that the experimenter either measures or manipulates.

Why is it important to use operational definitions?

Clearly, researchers must define variables in precise ways that reflect the methods used to assess them. They do this by using **operational definitions**, which *quantify* the variables in order to measure them. For example, if you choose to study how coordination is affected by alcohol, how will you quantify "coordination" in order to judge whether it is affected by alcohol? One option might be to measure how easily people can touch their fingers to their noses with their eyes closed. In this case, the operational definition for coordination could be the number of inches by which people miss. This concrete definition would help other researchers know precisely what is being measured, which would then allow them to replicate the research.

variable Something in the world that can be measured and that can vary.

operational definition The quantification of a variable that allows it to be measured.

An Experiment Involves Manipulating Conditions

In experimental research, the investigator has maximal control over the situation. An **experiment** is defined as a study wherein variables are both measured and manipulated. The basic idea is that one variable is manipulated in order to examine how it affects a second variable. In studying how alcohol intoxication affects people's ability to drive, you could manipulate whether participants were intoxicated and then measure their driving performance. The *conditions* of the experiment refer to the experimental group to which a participant is assigned by the researcher. Within this context, what is manipulated is referred to as the **independent variable**, and what is measured is referred to as the **dependent variable**. That is, the conditions of the experiment refer to the different levels of the independent variable (such as impaired or sober) to which participants are assigned. In this example, driving performance is the dependent variable since it *depends* on the level of the independent variable. Thus, for example, to study whether alcohol increases risky driving decisions, researchers assigned participants to independent variable conditions in which they consumed a drink that either contained alcohol or did not contain alcohol. A second independent variable in this study was whether participants were led to believe that the drink contained alcohol or did not. The dependent measure was how they performed on a driving simulator. Those participants who had consumed alcohol made more risky driving decisions, and this occurred whether or not they believed they had consumed alcohol (Burian, Hensberry, & Liguori, 2003).

The benefit of experiments is that they allow the researcher to study the *causal relationship* between two variables. If the independent variable (such as intoxication state) proves to influence the dependent variable (such as driving performance) in a systematic way, then the independent variable is assumed to be the cause of the change in the dependent variable. In making the decision about whether or not to perform an actual experiment, the crucial issue is whether it is desirable to show a causal relationship between two variables.

ESTABLISHING CAUSALITY A properly performed experiment depends on rigorous *control.* Control refers to the steps that are taken to minimize the possibility that anything other than the independent variable may have been affecting the outcome of the study. In experiments, a **confound** is anything affecting a dependent variable that may unintentionally vary between the different experimental conditions of a study. A properly performed experiment requires making sure that the only thing that varies is the independent variable. Control thus represents the foundation of the experimental approach, in that it allows the researcher to rule out *alternative explanations* for the observed data. The more confounds and thus alternative explanations that can be eliminated, the more certain we can be that the independent variable actually was responsible for producing the change (or effect) in the dependent variable.

For example, in the hypothetical study of alcohol and driving performance, what if a car with an automatic transmission was used to assess driving when participants were sober, but a car with a manual transmission was used to assess performance when participants were intoxicated? Given that manual transmissions require greater dexterity to operate than automatic transmissions, any apparent effect of intoxication on driving performance might actually have been caused by the change in the kind of car.

Other potential confounds include changes in the sensitivity of the measuring instruments, such as a systematic change in a scale so that it weighs things more heavily in one condition than another, or even changes in the time of year when the experiment is conducted. For this reason, researchers have to be vigilant in watching for

experiment A study that tests causal hypotheses by measuring and manipulating variables.

independent variable In an experiment, the condition that is manipulated by the experimenter to examine its impact on the dependent variable.

dependent variable In an experiment, the measure that is affected by manipulation of the independent variable.

confound Anything that affects a dependent variable that may unintentionally vary between the different experimental conditions of a study.

How is causation between two variables established?

potential confounds. Fortunately, however, controlling confounds is straightforward: change nothing but the independent variable in an experiment.

RANDOM ASSIGNMENT IS USED TO ESTABLISH EQUIVALENT GROUPS

One major potential confound in studies is any differences that exist between groups that are exposed to different conditions. Suppose, for instance, that you want to study the effect of electric shock on memory. Because you are concerned about the ethics of the study, you ask people how afraid they are of shock, and then you assign the high-fear participants to the no-shock condition and the low-fear participants to the high-shock condition. Any differences in the reaction observed may be due not to the shock but instead to differences in how participants are assigned to the conditions, which is known as *selection bias.* Selection bias also occurs when participants differ between conditions in unexpected ways. For instance, let's once again consider the intoxication and driving study. Suppose you have two experimental conditions, one in which people drink a small amount of alcohol and one in which they drink a large amount. What follows if the people assigned to the large-amount condition just happen to be heavier drinkers and are therefore less affected by alcohol? Indeed, some heavy drinkers develop such a tolerance to alcohol that they show few outward signs of intoxication even when they have blood levels of alcohol that would knock out or even kill a typical person (Chesher & Greeley, 1992). How do you know if the people in the different conditions of the study are equivalent? The problem is that you really can't. You could give them measures on drinking habits and so on, but you can never be sure that you have assessed all possible factors that may differ between the groups. In spite of this problem, the results from relatively uncontrolled studies can influence public policy, even in Canada (Mann et al., 2002). For example, changes were made in the driver's licence suspension law based on blood alcohol concentrations and their presumed relationship to driver fatalities without controlled experimental studies

The only way to ensure equivalency of groups is to use **random assignment**, in which each potential research participant has an equal chance of being assigned to any level of the independent variable. Thus, perhaps you flip a coin to determine whether people are in the control group or the treatment group. Now, of course, there are bound to be individual differences among participants, and it is possible that there are some differences between conditions on a variety of factors. But these differences will tend to average out when participants are assigned randomly so that the groups are equivalent *on average.* Random assignment helps balance out both known and unknown factors.

random assignment The procedure for placing research participants into the conditions of an experiment in which each participant has an equal chance of being assigned to any level of the independent variable.

correlational study A research method that examines how variables are naturally related in the real world, without any attempt by the researcher to alter them.

Correlational Designs Examine How Variables Are Related

Many times it is not possible to manipulate the variables of interest, and you have to deal with them as they naturally occur. A **correlational study** cannot show causation; it examines how variables are naturally related in the real world, without any attempt by the researcher to alter them. To study the effects of alcohol intoxication on behaviour, one option would be to sift through police records of alcohol-related accidents. Under this scenario, the goal would be to match the reported intoxication levels of drunken drivers with some measure of their driving performance, such as the severity of the drivers' accidents. A good example of this is a study in which researchers demonstrated that the more alcohol was available in different communities, the more likely it was that there were late night, single-vehicle accidents, a measure believed to indicate drinking and driving (Gruenewald et al., 1996). In this case you would have data that allow you to compare how intoxication *might* have affected driving performance, but, unlike in the experiment scenario, you haven't actually *shown* that alcohol affected driving performance. To see why, let's examine the potential problems with correlational studies that prevent researchers from drawing causal conclusions.

In correlational studies, it is always possible that some extraneous factor is responsible for the apparent relation between your variables. This **third-variable problem** occurs when you cannot directly manipulate the independent variable and therefore cannot be confident that it was the actual cause of differences in the dependent variable. For instance, it is possible that people who were really stressed in their daily lives were more likely both to drink regularly and to be distracted while driving. Thus, the cause of both drinking and bad driving is the third variable, stress. The third variable is a problem for all correlational studies. Sometimes the third variable is obvious. For instance, if you were told that the more churches there were in a town, the greater the rate of crime, you would not be tempted to say that churches cause crime, since obviously the population size of a city determines both the number of churches and the frequency of crime. But sometimes the third variables are not so obvious and may not even be identifiable. For instance, you have heard that smoking causes cancer. For ethical reasons researchers can't manipulate smoking levels in humans to see who gets cancer. Suppose that some genetic defect makes people more likely to smoke and also more likely to get cancer. Indeed, recent evidence indicates that a particular gene predisposes some smokers to developing lung cancer (Paz-Elizur et al., 2003). Although this may seem unlikely, it is possible, and the simple possibility of a third-variable explanation rules out concluding a causal relationship. Indeed, tobacco companies have long used this sort of logic to cast doubt on the claims that smoking causes cancer. Although the overwhelming evidence suggests that smokers die sooner than nonsmokers, and other research shows that tobacco causes cancer in animals, it is always possible that some unidentified third variable is responsible for the relationship between smoking and cancer in humans.

third-variable problem When the experimenter cannot directly manipulate the independent variable and therefore cannot be confident that another, unmeasured variable is not the actual cause of differences in the dependent variable.

What are some of the problems with correlational studies?

Another problem with correlational studies is in knowing the direction of the cause-effect relation between variables. Suppose some researchers took a survey and found that people who reported sleeping a great deal also reported having a lower level of stress. Does reduced stress lead to longer and better sleep, or does sleep reduce people's stress levels? In this case, both scenarios seem equally plausible, an ambiguity that is known as the *directionality problem.* Without manipulating the frequency of sleep or the degree of stress as an independent variable and measuring the other as a dependent variable, we can never conclude with scientific certainty that one variable *caused* the other to change. The direction of causal relationship, if any, remains ambiguous.

Correlational studies are widely used in psychological science because they provide important information about how variables are naturally related. Some questions cannot be studied via an experiment because manipulating an independent variable would be unfeasible, unethical, or both. In this case, you need to conduct correlational research. Knowing the relation between variables allows us to make predictions. For instance, correlational research has established a relation between complexion and risk of skin cancer from sun exposure. Complexion cannot be manipulated, but knowing that people with fair skin are especially at risk for being harmed by the sun allows us to warn them to be especially vigilant with sunscreen. Similarly, correlational research has identified a strong relation between depression and suicide, and therefore clinical psychologists often assess symptoms of depression in order to determine suicide risk. Indeed, most of the research on psychopathology uses the correlational method, since it is unethical to try to induce mental disorders in people to study the effects. Typically, researchers who use the correlational method use statistical procedures to try to rule out potential third variables. By showing that a relation between two variables holds even when potential third variables are taken into account, researchers can be more confident that the relationship is meaningful. In the real world, events *do* happen together, and therefore understanding their relationship allows us to predict one from the other.

Thinking Critically

Does Smoking Marihuana Cause Heart Attacks?

Understanding the limitations of correlational research and how confounds might affect the outcomes of studies is important when dealing with the increasing amount of scientific information about ourselves and our world. Knowing how to evaluate such information is an indispensable critical thinking skill. What should we believe? What should we ignore? These questions are crucial because we often make important decisions for ourselves and for society in direct response to conclusions we draw from science-based research. The practical benefit of thinking critically about the methods of psychological science is that doing so provides an invaluable basis for you to be an informed consumer of scientific information.

While medical use of marihuana continues to be debated in the United States, in Canada the Marihuana Medical Access Regulations, implemented in 2001, permit access to marihuana for clearly defined conditions, such as severe complications of cancer, epilepsy, and multiple sclerosis. But people's opinions about such use of the drug, which affect public policy debates, depend largely on interpreting available scientific information about marihuana—both its positive and negative aspects. For example, studies in Canada continue to explore marihuana use and its psychological effects, such as whether it can trigger schizophrenia during adolescence and young adulthood (Hall & Degenhardt, 2006). In another case, on March 3, 2000, the *New York Times* published an article that began: "In what is believed to be the first documented link between smoking marijuana and heart attacks, a study has found that a middle-aged person's risk of heart attack rises nearly fivefold in the first hour after smoking marijuana." The conclusion is clear: Middle-aged people are at increased risk for a heart attack if they smoke marihuana. But is the conclusion valid?

According to the article, the researchers who conducted the study took brief case histories from 3,882 middle-aged heart attack patients. Of these, 124 were found to have been regular users of marihuana. Of these 124, 37 had smoked within the 24 hours preceding their attack, and 9 had smoked within the actual hour. On the basis of this evidence, the researchers suggested that there is a significantly increased risk of heart attack within the first several hours after smoking marihuana. Let's consider several important issues regarding these data—and the conclusions that we can draw.

First, the study looked only at people who had had heart attacks, which is a nonrandomized sampling technique. As a result, the findings don't pertain to *all* middle-aged people. Rather, the results generalize only to middle-aged people who have had heart attacks *and* who smoke marihuana. To assess the risk of marihuana-induced heart attacks in all middle-aged people, we would need to compare the rates of heart attacks in the middle-aged population that smokes marihuana and in the middle-aged population that doesn't. If the smoking population shows a greater rate of heart attacks, *then* we should advise middle-aged people to think twice before lighting up. But even then, this would be a correlational rather than a true experimental design, and causality could not be established.

Second, we need to consider whether factors other than marihuana may have influenced the rate of heart attacks for marihuana smokers. Did these patients do anything else near the time of the heart attack that may have contributed to—or fully caused—the cardiac event? One possibility might be that the heart attack victims had gotten high prior to engaging in sex or exercise, two activities that can contribute to the occurrence of attacks. These possible confounds are examples of third-variable problems that again prevent any conclusions about causality. However, the *Times* article does not make it clear whether or how the researchers treated this concern. As a result, we again lack necessary information, and an informed reader will have serious doubts about the link between marihuana and heart attacks.

Critical thinkers might consult the original article, since media can distort the results of studies. The study appeared in the scientific literature following its report in the *New York Times,* and it indicates that the researchers did their best to account for other behaviours in the sample, such as using other drugs or having sex (Mittleman, Lewis, Maclure, Sherwood, & Muller, 2001). They conclude that the effects are real, even controlling for other potential causes. However, significant problems with sampling remain. The marihuana smokers were much younger (44 vs. 62), more likely to be male, more out of shape, and more likely to smoke cigarettes. Could other lifestyle factors in addition to drug use—such as diet, exercise, socioeconomic status, and regular physician checkups—be associated with heart disease? What does this finding tell us about the cardiac risks of smoking marihuana for adults in general? ■

Descriptive Studies Observe and Classify Behaviour

Descriptive studies, sometimes called observational studies because of the manner in which the data are typically collected, involve observing and noting behaviour in order to provide a systematic and objective analysis. For instance, an observer might watch the types of foods that people eat in cafeterias, measure the time that people spend talking during an average conversation, or evaluate the mating behaviour of animals within a group. Some studies may rely on observing behaviour at regular time intervals, spanning durations from as short as seconds to as long as years. In this manner, the researcher can keep track of what a subject is doing at each point of time, and behaviours can be studied that may take years to unfold, such as tracking the job histories of college graduates.

There are two basic types of descriptive studies. In **naturalistic observation**, the observer is apart from and makes no attempt to alter or change the situation. In contrast, in **participant observation**, the researcher is involved in the situation. An example of the latter was conducted by social psychologists who joined a doomsday cult to see how the cult members would respond when the world did not end (which they correctly assumed would not happen). There are problems with participant observation, especially related to the observer's losing objectivity and the participants' changing their behaviour if they know they are being observed. Thus, participant observers need to try to keep their objectivity and to do what they can to minimize their impact on the situation.

Descriptive techniques are especially valuable in the early stages of research, when researchers are simply trying to see whether a phenomenon exists. From initial observations, researchers can design ways to study the behaviour in a more systematic

descriptive study A research method that involves observing and noting the behaviour of people or other animals in order to provide a systematic and objective analysis of behaviour.

naturalistic observation A passive descriptive study in which observers do not change or alter ongoing behaviour.

participant observation A type of descriptive study in which the researcher is actively involved in the situation.

How does an experiment differ from a descriptive study?

fashion. The power and value of even the simplest observations should not be underestimated. For example, while studying the behaviour of finches in the Galápagos Islands, Charles Darwin noted that the birds had a number of different beak types, and moreover, that how a finch foraged for food depended on its type of beak. Birds with short, large beaks mostly fed on large seeds out in the open, while birds with thin, long beaks tended to search for food in the crevices and cracks of the rocky shoreline. From these simple observations of finches and their behaviour, Darwin surmised that, over many generations within a species, both the body and the behaviour of animals adapt or change in order to take maximum advantage of the resources available in the environment. Thus the theory of evolution had its foundations in simple observations.

REVIEWING the Principles | What Are the Types of Studies in Psychological Research?

There are three main types of studies in psychological research: experiments, correlational studies, and descriptive studies. In experiments, researchers manipulate the independent variable to study how it affects the dependent variable. An experiment allows a researcher to establish a causal relationship between the independent and dependent variables and to avoid the directionality problem when trying to understand how one variable might affect another. Therefore, experiments depend on control, so that the only thing changing is the independent variable. Not all questions can be addressed by an experiment. Alternative research designs include observational or correlational studies, in which the researcher examines behaviour as it naturally occurs. Although these latter types of studies are useful for describing and predicting behaviour, they do not allow the researcher to assess causality.

What Are the Data-Collection Methods of Psychological Science?

Once the researcher has established the best design for conducting the study, the next task is to choose a *method* to collect the data. A fundamental principle of psychological research is that the question itself dictates the appropriate method to use for collecting data. In short, you start with an appropriate, theory-based question, as discussed above, and then ask yourself: What data do I need to collect to best answer my question? The first issue in selecting a data-collection method is to understand what level of analysis a particular question is addressing. Recall from Chapter 1 the three major research categories that span the levels of analysis: social, individual, and biological. You then select data-collection methods that are appropriate for questions at that level.

Researchers studying different levels of analysis tend to use different types of methods to collect data. For instance, at the social level, researchers often collect data by observing people and seeing how they interact with each other. At the individual level, researchers question participants or use indirect assessments, such as seeing how quickly participants respond to a particular question or whether they accurately discriminate between different sizes of objects. At the biological level, researchers use techniques to measure bodily and brain processes in animals or humans. For instance, they might use brain imaging to examine how people respond to observing pictures of scary objects. Researchers often use a variety of data-collection methods depending on the scientific question they are investigating. Let's turn now to the major methods used to collect data to study the mind, brain, and behaviour.

Observing Is an Unobtrusive Strategy

Observational techniques involve systematic assessment and coding of overt behaviour, such as watching people's gestures during social interaction or coding the eating or sexual behaviour of animals that are injected with drugs that affect brain function. As the name implies, these techniques are the major data-collection method for descriptive studies, but they can be used in experiments and correlational designs as well.

observational technique A research method of careful and systematic assessment and coding of overt behaviour.

Using observational techniques to collect data requires you to make at least three decisions. First, should the study be conducted in the laboratory or in the natural environment? At issue is the extent to which you are interested in behaviour as it occurs in the real world, and the possibility that the laboratory setting will lead to artificial behaviour (Figure 2.3).

Second, how should the data actually be collected—as a written description of what was seen, or as a running tally of prespecified categories of behaviour? For example, suppose you have a theory that people greet friends and family more effusively at the airport than at the train station. You could operationally define different categories of "effusive" greetings, such as "hugging," "kissing," "hand shaking," and the like. Then, as each episode of greeting is observed at the gate or platform, you could mark down the proper categories on a tally sheet. Using preestablished categories is usually preferred for more objectivity. However, badly chosen categories

2.3 Observational Technique Primate behaviour is often observed in natural settings. Here Jane Goodall observes a family of chimpanzees. Animals are more likely to act naturally when they are in their native habitat, relative to animals in captivity.

may mean that you miss important behaviour. Furthermore, the choice of which type of description is preferable may not always be clear.

Third, should the observer be visible? The concern here is that observation might alter the behaviour being observed. This effect is known as **reactivity**. For instance, people may feel compelled to try to make a positive impression. A classic example of this happened when researchers tried to investigate the effects of workplace conditions, such as lighting, on productivity at the Hawthorne electric plant in Cicero, Illinois (U.S.A.), between 1924 and 1932. The workers, who knew they were being observed, increased productivity to impress the researchers and their supervisors, thereby obliterating the effect of interest to the researchers. The *Hawthorne effect* refers to changes in behaviour that occur when people know that others are observing them. In general, observation should be as unobtrusive as possible and behaviour should be allowed to flow naturally (Adair, 1984).

2.4 Robert Rosenthal Social psychologist Rosenthal's studies of experimenter expectancy effects showed that subtle cues given off by the experimenter can affect how research participants behave.

How might culture influence observer bias?

OBSERVER BIAS In conducting observation research, it is important to be aware of the possibility of **observer bias**, which refers to systematic errors in observation that occur due to an observer's expectations. For instance, if observers are coding the facial expressions of men and women, they may be more likely to rate female expressions as indicating sadness because they believe that men don't show sadness. Likewise, the same level of assertiveness may be rated more strongly for females than for males because it might not conform to expectations about how men and women typically behave. This can be especially problematic if cultural norms favour inhibiting or expressing certain behaviours. There is even evidence that observer bias can lead to actual changes in the behaviour of the people or animals being observed, which is known as the **experimenter expectancy effect**. For example, in one classic study conducted in the 1960s by Robert Rosenthal (Figure 2.4), college students trained rats to run a maze. Half of the students were told in advance that their rats were genetically bred to be really good at running mazes, whereas the other group was not given this expectation. In reality, there were no initial differences between the groups of rats. Nonetheless, those trained by the students who believed that their rats were bred to be fast learners actually did learn the task more quickly. Thus, the students' expectations might have altered how they treated their rats, which in turn influenced the speed at which the rats learned. To protect against experimenter expectancy effects, it is best if the person running the study is *blind* to, or unaware of, the hypotheses of the study. It is also desirable to have multiple observers to protect against idiosyncratic coding. However, the significant challenge with multiple observers is to minimize variability in how each observer catalogs or interprets the behaviours encountered during observation.

reactivity The effect that occurs when the knowledge that one is being observed alters the behaviours being observed.

observer bias Systematic errors in observation that occur due to an observer's expectations.

experimenter expectancy effect Actual change in the behaviours of the people or animals being observed that is due to observer bias.

Asking Takes a More Active Approach

If observation is an unobtrusive approach for studying behaviour, *asking* people about themselves—what they think, why they act, and how they feel—is a much more interactive way of collecting data. Asking-based methods include surveys, interviews, questionnaires, and self-reports; all are aimed at getting people to divulge information about themselves. The type of information sought ranges from demographic facts (e.g., ethnicity, religious affiliation) to past behaviours, personal attitudes, beliefs, and so on. Have you ever used an illegal drug? Should people who drink and drive be jailed for a first offense? Are you comfortable sending food back to the kitchen in a restaurant when there is a problem? These questions require people to recall certain events from their lives or reflect on their mental or emotional states. For example, to study whether people who start drinking earlier are more likely to be involved in alcohol-related car accidents, Ralph Hingson and colleagues (2002) interviewed nearly 43,000 people and asked them the age when they started drinking, whether they drove after drinking too much, and whether they had ever had an accident because of their drinking. The researchers found strong evidence that early drinking was related to both driving and crashing while drinking. Although 18 percent of those who started drinking after age 21 reported driving while drinking, more than 50 percent of those who started drinking before age 16 reported doing so. A similar pattern occurred for reports of car accidents.

A critical issue in asking-based research concerns how to frame the questions used in the study, as there are several different options. *Open-ended questions* allow respondents to answer in as much detail as they feel is appropriate. In contrast, *closed-ended questions* require respondents to select among a fixed number of options, as in a multiple-choice exam (Figure 2.5). Ultimately, the researcher decides what style of question to use in relation to what will provide the most appropriate information for the hypothesis.

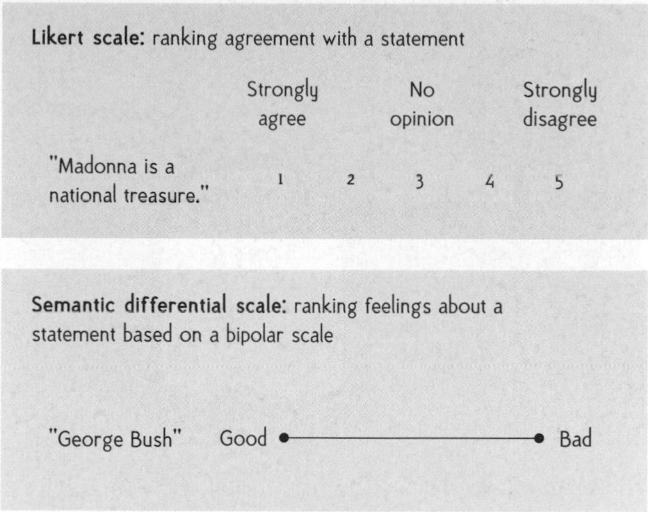

Likert scale: ranking agreement with a statement

	Strongly agree		No opinion		Strongly disagree
"Madonna is a national treasure."	1	2	3	4	5

Semantic differential scale: ranking feelings about a statement based on a bipolar scale

"George Bush" Good •————————• Bad

2.5 Quantifying Responses Examples of two different ways to quantify people's responses by offering a fixed number of options.

Asking-based methods have a number of strengths and weaknesses, but they depend on whether respondents are simply given a questionnaire or survey to fill out, or whether each person in the study is actually interviewed by a researcher. **Self-report methods** such as questionnaires can be used to gather data from a large number of people. They can be mailed out to the population of interest or handed out in appropriate locations. They are easy to administer, relatively cheap and cost-efficient, and can collect a great deal of data in a relatively short period of time. In contrast to self-reports, interviews can be used with great success for groups that cannot be studied via surveys. Furthermore, interviewing people gives the researcher the opportunity to explore new lines of questioning, since the answers a respondent gives to initial questions may inspire avenues of inquiry that were not originally planned.

Why are anecdotal reports limited in what they can tell us?

An important issue to be considered in both self-reports and interviews is how to select people for inclusion in the study. Psychological scientists typically want to know that their findings *generalize,* or apply, to people beyond those individuals who are in the study. For instance, they might wish to know that their results generalize to all women, or to all college students, or to students who belong to sororities and fraternities, or to everyone. The group of interest is known as the *population,* whereas the subset of people who are studied are known as the *sample. Sampling* is the process by which people are selected from the population to be members of the sample. Because it is desirable that the sample be representative of the population, researchers often use *random sampling,* in which each member of the population has an equal chance of being chosen to participate.

SELF-REPORT BIAS A problem common to all asking-based methods is that people often introduce biases into their answers. In particular, people may be hesitant to reveal personal information that casts them in a negative light. Consider the question, "How many times have you lied in order to get something that you wanted?" Although most people have lied at some point in their lives to obtain a desired outcome or object, few of us want to admit this to total strangers. At issue is the extent to which questions produce **socially desirable responding**, or "faking good," in which the person responds in a way that is most socially acceptable. Imagine having a middle-aged interviewer ask you to describe intimate aspects of your sexual life. Many of us would find talking about this quite embarrassing and therefore would not be very forthcoming. Moreover, even when respondents may not be purposefully answering incorrectly,

self-report method A method of data collection in which people are asked to provide information about themselves, such as in questionnaires or surveys.

socially desirable responding When people respond to a question in a way that is most socially acceptable or that makes them look good.

their answers may reflect less than accurate self-perceptions. In other words, research has shown that at least within North American culture, people have the tendency to describe themselves in especially positive ways, and in many cases this occurs because people believe things about themselves that might not actually be true. For instance, most people believe they are better than average drivers.

This tendency to express positive things about oneself is especially common in Western cultures, such as those of North America and Europe, but is less pronounced in Eastern cultures, such as Korea and Japan (as you will learn in Chapter 15). How questionnaires are worded is especially important in conducting cross-cultural research. The researcher has to be sensitive not only to any language differences among cultures, but also to any cultural norms that might shape how people answer questions. For example, Western cultures tend to prize individuality and speaking one's mind, whereas Eastern cultures favour modesty and not wanting to stand out from the group (Heine, 2003). Also, some words do not translate easily into other languages, such as *schadenfreude,* the German word for taking pleasure in another person's misfortune. Thus, the researcher would have to explain the concept to other cultural groups that lack a single word to describe this emotion. Any differences that emerge among cultures may reflect differences in language, in participants' willingness to report things about themselves publicly, or in ways they use numbers in measurement scales; or they may reflect genuine differences between the cultures. Therefore, the central challenge for cross-cultural researchers is to refine their measurements to rule out alternative explanations.

Case Studies Examine Individual Lives

case study A research method that involves the intensive examination of one person.

A **case study** involves the intensive examination of one person, typically one who is unusual. In psychology, case studies are most commonly conducted on those who have a brain injury or psychological disorder. Case studies of those with brain injuries have provided a wealth of evidence about which parts of the brain are involved in various psychological processes. In one case, a man who was accidentally stabbed through the middle part of the brain with a fencing foil lost the ability to store new memories, and case histories of individuals with damage to the front portions of their brains reveal consistent difficulties with inhibiting impulsive behaviours. Research at the University of Manitoba examined patients with multiple personality disorder at four different centres (Ross, Miller, Reagor, Bjornson, Fraser, & Anderson, 1990). The emerging patient clinical profile from all four centres exhibited similar characteristics, including 91.5 percent reporting a history of childhood physical and/or sexual abuse. Other similarities included somatic symptoms, borderline personality disorder criteria, and extrasensory experiences. The results of these case studies indicate that multiple personality disorder has a stable, consistent set of features. One famous case study in clinical psychology involved "Sybil," who apparently had *multiple personality disorder,* a rare condition in which a person appears to have more than one distinct personality. These multiple personalities were viewed as Sybil's way of coping with severe abuse she experienced during childhood. You will learn more about this disorder in Chapter 13.

The major problem with clinical case studies such as Sybil's is that it is difficult to know whether the researcher's hypothesis about the cause of the psychological disorder is correct. The researcher has no control over the person's life and is forced to make assumptions about the effects of various life events. Thus, the interpretation of case studies is often very subjective. Similarly, understanding the effects of brain injury or disease requires the assumption that apparent mental deficits were caused by the damage to the brain. Although many times such conclusions seem reasonable, especially if friends or family report that the person was different before the brain damage, the lack of control over the situation makes it difficult to draw firm conclusions. For instance, in 1966 a man named Charles Whitman climbed a tower at the University of Texas and began firing a rifle at passersby, killing 14 and wounding many more in a 90-minute spree. The night before, he had killed his wife and mother and had composed a letter trying to explain his actions. His case history, including his visits to a psychiatrist and the letter he wrote, indicate that Whitman might have had a brain tumour, and an autopsy did turn up evidence consistent with this assessment. But was a brain tumour responsible for Whitman's murderous actions? We will never know.

Profiles in Psychological Science

What Makes Killers Kill?

The subjective nature of case studies means that we need to be cautious in deciding how to interpret them. One of the hallmarks of critical thinking is to avoid relying on anecdotal cases that seem to prove a point, such as when someone proclaims, "I know smoking doesn't cause cancer because my uncle smokes and he is still alive." Anecdotal cases can be exceptions to general rules of behaviour and therefore must be critically examined. However, if you have a good theory and can accumulate evidence from multiple case studies, then the conclusions that are drawn are more persuasive. Let's look at how such an approach was used to test a novel hypothesis about why people become murderers.

Lee Boyd Malvo was 17 years old when he began killing people. His participation in the 2002 Washington, D.C.-area sniper killings came as a tremendous surprise to teachers and school officials who had known him years earlier and described him as a polite, hard-working, and intelligent student. Indeed, his grade-school teacher Beverly Clark described his personality as striking: "You look at the kid and you know he's cut out to be something great in life" (Kovaleski & Sheridan, 2003, p. C.01). How did this religious and studious young boy become a mass murderer?

In May 1998, Kip Kinkel shot and killed his parents. Earlier in the day, the 15-year-old had been arrested for purchasing a stolen gun at school. His father picked him up at the police station and then berated Kip for bringing shame on the family. A few hours later, Kip approached his father from behind in the kitchen and shot him in the head, using one of the father's many guns. He then waited two hours for his mother to come home, at which point he shot her repeatedly: "I shot her again so she wouldn't know that I killed her. . . . I loved my mom." The next morning he walked into the cafeteria at Thurston High School in Springfield, Oregon, and sprayed the room with semi-automatic gunfire, killing two students and wounding 25 others.

According to the neurologist Jonathan Pincus, both Malvo and Kinkel are typical of murderers, albeit younger than most (Figure 2.6). Pincus has examined more than 150 murderers, including Kinkel, and has found a combination of early childhood abuse, psychiatric disorders such as paranoia, and frontal-lobe dysfunction in more than 90 percent of these violent killers. None of these problems appears to be sufficient on its own, but when all three combine in an individual, serious trouble often results. Kinkel fits this pattern well. His parents were verbally abusive, and he displayed many mental health problems, such as depression and paranoid thoughts; apparently he heard voices telling him he was worthless. These same voices

commanded him to kill his parents and shoot at students in his school. Importantly, brain imaging showed evidence of abnormalities in several brain regions, especially the frontal lobes. This is relevant because the frontal lobes are crucial in helping people inhibit their impulses and control their behaviour. The cause of Kinkel's abnormalities is not known, but the consequences of having impaired frontal lobes have been well established in psychological research.

The immaturity of frontal lobes in adolescent killers raises perplexing policy issues regarding punishment; should they be treated as adults or as children? The psychologist Lawrence Steinberg and his colleagues (e.g., Steinberg & Scott, 2003) argue that adolescent cognitive capacities for self-control and decision making are not fully developed and lessen the extent to which adolescents can be held responsible for criminal behaviours. Making plans or decisions, thinking about their consequences, controlling impulses, and resisting social pressure all are functions of the frontal lobes. Although scientists once believed that the frontal lobes were pretty well developed in early childhood, overwhelming evidence from the past decade has established that frontal brain regions continue to develop well into early adulthood (Giedd, 2004). This means that adolescents are typically more impulsive and more easily influenced by others than are adults. Indeed, although laboratory research indicates that adolescents perform adequately on decision-making tasks, even comparable to adults, the sterile lab environment does not reflect the real world. This is especially true in situations in which crimes are committed, when decisions are made under great emotional arousal and are often accompanied by strong pressure from peers or accomplices. There is considerable evidence that Lee Malvo was coerced, even brainwashed by his father figure, John Allen Muhammad. Malvo's lawyer based his closing argument on psychological evidence suggesting that immature brain development makes adolescents less culpable for their crimes. The American jury apparently was persuaded by this science, since it did not impose the death penalty, as another jury had for the older Muhammad. ■

2.6 Lee Boyd Malvo, Kip Kinkel, and Kimveer Gill Psychologists think that immature brain development, in combination with other factors, contributes to violent behaviour. Lee Boyd Malvo and Kip Kinkel were both convicted of murder in their teens. Kimveer Gill went on a shooting rampage that killed one woman and injured 19 other people at Dawson College in Montreal.

Response Performance Measures Stimulus Processing

In Chapter 1 we noted that Wilhelm Wundt established the first laboratory in psychology in 1874. Wundt and his students pioneered many of the methods for studying how the mind works by examining how people respond on psychological tasks. The typical goal of measuring **response performance** is to quantify how a perceptual or cognitive process responds to a specific stimulus. Specifically, researchers infer how a stimulus is processed from how a person responds to it. Response performance is quantified in three basic ways. First, a researcher can measure **reaction time**, the speed of a response—which has been the workhorse method of cognitive psychology. The interpretation of reaction times is based on the idea that processing in the brain

response performance A research method in which researchers quantify perceptual or cognitive processes in response to a specific stimulus.

reaction time A quantification of performance behaviour that measures the speed of a response.

red blue green red blue yellow red blue

blue red green yellow red blue green red

green yellow red yellow blue red green blue

2.7 The Stroop Effect Quickly name the colour of ink that each of these words is printed in. Name the colour for each word in the row as quickly as you can. Notice how it took you longer to name the colours for the last row. This is called the Stroop effect, in which it takes longer to name the colour of words printed in conflicting colours. The interference from your automatic reading of the word slows you down.

takes time; the more processing a stimulus requires, the longer the reaction time to that stimulus will be. By manipulating what a subject has to do with a stimulus and measuring reaction times, researchers can gain much information regarding the different operations involved in stimulus processing. For example, suppose you have a theory that it takes longer to determine the shape of an object than it does to determine its size. Your theory generates the hypothesis that reaction times should be longer when people are asked to discriminate the shape of an object compared with when they have to discriminate the size of the object. Why? The assumption is that the longer the process takes, the longer the corresponding reaction time will be. An example of a reaction-time task is presented in Figure 2.7.

Researchers can also measure *response accuracy,* which usually concerns how well a stimulus was perceived. For example, does paying attention to a stimulus actually improve its perception? One way to study this in the visual domain would be to ask participants to pay attention to just one side of a computer screen while keeping their eyes focused on the centre of the screen. We can then present a stimulus requiring a discrimination response (to a question such as, Was the shape a hexagon or an octagon?) to either the attended or the unattended side of the screen. If the accuracy of responses is greater for stimuli presented on the attended side, then the implication is that attention is improving the perception of the stimulus.

Finally, researchers can measure response performance by asking people to make *stimulus judgments* regarding the different stimuli with which they are presented. Typical examples would be asking subjects to indicate whether a faint stimulus was noticed, or asking them to compare two objects and judge whether they are the same on some parameter, such as colour, size, or shape (Figure 2.8). The benefit of using response performance as a methodology is that it can be a relatively simple way to study cognition and perception.

2.8 Stimulus Judgments A common method for studying questions at the cognitive/perceptual level is to ask participants to make judgments about stimuli. In this case, the participant is required to decide whether the two letters on the screen are the same or different.

Body and Brain Activity Can Be Directly Measured

The activity of the body and the brain can be directly measured in different ways. For instance, it is now well established that certain emotional states influence the body in predictable ways. When people are frightened their muscles become tense, and their hearts start beating faster. Other bodily systems influenced by mood and mental states include blood pressure, blood temperature, rate of perspiration, pupil size, breathing rates, and so on. These various measures are examples of **psychophysiological assessment**, in which researchers examine how changes in bodily functions are associated with behaviour or mental state. For example, police investigators often use *polygraphs,* popularly known as lie detectors, which assess these bodily states, under the assumption that people who are lying are experiencing more arousal and therefore are more likely to show physical signs of stress. It should be noted that the correspondence between bodily response and mental state is not perfect. For instance, some stressful events lead to increased heart rate whereas others

psychophysiological assessment A research method that examines how changes in bodily functions are associated with behaviour or mental state.

lead to reduced heart rate. Nonetheless, these methods allow researchers to study bodily responses that circumvent the problems inherent in self-report, such as people being unwilling or unable to report on their thoughts or feelings.

electrophysiology A method of data collection that measures electrical activity in the brain.

ELECTROPHYSIOLOGY **Electrophysiology** is a method of data collection that measures electrical activity in the brain to see how it is related to cognitive and perceptual tasks. A researcher fits electrodes onto the participant's scalp; they act almost like small microphones to pick up the electrical activity of the brain (Figure 2.9). The device that measures brain activity is an *electroencephalogram,* or *EEG,* which provides a record of overall brain activity. This method is useful because different behavioural states produce predictable EEG patterns. As we will explore in Chapter 4, the EEG shows specific and consistent patterns as people fall asleep, and it reveals that the brain is very active even when the body is at rest, especially during dreams. As a measure of specific cognitive states, the EEG is limited because the recordings reflect all brain activity and therefore are too noisy to isolate specific responses to particular stimuli. A more powerful method of examining how brain activity changes in response to a specific stimulus involves conducting many trials and averaging across the trials. Researchers are thus able to observe patterns associated with specific events, hence the name *event-related potential* or *ERP.* The real advantage of the EEG and the ERP is not that they provide information about *where* in the brain an electrical event comes from, but that they determine exactly *when* it happens. They provide maps of approximate origins of various electrical signals and give insight into how the brain processes information on the basis of which regions are activated and when.

Electrophysiological measures can be used with other methods to assess complex behaviours, such as drinking and driving. Working across levels of analysis, researchers have measured EEGs and ERPs in relatives of alcoholics and found that their brain responses appear to be muted overall (Bierut et al., 2002). Interestingly, the researchers have linked these brain responses to genes that are believed to predispose people to alcoholism. Over the past decade, researchers have made significant advances in the use of genetic methods to predict alcoholism, and although there is unlikely to be a specific gene, we now have compelling evidence that many

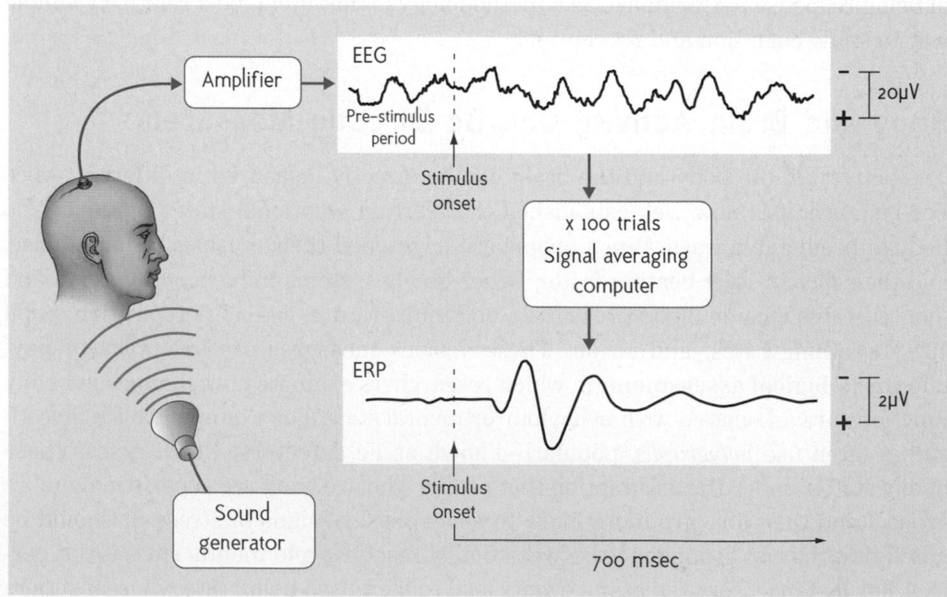

2.9 Measuring the Brain's Electrical Activity The relatively small electrical responses to specific events can be observed only by averaging the EEG traces over a series of trials. The large background oscillations of the EEG trace make it impossible to detect the evoked response to the sensory stimulus from a single trial. However, averaging across tens or hundreds of trials removes the background EEG and leaves the event-related potential (called ERP).

aspects of alcoholism have a strong genetic component. For instance, evidence indicates that children of alcoholics report needing to consume much greater quantities of alcohol than most people before feeling alcohol's effects and are therefore much more likely to report drinking large amounts of alcohol (Schuckit et al., 2001). In cases of drunk driving, some individuals may consume large amounts of alcohol to obtain subjective effects, and these large amounts of alcohol may interfere with their motor skills and reaction times without their awareness. This suggests that some individuals may fail to notice that they are impaired in their motor and cognitive abilities because they do not feel as though alcohol has had much effect.

BRAIN IMAGING The electrical activity of the brain is associated with changes in the flow of blood as it carries oxygen and nutrients to the active brain regions. Brain-imaging methods measure such changes in the flow, and by keeping track of these changes, researchers can monitor which brain areas are active during a study. Imaging is a powerful tool for uncovering where different systems reside in the brain and the manner in which different brain areas interact in order to process information. For example, research has shown that certain brain regions become active whenever people look at pictures of faces, whereas other brain regions are active when people try to understand what other people are thinking. In another example of drinking and driving, one study found that alcohol leads to a one-third reduction in activity in areas of the brain concerned with vision, which suggests that alcohol may impair drivers by directly affecting their sight (Levin et al., 1998). We will discuss the two major imaging technologies.

Positron emission tomography, or **PET**, was developed in the late 1980s. It is the computer-aided reconstruction of the brain's metabolic activity through the use of a relatively harmless radioactive substance injected into the bloodstream. The research participant lies in a special scanner that detects the radiation, and a three-dimensional map of the density of radioactivity within the participant's brain is produced (Figure 2.10). This map is useful because as the brain performs a mental task, blood flow increases to the most active regions, leading to more emitted radiation. The amount of radioactivity emitted by a brain region roughly corresponds to the amount of electrical activity in the local neurons. By scanning participants as they perform some psychological task (for example, looking at pictures of faces expressing fear), researchers obtain a map of brain metabolic activity during the task. However, since the entire brain is extremely metabolically active all of the time, scans must also be made while the participant performs another, closely related task (for example, looking at pictures of faces with neutral expressions). By subtracting one image from the other, experimenters obtain a "difference image" of which brain regions are most active during the task in question. In this way, regions of the brain can be correlated with specific mental activities.

positron emission tomography (PET) A method of brain imaging that assesses metabolic activity by using a radioactive substance injected into the bloodstream.

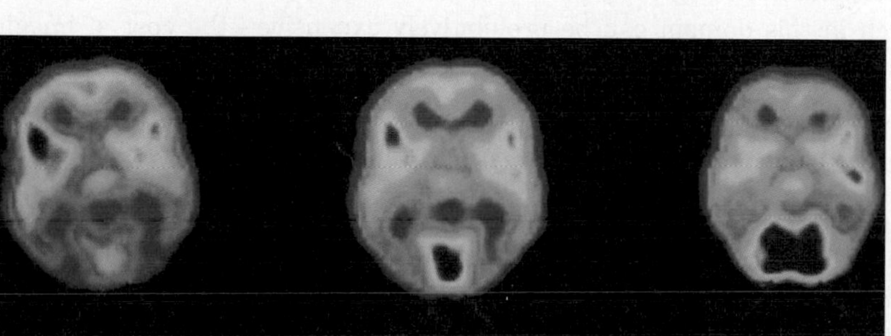

2.10 Positron Emission Tomography (PET) A PET scan of blood flow in a subject's brain under different conditions.

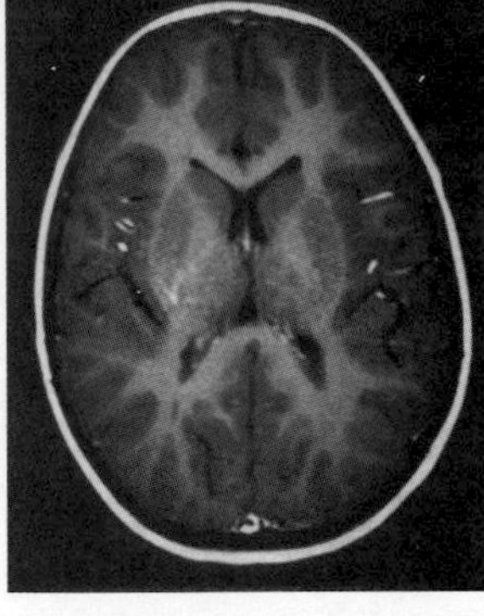
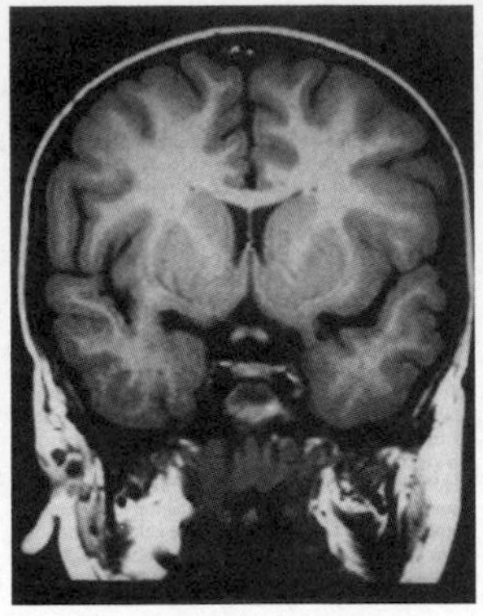
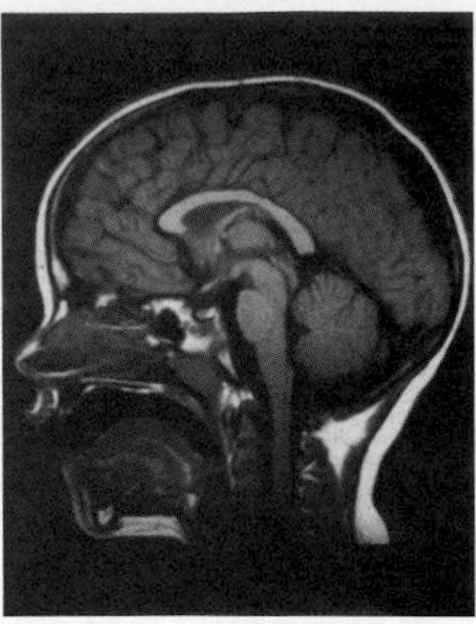

2.11 Magnetic Resonance Imaging (MRI) MRI provides very high quality images of intact human brains.

Magnetic resonance imaging, or **MRI**, is the newest and perhaps most powerful of imaging techniques. It relies on the fact that protons contained in hydrogen, a major component of water and fat, behave like tiny magnets. In MRI, which is not dangerous, a research participant lies in a scanner that produces a powerful magnetic field. The field causes the protons in the participant's brain to line up with it, just as a magnetized needle turns to point north. Radio waves are then sent through the magnetic field and cause the protons to briefly align to a different orientation. As they return to the direction of the MRI magnetic field, energy is released in a form that can be measured by detectors surrounding the head. Since the protons in fat and water release this energy differently, researchers can discriminate between them, which results in a high-resolution image of the brain (Figure 2.11).

MRI images are extraordinarily valuable for determining the location of brain damage, for example, but they can be even more profitable when used to create an image of the working brain. This is what **functional magnetic resonance imaging**, or **fMRI**, can do; like PET, it makes use of the brain's blood flow to map its activity, scanning the participant during the performance of several tasks. Whereas PET works by measuring the flow of blood directly, fMRI works by assessing changes in the oxygen level of blood, which is an indirect measure of blood flow. In all brain-imaging methods, including fMRI, the tasks used should differ in only one way, which reflects the particular mental function of interest. The images are then compared to examine differences in blood flow and therefore brain activity (Figure 2.12).

The value of measuring brain activity is that clear links can be made between actual activity in the brain and aspects of personality, social behaviour, cognition, and the like. Thus, many different types of researchers are now using brain imaging. For instance, both cognitive and social psychologists have used fMRI to study how people interpret facial expressions. Brain imaging is also widely used to help diagnose psychological disorders, especially those that may result from brain injury or disease. In particular, it can be used to assess the relationship between epilepsy and brain structure disintegration (Bernasconi, Natsume, & Bernasconi, 2005). However, research in this domain can be prohibitively expensive—the cost of imaging machines, their upkeep, and their attendant personnel runs in the millions of dollars.

magnetic resonance imaging (MRI) A method of brain imaging that produces high-quality images of the brain.

functional magnetic resonance imaging (fMRI) An imaging technique used to examine changes in the activity of the working human brain.

Research with Animals Provides Important Data

Throughout the history of psychological science, many of the most important research findings have been obtained by studying the behaviour of animals. For instance, watching animals—usually rats—run through mazes or press levers to earn rewards led to the development of many of the principles of learning. Indeed, Pavlov's observa-

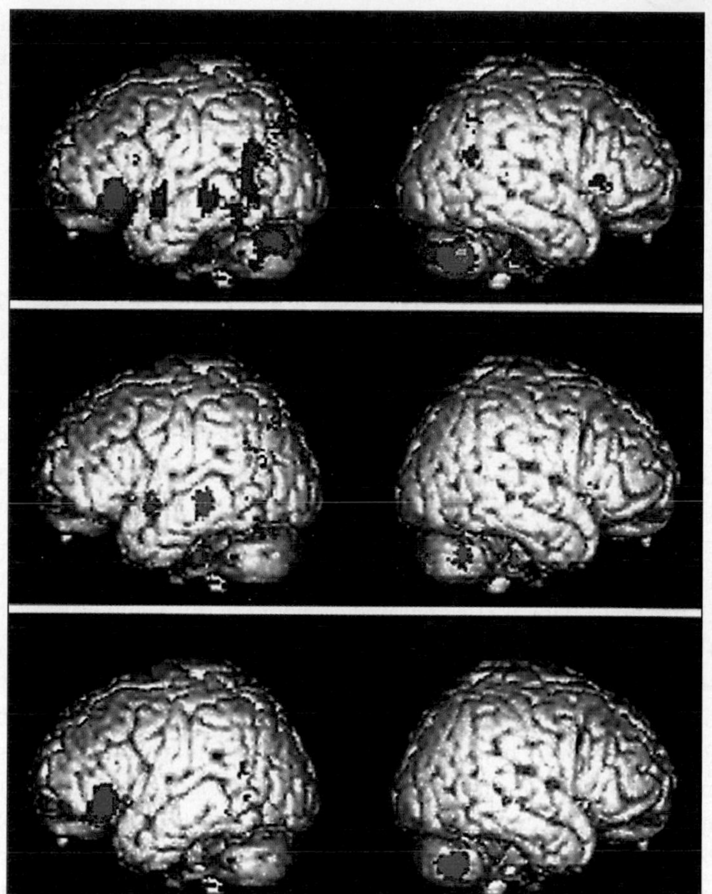

2.12 Functional Magnetic Resonance Imaging (fMRI) Activity maps showing changes in blood oxygenation, obtained by fMRI while the subject performed a task (and subtracted from another "control" image), superimposed on higher-resolution MRI images taken of the same brain. This procedure allows precise determination of which brain areas are active during different mental activities.

tion of a salivating dog inspired John B. Watson to launch the behaviourist movement (see Chapter 6). One of the central assumptions underlying Watson's behaviourism was that humans are subject to the same laws of nature as other animals. Although the behaviours displayed by humans might seem more complex than those of rats, the forces that control the behaviours of rats and humans are in many ways the same.

Besides in learning studies, animals have also been used in research that examines the physiological basis of behaviour. For example, psychological scientists manipulate levels of a certain chemical in the brain or body and then observe the effects on behaviour. Another method involves selectively removing specific brain regions and observing whether behaviour is impaired. Researchers also use electrophysiological methods to measure the activity of single neurons in the brains of animals. This method has helped us begin to uncover the different kinds of information single neurons code in different parts of the brain, something that can't be done by recording from the scalp. Recall the study by Hubel and Wiesel in which the stuck slide led to the serendipitous finding that cells in a cat's primary visual cortex respond to moving lines. Without the ability to measure directly the activity of single neurons in animals, the questions raised by Hubel and Wiesel could not have been answered.

Psychological scientists working at the biological level of analysis manipulate genes directly to examine their effects on behaviour. For instance, *transgenetic* mice are produced by inserting or manipulating the genes in a developing embryo, for example, by inserting strands of foreign DNA. The inserted gene is integrated into every cell of the body. As you will read in greater detail in Chapter 3, specific natural genes can also be targeted for genetic manipulations. Researchers can delete genes to eliminate their effects or replace genes in other locations to enhance their effects. These genes are known respectively as *knock-outs* or *knock-ins.* There has been rapid growth in the use of transgenetic and knock-out/knock-in models to study psychological activity, with the hope that these methods will help in the development of new therapies for psychological disorders.

There Are Ethical Issues to Consider

How do ethical issues constrain psychological research?

When scientists select a method of research, it is imperative that they make decisions with full knowledge of the ethical issues involved. Are the participants being asked to do something unreasonable? Is there a risk of physical or emotional harm from the study? Some ethical concerns are specific to the kind of method used, while other concerns apply across all methods. Therefore, to ensure the well-being of study participants, all colleges and universities have strict guidelines in place regarding both human- and animal-based research. **Institutional Review Boards (IRBs)** consisting of trained scholars, administrators, and legal advisers are given the task of reviewing proposed research to ensure that it meets the accepted standards of science. In Canada, IRBs operate according to the regulations set out by the Canadian Psychology Association (1992).

Institutional Review Boards (IRBs) Groups of people responsible for reviewing proposed research to ensure that it meets the accepted standards of science and provides for the physical and emotional well-being of research participants.

One of the more prominent ethical concerns with research is the reasonable expectation of *privacy.* If behaviour is going to be observed, is it okay to observe people without their knowledge? This question obviously depends on what sorts of behaviours one might be looking at. If the behaviours occur in public, then the concerns may be less than if one is interested in behaviours that are considered more private. For example, it would be okay to observe the behaviour of couples saying good-bye in an airport, as this is a public place, but it would be inappropriate to examine the private mating behaviours of couples—at least without their knowledge. The concern over privacy is compounded by the growing technology available for remotely monitoring people. Although someone might like to compare the behaviour of men and women in public bathrooms, would it be okay to install a discreet video camera to monitor people in restrooms? Likely not.

When asking people for information, are there topics that should not be raised because they may be insensitive or inappropriate? For example, many researchers would like to understand how physical and emotional trauma affect people in the months and years after a traumatic event. Although exploring such issues is necessary to develop strategies for overcoming anguishing and difficult experiences, researchers exploring such topics must *always* remain sensitive to how their line of questioning is affecting the individuals they are studying.

No matter what kind of method is employed, it is critical to consider who will have access to the data that have been collected. Participant *confidentiality* should always be carefully guarded so that personal information is not publicly linked to the behavioural findings of the study. Often the quality and accuracy of data directly depend on how confident the subjects are that their responses will be kept confidential, especially when emotionally or legally sensitive topics are involved. For extremely sensitive topics, it is preferable that the participants' responses be anonymous.

INFORMED CONSENT Research with human participants is a partnership based on mutual respect and trust. People who volunteer for psychological research, whether compensated or not, have the right to know what will happen to them during the course of the study. Ethical standards require providing people with all relevant information that would affect their willingness to become participants. This **informed consent** means that participants make a knowledgeable decision to participate. Typically, informed consent is obtained in writing; in the case of observational studies of public behaviour, to protect privacy, individuals are treated anonymously. Minors, the intellectually incapacitated, and the mentally ill cannot legally provide informed consent, and therefore the permission of a legal guardian is necessary.

informed consent A process in which people are given full information about a study, which allows them to make a knowledgeable decision about whether to participate.

It is not always possible to inform participants fully about details of a study. Sometimes researchers must use *deception,* which involves either misleading participants about the goals of the study or not fully revealing what will take place, because knowing the specific goals of the study may alter the participants' behaviour, thereby rendering the results meaningless. Deception is used only when other methods are not appropriate, when the goals of the research are important, or when the deception does not involve situations that would strongly affect the people's willingness to participate. If deception is used, a careful *debriefing,* or explanation of the study after its completion, must take place to fully inform participants of the goals of the study and the need for deception, and to eliminate or counteract any negative effects produced by the deception.

RELATIVE RISKS OF PARTICIPATION Another ethical issue is the *relative risk* to mental or physical health. Researchers must always be aware of what they are asking participants to do in the name of data collection. People cannot be asked to endure unreasonable amounts of pain or discomfort, either from stimuli or from the manner in which data measurements are taken. Fortunately, in the vast majority of research being conducted, these types of concerns are not an issue. However, although risk is low, researchers have to think carefully about the potential for risk to specific participants. For example, fMRI involves placing participants in powerful magnetic fields. As a result, participants must be carefully screened to ensure they have no metal in or on their bodies, even if it's only a tattoo with metal-based inks. Again, it is important to note that all research now conducted at colleges and universities must be approved by an IRB that is familiar with the various rules and regulations that protect research participants from possible physical or psychological harm. Most IRBs look at the relative trade-off between risk and benefit. There is much to be gained from the scientific enterprise, and sometimes it requires asking participants to expose themselves to some risk in order to obtain important findings. The *risk-benefit ratio* asks whether the research is important enough to warrant placing participants at risk. Poorly designed studies have little benefit to science, and therefore no element of risk is justified by them. Thus, scientists have an ethical obligation to conduct carefully designed research.

RESEARCH WITH ANIMALS Many of the major medical and scientific discoveries of the past century have involved the use of animals as research subjects. The use of animals allows researchers to examine biological mechanisms that cannot be studied in humans. Obviously, animal research should be undertaken only when the potential benefits outweigh the costs and when the studies are well designed. In addition, researchers have a responsibility to treat the animals as humanely as possible. Indeed, government regulations require a careful monitoring of animals used in research, and potential animal studies are screened for problems before they are conducted. Fortunately, the majority of those who conduct animal research are compassionate and do their best to minimize stress or discomfort in the animals they study. In Canada, all research with animals is conducted under the guidelines and directives of the Canadian Council on Animal Care (1993).

REVIEWING **the Principles** | What Are the Data-Collection Methods of Psychological Science?

There are five basic data-collection methods in psychological science; the choice of which to use is generally dictated by the research question. First, we can observe behaviour as it is taking place and either write down general descriptions of the behaviours or check off a tally sheet of prespecified behaviour categories. Second, we can ask people for information about their thoughts, feelings, and preferences by using surveys, questionnaires, interviews, and self-reports. Third, we can measure how quickly and accurately people respond to a stimulus. Fourth, we can directly measure electrical activity and blood flow in the brain, techniques that are increasingly being combined with the other three methods. Finally, researchers can use animal models in which genes, neurochemicals, or brain structures are altered to study the effects on behaviour. Regardless of the method chosen, it is essential always to consider the ethical consequences associated with the method of data collection; attention must be given to the relative risks and benefits associated with the study.

How Are Data Analyzed and Evaluated?

How do we ensure that our measurements are accurate?

Up to this point, we have examined the essential elements of scientific inquiry in psychology. In particular, we have examined how to frame an empirical question within the realm of theories, hypotheses, and research; discussed what type of study to run; and looked at the different methods of collecting data. This section focuses more closely on the issue of data, such as the characteristics that make for good data and the statistical procedures that are used to analyze them.

Good Research Requires Valid, Reliable, and Accurate Data

If data are to properly answer a hypothesis, they must actually address it. **Validity** refers to whether the data you collect address your question; valid data provide clear and unambiguous information from which to evaluate the theory or hypothesis. Suppose a theory predicts that children who are physically abused by their parents are more likely than nonabused children to use drugs in high school. The obvious hypothesis generated by this theory is that if we compare the childhoods of high school drug users and non–drug users, we should find that a greater percentage of the drug users were abused by their parents. In this case, examining the medical records of all the students might be one way to document whether abuse had occurred. Regardless of whether the hypothesis is supported, the data would be considered valid on the reasonable assumption that in children, abuse often leads to suspicious medical problems. The key here is that the validity of data depends on the question being studied. Data that might be invalid for one question may be perfectly valid for a different question.

validity The extent to which the data collected address the research hypothesis in the way intended.

reliability The extent to which a measure is stable and consistent over time in similar conditions.

Another important aspect of data is **reliability**, or the stability and consistency of a measure over time. If the measurement is reliable, the data collected will not vary because of changes over time in the measurement device. Suppose that you have a theory that people in their twenties are more likely to channel surf than are people in their fifties. To test your theory you would need to study television-watching behaviour and, in particular, the average length of time people stay tuned to each station. One option for measuring the duration of each channel stay would be to have an observer use a handheld stopwatch. However, there will likely be some variability in when the stopwatch is started and stopped, relative to when the surfer actually changes channels. As a consequence, the data in this scenario would be relatively unreliable in comparison to data collected by a computer that was linked to the television remote.

What kinds of statistics are used to analyze data?

A third and final characteristic of good data is **accuracy**, or the extent to which the measure is free from *error.* While this may seem obvious, the problem is that although a measure may be both valid and reliable, this does not guarantee that the measure is *accurate.* The way psychological scientists think about this problem is by turning it on its head and asking, How do errors creep into a measure? There are two basic types of error, *random* and *systematic.* Take the channel-surfing study. The problem with using a stopwatch to measure the duration of each channel stay is that each measurement will tend to either overestimate or underestimate the actual duration (because of human error). This is known as a *random error,* because although an error is introduced into each measurement, the actual value of the error is different each time. But suppose that the stopwatch has a glitch, such that it always overstates the actual time measured by two seconds. This is known as a *systematic error,* because the amount of error introduced into each measure is a constant. Generally, systematic error is more problematic than random error because the latter tends to average out over time and therefore is less likely to lead to inaccurate results (Figure 2.13).

accuracy The extent to which an experimental measure is free from error.

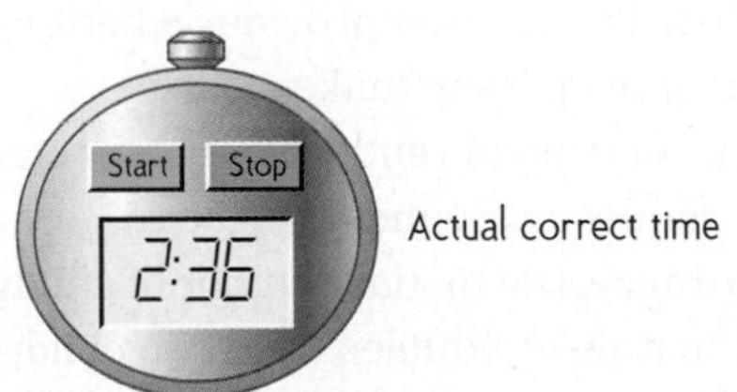

Random error: notice that the final average is correct

Systematic error: notice that the final average is incorrect

2.13 Systematic versus Random Errors Systematic errors occur when the measurement has the same degree of error on each occasion, whereas random errors occur when the degree of error varies each time.

central tendency A measure that represents the typical behaviour of the group as a whole.

mean A measure of central tendency that is the arithmetic average of a set of numbers.

median A measure of central tendency that is the value in a set of numbers that falls exactly in the halfway point between the lowest and highest values.

mode A measure of central tendency that is the most frequent score or value in a set of numbers.

variability In a set of numbers, how widely dispersed the values are from each other and from the mean.

standard deviation A statistical measure of how far away each value is on average from the mean.

Descriptive Statistics Provide a Summary

The first step in evaluating data is simply to inspect the raw values. This allows researchers to familiarize themselves with the data and to detect possible errors in data recording. For instance, you would assess whether any of the responses seemed especially unlikely, such as finding someone with a blood alcohol content of 50 percent. Once you are satisfied that the data make sense, you summarize the basic patterns using *descriptive statistics,* which provide an overall sense of the results of the study, such as how people performed in one condition compared to another. The simplest descriptive statistic is the **central tendency**, which is a single value that describes a typical response, or the behaviour of the group as a whole. Perhaps the most intuitive measure of central tendency is the **mean**, which is the arithmetic average of a set of numbers. The class average on an exam is an example of a mean score. Consider the hypothetical study of alcohol and driving performance. A basic way to summarize the data would be to calculate the mean driving performance across all participants for when they were (1) sober and (2) intoxicated. If alcohol affects driving, then we would expect to see a difference in the mean between sober and intoxicated driving performance.

A second measure of central tendency is the **median**, which is the value in a set of numbers that falls exactly halfway between the lowest and highest values. For instance, if you received the median score on a test, then half the people in your class scored lower than you and half the people in the class scored higher. Sometimes researchers will summarize data using a median instead of a mean because if one or two numbers in the set are dramatically larger or smaller than all the others, the mean will give either an inflated or deflated summary of the actual "average." This occurs when we try to examine average income per adult. Perhaps only 50 percent of Canadians make more than $30,000 per year, but some make so much more that the *mean* income is much higher. The median provides a better estimate of how much money the average person makes.

A third measure of central tendency encountered in psychological research is the **mode**, or the most frequent score or value in a set of numbers. For instance, the modal number of children in a family is two, which means that most families have two children. Examples of all three central tendency measures are shown in Figure 2.14.

In addition to measures of central tendency, another important characteristic of data is the **variability** in a set of numbers, or how widely dispersed the values are about the mean. For instance, the *range* tells you the distance between the largest and smallest value, which is often not of much use because it only uses two scores. The most common measure of variability is the **standard deviation**, which refers to how far away each value is, on average, from the mean. For instance, if the mean of an exam is 75 percent and the standard deviation is 5, this tells you that most people scored between 70 and 80 percent on the exam.

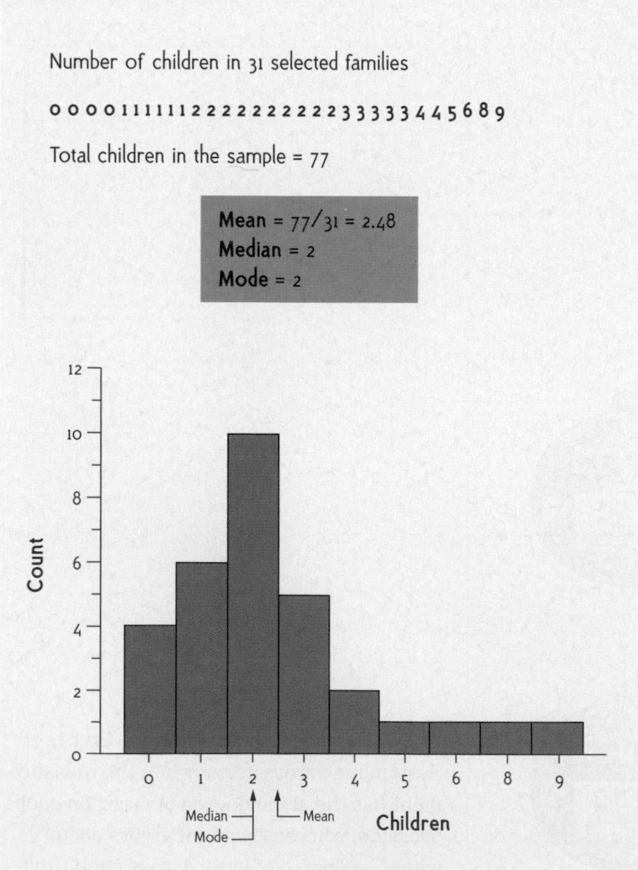

2.14 Central Tendency Measures The mean, median, and mode are different measures of central tendency; each reveals important information on a set of numbers. The graph here shows that for some data sets, all three measures can be the same; however, for other data sets, the values may be quite different.

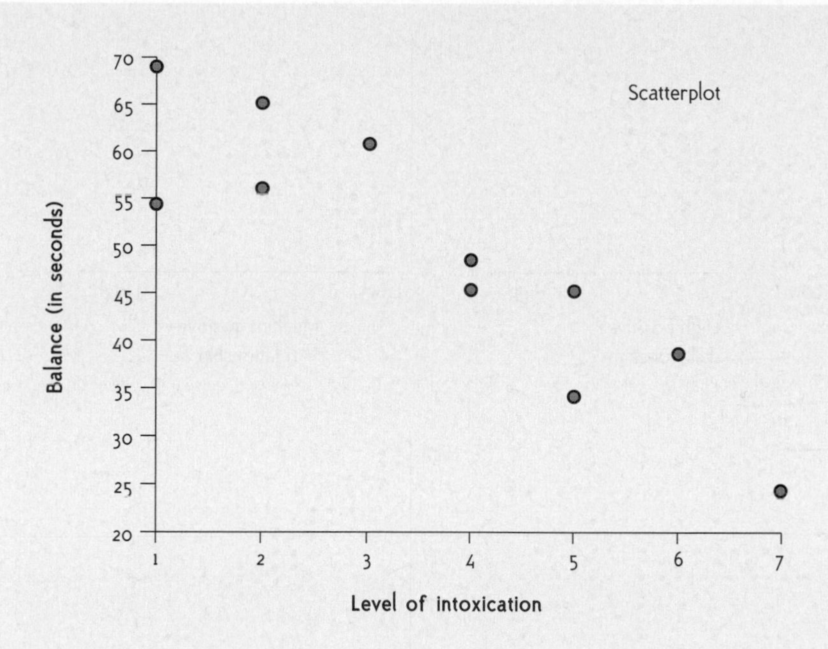

2.15 Scatterplots Scatterplots illustrate the relationship between two variables. Note that it appears that the more intoxicated people are, the fewer seconds they can maintain their balance.

Correlations Describe the Relationships between Variables

The descriptive statistics discussed thus far pertain to summarizing the central tendency and variability in a set of numbers. Descriptive statistics can also be used to summarize how two variables relate to each other. As shown in Figure 2.15, the first step in comparing two variables is to create a graph known as a *scatterplot,* which provides a convenient picture of the data. A scatterplot represents each participant's measure on two different variables—in this case, the level of intoxication—on an eight-point scale (with higher numbers meaning more intoxicated) and the ability to balance on one leg as measured in seconds. By looking at the scatterplot we can determine that the people who were more intoxicated also were less able to maintain balance.

correlation A statistical procedure that provides a numerical value, between +1 and −1, indicating the strength and direction of the relation between two variables.

In analyzing the relationship between two dependent variables, researchers are not limited to creating a scatterplot. They can also compute a **correlation**, which is a descriptive statistic that provides a numerical value (between +1.0 and −1.0) indicating the strength of the relationship between two variables. If two variables have a strong relationship, knowing how people measure on one variable becomes a good predictor of how they might measure on the other variable.

What signifies a strong relationship? Consider the different scatterplots in Figure 2.16 (on p. 66). Two variables have a *negative correlation*: as one *increases* in value, the other *decreases* in value. For example, as people become more intoxicated, they are less able to balance on one foot. In terms of values, a correlation of −1.0 indicates a perfect negative correlation. Two variables can also have a *positive correlation,* in which the variables increase or decrease together—for example, taller people often weigh more than shorter people. In this case, the correlation is positive, and a perfect positive correlation is indicated by a value of +1.0. If two variables show no apparent relationship, then the value of the correlation will be a number close to zero.

"I think you should be more explicit here in step two."

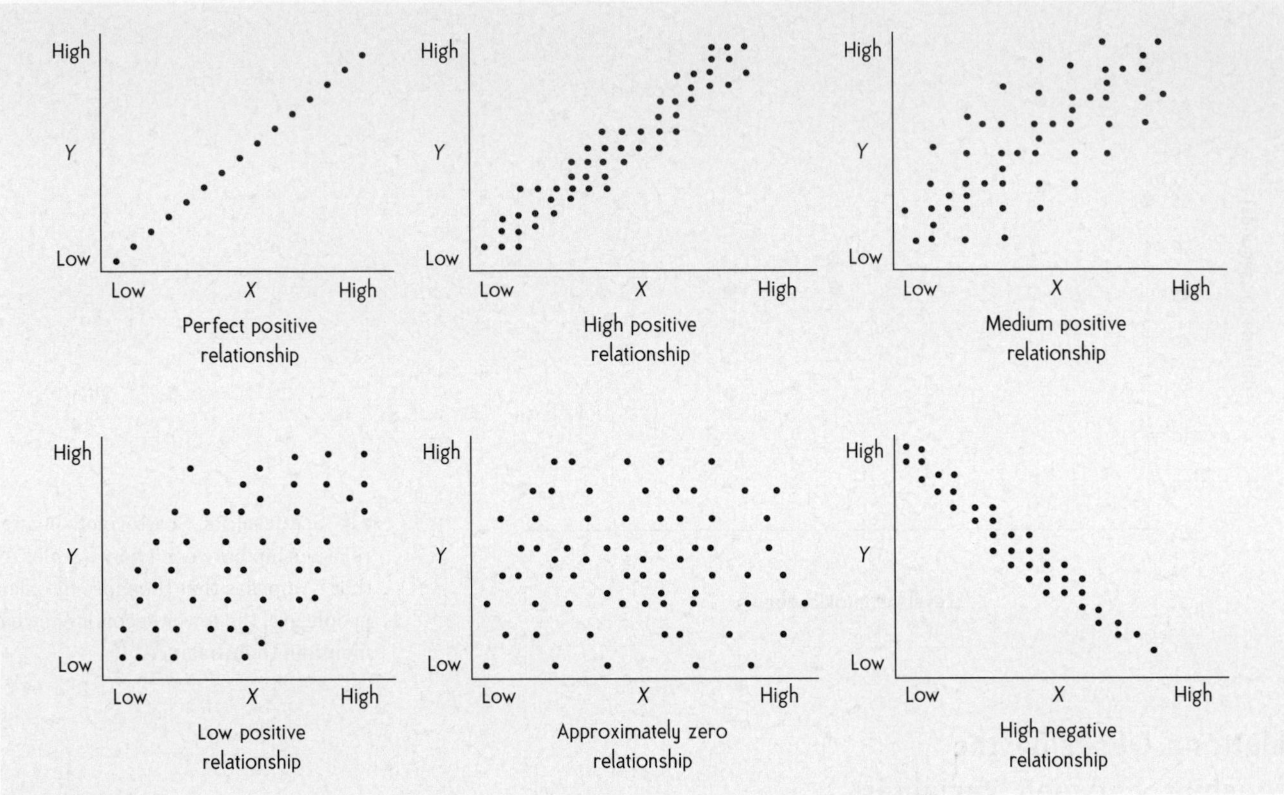

2.16 Correlation Correlations can have different values, which in turn reveal different kinds of relationships between two variables. The greater the scatter of values, the lower the correlation. A perfect correlation occurs when all of the values fall on a straight line.

Inferential Statistics Predict Chance Results

inferential statistics A set of procedures used to make judgments about whether differences actually exist between sets of numbers.

Whereas researchers use descriptive statistics to summarize data sets, they use **inferential statistics** to decide whether differences actually exist between different sets of numbers. For instance, suppose you find that the mean driving performance for intoxicated drivers is lower than the mean driving performance for sober drivers. How different do these two means need to be in order for you to conclude that drinking alcohol does in fact reduce people's ability to drive?

To answer this kind of question, you can use inferential statistics to compute the probability that a difference between means was actually due to chance rather than a causal effect of one variable on another. How does this work? Pretend for a moment that intoxication doesn't influence driving performance. Even so, if you go out and measure the driving performance of sober and drunken drivers, just by chance there will be some degree of variability in the mean performance of the two groups. Most of the time the difference in means will be small, but sometimes the difference may happen to be large. The key is that if alcohol does not actually affect driving performance, then the *probability* of showing a large difference between the two means is relatively small. The principle is the same as for how many heads are obtained when flipping a coin ten times: on average, the number will usually be five or close to it, but every now and then we may get either no heads or ten heads, just by chance.

Therefore, when you are comparing two means, inferential statistics tell you how probable the outcome would be if in fact there were no actual difference between the variables measured. If the probability is small enough, then you can feel reasonably certain that your result was more likely due to an actual difference (or an effect of one variable on another) rather than due simply to chance. How small is small enough? Although the question has long been contentious, a 5 percent chance of concluding there is a difference, when there is really not, tends to be the most liberal value encountered in psychological research. Regardless, the critical thing when deciding to use inferential statistics is to determine beforehand what probability level will be accepted as "small enough." If the results of the study fall below this predetermined probability value, then the difference in means can be labeled as *statistically significant*. If not, the alternative is to report the effect as *nonsignificant,* meaning that the observed difference is likely due to chance.

Psychological Science *in Action*

Understanding Consumers

Have you ever wondered why the managers of grocery stores put daily necessities like milk and eggs at the back of the store and sweets like candy near the checkout stand? They do so because it increases sales. Forcing people to walk to the back of the store ensures that they pass many items along the way, perhaps remembering their need to purchase one or two of them. Nearer to the cash register are items that people buy on impulse. The study of how people relate to the products and services they buy is a growing field known as *consumer psychology*: using the principles and methods of psychological science to understand how people spend their money.

Consumer psychologists receive training in research methods and often complete advanced coursework in topics such as marketing, advertising, and management. They work in a variety of settings, such as advertising agencies, automobile manufacturers, airlines, food manufacturers, and universities. For instance, a consumer psychologist might study how someone's personality predicts whether they will purchase one product over another. This information can then be used to target certain types of people with advertising. Because psychological scientists understand a great deal about how the mind works, they can also help devise strategies to increase sales. There are a number of ways that consumers can be influenced to purchase products, such as being invited to try a product at home without any obligation to purchase it—people come to think of the product as their own and are unlikely to return it.

A variety of methods are used to understand consumer behaviour. Researchers use observational studies to watch how people interact with products, such as observing children playing with toys. Consumers are also surveyed about their attitudes, sometimes with other people in focus groups in which the manufacturer is present as people discuss their thoughts and feelings about a product. Interestingly, simply asking questions can affect subsequent consumer behaviour. For instance, do you plan to buy a Krispy Kreme doughnut today? According to research by Patti Williams and colleagues (2004), simply being asked that question, no matter how you answer it, makes you more likely to actually buy a doughnut. So, go ahead and have one (any brand you like—preferably Tim Horton's, of course). You can blame it on us.

One of the most common methods of consumer research is to assess people's memory for products that have been advertised. For example, researchers have

2.17 Understanding Consumer Psychology Consumer psychologists use the methods of psychological science to understand how and why people spend their money. The results of their research have had a big effect on how advertisers devise sales strategies and target their advertising.

studied what type of advertisements are most likely to capture attention and be remembered. In one study, researchers created a magazine that contained a number of interesting articles as well as advertising. The researchers were interested in comparing memory for ads that used pictures to those that used only verbal descriptions. Participants were told that the study was about either the quality of the articles or the quality of the ads. Research participants were much more likely to remember the ads that contained pictures, especially when they had not been instructed to pay attention to the ads (McQuarrie & Mick, 2003). Perhaps a picture really is worth a thousand words.

As the methods of psychological science have evolved, so has their use in consumer research (Figure 2.17). For instance, some years ago the popular Pepsi Challenge ad campaign showed people taking a blind taste test of Coca-Cola against Pepsi. Pepsi was usually the winner. But Coke outsells Pepsi in most countries around the world, and if they know which brands they are tasting, Coke drinkers prefer Coke. How does knowing the brand affect taste perception? In a recent brain-imaging study, participants tasted Pepsi or Coke while they were in the brain scanner, but they were told the brand only on some trials (McClure et al., 2004). In a previous session, the participants had made blind taste ratings on Coke and Pepsi. Interestingly, about half preferred Coke and half Pepsi, and their blind ratings did not match their stated preference. In the scanner, an area of the brain that has been found to be involved in rewarding experiences was more active when participants tasted the drink that they preferred in the behavioural tasks. It was not more active when they tasted their stated preference. But when the products were labeled, brain activity changed when participants drank Coke. When they knew it was Coke, a number of other brain regions, many of them involved in emotion and decision making, became active. Thus, brand preferences are determined by not only our sensory experience, but also our beliefs and expectations regarding the product.

The use of techniques such as imaging to study consumer behaviour has been labeled neuromarketing. The idea that corporations can peek inside your head to study your purchasing habits makes some people nervous. But marketing specialists argue that people are not aware of many of the factors that influence buying, so they can't report them during focus groups or other traditional data-collection methods. They also point out that these methods will help businesses understand how people decide which products they prefer, which ultimately will benefit consumers. As with psychological science, the ultimate goal of consumer research is to understand the factors that guide behaviours that are important to our everyday lives. ■

REVIEWING the Principles | How Are Data Analyzed and Evaluated?

Data analysis begins with descriptive statistics, which summarize the data. Measures of central tendency indicate statistical averages across sets of numbers, whereas standard deviations and variances indicate how widely numbers in a set are distributed about an average. Correlations in turn are used to describe the relationship between two variables: positive, negative, or none. Inferential statistics are used to decide whether the results of a study were due to an effect of one variable on another or whether the results were more likely due to chance.

Conclusion

Throughout this chapter we have emphasized the major issues involved in designing and conducting research in psychological science. However, the ideas we have discussed have another important application as well: evaluating research. Understanding the quality of research is important whether you go on to conduct experiments yourself or not. Each day the media report some new major finding, such as the link between smoking marihuana and heart attacks that was discussed earlier. Should you believe a report and perhaps change how you lead your daily life as a result, or should you ignore new and potentially important data, because they came from a flawed study? To make educated decisions in this domain requires understanding how good psychological science is performed, which in turn requires the use of critical thinking.

So what determines good science? A number of factors need to be considered. Is there a theoretical basis for the work? Were the method and level of analysis appropriate for the question of interest? Does the study have adequate operational definitions of the variables involved? Are the researchers presenting their results as showing a causal relationship between two variables even though an experiment was not performed? If an experiment was performed, was it carefully designed and well controlled, or are there potential confounds that might have been overlooked? Do the results of the study apply to everyone or just to the population (e.g., the elderly, teenage males) that was actually studied? These fundamental questions underscore the necessity of being a critical, well-informed research evaluator. If you can't answer these questions when reading about a study, then you can't properly evaluate whether you should believe the results or how those results have been interpreted.

Quality research stems from both sound methodology *and* good questions. What defines a good question? Although it may be a subjective issue, there are certainly things to think about. Does the question address something fundamental or something relatively minor? If you obtain the results you predict, what impact will it have on the specific field of study? Will the results establish new understanding between different levels of analysis, or will they simply confirm a minor aspect of an old and arcane theory? We can't all be doing research that places us in the running for a Nobel prize, but at the same time, we shouldn't shy away from focusing our efforts on important rather than trivial issues.

Summarizing the Principles of Research Methodology

What Is Scientific Inquiry?

1. **Scientific questions can be objectively answered:** Scientific inquiry relies on objective methods and empirical evidence to answer testable questions.

2. **The empirical process depends on theories, hypotheses, and research:** Interconnected ideas or models of behaviour (theories) yield testable predictions (hypotheses), which are tested in a systematic way (research) by collecting and evaluating evidence (data). Unexpected (serendipitous) discoveries sometimes occur.

What Are the Types of Studies in Psychological Research?

3. **An experiment involves manipulating conditions:** In experiments, researchers control the variations in the conditions that the participant experiences (independent variables) and measure the outcomes (dependent variables), which lead to an understanding of causality.

4. **Correlational designs examine how variables are related:** Correlational studies are used to examine how variables are naturally related in the real world but cannot be used to establish causality or the direction of the relationship.

5. **Descriptive studies observe and classify behaviour:** Researchers observe and describe naturally occurring behaviours in order to provide a systematic and objective analysis.

What Are the Data-Collection Methods of Psychological Science?

6. **Observing is an unobtrusive strategy:** Data collected by observation must be clearly defined and systematically collected. Bias in the data may occur because the participant is aware they are being observed or because of the expectations of the observer.

7. **Asking takes a more active approach:** Using surveys, questionnaires, and interviews, we can directly ask people

about their thoughts and behaviours. These self-report data may be biased by the respondents' needs to present themselves in a particular way.

8. **Case studies examine individual lives:** An intensive examination of an individual is useful for examining unusual participants and research questions. But interpretation of case studies can be subjective.

9. **Response performance measures stimulus processing:** Measuring reaction times and accuracy, and asking people to make stimulus judgments are methods used to examine how people respond to psychological tasks.

10. **Body and brain activity can be directly measured:** Electrophysiology (EEG and ERP) measures the brain's electrical activity. Brain imaging is done using positron emission tomography (PET), magnetic resonance imaging (MRI), and functional magnetic resonance imaging (fMRI).

11. **Research with animals provides important data:** Animal research provides useful, albeit simpler, models of behaviour and genetics.

12. **There are ethical issues to consider:** Ethical research is governed by a variety of principles, which ensures fair and informed treatment of participants.

How Are Data Analyzed and Evaluated?

13. **Good research requires valid, reliable, and accurate data:** Data must be meaningful (valid) and its measurement consistent and stable (reliable) and accurate.

14. **Descriptive statistics provide a summary:** Measures of central tendency and variability are used to describe data.

15. **Correlations describe the relationships between variables:** A correlation is a descriptive statistic that describes the strength and direction of the relationship between two variables.

16. **Inferential statistics predict chance results:** Inferential statistics evaluate the variability and difference between sets of data to determine if they are due to chance or truly represent real change.

Applying the Principles

1. Philosophers, theologians, and psychological scientists are concerned with human behaviour. What separates psychological science from other areas of inquiry, such as philosophy and theology?

_____ a) Philosophers only ask "why" and do not seek the answer.
_____ b) Psychological science deals with theories and the others deal with facts.
_____ c) Philosophy and theology do not require scientific objectivity.
_____ d) Psychological science is a subjective science.

2. A hypothesis states that television violence causes children to be violent. A researcher tests this hypothesis by using a television-viewing behaviour questionnaire and then observing how children behave during recess at school. This type of study is

_____ a) correlational and therefore cannot establish if TV causes violence, because of a potential third variable.
_____ b) experimental because the researcher has control over the setting in which the behaviour is observed.
_____ c) descriptive, only because there is observation involved.
_____ d) able to clearly establish that viewing violence on television causes violence at school.

3. In Gordon's introductory psychology class, students earn credit for participating in research. Gordon takes part in three research activities. In the first, he completes a questionnaire that asks about racial attitudes. Next, he is shown a video of various people (of different ethnicities) at their jobs and later measured on his speed and accuracy in matching the inividuals with their occupations. Finally, physiological measures (i.e., heart rate and brain activity) are recorded while he views screen images of individuals. Which type of data might have the most bias?

_____ a) the questionnaire response
_____ b) the speed and accuracy measures
_____ c) the physiological measures
_____ d) all are unbiased measures

4. A researcher is investigating the likelihood of success of entering college freshmen. She has existing data about the accuracy of GPA and SAT scores. She next uses a variety of measures, including GPA and SAT scores, and examines whether or not students complete their undergraduate degrees. Her research reveals that GPA has more predictive value of college success than SAT scores. She concludes

that SAT scores in relationship to predicting college success are

_____ a) reliable and valid measures.
_____ b) stable but unreliable measures.
_____ c) reliable but not valid measures.
_____ d) valid but not reliable measures.

ANSWERS: *1. c 2. a 3. a 4. c*

Key Terms

accuracy, p. 63
case study, p. 52
central tendency, p. 64
confound, p. 43
correlation, p. 65
correlational study, p. 44
data, p. 41
dependent variable, p. 43
descriptive study, p. 47
electrophysiology, p. 56
experiment, p. 43
experimenter expectancy effect, p. 50
functional magnetic resonance imaging (fMRI), p. 58
hypothesis, p. 41
independent variable, p. 43
inferential statistics, p. 66
informed consent, p. 61
Institutional Review Boards (IRBs), p. 60
magnetic resonance imaging (MRI), p. 58
mean, p. 64
median, p. 64
mode, p. 64
naturalistic observation, p. 47
observational technique, p. 49
observer bias, p. 50
operational definition, p. 42
participant observation, p. 47
positron emission tomography (PET), p. 57
psychophysiological assessment, p. 55
random assignment, p. 44
reaction time, p. 54
reactivity, p. 50
reliability, p. 62
replication, p. 40
research, p. 41
response performance, p. 54
self-report method, p. 51
socially desirable responding, p. 51
standard deviation, p. 64
theory, p. 41
third-variable problem, p. 45
validity, p. 62
variability, p. 64
variable, p. 42

Further Readings

Fossey, D. (1988). *Gorillas in the mist.* New York: Houghton Mifflin.

Huff, D. (1993). *How to lie with statistics.* New York: Norton.

Martin, P., & Bateson, P. (1993). *Measuring behaviour: An introductory guide.* New York: Cambridge.

Stanovich, K. E. (2003). *How to think straight about psychology* (7th ed.). Boston: Allyn &Bacon.

Zechmeister, J. S., Zechmeister, E. B., & Shaughnessy, J. J. (2000). *Essentials of research methods in psychology.* Boston: McGraw-Hill.

Further Readings: A Canadian Presence in Psychology

Adair, J. G. (1984). The Hawthorne effect: A reconsideration of the methodological artifact. *Journal of Applied Psychology, 69*, 334–345.

Bernasconi, N., Natsume, J., & Bernasconi, A. (2005). Progression in temporal lobe epilepsy: Differential atrophy in mesial temporal structures. *Neurology, 65*, 223–228.

Canadian Psychological Association. (1992). *Canadian code of ethics for psychologists, 1991.* Old Chelsea, Quebec.

Hall, W., & Degenhardt, L. (2006). What are the policy implications of the evidence on cannabis and psychosis? *Canadian Journal of Psychiatry, 51*, 566–574.

Mann, R. F., Smart, R. G., Stoduto, G., Beirness, D., Lamble, R., & Vingilis, E. (2002). The early effects of Ontario's administrative driver's licence suspension law on driver fatalities with a BAC > 80 mg%. *Canadian Journal of Public Health, 93*, 176–180.

Olfert, E. D., Cross, B. M., & McWilliam, A. A. (Eds.). (1993). *Guide to the care and use of experimental animals* (2nd ed., vols. 1 & 2). Ottawa, Ontario: Canadian Council on Animal Care.

Investigating Our Genetic Heritage

A visitor at a museum examines a digital representation of the human genome. As technology has advanced, psychological scientists have developed more sophisticated tools to study the biological bases of mind and behaviour. Recent research has focused less on whether specific genes cause specific outcomes, and more on how genes are expressed—turned "on" and "off"—by environmental factors. Changing the expression of even a single gene can have remarkable effects on complex behaviours such as social interaction.

CHAPTER

3

Genetic and Biological Foundations

They were a happy couple in their early thirties, with three healthy young children. When their infant son died during a routine surgical procedure, they understandably were devastated. Although many people in this situation might have wanted to have another child, this couple did not; they wanted to bring their dead son back to life. To do so, they decided to finance—using money from a malpractice settlement with the hospital where he died—an attempt to clone their son. Welcome to the twenty-first century.

The idea of human cloning moved from the realm of science fiction to remote plausibility with the announcement in 1997 that Scottish researchers had created a successful sheep clone named Dolly. If sheep could be cloned, why not humans? Aghast at the idea, many governments around the world

Outlining the Principles

What Is the Genetic Basis of Psychological Science?

- Heredity Involves Passing along Genes through Reproduction
- Genotypic Variation Is Created by Sexual Reproduction
- Genes Affect Behaviour
- Social and Environmental Contexts Influence Genetic Expression
- Gene Expression Can Be Modified

How Does the Nervous System Operate?

- Neurons Are Specialized for Communication
- Action Potentials Cause Neuronal Communication
- Changes in Electrical Potential Lead to Action
- Action Potentials Spread along the Axon
- Neurotransmitters Bind to Receptors across the Synapse

How Do Neurotransmitters Influence Emotion, Thought, and Behaviour?

- Acetylcholine Affects Motor Control and Mental Processes
- Monoamines Are Involved in Affect, Arousal, and Motivation
- Amino Acids Are General Inhibitory and Excitatory Transmitters in the Brain
- Peptides Modulate Neurotransmission

How Are Neural Messages Integrated into Communication Systems?

- The Central Nervous System Consists of the Brain and Spinal Cord
- The Peripheral Nervous System Includes the Somatic and Autonomic Systems
- The Endocrine System Communicates through Hormones
- Actions of the Nervous System and Endocrine System Are Coordinated

banned research on human cloning. However, in 2001 separate teams of researchers confirmed that they were actively attempting to clone a human using the methods pioneered in Scotland. Ethicists, religious leaders, and scientists are locked in debate about whether human cloning is viable and ethical, with most agreeing that it is neither. In the midst of this debate, psychological scientists can't help wondering what cloning might tell us about what it means to be a human being.

In the cloning debate, it is important to separate efforts to clone human cells to be used for basic medical research from reproductive cloning. Most scientists want to use cloning to study genetic processes, not to grow a human being. In England, the team that cloned Dolly received permission to conduct human cloning research to study amyotrophic lateral sclerosis (ALS), known in North America as Lou Gehrig's disease. Although ALS usually leads to death within five years, some people with the disease, such as the physicist Stephen Hawking, live much longer. These researchers want to take a cell from an ALS patient and, through cloning, create embryonic stem cells. From these, nerve cells can be extracted and the genes responsible for ALS can be studied. This technique could be a valuable research method for understanding many genetic outcomes, including psychological disorders. The approach is completely different from using cloning to create a human being, which is unlikely to occur any time soon. But let's imagine for a second that you could clone a person, because that allows us to consider fundamental and profound questions about what it means to be a human being. What if we could clone a person, so that the clone and the person contained identical genes? Would the clone be the same person? Are we so hard-wired that our clone would share our favourite foods, our work habits, and even our ethical views of cloning?

Nature already clones people, in the form of identical twins. As you will read in this chapter, such twins have many remarkable similarities, even if they are raised apart. But would "clones" be as similar to their gene donors as identical twins are to each other? There are many reasons to believe they would not. Consider the infant boy whom the young couple wants to recreate, assuming again that it was even possible. The clone's natural environment would be similar but not identical to that of his predecessor. Among other things, the clone would have a different uterine environment and would be raised by parents who are now older and who have undergone a major transformative experience; he would also be interacting with sib-

1817

Parkinson's Disease The physician James Parkinson identifies this major neurological disorder, which affects muscular control and results from depletion of the dopamine system.

1866

Mendel's Peas Gregor Mendel, a monk, conducts studies on heredity in pea plants. His results provide the basis of genetics, including the idea of dominant and recessive genes.

1870s

Golgi Stain Italian physician Camillo Golgi invents what becomes known as the "Golgi stain," a method of tissue preparation that allows the examination of single neurons.

1880s

Neuron Doctrine The Spanish anatomist Santiago Ramón y Cajal uses the Golgi stain to chart the microscopic anatomy of different regions of the brain. He correctly argues that the nervous system is composed of distinct cells.

lings who are older. Moreover, even though identical twins grow up to be quite similar in many respects, they still differ from each other in multiple ways, as we'll see in our discussions on twin studies. There would be similarities, but also substantial differences. This clone would be a unique individual, just like his predecessor. Thus, much of the debate over the morality of cloning should be muted by Mother Nature's near-perfect clones, identical twins. Science can only hope to improve on the "identicalness" of identical twins by cloning two or more individuals from the same cell donor and developing them in the same environment at the same time.

As a thought experiment, human cloning raises issues of great interest to psychological scientists. Would a cloned child develop at the same rate as the first child, or might his development be affected by the parents' expectations? How much of who we are is determined by our genes and how much by the environment? If two clones were raised in different cultures, how different might they be? What if they were raised ten years apart, and therefore exposed to different cultural environments? Every person is unique. What determines this uniqueness? Indeed is human cloning ethical (Caulfield, 2003)?

For many years, geneticists focused on whether people possessed certain types of genes, such as genes for psychological disorders or even genes for intelligence. The emphasis was on the extent to which human traits were genetically hard wired, a formulation that left little room for environmental factors. Recent research is approaching the study of genes by trying to understand not only the possession of certain genes, but also gene expression, how genes are turned "on" and "off." The major lesson emerging from this research is that the environment plays an important role in how genes are expressed and therefore in how they influence mind and behaviour. Although there is evidence that people might possess specific genes that have specific effects, it is the expression of those genes ("turning on and off") as much as mere possession that makes us who we are. This chapter begins our examination of the genetic and biological foundations of psychological science. To understand fully what makes us who we are, we need to understand how basic physiological processes affect behaviour, thought, and emotion. We also need to understand how those physiological processes are shaped by the environment—how nurture influences nature. Along the way, we will look at some of the exciting new genetic research that is reshaping the way we think about the human mind and behaviour. ■

1897

The Synapse Sir Charles Sherrington proposes that neurons communicate with each other across the synapse through electrical means. A few years later, Thomas Renton Elliott suggests that synapses are chemical rather than electrical.

1921

Neurotransmission Is Chemical Otto Loewi shows that neurons communicate through the release of neurotransmitters, chemicals released from one neuron that signal other neurons. He is awarded a Nobel prize.

1936

First Neurotransmitter Identified Sir Henry Dale identifies acetylcholine as the neurotransmitter that operates in the control of muscles.

1950s

Sodium-Potassium Pump Alan Hodgkin and Andrew Huxley show that there are active components within the neuron that stabilize the concentration of sodium and potassium, the main chemicals involved in action potentials.

1950s

Pharmacology Researchers show that chemicals that bind to neurotransmitter receptors can be used in therapies for various forms of mental illness, setting the stage for modern-day drug therapy.

Over the past three decades, scientific understanding of the genetic and physiological foundations of psychological activity has increased dramatically. As technology has advanced, sophisticated tools have been developed to explore the biological bases of mind and behaviour. Researchers can now examine DNA to predict who will develop specific disorders and to understand how certain diseases are passed from one generation to the next. Researchers have also identified genes that predispose people to be, among other things, outgoing, sociable, and intelligent. Most recently, researchers have developed techniques that allow them to turn specific genes on and off. Remarkably, changing the expression of a single gene can have profound effects even on complex behaviours such as social interaction (Insel & Young, 2001). Research at the biological level of analysis has led to great strides in understanding the various biological processes that occur within the brain and body to produce experience, thought, emotion, memory, and behaviour. By building across the levels of analysis, we plan to demonstrate how psychological scientists conduct research to understand different questions related to mind, brain, and behaviour. This chapter examines psychological activity at the genetic and neurochemical levels; in Chapter 4 we will describe the functions of various brain regions at the brain-systems level of analysis.

What Is the Genetic Basis of Psychological Science?

One of the major outgrowths of the new biological revolution occurred in February 2001, when two groups of scientists published separate articles detailing the results of the first phase of the *Human Genome Project,* an international research effort to map the entire structure of human genetic material. This achievement represents the coordinated efforts of hundreds of scientists around the globe. To understand their goals it is important to have a basic understanding of genetic processes.

Within nearly every cell in the body is the genome for making the entire organism: It is the master recipe that provides detailed instructions for everything from how to grow a gall bladder to where to place the nose on your face. Whether a cell becomes part of a gall bladder or a nose is determined by which genes are turned "on" or "off" within the cell, and this is determined by cues from outside the cell. The genome provides the option, and the environment determines which option is taken (Marcus, 2004). As you will learn, recent discoveries about how genes are turned on or off have provided important insights into not only how physical bodies develop but also how the mind works. The term *genetics* is typically used to describe how characteristics, such as height, hair colour, and weight, are passed along through inheritance, but it also refers to those processes that are involved in turning genes off and on.

1974

Endogenous Opiates Scottish researchers Hans Kosterlitz and John Hughes discover that endorphins are important for reward and for the natural control of pain.

1977

Releasing Hormones Roger Guillemin, Andrew Shally, and Rosalyn Yalow share a Nobel prize for their work showing the existence of releasing factors that regulate hormonal function.

1980s

Behavioural Genetics of Personality Researchers at the University of Minnesota study identical twins who have been raised apart, demonstrating that genetics plays an important role in human personality.

1998

Gene Silencing Researchers discover that modified RNA causes a shutdown in gene expression, effectively silencing genetic instructions. This discovery might allow treatments for genetic disorders.

2001

Human Genome Project An international collaboration of genetics researchers maps the basic gene sequence of human DNA, setting the stage for the more complex task of figuring out how genes interact to produce illness and influence behaviour.

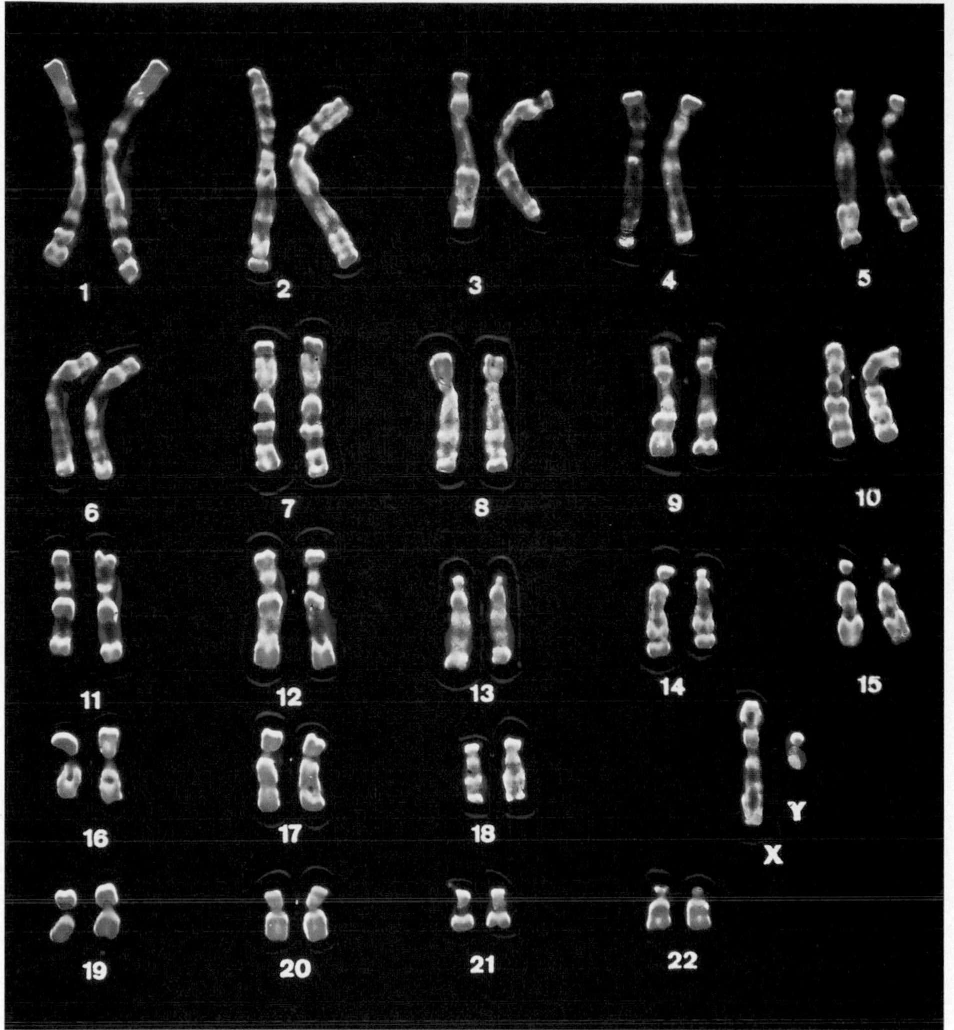

3.1 Human Chromosomes Each person has 23 pairs of chromosomes. Each parent contributes one chromosome to each pair. Males and females differ on the 23rd chromosome pair, with males having an X and a Y chromosome (shown) and females having two X chromosomes (not shown).

Within each cell are **chromosomes**, which are structures made up of genes. The typical human has 23 pairs of chromosomes, half of each pair coming from each parent (Figure 3.1). Genes in turn are components of a substance called *DNA* (*deoxyribonucleic acid*), which consists of two intertwined strands of molecules of sugar, phosphate, and nitrogen. The precise sequence of these molecules along the DNA strand specifies an exact instruction, through the production of *RNA* (*ribonucleic acid*), to manufacture distinct proteins. Proteins, of which there are thousands of different types, are the basic chemicals that make up the structure of cells and direct their activities. A **gene**, then, is a segment of DNA that is involved in producing proteins that carry out specific tasks. The process by which the gene produces RNA and then protein is known as *gene expression,* in that the gene is "switched on." The gene has not only recipes for how to produce specific proteins, but also instructions for when to produce them; the environment determines all of this. For example, a certain species of butterfly becomes colourful or drab depending on the season in which it is born. The environment during its development likely causes a gene sensitive to temperature to be expressed, or switched on (Marcus, 2004). Genes provide options, and the environment determines which option is taken. Each cell in the human body contains the same DNA, but cells become specialized for different tasks or for having

chromosomes Structures within the cell body that are made up of genes.

gene The unit of heredity that determines a particular characteristic in an organism.

different effects when specific genes are expressed. Gene expression not only determines the physical makeup of the body but also remains important throughout life and is involved in all psychological activity. It is gene expression that allows us to learn, to see, and to fall in love.

The first step of the Human Genome Project was to map out the entire structure of DNA—in other words, to identify the precise order of molecules that makes up each of the thousands of genes on each of the 23 pairs of human chromosomes. One of the most striking findings from the Human Genome Project is that there are only around 30,000 genes, not many more than are found in a fly (13,000), a worm (18,000), or even a plant (26,000). This suggests that the complexity of humans is not due simply to our possessing a large number of genes, but more likely due to subtleties in the way that genes are expressed and regulated (Baltimore, 2001). Now that the initial map of the human genome is almost complete, scientists are directing their efforts to understanding how genes influence mind and behaviour. The eventual goal of the project is to understand how genes interact with each other to produce illness. By understanding how genes work, researchers can develop methods to alter gene function to cure various ailments. At the same time, genetic research is providing a compelling new understanding of the biological basis of psychological activity.

Heredity Involves Passing along Genes through Reproduction

The first clues to the mechanisms responsible for heredity were discovered by the monk Gregor Mendel around 1866. Mendel developed an experimental technique called *selective breeding* for studying genetics. At the monastery where Mendel lived, the study of plants had a long history. By using selective breeding, Mendel could strictly control which plants were bred with which other plants.

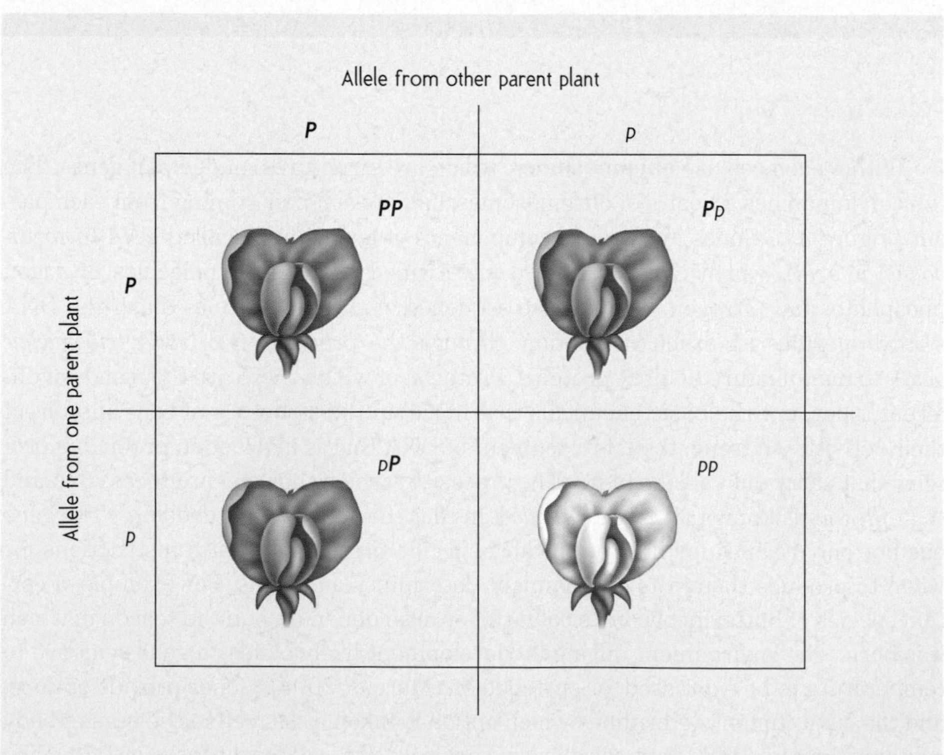

3.2 Cross-Bred Pea Blossoms In cross-breeding, when purple is the dominant gene (indicated ***P***) and white is the recessive gene (indicated *p*), there are four possible genotypes, and three of the four phenotypes will be purple.

Consider a simple study by Mendel. He chose peas that had either only purple flowers or only white flowers. Mendel then cross-pollinated the two types of plants to see which colour flowers would be produced. Mendel found that the first generation of pea offspring tended to be completely white or completely purple. If he had stopped there, he would never have discovered the basis of heredity; fortunately, he then allowed the plants to self-pollinate into a second generation. This second generation revealed an interesting pattern. Of the hundreds of pea plants, around 75 percent had purple flowers and 25 percent had white flowers. This three-to-one ratio repeated itself in additional studies, and it also turned out to be true for other characteristics, such as pod shape. From this pattern, Mendel deduced that there must be discrete units, now referred to as genes, that exist in two versions (e.g., white and purple). These two versions of a gene are known as *alleles.* According to Mendel, if the alleles differ (such as in cross-breeding), one of them is dominant and the other is recessive. **Dominant genes** are expressed (become apparent) whenever they are present, whereas **recessive genes** are expressed only when they are matched with a similar gene from the other parent. Thus, because white flowers were recessive, they occurred only when the gene for purple flowers was not present, which happened in only one of the four combinations of white and purple genes (see Figure 3.2).

dominant gene A gene that is expressed in the offspring whenever it is present.

recessive gene A gene that is expressed only when it is matched with a similar gene from the other parent.

genotype The genetic constitution determined at the moment of conception.

phenotype Observable physical characteristics that result from both genetic and environmental influences.

GENOTYPE AND PHENOTYPE The existence of dominant and recessive genes means that not all genes are outwardly expressed. The **genotype** is the genetic constitution of an organism, the actual genetic makeup that is determined at the moment of conception. The **phenotype** is the observable physical characteristics of an organism that result from both genetic and environmental influences. So, for instance, in Mendel's experiments two purple flowers would have the same phenotype, but they might differ in genotype, in that one could have two dominant genes for purple and another could have one dominant purple gene and one recessive white gene. The environment can also affect the phenotype; for instance, good nutrition leads to increased physical size and a suntan changes the colour of the skin. An excellent example of environmental influence on the phenotype is *phenylketonuria* (*PKU*), a disorder in which infants are unable to break down an enzyme (phenylalanine) contained in dairy and other products, such as the sweetener aspartame. This rare genetic disorder can lead to severe brain damage. Fortunately, providing a bland diet that is low in phenylalanine until the child passes critical stages of neuronal development greatly helps to reduce brain damage. The phenotype, then, is modified by diet.

POLYGENIC EFFECTS Mendel's flower-colour experiments dealt with single-gene characteristics. These are traits that appear to be determined by only one gene. But when there is a range of variability within a population for certain characteristics, such as height, this indicates that the characteristic is *polygenic,* that is, influenced by many genes, as well as the environment. Most human traits and diseases are polygenic.

Genotypic Variation Is Created by Sexual Reproduction

Although they have the same parents, siblings differ from each other in many ways, such as eye colour, height, and personality. This occurs because each has a specific combination of genes that differ in part due to random cell division that occurs prior to reproduction. Reproductive cells from each parent divide to produce *gametes,* egg and sperm cells, that contain only one half of each pair of chromosomes. The sperm and egg cells then combine during fertilization. One half of each parent's chromosomes join in a random fashion. The fertilized cell, known as a *zygote,* contains 23 pairs of chromosomes, one half of each pair from the mother and the other half from the father. The two chromosomes in the 23rd pair are the *sex chromosomes,* denoted *X* and *Y* due to their shape. Females have two X chromosomes, whereas males have one X and one Y chromosome. Calculating all possible combinations of the 23 chromosomes indicates that this can result in some 8 million different outcomes. Further, since both mother and father are themselves one of 8 million possible outcomes, the zygote is one of 64 trillion (8 million × 8 million) possible combinations. The net outcome is that a unique genotype is created at the moment of conception, and this accounts for the genetic variation in the human species.

What determines the genetic uniqueness of individuals?

The zygote grows when the chromosome duplicates and then the cell divides into two new cells with an identical chromosome structure. Cell division is the basis of the life cycle and is responsible for growth and development. Sometimes there are errors in this process, which lead to *mutations,* most of which are benign and have little influence. Occasionally, a genetic mutation may produce a selective advantage or disadvantage in terms of survival or reproduction. The evolutionary significance of these changes in adaptiveness is complex, but mutations that lead to abilities or behaviours that are advantageous to an organism may spread through the gene pool because those who carry the gene are more likely to survive and reproduce. For instance, consider *industrial melanism,* in which areas of the world with heavy soot or smog tend to have moths and butterflies that are darker in colour. Prior to industrialization, darker insects were more likely to be spotted against pale backgrounds. Any mutation that led to darker colouring was quickly eliminated through natural selection. But, with industrialization, pollution led to a darkening of trees and buildings, and therefore darker-coloured insects became more adaptive.

You may wonder how genetic mutations lead to disease and why they remain in the gene pool. For instance, *sickle-cell disease* is a genetic disorder that mostly affects people of African ancestry and alters how oxygen is processed in the bloodstream; it can lead to pain, physical damage, and anemia. The gene for the disease is recessive, so that most carriers have healthy phenotypes in spite of genotypes that contain the disease. Thus, only those who inherit recessive genes from *both* parents will develop the disease. Recessive genes do not interfere with health in most people, which allows them to survive in the gene pool. In addition, the gene for sickle-cell disease also increases resistance to malaria, and therefore in environments in which malaria is prevalent, such as certain regions in Africa, the sickle-cell gene has some benefit. In contrast to recessive genes, most dominant gene disorders are lethal for most of their carriers and therefore do not last in the gene pool.

Psychological Science *in Action*

Revealing and Challenging Genetic Destiny

Unlike disorders carried on recessive genes, which do not typically affect their carriers, most dominant gene disorders are lethal for their carriers and therefore do not last in the gene pool. One notable exception is Huntington's disease, a neurological disorder that results in mental deterioration and abnormal body movements. This disorder escapes elimination from the gene pool because the genes for it do not become expressed until later in life, so that many of those with Huntington's have children before they realize they have a genetic disorder. Because the gene is dominant, if a parent has it, each of his or her children has a 50–50 chance of also developing the disorder. The early symptoms of Huntington's disease strike when people are middle aged, typically around 40, but sometimes much earlier. The first signs are restless and involuntary movements of whole limbs or major portions of limbs. The afflicted person begins to walk in a jerky fashion, often stumbling, and eventually loses control of all movement, including the ability to walk, write, or speak. The individual also experiences emotional and personality changes, such as extreme anxiety and depression. Suicide is not uncommon among those in the early stages of the disease. No cure for Huntington's has yet been found. All those who have the disease die within ten years or so after its onset. Those who have relatives with Huntington's spend a good part of their lives wondering whether they will develop symptoms and whether it is safe for them to have children.

Nancy Wexler (Figure 3.3), a clinical psychologist, dedicated herself to finding a genetic marker for the disorder after her mother died of the disease in 1978. Her mother's illness shaped many aspects of Wexler's academic life. Her doctoral dissertation focused on people like herself who were at risk for developing disorders. Being at risk seems to have affected all aspects of her life; she has never married and never had children, not wanting to expose them to risks of inheriting the disease. Instead, as a professor of neuropsychology at Columbia University, she focuses her time and energy on trying to unlock the genetic mysteries of Huntington's. Over the last two decades, her collaborative research has identified the gene responsible for the disease as well as how this gene leads to destruction of the brain.

The discovery of the genetic marker for Huntington's happened in 1981 after Wexler studied a large extended family in Venezuela in which many had the disorder. Over the next decade, Wexler and her colleagues identified the defective gene itself. Since that time researchers have discovered that when the gene for Huntington's disease is expressed it creates a protein that attacks the brain, producing the symptoms of the disease and eventually causing death. Wexler's team, along with scientists in many other countries, is now searching for ways to interrupt this genetic process. As you will read, the use of knock-out gene models was not an option, because the gene involved in Huntington's serves other vital brain functions. However, the recent discovery that a form of RNA can be used to silence gene expression holds tremendous promise for controlling the disease. Most recently, Wexler and her colleagues have found that the environment has a strong influence on when the Huntington's gene is expressed, and therefore when people first develop the disorder. They are now working to identify which specific factors in the environment are involved (Wexler et al., 2004).

Thanks to Wexler and her colleagues, relatives of those with Huntington's can now take a genetic test to determine whether they will develop the disease. Of course, the decision to have the test is difficult, since there is still no cure. Moreover, there are ethical issues

3.3 Nancy Wexler Nancy Wexler is seen here working with a patient. After her mother's death from Huntington's disease, Wexler helped develop a genetic test to determine if people are at risk for developing the disease.

related to genetic testing, such as whether the information should be available to employers and insurance companies. There are personal issues, too. If you tested positive when you were 20, how might the knowledge affect your occupational and relationship goals? Interestingly, although she played an integral role in developing the genetic test, Nancy Wexler has never revealed whether she took it. Now approaching her sixties, Wexler shows no signs of the disease, and her chances of developing it diminish each year. If you were at risk, would you want to take the test? How might being at risk affect your decision to get married or have children? ■

Genes Affect Behaviour

What determines the kind of person you are? What factors make you more or less bold, intelligent, or able to read a map? All of these are influenced by the interaction of your genes and the environment in which you were raised. The study of how genes and environment interact to influence psychological activity is known as *behavioural genetics*. Behavioural genetics has made important contributions to the biological revolution, providing information about the extent to which biology influences mind, brain, and behaviour.

Any research that suggests that abilities to perform certain behaviours are based in biology is controversial. Who wants to be told that there are limitations to what you can achieve based on something that is beyond your control, such as your genes? It is easy to accept that genes control physical characteristics such as sex, race, eye colour, and predisposition to diseases such as cancer, alcoholism, or migraines. But can genes also determine whether people will get divorced, how smart they are, or what career they're likely to choose? A concern of psychological scientists is the extent to which all of these characteristics are influenced by nature and nurture, by genetic makeup and the environment. Increasingly, science indicates that genes lay the groundwork for many human traits. From this perspective, people are born essentially like undeveloped photographs: The image is already captured, but the way it eventually appears can vary based on the development process. However, the basic picture is there from the beginning.

BEHAVIOURAL GENETICS METHODS Siblings are different, even when they are raised in the same household; this could be because they do not share identical genes. Remember that each individual is one among eight million possible outcomes (Figure 3.4). But there are also subtle differences in the environment, both within and outside the household. Siblings have different birth orders, they have different friends and teachers, and perhaps they are treated differently by parents. It is difficult to know what causes the similarities and differences between siblings because some of the genes are the same and often most of the environment is the same. Therefore, behavioural geneticists use two basic methods to assess the degree to which traits are inherited: twin studies and adoption studies.

Twin studies compare similarities between different types of twins to determine the genetic basis of specific traits. **Monozygotic twins**, also called identical twins, are the result of one zygote dividing into two, each having the same chromosomes and the genes they contain. **Dizygotic twins**, sometimes called fraternal or nonidentical twins, are the result of two separately fertilized eggs and are no more related than any other pair of siblings. The process of fertilization occurs separately to produce the two zygotes, which then develop independently. To the extent that monozygotic twins are more similar on a given dimension than dizygotic twins, the increased similarity is considered most likely due to genetic influence. Of course,

monozygotic twins Twin siblings who result from one zygote splitting in two, and therefore share the same genes (i.e., identical twins).

dizygotic twins Twin siblings who result from two separately fertilized eggs (i.e., fraternal twins).

3.4 Two Couples, Four Children Shown in the top row are two couples. Below are four children. Try to match which child comes from which parents. The answer is revealed at the bottom of the page.*

even identical twins do not have the exact same environment (and in rare circumstances might even have different genes due to random mutations), and therefore they have different phenotypes, but they are typically more similar than dizygotic twins, who differ both in genotype and phenotype.

Adoption studies compare the similarities between biological relatives and adoptive relatives. Adopted siblings have similar home environments but different genes. Therefore, the assumption is that similarities among adopted siblings have more to do with environment than genes. Interestingly, it turns out that growing up in the same household has relatively little influence on many traits, such as personality. Indeed, after we control for their genetic similarity, even biological siblings raised in the same household are no more similar than two strangers plucked at random off the street. This point will be examined in greater detail in the development and personality chapters, but for now it is sufficient to understand the basic logic of the different types of studies.

How do studies of twins help us understand genetics?

One interesting behavioural genetic study is to compare monozygotic twins who have been *raised together* or *raised apart*. This would seem the ideal way to test the relative contributions of genes and the environment. Thomas Bouchard and his colleagues at the University of Minnesota have identified more than 100 pairs of identical and nonidentical twins who were either raised together or raised apart (1990; Figure 3.5). They examined a variety of traits and characteristics, including intelligence, personality, sense of well-being, achievement, alienation, and aggression. The general finding was that identical twins were likely to be similar, whether raised together or not. The "Jim Twins" are perhaps the most famous case study to emerge from the Minnesota Twin Project. How is it possible that two boys separated from birth and raised by different families had the same first name, had each married and divorced women named Linda, had each remarried women named Betty, had named their sons James Alan and James Allen, and had each named their dogs Toy? In addition, both were part-time law-enforcement officers who drove Chevrolets and vacationed in Florida. Just to complete the circle, they were the same height and weight, chain-smoked the same brand of cigarettes, and drank the same brand of

*The boy and the girl on either end of the bottom row are children of the parents at the top right. The two girls in the middle of the bottom row are children of the parents at top left.

3.5 Gerald Levey and Mark Newman Participants of Dr. Bouchard's study, Levey and Newman are identical twins who were separated at birth. Reunited at age 31, they discovered that they were both firefighters and had similarities in personality traits.

beer. Although no one suggests that there are genes for naming dogs Toy or for marrying and divorcing women named Linda, the obvious similarities in many aspects of the Jim Twins' lives point to the importance of genetic influences in shaping personality and behaviour.

Some critics have argued that most of the adopted twins were raised in relatively similar environments, in part because adoption agencies try to match the child to the adoptive home. However, this argument doesn't explain the identical twins Oskar Stohr and Jack Yufe (Bouchard et al., 1990). Oskar was raised a Catholic in Germany and even joined the Nazi party. Jack was raised a Jew and lived for a while in Israel. What twins could be more different? Yet when they met at an interview, they were wearing similar clothes, exhibited similar mannerisms, and had the same odd habits, such as flushing the toilet before using it, dipping toast in coffee, storing rubber bands on their wrists, and surprising people by sneezing in the elevator. Critics have suggested that nothing more than coincidence is at work in these case studies, but it is clear that twins share similarities other than just behaviour quirks, such as intelligence and personality. Indeed, some evidence suggests that twins raised apart are more similar than twins raised together. This would likely occur if parents encouraged individuality in twins raised together by emphasizing different strengths and interests. In effect, the parents would actively create a different environment for each twin. Continued research using the methods of behavioural genetics is needed to provide additional insight into the relative influence of heredity and environment on individual behaviour.

heritability A statistical estimate of the fraction of observed measure of the overall amount of difference among people in a population that is caused by differences in heredity.

UNDERSTANDING HERITABILITY *Heredity* is the transmission of characteristics from parents to offspring by means of genes. A term that is often confused with heredity, but that means something else altogether, is **heritability**, which is a statistical estimate of the portion of observed variation in a population that is caused by differences in heredity. *Variation* is the measure of the overall amount of difference among people. This variation is estimated based on studies of twins and other methods.

Why is heritability lower when the population is more diverse?

It is important to understand that heritability refers to differences in a certain trait among individuals, not to the trait itself. If a trait such as height has a heritability of .60, it means that 60 percent of height variation among individuals is genetic, not that you get 60 percent of your height from genetics and 40 percent from your environment. Heritability refers to populations, not to individuals. The heritability for a trait depends on variability within a population. For instance, almost everyone has two legs, and more people lose legs through accidents than are born without them. Thus, the heritability value for having two legs is nearly zero, in spite of the obvious fact that the human genome contains instructions to grow two legs. Herein lies a key lesson: Estimates of heritability are concerned only with the extent that people differ in terms of their genetic makeup. For example, how shy you are compared to others is determined in part by genes, and about half of the variability in shyness among different people is determined by genetic variation. Thus, the heritability for shyness is approximately 50 percent. It's important to point out that although there is genetic variation among us, the vast majority of genes work nearly identically in all people, and, indeed, across many species of other animals.

It is also important to point out that the population used to estimate total variation can affect the estimate obtained for heritability. Typically, the more diverse the population, the lower the estimate of heritability. This occurs because of the increased variability that comes from diversity and because the estimates of the genetic variation do not consider such diversity. The important point is that heritability is an estimate that is not precise and that can be affected by a number of factors. Still, it helps behavioural geneticists in clarifying the interaction between the environment and genes, so that they can understand the circumstances in which genes operate.

Social and Environmental Contexts Influence Genetic Expression

Let's examine how social and environmental contexts can influence genetic processes. For instance, Americans of African ancestry are especially prone to hypertension. According to some researchers, the dismal conditions of the slave trade may have led to a change in the genetic constitution of African Americans (Wilson & Grim, 1991). As many as one in four slaves died during transportation, and approximately the same number died within the first few years of arrival in America. Wilson and Grim argue that the horrific conditions imposed a strong selection factor in determining the type of individual who was able to survive. Those slaves who were able to maintain high blood pressure on low levels of salt might have been more likely to survive. In contemporary times, however, with salt readily available, a greater sensitivity to salt may cause African Americans to become salt hypertensive. The theory is provocative and difficult to test, but it highlights one possible way in which social context can affect genetic constitution.

Another strong example of how social environment interacts with gene expression is found in a longitudinal study of criminality conducted by Avshalom Caspi and his colleagues. These researchers followed a group of more than 1,000 New Zealanders from their births in 1972–1973 until adulthood, collecting enormous amounts of information about the participants and their lives each few years. When the participants were 26 years old, the investigators were able to examine which factors predicted who became a violent criminal. Prior research had demonstrated that children who are mistreated by their parents are more likely to become violent offenders. But not all mistreated children become violent. These researchers wanted to know why. They developed the hypothesis that differences in a gene that controls a certain enzyme (monoamine oxidase, or *MAO*) may be important in determining susceptibility to the effects of maltreatment. You will read about MAO in Chapter 13; for now it is sufficient to know that low levels of MAO have been implicated in aggressive behaviours. The gene for MAO comes in two forms, one of which leads to higher levels of MAO and one of which leads to lower levels. Caspi and colleagues (2002) found that boys with the low-level MAO gene appeared to be especially susceptible to early childhood maltreatment. Those children who were mistreated and had the low-activity MAO gene were much more likely to be convicted of a violent crime than those with the high-activity gene. Indeed, although only 1 in 8 boys was mistreated and had the low MAO gene, they were responsible for nearly half of all violent crimes committed by the group (Figure 3.6). The New Zealander study is a good example of how the combination of nature and nurture can work together to affect human behaviour—unfortunately, in this case, violent behaviour.

3.6 MAO Activity and Violent Crimes Those who had the gene for low MAO activity were much more likely than others to have been convicted of violent crimes if they had been maltreated. The effects of maltreatment had less influence on those with the high MAO gene.

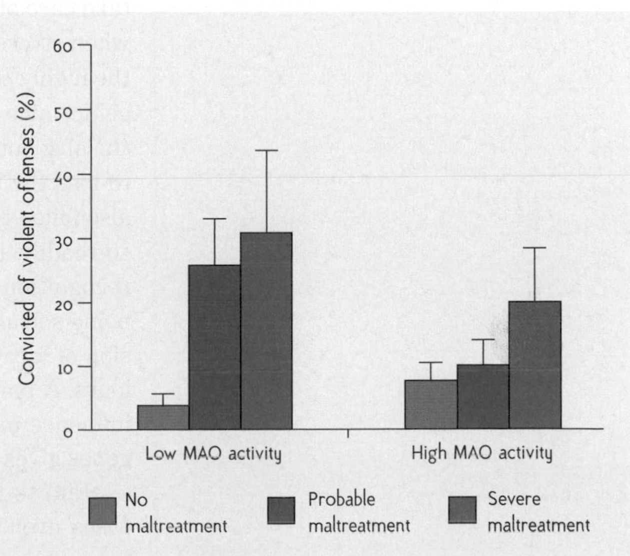

Finally, there is also evidence that genes and the social context interact to affect the phenotype. Sandra Scarr (Scarr & McCarthy, 1983) proposed a theory of development that stresses the interactive nature of genes and environment. According to Scarr, early environments influence young children, but children's genes also influence the experiences they receive. For instance, children exposed to the same environment interpret and react to it in different ways. Some children who are teased withdraw; others shrug it off without concern. Similarly, different sorts of children evoke different responses from others. A well-mannered, cuddly child cues more nurturing from parents and others than a baby who is irritable and fussy. Finally, as children become older they can choose their social situations. Some children prefer to spend time outdoors engaged in vigorous physical pursuits; others prefer the tranquility of a good book in a comfortable setting. Thus, genes predispose certain behaviours that elicit different responses, and these subsequent interactions then shape the phenotype. Because genes and social contexts interact, it can be very difficult to separate their independent effects.

Gene Expression Can Be Modified

What can we learn and ultimately achieve through the manipulation of our genes? Throughout this section, we have highlighted the new emphasis on gene expression—what causes genes to turn on and off—in addition to the heritability of particular traits or behaviours. In Chapter 2 we described methods that researchers use to study gene expression. A variety of gene-manipulation techniques are employed to enhance or reduce the expression of genes, or even to insert genes into a different animal species. Animals with genetic modifications are then compared with animals without such modifications to test theories about the genes' function. For instance, genes can be removed from or disrupted within the genome to create knock-outs; if the gene is important for a specific function, then disrupting it should interfere with that function. These gene-manipulation techniques have led to dramatic increases in understanding how gene expression influences mind and behaviour.

One remarkable finding so far is that changing even a single gene can lead to dramatic changes in behaviour (Wahlsten, 1999). Using gene-manipulation techniques, researchers have created anxious mice, hyperactive mice, mice that can't learn or remember, mice that groom themselves to the point of baldness, mice that fail to take care of their offspring, and even mice that progressively increase alcohol intake when stressed (Marcus, 2004; Ridley, 2003). Consider a study in which a gene from the highly social prairie vole was inserted into normally antisocial mice. These transgenic mice exhibited social behaviour more typical of prairie voles than of mice (Insel & Young, 2001). Another study found that knocking out specific genes led mice to lose the ability to remember mice they had previously encountered. These mice also failed to investigate new mice placed in their cages when normal mice would do so readily. In essence, knocking out one gene led to multiple impairments in social recognition (Choleris et al., 2003). This indicates not that there is a specific gene for being social, but rather that changing the expression of one gene leads to the expression of a cavalcade of other genes, which ultimately influences even complex behaviours. A major lesson from recent research is that genes seldom work in isolation to influence mind and behaviour; rather, the complex interaction among thousands of genes gives rise to the complexity of human experience (Suzuki et al., 1989).

Efforts to manipulate genes in humans have so far proven difficult and in some cases even dangerous. But there is new hope for altering gene expression with a method that focuses on RNA rather than on DNA. Recall that gene expression involves

DNA sending instructions via RNA to create specific proteins. Although RNA was once viewed as little more than a messenger, recent dramatic discoveries have indicated that RNA might be more important than ever imagined. These findings are so amazing that the most influential research journal, *Science,* labeled them the number one scientific breakthrough for the year 2002, and the researchers received the 2006 Nobel Prize for this work. It began in 1998, when researchers discovered that injecting a slightly modified form of RNA into worms led to a dramatic shutdown in gene expression in the very genes that were associated with producing that RNA in the first place (Fire et al., 1998). The modified RNA is broken down by other gene expression processes into tiny segments called small RNA. This process, now referred to as *RNA interference* (RNAi), prevents gene expression by silencing the DNA instructions to make proteins. Researchers view the potential of RNAi as astounding, and the race is on among various laboratories around the world to use RNAi as therapy for a variety of diseases and disorders ranging from cancer, to AIDS, to schizophrenia (Shankar, Manjunath, & Lieberman, 2005). Recall Nancy Wexler, the psychologist who has devoted her life to understanding Huntington's disease. Her team has discovered that the gene for Huntington's produces a protein that attacks the brain. We already know that knock-out methods will not work to prevent Huntington's because the gene serves many important functions. But what if RNAi could be used to silence the production of the specific brain-damaging protein? Indeed, using a mouse model of Huntington's disease, researchers recently have demonstrated that RNAi can dramatically reduce signs of brain damage and improve behavioural symptoms of the disease (Harper et al., 2005). The hope of being able to control such gene expression opens the possibility of bold new advances in our understanding of how genes influence mind and behaviour.

REVIEWING the Principles | What Is the Genetic Basis of Psychological Science?

Human behaviour is influenced by genetic processes. People inherit both physical characteristics and personality traits from their parents. Only recently have scientists developed the tools to measure genetic processes and the roles that various genes play in psychological activity. The Human Genome Project has succeeded in mapping the basic sequence of DNA, but a great deal more needs to be accomplished before that information can be translated into medical treatments or a greater understanding of individual differences among people. Within the last few years, researchers have begun to focus more on how and when genes are expressed in addition to the heritability of particular traits. Methods have been developed to enhance or interrupt gene expression, and most recently scientists are looking at how modified RNA can be used to silence gene expression entirely.

How Does the Nervous System Operate?

Many of the genes in the human genome direct the development and ongoing functions of the *nervous system,* a communication network that serves as the foundation for all psychological activity. Comprising billions of specialized cells known as nerve cells, the nervous system takes in a variety of information from the external world, evaluates that information, and subsequently produces behaviours or makes bodily adjustments to adapt to the environment. At the biological level of analysis, the essence of life—from smelling a rose, to thinking, to moving your big toe—comes down to communication among the billions of nerve cells throughout the body. So how do nerve cells operate? How do they communicate with each other? How do communicating networks function to produce behaviours, thoughts, and feelings? Before we get to the workings of the nervous system as a whole, we begin by describing how cells within this system communicate with each other to produce psychological activity.

Neurons Are Specialized for Communication

neuron The basic unit of the nervous system that operates through electrical impulses, which communicate with other neurons through chemical signals. Neurons receive, integrate, and transmit information in the nervous system.

dendrites Branchlike extensions of the neuron that detect information from other neurons.

cell body In the neuron, where information from thousands of other neurons is collected and processed.

axon A long narrow outgrowth of a neuron by which information is transmitted to other neurons.

Neurons, the basic units of the nervous system, are cells that specialize in communication. Neurons differ from most other cells because they are excitable: They operate through electrical impulses and communicate with other neurons through chemical signals. They have three functions: to take in information from neighboring neurons (reception), integrate those signals (conduction), and pass signals to other neurons (transmission).

Neurons come in a wide assortment of shapes and sizes, but they typically share four structural regions that assist the neuron's communication functions (Figure 3.7). The first region is the **dendrites**—short outgrowths that increase the neuron's receptive field. These branchlike appendages detect chemical signals from neighboring neurons.

The second region of the neuron is the **cell body**, where information from thousands of other neurons is collected and integrated. The cell body is also the site of metabolism and genetic action. Like all cells in the body, neurons contain DNA, the recipe for action for the neuron. Once the incoming information has been integrated in the cell body, electrical impulses are transmitted along a long narrow outgrowth known as the **axon**. Axons, the third region of the neuron, vary tremendously in

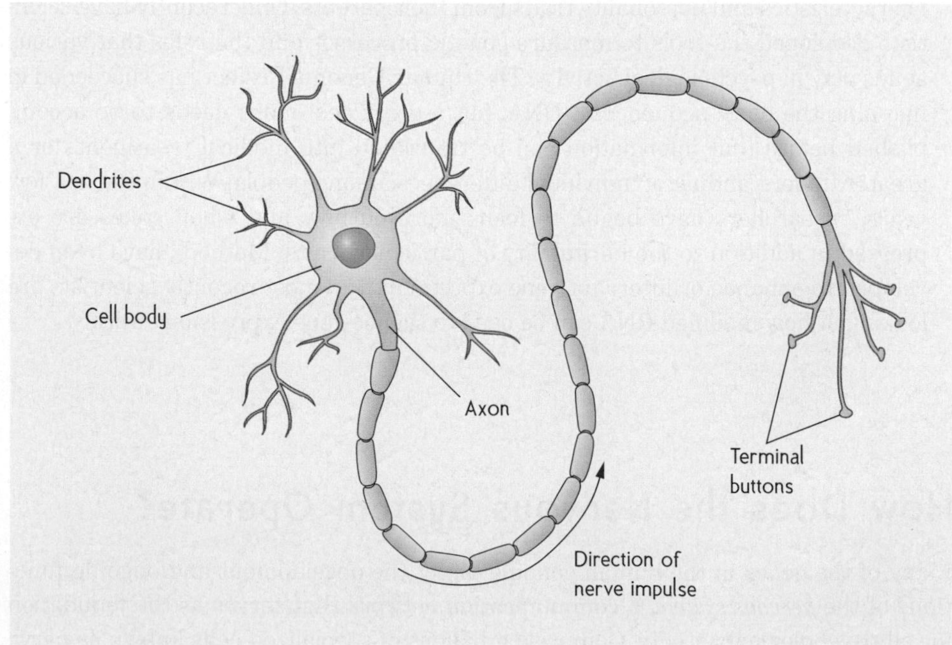

3.7 A Neuron Messages are received by the dendrites, processed in the cell body, transmitted along the axon, and sent to other neurons via chemical substances released from the terminal buttons.

length, from a few millimeters to more than a meter. The longest axons stretch from the spinal cord to the big toe. You have probably heard the term *nerve,* as in a pinched nerve. Used in this context, a nerve refers to a bundle of axons that carry information between the brain and the body.

Terminal buttons, small nodules at the ends of axons, are the fourth region of the neuron. Terminal buttons receive the electrical impulses and release chemical signals from the neuron to an area called the synapse. The **synapse** is the site for chemical communication between neurons. Chemicals leave one neuron, cross the synapse, and then pass signals along to the dendrites of other neurons.

The boundary of a neuron is defined by its membrane, a double layer of fatty molecules called *lipids.* As you will see, the membrane plays an important role in communication between neurons by regulating the concentration of electrically charged molecules that are the basis of the neuron's electrical activity.

terminal buttons Small nodules at the ends of axons that release chemical signals from the neuron to an area called the synapse.

synapse The site for chemical communication between neurons.

sensory neurons One of the three types of neurons, these afferent neurons detect information from the physical world and pass that information along to the brain.

motor neurons One of the three types of neurons, these efferent neurons direct muscles to contract or relax, thereby producing movement.

interneurons One of the three types of neurons, these neurons communicate only with other neurons, typically within a specific brain region.

TYPES OF NEURONS The three basic types of neurons are *sensory neurons, motor neurons,* and *interneurons.* **Sensory neurons** detect information from the physical world and pass that information along to the brain, usually via the spinal cord. You know from hitting your funny bone that sensory neurons can transmit fast-acting signals that trigger a nearly instantaneous bodily response and sensory experience. Sensory neurons are often called *afferent* neurons, because they send their signals from the body to the brain. Signals that travel from the brain to the body are known as *efferent* (Figure 3.8).

Motor neurons direct muscles to contract or relax, thereby producing movement. Motor neurons are efferent neurons, in that signals *from* the brain are transmitted *to* the body. **Interneurons** communicate within local or short-distance circuits. That is, interneurons integrate neural activity within a single area rather than transmitting information to other brain structures or to the body organs.

Together, sensory and motor neurons control movement. For instance, when you use your pen your brain sends a message via motor neurons to the muscles in your fingers to move in specific ways. Receptors in your skin and muscles send back

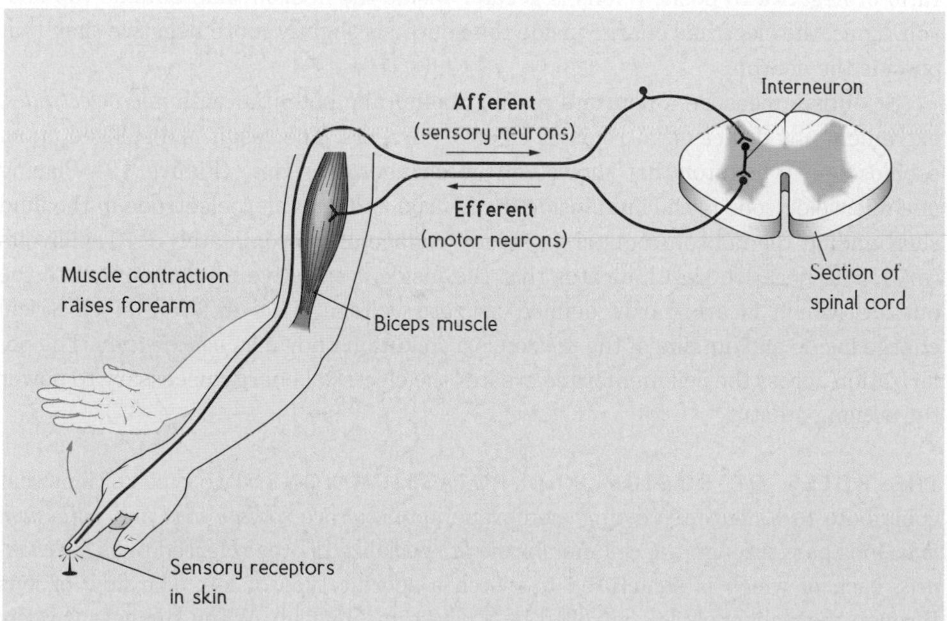

3.8 Afferent and Efferent Signals Receptors send afferent signals to the brain for processing. An efferent signal is then sent from the brain to the body via the spinal cord to produce a response.

messages through sensory neurons that help to determine how much pressure is needed to hold the pen. The nerves that provide information from muscles are referred to as *somatosensory,* which is the general term for sensations experienced from within the body.

Even the tiniest actions require the integration and coordination of multiple brain and body systems. Writing requires brain mechanisms that initiate the desire to use the pen, motor messages that direct the arm, hand, and fingers to reach for and hold the pen, and somatosensory messages that provide feedback necessary for successful use of the pen. If any of the components fail to work, you will have great difficulty even writing your name.

Complex networks of thousands of neurons sending and receiving signals are the functional basis of all psychological activity. Although the actions of single neurons are simple to describe, human complexity is a result of billions of neurons, each making contact with tens of thousands of other neurons. Neurons don't communicate randomly or arbitrarily; they selectively communicate with other neurons to form circuits or *neural networks.* These networks develop through maturation and experience, forming permanent alliances among groups of neurons.

Action Potentials Cause Neuronal Communication

How do nerve cells operate?

Neuronal communication depends on the ability of the neuron to respond to incoming stimulation by changing electrically and subsequently passing along signals to other neurons. An **action potential**, also called *neuronal firing,* is the electrical signal that passes along the axon and causes release of chemicals that transmit signals to other neurons. Let's examine some of the factors that contribute to the firing of an action potential.

action potential The neural impulse that passes along the axon and subsequently causes the release of chemicals from the terminal buttons.

resting membrane potential The electrical charge of a neuron when it is not active.

THE RESTING MEMBRANE POTENTIAL IS NEGATIVELY CHARGED When not active, the inside and outside of a neuron differ electrically because of the balance of *ions* in the intracellular fluid. This electrical difference is referred to as the **resting membrane potential**. As with all salty solutions, the salts in intracellular fluid consist of particles called *ions* that have negative or positive charges. The ratio of negative to positive ions is greater inside the neuron than outside the neuron. Thus, the electrical charge inside the neuron is slightly more negative than that outside the neuron.

Scientists measure a neuron's resting membrane potential with *microelectrodes,* extremely tiny devices that register electric currents. The result is displayed on an oscilloscope, a monitor that shows voltage changes over time (Figure 3.9). Placing one microelectrode in the fluid inside the neuron and one microelectrode in the fluid surrounding the neuron registers a steady voltage of approximately −70 millivolts (mV). The negative sign indicates that the inside is negative in comparison to the outside, which is arbitrarily defined as zero voltage. This differential electrical charge inside and outside of the neuron is a condition known as *polarization.* The polarization across the cell membrane creates the electrical energy necessary to power the action potential.

THE ROLES OF SODIUM AND POTASSIUM IONS Two types of ions that contribute to a neuron's resting membrane potential are *sodium ions* and *potassium ions.* Ions pass through the cell membrane at specialized pores referred to as *ion channels,* each of which is structured to match a specific type of ion. The flow of ions through their channels is controlled by a gating mechanism. When the gate is open,

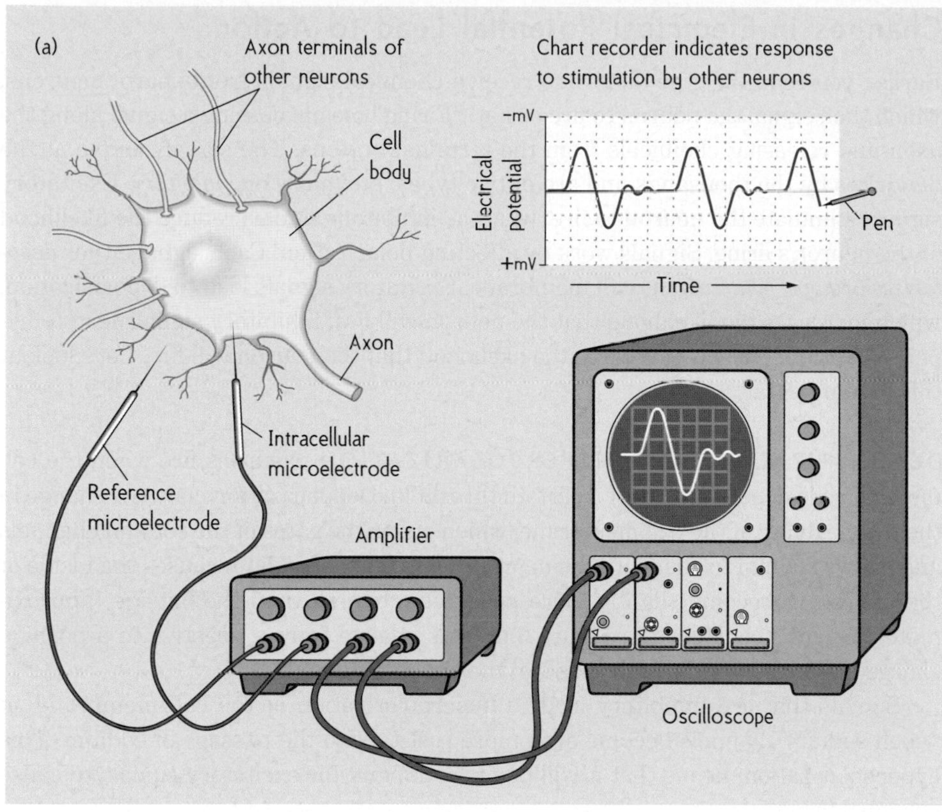

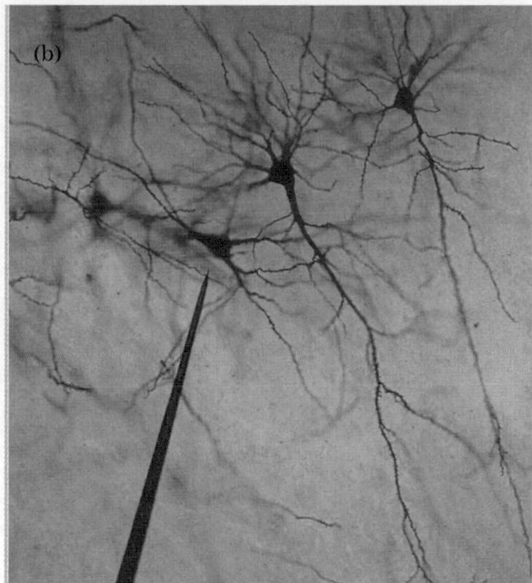

3.9 Resting Membrane Potential (a) The neuron's resting membrane potential can be recorded by placing one microelectrode in the fluid inside the neuron and one microelectrode in the fluid surrounding the neuron and measuring the difference in voltage, approximately −70 millivolts (mV). (b) A microelectrode being placed into a neuron.

ions flow freely past, but when closed the gate prevents passage of ions through the cell membrane (Figure 3.10). The flow of ions also is affected by the fact that the cell membrane is *selectively permeable*—it allows some types of ions to cross more easily than others, much like a bouncer at an exclusive nightclub. Partially as a result of the selective permeability of the cell membrane, there is more potassium inside the neuron than sodium, which contributes to polarization.

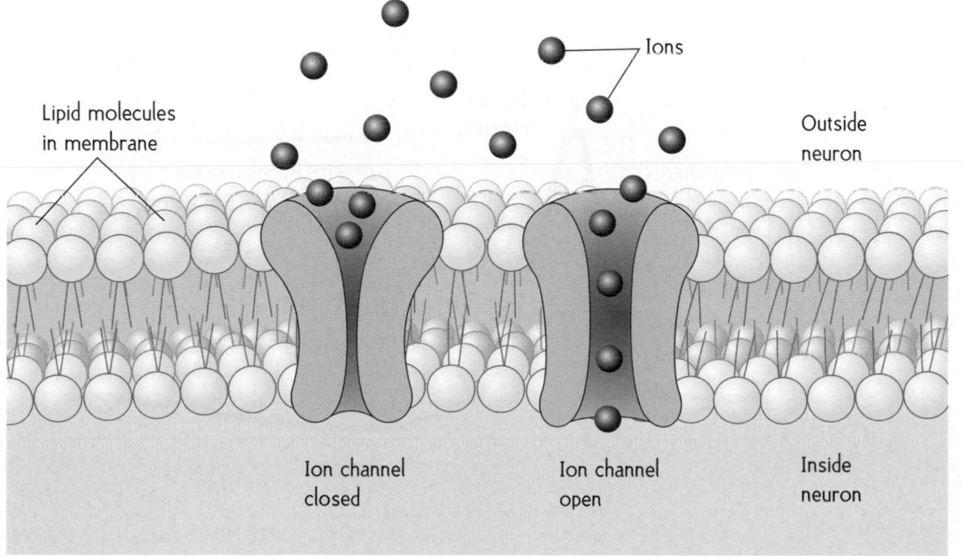

3.10 Cell Membrane Permeability The cell membrane is selectively permeable. Ions can travel into the neuron when ion channels are open but not when ion channels are closed. The channels are specialized for specific ions. Thus, the sodium channel controls only the passage of sodium.

Changes in Electrical Potential Lead to Action

Earlier you read that the dendrites receive chemical signals from nearby neurons, which then signal the neuron to fire or not. "Firing" means passing a signal along the axon and releasing chemicals from the terminal buttons. The signals arrive at the dendrites by the thousands and are of two types: *excitatory* or *inhibitory.* Excitatory signals stimulate the neuron to fire, whereas inhibitory signals reduce the likelihood of the neuron's firing. Signals work by affecting polarization. Each signal either *depolarizes* or *hyperpolarizes* the cell membrane. Excitatory signals lead to depolarization, which increases the likelihood that the neuron will fire; inhibitory signals lead to hyperpolarization, which decreases the likelihood that the neuron will fire. Let's look at this in more detail.

DEPOLARIZATION AND HYPERPOLARIZATION Neurons fire when the cell membrane is depolarized past a certain threshold. Depolarization causes a change in the permeability of the cell membrane, which opens the gates of the sodium channels and allows sodium to rush into the neuron. This influx of sodium causes the inside of the neuron to become slightly more positively charged than the outside (approximately 50 mV), as shown in Figure 3.11. This change from a negative to a positive charge inside the neuron is the basis of the action potential.

Signals that are inhibitory lead to hyperpolarization of the cell membrane, in which sodium channels become even more resistant to the passage of sodium. This hyperpolarization means that it will be more difficult for excitatory signals to cause neuronal firing.

Action Potentials Spread along the Axon

Inhibitory and excitatory signals received by the dendrites are integrated within the neuron. If the total amount of depolarization surpasses the neuron's threshold, which occurs when there are more excitatory than inhibitory signals, an action potential is generated. When the neuron fires, the depolarization of the cell membrane moves along the axon like a wave, which is called *propagation.* Sodium rushing through its ion channels leads to further depolarization, causing adjacent sodium channels to

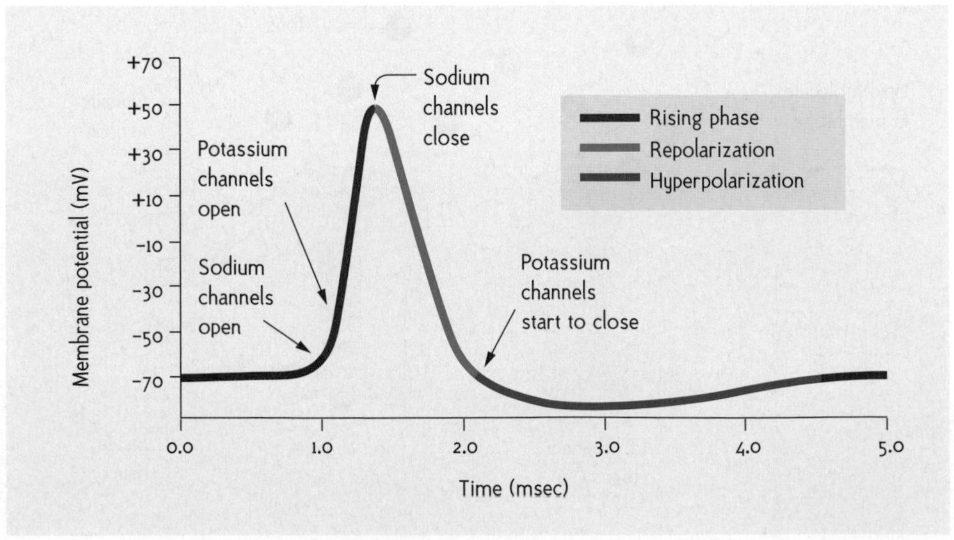

3.11 Membrane Potential In the resting state, sodium ion channels are closed. Depolarization opens the sodium channels, which allows sodium to flow into the neuron and changes the electrical charge of the neuron.

open. Thus, like toppling dominoes, sodium ion channels open successively along the cell body to the end of the axon. At any point along the axon, the sudden influx of sodium repels the potassium ions, which are forced out of the neuron. This influx of sodium and outflow of potassium continues until a state of near equilibrium is reached, at which point other forces take over to shut the sodium ion channel gates and stop the influx of sodium. All this occurs within about 1/1,000 of a second.

ABSOLUTE AND RELATIVE REFRACTORY PERIODS Once the gating mechanism stops the flow of sodium into the neuron, potassium stops leaving. However, this takes 1 to 2 milliseconds, during which a decreased concentration of potassium in the cell body momentarily creates a state of hyperpolarization. During this brief period, known as the *absolute refractory period,* it is impossible for the neuron to fire, which keeps the action potential from repeating up and down the axon in a kind of ripple effect (McCormick, 1999). As each section of the axon depolarizes, the preceding section enters its absolute refractory state. Hence, under natural conditions, action potentials move along the axon in only one direction. The signal can't travel backward because the preceding axonal section is in its refractory period and its sodium ion channels are blocked.

As potassium stops exiting the neuron and the cell membrane is returning to its resting membrane potential, the neuron can fire, but only in response to an especially strong signal, such as might occur if the neuron were bombarded by excitatory signals. This brief period of time is called the *relative refractory period.* Until the neuron returns to its resting potential, it is somewhat hyperpolarized and requires a stronger signal to become depolarized.

ALL-OR-NONE PRINCIPLE Any one signal received by the neuron has little influence on whether it fires. Normally, the neuron is barraged by thousands of inhibitory and excitatory signals, and its firing is determined by the number and frequency of those signals. The firing criterion is the total electrical voltage produced by the signals at a given time. If the sum of the depolarizations and hyperpolarizations leads to a drop in voltage that exceeds the neuron's firing threshold, then an action potential is generated. It is as if you decided to change your major from English to psychology because five of the eight friends you asked said to go for it.

The firing of the neuron is all or none—a neuron cannot partially fire. The **all-or-none principle** dictates that a neuron fires with the same potency each time, but at intervals of different frequency depending on the strength of stimulation. Suppose you are playing a video game in which you fire missiles by pressing a button. Every time you press the button, a missile is launched at the same velocity as the last. It makes no difference how hard you press the button. But if you keep your finger on the button, additional missiles fire in rapid succession. Thus, the strong stimulus—your finger holding down the button—controls the firing frequency.

all-or-none principle The principle whereby a neuron fires with the same potency each time, although frequency can vary; it either fires or not, it cannot partially fire.

myelin sheath A fatty material, made up of glial cells, that insulates the axon and allows for the rapid movement of electrical impulses along the axon.

THE MYELIN SHEATH The propagation of action potentials along the axon happens quickly, which permits the fast and frequent adjustments required for coordinating motor activity. However, the action potential moving along the axon by depolarizing adjacent sodium channels does not occur rapidly enough to explain the precision of fine motor movements. So what permits rapid firing? Signals can travel quickly down the axon because it is insulated by a **myelin sheath**, a fatty material that encases axons (Figure 3.12). Made up of *glial cells* (Greek, "glue"), myelin, like the plastic tubing around wires in an electrical cord, insulates sections of the axon and facilitates the passage of electrical impulses.

3.12 The Myelin Sheath and the Nodes of Ranvier The myelin sheath surrounds the axon, except at the nodes of Ranvier. The electrical signal jumps from node to node.

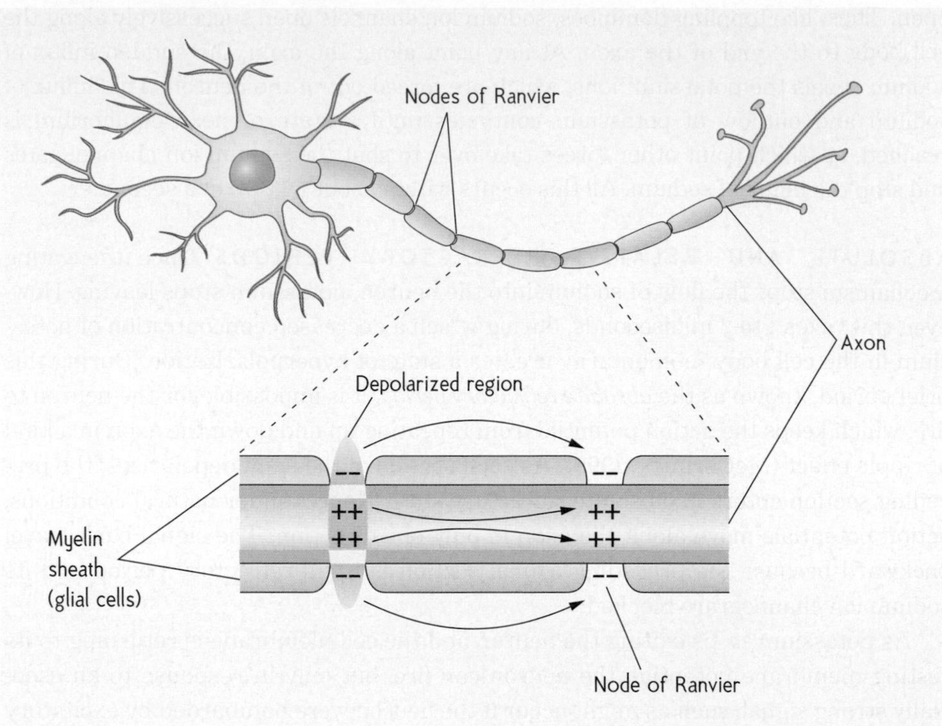

nodes of Ranvier Small gaps of exposed axon, between the segments of myelin sheath, where action potentials are transmitted.

As shown in Figure 3.12, the myelin sheath grows along axons in short segments. Between these segments are small gaps of exposed axon known as the **nodes of Ranvier**, named for Louis A. Ranvier, the researcher who first described them. It is only at these unmyelinated gaps that ion channels can transmit action potentials. Thus, action potentials skip quickly along the axon, pausing only briefly at each node of Ranvier to get recharged. This method of conduction, known as *saltatory conduction* (from Latin *saltare,* "to jump"), enables signals to move through the nervous system extraordinarily fast. Deterioration of the myelin sheath leads to *multiple sclerosis* (*MS*), an especially tragic neurological disorder that mostly affects young adults. The symptoms result from decay of the myelin sheath surrounding axons. The earliest symptoms are often numbness in the limbs and blurry vision. Since the myelin insulation helps messages move quickly along axons, demyelination slows down neural impulses. The axons essentially short-circuit, and normal neuronal communication is interrupted. Motor actions become jerky, and victims lose the ability to coordinate motor movements. Over time, movement, sensation, and coordination are severely impaired. As the myelin sheath disintegrates, axons are exposed and may themselves start to break down. The term *sclerosis* means "hardening," which is noticeable from the scarring in the brains of patients with MS, as shown in Figure 3.13.

3.13 The Brain of an MS Patient The arrows point to areas of sclerosis.

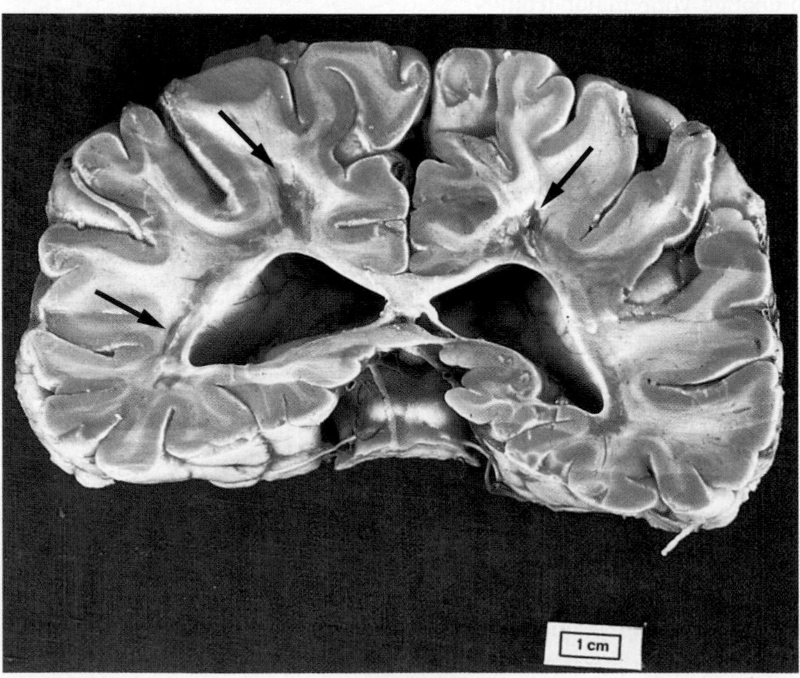

This progressive disease appears to have both genetic and environmental causes. Evidence for a genetic influence includes that it is more common among identical twins than nonidentical twins and that it is more common among Caucasians than Asians. However, it is also more common in cold climates than in warm climates. Although the connection between cold weather and MS is unclear, growing up in a cold climate seems more relevant than moving to one as a young adult; moving from a cold climate to a warmer one after childhood does not seem to reduce the risk of developing MS. Some theories suggest that MS may be an autoimmune disorder in which the body views myelin as an intruder, triggering an immune reaction that attacks and kills the myelin sheath. This autoimmune disorder may have its origins in a slow-acting infection that is contracted early in childhood. There is currently no known cure for multiple sclerosis.

synaptic cleft The small space between neurons that contains extracellular fluid.

neurotransmitter A chemical substance that carries signals from one neuron to another.

receptors In neurons, specialized protein molecules on the postsynaptic membrane that neurotransmitters bind to after passing across the synaptic cleft.

Neurotransmitters Bind to Receptors across the Synapse

Neurons do not touch one another; they are separated by a small space known as the **synaptic cleft**, which is the site of chemical communication between neurons. Action potentials cause neurons to release from their terminal buttons chemicals that travel across the synaptic cleft and are received by the dendrites of other neurons. The neuron that sends the signal is called *pre*synaptic, and the one that receives the signal is called *post*synaptic.

How do nerve cells communicate with each other to influence mind and behaviour?

How do these chemical signals work? Inside the terminal buttons are small packages, or *vesicles,* that contain chemical substances known as neurotransmitters. The term **neurotransmitter** is a generic word used for chemical substances that carry signals across the synaptic cleft. After an action potential travels to the terminal button, it causes the vesicles to spill their neurotransmitters into the synaptic cleft. These neurotransmitters then spread across the synaptic cleft and attach themselves, or *bind,* to receptors on the postsynaptic neuron (Figure 3.14).

Receptors are specialized protein molecules. The binding of neurotransmitter to receptor causes ion channels to open, which changes the membrane potential at that location, thus affecting the probability that the neuron will fire. If a neurotransmitter binds with a receptor and depolarizes the membrane, it is excitatory and increases the likelihood that the receiving neuron will fire. By contrast, if the neurotransmitter's binding hyperpolarizes the membrane, it is inhibitory and makes the receiving neuron less likely to fire.

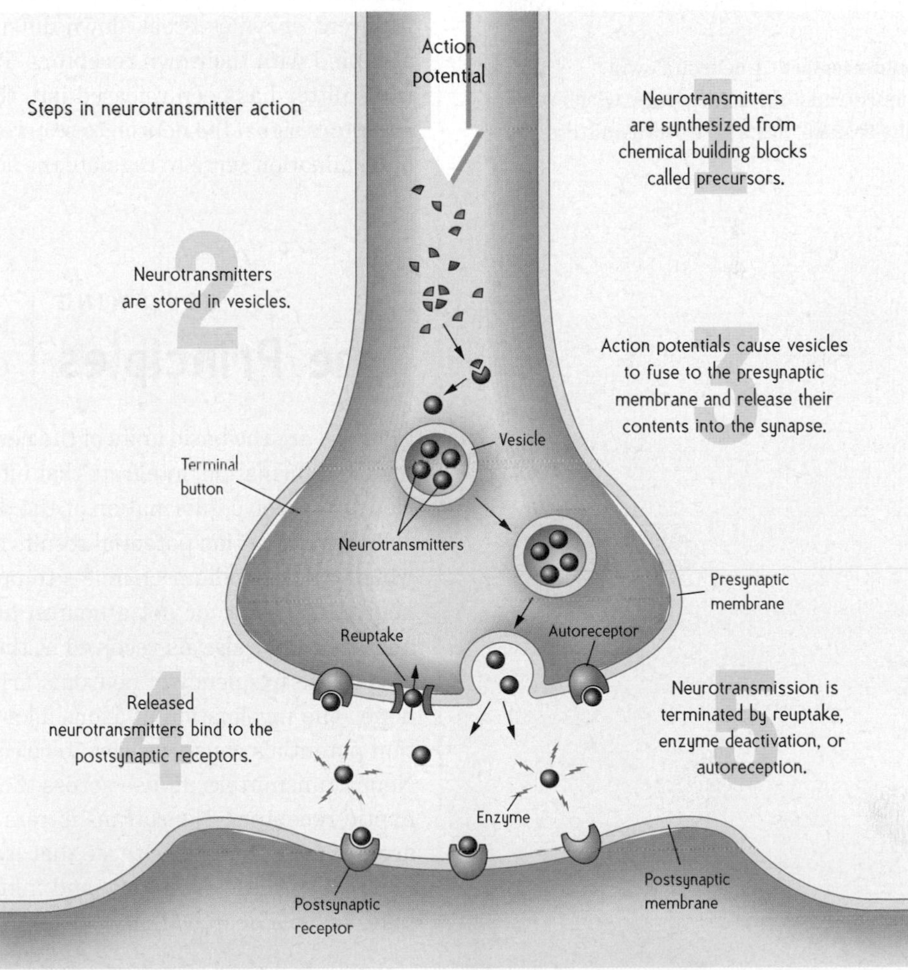
Steps in neurotransmitter action
Action potential
1
Neurotransmitters are synthesized from chemical building blocks called precursors.
2
Neurotransmitters are stored in vesicles.
3
Action potentials cause vesicles to fuse to the presynaptic membrane and release their contents into the synapse.
Vesicle
Terminal button
Neurotransmitters
Presynaptic membrane
Reuptake
Autoreceptor
4
Released neurotransmitters bind to the postsynaptic receptors.
5
Neurotransmission is terminated by reuptake, enzyme deactivation, or autoreception.
Enzyme
Postsynaptic receptor
Postsynaptic membrane

3.14 **Neurotransmitters** An overview of how neurotransmitters work.

RELATIONSHIP BETWEEN RECEPTORS AND NEUROTRANSMITTERS In much the same way as a lock opens only with the correct key, each receptor can be influenced by only one type of neurotransmitter. Yet drugs and toxins can mimic neurotransmitters and bind with their receptors as if they were the real thing. Addictive drugs such as heroin and cocaine, for example, have their effects because they are structurally similar to naturally occurring neurotransmitters; the receptors cannot differentiate between the ingested drug and the real neurotransmitter. Through knowledge of how neurotransmitters bind with receptors, researchers have been able to develop effective pharmacological therapies for many psychological and medical disorders.

TERMINATING SYNAPTIC TRANSMISSION The three events that terminate the influence of transmitters in the synaptic cleft are *reuptake, enzyme deactivation,* and *autoreception.* Unless one of these processes occurs, once a neurotransmitter is released into the synaptic cleft it continues to bind with receptors and blocks new signals from getting through. This would lock up the brain and prevent new activity. For change to occur, the signals have to be interrupted. The most common termination process is **reuptake**, when the neurotransmitter molecules are taken back into the presynaptic terminal buttons. The cycle of reuptake and release repeats continuously. An action potential prompts terminal buttons to release the transmitter into the synaptic cleft and then take it back for recycling. Another process that terminates the actions of a neurotransmitter occurs when an enzyme destroys the transmitter substance in the synaptic cleft, which is appropriately called **enzyme deactivation**. Different enzymes break down different neurotransmitters. Neurotransmitters can also bind with their own receptors. These **autoreceptors** monitor how much neurotransmitter has been released into the synapse. When excess is detected, the autoreceptors signal the neuron to stop releasing the neurotransmitter. All three methods of termination serve to regulate the activity of neurotransmitters in the synapse.

reuptake The process whereby the neurotransmitter is taken back into the presynaptic terminal buttons, thereby stopping its activity.

enzyme deactivation The process whereby the neurotransmitter is destroyed by an enzyme, thereby terminating its activity.

autoreceptors A neuron's own neurotransmitter receptors, which regulate the release of the neurotransmitters.

REVIEWING the Principles | How Does the Nervous System Operate?

Neurons are the basic units of the nervous system. The neuron's primary task is to take in information, integrate that information, and pass a signal to other neurons. Neurons receive information at the dendrites and process that information in the cell body. An action potential results from the depolarization of the cell membrane, which causes sodium channels to open and allows sodium ions to rush into the neuron. Whether or not a neuron fires depends on the integration of inhibitory and excitatory signals received at the dendrites. Summation of excitatory signals affects the frequency of neuronal firing but not the strength—neurons fire all-or-none. The myelination of axons allows action potentials to propagate rapidly. Action potentials cause vesicles to release neurotransmitters into the synaptic cleft. Neurotransmitters diffuse across the synaptic cleft and bind with specific postsynaptic receptors. Neurotransmitters that depolarize the postsynaptic membrane are excitatory, whereas those that hyperpolarize the postsynaptic membrane are inhibitory. These excitatory and inhibitory signals are terminated through reuptake, enzyme deactivation, and autoreception.

How Do Neurotransmitters Influence Emotion, Thought, and Behaviour?

Before the 1970s, most researchers believed that communication in the brain took place through the actions of just five or so narrowly defined neurotransmitters. Now we know that more than 60 chemicals transmit information in the brain and body, and that different transmitters are responsible for influencing emotion, thought, and behaviour.

How do drugs alter neural activity?

As you have learned, all neurotransmitters act to enhance or inhibit action potentials, by depolarizing or hyperpolarizing postsynaptic cell membranes. Many substances, such as drugs and toxins, can alter the actions of neurotransmitters in several ways. This fact is important because much of what we know about neurotransmitters has been learned through the systematic study of the effects of drugs and toxins on emotion, thought, and behaviour. For instance, drugs and toxins can alter how the neurotransmitter is synthesized, they can raise or lower the amount of neurotransmitter released from the terminal buttons, and they can change the way the neurotransmitter is deactivated in the synaptic cleft by blocking reuptake or preventing enzyme deactivation. Drugs that enhance the actions of neurotransmitters are known as **agonists**; drugs that inhibit action are **antagonists** (see Figure 3.15).

agonist Any drug that enhances the actions of a specific neurotransmitter.

antagonist Any drug that inhibits the action of a specific neurotransmitter.

Researchers often inject neurotransmitter agonists or antagonists into the brains of animals to assess the behavioural effects of a neurotransmitter. For instance,

Agonistic drug effects
Antagonistic drug effects
Action potential
Drugs increase the synthesis of neurotransmitters.
Drugs block the synthesis of neurotransmitters.
Drug molecule
Drugs increase the release of neurotransmitters.
Drugs block the release of neurotransmitters.
Vesicle
Drugs bind to autoreceptors and block their inhibitory effect.
Drugs activate autoreceptors so that they inhibit release of neurotransmitters.
Neurotransmitter
Drugs block the deactivation or reuptake of drugs from the synapse.
Reuptake
Auto-receptor
Drugs destroy the neurotransmitter in the synapse.
Drugs bind to postsynaptic receptors and either activate them or increase the effect of the neurotransmitter.
Drugs bind to postsynaptic receptors thereby blocking neurotransmitter binding.
Enzyme
Postsynaptic receptor

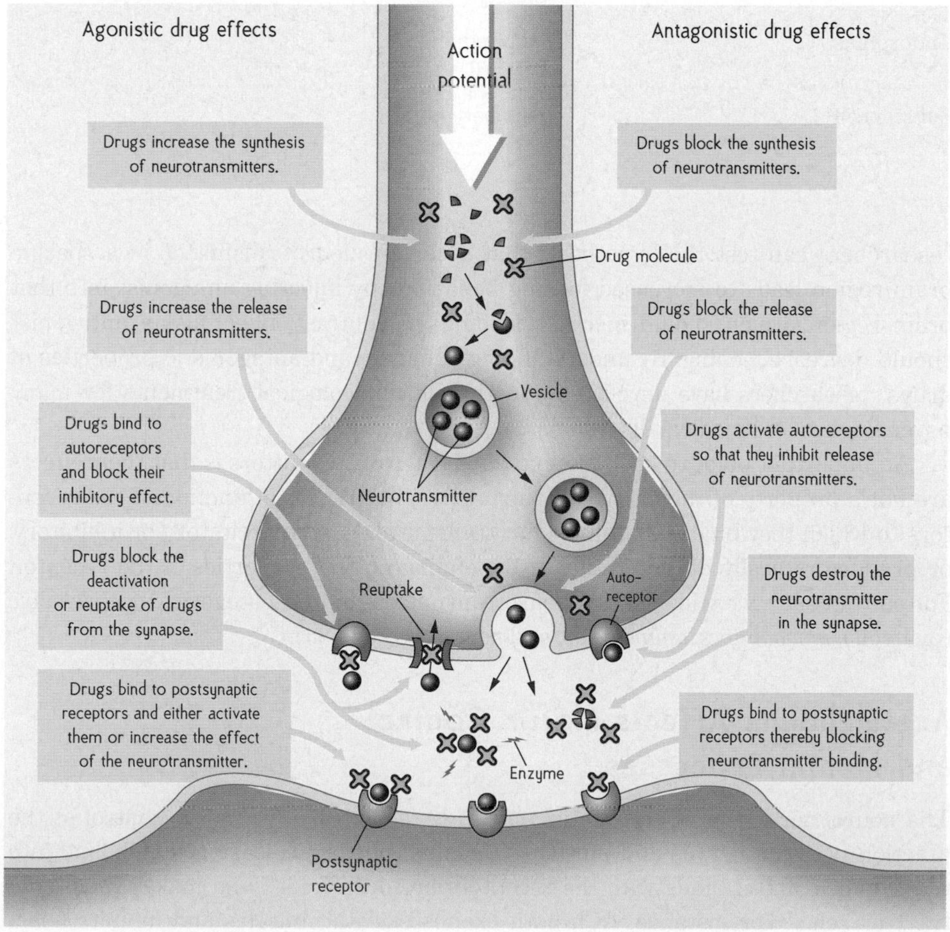

3.15 How Drugs Work The typical mechanisms of drug action.

TABLE 3.1 Common Neurotransmitters and Their Major Functions

NEUROTRANSMITTER	FUNCTIONS
Acetylcholine	Motor control over muscles Learning, memory, sleeping, and dreaming
Monoamines	
Norepinephrine	Arousal and vigilance Eating behaviour
Dopamine	Reward and motivation Motor control over voluntary movement
Serotonin	Emotional states and impulsiveness Dreaming
Amino Acids	
GABA (gamma-aminobutyric acid)	Inhibition of action potentials Anxiety and intoxication
Glutamate	Enhances action potentials Learning and memory
Peptide Modulators	
CCK (cholecystokinin)	Learning and memory Satiety
Endorphins	Pain reduction Reward
Substance P	Pain perception

researchers can test the hypothesis that a certain neurotransmitter in a specific brain region leads to increased feeding behaviour by injecting an agonist into that brain region, which should increase feeding; in contrast, injecting an antagonist should decrease feeding. By understanding agonistic and antagonistic properties of drugs, researchers have developed effective pharmacological treatments for many psychological and medical disorders.

An important point to keep in mind about neurotransmitters is that their effects are not a property of the chemicals themselves, but rather a function of the receptors to which they bind. The same neurotransmitter can be excitatory or inhibitory, or produce radically different effects, depending on the properties of the receptor. For convenience, we will describe neurotransmitters in four categories: *acetylcholine* (by itself), *monoamines, amino acids,* and *peptides* (Table 3.1).

Acetylcholine Affects Motor Control and Mental Processes

acetylcholine (ACh) The neurotransmitter responsible for motor control at the junction between nerves and muscles; also involved in mental processes such as learning, memory, sleeping, and dreaming.

The neurotransmitter **acetylcholine** (ACh) is responsible for motor control at the junction between nerves and muscles. Terminal buttons release acetylcholine into the synapse, and it binds with the receptors on muscle cells and makes them contract or relax. For instance, ACh both excites skeletal muscles and inhibits heart muscles. Whether its effects will be inhibitory or excitatory depends on the receptors. This difference reinforces the rule that the receptor controls the process.

Acetylcholine is also involved in complex mental processes such as learning, memory, sleeping, and dreaming. Because ACh affects memory and attention, drugs that are ACh antagonists inhibit its activity and can cause temporary amnesia. In a similar way, Alzheimer's disease, a condition characterized primarily by severe memory deficits, is associated with diminished ACh functioning (Geula & Mesulam, 1994). Drugs that are ACh agonists may enhance memory and decrease other symptoms, although so far there has been only marginal success in the pharmacological treatment of Alzheimer's dementia (Dunnett & Fibiger, 1993).

Toxins that mimic ACh can bind to its receptors and cause temporary paralysis. For example, curare, an herbal poison used by South American Indians, competes with ACh for receptor binding and inhibits the mechanisms that produce muscle movement. Administration of curare paralyzes the limbs and lungs of its victims, asphyxiating them. Botulism, a form of food poisoning, inhibits the release of ACh from terminal buttons in a similar fashion and leads to trouble with breathing, difficulty in chewing, and often death. Because of its ability to paralyze muscles, botulism is now used in cosmetic surgery in smaller, much less toxic doses. Physicians inject botulism into the eyebrow region, paralyzing muscles that produce some wrinkles. Because the effects wear off over time, a new dose of botulism needs to be injected every two to four months.

NICOTINE One substance that excites ACh receptors is nicotine, the addictive drug found in tobacco leaves. Given the role of ACh in memory and sleeping, it may not surprise you that the ingestion of nicotine has several cognitive effects. For instance, people who wear the nicotine patch often have extremely vivid dreams, probably because of the activation of ACh neurons known to affect the stage of sleep associated with dreaming. Studies demonstrate that smoking a cigarette can heighten attention, improve problem solving, and facilitate memory (Warburton, 1992). But before you rush out to buy a pack, some evidence suggests that nicotine may produce these effects only for current smokers (Heishman, Taylor, & Henningfield, 1994). When nonsmokers are given nicotine gum, they perform more poorly on cognitive tasks, as do former smokers who have not had nicotine for a while. Thus, smoking or chewing tobacco alters how the mind works, and over time users suffer deficits in cognitive performance *unless* they ingest nicotine. This is one of the many reasons why giving up nicotine is so difficult; recent ex-smokers find it hard to think or concentrate.

Monoamines Are Involved in Affect, Arousal, and Motivation

The **monoamines** are a group of neurotransmitters synthesized within the neuron from single amino acids—*mono* means "one." Although the monoamines are involved in a variety of psychological activities, their major functions are to regulate states of arousal and affect (feelings) and to motivate behaviour. Earlier we noted that men who had a gene that produced low levels of monoamine oxidase (MAO) and who had been mistreated as children were much more likely to be convicted of violent crimes. MAO is an enzyme that interrupts the activity of all monoamines. The four monoamines are *epinephrine, norepinephrine, serotonin,* and *dopamine.*

EPINEPHRINE AND NOREPINEPHRINE **Epinephrine** is found primarily in the body, although small amounts are in the brain. It was initially called *adrenaline*

monoamines A group of neurotransmitters synthesized from a single amino acid that are involved in a variety of psychological activities.

epinephrine A monoamine, found primarily in the body, which causes a burst of energy after an exciting event.

norepinephrine A monoamine neurotransmitter involved in states of arousal and vigilance.

serotonin A monoamine neurotransmitter important for a wide range of psychological activity, including emotional states, impulse control, and dreaming.

dopamine A monoamine neurotransmitter involved in reward, motivation, and motor control.

Parkinson's disease A neurological disorder that seems to be caused by dopamine depletion, marked by muscular rigidity, tremors, and difficulty initiating voluntary action.

and is the basis for the phrase "adrenaline rush," a burst of energy caused by its release in the body. **Norepinephrine** is involved in states of arousal and alertness. Norepinephrine is especially important for vigilance, a heightened sensitivity to what is going on around you. Norepinephrine appears to be useful for fine-tuning the clarity of attention; it inhibits responsiveness to weak synaptic inputs and strengthens or maintains responsiveness to strong synaptic inputs.

SEROTONIN **Serotonin** is a monoamine neurotransmitter involved in many different behaviours; it is especially important for emotional states, impulse control, and dreaming. Low levels of serotonin are associated with sad and anxious moods, food cravings, and aggressive behaviour. Drugs that block the reuptake of serotonin are now used to treat a wide array of mental and behavioural disorders, including depression, obsessive-compulsive disorders, eating disorders, and obesity (Tollesfson, 1995).

The drug LSD is structurally similar to the neurotransmitter serotonin; when ingested, LSD enters the brain and binds with serotonin receptors. When LSD binds with serotonin receptors normally involved in dreaming, it causes hallucinations, a dreamlike state of mind people experience while they are awake.

DOPAMINE **Dopamine** serves many significant brain functions, especially motivation and motor control. Many theorists believe dopamine is the primary neurotransmitter that communicates which activities may be rewarding. Eating when hungry, drinking when thirsty, or having sex when aroused all lead to activation of dopamine receptors and therefore are experienced as pleasurable. At the same time, dopamine activation is involved in motor control and planning, thereby guiding behaviour toward objects and experiences that will lead to additional reward. One theory of drug addiction is that certain drugs are dopamine agonists; people self-administer these drugs because they activate dopamine receptors and cause pleasure (Wise & Rompre, 1989). For example, cocaine blocks the reuptake of dopamine into presynaptic vesicles and allows dopamine to have a longer-lasting effect on postsynaptic receptors, leading to heightened arousal and feelings of euphoria. Rats quickly learn to self-administer cocaine, which supports the idea that it has rewarding qualities.

Dopamine also is involved in controlling voluntary muscle movements. Research in the past few decades implicates dopamine depletion in **Parkinson's disease**, a neurological disorder marked by muscular rigidity, tremors, and difficulty initiating voluntary action. With Parkinson's disease, the dopamine-producing neurons slowly die off. In the later stages of the disorder, people suffer from cognitive and mood disturbances. Parkinson's is a slow and degenerative disease that eventually leads to death.

From a treatment standpoint, drugs that enhance dopamine production can compensate for the lack of dopamine-producing neurons. For instance, injections of one of the chief building blocks of dopamine, L-DOPA, help the surviving neurons produce more dopamine. When used to treat Parkinson's disease, L-DOPA often produces a remarkable, though temporary, recovery. However, L-DOPA does nothing to stop the underlying disease that is destroying dopamine neurons in the first place. Fewer and fewer neurons work harder and harder to provide enough dopamine for fine motor movement; eventually, too many neurons die, and L-DOPA's effect diminishes.

A promising development in Parkinson's research is the transplanting of fetal tissue into human brains in the hope that the new fetal cells will produce dopamine. One of the first patients to undergo fetal neural transplantation, Donald Wilson, regained the ability to walk and was able to return to his hobby of woodworking. Brain scans showed increased release of dopamine one year following implantation (Figure 3.16).

Although case studies indicate that fetal transplants have promise for treating Parkinson's disease, clinical studies using random assignment have not found large differences between patients receiving fetal-cell transplants and those undergoing sham surgery, which mimics the surgery but does not involve transplantation (Olanow et al., 2003). Nonetheless, these methods are still being developed, and researchers are continuing to explore how fetal-cell transplants might be used to treat brain disorders. As discussed in Chapter 4, there is also hope that implanting stem cells, which might be more potent than fetal cells, might aid in the treatment of disorders such as Parkinson's.

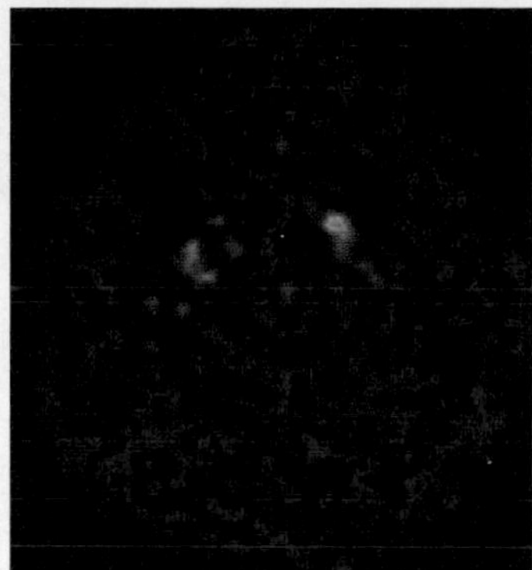

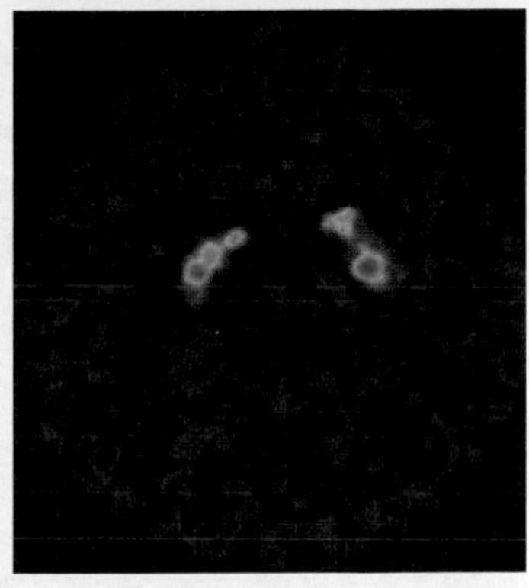

3.16 Dopamine Activity PET scans showing dopamine activity before (left) and after (right) fetal transplant surgery. This increase suggests that the transplant was successful and that the new neurons are producing dopamine.

Profiles in Psychological Science

Frozen Addicts

Although the cause of Parkinson's disease is not known, there is evidence that ingestion of certain drugs and toxins can produce symptoms that mimic the disease, as was dramatically demonstrated in the mysterious case of the "frozen addicts." Every now and then a patient baffles emergency-room physicians, challenging their years of medical training and practical experience. In July 1982, six patients ranging in age from 26 to 42 showed up at hospitals in San Jose, California, with symptoms unlike any the ER physicians had previously encountered. The patients' eyes were open and they appeared to be conscious; their limbs were stuck in place and their faces were frozen into hideous masks. Unable to speak, the six could not give any clues about the cause of their condition.

That all six patients had the same unusual symptoms could mean they had been exposed to a common toxic substance, but interviews with friends and family allowed the doctors to dismiss that possibility. The interviews did reveal, however, that each was a frequent heroin user. Still, the doctors remained uncertain about the role of heroin in their condition, since none of them bore any resemblance to the usual heroin overdose victim. Heroin depresses the body's nervous system, so a typical overdose is associated with lethargy and lax muscles; it was obvious that they had not taken real heroin. This was confirmed through interviews with their friends, who told of injecting what they believed to be heroin but experiencing sensations and bodily movements that were nothing like what they had come to expect from the drug. Most heroin users report a warm glow that comes almost immediately after injection; the warmth spreads throughout the body and leads to orgasmic sensations. Those who took the tainted heroin felt a stinging, burning sensation at the site of the injection, their vision became blurry, they experienced a metallic taste in the backs of their mouths, and their limbs starting jerking uncontrollably. This was certainly not the sought-after high associated with intravenous heroin injection. Drug users are accustomed to unanticipated side effects, since most street drugs are mixtures of narcotics and a wide array of fillers, some of which are poisons, used by dealers to increase profits.

The patients' symptoms persisted over many days. Doctors began to suspect brain damage, but no signs of it could be identified using CT scans or MRI. The six

3.17 MPTP and Parkinson's Disease These people developed symptoms of Parkinson's disease after ingesting the substance MPTP, which they mistakenly believed to be a synthetic form of heroin.

had undoubtedly ingested the same drug, but what was the drug and why was it causing these symptoms? William Langston, a brain scientist from Stanford University, examined the patients and was struck by the similarity between their symptoms and those of patients with Parkinson's disease. The six patients (five of whom are shown in Figure 3.17) were unlikely candidates for Parkinson's disease, which rarely affects people under age 45. And the chance that it would strike six young people from the same basic area at the same time is outside the realm of possibility. Dr. Langston suspected that the six might have injected a synthetic drug that was supposed to act like heroin, and that this drug had led to a condition similar to Parkinson's. His diagnosis turned out to be correct. The drug (later called MPTP) taken by these patients damaged a region of the brain that is normally involved in communicating with muscles to produce voluntary movements. The drug MPTP is now used to study Parkinson's disease, since it can be used to create animal models of the disease.

Whatever happened to the frozen addicts? Dr. Langston has tried a variety of experimental procedures on them over the two decades since their poisoning, including implanting new dopamine neurons in their brains, which produced modest improvement. Of the original six, some have died, one is in jail, and only one is completely able to take care of herself. In spite of this sad outcome, the finding that MPTP damages specific brain regions has allowed scientists to explore methods that might cure Parkinson's disease or at least delay the development of its symptoms in those with the disease. ■

Amino Acids Are General Inhibitory and Excitatory Transmitters in the Brain

For much of the past century, the monoamines and acetylcholine were considered to be the only neurotransmitters. In the past 40 years, however, it has become apparent that many other substances are also involved in neuronal communication. For instance, although the monoamines are made up of amino acids, some amino acids serve as neurotransmitters all by themselves. More recent research has established important roles for amino acids in general levels of inhibition and activation of the nervous system. Here we will look at two amino acids, *GABA* and *glutamate.*

GABA (gamma-aminobutyric acid) The primary inhibitory transmitter in the nervous system.

GABA **GABA** (gamma-aminobutyric acid) is the primary inhibitory transmitter in the nervous system and works throughout the brain to hyperpolarize postsynaptic membranes. Without the inhibitory effect of GABA, synaptic excitation might get out of control and spread through the brain in a reverberating circuit. Epileptic seizures may happen because of an abnormality in GABA functioning (Upton, 1994).

Drugs that affect the GABA system are widely used to treat anxiety disorders. For instance, people with nervous disorders commonly use benzodiazepines, which include drugs such as Valium. Benzodiazepines assist GABA in binding with one of its primary receptors, an action that inhibits neuronal connections and reduces symptoms of anxiety. These drugs also treat insomnia and other stress-related disorders. Ethyl alcohol—the type that people drink—has similar effects on GABA receptors, which is why alcohol typically is experienced as relaxing. GABA reception also may be the primary mechanism by which alcohol interferes with motor coordination. Drugs that block the effects of alcohol on GABA receptors also prevent alcohol intoxication. However, drugs that prevent the effects of alcohol are not used to treat alcoholics because reducing the symptoms of being drunk could easily lead to even greater alcohol abuse.

GLUTAMATE **Glutamate** is the primary excitatory transmitter in the nervous system and is involved in fast-acting neuronal transmission throughout the brain. It opens sodium gates in postsynaptic membranes and causes depolarization. Glutamate receptors aid learning and memory by strengthening synaptic connections (Cain, 1998).

In addition to being released by neurons, glutamate is also released by *astrocytes,* which are a type of glial cell. Until very recently, glial cells were believed to play primarily a support role in neuronal communication, such as being the building blocks of the myelin sheath. However, within the last few years it has become apparent that astrocytes play an important role in the nervous system. The release of neurotransmitters into the synapse causes a change in the activity of *calcium ions* in astrocytes, which causes the release of glutamate into the synapse. In this way, astrocytes contribute to the activity of neurons, such as by modifying their levels of polarization. There are at least three times as many glial cells as there are neurons, and unlike neurons, they are reproduced constantly. Philip Haydon (2001) has described astrocytes as being analogous to stagehands and directors in the theater; they may not be centre stage, but they are nonetheless important for the performance. How this alternative mode of communication within the nervous system affects higher brain processes, such as psychological activity, is currently unknown.

glutamate The primary excitatory transmitter in the nervous system.

peptides Chains of two or more amino acids found in the brain and the body; they can act like classic neurotransmitters or modify the quality of the neurotransmitter with which they are released.

cholecystokinin (CCK) The peptide found in highest concentration in the cerebral cortex; it plays a role in learning and memory, pain transmission, and exploratory behaviour.

Peptides Modulate Neurotransmission

Peptides are chains of two or more amino acids that are found in the brain and the body. Often, neurotransmitters and peptides are released simultaneously into the synaptic cleft, and the peptide modifies the effect of the neurotransmitter with which it is paired. Some peptides, for example, prolong or shorten the action of neurotransmitters; others influence postsynaptic receptors. An analogy of the effects of peptides on neurotransmitters is the sound controls on a stereo. Peptide *modulation* can turn the volume up or down and alter the tone, but it cannot change the tune. As such, peptides help to explain the vast subtleties and nuances of human experience. There are more than 30 known peptides that act as neurotransmitters or modulators. Three examples are *cholecystokinin* (*CCK*), *endorphins,* and *Substance P.*

CCK **Cholecystokinin (CCK)** is the peptide found in highest concentration in the cerebral cortex. It plays a role in learning and memory, pain transmission, and exploratory behaviour. The administration of CCK triggers panic attacks in people who suffer from panic disorder, creating intense anxiety, a feeling of impending doom, and often feelings of suffocation (Bourin, Baker, & Bradwejn, 1998). CCK may also contribute to social anxiety. In one experiment, two rats are placed together in a novel or brightly lit setting. The amount of time they interact, grooming and sniffing, measures their social behaviour, which usually declines under stress. Administration of CCK antagonists leads to more social interaction whereas administration of CCK agonists has the opposite effect (Woodruff & Hughes, 1991). CCK is also found in the gastrointestinal system, where it promotes *satiety,* the feeling of fullness that terminates feeding. Administration of CCK before a meal leads people to feel full more quickly and decreases the amount they eat (Woods & Stricker, 1999). As you might imagine, research is currently under way to see whether CCK can be used to help treat human obesity.

endorphins Peptides involved in natural pain reduction and reward.

substance P A peptide that acts as a neurotransmitter and is involved in pain perception.

ENDORPHINS **Endorphins** are peptides involved in natural pain reduction and reward. In the early 1970s Candace Pert and Solomon Snyder established that opiate drugs such as heroin and morphine bind to receptors in the brain, which led to the discovery of naturally occurring substances that bind to those sites. These substances were called endorphins (short for "endogenous morphine").

Endorphins are part of the body's natural defense against pain. Pain is useful because it signals that an animal is hurt or in danger and therefore should try to escape or withdraw, but pain can also interfere with adaptive functioning. If pain prevented animals from engaging in behaviours such as eating, competing, and mating they would fail to pass along their genes. The analgesic effects of endorphins help animals perform these behaviours even when they are in pain. In humans, administration of drugs that bind with endorphin receptors, such as morphine, reduces the subjective experience of pain. People still feel pain but report detachment; they know about the pain but do not experience it as aversive (Foley, 1993). Apparently morphine alters the way pain is experienced rather than blocking the nerves that transmit pain signals.

Endorphins may account for the *placebo effect.* A placebo is a neutral substance, such as water, that has no pharmacological effect. Yet people being treated for pain with a placebo might report relief just because they expect relief. In a classic report published in the 1950s, 4 out of 10 surgical patients reported satisfactory relief from pain after the injection of salt water (Lasagna, Mosteller, von Felsinger, & Beecher, 1954). Placebo effects are real; administration of a placebo leads to physiological changes in heart rate and digestion. When people take a medication and expect to gain relief, mechanisms release endorphins, which confirm their expectations.

Endorphins are associated with euphoric moods; this may explain why drugs such as heroin or morphine, which mimic endorphins by binding with their receptors, are so addictive. Endorphins may also explain "runner's high," the experience that runners get when they push their bodies beyond endurance, and pain turns to pleasure. One theory is that the body produces endorphins to cope with anticipated pain; when pain fails to materialize or is less than expected, the result is pleasure. Skydiving and other high-risk activities may be pleasurable for similar reasons (Jones & Ellis, 1996).

SUBSTANCE P **Substance P** is another peptide that functions as a neurotransmitter and is involved in pain perception. This mysterious-sounding stuff was first identified in 1931 by von Euler and Gaddum, who referred to it in their notes simply by the initial "P," and this simple name stuck. Substance P helps transmit signals about pain to the brain. Probably the best evidence for it can be found at your local Mexican restaurant, where you can conduct your own experiment. Chili peppers, especially jalapeño peppers, contain a substance known as capsaicin, which activates sensory neurons and leads to the release of Substance P in the brain. The release of Substance P makes your tongue and mouth burn, your eyes water, and your hand reach for the nearest pitcher of water—a bad idea, because water spreads capsaicin around and releases more Substance P, which only intensifies the pain.

Over-the-counter liniments also contain capsaicin; when rubbed on your skin they activate sensory neurons that transmit pain signals to the brain, causing a burning sensation. Used in sufficient quantity, the capsaicin in these ointments depletes reserves of Substance P, preventing the neurons from transmitting other pain messages to the brain. This analgesic effect lasts until more Substance P is produced.

REVIEWING
the Principles | How Do Neurotransmitters Influence Emotion, Thought, and Behaviour?

Neurotransmitters are chemical substances that carry signals across the synaptic cleft and act to enhance or inhibit action potentials by depolarizing or hyperpolarizing postsynaptic cell membranes. Substances that enhance the actions of neurotransmitters are known as agonists; those that inhibit action are antagonists. Acetycholine is involved in motor movement as well as complex thought. The monoamines are involved in arousal and motivation, such as dopamine's role in reward. GABA and glutamate are related to general inhibiion and excitation. The number of known substances that act as neurotransmitters is now over 60 and growing. In addition to the classic neurotransmitters, other amino acids and peptides also affect neurotransmission.

How Are Neural Messages Integrated into Communication Systems?

How does the brain communicate with the body?

A number of communication systems operate together to regulate all psychological activity. The most important communication system is the nervous system, made up of networks of neurons throughout the body. The nervous system is divided into two functional units: the **central nervous system (CNS)**, which consists of the brain and spinal cord, and the **peripheral nervous system (PNS)**, which consists of all other nerve cells in the body. The two systems are anatomically separate, but their functions are highly interdependent. The PNS transmits a variety of information to the CNS, which organizes and evaluates that information and then directs the PNS to perform specific behaviours or make bodily adjustments. In addition, the CNS receives information from the *endocrine system,* which uses a different mode of communication. This section describes the interaction of the nervous system and the endocrine system in the production of psychological activity.

central nervous system (CNS) The brain and spinal cord.

peripheral nervous system (PNS) All nerve cells in the body that are not part of the central nervous system. The PNS includes the somatic and autonomic nervous systems.

The Central Nervous System Consists of the Brain and Spinal Cord

Nearly all the functions of the CNS are performed by the brain, and there has recently been dramatic growth in scientific knowledge about how the brain works. For our purposes of understanding how neural messages are integrated into communication systems, remember that behaviour and mental activity are produced within specific locations in the brain. For example, some brain regions are involved with emotional states, others with the ability to see, and others with the ability to speak. The brain has developed specialized mechanisms over millions of years to solve problems related to survival and reproduction. Using methods such as brain imaging, psychological scientists have been able to map the functions of many structures in the brain, and you will learn about these in the next chapter.

The other part of the CNS is the spinal cord. Although the spinal cord is capable of reflex action, its main job is to receive sensory signals from the body and transmit them to the brain, and to receive signals from the brain and relay them to the body to control muscles, glands, and internal organs. When the spinal cord is damaged, the brain loses both sensory input from and control over the body. The severity of feeling loss and paralysis depends on where the spinal cord is damaged. The higher up on the spine the damage is, the greater the number of nerve connections between the brain and the body that are severed.

The CNS is separated from the rest of the body by the *blood-brain barrier,* which refers to the selectively permeable nature of blood vessels throughout the CNS that prevents certain toxins and poisons in the blood from entering the brain or spinal cord. The extent to which chemical substances have an influence on the brain depends on the ease with which they pass through the blood-brain barrier. Many drugs, such as cocaine and heroin, cross the blood-brain barrier easily and therefore are able to bind with receptors and affect neurotransmitter functioning.

The Peripheral Nervous System Includes the Somatic and Autonomic Systems

somatic nervous system A major component of the peripheral nervous system, which transmits sensory signals to the CNS via nerves.

autonomic nervous system (ANS) A major component of the peripheral nervous system, which regulates the body's internal environment by stimulating glands and by maintaining internal organs such as the heart, gall bladder, and stomach.

The PNS has two primary components, which themselves are referred to as nervous systems: the *somatic nervous system* and the *autonomic nervous system* (Figure 3.18). The **somatic nervous system** transmits sensory signals to the CNS via nerves (bundles of axons). Specialized receptors in the skin, muscles, and joints send sensory information to the spinal cord, which relays it to the brain. In addition, signals are sent from the CNS to muscles, joints, and skin to initiate, modulate, or inhibit movement.

The second major component of the PNS is the **autonomic nervous system (ANS)**, which regulates the body's internal environment by stimulating glands (such as sweat glands) and by maintaining internal organs such as the heart. Nerves in the ANS carry *somatosensory* signals to the CNS, providing information about, for example, the fullness of your stomach or bladder, and how anxious you feel.

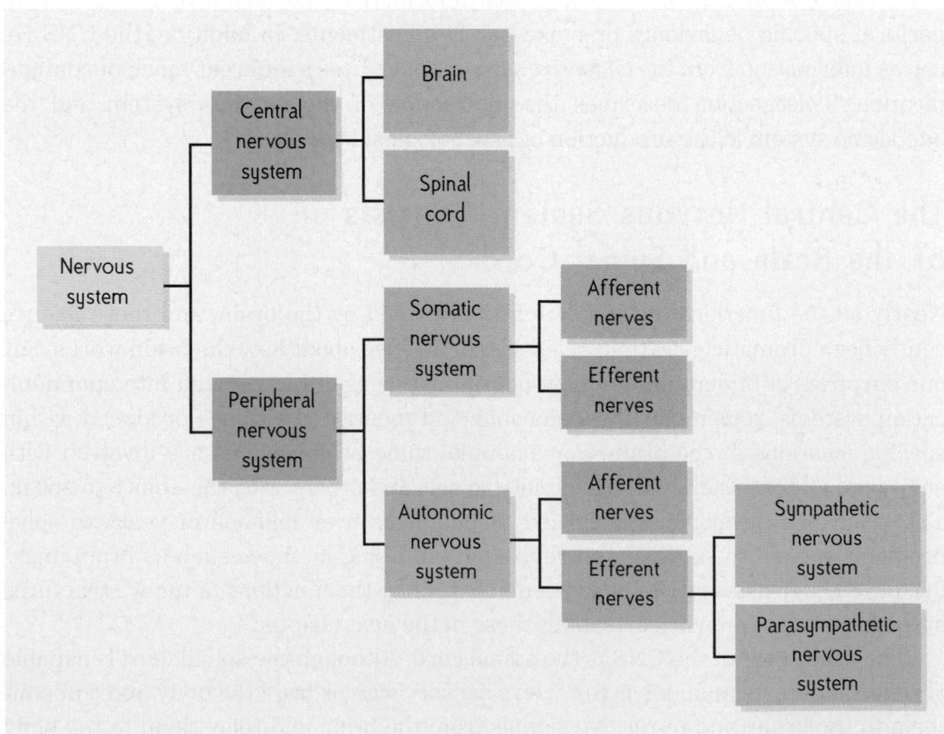

3.18 The Nervous System The major divisions of the nervous system.

SYMPATHETIC AND PARASYMPATHETIC DIVISIONS Two types of signals travel from the CNS to organs and glands in the PNS. To understand them, imagine that as you are studying, a fire alarm goes off. In the second after hearing the alarm, signals have been sent out to parts of your body to tell it to prepare for action: blood flows to skeletal muscles, epinephrine is released to increase heart rate and blood sugar, your lungs start taking in more oxygen, you stop digesting food to conserve energy, your pupils dilate to maximize visual sensitivity, and you start perspiring to keep cool. These actions are the result of the **sympathetic division** of the autonomic nervous system, which prepares the body for action. Should there be a fire, you will be physically prepared to flee. As often happens, the fire bell is a false alarm. Now your heart returns to its normal steady beat, your breathing slows, you resume digesting food, and you quit perspiring. This return to a normal state results from the action of the **parasympathetic division** of the ANS (Figure 3.19). Parasympathetic signals return your body to a resting state after sympathetic activation. Most of your internal organs are controlled by inputs from sympathetic and parasympathetic systems. The more aroused you are, the greater the dominance of the sympathetic system.

It doesn't take a fire alarm to activate your sympathetic nervous system. When you meet someone you find attractive, your heart starts beating quickly, you start

sympathetic division of ANS A division of the autonomic nervous system that prepares the body for action.

parasympathetic division of ANS A division of the autonomic nervous system that returns the body to its resting state.

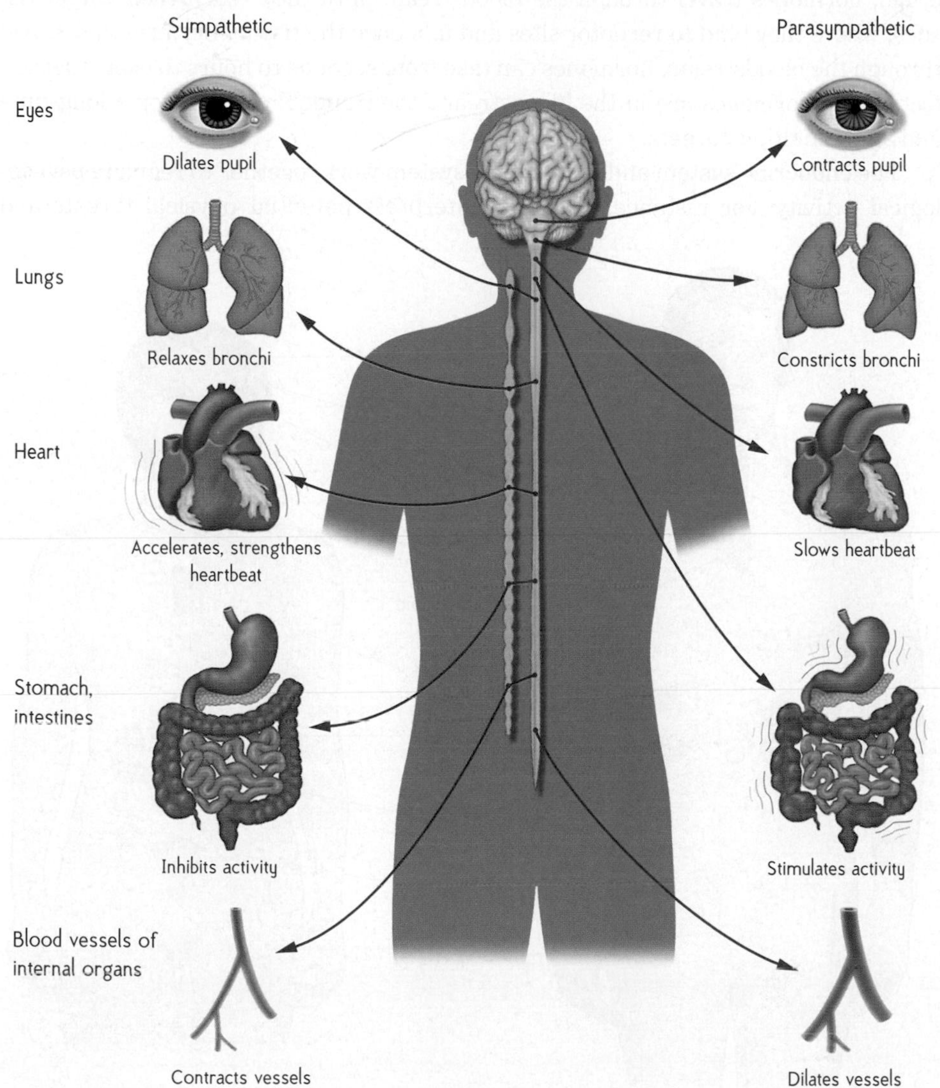

3.19 The Sympathetic and Parasympathetic Systems The sympathetic division of the nervous system prepares the body for action, whereas the parasympathetic returns it to a resting state.

perspiring, you may start breathing heavily, and although you may not know it, your pupils widen. These are all signs of sexual arousal, which relies on activation of the sympathetic division of the ANS, and they provide nonverbal cues during social interaction.

The sympathetic nervous system is also activated by psychological states such as anxiety or unhappiness. Certain people worry a great deal or cannot cope with stress; their are bodies in a constant state of arousal. Chronic activation of the sympathetic nervous system is associated with medical problems that include ulcers, heart disease, and asthma (Davison & Pennebaker, 1996).

The Endocrine System Communicates through Hormones

endocrine system A communication system that uses hormones to influence thoughts, behaviours, and actions.

hormones Chemical substances, typically released from endocrine glands, that travel through the bloodstream to targeted tissues, which are subsequently influenced by the hormone.

Like the nervous system, the **endocrine system** is a communication system that influences thoughts, behaviours, and actions. The main distinction between the two is their mode of communication: The endocrine system uses hormones, whereas the nervous system uses electrochemical signals. **Hormones** are chemical substances released into the bloodstream by *endocrine glands,* ductless glands such as the pancreas, thyroid, and testes or ovaries (Figure 3.20). Once released by a gland or an organ, hormones travel through the bloodstream until they reach their target tissues, where they bind to receptor sites and influence the tissue. Because they travel through the bloodstream, hormones can take from seconds to hours to exert their effects; once hormones are in the bloodstream, their effects can last for a long time and affect multiple targets.

The endocrine system and the nervous system work together to regulate psychological activity. For instance, the brain interprets potential physical threats and

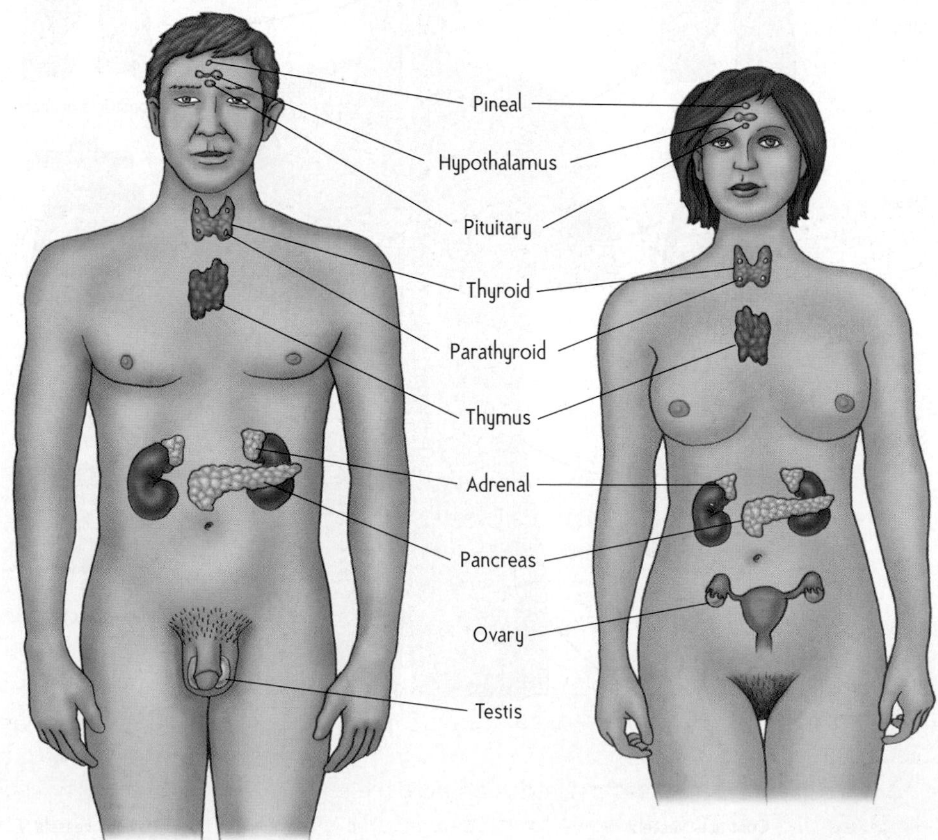

3.20 Glands The major endocrine glands.

directs action in the endocrine system to prepare for battle or to deal with possible injury.

"You have been charged with driving under the influence of testosterone."

EFFECTS OF HORMONES ON SEXUAL BEHAVIOUR An example of hormonal influence is evident in sexual behaviour. The main endocrine glands in sexual behaviour are the **gonads**, which for males are testes and for females are ovaries. The two major gonadal hormones are identical in males and females, although the quantity differs: *androgens* such as testosterone are more prevalent in males, whereas *estrogens* such as estradiol are more prevalent in females. Gonadal hormones influence both the development of secondary sex characteristics and adult sexual behaviour.

For males, successful sexual behaviour depends on having at least a minimum amount of testosterone. Surgical removal of the testes diminishes the capacity for developing an erection and lowers sexual interest. Yet a castrated male who receives an injection of testosterone will be able to perform sexually. Do injections of testosterone increase sexual behaviour among healthy men? The answer is no, which implies that as long as a minimum amount of testosterone is available, the healthy male will be able to perform (Sherwin, 1988).

The influence of gonadal hormones on females is much more complex. In many nonhuman animals there is a finite period called *estrus* when the female is sexually receptive and fertile. During estrus the female displays behaviours designed to attract the male. Surgical removal of the ovaries terminates these behaviours and the female is no longer receptive. However, injections of estrogen reinstate estrus. What does this mean for women? Apparently not very much. The sexual behaviour of women is not particularly linked to their menstrual cycle, and surgical removal of the ovaries has a minimal effect on sexual interest or behaviour (Dennerstein & Burrows, 1982). Moreover, sexual behaviour in women may have more to do with androgens than estrogens (Morris, Udry, Khan-Dawood, & Dawood, 1987). Women with higher levels of testosterone report greater interest in sex, and injections of testosterone increase women's sexual interest after surgical removal of the uterus (Sherwin, 1994).

How does hormonal communication differ from neurotransmission?

gonads The main endocrine glands involved in sexual behaviour: in males, the testes; in females, the ovaries.

Thinking Critically 

Does Taking Steroids Lead to Violence?

Imagine that you have been asked to serve on a jury deliberating the case of 16-year-old Jamie Fuller, who in August 1991 killed his 14-year-old girlfriend, Amy Carnevale. Fuller, who reportedly was angry that Amy had gone to the beach the day before with some other boys, repeatedly stabbed her and then dumped her body in a pond. Those are the facts. But why did he kill her?

According to the prosecutors, Fuller was a troubled youth who had planned the murder for months and discussed it with his friends. Afterward, he even bragged about killing her. According to the defense attorneys, however, Fuller had been taking steroids at the time of the crime, and the effects of steroids caused him to become enraged and lose control. Let's examine some of the issues involved in the defense claim.

According to the defense, steroids can produce violent outbursts in those who take them, so-called "roid rage." The particular steroid that Fuller ingested was a synthetic form of the male sex hormone testosterone, which enhances muscle growth. Although steroids are banned in most sports, some athletes use them to become stronger, faster, and more competitive. Steroid use is reputed to be widespread in some sports, such as weightlifting, but it is apparently on the rise in others, such as baseball, and even among high school athletes who want to be more muscular. There is little doubt that steroids can enhance performance, but can they create killers?

Given that the steroid that Fuller was taking was a synthetic form of testosterone, is there any evidence that testosterone itself is associated with aggression? Let's critically examine the evidence. Males have more testosterone than females, and males carry out the vast majority of aggressive and violent acts in this world. Boys play more roughly than girls at an early age. They become especially aggressive during early adolescence, a time when their levels of testosterone rise tenfold (Mazur & Booth, 1998). But this could just as easily be because increases in testosterone coincide with other maturational changes that promote aggression, such as physical growth.

The overall picture suggests only a modest correlation between testosterone and human aggression. Particularly aggressive men, such as violent criminals, and particularly physical athletes, such as hockey players, have been found to have higher levels of testosterone than other males (Dabbs & Morris, 1990). This relationship, however, is not large, and it is unclear how testosterone is linked to greater aggressiveness. For instance, it may be that testosterone changes are the result of rather than the cause of aggressive behaviour. A number of studies have shown that testosterone rises just prior to athletic competition; it remains high for the winners of competitive matches and drops lower for the losers. Even those who simply watch a competition can be affected. Testosterone increased in Brazilian television viewers who watched Brazil beat Italy in the 1994 World Cup soccer tournament, but decreased for Italian television viewers (Bernhardt, Dabbs, Fielden, & Lutter, 1998). This suggests that testosterone might not play a direct role in aggression, but rather may be related to dominance, the results of having greater power and status (Mazur & Booth, 1998). If this is true, the question then is how feelings of dominance are related to aggression.

It remains possible that taking large doses of some steroids might increase testosterone to such a level that it produces an extreme need for dominance and control, thereby provoking violent behaviour when a male is challenged. Jamie Fuller had been a skinny, introverted teen who often was picked on by bigger kids. After he took steroids for many months, his appearance and personality changed dramatically: he gained some 30 pounds of muscle and became more socially assertive, as well as increasingly short-tempered and belligerent. According to testimony by the Harvard psychiatrist Harrison Pope, steroids caused these changes in Jamie Fuller, which should mitigate his guilt in the murder of Amy Carnevale. Pope claims there have been a number of case studies of steroid users becoming extremely violent. Athletes themselves report becoming more aggressive and even committing violent crimes after steroid use. In self-report studies, those who admit taking large quantities of steroids also report experiencing grandiose beliefs and frenzied mental states and engaging in reckless and aggressive behaviour (Figure 3.21). Should we believe these anecdotal reports? Isn't it possible that the athletes are simply repeating what they have heard on television or that they are using this possible link to excuse their aggressive actions?

It is important to remember that correlation does not prove causation. In self-report studies, researchers cannot determine whether it is the steroids or some other factor that is responsible for aggressive outbursts. Perhaps the social environment of athletes encourages aggressive behaviour; emotionally intense practices and demanding coaches can't be discounted. Alternatively, perhaps male athletes who are susceptible to violence or who have other psychological problems are more likely to use steroids to enhance athletic performance (Yates, Perry, & Murray, 1992). A study by Pope and his colleagues found that weightlifters who used steroids were much more likely to be substance abusers, as was Fuller, than those who had never taken steroids (Kanayama, Pope, Cohane, & Hudson, 2003). Which causes which?

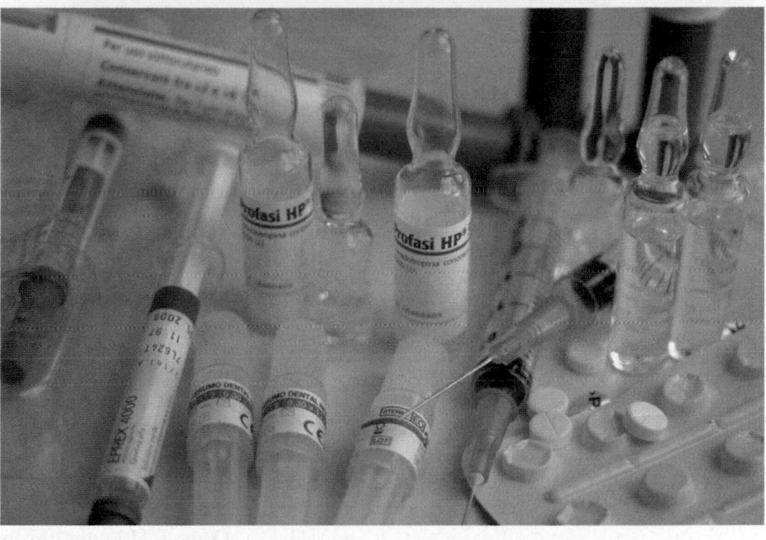

3.21 **"Roid Rage"** Some athletes report that they become more aggressive and violent after taking steroids, but psychological research has been unable to determine the exact relationship between steroids and aggression.

An effective research strategy to understand the effect of steroids is to administer randomly either steroids or placebos to research participants, without anyone knowing which participants received which, and then examine changes in mood and behaviour. Pope and his colleagues (2000) conducted such a study and found that those subjects who took steroids experienced overall increases in aggression and frenzied emotions. But they also found that these effects were highly variable, with 16 percent of the participants showing quite dramatic changes and the remainder showing minimal effects. One limitation of this kind of research is that it is unethical to give research participants the massive doses of steroids that athletes often take, since serious health consequences are associated with such use. Maybe if higher doses were used, all of the participants would have shown increased aggression. It is impossible to know. But the preponderance of evidence suggests that chronic high doses of steroids may promote aggressive outbursts, at least among some users. Still, is this a sufficient explanation of why Jamie Fuller killed Amy Carnevale? Obviously there are many details about the crime that you do not know. But how would knowing about Fuller's steroid use affect your willingness to convict him? Do you think his changes in personality and aggressiveness were caused by steroid use?

The actual trial jury didn't accept the steroid defense; Fuller was convicted and sentenced to life in prison. What evidence would you require in order to determine whether steroids could ever be responsible for violent behaviour? ■

Actions of the Nervous System and Endocrine System Are Coordinated

Throughout this chapter, we have described communication systems that link neurochemical and physiological processes to behaviours, thoughts, and feelings. These systems are fully integrated and interact to facilitate survival. The nervous and endocrine systems use information from the environment to direct behavioural responses that are adaptive. Ultimately, the endocrine system is under the control of the central nervous system. External and internal stimuli are interpreted in the brain, and signals are sent from the brain to the endocrine system, which then initiates a variety of effects on the body and on behaviour.

Most of the central control of the endocrine system is accomplished by a small brain structure called the *hypothalamus,* which is located just above the roof of the mouth. As you will learn in the next chapter, the hypothalamus plays an extremely important role in behaviours related to survival and reproduction, such as feeding and sex. How does this central control work? At the base of the hypothalamus is the **pituitary gland**, which controls the release of hormones from the rest of the endocrine glands. Based on some sort of neural activation, a *releasing factor* is secreted from the hypothalamus. This releasing factor causes the pituitary to release a specific

pituitary gland Located at the base of the hypothalamus, the gland that sends hormonal signals that control the release of hormones from endocrine glands.

hormone, which then travels to endocrine sites throughout the body. Once the hormone is at the target sites, it touches off the release of other hormones, which subsequently affect bodily reactions or behaviour.

It is fascinating how finely tuned this integration can be. For example, consider the case of physical growth. *Growth hormone (GH)* is an extract from the pituitary gland that affects bone, cartilage, and muscle tissues, helping them to grow or regenerate after injury. External administration of GH can increase body size, something known since the 1930s. A common and effective therapy for children with medical conditions such as dwarfism is a synthetic version of GH. How does it work? To build body tissues, GH requires dietary protein. A lack of protein in the diet is associated with a lack of normal growth.

GH is released in bursts throughout the day. Its release is triggered by the *growth hormone releasing factor (GRF)*. GRF neurons are connected to an area of the hypothalamus that is involved in sleep-wake cycles. Thus, the bursts of GH are controlled by the body's internal clock.

Once GH is available, it needs protein. GRF also plays a role in the control of feeding behaviour. Research at the University of Toronto shows that injections of GRF into the area of the brain that controls the body clock increase consumption of protein but not fats or carbohydrates (Dickson & Vaccarino, 1994). Consider the whole picture: GRF selectively stimulates eating protein, perhaps by making protein especially enjoyable. At the same time, it also releases GH, which relies on the higher intake of dietary protein to help build strong bones and muscles. Hence, the CNS, PNS, and endocrine system are clearly integrated to ensure that behaviours provide the body with substances it needs for survival at the times they are required.

REVIEWING the Principles | How Are Neural Messages Integrated into Communication Systems?

The central nervous system, consisting of the brain and spinal cord, attends to the body and the environment, initiates actions, and directs the peripheral nervous system and endocrine system to respond appropriately. All three systems use chemicals to transmit their signals, but transmission in the nervous system occurs across synapses whereas transmission in the endocrine system uses hormones that travel through the bloodstream. The hypothalamus controls the endocrine system by directing the pituitary to release specific hormones. The various communication systems are integrated and promote behaviour that is adaptive to the environment.

Conclusion

The human body is a biological machine, produced by a genome that has been shaped by millions of years of evolution. Some 30,000 genes provide instructions for producing proteins, and this modest process builds a brain that thinks, feels, and acts. The biological revolution has revealed that these genes not only build the common structures of the human body, but they are also partially responsible for differences among individuals, such as physical appearance, personality, and mental abilities. So far, we only have a minimal understanding of the complexity of our mental lives at the genetic level of analysis. Only now are we beginning to understand the relationship between our environments and our genes. There are more neuronal connections in the human brain than there are stars in our galaxy (Kandel, Schwartz, & Jessell, 1995), and the messages that are transmitted are modulated by subtle variations in the actions of dozens of chemicals in the synapses. Not only do we need to understand how these mechanisms work, but we also need to know how our environment causes our genes to "turn on and off." Understanding the inner workings of genetic expression may well lead to profound insights into what makes us human and how our minds work.

Summarizing the Principles of Genetic and Biological Foundations

What Is the Genetic Basis of Psychological Science?

1. **Heredity involves passing along genes through reproduction:** The Human Genome Project has mapped the approximately 30,000 genes that make up the 23 chromosomal pairs in humans. Variations of genes (alleles) are either dominant or recessive. The genome represents the genotype and the observable characteristics are the phenotype. Many characteristics are polygenetic.
2. **Genotypic variation is created by sexual reproduction:** Because one half of each chromosome comes from each parent and is randomly joined, there is enormous variation in the genome of the resulting zygote. Mutations also give rise to variations.
3. **Genes affect behaviour:** Behavioural geneticists are able to examine the similarity and variation in shared characteristics in a population, based on genetic and environmental similarity; this assessment is termed heritability.
4. **Social and environmental contexts influence genetic expression:** Though genes guide the development of characteristics, their final expression is a complex interaction between genetic makup and environmental context.
5. **Gene expression can be modified:** Genetic manipulation has been achieved in other mammals (e.g., mice) but has proven difficult in humans. However, studies focusing on RNA offer new hope.

How Does the Nervous System Operate?

6. **Neurons are specialized for communication:** Neurons are the basic building blocks of the nervous system that receive and send chemical messages. The structure of all neurons is the same but vary by function and location in the nervous system.
7. **Action potentials cause neuronal communication:** Changes in the electrical charge of neurons results in an action potential (neural firing). This is the means of communication within networks of neurons.
8. **Changes in electrical potential lead to action:** Neural firings either continue the message to other neurons or halt it. Neurons fire due to a change in the electrical balance of positive and negative sodium and potassium ions inside and outside the neuron.
9. **Action potentials spread along the axon:** The dendrites receive a signal that spreads the changes in the exchange of positive and negative ions to reach a threshold, which generates an action potential. The firing of a neuron is all or none—a neuron does not partially fire. This electrical signal travels along the axon. The myelin insulation on the axon speeds transmission.
10. **Neurotransmitters bind to receptors across the synapse:** Neurons do not touch; they release chemicals (neurotransmitters), which bind with receptors of other neurons thus changing the charge in that neuron. The effect of neurotransmitters is halted by a variety of actions.

How Do Neurotransmitters Influence Emotion, Thought, and Behaviour?

11. **Acetylcholine affects motor control and mental processes:** Acetylcholine acts on neurons directing muscle movement.
12. **Monoamines are involved in affect, arousal, and motivation:** Epinephrine and norepinephrine are involved in arousal and alertness. Serotonin is involved in impulse control, emotion, and dreaming. Dopamine affects motor control and motivation.
13. **Amino acids are general inhibitory and excitatory transmitters in the brain:** The amino acid GABA is the transmitter for inhibiting neuronal transmission. Glutamate does the opposite; it is involved in neuronal transmission and aids learning and memory.

14. **Peptides modulate neurotransmission:** Peptides change the effect of other transmitters. Cholecystokinin (CCK), endorphins, and substance P affect satiation and pain.

How Are Neural Messages Integrated into Communication Systems?

15. **The central nervous system consists of the brain and spinal cord:** The central nervous system is made up of the structures and functions of the brain and spinal cord.

16. **The peripheral nervous system includes the somatic and autonomic systems:** The structures and functions that regulate the body's internal environment is the autonomic system, which has the complementary division of the alarm response (sympathetic) and the return-to-normal-state response (parasympathetic). The somatic system relays sensory information.

17. **The endocrine system communicates through hormones:** Endocrine glands and organs produce and release chemical substances, which travel to tissues in the body through the bloodstream and influence a variety of processes, including sexual behaviour.

18. **Actions of the nervous system and endocrine system are coordinated:** The endocrine system is controlled by the central nervous system. Most of the central control of the endocrine system occurs through the action of the hypothalamus and the pituitary gland, which controls the release of hormones from the rest of the endocrine glands.

Applying the Principles

1. You've just made two new friends at school today, Jack and Jim, who happen to be identical twins. They look physically alike, but you're more curious about whether they will behave exactly alike. Because of what you know about the role of genes and the environment in shaping the psychological characteristics of an individual, you would expect that Jack and Jim will be

_____ a) identical in most characteristics.
_____ b) very different in all characteristics if their parents make a special effort to treat them differently.
_____ c) different only if separated and raised in different families.
_____ d) very similar in most characteristics.

2. As you read this question, how is your nervous system responding?

_____ a) The neurons in the brain plug into each other and fire all at once.
_____ b) An action potential causes the release of neurotransmitters, which affect whether other neurons fire.
_____ c) The strength of the action potential changes as you understand the words.
_____ d) A different chemical messenger is released for every neuron that fires and the sum of these chemical messages is the response.

3. As you sit in a large lecture hall listening to the professor, your ability to stay alert and effectively remember the material is due to

_____ a) the action of only one class of neurotransmitters.
_____ b) the actions of a variety of different classes of neurotransmitters.
_____ c) the actions of peptides (for memory response) and neurotransmitters (for arousal).
_____ d) the continuous availability of excitatory neurotransmitters for memory and inhibitory neurotransmitters to suppress sleep.

4. On the television show *24,* Jack Bauer, played by Kiefer Sutherland, spends an entire day in unbelievably stressful situations as an agent for the Counter Terrorism Unit. Throughout Jack's day, his _______ controls his planning; his _______ executes behaviours such as running and shooting; and his _______ releases chemicals into the bloodstream to respond to stress.

_____ a) PNS; CNS; endocrine system
_____ b) endocrine system; CNS; PNS
_____ c) CNS; PNS; endocrine system
_____ d) CNS; endocrine system; PNS

ANSWERS: 1.d 2.b 3.d 4.c

Key Terms

acetylcholine (ACh), p. 98
action potential, p. 90
agonist, p. 97
all-or-none principle, p. 93
antagonist, p. 97
autonomic nervous system (ANS), p. 106
autoreceptors, p. 96
axon, p. 88
cell body, p. 88
central nervous system (CNS), p. 105
cholecystokinin (CCK), p. 103
chromosomes, p. 77
dendrites, p. 88
dizygotic twins, p. 82
dominant gene, p. 79
dopamine, p. 100
endocrine system, p. 108
endorphins, p. 104
enzyme deactivation, p. 96
epinephrine, p. 99
GABA (gamma-aminobutyric acid), p. 102
gene, p. 77
genotype, p. 79
glutamate, p. 103
gonads, p. 109
heritability, p. 84
hormones, p. 108
interneurons, p. 89
monoamines, p. 99
monozygotic twins, p. 82
motor neurons, p. 89
myelin sheath, p. 93
neuron, p. 88
neurotransmitter, p. 95
nodes of Ranvier, p. 94
norepinephrine, p. 100
parasympathetic division of ANS, p. 107
Parkinson's disease, p. 100
peptides, p. 103
peripheral nervous system (PNS), p. 105
phenotype, p. 79
pituitary gland, p. 111
receptors, p. 95
recessive gene, p. 79
resting membrane potential, p. 90
reuptake, p. 96
sensory neurons, p. 89
serotonin, p. 100
somatic nervous system, p. 106
substance P, p. 104
sympathetic division of ANS, p. 107
synapse, p. 89
synaptic cleft, p. 95
terminal buttons, p. 89

Further Readings

Carlson, N. R. (2003). *Physiology of behavior* (8th ed.). Needham Heights, MA: Allyn & Bacon.

Gazzaniga, M. S., Ivry, R. B., & Mangun, G. R. (2002). *Cognitive neuroscience: The biology of the mind* (2nd ed.). New York: Norton.

Kandel, E. R., Schwartz, J. H., & Jessell, T. M. (2000). *Principles of neural science* (4th ed.). New York: McGraw-Hill.

Marcus, G. (2004). *The birth of the mind.* New York: Basic Books.

Marshall, L. H., & Magoun, H. W. (1998). *Discoveries in the human brain: Neuroscience prehistory, brain structure, and function.* Totowa, NJ: Humana Press.

Pinel, J. P. J. (2005). *Biopsychology* (6th ed.). Needham Heights, MA: Allyn & Bacon.

Ridley, M. (2003). *Nature versus nurture: Genes, experience, and what makes us human.* New York: HarperCollins.

Zigmond, M. J., Bloom, F. E., Landis, S. C., Roberts, J. L., & Squire, L. R. (2002). *Fundamentals of neuroscience* (2nd ed.). San Diego, CA: Academic Press.

Further Readings: A Canadian Presence in Psychology

Cain, D. P. (1998). Testing the NMDA, long-term potentiation, and cholingergic hypotheses of spatial learning. *Neuroscience and Biobehavioral Reviews, 22*, 181–193.

Caulfield, T. (2003). Human cloning laws, human dignity and the poverty of the policy making dialogue. *BMC Medical Ethics, 4*, E3.

Dunnett, S. B., & Fibiger, H. C. (1993). Role of forebrain cholinergic systems in learning and memory: Relevance to the cognitive deficits of aging and Alzheimer's dementia. *Progress in Brain Research, 98*, 413–420.

Suzuki, D. T., Griffiths, A. J. F., Miller, J. H., & Lewontin, R. C. (1989). *An introduction to genetic analysis* (4th ed.). New York: Freeman Press.

Wahlsten, D. (1999). Single gene influences on brain and behaviour. *Annual Review of Psychology, 50*, 599–624.

Wise, R. A., & Rompre, P. P. (1989). Brain dopamine and reward. *Annual Review of Psychology, 40*, 191–225.

Focusing on the Physical Brain

Former heavyweight boxing champion Muhammad Ali and Canadian actor Michael J. Fox, both of whom suffer from Parkinson's disease, pose for the cameras before the start of a congressional subcommittee hearing on Parkinson's. Specific parts of the brain control specific mental and physical capabilities. Parkinson's patients, for example, have difficulty learning tasks that require repetition because of damage to their basal ganglia, which is crucial for planning and producing movement.

CHAPTER

4

The Brain and Consciousness

It was the 1940s, and a patient lay on the operating table, fully conscious, with part of her skull temporarily removed to expose the surface of her brain. Her surgeon at the Montreal Neurological Institute, Dr. Wilder Penfield (Figure 4.1), delicately touched a small electrode to her brain, and the patient announced that she had the sudden experience *"of being in her kitchen listening to the voice of her little boy who was playing outside in the yard. She was aware of the neighborhood noises, such as passing motor cars, and understood that might mean danger to him."*

The patient suffered from *epilepsy,* the debilitating affliction in which *seizures,* uncontrolled "storms" of electrical activity, begin in some part of the brain and spread throughout much of it, often causing violent, life-threatening convulsions of the entire body. As a last resort, she had agreed to undergo surgery to try to find and remove the part of her brain in which the seizures began.

Outlining the Principles

What Are the Basic Brain Structures and Their Functions?

- The Spinal Cord Is Capable of Autonomous Function
- The Brainstem Houses the Basic Programs of Survival
- The Cerebellum Is Essential for Movement
- Subcortical Structures Control Basic Drives and Emotions
- The Cerebral Cortex Underlies Complex Mental Activity

How Does the Brain Change?

- The Interplay of Genes and the Environment Wires the Brain
- The Brain Rewires Itself throughout Life
- The Brain Can Recover from Injury

How Is the Brain Divided?

- The Hemispheres Can Be Separated
- The Separate Hemispheres Can Be Tested
- The Hemispheres Are Specialized
- The Mind Is a Subjective Interpreter

Can We Study Consciousness?

- Definitions of Consciousness Allow Its Empirical Study
- Unconscious Processing Influences Awareness
- Awareness Has Many Seats in the Brain

What Is Sleep?

- Sleep Is an Altered State of Consciousness
- Sleep Is an Adaptive Behaviour
- Sleep and Wakefulness Are Regulated by Multiple Neural Mechanisms
- People Dream while Sleeping

4.1 **Wilder Penfield** In 1934, Penfield opened the Montreal Neurological Institute, the first of its kind.

Penfield was electrically stimulating points on the surface of her brain in an effort to set off the beginning of a seizure, to determine exactly which part of the brain should be excised (Figure 4.2). He was also mapping out the functions of specific brain areas to determine which ones could be removed without damaging the patient's ability to speak. The small electric current interfered with the local area of brain tissue that it flowed through, temporarily deactivating it. If the patient stopped speaking during one of these stimulations, Penfield would know that that area was vital to speech and should not be removed.

Interestingly, the current, while *deactivating* the region in the immediate vicinity of the electrode, *reactivated* more distant brain regions that were connected to neurons in the stimulated area. The result was the vivid reawakening of specific memories in the patient.

Although Penfield repeated this demonstration many times, it is now believed that the specific memory phenomena may have resulted from the patients' disorders, thus not necessarily reflecting a normal brain. However, electrical inductions of experiences, sensations, movements, emotions, and even beliefs, have been reproduced countless times since then. These physical reawakenings of mental events demonstrate the physical nature of the mind. Far from existing as an incomprehensible substance distinct from the physical world, the mind is very much a part of it, consisting of the electrical and chemical activity of our brains. ■

You learned in the last chapter that mental activities and behaviours are produced by biological processes within the brain, such as the action of nerve cells and associated chemical reactions. In this chapter we will make connections across levels of

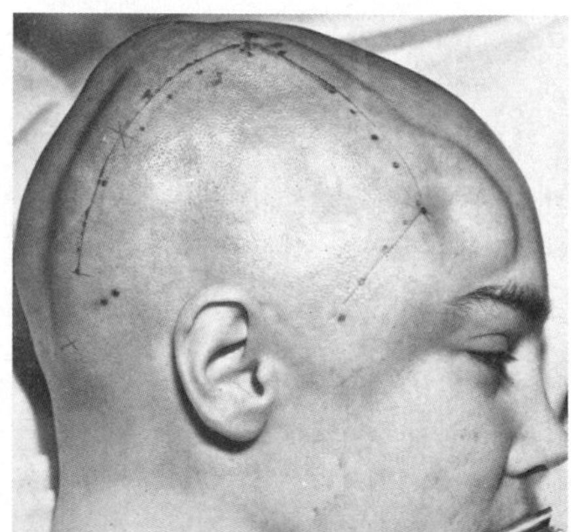

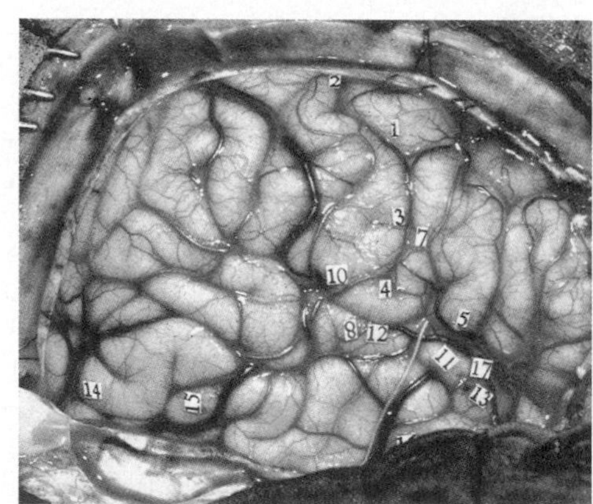

4.2 **Direct Stimulation** One of Wilder Penfield's patients undergoing direct stimulation of the brain. Left: the patient immediately before surgery. A local anesthetic has been applied to her scalp; Penfield's patients remained fully conscious throughout the procedures. Right: the exposed surface of her cortex. Numbered tags denote electrically stimulated locations.

Fourth century B.C.E.

Locating the Mind Although Greek philosophers such as Aristotle believed the mind was located in the heart, Hippocrates describes the brain as the location of the mind, a view supported in the second century C.E. by the influential Roman physician Galen.

1650s

Mind-Body Dualism French philosopher René Descartes proposes a dualistic theory of the mind and body in which the body's movements are controlled by mechanical reflexes interacting with a nonphysical soul located in the brain's pineal gland.

1800s

Phrenology German physiologist Franz Joseph Gall develops phrenology, arguing that personality traits and mental abilities are housed in distinct areas of the brain, and can be assessed by measuring the external dimensions of the skull.

1848

Frontal Lobe Pathology In Vermont, railroad worker Phineas Gage is accidentally lobotomized by an iron rod. Major changes are observed in his personality and impulse control, but he is otherwise unimpaired.

analysis, from the biological level to the level of individual personal experience. We will consider a number of intriguing questions about how the brain enables the mind: How do different regions of the brain contribute to our mental experience of the world? What role do genes and the environment play in shaping individual brains? Can the brain recover following injury? Following this physical consideration of the brain, we will probe its most distinguishing capacity—the ability to have a sense of consciousness. The coordinated action of a number of different brain regions contributes not only to consciousness, but also to variations in conscious experience, from being highly aware of our surroundings to being fast asleep and dreaming. Throughout it all, the brain is active, functioning at every moment to keep us alive as it processes information from the external world—and in doing so creating our mental worlds.

How does the brain enable the mind?

The brain wasn't always recognized as the home of the mind. The Egyptians, for example, viewed the heart as more important; they elaborately embalmed the hearts of their deceased, but the brains they simply threw away. The heart was to be weighed in the afterlife to determine the deceased's fate. But in the ensuing centuries, there was a growing recognition that the brain was essential for normal mental functioning. Much of this came from observing those with brain injuries—at least since the time of the Roman gladiators it was apparent that a blow to the head often produced disturbances in mental activity, such as unconsciousness or loss of the ability to speak.

phrenology An early method of assessing personality traits and mental abilities by measuring bumps on the skull.

By the beginning of the nineteenth century, anatomists had a reasonably good understanding of the basic structure of the brain. But debates raged over how the brain worked to produce mental activity. Did different parts do different things or did the entire brain act in unison? In the early nineteenth century, Franz Gall and Johann Spurzheim proposed their theory of *phrenology,* based on the idea that the brain operates on a principle of functional localization. **Phrenology** is the practice of assessing personality traits and mental abilities by measuring bumps on the human skull (Figure 4.3). Although an influential theory in its day, its validity could not be tested scientifically. One of its most vigorous critics was French scientist Pierre Flourens, who believed that the brain functions as a single unit, with all parts working together. To test his theory, Flourens conducted experiments in the 1820s on birds and rodents, in which he removed brain regions and observed the effects on behaviour. In the first half of the twentieth century, the psychologist Karl Lashley used a similar method to identify the places in the brain where learning occurred. Unlike Flourens, who did not believe in brain specialization, Lashley conceded that specific brain regions were involved in motor control and sensory experiences. Yet Lashley believed that all parts of the cortex contributed equally to mental abilities, such as problem solving and memory, a concept called *equipotentiality.* Today, his theory has been discredited, and we now know that the brain consists of a patchwork of highly specialized areas.

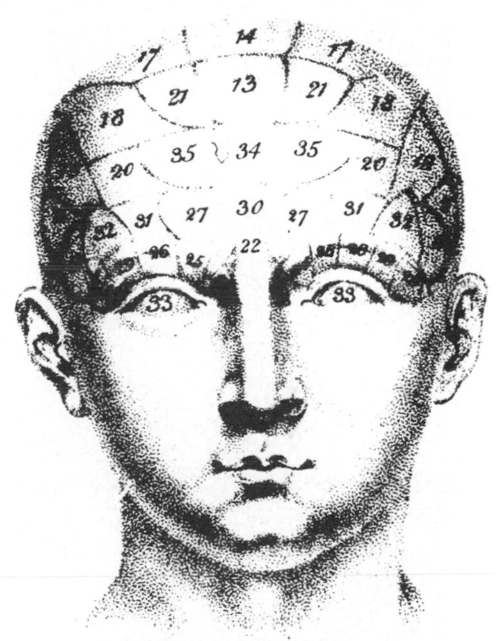

4.3 Phrenology One of Johann Spurzheim's phrenological maps of the skull. Each numbered region corresponds to a different characteristic.

1861
Broca's Area French physician Paul Broca presents the first widely accepted and rigorously substantiated assignment of a specific function to a particular cortical area. "Broca's area" is still recognized today as important for speech.

1920s
Equipotentiality American experimental psychologist Karl Lashley removes pieces of a rat's cortex in an effort to locate the memory trace. He wrongly concludes that much of the cortex can contribute equally to most mental functions.

1940s
Stimulating the Awake Brain Canadian neurologist Wilder Penfield experiments with electric current applied to different brain areas to examine the functions of these areas. He later founds the Montreal Neurological Institute.

1949
Reticular Formation and Arousal Giuseppe Moruzzi and Horace Magoun discover that stimulating the reticular formation in the brainstem leads to arousal, a state of heightened physical and brain activity.

1950s
Recording Neurons David Hubel and Torsten Wiesel make the first recordings from individual neurons in the cerebral cortex, characterizing the response properties of neurons in a cat's primary visual cortex using visual displays.

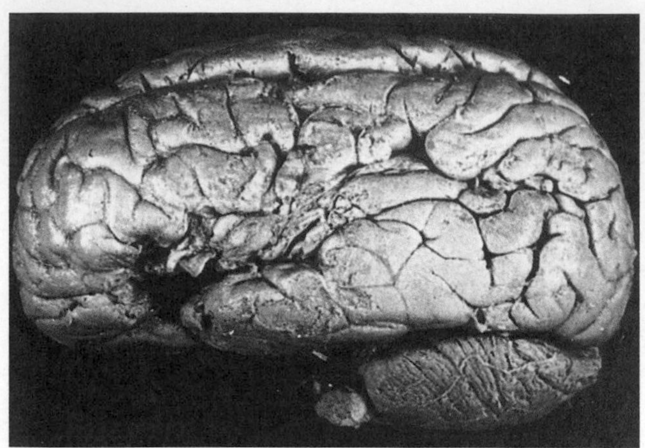

4.4 Monsieur Leborgne's Brain Paul Broca studied the brain of Monsieur Leborgne and identified the lesioned area as crucial for speech production.

The first strong evidence that the brain contains specialized functions came from the work of French scientist Paul Broca (Finger, 1994). In 1861, Broca performed an autopsy on Monsieur Leborgne. Before his death, Leborgne was a patient who had lost the ability to say anything other than the word "tan" but could still understand language. When he examined Leborgne's brain, Broca found substantial damage to the front left side caused by a large lesion (Figure 4.4). This observation led him to conclude that this particular region was important for speech. Broca's theory has survived the test of time. This left frontal region became known as **Broca's area**, and it has since been repeatedly confirmed to be crucial for the production of language.

The debate over whether psychological processes are located in specific parts or distributed throughout the brain continued so long, in part, because until fairly recently researchers have not had methods for studying ongoing mental activity in the working brain. The invention of brain-imaging methods in the late 1980s changed that swiftly and dramatically. Since then there has been an explosion of research, cutting across various levels of analysis, linking specific brain areas with particular behaviours and mental processes. These methods have allowed psychological scientists to examine mental states, such as consciousness, that were previously viewed as too subjective to be studied. Although philosophers have long debated about what it means to be conscious of something, psychologists avoided the topic because of the difficulty in measuring consciousness objectively in the laboratory. Our new ability to watch the working brain in action has helped us to learn a great deal about how the brain works. Let's now consider the fascinating question of how mind and brain are connected.

What Are the Basic Brain Structures and Their Functions?

In this section, we explore how the mind is adaptive, which is one of this text's four major themes. The first nervous systems were probably little more than a few specialized cells with the capacity for electrical signaling. Today's brain is best viewed as a collection of interacting neuronal circuits that have accumulated and developed throughout human evolution. Through the process of adapting to the environment, the brain has evolved specialized mechanisms to regulate breathing, food intake, sexual behaviour, and bodily fluids, as well as sensory systems to aid in navigation and assist in recognizing friends and foes.

As discussed in Chapter 3, the nervous system is involved in almost every aspect of an organism's maintenance, regulation, and behaviour. It is composed of a vast number of interacting brain circuits ranging from those controlling the contractions

Broca's area The left frontal region of the brain that is crucial to the production of language.

1953
REM Sleep Nathaniel Kleitman, Eugene Aserinsky, and William Dement note that changes in the electrical activity of the brain are associated with rapid eye movements. They believe that these REM periods are the physiological basis of dreaming.

1960s
Split Brains American psychobiologist Roger Sperry and his student Michael Gazzaniga conduct research on patients who have had their cerebral hemispheres disconnected to treat epilepsy. They find that the hemispheres can function independently.

1980s
Brain Imaging Invention of PET imaging and functional MRI enables visualization of the activity of the working human brain. Brain imaging allows researchers to examine subjective mental states, such as emotion and consciousness.

1998
Neurogenesis Researchers such as Fred Gage and Elizabeth Gould find evidence that new neurons are produced in the adult brain, and that this process is affected by events in the environment.

1990s–2000
Stem Cells Researchers accumulate evidence that stem cells might be useful for treating injuries and diseases of the central nervous system, including paralysis and Parkinson's disease.

of the intestines, to those allowing a child to play with a dog, to those allowing college students to choose their majors. All of these are orchestrated by parts of the brain that are as different in their structure and organization as the roles they fulfill. Some of these roles have remained essentially the same throughout our evolution; the neural circuits responsible for such basics as breathing have changed correspondingly little. But although the human nervous system has a fundamental layout shared with all other vertebrates, it has developed an impressive elaboration of structures responsible for our enormous capacities for communication and thought.

Figure 4.5 shows some of the basic structures of the central nervous system that we will now consider.

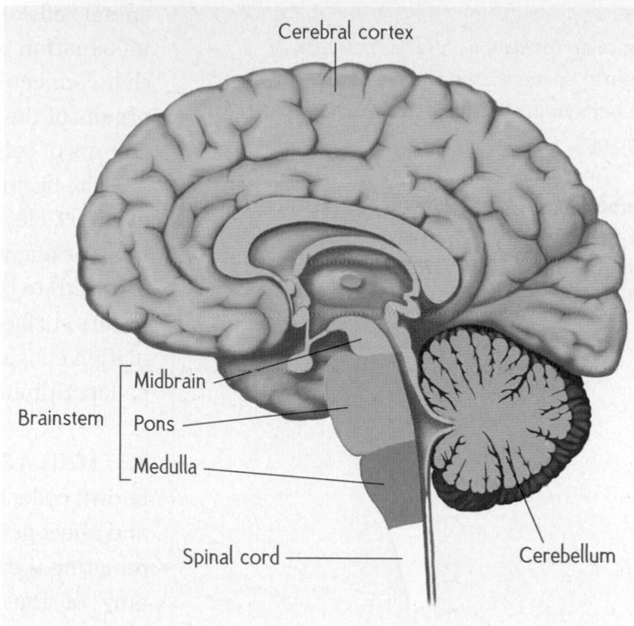

4.5 The Human Central Nervous System

The Spinal Cord Is Capable of Autonomous Function

The **spinal cord** is a rope of neural tissue that runs inside the hollows of the vertebrae, from just above the pelvis up into the base of the skull. It is segmented, with each segment marked by its own pair of spinal nerves emerging from the sides of the cord and communicating information to and from the rest of the body. In cross section, the cord is seen to be composed of two distinct tissue types: the **gray matter**, which is dominated by the cell bodies of neurons, and the **white matter**, which consists mostly of axons and the fatty sheaths that surround them. Gray and white matter are clearly distinguishable throughout the brain as well. Sensory information from the body enters the spinal cord and is passed up to the brain. However, besides relaying information, the spinal cord is able to take action on its own.

STRETCH REFLEX The spinal cord handles one of the simplest behaviours, the *spinal reflex*. This is the conversion of sensation into action by a handful of neurons and the connections between them. As an example, consider the stretch reflex. When the tendon attached to your kneecap is tapped with a rubber hammer, the leg gives a little reflexive kick. This reflex is present throughout the skeletal musculature and functions to maintain the positions of the joints under varying loads. It works as follows: All muscles have stretch receptors inside them to sense changes in length. These receptors are actually the dendritic tips of receptor neurons whose cell bodies are located in the spinal cord. Stretching the muscle causes the receptor neurons connected to it to fire. The receptor neurons' axons enter the spinal cord and transmit their signal directly to motor neurons, which lead back out to the same muscle. This closes the loop: Stretching the muscle causes the stretch receptor neurons to fire, which causes the motor neurons to increase their firing, which contracts the muscle.

The Brainstem Houses the Basic Programs of Survival

The spinal cord continues up into the base of the skull, thickening and becoming more complex as it transforms into the **brainstem**, which houses the most basic programs of survival, such as breathing, swallowing, vomiting, urination, and orgasm. Since the brainstem is also simply the spinal cord continued up into the head, it performs functions for the head similar to those that the spinal cord performs for the rest of the body. A whole complement of reflexes is housed here, analogous to the

spinal cord Part of the central nervous system, a rope of neural tissue that runs inside the hollows of the vertebrae from just above the pelvis and into the base of the skull.

gray matter A segment of the spinal cord that is dominated by the cell bodies of neurons.

white matter A segment of the spinal cord that consists mostly of axons and the fatty sheaths that surround them.

brainstem A section of the bottom of the brain that houses the most basic programs of survival, such as breathing, swallowing, vomiting, urination, and orgasm.

reticular formation A large network of neural tissue within the brainstem involved in behavioural arousal and sleep-wake cycles.

cerebellum A large convoluted protuberance at the back of the brainstem that is essential for coordinated movement and balance.

spinal reflexes; gagging is one example. Just as the spinal cord has nerves that carry information to and from the skin and muscles of the body, the brainstem has nerves that connect it to the skin and muscles of the head, as well as to the specialized sense organs of the head such as eyes and ears. These nerves each have distinct, dedicated clumps of cells within the brainstem that handle their needs.

The brainstem uses the reflexes of the spinal cord to produce useful behaviour. For example, electrically stimulating a part of the brainstem can cause an anesthetized animal (or one whose spinal cord and brainstem have been disconnected from the rest of its brain) to begin walking, very much as Wilder Penfield's stimulation of a point on the brain's surface initiated a memory in his patient. Increasing the frequency of the stimulation causes the animal to go from a walk to a trot, and increasing it still more, to a gallop. Stimulating another area will cause the animal to turn to one side.

RETICULAR FORMATION The brainstem also contains networks of neurons, known collectively as the **reticular formation**, that project up into the cerebral cortex and affect general arousal. The reticular formation is also involved in inducing and terminating the different stages of sleep, as you will read later in this chapter. The autonomy of the brainstem can be dramatically illustrated by severing an animal's brainstem from the entire brain above it, including its entire cerebral cortex. Cats that receive this treatment can still walk around and direct attacks at noises; if they then find themselves holding on to food, they will eat it. Some cases have been reported of humans born without cerebral cortices, and their behaviours are extremely basic and reflexive. Such infants tend not to develop normally and also do not tend to survive.

The Cerebellum Is Essential for Movement

The **cerebellum** (Latin, "little brain") is a large protuberance connected to the back of the brainstem (Figure 4.6). Its size and convoluted surface make it look like a supplementary brain. Lesions to different parts of the cerebellum produce very different effects. Its cellular organization, however, appears to be identical throughout. This suggests that the cerebellum is performing identical operations on all of its inputs, with the different effects resulting from the differences in origin, and destination, of the information.

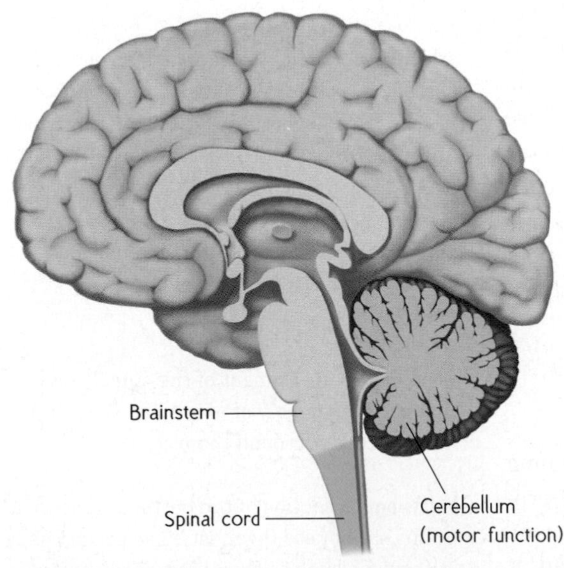

4.6 The Cerebellum The cerebellum, at the back of the brainstem, and the spinal cord.

The cerebellum is extremely important for proper motor function. Damage to the little nodes at the very bottom causes head tilt, balance problems, and a loss of smooth compensation of eye position for movement of the head. Damage to the ridge that runs up its back affects walking. Damage to the bulging lobes on either side causes a loss of limb coordination. The most obvious role of the cerebellum is in motor learning. It seems to be "trained" by the rest of the nervous system and operates independently and unconsciously. The cerebellum allows us to ride a bicycle effortlessly while we think about what we'll have for lunch. Functional imaging, however, indicates a broader role for the cerebellum, suggesting that it is involved in a variety of cognitive processes, including such things as making plans, remembering events, using language, and experiencing emotion. Indeed, during imaging studies, researchers have observed activation of the cerebellum when people either experience a painful stimulus or observe their loved ones receiving the same painful stimulus, suggesting that it is involved somehow in the experience of empathy (Singer et al., 2004).

Patients with disorders of the cerebellum typically have symptoms of *ataxia,* which involves clumsiness and a loss of motor coordination. A recent examination of such patients found that many of them also were impaired on cognitive tasks and showed a lack of normal emotional responses (Schmahmann & Sherman, 1998). There is also emerging evidence that patients with damage to the cerebellum experience changes in personality and display greater than expected symptoms of mental disorders.

Subcortical Structures Control Basic Drives and Emotions

Above the brainstem and cerebellum is the *forebrain,* which consists of the two cerebral hemispheres. Looking at the brain from the outside, the most noticeable feature of the forebrain is the *cerebral cortex,* which we will discuss in the next section. Below this are the *subcortical* regions, so named because they lie under the cortex. Subcortical structures that are important for understanding psychological functions include the hypothalamus, thalamus, hippocampus, amygdala, and the basal ganglia. Many of these structures are sometimes referred to as being part of the *limbic system*—the term *limbic* means "border" and can be thought of as separating the evolutionarily older (brainstem and cerebellum) and newer (cerebral cortex) parts of the brain. Many brain structures in the limbic system are especially important for controlling basic drives (such as eating and drinking) and emotions.

HYPOTHALAMUS The **hypothalamus** is the master regulatory structure of the brain and is indispensable to the organism's survival. It is responsible for regulating the vital functions—body temperature, bodily rhythms, blood pressure and glucose level—and, to these ends, impelling the organism by such fundamental drives as thirst, hunger, aggression, and lust (as will be described in Chapter 9).

The hypothalamus is one of the most vital regions of the brain. It receives input from almost everywhere and projects its influence, directly or indirectly, to almost everywhere. Through its projections to the rest of the brain, the hypothalamus induces motivational drives and the behaviours to satisfy them. Through its projections to the spinal cord, it governs much of the function of the internal organs. As discussed in Chapter 3, it controls the *pituitary gland,* the "master gland" of the body, which by releasing hormones into the bloodstream controls all other glands and governs such major processes as development, ovulation, and lactation.

The hypothalamus governs sexual and reproductive development and behaviour. It is one of the only places in the human brain where clear differences exist between men and women, due to early hormonal influences during nervous system development. Female rats exposed to high levels of testosterone while in the womb develop hypothalamic organization that is more typical of males—so-called *fetal masculinization.* Differences in hypothalamic structure may influence sexual orientation. Using postmortem methods, LeVay (1991) found that the anterior hypothalamus was only half as large in homosexual men as in heterosexual men. In fact, the size of this area in homosexual men was comparable to its size in heterosexual women. The implications of this finding are discussed more fully in Chapter 11.

THALAMUS The **thalamus** is the gateway to the cortex: Almost all incoming sensory information must go through the thalamus before reaching the cortex. The only exception to this rule is the sense of smell, the oldest and most fundamental of the senses; it has a direct route to the cortex, bypassing the thalamus. During sleep, the thalamus shuts the gate on incoming sensations while the brain rests. The thalamus appears to play a role in attention as well.

hypothalamus A small brain structure that is vital for temperature regulation, emotion, sexual behaviour, and motivation.

thalamus The gateway to the brain that receives almost all incoming sensory information before it reaches the cortex.

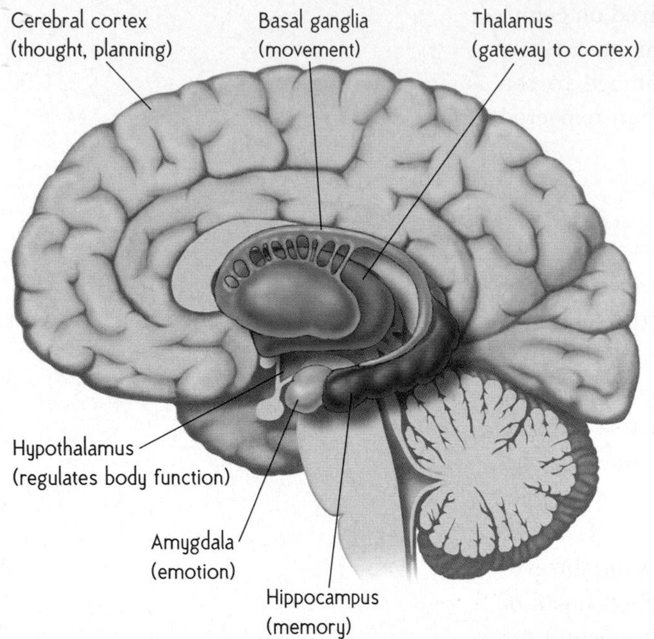

4.7 The Cerebral Hemisphere One of the two cerebral hemispheres, containing the thalamus, hippocampal formation, amygdala, basal ganglia, hypothalamus, and cerebral cortex.

HIPPOCAMPUS AND AMYGDALA Two subcortical structures are essential for memory and emotions, respectively: the hippocampus (Greek, "sea horse," after its distinctive shape) and the amygdala (Latin, "almond") (Figure 4.7).

The **hippocampus** plays an important role in the storage of new memories. It seems to do this by creating new interconnections within the cerebral cortex with each new experience. We noted earlier that Karl Lashley had failed to find the location of the memory trace by removing parts of rats' cerebral cortices. Had he damaged their hippocampal formations as well, his results would have been quite different.

Consistent with its role in memory formation, the hippocampus has recently been shown to change in size with increased use. One hypothesis suggests that the hippocampus may be involved in how we remember the arrangement of places and objects in space, such as how streets are laid out in a city or how furniture is positioned in a room. The best study to support this theory focused on taxi drivers in London, England. London taxi drivers were chosen because in order to acquire a commercial licence drivers must take an exam testing their knowledge of the streets of London, which suggests that they should have greater navigational expertise than the population at large. Maguire and colleagues (2003) found that a region of the hippocampus was much larger in the taxi drivers than in most other London drivers. Is this the result of experience or are those with a large hippocampus more likely to drive a taxi? The researchers also found that the volume of gray matter in this region of the taxi drivers' brains was highly correlated with the number of years of experience as a taxi driver. This suggests that the hippocampus increases its volume to store more accurate and larger representations of the spatial world. Thus, it changes with experience.

The **amygdala**, located immediately in front of the hippocampus, serves a vital role in our learning to associate things in the world with emotional responses: an unpleasant food with disgust afterward, for example. To this end, its connections with the cerebral cortex and the hypothalamus allow it to supplement the more primitive emotional pathways with new associations. The amygdala thus enables the organism to overrule instinctive responses by connecting memories of things to the emotions they engender. The amygdala also intensifies memory during times of emotional arousal. A frightening experience can be seared into our memories for life.

The amygdala plays a special role in responding to stimuli that elicit fear. Affective processing of frightening stimuli in the amygdala is a hard-wired circuit that has developed over the course of evolution to protect animals from danger. The amygdala is also involved in evaluating the emotional significance of facial expressions (Adolphs et al., 2005). Imaging studies have found that the amygdala activates especially strongly in response to fearful faces (Whalen et al., 2005).

Recent neuroimaging investigations have also implicated the amygdala in the processing of more positive emotions, including sexual arousal. Stephan Hamann and colleagues (2004) found that activity within the amygdala increased when both men and women viewed sexually arousing stimuli, such as short film clips of sexual activity or pictures of opposite-sex nudes, but that such activity was markedly higher in men. It has long been observed that males are more responsive than females to visual sexual stimuli; this study suggests that the amygdala may play a role in that greater responsiveness.

The heightened responsiveness of the hippocampus and amygdala is further seen in their unique sensitivity for becoming clinically "over-excited" and developing epileptic seizures (Goddard et al., 1969). Indeed, it is not well-known that epilepsy is

hippocampus A brain structure important for the formation of certain types of memory.

amygdala A brain structure that serves a vital role in our learning to associate things with emotional responses and for processing emotional information.

our most common neurological disorder (1–3% of the population), which is often difficult to control. This fact has resulted in the employment of radical treatments to stop it (see Penfield & Jasper, 1954), as described in our opening vignette.

basal ganglia A system of subcortical structures that are important for the initiation of planned movement.

cerebral cortex The outer layer of brain tissue that forms the convoluted surface of the brain.

occipital lobes A region of the cerebral cortex at the back of the brain that is important for vision.

THE BASAL GANGLIA The **basal ganglia** are a system of subcortical structures crucial for planning and producing movement. They receive input from the entire cerebral cortex and project to the motor centres of the brainstem and, via the thalamus, back to the cortex's motor-planning area. Damage to the basal ganglia can produce symptoms ranging from the tremors and rigidity of Parkinson's disease to the uncontrollable jerky movements of Huntington's disease. There is evidence that damage can impair the learning of movements as well, because the basal ganglia may be involved in the learning of habits, such as automatically looking for cars before you cross the street. Indeed, Parkinson's patients have trouble learning tasks that require routine actions.

One structure in the basal ganglia, the *nucleus accumbens,* provides a good example of how our environment interacts with our brains. We know from research that the nucleus accumbens is important for experiencing reward. As you will read in Chapter 6, nearly everything that you do that is pleasurable, from eating chocolate to looking at an attractive person, is associated with activation of dopamine neurons in the nucleus accumbens. One recent brain-imaging study found that viewing fancy, expensive sports cars led to greater activation of the nucleus accumbens in men than did viewing less attractive, small economy cars. Because we know that society values expensive objects, this finding and others suggest that cultural beliefs indicate which objects are desirable, and the more desirable objects are, the more they activate basic reward circuitry in our brains (Erk, Spitzer, Wunderlich, Galley, & Walter, 2002).

The Cerebral Cortex Underlies Complex Mental Activity

The **cerebral cortex** is the outer layer of the cerebral hemispheres and gives the brain its distinctive appearance (the word *cortex* is Latin for "bark"). In humans, the cortex is relatively enormous—the size of a sheet of newspaper—and folded in against itself many times so as to fit within the skull, which gives it its wrinkled appearance. The cerebral cortex is the site of all thoughts, detailed perceptions, and consciousness—in short, everything that makes us human. The cortex allows us to learn fine distinctions and intricate details of the outside world, as well as permitting complex behaviours and conferring the ability to *think* before we act. The cortex is the source of our complex culture and communication, allowing us to understand other people and to follow the rules of society. Each hemisphere has four "lobes": the *occipital, parietal, temporal,* and *frontal* (Figure 4.8). A massive bridge of millions of axons, called the *corpus callosum,* connects the two hemispheres and allows information to flow between them.

The **occipital lobe** is almost exclusively devoted to the sense of vision. It is divided into a multitude of different visual areas, of which by far the largest is the *primary visual cortex.* This is the major destination for visual information. Typically for the cerebral cortex, the information is represented in a way that preserves spatial relationships: that is, the visual image, relayed from the eye, is "projected" more or less faithfully onto the primary visual cortex. Two objects near to one another in a visual image, then, will activate neurons that are near to one another in the primary

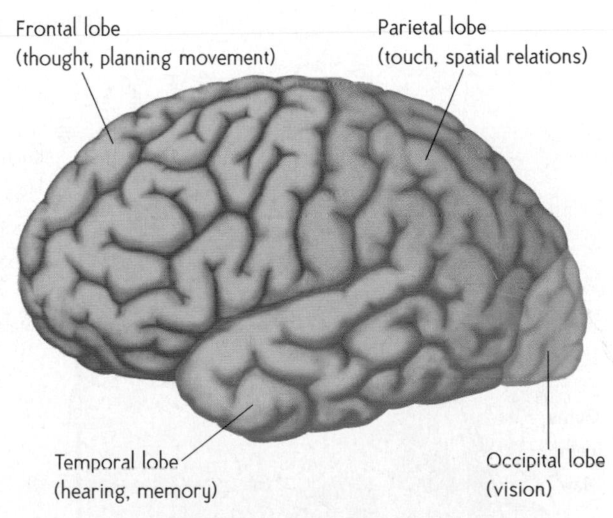

4.8 The Lobes of the Cerebral Hemispheres
The parietal, occipital, temporal, and frontal lobes.

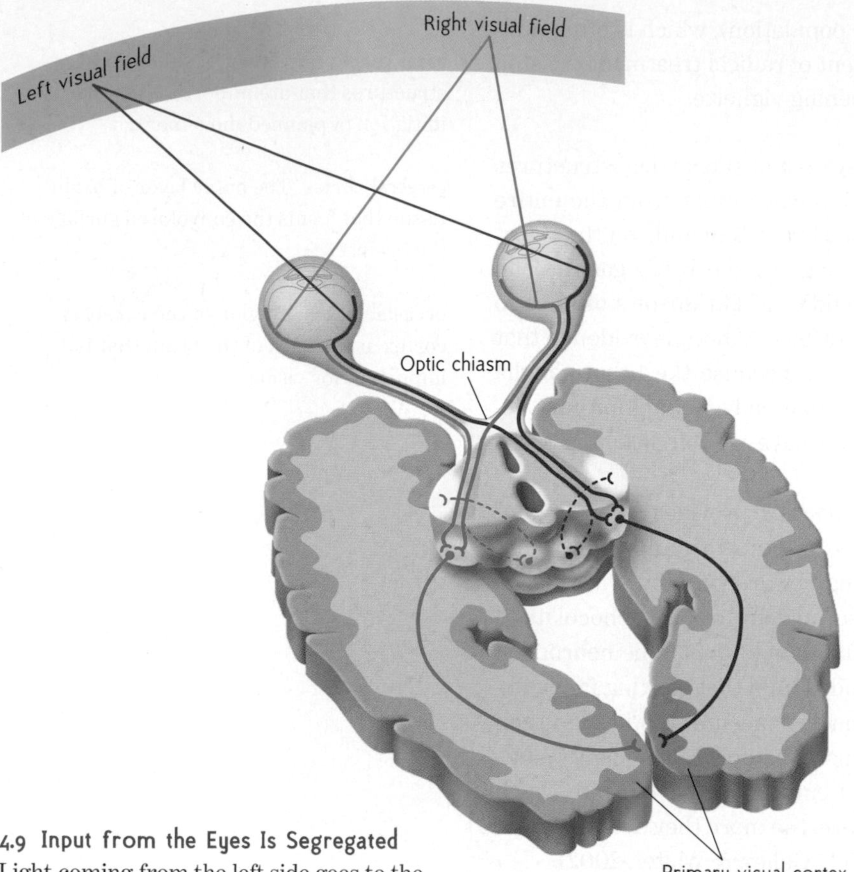

4.9 Input from the Eyes Is Segregated Light coming from the left side goes to the right hemisphere; that coming from the right side goes to the left hemisphere.

visual cortex. This is another way of saying that nearby neurons in the cortex tend to have similar jobs. Each hemisphere takes half of the information: The left hemisphere gets the information from the right side of the visual world, the right hemisphere gets that coming from the left side (Figure 4.9). Surrounding the primary visual cortex is a patchwork of secondary visual areas that process various attributes of the visual image, such as its colour, motion, and forms.

The **parietal lobe** is partially devoted to the sense of touch: It contains the *primary somatosensory* (Greek, "bodily sense") *cortex,* a strip running from the top of the brain down the side. Again, the labour is divided between the left and right cerebral hemispheres: The left hemisphere receives touch information from the right side of the body; the right hemisphere receives information from the left side of the body. This information is also represented so that sensations on the fingers are near ones that respond to sensations on the palm, and so on. The result is a distorted representation of the entire body covering the primary somatosensory area: the so-called *somatosensory homunculus* (Greek, "little man"; Figure 4.10). The homunculus is distorted because the more sensitive areas of the body, such as the face and fingers, have much more cortical area devoted to them.

One of the most fundamental jobs of the mind is to represent the spatial relationships between ourselves and objects around us. It has become clear that this is a specialty of the parietal lobe. A quite common result of a stroke or other damage to the right parietal region is what is called *hemineglect:* patients' failure to notice anything on their left side. If two objects are held up before them, they will see only the one on the right. Asked to draw a simple object, they will draw only its right half (Figure 4.11). Looking in a mirror, such patients will shave or put makeup on the right side of their faces only, not noticing the left side at all.

Work with stroke patients has demonstrated that the neglect extends to the inside world. The Italian neurologists Eduardo Bisiach and Claudio Luzzatti (1978) asked "neglect" patients to close their eyes and imagine themselves standing at one end of Milan's Piazza del Duomo, a public square well known to all of them. Asked to describe the square, the patients talked about the landmarks in great detail, but only those on the right-hand side of their imagined direction of gaze! If they were asked to imagine themselves standing at the other end of the square, they described the landmarks they had previously omitted. Evidently,

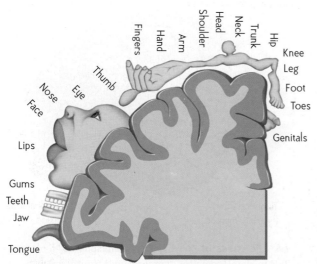

4.10 The Somatosensory "Homunculus" The cortical representation of the body surface is organized in a strip that runs down the side of the brain. Connected areas of the body tend to be represented next to each other in the cortex, and more sensitive skin regions have more cortical area devoted to them.

losing a cortical area that processes a certain kind of information prevents even the imagination or memory of whatever that cortical area was responsible for processing, a conclusion that we will reconsider when we discuss consciousness later in this chapter.

The **temporal lobes** contain the *primary auditory cortex,* an area for hearing analogous to the primary visual and somatosensory cortices, as well as secondary auditory areas that further process what we hear, including, in the left hemisphere, the decoding of words and sentences. The temporal lobes also contain more specialized visual areas for recognizing detailed objects such as faces. The temporal lobes are also critical for memory, containing the hippocampal formation and amygdala, discussed earlier; it was the temporal lobes that Dr. Wilder Penfield electrically stimulated to reawaken his patients' past experiences.

At the intersection of the temporal and occipital cortices is the *fusiform face area.* Its name comes from the fact that this area is much more active when people look at faces than when they look at other things. In contrast, other regions of the temporal cortex are more activated by objects, such as houses or cars, than by faces. Damage to the fusiform face area region can cause specific impairments in recognizing people, but not in recognizing other objects. There has been vigorous debate about whether this region really is specific to faces, or whether it is involved in any task in which an expert has to classify objects (such as a guitar enthusiast knowing the differences between a Stratocaster and a Telecaster). Although the debate continues, recent evidence suggests that this region is more important for faces than for general expertise (Rhodes, Byatt, Michie, & Puce, 2004).

The **frontal lobes** are essential for planning and movement. The rearmost portion of the frontal lobes is the *primary motor cortex.* Instead of responding to sensations coming from the body, though, the primary motor cortices send information to it: They project directly to the spinal cord to move the muscles of the body. Just as for the sensory areas, the motor cortex's responsibilities are divided down the middle of the body: The left hemisphere controls the right arm, for example, whereas the right hemisphere controls the left arm. The primary motor cortex also is supplemented by several auxiliary motor areas responsible for more complex movements.

The rest of the frontal lobes, not directly responsible for movement, is collectively termed the **prefrontal cortex**. The prefrontal cortex, occupying about 30 percent of the brain in humans, is indispensable for rational, directed activity. Parts of the prefrontal cortex are responsible for directing and maintaining attention, keeping ideas in mind while distractions bombard us from the outside world, and developing and acting on plans.

It has long been thought that what makes humans unique in the animal kingdom is our extraordinarily large prefrontal cortex. Recent evidence, however, indicates that what separates humans from other animals is not how much of the brain the prefrontal cortex occupies, but rather the complexity and organization of its neuronal circuits—the way it is put together (Bush & Allman, 2004; Schoenemann, Sheehan, & Glotzer, 2005). The prefrontal cortex is especially important for many aspects of human social life, such as understanding what other people are thinking, behaving according to cultural norms, and even contemplating our own existence. These are all concerns that do not exist or have a very limited presence in other species of animals. As we think about how our minds and environments interact, it can be said that the prefrontal cortex helps define who we are as people.

The area in the centre of the prefrontal cortex just behind the eyes, called the *orbitofrontal cortex,* is especially important for personality, emotion, and impulse control. It is often considered to be part of the limbic system. People with damage to the orbitofrontal cortex often are easily distracted and engage in improper social

4.11 Hemineglect A drawing made by a hemineglect patient, omitting much of the left side of the flower.

parietal lobes A region of the cerebral cortex lying in front of the occipital lobes and behind the frontal lobes that is important for the sense of touch and the spatial layout of an environment.

temporal lobes The lower region of the cerebral cortex that is important for processing auditory information and also for memory.

frontal lobes The region at the front of the cerebral cortex concerned with planning and movement.

prefrontal cortex A region of the frontal lobes, especially prominent in humans, important for attention, working memory, decision making, appropriate social behaviour, and personality.

behaviour. It is as if these people are oblivious to how they are being evaluated by others. In the early part of the twentieth century, deliberately damaging the frontal lobes was established as a means of treating mental patients. This procedure, known as *lobotomy*, left many patients emotionally flat and lethargic, which made them much easier to deal with in mental hospitals.

Profiles in Psychological Science

Phineas Gage

We have learned a great deal of what we know about the functioning of different brain regions through the careful study of people whose brains have been damaged by disease or injury. Perhaps the most famous historical example of brain damage is the case of Phineas Gage. In 1848, Gage was a 25-year-old foreman on the construction of Vermont's Rutland and Burlington Railroad. One day he dropped his tamping iron on a rock, which ignited some blasting powder. The resulting explosion drove the iron rod—over a yard long and an inch in diameter—into his cheek, through his frontal lobes, and out through the top of his head (Figure 4.12). Gage was still conscious as he was hurried back to town on a cart. Able to walk, with assistance, upstairs to his hotel bed, he wryly remarked to the awaiting physician, "Doctor, here is business enough for you," and said he expected to return to work in a few days. In fact, Gage lapsed into an unconscious state for two weeks. His condition steadily improved subsequently, though, and he recovered remarkably well, at least physically.

Unfortunately, the accident led to major personality changes. Whereas the old Gage had been regarded by his employers as "the most efficient and capable" of workers, the new Gage was not. As one of his doctors later wrote: "The equilibrium or balance, so to speak, between his intellectual faculties and animal propensities seems to have been destroyed. He is fitful, irreverent, indulging at times in the grossest profanity . . . impatient of restraint or advice when it conflicts with his desires. . . . A child in his intellectual capacity and manifestations, he has the animal passions of a strong man." In sum, Gage was "no longer Gage."

4.12 Phineas Gage A photo of Gage's death mask next to his skull. Analysis of Gage's skull provided the basis for the first modern theories of the role the prefrontal cortex plays in personality and self-control.

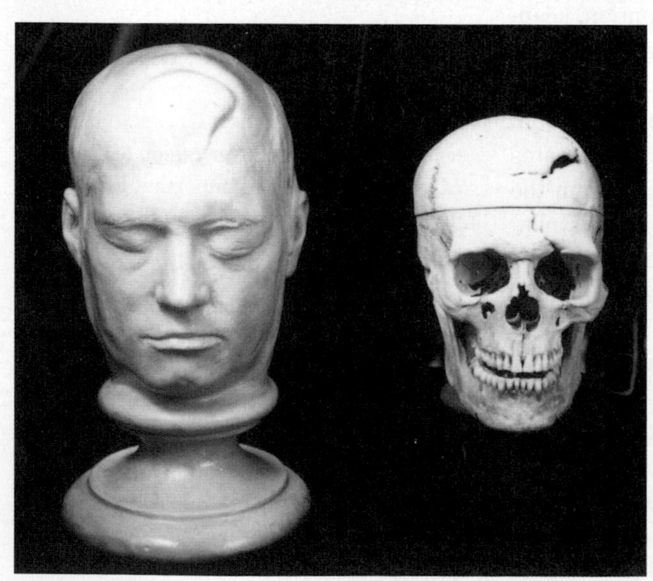

Unable to get his foreman's job back, Gage exhibited himself in various New England towns and at the New York Museum (owned by P. T. Barnum), worked at the stables of the Hanover Inn at Dartmouth College, and drove coaches and tended horses in Chile. After a decade, his health began to decline, and in 1860 he started having epileptic seizures and died within a few months.

Gage's recovery was initially used to argue that the brain worked in a uniform fashion and that the remaining brain had the ability to take over the work of the damaged parts. However, the medical community eventually recognized that Gage's psychological impairments were severe, suggesting that some areas of the brain did indeed have specific functions. The case of Phineas Gage provided the basis for the first modern theories of the roles of the prefrontal cortex in personality and self-control. Reconstruction of the injury through examination of Gage's skull makes it clear that the orbitofrontal cortex was the area most damaged by the tamping rod (Damasio, Grabowski, Frank, Galaburda, & Damasio, 1994). Recent studies of patients with similar injuries reveal that this prefrontal region is particularly concerned with social phenomena, such as following social norms, understanding what other people are thinking, and feeling emotionally connected to others. People with damage to this region do not typically have problems with memory or general knowledge, but they often have profound disturbances in their ability to get along with others. ■

REVIEWING **the Principles** | What Are the Basic Brain Structures and Their Functions?

The different parts of the nervous system all have essential roles. The spinal cord is involved in basic movement and reflexes. The brainstem serves survival functions, such as breathing and heart rate. At the back of the brainstem is the cerebellum, a workhorse that learns routine habits of movement and maybe thought. The midbrain integrates motor and sensory information. Beneath the cortex, the thalamus serves as a way station through which sensory information travels; the hypothalamus regulates bodily systems and controls the hormonal system; the basal ganglia aid in motor planning and habit learning; the hippocampus is involved in memory; and the amygdala influences emotional states, especially fear. Finally, the cerebral cortex is the outer surface of the brain, divided into occipital, parietal, temporal, and frontal lobes. Each lobe serves specific functions; the frontal lobe is essential for higher-level thought and social behaviour.

How Does the Brain Change?

Despite the great precision and specificity of its connections, the brain is extremely malleable. Over the course of development, after injury, and throughout our constant stream of experience, the brain is continually changing, a property known as **plasticity**. We will see that the brain can reorganize based on which parts of it are underused or overused. Determining the nature of these changes, and the rules that they follow, is providing major insights into the mind, and is a direct outgrowth of the biological revolution that is energizing the field.

plasticity A property of the brain that allows it to change as a result of experience, drugs, or injury.

The brain follows a predictable development pattern, with different structures and abilities progressing at different rates and maturing at different points in life. Reptiles hatch from their leathery eggs ready to go; human infants sleep and continue to grow and develop their brains, actively rewiring them, in major ways, for many years. In addition, our brains' connections are refined and retuned with every experience of our lives. Brain plasticity research promises to reveal great insights into the interactive nature of biological and psychological influences on our behaviour, drawing on exciting research that crosses all levels of analysis.

The Interplay of Genes and the Environment Wires the Brain

As you read in Chapter 3, there are complex interactions between genes and the environment. The brain's development follows set sequences programmed in the genes: Babies' vision develops before their ability to see in stereo, for example, and

Are there critical periods during brain development?

the prefrontal cortex is not anatomically fully mature until early adulthood. But even with these meticulously specified genetic instructions, environment plays a major role. The behaviour of the genes themselves is utterly dependent on their environment. Different genes are expressed in different cells, and which genes are expressed, and to what extent, is determined by the environment. The environment does not just affect the products of our DNA's activity; it affects the DNA's activity itself (Liu et al., 2002).

CHEMICAL SIGNALS GUIDE GROWING CONNECTIONS In the developing embryo, new cells receive signals from their surroundings that determine what type of cells they will become. Cells release chemical signals to each other that act directly on their genetic material to determine which of the millions of instructions contained in the genetic code will be executed.

If cells from one part of an embryo are surgically transplanted to another part, the result depends on how far along the process has gotten. Tissue that is transplanted early on completely transforms into whatever type is appropriate for its new location. But as time passes, cells become more and more committed to their identities, so that transplanting them results in disfigured organisms. In the case of neural tissue, cells transplanted early take on the identity appropriate to their new location, and the organism develops normally. Cells transplanted later, however, grow connections as if they hadn't been moved; that is, the cells connect to areas that would have been appropriate if they had remained in their old location.

This predestination of connections from cells that have had time to determine their identities exposes the preprogrammed nature of brain wiring. The connections of the brain are produced largely by the growing axons' detection of particular chemicals that tell them where to go and where not to. The broad brushstrokes of the brain are laid out by this chemical specificity: Neurons in one region are looking for particular chemicals, and neurons in another are producing them. Axons from the first region grow steadily toward or away from increasing or decreasing concentrations of these signaling chemicals—so-called *chemical gradients*. Following the gradients of some chemicals guides axons on long trips to their destination areas, whereas the concentration of other chemicals governs their branching patterns once they're there. This method lays down the connections between all brain structures and creates the specific connections between areas within the cerebral cortex.

EXPERIENCE FINE-TUNES NEURAL CONNECTIONS Even though the major connections are established by chemical gradients, the detailed connections are governed by experience. If a cat's eyes are sutured closed at birth, depriving it of visual input, the maps in its visual cortex fail to develop properly; when the sutures are removed weeks later, the cat is blind (Wiesel & Hubel, 1963). Evidently the ongoing activity of the visual pathways is necessary to refine the map enough for it to be useful. In general, such plasticity has **critical periods**, times in which certain experiences must occur for development to proceed normally. Adult cats that are similarly deprived do not lose their sight.

critical period The time in which certain experiences must occur for normal brain development, such as exposure to visual information during infancy for normal development of the brain's visual pathways.

A few years later, it was discovered that rats raised in "enhanced" environments with lots of toys and obstacles grew up to have bigger brains than those raised in normal laboratory conditions (Rosenzweig, Bennett, & Diamond, 1972). However, consider the "normal" conditions: essentially, featureless boxes with bedding in the bottom. The reality may be that the "enhanced" conditions were simply an approximation of rat life in the wild, allowing normal rat development, while the mental and social deprivation caused by the lab cage environment caused atrophy in the rats' unused brains. Nonetheless, this experiment does demonstrate at a minimum that environment is important for normal development. We will consider enriched environments and their effects on gene expression in Chapter 8.

The Brain Rewires Itself throughout Life

Although plasticity decreases with age, the brain retains the ability to rewire itself throughout life. This is the biological basis of learning.

CHANGE IN THE STRENGTH OF CONNECTIONS UNDERLIES LEARNING

With every moment of life, we gain new memories: experiences and knowledge instantaneously acquired that can later be recalled and habits that gradually form. All these forms of memory consist of physical changes in the brain.

It is now widely accepted by psychological scientists that these changes are most likely not in the brain's gross wiring or arrangement, but simply in the strength of preexisting connections. One possibility is that two neurons firing at the same time strengthens the synaptic connection between them, making them more likely to fire together in the future, and that, conversely, *not* firing at the same time tends to weaken two neurons' connection. Known technically as *Hebbian learning* (discussed in Chapter 6), this theory is summarized by the catchphrase "Fire together, wire together," and it is consistent with a great deal of experimental evidence and many theoretical models. It accounts for both the "burning in" of an experience—a pattern of neuronal firing is made more likely to recur, leading us to recall an event—and for the ingraining of habits: Merely repeating a behaviour makes us tend to perform it automatically (Hebb, 1949).

Another possible method of plasticity is the growth of entirely new connections; this takes much longer but appears to be a major factor in recovery from injury. Until very recently, it was believed that, uniquely among bodily organs, adult brains produced no new cells—presumably to avoid obliterating the connections that had been established in response to experience. However, Fred Gage and his colleagues at the Salk Institute demonstrated that new neurons *are* produced in the adult brain, a process known as *neurogenesis* (Eriksson et al., 1998). There appears to be a fair amount of neurogenesis in the hippocampus. Recall that memories are retained within (or at least require) the hippocampus initially but are eventually transferred to the cortex, with the hippocampus being continuously overwritten. It may be that lost neurons can be replaced without disrupting memory.

How might environment affect brain development in young children?

Princeton University psychologist Elizabeth Gould and her colleagues have demonstrated that environmental conditions play an important role in neurogenesis. For example, they have found in their tests with rats, shrews, and marmosets that stressful experiences, such as being confronted by strange males in their home cages, interfere with neurogenesis both during development and adulthood (Gould & Tanapat, 1999). When animals are housed together, they typically form dominance hierarchies that vary along social status. Dominant animals—those who possess the highest social status—show greater increases in new neurons than do subordinate animals (Kozorovitskiy & Gould, 2004). This demonstrates that the social environment can exert important effects on brain plasticity. Although we are at the earliest stages of understanding this dynamic process, it is possible that neurogenesis underlies the entire basis of neural plasticity. Another implication is that neurogenesis may be able to reverse the loss of neurons and slow down the rate of decline as we age. Imagine how quality of life would improve for all of us if our minds remained nimble throughout our lives.

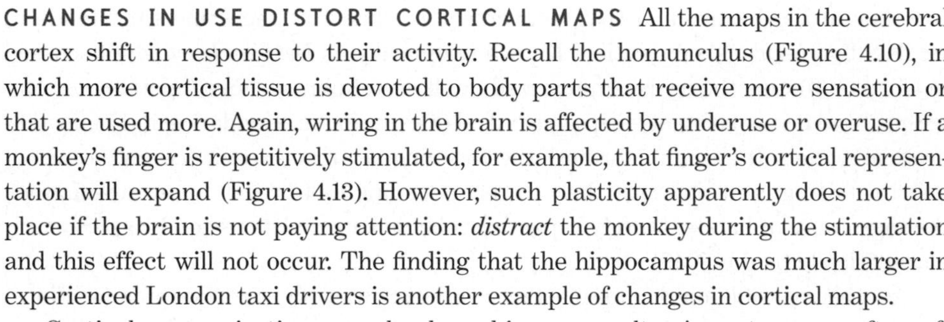

4.13 Skin Regions Changes in cortical representations of skin regions in monkeys that were trained or not trained to detect certain touch stimulation. Repeated training led to an increased representation in cortical areas (as shown in a). In (b), the filled circles indicate where the fingers were touched for the trained (blue) and untrained (green) fingers. The circles show the regions of the finger that became sensitive to the stimulation. More of the finger becomes sensitive to the stimulation after training.

Does the brain rewire itself during learning, aging, and repair?

CHANGES IN USE DISTORT CORTICAL MAPS All the maps in the cerebral cortex shift in response to their activity. Recall the homunculus (Figure 4.10), in which more cortical tissue is devoted to body parts that receive more sensation or that are used more. Again, wiring in the brain is affected by underuse or overuse. If a monkey's finger is repetitively stimulated, for example, that finger's cortical representation will expand (Figure 4.13). However, such plasticity apparently does not take place if the brain is not paying attention: *distract* the monkey during the stimulation and this effect will not occur. The finding that the hippocampus was much larger in experienced London taxi drivers is another example of changes in cortical maps.

Cortical reorganization can also have bizarre results. Amputees are often afflicted with *phantom limbs,* the intense sensation that the amputated body part still exists (Melzack et al., 1997). Some phantom limbs are experienced as moving normally, being used to gesture in conversation as if they really existed, for example, whereas some are frozen in position. Unfortunately, phantom limbs are often accompanied by sensations of pain, which may result from the misgrowth of the severed pain nerves at the stump, with the pain being interpreted by the cortex as coming from the place those nerves originally came from (Figure 4.14). This would suggest that the brain had *not* reorganized in response to the injury, and that the cortical representation of the arm was still intact. However, V. S. Ramachandran of the University of California, San Diego, discovered that some of these people, when their eyes were closed, perceived a touch on the cheek as if it were on their missing hand! Apparently, the hand is represented next to the face in our somatosensory homunculus. Following the loss of the limb, the patient's unemployed cortex assumed to some degree the function of the closest group, which was to represent the skin of the face. Touching the face then activated these neurons. Somehow, the rest of the brain had not kept pace with the somatosensory area enough to figure out these neurons' new job.

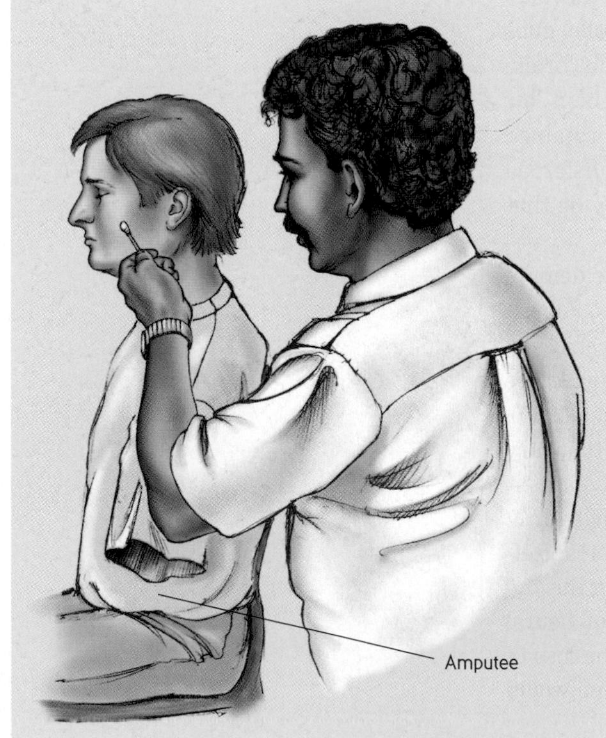

4.14 Cortical Remapping Following Amputation A cotton swab touching the subject's cheek was felt as touching his missing hand.

Psychological Science *in Action*

Manipulating Objects with the Mind

Let's consider an example of how brain plasticity might allow researchers to improve the lives of those with spinal injuries. In our daily interactions with the world, we use our arms and hands to manipulate objects around us, as we type on keyboards, play computer games with joysticks, change television channels with remotes, and control our cars with steering wheels. Imagine being able to manipulate these objects simply by thinking about using them. How amazing this would be for the many Canadians who do not have use of their arms or legs because of spinal cord injuries or disease, who could perhaps be taught to control prosthetic limbs with their minds. In groundbreaking research that provides hope for those with paralysis, neuroscientists at Duke University have taught monkeys how to control a robot arm using only their thoughts (Carmena et al., 2003). As the media have quipped, "Monkey see, robot do."

How did the researchers accomplish this remarkable feat? The first step was to implant small electrodes in the frontal and parietal lobes of two rhesus monkeys and then record impulses from those electrodes while the monkeys learned to play a simple computer game. These brain areas were chosen because they are known to be involved in controlling complex movement. In the computer game, the monkeys used a joystick that allowed them to perform actions such as steering a cursor and squeezing a trigger to achieve a goal, which involved moving a robotic arm that they could see but not touch. By examining the recordings from various brain regions, the research team was able to determine which patterns of brain activity led to specific motor actions. At this point, scientists unplugged the joystick so that it no longer controlled the movement of the robotic arm and then made movements of the robotic arm directly dependent on the pattern of neuronal firing from the monkeys' motor cortices. At first, the monkeys moved their arms wildly as if controlling the joystick, but it was not their arm movements that controlled the robot; it was their thoughts. "The most amazing result, though, was that after only a few days of playing with the robot in this way, the monkey suddenly realized that she didn't need to move her arm at all," says the lead neuroscientist, Miguel Nicolelis, describing one of the experiment's subjects (Figure 4.15). "Her arm muscles went completely quiet, she kept the arm at her side and she controlled the robot arm using only her brain and visual feedback. Our analyses of the brain signals showed that the animal learned to assimilate the robot arm into her brain as if it were her own arm" ("Gazette: Monkeys Move Matter, Mentally," 2004).

This experiment reveals an amazing property of brain plasticity: The brain learns to use tools as natural extensions of our physical bodies. Watch young children who have learned to play computer games; they aren't thinking about using the mouse to control the action on the monitor—the mouse becomes part of their physical being. In many ways, this is no different from the way our brains control our biological appendages; sensory information is received by the brain, and motor commands control physical action. When you drive, your arms don't control the steering wheel; your brain turns the wheel. Researchers have been able to tap into the motor command signals from the brain and use those signals to control an external device.

Is the brain activity identical in the two situations? According to Nicolelis, it is not. When the researchers removed the joystick, the physiological properties of the

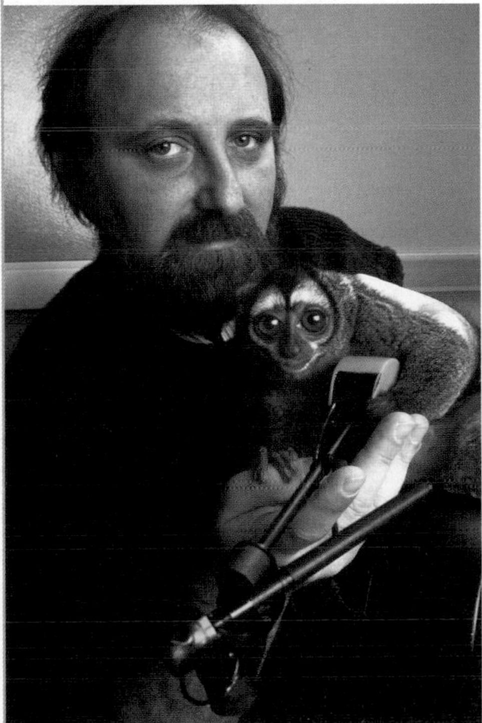

4.15 Advances in Motor Control Neurobiologist Miguel Nicolelis with one of the monkeys used in his study of the brain's motor command signals. Dr. Nicolelis's research revealed new information on brain plasticity and the brain's capabilities to adapt.

neurons changed immediately, and they changed a second time when the monkeys were once again required to use the joystick. The methods used in this study will allow neuroscientists to study brain plasticity, because they can modify subtle aspects of the robot arm (such as building in a short delay between thought and action) and watch how the brain adapts.

How soon might this research help those who are paralyzed? A team of researchers led by John Donoghue at Brown University, in collaboration with a private company, received permission from the U.S. Food and Drug Administration to implant a small computer chip into five quadriplegic individuals. The chip reads signals from 20 to 100 neurons in the motor cortex. The first human to benefit from this technology was 25-year-old Matthew Nagle, who had been paralyzed from the neck down following a knife attack in 2001. The implanted device, called BrainGate, has allowed him to use a remote control to perform physical actions such as change television channels and open and close curtains, entirely with his thoughts. As in the Duke research, this firing pattern is translated into signals that control the movement of a robotic arm. The longer-term goals are to allow paralyzed individuals to control prosthetic devices that will assist them in all aspects of daily living. Possibly within the next few years we will witness quadriplegics doing things that are now unimaginable, and in doing so we will learn a great deal about how brain plasticity allows the mind to control behaviour. ■

The Brain Can Recover from Injury

The converse of the brain's reorganization in response to overuse or underuse is its reorganization in response to brain damage. Following a lesion in the cortex, the surrounding gray matter assumes the function of the damaged area, with the map distorting everywhere around it to recover the lost capability, like local businesses scrambling to pick up the customers of a newly bankrupt store. Some of this remapping seems to occur immediately and to continue for years. Such plasticity involves all levels of the nervous system, from the cortex down to the spinal cord.

This reorganization is much more prevalent in children than adults, in accordance with the critical periods of normal development. Young children afflicted with severe and uncontrollable epilepsy sometimes undergo "radical hemispherectomy": the surgical removal of one entire cerebral hemisphere. This procedure is not possible in adults because of the inevitable and permanent paralysis and loss of function that results. Young children, however, eventually regain almost complete use of the initially paralyzed arm and leg, as well as the other functions of the lost hemisphere: These are simply taken up by the remaining one. But even in adults, recovery from brain injury can be quite dramatic. Stroke patients with a paralyzed arm, for example, very often recover its use within a few months.

Finally, one of the most exciting areas of current neurological research is the transplantation of stem cells into the brain to repair damage. *Stem cells* are "master" cells that not only are able to regenerate themselves, but also have the capacity to develop into any type of tissue, such as muscles or nerve cells. This procedure, using stem cells typically obtained from human fetal tissue, is beginning to be explored as a possible treatment for degenerative diseases such as Parkinson's or Huntington's, and also for strokes. The significant challenge is to get the newly introduced cells to make the proper connections and regrow the damaged circuits. These techniques are still in their earliest days, and they pose obvious ethical dilemmas, but they may well soon become very important treatment methods.

Thinking Critically

Is Stem Cell Research Ethical?

One of the most exciting areas of current neurological research is the transplantation of stem cells into the brain to repair damage. It is also extremely controversial. Let's critically examine the promise and the controversy.

When the actor Christopher Reeve was paralyzed in a horseback riding accident, he vowed to walk again some day. Until recently, such vows would be dismissed as wishful thinking, since paralysis from spinal injuries was believed to be permanent. Within the last decade, however, the discovery of stem cells has provided new hope for the treatment of many diseases. *Stem cells* are special "master" cells that have the ability to develop into any type of tissue, such as muscles or nerve cells. They have another amazing property: They can renew themselves through cell division, perhaps indefinitely. The hope is that injecting stem cells into damaged tissue will allow the stem cells to repair the damage and restore function. Findings from animal studies provide tantalizing evidence that stem cells might reverse the effects of spinal injuries, repair damage following heart attacks, cure diabetes, and provide effective treatments for many neurological disorders. Brain diseases such as Huntington's, Parkinson's, and Lou Gehrig's disease may soon go the way of polio and smallpox and become diseases of the past.

Using stem cells to cure diseases seems an obvious choice, but some ethicists and religious leaders have raised moral objections to stem cell research (Figure 4.16). Their concerns revolve around the obtaining of stem cells from embryos because the process actually destroys the embryos. They argue that in destroying embryos, we are destroying life. Stem cells can be harvested from various sources, but scientists prefer embryonic stem cells because they are the purest and therefore likely to be most useful for medical treatment. There is also a potentially huge supply of embryos from which

4.16 The Stem Cell Controversy Stem cell research is a highly controversial and divisive subject. Stem cell researchers, such as cell biologist Lesley Young from the UK Stem Cell Bank in London (left), work to develop treatments that could save millions of lives. Political and religious leaders, such as U.S. President George W. Bush and the late Pope John Paul II (right), have publicly criticized such research on moral and ethical grounds.

4.17 Stem Cell Potential Jean Chretien, former prime minister of Canada, is seen with Christopher Reeve, actor and founder of the Christopher and Dana Reeve Paralysis Resource Center, during a visit to the Ottowa Rehabilitation Centre in 2000. Reeve and Chretien met to discuss new developments for paralysis cases, such as advancements with stem cells.

stem cells can be obtained. When people undergo in vitro fertilization, dozens of embryos are created. Once the treatment is successful, the leftover embryos are destroyed or frozen for later use. Researchers want to use these "spare" embryos to obtain stem cells.

But are embryos people? At the heart of the ethical debate is whether these clumps of 100 or so cells really deserve the same degree of respect and protection given to adults and children. Are we humans from the moment of fertilization, or do we become humans with the development of a nervous system capable of sensation and conscious experience? Recognizing Canadians' diverse views on *in vitro* human embryonic research, the Assisted Human Reproduction Agency of Canada was created; this agency ensures through clear guidelines that all research involving the use of the *in vitro* human embryo is carefully regulated. The government of the United Kingdom considered the available evidence and decided that embryos can be used as long as they have not developed beyond 14 days. In the first 14 days following fertilization, the embryo can go through many changes, such as splitting into twins or recombining back into one embryo, that make it difficult to accord to it the status of an individual human being. From this perspective, conception of a person does not happen in an instant but occurs over 14 days. Moreover, during this period there is no trace of a nervous system, and the possession of a nervous system is considered an important determinant of whether someone is alive. Following birth, a person whose brain stops working is considered dead for all practical purposes, even if other bodily systems function. The question of when life begins can be addressed from scientific, religious, or other philosophical perspectives; the implications of how this question is addressed for stem cell research are enormous.

In the United States, President George W. Bush's administration enacted rules in 2001 that severely restrict the use of federal research grant dollars to fund stem cell research. This raises other important ethical questions. Should moral values be considered in federal funding decisions? Is it ethical to block the development of treatments that might save millions of lives? Should political leaders be allowed to shape medical research according to their religious beliefs—beliefs that others might not share? In response to the federal restrictions, a number of private companies and universities, such as Harvard and Stanford, have begun their own stem cell initiatives. Research conducted by private industry also raises critical questions, such as whether the findings will be made available to other scientists and whether treatments that are developed will be available to the public.

History has taught us that science marches on, one way or the other. Even if stem cell research is banned in the United States, or dramatically slowed down, other countries, such as Canada, Great Britain, and South Korea, will forge ahead, and a more practical set of issues will emerge. If stem cells and the treatments that emerge from the research are not available in the United States, those with limited economic means could die from diseases while wealthier Americans could be treated in countries that have active stem cell programs. Perhaps someone like Christopher Reeve (Figure 4.17) could have raised millions of dollars to receive treatment in another country to restore his ability to walk. Could the average American? More important, should Americans make it impossible for someone like Christopher Reeve to walk again? Sadly, Reeve died in October 2004 and did not live to see the promise of stem cell research fulfilled. But, should Americans make it impossible for someone with Huntington's disease to live a full and meaningful life? How should ethical values shape research? Should one's government, its leaders or their beliefs place limits on research into stem cell possibilities? ■

REVIEWING

the Principles | How Does the Brain Change?

Though neural connections are intricate and precise, they are malleable. The human genome is the blueprint for normal development, but it is affected by environmental factors, such as injury or sensory stimulation or deprivation. During development, and throughout our lifetimes of learning, the circuitry is reworked and updated. Reorganization also occurs following brain injury, with recovery ability far greater in children than in adults.

corpus callosum A fibre of axons that transmits information between the two cerebral hemispheres of the brain.

split brain A condition in which the corpus callosum is surgically cut and the two hemispheres of the brain do not receive information directly from each other.

How Is the Brain Divided?

Does each half of the brain possess the same capacities?

As we have seen, studying humans whose brains were damaged by accidents or surgically altered has been a rich source of insight into the mechanics of the mind. Many of these surgeries were attempts to treat epilepsy by specifically removing the part of the brain in which the seizures begin. Another strategy, pioneered in the 1940s and still practiced on occasion when other interventions have failed, is to cut connections within the brain to try to isolate the site of seizure initiation, so that a seizure that begins there will be less able to spread throughout the cortex. This surgical procedure has provided many important insights into the basic organization of each hemisphere of the brain and the specialized functions they perform. But this procedure also raises another interesting question: "If you split the brain, do you split the mind?"

The Hemispheres Can Be Separated

The only connection between the hemispheres that may readily be cut without damaging the gray matter itself is the massive fibre bundle called the **corpus callosum** (Figure 4.18). To treat epilepsy, the corpus callosum may be cut only partway through, sparing some of the connecting axons, or it may be completely severed. Most epilepsy patients, and all of those on whom the initial operations were performed and who provided the first striking results, have had the corpus callosum completely cut. This leaves the two halves of the forebrain almost completely isolated from each other, and the two hemispheres connected only at the back of the brainstem, hence only indirectly to each other. This condition is known as the **split brain**.

Perhaps the most obvious thing about the split-brain patients after their operations was how very normal they were. Unlike patients following other types of brain surgery, split-brain patients had no major problems that were immediately apparent. In fact, some early investigation suggested that the surgery had not affected the patients in any discernible way. They could walk and talk normally, think clearly, and interact socially.

However, work with animals had shown that this could not be the full story. One of the authors (M.S.G.), working with the eventual Nobel laureate Roger Sperry, conducted a series of tests on the first split-brain participants. The results were stunning: Just as the brain had been split in two, so had the mind!

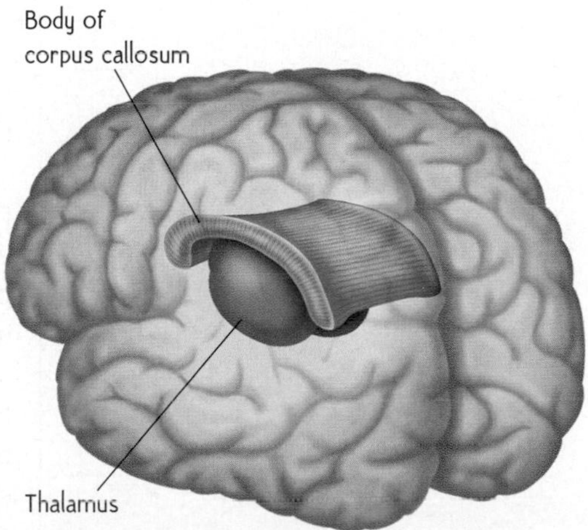

4.18 The Corpus Callosum A massive bundle of millions of axons connect the two cerebral hemispheres.

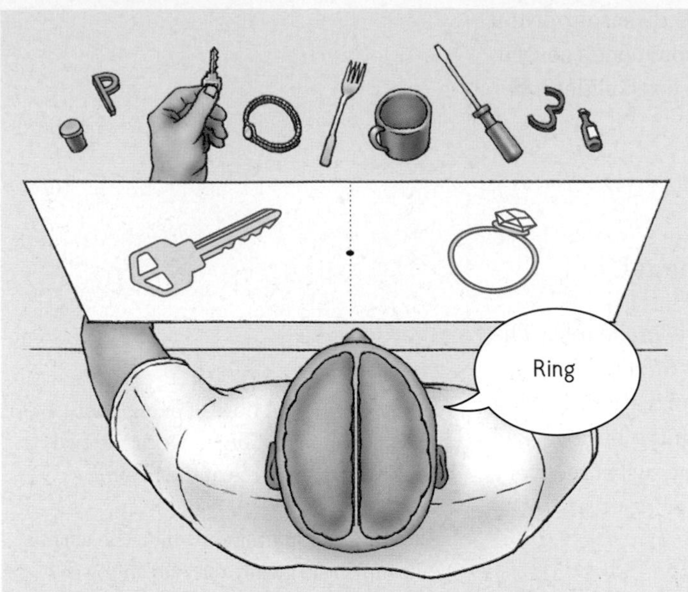

4.19 An Experiment with a Split-Brain Subject When asked what he sees, the left hemisphere responds that he sees a "ring," because that is what is shown on the right side of the screen. The right hemisphere sees the left side of screen, and although not able to verbalize what it sees, it can pick the correct object using the left hand.

The Separate Hemispheres Can Be Tested

Recall from earlier in this chapter that images from the left side of the visual field go to the right hemisphere, and those from the right side go to the left. Remember also that the left hemisphere controls the right hand, and the right hemisphere controls the left hand. These divisions allow researchers to provide information to and get information from only a single hemisphere at a time.

Figure 4.19 shows a typical experimental setup. The split-brain patient sits in front of a screen, staring at a central dot. Two images are flashed simultaneously, one on each side of the dot. Because each hemisphere sees only the contents of the opposite side of the visual field, the right hemisphere sees the image on the left, and the left hemisphere the image on the right.

It has long been known that the left hemisphere is dominant for language in most people. If a split-brain patient, having just been flashed two pictures in the manner just described, is asked to report what was shown, he or she will announce that only one picture was shown and will describe the one on the right. Why is this? Because the left hemisphere, with its control over speech, saw only the picture on the right side. The mute right hemisphere (or "right brain"), having seen the picture on the left, is unable to articulate a response. It can be shown, though, that the right brain indeed saw the picture. If the picture on the left was, for example, of a spoon, the right hemisphere can easily pick out an actual spoon from a selection of objects, using, of course, the left hand. The left hemisphere has no knowledge whatsoever of what the right one saw. Splitting the brain, then, produces two half brains, each with its own independent perceptions, thoughts, and consciousness!

The Hemispheres Are Specialized

Further explorations revealed much more about the division of labour within the brain. In all the patients studied, the left hemisphere was far more competent at language than the right, so much so that in most patients the right hemisphere had no discernable language capacity at all. In some patients, though, the right hemisphere did have some rudimentary language comprehension. Interestingly, such right-hemisphere language capabilities tend to improve in the years following the split-brain operation, presumably as the right hemisphere attains communication skills that were unnecessary when the hemisphere was fully connected to the fluent left brain.

The right hemisphere, however, has its own competencies, which complement those of the left. The left brain is generally hopeless at spatial relationships. In one experiment, a split-brain participant is given a pile of blocks and a drawing of a simple arrangement in which to put them—for example, a square. If the participant is using his or her left hand, controlled by the right hemisphere, the blocks are arranged effortlessly. If, however, the left brain is doing the arranging via the right hand, the result is a meandering, incompetent attempt. During such a dismal performance, the idle left hand will attempt to slip in and help out, as the right brain presumably looks on in frustration!

Split-brain patients have provided insight into not only the functional specializations of the left and right sides, but also the organization within each hemisphere. When the corpus callosum was not cut along its entire length, the relative contributions of its different ends to the transfer of different kinds of information could be explored. Figure 4.20 shows the reaction of one such patient, who had enough right-hemisphere language capability to comprehend words flashed in the left visual field. His operation was done in two parts, with the back half of the corpus callosum cut first. He was tested following the first operation. Since the visual cortex occupies the back of the brain, the cut rear half of the corpus callosum prevented any direct transfer of visual information. The left brain, then, was unable to see the word "knight." Fascinatingly, though, the left brain was nevertheless able to visualize what the right brain was thinking about! Evidently, more "meaningful" information *could* be transferred, through the axons in the front half of the corpus callosum connecting the two front halves of the brain.

The Mind Is a Subjective Interpreter

Another interesting dimension to the relationship between the brain's hemispheres is how they work together to reconstruct our experiences. This can be demonstrated by asking a disconnected left hemisphere what it thinks about previous behaviour that has been produced by the right hemisphere. In one such experiment, different images are flashed simultaneously to the left and right visual fields and the patient is asked to point with both hands to the pictures that seem most related to the images on the screen (Figure 4.21). As one example, a picture of a chicken claw was flashed to the left hemisphere and a picture of a snow scene to the right hemisphere. In response, the left hemisphere directed the right hand to point to a picture of a chicken head, and the right hemisphere pointed the left hand at a snow shovel. The participant was then asked why he chose

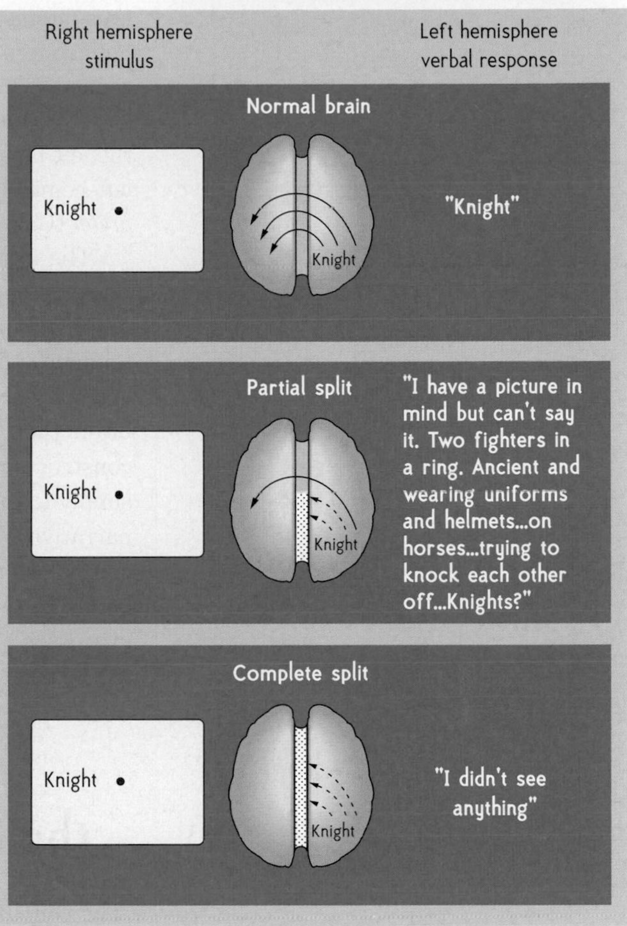

4.20 Information Transfer in the Brain Different kinds of information are transferred by different parts of the corpus callosum. If the back half is severed but the front half left intact, meaning transfers from one hemisphere to the other, but visual information does not.

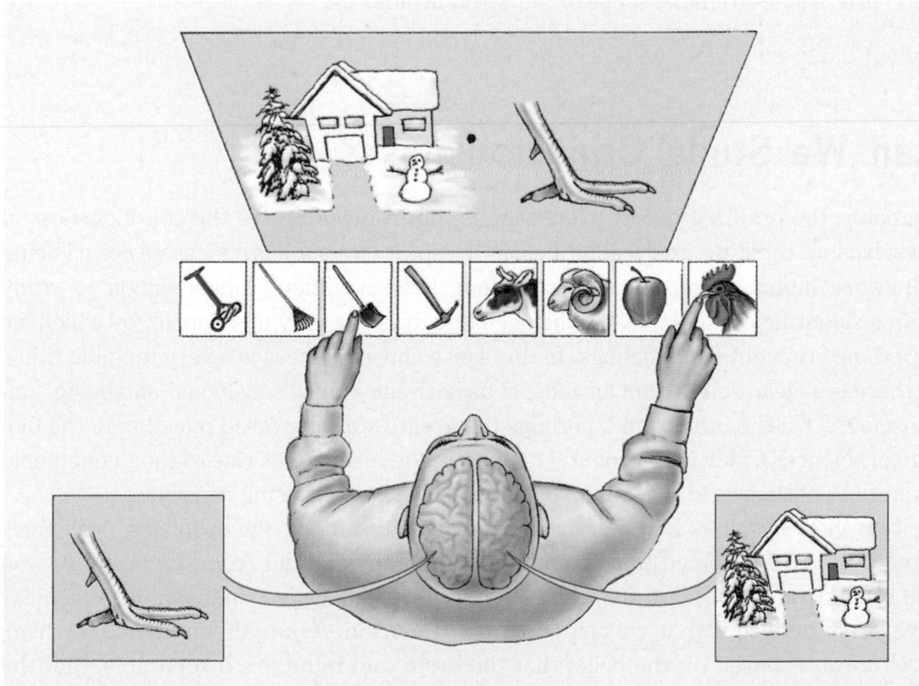

4.21 The Left-Brain Interpreter Mechanism The left hemisphere attempts to explain the behaviour of the right hemisphere on the basis of limited information.

those items. Clearly, the speaking left hemisphere could have no idea what the right hemisphere had seen. However, the patient (or rather his left hemisphere) calmly replied, "Oh, that's simple. The chicken claw goes with the chicken, and you need a shovel to clean out the chicken shed." The left hemisphere had evidently interpreted the left hand's response in a manner consistent with the left brain's knowledge. This left-hemisphereic propensity to construct a world that makes sense is called the *interpreter* (Gazzaniga, 2000).

The interpreter strongly influences the way we view and remember the world. Shown a series of pictures that form a story and asked later to choose which of another group of pictures they had seen previously, normal participants have a strong tendency to falsely "recognize" pictures that are consistent with the theme of the original series and to reject those that are inconsistent with the theme. The left brain, then, tends to "compress" its experience into a comprehensible story and reconstructs remembered details based on the gist of the story. The right brain seems simply to experience the world and remember things in a manner less distorted by narrative interpretation. Given the finite capacity of the brain, the advantages of compression seem clear, though it appears the right hemisphere may check the left hemisphere's potentially unwarranted speculations. Many of our thoughts reflect the processes of the interpreter. Thus, our sense of consciousness, to which we turn in the next section, is influenced greatly by our left hemisphere interpreter.

REVIEWING

the Principles | How Is the Brain Divided?

The corpus callosum joins the two cerebral hemispheres. For epilepsy patients, the corpus callosum is cut in order to treat the disease. Research on split-brain patients shows that the left hemisphere is dominant for language whereas the right hemisphere is dominant for spatial relationships. In split-brain patients each hemisphere seems to have an independent consciousness. The left-hemisphere interpreter tries to make sense of the world around us.

Can We Study Consciousness?

Arguably, the brain's greatest expression is that it gives rise to the consciousness of ourselves as thinking and feeling beings. Despite our subjective *experience* of being self-aware humans, consciousness has long been a difficult phenomenon to study from a scientific perspective. On the one hand, it is the very mechanism by which we experience thoughts and feelings. In this sense consciousness is a very tangible thing, as there is a clear delineation for most of us between being "conscious" and being "unconscious." On the other hand, perhaps the greatest unanswered question in the biological sciences today is how neural activity in the brain gives rise to the phenomenal awareness of the world that we associate with the state of being conscious.

Consciousness has long been a topic of debate among philosophers, ever since René Descartes asserted that the mind is physically distinct from the brain. As you will recall from Chapter 1, this view of consciousness, called dualism, posits a clear separation between what we call mind and the brain. As an alternative to dualism, *materialism* is based on the belief that the brain and mind are inseparable, that the

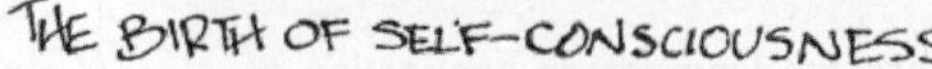

brain directly enables the mind. According to materialism, it is the activity of neurons in the brain that produces the contents of consciousness, such as cognition and perception. When we drink too much, our decision making becomes impulsive and confused because alcohol is affecting the synaptic transmissions between neurons. When we have strokes or traumatic brain injuries, we may lose our ability to feel empathy because brain tissue has been damaged. Therefore, we can study consciousness only through studying the brain.

Definitions of Consciousness Allow Its Empirical Study

As discussed in Chapter 2, the methods of psychological science dictate that in order to study a phenomenon, we need to be able to define operationally exactly what it is we're studying. In this regard, consciousness has proven difficult to define in scientifically practical terms. In recent years the solution that researchers have converged on is to break down consciousness into more basic issues that are easier to study, which include:

Subjectivity: This is the unique perspective each of us has on our own conscious experience. What we see, hear, feel, and think can be discussed with other people, but the actual *experience* of these things can never be shared in a manner resembling our own internal awareness of them. How can we explain to others what the colour red looks like to us? These aspects of consciousness—the properties of our subjective experience—are referred to as **qualia**, and perhaps the most difficult aspect of a scientific study of consciousness concerns understanding how qualia arise from neural activity (Figure 4.22).

qualia The properties of our subjective, phenomenological awareness.

Access to Information: Mental experience presupposes that we have knowledge of the contents of our consciousness, at least when we are in a normal, unaltered mind state. However, that content is the by-product of the actions of a number of different

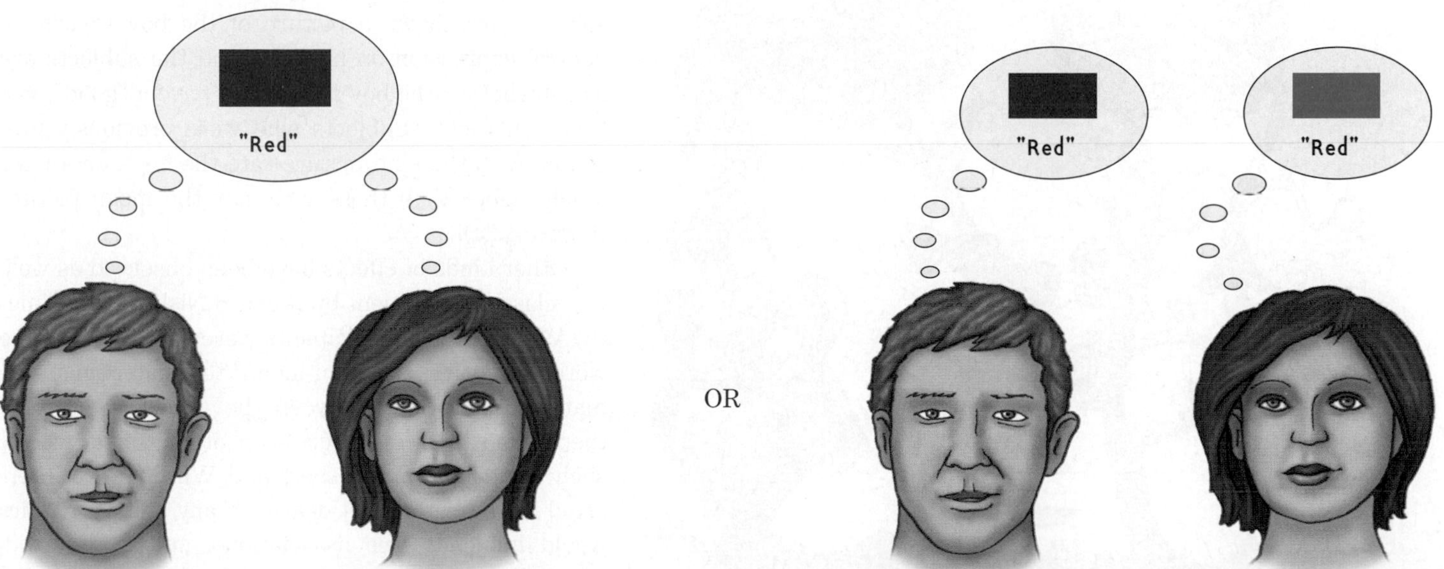

4.22 Seeing Red One of the difficult questions of consciousness is how to account for "qualia," or the phenomenological percepts of the world. For instance, does the colour red appear the same to everyone who has normal colour vision?

What are the elementary properties of consciousness?

unconscious Those processes that are outside the realm of conscious awareness.

underlying processes, of whose workings we have no direct knowledge. We see the products of the visual system, not the computations upon which they are based. What does this mean? Information processing in the brain can be divided into two classes: processes that are accessible to consciousness and processes that are not. The processes that we are not conscious of we label **unconscious**, or outside our mental awareness.

A Unitary Experience: Consciousness brings together the fruits of our sensory systems into a unified phenomenal experience that remains continuous over time. Our awareness of the world is based on a melding of these sensations into a single, multimedia event.

The goal of breaking down consciousness is to facilitate empirical study. For instance, now that we have defined qualia, we may be able to design studies that begin to investigate what parts of the brain are active and what parts are inactive when experiencing specific qualia. We now turn to several different definitions of consciousness and the different empirical approaches they have spurred.

Unconscious Processing Influences Awareness

One of the properties of consciousness is the idea that we are aware of some mental processes and not aware of others. Over the last several decades, one of the more fruitful areas of consciousness research has explored different ways in which unconscious or *subliminal perception* can influence cognition. Subliminal perception refers to stimuli that our sensory systems respond to but that because of their short duration or subtle form never reach the threshold of entering into consciousness (Figure 4.23).

4.23 **Seeing the Subliminal** This design contains a subliminal message that is dramatically obvious when you discover it. Can you find it?

THE CASE FOR UNCONSCIOUS INFLUENCE There is no doubt that stimuli can influence our thoughts and actions even though they stay outside the realm of consciousness (Gladwell, 2005). One often-used illustration of this point involves priming test subjects by flashing before them a picture of a boy who appears either mean or kind. Subjects remain unaware that they have been shown the initial picture. For the main test, subjects are shown a picture of the boy wearing a neutral expression on his face, and the subjects are asked whether the boy appears to have a "good" or a "bad" character. Subjects who were previously presented with the kind picture rate the boy's character much higher than those who saw the mean picture (Figure 4.24).

Other kinds of effects have been observed as well. In a classic experiment by Richard Nisbett and Timothy Wilson (1977), participants were asked to examine word pairs such as "ocean-moon" that had obvious semantic associations between the words. They were then asked to free associate on other, single words, such as "detergent." Nisbett and Wilson were interested in finding to what degree, if any, the word pairs would influence free associations, and if they did, whether the participants would have any awareness that the pairs had indeed influenced their thoughts. What they found was that when given the word "detergent" after the word pair "ocean-moon," participants' typical free association was the word "Tide." Interestingly, they would usually give reasons such as "My mom used Tide when I was a kid"; they had no awareness that they were being influenced by the word pairs. Here again we see the effects of the left hemisphere interpreter at work, making sense of a situation and providing a plausible explanation for cognitive events when complete information is not available. What this reveals is that we are frequently unaware of the different things that can affect our decisions about what we say and do. Similar effects underlie the classic *Freudian slip*, in which unconscious thoughts are suddenly expressed at inappropriate times, often to inappropriate people.

Another example comes from the work of John Bargh and his colleagues (1996), who had subjects make sentences out of words that were jumbled up. Some of the words were associated with the elderly. For instance, the subjects were asked to construct a sentence using "old," "Florida," and "wrinkles." After research participants had made up a number of such sentences, they were told the experiment was over. But instead, what Bargh and his colleagues were interested in was whether the unconscious activation of beliefs about old people would influence the participants' behaviour. They found that participants primed with the elderly stereotype walked much more slowly down the hall than those who performed a sentence task that did not activate beliefs about elderly people. When questioned later, the participants had no awareness that the concept of elderly had been activated or that it had changed their behaviour. Other researchers have obtained similar findings. For instance, Ap Dijksterhuis and Ad van Knippenberg (1998) at Nijmegen University in the Netherlands found that people better answered trivial questions when they were unconsciously presented with information about "professors" than when they were presented with subliminal information about "soccer hooligans." Even how hard people work can be influenced by unconscious cues. Students who receive incidental information that emphasizes "achievement" perform better than those who do not have achievement primed. What these findings indicate is that much of our behaviour occurs without our constant awareness or intention (Ferguson & Bargh, 2004).

4.24 Subliminal Perception If asked to judge the character of the boy in panel B, subliminal perception of panel A1 beforehand will bias people to give a negative report. Alternatively, subliminal perception beforehand of panel A2 will bias people toward a positive report.

THE ILLUSION OF CONSCIOUS WILL Many of our behaviours occur automatically and without much conscious awareness. You walk along without thinking about how to move your legs or swing your arms, you drive home without paying much specific attention to what you are doing, and you engage in breezy conversation even though you are not consciously aware of how you know what to say or how to say it. On other occasions, however, we consciously choose to engage in a specific behaviour—we feel that we willed the action. This usually happens when we are aware of the action before we perform it. In his book *The Illusion of Conscious Will* (2002), Harvard social psychologist Daniel Wegner presents examples of situations in which people have a feeling of will. According to Wegner, our sense of will occurs in part because brain activity causes both the feeling of will and the action that is associated with the feeling of will, even when they did not cause the action. You may have thought that you are thirsty and then have a drink of water. Did your thought cause you to drink? Or did feeling thirsty lead to both the thought of drinking and the action? When letters are chosen on a Ouija board, is it because of supernatural forces or because your fingers are responding to unconscious choices in your mind? According to Wegner, even though people using a Ouija board do not experience a conscious sense of will, it can be shown that they are in fact providing the answers.

Awareness Has Many Seats in the Brain

How does the brain give rise to awareness?

As we have stated, perhaps the greatest challenge in the psychological sciences today is to understand how neural activity gives rise to the conscious awareness we associate with cognition and perception. However, to approach this issue, we need to ask: What exactly is *awareness*? Although there are a host of definitions, they all converge on the notion that to be aware of information is to be able to report that information is being—or has been—perceived. Based on this general definition, a great deal of research over the last several decades has focused on identifying the brain areas involved in *awareness*. By studying awareness in individuals with damage to specific regions of the brain, researchers hope to link selective losses of awareness—that is, awareness for specific forms of information—to the damaged brain areas.

blindsight A condition in which people who are blind have some spared visual capacities in the absence of any visual awareness.

BLINDSIGHT One of the most fascinating areas of study in cognitive neuroscience has concerned the neural basis of visual awareness. In particular, researchers have focused on the phenomenon of **blindsight**, a condition in which people suffer blindness due to damage to their visual cortex but continue to have some visual capacities in the absence of any visual awareness. Although blindsighted patients are exceedingly rare, they typically have loss of vision in only a small portion of their visual field. Researchers have discovered that if these patients are presented with a stimulus in their blind field, they can unconsciously perceive important aspects of the stimulus. For example, patients might be presented with a moving dot in their blind spot, and their task is to indicate in which direction the dot is moving. The typical scenario will have patients declaring that they have seen nothing and thus have nothing to report. However, when pressed to guess the direction of motion, more often than not patients will guess correctly.

A fascinating example of blindsight was recently found in a 52-year-old physician who had become blind following two consecutive strokes that destroyed the primary visual cortices in both hemispheres of his brain (Pegna, Khateb, Lazeyras, & Seghier, 2005). Although there was nothing wrong with his eyes, the visual regions of his brain were unable to process any information they received from them. So although alert and aware of his surroundings, the patient reported being unable to see anything, even the presence of intense light. Visual information also goes to other brain regions, such as the amygdala. As you will learn in Chapter 10, one theory suggests that the amygdala processes visual information very crudely and quickly, which helps identify potential threats. The amygdala, for example, becomes activated in imaging studies when people observe subliminal presentations of faces expressing fear (Whalen et al., 2005). When our blind patient was shown a series of faces and simply asked to guess the expression, although he had no sense of having seen an object, he was able to correctly identify the expression at a much better than chance level. He did not respond to other stimuli (such as shapes, animal faces, or other scary stimuli). When this patient was placed in a brain scanner, his amygdala became activated when presented with emotional faces but not to faces with neutral expressions. Thus, his amygdala may well have been processing the emotional content of the faces, in spite of his lack of awareness of seeing anything at all. This raises intriguing questions of whether the patient is "seeing" the faces, as well as how visual information reaches the amydala when primary visual areas are damaged.

Many of our responses to events around us occur without our conscious awareness of them.

NEURONAL WORKSPACE How does activity within the brain relate to awareness? An influential model of consciousness known as the *neuronal workspace model* posits that consciousness arises as a function of which brain regions are active. When one region is more highly activated than another, the image that reaches consciousness is the one associated with that region's functional activity. A study by Frank Tong and colleagues (1998) looked at the relationship between consciousness and neural responses in the brain. Participants were shown images with houses superimposed on faces (Figure 4.25). The researchers found that the fusiform face area increased activity when participants reported seeing a face, but that when participants reported seeing a house, temporal cortex regions associated with object recognition became active. This finding suggests that the neural correlates of awareness depend on the activity of brain regions that process certain types of sensory information. In the case of rivalry between faces and houses, if the fusiform face area is more highly activated than areas associated with objects, the participant's conscious experience is that of seeing a face, rather than a house.

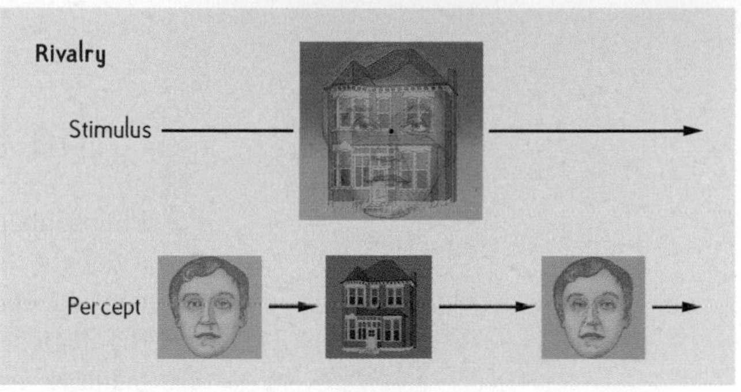

4.25 Consciousness and Neural Responses The picture at the top can be viewed as a face or a house; how it is perceived activates different brain regions.

Work with brain-injured patients reveals a pattern that supports the neuronal workspace model, in which the output of active brain regions is experienced as conscious awareness. If one of these systems is damaged or not working, then there is no output to consider and nothing is noted as being wrong. This can be seen with the phenomenon of hemineglect discussed earlier in the chapter. People with hemineglect are not aware that part of their visual world is missing. In one patient's words: "I knew the word 'neglect' was a sort of medical term for whatever was wrong but the word bothered me because you only neglect something that is actually there, don't you? If it's not there, how can you neglect it?" (Halligan & Marshall, 1998). Cooney and Gazzaniga (2003) explain this phenomenon by arguing that our left hemisphere interpreter can make sense only of information that is available, so even though we might find the behaviour of the hemineglect patients bizarre, they see the state of their world as perfectly normal. Studies like these show us that the experience of the patient with hemineglect does not include awareness of the deficit, which supports the idea that consciousness arises as a result of the brain processes that are active at any point in time.

The importance of the neuronal workspace model is that it demonstrates that there appears to be no single area in the brain responsible for general "awareness." Rather, different areas of the brain deal with different types of information, and each of these systems in turn is responsible for awareness of that type of information (Figure 4.26). Taken from this perspective, consciousness can be viewed as the mechanism that is actively "aware" of information, and which serves the important role of prioritizing what information we need or want to deal with at any moment.

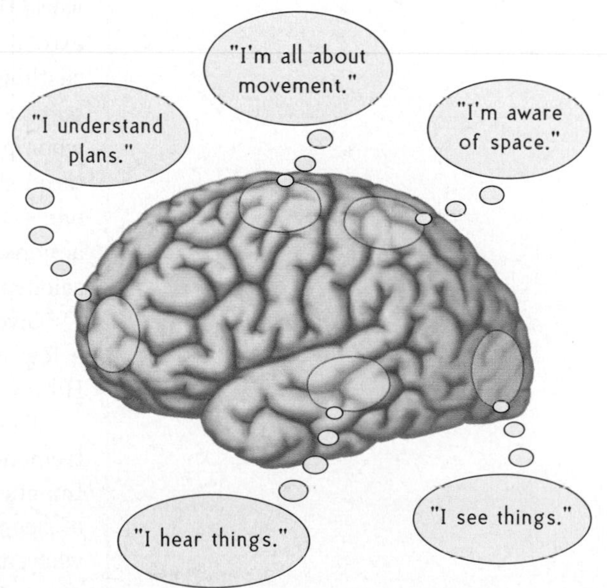

4.26 Areas of Awareness The central theme emerging from cognitive neuroscience is that awareness of different aspects of the world is associated with functioning in different parts of the brain.

REVIEWING

the Principles | Can We Study Consciousness?

Philosophers have long debated whether the mind and brain are separate entities, a debate between dualist and materialist beliefs. However, a scientific study of consciousness is consistent with the materialist view, and researchers have begun the critical task of defining the different elements of consciousness in order to facilitate empirical investigations into its nature. One main area that has been explored concerns the degree to which unconscious processes influence the contents of consciousness. In a more neuroscientific vein, researchers have also explored how awareness of information changes with damage to the brain, such as with blindsight. Results from this domain of study have converged on the theory that different parts of the brain are responsible for awareness of different types of information, findings that challenge the more traditional idea that awareness has a single centre in the brain.

What Is Sleep?

At regular intervals the brain does a strange thing—it goes to sleep. A common misperception is that the brain shuts itself down during sleep, and is no longer processing information from the external world; nothing could be further from the truth. As you will discover, many regions of the brain are more active during sleep than during wakefulness. It is even possible that complex thinking, such as working on difficult problems, occurs while people are asleep. In one study, people had to learn a task at which they slowly got better with practice. Unbeknownst to participants, there was a hidden rule that once discovered led to much better performance. The researchers found that people were more likely to figure out the rule after sleep than after wakefulness, independent of the time of day (Wagner, Gals, Haider, Verleger, & Born, 2004).

Rather than the brain shutting off, it's the conscious experience of the outside world that is apparently turned off during sleep. Even then people are aware to some extent of their surroundings, as when a sleeping mother is aware of her baby rustling in his crib. Many psychological scientists view sleep as an *altered state of consciousness,* one sharing many properties with other altered mental states, such as when people consume drugs or alcohol, meditate, or are under hypnosis. These altered states are associated with unusual subjective experiences, diminished or enhanced levels of self-awareness, and often disturbances in control over physical actions. In this section we focus on sleep as the prototypical altered state of consciousness (we discuss other altered states throughout the text).

Given that all animals sleep, and that people can't go without sleep for more than a few days, most researchers believe that sleep serves some biological purpose. However, as you will see, it isn't clear exactly *why* sleep is important.

The average person sleeps around eight hours per night, although there is tremendous variability, both in terms of individual differences and in terms of age. Infants sleep most of the day, whereas elderly individuals may need only a few hours of sleep per night. Some adults report needing 9 or 10 hours of sleep to feel rested, whereas others report needing only an hour or two a night. One 70-year-old retired nurse, Miss M., reported sleeping only an hour a night. Perhaps, like the researchers who studied Miss M., you find this hard to believe. But after spending two sleepless nights in a sleep laboratory, apparently because of the excitement, she slept for only 99 minutes on the third night, awaking refreshed, cheerful, and full of energy (Meddis, 1977). Imagine having all those extra hours of spare time!

Sleep Is an Altered State of Consciousness

When you are asleep, your brain is still processing information. People who sleep with pets or children tend not to roll over and smother them. Most people also don't fall out of bed while sleeping, indicating that the brain is still aware of the environment, such as the relative position of the edge of the bed. Indeed, even though you are not conscious when asleep, your mind is at work, analyzing potential dangers, controlling body movements, and shifting body parts to maximize comfort. The difference between sleep and awake states has as much to do with conscious experience as biological processes.

Before the discovery of objective methods to assess brain activity, most people believed that the brain went to sleep along with the rest of the body. However, the invention of the electroencephalogram (EEG), a machine that measures the electrical activity of the nervous system, revealed that a great deal goes on in the brain during sleep. The fact that there are different psychophysiological states during sleep has been described as one of the first major discoveries of neuroscience (Hobson, 1995). When people are awake, the neurons in their brains are extremely active, as evidenced by short, frequent, desynchronized brain signals known as *beta waves* (Figure 4.27). When people close their eyes and relax, brain activity slows down and becomes more synchronized, a pattern that produces *alpha waves.*

STAGES OF SLEEP Sleep occurs in stages, as evidenced by changes in EEG readings. As you drift off to sleep, you enter stage 1, characterized by *theta waves,* from which you can be easily aroused. Indeed, if you are awakened you will probably deny that you were sleeping. In this light sleep you might see fantastical images, such as geometric shapes, or have the sensation that you are falling or that your limbs are jerking. As you progress to stage 2, your breathing becomes more regular and you become less sensitive to external stimulation. You are now really asleep. Interestingly, although the EEG continues to show theta waves, there are occasional bursts of activity known as *sleep spindles* and large waves called *k-complexes.* Some researchers believe that these are signals of brain mechanisms involved with shutting out the external world and keeping people asleep (Steriade, 1992). Abrupt noise can trigger k-complexes, and as

4.27 Brain Activity during Sleep Patterns of electrical brain activity during different stages of sleep.

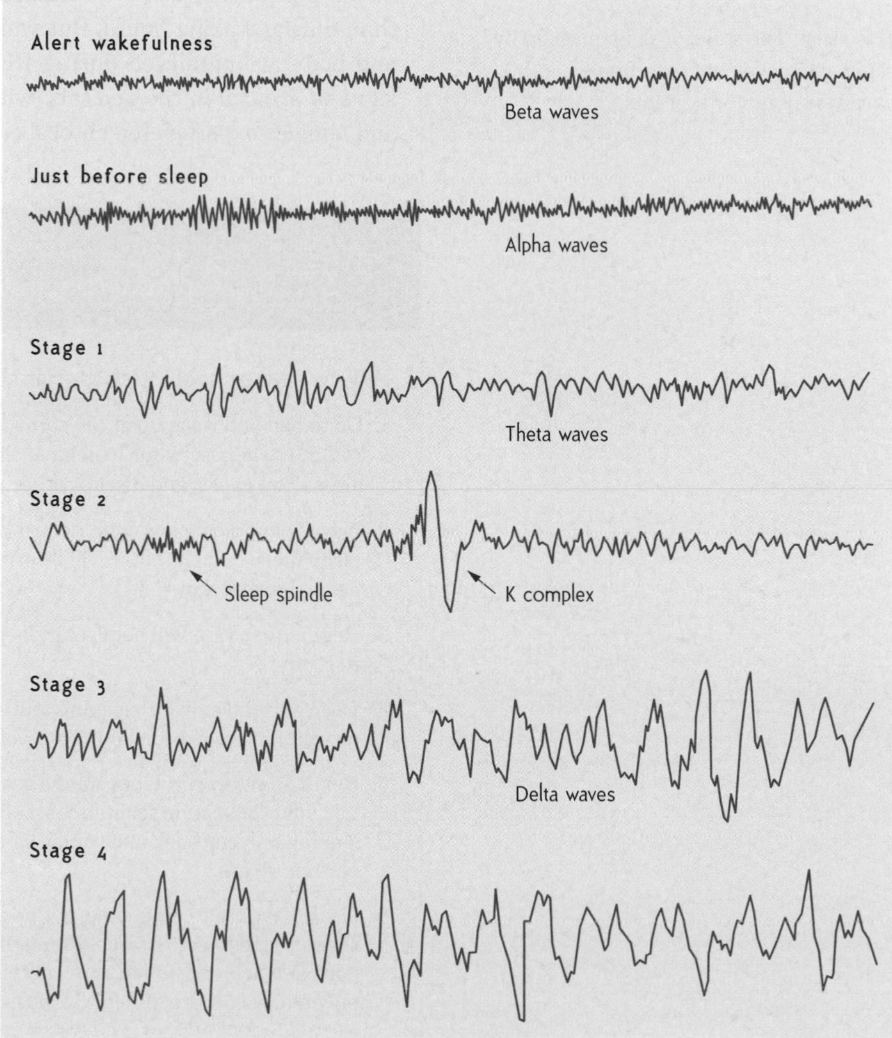

people age they show fewer sleep spindles and sleep much more lightly. This is interesting because it suggests that the brain actually has to work to keep you sleeping. For some people, getting to sleep and staying asleep is difficult. **Insomnia** is a sleep disorder in which people's mental health and ability to function are compromised by their inability to sleep. Ways to deal with insomnia are listed in Table 4.1.

How would you define sleep?

The progression to deep sleep occurs through stages 3 and 4, which are marked by large, regular brain patterns referred to as *delta waves.* This period of sleep is often referred to as *slow-wave sleep,* reflecting the presence of delta waves. It is extremely hard to wake people up if they are in slow-wave sleep, and when they do wake up they are often very disoriented. However, people still process some information in stage 4. Parents can be aroused by crying children but blissfully ignore sirens or traffic noise. The mind continues to evaluate the environment for potential danger.

REM SLEEP After about 90 minutes of sleep, a very peculiar thing happens. The sleep cycle reverses, returning to stage 3 and then to stage 2. At this point, the EEG suddenly indicates a flurry of beta-wave activity that usually indicates an alert, awake mind. At the same time, the eyes start darting back and forth rapidly beneath closed eyelids. This is called **REM sleep** for the *rapid eye movements* that occur during this stage. It is sometimes called *paradoxical sleep* because of the paradox of a sleeping body with an activated brain. Indeed, some neurons in the brain, especially in the occipital cortex and brainstem regions, are more active during REM sleep than during waking hours. But while the brain is active, most muscles in the limbs and body are paralyzed during REM episodes. At the same time, the body shows signs of arousal in the genitals, with most males of all ages developing an erection and females experiencing clitoral engorgement.

insomnia A disorder characterized by an inability to sleep.

REM sleep The stage of sleep marked by rapid eye movements, dreaming, and paralysis of motor systems.

TABLE 4.1 Sleeping through the Night

HOW TO DEVELOP GOOD SLEEPING HABITS

1. Go to bed and wake up at the same time every day, including weekends. Establish a routine to help set your biological clock. Changing the time you wake up each day can alter sleep cycles and disrupt other physiological systems.
2. Never have alcohol or caffeine just before going to bed. Alcohol might help you get to sleep more quickly, but it will interfere with your sleep cycle and cause you to wake up early the next day.
3. Regular exercise will help your sleep cycles, but do not do it immediately before going to sleep.
4. Use your bed only for sleeping and sex. Do not spend time in your bed reading, eating, or watching television. You want your mind to associate your bed with sleeping.
5. Relax. Don't worry about the future. Have a warm bath or listen to soothing music. Learning relaxation techniques, such as imagining you are on the beach, with the sun shining on your back and radiating down your hands, may assist in dealing with chronic stress.
6. If you have trouble sleeping, get up and do something else. Don't force yourself to lie there trying to get to sleep. Remember that one sleepless night won't affect your performance very much and worrying about how you will be affected by not sleeping only makes it more difficult to sleep.

The psychological significance of REM sleep is that about 80 percent of the time that people are awakened during REM sleep they report dreaming, compared to less than half of the time during non-REM sleep (Solms, 2000). As you will see below, the dreams themselves are quite different during the two types of sleep.

Over the course of the night, the sleep cycle repeats, with progression from slow-wave sleep to REM sleep and then back to slow-wave sleep. As morning approaches, the sleep cycle becomes shorter and relatively more time is spent in REM sleep (Figure 4.28). People also briefly awaken many times during the night but do not remember these awakenings in the morning. As people age, they sometimes have more difficulty going back to sleep after awakening.

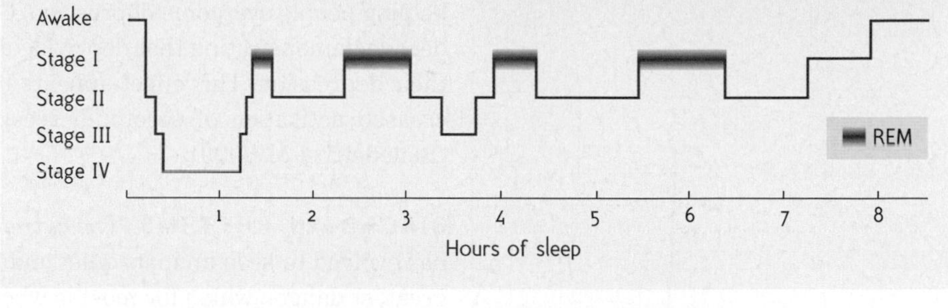

4.28 Stages of Sleep Stages of sleep over the course of the night.

Sleep Is an Adaptive Behaviour

Why do we sleep? It can be dangerous to tune out the external world. If nothing else, it might seem like a huge waste of time that you could spend in more productive ways. But people cannot override indefinitely the desire to sleep; the body shuts down whether we like it or not. Sleep must do something important, because nearly all animals sleep. Some animals have peculiar sleeping styles. Some species of dolphins have *unihemispherical sleep,* in which the cerebral hemispheres take turns sleeping! Researchers do not yet know exactly why animals sleep, but there are three general explanations that are used to describe the adaptiveness of sleep: restoration, circadian cycles, and facilitation of learning.

RESTORATION AND SLEEP DEPRIVATION The *restorative theory* of sleep emphasizes that the brain and body need to rest and that sleep allows the body to repair itself. Indeed, growth hormone is released during deep sleep, and one of its functions is to facilitate repair of damaged tissue. Additional evidence that sleep is a time of restoration is that people who have engaged in vigorous physical activity, such as running marathons, seem to sleep longer. But people sleep even if they spend the day being physically inactive. In addition, it appears that sleep allows the brain to replenish glycogen stores and strengthen the immune system (Hobson, 1999).

Numerous laboratory studies have examined the effects of sleep deprivation on physical and cognitive performance. Surprisingly, most studies find that two or three days of deprivation have little effect on strength, athletic ability, or cognitive performance on complex tasks. Performing boring or mundane tasks when sleep deprived, however, is nearly impossible. A brain-imaging study of sleep-deprived people found increased activation of the prefrontal cortex, suggesting that some brain regions may compensate for the effects of deprivation (Drummond et al., 2000). Over long periods, however, sleep deprivation eventually causes problems with mood and cognitive performance. Indeed, studies using rats have found that extended sleep deprivation compromises the immune system and leads to death.

People who suffer from chronic sleep deprivation, such as many college students, may experience lapses in attention, reduced short-term memory, and **microsleeps**, in which they fall asleep during the day for brief periods of time ranging from a few seconds to as long as a minute. These microsleeps may lead to disastrous results if the person is driving a car or performing a dangerous task.

microsleeps Brief, unintended sleep episodes, ranging from a few seconds to a minute, caused by chronic sleep deprivation.

Interestingly, sleep deprivation might serve one very useful purpose, which is helping people overcome depression. Consistent evidence has emerged over the past decade demonstrating that depriving depressed people of sleep sometimes alleviates their depression. This effect appears to occur because sleep deprivation leads to increased activation of serotonin receptors, as do drugs used to treat depression (Benedetti et al., 1999).

circadian rhythms The regulation of biological cycles into regular patterns.

CIRCADIAN RHYTHMS The *circadian rhythm* theory of sleep proposes that sleep has evolved to keep animals quiet and inactive during times of the day when there is greatest danger, which for most is when it is dark. Physiological and brain processes are regulated into regular patterns known as **circadian rhythms** (*circadian* roughly translates to "about a day"). Body temperature, hormone levels, and sleep-wake cycles are all examples of circadian rhythms, which operate like biological clocks. Circadian rhythms are themselves controlled by cycles of light and dark, although animals continue to show these rhythms when light cues are removed.

According to the circadian rhythm theory, animals need only so much time in the day to accomplish the necessities of survival, and it is adaptive to spend the remainder of the time inactive, preferably hidden away. Accordingly, the amount an animal sleeps depends on how much time it needs to obtain food, how easily it can hide, and how vulnerable it is to attack. Small animals sleep a great deal, whereas large animals vulnerable to attack, such as cows and deer, sleep little. Large animals that are not vulnerable, such as lions, also sleep a great deal. After a fresh kill, a lion may sleep for days on end. Humans, who depend greatly on vision for survival, adapted to sleeping at night, when the lack of light put them in possible danger.

FACILITATION OF LEARNING It has been proposed that sleep may be important because it is involved in the strengthening of neuronal connections that serve as the basis of learning. The general idea is that circuits that have been wired together during the waking period are consolidated, or strengthened, during sleep (Wilson & McNaughton, 1994). Robert Stickgold and colleagues (2000) conducted a study in which they required participants to learn a complex task. They found that participants improved at the task only if they had slept for at least six hours following training. Both slow-wave sleep and REM sleep appeared to be important for learning to take place. The researchers argued that learning the task required neuronal changes that normally occur only during sleep. Indeed, some evidence indicates that students have more REM sleep during exam periods, when it might be expected that greater consolidation of information is taking place (Smith & Lapp, 1991).

The argument that sleep, especially REM sleep, promotes development of brain circuits for learning is also supported by the changes in sleep patterns that occur over the life course. Infants and the very young, who learn an incredible amount in a few short years, sleep the most and also spend the most time in REM sleep.

Sleep and Wakefulness Are Regulated by Multiple Neural Mechanisms

Multiple neural mechanisms are involved in producing and maintaining circadian rhythms and sleep. For instance, the biological clock is located in the suprachiasmatic nucleus (SCN) of the hypothalamus. Light-sensitive photoreceptors in the eye send signals to the SCN. These photoreceptors are different from those that provide visual information, and they may even work for people who are otherwise blind. Individual neurons within the SCN seem to be the neural basis of the biological clock, as they show a peak in firing activity about once a day (Welsh, Logothetis, Meister, & Reppert, 1995). The SCN also signals a tiny structure called the *pineal gland* (Figure 4.29) to secrete melatonin, a hormone that travels through the bloodstream and affects various receptors in the body and the brain. Bright light suppresses production of melatonin, whereas darkness triggers its release. It has recently been noted that taking melatonin can help people cope with jet lag and shift work, both of which interfere with circadian rhythms. Taking melatonin also appears to help people get to sleep, although it is presently unclear why this happens.

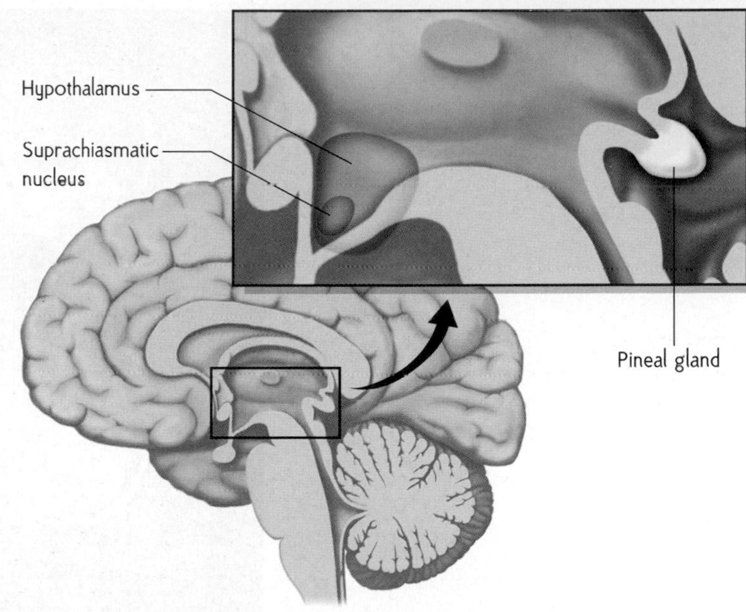

4.29 The Biological Clock The suprachiasmatic nucleus houses the biological clock. It signals the pineal gland to secrete melatonin, which affects bodily states related to being tired.

BRAINSTEM AND AROUSAL Sleep involves alterations in states of arousal, thereby implicating brain mechanisms that produce aroused states. In 1949, Giuseppe Moruzzi and Horace Magoun found that stimulating the *reticular formation* in the brainstem leads to increased arousal in the cerebral cortex. If you cut the fibres from the reticular formation to the cortex, animals fall asleep and stay asleep until they die. Accordingly, Moruzzi and Magoun proposed that low levels of activity in the reticular formation produce sleep, and high levels lead to awakening.

It is now known that multiple regions within the reticular formation participate in the control of sleep-wake cycles. For instance, one region sends the neurotransmitter norepinephrine to many regions of the brain, thereby increasing cortical arousal. Research has demonstrated that this region is highly active when animals are awake, less active during non-REM sleep, and almost completely inactive during REM sleep (Figure 4.30). The system is also involved in sleep.

What triggers sleep? There is some evidence that a small area of the forebrain, just in front of the hypothalamus, is involved in inducing non-REM sleep. Neurons in this region become more active during non-REM sleep, and lesioning this region leads to insomnia. Once this area is activated, inhibitory signals are sent to the reticular formation, thereby reducing arousal and triggering sleep.

REM SLEEP During REM sleep, various neural processes are activated, of which some lead to paralysis of motor systems whereas others lead to activation of mental circuits related to motivational states. For instance, neurons in the *pons,* a region of the brainstem, send signals to the spinal cord that block movement during REM sleep. Surgical lesioning of the pons causes animals to become very active while in REM sleep, as if acting out scenes from their dreams. People who have strokes that damage the brainstem sometimes develop *REM behaviour disorder,* in which the normal muscle paralysis that accompanies REM is absent. Unfortunately for their sleeping partners, these individuals can become quite violent during REM episodes. In contrast, people with *cataplexy* experience REM muscle paralysis while awake; their muscles can suddenly become flaccid and they collapse to the floor.

REM sleep is triggered by acetylcholine neurons in the pons. In the minute before REM episodes, these neurons become increasingly active, eventually producing

4.30 Neuronal Activity Activity of neurons in one regional reticular formation during wakefulness and stages of sleep.

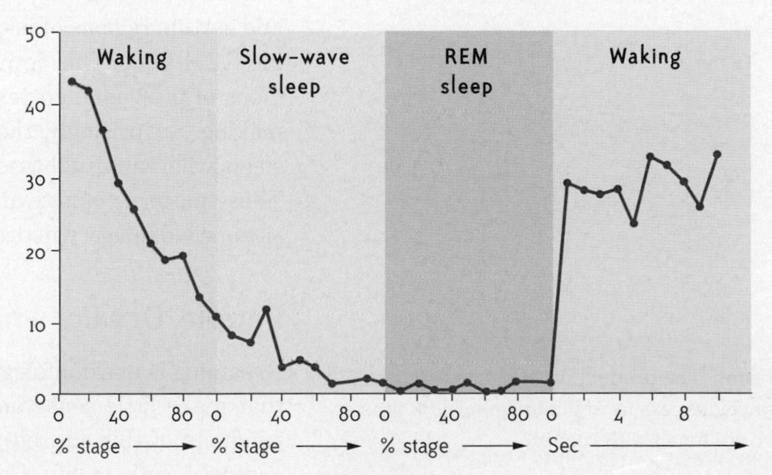

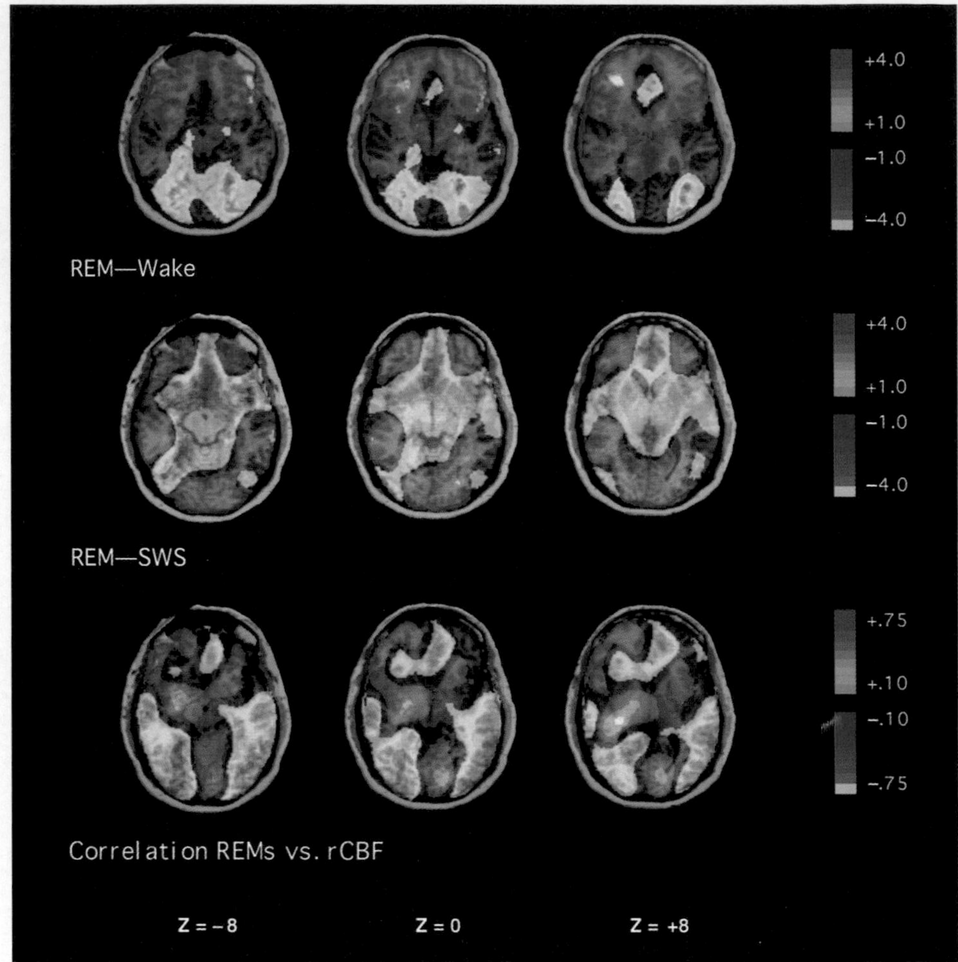

4.31 PET Scans Patterns of activation using PET, comparing REM to slow-wave sleep and waking. Note that there is greater activation (indicated by yellow and red) in limbic regions during REM than during slow-wave sleep. There is also less activation (indicated by purple) of much of the prefrontal cortex.

REM. Signals from this region are transmitted to the thalamus and the occipital lobes and also appear to trigger the eye movements that give REM sleep its name.

Brain-imaging studies show activation of limbic structures, such as the amygdala and certain regions of prefrontal cortex (i.e., the middle region behind the eyes), during REM sleep. This activation occurs along with a lessening of activity of other regions of prefrontal cortex, areas typically involved in rational thought and decision making. Surprisingly, the primary visual areas seem to be inhibited during REM sleep, whereas visual association areas are activated (Braun et al., 1998; Figure 4.31). These patterns of activation may produce one of the hallmark associations of REM sleep, vivid imagery in dreaming.

People Dream while Sleeping

dreams The product of an altered state of consciousness in which images and fantasies are confused with reality.

Dreaming is one of life's great mysteries, as the mind conjures up images and stories that make little sense and sometimes scare the dreamer awake. **Dreams** are the products of this altered state of consciousness in which images and fantasies are confused with reality. Usually, people realize they were dreaming only when they wake up. Some people claim not to remember their dreams, but unless a person has a specific brain injury or is taking medication, everyone dreams. Indeed, the average person spends six full years of his or her life dreaming. People who have trouble remembering their dreams can be taught to do so, mostly by trying to write down their dreams as soon as they wake up.

Dreams occur both in REM and non REM sleep, although the contents of dreams differ in the two types of sleep. Non-REM dreams are often very dull, such as deciding what clothes to wear or thinking about taking notes in class. In contrast, REM dreams are often bizarre. They involve intense emotion, visual and auditory hallucinations (but rarely taste, smell, or pain), illogical content, and an uncritical acceptance of events. Rules of time and space and physical laws are ignored in dreams.

When researchers first noticed rapid eye movements in the 1950s, they initially believed that dreams were mainly a product of REM sleep. Initially, non-REM dreams were dismissed as trivial or as based on faulty recollection. However, it is now clear that REM sleep can occur without dreaming and dreams can occur without REM sleep. Moreover, REM and dreaming appear to be controlled by different neural signals (Solms, 2000). Activation of different brain regions during REM and non-REM sleep may be responsible for their different types of dreams. The content of REM dreams is a result of the activation of brain structures associated with motivation, emotion, and reward along with the visual association areas. This closed loop allows the brain's emotion centres and visual association areas to interact without self-awareness, reflective thought, or input from the external world.

manifest content The plot of a dream; the way a dream is remembered.

latent content What a dream symbolizes, or the material that is disguised in a dream to protect the dreamer.

activation-synthesis hypothesis A theory of dreaming that proposes that neural stimulation from the pons activates mechanisms that normally interpret visual input.

WHAT DO DREAMS MEAN? Sigmund Freud, in one of the first major theories of dreams, argued that dreams contain hidden content that represents unconscious conflicts. According to Freud, the **manifest content** is the dream the way the dreamer remembers it, whereas the **latent content** is what the dream symbolizes, or the material that is disguised to protect the dreamer. However, there is virtually no support for Freud's idea that dreams represent hidden conflicts or that objects in dreams have special symbolic meanings. What is true is that daily life experiences influence the content of dreams. That is, you may be especially likely to have anxiety dreams while studying for exams. The dreams themselves have a thematic structure, in that they unfold as events or stories rather than as jumbles of disconnected images, but there is apparently no secret meaning to the way the story is told.

What do dreams really mean?

ACTIVATION-SYNTHESIS HYPOTHESIS The sleep researcher Alan Hobson proposed an influential theory that has dominated thinking about dreaming for the past two decades. Hobson's **activation-synthesis hypothesis** proposes that neural stimulation from the pons activates mechanisms that normally interpret visual input. The sleeping mind tries to make sense of random neuronal firing by synthesizing apparent activity in visual and motor neurons with stored memories. From this perspective, then, dreams are *epiphenomenal*—they are the experienced side effects of mental processes. Hobson and his colleagues (2000) have recently revised the activation-synthesis model to take into account recent findings in cognitive neuroscience. For instance, they have included amygdala activation as the source of the emotional content of dreams, and they propose that deactivation of the frontal cortices contributes to the delusional and illogical aspect of dreams. By its nature, the activation-synthesis hypothesis is more concerned with REM than non-REM dreams.

"Look, don't try to weasel out of this. It was my dream, but you had the affair in it."

EVOLVED THREAT-REHEARSAL STRATEGIES Antti Revonsuo (2000) has proposed an evolutionary account wherein dreams simulate threatening events to allow people to rehearse coping strategies. Although Revonsuo does not believe that all dreams perform this function, he does suggest that dreams that provided people with solutions to adaptive problems helped them to survive and reproduce; thus dreaming is a result of natural selection. That the majority of dreams reported by people involve negative emotions, such as fear and anxiety, supports the evolved threat-rehearsal theory. Dreaming is associated with the activation of limbic structures, such as the amygdala, that are also activated during real threat encounters. Moreover, people tend to dream about real threats in their lives and to have nightmares about past traumas for long periods of time.

REVIEWING the Principles | What Is Sleep?

All animals experience sleep, an altered state of consciousness in which the sleeper loses most contact with the external world. Sleep has a number of stages that can be identified by different patterns on EEG recordings. There is a basic distinction between non-REM and REM sleep, and different neural mechanisms are responsible for producing each type, although the brainstem figures prominently in the regulation of sleep-wake cycles. Dreams occur in REM and non-REM sleep, although the content of those dreams differs. This may be due to differential activation of brain structures associated with emotion and cognition. Although a number of theories have been proposed to explain sleeping and dreaming, their biological function is currently unclear.

Conclusion

The brain is a unique and remarkable set of structures that have collectively evolved to control the organism as it goes about the business of survival and reproduction. In humans, this evolutionary adaptiveness has led to the development of complex communication, culture, and thought. Our knowledge about the brain has expanded rapidly and dramatically over the past few decades, drawing on research from all the levels of analysis. For more than 100 years scholars debated whether psychological processes were located in specific parts of the brain or distributed throughout it. Along the way, they also built a formidable base of knowledge and principles. It was not until the 1990s, with the results of brain-imaging studies, among other methods, that it became clear that many mental processes are localized to specific regions of the brain that work together to produce behaviour and mental activity. Research on the brain has revealed many fascinating phenomena, such as phantom limbs, split brains, and a left brain interpreter that struggles to make sense of the world. Brain-imaging techniques have also allowed psychological scientists to begin investigating consciousness, the personally familiar but scientifically elusive mental process that in some sense defines our entire experience with the world. Similarly, being able to watch the working brain has taught us an incredible amount about altered states of consciousness such as sleep and dreaming. Although much remains to be discovered, scientists using the tools of the biological revolution are making great and rapidly accelerating progress in understanding how the circuits of the brain enable the mind.

Summarizing the Principles of the Brain and Consciousness

What Are the Basic Brain Structures and Their Functions?

1. **The spinal cord is capable of autonomous function:** The spinal cord is a bundled rope of neurons that responds to sensory inputs and directs muscle responses (reflexes).
2. **The brainstem houses the basic programs of survival:** The top of the spinal cord at the base of the skull forms the brainstem, which is involved in basic functions and general arousal.
3. **The cerebellum is essential for movement:** The cerebellum, the bulging structure connected to the back of the brainstem, is essential for balance and movement.
4. **Subcortical structures control basic drives and emotions:** The subcortical structures play a key part in psychological functions because they control vital functions (the hypothalamus), sensory relay (the thalamus), memories (the hippocampus), emotions (the amygdala), and planning and producing movement (the basal ganglia).
5. **The cerebral cortex underlies complex mental activity:** The lobes of the cortex play specific roles in controlling vision (occipital), touch (parietal), hearing and speech comprehension (temporal), and planning and movement (frontal).

How Does the Brain Change?

6. **The interplay of genes and the environment wires the brain:** Chemical signals influence the growth and function of cells. Environmental experiences, especially during critical periods, are necessary for cells to develop properly and for them to make more detailed connections.
7. **The brain rewires itself throughout life:** Although plasticity decreases with age, the brain retains the ability to rewire itself throughout life. This ability is the biological basis of learning.
8. **The brain can recover from injury:** The brain can reorganize the function of areas in response to brain damage, although this capacity decreases with age. Transplanted stem cells may be able to grow to replace damaged neurons.

How Is the Brain Divided?

9. **The hemispheres can be separated:** A bundle of axons (the corpus callosum) connects the two sides of the brain; the connection can be cut, resulting in two independently functioning sides.
10. **The separate hemispheres can be tested:** Testing of split-brain patients has revealed the functions of each hemisphere.
11. **The hemispheres are specialized:** The left hemisphere is primarily responsible for language and the right hemisphere is competent in images and spatial relations.
12. **The mind is a subjective interpreter:** Our left hemisphere interprets and strives to make sense of our experiences, influencing the way we view and remember the world.

Can We Study Consciousness?

13. **Definitions of consciousness allow its empirical study:** To better study consciousness, researchers have divided the topic into basic issues, including subjectivity, access to information, and the unitary aspect of experience.
14. **Unconscious processing influences awareness:** Research findings indicate that much of our behaviour occurs automatically, without our constant awareness.
15. **Awareness has many seats in the brain:** Blindsight demonstrates visual ability without awareness. The neuronal workspace model of consciousness demonstrates how awareness depends on activity in a variety of different cortical areas.

What Is Sleep?

16. **Sleep is an altered state of consciousness:** The stages of sleep vary by brain and respiration actitivity. REM sleep activates the brain and produces body paralysis and genital stimulation.
17. **Sleep as an adaptive behaviour:** Sleep restores the body and circadian rhythms control changes in body function and sleep. Learning is consolidated during sleep.
18. **Sleep and wakefulness are regulated by multiple neural mechanisms:** The brainstem structures are involved in arousal (the reticular formation) and REM sleep (the pons).
19. **People dream while sleeping:** REM and non-REM dreams involve activation of different areas of the brain. Sigmund Freud thought dreams revealed unconscious conflicts. The activation-synthesis hypothesis posits that dreams are side effects of brain activity. Antti Revonsuo has theorized that dreaming is adaptive.

Applying the Principles

1. You are riding your mountain bike, listening to your iPod, and thinking about your plans for spring break, when suddenly you are frightened by a barking dog that starts chasing you. What parts of the central nervous system would be primarily responsible for each of these underlined behaviours?

   ____ a) cerebellum; temporal lobes; prefrontal cortex; amygdala
   ____ b) reticular formation; parietal lobe; occipital lobe; hippocampus
   ____ c) frontal lobes; temporal lobes; cerebellum; amygdala
   ____ d) occipital lobe; amygdala; prefrontal cortex; cerebellum

2. In your sociology class, while watching a video that deals with the conditions of infants in institutional care in Romania, you are struck by the limited interaction the children have with others, and by the complete lack of stimulating materials in their environment. The children appear listless and withdrawn. Based on what you have learned about the development of the brain, you correctly conclude that

   ____ a) their brains will not develop, leaving them like newborns.
   ____ b) their development will be significantly slowed, since the brain is "wired" by experience rather than genetics.
   ____ c) their brain development can catch up later in life, when they are placed in more stimulating environments.
   ____ d) their brains will develop normally since genes, not experiences, control development.

3. You are right-handed, but after breaking your right hand you are forced to use your left hand for several months. Initially you are able to do very little with your left hand, but with practice, you develop adequate skills. This is an example of

   ____ a) the human brain's preference that most motor skills be done by the right hand.
   ____ b) the simultaneous damage that occurs in the left cortex when the right hand is damaged.
   ____ c) the opposite-side control of the body by the brain and the ability of each hemisphere to specialize and to assume new functioning.
   ____ d) the left side of your brain, which controls the left hand, rewiring itself for the new skills.

4. You stayed up studying all night right before an exam on formal logic for your philosophy class. When you get back the graded exam, you discover that you performed poorly compared to your roommate, who had hit the sack at midnight. Which of the following reasons might account for this difference in performance?

   ____ a) One of the roles of sleep is consolidation of memories and learning.
   ____ b) The horizontal position during sleep enabled a beneficial blood flow pattern for the brain.
   ____ c) The lack of sleep made it hard for you to have enough motor control to write your exam responses.
   ____ d) The lack of sleep made your mind drift into REM activity, thus interfering with conscious thought.

ANSWERS: *1.a 2.b 3.c 4.a*

Key Terms

activation-synthesis hypothesis, p. 153
amygdala, p. 124
basal ganglia, p. 125
blindsight, p. 144
brainstem, p. 121
Broca's area, p. 120
cerebellum, p. 122
cerebral cortex, p. 125
circadian rhythms, p. 150
corpus callosum, p. 137
critical period, p. 130
dreams, p. 152
frontal lobes, p. 127
gray matter, p. 121
hippocampus, p. 124
hypothalamus, p. 123
insomnia, p. 148
latent content, p. 153
manifest content, p. 153
microsleeps, p. 149
occipital lobes, p. 125
parietal lobes, p. 126
phrenology, p. 119
plasticity, p. 129
prefrontal cortex, p. 127
qualia, p. 141
REM sleep, p. 148
reticular formation, p. 122
spinal cord, p. 121
split brain, p. 137
temporal lobes, p. 127
thalamus, p. 123
unconscious, p. 142
white matter, p. 121

Further Readings

Brodal, P. (1998). *The central nervous system.* Oxford, U.K.: Oxford University Press.

Finger, S. (1994). *Origins of neuroscience.* Oxford, U.K.: Oxford University Press.

Gazzaniga, M. S., Ivry, R. B., & Mangun, G. R. (2002). *Cognitive neuroscience: The biology of the mind* (2nd ed.). New York: Norton.

Gladwell, M. (2005). *Blink.* New York: Little, Brown.

Kolb, B., & Whishaw, I. Q. (1998). *Fundamentals of human neuropsychology.* New York: Freeman.

McConnell, S., Roberts, J., Spitzer, L., Zigmond, M., Squire, L., & Bloom, F. (2002). *Fundamental neuroscience* (2nd edition). San Diego, CA: Academic Press.

Ramachandran, V. S., & Blakeslee, S. (1998). *Phantoms in the brain: Probing the mysteries of the human mind.* New York: William Morrow.

Sacks, O. (1985). *The man who mistook his wife for a hat.* New York: Summit Books.

Wegner, D. M. (2002). *The illusion of conscious will.* Cambridge, MA: MIT Press.

Further Readings: A Canadian Presence in Psychology

Cheeseman, J., & Merikle, P. M. (1986). Distinguishing conscious from unconscious perceptual processes. *Canadian Journal of Psychology, 40,* 343–367.

Goddard, G. V., McIntyre, D. C., & Leech, C. K. (1969). A permanent change in brain function resulting from daily electrical stimulation. *Experimental Neurology, 25,* 295–330.

Hebb, D. O. (1949). *The organization of behavior.* New York: Wiley.

Kolb, B., & Whishaw, I. Q. (1998). Brain plasticity and behavior. *Annual Review of Psychology, 49,* 43–64.

Liu, D., Diorio, J., Day, J. C., Francis, D. D., & Meaney, M. J. (2002). Maternal care, hippocampal synaptogenesis, and cognitive development in rats. *Nature Neuroscience, 3,* 799–806.

Melzack, R., Israel, R., Lacroix, R., & Schultz, G. (1997). Phantom limbs in people with congenital limb deficiency or amputation in early childhood. *Brain, 120,* 1603–1620.

Penfield, W., & Jasper, H. (1954). *Epilepsy and the functional anatomy of the human brain.* Boston: Little, Brown.

Smith, C. A., & Lapp, L. (1991). Increases in number of REMs and REM density in humans following an intensive learning period. *Sleep, 14,* 325–330.

Steriade, M. (1992). Basic mechanisms of sleep generation. *Neurology, 42* (Suppl.), 9–18.

Sensation and Perception

Sensation and perception bridge the physical and psychological worlds. The museum visitors pictured here experience stimuli (e.g., the colours of the painting, the sounds of the crowds in the gallery room), which are first detected by sensory organs (such as the eyes or ears) before they are processed by the brain to produce a conscious, internal experience. The way in which our sensory organs convert information about the physical environment into signals that the brain can understand is an adaptive system developed over the course of human evolution.

CHAPTER

5

Sensation, Perception, and Attention

Imagine yourself to be Helen Keller at the age of six. Ever since you were 19 months of age, your life has proceeded in complete darkness and perpetual silence. For you the world exists only through touch, smell, and taste. You can recognize your parents by the feel of their skin and clothing, and you can tell where you are by smell and by touching objects that are around you. But you are otherwise completely isolated. You realize that others can communicate, but you are locked out. You are so enraged and frustrated by your mental prison that you throw daily tantrums. Your parents seek assistance from Alexander Graham Bell, the inventor of the telephone who has worked with deaf children, and he puts them in touch with the Perkins school for the blind in Watertown, Massachusetts. Through the school, they hire a teacher, Anne Sullivan, to teach you to communicate with signs. She makes strange sequences of hand motions in your hands. You mimic them, but they make no sense. Then one day she runs cool well water

Outlining the Principles

How Do We Sense Our Worlds?

- Stimuli Must Be Coded to Be Understood by the Brain
- Psychophysics Relates Stimulus to Response

What Are the Basic Sensory Processes?

- In Gustation, Taste Buds Are Chemical Detectors
- In Smell, the Nasal Cavity Gathers Particles of Odour
- In Touch, Sensors in the Skin Detect Pressure, Temperature, and Pain
- In Hearing, the Ear Is a Sound-Wave Detector
- In Vision, the Eye Detects Light Waves

What Are the Basic Perceptual Processes?

- Perception Occurs in the Brain
- Object Perception Requires Construction
- Depth Perception Is Important for Locating Objects
- Size Perception Depends on Distance Perception
- Motion Perception Has Both Internal and External Cues
- Perceptual Constancies Are Based on Ratio Relationships

How Does Attention Help the Brain Manage Perceptions?

- Visual Attention Is Selective and Serial
- Auditory Attention Allows Selective Listening
- Selective Attention Can Operate at Multiple Stages of Processing

5.1 Keller and Sullivan Helen Keller (left) with her teacher, Anne Sullivan.

over one hand while spelling w-a-t-e-r in the other. You finally make the connection. You then grab some dirt and ask for that to be spelled. By evening you have learned your first 30 words, and have begun a life of passionate learning and social activism (Figure 5.1).

Now imagine that you are not only blind and deaf but also unable to experience touch, or smell, or taste. You still feel hunger and other bodily sensations, such as fatigue, but for you there is no world beyond this. You have no way of knowing about the existence of other people or an environment outside your body. You are truly alone and no one can reach you, and you can reach no one. What can you do without perception?

Perception is the only bridge we have to the world. Without it we would be prisoners in the loneliest of solitary confinements. In order to perceive the world, we rely on information provided by our sense organs—our eyes, ears, skin, nose, and tongue. Each of these organs is sensitive to different physical stimuli, and each contributes different information. Ultimately, our perceptual representation of the environment is limited by the kinds of stimuli to which we are sensitive, and by the limits of our sensory systems in responding to those stimuli. And yet, we are sensitive to only a tiny fraction of the stimulus energy surrounding us. What we call light is far less than 1 percent of the electromagnetic spectrum, which also includes radio and television waves, x-rays, ultraviolet and infrared light, cosmic rays, and microwaves. We cannot sense magnetic fields, for example, or experience the world with sonar, as do bats and dolphins, or see the polarization of light, as do certain insects, or experience objects at a distance using electric fields, as do some fish. Even with fully functioning sensory organs we are blind to the vast majority of energy, events, and information surrounding us. If we are blind to most of what surrounds us, why does this not bother us? How would we know we are blind? Unless we have instruments, such as radios and televisions, that can convert these "invisible" forms of energy into forms that we can detect, it is for us as if this energy and information did not exist. Yet evolution has bestowed on us a system that allows us to flourish in our particular environment, providing rich sensory experiences of the world around us. ■

sensation How sense organs respond to external stimuli and transmit the responses to the brain.

perception The processing, organization, and interpretation of sensory signals that result in an internal representation of the stimulus.

Psychological scientists often divide the way we experience the world into two distinct phases: sensation and perception. **Sensation** refers to how our sense organs respond to and detect external stimulus energy (lights, air vibrations, odours, and so

c. 1500
Art and Perception Leon Alberti describes the use of linear perspective and other depth cues to create the impression of depth in representational art.

1672
The Nature of Light Sir Isaac Newton performs his prism experiment, demonstrating that white light is in fact composed of a mixture of the spectral colours.

1709
The Empiricist View of Perception In his *New Theory of Vision* George Berkeley outlines the Empiricist view that perception is learned through experience with the world.

1838
The Stereoscope Charles Wheatstone invents the stereoscope and documents the role of binocular vision in spacial perception.

on), and how those responses are transmitted to the brain. **Perception** refers to the brain's further processing of these detected signals that ultimately results in an internal representation of stimuli and a conscious experience of a world. Whereas the essence of sensation is detection, the essence of perception is construction of useful and hopefully meaningful information about the environment. For example, a green light emits photons that are detected by specialized neurons in the eyes, which transmit signals to the brain (sensation). The brain processes those neural signals and the observer experiences a green light (perception).

The study of sensation and perception, our focus in this chapter, is the study of those bodily systems that convert stimulus energy into useful information. We will study how various types of stimulus energy are detected by the sense organs, how the brain constructs information about the world on the basis of what has been detected, and how we use this constructed information to guide ourselves through the world around us. Many times, perception is based on our prior experience, which shapes our expectations during identification of sensory experiences. Thus, you are unlikely to see a blue, apple-shaped object as an apple, because you know from past experience that apples are not blue. An important lesson that will emerge in this chapter is that our perception of the world does not work like a tape recorder or camera, faithfully and passively capturing the physical properties of stimuli we encounter. Rather, what we *see* or *hear* is the result of brain processes that actively construct perceptual experiences to allow adaptation to the environment. This system can get the details wrong, such as filling in information that doesn't exist, but it does so in an intelligent and efficient way that produces a meaningful understanding of what is going on in the world around us.

How is information from stimuli in the world transformed into neural activity in the brain?

How Do We Sense Our Worlds?

Our sensory organs gain information about the environment by converting forms of physical energy into signals that the brain can understand. This system has been shaped over the course of evolution to solve adaptive problems, such as providing information on potential dangers or assisting in the identification of items that are vital for survival, such as edible foods and fresh water. Because they faced different adaptive challenges, each animal species is sensitive to different types of physical energy. Animals that hunt at night have poor vision but superb hearing. Unlike humans, many insects can perceive ultraviolet light, which is useful because such light reveals their primary food source. This is important to keep in mind as we focus mainly on human sensation and perception. We sense and perceive energy information that has solved adaptive problems for humans over the course of evolution.

1860
Psychophysics Gustav Theodor Fechner publishes *Elements of Psychophysics*, laying the foundations for the systematic investigation of sensory processes.

1867
Searching for Biological Foundations Hermann von Helmholtz publishes his seminal *Handbook of Physiological Optics*, integrating the physics, physiology, and psychology of visual perception.

1907
Birth of the Gestalt School Max Wertheimer's experiments with Phi motion mark the beginning of the Gestalt movement in perceptual psychology.

1950
Sensation and the Brain Wilder Penfield and Theodore Rasmussen publish *The Cerebral Cortex of Man*, describing cortical stimulation experiments during neurosurgery that evoke sensory experiences.

1961
A Theory of Hearing Georg von Békésy is awarded the Nobel prize for his research into the mechanisms of hearing. His theory articulates how sound waves are transformed into neuronal information.

Stimuli Must Be Coded to Be Understood by the Brain

How is energy converted into sensation?

The way our sensory organs translate a stimulus's physical properties into neural impulses is called *sensory coding*. Different features of the physical environment are coded by different patterns of neural impulses. Thus, a green stoplight will be coded by a particular pattern of neural responses in the retina of the eye; when the hand touches a hot skillet, other neurons will signal pain; even when you feel the most intense of physical pleasures, that is still due entirely to neurons firing away in your brain. Recall that receptors are specialized neurons in the sense organs that pass impulses to connecting neurons when they receive some sort of physical or chemical stimulation. This process is called **transduction**. After transduction in the receptors, connecting neurons in the sense organs transmit information to the brain in the form of neural impulses. As you will recall from the last chapter, most sensory information first goes to a structure in the middle of the brain called the thalamus. Neurons in the thalamus then send information to the cortex, where incoming neural impulses are interpreted as sight, smell, sound, touch, or taste. Table 5.1 lists the stimuli, receptors, and pathways for each major sensory system. We will address the issue of how the brain interprets these impulses later in the chapter when we discuss perception.

transduction A process by which sensory receptors produce neural impulses when they receive physical or chemical stimulation.

Sensations refer to the transduced messages that are carried by nerve impulses. Sensory coding can be divided into two categories: quantitative and qualitative. Coding for *quantitative* factors, such as intensity, brightness, and loudness, is often indexed by the neural firing frequency (Figure 5.2). The higher the firing frequency, the brighter or louder the stimulus. The number of neurons triggered by the stimulus also contributes to quantitative coding in that more intense stimuli tend to recruit more neurons.

Although quantitative sensory coding is useful for some dimensions of a sensation, it is less useful for others, such as colour or taste. For these tasks, the brain relies on *qualitative* sensory coding. Qualitative coding is possible because different sensory receptors respond to different qualities of a stimulus. The simplest form of qualitative sensory coding would be to have a dedicated receptor type for every possible stimulus. So, for example, one class of visual receptors might respond only to red light, another class might respond only to purple light, and so on. Obviously, for this to be strictly true we would need to have an enormous number of different receptor types for each sensory modality. In most sensory systems—with the possible exception of olfaction (the sense of smell)—receptors provide what is called *coarse coding,* in which sensory qualities are coded by only a few receptors, each of which responds to a broad range of stimuli. Only by comparing and integrating activity across the whole range of receptors do we compute the final percept.

5.2 Neural Firing Frequency A more intense stimulus (e.g., brighter light) leads to a greater frequency of neural responses during an identical time period.

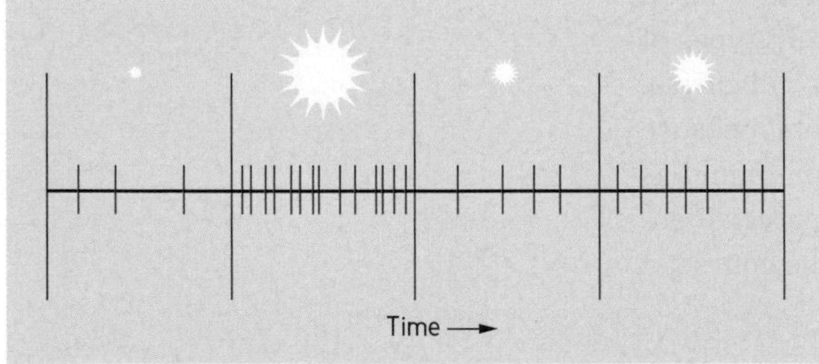

1962
Exploring the Neuronal Basis of Vision David Hubel and Torsten Wiesel describe the functional anatomy of the primary visual cortex of cats. They receive a Nobel prize for this work in 1981.

1966
Evolutionary Aspects of Perception James J. Gibson outlines his theory of direct perception proposing that perceptual systems must be understood in terms of the problems they evolved to solve.

1982
Birth of Computational Vision David Marr publishes his monograph *Vision,* emphasizing the computational approach to the study of visual perception.

1987
Feature Integration Theory Anne Treisman investigates aspects of conscious and unconscious visual perception. She articulates the role of attention in perceiving objects.

1990s–2000s
Perception as Construction Vision scientists increasingly emphasize that the brain actively constructs our perception of the world.

TABLE 5.1 The Stimuli, Receptors, and Pathways for Each Sense

SENSE	STIMULUS	RECEPTORS	PATHWAY TO THE BRAIN
Hearing	Sound waves	Pressure-sensitive hair cells in cochlea of inner ear	Auditory nerve (8th cranial nerve)
Vision	Light waves	Light-sensitive rods and cones in retina of eye	Optic nerve (2nd cranial nerve)
Touch	Pressure on the skin	Sensitive ends of touch neurons in skin	Trigeminal nerve (5th cranial nerve) for touch above the neck. Spinal nerves for touch elsewhere.
Pain	Wide variety of potentially harmful stimuli	Sensitive ends of pain neurons in skin and other tissues	Trigeminal nerve (5th cranial nerve) for pain above the neck. Spinal nerves for pain elsewhere.
Taste	Molecules dissolved in fluid on the tongue	Taste cells in taste buds on the tongue	Portions of facial, glossopharyngeal, and vagus nerves (7th, 9th, and 10th cranial nerves)
Smell	Molecules dissolved in fluid on mucous membranes in the nose	Sensitive ends of olfactory neurons in the mucous membranes	Olfactory nerve (1st cranial nerve)

Psychophysics Relates Stimulus to Response

Your perceptual experience is constructed from information detected by your sense organs. For over a century, psychological scientists have tried to understand the relationship between the physical properties of the world and how we sense or perceive them. This is the focus of *psychophysics,* which examines our *psychological* experiences of *physical* stimuli. Developed during the nineteenth century by Gustav Fechner, psychophysics assesses such things as how much physical energy is required for our sense organs to detect that energy and how much change is required before we notice that change. To test this, researchers present very subtle changes in the stimuli and observe how people respond. Through careful study, researchers are able to test the limits of the human sensory systems (Norwich & Wong, 1997). As you will see, our sensory systems are remarkably sensitive.

SENSORY THRESHOLDS Our sensory organs constantly acquire information from the environment, much of which we do not notice. How much of a physical energy source needs to be present before it is detected by the sensory organs? The **absolute threshold** is the minimum intensity of stimulation that must occur before we can experience a sensation. For instance, the absolute threshold for hearing is the faintest sound a person can hear. How loudly must someone in the next room whisper for you to be able to hear it? In this case the absolute threshold for auditory stimuli would be the quietest whisper that you could hear. Because of this, the absolute threshold is defined as the stimulus intensity that is detected above chance (Table 5.2 lists some approximate minimum stimuli for various senses).

absolute threshold The minimum intensity of stimulation that must occur before one can experience a sensation.

TABLE 5.2 | Approximate Minimum Stimulus for Each Sense

SENSE	MINIMUM STIMULUS
Vision	A candle flame seen at 30 miles on a dark, clear night
Hearing	The tick of a clock at 20 feet under quiet conditions
Taste	One teaspoon of sugar in 2 gallons of water
Smell	One drop of perfume diffused into the entire volume of six rooms
Touch	The wing of a fly falling on your cheek from a distance of 1 centimeter

difference threshold The minimum amount of change required in order to detect a difference between intensities of stimuli.

A **difference threshold** is the *just noticeable difference* between two stimuli—the minimum amount of change required in order for us to detect a difference. If a friend is watching a television show while you are reading, and the show goes to a commercial that is louder in volume, you might look up, noticing that something has changed. The minimum change in volume required for you to be able to detect a difference would be the difference threshold. It turns out that the difference threshold increases as the stimulus becomes more intense. Pick up a one-ounce letter and a two-ounce letter, and you will easily detect the difference. But pick up a five-pound package and one weighing one ounce more, and the difference is harder to discern. This principle, called *Weber's law* after the researcher who first described it, states that the size of a just noticeable difference is based on a relative proportion of difference rather than a fixed amount of difference. Imagine an exam in which you score 6 out of 10. This feels considerably different from scoring 96 out of 100, even though both differ by only four wrong answers. What is important in determining the difference threshold is the percentage, not the absolute size of the difference.

How do we make judgments when stimuli are ambiguous?

SIGNAL-DETECTION THEORY Classical psychophysics was based on the idea of a *threshold;* you either see something or not, depending on whether the intensity of the stimulus is above or below the sensory threshold. As research progressed, however, it became clear that early psychophysicists had ignored an important variable: human judgment. Researchers began to realize that the concept of an absolute threshold was flawed. People sometimes believed they saw a weak stimulus when there was none, and sometimes failed to detect a stimulus that was presented. Such observations led to the formulation of *signal-detection theory,* which states that detecting a stimulus requires making a judgment about its presence or absence, based on a subjective interpretation of ambiguous information. This is very much the daily task of radiologists who scan medical images to detect early signs of cancer. Even after years of training and experience, it can be difficult to judge whether an abnormality in the image is likely to be cancerous or not. The radiologist's knowledge of the patient (e.g., the patient's age and sex, a family history of cancer) will likely affect this judgment; so, of course, will such factors as the radiologist's level of motivation and attention. Moreover, the consequences of their judgments weigh on the radiologists. Being wrong could mean missing a fatal cancer or, conversely, causing unnecessary and potentially dangerous treatment. All of the factors run through the radiologist's mind while the decision is being made.

Signal-detection theory involves a series of trials in which a stimulus is presented on some trials and not on others. The task of the research participant is to

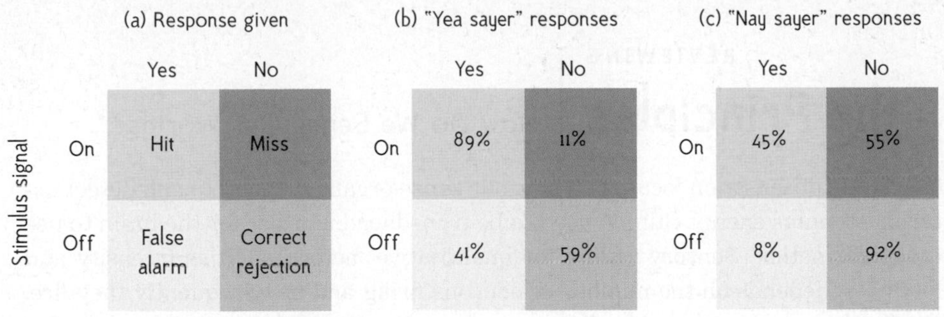

(a) Response given

Stimulus signal	Yes	No
On	Hit	Miss
Off	False alarm	Correct rejection

(b) "Yea sayer" responses

	Yes	No
On	89%	11%
Off	41%	59%

(c) "Nay sayer" responses

	Yes	No
On	45%	55%
Off	8%	92%

5.3 Payoff Matrices for Signal-Detection Theory There are four possible outcomes when a subject is asked whether something occurred during a trial (matrix a). Those who are biased toward reporting a signal tend to give the responses in matrix b; those who are biased toward denying that a signal occurred tend to respond according to the percentages in matrix c.

state whether or not they observed the stimulus. Any given trial in which people are asked to judge whether an event is present can have one of four outcomes (Figure 5.3). If the signal is present and the observer detects it, the outcome is a *hit*. A failure to detect a true signal is termed a *miss*. If the observer erroneously "detects" a stimulus that wasn't there, the outcome is a *false alarm*. Finally, if the stimulus is not presented and the observer denies having seen it the outcome is a *correct rejection*. The observer's sensitivity to the stimulus is usually computed by comparing the hit rate with the false-alarm rate—thus correcting for any bias the observer might bring to the testing situation. *Response bias* refers to a participant's tendency to report detecting the stimulus on ambiguous trials. In some circumstances, an observer will be strongly biased against making a response and will need a lot of evidence that the signal is present. In other conditions, the same observer will need only a small amount of evidence. For example, if you were a radiologist checking a CAT scan for signs of a brain tumour, you might be extra cautious about accepting any abnormality as a signal (i.e., a tumour), since your response could lead to drastic and dangerous neurosurgery. However, if you were examining an x-ray image for signs of a broken bone, you might be more willing to make a positive diagnosis, since treatment—although it is uncomfortable—doesn't endanger the life of the patient. Figure 5.3 illustrates how bias can influence responses to ambiguous stimuli. People's expectations often influence the extent to which they are biased. So, for instance, a soldier who is expecting an imminent attack is likely to err on the side of responding, perhaps having many false alarms. This is another example of how higher-level processes in the brain, such as beliefs and expectancies, influence how sensations from the environment are perceived.

SENSORY ADAPTATION In terms of sensing the world, your response to a stimulus changes over time. Imagine that you are studying in the library when work begins at a nearby construction site. When the equipment starts up, the sound seems particularly loud and disturbing. After a few minutes, however, you hardly notice the noise; it seems to have faded into the background. This is an example of what researchers call **sensory adaptation**. Researchers have often noticed that observers' sensitivity to stimuli decreases over time. If a stimulus is presented continuously, the responses of the sensory systems that detect it tend to diminish over time. One way to think about this is to consider that sensory systems are tuned to detect change in the environment. When some aspect of the environment changes, it is important for us to be able to detect it; it is less critical to keep responding to unchanging stimuli. Note that when a continuous stimulus stops, there is usually a large response as well. If the construction noise suddenly halted, you would likely notice the silence. As you will see later in this chapter, researchers often take advantage of sensory adaptation to explore the nature of sensory systems.

sensory adaptation When an observer's sensitivity to stimuli decreases over time.

REVIEWING

the Principles | How Do We Sense Our Worlds?

The study of sensation focuses on how our sense organs respond to and detect external stimulus energy. Stimuli need to be transduced in order for the brain to use that information. Sensory coding for quantitative factors, such as intensity and loudness, depends on the number of neurons firing and how frequently they fire. Qualitative aspects, such as colour or bitterness, are coded by the integration of activation across specific receptors. The development of psychophysical methods allowed psychological scientists to study psychological reaction to physical events. Psychophysical methods can be used to determine thresholds for detecting events and for noticing change. These thresholds can be influenced by situational factors and by biases in human judgment.

What Are the Basic Sensory Processes?

How does information about the world get into the brain? Helen Keller's perceptual experience was restricted because her sensory systems were damaged. The neurons in her brain that construct visual and auditory perceptions did not receive any sensory signals. As you learned in Chapter 3, no neuron beyond the sensory organs responds directly to events in the world. All that a neuron in the brain receives is signals from other neurons. In other words, a neuron in the brain does not respond to events in the world, it responds to its input from other neurons. As a result, you could say that your experience is a construction of your brain and resides inside your skull, despite the illusion that your experience is of objects or events out in the space around you. Neurons are talking to neurons in total darkness, and yet from this emerges our conscious experience of the outside world. How is this information transmitted to the brain? In this section, we will review the first step in this process, how stimuli are detected and sent to the brain for each of the five primary senses: taste, smell, touch, hearing, and vision.

In Gustation, Taste Buds Are Chemical Detectors

taste buds Sensory receptors that transduce taste information.

The job of our sense of taste, or *gustation,* is to keep poisons out of our digestive systems while allowing good food in. The stimuli for taste are chemical substances from food that dissolve in saliva, though how these stimuli work is still largely a mystery. The taste receptors are part of the **taste buds** of the tongue and mouth. People differ greatly in how many taste buds they have, ranging from 500 to 10,000, mostly located on the tongue, but also spread throughout the mouth and even the throat. Each taste bud has about 50 receptor cells. *Microvilli*—short, hairlike structures at the tip of each taste bud—come into direct contact with saliva. When stimulated, they send electrical signals to a brainstem region called the medulla and from there to the thalamus and cortex, which ultimately produces the experience of taste.

We have several different kinds of taste receptors. These taste buds each respond most strongly to one of four "primary" taste sensations: sweet, sour, salty, and bitter. Every taste experience is composed of a mixture of these four qualities. This experience reflects a general principle of sensory experiences, in that a near-infinite variety of perceptual experiences arises from the activation of unique combinations of a small number of receptor types using coarse coding. Although it was once believed that different regions of the tongue are more sensitive to certain tastes, we now know that the different taste buds are spread relatively uniformly throughout the tongue and mouth (Lindemann, 2001).

It is not just taste that affects how much you like a certain type of food. You know from having a cold that taste relies heavily on the sense of smell—if your nasal passages are blocked, food seems tasteless. In addition, the texture of the food matters a great deal; whether a food is soft or crunchy, creamy or granular, tender or tough—all these textures affect the sensory experience, as does the extent to which food causes discomfort, such as occurs with spicy chilies. The entire taste experience occurs in your brain, not in your mouth. It is the integration of these various sensory signals that gives us the experience of taste. The way these various aspects of food are integrated may account in part for why people differ in which foods they like. Our taste experiences are also influenced by the culture in which we live. These cultural beliefs shape how we experience the sensory signals that arrive from our taste buds.

Some people experience especially intense taste sensations, which is largely determined by genetics. These individuals, known as *supertasters,* are highly aware of flavours and textures, and are also more likely to feel pain when eating very spicy foods (Bartoshuk, 2000). First identified by their extreme dislike of bitter substances, supertasters have nearly six times as many taste buds as normal tasters. Although you might think that it would be enjoyable to experience intense tastes, it turns out that many supertasters are especially picky eaters because so many tastes seem overwhelming to them. When it comes to sensation, more isn't necessarily better.

In Smell, the Nasal Cavity Gathers Particles of Odour

The sense of smell, or *olfaction,* has the most direct route to the brain of all of the senses, but it may be the least understood. Like taste, it involves sensing chemicals from outside the body. We smell something when odourous particles pass into the nose and the upper and back portions of the nasal cavity. There they come into contact with the *olfactory epithelium,* a thin layer of tissue embedded with olfactory receptors; the particles dissolve in the solution that surrounds the epithelium and cause a reaction that triggers chemical receptors. These nerve impulses convey information to the **olfactory bulb**, the brain centre for smell, just below the frontal lobes (Figure 5.4). Unlike other sensory information, from here smell signals go directly to other brain areas, initially bypassing the thalamus. A recent imaging study found that regions in the prefrontal cortex process information about whether a smell is pleasant or aversive, whereas the intensity of the smell is processed by the amygdala (Anderson et al., 2003). Given that these regions are involved in the experience of emotion and also contribute to memory formation, it is perhaps not surprising that olfactory stimuli can evoke powerful memories and feelings. For many people, a wisp of the aroma of certain holiday foods cooking, the smell of bread baking, or the fragrance of a familiar perfume can all generate fond childhood memories.

olfactory bulb The brain centre for smell, located below the frontal lobes.

Researchers have identified thousands of different receptors in the olfactory epithelium, each responsive to a different chemical group. It remains unclear exactly how these receptors encode different smells. One possibility is that each receptor type is uniquely associated with a specific odour, so that there is a receptor that encodes only, for example, the scent of a rose. Another possibility is that odours will each stimulate several receptors, and that the pattern of activation across several receptor types will determine the olfactory percept (Lledo, Gheusi, & Vincent, 2005).

The human sense of smell is vastly inferior to that of many animals, such as dogs. This may be, in part, because we rely most heavily on vision and other senses and

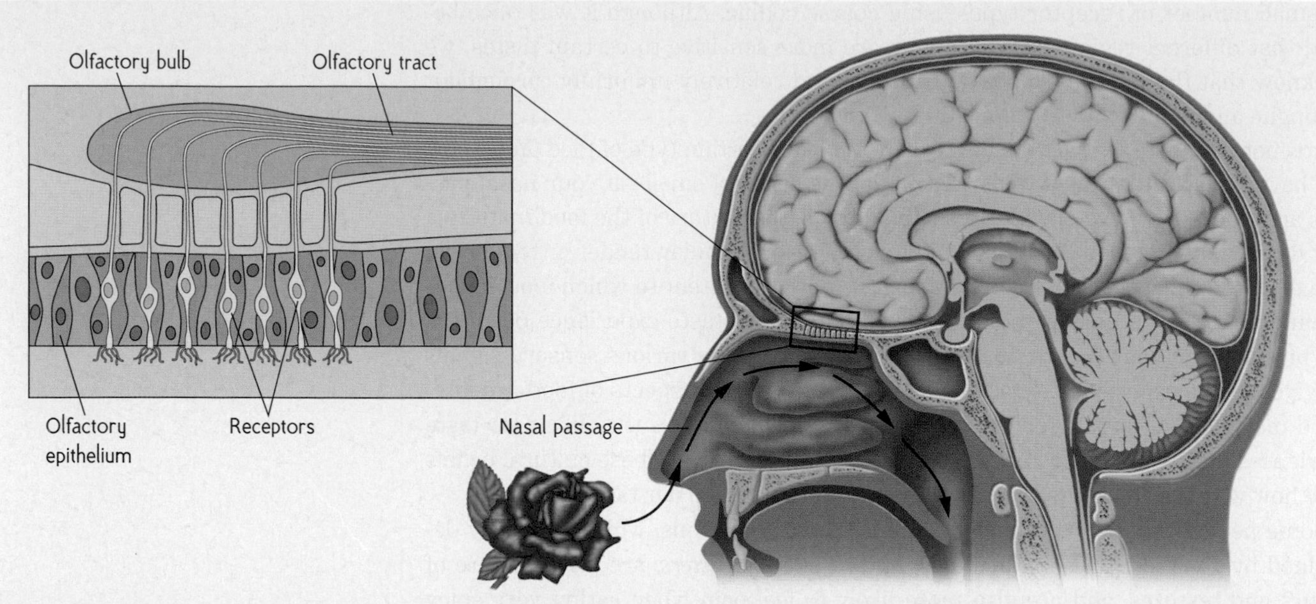

5.4 The Olfactory System Once an odour particle enters the nose, it activates receptors in the olfactory epithelium, which convey odour information to the olfactory bulb. Smell signals are first sent to regions of the prefrontal cortex and the amygdala, and then to the thalamus.

fail to pay much attention to smell. For Helen Keller, who did not have visual or auditory sensations, smell was what she called a "potent wizard" that carried her through life. She was able to smell oncoming storms well before others even noticed the smell of newly fallen rain. What is that smell, anyway? Although quite a few odours are familiar to many people, such as the smell of freshly baked chocolate-chip cookies, they are hard to describe in words.

Humans do not have the olfactory capabilities of other animals, but this doesn't mean that smell is not important to humans overall, as evidenced by the vast sums of money spent on deodourants and mouthwash, at least in Western cultures. Moreover, certain fragrances can lead to powerful changes in mood or evoke specific memories. Indeed, when contrasting the importance of senses, we probably underestimate how important smell is to us in our daily lives.

The sense of smell is also involved in an important mode of communication. **Pheromones** are chemicals released by animals—including probably humans—that trigger physiological or behavioural reactions in other animals. These chemicals do not elicit "smells" we are conscious of, but they are processed in a similar manner to olfactory stimuli. Specialized receptors in the nasal cavity respond to the presence of pheromones. Pheromones play a major role in sexual signaling in many animal species and may affect humans in similar ways, as we will discuss in Chapter 9. For example, pheromones may explain why the menstrual cycles of women who live together tend to synchronize (McClintock, 1971).

pheromones Chemicals released by animals and humans that trigger physiological or behavioural reactions in other members of the same species.

haptic sense The sense of touch.

In Touch, Sensors in the Skin Detect Pressure, Temperature, and Pain

The **haptic sense**, commonly referred to as touch, conveys sensations of pain, temperature, and pressure. Anything that makes contact with our skin provides *tactile stimulation,* which gives rise to an integrated experience of touch. Haptic receptors are sensory neurons that terminate in the outer layer of skin. Their long axons enter the central nervous system by way of spinal or cranial nerves. Some of the receptors for pressure are nerve fibres at the bases of hair follicles that respond to movement in the hair. Other receptors are capsules in the skin that respond to continued vibration, sudden movements, and steady pressure. In terms of temperature, there appear to be separate receptors for hot and cold, although both can be triggered at the same time by intense stimuli. The simultaneous activation of hot and cold receptors can produce strange sensory experiences, such as a feeling of wetness. The integration of these various signals produces haptic experiences. For instance, stroking multiple pressure points can produce a tickling sensation, which can be pleasant or unpleasant depending on the mental state of the person being tickled. This raises the age-old question of why you can't tickle yourself. Indeed, recent imaging research finds that the brain areas involved in touch sensation respond less to self-produced tactile stimulation than to external tactile stimulation (Blakemore, Wolpert, & Frith, 1998).

TWO TYPES OF PAIN Pain is part of a warning system that stops you from continuing activities that may harm you. Whether the message is to remove your hand from a hot burner or to stop running when you have damaged a tendon, pain signals that you should quit doing whatever you are doing. Children born with a rare genetic disorder that leaves them insensitive to pain usually die young, no matter how carefully they are supervised. They simply do not know how to avoid activities that harm them or to report when they are not feeling well (Melzack & Wall, 1982). As with other sensory systems, the actual experience of pain is created by the brain. For instance, as you recall from Chapter 4, people with amputated limbs sometimes feel phantom pain in the nonexistent limb. The person really feels pain, but the pain occurs because of a misinterpretation of neural activity (Melzack, 1992).

Can we override pain?

Most experiences of pain result from damaging the skin, which activates haptic receptors. The nerve fibres that convey pain information are thinner than those for temperature and touch and are found in all body tissues that sense pain: skin, muscles, membranes around bones and joints, organs, and so on. Two kinds of nerve fibres have been identified for pain: fast fibres for sharp, immediate pain, and slow fibres for chronic, dull, steady pain. An important distinction between these fibres is the myelination of their axons that travel from the pain receptors to the spinal cord. Recall from Chapter 3 that myelination speeds up neural signals; fast-acting axons are myelinated, whereas the axons of slow-acting receptors are not. Think of when you touch a hot pan. You feel two kinds of pain: a sharp, fast, localized pain at the point where your skin touches the pan, followed by a slow, dull, more diffuse burning pain. The fast-action receptors are activated by strong physical pressure and temperature extremes, whereas the slow-acting receptors are activated by chemical changes in tissue when skin is damaged. From an adaptive perspective, fast pain leads to recoil from harmful objects and therefore is protective, whereas slow pain keeps us from using the affected body part and therefore helps in recuperation.

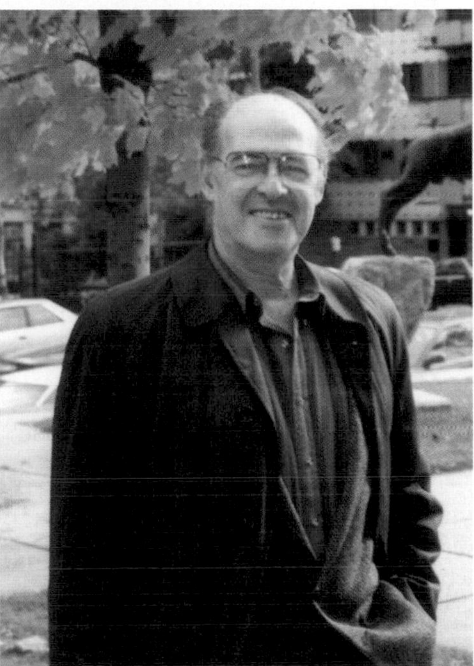

5.5 Richard Melzack A professor at McGill University in Montreal, Quebec, Melzack has produced much of the world's most groundbreaking work on pain. In 1965, he produced one of the most influential papers ever written on the subject, revolutionizing the way clinicians think about the topic of pain.

GATE-CONTROL THEORY The brain regulates the experience of pain, sometimes producing it, sometimes suppressing it. The *gate-control theory of pain,* formulated by Ronald Melzack (Figure 5.5) and Patrick Wall (1982) at McGill University in Montreal, states that in order for us to experience pain, pain receptors must be activated and a neural "gate" in the spinal cord must allow these signals through to the brain. One way to close the gate is to stimulate other haptic receptors, which overwhelms the signals from pain receptors. This is why rubbing an aching muscle helps to reduce pain or why scratching an itch is so satisfying. Experienced nurses often vigorously rub the skin surrounding the injection site, which overwhelms the haptic signals and reduces the sting of the needle. A number of cognitive states can also close the gate, such as when people are distracted by more pressing concerns. Athletes sometimes

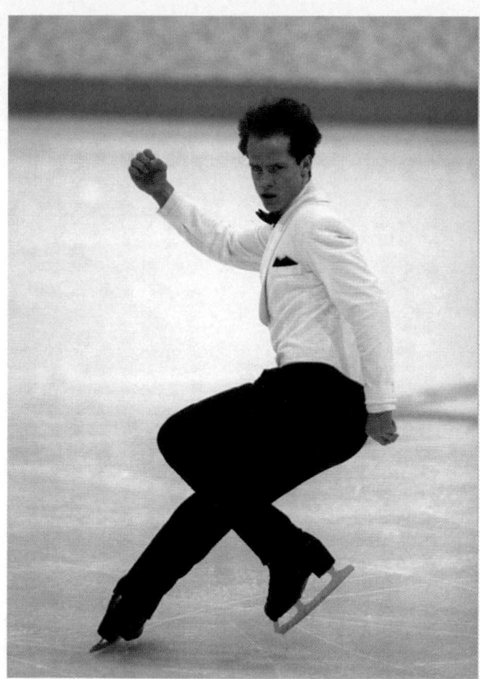

5.6 Kurt Browning Despite hitting the ice hard in the middle of his short program, Browning was still able to finish his performance in the 1994 Olympic games in Lillehammer, Norway, ultimately finishing fifth.

can play through pain because of their intense focus on the game (Figure 5.6). An annoying insect bite bothers us more when we are trying to sleep and have few other distractions. Conversely, some mental processes seem to open up the pain gates even wider, such as worrying about or focusing on the painful stimulus.

An important aspect of the gate-control theory is the role of the brain in controlling the gate. One region of the midbrain influences whether the gate is open or shut. For instance, this midbrain region can block the gate and keep pain signals from reaching the cortex, such as when athletes play through pain. Painkillers such as morphine stimulate this region, making the gate close and blocking pain (Basbaum & Fields, 1984). Recall from Chapter 3 that naturally occurring substances, called endorphins, block pain. Endorphins are believed to act on this midbrain region in the same way. With rats, if this region is stimulated during abdominal surgery, the rats do not seem to feel pain (Reynolds, 1969). Indeed, it is possible that activation of this region is what allows acupuncture to reduce feelings of pain.

In Hearing, the Ear Is a Sound-Wave Detector

Hearing, or audition, is second only to vision as a source of information about the world. It is not only a mechanism for determining what is happening in the environment, but it also provides a medium for spoken language. The stimulus for hearing is the displacement of air molecules caused by changes in air pressure. The pattern of the changes in air pressure through time is called a **sound wave**. The *amplitude* of the wave determines its *loudness,* with higher amplitude perceived as louder. The *frequency* of a sound wave determines its *pitch,* with higher frequencies perceived as higher in pitch. The frequency of sound is measured in vibrations per second, called *hertz* (Hz). Humans can detect sound waves with frequencies from about 20 Hz to 20,000 Hz. Most sounds are much more complex than a simple sine wave. However, since the pattern of compression and expansion that describes any complex sound can be represented by a unique combination of simple sine waves with different frequencies and amplitudes, we can usually predict the response of the auditory system to a complex sound from the way it responds to the component tones. Once again, what you hear, the sensory experience, occurs within your brain as it integrates the different signals provided by various sound waves. Thus, if a tree falls and there is no one around to hear it, it might cause changes in air pressure, but unless those changes are registered in a brain, there is no sound.

Our ability to hear is based on the intricate interactions of various regions of the ear, which transduce sound waves into brain activity, producing the sensation of meaningful sound. Figures 5.7 and 5.8 show the structures of the ear and a cross section illustrating how these structures convert sound waves to neural signals. Air-pressure changes that produce sound waves arrive at the **outer ear** and travel down the auditory canal to the **eardrum**, a membrane stretched tightly across the canal, marking the beginning of the *middle ear.* Changes in air pressure make the eardrum vibrate. These vibrations are transferred to three tiny bones called **ossicles**—commonly called the hammer, anvil, and stirrup. The ossicles transfer the vibrations of the eardrum to the *oval window,* a membrane of the cochlea. The **cochlea**, or inner ear, is a fluid-filled tube that curls into a snail-like shape. The ossicles mechanically amplify the vibrations, so that when they reach the oval window from the eardrum these vibrations are about 1,000 times greater in pressure.

Running through the centre of the cochlea is the thin *basilar membrane* that divides the cochlea into three chambers, two outer ducts and an inner duct. The vibrations of the oval window create pressure waves in the fluid of the inner ear, and these

sound wave The pattern of the changes in air pressure through time that results in the percept of a sound.

outer ear The structure of the ear at which sound waves arrive.

eardrum (tympanic membrane) A thin membrane, which sound waves vibrate, that marks the beginning of the middle ear.

ossicles Three tiny bones, the malleus (hammer), incus (anvil), and stapes (stirrup), in the middle ear that transfer the vibrations of the eardrum to the oval window.

cochlea (inner ear) A fluid-filled tube that curls into a snail-like shape. The cochlea contains the basilar membrane, which in turn contains auditory receptor cells called hair cells. These transduce the mechanical energy of the sound wave into neural impulses.

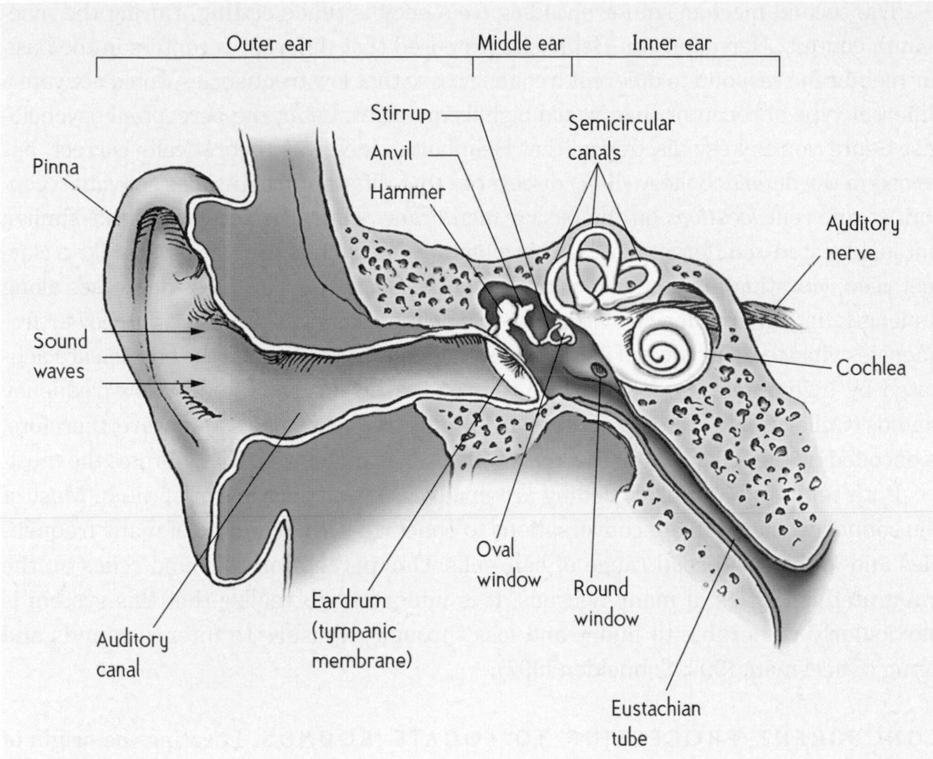

5.7 The Human Ear Sound waves enter the auditory canal, causing the eardrum to vibrate. These vibrations are carried along as waves in the liquid of the inner ear.

waves stimulate the *hair cells* located on the surface of the basilar membrane. These hair cells are primary auditory receptors; the oscillations of the basilar membrane prompt the hair cells to generate action potentials. Thus, the mechanical signal of the oscillations produced by sound waves is converted into a neural signal that travels down the auditory nerve to the brain.

temporal coding A mechanism for encoding low-frequency auditory stimuli in which the frequency of the sound wave is encoded by the frequency of firing of the hair cells.

TIME AND PLACE CODING FOR PITCH How does the firing of auditory receptors signal different frequencies of sound, such as high notes and low notes? In other words, how is pitch coded by the auditory system? Two mechanisms for encoding the frequency of an auditory stimulus operate in parallel in the basilar membrane: temporal coding and place coding.

Temporal coding is used to encode relatively low frequencies, such as the sound of a tuba. The firing rates of cochlear hair cells match the frequency of the pressure wave, so that a 1,000 Hz tone causes hair cells to fire 1,000 times per second. Physiological research has shown that this strict matching between the frequency of auditory stimulation and firing rate of the hair cells can occur only for relatively low frequencies—up to about 4,000 Hz. At higher frequencies, temporal coding can be maintained only if hair cells fire in volleys, in which different groups of cells take turns firing, so that the overall temporal pattern matches the sound frequency.

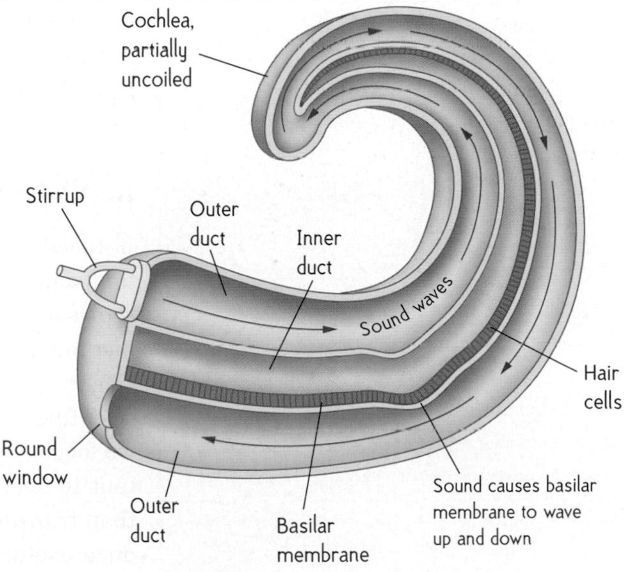

5.8 Path of Transduction in the Inner Ear When sound waves hit the fluid of the inner ear, the fluid causes the basilar membrane to move up and down, activating electrical potentials in the hair cells (the receptor cells for hearing).

place coding A mechanism for encoding high-frequency auditory stimuli in which the frequency of the sound wave is encoded by the location of the hair cells along the basilar membrane.

The second mechanism for encoding frequency is **place coding**. During the nineteenth century, Hermann von Helmholtz proposed that different receptors in the basilar membrane respond to different frequencies, so that low frequencies would activate a different type of receptor than would high frequencies. Later, the perceptual psychologist Georg von Békésy discovered that Helmholtz's idea was theoretically correct, but wrong in the details. Békésy (1957) discovered that different frequencies activate receptors at different *locations* on the basilar membrane. Thus, the receptors were similar, but just located in different places. This membrane responds to sound waves like a clarinet reed, vibrating in resonance with the sound. Because its stiffness decreases along its length, higher frequencies vibrate better at the membrane's base, while lower frequencies vibrate more toward its tip. Thus, hair cells at the base of the cochlea are activated by high-frequency sounds; hair cells at the tip are activated by low-frequency sounds (Culler, Coakley, Lowy, & Gross, 1943). The frequency of a sound wave, therefore, is encoded by the receptors on the area of the basilar membrane that vibrates the most.

Both temporal and place coding are involved in our perception of pitch. Most of the sounds we hear—from conversations to concerts—are made up of many frequencies and activate a broad range of hair cells. Our perception of sound relies on the integrated activities of many neurons. It is important to realize that this system is particularly vulnerable to injury and loss through exposure to intense sounds and aging (Cheesman, 1997; Schneider, 1997).

How do we know where sound is coming from?

CONCURRENT PROCESSING TO LOCATE SOUNDS Locating the origin of a sound is another significant problem in auditory perception. In audition, the sensory receptors do not code where in the world events occur. Instead, the brain integrates the different sensory information coming from each of our two ears. Much of our understanding of auditory localization has come from examining barn owls, which use a fine-tuned sense of hearing to locate their prey. These nocturnal animals can locate a mouse in a dark laboratory using their sense of hearing alone. Barn owls use two cues to locate a sound: the difference in timing between its arrival in each ear, and the difference in its intensity in the ears. Unless a sound is coming from a point exactly in front or in back of the owl, it will reach one ear before the other. Likewise, if sound comes from the right, it will be softer on the left because the head acts as a barrier. These differences in timing and magnitude are minute—but not too small for the owl's brain to detect and act on. The neurobiologist Mark Konishi (1993) has discovered that separate neural pathways process the timing cue and the intensity cue. Whether similar neural mechanisms are responsible for sound localization in humans is unknown, but the concept that such functions are likely to be segregated is a key to understanding how and what we perceive.

In Vision, the Eye Detects Light Waves

If knowledge is acquired through our senses, then vision is by far our most important source of knowledge. Consider how much of what we know comes from what we see. Is a place safe or dangerous? Does that person look friendly or hostile? Even our metaphors for knowledge and understanding are often visual: "I see," "the answer is clear," "I'm fuzzy on that point." It should come as no surprise, then, that most of the scientific study of sensation and perception is concerned with vision. In this section we focus (so to speak) on how energy is transduced in the visual system. It is important to keep in mind, however, that what is commonly called *seeing* is much more than transducing energy; it is an intelligent process that occurs in the brain to produce useful information about the environment. In his book *The Thinking Eye, the*

Seeing Brain, James Enns (2005) of the University of British Columbia notes that very little of what we call seeing actually takes place in the eyes; rather, what we see results from active constructive processes that occur throughout much of the brain to produce our visual experiences. When you open your eyes and look around, nearly half your brain springs into action to make sense of the energy arriving in your eyes. It seems so effortless and automatic that we often take it for granted, but our ability to see results from a cascade of neural processes that occur when light hits our eye. Of course, although the brain sees, it does so based on sensory signals from the eyes. If the eyes are damaged, then the sensory system fails to process new information. Here we examine how this energy is processed in the visual system.

The human eye works like a rather crude camera, focusing light to form an image on the retina (Figure 5.9). Light first passes through the *cornea,* a thick, transparent outer layer of the eye. The cornea focuses incoming light in a process called refraction. Light rays then enter and are bent farther inward by the lens, which focuses the light to form an image on the **retina**, the inner surface of the back of the eyeball. Although more refraction (bending of light) happens at the cornea than at the lens, the lens is adjustable, whereas the cornea is not. The *pupil,* a small opening in the front of the lens, contracts or dilates to alter how much light enters the eye. The *iris,* an opaque, circular muscle, controls the size of the pupil and gives eyes their colour. Behind the iris, muscles change the shape of the lens—flattening it to focus on distant objects and thickening it to focus on closer objects. This is called **accommodation.** The lens and cornea work together to collect and focus light rays reflected from an object to form an inverted image of the object on the retina.

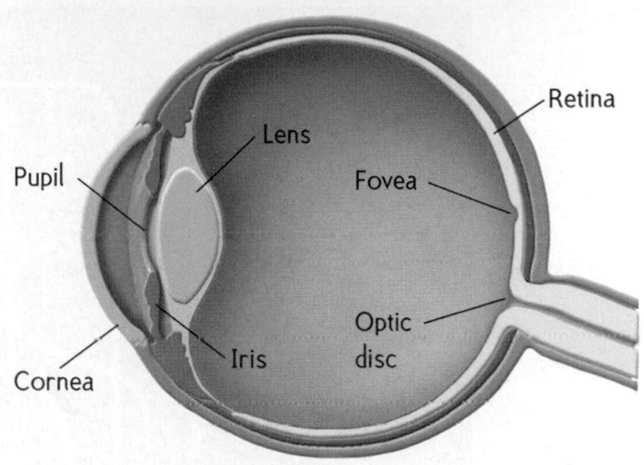

5.9 The Eye The cornea is a clear protective layer covering the lens, which focuses images on the retinal surface. Receptors in the retina send information to the visual cortex via the optic nerve.

RODS AND CONES The retina has two types of receptor cells: rods and cones, so called because of their distinctive shapes. Rods respond at extremely low levels of illumination and are primarily responsible for night vision; they do not support colour vision, and they are poor at resolving fine detail. This is why everything on a moonless night appears as shades of gray. Cones, in contrast, are less sensitive to low levels of light; they are primarily responsible for vision under high illumination, and for colour and detail. Within the rods and cones, light-sensitive chemicals called *photopigments* initiate the transduction of light waves into electrical neural impulses.

There are approximately 120 million rods and 6 million cones in each retina. Cones are densely packed in a small region near the centre of the retina called the **fovea**. Although cones are spread throughout the remainder of the retina (except in the blind spot, as you will see shortly), they become increasingly scarce near the outside edge (Figure 5.8). Rods are all located at the edges of the retina; none are located in the fovea. You can prove this to yourself at night by looking at a very dim star on a moonless night. If you look directly at the star it appears to vanish, because its light falls on the retina, where there are no rods. However, if you look just to the side of the star, it becomes visible again, because now its light is falling just outside the fovea, where there are rods.

retina The thin inner surface of the back of the eyeball. The retina contains the photoreceptors that transduce light into neural signals.

accommodation A process by which muscles change the shape of the lens by flattening it to focus on distant objects or by thickening it to focus on closer objects.

fovea The centre of the retina where cones are densely packed.

TRANSMISSION FROM THE EYE TO THE BRAIN Shine a light in someone's eyes so that you can see their retina, and you are in fact looking at the only part of their brain that is visible to us from outside the skull. It is the one part of the central nervous system that is located where you can see it. The generation of electrical signals by the photoreceptors in the retina begins the visual process. Immediately after light is transduced into neural impulses by the rods and cones, other cells in the retina (*bipolar, amacrine, horizontal;* see Figure 5.10) perform a series of sophisticated computations on those impulses that help the visual system process incoming information. The outputs from these cells converge on about 1 million retinal *ganglion* cells. These cells are the first in the visual pathway to generate action potentials.

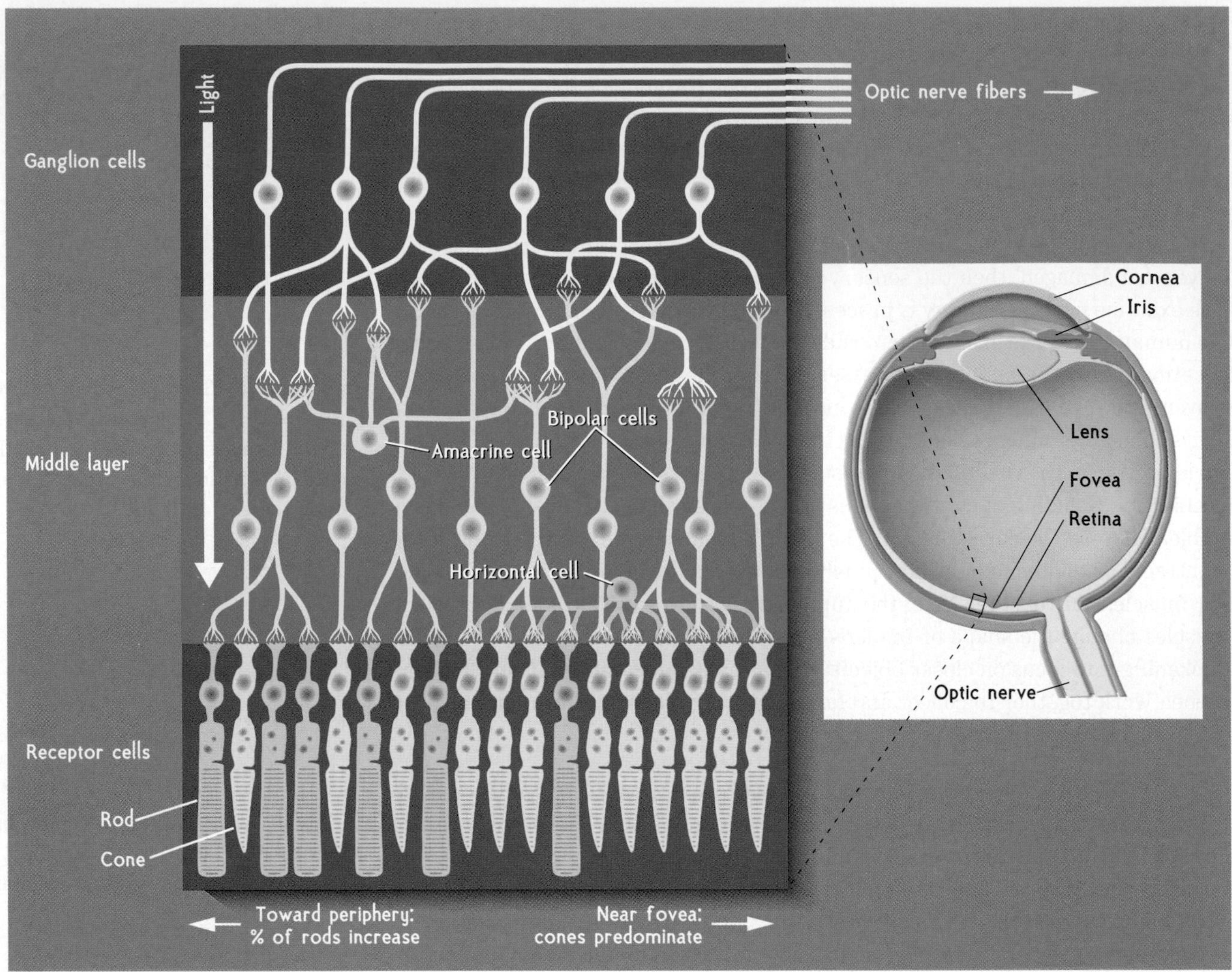

5.10 A Schematic Cross Section of the Retina Light passes through the cornea and is focused on the retina by the lens. It passes through the ganglion cell layer and middle layer before it is transduced into neural impulses by the receptor cells. There are two types of receptors in the retina: rods and cones. The cells in the middle layer transduce the neural impulses, which form a synapse with the ganglion cells. The ganglion cells, in turn, transmit neural impulses to the brain.

A surprising fact about the structure of the eye is that ganglion and other retinal cells and blood vessels lie in front of the cones and rods, potentially blocking light from being detected. Try this: hold a penlight near the outside corner of one eye (being careful not to touch your eye or burn yourself) and gently wiggle this light back and forth while looking at a blank wall. If you do this, you may be able to see shadows cast by the blood vessels in your eye. These will typically look like a large river delta as seen from the sky. The brain normally discounts these shadows. But when light enters the eye from a sharp angle, shadows are cast on cones that usually do not have this shadow cast on them. The brain fails to discount these shadows, and they are seen as if they were "out there" in the world.

Because the ganglion cells are inside the eye, they must leave it in order to send their signals to higher areas of the brain. Axons from ganglion cells are gathered into a bundle called the *optic nerve* that exits the eye at the back of each retina like a cable leaving a room. The point at which the optic nerve exits the retina has no rods or cones at all, resulting in a blind spot in each eye that is about the size of a fist held at arm's length. This blind spot can be isolated, as demonstrated in Figure 5.11, but we are generally not aware of it. Each optic nerve carries information to the central nervous system. The optic nerve splits into two parts that cross at the *optic chiasm*, causing all information from the left side of visual space (i.e., everything visible to

5.11 The Blind Spot The optic nerve creates the blind spot we all have—a small point at the back of the retina. To find your blind spot, hold this book in front of you and look at the dot, closing your left eye. Move the book toward and away from your face until the rabbit disappears. You can repeat this for your right eye by turning the book upside down.

the left of the point of gaze) to be projected to the right hemisphere of the brain and vice versa, as shown in Figure 5.12. The first synapse of the majority of ganglion cells lies within the visual areas of the thalamus, and visual information is transmitted from there to cortical areas at the back of the head called the *primary visual cortex* (see Figure 5.12). The pathway from the retina to this region carries all the information that we consciously experience as seeing.

THE DETECTION OF VISUAL PRIMITIVES Sensory neurons are generally picky about what they respond to in the world. Scientists say that sensory neurons have a particular "tuning." This can be understood by thinking of a tuning fork tuned to, say, middle C. If the air around this tuning fork vibrates at the frequency of middle C, then this will make our tuning fork vibrate, resonating with the energy around it. However, if the air vibrates at a different frequency, our tuning fork will not resonate. Similarly, visual sensory neurons can be thought to be tuned to particular types of information in light. Some visual neurons respond best to particular colours, shape orientations, or directions of motion. This tuning property of a given visual neuron specifies its receptive field. A **receptive field** is the population of sensory receptors that influences activity in a sensory neuron, the region within which the cell

receptive field The region of visual space to which neurons in the primary visual cortex are sensitive.

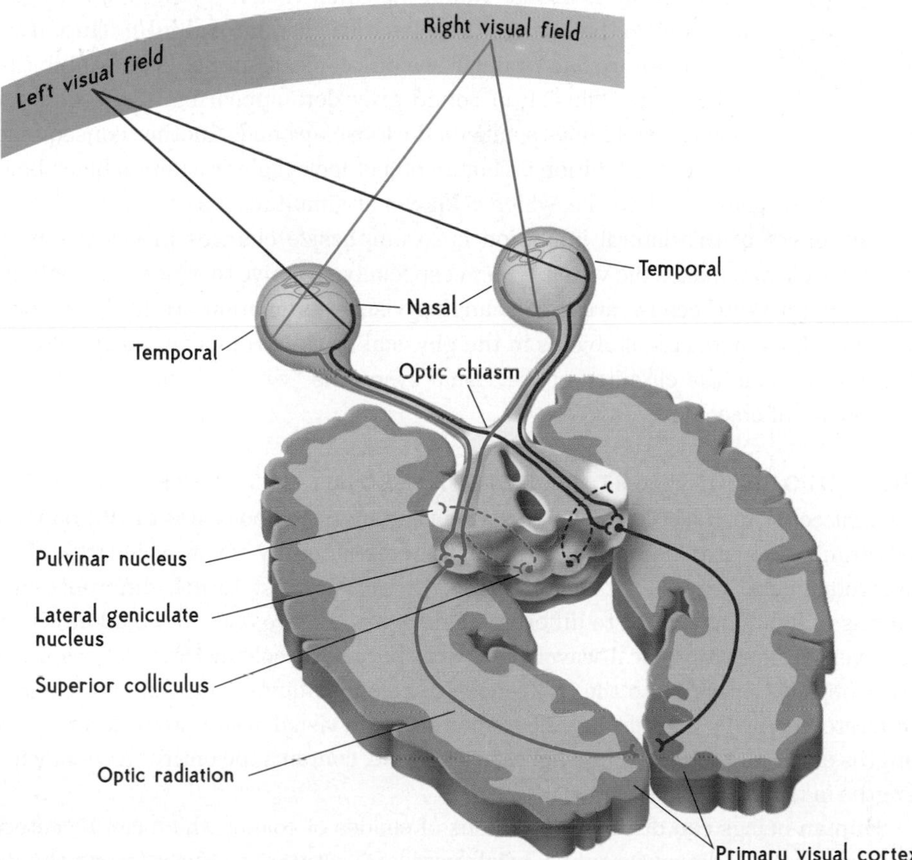

5.12 Visual Pathways The pathways along which information from the left visual field projects to the right visual cortex and the information from the right visual field projects to the left visual cortex.

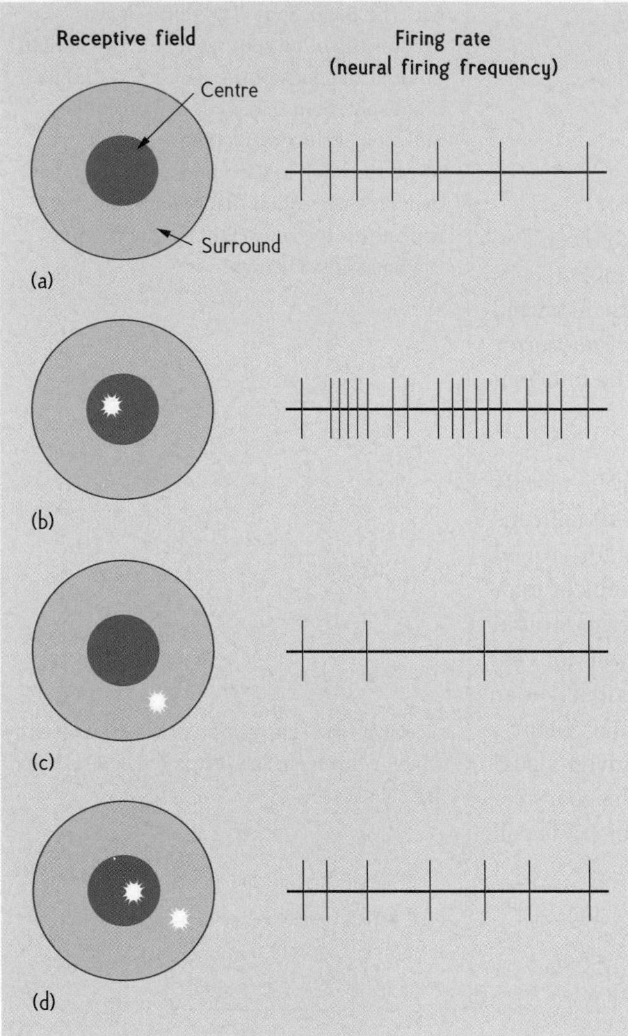

5.13 Receptive Fields A typical receptive field consists of a centre and a surround. When there is no light the cell fires at its baseline rate (a). When a light is shined in the centre, the neural firing frequency increases (b); but when a light is shined in the surround, the cell decreases its firing (c). When a light shines in both the centre and the surround, the cell's firing rate is similar to its baseline rate (d).

lateral inhibition A visual process in which adjacent photoreceptors tend to inhibit one another.

responds to a given stimulus. As in the tuning analogy, it responds best to some stimuli. Receptive fields are found for many sensory systems; the receptors can be in a region of skin or a set of photoreceptors in the retina. Since the eye projects an image of the visual scene onto the retina, *visual receptive fields* can be thought of as being located on a specific region of the retina or a specific location in visual space. To understand the nature of a receptive field, consider one of the most common types, consisting of a centre region and a surrounding region (see Figure 5.13). In this type of receptive field, light that is directed towards the centre region causes the cell to become more active, whereas light directed toward the surrounding region inhibits cell firing.

The processing of form information begins in the retina. Ganglion cells have receptive fields that make them especially sensitive to edges (abrupt changes in brightness or colour) while at the same time relatively insensitive to uniform regions, such as large areas of the same colour or brightness. In effect, the retina filters out uniform regions because they convey little useful information. The information that the retina transmits to the brain for further processing is a compressed version of the image that emphasizes information about borders.

LATERAL INHIBITION Another process that contributes to edge detection occurs because adjacent photoreceptors in neighboring regions of the retina tend to inhibit one another. If a cell is stimulated it sends information up to the brain, but it also sends information to its neighboring neurons, inhibiting their activity. The physiological mechanism that causes this effect is called **lateral inhibition**. If you look at Figure 5.14, you will see one consequence of lateral inhibition, which is the illusion of round grey dots appearing at the intersections of white lines against a dark background. Another consequence of lateral inhibition is that an object looks lighter againt a black background than against a white one, which is known as simultaneous contrast.

The effect of this lateral inhibition is to emphasize changes in visual stimuli, which once again makes the visual system especially sensitive to edges and contours. This is important because areas of changing visual stimulation are likely to correspond to the boundaries of objects in the physical world. From a very early stage of processing, then, the circuitry of the visual system is "wired" to make finding the boundaries of objects easier.

THE COLOUR OF LIGHT IS DETERMINED BY ITS WAVELENGTH Visible light consists of electromagnetic waves ranging from about 400 to 700 nanometers (nm) in length. The colour of light is determined by the wavelengths of the electromagnetic waves that it comprises. In the simplest terms, different wavelengths of light correspond to different colours (see Figure 5.15). It would be wrong, however, to equate physical wavelength with perceived colour. How the brain converts physical energy into the experience of colour is quite complex, and it can be understood only by considering the response of the visual system to different wavelengths at the same time. White light, for example, contains the entire range of wavelengths in the visible spectrum.

Human beings can distinguish millions of shades of colour, which can be categorized along three dimensions: hue, brightness, and saturation. *Hue* refers to the dis-

5.14 The Hermann Grid The Hermann grid is a demonstration of lateral inhibition at work. Look at the figure as a whole and you will immediately see darkened spots at the intersections of the white lines. However, if you look directly at the intersections, or cover all but one row of squares, you see that the dark spots are illusory. What is happening? Receptors coding information from the white lines are inhibited by their neighbors on two sides. Those receptors that code information from the intersections, however, are inhibited from four sides, so they respond less vigorously. This makes it look as though the intersections are darker than the lines.

tinctive characteristics of a colour that place it in the spectrum; hue depends primarily on the light's wavelength when it reaches the eye. *Brightness* relates to the perceived intensity or luminance of a colour, which is determined chiefly by the total amount of light reaching the eye. One should be careful not to confuse brightness—a physical dimension—and lightness—a psychological dimension. The *lightness* of an area is determined by its brightness relative to its surroundings. Because of simultaneous contrast, two grays with the same brightness can differ in lightness, depending on the surrounding levels of brightness. Although lightness is a more important variable than brightness for describing visual appearance, it has no simple physical correlate. The third dimension, *saturation,* refers to the purity of a colour. Saturation varies according to the mixture of wavelengths present in a stimulus. Pure spectral colours have only a single wavelength, whereas pastels have a mixture of many wavelengths.

SUBTRACTIVE COLOUR MIXING Colour is determined not just by wavelength but also by the mixture of wavelengths (or spectral pattern) of a stimulus. There are two ways to produce a given spectral pattern: *subtractive* and *additive* mixture of wavelengths. Here we consider the subtractive process. In mixing, say, paints, the mixture occurs within the stimulus itself and is a physical process. This is an example of **subtractive colour mixing**. Subtractive mixing of paints occurs because colours are determined by pigments—chemicals on objects' surfaces that absorb different wavelengths of light and prevent them from being reflected to the eye. The colour of

subtractive colour mixing A way to produce a given spectral pattern in which the mixture occurs within the stimulus itself and is actually a physical, not psychological, process.

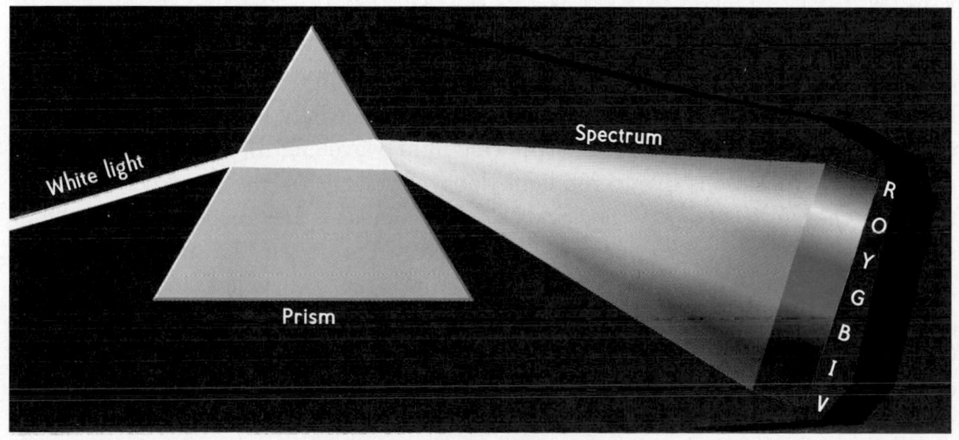

5.15 The Colour Spectrum A prism breaks apart light rays into the spectrum of colours: red, orange, yellow, green, blue, indigo, and violet (ROYGBIV).

How do we see colour?

the pigment is determined by the wavelengths that are reflected. Thus blue paint contains pigments that absorb only long wavelengths (yellow and red); the paint appears blue because it reflects the shorter waves. Yellow pigments absorb shorter (blue) and longer (red) wavelengths, but they reflect wavelengths in the central yellow region of the visible spectrum. If we mix blue and yellow paint we get green because the yellow pigment absorbs the blue wavelengths and the blue pigment absorbs the red and yellow wavelengths. What remains is the wavelength corresponding to green. Thus the pigments mix to make green by subtraction. Red, yellow, and blue are the *subtractive primary colours.* Mix all three together and you get black, because together these pigments absorb nearly all the colours of the visible spectrum.

additive colour mixing A way to produce a given spectral pattern in which different wavelengths of lights are mixed. The percept is determined by the interaction of these wavelengths with receptors in the eye and is a psychological process.

ADDITIVE COLOUR MIXING When lights of different wavelengths are mixed, what you see is determined by the interaction of these wavelengths within receptors in the eye and is a psychological process. This is called **additive colour mixing**. Additive colour mixing is a technique known to stage lighting designers. For them, red, yellow, and blue are not the primary colours because they can aim red and green lights at the same point on a stage and create a yellow light. In fact, almost any colour can be created by combining just three wavelengths, so long as one is from the long-wave end of the spectrum (red), one is from the middle (green-yellow), and one is from the short end of the spectrum (blue-violet). This phenomenon is called the *three primaries law of colour.* Note that the exact colours of the three primaries in additive colour mixing are more or less arbitrary. However, for reasons that will become clear soon, most psychologists consider the *additive primary colours* to be red, green, and blue. Whereas mixing red, green, and blue paint yields black paint, mixing red, green, and blue light yields white light. Isaac Newton discovered that white light is made up of many colours of light. You can see this for yourself with a prism (see Figure 5.15). Because different wavelengths of light bend (refract) at different angles when they pass through a prism, white light entering a prism leaves it with all the colours of a rainbow. Indeed, rainbows form in the sky because tiny water droplets in the air function as prisms, refracting sunlight in different directions.

EXPLAINING THE PHENOMENA OF COLOUR VISION Whether colour is created by subtracting hues from white light or mixing colours together, the result is the same: a certain combination of wavelengths striking the retina. However complicated this pattern is, it can be replaced with light of a single wavelength to yield the same perceived colour.

If different wavelengths correspond to different colours, how is it that combining multiple wavelengths can create unique colours? This occurs because of the way light is transduced into neural impulses in the retina. Colour vision begins in the retinal cone cells, which transduce light into neural impulses. There are actually three distinct types of cones in the retina, each of which responds best to a different wavelength of light. One type of cone is most sensitive to blue light (short wavelength), another is most sensitive to green light (medium wavelength), and the remaining population is most sensitive to red light (long wavelength). The three cone populations are therefore called "S," "M," and "L" cones because they respond maximally to short, medium, and long wavelengths, respectively. The colour of a stimulus is determined by how much of each cone type it activates. Thus, yellow light looks yellow because it stimulates the L and M cones about equally and hardly stimulates the S cones at all. Similarly, we can create yellow light from red and green light because each stimulates the corresponding cone population. As far as the brain can tell, there is no difference between yellow light and a combination of red and green lights.

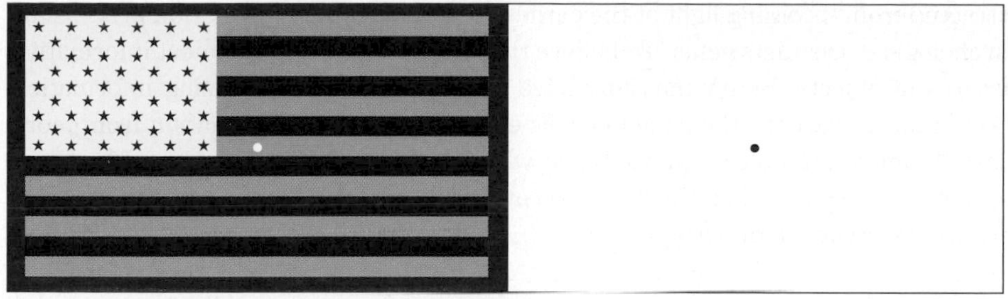

5.16 Afterimage Stare at the dot in the middle of the flag for at least 30 seconds. Then look at the dot in the blank space to the right. Because your receptors have adapted to the green and yellow in the first image, the afterimage appears in the complementary colours of red and blue.

Ultimately, our perception of different colours is determined by the ratio of activity among the three cone receptors.

The initial coding of colour information by just three types of retinal cone cells is one of the most important discoveries in the scientific study of visual perception. It illustrates how an essentially limitless variety of colours can be encoded by a small number of receptors. We've seen this integration property with other sensory systems. There are aspects of colour vision, however, that are not predicted by the existence of three types of cones in the retina. For example, some colours seem to be "opposites" in some sense (Hering, 1878/1964). When we stare at a red image for some time, we see a green afterimage when we look away (and vice versa). Likewise, when we stare at a blue image for some time, we see a yellow afterimage when we look away (and vice versa) (Figure 5.16). We also have trouble visualizing certain colour mixtures. For instance, it is easier to imagine reddish yellow or bluish green than reddish green or bluish yellow.

These phenomena cannot be explained by the responses of the different cones in the retina. To account for them, we must turn to the next stage of visual processing, in the retinal ganglion cells. As you have seen, the information from the cones converges on ganglion cells in the retina, some of which are sensitive to the colour of a stimulus, and some of which are not. One class of colour-sensitive ganglion cells receives excitatory input from L cones but is inhibited by M cones (or vice versa). These cells create the perception that red and green are "opposites." Other ganglion cells are excited by input from S cones, but inhibited by both L- and M-cone activity (or vice versa). These create the perception that blue and yellow are opposites.

A typical receptive field for a "colour-opponent" ganglion cell is shown in Figure 5.17. A ganglion cell with this receptive field structure will respond well when red light is shined in its centre, but will respond less when red light is shined in both its centre and surround. It will respond even less when green light is shined in its inhibitory surround. Cells with colour-opponent receptive fields respond most to colour boundaries, and less to uniform regions of colour. Edge information is therefore

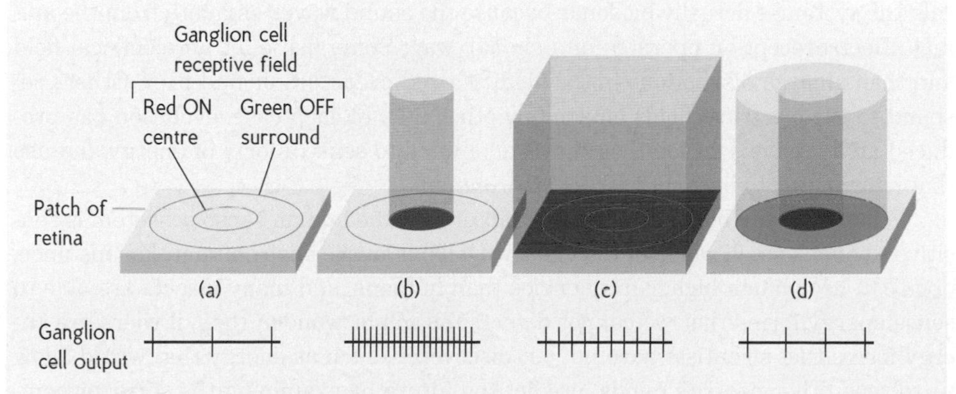

5.17 Colour-Opponent Ganglion Cell A ganglion cell with this receptive field structure will respond well when red light is shined in its centre, but will respond less when red light is shined in both its centre and surround. It will respond even less when green light is shined in its inhibitory surround.

extracted from incoming light at the earliest stages of visual information processing. Psychological scientists generally believe that edges are among the most informative features of objects. By emphasizing edge information and minimizing information about uniform regions, the ganglion cells efficiently process information from more than 100 million visual receptors. Later, when we examine the nature of visual perception, we will see that the brain recovers information about uniform regions through constructive processes.

Thinking Critically

Is There a Sixth Sense?

Now that we have covered the primary senses, you might wonder whether we have additional senses beyond these five. For instance, can we perceive the future? The term *sixth sense* refers to the "unexplainable" feeling that something is about to happen. Before we consider the case for whether we can perceive information beyond ordinary sensory information—so-called extrasensory perception, or ESP—we note that we have many sensory systems that provide information about the world. In actuality, as Howard Hughes (Figure 5.18) points out in his book *Sensory Exotica,* humans have several internal sensory systems in addition to the five primary ones we've discussed (Hughes, 2000). Subtle sensory systems send messages to our brains about things like our blood pressure or blood glucose levels. The kinesthetic sense, which some researchers group with the touch senses, refers to sensations we gather from receptors in muscles, tendons, and joints that pinpoint the position and movements of our limbs and body in space. This helps us to coordinate voluntary movement and is invaluable in avoiding injury. The vestibular senses use data from receptors in the semicircular canals of the inner ear. These canals contain a liquid that moves when the head moves, bending hair cells at the ends of the canal. The bending generates nerve impulses that inform us of the head's rotation and is thus responsible for a sense of balance. It explains why inner-ear infections make you dizzy, and why standing up quickly can give you a "head rush." The experience of being seasick or carsick results in part from conflicting signals arriving from the vestibular system and the visual system.

More exotic sensory systems are also being studied in animals, such as sonar senses and electroreception (senses based on electrical fields). Our understanding of sensory systems that use sonar for navigation comes primarily from studying bats and dolphins. These animals produce a call and then respond to the echoes of that call; the system is actually biosonar because the sound waves emanate from the animal. Electroreception operates in a similar way: Some fish emit an electrical field and then analyze disruptions in the field to avoid predators or find prey. Others respond to the electrical fields emitted by other fish. In each case, evolution has produced an adaptive solution that allows an animal to sense a form of energy and use that information to interact with its environment.

We noted at the beginning of this chapter that the human sensory system is sensitive to only a small range of the energy available in the environment. For instance, dogs can hear much higher frequencies than humans, and many insects are able to sense energy forms that we cannot detect. You might wonder, then, if there are energy forms that scientists have not yet discovered, such as energy that would allow us to read other people's minds, predict the future by examining the stars, or communicate with ghosts. Many reports of ESP are based on anecdotal cases. Psychological scientists tend to reject such claims because they are difficult to prove and because ordinary thought processes can explain many of them. For instance, if you see a couple fighting all the time, you might predict accurately that they will break up, but that doesn't make you a psychic. Much of social perception requires us to be sensitive to the subtle cues that guide behaviour in a situation. Although this form of perception is beyond our basic senses, it cannot be considered ESP.

Perhaps the best evidence for ESP was obtained by the Cornell social psychologist Daryl Bem and his collaborator Charles Honorton. In these studies, a "sender" in a soundproof booth focuses on a randomly generated image and a receiver in another room tries to sense the sender's imagery. The receiver is then asked to choose among four alternatives, one of which is correct. By chance, the receivers should be correct 25 percent of the time, but across 11 studies, Bem and Honorton (1994) found that receivers were right about 33 percent of the time. Is this evidence of ESP? Many psychological scientists say that other factors in the experiments might have affected the results. A statistical review of many such studies found little support for ESP (Milton & Wiseman, 2001). Moreover, numerous scientific organizations and government agencies have reviewed decades of research and have concluded that no such phenomenon exists.

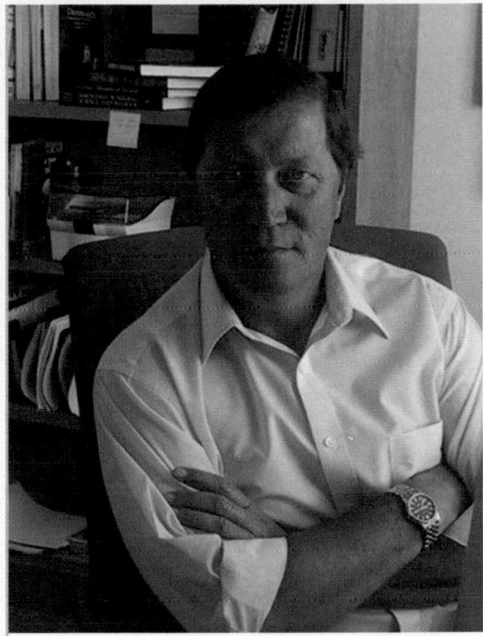

5.18 Howard Hughes Hughes describes several sensory systems in humans and other animals in addition to the five primary systems of touch, taste, smell, sight, and hearing.

Perhaps you are a believer because you've seen a stage performer accomplish amazing feats. Be warned: Many magic tricks are designed to fool your basic senses; they can be viewed as elaborate examples of other visual illusions you will encounter in this chapter. They rely on the fact that our sensory system is far from perfect, and that our expectations shape much of what we perceive. These are tricks, not evidence of psychic abilities. But does any of this prove that ESP does not exist? An important quality of critical thinking is to keep an open mind and not to conclude that something doesn't exist just because it hasn't yet been demonstrated. The only reasonable thing to conclude is that the evidence for ESP is currently weak and that healthy skepticism demands better evidence.

What kinds of evidence would it take to convince you that ESP does or does not exist? What social factors might psychics use to *read* information from their clients? What might explain how some dogs can be trained to predict when their owners will have an epileptic seizure? ■

REVIEWING the Principles | What Are the Basic Sensory Processes?

Each of our sensory systems has receptors that respond to different physical or chemical stimuli by transducing them into some pattern of brain activity. Typically, different receptors respond to different types of stimuli, and most sensory systems use coarse coding to integrate signals from these different receptors into an overall sensation. This allows a relatively small number of receptors to code a wide variety of stimuli. For example, the visual system can interpret the entire range of colours with only three cone types. These various sensory receptors help the perceptual system receive important information that assists in solving adaptive problems. Sensory information, although obtained from the outside world, is processed entirely in the brain to produce sensory experience through perception.

What Are the Basic Perceptual Processes?

The perceptual system is stunningly intelligent in its ability to guide us through the world around us. Every minute, our brains make millions of calculations—all in milliseconds—producing a coherent experience of our environment. For instance, as you stare at a computer screen, you are aware of one image, not the thousands that dance across your retina to create that constant, static scene. What you perceive, then, is vastly different from the pattern of stimulation your retina is taking in. How does the brain extract a stable representation of the world from the information the senses provide? This is what perception research is about.

If we were aware every moment of what our brain was doing, we would be paralyzed by the overload of information. Most of the computations the brain performs never reach our conscious awareness—only important new outcomes do. Perceptual psychology has drawn on many disciplines to understand how we represent our world. They have made many advances by noting how perception is disrupted following some injury to the brain, working backward to infer how the intact brain processes information. In addition, modern psychophysicists have developed clever techniques to describe the perceptual functions of the brain. Areas as diverse as art, computer science, music, philosophy, anatomy, and physiology have also informed our understanding of perception. Recall our discussion of music perception in Chapter 1, in which we saw how music can be used to explore various aspects of perception, from how genes predispose perfect pitch to how music can affect our moods. It is truly an interdisciplinary study that crosses the levels of analysis.

Perception Occurs in the Brain

So far, you have seen how sensory stimuli are transduced into electrical impulses and transmitted to the brain via nerves. These electrical impulses are all that the brain has to work with to create our rich variety of perceptual experiences. With the exception of olfaction, all sensory information is relayed to the brain via the thalamus. From the thalamus, information from each sense is projected to specific regions of the cerebral cortex, called *primary sensory areas* (Figure 5.19). It is in these areas that the perceptual process begins in earnest. This section considers perceptual processing in audition, touch, and vision. The brain regions involved in taste and smell are less well understood, and therefore we will not consider them here.

HEARING Auditory neurons in the thalamus extend their axons to the *primary auditory cortex* (called A1, for the *first* auditory area) in a region of the temporal lobe.

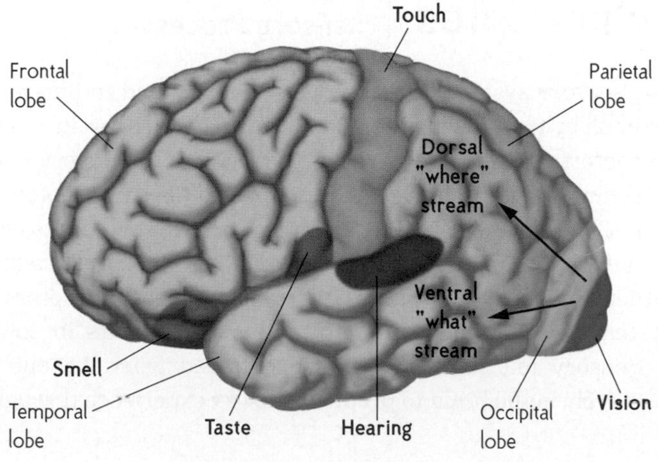

5.19 Primary Sense Regions These are the primary sense regions for the five primary senses. From the primary visual cortex, the dorsal stream codes for "where" objects are, whereas the ventral stream codes for "what" objects are.

Single-cell recording experiments have shown that neurons in A1 code the frequency (or pitch) of auditory stimuli. These studies have also revealed that A1 has a *tonotopic organization.* That is, neurons at the rear end of A1 respond best to lower frequencies, such as the sound of a fog horn, whereas neurons at the front end respond best to higher frequencies, such as a train whistle. Thus, the encoding of frequency in A1 is primarily by location. Several secondary auditory areas in the temporal and parietal lobes surround A1. Although we have little information about the organization of these secondary areas, it appears that they are also organized tonotopically.

"Great! O.K., this time I want you to sound taller, and let me hear a little more hair."

TOUCH Touch information from the thalamus is projected to the *primary somatosensory cortex* (called S1) in the parietal lobe. In a classic series of studies of patients undergoing brain surgery, described in Chapter 4, the neurosurgeon Wilder Penfield discovered that electrical stimulation of S1 could evoke the sensation of touch in different regions of the body. Penfield found that neighboring body parts tended to be represented next to one another in the primary somatosensory cortex. That is, S1 has a *somatotopic* organization: The body is effectively mapped out according to physical proximity, such that connected areas of the body tend to be next to each other in the cortex. Remember that the drawing of the homunculus from Chapter 4 (see Figure 4.10) showed that more sensitive body parts have relatively larger amounts of S1 dedicated to them. The most sensitive regions of the body, such as lips and fingers, have a great deal of cortical tissue devoted to them. Other areas, such as the back or the calves, have very little. You can confirm this by having someone touch your back at adjacent locations. How far apart do the two points have to be before you notice that they are different? It turns out that this distance is much larger for the back than for, say, the cheek. The cheek has many more receptors in the skin and much more cortex devoted to it than does the back.

VISION The study of perception has focused to a large extent on the visual cortex and the multiple areas in which the retinal image is processed. The complexity of visual perception is underscored by the amount of cortical real estate that is dedicated to processing visual information. Some estimates suggest that up to half of the cerebral cortex may participate in visual perception in some way. By now you are aware that the **primary visual cortex**, also called area **V1**, is in the occipital lobe of the brain. The neuronal pathway from the retina to the occipital lobe preserves spatial relationships, so that adjacent areas of the retina correspond to adjacent areas in V1. This systematic ordering is known as **retinotopic organization**. The more forward we go along V1, the more the peripheral (away from the centre) portion of the visual field is represented.

primary visual cortex (V1) The largest area in the occipital lobe, where the thalamus projects the image.

retinotopic organization The systematic ordering of the neuronal pathway from the retina to the occipital lobe; this organization preserves spatial relationships, so that adjacent areas of the retina correspond to adjacent areas in the primary visual cortex.

Early cortical processing consists of a rough description of various image properties and their locations within the scene. This description is a processed version of the original retinal image, but it is still a long way from identification of the three-dimensional structure of the visible world; a great deal of additional computation is required. A keystone in thinking about the neural mechanisms of visual perception is the concept of *hierarchical processing* of the details of the visual image. A widely held view is that this processing occurs in a number of stages, the first of which performs an analysis of the retinal image by extracting different elementary features (called *primitives*), such as lines, angles, colours, curves, and motion. It has been argued that relatively independent modules that specialize in extracting and interpreting

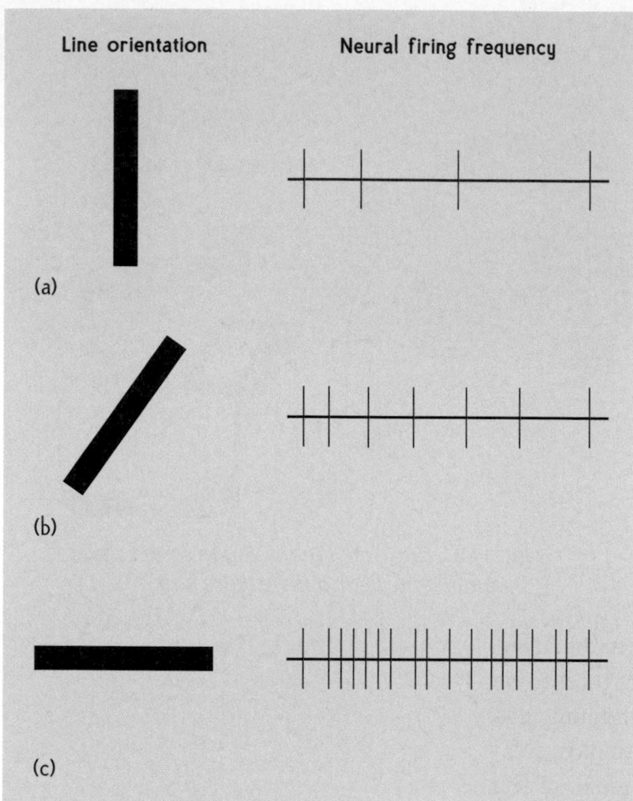

5.20 Simple Cells This neuron responds minimally to the vertical bar (a) and maximally to the horizontal bar (c). A slanted bar produces an in-between response (b).

particular classes of visual information may process different primitive features.

In the 1960s, the neurophysiologists David Hubel and Thorsten Wiesel (1962) began exploring the properties of neurons in V1 by recording activity from single cells—work for which they were awarded a Nobel Prize. They discovered that some neurons in the primary visual cortex respond more to lines of particular orientations; for example, some neurons increase their firing rate when a vertical line segment is presented in their receptive field (Figure 5.20). The firing rate of these cells—termed *simple cells* by Hubel and Wiesel—decreases as the orientation of the line segment is rotated away from the preferred orientation. Further studies by Hubel and Wiesel and others have also found neurons that specialize in detecting the ends of lines, corners, and colours, as well as more complex visual features.

WHAT VERSUS WHERE One of the core discoveries of the past few decades has been that neurons in different parts of the brain tend to have different types of receptive fields. Besides V1, visual information is processed in a cascade of other visual areas many of which are also organized retinotopically. These visual areas appear to process specific aspects of a visual stimulus, such as colour or motion. One important theory proposes that visual areas beyond V1 form two parallel processing streams—a lower ventral pathway that includes occipital and temporal lobe regions, and a higher dorsal pathway involving occipital and parietal lobe regions (see Figure 5.19). The *ventral stream* appears to be specialized for object perception and recognition, such as colour and shape, whereas the *dorsal stream* seems to be specialized for spatial perception—determining where an object is and relating it to other objects in a scene. These two processing streams therefore are known as the *"what" pathway* and the *"where" pathway* (Ungerleider & Mishkin, 1982). Subsequent studies using brain imaging of humans have confirmed that brain regions in the dorsal stream are activated by tasks that require decisions about spatial relationships between objects, whereas regions in the ventral stream are activated by tasks that require identifying objects.

Damage to certain regions of visual cortex provides support for distinguishing between ventral and dorsal streams. Consider the case of D. F., described by Goodale and Milner (1992) at the University of Western Ontario, who at 34 years of age suffered carbon monoxide poisoning that damaged her visual system, in particular regions involved in the "what" pathway. She was no longer able to recognize the faces of her friends and family, common objects, or even a drawing of a square or circle. She could recognize people by their voices, however, and if you placed objects in her hands, she could tell you what they were. Her condition—*object agnosia,* the inability to recognize objects—is striking in what she can and cannot do. When presented with a drawing of, say, an apple, she cannot identify or reproduce it. But if asked to "draw an apple," she can do so from memory. Despite major deficits in object perception, she can use visual information about the size, shape, and orientation of objects to control visually guided movements—her "where" pathway appears intact. For instance, she can walk across a room and step around things adeptly. She can reach out and shake your hand. Most confounding, in laboratory tests, she can reach out and grasp a block, with the exact right distance between her fingers, even though she cannot tell you what she is going to pick up, or how big it is. Thus, her conscious perception of objects is impaired—she has no awareness that she is taking in any visual information about objects she sees. However, other aspects of visual processing are unaffected. The intact regions of her visual cortex allow her to use information about the size and location of objects despite her lack of awareness. These systems operate independently to help us understand the world around us.

VISUAL FILLING-IN The central characteristic of perception is the construction of information about the world based on sensory information. A process known as *filling-in* offers a fascinating example of the constructive nature of perception. Earlier in this chapter you learned that the information sent from the ganglion cells to V1 is compressed in the sense that it emphasizes information about brightness and colour boundaries and deemphasizes information about regions that are more uniform, such as large swaths of similar colour. Yet our perception includes the whole scene, with an equal emphasis on edges and areas that are more uniform. When you look at a wall you do not only see the corners, you notice the vast expanse of the wall inside its edges. Most perceptual psychologists assume that the early visual areas, such as area V1, actually fill in the constant region information that was thrown out at the level of the ganglion cells. An example of perceptual filling-in is your blind spot, as you might have discovered in Figure 5.11. When the gap falls on your blind spot, your brain fills in the portion of the area that it assumes is missing because of the blind spot. Note that filling-in happens automatically. It is "cognitively impenetrable" because your thought cannot change the perceptual experience. Even though you know that there is a gap out there in the world, you cannot help but see a continuous line when the image of the gap falls on your blind spot (Pylyshyn, 1999). Now that we have seen that perception occurs in the brain, let's see how we actually construct our perceptual worlds.

Object Perception Requires Construction

How do we identify objects?

The neural computations required for object perception begin early in visual processing. We have already seen how lateral inhibition among neurons in the retina helps to accentuate areas of changing stimulation—which are likely to correspond to the boundaries of objects. In addition, the work of Hubel and Wiesel on the properties of neurons in the primary visual cortex strongly suggests that one of the most important roles for V1 is extracting the edges and contours that define the boundaries of objects. Thus it appears that one of the first steps in processing a form is encoding the features that compose it.

The psychologist Anne Treisman and her colleagues developed a clever technique to reveal how these feature detectors can be used (Treisman, 1988). They employed a **visual search task** in which an observer tries to detect a target stimulus among an array of identical distracter stimuli. For example, the display might consist of one vertical line in a crowd of horizontal ones. The observer's job was to respond as soon as he or she saw the target. Treisman and her coworkers found that, for simple stimuli, subjects took no more time to find the target when it was buried in a large number of distracters than when it appeared with only one or two. They termed this phenomenon **pop-out**. The experimenters concluded that these simple scene elements are processed all at once, with oddballs standing out automatically from the crowd. These and other behavioural studies suggest that a vast amount of low-level visual information is automatically detected at the same time. We will see later that Treisman's visual search tasks have provided important information about how we pay attention to objects around us.

visual search task An experiment used to study form perception, in which an observer tries to detect a target stimulus among an array of distracter stimuli.

pop-out The phenomenon whereby, when simple stimuli are used, subjects take the same amount of time to find the target, whether there are a few or many distracters.

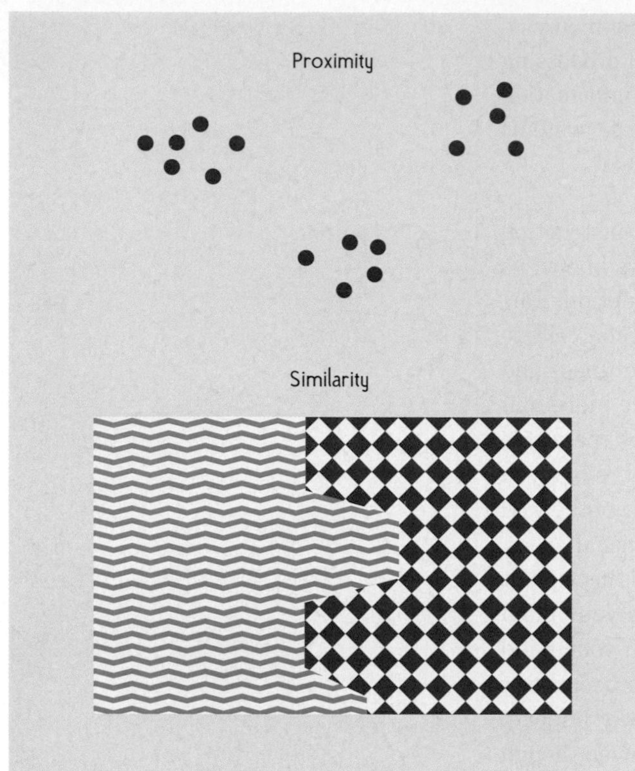

5.21 **Proximity and Similarity** Proximity and similarity are two powerful organizing principles identified by Gestalt psychologists.

GESTALT PRINCIPLES OF PERCEPTUAL ORGANIZATION

So what do we do with the information we take in about an object's features? How do we organize that information? As you will recall from Chapter 1, in the years prior to World War I, psychologists in Germany and North America began theorizing that perception is more than the result of accumulating sensory data. This *Gestalt school* of perceptual psychology, founded by Max Wertheimer along with the experimental psychologists Wolfgang Köhler and Kurt Koffka, contributed much to the study of perception and other fields of psychology. The Gestalt psychologists believed that our perceptions are different from the sum of their constituent sensations. The German word *gestalt* means "shape" or "form," but as used in psychology it means "organized whole." Gestalt psychology holds that our brains use certain built-in (innate) organizing principles to organize sensory information; that is why we perceive, say, "car" as opposed to "metal, tires, glass, door handles, hubcaps, fender." The founders of Gestalt psychology postulated a series of laws to explain how perceived features of a visual scene are grouped into organized wholes.

PROXIMITY AND SIMILARITY Two of the most important Gestalt laws are the principles of proximity and similarity. The *principle of proximity* states that the closer two figures are to each other, the more likely we are to group them together and see them as part of the same object. The principle of similarity is illustrated by the *Sesame Street* song/game "One of These Things Is Not Like the Others." We tend to group figures according to how closely they resemble each other, whether it is in shape, colour, or orientation (Figure 5.21). Both of these principles tend to clump elements of the visual scene into clusters, enabling us to consider them as a whole rather than as their individual parts. For example, we often perceive a flock of birds as a single entity because all the elements—the birds—are similar and in close proximity.

THE "BEST" FORMS Other organizing principles of Gestalt psychology describe how we perceive a form's features. *Good continuation* is the tendency to interpret intersecting lines as being continuous, rather than as changing direction radically. Good contour continuation appears to play a role in completing objects behind *occluders.* For example, in Figure 5.22 the bar appears to be completely behind the

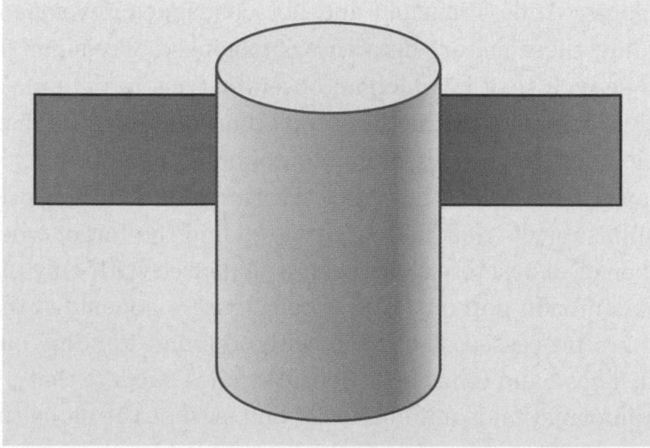

5.22 **Bar Behind Occluders** We perceive the bar to be continuous.

occluder. However, good continuation may operate over more complex representations than contour information. For example, the apparently very long cat in Figure 5.23 appears to wrap around the pole, yet there are no continuous contours permitting this completion to take place. Completion behind the occluder must happen at a level where three-dimensional surfaces are completed. *Closure* refers to the tendency to complete figures that have gaps (Figure 5.24).

A phenomenon that illustrates several Gestalt principles is *illusory contours,* which we perceive even though they don't exist. Illusory contours appear when stimulus configurations suggest that contours ought to be present—for example, when depth transitions are implied or useful in interpreting a pattern (Figure 5.25).

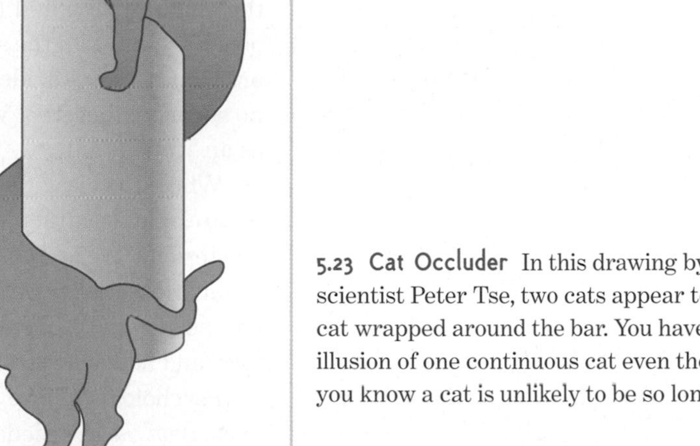

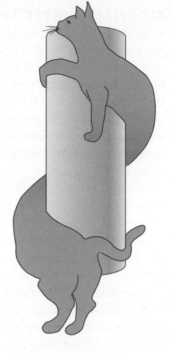

5.23 Cat Occluder In this drawing by vision scientist Peter Tse, two cats appear to be one cat wrapped around the bar. You have the illusion of one continuous cat even though you know a cat is unlikely to be so long.

FIGURE AND GROUND One of our most basic organizing principles is distinguishing between figure and ground. A classic illustration of this is the *reversible figure illusion,* an example of which is in Chapter 1 (see Figure 1.13). In the figure, we see either a full face or two faces looking at each other—but not both at the same time. This illusion shows that our visual perceptual system divides scenes into figure and ground. When we identify a figure, we assign the rest of the scene as the background. In reversible figures, the "correct" assignment of figure and ground is ambiguous, so they periodically switch back and forth as the visual system strives to make sense of the stimulation. This illustrates the dynamic and ongoing nature of visual perception.

In Chapter 1, we described the work of Richard Nisbett and his colleagues (2001), who have demonstrated that there are cultural differences in how people from the East and West perceive the world. Westerners focus on single elements in the forefront, whereas those from Eastern cultures focus on the entire scene more holistically. From this perspective, Westerners are more likely to extract figure from ground, whereas Easterners are more likely to be influenced by both the figure and the ground.

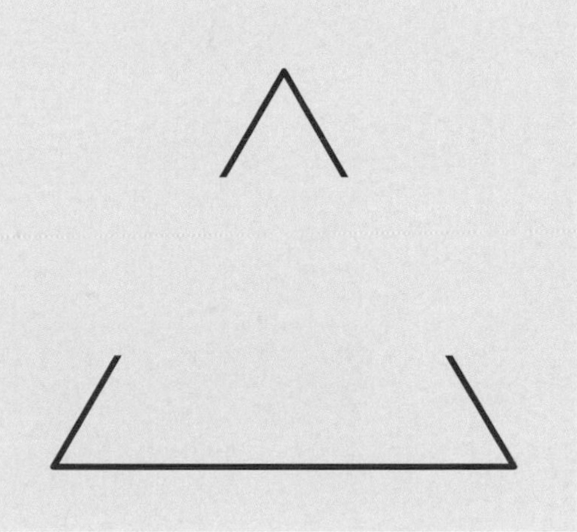

5.24 Closure We tend to complete figures even when gaps exist.

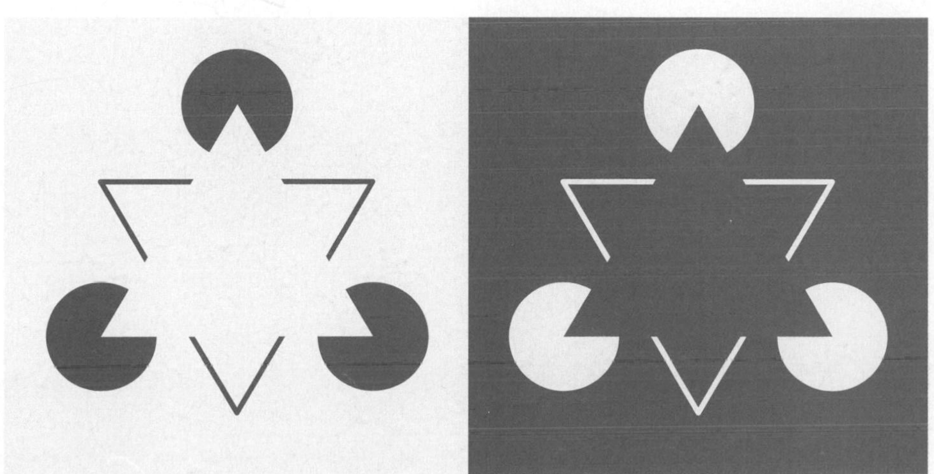

5.25 Illusory Contours Related to closure, we tend to see contours even when they don't exist.

RECOVERING SHAPE FROM IMAGE CUES In visual perception, an object's form (shape) seems to be the most salient cue for the brain to identify it. A square of any size or colour is still a square. The ability to recognize objects by their shape develops at an early age. The psychologist Barbara Landau studied 3-year-olds and found that when scientists showed the children a shape and told them it was a "dax," the children accepted this definition (Landau, 1994). When they were shown other objects, some with the same shape as the "dax" but made of different materials, sizes, or colours, they still identified each object of the same shape as a "dax." It is perhaps no surprise, then, that we recognize a Granny Smith, a Red Delicious, or a Macintosh as an "apple."

While it is clear that form is the most important cue for object recognition, how we are able to extract an object's form from the image on our retina is still somewhat mysterious. We are able to recognize objects from different perspectives and in unusual orientations. We generally have little trouble telling where one object ends and another begins. Consider how we know that a horse and rider are two distinct beings, and not some strange animal.

Psychologists have described several shape-formation models. One of the main ones, *geon theory* (Biederman, 1987; see also Marr, 1982), is built on the idea that objects can be represented as an assemblage of primitive parts called *geons,* which is an abbreviation of the term "geometric ions." Geons are constructed by sweeping a basic two-dimensional shape like a circle or square through space to define a volume. This theory claims that all objects can be represented as a combination of geons, which function as a sort of shape alphabet from which more complex objects can be constructed (Figure 5.26). Some researchers have objected to this theory because they feel that a solution to the problem of shape formation should not be limited to combinations of primitive volumes. They pointed out that many shapes lack a distinct form entirely, such as the surface of the ocean, stalagmites, or swirls of smoke.

Even if the visual system does not use a geon description for recognition, it is likely that some other similar process is used. In general, the visual system attempts to recover the internal properties of objects (surface reflectance, material substance, three-dimensional shape) because these are more or less constant, whereas external properties (lighting, shading, shadows, distance, orientation) are constantly changing.

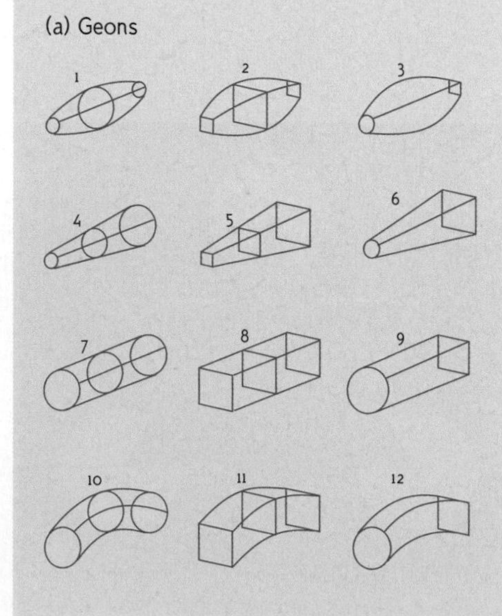

5.26 Geons The geon theory proposes that object recognition is based on identifying patterns of geons. (a) Twelve geons that differ by shape and the extent to which the lines are parallel. (b) A set of common objects with their component geons marked, using the geons from panel a.

Both internal and external information can be derived from an image, and probably both types are stored and used for various tasks, including recognition. Once even a part of an unfamiliar object has been recognized, the object's representation in memory can be used to provide information about other, still unrecognized parts. For example, in the image shown in Chapter 1 (Figure 1.12), it is hard to discern a Dalmatian standing among many black spots scattered on a white background because the part of the image corresponding to the dog lacks contours that define the edges of the dog, and the dog's spotted texture resembles that of the background. Many observers find that they first recognize one part of the dog, say the head, which then makes the whole dog's shape apparent. This is an example of how our model of a dog's shape in memory informs shape processing.

BOTTOM-UP AND TOP-DOWN INFORMATION PROCESSING How do we assemble the information about parts into a perception of a whole object? Most models of pattern recognition are hierarchical, using **bottom-up processing**. This means that data are relayed from lower to higher levels of processing. But perception is actually a combination of both bottom-up and **top-down processing**, in which information at higher levels of processing can influence lower, "earlier" levels in the processing hierarchy, such as helping us perceive the Dalmatian's body once we recognize its head. One illustration of this is the effect of context on perception: What we expect to see influences what we perceive. Consider, for example, the incomplete letter in Figure 5.27; the two lines in the centre of each word are perceived as either "H" or "A" depending on which interpretation would make sense in the context of the word.

How do expectations affect perception?

THE CAT

5.27 Context Context plays a role in object recognition.

If expectations are faulty, it can lead to faulty perceptions. On November 28, 1979, Air New Zealand Flight 901 crashed into the slopes of Mount Erebus on Ross Island in Antarctica, with the loss of all 257 passengers and crew. Several factors contributed to this disaster. The aircraft's flight computer had been incorrectly programmed, so the plane was badly off course. In addition, the pilot had descended below the minimum altitude allowed for the flight. These factors don't explain, however, why the flight crew failed to notice the 12,000-foot volcano looming in front of them until moments before impact.

Psychologists testifying at the commission of inquiry offered a possible, if startling, explanation—the pilots saw what they expected to see. One of the unique hazards of Antarctic aviation is "white out," in which the sky and the snow-covered terrain appear to merge, and pilots are unable visually to distinguish the ground or the horizon. The pilots believed they were hundreds of miles away, flying over the Ross Ice Shelf, so they had no expectation that there would be mountains anywhere near their flight path. The psychologists argued that the few visual cues available to the pilots were sufficiently consistent with what they expected to see that in their minds their expectations were confirmed. Since there appeared to be no danger, the pilots decided—fatally—to reduce altitude to give the passengers a better view of the spectacular Antarctic landscape. The combination of an unusually sparse visual environment and the pilots' own beliefs had conspired to fool their visual systems into seeing terrain that wasn't there—and failing to see the mountain that was. This story conveys the importance of expectation in determining what we perceive. Recall the killing of Amadou Diallo described at the beginning of Chapter 1. The police officers expected Diallo to pull out a gun possibly in part because of stereotypical expectations, as well as the context they were in. We will see in Chapter 15 that the chances of an object being perceived as a gun rather than as a tool are higher when the object is held by a black person than when it is held by a white person (Payne, 2001).

bottom-up processing A hierarchical model of pattern recognition in which data are relayed from one processing level to the next, always moving to a higher level of processing.

top-down processing A hierarchical model of pattern recognition in which information at higher levels of processing can also influence lower, "earlier" levels in the processing hierarchy.

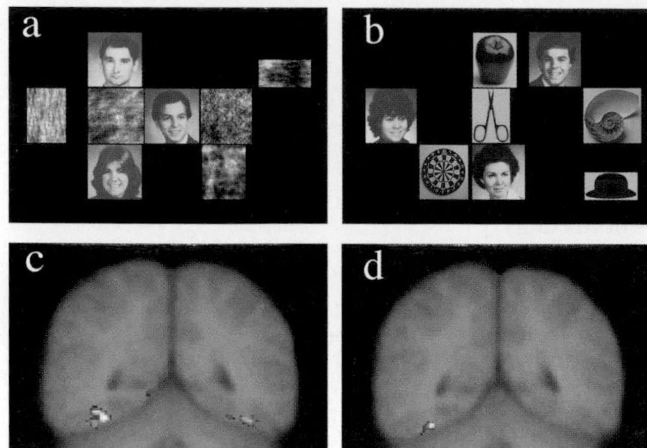

5.28 Perceiving Faces Faces presented within an array of (a) nonobjects or (b) objects activated the fusiform gyrus. This occurred more strongly for the right hemisphere (c and d), especially when faces were presented among objects (d). Note that the right side of the brain (in c and d) is presented on the left side of the image.

FACE PERCEPTION One special class of object that the visual system cares about is faces, and psychological scientists have studied face perception across multiple levels of analysis, from how different brain regions seem to be especially responsive to anything that resembles a face, to how people determine a face's identity (i.e., who it is), to how various characteristics of the face determine whether we find it attractive. As highly social animals, humans have a well-developed ability to perceive and interpret facial expressions. At the biological level of analysis, faces are so important that certain brain regions appear to be dedicated solely to perceiving them. As part of the "what" stream, certain cortical regions, and even specific neurons, seem to be specialized to perceive faces. For example, neurons in the temporal lobes of monkeys became active only when the monkeys are looking at faces (Gross, Rocha Miranda, & Bender, 1972; Perrett, Mistlin, & Chitty, 1987). A number of brain-imaging studies have identified an area of the right hemisphere that seems particularly active during the observation of faces. As mentioned in Chapter 4, researchers such as Gregory McCarthy, James Haxby, and Nancy Kanwisher, working in separate laboratories, have found that a region of the *fusiform gyrus* may be specialized for perceiving faces (see Figure 5.28; Grill-Spector, Knouf, & Kanwisher, 2004; McCarthy, Puce, Gore, & Allison, 1997). Indeed, this brain area responds most strongly to upright faces, as we would perceive them in the normal environment (Kanwisher, Tong, & Nakayama, 1998).

Although some have argued that faces are simply a specialized category of object, there are people who have selected deficits in the ability to recognize faces (a condition known as *prosopagnosia*) but not in the ability to recognize other objects (Farah, 1996). In general, it appears that some areas of the brain, especially the fusiform gyrus, are important for the identification of faces, whereas other areas are sensitive to changes in faces, such as in facial expression and gaze direction. The emotional significance of a face appears to activate the amygdala, which is involved in calculating the potential danger of objects (Adams, Gordon, Baird, Ambady, & Kleck, 2003; Adolphs, 2003). Overall, the evidence suggests that many brain regimes contribute to the perception of faces, which aids in survival.

Interestingly, people have a surprisingly hard time recognizing faces that are upside down. We are much worse at this than we are at recognizing other inverted objects. People have difficulty with inverted faces, especially unknown faces, because such faces interfere with how people perceive the relations between various facial features (Hancock, Bruce, & Burton, 2000). For instance, if the eyebrows are bushier—the change is obvious if the face is upright, but not detectable when the face is inverted. One interesting example of the perceptual difficulties associated with inverted faces is evident in the Thatcher illusion (Thompson, 1980). When you view the faces upside down (see Figure 5.29), they both look pretty normal. However, if you turn your textbook upside down you will notice that one of the faces looks grotesque. Inversion of the whole face interferes with the perception of the individual components. This implies that we pay most attention to the eyes and mouth, and as long as they are in the right orientation the rest of the face appears normal, even if it actually is not.

5.29 Thatcher Illusion These two inverted pictures of former British prime minister Margaret Thatcher look normal. Turn your book upside down to reveal a different perspective.

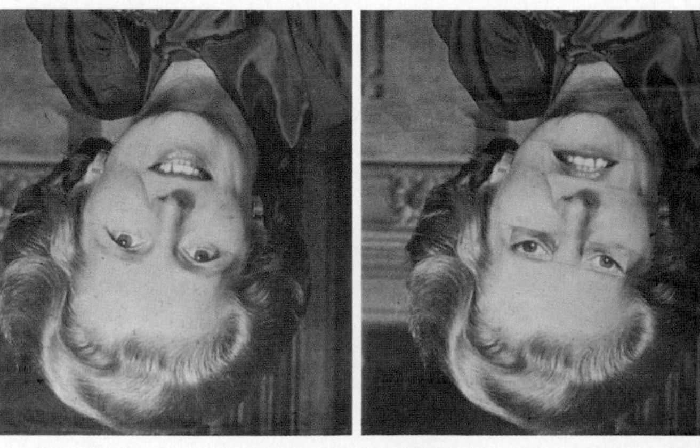

At the social level, people are better at recognizing members of their own ethnic group than members of other ethnic groups. There is some truth to the old saying "They all look

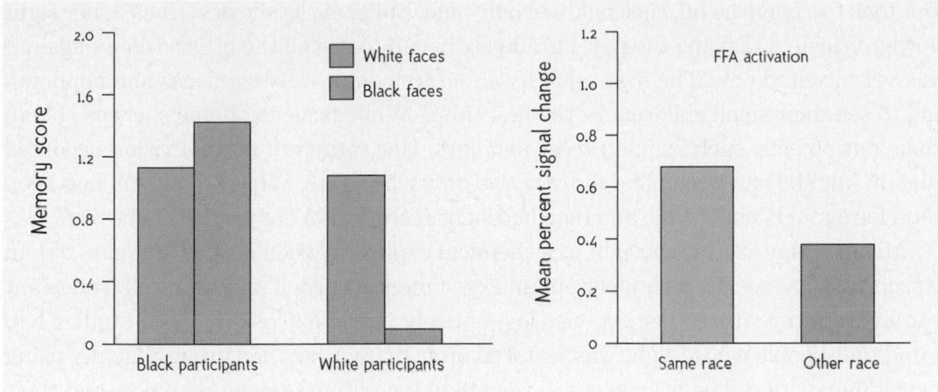

5.30 Face Recognition Both whites and blacks showed better memory for same-race faces than for other-race faces, but this effect was only significant for whites. In both groups, there was greater activation of the fusiform face area (FFA) for same-race faces than for other-race faces. This response might indicate greater exposure to same-race faces, especially for whites.

alike," but the saying applies to all groups. This may occur because people have more exposure to members of their own ethnicity (Gosselin & Larocque, 2000). From this Canadian study of facial morphology and children's categorization of facial expressions of emotions, Gosselin and Larocque were able to determine that Asian faces were categorically more expressive of fear and surprise, whereas Caucasian faces were categorically more expressive of disgust. Indeed, in Canada, where whites greatly outnumber Asians, whites are much better at recognizing white faces than Asian faces (Ng & Lindsay, 1994). An interdisciplinary team of social and cognitive psychologists from Stanford University sought to examine the neural correlates of this same-race effect (Golby, Gabrieli, Chiao, & Eberhardt, 2001). They asked white and black participants to remember pictures from both groups along with pictures of objects (antique radios) as a control comparison. They found that there was a significantly greater activation in a region of the fusiform gyrus for same-race faces compared to other-race faces, especially for whites, and that these activations were associated with better recognition of those faces (Figure 5.30). One possible explanation for why this effect appears to be smaller for blacks is that they are in the minority and therefore have more outgroup exposure than do whites. Thus, cultural context may affect the perception of faces, as well as other objects.

Why might we possess specialized mechanisms for perceiving faces?

Profiles in Psychological Science

Regaining Vision by Fixing the Eyes

People see with their brains, not with their eyes (Enns, 2005). You can prove this to yourself by closing your eyes and thinking about a familiar object. For example, try to picture in your mind the head of a lion. Can you see its mane? What colour is it? Most people can "see" an image, even with their eyes closed. Most people who become blind have lost sensory input from the retina, yet they can still remember what objects look like and visualize them. What happens when we repair the visual system? Does perception return to the way it was before? With advances in medical treatment, many people who became blind due to illness are now regaining their sight. For instance, consider the case of Nogaye Gueye, a 58-year-old woman from Senegal (Figure 5.31) who had lost her sight due to cataracts. Volunteer surgeons from Sight Savers International conducted a 20-minute procedure that restored her vision. In a BBC interview, she describes not wanting to be too excited in case the surgery did not work. Indeed, when the doctors

5.31 Advances in Treatment Nogaye Gueye, who had been blind for two years, regained her vision following a 20-minute cataract-removal surgery.

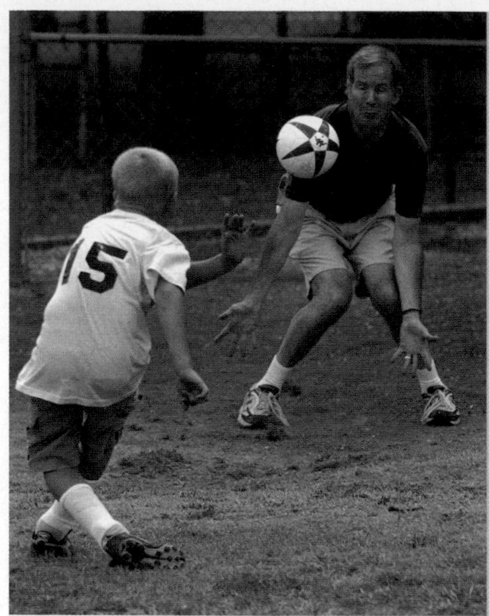

5.32 **Recovering Vision** Michael May, pictured here playing soccer with his son, became blind at age three; his vision in one eye was partially restored by corneal and stem cell implants.

first took the bandage off, she could see only blue. But then, as she describes it, her sight slowly returned: "During the day, I gradually began to see all the other colours again. I was very excited now. The first thing I wanted to look at was my nieces and nephews, and to see their small children for the first time." While traveling home she was able to make out objects, such as palm trees and cars. Her return to normal vision occurred quickly, but she had been blind for only two of her 58 years. What if a person had been blind for a much longer time and then had sight restored? What would that be like?

Michael May lost his eyesight in a chemical explosion when he was 3½ years old. In March 2000, at age 43, he underwent an experimental cornea and stem cell transplant into his right eye (the other eye was lost entirely in the explosion). Prior studies had found that people whose sight was restored after many years had great difficulty using visual information. The psychological scientists Ione Fine, Donald MacLeod, and their colleagues followed May's progress and tested his visual capacities (Fine et al., 2003). Two years after the surgery, May (Figure 5.32) was able to detect colour and motion, but he still had great difficulty with other visual tasks, especially those that required higher-level constructive processes, such as using depth cues and identifying objects. Indeed, he had special difficulty recognizing faces; he could not recognize his wife or children by sight alone, and he had a hard time even judging gender from faces. A brain-imaging study showed that brain areas that normally react robustly to faces and objects were not at all active for May. But the brain regions that process motion reacted similarly to the same brain regions in typical research participants. Motion, however, is still a strange experience for him. May was an expert skier, having learned through verbal instructions. After the surgery he skied with his eyes closed because the overwhelming sensation of motion gave him a frightening sense that he was going to crash. He also became more nervous about crossing roads, which he used to do without much thought. So was having his sight restored a good or bad experience for May? According to May, not a day goes by that he doesn't appreciate the return of his sight.

The case of Michael May provides some interesting insights about visual perception. The brain is fine-tuned through experience; if a cortical region is not used in perception, it ceases to develop normally. Perhaps most interesting is how his brain had adapted to his loss of vision, so that he was highly successful in navigating his environment using his other senses. Regaining his sight added new joys, but his way of interacting with the world was thrown into turmoil by the addition of the novel experience of vision. Even though progress is slow, May continues to develop better vision with hard work. This demonstrates that brain plasticity continues well into adulthood. It will be interesting to observe May's progress over the coming years as his brain continues to learn how to use effectively the sensory cues it receives. ■

Depth Perception Is Important for Locating Objects

One of the most important tasks for our visual systems is to locate objects in space. Without this capacity we would find it difficult to navigate in and interact with the world. One of the most enduring problems in psychological research is how we are able to construct a three-dimensional mental representation of the visual world on the basis of two-dimensional retinal input. The fact that we can see depth in a photograph illustrates this point. A three-dimensional array of objects creates exactly the same image on the retina as a photograph of the array of objects. Despite this inherent ambiguity, we seldom have trouble understanding the arrangement of objects in the world. We actually make use of the ambiguous nature of depth perception when we look at photographs, movies, and television images. We are able to perceive depth in these two-dimensional patterns because the brain is able to apply the same rules or mechanisms that it uses to work out

binocular depth cues Cues of depth perception that arise from the fact that people have two eyes.

monocular depth cues Cues of depth perception that are available to each eye alone.

binocular disparity A cue of depth perception that is caused by the distance between a person's eyes.

the spatial relations between objects in the (three-dimensional) world. To do this, the brain must rapidly and automatically exploit certain prior assumptions it has about the nature of the relationship between two-dimensional image cues and three-dimensional world structure. To understand this process, let us begin by considering the cues that are available to help the visual system perceive depth. These depth cues can be divided into those that arise from the fact that we have two eyes, called **binocular depth cues**, and those that are available to each eye alone, called **monocular depth cues**.

BINOCULAR DEPTH PERCEPTION One of the most important cues to depth perception is the **binocular disparity** (or *retinal disparity*) caused by the distance between your two eyes. Because each eye has a slightly different view of the world, your brain has access to two different but overlapping retinal images. The brain uses the disparity between these two retinal images to compute distances to nearby objects (see Figure 5.33).

Try this simple demonstration: hold your finger out in front of you and close first one eye and then the other. Your finger appears to move because each eye, due to its position relative to the object in question, has a unique retinal image.

The ability to determine the depth of an object on the basis of its different projections to each eye is called stereoscopic vision. In 1838 the physicist and inventor Charles Wheatstone devised the stereoscope, a device that enables a viewer to perceive depth by presenting a pair of two-dimensional pictures—each taken from a slightly different perspective. Wheatstone used this device to demonstrate that depth perception was influenced by binocular disparity, and thus that stereoscopic vision was an important cue for the depths of objects. In 1849 David Brewster, the Scottish physicist who invented the kaleidoscope, discovered the autostereogram—an optical illusion in which you can turn specially designed pairs of two-dimensional images into three-dimensional images by looking at them in an unfocused manner (see Figure 5.34).

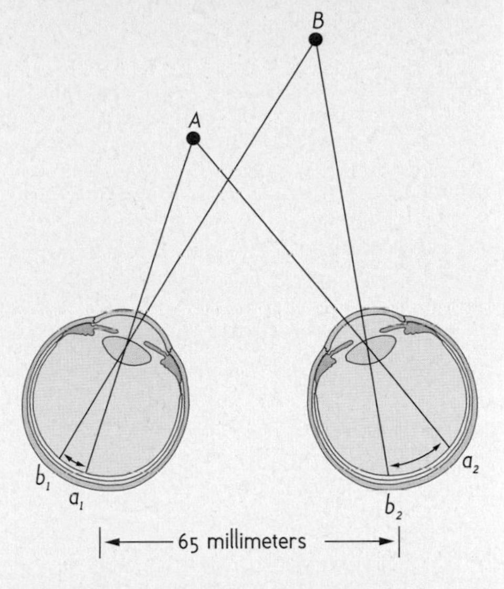

5.33 Binocular Disparity Our two eyes cause us to see every object from two distinct vantage points, resulting in two slightly different retinal images. The distance between the images for objects A and B is different for each eye. This disparity is an important cue for depth.

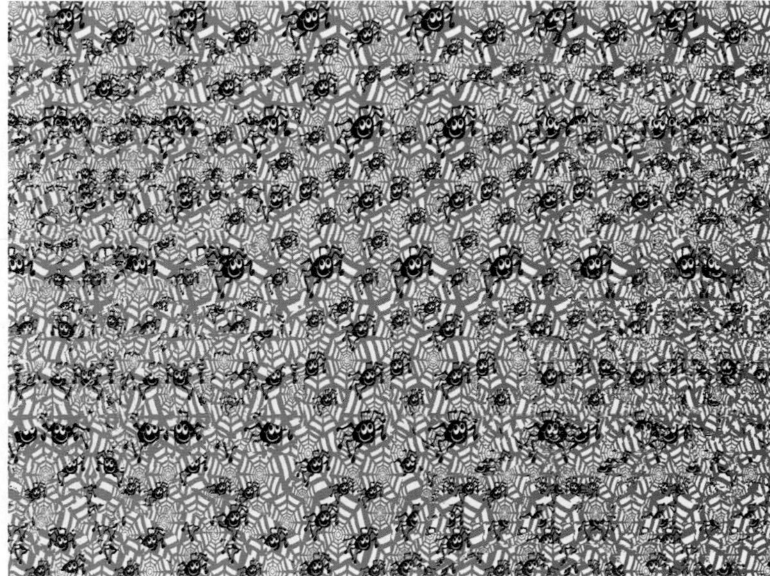

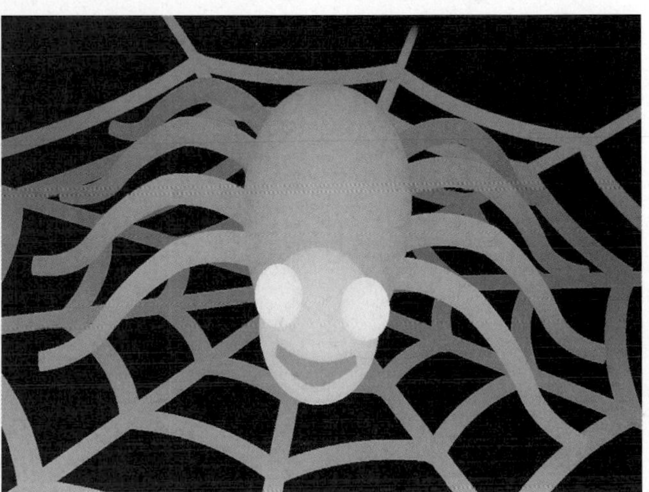

5.34 Autostereogram Hold the picture on the left close to your eyes and stare straight ahead. Try to relax your eyes and let them look through the book to an imaginary point in the distance. Slowly move the book away from you, still fixating on that imaginary point. Try to allow the doubled images to fall on top of each other and keep them there. The result of superimposing the images should be a three-dimensional effect. The picture on the right is the embedded image. If you don't see it in a few minutes, stop and try again later.

MONOCULAR DEPTH PERCEPTION Although binocular disparity is an important cue for depth perception, it is useful only for objects that are relatively close to us. Furthermore, we can still perceive depth with one eye closed. Monocular depth cues allow us to do this. Because artists routinely use these cues to create a sense of depth, they are also called pictorial depth cues. Leonardo da Vinci first identified many of these cues, which include:

- *occlusion:* A near object occludes (blocks) an object that is farther away.
- *relative size:* Far-off objects project a smaller retinal image than close objects.
- *familiar size:* We know how large familiar objects are, so we can tell how far away they are from the size of their retinal images.
- *linear perspective:* Parallel lines appear to converge in the distance.
- *texture gradient:* There is a continuous change in uniformly textured surfaces. As a surface recedes, its texture becomes denser, as shown in Figure 5.35.
- *position relative to horizon:* All else being equal, objects below the horizon that appear higher in the visual field are perceived as being farther away. Objects above the horizon that appear lower in the visual field are perceived as being farther away.

MOTION CUES FOR DEPTH PERCEPTION Motion is also a cue for depth. *Motion parallax* is the relative movement of objects at varying distances from the observer. Imagine sitting in a moving car and watching the scenery. Near objects seem to pass quickly, far objects more slowly (Figure 5.36). If we fixate on a point farther away, such as the moon, it appears to match our speed. If we fixate on an object at an intermediate distance, anything closer moves opposite our direction relative to that object, whereas anything farther moves in our direction relative to the object. Motion cues such as these help the brain to calculate which objects are closer and which are farther away.

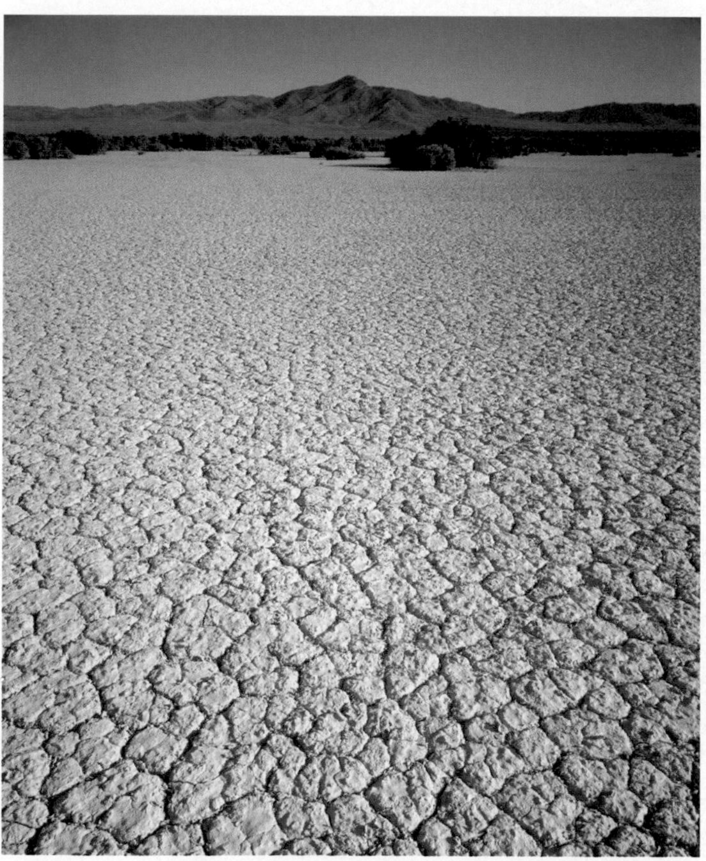

5.35 Texture Gradients Texture gradient is another important depth cue. Uniformly textured surfaces appear denser as objects recede. When texture disappears, it indicates a drop-off.

5.36 Motion Parallax Near objects seem to pass us more quickly in the opposite direction of our movement. Objects farther away seem to move more slowly.

Size Perception Depends on Distance Perception

The size of the retinal image of an object depends on the distance of the object from the observer; the farther away it is, the smaller its retinal image. In order to determine the size of an object, then, the visual system needs to know how far away it is. Most of the time enough depth information is available for the visual system to work out the distances of objects, and thus how big they are. However, in some circumstances size perception fails, and objects look bigger or smaller than they really are (see Figure 5.37). These optical illusions arise when normal perceptual processes result in an incorrect representation of the distal stimulus. Optical illusions are some of the favourite tools perceptual scientists have for understanding the way the brain uses information. Researchers rely on these tricks to reveal automatic perceptual systems that, in most circumstances, result in accurate perception. Optical illusions rely on depth cues to fool us into seeing depth when it is not really there. There are many such illusions. We will concentrate on just three: Ames boxes (also called Ames rooms), the Ponzo illusion, and the moon illusion.

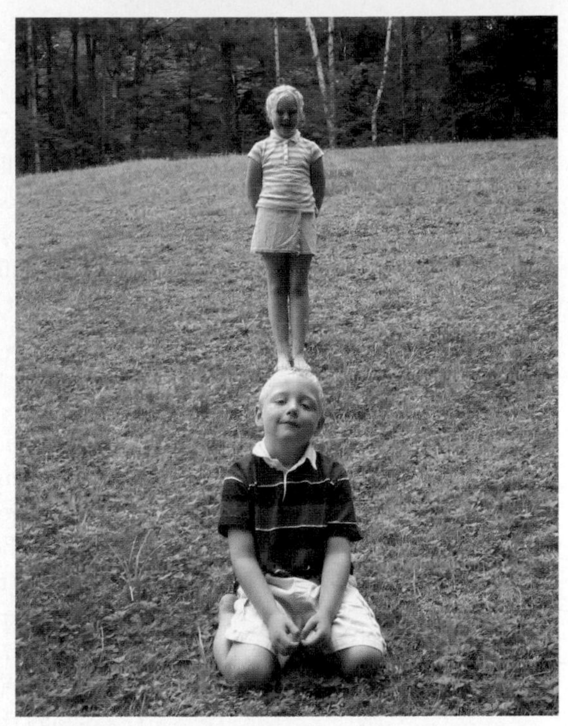

5.37 Distance Perception This picture was taken by 14-year-old babysitter Rebecca Robinson, who captured what appears to be a tiny Sarah Heatherton standing on James Heatherton's head. This illusion occurs because the photo does not capture the hill on which Sarah is standing and so fails to convey depth information.

AMES BOXES Ames boxes, crafted in the 1940s by Adelbert Ames, a painter turned psychologist, are a powerful example of depth illusions. His boxes elaborated on the Victorian trick in which a person looking through a peephole would see a furnished room and then open the door and find the room empty—with a dollhouse room nailed to the door. He built rooms that played with linear perspective and other distance cues. One such perspective illusion was a room that made a far corner appear to be the same distance away as a near corner. Normally, a nearby child projects a larger retinal image than a child farther away but does not appear larger because the perceptual system takes depth into account when assessing size. If the depth cues are wrong, so that the child appears to be farther away than he is—as in the Ames box—the disproportionate size of his image on your retina makes him look huge (see Figure 5.38).

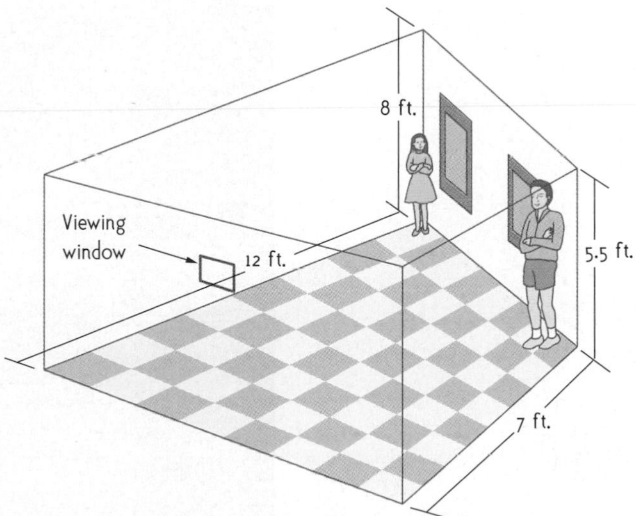

5.38 An Example of the Ames Box Ames played with depth cues to create size illusions. He made a diagonally cut room appear rectangular by using crooked windows and floor tiles. When one child stands in a near corner and another (of similar height) stands in a far corner, the room creates the illusion that they are both equidistant from the viewer; therefore the closer child appears as a giant compared to the child farther away. The Ames room can be experienced by viewers at the Canada Science and Technology Museum in Ottawa, Ontario.

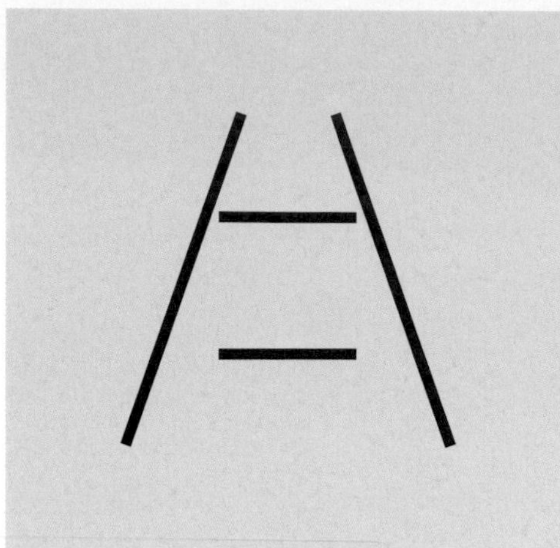

5.39 The Ponzo Illusion A classic size illusion caused by misleading cues of depth perception. The two horizonal lines appear to be different sizes because of the law of perspective that tells us that parallel lines converge in the distance. In fact, the two lines are the same length.

THE PONZO ILLUSION The Ponzo illusion, shown in Figure 5.39, was first described by Mario Ponzo in 1913 and is another classic example of a size-distance illusion. The common explanation for this illusion is that monocular depth cues make the two-dimensional figure seem three-dimensional (Rock, 1984). As noted earlier, parallel lines appear to converge in the distance. The two lines drawn to look like a railroad track receding in the distance trick your brain into thinking that the lines are parallel. Therefore, you perceive the two parallel lines in the centre as if they are at different distances, and thus different in size when they actually are the same exact size. This shows how much we rely on depth perception to gauge size—your brain defaults to using depth cues even when there is no depth present.

THE MOON ILLUSION You may have noticed that a full moon looks much larger when it is near the horizon than when it is overhead (Figure 5.40). This is an illusion—the moon (obviously) remains the same size and distance from the earth, and the image of the moon on your retina is the same size whether it is on the horizon or overhead. The most common explanation for this illusion is that when the moon is near the horizon, several depth cues indicate that it is really very far away. When the moon is overhead, no such cues are available so the moon looks as if it is closer to the earth. The logic of this explanation is similar to that offered for the Ponzo illusion. The horizon moon looks farther away than the overhead moon, yet since they are really the same distance away, they create identical images on the retina. The only way for the brain to reconcile this discrepancy is to assume that the horizon moon is larger than the overhead moon. The mechanism responsible for the apparent difference in distance between horizon and overhead moon is still disputed, but all current theories of the moon illusion agree that perceived distance is the source.

What causes perceptual illusions?

Many perceptual psychologists believe that illusions reveal the operation of the mechanisms that help our visual systems determine the sizes and distances of objects

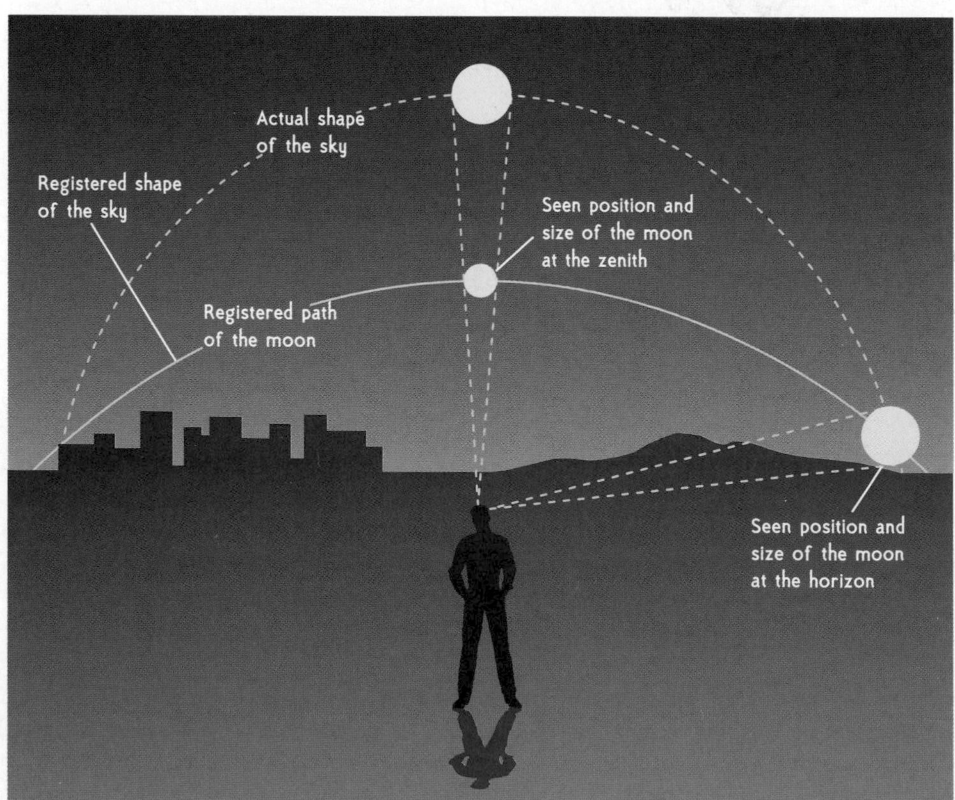

5.40 The Moon Illusion The moon looks larger when it is near the horizon than when it is overhead, due to various visual cues.

in the visual environment. In doing so, illusions illustrate the interdependence of size and distance perception in how we form accurate representations of the three-dimensional world.

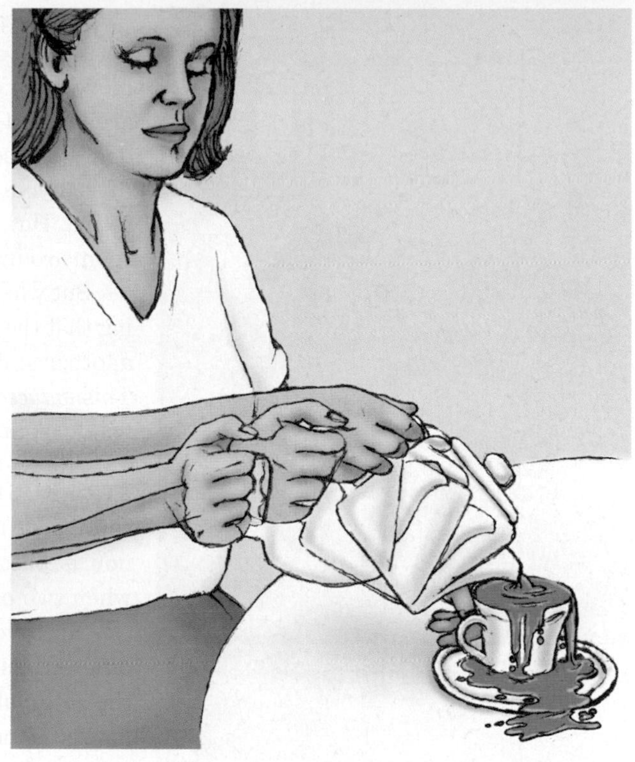

5.41 **Motion Blindness** Patients with motion blindness (akinetopsia) see the world as a series of snapshots rather than as a moving image. Even pouring a cup of tea becomes difficult because by the time they see the cup as full, it is overflowing.

Motion Perception Has Both Internal and External Cues

We know how motion can cue depth perception, but how does our brain perceive motion? One answer is that we have neurons that are specialized for detecting movement—they fire when movement occurs. But how does the brain then know what is moving? If you look out a window and see a car driving past a house, how does your brain know that the car is moving and not the house? Consider the dramatic case of M.P., a German woman who developed selective loss of motion perception following damage to secondary visual areas of the brain. She saw the world as a series of snapshots rather than as a moving image (Zihl, von Cramon, & Mai, 1983). Pouring tea, M.P. would see the liquid frozen in air and be surprised when her cup overflowed (Figure 5.41). Before crossing a street, she would spot a car far away, but when she tried to cross it would be right in front of her. M.P.'s unique deficit has been termed *akinetopsia*—the inability to perceive motion. She could perceive objects and colours, but why couldn't she perceive continuous movement? Unraveling mysteries such as these is the goal of researchers who study motion perception. We will consider three phenomena that offer some insight into how the visual system perceives motion. These are motion aftereffects, compensation for head and eye motion, and stroboscopic motion perception.

MOTION AFTEREFFECTS Motion aftereffects occur when you gaze at a moving image for a prolonged period and then look at some stationary scene. You experience a momentary illusion of seeing the new scene moving in the opposite direction from the moving image. This is also called the *waterfall effect,* because if you stare at a waterfall and then turn away, the rocks and trees will seem to move upward for a moment. Aftereffects are strong evidence that motion-sensitive neurons exist in the brain. The theory behind this illusion combines the phenomenon of sensory adaptation with neural specificity. The visual cortex has neurons that respond to movement in a given direction. When you stare at a moving stimulus for long enough, these direction-specific neurons begin to adapt to the motion, becoming less sensitive, or fatigued. If the stimulus is suddenly removed, the motion detectors that respond to all the other directions are more active than the fatigued motion detectors. This results in the impression that a new scene is moving in the other direction.

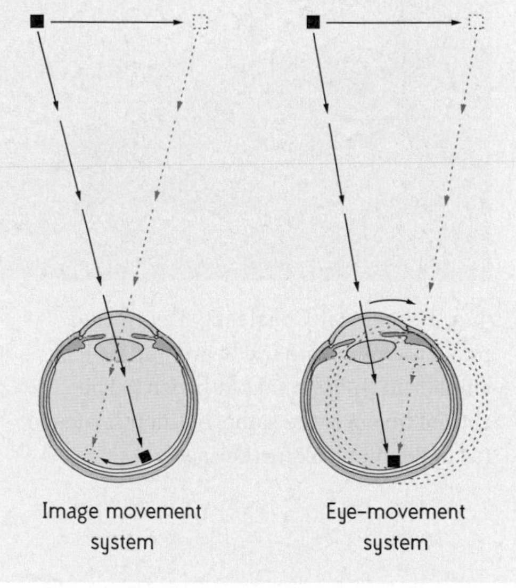

5.42 **Perceiving Movement** There are two ways the visual system detects movement of objects: (1) if the eye is fixed, the image moves across the retina; (2) if the eye moves to follow the object, the retinal image stays in place.

COMPENSATORY FACTORS The fact that we have motion-sensitive neurons doesn't completely explain motion perception. How do you know, for instance, whether an object is moving or whether you, or your eyes, are moving? Images move across your retina all the time and you don't always perceive them as moving. Each slight blink or movement of the eye creates a new image on the retina. Why is it that every time you move your eye, or your head, the images you see don't jump around? One explanation is that the brain calculates objects' perceived movement by monitoring the movement of the eyes or head as they track a moving object. In addition, motion detectors track the motion of an image across the retina, receptors in the retina firing one after the other (see Figure 5.42).

5.43 **Perceptual Constancy** Depth cues make the man with the beard and tie look smaller in (b) than (a) but he isn't. The retinal image is the same, but depth cues in (a) make him look further away.

perceptual constancy People correctly perceive objects as constant in their shape, size, colour, and lightness despite raw sensory data that could mislead perception.

Another explanation for movement perception is the *frame of reference effect:* The perceptual system establishes a stable frame of reference and uses it to assess the movement of other objects. Suppose you are sitting in the library reading, and around you are dozens of other students studying. While you are watching one student, he rises and moves down two seats. Since you were watching him move, you moved your head, or at least your eyes, prompting an array of images across your retina. However, your brain knows that it was the student who moved, and not the room around him, because it had established a frame of reference.

But when that frame of reference is the wrong one, you can be tricked into thinking that the wrong object is moving. Say you are on a train looking out the window at another stationary train. That train starts to move—and you think you are moving. This *induced movement illusion* also happens when you look up at the sky on a cloudy, windy night and it seems as if the moon is moving through the clouds.

STROBOSCOPIC MOVEMENT Movies are made up of still-frame images, presented one after the other to create the illusion of a "motion picture." This phenomenon is based on a perceptual illusion called *stroboscopic movement,* which occurs when two or more slightly different images are presented in rapid succession. The Gestalt psychologist Max Wertheimer conducted experiments in 1912 by flashing, at different intervals, two vertical lines placed close together. He discovered that when the interval was less than 30 milliseconds, subjects thought the two lines were flashed simultaneously. When the interval was greater than 200 milliseconds they saw two lines being flashed at different times. Between those times movement illusions occurred. When the interval was about 60 milliseconds the line appeared to jump from one place to another; at slightly longer intervals, the line appeared to move continuously—this is called *phi movement.*

Perceptual Constancies Are Based on Ratio Relationships

As we discussed earlier, illusions are inaccurate representations of stimuli. The opposite situation is called **perceptual constancy**, in which we correctly perceive objects as constant despite sensory data that could lead us to think otherwise. Consider the profile of the Queen's head on a Canadian penny. At first glance, it seems like a perfect representation of the Queen's head. But if you were to meet someone with as flat a head as this, you might be shocked. The reason a profile can provide such a good likeness is that the relative positions and orientations of features on the head are the same as they would be for an actual head. Similarly, how does the brain know that a person is six feet tall when the retinal image of that person changes size according to how near or far he is (Figure 5.43)? How does the brain know that snow is white and a tire is black, even when snow at night or a tire in bright light might send the same luminance cues to the retina? For the most part, changing the angle, distance, or illumination of an object does not change our perception of that object's size, shape, colour, or lightness. But to perceive any of these four *constancies,* we need to understand the relationship between at least two factors. For instance, for *size constancy* we need to know how far away the object is from us. For *shape constancy* we need to know from what angle we're seeing the object. For *colour constancy* we need to compare the wavelengths of light reflected from the object with those from its background. Likewise, for *lightness constancy* we need to know how much light is being reflected from the object and from its background. In each case, the brain is computing a ratio based on the relative magnitude rather than relying on the absolute magnitude of each sensation. The ability of the perceptual system to make relative judgments is what allows constancy to be maintained across a variety of perceptual contexts. Although the precise mechanisms of the constancies are unknown, they illustrate that perceptual systems are tuned to detect change from some baseline condition, and not just to respond to sensory inputs.

How do perceptions remain constant when sensory information changes?

HELMHOLTZ'S VS. GIBSON'S DIRECT-PERCEPTION THEORIES Broadly speaking, there are two approaches to the study of how we arrive at these ratio relationships. Classical perceptual theory, which originated with the nineteenth-century works of Hermann von Helmholtz, assumes that perceptual experience and perhaps innate knowledge provide the information about the ratio relationships that determine constancies. Thus, sensations create impulses to the brain, and the brain then interprets those impulses according to what it has learned in the past or had hard wired into it over the course of evolution. Earlier in this chapter we pointed out that the visual system must have prior assumptions in place about the relationship between two-dimensional retinal activation and the nature of events and objects in the three-dimensional world in order to be able rapidly to construct a representation of the world from the retinal image. These prior assumptions allow the brain to infer the correct state of the world from highly ambiguous retinal information. In this view, visual perception is always based on *unconscious inferences:* The brain is actively "connecting the dots" provided to it by sensation via automatic and unconscious Gestalt grouping procedures and prior assumptions that it has learned or inherited.

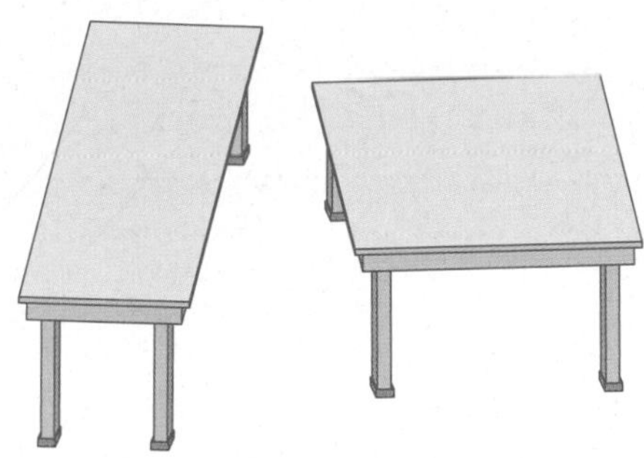

5.44 The Tabletop Illusion This unbelievable tabletop illusion of Roger Shepherd's demonstrates the automatic perceptual processes of the brain. Even when we know the two surfaces are the same, we cannot make ourselves "see" it.

Studying how illusions work, many perceptual psychologists have come to believe that the brain has built-in prior assumptions that influence perceptions. The vast majority of visual illusions appear to be beyond our conscious control—we cannot make ourselves not see illusions. As an example, look at the two tables in Figure 5.44. Believe it or not, the tops of the tables are the exact same size and shape, not just the same surface area. Our brains will not allow us to accept that fact, however. Even if you trace one and place it on top of the other, your brain still cannot see it. The reason is that you have automatic perceptual processes that use perspective cues to tell you the size and shape of the tables are different.

Beginning in the 1940s, the perceptual psychologist James J. Gibson offered an alternative to classical perceptual theory, *direct perception.* Gibson (1966) approached the study of perception from an evolutionary perspective. He believed that to understand visual perception, the most important question to ask is: What is vision good for? In the most basic sense, the function of vision is to identify and locate predators and prey, find mates, avoid cliffs, and so on. The idea that our brains would have to rely on inference, on learning and memory, to form concepts that were basic to survival didn't make sense to him. In his view, the visual system is not built to enable us to see an exact copy of the real world; it is built to interpret cues that maximize its function. His direct-perception theory, then, stated that stimuli already have enough information for us to perceive them—we don't need additional memories or calculations to understand the sensory data. From this perspective, the brain does not have to construct representations of the three-dimensional world, it just needs to extract useful information from light. Whereas Helmholtz might have said that your brain effectively adds information that is not present in the retinal image when it "connects the dots" provided by sensation, Gibson simply rejected the notion that perception involves the construction of any information in addition to what is already available in the light itself.

Modern perceptual psychology has synthesized the ideas of Gibson and Helmholtz with Gestalt insights. Most perceptual psychologists now accept that the brain makes rapid inferences on the basis of prior assumptions about the relationship of the retinal image to the three-dimensional world, as Helmholtz first argued. Many of these inferences appear to arise from the global grouping procedures first discussed by the Gestalt psychologists. Yet there appears to be truth in Gibson's notion that much of perception is inaccurate but adequately useful. The big puzzle that occupies the minds of modern psychological scientists is unraveling the nature of the connection between electrochemical activity within neuronal circuits and the complex information processing that culminates in our perception of the world.

REVIEWING the Principles | What Are the Basic Perceptual Processes?

The perceptual system is stunningly intelligent. It take ambiguous sensory information and constructs rich and meaningful experiences that allow us to navigate the world around us. All perception takes place in the brain, in which a variety of perceptual processes take incoming sensations and construct them into percepts. Information first arrives in primary sensory regions, such as V1 for vision and A1 for audition, but multiple brain regions contribute to our unified perceptual experience. The perceptual system uses cues from the environment to help interpret sensory information. For instance, visual cues help provide information about what objects are and where they are located, as well as other cues that provide information about depth and motion. Contemporary theorists emphasize that perceptions are not faithful reproductions of the physical world, but rather are constructed based on multiple processes that allow us to taste, smell, touch, hear, and see.

How Does Attention Help the Brain Manage Perceptions?

The brain is constantly receiving input from an enormous number of sources. How does the brain combine all this information into coherent, usually correct, perceptions? And how do we manage all of the sensory information that constantly bombards us? The study of attention is the study of how the brain selects which sensory stimuli to discard and which to pass along to higher levels of processing. Attention is important to our ability to function in our daily lives, as we try to focus attention on the task at hand and ignore all of the things that might distract us. Even having a decent conversation requires paying attention to the speaker, which might be difficult when the person has unusual facial features (such as a wandering eye) or if there is food hanging from the chin. In these cases, the unusual features capture our attention and make it difficult to comprehend what the person is saying. How does the brain know which features of the environment are most important and therefore are most deserving of our attention? In this section we consider the basic principles of how human attention works.

Visual Attention Is Selective and Serial

Can we perceive something without attending to it?

As you learned earlier, the psychologist Anne Treisman proposed that we identify "primitive" features automatically (colour, shape, orientation, etc.). Her theory of the automatic recognition of these features has been a major development in the study of attention. Treisman proposed that different visual features of objects are analyzed by separate systems. These systems process information in parallel (at the same time), and we can attend selectively to one feature by effectively blocking the further processing of the others (Treisman & Gelade, 1980). Recall that in Treisman's visual search tasks (called feature search tasks), participants were looking for targets that differed in only one feature, and that these would pop out immediately, regardless of the number of distracters.

A very different picture emerges, however, if the participants are looking for a target that is harder to discriminate from the distracters. In conjunction search tasks, observers are asked to search for a target with two or more features. In Figure 5.45, for example, the subject might be asked to search for a green T among red Ts and green Xs. In this case the target can be distinguished from the distracters only by the conjunction of its features—it shares colour with one half of the distracters and shape with the other half. In contrast to the immediate pop-out found in feature search, the presence of distracters markedly affected the participants searching for conjunctions. The time it takes to find the target increases linearly with the number of distracters—suggesting that subjects labouriously consider each distracter one by one.

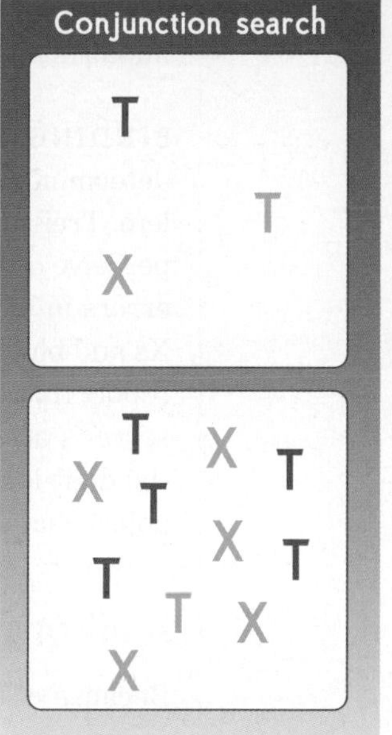

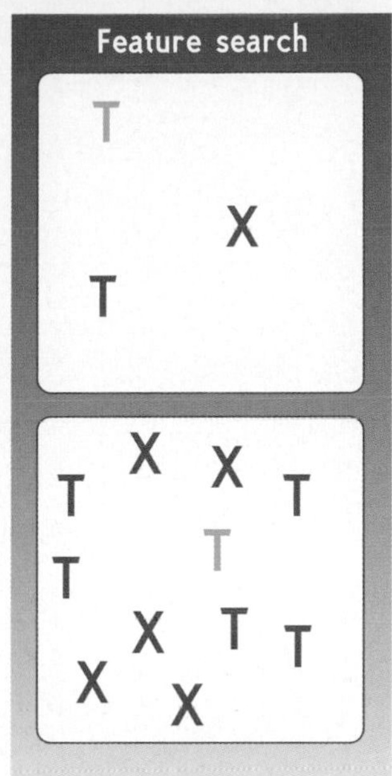

5.45 Visual Search Tasks In visual search tasks, subjects are asked to find the target—the thing unlike the others. In feature searches, in which the target has only one feature (colour or shape) that differs from the distracters, this is simple. When the target has more than one feature that differs (for instance, the green T among red T's and green X's), it takes longer.

These studies led Treisman to propose a two-stage theory of visual-image processing. According to this theory, visual processing begins with a rapid, parallel extraction of elementary features, whose distribution the visual system is simply mapping. Separate maps are created for each feature type—one for colour, one for form, and so forth. The second stage of processing is slower and more effortful as this visual system combines the features from the various maps to form objects. This stage is not automatic—it requires the deployment of attention. Because of this, only selected regions of the feature maps get bound together. In feature searches, the average time it takes viewers to respond to targets is not affected by the number of distracters. This is the hallmark of preattentive searches (searches that do not require attention); the entire array is scanned in parallel, and the critical feature is rapidly registered if it is present. Conjunction searches, however, require attention and are processed serially, so the search time lengthens as more distracters are included.

SEGREGATION OF VISUAL FUNCTIONS Other evidence suggests that elementary visual processes proceed in parallel and are integrated relatively late in processing. For example, consider the versions of the Ponzo illusion shown in Figure 5.46. In the left red-on-green version, the illusion remains, but in the version on the right the illusion is much less effective. This is because in the right version the red and green have been set to the same approximate luminance, such that the figure and background are distinguished only by their colour; in contrast, the left version has both colour and brightness differences. Colour and depth are processed in separate pathways in the visual cortex, so the depth information is lost in the right version. Without depth information, the parallel lines appear equal in length.

Cases in which people suffer visual deficits as a result of brain injuries also support the view that humans have distinct perceptual systems for processing different visual features. For instance, colour-blind people do not have problems with depth and texture perception—they can still see and recognize objects. The akinetopsia patient described

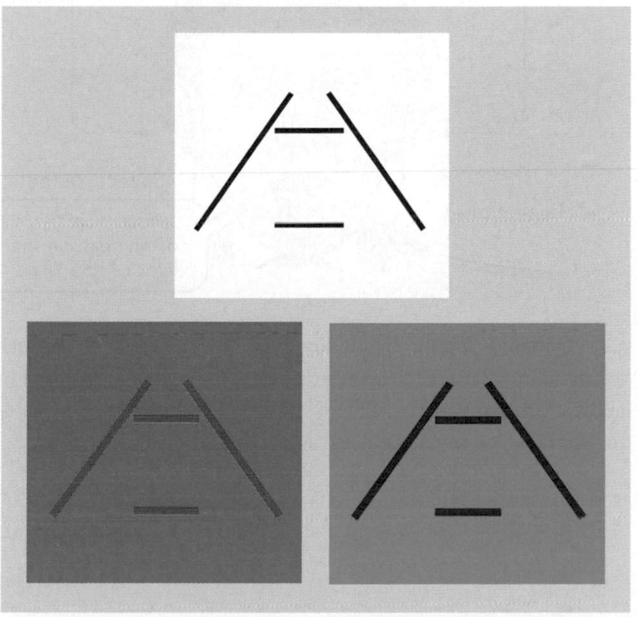

5.46 The Ponzo Illusion The Ponzo illusion is effective in all but the third panel. When the background and lines are the same brightness, depth cannot be perceived. This indicates that visual processing of brightness and depth are separate.

earlier apparently has normal processing of colour and form, despite her profound impairment in processing motion.

BINDING If the brain processes features automatically and separately, how does it determine which features go with which objects? This is known as the binding problem. Treisman has shown that attention helps integrate features so that we correctly perceive objects. When a normal observer is given too many cues to pay attention to, errors in binding occur. For example, in tests in which people are briefly shown red Xs and blue Os and are asked to name a digit presented at the same time, they often report that they saw a blue X or red O. These illusory events happen because the observer's attention was divided and overextended between the coloured letters and the digit-identification task. As a result, features from objects recombine to form an object that doesn't exist. The binding system breaks down.

Auditory Attention Allows Selective Listening

What captures attention?

Because of attention, it is hard to do two tasks at the same time, especially if they rely on the same mechanisms. We easily can listen to music and drive at the same time but find it hard to listen to two conversations at once. Driving and listening to the radio at the same time can even be hazardous depending on what you're listening to—a sports broadcast may engage your visual system to imagine the game in progress, diverting attention away from the visual cues on the road ahead.

The **cocktail party phenomenon**, so called by the British psychologist E. C. Cherry (1953), refers to how you can focus on a single conversation in the midst of a chaotic cocktail party, yet a particularly pertinent stimulus, such as hearing your name mentioned in another conversation, can capture your attention. While proximity and loudness contribute, your selective attention can also determine which conversation you hear. Imagine you are having dinner with a friend at a crowded restaurant. Suddenly the conversation at the next table piques your interest. If you really want to hear that conversation, you can focus your attention on it rather than on what your (closer and therefore louder) friend is saying. When your friend notices the blank look on your face and protests, "You're not listening to me!" chances are you won't be able to tell your friend what he or she said—but you can relate the drama at the next table.

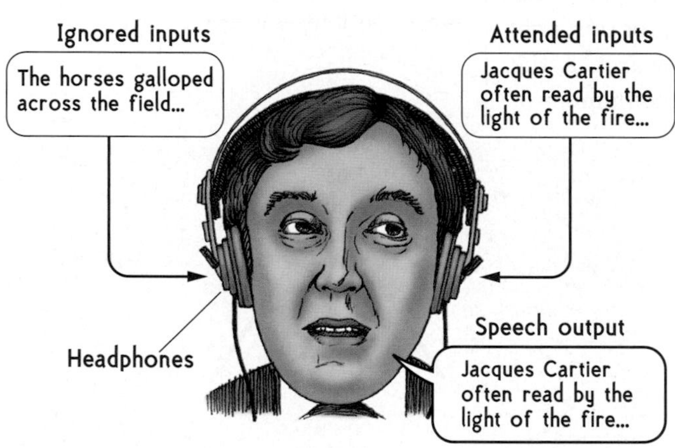
Ignored inputs
The horses galloped across the field...
Attended inputs
Jacques Cartier often read by the light of the fire...
Headphones
Speech output
Jacques Cartier often read by the light of the fire...

5.47 Listening Inputs In shadowing experiments, subjects receive different auditory messages in each ear but are asked to "shadow" or say aloud only one.

Cherry developed selective-listening studies to examine this aspect of attention. He used a technique called shadowing, in which a participant wears headphones, with one ear receiving one message, and the other ear receiving a completely different message. The person is asked to attend to one of the two messages and "shadow" it by repeating it aloud. In this situation, the subject will usually notice the unattended sound (the message given to the other ear), but will have no idea about the content of that message (Figure 5.47).

Selective Attention Can Operate at Multiple Stages of Processing

The psychologist Donald Broadbent (1958) developed the **filter theory** to explain selective-listening findings. He assumed that we have a limited capacity for sensory information and thus screen incoming information, only letting in the most important. In this model, attention is like a gate that opens for important information and closes for information to be ignored. But can we really close the gate for irrelevant information? When and where is the gate closed?

Broadbent's filter theory is an example of **early selection**, which is the theory that we can choose which stimuli we will attend to even before we process their basic features. In visual processing we could choose to ignore features such as colour or form and process only those features that we had decided would be important for the task at hand. Accordingly, attention effectively filters out certain features at an early stage so that they never get processed.

However, some studies reveal that "unattended" visual information is processed to at least some extent. **Late selection** theories assume that we take in sensory information, process it, and then select which aspects of the stimuli should be attended. The attention stage occurs when the perception enters our conscious awareness. Several studies of selective listening found that although a subject could not repeat an unattended message, he or she still processed its contents. Fascinating studies demonstrate that if the unattended message is relevant, we automatically use the information. For instance, subjects attended to the phrase "they threw stones at the bank yesterday," while at the same time either ear was presented with one of two words: "river" or "money." Afterward, subjects could not report which of these two words they heard; however, if they had been presented with "river," they interpreted the sentence to mean that someone had thrown stones at a riverbank. If they had been presented with the word "money," they interpreted the sentence to mean someone had thrown stones at a financial institution (MacKay, 1973). Thus, they extracted meaning from the word even though they were not aware of it.

It now seems clear that attentional selection is not an all-or-none process that eliminates unattended stimuli, as proponents of early-selection theories argued. Nor is unattended information fully processed before selection occurs, as suggested by late-selection theories. Rather, most modern theories of attention suggest that selective attention can operate at multiple stages of perceptual processing. A variety of studies has shown that the processing of attended stimuli is enhanced relative to that of unattended stimuli, but that unattended stimuli are still processed. Treisman proposed that unattended information is not "gated" completely but instead is simply reduced. Broadbent, the major proponent of the filter theory, agreed and amended his view.

cocktail party phenomenon The ability to focus on a single conversation in the midst of a chaotic cocktail party or other similarly noisy situation.

filter theory A theory that people have a limited capacity for sensory information and thus screen incoming information, letting in only the most important.

early selection theory A theory that we can choose the stimuli to which we will attend before we process their basic features.

late selection theory A theory that people take in sensory information, process it, and then select which aspects of the stimuli should be attended after processing.

Psychological Science *in Action*

Making the Roads Safer for All

Attention is important to many aspects of our daily lives and crucial for others, such as driving. Failing to pay attention is a major cause of car accidents, which are a serious problem around the globe, injuring or killing millions. Accidents have two main causes: bad road conditions and bad drivers. Both of these risk factors are ultimately related to perception. Thus, psychological scientists conduct research on the conditions that produce unsafe roads or on the types of people who are likely to have accidents (e.g., young drivers or very elderly drivers). For example, research has found that younger drivers tend to spend much longer looking at road hazards (Chapman & Underwood, 1998), which means they are less likely to notice other hazards and therefore more likely to crash. Other research shows that drivers have greater difficulty driving at night because of low illumination (Leibowitz, 1996). Because people are not aware of this degradation of their perceptual abilities, they often drive as

5.48 Making the Roads Safer In Canada, chevrons help drivers better perceive the distance between their cars and the cars in front of them. By keeping the proper number of chevrons behind another vehicle, a driver is simultaneously a safe distance back and free to concentrate on other potential road hazards that may occur, reducing the human-factor involved in driving.

quickly at night as they do in the day, the result being an increased likelihood of crashing. Due to the effects of aging on the visual system, older drivers are likely to experience difficulty at night or when contrast is reduced, as in rain or fog.

Ultimately, many accidents are due to limitations of perception and attention. For instance, road glare can overwhelm the visual system. People can fail to notice sharp turns, or they can become confused by traffic signs or strange intersections. Human-factors researchers are psychologists who help road designers build safer roads and car makers build safer cars. On busy highways in Canada, for example, there are markers on the road at fixed distances that help drivers calculate how much space to leave between their car and the car in front of them (Figure 5.48). Signs advise drivers to keep two "chevrons" in distance, which works no matter how fast you are driving. As drivers exit from highways, some roads contain horizontal lines that get closer together near the end of the exit. This produces an illusion of speed that encourages drivers to slow down even more. Car makers also have worked with psychologists to build safer cars with new features, such as daytime running lights, larger and more prominent brake lights, and even sonar that warns drivers that they are too near to objects, and adaptive cruise control that keeps them multiple car lengths from vehicles in front of them.

Some psychologists focus on bad drivers, such as those who eat, talk, read, put on their makeup, or are impaired while driving. In Canada, 3,124 deaths resulted from alcohol-related crashes in 2003 (Traffic Injury Research Foundation of Canada, 2005), and car accidents remain the leading cause of death among Canadian youth (Mayhew, Singhal, Simpson, & Beirness, 2004). Paying attention is an important aspect of safe driving. Indeed, beyond impaired driving, being distracted or inattentive is one of the major causes of car accidents. As we saw earlier, attending to multiple things at the same time is difficult, as attention has a limited capacity. Simulated-driving experiments have shown that talking on a cell phone is especially dangerous, impairing driving skills nearly as much as alcohol. For instance, talking on a cell phone interferes with drivers' braking reactions to cars in front of them, possibly because talking impairs visual processing (Strayer, Drews, & Johnston, 2003). Interestingly, researchers from Spain have demonstrated that conversations in general may be dangerous. They found that talking either on a cell phone or to a passenger impaired visual processing in real traffic conditions (Recarte & Nunes, 2003).

Even when stressed, drivers need to be able to pay attention. One group that is especially likely to encounter stressful driving situations is police officers, who sometimes need to pursue fleeing suspects. Police chases have come under increased public scrutiny because of their risks to the general public. Some 40 percent of police chases end in crashes, with about 1 percent ending in a fatality (Hill, 2002). Psychologists help to train police officers to control their emotions while driving, in part by recognizing the adrenaline rush they feel as they start the chase. Police officers are also trained to pay attention to possible hazards during a chase. One study in England examined how police respond to the visual demands made on them during a chase (Crundall, Chapman, Phelps, & Underwood, 2003). The researchers found that police officers were more likely than other drivers to scan the horizon with quick visual fixations, allowing them to attend to more of the road and possible hazards.

The various types of research that we've just considered provide excellent examples of how psychological research in the laboratory can be applied to help solve real-world problems. At the same time, working in applied settings often enriches the questions that psychological scientists address. The goal of psychological research, as we mentioned in Chapter 1, is to explain the most important human behaviours in real-life contexts. Applying principles developed in perceptual psychology can help make the roads safer for all. ■

REVIEWING

the Principles | How Does Attention Help the Brain Manage Perceptions?

Attention is how the brain selects which sensory stimuli to ignore, and which to pass along to higher levels of perceptual processing. Visual search tasks indicate that we process visual information about primitive features along parallel path ways. Recombining information from these pathways may require attention. A key aspect of attention is that it is selective; we have the ability to choose the stimuli to which we attend. Despite this, we can still process some information contained in sensory stimuli to which we are not consciously attending. It is now clear that attention can operate on multiple stages of perceptual processing, and that unattended stimuli are reduced rather than eliminated from further processing.

Conclusion

Perceptual psychologists seek to understand how elementary sensations are translated into conscious perception. Our entire experience of the world relies on this process. Although it seems effortless, perceiving relies on numerous active processes that work together to construct our experience. Our senses gather information from relatively limited sources of physical energy. We hear few of the available sound waves and see only a small portion of the electromagnetic spectrum. But our experience does not feel limited, because these perceptual systems allow us to perform the tasks necessary for our survival and reproduction. After over a century of research, science still cannot state definitely how perception works. No single theory or model has yet successfully accounted for more than a fraction of perceptual phenomena. Despite this, the workings of our perceptual systems have begun to yield to the concerted efforts of psychological scientists working at many levels of analysis. With the advent of new technologies for studying the brain, these efforts have increased in their urgency. Perhaps the twenty-first century will finally yield answers to two of humankind's most enduring questions: How do we know our world? And how are mental events realized in the physical events of the brain?

Summarizing the Principles of Sensation, Perception, and Attention

How Do We Sense Our Worlds?

1. **Stimuli must be coded to be understood by the brain:** Stimuli reaching the receptors are converted to neural impulses through the process of transduction.
2. **Psychophysics relates stimulus to response:** By studying how people respond to different sensory levels, scientists can determine thresholds and perceived change (signal-detection theory). Our sensory systems are tuned to adapt to constant levels of stimulation and detect changes in our environment.

What Are the Basic Sensory Processes?

3. **In gustation, taste buds are chemical detectors:** The gustatory sense uses taste buds to respond to the chemical substances producing basic sensations of sweet, sour, salty, and bitter. The amount and concentration of taste buds vary individually.
4. **In smell, the nasal cavity gathers particles of odour:** Receptors in the olfactory epithelium respond to chemicals and send signals to the olfactory bulb in the brain. Pheromones are particular chemical signals linked to physiological responses in animals.

5. **In touch, sensors in the skin detect pressure, temperature, and pain:** The haptic sense relies on tactile stimulation to activate receptors for temperature, sharp and dull pain, and other sensations. Neural "gates" in the spinal cord also control pain.

6. **In hearing, the ear is a sound-wave detector:** The size and shape of sound waves activate hair cells in the inner ear. The receptors respond depending on frequency of the sound waves, timing, and the location of the activated receptors. Having two ears allows us to locate the source of a sound.

7 **In vision, the eye detects light waves:** Receptors (rods and cones) in the retina detect different forms of light waves. The lens helps the eye focus the stimulation on the retina for near versus far objects. Colour is determined by wavelengths of light activating certain types of cones, by the absorption of wavelengths by objects, or the mixing of wavelengths of light.

What Are the Basic Perceptual Processes?

8. **Perception occurs in the brain:** The primary auditory cortex handles hearing. Touch is handled by the primary somatosensory cortex. Vision results from a complex series of events in a variety of areas of the brain but primarily in the occipital lobe.

9. **Object perception requires construction:** By using the Gestalt principles of stimulus organization, we are able to perceive our world. We use cues about similarity, proximity, form, figure and background properties, and shading. Perception involves the dual processes: bottom-up (sensory information) and top-down (brain organization).

10. **Depth perception is important for locating objects:** The pattern of stimulation of an object on each of the two retinas (binocular) informs the brain about depth. Pictorial (monocular) cues use information from the appearance of objects relative to their surroundings to inform about depth, as well as relative motion.

11. **Size perception depends on distance perception:** Illusions of size can be created when the retinal size conflicts with the known size of objects in the visual field, such as with the Ames, Ponzo, and moon illusions.

12. **Motion perception has both internal and external cues:** Motion detectors in the cortex respond to stimulation. The perceptual system establishes a stable frame of reference and relates object movement to it. Intervals of stimulation of repeated objects give the impression of continuous movement.

13. **Perceptual constancies are based on ratio relationships:** Hermann von Helmholz felt that experience provides ratio information about objects in their surroundings to achieve constancy. James Gibson felt the information was a primary, evolutionarily based aspect of perception. The contemporary view is a blend of the two theories.

How Does Attention Help the Brain Manage Perceptions?

14. **Visual attention is selective and serial:** We process the elements and features of visual stimulus simultaneously. Studies of brain-injured patients reveal that different areas of the brain process different information.

15. **Auditory attention allows selective listening:** The cocktail party phenomenon demonstrates how we readily shift attention to relevant auditory information.

16. **Selective attention can operate at multiple stages of processing:** The debate on selective attention centres on when information is filtered or passed on for further processing.

Applying the Principles

1. In grade school, most of us were given a hearing test. We wore headphones and were presented with tones in one ear and told to raise a hand whenever we heard a tone. Some of the time you may have raised your hand even though you were not certain if you heard something. Your uncertainty was the result of the _______ principle of signal-detection theory; and the lowest determined level of your hearing ability was your _______.

    ____ a) response and signal; difference threshold
    ____ b) response and noise; minimal threshold
    ____ c) noise and response bias; difference threshold
    ____ d) noise and response bias; absolute threshold

2. While studying in her room, Michiko hears a song she likes playing on her roommate's stereo in the living room. She then goes into the living room, turns up the stereo, and turns up the halogen lamp in order to read the CD case. What properties of the stimuli has she changed?

____ a) light- and sound-wave frequency
____ b) light- and sound-wave amplitude
____ c) light-wave frequency; sound-wave amplitude
____ d) light-wave amplitude; sound-wave timbre

3. Driving in the country, you notice in the distance something that appears to be a hawk on the side of the road. As you get closer to it, you realize it is only a bent county road marker. This change in visual perception is an example of

____ a) the basic uncertainty of our visual experience.
____ b) how motion affects depth perception.
____ c) how our brain engages in an ongoing process of organization and identification of sensory information.
____ d) how our sensory system picks up the wrong information.

4. In the movie *At First Sight,* actor Val Kilmer plays a man who, blind from birth, has his eyesight restored in adulthood and finds that he has incredible difficulty making sense of the visual information that bombards him. This scenario demonstrates

____ a) the importance of attention in selecting which information to process.
____ b) that seeing is solely a function of the visual system.
____ c) that older people process visual information differently.
____ d) that if you don't see something it is because the sensory system didn't process it.

ANSWERS: *1.d 2.b 3.c 4.a*

Key Terms

absolute threshold, p. 163
accommodation, p. 173
additive colour mixing, p. 178
binocular depth cues, p. 193
binocular disparity, p. 193
bottom-up processing, p. 189
cochlea (inner ear), p. 170
cocktail party phenomenon, p. 202
difference threshold, p. 164
eardrum (tympanic membrane), p. 170
early selection theory, p. 203
filter theory, p. 202
fovea, p. 173
haptic sense, p. 168
lateral inhibition p. 176
late selection theory, p. 203
monocular depth cues, p. 193
olfactory bulb, p. 167
ossicles, p. 170
outer ear, p. 170
perception, p. 161
perceptual constancy, p. 198
pheromones, p. 168
place coding, p. 172
pop-out, p. 185
primary visual cortex (V1), p. 183
receptive field, p. 175
retina, p. 173
retinoptic organization, p. 183
sensation, p. 160
sensory adaptation, p. 165
sound wave, p. 170
subtractive colour mixing, p. 177
taste buds, p. 166
temporal coding, p. 171
top-down processing, p. 189
transduction, p. 162
visual search task, p. 185

Further Readings

Enns, J. (2005). *The thinking eye, the seeing brain.* New York: Norton.

Gazzaniga, M. S., Ivry, R. B., & Mangun, G. R. (2002). *Cognitive neuroscience: The biology of the mind* (2nd ed.). New York: Norton.

Gibson, J. J. (1979). *The ecological approach to visual perception.* Boston: Houghton Mifflin.

Hoffman, D. D. (1998). *Visual intelligence.* New York: Norton.

Hughes, H. C. (2000). *Sensory exotica.* Cambridge, MA: MIT Press.

Kandel, E. R., Schwartz, J. H., & Jessell, T. M. (2002). *Essentials of neural science and behavior.* New York: Appleton & Lange.

Rock, I. (1984). *Perception.* New York: Scientific American Books.

Further Readings: A Canadian Presence in Psychology

Cheesman, M. F. (1997). Speech perception by elderly listeners: Basic knowledge and implications for audiology. *Journal of Speech-Language Pathology and Audiology, 21,* 104–119.

Dodwell, P. C., & Humphrey, G. K. (1990). A functional theory of the McCollough effect. *Psychological Review, 97,* 78–89.

Goodale, M. A., & Milner, A. D. (1992). Separate visual pathways for perception and action. *Trends in Neuroscience, 15,* 22–25.

Gosselin, P., & Larocque, C. (2000). Facial morphology and children's categorization of facial expressions of emotions: A comparison between Asian and Caucasian faces. *Journal of Genetic Psychology, 161,* 346–358.

Melzack, R. (1992). Phantom limbs. *Scientific American, 266,* 120–126.

Melzak, R., & Wall, P. D. (1982). *The challenge of pain.* New York: Basic Books.

Ng, W., & Lindsay, R. C. (1994). Cross-race facial recognition: Failure of the contact hypothesis. *Journal of Cross-Cultural Psychology, 25,* 217–232.

Norwich, K. H., & Wong, W. (1997). Unification of psychophysical phenomena: The complete form of Fechner's Law. *Perception and Psychophysics, 59,* 929–940.

Schneider, B. (1997). Psychoacoustics and aging: Implication for everyday listening. *Journal of Speech-Language Pathology and Audiology, 21,* 11–124.

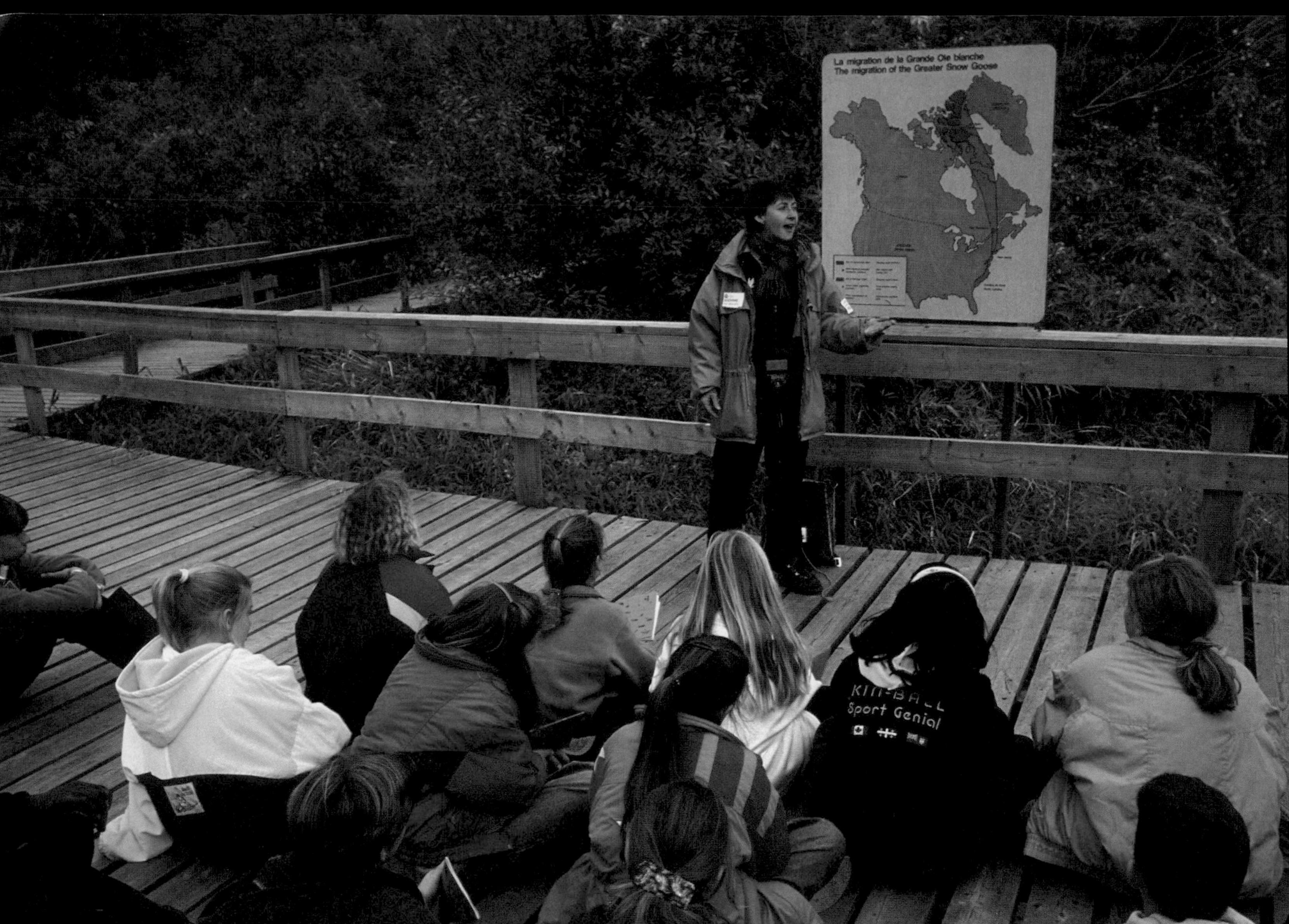

Learning through Conditioning

Children receive higher grades and positive reinforcement for studying and paying attention in class. In this way, children learn how to become good students. Many teachers in Canada find new and interesting ways to keep their students motivated, including classes taught with nontraditional activities such as nature walks.

CHAPTER

6

Learning and Reward

The person who arguably had some of the greatest influence on contemporary psychological science started out as a novelist. Burrhus Frederick (B. F.) Skinner was inspired to pursue writing during his senior year in college after the poet Robert Frost read some of his work and provided encouraging comments. After many months spent toiling away in his parents' attic, with measly results, Skinner began to contemplate alternative careers. He originally had wanted to be a novelist so that he could explore big questions about the human condition, such as understanding why people act the way they do. Skinner, known as "Fred" to his friends, had always been a bit socially awkward and aloof and did not have great insight into other people's emotions. Unfortunately, the process of writing did not provide additional insights for him. As a novelist, he could describe people's actions, but this did not help him gain the true understanding

Outlining the Principles

How Did the Behavioural Study of Learning Develop?

- Behavioural Responses Are Conditioned
- Phobias and Addictions Have Learned Components
- Classical Conditioning Involves More Than Contiguity

How Is Operant Conditioning Different from Classical Conditioning?

- Reinforcement Increases Behaviour
- Both Reinforcement and Punishment Can Be Positive or Negative
- Operant Conditioning Is Influenced by Schedules of Reinforcement
- Biology and Cognition Influence Operant Conditioning
- The Value of Reinforcement Follows Economic Principles

How Does Watching Others Affect Learning?

- Learning Can Be Passed On through Cultural Transmission
- Learning Can Occur through Observation
- Animals and Humans Imitate Others

What is the Biological Basis of Reward?

- Self-Stimulation Is a Model of Reward
- Dopamine Signals Reward

How Does Learning Occur at the Neuronal Level?

- Habituation and Sensitization Are Simple Models of Learning
- Long-Term Potentiation Is a Candidate for the Cellular Basis of Learning
- Learning Can Be Simulated by Computerized Neural Networks

of people that he was seeking. To pursue his intellectual interests in the human condition, particularly in human behaviour, Skinner began to contemplate a career in psychology, though he had not taken a single psychology course in college. He became convinced that psychology was his calling after reading John B. Watson's *Behaviorism* (1924), and the novelist H. G. Wells's 1927 *New York Times Magazine* article in which he expressed admiration for the work of Ivan Pavlov, a Russian physiologist. Increasingly, the perspective of the behaviourists made sense to Skinner. Having made his decision, he enrolled in graduate school at Harvard University and began his explorations of human behaviour. As they say, the rest is history—this time, to a large extent, a major part of the history of psychological science.

Skinner really came into his own at Harvard, in spite of the fact that he shared few interests with the faculty. Many faculty members were concerned about his apparent disdain for their efforts to analyze the inner mind through introspection, an approach that was common at Harvard at the time. Because of the influence of Pavlov and Watson on this thinking, Skinner became obsessed with the idea that he could dramatically change an animal's behaviour simply by providing incentives to the animal for performing the behaviour. For the next half century, Skinner conducted systematic studies of animals, often pigeons or rats, to discover the basic rules of learning. The importance of Skinner's work cannot be underestimated: Skinner discovered many of the most important principles that shape animal and human behaviour. These principles are as relevant today as they were more than 50 years ago. As you'll see later in the chapter, we typically refer to this kind of approach to learning as operant conditioning and when it's applied to people as behaviour modification.

Through his groundbreaking work, B. F. Skinner came to be radical in his beliefs about behaviourism, dismissing the importance of mental states and questioning philosophical concepts such as free will. He believed that the application of basic learning principles could create a better, more humane world for all, one that was free of poverty and violence. His best-selling novel *Walden Two* depicts a utopia in which children are raised only with praise and incentives, never with punishment. In his fictional community, behavioural engineers use principles of learning to create a world in which all people are well satisfied. *Walden Two* has inspired many people over the past 50 years; entire communities, such as Los Horcones in northern Mexico and Twin Oaks in rural Virginia, live according to the principles discovered in Skinner's research.

1690
The Mind as a Blank Slate Building on ideas that can be traced back to Aristotle, the philosopher John Locke argues that all knowledge is acquired through contact with the environment.

1898
The Law of Effect Edward Thorndike proposes the Law of Effect, which shows that behaviours that lead to satisfying outcomes are likely to be repeated.

1906
Learning from Association Russian physiologist Ivan Pavlov provides the first empirical demonstration of simple associative learning, such as a dog associating the sound of a bell with the arrival of food.

1920s
The Rise of Behaviourism John B. Watson proposes the theory of behaviourism, which emphasizes that all behaviour is determined by external forces and that only behaviours that can be manipulated and observed are of value to psychological science.

Skinner also used the ideas established through his research to raise his own children, especially in emphasizing reward over punishment. This was sometimes misinterpreted, as typified by the "baby in the box" incident. For his infant daughter Deborah, Skinner developed a sleeping chamber that kept an optimal temperature and provided a continuous supply of fresh linens. Unfortunately, the media portrayed this as the workings of a cold, mad scientist cruelly mistreating his child by locking her in a box, and rumors circulated that she was depressed and had killed herself. Nothing could be further from the truth. Deborah is an artist who is alive and well, and both daughters had excellent relationships with their father until his death in 1990.

B. F. Skinner's work has been enormously influential throughout society, from classrooms to clinics and beyond. During World War II, the U.S. military funded Skinner to conduct research on how animals might be trained to perform tasks such as guiding missiles. Although none of Skinner's animals were ever used as weapons, Skinner was able to demonstrate that animals could be trained to perform tremendously complex tasks, all via the selective administration of rewards for engaging in specific behaviours. These techniques continue to be used today, as in training dogs to serve as guides for people with handicaps; to sniff out drugs, bombs, or cadavers; or even to sense oncoming seizures in people with epilepsy. Early in the Iraq war, specially trained dolphins were used by the U.S. Navy to search for explosives in the Persian Gulf. Not surprisingly, these techniques are equally effective with humans. Skinner was especially interested in education, and many of his findings have been translated directly into the classroom, for example, the importance of self-paced computerized instruction. Many of the methods used to help people with mental limitations and disorders lead healthy and productive lives are also based on the behavioural principles discovered by Skinner and other psychological scientists.

In this chapter, you will find out what Skinner and a number of other learning theorists have discovered about how learning takes place. This material is central to what psychology has contributed to our understanding of human and animal behaviour and affects all the other major areas of psychology. It is therefore a cornerstone for a number of the foundational principles of psychological science. We will see how these findings have been used to improve quality of life and to train humans and animals to learn new tasks. We will also see how contemporary psychological scientists are discovering the brain mechanisms that support this learning and what new theories are emerging to explain how we learn. ■

1930s
Learning from Reinforcement B. F. Skinner develops the theory of operant conditioning, which states that behaviours are controlled by reinforcement or punishment. The widely used method of behaviour modification is based on this theory.

1940s
Cognitive Learning Approach Edward Tolman, among others, sets the groundwork for the cognitive revolution in psychology by arguing that learning is not the change in behaviour as such but is instead the acquisition of new knowledge or cognitions.

1948
Learning by Synaptic Alteration The Canadian psychologist Donald Hebb proposes that learning results from alterations in synaptic connections. Psychological science has since shown the basic tenets of this proposal to be accurate.

1960s
Social Learning Theory Concerned that behaviourism is ignoring the social context of learning, Albert Bandura shows that people learn behaviours by observing whether others are rewarded or punished for performing those behaviours.

1960s
Learning as an Evolved Mechanism John Garcia argues that certain types of learning involve special mechanisms that have evolved through natural selection.

learning An enduring change in behaviour that results from experience.

Learning is a relatively enduring change in behaviour that results from experience. It occurs when organisms benefit from experience so that their future behaviour is better adapted to the environment. The ability to learn is crucial for all animals. To survive, animals need to learn such things as which types of foods are dangerous, when it is safe to sleep, and which sounds indicate potential dangers. Learning is central to almost all aspects of human existence, from learning such basic abilities as walking and speaking, to even more complex abilities, such as flying airplanes, performing surgery, or maintaining intimate relationships. Learning also shapes many aspects of daily life, from clothing and musical tastes, to cultural rules for how close we stand to each other, to our values about the environment.

The focus in this chapter is on how learning occurs. The essence of learning is in understanding how events are related, such as how you associate going to the dentist with being in pain or how you associate working with getting paid. These associations develop through a process known as *conditioning,* in which environmental stimuli and some sort of behavioural response become connected. We will see that psychologists have tended to focus on two types of conditioning. The first, known as *classical conditioning,* occurs when we learn that two types of events go together, such as walking in the rain and getting wet. The second, known as *operant conditioning,* occurs when we learn that a behaviour leads to a particular outcome, such as that studying leads to better grades. This is the type of learning that was of greatest interest to B. F. Skinner. We will also consider other types of learning, such as by observing others—for example, learning about new fashions by paying attention to what movie actors and rock stars wear.

The study of learning has been central to psychological science since the early nineteenth century and is an excellent example of how the principles of psychological science are cumulative. As we read about the major learning theories, we'll see how great theorists such as Thorndike, Watson, Skinner, Hebb, Bandura, and Kandel were responding to and expanding on the ideas of their predecessors in an effort to understand the complexities of how people and animals learn. Let's now look at the different ways that we learn and the most recent efforts to understand the biological underpinnings of learning.

6.1 John B. Watson Watson was the founder of behaviourism and one of its most ardent supporters.

How Did the Behavioural Study of Learning Develop?

The rise of learning theory in the early twentieth century was due in part to the dissatisfaction among some psychologists with the widespread use of verbal reports to assess mental states. At the time, Freudian ideas were at the heart of psychological theorizing. To assess the unconscious mental processes that they believed were the

1972
Expectancies and Learning Robert Rescorla and Allan Wagner propose that learning occurs when organisms develop expectancies that allow them to predict future events.

1973
Long-Term Potentiation Working in Norway, Terje Lømo and Tom Bliss find that rapid stimulation of a synapse strengthens the synaptic connection. This potentiation is now known to be an important cellular process underlying learning.

1980s
The Neural Basis of Learning Eric Kandel studies the cellular and molecular mechanisms of learning in a type of invertebrate creature. This work earns him a Nobel Prize in 2000.

1990s
Computer Models of Neural Networks Superfast modern computers allow psychological scientists to develop neural-network models of learning.

2000s
Mirror Neurons Giacomo Rizzolatti and colleagues discover that neurons involved in performing an action become activated when monkeys simply observe others performing that action. These mirror neurons might support observational learning.

primary determinants of behaviour, Freud and his followers used verbal report techniques such as dream analysis and free association. The American psychologist John B. Watson (Figure 6.1) scorned any form of psychological enterprise that focused on things that could not be observed directly, such as people's mental events. He argued that Freudian theory was unscientific and ultimately meaningless. According to Watson, overt behaviour was the only valid indicator of psychological activity. Although he acknowledged that thoughts and beliefs existed, he denied that they could be studied using scientific methods.

Watson founded the school of behaviourism, which was based on the belief that animals and humans are born with the potential to learn anything. Based on philosopher John Locke's idea of *tabula rasa* (Latin, "blank slate"), which states that infants are born knowing nothing and that all knowledge is acquired through sensory experiences, behaviourism stated that the environment and its associated effects on organisms were the sole determinants of learning. Watson felt so strongly about the preeminence of the environment that he issued the following bold challenge: "Give me a dozen healthy infants, well formed, and my own specified world to bring them up in and I'll guarantee to take any one at random and train him to become any type of specialist I might select—doctor, lawyer, artist, merchant-chief, and yes, even beggar-man and thief, regardless of his talents, penchants, tendencies, abilities, vocations and race of his ancestors" (Watson, 1924, p. 82).

6.2 Ivan Pavlov Pavlov conducted groundbreaking work on classical conditioning.

Behavioural Responses Are Conditioned

Watson had an incredible influence on the study of psychology in North America. Behaviourism was the dominant psychological paradigm well into the 1960s, and it influenced the methods and theories of every area within psychology. Watson developed his ideas about behaviourism after he read the work of the Russian physiologist Ivan Pavlov (Figure 6.2), a distinguished scientist who had won a Nobel Prize for his work on the digestive system. Pavlov was interested in the *salivary reflex,* the automatic and unlearned response that occurs when the stimulus of food is presented to hungry animals, including humans. For his work on the digestive system, Pavlov had created an apparatus that collected saliva from dogs (Figure 6.3). He placed various types of food into a dog's mouth and measured differences in salivary output. Like so many major scientific advances, Pavlov's contribution to psychology started with a simple observation. One day he realized that the dogs were salivating well before they tasted the food. Indeed, the dogs started salivating the moment that a lab technician walked into the room, or whenever they saw the bowls that usually contained food. The genius of Pavlov was in recognizing that this behavioural response was a window to the working mind. Unlike inborn reflexes, salivation at the sight of a bowl is not automatic and therefore must have been acquired through experience. The insight that dogs could associate bowls and lab technicians with food led Pavlov to devote the rest of his life to studying the basic principles of learning.

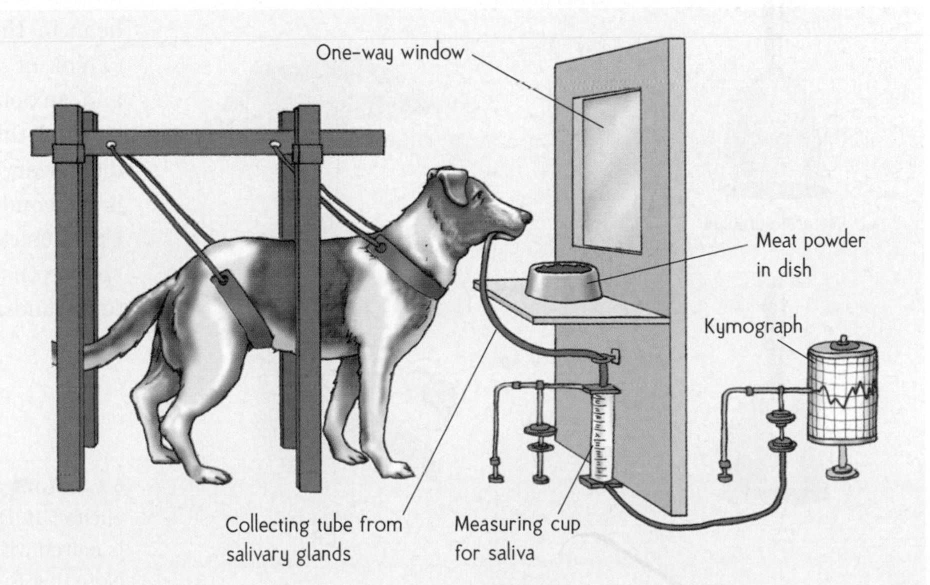

6.3 Pavlov's Apparatus for Collecting Saliva The dog was presented with a bowl containing meat. A tube from the salivary glands carried the saliva to a container that measured the amount of salivation.

PAVLOV'S EXPERIMENTS In a typical Pavlovian experiment, a *neutral stimulus* unrelated to the salivary reflex, such as a ringing bell, is presented together with a stimulus that reliably produces the reflex, such as food. This pairing, known as a *conditioning trial,* is repeated a number of times; then, on *critical trials,* the bell sound is presented alone and the salivary reflex is measured. Pavlov found that under these conditions, the sound of the bell on its own produced salivation. This type of learning is now referred to as *classical conditioning* or *Pavlovian conditioning.* **Classical conditioning** occurs when a neutral object comes to elicit a reflexive response when it is associated with a stimulus that already produces that response.

How do events become associated?

Pavlov called the salivation elicited by food the **unconditioned response (UR)**, because it occurred without any prior training. Unconditioned responses are unlearned, automatic behaviours, such as any simple reflex. Similarly, the food is referred to as the **unconditioned stimulus (US)**. In the normal reflex response, the food (US) leads to salivation (UR). Because the ringing bell produces salivation only after training, it is called the **conditioned stimulus (CS)**; it stimulates salivation only after learning takes place. The salivary reflex that occurs when only the conditioned stimulus is presented is known as the **conditioned response (CR)**; it is an acquired response that is learned. Note that both the unconditioned and the conditioned responses are salivation, but they are not identical: The bell produces less saliva than does the food. The conditioned response usually is less strong than the unconditioned response. The process of conditioning is outlined in Figure 6.4.

Consider this common reaction. Watching a frightening scene in a movie, such as someone being attacked, makes you feel tense, anxious, and perhaps even disgusted. In this scenario, the frightening scene, the stimulus, and your reaction, the response, are unconditioned—they occur naturally. Now imagine a piece of music that doesn't initially have much effect on you but that you hear in the movie just before each frightening scene. (Think of *Jaws*!) Eventually you will begin to feel tense and anxious as soon as you hear the music. You have learned that the music, the conditioned stimulus, predicts scary scenes, and therefore you feel anxious, which is the conditioned response. As in the Pavlov studies, the CS (music) produces a somewhat different emotional response than the US (the scary scene), perhaps less intense and more a feeling of apprehension than of being

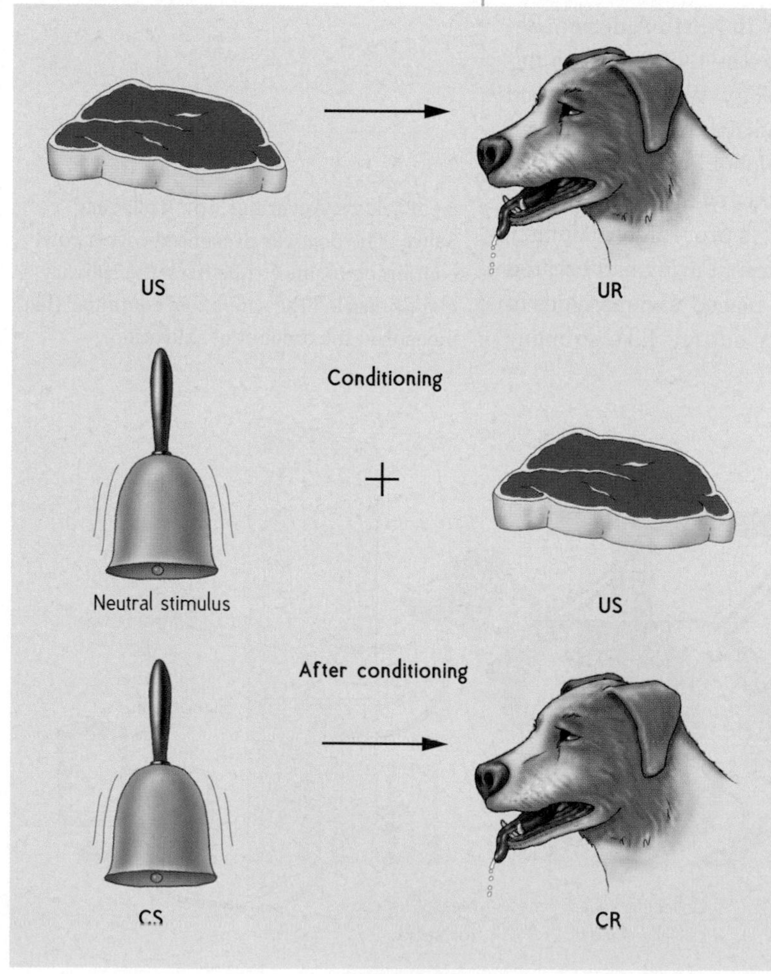

6.4 Before and After Conditioning Initially, the US elicits UR. During acquisition, the formerly neutral CS is paired with the US. Eventually the CS elicits CR. Note that the UR is typically stronger than the CR.

afraid or disgusted. If you later hear this music in a different context, for example, on the radio, you will again feel tense and anxious even though you are not watching a movie. You have been classically conditioned to be anxious when you hear the music.

ACQUISITION, EXTINCTION, AND SPONTANEOUS RECOVERY Like many other scientists at the time, Pavlov was greatly influenced by Darwin's *On the Origin of Species.* Pavlov believed that conditioning was the basis for how animals learn to adapt to their environments. By learning to predict what objects bring pleasure or pain, animals acquire new behaviours that are adaptive to the environment. For instance, let's suppose that each time it rains, a delicious and nutritious plant blooms. An animal that learns this association will seek out this plant each time it rains. **Acquisition**, the initial learning of a behaviour, is the gradual formation of an association between the conditioned and unconditioned stimuli. From his research, Pavlov concluded that the critical element in the acquisition of a learned association is that the stimuli occur together in time, which is referred to as *contiguity.* Subsequent research has shown that the strongest conditioning actually occurs when there is a very brief delay between the CS and the US (Jenkins et al., 1978; Macintosh & Honig, 1969; Spetch et al., 1981). Thus, you will develop a stronger conditioned response to a piece of music if it comes just before a scary scene than if it occurs during the scary scene. As you will see later, the role of the music in predicting the frightening scene is an important part of classical conditioning.

Once a behaviour is acquired, how long does it persist? For instance, what if the animal sought out the tasty blooms following rain, but they stopped appearing? Animals sometimes have to *unlearn* previous associations when they are no longer adaptive. Let's look at how this happens. After conditioning, the bell (CS) leads to salivation (CR) because the animal learns to associate the bell with the food (US). If the bell is presented many times and food does not arrive, the animal learns that the bell is not a good predictor of food, and therefore the salivary response gradually disappears. This process is known as **extinction**. The conditioned response is *extinguished* when the conditioned stimulus no longer predicts the unconditioned stimulus (Figure 6.5).

But suppose the delicious plant blooms only during a certain time of the year. The adaptive response is to check back once in a while to see if the plant blooms following rain. In the lab, an analogous situation occurs when the conditioned stimulus is presented a long time after extinction. Sounding the bell will once again produce the conditioned response of salivation (see Figure 6.5). This **spontaneous recovery**, in which the extinguished CS again produces a CR, is temporary and will quickly fade unless the CS is again paired with the US. Even a single pairing of the CS with the US will reestablish the CR, which will then again diminish if CS-US pairings do not continue. Thus, extinction inhibits but does not break the associative bond. It is a new form of learning that overwrites the previous association; what is learned is that the original association no longer holds true (Bouton, 1994; Bouton, Nelson, & Rosas, 1999).

GENERALIZATION, DISCRIMINATION, AND SECOND-ORDER CONDITIONING In any learning situation, hundreds of possible stimuli can be associated with the unconditioned stimulus to produce the conditioned response. How does the brain determine which of these stimuli is relevant? For instance, suppose we classically condition a dog so that it salivates (CR) when it hears a 1,000 Hz (hertz) tone. After the CR is established, tones that are similar to 1,000 Hz will also produce it, but the farther the tones are from 1,000 Hz, the less the dog will salivate (Figure 6.6). **Stimulus generalization** occurs when stimuli that are similar but not identical to the CS

Why might a delay strengthen conditioning?

classical conditioning A type of learned response that occurs when a neutral object comes to elicit a reflexive response when it is associated with a stimulus that already produces that response.

unconditioned response (UR) A response that does not have to be learned, such as a reflex.

unconditioned stimulus (US) A stimulus that elicits a response, such as a reflex, without any prior learning.

conditioned stimulus (CS) A stimulus that elicits a response only after learning has taken place.

conditioned response (CR) A response that has been learned.

acquisition The gradual formation of an association between the conditioned and unconditioned stimuli.

extinction A process in which the conditioned response is weakened when the conditioned stimulus is repeated without the unconditioned stimulus.

spontaneous recovery A process in which a previously extinguished response reemerges following presentation of the conditioned stimulus.

stimulus generalization Occurs when stimuli that are similar but not identical to the conditioned stimulus produce the conditioned response.

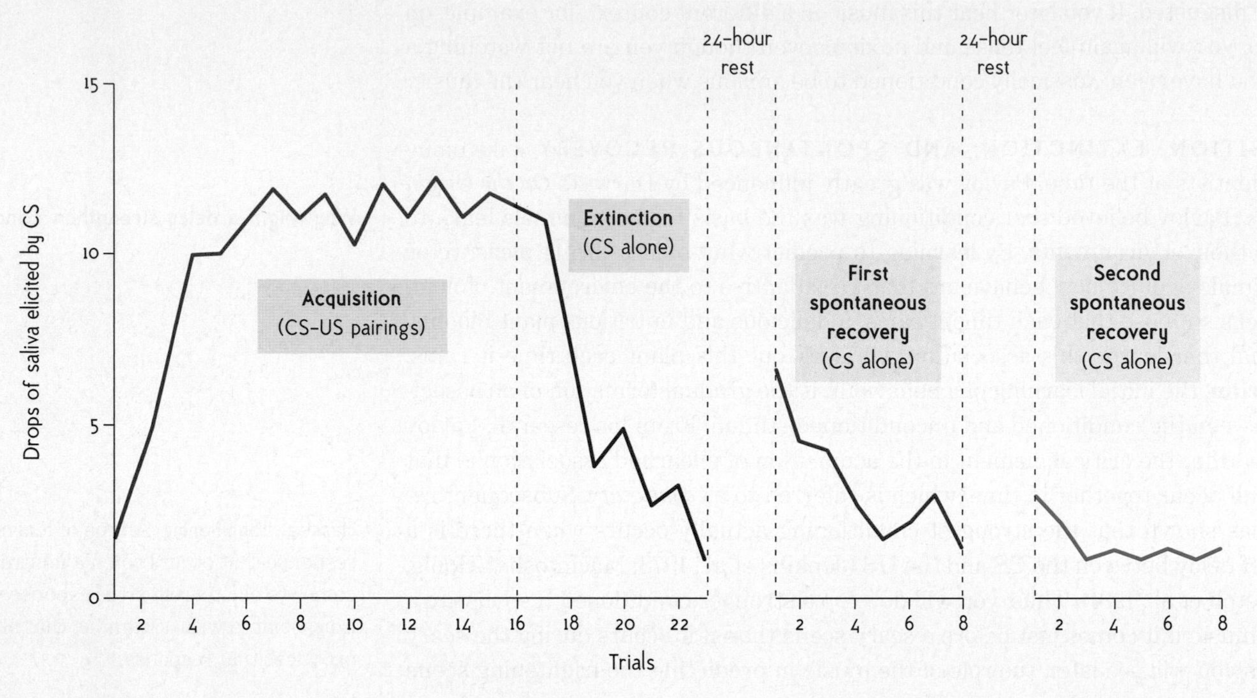

6.5 Extinction During acquisition, the CS-US pairings lead to increased learning such that the CS can produce the CR. However, if the CS is presented without the US, eventually the CR extinguishes. Later, if the CS is presented alone it will produce a weak CR, known as spontaneous recovery. This CR will quickly extinguish if the US does not appear.

produce the CR. Generalization is adaptive because in nature the CS is seldom experienced repeatedly in an identical fashion. Slight differences in variables such as background noise, temperature, and lighting lead to slightly different perceptions of the CS, so animals learn to respond to variations in the CS.

Of course, there are limits to generalization. Sometimes it is important for animals to distinguish among similar stimuli. For instance, two species of plants may look similar, but one might be poisonous. **Stimulus discrimination** means that animals learn to differentiate between two similar stimuli if one is consistently associated with the unconditioned stimulus and the other is not. In various studies, Pavlov and his students demonstrated that dogs could learn to make very fine distinctions between similar stimuli, such as subtle differences in shades of gray.

Sometimes a conditioned stimulus does not become directly associated with an unconditioned stimulus, but rather with other stimuli that themselves are associated with the US, a phenomenon known as *second-order conditioning.* In one of Pavlov's early studies, a CS-US bond was formed between a tone and food so that the tone (CS) led to salivation (CR). Following this conditioning, a second training session was conducted in which a black square was repeatedly presented at the same time as the tone. There was no US during this phase of the study. Following many trials, the black square was presented alone and it also produced salivation. Second-order conditioning helps to account for the complexity of learned associations, especially among

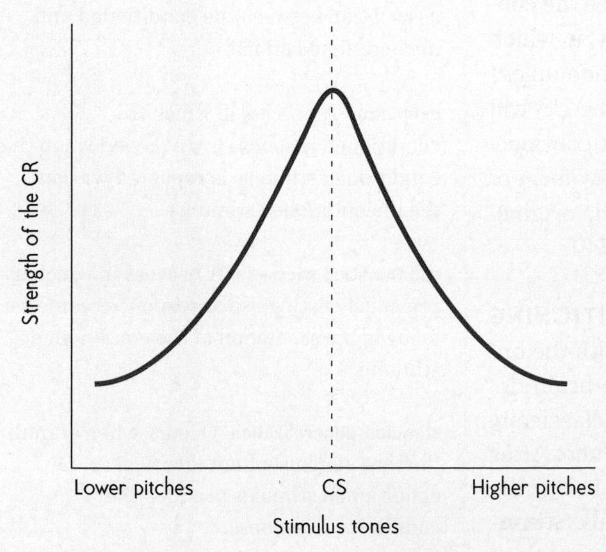

6.6 Generalization and Discrimination Maximal salivation occurs at the CS of 1,000 Hz, but stimuli that are similar to the CS also produce the CR. The strength of the CR diminishes as the tone diverges from 1,000 Hz.

people. For instance, money is usually just paper or cheap metal, but it means something entirely different to people. It is the association between paper and what it buys that makes us want to have it. As we will see later, second-order conditioning is why money is rewarding for those who receive it. Similarly, when we see our favourite actor drinking a particular brand of soda we develop more positive attitudes about that soda, which is why advertisers often seek out celebrities to endorse products. Even though it has a powerful influence over many of our beliefs and attitudes, most second-order conditioning occurs implicitly, without our awareness or intention.

Phobias and Addictions Have Learned Components

Classical conditioning helps explain a number of behavioural phenomena, including phobias and addictions.

stimulus discrimination A learned tendency to differentiate between two similar stimuli if one is consistently associated with the unconditioned stimulus and the other is not.

phobia An acquired fear that is out of proportion to the real threat of an object or a situation.

PHOBIAS AND THEIR TREATMENT **Phobia** is an acquired fear that is out of proportion to the real threat, as you will learn more about in Chapter 13. Common phobias include fear of heights, dogs, insects, snakes, and the dark. According to classical-conditioning theory, phobias develop through the generalization of a fear experience, as when a person who is stung by a wasp develops a fear of all flying insects.

Animals can be classically conditioned to fear neutral objects, a process known as *fear conditioning*. For example, if an animal is repeatedly presented with a flash of light followed by a moderately painful electric shock, it will soon display physiological and behavioural responses indicating fear, such as change in heart rate, whenever it sees a flash of light. As you will read later in this chapter, psychological scientists now know a great deal about the biological basis of fear conditioning. Over the last decade, researchers have made enormous progress in understanding how the brain learns to fear an object or a situation, how these learned fears can help the animal learn to avoid danger, and how the experience of fear can strengthen learning for important events. The most important brain structure for fear conditioning is the amygdala. Without an amygdala, there is no fear conditioning.

Techniques from classical conditioning have been valuable for developing behavioural therapies to treat phobias. For instance, exposing people to small doses of the feared stimulus while having them engage in a pleasurable task, called *counterconditioning,* can help people overcome their fears. Joseph Wolpe, a behavioural therapist, developed a formal treatment based on counterconditioning known as *systematic desensitization* (Wolpe, 1997). Patients are taught how to relax their muscles, and when they are able to do so, they are asked to imagine the feared object or situation while continuing to use the relaxation exercises. Eventually the person is exposed to the feared stimulus while relaxing. The general idea is that the $CS \rightarrow CR_1$ (fear) connection can be broken by developing a new $CS \rightarrow CR_2$ (relaxation) connection. The behavioural treatment of phobias has proven to be very effective.

DRUG ADDICTION Classical conditioning also plays an important role in drug addiction. Conditioned drug effects are common. For example, the smell of coffee can become a conditioned stimulus, leading coffee drinkers to feel activated and aroused. Likewise, because the needle becomes a CS, heroin addicts will sometimes inject themselves with water, when heroin is not available, to reduce their cravings. These learned associations demonstrate the power of conditioning and the potential for problems. When former heroin addicts are exposed to environmental cues associated with their drug use, they often experience cravings and physiological sensations similar to those they experienced during *withdrawal,* the unpleasant state of

6.7 **Theoron Fleury** Fleury has struggled with substance abuse throughout his NHL career; he first entered the NHL Behavioral and Substance Abuse Program in March 2001. Because of relapses, Fleury has not played a game in the NHL since the 2002–2003 season; he currently plays for the Coors Belfast Giants of the Elite Ice Hockey League in the United Kingdom.

tension and anxiety that occurs when addicts stop using drugs. Thus, addicts who quit using drugs in treatment centres often relapse when they return to their old environments (Figure 6.7). They experience conditioned withdrawal effects.

In the laboratory, presenting heroin or cocaine addicts with cues associated with drug ingestion leads to cravings and a variety of physiological responses associated with withdrawal, such as changes in heart rate and blood pressure. Brain-imaging studies have found that such cues lead to activation of the prefrontal cortex and various regions of the limbic system (Figure 6.8; Childress et al., 1999; see also Maas et al., 1998). Recall from Chapter 4 that the limbic system is involved in emotion and motivation. The role of the limbic system in addiction will be discussed in greater detail in Chapter 9.

The learning theorist Shepard Siegel at McMaster University has conducted fascinating research showing that drug-tolerance effects are specific to certain situations. *Tolerance* is a process by which addicts need more and more of a drug to experience the same effects. According to research by Siegel, tolerance effects are greatest when the drug is taken in the same location as previous drug use. Tolerance can be so great that addicts regularly use drug doses that would be fatal for the inexperienced user. Conversely, Siegel's findings imply that if addicts take their usual large doses in a novel setting they are more likely to overdose (Siegel, 1984; Siegel, Hinson, Krank, & McCully, 1982).

Profiles in Psychological Science

Albert B.

An early case study demonstrating the role of classical conditioning in the development of phobias was John B. Watson's teaching of an infant named Albert B. to fear neutral objects. It is important to note Watson's motives for conditioning Albert. At the time, 1919, the prominent theory of phobia was based on Freudian ideas about unconscious repressed sexual desires. Watson thought that Freudian ideas were unscientific and unnecessarily complex. He proposed that phobias could be explained by simple learning principles, such as classical conditioning. To prove his point, Watson talked a woman he knew into having her son take part in a learning study. The child was chosen because he was emotionally stable and Watson believed the experiment would cause him little harm. When Albert was nine months of age, Watson and his lab assistant, Rosalie Rayner, presented him with a variety of neutral objects, including a white rat, a rabbit, a dog, a monkey, costume masks, and a ball of white wool. Albert showed a natural curiosity regarding these items, but displayed no overt emotional response.

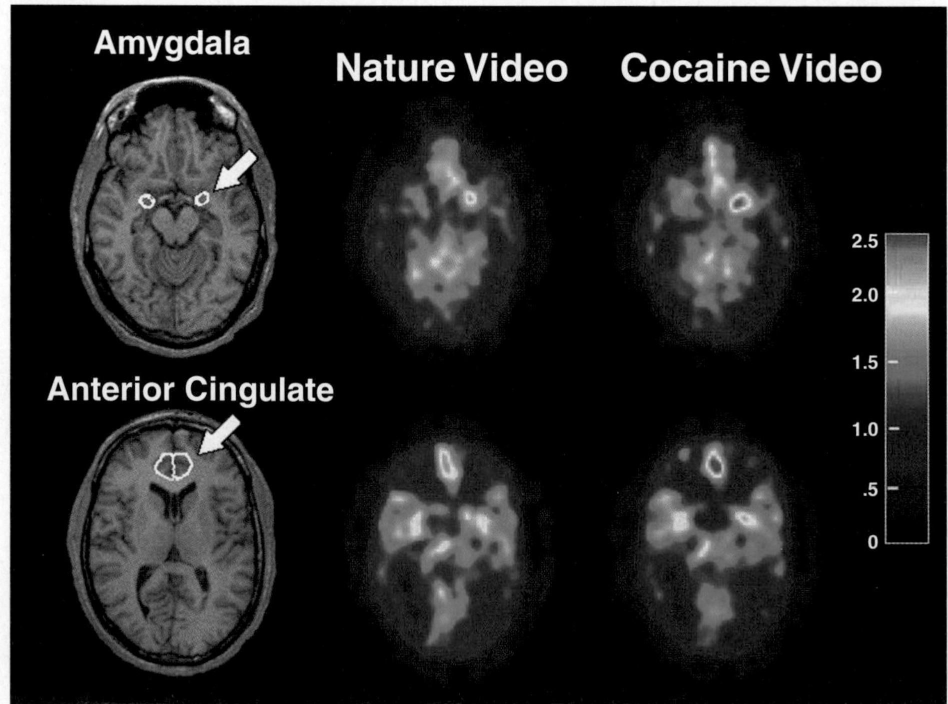

6.8 PET Scans Showing Activation of Abstinent Limbic System Structures Cocaine addicts were shown videos of nature scenes or cocaine cues. Areas with greatest activation are shown in red.

6.9 Albert B. John B. Watson and Rosalie Rayner conducted a conditioning experiment on their infant subject, Albert B., in which they taught him to fear neutral objects.

When Albert was eleven months old, Watson and Rayner began the conditioning trials. This time, as they presented the white rat and Albert reached for it, Watson smashed a hammer into an iron bar, producing a loud clanging sound (Figure 6.9). The sound scared the child, who immediately withdrew and hid his face. Watson did this a few more times at intervals of five days until Albert would whimper and cringe when the rat was presented alone. Thus, the US (smashing sound) led to a UR (fear), and eventually the pairing of the CS (rat) and US (smashing sound) led to the rat producing fear (CR) on its own. The fear response generalized to other stimuli that Watson had presented along with the rat at the initial meeting, including the rabbit and the ball of wool, and Albert eventually became frightened of them all. Even a Santa Claus with a white beard produced a fear response. Thus, classical conditioning was shown to be an effective method of inducing phobia.

Although Watson had planned to conduct extinction trials to remove the learned phobias, Albert's mother removed him from the study before Watson could conduct those trials. Whatever became of Albert? No one knows. Watson's conditioning of Albert has long been criticized as unethical, and nowadays it is unlikely that an ethics committee would approve such a study. It is only fair to note that Watson had developed detailed plans to change the conditioning so that Albert no longer feared the items used in the study. For instance, he described a method of continually presenting the objects until the response habituated, while at the same time "reconditioning" the feared objects with more pleasant things, such as candy. Behavioural methods such as these have since been found to be effective for treating phobias. ■

Classical Conditioning Involves More Than Contiguity

Pavlov's original explanation for classical conditioning was that any two events presented at the same time, in contiguity, would produce a learned association. Pavlov and his followers believed that the strength of the association was determined by factors such as the intensity of the conditioned and unconditioned stimuli, with greater intensity associated with increased learning. In the mid-1960s a number of challenges suggested that some conditioned stimuli were more likely to produce learning than others, and that contiguity was not sufficient to create CS-US associations.

EVOLUTIONARY SIGNIFICANCE According to Pavlov, any object or phenomenon could be converted into a conditioned stimulus during conditioning trials. Thus, a light, tone, colour, or odour could be associated with the unconditioned stimulus. The idea that all stimuli are equally capable of producing conditioning is known as *equipotentiality.* However, it appears that not all stimuli are equally potent. Research conducted by the psychologist John Garcia and his colleagues showed that certain pairings of stimuli are more likely to become associated than others. For instance, animals that are given poison in their food quickly learn to avoid the tastes or smells associated with the food (Garcia & Koelling, 1966). In these cases, just a few trials are enough to produce a long-lasting avoidance of the food item that was poisoned. These conditioned food aversions are easy to produce with taste or smell, but very difficult to produce with light or sound. This makes sense, since taste and smell are the main cues that guide eating behaviour in animals. From an evolutionary viewpoint, animals that quickly associate a certain flavour with illness, and therefore avoid that flavour, are more likely to survive and pass along their genes.

Why are some associations easier to learn than others?

Other research has shown that it is easier to condition monkeys to fear snakes than to fear objects such as flowers or rabbits (Cook & Mineka, 1989). Psychologist Martin Seligman (1970) argued that animals are genetically programmed to fear specific objects, which he refers to as **biological preparedness**. Preparedness helps to explain why animals tend to have phobias of things that have potential danger (e.g., snakes, fire, heights) rather than objects that pose little threat (e.g., flowers, shoes, babies).

biological preparedness The idea that animals are biologically programmed to learn to fear specific objects.

At the most general level, contemporary researchers are interested in how classical conditioning helps animals learn adaptive responses (Hollis, 1997; Shettleworth, 2001). The adaptive value of a particular response varies according to the animal's evolutionary history. For example, taste aversions are easy to condition in rats, but difficult to condition in birds because they rely more on vision than taste to select food. However, birds quickly learn to avoid a visual cue that has been associated with illness. Rats freeze and startle if the CS is auditory, but rise on their hind legs when the CS is a visual cue (Holland, 1977). These differences in learned adaptive responses may reflect the meaning of, and potential danger associated with, auditory and visual stimuli in the environment.

THE COGNITIVE PERSPECTIVE Prior to the 1970s, most animal learning theorists were concerned only with observable stimuli and responses. In the past three decades, learning theorists have placed a greater emphasis on trying to understand the mental processes that underlie conditioning. An important principle that has emerged is that classical conditioning is a means by which animals come to *predict* the occurrence of events. This rise in consideration of mental processes, such as prediction and expectancy, is referred to as the cognitive perspective on learning (Hollis, 1997).

Robert Rescorla (1966) conducted one of the first studies highlighting the role of cognition in learning. He argued that for learning to take place, the conditioned stimulus needs to be an accurate predictor of the unconditioned stimulus. For instance, a stimulus that occurs *before* the US is more easily conditioned than one that comes *after* it. Even though the two are both contiguous presentations with the US, the first stimulus is more easily learned because it predicts rather than comes after the US. Indeed, as mentioned earlier, across all conditioning situations, some delay between the CS and the US is optimal for learning. The length of delay varies depending on the nature of the conditioned and unconditioned stimuli. For instance, eyeblink conditioning occurs when a sound

"Oh, not bad. The light comes on, I press the bar, they write me a check. How about you?"

(CS) is associated with a puff of air blown into the eye (US), which leads to a blink. Optimal learning for eyeblink conditioning is measured in milliseconds. By contrast, conditioned taste aversions often take many hours, since the ill effects of consuming poisons often take time to occur.

In 1972, Robert Rescorla and his colleague Allan Wagner published a cognitive model of classical conditioning that profoundly changed our understanding of learning (Rescorla & Wagner, 1972). The **Rescorla-Wagner model** states that the strength of the CS-US association is determined by the extent to which the US is unexpected or surprising. The greater the surprise of the US, the more effort an organism puts into trying to understand its occurrence so that it can predict it more accurately in the future. The end result is greater classical conditioning of the surprising event (CS) that predicted the US. According to the model, conditioning is a process by which organisms learn to expect the unconditioned stimulus based on the conditioned stimulus. Your pet wags its tail and runs around in circles when you start to open a can of food because it expects that it will soon be fed. According to Rescorla and Wagner, your pet has developed a mental representation in which the sound of the can opener (CS) predicts the appearance of food (US). When the US occurs unexpectedly, the animal attends to events in the environment that might have produced it. For example, say you use a manual can opener because the electric one is broken. Your pet finds itself presented with the food even though it did not hear the can being opened. Your pet soon will learn to anticipate being fed by the wrist movements you make with the new opener.

Rescorla-Wagner model A cognitive model of classical conditioning that states that the strength of the CS-US association is determined by the extent to which the unconditioned stimulus is unexpected.

Why does learning occur more readily for novel events?

Consistent with the Rescorla-Wagner model, novel stimuli are more easily associated with the unconditioned stimulus than are familiar stimuli. For example, dogs can be conditioned more easily with the novel smell of almonds than with smells more familiar to their environments, like dog biscuits. Once learned, a conditioned stimulus can prevent the acquisition of a new conditioned stimulus, a phenomenon known as the *blocking effect*. For example, a dog that has acquired the smell of almonds (CS) as a good predictor of food (US) does not need to look for other predictors. Furthermore, a stimulus that is associated with a CS can act as an *occasion setter* or trigger for the CS (Schmajuk, Lamoureaux, & Holland, 1998). For example, a dog can learn that the smell of almonds predicts food only when the smell is preceded by a sound or a flash of light and not at other times. The smell or light indicates whether or not the association between the smell of almonds and food is active.

REVIEWING the Principles | How Did the Behavioural Study of Learning Develop?

Behaviourism, founded by John B. Watson, focused on observable aspects of learning. Pavlov developed the classical-conditioning theory to account for the learned association between neutral stimuli and reflexive behaviours. Conditioning occurs when the conditioned stimulus becomes associated with the unconditioned stimulus. For learning to occur, the conditioned stimulus needs to be a reliable predictor of the unconditioned stimulus, not simply contiguous. A cognitive model that accounts for most conditioning phenomena is the Rescorla-Wagner model, which states that the amount of conditioning is determined by the extent to which the unconditioned stimulus is unexpected or surprising.

How Is Operant Conditioning Different from Classical Conditioning?

How is behaviour shaped by rewards and punishments?

Classical conditioning is a relatively passive process in which subjects associate events that happen around them. This form of conditioning does not account for the many times that one of the events occurs because of some action on the part of the subject. We don't sit idly waiting for food to be presented to us; we go out and purchase it. Our behaviours often represent a means to an end. We buy food to eat it, we study to get good grades, and we work to receive money. Thus, many of our actions are *instrumental*; they are done for a purpose. We learn to behave in certain ways in order to be rewarded, and we avoid behaving in certain ways in order not to be punished; this is called *instrumental conditioning* or *operant conditioning*. B. F. Skinner, the psychologist most closely associated with this type of learning, selected the term *operant* to express the idea that animals operate on the environment to produce an effect. **Operant conditioning** is the learning process in which the consequences of an action determine the likelihood that it will be performed in the future.

The study of operant conditioning began in the basement of the psychologist William James's house in Cambridge, Massachusetts. A young graduate student working with James, Edward Thorndike, had been influenced by Darwin and was studying whether nonhuman animals showed signs of intelligence. As part of his research, Thorndike built an apparatus called a puzzle box, a small cage with a trapdoor (Figure 6.10). The trapdoor would open if the animal performed a specific action, such as pulling a string. Thorndike placed food-deprived animals, usually cats, inside the puzzle box to see if they could figure out how to escape. To motivate the cats, he would place food just outside the box. When a cat was first placed in the box, it usually engaged in a number of nonproductive behaviours in an attempt to escape. After 5 to 10 minutes of struggling, the cat would *accidentally* pull the string and the door would open. Thorndike would then return the cat to the box and repeat the trial. Thorndike found that the cats would pull the string more quickly on each trial, until they soon learned to escape from the puzzle box within seconds. From this line of research Thorndike developed a general theory of learning, known as the **law of effect**, which states that any behaviour that leads to a "satisfying state of affairs" is more likely to occur again, and that those that lead to an "annoying state of affairs" are less likely to occur again.

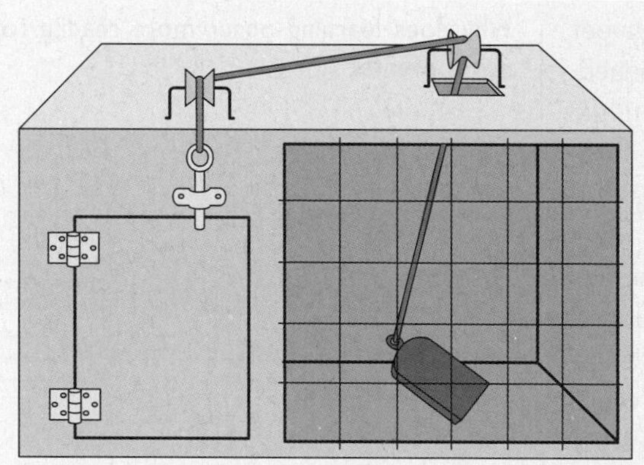

6.10 The Puzzle Box Thorndike used a primitive puzzle box, such as the one depicted here, to assess learning in cats.

operant conditioning A learning process in which the consequences of an action determine the likelihood that it will be performed in the future.

law of effect Thorndike's general theory of learning, which states that any behaviour that leads to a "satisfying state of affairs" is more likely to occur again, and that those that lead to an "annoying state of affairs" are less likely to recur.

reinforcer A stimulus following a response that increases the likelihood that the response will be repeated.

Reinforcement Increases Behaviour

Thirty years after Thorndike, another Harvard graduate student in psychology, B. F. Skinner (Figure 6.11), developed a more formal learning theory based on the law of effect. As you read in the beginning of this chapter, Skinner had been greatly influenced by John B. Watson and shared his philosophy of behaviourism. He therefore objected to the subjective aspects of Thorndike's law of effect: states of "satisfaction" are not observable empirically. Skinner coined the term *reinforcer* to describe events that produced a learned response. A **reinforcer** is a stimulus that occurs following a response that increases the likelihood that the response will be repeated. Skinner believed that behaviour, from studying to eating to driving on the proper side of the road, occurs because it has been reinforced.

THE SKINNER BOX Skinner developed a simple device for assessing operant conditioning called the *Skinner box,* a small chamber or cage in which a lever (or response key) is connected to a food or water supply. An animal, usually a rat or a pigeon, is placed in the box; when it presses the lever, food or water becomes available. Skinner's earlier research used mazes in which a rat had to take a specific turn to get access to the reinforcer. After the rat completed the trial, Skinner had to get the rat and return it to the beginning. He developed the Skinner box, or *operant chamber,* as he called it, basically because he got tired of constantly fetching the rats. With the Skinner box, the rats could be exposed to repeated conditioning trials without the experimenter having to do anything but observe. Skinner later built mechanical recording devices that allowed trials to be conducted without the experimenter even being present.

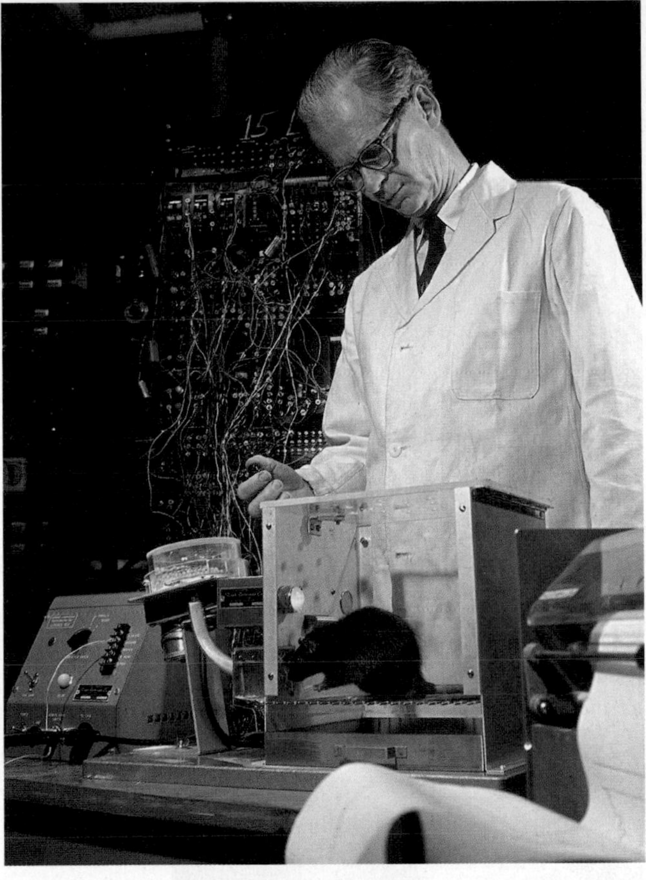

6.11 B. F. Skinner Skinner (seen here demonstrating the Skinner box) was one of the most influential psychologists of all time.

SHAPING With little to do in a Skinner box, the animal typically presses the lever sooner rather than later. One of the major problems encountered with operant conditioning outside the Skinner box is that you need to wait until the animal emits an appropriate response before you can provide the reinforcer. Let's say you are trying to teach your dog to roll over. Rather than waiting for your dog to spontaneously perform this action, you can use an operant conditioning technique to teach your dog to do so. The process, called **shaping**, involves reinforcing behaviours that are increasingly similar to the desired behaviour. You initially reward your dog for any behaviour that even slightly resembles rolling over, such as lying down. Once this behaviour is established you become more selective in which behaviours you reinforce. Reinforcing *successive approximations* will eventually produce the desired behaviour, as your dog learns to discriminate which behaviour is being reinforced (Figure 6.12). Shaping is a powerful procedure that can condition animals to perform amazing feats, such as pigeons playing table tennis, dogs playing the piano, and pigs doing housework such as picking up clothes and vacuuming. Shaping has also been used to teach mentally ill people appropriate social skills, autistic children language, and mentally retarded individuals basic skills such as dressing themselves. More generally, parents and educators often use subtle forms of shaping to encourage appropriate behaviour in children, such as praising children for their initial—often illegible—attempts at handwriting.

REINFORCERS CAN BE CONDITIONED The most obvious reinforcers are those that are necessary for survival, such as food or water. Those that satisfy biological needs are referred to as **primary reinforcers**. From an evolutionary standpoint, the learning value of primary reinforcers makes a great deal of sense, since organisms that repeatedly perform behaviours reinforced by food and water are more likely to survive and pass along their genes. But many apparent reinforcers do not directly satisfy basic biological needs. For instance, as we noted already, money is only pieces of metal or paper, but many people work hard to receive it. Likewise, a compliment, a hug from a friend, or an "A" on a paper can be reinforcing. Events or objects that serve as reinforcers but that do not satisfy biological needs are referred to as **secondary reinforcers**. As mentioned earlier in this chapter, secondary reinforcers are established through classical conditioning. We learn to associate a neutral stimulus, such as paper money (CS), with, for example, food (US). Of course, money also takes on other meanings, such as power and security, but the essential point is that a neutral object becomes meaningful through its association with unconditioned stimuli.

shaping A process of operant conditioning that involves reinforcing behaviours that are increasingly similar to the desired behaviour.

primary reinforcers Reinforcers that are innately reinforcing, such as those that satisfy biological needs.

secondary reinforcers Events or objects that become reinforcers through their repeated pairings with primary reinforcers.

6.12 Shaping The process of shaping can be used to train extraordinary behaviours. Here, a trained dog water skis for a boat show.

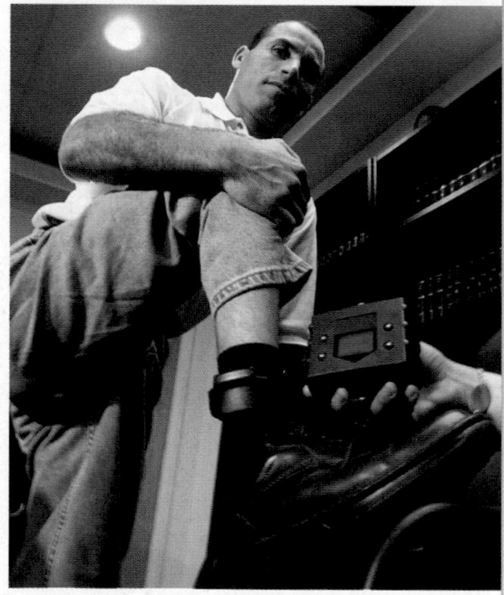

6.13 Punishment as a Deterrent This man was stripped of his freedom after Canadian authorities thwarted his plot to commit an act of terrorism against the country.

How does negative reinforcement differ from positive punishment?

REINFORCER POTENCY Some reinforcers are more powerful than others. An integrative theory of reinforcement was proposed by the psychologist David Premack, who theorized that the value of a specific reinforcer could be determined by the amount of time an organism engages in that behaviour when free to choose anything. For instance, you could observe that most children choose to spend more time eating ice cream than spinach, indicating that ice cream is more reinforcing. One great advantage of Premack's theory is that it can account for individual differences in values. Undoubtedly some people prefer spinach to ice cream; therefore, spinach serves as a more potent reinforcer for them.

A logical extension of Premack's theory is that a more valued activity can be used to reinforce the performance of a less valued activity, which is referred to as the *Premack principle.* Parents use the Premack principle all the time: "Eat your spinach and then you'll get dessert"; "Finish your homework and then you can go out."

Both Reinforcement and Punishment Can Be Positive or Negative

Reinforcement and punishment have opposite effects on behaviour. Whereas reinforcement increases behaviour, punishment decreases behaviour. Formally, **punishment** is the process by which the consequences of an action reduce the likelihood that the action will be repeated. For example, giving a rat an electric shock for pressing a lever will decrease the number of times that it presses the lever. Both reinforcement and punishment can be positive or negative.

POSITIVE AND NEGATIVE REINFORCEMENT **Positive reinforcement** increases the probability of a behaviour's being repeated by administration of a pleasurable stimulus. Positive reinforcement often is referred to as *reward.* Behaviours that are rewarded increase in frequency, such as working hard because of praise or money. In contrast, **negative reinforcement** increases behaviour through the *removal* of an aversive stimulus. For instance, a rat is negatively reinforced when required to press a lever to turn off an electric shock. Note how negative reinforcement is different from punishment, in which the rat receives a shock for pressing the lever. Negative reinforcement is not punishment. Reinforcement—positive or negative—*increases* whereas punishment *decreases* the likelihood of a behaviour. Negative reinforcement is quite common in everyday life. You close the door to your room to shut out noise. You change the channel to avoid watching an awful program. In each case, you are trying to escape or avoid an unwanted stimulus, which is the negative reinforcer.

POSITIVE AND NEGATIVE PUNISHMENT Punishment reduces behaviour, but it can do so through positive or negative means. **Positive punishment** decreases the probability of a behaviour's recurring by administration of an averse stimulus. A rat's getting a shock for pressing a lever is an example of positive punishment. **Negative punishment** decreases the probability of a behaviour by removal of a pleasurable stimulus (Figure 6.13). Youths whose driving privileges are revoked for speeding are less likely to speed the next time they get behind the wheel. Although losing driving privileges is a form of negative punishment, getting a speeding ticket is a form of positive punishment. Figure 6.14 provides an overview of positive and negative reinforcement and punishment.

EFFECTIVENESS OF PARENTAL PUNISHMENT Parents who want their children to behave may use punishment as a means of discipline. But many contempo-

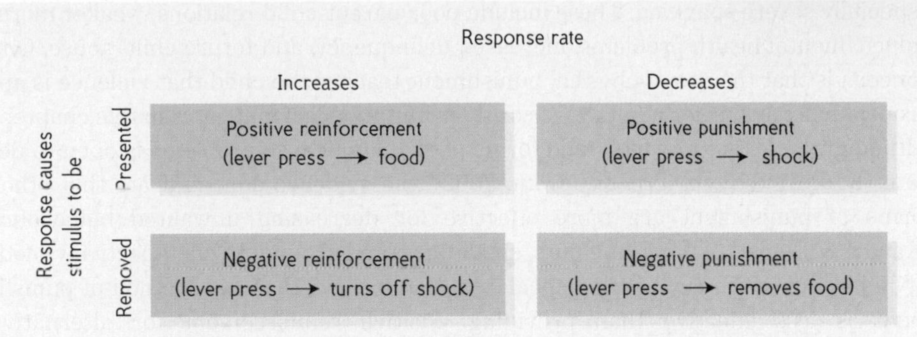

6.14 An Overview of Negative and Positive Reinforcement and Punishment

rary psychologists believe that punishment is often applied ineffectively and that it may have unintended and unwanted consequences. Research has shown that for punishment to be effective, it must be reasonable, unpleasant, and applied immediately so that the relationship between the unwanted behaviour and the punishment is clear (Goodall, 1984; O'Leary, 1995). Also, it must be applied only in direct response to unwanted behaviour. But experience shows us that there is considerable potential for confusion here. Sometimes punishment is applied following a desired action, as when a student is punished after admitting to cheating on an exam. The student may then associate the punishment with being honest rather than with the original offense; the result may be that the student learns not to tell the truth. As Skinner once pointed out, one thing that people learn from punishment is how to avoid it; they learn not to get caught rather than how to behave appropriately.

Punishment can also be problematic because it can lead to negative emotions, such as fear or anxiety, which may become associated, through classical conditioning, with the person who administers the punishment. A child may learn to fear a parent or teacher rather than stop the undesired behaviour, and this may damage the long-term relationship between adult and child (Gershoff, 2002). Another potential problem with punishment is that it often fails to offset the reinforcing aspects of the behaviour. In real life, any behaviour can be reinforced in multiple ways. For instance, thumb sucking may be reinforced because it makes a child feel good and provides relief from negative emotions and hunger. The threat of punishment may not be sufficient to offset the rewards of thumb sucking, though it may reinforce secrecy. For these and other reasons, most psychologists agree with Skinner's recommendation that reinforcement be used rather than punishment. A child who receives compliments for being a good student is likely to show better academic performance than one who is punished for doing poorly. After all, reinforcing a behaviour tells the child what to do, whereas punishing a student for poor grades doesn't provide information about what the child should do to improve.

One form of punishment that most psychological scientists believe is especially ineffective is physical punishment, such as spanking. However, spanking is very common, with 80 percent of Canadians reporting they have been spanked (MacMillan et al., 1999). As noted by Kazdin and Benjet (2003), beliefs about the appropriateness of spanking involve religious beliefs and cultural views, as well as legal issues. In 2004, the Supreme Court of Canada upheld a law that allows parents and teachers to spank children, at least those under age 12. Many countries (e.g., Austria, Denmark, Israel, and Italy) have banned corporal punishment in homes or schools, and even the United Nations has passed resolutions discouraging it. Indeed, some researchers have provided evidence of numerous negative outcomes associated with spanking,

punishment A stimulus following a response that decreases the likelihood that the response will be repeated.

positive reinforcement The increase in the probability of a behaviour's being repeated following the administration of a pleasurable stimulus, referred to as a reward.

negative reinforcement The increase in the probability of a behaviour's being repeated through the removal of an aversive stimulus.

positive punishment Punishment that occurs with administration of a stimulus that decreases the probability of a behaviour's recurring.

negative punishment Punishment that occurs with removal of a stimulus that decreases the probability of a behaviour's recurring.

especially severe spanking. These include poor parent-child relations, weaker moral values, mental health problems, increased delinquency, and future child abuse. One concern is that the use of physical punishment teaches the child that violence is appropriate behaviour for adults. (We consider imitation learning later in this chapter.) Although the extent to which mild forms of spanking cause problems is open to debate (Baumrind, Larzelere, & Cowan, 2002), the evidence does indicate that other forms of punishment are more effective for decreasing unwanted behaviours (Kazdin & Benjet, 2003). Time-outs, small fines, and grounding can effectively modify behaviour. Yet many psychological scientists believe that any method of punishment is less effective than providing positive reinforcement for alternative behaviours. By rewarding the behaviours they wish to see, parents are able to increase those behaviours while building more positive bonds with their children.

Operant Conditioning Is Influenced by Schedules of Reinforcement

Which reinforcement schedule would you use if you wanted a behaviour to continue over time?

How often should a reinforcer be given? For fast learning, you might wish to reinforce the desired behaviour each time it occurs, referred to as **continuous reinforcement**. In the real world, behaviour is seldom reinforced continuously. Animals do not find food each time they look for it, and people do not receive praise each time they behave in an acceptable fashion. Most behaviour is reinforced intermittently, which is referred to as **partial reinforcement**. The effect of partial reinforcement on conditioning depends on the reinforcement schedule.

continuous reinforcement A type of learning in which the desired behaviour is reinforced each time it occurs.

partial reinforcement A type of learning in which behaviour is reinforced intermittently.

ratio schedule A schedule in which reinforcement is based on the number of times the behaviour occurs.

interval schedule A schedule in which reinforcement is based on a specific unit of time.

fixed schedule A schedule in which reinforcement is consistently provided upon each occurrence.

variable schedule A schedule in which reinforcement is applied at different rates or at different times.

partial-reinforcement extinction effect The greater persistence of behaviour under partial reinforcement than under continuous reinforcement.

RATIO AND INTERVAL SCHEDULES Partial reinforcement can be administered according to the number of behavioural responses or the passage of time. For instance, factory workers can be paid by the piece (behavioural response) or by the hour (passage of time). In a **ratio schedule**, reinforcement is based on the number of times the behaviour occurs, such as reinforcing every third or tenth occurrence of the behaviour. With **interval schedules**, reinforcement is based on a specific unit of time, such as once every minute or hour. In general, ratio reinforcement leads to greater responding than does interval reinforcement. Factory workers who are paid by the piece are usually more productive than those paid by the hour, especially if there are incentives for higher levels of productivity.

FIXED AND VARIABLE SCHEDULES Partial reinforcement also can be given on a *fixed* or *variable* schedule. In a **fixed schedule**, the reinforcer is consistently given following a specific number of occurrences or after a specific amount of time. Whether factory workers are paid by the piece or by the hour, they usually are paid according to a fixed rate, earning the same for each piece or for each hour. The rate of reinforcement is entirely predictable. In **variable schedules** of reinforcement, the reinforcer is applied at different rates or at different times. The responder does not know how many behaviours need to be performed or how much time needs to pass before reinforcement occurs. Variable reinforcement occurs when salespeople are paid only when a customer agrees to purchase a product.

The influence of reinforcement schedules on behaviour can be observed in the study habits of students. Ideally, students will spread their studying out over the entire term, but the reality is that when exams are scheduled at fixed points, students tend to study a great deal just before the exam and not at all afterward. It is typical of *fixed-interval reinforcement* that the behaviour is performed only when it is time for the reinforcer to be administered. When exams are on a *variable-interval schedule*—say, with surprise quizzes—students tend to study more often and more consistently. The patterns of behaviour typically observed under different schedules of reinforcement can be seen in Figure 6.15.

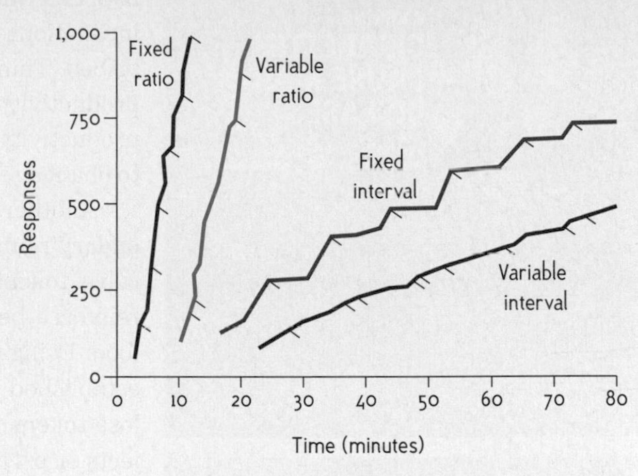

6.15 Learning Curves These curves show typical learning under different schedules of reinforcement. The steeper the line, the higher the response rate. The slash marks on the lines indicate when reinforcement is given. Note that ratio reinforcement leads to the highest rate of response.

BEHAVIOURAL PERSISTENCE The schedule of reinforcement also affects the persistence of behaviour. Continuous reinforcement is highly effective for acquiring a behaviour, but if the reinforcement is stopped, the behaviour extinguishes quickly. For instance, typically when you put money in a vending machine it gives you a product. If it fails to do so, you will quickly stop giving it your money. In contrast, a slot machine pays out on a *variable-ratio schedule.* People continue to pour money into slot machines even if the machine rarely reinforces their behaviour. The **partial-reinforcement extinction effect** describes the greater persistence of behaviour under partial reinforcement, due to the difficulty of detecting a lack of reinforcement. During continuous reinforcement, it is easy to detect when reinforcement has ceased. But if the behaviour is reinforced only some of the time, the subject needs to repeat the behaviour comparatively more times to detect the absence of reinforcement. Thus, the less frequent the reinforcement during training, the greater the resistance to extinction. To condition a behaviour that you wish to persist, reinforce it continuously during early acquisition and then slowly change to partial reinforcement. Parents naturally follow this strategy in teaching their children toilet training and other behaviours.

Psychological Science *in Action*

Behaviour Modification

Parents, teachers, and animal trainers widely use operant-conditioning strategies. *Behavioural modification* refers to using the techniques of operant conditioning to eliminate unwanted behaviours and replace them with desirable ones. The general rationale behind behaviour modification is that most unwanted behaviours are learned and therefore can be unlearned (Figure 6.16). Through conditioning principles, people can be taught, for example, to be more productive at work, to save energy, and to drive more safely. These techniques can even be used to train children with profound learning disabilities to communicate and interact with their classmates. As we will see in Chapter 14, operant techniques are also effective for treating psychological conditions such as depression and anxiety disorders.

Because they are so successful, many businesses have embraced the use of behaviour modification in the workplace. Consider the classic case of Emery Air Freight, a large company in the shipping industry. In the 1970s, Emery was losing a great deal of money because workers were not filling containers to capacity before shipping. Management instituted a program in which workers were monitored and rewarded for filling the containers completely. Within a matter of days the percentage of full containers more than doubled, saving the company millions of dollars. Other companies quickly took notice, and many developed programs—such as profit sharing, stock options, and bonuses—to reward productivity. These programs have to be used carefully, however, because if rewards are not given properly they can backfire, as

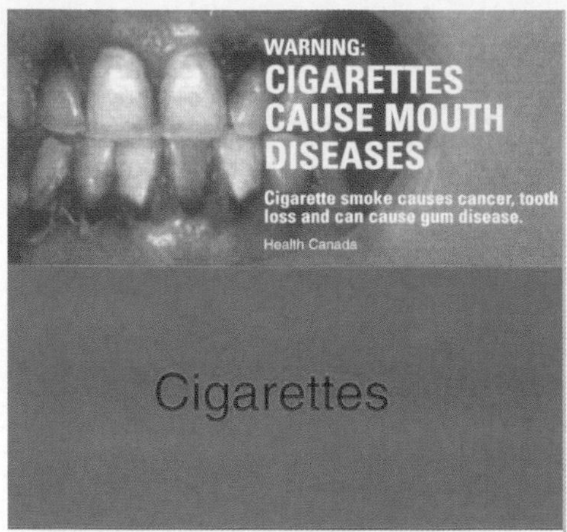

6.16 Behaviour Modification Behaviour modification can come in many forms and from many sources. Here, the Canadian government tries to stop its citizens from acting in ways harmful to themselves by posting graphic advertisements depicting the physical side effects of smoking. What kind of positive reinforcement might help a smoker eliminate smoking?

happens when workers become overly competitive, when rewards undermine natural inclinations to work hard, or when rewards make people feel as if they are being bribed. Think about what you know regarding how schedules of reinforcement affect productivity. If a company used a reward program on a fixed schedule, it is likely that productivity would decline quickly when the rewards ceased. Thus, companies need to change rewards and offer them on a variable schedule to maintain productivity.

Another widespread behaviour-modification method uses the principles of secondary reinforcement. Chimpanzees can be trained to perform tasks in order to receive tokens, which they can later trade for food. The tokens themselves thus reinforce behaviour and the chimps work as hard to obtain the tokens as they do food. Using similar principles, many prisons, mental hospitals, and classrooms have established *token economies* in which people earn tokens for completing tasks and lose tokens for behaving badly. These tokens can be saved and later traded for objects or privileges. So, for instance, teachers can provide tokens to students for obeying class rules, turning in homework on time, and helping others. At some future point the tokens can be exchanged for rewards, such as extra recess time or fun activities. Token economies are often established in mental hospitals to encourage good grooming and appropriate social behaviour and to discourage bizarre behaviours. The use of a token economy not only reinforces participants, but also gives them a sense of control over their environment.

You can use the principles of behaviour modification to change your own behaviour. To do this you must first select some target behaviour you wish to change, such as swearing, not spending enough time studying or exercising, or spending too much time watching television—anything, as long as it is specific and you have a realistic goal for changing it. You then need to monitor the behaviour to see how often you do it (or not do it), such as how many hours per day you spend productively studying. Simply noting the behaviour is likely to move you toward your goal, since you will be more conscious of it, but keeping careful track will also allow you to assess your progress. By monitoring, you can also identify potential triggers of an unwanted behaviour, such as time pressure or stress. Knowing what the triggers are will allow you to develop plans or strategies for dealing with them. Now you are ready to take charge. The first thing to do is pick a good reinforcer, something that you really want and that is reasonably attainable, such as going out to a movie, having a nice meal at a restaurant, or hanging out with friends. But remember: If you don't behave appropriately, you don't get the reinforcer! Over time, as you become more successful, phase out the reinforcer so that you are performing the behaviour out of habit. Once you are used to exercising regularly, for example, you will do it regularly and it may even become reinforcing on its own. Give it a try—you might amaze yourself with the power of behaviour modification. ■

Biology and Cognition Influence Operant Conditioning

Behaviourists such as B. F. Skinner believed that all behaviour could be explained by straightforward conditioning principles. Recall that Skinner's *Walden Two* describes a utopia in which all of society's problems are solved through operant-conditioning principles. In reality, however, we know there are limits to explaining human behaviour through schedules of reinforcement. Biology places constraints on learning, and reinforcement does not always have to be present for learning to take place.

BIOLOGICAL CONSTRAINTS Although behaviourists believed that any behaviour could be shaped through reinforcement, it is now apparent that animals have a hard time learning behaviours that run counter to their evolutionary adaptation. A good example of biological constraints was obtained by the Brelands, a husband-and-wife team of psychologists who used operant-conditioning techniques to train animals for commercials (Breland & Breland, 1961). They discovered that many of their animals refused to perform certain tasks they had been taught. For instance, they trained a raccoon to place coins in a piggy bank, but, although the raccoon initially learned the task, it eventually refused to deposit the coins. Instead, it stood over the piggy bank and briskly rubbed the coins in its paws. The rubbing behaviour was not reinforced; it actually delayed reinforcement. One explanation for the raccoon's behaviour is that the task was incompatible with innate adaptive behaviours. The raccoon associated the coin with food and treated it the same way: rubbing food between the paws is hardwired for raccoons. Along similar lines, pigeons can be trained to peck at keys to obtain food or secondary reinforcement, but it is difficult to get them to peck at keys to avoid electric shock. However, they can learn to avoid shock by flapping their wings because it is their natural means of escape. Psychologist Robert Bolles argues that animals have built-in defense reactions to threatening stimuli (Bolles, 1970). Conditioning is most effective when the association between the behavioural response and the reinforcement is similar to the built-in predispositions of the animal.

The evolutionary perspective views the brain as a compilation of different domain-specific modules, each responsible for different cognitive functions. This suggests that learning is the result of many unique mechanisms that solve individual adaptive problems, not the result of general learning mechanisms (Rozin & Kalat, 1971; Shettleworth, 2001). As the evolutionary psychologist Randy Gallistel (2000) argues, people readily accept that the lungs are adapted for breathing and the ear is adapted for hearing. No one would argue that the lungs could be used to hear, or the ears to breathe. There is no general "sensing organ." Thus, it makes equal sense to postulate that a variety of learning mechanisms have evolved to solve specific problems. Consider an ant that leaves its nest to forage for food. Typically, the ant takes a circuitous and wandering path until it finds food, at which point it takes a direct path back to the nest, even over unfamiliar territory. This differs from what would be expected from traditional models of learning. The ant has never been rewarded for following the path, and there has been no classically conditioned association between the unfamiliar environmental objects and the most direct path back to the nest. Instead, mental processes that compute small changes in distance and direction provide a solution for the most direct path home. Gallistel's point is that learning consists of specialized mechanisms (rather than universal mechanisms) that solve the adaptive problems faced by animals in their environments.

ACQUISITION-PERFORMANCE DISTINCTION Another challenge to the idea that reinforcement is responsible for all behaviour is that learning can take place without reinforcement. Edward Tolman, an early cognitive theorist, argued that reinforcement has more impact on performance than on learning. At the time, Tolman was conducting experiments in which rats had to learn to run through complex mazes to obtain food. Tolman believed that the rats developed **cognitive maps**, spatial representations of the maze that helped them learn to quickly find the food. To test his theory, Tolman and his students studied three groups of rats whose task was to travel through a maze to a "goal box" containing the reinforcer, usually food. The first group did not receive any reinforcement, and their performance was quite poor. With no food in the goal box, the rats wandered through the maze. A second group was reinforced on every trial, and they learned to find the goal box quickly. The critical group was the third group, which was not reinforced for the first ten tri-

cognitive map A visual/spatial mental representation of the environment.

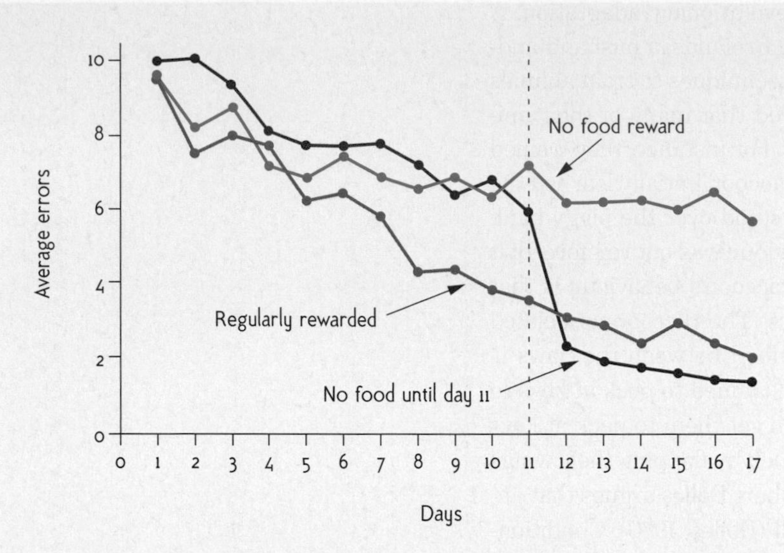

6.17 Reinforcement Rats that were regularly reinforced for correctly running through a maze showed improved performance over time compared to rats that did not receive reinforcement. Rats that were not reinforced for the first ten trials but were reinforced thereafter showed an immediate change in performance, indicating that they had learned a path through the maze, but that their learning was not revealed, or was latent, until it was reinforced. (Note that between days 11 and 12 their average number of errors decreased dramatically.)

als but was then reinforced on subsequent trials. Tolman found that this group showed an amazingly fast learning curve, so that they immediately caught up to the group that had been continuously reinforced (Figure 6.17). This result implies that the rats had learned a cognitive map of the maze and could use it when reinforcement became available. Tolman used the term **latent learning** to refer to learning that takes place in the absence of reinforcement.

Another form of learning that takes place in the absence of reinforcement is *insight learning*—a form of problem solving (discussed in Chapter 8) in which a solution suddenly emerges after a period of inaction or following contemplation of the problem. You probably have had this sort of experience, in which you mull over a problem for a while and then suddenly know the answer. Reinforcement does not provide an adequate account of the process of insight learning, although it does predict whether the behaviour is subsequently repeated.

The Value of Reinforcement Follows Economic Principles

A relatively new approach to understanding operant conditioning considers the value of reinforcement in the context of basic economic principles, such as supply and demand. That which is in short supply is typically valued more, and therefore is a more potent reinforcer. Sometimes the economic considerations are more complicated. Which would you prefer, $100 today or $1,000 next year? Although it might appear that the larger payment would be more reinforcing, its value is discounted because there is a significant delay before it is received. But what if you really needed money right now? In that case, waiting has additional costs. Indeed, adults with lower incomes discount future payments more steeply than those with large incomes (Green, Myerson, Lichtman, Rosen, & Fry, 1996). At the heart of the behavioural economics approach is the idea that people and other animals often need to make choices between reinforcers. The personal worth of the reinforcer is affected by how long one has to wait to receive it as well as the likelihood that the payoff will actually be received. This approach has provided insights into a variety of behaviours, especially those associated with addiction. Warren Bickel and his colleagues (1999; Madden, Petry, Badger, & Bickel, 1997) found that both smokers and heroin addicts discounted future rewards more greatly than did nonaddicts, which may contribute to their problems with impulsivity and self-control. The issue of delaying immediate gratification to obtain long-term rewards is discussed more fully in Chapter 9.

Behavioural choice is also implicated in studies of animal foraging. Where animals choose to eat is dependent on the likelihood that food will be present, the energy costs associated with obtaining the food, and the risks associated with predators (Shettleworth, 2001). Should an animal eat all the berries in one patch and then search for a new patch, or should it save some for later and search for more patches? The new patches might have berries, but they might not. Researchers who study animals in their natural habitats find that animals are highly sensitive to the relative rates of reinforcement among different patches. The time they spend eating in one patch is influenced by the rate of reinforcement, the relative rate of reinforcement in other patches,

latent learning Learning that takes place in the absence of reinforcement.

and the time it takes to travel to those other patches. *Optimal foraging theory* describes how animals in the wild choose to provide their own schedules of reinforcement. It predicts that animals will sometimes act in apparently risky fashion. Consider an animal faced with the choice of feeding from two locations that have the same average amount of food, but one is much more variable, sometimes having no food and sometimes having lots. If an animal is facing starvation, its best chance for survival is to follow the apparently risky strategy of foraging in the more highly variable location. Animals also learn to vary how they eat according to the likelihood of being attacked by predators. Research at the University of Lethbridge in Alberta found that rats eat very quickly when in the dark, especially when they are food deprived (Wishaw, Drigenberg, & Comery, 1992). Although eating quickly is optimal for consuming large amounts of food, it is not optimal for digestion. When in the light, rats eat more slowly, but they make frequent scans of the environment to check for predators. These variations reflect the fact that even a simple behaviour such as eating requires a number of calculations to detect the costs and benefits associated with different behavioural options. The choices made by animals reflect specialized learning capacities that are adaptive to the environment.

REVIEWING the Principles | How Is Operant Conditioning Different from Classical Conditioning?

Whereas classical conditioning involves the learned association between two events, operant conditioning involves learning the association between a behaviour and its consequences. Developed initially by B. F. Skinner, operant conditioning explains why some behaviours are repeated and others are not. The application of reinforcement increases the likelihood that a behaviour will be repeated, whereas punishment reduces that likelihood. The extent to which a reinforcer increases behaviour depends on whether it is positive or negative, its relative value, and on how it is provided. Although Skinner was confident that operant conditioning could ultimately explain all behaviour, it has faced a number of challenges. Chief among these is that it is difficult to change instinctive behaviours and that learning can take place in the absence of reinforcement. Modern learning theorists recognize the important role of cognitive processes and biological constraints on behaviour. Models based on economic theory have become increasingly useful for understanding how animals choose among reinforcers.

How Does Watching Others Affect Learning?

What is learned from observing others?

Suppose that you were trying to teach someone to fly an airplane. How might you apply learning principles discussed in this chapter to accomplish your goal? Reinforcing arbitrary correct behaviours obviously would be a disastrous strategy for training aspiring pilots. Similarly, teaching people to play football, eat with chopsticks, or perform complex dance steps requires more than simple reinforcement. For many behaviours, we learn not by doing but by observing the actions of others. We learn things such as social etiquette through observation, and we learn whether to be anxious in a situation by watching to see if others are anxious. Similarly, we acquire many of our attitudes about politics, religion, people, and objects from outside sources, such as parents, peers, teachers, and the media.

Learning Can Be Passed On through Cultural Transmission

Although humans around the world belong to the same species and share the vast majority of genes, there is enormous cultural diversity in what people think and how they behave. Would you be the same person if you were raised in a small village in China, or the jungles of South America, or the mountains of Afghanistan? Probably not, since your religious beliefs, personal values, and even your musical tastes are shaped by the culture in which you are raised. The term that evolutionary psychologists use for the transmission of cultural knowledge is **meme**. Memes are analogous to genes, in that they are selectively passed on from one generation to the next, with some, such as fads, dying out quite quickly.

meme The knowledge transferred within a culture.

observational learning Learning that occurs when behaviours are acquired or modified following exposure to others performing the behaviour.

A good example of cultural transmission of knowledge can be found in the case of Imo the monkey. In the 1950s, researchers who were studying monkeys in Japan threw some sweet potatoes onto the sandy beach for the monkeys to eat. Imo developed the habit of washing her sweet potato in the ocean to get the sand off. Within a short time, other monkeys copied Imo and soon potato-washing behaviours were picked up by many monkeys all over the island. This behaviour has continued to be passed along from one generation to the next through social learning, and monkeys can still be observed washing their potatoes to this day (Dugatkin, 2004). We discuss this further in Chapter 16. Unlike natural selection, which typically occurs slowly over thousands of years, memes can spread quickly, such as the worldwide adoption of surfing the Internet. Although memes can be conditioned through association or reinforcement, many memes are learned from watching the behaviour of others.

Learning Can Occur through Observation

Observational learning, which occurs when behaviours are acquired or modified following exposure to others performing the behaviour, is a powerful adaptive tool both for humans and for other animals. Animals learn that food is safe by watching what their parents eat, and they learn to fear dangerous objects by watching their parents avoid those objects. Similarly, children acquire beliefs and basic skills by observing their parents and other role models. Any parent will tell you that young children are sponges, absorbing everything that goes on around them. They learn by watching as much as by doing.

6.18 Albert Bandura Bandura conducted important research showing that people could learn through observation.

BANDURA'S OBSERVATIONAL STUDIES The most thorough work on observational learning was conducted by Canadian psychologist Albert Bandura (Figure 6.18) in the 1960s. In a now-classic series of studies, preschool children were shown a film of an adult playing with a large inflatable doll called Bobo (Figure 6.19). In the film, the adult either played quietly with Bobo or attacked the doll furiously, whacking it with a mallet, socking it in the nose, and kicking it around the room. When the children were later allowed to play with a number of toys, including the Bobo doll, those who had seen the more aggressive display were more than twice as likely to act aggressively toward the doll (Bandura, Ross, & Ross, 1961). These results suggest that exposure to media violence may increase the likelihood that children will act aggressively.

OBSERVATIONAL LEARNING OF FEAR Observational learning also occurs in animals. For example, psychologist Susan Mineka explored whether rhesus monkeys could develop a phobia of snakes by observing other monkeys react fearfully to them. She had noticed that monkeys raised in the laboratory do not fear snakes, whereas

6.19 Scenes from the Classic Bobo Doll Studies After viewing the adult act aggressively (top row), children (bottom two rows) imitated what they had seen.

those raised in the wild have an intense fear of them. Mineka and her colleagues set up an experiment with two groups of monkeys, one group reared in the wild and one group reared in the laboratory. The monkeys were required to reach beyond a clear box that contained either a snake or a neutral object to obtain food. When a snake was in the box, not only did the wild-reared monkeys not touch the food, but they also showed signs of distress, such as clinging to their cages and making threatening faces (Figure 6.20). The laboratory-raised monkeys reached past the box whether it contained a snake or not, and they showed no overt signs of fear. Mineka then showed the laboratory-raised monkeys the fearful response of the wild monkeys to see if it would affect their reaction to the snake. The laboratory monkeys quickly developed a fear of the snakes, and this fear was maintained over a three-month period (Mineka, Davidson, Cook, & Keir, 1984).

6.20 The Fear Response In Susan Mineka's experiment, two sets of monkeys had to reach past a clear box to get food. When the clear box contained a snake, the wild-reared monkeys refused to reach across the box. By observing the actions of those monkeys, the laboratory-reared monkeys, who did not originally fear snakes, learned to become afraid of them. This suggests that fears can be learned through observation.

TEACHING THROUGH DEMONSTRATION Because humans can learn through observation, they can be taught many complex skills through demonstration. For instance, parents use slow and exaggerated motions to show their children how to tie their shoes. Do animals other than humans teach their offspring? Although the idea is somewhat controversial, there do appear to be instances in which animals teach their offspring through demonstration (Caro & Hauser, 1992). For instance, cheetahs appear to teach their young to hunt by injuring rather than killing their prey, which makes it easier for the cubs to knock down the prey. One chimpanzee who learned sign language (which you will read about in Chapter 11) was observed trying to demonstrate the proper use of signs to another chimpanzee.

6.21 ***Thank You for Smoking*** Director Jason Reitman (left) and actor Cameron Bright pose for the press at the premiere of their 2005 movie. Though they are smoking candy cigarettes, the image is nonetheless striking. Bright looks to be a happy and successful 13-year-old movie star with a cigarette in his mouth, proudly touting it for the cameras. What affect might this have on other teenagers?

Animals and Humans Imitate Others

Animals and humans readily imitate the actions of others. In one study, pigeons observed other pigeons being reinforced at a feeder when they either stepped on a bar or pecked at the feeder directly. When the observing pigeons were themselves placed before the feeder, they tended to use the same technique they had seen (Zentall, Sutton, & Sherburne, 1996). Within a few days of birth, if not hours, human newborns will imitate facial expressions they observe, and they will continue to imitate gestures and other actions as they mature, just like the monkeys who copied the potato-washing behaviour of Imo the monkey.

The imitation of observed behaviour is commonly referred to as **modeling**, in that humans or animals reproduce the behaviours of *models*—those being observed. Modeling in humans is influenced by a number of factors. In general, we are more likely to imitate the actions of models who are attractive, have high status, and are somewhat similar to ourselves. In addition, modeling will be effective only if the observer is physically capable of imitating the behaviour. Simply watching Tiger Woods blast 300-yard drives does not mean we could do so if handed a golf club.

The influence of models on behaviour often occurs implicitly, without our being aware that our behaviour is being altered. Indeed, who wants to admit that we change the way we speak or dress to act like some celebrity? But the evidence is overwhelming that we imitate what we see in others, especially those we admire. Adolescents whose favourite actors smoke in the movies are much more likely to smoke themselves (Tickle, Sargent, Dalton, Beach, & Heatherton, 2001; see Figure 6.21), and the more that adolescents observe smoking in movies, the more positive their attitudes about smoking become and the more likely they are to begin smoking (Dalton et al., 2003). This effect is strongest among children whose parents do not smoke, which may be because what these children learn about smoking comes completely through the media, which tend to glamorize the habit. Smokers in movies are often presented as attractive, healthy, and wealthy, which does not really reflect the typical smoker. It isn't that adolescents watch a movie and decide to smoke, but rather that the images of smokers as mature, attractive, and cool—things that teenagers want to be—shape adolescent attitudes about smoking and subsequently lead to imitation.

VICARIOUS REINFORCEMENT Another factor that determines whether observers imitate a model is whether the model is reinforced for performing the behaviour. In another study, Bandura and his colleagues showed children a film of an adult aggressively playing with a Bobo doll, but this time the film had one of three different endings (Bandura, Ross, & Ross, 1963). A control condition showed no consequences for the model's behaviour, while the second film showed the adult being rewarded by being given candy and praise, and the third showed the adult being spanked and verbally reprimanded. When subsequently allowed to play with the Bobo doll, the children who observed the model being rewarded were much more likely to be aggressive toward the doll than those children in the control group. In contrast, those who saw the model being punished were less likely to be aggressive than the control group. Does this mean that those children did not learn the behaviour? No. When the children later were offered small gifts to perform the model's actions, all of them could do so reliably. It is important to distinguish between the *acquisition* of a behaviour and its *performance*. All of the children learned the behaviour, but only those who saw the model being rewarded actually performed the behaviour. **Vicarious learning** occurs when people learn about the consequences of an action by observing others being rewarded or punished for performing the action.

modeling The imitation of behaviour through observational learning.

vicarious learning Learning that occurs when people learn the consequences of an action by observing others being rewarded or punished for performing the action.

MIRROR NEURONS What happens in the brain during imitation learning? An intriguing study found that special neurons in the brain called **mirror neurons** become activated when a monkey observes another monkey reaching for an object (Rizzolatti, Fadiga, Gallese, & Fogassi, 1996). These mirror neurons are the same neurons that would be activated if the monkey performed the behaviour itself. Mirror neurons are especially likely to become activated when monkeys observe a target monkey engaging in movement that has some goal, such as reaching out to grasp an object. Neither the sight of the object on its own nor the mere sight of the target monkey leads to activation of these mirror neurons.

mirror neurons Neurons in the premotor cortex that are activated during observation of others performing an action.

Brain-imaging techniques have identified similar mirror neurons in humans (Rizzo-latti & Craighero, 2004). Thus, every time you observe another person engaging in an action, similar neural circuits are firing in both your brain and in the other person's. The function of mirror neurons is currently open to debate. This system may serve as the basis of imitation learning, but note that the firing of mirror neurons does not lead to imitative behaviour in the observer. This has led some theorists to speculate that mirror neurons help us explain and predict the behaviour of others. In other words, they allow us to step into the shoes of those we observe so that we can better understand their actions.

Humans also have mirror neurons for mouth actions, and these are stimulated when observers see a mouth move in a way that is typical of chewing or speaking (Ferrari, Gallese, Rizzolatti, & Fogassi, 2003). This has led to speculation that mirror neurons are important not only for imitation learning, but also for the human ability to communicate through language. Mirror neurons may be a brain system that creates a link between the sender and receiver of a message. Rizzolatti and Arbib (1998) proposed that the mirror neuron system evolved to allow language. Their theory relies on the idea that speech evolved mainly from gestures, and indeed people readily understand many nonverbal behaviours, such as waving or thrusting a fist in the air. If their theory is true, then speech sounds came to represent gestures, and this suggests a direct link between words that describe actions and the very actions themselves. Indeed, there is evidence that listening to sentences that describe actions activates the same brain regions that are active when those actions are observed (Tettamanti et al., 2005). Even reading words that represent actions leads to brain activity in relevant motor regions, such as the word *lick* activating brain regions that control tongue movements (Hauk, Johnsrude, & Pulvermüller, 2004).

Thinking Critically

Does Watching Violence on Television Cause Aggression?

The average television set in North America is on for five or six hours per day, and young children often spend more time watching television than doing any other activity, including schoolwork (Roberts, 2000). There is a great deal of violence in television programming. The average child witnesses more than 100,000 violent acts on TV before the end of elementary school; you probably witnessed over 18,000 murders on TV before you started college. What effect does this massive exposure to violence have on behaviour? Do we imitate the violence we see on television?

Three decades of research have linked the viewing of violence with imitative violence, aggressive behaviour, acceptance of violence as a solution, increased feelings of hostility, and willingness to deliver painful stimuli to others. Some researchers

6.22 What Causes Violent Behaviour? Many scientists believe there is a direct correlation between exposure to violent television shows or video games and a tendency to aggressive behaviour. In the popular TV series *24*, starring Kiefer Sutherland, explosions, fighting, and the use of firearms are regular occurrences. Does this affect viewers? Perhaps, but it is important to note that generalization based on the interpretation of specific studies is problematic, as correlation does not prove causation.

have conducted controlled laboratory studies in which they randomly assign children to watch either violent or neutral film clips and then measure subsequent aggressive behaviour. In general, compared to those who watch neutral clips, children who watch the violent ones are more likely to act in an aggressive manner during subsequent tasks. A study by Leonard Eron and his colleagues found that television viewing habits at age 8 predicted violent behaviour and criminal activity at age 30 (Eron, 1987). The average effect size (the magnitude of the association between independent and dependent variables), determined by statistically averaging across the hundreds of studies, is nearly as large as that linking smoking and cancer, and is larger than the effect size for condom nonuse and HIV transmission or for lead exposure and IQ in children (Bushman & Anderson, 2001).

There are, however, a number of problems with the studies on this topic. The University of Toronto social psychologist Jonathan Freedman (1984) notes that many of the so-called aggressive behaviours displayed by children could be interpreted as playful rather than aggressive. A more serious concern is whether the studies generalize to the real world. Viewing a violent film clip in a lab is unlike watching TV in one's living room. The film clips used in studies are often brief and extremely violent, and the child watches them alone. In the real world, violent episodes are interspersed with nonviolent material, and children often watch them with others who may buffer the effect.

Even the longitudinal studies that assess childhood television watching and later violent behaviour fail to prove satisfactorily that TV caused the behaviour. It could easily be that extraneous variables, such as personality, poverty, or parental negligence, have an effect on both television viewing habits and violent tendencies. After all, not all of those who view violence on television become aggressive later in life. Perhaps those who watch excessive amounts of TV, and who therefore have fewer opportunities to develop social skills, act aggressively. Correlation does not prove causation. Only through careful laboratory studies in which participants are randomly assigned to experimental conditions can we determine causality. Obviously, it isn't practical to assign children randomly to different types of media, and it is ethically questionable to expose any children to violence if we think it will make them more aggressive.

In spite of the problems with interpreting specific studies, most scientists believe that there is a direct relation between exposure to violence and aggressive behaviour. Indeed, a recent joint statement by professional groups representing psychologists and pediatricians concluded that a plethora of studies "point overwhelmingly to a causal connection between media violence and aggressive behaviour in some children" (Joint Statement, 2000, p. 1).

How might media violence promote aggression in children? One possibility is that exposure to massive amounts of violence in movies, which misrepresents the prevalence of violence in real life, leads children to believe that violence is both common and inevitable (Figure 6.22). Because few people are punished for acting violently in movies, children may come to believe that such behaviours are justified (Bushman & Huesmann, 2001). That is, the portrayal of violence in movies teaches children questionable social scripts for solving personal problems. By mentally rehearsing the script, or observing the same script enacted many times in different movies, a child might come to believe that engaging in brutality is an effective way to solve problems and dispense with annoying people (Huesmann, 1998).

What do you think? How much violence have you seen in movies or on television? Do you think it affected you? Should it be banned even if it only affects some children? What might parents do if they are concerned about what their children watch on television? ■

REVIEWING the Principles | How Does Watching Others Affect Learning?

Thanks to psychological research, we now know that much of behaviour is learned by observing the behaviour of others. Children learn language, social skills, and political attitudes from observing their parents, peers, and teachers, and we teach complex skills, such as surgery and driving, by demonstration. Nonhuman animals also learn through observation, for example, which food is safe to eat and which objects should be feared. People imitate models that are attractive, high status, and somewhat similar to themselves. Modeling is more likely to occur when the model has been rewarded for the behaviour, and less likely when the model has been punished. Vicarious conditioning influences whether a behaviour is performed, but not whether it is learned. It is possible that mirror neurons, which fire when a behaviour is observed, may be the neural basis of imitation learning.

What Is the Biological Basis of Reward?

Although people often use the term *reward* as a synonym for positive reinforcement, Skinner and other traditional behaviourists defined reinforcement strictly in terms of whether it increased behaviour. They were relatively uninterested in *why* it increased behaviour. Indeed, they carefully avoided any speculation about whether subjective experiences had anything to do with behaviour, since they believed that mental states were impossible to study empirically. The biological revolution, however, has begun to provide new insights into how we learn. In this section we examine the biological basis of reinforcement and its application to understanding conditioned behaviours.

Self-Stimulation Is a Model of Reward

One of the earliest discoveries pointing to the role of neural mechanisms in reinforcement came about because of a small surgical error. In the early 1950s, Peter Milner and James Olds at McGill University were testing whether electrical stimulation to a specific brain region would facilitate learning. To see whether the learning they observed was caused by the activity of the brain or by the aversive qualities of the electrical stimulus, Olds and Milner administered electrical stimulation to the brains of rats only while the rats were in one specific location in the cage. The logic was that if the application of electricity was aversive, the rats would selectively avoid that location. Fortunately for science, they administered the shocks to the wrong part of the brain, and instead of avoiding the area of the cage associated with electrical stimulation, the rats quickly came back, apparently looking for more.

PLEASURE CENTRES Olds and Milner then set up an experiment to see whether rats would press a lever to self-administer shock to specific sites in their brains, a procedure subsequently referred to as **intracranial self-stimulation** (ICSS) (Figure 6.23). The rats self-administered electricity to their brains with gusto, pressing the lever

intracranial self-stimulation (ICSS) A procedure in which animals are able to self-administer electrical shock to specific areas of the brain.

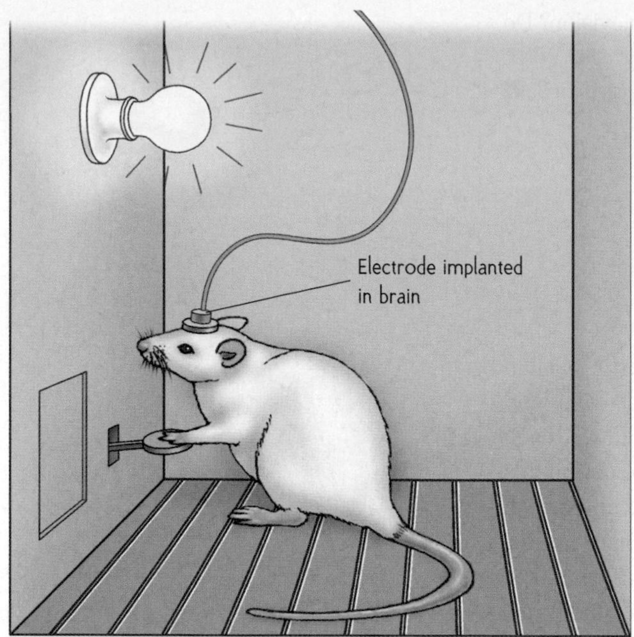

6.23 Intracranial Self-Stimulation A rat will self-administer electricity to pleasure centres in the brain.

hundreds of times per hour (Olds & Milner, 1954). Olds and Milner referred to brain regions that support ICSS as *pleasure centres.* Although behaviourists might have objected to the term *pleasure,* it was clear to everyone that ICSS was a powerful reinforcer. In one experiment, rats that had been on a near-starvation diet for ten days were given a choice between food and the opportunity to administer ICSS. They chose the electrical stimulation more than 80 percent of the time! Deprived rats also chose electrical stimulation over water or receptive sexual partners; they even crossed a painful electrified grid in order to receive ICSS. Rats will continue intracranial self-stimulation until they collapse from exhaustion. Monkeys tested in similar studies have been found to press a bar for electrical stimulation up to 8,000 times per hour (Olds, 1962).

ELECTRICAL STIMULATION AND NATURAL REINFORCEMENT Most psychologists believe that ICSS acts on the same brain regions as those activated by natural reinforcers, such as food, water, and sex. Electrical stimulation applied to pleasure centres in rats elicits naturally motivated behaviours, such as feeding, drinking, and copulating with an available partner. Also, depriving an animal of food or water leads to increased ICSS, which is taken to indicate that the animal is trying to obtain the same experience associated with natural reward. Finally, the neural mechanisms underlying ICSS and natural reward appear to use the same neurotransmitter system, namely dopamine. In terms of operant conditioning, this evidence suggests that dopamine serves as the neurochemical basis of positive reinforcement.

Dopamine Signals Reward

Recall from Chapter 3 that the neurotransmitter dopamine is involved in motivation and emotion. It also appears to be crucial for positive reinforcement. Research over the past 50 years has shown that dopamine plays an important role in the experience of reward (Wise & Rompre, 1989). For instance, intracranial self-stimulation activates dopamine receptors; interfering with dopamine eliminates self-stimulation as well as other naturally motivated behaviours, such as feeding, drinking, and copulating.

Why are certain experiences rewarding?

NUCLEUS ACCUMBENS ACTIVATION The *nucleus accumbens* is a subcortical brain region that is part of the limbic system. The experience of pleasure usually results from activation of dopamine neurons in the nucleus accumbens. For example, enjoying food depends on dopamine activity. Hungry rats that are given food have increased release of dopamine in the nucleus accumbens, and the greater the hunger, the greater the release of dopamine (Rolls, Murzi, Yaxley, Thorpe, & Simpson, 1980). Food tastes better when you are hungry and water is more rewarding when you are thirsty because more dopamine is released under deprived than nondeprived conditions. Even the enjoyable activity of looking at funny cartoons activates the nucleus accumbens (Mobbs, Greicius, Abdel-Azim, Menon, & Reiss, 2003).

In operant conditioning, the release of dopamine sets the value of a reinforcer. Drugs that block the effects of dopamine disrupt operant conditioning. For instance, Jim Stellar and his colleagues taught rats to run a maze in order to receive electrical stimulation, but when they injected a dopamine blocker into the rats, they had to turn up the electrical current in order to get the rats to run the maze (Stellar, Kelley, & Corbett, 1983). The blocker decreased the value of the electrical-stimulation reinforcement. As you might expect, drugs that enhance the activation of dopamine, such as cocaine and amphetamine, increase the reward value of stimuli.

SECONDARY REINFORCERS ALSO RELY ON DOPAMINE Natural reinforcers appear to signal reward directly through the activation of dopamine receptors in the nucleus accumbens. But what about secondary reinforcers, such as money or good grades? Through a process of classical conditioning, neutral stimuli that fail to stimulate a release of dopamine at first do so readily after they are paired with unconditioned stimuli. In one study, monkeys were presented with a trapdoor that opened occasionally. The door opening did not activate dopamine activity. The experimenters then placed apples in the doorway, such that the door opening was associated with the unconditioned stimulus of food. After many conditioning trials, the door opening led on its own to increased activation of dopamine (Ljungberg, Apicella, & Schultz, 1992). Thus, the sight of a loved one, or getting a good grade or a paycheck, may be conditioned to produce dopamine activation. As mentioned earlier, money is an excellent example of a secondary reinforcer, and indeed, anticipated monetary rewards have been found to activate dopamine systems (Knutson, Fong, Adams, Varner, & Hommer, 2001). Drugs that block the effects of dopamine appear to block the rewarding qualities of most activities. Individuals with *Tourette's syndrome,* a disorder of motor control, are given dopamine blockers to control their involuntary body movements. These individuals often have trouble staying on their drug regimens because they feel the drugs rob them of life's enjoyment. We will discuss this in greater detail in Chapter 14.

DRUG USE AND ADDICTION People take drugs to feel good. Therefore, it should come as no surprise that most drugs activate dopamine receptors. As with intracranial self-stimulation, food, water, and sex, most drugs that people use and sometimes abuse—including alcohol, nicotine, heroin, cocaine, and cannabis—are associated with increased dopamine activation in the nucleus accumbens. These drugs either increase dopamine release or prevent the normal termination of its neural signal, thereby prolonging its effects. Over time, the continued use of drugs can lead to addiction; most researchers have concluded that the release of dopamine is a necessary condition for reinforcement and drug addiction (Koob, 1999). The potential for addiction to a specific drug is determined by the speed of its rewarding effects. For instance, people are more likely to become addicted to crack cocaine, which they smoke, than to powdered cocaine, which they sniff, because cocaine arrives at the brain faster through smoking. Of course, people become addicted to drugs for many other reasons, such as the social environment, and we will consider these in greater detail in Chapter 9.

REVIEWING the Principles | What Is the Biological Basis of Reward?

Although the behaviourists avoided any reference to internal mental states, it is clear that positive reinforcement generally works because it provides the subjective experience of pleasure. The neural basis of this reinforcement is the release of dopamine. Engaging in naturally motivated behaviours, self-administering electricity to the brain, and taking drugs or alcohol all lead to increased activation of dopamine neurons in the nucleus accumbens, which is associated with the experience of pleasure. Understanding the neural basis of reinforcement helps to explain complex conditioning, such as the development of addiction.

6.24 Donald Hebb Hebb proposed that learning results from changes in synaptic connections: *Cells that fire together wire together.*

How Does Learning Occur at the Neuronal Level?

From a biological perspective, scientists have long believed that learning involves relatively permanent changes in the brain that result from exposure to environmental events. The roots of this idea can be traced back to a number of scientists, including the German researcher Richard Semon, who in 1904 proposed that memories are stored through changes in the nervous system. Semon called the storage of learned material an *engram,* a term later popularized by the eminent psychologist Karl Lashley. In 1948, the Canadian psychologist Donald Hebb at McGill University (Figure 6.24) proposed that learning results from alterations in synaptic connections. According to Hebb (1949), when one neuron excites another, some change takes place such that the synapse between the two becomes strengthened. Subsequently, the firing of one neuron becomes increasingly likely to cause the firing of the other. Although there have been a number of different interpretations of Hebb's postulate, most of them can be summed up as "cells that fire together wire together." Hebb did not have the technology to examine whether his hypothesis was true or not, but we now know that his basic theory was correct.

Habituation and Sensitization Are Simple Models of Learning

What is the neural basis of learning?

What happens at the synapse that leads to learning? One answer is found in research using simple invertebrates such as the *aplysia,* a small marine snail that eats seaweed (Figure 6.25). The aplysia is an excellent species to use to study learning because it has a relatively small number of neurons, some of which are large enough to be seen without a microscope (Kandel, Schwartz, & Jessell, 1995). The neurobiologist Eric Kandel and his colleagues have used the aplysia to study the neural basis of two types of simple learning: *habituation* and *sensitization.* As a result of this research, Kandel received a Nobel Prize for medicine in 2000.

Habituation is a decrease in behavioural response following repeated exposure to nonthreatening stimuli. When an animal encounters a novel stimulus, it pays attention to it, behaviour known as an *orienting response.* If the stimulus is neither harmful nor rewarding, the animal learns to ignore it. We constantly habituate to meaningless events around us. For instance, sit back and listen. Perhaps you can hear a clock, or a computer fan, or your roommates playing music in the next room. You didn't really notice these sounds in the background before because you had habituated to them. Habituation in the aplysia can be demonstrated quite easily by repeatedly touching it. The first few touches cause it to withdraw its gills, but after about 10 touches it quits responding, and this lack of response lasts about 2 to 3 hours. Repeated habituation trials can lead to a state of habituation that lasts several weeks.

6.25 Simple Models of Learning The aplysia is a marine invertebrate that is used to study the neurochemical basis of learning.

Sensitization is an increase in behavioural response following exposure to a threatening stimulus. For instance, imagine that while you are studying you smell burning. You are unlikely to habituate to this smell. You might focus even greater attention on your sense of smell in order to assess the possible threat of fire, and you will be highly vigilant for any indication of smoke or flames. In general, sensitization leads to heightened responsiveness to other stimuli. Giving a strong electrical shock to the tail of the aplysia leads to sensitization. Following the shock, a mild touch anywhere on the body will cause the aplysia to withdraw its gills.

Kandel's research on aplysia has shown that alterations in the functioning of the synapse lead to both habituation and sensitization. For both types of simple learning, presynaptic neurons alter their release of neurotransmitters. A reduction in neurotransmitter release leads to habituation; an increase in neurotransmitter release leads to sensitization. Knowing the neural basis of simple learning gives us the building blocks to understand more complex learning processes.

habituation A decrease in behavioural response following repeated exposure to nonthreatening stimuli.

sensitization An increase in behavioural response following exposure to a threatening stimulus.

long-term potentiation (LTP) The strengthening of a synaptic connection so that postsynaptic neurons are more easily activated.

Long-Term Potentiation Is a Candidate for the Cellular Basis of Learning

To understand learning in the complex mammalian brain, researchers have investigated a phenomenon known as *long-term potentiation.* The word *potentiate* means to strengthen, to make something more potent. **Long-term potentiation (LTP)** is the strengthening of the synaptic connection so that postsynaptic neurons are more easily activated. To demonstrate LTP, researchers first establish the extent to which electrically stimulating one neuron leads to an action potential (the level of stimulation needed to activate or "fire") in a second neuron. They then provide intense electrical stimulation to the first neuron, perhaps giving it a hundred pulses of electricity in five seconds. Finally, a single electrical pulse is readministered to measure the extent of activation of the second neuron. As you can see in Figure 6.26, LTP occurs when the intense electrical stimulation increases the likelihood that stimulating one neuron leads to an action potential in the second neuron. Whereas habituation and sensitization in aplysia are due to changes in neurotransmitter release from the presynaptic neuron, LTP results from changes in the postsynaptic neuron that make it more easily activated. There is also evidence that a similar process can weaken synaptic connections, a situation referred to as *long-term depression.*

A number of lines of evidence support the idea that long-term potentiation may be the cellular basis for learning and memory (Beggs et al., 1999). For instance, LTP effects are most easily observed in brain sites known to be involved in learning and memory, such as the hippocampus. Moreover, the same drugs that improve memory also lead to increased LTP, and those that block memory also block LTP. Finally, behavioural conditioning produces neurochemical effects that are nearly identical to LTP.

The process of long-term potentiation also supports Hebb's contention that learning results from a strengthening of synaptic connections that fire together. Hebb's rule can be used

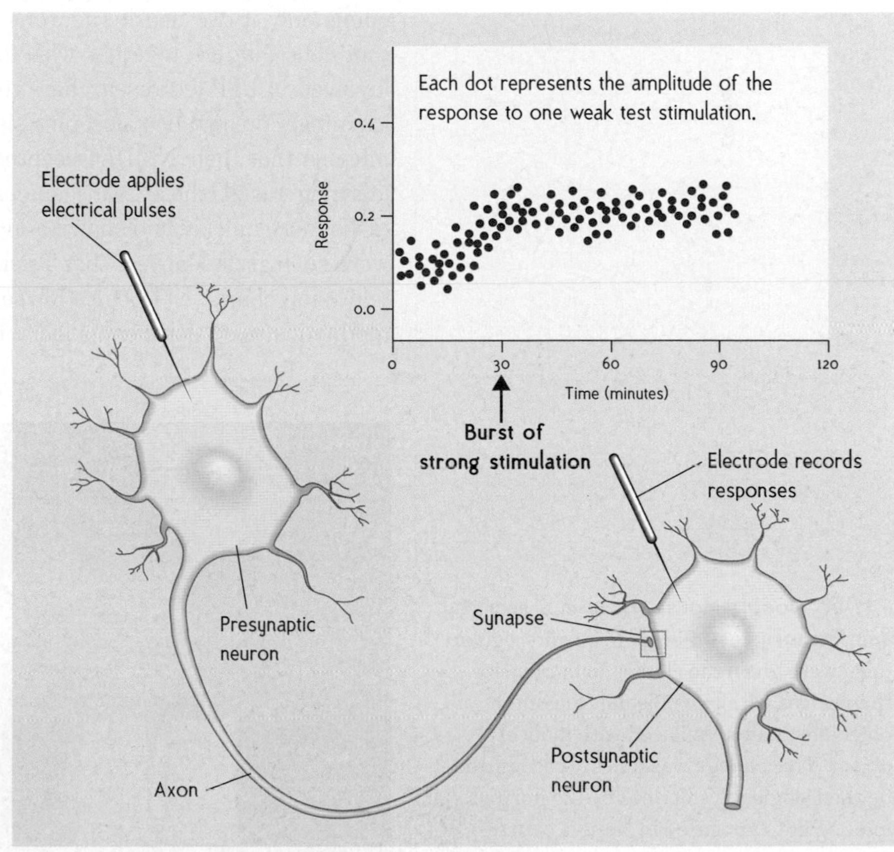

6.26 Electric Pulses in Neurons A presynaptic neuron is given a brief electrical pulse, which causes a slight response in the postsynaptic neuron. Applying intense and frequent electricity leads to a greater response of the postsynaptic neuron. When a single brief pulse is subsequently applied, it results in a greater effect on the postsynaptic neuron than occurred originally.

to explain a variety of learning phenomena, including classical conditioning. Neurons that signal the unconditioned stimulus are active at the same time as those that signal the conditioned stimulus. Over repeated trials, the synapses that connect these two events become strengthened, so that when one fires, the other fires automatically, producing the conditioned response.

FEAR CONDITIONING Although LTP was originally observed in the hippocampus, there is recent evidence that fear conditioning may induce LTP in the amygdala. In a typical fear-conditioning study, a rat is classically conditioned to produce a fear response to an auditory tone. Electric shock follows the tone, and eventually the tone produces fear responses on its own, which include specific physiological and behavioural reactions (Annau & Kamin, 1961). One interesting response, *freezing*—standing still—is observed in many species, including humans. If you have seen video footage from the bombing at the Atlanta Summer Olympics in 1996, you probably noticed that most people froze in a crouch for the first few seconds after the bomb exploded. Immediately keeping still might be a hard-wired response that helped animals deal with predators, which often are attracted by movement (LeDoux, 2002). There is substantial evidence that the amygdala is crucial for fear conditioning. If one particular part of the amygdala is removed, animals are unable to learn that shock follows the tone (Davis, 1997). Joseph LeDoux and his students have demonstrated that both auditory fear conditioning and artificial LTP induction lead to similar changes in amygdala neurons, suggesting that fear conditioning might produce long-lasting learning through the induction of LTP (Rogan, Stäubli, & LeDoux, 1997).

LTP AND THE NMDA RECEPTOR Researchers have made considerable progress over the last decade in understanding how LTP works. A special receptor, called the *NMDA* receptor, is required for LTP. This receptor has a special property: It opens only if two nearby neurons fire at the same time, supporting the Hebb rule that cells that fire together wire together. The findings that NMDA receptors were involved in LTP led researchers to examine genetic processes that might influence learning. For instance, Joseph Tsien at Princeton University modified the genes in mice so that their NMDA receptors were more efficient. When tested in standard learning tasks, these transgenic mice performed amazingly well, learning novel tasks more quickly and showing increased fear conditioning (Tsien, 2000). The mice were such great learners that Tsien named them "Doogie mice" after the prime-time television character Doogie Howser, a boy doctor (Figure 6.27). Might we be able to modify genes so that people learn more quickly? This fascinating question raises all

6.27 A "Doogie Mouse" Used in a Learning and Memory Test Doogie mice and regular mice were given the chance to familiarize themselves with two objects. One of the objects was later replaced with a novel object. This change was quickly recognized by the Doogie mice but not by the normal mice. (Photo appeared in *Nature,* courtesy of Princeton University.)

sorts of ethical issues, but various pharmaceutical companies are exploring drugs that might enhance the learning process by manipulating gene expression or activating NMDA receptors. If successful, such treatments might prove valuable for treating patients with diseases such as Alzheimer's. This is an especially active area of research, and we will likely continue to hear of many discoveries about how genes, neurotransmitters, and the environment interact to produce learning.

Learning Can Be Simulated by Computerized Neural Networks

Psychological scientists have developed computer models of neural networks to understand how the brain learns. The computer is a widely used metaphor for the working mind. Both the brain and the computer receive information as input, process that information, and create output in the form of actions or behaviours. Computerized models of learning are often referred to as *connectionist models* because they are roughly based on the idea that neurons are connected or associated with each other. Instead of neurons, computer models have units, which are connected with each other in the computerized neural network.

A simple connectionist model is shown in Figure 6.28. Note that this model has three layers of units: It takes in information through the *input units,* processes that information in the *hidden units,* and generates an action through the *output units.* Just as one neuron can pass a message to another neuron to fire or not to fire, the computer units also can be excitatory or inhibitory, and the strength of their connections is modified through learning.

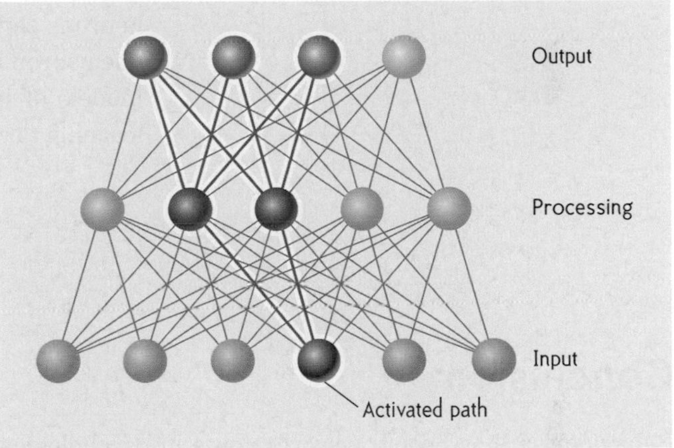

6.28 Neural Networks A simple network with three layers of units.

RULES FOR LEARNING The rule by which connectionist models operate is based on the degree to which a model's output matches the input. For instance, suppose you are trying to learn someone's name. When you see her face, you want to retrieve the correct name. When the output (i.e., the retrieved name) fails to match the input (i.e., the relevant face), you adjust the strength of association between the units so that the output becomes more similar to what it is supposed to be. For example, perhaps you recall the name as Karen when it is actually Kathy. To rectify the error, your neural network strengthens the association between the person's face and "Kathy" and weakens the association between her face and "Karen." After many trials, you call Kathy by her correct name. Subsequently, the connection between the person's face and the name Kathy is strengthened each time you associate them.

How connectionist models "learn" is similar to the Rescorla-Wagner model of classical conditioning discussed earlier. In both models, the greater the deviation from expectancy, the more learning takes place. If Kathy's face triggered an output of the name "Jake," you would need to learn a lot more! However, once the input matches the appropriate output, relatively little further learning takes place.

THE PARALLEL-DISTRIBUTED PROCESSING MODEL One of the best-known connectionist models of learning is the *parallel-distributed processing (PDP)* model developed by David Rumelhart and James McClelland. An important feature of the PDP model is that it is based on how neural networks actually operate. For instance, it views neuronal connections as richly distributed throughout the brain and it involves *parallel* processing, in which everything happens at the same time. This is an advance over serial models, in which learning was thought to occur in a fixed order, like the sequence of operations in a computer program.

The human brain is massively parallel, with millions of synapses within neural networks firing at the same time. In the PDP model, each unit in the network can be connected to all other units, and each is activated or inhibited by other units in the network. Essentially, each unit adds up the various inhibitory and excitatory messages from the other units, and it becomes activated when this sum passes a critical threshold. In this way, units in the PDP model operate as neurons do when they integrate information from other neurons to determine whether they will fire.

REVIEWING the Principles | How Does Learning Occur at the Neuronal Level?

Researchers are rapidly identifying the neurophysiological basis of learning. Much of what has been learned supports Hebb's theories that neurons that fire together wire together. Kandel's work on aplysia has shown that habituation and sensitization, two simple forms of learning, occur through alteration in neurotransmitter release. The discovery of long-term potentiation shows that intense stimulation of neurons can strengthen synapses, increasing the likelihood that the activation of one neuron will increase the firing of other neurons in the network. Computerized models of neural networks have helped make predictions about how complex learning takes place.

Conclusion

Behaviourism has been a powerful force in psychological science since early in the twentieth century. The shift from subjective to empirical methods established psychology as a science. A renewed interest in mental processes eventually led many to abandon the strict principles of behaviourism, but basic conditioning and learning processes are still foundational principles in understanding mind and behaviour. Psychological scientists use learning principles in studies across all levels of analysis, from synaptic connections in aplysia (cellular) to the cultural transmission of morals and values (cultural). With all its competing theories and sharing of ideas, the field of learning also serves as an excellent example of how psychological research is based on cumulative principles. We still see that dynamic in operation today. The principles of classical and operant conditioning are basic methodologies used by all neuroscientists to study brain mechanisms. In turn, recent advances in neuroscience techniques resulting from the biological revolution have allowed for a more complete understanding of learning processes. Learning occurs in the brain, and our relatively new access to brain-imaging techniques is reigniting interest in basic conditioning processes. Understanding the neurochemical basis of learning may help us to develop more effective treatments for a variety of psychological problems, including phobias and drug addictions.

Summarizing the Principles of Learning and Reward

How Did the Behavioural Study of Learning Develop?

1. **Behavioural responses are conditioned:** Pavlov established the principles of classical conditioning, a process that occurs when associations are made between stimuli and responses. This type of learning is reflex based and has many measurable aspects with regard to acquisition, discrimination, generalization, and extinction. Some emotional responses are learned through conditioning.

2. **Phobias and addictions have learned components:** Phobias are learned fear associations. Similarly, addiction involves a conditioned response, which results in withdrawal and tolerance.

3. **Classical conditioning involves more than contiguity:** Not all stimuli are equally potent in producing conditioning. Animals are biologically prepared to make connections between stimuli that are potentially dangerous. Animals are also predisposed to form predictions to enhance survival.

How Is Operant Conditioning Different from Classical Conditioning?

4. **Reinforcement increases behaviour:** Positive consequences of a behaviour will likely strengthen and reinforce the behaviour. Shaping is a procedure where successive approximations of a behaviour are reinforced, which leads to the desired behaviour. Reinforcers are primary (those that satisfy biological needs) and secondary (those that do not directly satisfy biological needs).

5. **Both reinforcement and punishment can be positive or negative:** Positive reinforcement or punishment means the delivery of a stimulus after the response; negative reinforcement or punishment means the removal of a stimulus after the response. Positive and negative reinforcements increase the likelihood of a behaviour; positive and negative punishments decrease the likelihood of the behaviour.

6. **Operant conditioning is influenced by schedules of reinforcement:** Reinforcement is delivered at a fixed rate or a variable rate that depends on the number (ratio) or time (interval) of responses.

7. **Biology and cognition influence operant conditioning:** The biological makeup of an organism restricts the types of behaviours it can learn. Latent learning refers to learning that takes place in the absence of reinforcement.

8. **The value of reinforcement follows economic principles:** Animals are predisposed to calculate the relative cost and benefit of behaviours, thus aiding their survival.

How Does Watching Others Affect Learning?

9. **Learning can be passed on through cultural transmission:** Memes (knowledge transferred within a culture) are analogous to genes, in that behaviours are selectively passed on from generation to generation.

10. **Learning can occur through observation:** Observational learning is a powerful adaptive tool. Humans and other animals learn by watching others' behaviours and the consequences.

11. **Animals and humans imitate others:** Modeling occurs when the behaviour of another is reproduced. Vicarious learning occurs when we see other people's behaviours reinforced or punished. Mirror neurons are activated when we watch a behaviour, the same as when we actually perform the behaviour.

What Is the Biological Basis of Reward?

12. **Self-stimulation is a model of reward:** The brain has specialized centres, which produce pleasure when stimulated. Behaviours that activate these centres are reinforced.

13. **Dopamine signals reward:** The nucleus accumbens (a part of the limbic system) has dopamine receptors, which are activated by pleasurable behaviours. Through conditioning, secondary reinforcers can also activate dopamine receptors. Drugs also increase habituation, which can lead to addiction.

How Does Learning Occur at the Neuronal Level?

14. **Habituation and sensitization are simple models of learning:** Repeated exposure to a stimulus results in habituation, a decrease in behavioural response. Sensitization is an increase in behavioural response to a new threatening stimulus.

15. **Long-term potentiation is a candidate for the cellular basis of learning:** Synaptic connections are strengthened when neurons fire together. This occurs in the hippocampus and, in fear responses, in the amygdala. The receptor NMDA is required for long-term potentiation.

16. **Learning can be simulated by computerized neural networks:** The connectionist model of neural networks is based on the idea that neurons are connected, which allows the brain to receive information (through input units), process the information (in hidden units), and generate an action (through output units). The strength of the connections is modified through learning. The parallel-distributed process (PDP) model is an advanced connectionist model that assumes learning involves parallel processing—it does not occur in a fixed order.

Applying the Principles

1. Imagine that in your experience, every time you saw a dog, it barked and frightened you; or each time you ate pepperoni pizza, you became ill; or perhaps your favourite grandmother always wore the same perfume. Now, whenever you see any kind of dog, it frightens you; the sight of a pizza makes you queasy; and the smell of that perfume always makes you smile. In all of these cases your response is due to

    ____ a) your ability to associate two different stimuli just because they happen together.
    ____ b) your ability to form contingencies between stimuli.
    ____ c) your intelligence.
    ____ d) your ability to override your reflexes with thought.

2. When your statistics instructor begins to sum up her main points and closes her lecture notes, you and the other students begin noisily packing your bookbags to get ready to leave. The instructor falls silent and glares at the class until the room is quiet again. What principles of learning are occurring with regard to your behaviours?

    ____ a) classical conditioning for both packing up and being quieted because two behavioural stimuli are being associated
    ____ b) operant conditioning with positive reinforcement for packing up and negative reinforcement for being quieted
    ____ c) classical conditioning for packing up; operant conditioning with negative reinforcement for being quieted
    ____ d) operant conditioning with negative reinforcement for the packing up; classical conditioning for being quiet

3. Amber is at a party, munching on some snacks while talking to Joe. Her friend Lisa, who is standing behind Joe, catches Amber's eye and brushes the right side of her face to signal that Amber has food on her cheek. Amber quickly raises a hand and brushes her cheek. What types of learning behaviours are occurring?

    ____ a) various conditioning responses in which Amber mimics Lisa
    ____ b) observational learning when Amber first brushes the left side of her face to mimic Lisa, and the likely firing of mirror neurons
    ____ c) observational learning that relies on species-specific responses
    ____ d) Amber is displaying a simple reflex response

4. German chocolate cake is your favourite dessert. You have many happy memories of eating a freshly made slice in your grandmother's kitchen after school. Your partner, knowing this, makes you a wonderful four-course birthday dinner with German chocolate cake for dessert prepared from your grandmother's recipe. At the biological level of analysis, what response in the nucleus accumbens of your brain may account for a less pleasurable experience than you had in the past?

    ____ a) the release of excess dopamine because you are older, which means you can't experience pleasure the same way
    ____ b) the communication of this area of the limbic system with the frontal cortex causing guilt for eating such a rich dessert
    ____ c) the big dinner has made you less hungry for dessert, so there is less dopamine activation
    ____ d) an activation of dopamine receptors due to thirst while consuming the cake results in a less pleasurable experience

5. Imagine that you've just bought a car. Initially, the key sticks in the lock and you have trouble opening the door, but after a while, the key goes in smoothly and you are able to unlock all the doors in a single turn of the key. This is similar to the effects of

    ____ a) long-term potentiation on the neuron.
    ____ b) dopamine on receptors.
    ____ c) long-term potentiation on the nucleus of the neuron.
    ____ d) genetic manipulation of neural receptors.

ANSWERS: 1.b 2.c 3.b 4.c 5.a

Key Terms

acquisition, p. 217
biological preparedness, p. 222
classical conditioning, p. 216
cognitive maps, p. 231
conditioned response (CR), p. 216
conditioned stimulus (CS), p. 216
continuous reinforcement, p. 228
extinction, p. 217
fixed schedule, p. 228
habituation, p. 242
interval schedule, p. 228
intracranial self-stimulation (ICSS), p. 239
latent learning, p. 232
law of effect, p. 224
learning, p. 214
long-term potentiation (LTP), p. 243
meme, p. 234
mirror neurons, p. 237
modeling, p. 236
negative punishment, p. 226
negative reinforcement, p. 226
observational learning, p. 234
operant conditioning, p. 224
partial reinforcement, p. 228
partial-reinforcement extinction effect, p. 229
phobia, p. 219
positive punishment, p. 226
positive reinforcement, p. 226
primary reinforcers, p. 225
punishment, p. 226
ratio schedule, p. 228
reinforcer, p. 224
Rescorla-Wagner model, p. 223
secondary reinforcers, p. 225
sensitization, p. 242
shaping, p. 225
spontaneous recovery, p. 217
stimulus discrimination, p. 218
stimulus generalization, p. 217
unconditioned response (UR), p. 216
unconditioned stimulus (US), p. 216
variable schedule, p. 228
vicarious learning, p. 236

Further Readings

Domjan, M. (2004). *The essentials of conditioning and learning.* Belmont, CA: Wadsworth.

Dugatkin, L. A. (2004). *Principles of animal behavior.* New York: Norton.

Kandel, E. R., Schwartz, J. H., & Jessell, T. M. (2000). *Principles of neural science* (4th ed.). New York: McGraw-Hill.

LeDoux, J. (2002). *Synaptic self.* New York: Viking.

McConnell, S., Roberts, J., Spitzer, L., Zigmond, M., Squire, L., & Bloom, F. (2002). *Fundamental neuroscience* (2nd ed.). San Diego, CA: Academic Press.

Further Readings: A Canadian Presence in Psychology

Annau, Z., & Kamin, L. J. (1961). The conditioned emotional response as a function of intensity of the UCS. *Journal of Comparative and Physiological Psychology, 54,* 428–432.

Hebb, D. O. (1949). *The organization of behavior.* New York: Wiley.

Jenkins, H. H., Barrera, F. J., Ireland, C., & Woodside, B. (1978). Signal-centered action patterns of dogs in appetitive classical conditioning. *Learning and Motivation, 9,* 272–296.

Kolb, B., & Whishaw, I. Q. (1998). Brain plasticity and behaviour. *Annual Review of Psychology, 49,* 43–64.

Macintosh, N. J., & Honig, V. R. (Eds.). (1969) *Fundamental issues in associative learning.* Halifax: Dalhousie University Press.

MacMillan, H. L., Boyle, M. H., Wong, M. Y., Duku, E. K., Fleming, J. E., & Walsh, C. A. (1999). Slapping and spanking in childhood and its association with lifetime prevalence of psychiatric disorders in a general population sample. *Canadian Medical Association Journal, 161,* 805–809.

Olds, J. (1962). Hypothalamic substrates of reward. *Psychological Review, 42,* 554–604.

Olds, J., & Milner, P. (1954) Positive reinforcement produced by electrical stimulation of the septal area and other regions of the rat brain. *Journal of Comparative and Physiological Psychology, 47,* 419–427.

Siegel, S. (1984). Pavlovian conditioning and heroin overdose. *Bulletin of the Psychonomic Society, 22,* 428–430.

Siegel, S., Hinson, R. E., Krank, M. D., & McCully, J. (1982). Heroin overdose death: Contribution of drug associated environmental cues. *Science, 216,* 436–437.

Spetch, M., Wilkie, D. M., & Pinel, J. P. J. (1981). Backward conditioning: A reevaluation of empirical evidence. *Psychological Bulletin, 89,* 163–175.

Stellar, J. R., Kelley, A. E., & Corbett, D. (1983). Effects of peripheral and central dopamine blockade on lateral hypothalamic self-stimulation. *Psychopharmacology, Biochemistry and Behavior, 18,* 433–442.

Wise, R. A., & Rompre, P. P. (1989). Brain dopamine and reward. *Annual Review of Psychology, 40,* 191–225.

Wishshaw, I. Q., Drigenberg, H. C., & Comery, T. A. (1992). Rats (rattus norvegicus) modulate eating speed and vigilance to optimize food consumption. *Journal of Comparative Psychology, 106,* 411–419.

How and What We Remember

In the 2000 film *Memento,* Leonard Shelby (played by Guy Pearce, above) suffers from severe short-term memory loss. In search of his wife's killer, he tattoos words onto his body to remind himself of what he's discovered. Decades of study have revealed a great deal about memory and its importance to survival.

CHAPTER

7

Memory

Imagine what the world would be like if you lost the ability to remember new experiences. You wouldn't be able to remember meeting people, or what you did last night, or even what you had for breakfast this morning. In a few minutes, you would not even remember having contemplated this problem. Such is the fate of a man who received brain surgery to relieve his epilepsy.

Anticonvulsive drugs had proven ineffective at controlling H.M.'s seizures, and therefore in September 1953 his doctors at the Montreal Neurological Institute performed a surgical technique in which they removed parts of his medial temporal lobes, including the hippocampus (Figure 7.1). The idea was to stop the seizures, which originated in the temporal lobes, from spreading throughout the brain. The surgery was successful in quieting the seizures, but

Outlining the Principles

What Are the Basic Stages of Memory?

- Sensory Memory Is Brief
- Short-Term Memory Is Active
- Long-Term Memory Is Relatively Permanent

What Are the Different Memory Systems?

- Explicit Memory Involves Conscious Effort
- Implicit Memory Occurs without Deliberate Effort

How Is Information Organized in Long-Term Memory?

- Long-Term Memory Is a Temporal Sequence
- Long-Term Storage Is Based on Meaning
- Schemas Provide an Organizational Framework
- Information Is Stored in Association Networks
- Retrieval Cues Provide Access to Long-Term Storage

What Brain Processes Are Involved in Memory?

- There Has Been Intensive Effort to Identify the Physical Location of Memory
- The Medial Temporal Lobes Are Important for Consolidation of Declarative Memories
- The Frontal Lobes Are Involved in Many Aspects of Memory
- Neurochemistry Influences Memory

When Do People Forget?

- Transience Is Caused by Interference
- Blocking Is Temporary
- Absentmindedness Results from Shallow Encoding
- Amnesia Is a Deficit in Long-Term Memory

How Are Memories Distorted?

- Flashbulb Memories Can Be Wrong
- People Make Source Misattributions
- People Make Bad Eyewitnesses
- People Have False Memories
- Repressed Memories Are Controversial
- People Reconstruct Events to Be Consistent

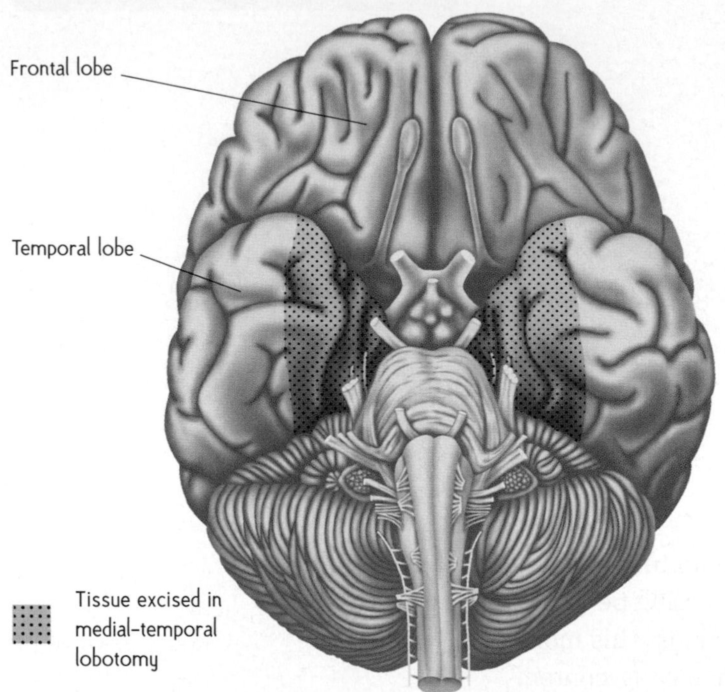

7.1 A Sketch of H.M.'s Brain The portions of the medial temporal lobe removed from H.M.'s brain are indicated by the shaded regions.

it also resulted in a most unfortunate side effect: H.M. lost the ability to form new long-term memories.

H.M.'s memory problems are profound. His world stopped in September 1953 when he was 27 years old. He can tell you about his childhood, explain the rules of baseball, and describe members of his family. According to neuropsychological testing, his IQ is slightly above average. Thus, his thinking abilities are perfectly fine, and he can hold a normal conversation as long as he is not distracted. Yet he cannot remember new information. Every moment is new and fresh. He never knows the day of the week, what year it is, or even his own age. The Canadian psychologist Brenda Milner, who has followed H.M.'s case for more than 40 years, and others who work with him continue to have to introduce themselves to him each time they meet. But he does seem to learn some new things . . . he just doesn't know it.

Most impressive is H.M.'s ability to learn new motor tasks. For instance, in one task he was required to trace the outline of a star while watching his hand in a mirror. This is a difficult task, and most people do poorly the first time they try it. H.M. was asked to trace the star 10 times on each of 3 consecutive days. As shown in Figure 7.2, H.M.'s performance improved over the 3 days, meaning that he had retained some information about the task; however, H.M. could not recall ever performing the task previously. H.M.'s ability to learn new motor skills allowed him to get a job at a factory, where he mounted cigarette lighters on cardboard cases. But he cannot give any description of the nature of his job or the place where he worked. The case of H.M. has contributed many clues to how memories are stored in the brain, and we will refer to his case throughout the chapter. ■

memory The capacity of the nervous system to acquire and retain usable skills and knowledge, allowing living organisms to benefit from experience.

modal memory model The three-stage memory system that involves sensory memory, short-term memory, and long-term memory.

In this chapter we are concerned with **memory**, the capacity of the nervous system to acquire and retain usable skills and knowledge, allowing organisms to benefit from experience. We remember millions of pieces of information, from the trivial to the vital. Our entire sense of self is that which we know from our memories, from our recollection of personal experiences. What kind of person are you? Are you shy or outgoing? To answer such questions we rely on our memories of past experiences. Yet as you will see, our memories are often incomplete, biased, and distorted. We are often surprised at how our memories for events differ vastly from those of others

1885
It Begins with Forgetting The psychologist Hermann Ebbinghaus studies how quickly people relearn nonsense syllables. He provides compelling evidence that forgetting occurs rapidly at first but then levels off over time.

1932
Reconstructive Memory Psychologist Frederic Bartlett suggests that human memory involves reconstruction and that people's memories are influenced by prior beliefs, challenging the view that memory is an objective recorder of experience.

1940s
Searching for the Engram While searching for the physical brain location of memory, called the engram, Karl Lashley concludes that the brain works as a whole to store memories.

1953
Amnesia and the Medial Temporal Lobes Brenda Milner publishes a landmark account of patient H.M., who underwent surgery for epilepsy and afterward experienced profound memory loss.

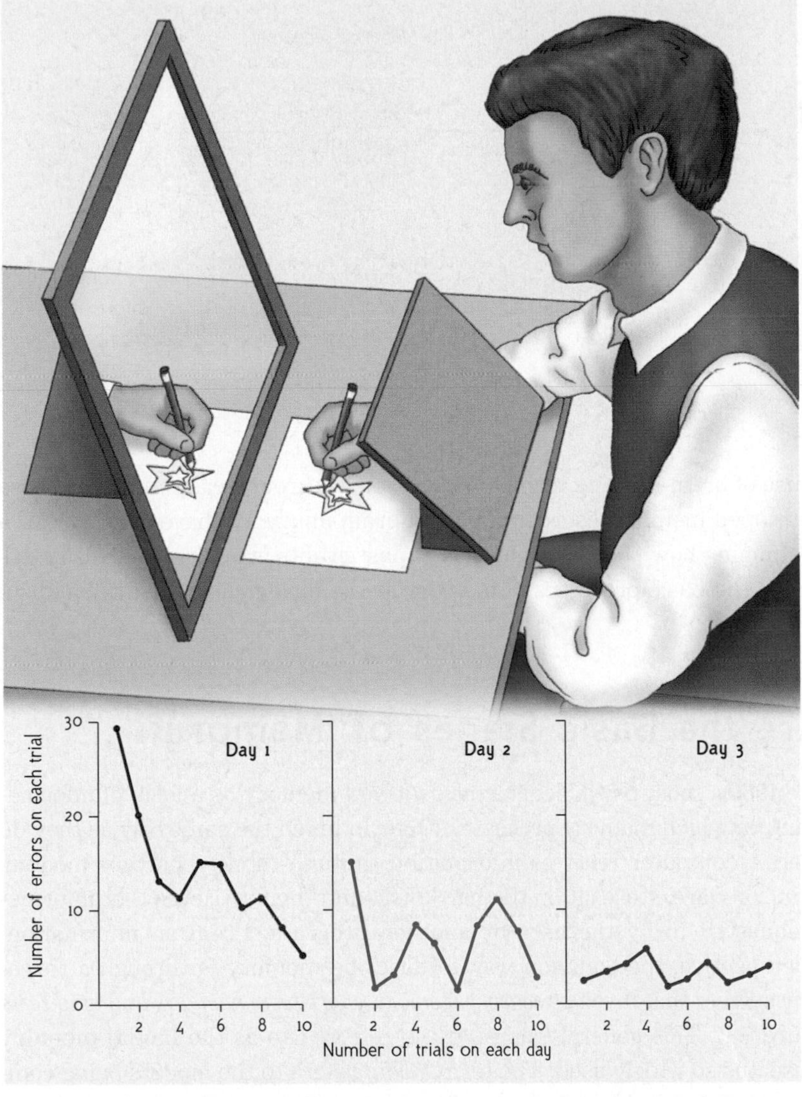

7.2 Tracing Task H.M. was required to trace a star figure while looking in a mirror. This is a difficult task, but H.M. improved in performance over time. However, he retained no conscious awareness that he had ever performed the task before.

who experienced the identical event. There are many sides to every story because each person stores and retrieves memory for the event in distinctive ways. Memory is not like a video camera that faithfully captures objective images. Rather, memory is a story that can be subtly altered through tellings and retellings.

In this chapter we explore the mental processes involved in acquiring and retaining knowledge. Beginning with a discussion of the different stages of memory, we then discuss how different types of information are represented in memories that persist over time. Knowledge about memory has accumulated rapidly over the last decade

1956

Memory Span George Miller, one of the intellectual founders of cognitive neuroscience, notes that short-term memory is limited, and demonstrates that people organize information into meaningful units, or chunks.

1960s

Stage Theories of Memory A number of memory researchers develop stage theories of memory. Most theories include a brief sensory memory buffer, short-term memory, and long-term memory.

1968–1970s

Neuropsychological Case Studies Researchers including Larry Squire, Stuart Zola-Morgan, Lawrence Weiskrantz, Elizabeth Warrington, and Tim Shallice report case studies of individuals with brain injury who have specific memory impairments.

1970s

Levels of Processing Fergus Craik, Robert Lockhart, and Endel Tulving emphasize that the way people process information determines how memory is stored and later retrieved. The more deeply people process information, the better they remember it.

1970s

Episodic Memory Endel Tulving introduces an important distinction between semantic memory of facts and knowledge and episodic memory of personal experiences.

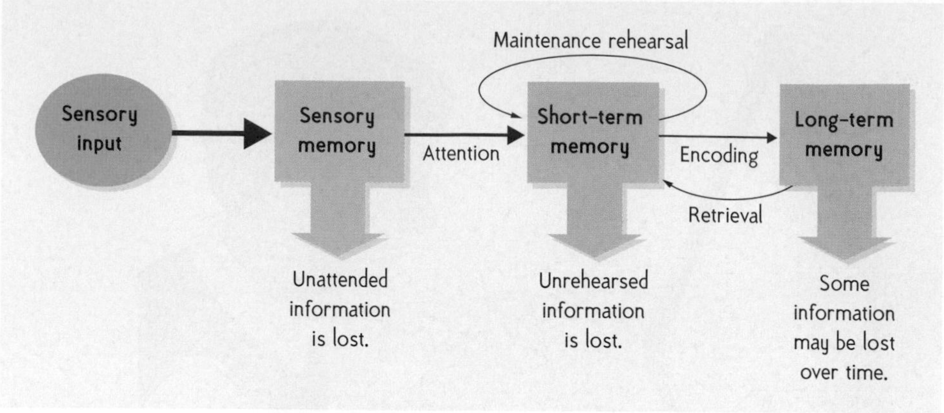

Maintenance rehearsal

Sensory input → Sensory memory → Attention → Short-term memory → Encoding → Long-term memory

Retrieval

Unattended information is lost.

Unrehearsed information is lost.

Some information may be lost over time.

7.3 Three-Stage Memory System The modal memory model serves as a useful framework for thinking about the basic stages of memory.

through the use of brain-imaging techniques, as well as through case studies of those who have developed memory disorders through brain injury. We have also learned a great deal examining how people's memories of past events are selectively distorted. We then explain the basic processes of memory at the biological, individual, and social levels of analysis.

What Are the Basic Stages of Memory?

Since the late 1960s, most psychologists have viewed memory as a form of information processing, in which memory processes occur in much the same way as they do in a computer. A computer receives information through the keyboard or modem, processes it in software, stores it on the hard disk, and then retrieves that information when requested to by the user or another program. From an information-processing perspective, the common way to describe memory is through a three-stage memory system that involves *sensory memory, short-term memory,* and *long-term memory* (Figure 7.3). This general framework is referred to as the **modal memory model** because it is so widely used. The term *modal* refers to the model's being common or standard. Proposed by Richard Atkinson and Richard Shiffrin in 1968, the modal model has dominated psychological thinking about memory, in spite of its being somewhat inaccurate and incomplete. For instance, many psychological scientists now believe that there are multiple memory systems that do not follow the three-stage sequence (we discuss these alternative views in the next section). However, the modal model is useful for introducing ideas about the memory system, and the vocabulary from this model remains widely used in memory research.

1970s
Animal Models of Memory Researchers such as Mortimer Mishkin, Patricia Goldman-Rakic, and Howard Eichenbaum develop memory experiments that can be performed by nonhuman animals.

1970s–1980s
Spreading Activation Models Network models suggest that memories emerge from associations between items based on semantic meanings. Activation of one element in the network leads to activation of other elements that share similar meanings.

1978
Ecological Realism Ulric Neisser emphasizes that researchers need to pay attention to how memory works in the real world, outside of laboratories.

1980s
Multiple Memory Systems A variety of memory processes are identified that involve separate but interacting brain systems, challenging the view that memory is a unitary entity.

1990s–2000s
Imaging and Memory Researchers use brain imaging to explore the neural basis of memory.

Sensory Memory Is Brief

Sensory information, such as lights, smells, and odours, leaves a trace on the nervous system for a split second and then vanishes. For instance, when you look at something and quickly glance away you can briefly picture the image and recall some of its details. When someone angrily proclaims, "you weren't listening to me," you can often repeat back the last few words the person spoke, even if you were thinking about something else. (Of course, this usually irritates the person further.)

This temporary sensory buffer is referred to as **sensory memory**. Visual sensory memory is also called *iconic memory,* whereas auditory sensory memory is also called *echoic memory.* George Sperling (1960) provided initial empirical support for sensory memory. In this classic experiment, three rows of letters were flashed on a screen for one-twentieth of a second, and then following various delay periods participants were asked to recall one of the rows. After very short delays people could perform this task quite well. But their performance became progressively worse with longer delay periods. Sperling concluded that the iconic memory persisted for about one-third of a second, after which the sensory-memory trace faded and was no longer accessible.

sensory memory Memory for sensory information that is stored briefly in its original sensory form.

According to many theorists, iconic and echoic memories allow us to experience the world as a continuous stream rather than in discrete sensations, much the same way that a movie projector plays a series of still pictures that follow each other closely enough to look like continuous action. When you turn your head, the scene passes smoothly in front of you rather than in jerky bits.

How are perceptual experiences transformed into memories?

Short-Term Memory Is Active

Information that is attended to is passed from sensory stores to **short-term memory (STM)**, a limited-capacity memory system that holds information in awareness for a brief period. Many contemporary researchers use the term *immediate memory* to emphasize the idea that this temporary buffer consists of our fleeting thoughts, feelings, and impressions of the world. A computer analogy for immediate memory is RAM, which can handle only a small amount of information compared with the vast amount stored in the computer's hard disk. The material in RAM is constantly replaced by new information, and it is lost forever if not saved.

short-term memory (STM) A limited capacity memory system that holds information in awareness for a brief period of time.

Short-term or immediate memory can hold information for no longer than about 20 seconds; it then disappears unless you actively prevent that from happening by thinking about the information or rehearsing it. For instance, when you call directory assistance for a telephone number, you repeat the number over and over again until you dial it. If the number is busy, or if there is some delay before you can dial it, you may forget the number and have to seek out directory assistance again (and again!). Similarly, suppose you are trying to remember some novel information, such as the three-letter string of consonants X, C, J. As long as you keep repeating the string over and over, you will keep it in short-term memory.

Typically, however, you are bombarded with other events that try to capture your attention, and you may not be able to stay focused. To simulate this, try again to remember X, C, J, but this time as you do it count backward in threes from the number 309. If you are like most people, you will find it difficult to remember the consonants after a few seconds of backward counting. This procedure demonstrates that the longer you spend counting, the less able you are to remember the consonant string. By 18 seconds of counting, people have extremely poor recall for the consonants. This occurs because of interference from previous items in STM.

MEMORY SPAN AND CHUNKING Why do new items interfere with the recall of older items? Short-term memory is a limited resource that can hold only so much information. The cognitive psychologist George Miller noted that STM is generally limited to about seven items, plus or minus two, which is commonly referred to as *memory span.* Some recent work suggests that memory span may be limited to as few as four units on average (Cowan, 2001). Memory span also varies among individuals. Indeed, some tests of intelligence use memory span as a measure of IQ.

Because STM is limited, you might expect people to have great difficulty remembering a string of letters such as

UTPHDUBCMASFUBAUWO.

These 20 letters would tax even the largest memory span. But, what if we organized the information into smaller, meaningful units? For instance,

UT-PHD-UBC-MA-SFU-BA-UWO.

Here we see that the letters can be separated to produce the names of universities or academic degrees. This organization makes them much easier to recall, for two reasons. First, memory span is limited to at most seven items, and probably fewer, but the items can be letters, numbers, words, or even concepts. Second, meaningful units are easier to remember than nonsense units. The process of organizing information into meaningful units is known as *chunking,* as in breaking down the information into chunks. The more efficiently you chunk information, the more you can remember. Master chess players are able to glance at a chess scenario for a few seconds and then later reproduce the exact arrangement of pieces (Chase & Simon, 1973). They are able to do this because they can chunk the board into a number of meaningful subunits based on their past experience with the game. Interestingly, if the game pieces are randomly placed on the chessboard so that the arrangement makes no "chess sense," experts are no better than novices at reproducing the board. In general, the greater your expertise with the material, the more efficiently you can chunk information, and therefore the more you can remember. The ability to chunk information efficiently relies on our long-term memory system, which we will discuss shortly.

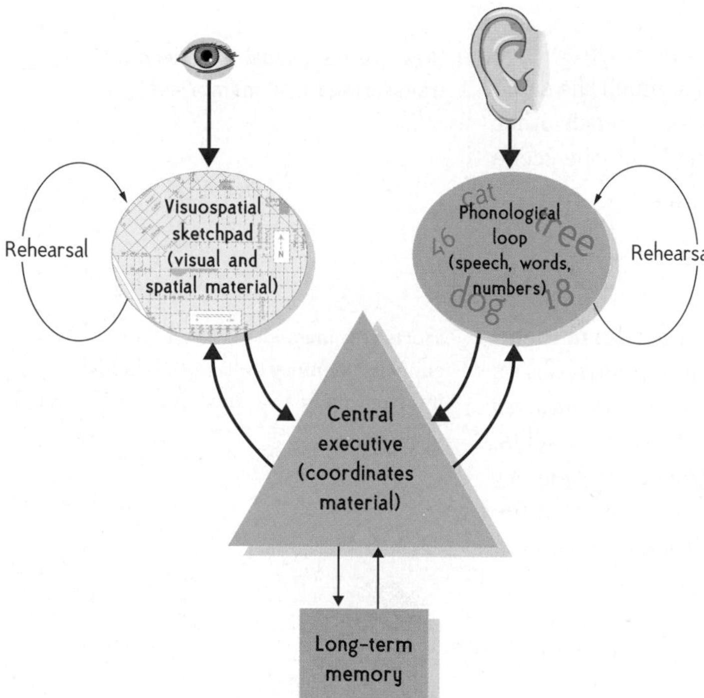

7.4 Baddeley's Working Memory System The working memory system developed by Alan Baddeley includes the central executive, phonological loop, and visuospatial sketchpad.

working memory An active processing system that keeps different types of information available for current use.

WORKING MEMORY The initial conception of short-term memory was that it was simply a buffer in which verbal information was rehearsed until it was stored or forgotten. It became apparent, however, that STM is not a single storage system but rather an active processing unit that deals with multiple types of information, such as sounds, images, and ideas. The British psychologist Alan Baddeley and his colleagues developed an influential model of a three-part active memory system that they called working memory (Figure 7.4). **Working memory** is an active processing system that keeps information available so that it can be used for activities such as problem solving, reasoning, and comprehension (Baddeley, 2001). For instance, H.M., the patient described in the beginning of this chapter, is able to keep track of a conversation as long as he stays actively involved in it. The three components of working memory are the *central executive,* the *phonological loop,* and the *visuospatial sketchpad.*

The *central executive* presides over the interactions between the subsystems and long-term memory; it's the boss. It encodes information from the sensory systems and then filters information that is sufficiently important to be stored in long-term memory. It also retrieves information from long-term memory as needed. The central executive relies on two subcomponents that temporarily hold auditory or visual information.

The *phonological loop* encodes auditory information and is active whenever you read, speak, or repeat words to yourself in order to remember them. You've probably noticed this "inner voice" that reads along as your eyes process written material. If you haven't, try to read the next sentence without "speaking" the words in your head. It is difficult to obtain meaning simply by scanning your eyes across the text. Evidence for the phonological loop is found in studies in which people are shown lists of consonants and asked to remember them. People tend to make errors with consonants that sound alike rather than those that look alike, for instance, misremembering a G as a T rather than a Q. Recall is also poorer when many words on a list sound the same compared to when they sound dissimilar, even when the latter words are related to each other in meaning. These examples suggest that words are processed in working memory by how they sound rather than by what they mean.

The *visuospatial sketchpad* processes visual information, such as objects' features and where they are located. Suppose while walking along you see a dog. The visuospatial sketchpad allows you to keep track of both where the dog is located and whether it is the sort of dog that one needs to be especially careful to track. The distinction between phonological loop and visuospatial sketchpad has been demonstrated by studying patients with specific brain damage. Patients with some types of brain injury might have great difficulty remembering spatial layouts but have little difficulty remembering words, whereas others show the exact opposite pattern. These sorts of research findings demonstrate that short-term memory consists of much more than simply an all-inclusive buffer.

Long-Term Memory Is Relatively Permanent

When people talk about their memories, they are usually referring to the relatively permanent storage of information known as **long-term memory (LTM)**. In the computer analogy, LTM is similar to information stored on a hard disk. Unlike computer storage, however, human LTM is nearly limitless. Long-term memory allows you to remember nursery rhymes from childhood, the meanings and spellings of rarely used words such as *aardvark,* and what you had for lunch yesterday.

DISTINGUISHING LONG-TERM MEMORY FROM SHORT-TERM MEMORY

Long-term memory is distinct from short-term memory in two important ways, duration and capacity. A controversy exists, however, as to whether LTM represents a truly different type of memory storage from STM. Initial evidence that LTM and STM are separate systems came from research at the individual level of analysis that required people to recall long lists of words. The ability to recall items from the list depended on order of presentation, with items presented early or late in the list remembered better than those in the middle. This better recall of early and late items is known as the **serial position effect** (Figure 7.5), which actually involves two separate effects. The **primacy effect** refers to the better memory people have for items presented first, whereas the **recency effect** refers to our better memory for the most recent items.

One explanation for the serial position effect relies on a distinction between STM and LTM. As people study the long list, they rehearse the earliest items the most,

long-term memory (LTM) The relatively permanent storage of information.

serial position effect The ability to recall items from a list depends on order of presentation, with items presented early or late in the list remembered better than those in the middle.

primacy effect In a list, the better memory for items presented first.

recency effect In a list, the better memory for words presented later in the list.

1.00
0.80
0.60
0.40
0.20
0.00
Probability of recall
Beginning
End
0 5 10 15 20 25
Primacy effect
Position of word in list
Recency effect

7.5 The Serial Position Effect After hearing a list of items, people typically remember more items from the beginning of the list (primacy effect) and the end of the list (recency effect) than from the middle of the list.

and that information is transferred into LTM. The last few items, by contrast, are in STM. The idea that primacy effects are due to LTM whereas recency effects are due to STM is supported by studies in which there is a delay between the presentation of the list and the recall task. Such delays interfere with the recency effect, but not the primacy effect, just as you would expect if the former involves STM. However, it is questionable to attribute the recency effect entirely to STM. After all, you probably remember your most recent class better than the classes you had prior to that, even though you are not holding that material in STM. Indeed, studies show a recency effect even for information presented over days and weeks, which makes it unlikely that the material is being maintained in STM. Thus, the recency effect on its own does not prove that STM and LTM are really different types of memory storage.

Perhaps the best support for the distinction between STM and LTM can be found at the biological level of analysis in case studies such as that of H.M. His STM system is perfectly normal and much of his LTM system is intact, but he is unable to transfer new information from STM into LTM. In a different case, a 28-year-old accident victim with damage to the left temporal lobe had extremely poor STM, with a span of only one or two items. However, he had perfectly normal LTM—he had a fine memory for day-to-day happenings and reasonable knowledge of past events (Shallice & Warrington, 1969). Somehow, in spite of the bottleneck in his STM, a normal flow of information was getting into LTM. These case studies demonstrate that LTM can be dissociated from STM. However, the two memory systems are

7.6 Long-term Memory and Rehearsal The process of learning lines, cues, and blocking for a play is a good example of LTM in use.

highly interdependent. For instance, to chunk information in STM, people need to form meaningful connections based on information stored in LTM.

WHAT GETS INTO LONG-TERM MEMORY Considering that we are bombarded with so much information and engaged in so many activities, it seems obvious that some type of filtering system or series of rules must constrain what goes into our LTM. Researchers have provided several possible explanations for this process. One possibility is that information enters permanent storage through rehearsal (Figure 7.6). When you study for exams you often go over the material many times to be sure that you have learned it. Memory researchers have even shown that *overlearning,* in which you keep rehearsing material that you already know pretty well, leads to improved memory, especially over longer periods of time. This is why studying that is spread out over time, known as *distributed practice,* is better remembered than material that is studied in a brief period of time, called *massed practice,* or cramming. The conclusion from many years of research is that studying for shorter periods of time, but spreading your study sessions out over several days or weeks, is the most efficient way to learn.

However, simply repeating or coming into contact with something many times does not mean that we develop long-lasting memory for the event or object. After all, sometimes we have extremely poor memory for objects that are highly familiar.

Another possibility is that only information that helps us adapt to the environment is typically transferred from short-term to long-term memory. Out of the billions of sensory experiences and thoughts we have each day, we want to store only information that is useful. Our memory that a penny is money is much more useful to us than remembering its specific features. By storing information that is *meaningful,* organisms can benefit from experience.

Evolutionary theory helps explain how we decide in advance what information will be useful. Memory allows us to use information in a way that assists in reproduction and survival. For instance, animals that can use past experiences to increase their chances of survival have a selective advantage over animals that fail to learn from past experiences. Recognizing a predator and remembering the escape route will help an animal avoid being eaten in the future. Accordingly, remembering which objects are edible, which people are friends and which are enemies, and how to get home are typically not challenging for people with intact memory systems.

Psychological Science *in Action* 

Strategies for Improving Memory

In this chapter we describe research indicating how and when people are more likely to remember or forget information. One of the central questions in psychological science is how people learn, and research over the past century has provided a wealth of information that is directly relevant to education. Much of what is found in psychological laboratories makes its way into classrooms, as teachers apply new techniques that have been found to enhance learning. Through research, psychological scientists have also developed specific strategies, called *mnemonics,* for improving memory. By using mnemonic strategies, most people are able to study and remember long lists of information. Let's examine some of the ways that findings from psychological science can help you study more effectively.

Practice. To become proficient you need to practice. The more times you repeat an action, the easier it is to do so on subsequent attempts. Motor skills, such as playing the piano, driving, and golf, become easier with practice.

Elaborate the Material. The deeper the level of processing, the more likely it is that you will remember the material later. This is another reason that critical thinking skills are important. Rather than just reading the material, think about its meaning and how it is related to other concepts. Try to organize the material in a way that makes sense to you, perhaps even putting the concepts into your own words. Indeed, making the material relevant to you is an especially good way to process material deeply, and therefore to remember it easily.

Overlearn. Rehearse material even after you think you have learned it. With the material in front of them, people are often overly confident that they will remember it later. But recognition is easier than recall. Test yourself by waiting a few hours after studying and then try to recall the material. Keep rehearsing until you can recall the material easily. Distributing your study over time rather than cramming will help you retain the information for longer periods of time. Three sessions of two hours each are much better for learning than one six-hour marathon.

Get Some Sleep. There is compelling evidence that sleep may help with the consolidation of memories and that disturbing sleep interferes with learning. How much sleep is necessary to promote learning and memory? A recent study found that a nap of 60 to 90 minutes is just as good as a full night of sleep (Mednick, Nakayama, & Stickgold, 2003). Chronic sleep deprivation, however, has its own costs that interfere with learning. Many students underestimate the importance of getting a good night's sleep.

"You simply associate each number with a word, such as 'table' and 3,467,009."

Use Verbal Mnemonics. How many days are there in September? Most people can readily answer this because of the old saying that begins "Thirty days hath September . . ." Children also learn early that it is "*i before e except after c*" and that "*weird is weird.*" By memorizing such phrases, we more easily remember things that are difficult to remember. Students have long used acronyms to remember information, such as HOMES to remember the great lakes (Huron, Ontario, Michigan, Erie, and Superior). In Chapter 12, you will see the acronym OCEAN will help you remember the major personality traits (openness to experience, conscientiousness, extraversion, agreeableness, and neuroticism). Advertisers often create slogans or jingles that rely on verbal mnemonics so that consumers can't help but remember them. Even complex ideas can be understood through simple mnemonics, such as *cells that fire together wire together* as a way to remember long-term potentiation, discussed in the last chapter as the brain mechanism responsible for learning.

Use Visual Imagery. Creating a mental image of material is especially good for memory. Memory researchers have developed specific strategies that rely on imagery to help people remember long lists of information. One, the *method of loci,* involves visualizing placing objects in familiar locations so that when you have to retrieve the material you simply imagine going back to that location. Other similar techniques involve learning lists of key words, or *pegs,* and then hanging new words onto those pegs by visualizing them together. Although these strategies may be useful for remembering long lists of items, they are typically not widely used in real life. For most of us, it is just as easy to keep lists of things we need on our computers or PDAs—or even to write important things down on a piece of paper (and don't forget where you put that!). ■

REVIEWING the Principles | What Are the Basic Stages of Memory?

The information-processing approach to memory occurs in three basic stages: sensory memory, short-term memory, and long-term memory. Sensory memory consists of brief traces on the nervous system that reflect perceptual processes. Material is passed from sensory memory to short-term memory, a limited buffer that holds information in awareness for a brief period of time. An influential model of STM is working memory, which involves a central executive, a phonological loop, and a visuospatial sketchpad. STM is limited to fewer than seven chunks of information, probably around four. The rules for chunking are determined by the meanings provided from long-term memory. LTM is a limitless, relatively permanent store. Only information that is somehow meaningful is stored in LTM. The distinction between STM and LTM is best established by case studies of those with impairments to one but not to the other.

What Are the Different Memory Systems?

Are there different types of memories, or are all memories essentially the same?

Until the last few decades, most cognitive psychologists thought long-term memory was a relatively unitary system. Memories were viewed as differing in terms of strength and accessibility, but generally they were considered to be of the same type. Cognitive psychologists such as Endel Tulving, Dan Schacter, and Larry Squire began to challenge this view in the late 1970s and early 1980s. They argued that memory is not just one monolithic entity, but rather a process that involves a number of interacting systems. Although the systems share a common function—to retain and use information—they encode and store different types of information in different ways. For instance, there are several obvious distinctions to be made between your memory for being able to ride a bicycle and your ability to recall what you ate for dinner last night or that the capital of Canada is Ottawa. The first task requires a behavioural component, an integration of motor and perceptual skills that you acquired over time during development. You are not consciously aware of your efforts to maintain balance or follow the basic rules of the road. By contrast,

7.7 Endel Tulving Tulving's most influential theory is his distinction of memories into two different types: episodic and semantic. Episodic memories are used to recall personal experiences, whereas semantic memories draw on knowledge of general facts and information. Thus, semantic memory is used to recall that Stephen Harper is the current prime minister of Canada, but episodic memory is used to remember shaking his hand.

explicit memory The processes involved when people remember specific information.

declarative memory The cognitive information retrieved from explicit memory; knowledge that can be declared.

episodic memory Memory for one's personal past experiences.

semantic memory Memory for knowledge about the world.

recalling a specific event or bit of knowledge has no behavioural component, and it sometimes requires a conscious effort to retrieve information from long-term memory.

Scientists do not yet agree on the number of human memory systems. For instance, some researchers have distinguished among memory systems based on how information is stored in memory, such as whether the storage occurs with or without deliberate effort. Other researchers have focused on the types of information stored, such as words and meaning versus particular muscle movements. Because the method of storage often differs depending on the type of information being stored, it is important to know whether reference is being made to process or content of memory. Understanding how different memory systems work has provided tremendous insight into memory, such as why it sometimes fails.

Explicit Memory Involves Conscious Effort

The most basic distinction between memory systems is between those for which we consciously remember and use information and those for which memory occurs without conscious effort or intention. **Explicit memory** involves the processes used to remember specific information. The information retrieved in explicit memory is known as **declarative memory**, which refers to cognitive information that can be brought to mind, that is, knowledge that can be declared. Many psychological scientists use the terms interchangeably, but explicit memory refers to the *process* of memory whereas declarative memory refers to the *content* of memory.

You use explicit memory when you try to recall what you had for dinner last night or what the word *aardvark* means. Declarative memories can involve words or concepts, visual images, or both. For instance, imagine earth's satellite. You are probably retrieving both the image and name of the moon. Most of the examples we have used so far in this chapter refer to explicit memories.

In 1972, Endel Tulving at the University of Toronto introduced a further distinction in explicit memory between episodic and semantic memory (Figure 7.7). **Episodic memory** refers to one's personal past experiences. **Semantic memory** represents one's knowledge of trivial or important facts independent of personal experience. For instance, people know the capitals of countries that they've never visited, and even those who have never played baseball know that three strikes mean you're out.

Evidence that episodic and semantic explicit-memory systems are separate can be found in cases of brain injury in which semantic memory is intact even though episodic memory is impaired. Such an instance was found in a group of British children who experienced brain damage during infancy or early childhood and developed poor memory for episodic information (Vargha-Khadem et al., 1997). They had trouble reporting what they had for lunch, what they were watching on television five minutes ago, or what they did during summer vacation. Their parents reported that the children had to be constantly monitored to make sure they remembered things such as going to school. Remarkably, these children attended mainstream schools and did reasonably well. Moreover, their IQs fell within the normal range. They learned to speak and read, and they could remember a great deal of facts. For instance, when given the question "Who is Martin Luther King Jr.?" one of the children responded, "An American; fought for Black rights, Black rights leader in the 1970 [*sic*]; got assassinated." These children, then, are capable of encoding and retrieving semantic information, but they do not remember how or when they learned the information.

Implicit Memory Occurs without Deliberate Effort

Implicit memory is the process by which people show without deliberate effort and without any awareness that they are remembering something. Implicit memory pervades daily experiences, such as brushing your teeth, or remembering how to get to class. Recall in Chapter 6 we described examples of classical conditioning; these are examples of implicit memory. You experience fear as you enter the dentist's office partially because of past associations (implicit memories) with pain. Implicit memory does not require attention. For instance, you may have had the experience while driving of realizing that you've been daydreaming and have no episodic memory of the past few minutes. However, during that time you *remembered* how to drive and where you were going. Implicit memory happens automatically and without deliberate effort.

An example of implicit memory is **procedural memory**, or *motor memory,* which involves motor skills, habits, and other behaviours that you employ to achieve a goal, such as coordinating muscle movements to ride a bicycle or following the rules of the road while driving. You remember to stop when you see a red light because you have learned to do so, and when you drive home, you follow a specific route without even thinking about it. Our procedural memories have an automatic, unconscious aspect to them. For instance, the next time you are walking try to think about each step involved in the process: first you lift your right foot off the ground and slightly bend the knee, then you extend the foot forward while transferring weight to the left leg, and so on. Most people find that consciously thinking about automatic behaviours, such as walking, interferes with the smooth production of those behaviours.

Implicit memory influences our lives in subtle ways. Many of our attitudes are formed through implicit learning. You might like someone because he or she reminds you of a favourite person, even if you are unaware of the connection. Advertisers rely on implicit memory to influence our purchasing decisions. Constant exposure to brand names makes us more likely to think of them when we buy a product.

At the social level of analysis, implicit attitude formation can even affect our beliefs about whether people are famous. Is Richard Shiffrin famous? Try to think for a second how you know him. You might recall that Richard Shiffrin was one of the psychologists who introduced the modal memory model (which might make him famous in scientific circles), or you might have known that you had read the name before but could not remember where, and therefore you assumed the person was famous. The psychologist Larry Jacoby at McMaster University called this the *false fame effect* (Jacoby, Kelley, Brown, & Jasechko, 1989). Jacoby had research participants read a list of made-up names as a test of pronunciation. The next day subjects completed an apparently unrelated study in which they were asked to read a list of names and decide whether each person was famous or not. Subjects misjudged some of the made-up names from the previous day as being famous. Why did this happen? Subjects knew they had heard the names before but couldn't remember where. Therefore, the implicit memory led them to assume the name was likely that of a famous person.

Implicit memory is also involved in **repetition priming**, the improvement in identifying or processing a stimulus that has previously been experienced. In a typical priming experiment, participants are exposed to a list of words and asked to do something, such as count the number of letters in the words. Following some brief delay, the participants are shown word fragments and asked to complete them with the first word that comes to mind. You can try this yourself. Count the letters in the following words: *appearance, chestnut, patent.* Now, complete the stems app_____, che_____,

implicit memory The process by which people show an enhancement of memory, most often through behaviour, without deliberate effort and without any awareness that they are remembering anything.

procedural memory A type of implicit memory that involves motor skills and behavioural habits.

repetition priming The improvement in identifying or processing a stimulus that has previously been experienced.

pat_____, with the first word that comes to mind. The participants in a typical experiment are much more likely to complete the fragments with the words they previously encountered, which were primed and therefore more easily accessible. In our example, you are much more likely to complete the stems as *appearance, chestnut,* and *patent* than, say, *application, cheese,* and *paternal.* This effect occurs even when participants cannot explicitly recall the words in the first task. Even many hours after viewing the primes, participants show implicit memory without explicit recall of the words.

REVIEWING **the Principles** | What Are the Different Memory Systems?

Memory researchers reject the idea that memory involves a single process or brain system. They now agree that there is a fundamental distinction between explicit and implicit memory systems. Explicit memory involves the conscious storage and retrieval of declarative memories, such as meanings of words or personal experiences. Implicit memory refers to an enhancement of memory without effort or awareness. Examples of implicit memory include procedural (or motor) memory, attitude formation, and repetition priming.

How Is Information Organized in Long-Term Memory?

Events that are sufficiently important to be remembered permanently need to be stored in a way that allows for later retrieval. Imagine if a video store just put each video or DVD onto a shelf, wherever there was empty space. What would happen if you wanted to find a certain movie? You would have to go through the entire inventory, film by film, until you encountered the right movie. Just as this system does not work for movies, it also does not work for memory. Rather, memories are stored in an organized fashion. In this section we examine the temporal and organizational principles of long-term memory.

Long-Term Memory Is a Temporal Sequence

Why is meaning important for long-term memory?

In our computer analogy, memory is a process that involves storing new information so that it is available when it is later required. Memory can be divided temporally into three processes: *encoding, storage,* and *retrieval.* In **encoding**, our perceptual experiences are transformed into representations, or *codes,* which are then stored. For instance, when your visual system senses a shaggy, four-legged animal and your auditory system senses barking, you perceive a dog. The concept of "dog" is a *mental representation* for a category of animals that share certain features, such as barking and fur. The mental representation for "dog" differs from that for "cat," even though the two are similar in many ways. You also have mental representations for complex things, such as ideas, beliefs, and the feelings of falling in love. We will discuss representations in more detail in Chapter 8.

encoding The processing of information so that it can be stored.

storage The retention of encoded representations over time that corresponds to some change in the nervous system that registers the event.

Whether simple or complex, information is stored in networks of neurons in the brain. **Storage** refers to the retention of encoded representations over time and corresponds to some change in the nervous system that registers the event. Stored representations are referred to as *memories.* Memories represent many different kinds of information, such as visual images, facts, ideas, tastes, or even muscle movements, such as memory for riding a bicycle. **Retrieval** is the act of recalling or remembering the stored information in order to use it. The act of retrieval often involves an explicit effort to access the contents of memory storage, such as when you try to remember the previous chapter or your fifth birthday. But many times you retrieve information implicitly, without any effort at all, as when you instantly remember the name of an acquaintance you encounter on the street. Thus, retrieval is involved in both explicit and implicit memory systems.

retrieval The act of recalling or remembering stored information in order to use it.

maintenance rehearsal A type of encoding that involves continually repeating an item.

elaborative rehearsal The encoding of information in a more meaningful fashion, such as linking it to knowledge in long-term memory.

schema A hypothetical cognitive structure that helps us perceive, organize, process, and use information.

Long-Term Storage Is Based on Meaning

Books are stored by Library of Congress numbers, video stores tend to organize movies by type, and telephone books are organized alphabetically. Memories, however, are not stored by category, type, or number. Memories are stored by meaning. In the early 1970s, the University of Toronto psychologists Fergus Craik and Robert Lockhart developed an influential theory of memory based on depth of elaboration. According to their *levels of processing* model, the more deeply an item is encoded, the more meaning it has and the better it is remembered. Craik and Lockhart proposed that different types of rehearsal lead to differential encoding. **Maintenance rehearsal** involves simply repeating the item over and over again. **Elaborative rehearsal** involves encoding the information in more meaningful ways, such as thinking about the item conceptually or deciding whether it refers to oneself. In other words, we elaborate on the basic information by linking it to knowledge from long-term memory.

How does this model work? Suppose you showed participants a list of words and then asked them to do one of three things. You could ask them to make simple perceptual judgments, such as, "Is each word printed in capital or small letters?" Or you could ask them to judge the acoustics of the word, such as, "Does the word _____ rhyme with boat?" Or you could ask them about the semantic meaning of the words, such as, "Does the word fit the sentence 'They had to cross the _____ to reach the castle'?" Once they had processed the information, you might later ask them to recall as many words as possible (Figure 7.8). You would find that words that were processed at the deepest level, based on semantic meaning, were remembered the best. At the biological (brain systems) level of analysis, brain-imaging studies have shown that semantic encoding activates more brain regions than shallow encoding (Kapur et al., 1994). This greater brain activity is associated with better memory.

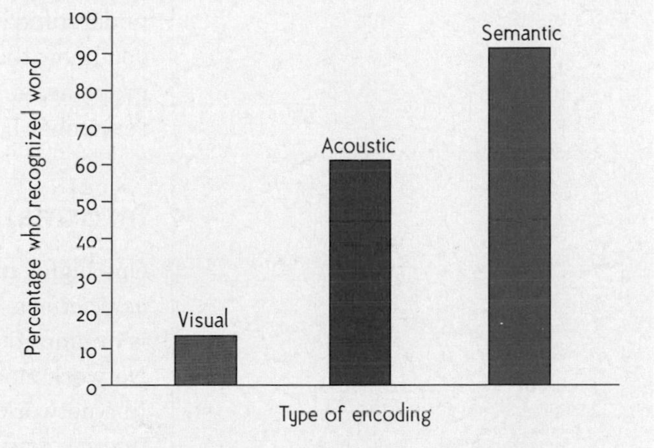

7.8 Encoding Participants are asked to consider a list of words: how the words are printed, how they sound, or what they mean. The more deeply the material is processed, the better it is remembered.

Schemas Provide an Organizational Framework

People store memories by meaning. But how is meaning determined? Earlier, we mentioned chunking as a good way to encode groups of items for easy memory. The more meaningful the chunks, the better they are remembered. Decisions about how to chunk information depend on previous organizational structures in long-term memory. **Schemas** are hypothetical cognitive structures that help us perceive, organize, process, and use information. They help us sort out incoming information and guide our attention to relevant features of the environment. People use their past memories and general knowledge about the world to shape incoming information. By doing so, they construct new memories. They fill in holes, overlook inconsistent information, and interpret meaning on the basis of past experience.

Existing schemas help us make sense of the world, but they can lead to biased encoding. In a classic demonstration conducted in the early 1930s, Frederick Bartlett asked British participants to listen to a Native American folk tale. After a 15-minute delay, Bartlett asked the participants to repeat the story exactly as they heard it. Bartlett found that the participants distorted the story a great deal, and they did so in a consistent way—they altered the story so that it made sense from their own cultural standpoint.

Schemas influence which information is stored in memory. Consider a study in which students read a story about a wild, unruly girl. Some of the participants were initially told that the girl was Helen Keller, whereas others were told it was "Carol Harris," a made-up name (Sulin & Dooling, 1974). One week later, the participants who were initially told the girl was Helen Keller were more likely to mistakenly report having seen the sentence "She was deaf, mute, and blind" than those who thought the story was about Carol Harris. Our schema for Helen Keller includes her various disabilities, and when we retrieve information about Helen Keller from memory, everything we know about her is retrieved along with the specific story we are trying to remember.

Read the following paragraph carefully.

> The procedure is actually quite simple. First arrange things into different bundles depending on make-up. Don't do too much at once. In the short run this may not seem important, however, complications easily arise. A mistake can be costly. Next, find facilities. Some people must go elsewhere for them. Manipulation of appropriate mechanisms should be self-explanatory. Remember to include all other necessary supplies. Initially the routine will overwhelm you, but soon it will become just another facet of life. Finally, rearrange everything into their initial groups. Return these to their usual places. Eventually they will be used again. Then the whole cycle will have to be repeated. (Bransford & Johnson, 1972)

How easy did you find this to understand? If you had to recall the sentences from this paragraph could you do so? It might surprise you to know that researchers presented this paragraph to college students and that those students found it easy to understand and relatively straightforward to recall. How is that possible? It is easy if you know that the paragraph describes washing clothes. Go back and read the paragraph again. Notice how your schema for doing laundry helps you to understand and remember how the words and sentences are connected to one another.

Information Is Stored in Association Networks

One highly influential set of theories of memory organization is based on *networks of associations.* We can trace back to Aristotle the idea that our knowledge of the world is organized so that things that naturally go together are linked together in storage. Network models of memory emphasize the links between semantically related items. In a network model proposed by Allan Collins and Elizabeth Loftus (Figure 7.9), distinctive features of an item are linked together in such a way as to identify it. Each unit of information in the network is known as a *node.* Each node is connected to many other nodes. When you look at a fire engine, for example, all of the nodes that represent features of a fire engine are activated, and the resulting pattern of activation gives rise to the knowledge that the object is a fire engine rather than, say, a cat. An important feature of network models is that activating one node increases the

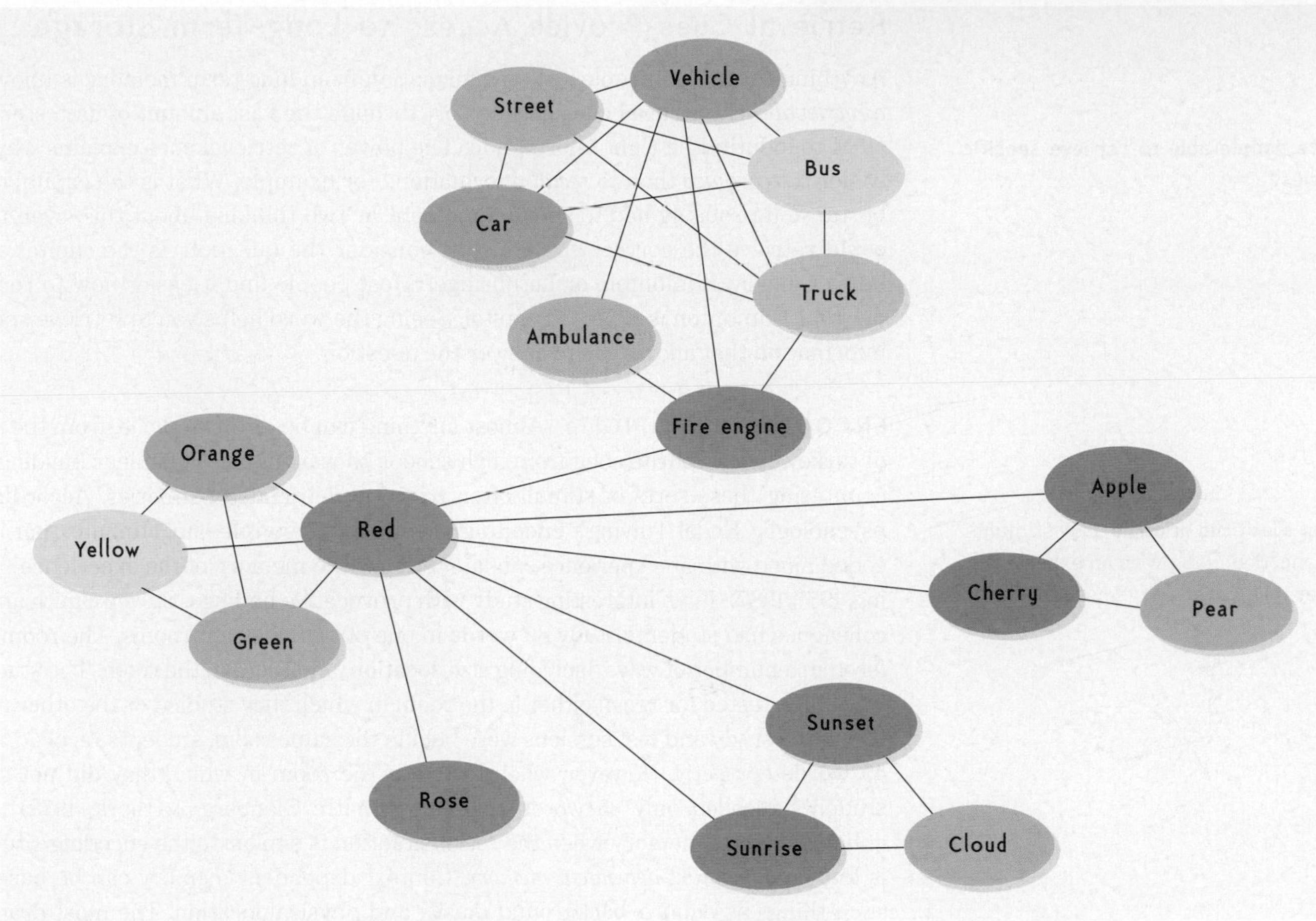

7.9 A Network of Associations A semantic network in which similar concepts are connected through their associations.

likelihood that closely associated nodes will also be activated. In Figure 7.9, the closer the nodes, the stronger the association between them and therefore the more likely that activating one will activate the other. Seeing a fire engine activates nodes that indicate other vehicles, so that once your fire-engine nodes are activated, you are much quicker to recognize other vehicles than, for instance, fruits or animals. The idea that activating one node increases the likelihood that associated nodes become active is a central tenet of *spreading activation* models of memory. According to these views, stimuli in working memory activate specific nodes in long-term memory, which increases the ease of access to that material, thereby facilitating retrieval.

The overall organization of associative networks is based on categories that are hierarchically structured, providing a clear and explicit blueprint for where to look for needed information. Given the vast amount of material in memory, it is truly amazing how quickly we can search for and obtain needed memories from storage. Each time you hear a sentence, you not only have to remember what all the words mean, but you also have to recall all relevant information that helps you understand the meaning. For this to occur, the information needs to be organized in some logical fashion. Imagine trying to find a specific file on a full 40-gigabyte hard disk by opening one file at a time. Such a method would be hopelessly slow. Instead, most computer disks are organized into folders, and within each folder are more specialized folders, and so on. This hierarchical storage system allows us to find needed files quickly.

Retrieval Cues Provide Access to Long-Term Storage

How are people able to retrieve specific memories?

Anything that helps people access information from long-term memory is known as a *retrieval cue.* Retrieval cues help us sort through the vast amount of data stored in LTM to identify the right information. The power of retrieval cues explains why it is easier to *recognize* than to *recall* information. For example, What is the capital of Alberta? You probably had to spend a moment or two thinking about this, even if you could retrieve the correct answer. Now consider the question, Is the capital of Alberta Calgary, Edmonton, or Lethbridge? Most people find it easier now to remember that Edmonton is Alberta's capital. Seeing the word helps you to retrieve specific information that allows you to answer the question.

encoding specificity principle Any stimulus that is encoded along with an experience can later trigger memory for the experience.

ENCODING SPECIFICITY Almost anything can be a retrieval cue, from the smell of turkey, to a favourite song from high school, to walking into a familiar building. Encountering these sorts of stimuli often triggers unintended memories. According to psychologist Endel Tulving's **encoding specificity principle**, any stimulus that is encoded along with an experience can later trigger the memory of the experience (Tulving, 1977, 1987). In an interesting study with provocative findings, Steven Smith and his colleagues had students study 80 words in one of two different rooms. The rooms differed in a number of ways, including size, location, and scent in the room. The students were then tested for recall either in the room in which they studied or the other room. When the study and test sessions were held in the same room, students recalled about 49 words correctly. However, when tested in the room in which they did not study, students recalled only 35 words correctly (Smith, Glenberg, & Bjork, 1978). Such enhancement of memory when the recall situation is similar to the encoding situation is known as *context-dependent memory.* Context-dependent memory can be based on such things as odours, background music, and physical location. The most dramatic demonstration showed that scuba divers who learned information underwater later tested better underwater than on land (Godden & Baddeley, 1975; see Figure 7.10).

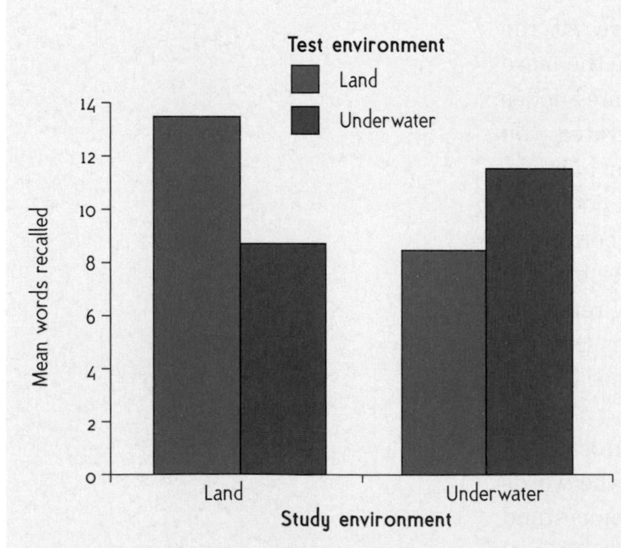

7.10 Context-Dependent Memory Mean number of words recalled in the Godden and Baddeley study (1975). Note that the best memory occurs for words that were studied and tested in the same environment.

STATE-DEPENDENT MEMORY Just as physical context can affect memory, so can internal cues, such as mood states or even inebriation. Enhancement of memory when there is a match between internal states during encoding and recall is known as *state-dependent memory.* Some of the research on this topic was inspired by the observation that alcoholics often misplaced important objects, such as paychecks, because they stored them in a safe place while they were drinking but could not remember where once they were sober. The next time they were drinking, however, they were able to remember where they had hidden the object. The psychological scientist Eric Eich and colleagues (1975) conducted a study of state-dependent memory using marihuana. Participants studied a test list either sober or high. Eich and colleagues found that memory was best when participants were tested in the same state in which they had studied. Note, however, that students recalled the information best when they were sober on both occasions. In a study that used alcohol, the worst performance was for students who studied when intoxicated and took the test sober. They did worse than students who studied sober and took the test intoxicated. Students who studied intoxicated and took the test intoxicated did much worse than students who were sober at both study and test (Goodwin, Powell, Bremer, Hoine, & Stern, 1969). Importantly, this phenomenon is not unique to humans. Indeed, rats will show a complete bidirectional dissociation in learning and recall between the normal state and, for example, a period of alcohol or barbiturate intoxication, or following a seizure (Overton, 1978; McIntyre & Reichert, 1971). State-dependent memory works because internal state is an additional retrieval cue that can facilitate the recovery of information from long-term memory.

REVIEWING the Principles | How Is Information Organized in Long-Term Memory?

Memory is composed of a temporal sequence of encoding, storage, and retrieval. Encoding and storage are organized so that information can be quickly retrieved. Human memory is stored according to meaning, and the more that meaning is elaborated at time of storage, the better the later memory will be. Shallow encoding does not lead to long-term storage. Schemas help people perceive, organize, and process information, which is then stored if processed deeply. Hierarchical networks of associated nodes provide semantic links between related items. Activation of a node spreads throughout the rest of its network. Retrieval cues help access stored information. Contextual cues and internal states can be retrieval cues.

What Brain Processes Are Involved in Memory?

In the first part of this chapter, we focused on the cognitive dimensions of memory, including types of memory and the kinds of information that are stored in memory. Now we need to ask another fundamental question: What role does biology play in the formulation of our memories? In this section, we focus on the biological level of analysis to examine brain structures involved in memory, as well as some of the neurochemical processes that affect memory. As an outgrowth of the biological revolution, there has been tremendous progress over the past two decades in understanding what happens in the brain when we store and retrieve memories. Studies of patients with brain injuries have provided the basic foundations for understanding the neural basis of memory, and brain-imaging techniques have permitted an understanding of memory processes in the healthy brain.

There Has Been Intensive Effort to Identify the Physical Location of Memory

Locating where in the brain memories are stored has been one of the central goals of memory researchers during the twentieth century. As you learned in Chapter 4, professor Karl Lashley spent much of his career trying to localize memory. He used the term *engram* to refer to the physical site of memory storage. Lashley first trained rats to run a maze. He then removed different areas of their cortices to test how much of the learning they retained. Lashley found that the size of the area removed rather than its location was most important in predicting retention. Thus, he concluded that memory is distributed throughout the brain rather than in any specific

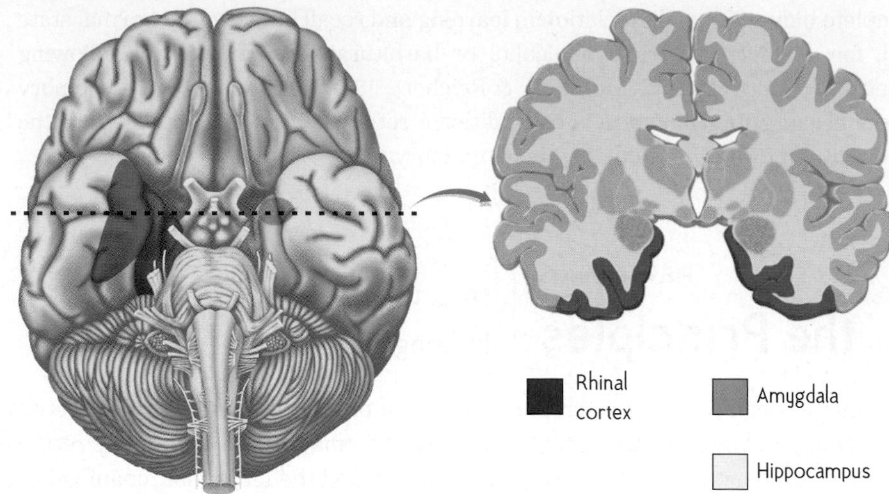

7.11 The Major Regions of the Medial Temporal Lobe

location, an idea known as *equipotentiality.* It turns out that Lashley was partially right, but also quite wrong.

It is now well established that memories are not stored in any one specific brain location. Rather, memories are stored in multiple regions of the brain and linked together through memory circuits, as proposed by Donald Hebb's idea that neurons that fire together wire together (see Chapter 6). This is not to suggest that all areas of the brain are equally involved in memory. As you'll see below, a great deal of neural specialization occurs, with different brain regions responsible for storing different aspects of information. Indeed, different memory systems, such as declarative and procedural memories, use different brain regions. Lashley's failure to find critical brain regions for memory is due to at least two factors. First, the maze task involved multiple sensory systems (such as vision and smell), so the rats could compensate for the loss of one by using other senses. Second, Lashley did not examine subcortical areas, which are now known to be important for the retention of memories.

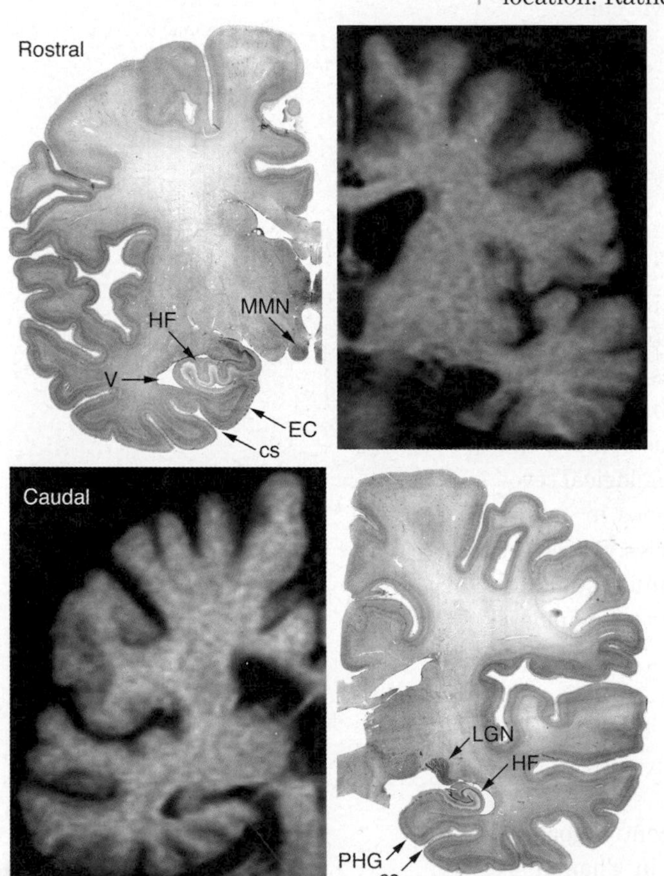

7.12 Images of H.M.'s Brain Compared to a Normal "Control" Brain Nissl-stained sections from a normal "control" brain (top left rostral, bottom right caudal) taken from the same level as H.M.'s MRI scans. Note the absence of the hippocampal formation (HF) and entorhinal cortex (EC) in the rostral MRI section of H.M.'s brain.

Over the past three decades, research at the biological (brain systems) level of analysis has identified a number of brain regions that contribute to learning and memory. For instance, we know from patient H.M. that the regions within the temporal lobes are important for the ability to encode new memories. It is clear that different brain regions are involved in different memory systems. You will see that the temporal lobes are important for declarative memory, but less important for implicit processes, such as motor learning, classical conditioning, and fear learning. We noted in Chapter 4 that the cerebellum plays a role in how motor actions are learned and remembered. Similarly, the basal ganglia, another brain region involved in motor actions, is important for classical conditioning, such as salivating at the sight of a tasty sandwich (Packard & Knowlton, 2002). One type of classical conditioning, fear learning, relies on the amygdala. You might recall from Chapter 6 that without an amygdala, animals cannot learn to fear objects that signal danger. In this section, we describe the most important brain systems involved in memory.

The Medial Temporal Lobes Are Important for Consolidation of Declarative Memories

The brain area that has been repeatedly identified as important for declarative memory is the middle section of the temporal lobes. Recall from Chapter 4 that Wilder Penfield, of the Montreal Neurological Institute, found that stimulating the temporal lobes led to reports of specific memories in some epilepsy patients. The medial temporal lobe consists of a number of structures relevant to memory, including the amygdala, hippocampus, and rhinal cortex, an area located around the front of the hippocampus (Figure 7.11). As was described earlier, H.M.'s brain surgery led to an inability to form new memories. Using brain imaging (Figure 7.12), Sue Corkin of MIT and her colleagues found that a large portion of H.M.'s hippocampus was removed, as well as a small section of nearby temporal lobe (Corkin, Amaral, Gonzalez, Johnson, & Hyman, 1997). It is now clear that damage to this region causes *anterograde amnesia,* which is the inability to store new explicit memories. However, it is *not* the case that the medial temporal lobes are the final repository of memory. After all, H.M. could remember things from before his surgery. Thus, damage to the medial temporal lobes interrupts storage of new material without impairing access to old material.

The transfer of contents from immediate memory into long-term memory occurs through a process known as **consolidation**. Consolidation results from changes in the strength of neural connections that support memory. Current thinking suggests that the medial temporal lobes are responsible for coordinating the strengthening between neuronal activations, but the actual storage most likely occurs in the particular brain regions engaged during perception, processing, and analysis of the material being learned. For instance, visual information is stored in the cortical areas involved in visual perception, whereas sound is stored in those areas involved in auditory perception. In one study, researchers exposed participants either to pictures of common objects (for example, dogs) or to their associated sounds (barking) and asked the participants to become very familiar with the stimuli. Two days later, while the participants underwent brain imaging, they were reexposed to the pictures or sounds and then asked to recall the stimuli as vividly as possible. Thus, some participants tried to remember how the objects looked, whereas others tried to remember how they sounded. The same areas that became activated when participants were exposed to the stimuli became reactivated when the participants vividly recollected those stimuli (Figure 7.13; Wheeler, Petersen, & Buckner, 2000). Thus, memory for sensory experiences involves reactivating the cortical circuits involved in perceiving them. The medial temporal lobes form links or pointers between the different sites of storage and direct the gradual strengthening of the connections between these links (Squire, Stark, & Clark, 2004).

consolidation A hypothetical process involving the transfer of contents from immediate memory into long-term memory.

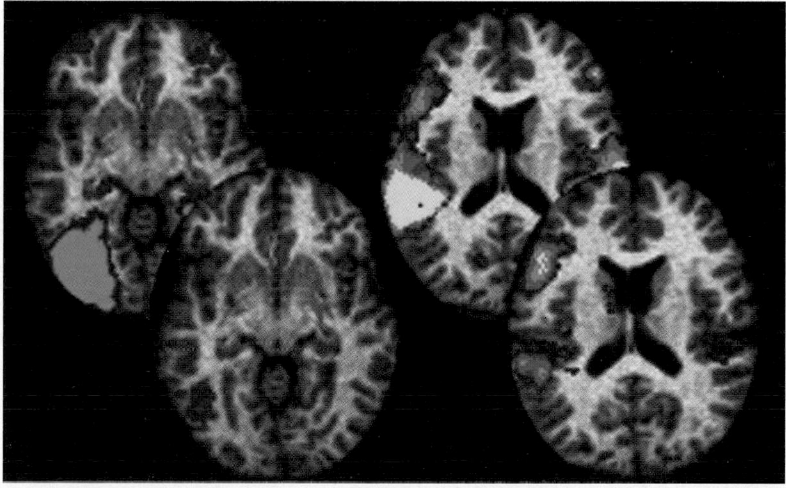

7.13 Brain Activation from Various Stimuli Four horizontally sliced brain images acquired using magnetic resonance imaging show regions of the brain active during perception of pictures (top left) and sounds (top right), and subsequently during retrieval of the same pictures (bottom left) and sounds (bottom right) from long-term memory. These data indicate that regions of sensory cortex are reactivated when we remember sensory-specific information.

It has been proposed that once memories are activated they need to be consolidated again in order to be stored back in memory, a process known as *reconsolidation* (LeDoux, 2002). This is sort of like taking a book out of the library—the book must be returned to the shelf for permanent storage. Evidence for reconsolidation can be found by administering drugs that interfere with memory storage after a memory has been activated (Debiec, LeDoux, & Nader, 2002). This suggests that memories for past

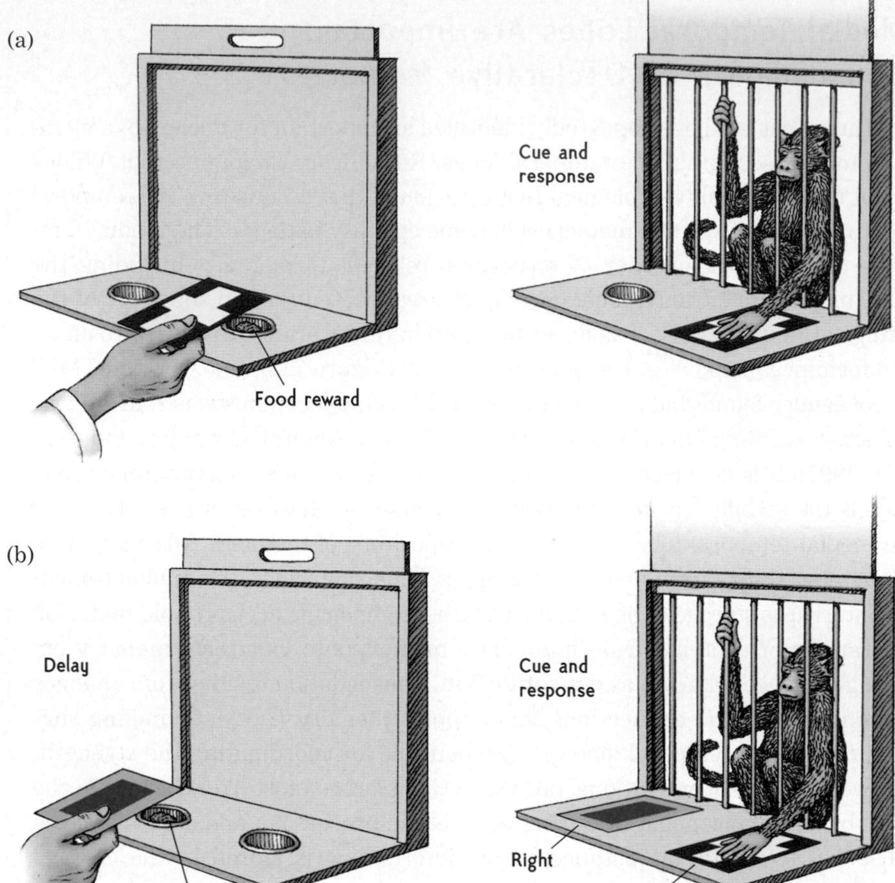

7.14 Mishkin's Animal Model of Memory
(a) The monkey is presented with a sample item, the red and yellow object. (b) After a delay, the monkey is presented with the same object and a new object. The monkey now has to select the new object, which implies it remembers the first object.

events can be affected by new circumstances once they are retrieved, so that the newly reconsolidated memories may differ from their original versions. In our library book analogy, this would be similar to what would happen if some pages were torn out of the book before it is returned. The book that is placed back on the shelf is slightly different from the one that was taken out; the information contained in those torn-out pages is no longer available for retrieval. The reconsolidation process repeats itself each time memory is activated and placed back in storage, which may explain why our memories for events can change over time. As you might imagine, this theory has received considerable attention, for it not only has implications for what it means to remember something but also opens up the intriguing possibility that bad memories could be erased by activating them and then interfering with reconsolidation.

Can old memories be changed by new circumstances?

LOCALIZING FUNCTION WITHIN MEDIAL TEMPORAL LOBES Which parts of the medial temporal lobes are most relevant to memory? Studies of primates with surgical lesions to the medial temporal lobes have given us a pretty good idea. In the 1970s, Mortimer Mishkin developed the first successful animal model of memory. In these studies, monkeys are placed in a box with a retractable door. In the first phase of the study, the monkey learns that a food reward is hidden under an object (Figure 7.14). Next, the door closes, then opens to reveal two objects: one that previously hid the reward and a new object. The rule is that the food reward is always under the novel object. This provides a good measure not only of learning, since the monkey has to learn the rule, but also of memory, since the monkey must remember which of the two objects was presented previously. The regions that are most important for this task are the hippocampus and surrounding rhinal cortex. Lesions to other areas of the medial temporal lobes do not interfere with this type of memory. Psychological scientists now view the hippocampus and surrounding regions as essential for declarative memory (Eichenbaum, 2001; Squire et al., 2004).

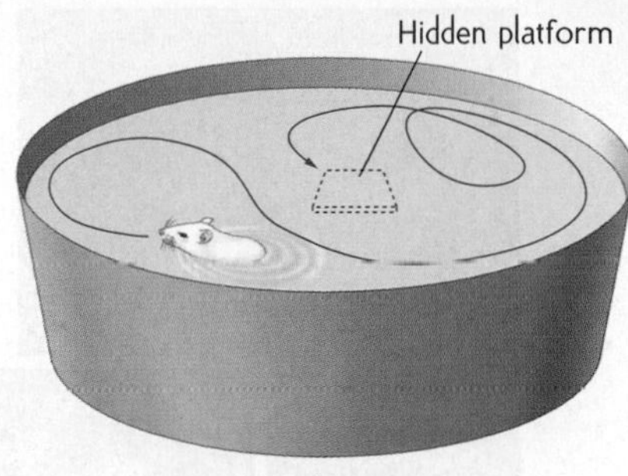

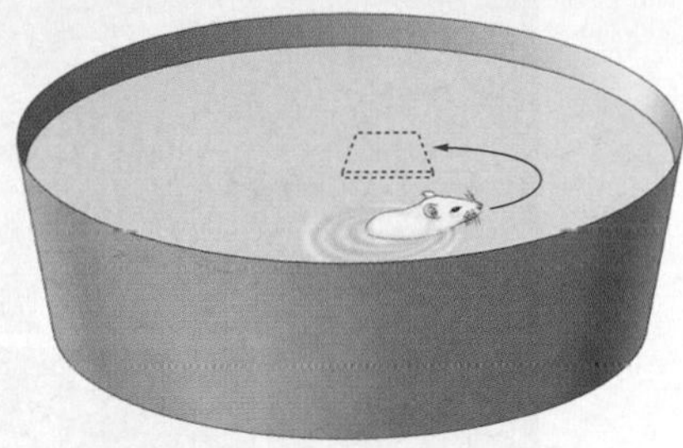

7.15 The Morris Water Maze Test A rat is placed in the Morris water maze and must find the hidden platform. After many trials, the animal learns where the platform is hidden and swims directly to it.

SPATIAL MEMORY Another important memory function of the hippocampus is **spatial memory**, which is memory for the physical environment and includes such things as direction, location of objects, and cognitive maps. A good example of this type of learning is the *Morris water maze test,* in which a rat is placed into a circular pool of murky water and learns to swim to an invisible platform that is just below the surface (Figure 7.15). Rats can learn this task quite readily, and they do so by using cues from the environment to remember where to find the platform. Rats with hippocampal damage are severely impaired at this task. The role of the hippocampus in spatial memory is supported by *place cells,* which are neurons that fire only when a rat returns to a specific location, such as one part of a maze. When rats are placed in novel environments, none of the place cells fire. But as the rat becomes familiar with the environment, its hippocampal place cells acquire a link to aspects of the surroundings. This suggests that the hippocampus helps animals to orient and find their way. Indeed, birds that rely on memory to find stored food, such as food-caching chickadees, have larger hippocampal regions than birds that do not store food. Patients with damage to the hippocampus also have trouble remembering the location of objects in pictures (Piggott & Milner, 1993; Rosenbaum et al., 2000), suggesting that the hippocampus is also important for spatial memory in humans.

spatial memory Memory for the physical environment that includes such things as location of objects, direction, and cognitive maps.

The Frontal Lobes Are Involved in Many Aspects of Memory

It has long been known that the frontal lobes are important to many aspects of memory, including episodic memory, working memory, spatial memory, time sequences (see below), and various aspects of encoding and retrieval (Wager & Smith, 2003). Extensive neural networks connect the prefrontal cortex with other brain regions involved in memory, such as the medial temporal areas. Thus, the frontal lobes work together with other brain regions to coordinate the encoding, storage, and retrieval of memory.

Patients with damage to prefrontal regions typically do not develop profound memory loss, but they often have great difficulty remembering the time sequence of events, such as which of two events happened first. Moreover, they often can't tell you where or when they learned information. The inability to recall the circumstances under which learning took place has been reported for elderly individuals, who often experience signs of frontal lobe deficits, as well as for young children, who have immature frontal lobes.

Brain-imaging studies have provided compelling evidence that the frontal lobes are crucial for encoding (Buckner, Kelley, & Petersen, 1999). Moreover, deeper encoding tasks are more likely to lead to frontal activation than are shallow encoding tasks. Frontal activation is also a particularly good predictor for which events are later remembered or forgotten (Brewer, Zhao, Glover, & Gabrieli, 1998). For instance, during a memory task researchers compared words that were later remembered with words that were later forgotten. As you can see in Figure 7.16, words that

7.16 Brain Activation in Memory Tasks This graph shows brain activation in the left prefrontal area during a memory task. Words that are later remembered elicit a larger response than words that are later forgotten.

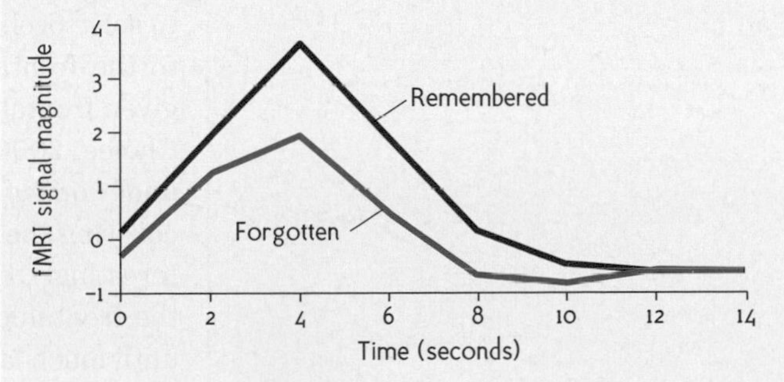

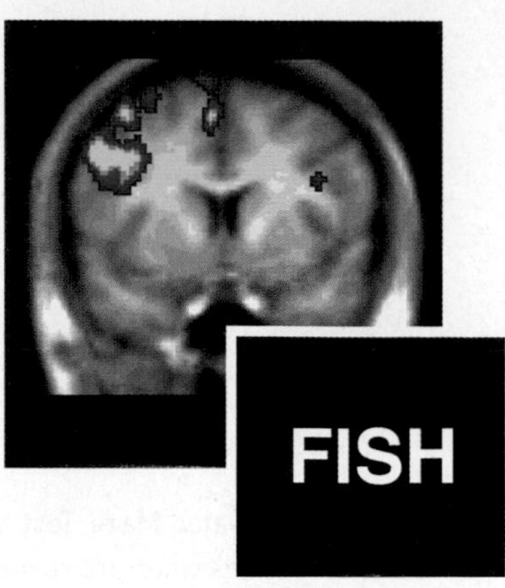

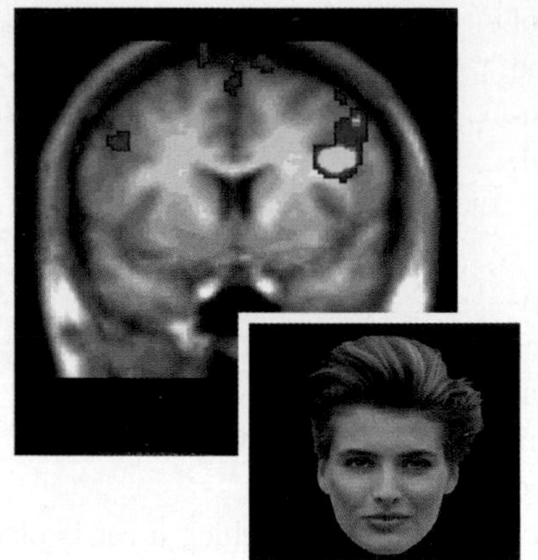
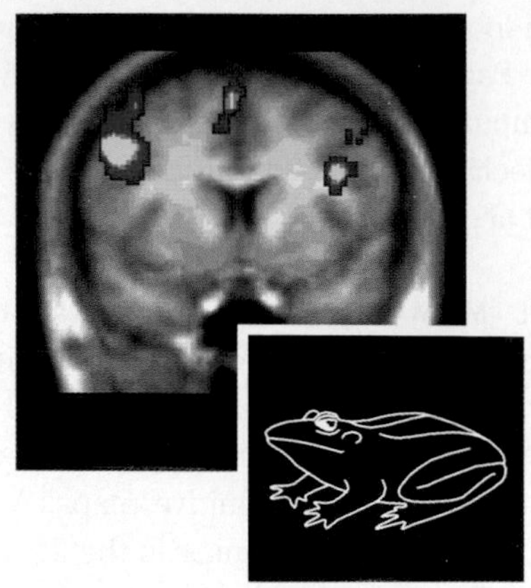
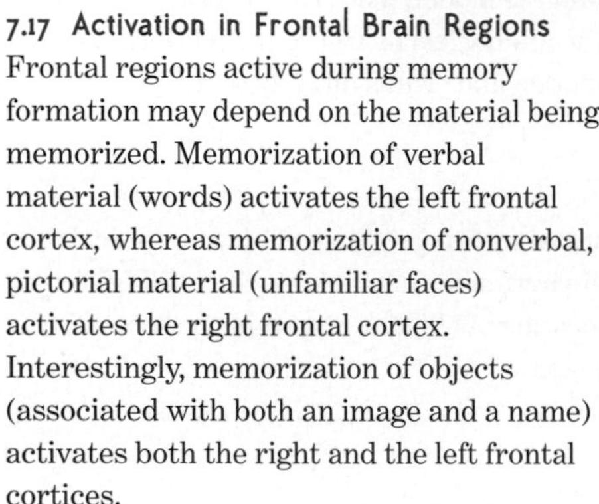

7.17 Activation in Frontal Brain Regions Frontal regions active during memory formation may depend on the material being memorized. Memorization of verbal material (words) activates the left frontal cortex, whereas memorization of nonverbal, pictorial material (unfamiliar faces) activates the right frontal cortex. Interestingly, memorization of objects (associated with both an image and a name) activates both the right and the left frontal cortices.

were later remembered were associated with stronger activation of the frontal lobes (Wagner et al., 1998). Interestingly, brain activity in regions that are involved in processing specific types of information is associated with better memory for that type of information (Paller & Wagner, 2002). For example, the medial prefrontal cortex is selectively active when people think about themselves. As a result, activity in this region predicts memory for information encoded about the self, but not for information encoded about others (Macrae, Moran, Heatherton, Banfield, & Kelley, 2004).

HEMISPHERICAL ASYMMETRIES IN ENCODING AND RETRIEVAL There is some evidence that encoding and retrieval occur in different hemispheres. The psychologist Endel Tulving and his colleagues at the Rotman Research Institute in Toronto analyzed a number of studies that used imaging to assess the encoding and retrieval of episodic experiences. When they looked at the brain-activation patterns, they found that the left frontal lobes were more active during encoding, whereas the right frontal lobes were more active during episodic retrieval (Habib, Nybert, & Tulving, 2003; Tulving, Kapur, Craik, Moscovitch, & Houle, 1994). They called this pattern the *hemispherical encoding retrieval asymmetry,* or *HERA*. William Kelley and his colleagues have argued that the pattern of left hemisphere activation might be due to the types of information used in the tasks. They argue that any stimulus that involves verbal information activates the left hemisphere, whereas objects that do not invoke verbal information (such as unfamiliar faces) activate the right hemisphere. Interestingly, nameable objects that contain both visual and verbal information (such as a picture of a frog; see Figure 7.17) activate both left and right frontal areas during encoding (Kelley et al., 1998). Thus, the asymmetry in memory applies more to the type of material studied than to memory processes themselves.

FRONTAL LOBES AND WORKING MEMORY So what do the frontal lobes do in memory? One hypothesis is that they play a role in working memory (Curtis & D'Esposito, 2003). Working memory holds information on line so that it can be used to solve problems, understand conversations, and follow plans. Patients with damage to the frontal areas often have difficulty following plans and goals, and monkeys given frontal lesions show impaired working memory (Goldman-Rakic, Scalaidhe, & Chafee, 2000). In one task, the monkey watches the experimenter hide a reward under one of two objects. After a delay, the monkey is allowed to reach for the object covering the reward. Monkeys with frontal lesions have difficulty with this task. Interestingly, human infants have trouble with a similar task. This further implicates the frontal lobes in working memory, since the frontal lobes are not fully matured until much later in human development (Diamond & Doar, 1989).

Frontal regions become active when information is being either retrieved from long-term memory into working memory or encoded from working memory into long-term memory. For example, when you enter the classroom, the frontal lobes oversee retrieval of stored knowledge from neural networks that are spread throughout cortical and subcortical locations. This information helps you remember what to do when you enter the room (sit down), who the person is at the front of the class (the professor), and the meaning of the sounds emanating from the professor's mouth. The frontal lobes and medial temporal lobes then work together to consolidate long-term declarative memory storage for the lecture. The hippocampus helps strengthen associations between the events in working memory so that new experiences are consolidated into permanent storage.

Neurochemistry Influences Memory

It is known that memory involves alterations in connections across synapses. As memories are consolidated, distributed networks of neurons become linked together. Research has shown that a variety of neurotransmitters can weaken or enhance memory. Collectively, these neurotransmitters are known as *memory modulators,* because they modulate, or modify, the storage of memory.

NEUROCHEMISTRY INDICATES MEANINGFULNESS OF STIMULI Earlier we pointed out that animals store only information that is meaningful. Memory modulation is the system that evolution has bestowed on organisms to help determine whether something is important. Important events lead to neurochemical changes that produce emotional experiences. An animal that outruns a predator has experienced a fear reaction that helps to stamp in avoidance of that predator. A child who eats a good-tasting food experiences a rewarding sensation that is easily remembered, as parents who have watched their preverbal children point at cookie jars can readily attest. Events that produce emotional reactions are especially likely to be stored in memory.

So what are the neurochemical signals that indicate that an experience is meaningful? When animals are engaging in minimally arousing tasks, little memory is later shown for those tasks. However, animals injected with epinephrine, which induces a state of arousal, show significant memory enhancement for trivial events. It is as if the jolt of epinephrine gives the message "Remember me!" Of course, epinephrine is in the periphery of the body, not in the brain, and so it does not have a direct effect on memory. Researchers initially believed that this effect occurs because epinephrine causes a release of glucose, which then enters the brain and influences memory storage. The role of glucose is also implicated by a study in which elderly participants were given a memory test after consuming lemonade with either glucose (sugar) or saccharine (a sugar substitute). Those who had glucose had better memory for the studied items the next day (Gold, 1987). More recent studies suggest that epinephrine enhances memory because of its effect on norepinephrine activity in the amygdala, as we will see shortly. Other drugs, such as opiates, alcohol, and surgical anaesthetics, may also interfere with memory through their effects on norepinephrine (McGaugh, 2002).

THE AMYGDALA AND THE NEUROCHEMISTRY OF EMOTION The amygdala is the prime candidate for the brain structure that controls the modulating effects of neurotransmitters on memory. It is the limbic structure closely tied to fear reactions, and it is located within the medial temporal lobe. Direct stimulation of the amygdala can enhance or impair memory, and damaging the pathways leading to or exiting it eliminates the effects of drugs on memory. As mentioned before, the modulation of memory occurs because of alterations in the activity of norepinephrine receptors in the amygdala. Any arousing event causes greater activity of norepinephrine receptors, which strengthens memory.

In humans, the amygdala is activated during recall of emotional film clips but not during recall of neutral film clips (Cahill et al., 1996). Interestingly, there appears to be a gender difference in this effect, such that emotional memory activates the right (but not left) amygdala in men and the left (but not right) amygdala in women (Cahill et al., 2001). Moreover, women have better memory than men for emotional events. For instance, Turhan Canli and colleagues (2002) showed men and women neutral and negative pictures and found that women not only reported greater emotional reactions to the negative pictures but also showed much better memory for those pictures. This may be because women process emotional information differently, such as rehearsing it and thinking about it more, or because they focus on different aspects of stimuli (Cahill, 2003). The effects of amygdala activity on memory are long lasting. Amygdala activity at encoding predicts increased memory for emotional pictures a year later (Dolcos, Labar, & Cabeza, 2005).

When people experience severe stress or emotional trauma—such as being in a serious accident, being raped, being in combat, or surviving a natural disaster—they often have negative reactions long after the threat has passed. In severe cases, people develop **posttraumatic stress disorder (PTSD)**, a serious mental health disorder that involves frequent and recurring unwanted thoughts related to the trauma, including nightmares, intrusive thoughts, and flashbacks. Those with PTSD often have chronic tension, anxiety, and health problems, and they may experience memory and attention problems in their daily lives. PTSD involves an unusual problem in memory, in that it is caused by the inability to forget. PTSD is associated with an attentional bias, such that people with PTSD are hypervigilant to stimuli associated with their traumatic event. For instance, soldiers with combat-induced PTSD show increased physiological responsiveness to pictures of troops, sounds of gunfire, and even words associated with combat. Exposure to stimuli associated with past trauma leads to activation of the amygdala (Rauch, van der Kolk, Fisler, & Alpert, 1996). It is as if the severe emotional event is "overconsolidated," burned into memory.

posttraumatic stress disorder (PTSD) A mental disorder that involves frequent nightmares, intrusive thoughts, and flashbacks related to an earlier trauma.

Knowing that neurochemistry can modify the formation of memory has led researchers to develop methods that may help people remember things they *want* to remember or forget things they *don't want* to remember.

Thinking Critically

Should Memory Be Altered by Drugs?

A paradox of memory is that many times we have trouble remembering things we want to remember, whereas we often have trouble forgetting things we want to forget. During an exam you might struggle to remember the answer to a question; at the same time you might feel anxious because you remember how it felt when you bombed on an exam in the past. The anxious memories might even be interfering with your ability to remember the relevant material. Or perhaps you witnessed a horrible car crash and you keep recalling gruesome details of the accident, causing you to relive your negative response over and over again. Wouldn't it be great if you could take a pill that would help you forget those unwanted memories? Equally desirable might be a pill that would help students better remember material for exams or aging baby boomers better remember where they left their car keys. We are now witnessing an explosion of drug therapies aimed at altering human memory capacities (Marshall, 2004). Let's consider some of these methods as well as their ethical implications.

In 1998, Eric Kandel, the Nobel laureate mentioned in Chapter 6, launched a company to develop drug treatments for learning and memory disorders associated with diseases such as Alzheimer's. Drawing on Kandel's own research into the molecular basis of memory, his company is developing drugs that alter gene expression to enhance learning and memory. Recall the transgenic "Doogie" mice that became fast learners. It is possible to alter gene expression, and indeed DNA itself, to produce better learners. But can we do the same to help people forget? Researchers at Emory University in Atlanta studied people who were trying to overcome their fear of heights through virtual reality training (Ressler et al., 2004). Those who received a drug that affected NMDA receptors (recall from Chapter 6 that this receptor is involved in long-term learning) responded much better than those given a placebo. In essence, the drug helped them to forget their fears. In other research, those given a type of beta blocker (drugs normally used to treat hypertension) right after experiencing a trauma had fewer emotional memories for that event months later (Pitman et al., 2002). Many other research efforts are underway to produce memory-altering drugs.

7.18 Altering Memories In the 2004 film *Eternal Sunshine of the Spotless Mind,* Joel Barish (played by Jim Carrey) tries to erase the painful memories of a recent breakup by undergoing an experimental procedure that eliminates unwanted memories. In real life, researchers struggle with the ethical implications of developing drugs that can alter memory.

On the surface, the ability to enhance wanted memories and eliminate unwanted ones seems like a good idea (Figure 7.18). As with many recent neuroscientific advances, however, the possibility of altering basic psychological functioning raises important issues. One is whether in some contexts memory enhancement might be like athletic performance enhancement obtained through steroids or other muscle-building supplements. Knowing that an athlete performed some amazing feat while using drugs diminishes our appreciation of that performance. How is competition meaningful unless the playing field is level? If students could take a drug that would help them do better on exams, how should their performances be graded?

A critical issue about using drugs to increase forgetting is whether taking the emotional sting out of life will make us less human. Clearly such drugs are called for in treating conditions such as PTSD, but what about taking a pill to erase the memory of a recent breakup? Feeling sorrow, embarrassment, and fear is part of being human. If we could give people a pill after a loved one dies so that they do not feel grief, would that be a good thing? We grieve because we love; does love change if we don't grieve? If we wiped out negative memories, would that diminish our appreciation of life's positive events? Even if beta blockers might reduce long-term problems for some people, should we routinely give those drugs after traumatic events, even before knowing which people will go on to develop unwanted and persistent memories?

Some have argued that any drug that alters normal cognitive processes is unnatural and thus unethical. But is pharmacological treatment of traumatic memories really any different from medical treatments for any disorder? Throughout this book, we discuss how neuroscientific discoveries are providing ways to better understand and treat a number of psychological disorders, such as schizophrenia and depression. New medications have helped many afflicted individuals. How does using medicine to treat those with mental illnesses differ from using medicine to improve memory among those with Alzheimer's or to help sexual assault victims forget their traumatic experiences? Or are we concerned only about the potential for abuse among those who do not really suffer from memory impairments? Perhaps the critical question to resolve is where we draw the line between treatment and enhancement. ■

REVIEWING the Principles | What Brain Processes Are Involved in Memory?

Research during the past thirty years has demonstrated that memories are encoded in distributed networks of neurons in relatively specific brain regions. It is now known that damage to the medial temporal lobes, especially the hippocampus or rhinal cortex, causes significant memory disturbances. These medial temporal regions are important for the consolidation of declarative memories into storage. The sites of memory storage are the brain structures involved in perception. The frontal lobes are especially important for working memory, and a variety of neurotransmitters modulate memory storage. Fear causes activation of the amygdala, which is associated with the strengthening of memories.

When Do People Forget?

forgetting The inability to retrieve memory from long-term storage.

Up until now we have been focusing on remembering information. But along the way we have noted that failures of memory are extremely common. **Forgetting** is the inability to retrieve memory from long-term storage. Forgetting is an everyday experience that is perfectly normal. Ten minutes after you see a movie you might remember plenty of details, but the next day you probably remember mostly the plot and the main characters. Years later you might remember the gist of the story, or you might not remember having seen the movie at all. We forget far more than we ever remember.

Most people bemoan forgetting, wishing that they could better recall information they study for exams, the names of childhood friends, or even the names of all the seven dwarfs. But imagine what life would be like if you couldn't forget. You would walk up to your locker and recall 10 or 20 different combinations. Consider the case of a Russian newspaper reporter who had nearly perfect memory. You could read him a tremendously long list of items, and as long as he could spend a few moments visualizing the items, he could recite them back, even many years later. But he was tortured by his condition and was eventually institutionalized. His memory was so cluttered with information that he had great difficulty functioning in normal society. Paradoxically, not being able to forget is as maladaptive as not being able to remember. It is therefore not surprising that memory tends to be best for meaningful and important points. We remember the forest rather than the individual trees. Normal forgetting helps us remember and use information that is important.

The study of forgetting has a long history in psychological science. Hermann Ebbinghaus, a German psychologist working at the individual level of analysis in the late nineteenth century, provided compelling evidence that forgetting occurs rapidly over the first few hours and days but then levels off. He used the so-called *methods of savings* to examine how long it took for people to relearn lists of nonsense syllables. Presumably, the more slowly people relearned the list, the greater evidence there was of forgetting. We've learned a great deal about forgetting over the last century. Harvard psychologist Daniel Schacter identifies what he calls *the seven sins of memory* (Schacter, 1999). These so-called sins are all too familiar for most people (see Table 7.1). They are also characteristics that are useful or perhaps even necessary for survival. Schacter argues that the seven sins of memory are by products of otherwise desirable aspects of human memory. The first three sins are related to forgetting: *transience, blocking,* and *absentmindedness* (we will consider some of the other sins later in this chapter).

TABLE 7.1 Seven Sins of Memory

ERROR	TYPE	EXAMPLE
Transience	Forgetting	Reduced memory over time, such as forgetting the plot of a movie.
Absentmindedness	Forgetting	Reduced memory due to failing to pay attention, such as losing your keys or forgetting a lunch date.
Blocking	Forgetting	Inability to remember needed information, such as failing to recall the name of a person you meet on the street.
Misattribution	Distortion	Assigning a memory to the wrong source, such as falsely thinking that Richard Shiffrin is famous.
Suggestibility	Distortion	Altering a memory because of misleading information, such as developing false memories for events that did not happen.
Bias	Distortion	Influence of current knowledge on our memory for past events, such as remembering our past attitudes as similar to our current attitudes even though they have changed.
Persistence	Undesired	The resurgence of unwanted or disturbing memories that we would like to forget, such as remembering an embarrassing faux pas.

SOURCE: Based on Schacter (2001).

Transience Is Caused by Interference

Memory **transience** refers to the pattern of forgetting over time, such as was obtained by Ebbinghaus in his studies of nonsense syllables. What causes transience? Many early theorists argued that forgetting was the result of *decay* of the memory trace in our nervous system. Indeed, there is some evidence that forgetting occurs for memories that are not used. However, research over the last few decades has established that most forgetting occurs because of *interference* from other information. There are two ways that additional information can lead to forgetting. In **proactive interference**, prior information inhibits our ability to remember new information. For instance, if you have a new locker combination each year, you may have difficulty remembering it because you keep recalling the old one. Indeed, because the physical context provides retrieval cues for that earlier combination, it becomes especially difficult to remember the new one while standing in front of your locker. In **retroactive interference**, new information inhibits our ability to remember old information. Now that you finally know your new locker combination, you may find that you have forgotten the old one. Figure 7.19 shows the typical experimental tests that have been used to demonstrate proactive and retroactive interference. In both cases, we forget because competing information displaces the information we are trying to retrieve.

Do people really forget things, and if so, why?

transience The pattern of forgetting over time.

proactive interference When prior information inhibits the ability to remember new information.

retroactive interference When new information inhibits the ability to remember old information.

blocking The temporary inability to remember something that is known.

Blocking Is Temporary

Blocking occurs when a person has a temporary inability to remember something that is known. Temporary blockages are common and frustrating, such as forgetting the name of a favourite CD, forgetting lines during a play, or forgetting someone's name

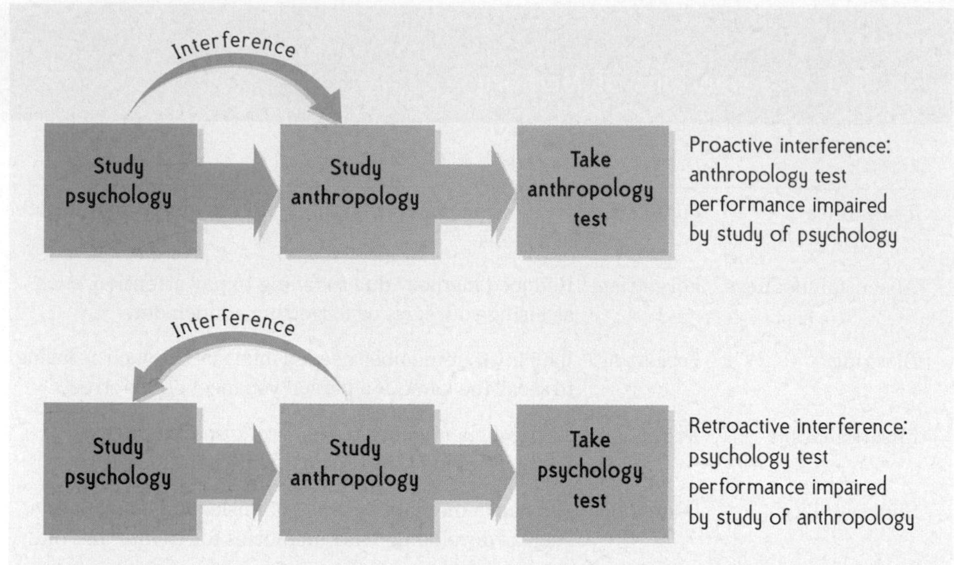

7.19 Proactive and Retroactive Interference Proactive interference occurs when material that is already known interferes with the ability to remember new information, whereas retroactive interference occurs when new material interferes with memory for old material.

tip-of-the-tongue phenomenon When people experience great frustration as they try to recall specific words that are somewhat obscure.

absentmindedness The inattentive or shallow encoding of events.

when you introduce them. Another good example of blocking is the **tip-of-the-tongue phenomenon**, first described by Roger Brown and David MacNeill, in which people experience great frustration as they try to recall specific words that are somewhat obscure. The researcher Alan Brown (no relation to Roger) has shown that tip-of-the-tongue can be reliably produced in the laboratory (Brown, 1991). For instance, when asked to provide a word that means "patronage bestowed on a relative, in business or politics," people often struggle. Or how about "an astronomical instrument for finding position"? Sometimes people know which letter the word begins with, how many syllables it has, and even what it sounds like, but they can't pull the precise word into working memory. Did you know the words were *nepotism* and *sextant*? Blocking often occurs because of interference from words that are similar in some way, such as in sound or meaning. These similar words often keep recurring as we try to remember the target word.

Absentmindedness Results from Shallow Encoding

Absentmindedness describes the inattentive or shallow encoding of events—for instance, forgetting where you left your keys, the name of a person you met five minutes before, or whether you took your vitamins this morning. The major cause of absentmindedness is failing to pay sufficient attention to details. Consider the fascinating phenomenon of "change blindness." Imagine that you are standing around and someone walks up to you and asks directions. While you are answering, two workers, holding a door, cut between you, so you momentarily lose sight of the person asking for directions. Do you think you would notice if the person standing in front of you was switched? When Daniel Simons and Daniel Levin conducted this study, only half the participants noticed the change (Simons & Levin, 1998). Simons and Levin refer to the inability to detect changes to an object or a scene as *change blindness.* Interestingly, in the original study, older people were especially likely not to notice a change in the person asking directions; college students were pretty good at noticing the change. Is this because older people are especially inattentive? Or is it that people process the gist of the situation rather than detail? Perhaps the older adults encoded the person as simply "a college student" and did not look for finer detail. To test this idea, Simons and Levin conducted an additional study in which college students were asked for directions by a construction worker. Sure enough, the college students failed to notice the replacement of one construction worker with another, proving the latter.

Amnesia Is a Deficit in Long-Term Memory

Sometimes we lose the ability to retrieve vast quantities of information from long-term memory. **Amnesia** refers to deficits in long-term memory that result from disease, brain injury, or psychological trauma. There are essentially two types of amnesia: retrograde and anterograde. In **retrograde amnesia,** people lose past memories for events, facts, people, or even personal information. Most portrayals of amnesia in the media are of retrograde amnesia, as when a character in a soap opera awakens from a coma and doesn't know who she is. By contrast, in **anterograde amnesia,** people lose the ability to form new memories. H.M. has a classic case of anterograde amnesia. He can remember old information about his past, but since his surgery he has lost the ability to form new memories. Interestingly, recent evidence suggests that H.M. has acquired some new semantic knowledge about people and events that have occurred since 1953. For instance, when he was given a list of people who became famous after 1953, H.M. was able to provide some information about them (O'Kane, Kensinger, & Corkin, 2004). Given the name Lee Harvey Oswald, H.M. described him as the man who "assassinated the president." This new learning may occur through extensive repetition of materials over a lengthy period of time.

Many cases of amnesia result from damage to the medial temporal lobes. Damage to other subcortical areas, such as around the thalamus, can also lead to amnesia. A common cause of this type of amnesia is *Korsakoff's syndrome,* a severe form of memory disturbance linked to chronic alcoholism. Long-term alcohol abuse can lead to vitamin deficiency that results in thalamic damage and, subsequently, amnesia.

Amnesia is a condition that can vary greatly. For instance, after getting hit in the head or losing blood flow to the brain, some people experience temporary memory loss that abates after a day or two. A serious blow to the head can cause a *concussion,* which involves a loss of consciousness that can range from moments to weeks. Trevor Rees-Jones, the bodyguard for Princess Diana, suffered a concussion during the car accident that claimed her life. He later was able to recall getting into the car but not the chase leading up to the crash. This is typical of amnesia following concussion. There is often a period of retrograde amnesia for events leading up to the incident and sometimes temporary anterograde amnesia. People visiting such patients in the hospital are often surprised that the person seems lucid during conversation but later fails to remember the visit. This occurs because working memory is not disrupted, but the memories are not consolidated to long-term memory.

amnesia Deficits in long-term memory that result from disease, brain injury, or psychological trauma.

retrograde amnesia The condition in which people lose past memories, such as memories for events, facts, people, or even personal information.

anterograde amnesia An inability to form new memories.

REVIEWING the Principles | When Do People Forget?

Forgetting is the inability to retrieve memory from long-term storage. The ability to forget is just as important as the ability to remember. Forgetting that occurs over time is often due to interference from competing stimuli. Blocking is a temporary inability to retrieve specific information, as exemplified by the tip-of-the-tongue phenomenon. It too is caused by interference. Absentmindedness is a result of shallow encoding, which occurs when people fail to pay sufficient attention to details. Amnesia is a deficit in long-term memory in which people forget past information (retrograde amnesia) or are unable to store new information (anterograde amnesia). Most amnesia is caused by injury to the brain.

How Are Memories Distorted?

What causes people to have memories that are distorted, false, or inaccurate?

Most people believe that human memory is a permanent storage, from which even minute and apparently forgotten details can be retrieved through hypnosis, truth drugs, or other special techniques. Research has clearly shown, however, that human memory is biased, flawed, and distorted. Memory illusions are common, as we described earlier in this chapter. However, not all memory distortions are so benign. In this section you will learn how the human memory systems provide a less than accurate portrayal of past events. Keep in mind, however, that simply because something doesn't work the way we think it works does not mean it is faulty. As Dan Schacter has argued, many of the seemingly flawed aspects of memory may be byproducts of mechanisms that are beneficial. Let's examine some of the most striking examples of the apparent fallibility of human memory.

Flashbulb Memories Can Be Wrong

flashbulb memories Vivid memories for the circumstances in which one first learned of a surprising, consequential, and emotionally arousing event.

Do you remember where you were when you found out about the terrorist attacks on the World Trade Center and the Pentagon? Some events cause people to experience what Roger Brown and James Kulik termed **flashbulb memories**, which are vivid memories for the circumstances in which one first learned of a surprising and consequential or emotionally arousing event (Figure 7.20). When Brown and Kulik interviewed people about their memory of the death of President John Kennedy in 1963, they found that people described their memories in highly vivid terms that contained a great deal of detail, such as who they were with, what they were doing or thinking, who told them or how they found out, and what their emotional reaction was to the event.

DO YOU REMEMBER WHERE YOU WERE WHEN YOU HEARD . . . ? Of course, the question is whether flashbulb memories are accurate. There is an obvious problem in conducting research on flashbulb memories—you have to wait for a "flash" to go off and then immediately conduct your study. The explosion of the space shuttle *Challenger* on January 28, 1986, at 11:38 EST provided a unique opportunity for research on this topic. Ulric Neisser and Nicole Harsch (1993) had 44 psychology students fill out a questionnaire the day the shuttle exploded and then tested their memory three years later. They found that the vast majority of students were incorrect about multiple aspects of the situation, and that only three students had perfect recall. However, other researchers have documented better memory for flashbulb experiences, such as where British participants were when they heard about Prime Minister Margaret Thatcher's resignation. Martin Conway and his colleagues (1994) showed that better memory for the flashbulb experience occurred among those who found the news surprising and felt that the event was important. Thus, students in the United Kingdom experienced stronger flashbulb memories for the Thatcher resignation than did students in the United States.

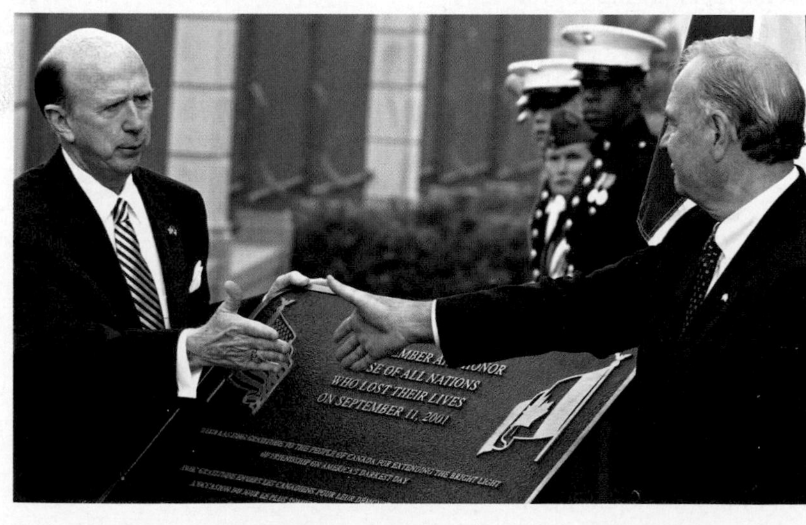

7.20 Flashbulb Memories Many Canadians report flashbulb memories for the destruction of the World Trade Center towers.

STRESS AND MEMORY REVISITED Although flashbulb memories are not perfectly accurate, this does not mean that they are less accurate than any other type of memory. It is indeed possible that flashbulb experiences are recalled more accurately than inconsequential or unsurprising events. To the extent that flashbulb memories contain an unusual number of incidental details, it is possible that this occurs because of the memory-modulating effects of stress hormones (Christianson, 1992). Thus, any event that produces a strong emotional response is likely to produce a vivid, although not necessarily accurate, memory. A second explanation is simply based on the well-known finding that distinctive events are recalled more easily than trivial events. This latter finding is known as the *von Restorff effect,* named after the researcher who first described it in 1933.

People Make Source Misattributions

Dan Schacter identified source misattributions as one of the "seven sins" of the memory system. **Source misattributions** occur when people misremember the time, place, person, or circumstances involved with a memory. A good example of this is the *false fame effect* described earlier, in which people mistakenly believe that someone is famous simply because they've previously encountered the person's name. Similarly, social psychologists have long known about the *sleeper effect,* in which strong arguments that are initially not very persuasive because they come from questionable sources become more persuasive over time. You would probably disbelieve a weekly tabloid if it claimed that scientists had discovered a way for people to learn calculus while sleeping. Yet over time you might remember the argument but fail to remember that the source was questionable; hence, you later believe that people can learn calculus while sleeping.

source misattributions Memory distortions that occur when people misremember the time, place, person, or circumstances involved with a memory.

cryptomnesia A type of misappropriation that occurs when people think they have come up with a new idea, yet have only retrieved a stored idea and failed to attribute the idea to its proper source.

CRYPTOMNESIA An intriguing example of source misattribution is **cryptomnesia**, which is when people think they've come up with a new idea, when really they have retrieved an old idea from memory but have failed to attribute the idea to its proper source (Macrae, Bodenhausen, & Calvini, 1999). Students who take verbatim notes while conducting library research sometimes experience the illusion that they've written the sentences, which can later lead to an accusation of plagiarism. Thus students need to be especially vigilant to indicate verbatim notes while they are taking them. George Harrison, the former Beatle, was sued because his song "My Sweet Lord" is strikingly similar to the song "He's So Fine," recorded in 1962 by the Chiffons. Although he acknowledged having heard the song before, he vigorously denied that he plagiarized it. He argued that with a limited number of notes, and an even smaller number of chord sequences in rock and roll, some overlap was inevitable. In a controversial verdict, the judge ruled against Harrison.

7.21 Thomas Sophonow In 1981, Thomas Sophonow was thrice convicted of strangling Barbara Stoppel in Winnipeg. After spending nearly four years in prison, Sophonow was acquitted and released from jail. In June 2000, Sophonow was exonerated of the crime based on DNA evidence. It was subsequently revealed that witnesses had been shown flawed photos and a lineup in which he stood a full head taller than anyone else. Is eyewitness really reliable?

People Make Bad Eyewitnesses

One of the most powerful forms of evidence is the eyewitness account. Early studies demonstrated that very few jurors are willing to convict an accused individual on the basis of circumstantial evidence alone. But add one person who says "That's the one!" and conviction rates shoot up, even if it is shown that the witness had poor eyesight or some other condition that raises questions about the accuracy of the testimony. The power of eyewitness testimony is troubling because it is so often in error (Figure 7.21). Indeed, Gary Wells and his colleagues studied 40 cases in which DNA evidence indicated that a person had been falsely convicted of a crime. In 36 of these cases, the person had been falsely identified by one or more eyewitnesses (Wells et al., 1998). No one knows this better than William Jackson (Figure 7.22), who served five years in an American prison after being wrongly convicted of a

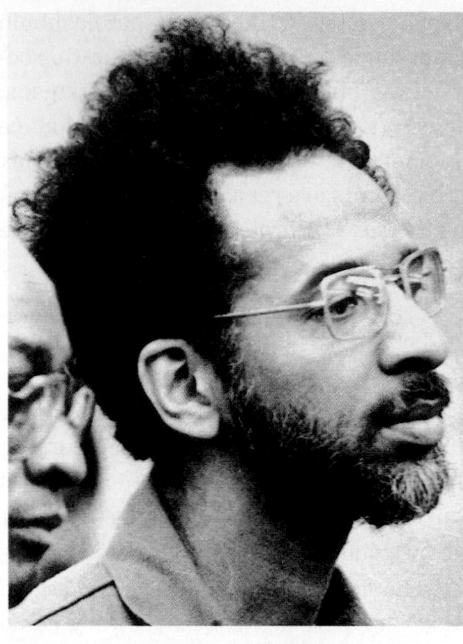
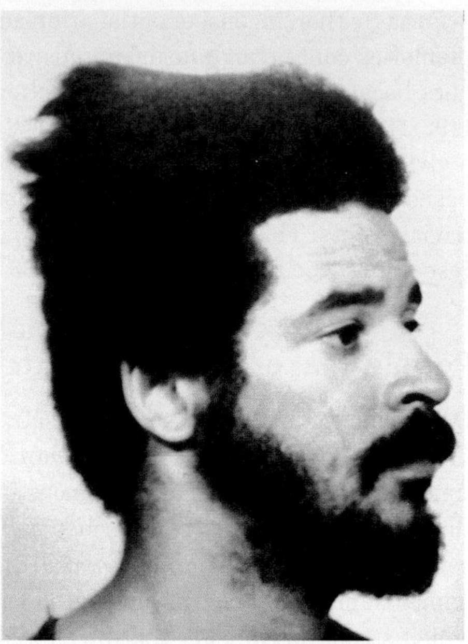

7.22 Eyewitness Accounts Can Be Unreliable William Jackson (left) served five years in prison because he was wrongly convicted of a crime based on the testimony of two eyewitnesses. The real perpetrator is shown on the right. Do these two men look the same to you?

crime because of the testimony of two eyewitnesses. Let's look at some of the memory processes that contribute to eyewitness misidentification.

CROSS-ETHNIC IDENTIFICATION At the social level of analysis, people are particularly bad at accurately identifying individuals of other ethnicities. This effect occurs for Caucasians, Asians, African Americans, and Hispanics. The phrase "they all look alike" appears to be true of all people who are not members of our own ethnic group. In one recent brain-imaging study conducted in the United States, both African American and Caucasian Americans showed better memory for same-race faces. The difference in performance by participants was associated with the degree of activity in the fusiform face area (Golby, Gabrieli, Chiao, & Eberhardt, 2001). As you recall from Chapter 4, this area responds more strongly to faces than to other objects. Apparently, the greater activation of this area to same-race faces accounts for the memory superiority for members of one's own racial group. One explanation for this effect is that people tend to have less frequent contact with members of other ethnicities; another is that people encode ethnicity according to gist rules of categorization, the way that older people in the change-blindness experiments saw "a college student" and students saw "a construction worker."

SUGGESTIBILITY AND MISINFORMATION Elizabeth Loftus (Figure 7.23) is one of the world's leading specialists in eyewitness memory. During the early 1970s, Loftus and her colleagues conducted a series of important studies demonstrating that people can develop biased memories when provided with misleading information—the "sin" of **suggestibility**. The general methodology of this research involves showing research participants an event and then asking them specific questions about it; the way the questions are worded alters the participants' memories for the event. For instance, in one experiment college students viewed a videotape of a car accident and then were asked, "How fast were the cars going when they ________ each other?" The following words were used in the question: contacted, hit, bumped, collided, or smashed. Loftus and John Palmer (1974) found that students estimated the cars to be traveling faster when the word "smashed" was used than when the other words were used. In a second study, groups of students saw videotapes of car accidents and were

suggestibility The development of biased memories when people are provided with misleading information.

then asked about seeing the cars either "smash" into or "hit" each other. One week later they were asked if they had seen broken glass on the ground, an event that did not happen. Nearly one-third of those who heard the term "smashed" falsely recalled having seen the broken glass, whereas very few of those who heard the term "hit" did.

Not only did participants report an altered view of the car accident, in other experiments Loftus and her colleagues demonstrated that people could be induced to report having seen things that did not exist, such as barns, through the use of such questions as "How fast was the white sports car going when it passed the barn while traveling along the country road?" Participants who were asked this question were more likely to report having seen the nonexistent barn than participants who were not asked this question.

You might be wondering whether these sorts of laboratory analogues are appropriate for studying eyewitness accuracy. After all, the sights and sounds of a real traffic accident impress the event on our awareness. Wouldn't our memory be better in the real world? Some evidence supports this idea. A study in Vancouver examined reports of those who had witnessed a fatal shooting (Yuille & Cutshall, 1986). All observers had been interviewed by the police within two days of the shooting. When the researchers went back months later, they found that eyewitness reports were highly stable, including memory for details. Indeed, it makes sense that true eyewitness accounts might be more vivid than those in the laboratory, given the memory modulation effect of stress hormones. However, the real world differs in a number of ways from a controlled laboratory setting, and it is not clear whether the stable memories are accurate. For instance, it could be that by retelling the story over and over again to friends, the eyewitnesses inadvertently developed stronger memories for inaccurate details.

7.23 Elizabeth Loftus Loftus has shown that memory is highly suggestible.

EYEWITNESS CONFIDENCE How good are observers such as jurors at judging whether eyewitnesses are accurate? Gary Wells and his colleagues conducted a number of studies at the University of Alberta and later at Iowa State University, and their general finding was that people cannot differentiate eyewitnesses who are accurate from those who are inaccurate. The problem is that eyewitnesses who are wrong are just as confident, often more confident, than eyewitnesses who are right. Eyewitnesses who report vivid details of all aspects of the scene are probably less credible than those who have poor memory for trivial details. After all, eyewitnesses to real crimes tend to be focused on the weapons or on the action—they fail to pay attention to minor details. Thus, strong confidence for minor details may be a cue that the memory is more likely to be inaccurate or perhaps even false. However, some people are just particularly confident, and whether they are correct or incorrect, jurors find them convincing.

People Have False Memories

Source amnesia occurs when a person shows memory for an event but cannot remember where he or she encountered the information. Think back to your earliest childhood memory. Most people cannot remember specific memories before three years of age. The absence of prior memories is known as *childhood amnesia,* and it may be due to an early lack of linguistic capacity as well as immature frontal lobes. But let's consider that first memory. How vivid is it? How do you know you are not remembering something you saw in a photograph, or a story related to you by family members? When people recall childhood events, the memories are often partial and hazy. How do you know if the memory is real?

source amnesia A type of amnesia that occurs when a person shows memory for an event but cannot remember where he or she encountered the information.

CREATING FALSE RECOGNITION Recently, some powerful demonstrations have shown that people can be misled falsely to recall or recognize events that did not happen. Read the following list of words, saying each one out loud: sour, candy, sugar, bitter, good, taste, tooth, nice, honey, soda, chocolate, heart, cake, tart, pie. Now put your book aside and try to write down as many of the words as possible.

Henry Roediger and Kathleen McDermott used this experiment to see whether they could create false memories among normal, healthy people (Roediger & McDermott, 1995). For instance, without looking back, which of the following words did you recall? Candy, honey, tooth, sweet, pie. If you recalled "sweet," then you have experienced a false memory. If you look back, you will see that "sweet" was not one of the original words, but note that all of the words are related to the word *sweet.* This basic paradigm is extremely reliable for producing false recollections. Moreover, people are often extremely confident in their judgments of previously having seen or heard the critical word.

Mike Miller and George Wolford have argued that responses in the Roediger-McDermott task are due to change in the criteria by which people are willing to say they recognize the words, therefore representing a response bias rather than something that could be called a false memory (Miller & Wolford, 1999). They make the important point that errors are common in many memory tasks, and we should be careful before concluding that memory illusions are "false." After all, no one ever said that participants who misjudged made-up names as famous (in the false fame effect discussed earlier) were having a false memory. Nonetheless, there is a growing body of evidence that false memories can be implanted (Loftus, 2003).

confabulation The false recollection of episodic memory.

CONFABULATION Some types of brain injury are associated with the false recollection of episodic memory, which is called **confabulation**. Morris Moscovitch of the University of Toronto has described confabulating as "honest lying" because there is no intent to deceive and the person is unaware of the falsehood of his or her story. Moscovitch (1995) provides an amazing example of confabulation. H.W. was a 61-year-old man who was the biological father of four children, all of them now grown. H.W. experienced severe frontal lobe damage following a cerebral hemorrhage. Here is part of the clinical interview:

Q. Are you married or single?
H.W. Married.
Q. How long have you been married?
H.W. About four months.
Q. How many children do you have?
H.W. Four. (He laughs.) Not bad for four months!
Q. How old are your children?
H.W. The eldest is 32, his name is Bob, and the youngest is 22, his name is Joe.
Q. How did you get those children in four months?
H.W. They're adopted.
Q. Does this all sound strange to you, what you are saying?
H.W. (He laughs.) I think it is a little strange.

Patients such as H.W. confabulate for no apparent purpose. They simply recall mistaken facts, and when questioned they try to make sense of their recollections by adding facts that make the story more coherent. Recall from Chapter 4 Gazzaniga's theory of the interpreter and how split-brain patients confabulate to make sense out of discrepant information fed to each cerebral hemisphere. A dramatic example of confabulation occurs in *Capgras syndrome,* in which patients have the delusional belief that their family members have been replaced by impostors. No amount of evidence can convince them that their siblings, parents, spouses, and children are real. Patients with Capgras often have damage to the frontal lobes and limbic brain regions.

Repressed Memories Are Controversial

7.24 Daniel Schacter Schacter's research has provided numerous insights into the everyday "sins" of human memory.

One of the most vitriolic debates over the past few decades has centred on repressed memories. On one side are some psychotherapists and patients who claim that long-repressed memories for traumatic events can resurface during therapy. Recovered memories of sexual abuse are the most common repressed memories reported, and in the early 1990s there was a rash of celebrity reports of early childhood sexual abuse.

On the other side are memory researchers, such as Elizabeth Loftus, who point out that there is little credible evidence indicating that recovered memories are genuine, or at least sufficiently accurate to be believable. Part of the problem is best summarized by the leading memory researcher Daniel Schacter (Figure 7.24): "I know that some traumatic events can be associated with temporary forgetting and subsequent memory recovery. I am convinced that child abuse is a major problem in our society. I have no reason to question the memories of people who have always remembered their abuse, or who have spontaneously recalled previously forgotten abuse on their own. Yet I am deeply concerned by some of the suggestive techniques that have been recommended to recover repressed memories" (Schacter, 1996, p. 251).

Schacter alludes to the frightening possibility that false memories for traumatic events were implanted by well-meaning but misguided therapists. There is convincing evidence that methods such as hypnosis, age regression, and guided recall can implant false memories. Likewise, a growing body of evidence from carefully controlled laboratory studies has demonstrated that children can be induced to remember events that did not occur (Ceci & Bruck, 1995). Indeed, one of the seven sins of memory identified by Schacter is *suggestibility,* which is illustrated by illusory memories that occur when people incorporate information provided by others into their own recollections. The Harvard psychologist Richard McNally (2003) has conducted research with people who had recovered memories of childhood abuse. In the laboratory, these individuals are more likely than control participants to develop false memories in general, as in the Roediger-McDermott method described earlier.

There have been a few infamous examples in which adults have accused their parents of abuse based on memories that they later realize were the products of therapy rather than reality. For instance, Diana Halbrook came to believe not only that she had been abused, but also that she had been involved in satanic ritualistic abuse and that she had herself killed a baby. When she expressed doubts to her therapist about the veracity of these events, the people in her "support" group and her therapist told her she was in denial and not listening to "the little girl" within. After all, all of the other members of the support group had recovered memories that they had been involved in satanic ritualistic abuse. After Diana left her therapy group she came to believe that she was neither abused nor a killer. It is interesting that so many people recover memories of satanic ritualistic abuse, since "though thousands of patients have 'remembered' ritual acts, not a single such case has ever been documented in the United States despite extensive investigative efforts by state and federal law enforcement" (Schacter, 1996, p. 269).

What is a reasonable person to conclude about repressed memories? The repressed-memory debate understandably involves strong and passionate beliefs on both sides. Research shows us that some therapeutic techniques seem especially likely to foster false memories, but it would be a mistake to simply dismiss all adult reports of early abuse. It is certainly possible that abuse occurs and is forgotten until some later period of time, and it is a disservice to actual victims of abuse to ignore their memories. It is interesting to note that in the latter half of the 1990s, the incidence of recovered memories fell dramatically. However, it is not clear whether this was because of less media attention to reports, or because people were less likely to seek therapy to uncover their past memories.

People Reconstruct Events to Be Consistent

The final example of memory distortion to consider is memory *bias,* in which people's memories for events are altered over time to be consistent with current beliefs or attitudes. One of psychology's greatest thinkers, Leon Festinger, when writing a tribute to his colleague Stanley Schachter, hit the nail on the head: "I prefer to rely on my memory. I have lived with that memory a long time, I am used to it, and if I have rearranged or distorted anything, surely that was done for my own benefit" (Festinger, 1987, p. 1).

Consider people who take study-skills courses. In spite of only modest evidence that such courses are beneficial, often because students fail to heed the advice they are given, most students who take them describe them as extremely helpful. How can something that often leads to only modest outcomes be so positively endorsed? To understand this, Michael Conway and Michael Ross randomly assigned University of Waterloo students to a genuine study-skills course or to a control group that received no special training. Students who took the real course showed few signs of improvement; in fact, their final-exam performance was slightly poorer than the control group's. Still they reported the study-skills program as helpful. Why? The experiment had one feature that allows us to understand what is going on. At the beginning of the course, participants were asked to rate their studying skills. At the end of the course they once again rated themselves *and* they were asked to recall how they had originally rated themselves. In describing their earlier ratings, students in the study-skills course recalled themselves as significantly worse than they actually had been at the beginning of the course, thereby "getting what they want by revising what they had" (Conway & Ross, 1984).

People tend to recall their past attitudes and beliefs as being consistent with their current attitudes and beliefs, often revising their memories when they have a change in attitude. People also tend to remember an event so that it casts them in a prominent or favourable light. We tend to exaggerate our contribution to group efforts, to take credit for success and blame failure on others, and to remember our successes more than our failures, as you will discover in Chapter 15.

Not only individuals bias their recollection of past events; societies do so as well. The collective memory of groups can seriously distort the past, as expressed in the Orwellian idea that we can change only the past, not the future. Consider that the histories of most societies tend to downplay the unsavory, immoral, and even murderous behaviours of their pasts; perpetrators' memories are generally shorter than victims'. Is there an objective reality? As Leon Festinger observed, we construct our memories for our own purposes. Our memory for events tends to be highly consistent with our personal beliefs.

Profiles in Psychological Science

False Confessions and Memories

On September 11, 2001, Abdallah Higazy was sleeping in a hotel room when he was awoken by the attack on the World Trade Center across the street. He was quickly evacuated but had to leave his possessions behind. A few months later he was allowed to retrieve his belongings, but one of them did not belong to him—a radio scanner capable of intercepting airline communications. The FBI immediately arrested and interrogated Higazy. During the interrogation, Higazy felt threatened and believed his family would be at risk if he did not tell the FBI agents what they wanted to hear. In an interview on the television program *60 Minutes* in early 2004, he reported that he felt it would be better for his family if he confessed to owning the radio, even though it did not belong to him, and so he confessed (Figure 7.25). He told *60 Minutes,* "It was either I go to prison, or I go to prison and something happens to my family. So I chose I go to prison." Higazy was eventually released because it turned out that the radio belonged to an airline pilot and that a security guard had been confused about where the radio was originally located.

This case has a reasonably happy ending. But how often do people confess to crimes that they did not commit? Many researchers believe it happens more frequently than you might think. Even more surprising is that many times people come to believe their false confessions (Kassin, 2005; Kassin, Goldstein, & Savitsky, 2003). Can people really believe they committed a crime when they did not? How easy is it for people to develop false memories? Let's examine the evidence.

7.25 False Confessions Abdallah Higazy was jailed during the investigation into the September 11, 2001 attacks on the World Trade Center because he falsely confessed to owning incriminating evidence. Higazy later told reporters he had lied because he feared that his family would be threatened if he didn't tell FBI agents what they wanted to hear.

Think back to when you were five. Do you remember getting lost in a mall and being found by a kind old man who returned you to your family? No? Well, what if your family told you about this incident, including how panicked your parents were when they couldn't find you? According to research by Elizabeth Loftus, you may well remember the incident, even if it didn't happen.

In an initial study, a 14-year-old named Chris was told by his older brother Jim, who was part of the study, about the "lost in the mall" incident. The context was a game called "Remember when . . . ." All of the other incidents narrated by Jim were true. Two days later, when asked if he had ever been lost in a mall, Chris began reporting memories of how he felt during the mall episode; within two weeks he reported the following:

> "I was with you guys for a second and I think I went over to look at the toy store, the Kay-bee toy and uh, we got lost and I was looking around and I thought, 'Uh-oh. I'm in trouble now.' You know. And then I . . . I thought I was never going to see my family again. I was really scared you know. And then this old man, I think he was wearing a blue flannel shirt, came up to me. . . . [H]e was kind of old. He was kind of bald on top. . . . [H]e had like a ring of gray hair . . . and he had glasses." (Loftus, 1993, p. 532)

You might wonder if there was something special about Chris that made him susceptible to developing false memories, but in a later study, Loftus and her colleagues used the same paradigm to assess whether they could implant false memories in 24 participants. Seven of the 24 participants falsely remembered events that had been implanted by family members who were part of the study. Note that it is unlikely that false memories could be created for certain types of unusual events, such as receiving an enema (Pezdek & Hodge, 1999). However, a study by Saul Kassin and Katherine Kiechel (1996) found that people could even be induced to make false confessions of minor transgressions that they didn't actually commit. Moreover, their research participants truly believed that they had committed the transgressions. It is well established that children are particularly suggestible and that false memories, such as getting their fingers caught in mousetraps and having to go to the hospital, can easily be induced in them. When you ask people to imagine an event happening, they create of mental image of that event; it is possible that they later confuse their memory of their imagination with reality, and they come to believe that the imagined event really happened. Essentially, people have a source-monitoring problem: Did the event really happen, or did they imagine it? To Chris, the memory of being lost in the mall became as real as other events from childhood.

What about when people confess to crimes they did not commit? Police interrogation can be brutal; many times suspects are drunk or on drugs or otherwise have memory problems. In some instances, the police officers tell the suspects that there are eyewitnesses or other evidence proving their guilt. In the United States, it is perfectly legal for police officers to make these false claims during interrogation. When confronted with evidence, suspects come to believe that it is possible that they actually committed the crime. It is apparent that with repeated questioning over many hours and extremely high stress levels, false memories can be formed. Ironically, *innocence* may put the innocent at risk because they fail to anticipate being accused of the crime, and so make decisions that are not in their best interests, such as not asking for a lawyer (Kassin, 2005). Over the last few years there have been a number of high-profile cases in which DNA analysis showed that a person who had confessed to a crime could not have been the perpetrator. Thus, just because someone confesses to a crime doesn't mean that he or she actually committed the crime. Sometimes confessions are motivated out of fear, such as Abdallah Higazy worrying about his family. But sometimes people confess even when they did not commit the crime because a false memory has been implanted. Confessions need to be treated skeptically unless there is corroborating evidence. Understanding the nature of false confessions may help the judicial system ensure that people are punished only for their genuine misdeeds. ■

REVIEWING the Principles | How Are Memories Distorted?

Memory is far from a true and faithful recorder of objective facts and events. Rather, memory often includes a number of biases, distortions, and outright lies. People tend to make poor eyewitnesses because human memory is better for gist than for detail. Yet people maintain an unjustified confidence in their personal memories, such as flashbulb memories. Memories can be distorted or even implanted by false information, and there is a general bias toward maintaining consistency in our memories.

Conclusion

The brain is a wonderfully evolved instrument that has allowed humans to develop language, build civilizations, and visit the moon. However, we can't always remember where we left our keys, the faces of people whom we've witnessed commit crimes, whether a person is famous, or details from a book we read last year. It needs to be emphasized that memory has evolved to solve certain problems related to survival, such as remembering our friends and foes. Memory did not evolve to remember phone numbers, textbook definitions, or how to drive a car, but it does a pretty good job at all of these. Memory is a system that provides an organism with the ability to retain information and skills that are useful to survival. As such, it works pretty well, considering the amount of information that bombards the nervous system. Finally, as one of the most dynamic areas of research in psychological science, memory highlights our four fundamental themes, in particular the way that researchers are working at various levels of analysis to unwrap the mystery of the mind.

Summarizing the Principles of Memory

What Are the Basic Stages of Memory?

1. **Sensory memory is brief:** Visual (iconic) and auditory (echoic) memories are maintained at the sensory memory stage long enough to ensure a continuous sensory experience.

2. **Short-term memory is active:** Immediate active memory is limited. Chunking reduces information into meaningful units that are easier to remember. The three components of working memory are the central executive, the phonological loop, and the visuospatial sketchpad.

3. **Long-term memory is relatively permanent:** Long-term memory (LTM) is the potentially indefinite storage of all memories. Meaningful memories are stored in LTM.

What Are the Different Memory Systems?

4. **Explicit memory involves conscious effort:** Explicit, declarative memories that we consciously remember include personal events (episodic memory) and knowledge (semantic memory).

5. **Implicit memory occurs without deliberate effort:** Procedural (motor) memories of how to do things automatically are implicit.

How Is Information Organized in Long-Term Memory?

6. **Long-term memory is a temporal sequence:** Memory processes include encoding, storage, and retrieval. Elaborate rehearsal involves encoding information in more meaningful ways, and results in better memory than maintenance (repetition) rehearsal.

7. **Schemas provide an organizational framework:** Cognitive structures of meaning (schemas) aid the organization of memories.

8. **Information is stored in association networks:** Networks of associations are formed by nodes of information, which are linked together and are activated by spreading activation.

9. **Retrieval cues provide access to long-range storage:** According to the encoding specificity principle, any stimulus that is encoded along with an experience can later trigger the memory of the experience. The context of the memory is also activated.

What Brain Processes Are Involved in Memory?

10. **There has been intensive effort to identify the physical location of memory:** Unable to locate a specific site of memory storage (engram), Lashley concluded, somewhat inaccurately, that memory is distributed throughout the brain (equipotentiality). Research has revealed that a number of specific brain regions contribute to learning and memory.

11. **The medial temporal lobes are important for consolidation of declarative memories:** The hypothetical process of consolidation of new memories involves changes in neural connections. The hippocampus and rhinal cortex are regions of the medial temporal lobe that are important for declarative memories. Hippocampal place cells aid spatial memory.

12. **The frontal lobes are involved in many aspects of memory:** Frontal lobe activation occurs with deep encoding tasks. The medial prefrontal region is active for memories about the self. The left frontal region is more active during encoding and the right more active during retrieval. The frontal lobes may be especially important for working memory.

13. **Neurochemistry influences memory:** A group of neurochemicals modulate the storage of memories. Epinephrine enhances memory. The amygdala is probably responsible for memory modulation through activity in its norepinephrine receptors.

When Do People Forget?

14. **Transience is caused by interference:** Forgetting over time occurs because of interference from both old and new information.

15. **Blocking is temporary:** The tip-of-the-tongue phenomenon occurs when there is trouble retrieving the right word, usually from interference from a similar word.

16. **Absentmindedness results from shallow encoding:** Inattentive or shallow processing causes memory failure.

17. **Amnesia is a deficit in long-term memory:** Injury and disease can result in amnesia, the inability to recall past memories (retrograde amnesia) or the inability to form new memories (anterograde amnesia).

How Are Memories Distorted?

18. **Flashbulb memories can be wrong:** The strong emotional response that attends flashbulb memories may modulate the strength of a memory and affect accuracy.
19. **People make source misattributions:** People misremember the source of a memory (source misattributions). Cryptomnesia refers to when people think a new idea is theirs, when they have only retrieved an old memory.
20. **People make bad eyewitnesses:** Poor eyewitness recall occurs particularly with regard to ethnic identification. Suggestibility leads to misinformation.
21. **People have false memories:** Immature frontal lobes cause childhood amnesia. False memories can be implanted. Confabulation can occur due to brain damage.
22. **Repressed memories are controversial:** Some therapeutic techniques can result in false repressed memories.
23. **People reconstruct events to be consistent:** People try to maintain a consistency between past memories and their current attitudes and knowledge.

Applying the Principles

1. Frequently, people complain about problems with their "short-term memory" because they can't remember something that happened over the last few days. How does the information-processing model of short-term memory help us understand the problem?

    ____ a) The problem is not short-term memory but rather retrieval from long-term memory.
    ____ b) The problem is not short-term memory but rather a malfunctioning working memory.
    ____ c) The problem is that sensory memory is not picking up the information to transfer to short-term memory.
    ____ d) The problem is with the central executive function of long-term memory.

2. When you are studying for an exam, you are working with memories that are ________. When you are recalling this information while taking an exam, you are using the ________ memory system and the ________ memory sytem to fill in your response on the answer sheet.

    ____ a) implicit; declarative; declarative
    ____ b) explicit; procedural; motor
    ____ c) declarative; explicit; implicit
    ____ d) episodic; implicit; explicit

3. Is long-term memory simply stored in the logical manner of a library? To test this, think of the word *red.* Very quickly your mind will recall a variety of different associations, (for example, fire truck; crayon; "seeing red"; "red, white, and blue"; "red-letter day"). This demonstrates what important aspect of long-term memory?

    ____ a) semantic networks that link different memories in a hierarchy of personal familiarity
    ____ b) schema networks that connect general concepts
    ____ c) long-term scripts that connect all the different memories in a story
    ____ d) implicit memory networks

4. You are driving at high speed on the freeway you take home every day. Suddenly, a deer leaps from the side of the road and you hit it, resulting in a scary crash. After the accident, you are flooded with detailed memories about that section of the freeway, and you are able to recall details you had never remembered of that road before. This demonstrates

    ____ a) how motor memories are permanently processed by the hippocampus.
    ____ b) the role of the amygdala in tagging information so that we remember it better.
    ____ c) the role of high speed in memory processing.
    ____ d) the way the frontal lobe only processes the most important and emotional aspects of a memory.

5. Aaron is studying for an exam when his roommate mentions that Aaron's mom had called and asked to have Aaron call her later that day. Throughout the rest of the day, Aaron remembers he must call someone, but can't remember whom. The next morning, Aaron's mother calls, irate that he never returned her phone call. His response is "I forgot." It is likely that his memory failure occurred because of which of the following?

    ____ a) absentmindedness
    ____ b) blocking
    ____ c) transience
    ____ d) misattribution

ANSWERS: *1.a 2.c 3.a 4.b 5.c*

Key Terms

absentmindedness, p. 280
amnesia, p. 281
anterograde amnesia, p. 281
blocking, p. 279
confabulation, p. 286
consolidation, p. 271
cryptomnesia, p. 283
declarative memory, p. 262
elaborative rehearsal, p. 265
encoding specificity principle, p. 268
encoding, p. 264
episodic memory, p. 262
explicit memory, p. 262
flashbulb memories, p. 282
forgetting, p. 278
implicit memory, p. 263
long-term memory (LTM), p. 257
maintenance rehearsal, p. 265
memory, p. 252
modal memory model, p. 252
posttraumatic stress disorder (PTSD), p. 276
primacy effect, p. 257
proactive interference, p. 279
procedural memory, p. 263
recency effect, p. 257
repetition priming, p. 263
retrieval, p. 265
retroactive interference, p. 279
retrograde amnesia, p. 281
schema, p. 265
semantic memory, p. 262
sensory memory, p. 255
serial position effect, p. 257
short-term memory (STM), p. 255
source amnesia, p. 285
source misattributions, p. 283
spatial memory, p. 273
storage, p. 264
suggestibility, p. 284
tip-of-the-tongue phenomenon, p. 280
transience, p. 279
working memory, p. 256

Further Readings

Bjork, E. L., & Bjork, R. A. (1996). *Memory.* San Diego, CA: Academic Press.

Gazzaniga, M. S. (2004). *The cognitive neurosciences* (3rd ed.). Cambridge, MA: MIT Press.

LeDoux, J. E. (2002). *Synaptic self.* New York: Viking.

Neath, I. (1998). *Human memory: An introduction to research, data, and theory.* Pacific Grove, CA: Brooks-Cole.

Schacter, D. L. (1996). *Searching for memory: The brain, the mind, and the past.* New York: Basic Books.

Schacter, D. L. (2001). *The seven sins of memory: How the mind forgets and remembers.* Boston: Houghton Mifflin.

Squire, L. R., & Kandel, E. R. (1999). *Memory: From mind to molecules.* New York: Scientific American Library.

Further Readings: A Canadian Presence in Psychology

Habib, R., Nybert, L., & Tulving, E. (2003). Hemispheric asymmetries of memory: The HERA model revisited. *Trends in Cognitive Sciences, 7*, 241–245.

Hebb, D. O. (1949). *The organization of behavior*. New York: Wiley.

McIntyre, D. C., & Reichert, H. (1971). State-dependent learning in rats induced by kindled convulsions. *Physiology & Behavior, 7*, 15–20.

Milner, B. (1972). Memory and the temporal regions of the brain. In K. H. Pribram and D. E. Broadbent (Eds.), *Biology of memory* (pp. 215–234). New York: Academic Press.

Moscovitch, M. (1995). Confabulation. In D. L. Schacter (Ed.), *Memory distortions: How minds, brains, and societies reconstruct the past* (pp. 226–251). Cambridge, MA: Harvard University Press.

Overton, D. A. (1978). Basic mechanisms of state-dependent learning. *Psychophamacological Bulletin, 14*, 67–68.

Piggott, S., & Milner, B. (1993). Memory for different aspects of complex visual scenes after unilateral temporal- or frontal-lobe resection. *Neuropsychologia, 31*, 1–15.

Rosenbaum, R. S., Priselac, S., Kohler, S., Black, S. E., Gao, F., Nadel, L., & Moscovitch, M. (2000). Remote spatial memory in the amnesic person with extensive bilateral hippocampal lesions. *Nature Neuroscience, 3*, 1044–1048.

Sherry, D. F., & Schacter, D. L. (1987). The evolution of multiple memory systems. *Psychological Review, 94*, 439–454.

Tulving, E. (1977). Context effects in the storage and retrieval of information in man. *Psychopharmacological Bulletin, 13*, 67–68.

Tulving, E. (1987). Multiple memory systems and consciousness. *Human Neurobiology, 6*, 67–80.

Tulving, E. (2002). Episodic memory: from mind to brain. *Annual Review of Psychology, 53*, 1–25.

Tulving, E. T., Kapur, S., Craik, F. I. M., Moscovitch, M., & Houle, S. (1994). Hemispheric encoding/retrieval asymmetry in episodic memory: Positron emission tomography findings. *Proceedings of the National Academy of Sciences, 91*, 2016–2020.

Wells, G. L., Small, M., Penrod, S., Malpass, R. S., Fulero, S. M., & Brimacombe, C. A. E. (1998). Eyewitness identification procedures: Recommendations for lineups and photospreads. *Law and Human Behavior, 22*, 603–647.

Yuille, J. C., & Cutshall, J. L., (1986). A case study of eyewitness memory of a crime. *Journal of Applied Psychology, 71*, 291–301.

Thinking and Intelligence

On July 12, 2005, Kyle MacDonald of Montreal decided he wanted a house. His goal was simple, yet complex: purchase a house, but only using one red paper clip. MacDonald traded the paper clip for a fish pen, the pen for a doorknob, doorknob for Coleman stove, and so on. How could such an unrealistic dream possibly become reality? MacDonald's stroke of genius was realizing that personal value and monetary worth are not necessarily equal. This enabled him to make intelligent decisions that ultimately lead to his success, and he moved into his new home in Kipling, Saskatchewan on September 2, 2006. MacDonald's story is a prime example of what psychologists already understand about intelligence: it is a combination of being able to learn quickly, make reasonable decisions, and adapt to diverse challenges.

CHAPTER 8

Thinking and Intelligence

In the aftermath of 9/11, it was impossible not to think of the deaths of the passengers in the hijacked planes and the office workers in the World Trade Center and the Pentagon. After the two planes struck the World Trade Center, footage of the results of the hijackings was repeatedly broadcast, resulting in vivid images that most of us will never forget. Although the events of that day are unparalleled, there is another tragic consequence of these highly publicized hijackings that has only rarely been discussed.

The psychologist Gerd Gigerenzer of the Max Planck Institute proposed that low-probability events that are publicized and have dire consequences, such as the deaths associated with the hijackings of 9/11, can result in fears he calls *dread risks* (Gigerenzer, 2004). These dread risks, although born of uncommon events, can have a profound impact on reasoning and decision making. For instance, after 9/11 many people, especially Americans, felt some trepidation about flying. In fact, in October, November, and December 2001 airline revenues dropped by 20 percent, 17 percent, and 12 percent compared

Outlining the Principles

How Does the Mind Represent Information?

- Mental Images Are Analogical Representations
- Concepts Are Symbolic Representations
- Schemas Organize Useful Information about the Environment

How Do We Make Decisions and Solve Problems?

- People Use Deductive and Inductive Reasoning
- Decision Making Often Involves Heuristics
- Problem Solving Achieves Goals

How Do We Understand Intelligence?

- Intelligence Is Assessed by Psychometric Tests
- General Intelligence Involves Multiple Components
- Intelligence Is Associated with Cognitive Performance
- Genes and Environment Both Influence Intelligence

to those same months in 2000. Although fewer people were flying after 9/11, estimates for miles driven in the United States during those months increased by almost 3 percent and the number of trips taken by Canadians the following year was greater (Statistics Canada, 2002). It seems that North Americans preferred to drive after 9/11.

For those who opted to drive, we can imagine that the images of the hijacked planes prompted an avoidance of being in a circumstance that could have the same horrible result. However, traveling by car is not without its risks. The number of people who die in car accidents every year far exceeds the number who die in airline disasters, much less in hijackings. Given that there were more people on the road after 9/11, was there an impact on U.S. traffic fatalities? The answer appears to be yes. The number of traffic fatalities in the three months following 9/11 was significantly above average. In fact, it is estimated that an additional 350 people may have died by avoiding flying during those months. The number of passengers and crew that died in all four hijackings on 9/11 was 266. Overall, it appears more Americans died by avoiding flying in the three months following 9/11 than died aboard airplanes on that fateful day. You can still see these effects four years later in North America. Airfares are at historic lows and gasoline prices are at historic highs, yet more gasoline is being consumed on a monthly basis than ever before, while air travel still hasn't reached its pre–9/11 mark. Surely some of this is due to the growing popularity of sport utility vehicles, but at least part of it reflects people's preference for driving rather than flying.

When reasoning about the right choice to make, humans do not always weigh the actual probabilities of different actions. In fact, we can be heavily influenced by a number of factors that might not be considered rational, such as how prominent an event or image is in our mind. Psychologists are starting to identify some of the typical biases that enter into reasoning and decision making. Gerd Gigerenzer suggests that we should start educating the public about these biases in decision making after highly unlikely, tragic events. Publicizing some of the typical biases in reasoning that dread risks can create might prompt people to reconsider choices that could result in additional negative consequences. ■

In the preceding two chapters we discussed how we learn and remember information. Once we have that information, what do we do with it? This chapter is concerned with how we use information in the form of thinking as well as what it means

1884
Defining Intelligence Sir Francis Galton opens his Anthropometric Laboratory, where people can measure the sensitivity of their sensory systems, the property that Galton proposes is the basis for intelligence.

1904
Spearman's G Charles Spearman proposed that one factor, known as general intelligence or g, underlies all mental abilities. Raymond Cattell subsequently differentiates fluid from crystallized general intelligence.

1905
Intelligence Quotient Alfred Binet and Théodore Simon of France develop the first intelligence test, which is later modified by the American Lewis Termin. Wilhelm Stern of Germany develops the concept of intelligence quotient, or IQ.

1925
Insight Wolfgang Köhler observes an ape, after long contemplation, join two sticks together in order to reach a banana, which suggests that animals are capable of insight.

to think intelligently. Our thoughts guide much of our behaviour as we solve problems, make decisions, and try to make sense of events going on around us. Sometimes our thought processes lead to great ideas and creative discoveries, but sometimes they lead to bad decisions and regret. Yet even in the face of all these thought processes and biases, some people seem to be better at using information than others, which we can describe as being intelligent. In this chapter we consider the nature of thought, from how we represent it in our minds, to how we use it to solve problems and make decisions in our daily lives, to explaining differences in intelligence among people. We consider the central issue of whether there is one monolithic thing we can call intelligence or whether it is best considered as reflecting a range of possible talents. We will see that there are several different ways to think about what it means to be intelligent. For instance, some people are superb at math; others are gifted in language. Some people excel at artistic endeavors or athletics; others are smart in matters of business and everyday affairs. Yet overall there does appear to be some form of general intelligence in which people vary. We will see that being able to use information intelligently matters more and more as society becomes more complex. We will also see that cultural and environmental factors play important roles in determining people's levels of intelligence.

An important idea to keep in mind is that for the most part our thinking is adaptive. For instance, we develop rules for making fast decisions because daily life demands it. In his 2005 book *Blink* (subtitled *The Power of Thinking without Thinking*), science writer Malcolm Gladwell makes the case that the ability to use information rapidly is a critical human skill. When we encounter a person on a street who might pose a threat, we might change our route to avoid that person. Being able to instantly size up whether a person is trustworthy allows us to avoid harm. This can happen without our conscious awareness. Recall from Chapter 4 that unconscious cognitive processes can affect thought and behaviour. We will see that this is also true for decision making and problem solving.

Sometimes snap judgments have important consequences. Gladwell describes a firefighter who was leading his colleagues in an effort to put out a kitchen fire. The fire did not respond in the expected way when they sprayed water on it. Feeling something was wrong, the firefighter quickly ordered everyone to leave the building, which was a good thing because the fire was under them in the basement and the kitchen floor on which they were standing collapsed moments after they left. If he had stopped to figure out the problem, he and his colleagues might have been killed. His intuitive decision saved him and the others. Sometimes, however, snap decisions can be disastrous—as in the case of the police shooting of Amadou Diallo, which we considered at the beginning of this book. Throughout this chapter we will see that just as thinking can solve important problems, it can also produce faulty outcomes. How do we use information intelligently so that it improves our lives?

1930s
Problem Solving Researchers such as Abraham Luchins, Norman Maier, and Karl Duncker study cognitive factors involved in problem solving. They find that the inability to see a problem in new ways can prevent people from finding solutions.

1947
Decisions Are Rational John von Neumann and Oskar Morgenstern publish the seminal book *Theory of Games and Economic Behavior*, which presents the central ideas underlying utility-based models of rational decision making.

1956
Decisions Are Less than Optimal Nobel Laureate Herbert Simon introduces the notion of *satisficing*, the idea that human decision making is based on finding approximations to statistically optimal solutions.

1970s
Genes and Intelligence Researchers such as Arthur Jensen and Richard Herrnestein propose that intelligence has a genetic component. They spark controversy by suggesting that racial differences in intelligence may be due to genetics.

1973
Mental Representation Through studies of how people mentally rotate objects, Roger Shepard shows that people form mental images, or representations, of objects.

How Does the Mind Represent Information?

cognition Mental activity such as thinking or representing information.

analogical representation A mental representation that has some of the physical characteristics of an object; it is analogous to the object.

symbolic representation An abstract mental representation that does not correspond to the physical features of an object or idea.

Do thoughts consist of images or words?

What form does information take in our minds? What do your thoughts seem like to you? Do you see your thoughts as images? Or do you hear them as spoken words within your head? Cognitive psychology was originally based on the notion that the brain *represents* information, and that the act of thinking—that is, **cognition**—is directly associated with manipulating these representations. In this way, we use representations to understand objects we encounter in our environment. Representations are all around us. A local road map represents the streets and avenues in your town or city; a menu represents the food options at a restaurant; a photograph might represent your mother; your description of the street you live on is also a representation. The challenge for cognitive psychologists is to understand the nature of these everyday mental representations. You don't need a map if you have a mental representation of the local streets, and although you may like a picture of your mother, you can describe what she looks like without it. But what are the characteristics of *mental* representations? Are they like a picture or map, but only in your mind? Or are they abstract, like language?

We use two basic types of representations every day, *analogical* and *symbolic*, which most often correspond to images and words. **Analogical representations** have some characteristics of (and are therefore analogous to) actual objects, such as maps reflecting the physical layout of geography or family trees indicating the degree of relationship between relatives. By contrast, **symbolic representations** most often are words or ideas, such as knowing that the word *violin* stands for a musical object. Notice that there is little correspondence between what a violin looks or sounds like and the letters that make up the word *violin*. Symbolic representations are abstract and do not have a relationship to physical qualities of objects in the world. Both types of representations are useful to us in understanding how we think because they form the basis of human thought, intelligence, and the ability to solve complex problems of everyday life. Let's consider the different ways in which mental representations are characterized and used.

Mental Images Are Analogical Representations

Put down your book for a minute and think about a lemon. Welcome back. What were you thinking? In our mind's eye, we often appear to *see* visual images, which happens without consciously trying to do so. It is difficult to think about a "lemon" without having some sort of image come to mind that resembles an actual lemon, with its yellow and somewhat waxy, dimpled skin. Not surprisingly, several lines of evidence support the notion that representations take on such picturelike qualities. First, in a famous set

1977
Script Theory Roger Schank and Robert Abelson propose that much of human behaviour is guided by scripts, which are schemas that describe everyday action sequences.

1979
Prospect Theory Daniel Kahneman and Amos Tversky propose the theory, based on 30 years of research, that decision makers (1) use points of reference, and (2) tend to give more weight to potential losses than to potential gains.

1983
Multiple Intelligences Howard Gardner expands the traditional definition of intelligence to recognize that people show intelligence in different ways.

1990s
Evolved Decision Making Gerd Gigerenzer argues that decision making is best understood by considering how humans have solved problems over the course of human evolution.

2000s
Intelligence and the Brain The frontal lobes are identified as being particularly important for fluid general intelligence, further supporting the link between intelligence and efficient executive functions, especially working memory.

of studies by Roger Shepard and colleagues in the early 1970s, participants were asked to view letters and numbers and to determine whether the given object was in its normal orientation or was a mirror image. The objects were presented in a variety of different rotated positions, such that sometimes the object was upright, sometimes it was upside down, and sometimes it was rotated somewhere in between (Figure 8.1). What Shepard and one of his colleagues found was that the length of time subjects took to determine whether an object was normal or a mirror image depended on its degree of rotation (Cooper & Shepard, 1973). The farther the object was rotated from the upright position, the longer the discrimination took, with the longest reaction time occurring when the object was fully upside down. From this evidence the researchers concluded that, in order to perform the task, participants mentally rotated representations or *images* of the objects in their heads in order to "view" the object in its upright position. Presumably, the farther the object was from the upright, the longer the task took, because more "rotation" of the representation was required. A related set of studies by Stephen Kosslyn and colleagues demonstrated a similar pattern using complex mental images (see Figure 8.2). But are all representations of objects analogical? Couldn't they also be simple representations based on factual knowledge about the world: lemons (1) are yellow and (2) have dimpled, waxy skin (Pylyshyn, 1984)?

Stephen Kosslyn, Martha Farah, and others have gone beyond Shepherd's original studies, using the tools of cognitive neuroscience to demonstrate that thoughts can indeed take the form of mental images. To make their case, Kosslyn and others have had to show that visual imagery is associated with activity in perception-related areas of the brain (that is, the primary visual cortex). If so, the argument goes, then these areas are likely responsible for providing the spatial aspects, such as size and shape, of analogical visual imagery. This is

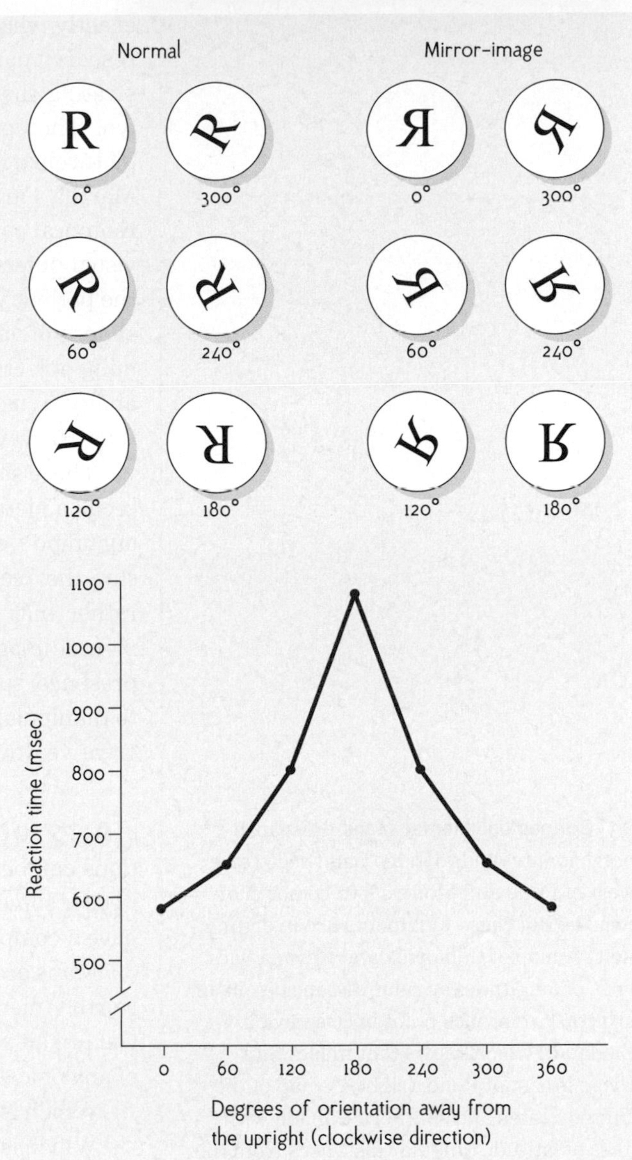

8.1 Mental Rotation The time it takes to determine whether the "R" is normal or a mirror image increases with the amount of rotation away from the upright letter.

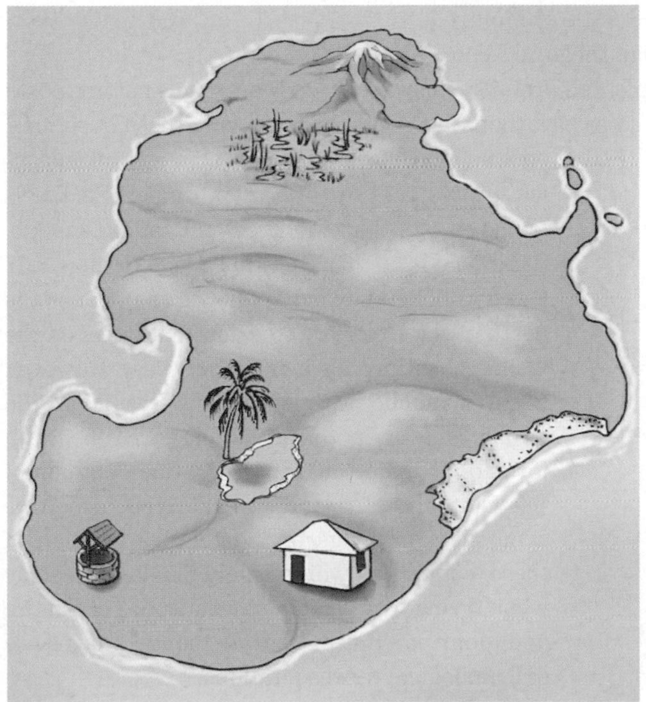

8.2 Kosslyn's Maps An example of the kind of maps used in Kosslyn's study. In this study, participants were required to mentally scan images of maps that they had committed to memory. The maps (of fictitious islands) included a number of different landmarks such as towns, lakes, and mountains. Participants were asked to visualize a landmark on the island and then, following a brief interval, to imagine a dot moving from this landmark to a second landmark located elsewhere on the map. The researchers found that the time it took to imagine shifting one's gaze between two points on the map increased with the actual distance between points.

exactly what Kosslyn and his colleagues found in a brain-imaging study in which research participants were asked to recall images they had just memorized, demonstrating that analogical representations activate the primary visual cortex (Kosslyn, Thompson, Kim, & Alpert, 1995). Additional evidence for the existence of picturelike imagery also comes from studies of brain-injured patients. In particular, Martha Farah and her colleagues (1988) examined a patient with damage to the temporal cortex, an area of the brain associated with processing the appearance of visual objects. Farah and colleagues found that, in comparison to normal subjects, the patient was deficient in calling up mental images of objects, such as animals and shapes of various U.S. states, but not at all impaired in spatial tasks, such as imagining objects rotating or comparing the relative locations of imagined objects. This ability to use spatial information is tied to the maturation of a child's nervous system (Harvey et al., 1986).

These studies show that when we retrieve information from memory, as when we recall a picture we recently saw in a newspaper, the representation of that picture in our mind's eye parallels the representation that was in our brain the first time we saw the picture. This doesn't mean the mental image is perfectly accurate, but rather that it corresponds generally to the physical object it represents. By using mental images, we are able to answer questions about objects that are not in our presence, such as when we visualize a lemon and can describe its colour. Being able to manipulate mental images allows us to think about our environment in novel and creative ways, which as we will see can help us solve problems.

LIMITS OF ANALOGICAL REPRESENTATIONS Although mental representations can be analogical, the range of knowledge we can represent in this way is limited. If something cannot be wholly perceived by our perceptual system, we cannot have a complete analogical representation of it.

Maps are an interesting case. It is possible to have a visual image of a map, say, of North America, but the landmass itself can't possibly be viewed (unless, of course, you are an astronaut hundreds of miles above earth). Mental maps involve a mixture of analogical and symbolic representations. Consider the following questions:

Which is farther north: Seattle or Montreal?

Which is farther east: San Diego or Reno?

If you are like most people, you probably answered that Montreal is farther north than Seattle and Reno is farther east than San Diego. In answering these questions, you may have formed a mental image, or analogical representation, of a map. In fact, Seattle is farther north than Montreal and, oddly enough, San Diego is farther east than Reno (Figure 8.3). Our symbolic knowledge tells us that Canada is north of the United States and that cities on the Pacific coast are always farther west. However, these symbolic representations result in the wrong answer in these instances. Even if you formed an analogical representation of a map of the United States and Canada, you also have symbolic knowledge of the relative location of these countries, which might lead you astray when trying to determine the latitudes of Seattle and Montreal. But inaccurate mental images are still analogical representations.

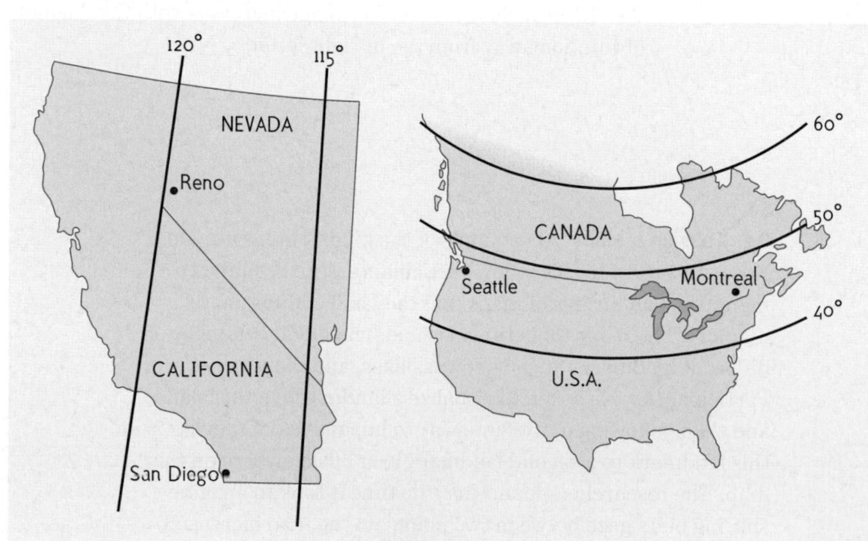

8.3 Conceptual Mental Maps Research participants tend to judge San Diego to be west of Reno and Montreal to be north of Seattle. But these judgments are in error. (left) A map of California and Nevada with lines of longitude (angular distance from an arbitrary reference point in Greenwich, England), which shows that in fact San Diego is east of Reno. (right) A map of the United States and southern Canada with lines of latitude (angular distances from the equator), which shows that Seattle is slightly north of Montreal.

Concepts Are Symbolic Representations

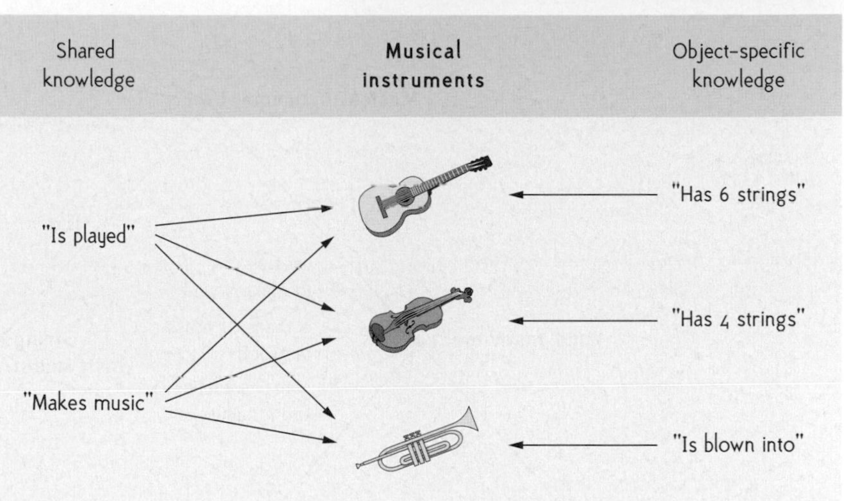

8.4 Categorization Shared knowledge helps us group objects into categories.

Much of our thinking reflects our general knowledge about objects in the world rather than simply their visual representations. Our symbolic representations consist of words and abstract ideas. They provide information. Picturing a lemon doesn't tell us what to do with it. But knowing that parts of it are edible helps us decide how to use it, for instance, to sprinkle its juice on a nice piece of fish. Thus, what you do with a lemon depends on how you think about it. One question of interest to cognitive psychologists is how we use knowledge about objects efficiently. Recall from Chapter 7 that our memory systems are organized so that we can quickly call up information as needed. The same principle is true for how we think about objects. For instance, what is a violin? Most people would respond that it is a musical instrument, a member of a group of objects that share common characteristics. Grouping things together based on shared properties, known as *categorization,* reduces the amount of knowledge we must hold in memory and is therefore an efficient way of thinking. For instance, if we understand the category "musical instruments"—objects that produce music when played—we can automatically apply this gross definition to all members of the category. As a result, we don't have to store this same bit of knowledge over and over for each musical instrument that we are aware of (Figure 8.4). However, we do have to store unique knowledge for each member of a category, such as "has six strings" for a guitar and "has four strings" for a violin.

Concept refers to a class or category that includes some number of individuals or subtypes. Concepts can be mental representatives of categories, such as musical instruments, fruits, or bachelors. Concepts can also be mental representations of relations, such as "smaller than" or "heavier than," as well as qualities or dimensions, such as "brightness" or "width." Concepts allow us to organize mental representations around common themes, ensuring that every instance of an object, a relation, or a quality does not need to be stored individually and allowing the abstract representation of knowledge with shared similar properties. How do we use concepts to understand the world?

How do we form concepts?

Consider the concept of a bachelor. It is likely that if asked you would be able to indicate, for most of your male acquaintances, whether or not each is a bachelor. The definition of *bachelor* is clear in the dictionary: an unmarried male. The notion that concepts are formed by defining attributes is called the **defining attribute model**. According to this model, each concept is characterized by a list of features that are necessary to determine if an object is a member of a category. For *bachelor*, the defining attributes would include "unmarried" and "male"; for musical instruments, attributes would include something that "can be played" and "makes music." These definitions provide a clear means of determining if an individual or object belongs to a category. Moreover, the concepts are organized hierarchically, such that they can be superordinate or subordinate to each other. For example, horns and stringed instruments would be subordinate categories of the superordinate category of musical instruments (Figure 8.5).

concept A mental representation that groups or categorizes objects, events, or relations around common themes.

defining attribute model The idea that a concept is characterized by a list of features that are necessary to determine if an object is a member of a category.

Although the defining attribute model is intuitively appealing, it fails to capture many key aspects of how we organize things in our heads. First, the model suggests

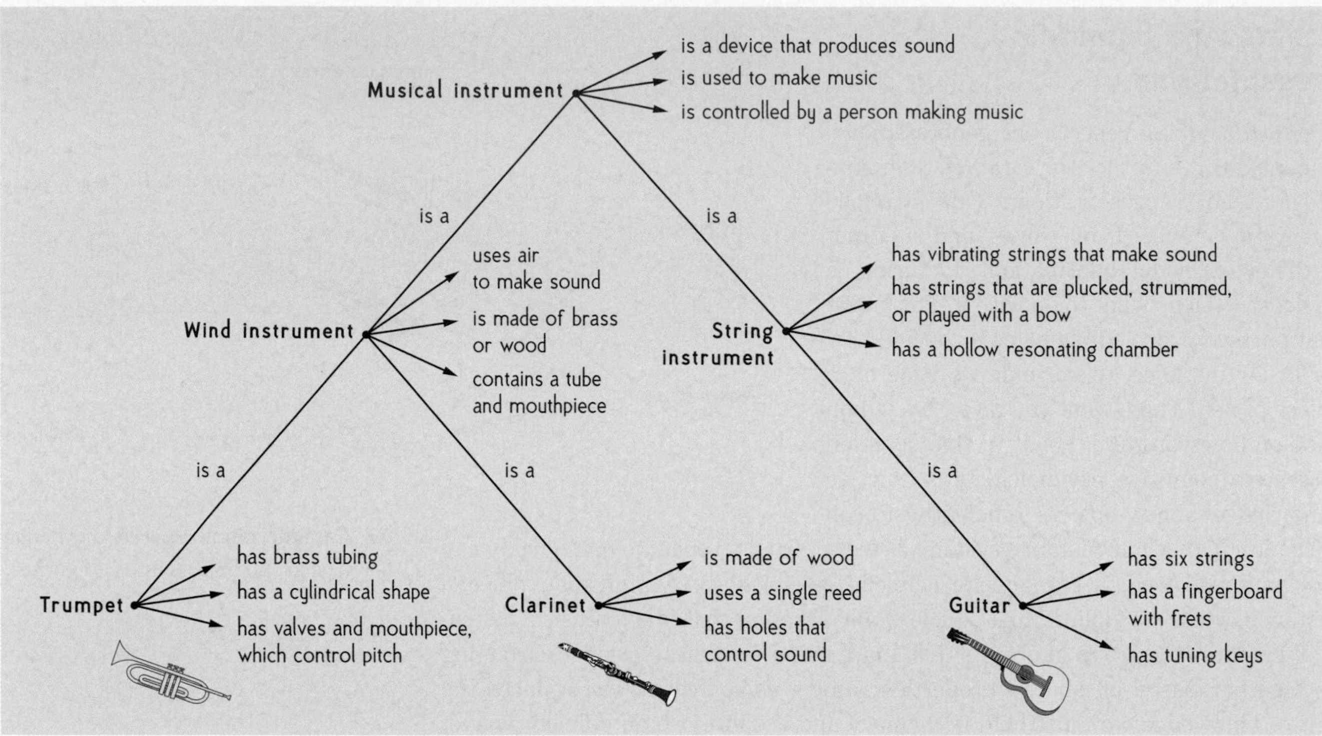

8.5 The Defining Attribute Model A schematic diagram of the defining attribute model.

that membership within a category is on an all-or-none basis, but in reality we often make exceptions in our categorizations, letting members into groups even if they do not have all the attributes or excluding them even if they have all the attributes. For instance, most people would include "can fly" as an attribute of *bird*. However, some birds, such as penguins, don't fly, yet they are still classified as birds. In addition, some people use spoons to play music. Even though they can be used to play music, it would be unusual to categorize spoons as a musical instrument. Second, the defining attribute model suggests that all attributes of a category are equally salient in terms of defining the given category. However, research demonstrates not only that some attributes are more important for defining membership than others, but that the boundaries between categories are much fuzzier and more indistinct than the defining attribute model suggests. For example, "has wings" is generally considered a clear attribute of a bird, whereas "is warm-blooded" does not as readily come to mind when we think of birds. Third, the model posits that all members of a category are equal in category membership—no one item is a better fit than any other. Consider, again, the concept of bachelor. According to the definition, a 16-year-old boy, a man who has been in a committed relationship for 25 years (but never married), and a man in his 20s who goes on dates a few nights a week would all be equal exemplars of the category *bachelor.* Do they seem the same to you?

prototype model An approach to object categorization that is based on the premise that within each category, some members are more representative than others.

To address the shortcomings of the defining attribute model, we often take a more flexible approach to object categorization that is based in **prototype models**. The basic premise of this model is that within each category, some members of a particular category are more representative or *prototypical* of that category than other members (Figure 8.6). A man in his 20s who dates a lot is more prototypical of a bachelor than an unmarried man in a 25-year committed relationship or a 16-year-old boy, although each fits the technical definition. The benefit of recognizing our tendency to view categories as having prototypical exemplars is that it closely resembles how we often organize our knowledge of objects. This tendency recognizes that not all members of a category will have the same attributes. Moreover, prototype models allow the boundaries between categories to be imprecise. For instance, tomatoes are often classified as a vegetable (used in salads), even though their attributes are also consistent with the fruit category (have seeds, are sweet).

Although prototype models allow for some additional flexibility in the representation of concepts, they do not always provide a clear indication of what a prototype representation would be like. Is a prototype the most common example of category? Is it a representation that all category members most resemble? Or is it a list of typical attributes? To address some of these concerns, *exemplar models* have been proposed. The premise behind exemplar models is that there is no single best representation of a concept. Instead, all of the examples, or *exemplars,* of category members form the concept. For example, your representation of dogs is made up of the hundreds of dogs you have encountered in your life. If you see a new animal in your yard, you would compare this new animal with the memories of other animals you have encountered. If this animal most closely resembles the dogs you have previously encountered (as opposed to the birds, squirrels, or other animals and objects), you might conclude it is a dog. Exemplar models assume that experience forms fuzzy representations of concepts because, in essence, there is no single representation. Exemplar models can account for the observation that some category members are more prototypical than others. Prototypical category members are simply those that we have encountered more often.

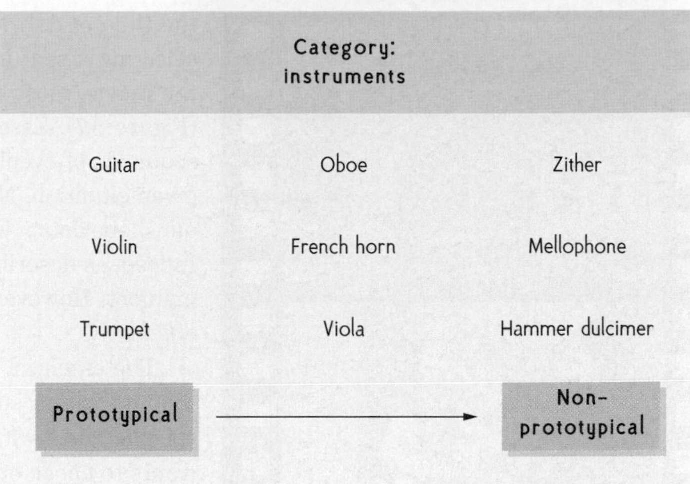

8.6 The Prototype Model Some items within a group or class are more representative (or prototypical) of that category than other members.

So far we know how we classify objects we encounter as well as how we represent those objects in the mind. Now we need to consider how we put this information to use in our daily lives.

Schemas Organize Useful Information about the Environment

How we think about the world extends well beyond a simple list of facts about the specific items we encounter every day. Rather, a whole different class of knowledge, called *schemas,* enables us to interact with the complex realities of our daily environments. Recall from Chapter 7 that schemas help us perceive, organize, and process information. As we find ourselves in different real-world settings, we draw on knowledge of what objects, behaviours, and events apply to a given setting in order to act appropriately. For example, at a casino blackjack table it is appropriate to squeeze in between the people already sitting down. However, if a stranger tried to squeeze into a group of people dining together in a restaurant, the reaction of the group would likely be quite negative. Obviously, this kind of knowledge regarding situations and social contexts is much different from the knowledge associated with object classification.

How does your schema for "student" affect your behaviour?

The basic view of researchers is that over time, we develop schemas about the different types of real-life situations that we encounter. One of the more prominent theories in this domain has focused on schemas about the sequences of events that occur in different situations. The researchers Roger Schank and Robert Abelson (1977) referred to these schemas about sequences as *scripts.* For example, "going to the movies" is a script that many people may have in common. First, we would expect to pay a fee to enter the movie, with the cost depending on age and, possibly,

How do scripts guide behaviour?

8.7 **Script Theory** This theory suggests that we tend to follow general scripts of how to behave in particular settings.

the time of day. Next, we might opt to buy some popcorn, candy, or soda prior to selecting a seat in the theater. Although quiet talking might be appropriate before the movie, most of us would expect talking to cease once the main feature begins (Figure 8.7). Essentially, scripts allow us to make a series of inferences about the sequence of events that arise in daily situations so that we know how to act in any given situation. Note that these scripts operate at the unconscious level in guiding our behaviour; we follow them without consciously knowing we are doing so. Indeed, as described in Chapter 7, sometimes our schemas can lead to distortions in memory. However, for the most part they help us interact with the external world efficiently.

The essential elements of schemas are that: (1) common situations have consistent attributes, such as a library's being quiet and having books; and (2) people have specific roles within the situational context, such as a librarian or a reader who wants to check out a book. Unfortunately, schemas and scripts can sometimes have unintended consequences, such as reinforcing sexist or racist beliefs (see Chapter 15). When asked to draw a "scientist," many young girls draw a picture of a male (Chambers, 1983). In *Blink,* Gladwell notes that until musicians began performing auditions behind screens, men were invariably chosen for principal positions because most conductors believed that women were incapable of playing as well as men. The conductors' schemas of women as inferior musicians interfered with their ability to rate auditioners objectively when they could observe the sex of the musicians.

Scripts dictate "appropriate" behaviours. What we view as appropriate is shaped by culture. Until recently, assuming it has changed, the script for going out on a date involved the male driving the car and paying for dinner. In the 1950s, the script for a black person getting on a bus in the southern United States involved going to the back of the bus. The schemas and scripts that children learn are likely to affect their behaviour when they are older. In one recent study, children aged two to six were asked to use props and dolls to act out a social evening for adults (Dalton et al., 2005). As part of the role play, each child selected items from a miniature grocery store, stocked with 73 different products, including beer, wine, and cigarettes, for an evening with friends. Surprisingly, two of the most common items purchased were alcohol (by 62 percent) and cigarettes (by 28 percent). Children whose parents smoked or consumed alcohol were four times more likely to select them. Children were asked about the items they chose, and it was clear that alcohol and cigarettes were included in the children's scripts for adult social life. One four-year-old girl who selected cigarettes exclaimed: "I need this for my man. A man needs cigarettes." These examples highlight the need for us to think critically about whether our automatic beliefs and actions reflect the values we wish to hold, which we consider in greater detail in Chapter 15. By recognizing the bias in evaluating female musicians, orchestras were able to address the problem by holding auditions with the musicians hidden behind screens and their names are withheld from the judges. Since these methods were instituted, the number of women in top U.S. orchestras has increased fivefold.

Given the potential problems with scripts and schemas, it is important to consider what they do for us. The adaptive value of schemas is that they minimize the amount of attention required to negotiate within a familiar environment. They also allow us to recognize and avoid unusual or dangerous situations. Mental representations in all forms assist us in using information about objects and events in adaptive ways. Being able to manipulate our mental representations—that is, to think about objects, events, and circumstances—allows us to take appropriate actions, make intelligent decisions, and function efficiently in our daily lives.

REVIEWING **the Principles** | How Does the Mind Represent Information?

Our thoughts consist of mental representations of objects and events in our environment. When we think of an object, we often bring to mind a visual image, which is an analogical representation of the object. It is not the same as the object; rather, the features of the representation correspond in some meaningful way to the features of the object. By contrast, symbolic representations do not correspond to the physical features of the object, but rather to our knowledge of the object. Much of our knowledge of the world is based on concepts, which are ways of classifying objects. Concepts are based on defining attributes, prototypes, and exemplars. Schemas assist in organizing information so that it is useful in our daily lives, such as understanding what script to follow in a restaurant.

How Do We Make Decisions and Solve Problems?

We have discussed the manner in which we represent and organize knowledge of the world in the mind. But how do we use knowledge to guide our daily actions? Throughout each day we make decisions, from what to eat for breakfast, to which clothes are most appropriate for the weather, to which route to take to work or school. We scarcely notice making many of these decisions. Sometimes decisions are much more consequential and require greater reflection, such as buying a house, selecting a college, or committing to a relationship partner. We also solve problems, such as figuring out a strategy for selecting the most appropriate car to meet our needs or identifying the best way to study for particular exams. Thinking enables us to do all of this.

One question that arises is whether *more* thinking is associated with better decisions. In *Blink,* Gladwell argues that decisions made quickly can often equal or even surpass decisions made following considerable reflection and deliberation. Yet this seems to fly in the face of many of our beliefs about the value of using our intellect. We are suspicious of quick decisions because we find it hard to believe that people can make good decisions without a rational consideration of all possible information. Throughout this section we will see that both conscious and unconscious mental processes are important for decision making and problem solving. As we learned in Chapter 4, much of the time we operate quite effectively on automatic pilot, with little conscious awareness of how we are interacting with the world around us. But we also sometimes need to make deliberative choices that require a more careful weighing of our options.

Why might being deliberative be better than making hasty decisions?

The ability to have rational thought and use it to guide decisions and actions is considered a fundamental characteristic of human cognition. People have pondered the nature of human reasoning since the days of the early Greek philosophers. In fact, Aristotle suggested that rational thought might be the defining characteristic that separates humans from other animals. In the last half century, cognitive psychologists have tested models derived from philosophical approaches of reasoning and decision making and have discovered that human behaviour, at times, diverges from what might be considered the most expedient or logical approach. Although more complete models encompassing how humans act outside the laboratory remain to be developed, exploring the work done so far provides an excellent window into understanding some of the most complex aspects of human cognition.

People Use Deductive and Inductive Reasoning

How do we evaluate evidence to draw conclusions? A police detective sifts through all the clues and tries to identify the right suspect. A psychological scientist analyzes data to see if they support or refute a given hypothesis. A parent listens to conflicting stories from two children and tries to figure out who broke the vase. Each situation requires *reasoning,* which refers to evaluating information, arguments, and beliefs in order to draw conclusions. Psychological scientists generally distinguish two classes of reasoning. *Inductive reasoning*, which we will discuss shortly, involves reasoning from the specific to the general, such as deciding that Vancouver is a friendly city because the five people you met from there were nice. *Deductive reasoning,* on the other hand, involves reasoning from the general to the specific, such as expecting a person from Vancouver you meet to be friendly. Both can be altered by psychopathology (Pelissier & O'Connor, 2002) as we will see in Chapter 14.

deductive reasoning A form of reasoning in which logic is used to draw a specific conclusion from given premises.

DEDUCTIVE REASONING In **deductive reasoning**, the task is to determine if a conclusion in a specific case can be drawn or *deduced* from a set of more general initial premises. Imagine that you are deciding on a place to go for dinner and your friend Bonnie tells you that the new Thai restaurant is excellent. Would you pick that restaurant? It obviously depends on whether you like Thai food and whether you think that Bonnie has good taste. If you do, then choosing her recommendation might be an excellent decision. Deductive reasoning uses logic to draw specific conclusions under certain assumptions, or *premises.* Deductive reasoning tasks are often presented in the form of syllogisms, which are logical arguments containing premises and a conclusion. Thus, in a *conditional syllogism*, the argument is in the form: *if A is true, then B is true.* If Bonnie has good taste, then the Thai restaurant she recommends will have delicious food. If you like Thai cuisine, then you will enjoy your meal. The conclusion of an argument is conditioned on whether its premises are true.

A common deductive reasoning task takes the form of a *categorical syllogism,* in which the logical argument contains two premises and a conclusion, which can be determined to be either valid or invalid. Categorical syllogisms take the form:

All A are B (Queen Elizabeth is a person)
All B are C (All people are mammals)
Therefore, all A are C (Therefore, Queen Elizabeth is a mammal)

This is an example of a valid conclusion in that it logically follows from the premises. The following syllogism would be considered invalid:

All A are B (All people from Vancouver are friendly)
Some B are C (Some friendly people like Thai Food)
Therefore, all A are C (Therefore, all people from Vancouver like Thai food)

In this case, the conclusion is invalid because there are some A that might not be C. After all, some friendly people might not like Thai food and it is therefore impossible to determine from the information provided if a person from Vancouver likes Thai food. Knowing a person is from Vancouver, even accepting the unlikely premise that all people from Vancouver are friendly, does not help us deduce their culinary preferences. Deductive reasoning allows the reasoner to determine the truth of a statement given the premises. It is possible, however, that the reasoner can come up with an incorrect but valid conclusion if the premises are incorrect. Consider the following:

Nothing is better than a piece of warm apple pie.
A few crumbs of bread are better than nothing.
Therefore, a few crumbs of bread are better than warm apple pie.

The ambiguity of the word *nothing* causes a logical error in this syllogism. This form of deductive reasoning can have other problems. Although we often assume that the principles of deductive reasoning should apply equally in all circumstances, research indicates that schemas about typical events and situations can influence performance on reasoning tasks (Klauer, Musch, & Naumer, 2000). Our reasoning is often influenced by our prior beliefs. For instance, if you were told that *all foods made of spinach are delicious* and that *the cake is made of spinach,* you could come to the valid conclusion that *the cake is delicious,* but you might have your doubts. In other words, it might be harder to agree that the statement *the cake is delicious* is valid if you know it is made of spinach—an ingredient not ordinarily used in baking cakes.

How do past experiences affect deductive reasoning?

INDUCTIVE REASONING Although in some situations we can determine the validity of a conclusion about a specific instance based on general premises, a more common reasoning problem is to determine general principles from specific instances. Suppose you arranged to meet a new friend for lunch, and your friend was late. In this one instance, you might conclude that special circumstances resulted in your new friend's tardiness. However, as you spend more time with this friend and arrange more occasions to get together, you notice that more often than not your friend is late. After a number of such instances, you might expect your friend to be late because you have *induced* the general conclusion that your friend is usually tardy. Determining general conclusions from specific instances is the basis of **inductive reasoning**. Using the scientific method to discover general principles relies on inductive reasoning. If a researcher has a hypothesis that students involved in school clubs perform better academically, she might select a random sample of students, half of whom participate in school clubs and half of whom do not, and assess their grade point average. Finding that the students who are in school clubs have a significantly higher GPA would lead to the general conclusion that, overall, students who participate in school clubs perform better academically. In other words, she would induce a general principle from the specific instances of the students in her experiment.

inductive reasoning A form of reasoning in which we develop general rules after observing specific instances.

As we discussed in Chapter 2, the scientific method dictates that scientists meet certain standards when inducing general principles from several specific instances. These standards are designed to guard against biases in inductive reasoning. In day-to-day life, however, we have no standards to guard against reaching inappropriate conclusions when we are reasoning about general principles from everyday circumstances. So, for instance, you might be considering buying a type of car that was highly rated by a consumer magazine. But when you mention your plans, a friend proclaims, "My uncle drove one of those and it broke down all the time." Although this single report should have little weight on the purchasing decision, some people are highly influenced by such anecdotal reports. Let's examine how people make decisions and how errors in reasoning can affect the quality of those decisions.

Decision Making Often Involves Heuristics

Closely linked to the ability to reason is the ability to decide. Reasoning allows the decision maker to form beliefs or conclusions. Decision making involves putting these beliefs into action by choosing among options. In order to decide, we must often reason about the options and weigh their relative value.

Research on decision making has been influenced by *normative* approaches and *descriptive* approaches. Historically speaking, normative models of decision making have viewed humans as optimal decision makers, whereas more recent descriptive models have tried to account for the tendencies humans have to misinterpret and misrepresent the probabilities underlying many decision-making scenarios. They focus on understanding how humans actually make decisions in everyday practice—decisions that often fail to comply with the predictions of "rational" behaviour. Moreover, as we noted, we need to make many decisions quickly, without time to consider all the possible pros and cons to each option. Thus, processes that allow us to make decisions quickly are often useful for dealing with real-world challenges.

Why do people not always make decisions rationally?

In 1947, John von Neumann and Oskar Morgenstern presented a normative model of how humans should make decisions—a model known as *expected utility theory*. The theory breaks down decision making into a computation of *utility,* an indication of overall value, for each possible outcome in a decision-making scenario. Expected utility theory proposes that decisions ultimately boil down to a consideration of possible alternatives, with people always choosing the most desirable alternative. To arrive at the most desirable alternative, we must first rank the alternatives in order of preference. This ordering of alternatives allows us to determine whether each alternative is more desirable, less desirable, or equally desirable compared to each competing alternative. But do we always choose the most desirable alternative?

One of the major findings in the field of cognition is that humans are far from perfectly rational. In the 1970s, Amos Tversky and Daniel Kahneman spearheaded descriptive research on decision making as they sought to examine the rules that people actually use in their everyday decisions. Because of the importance of reasoning and decision making in economic theory, this research was eventually awarded the Nobel Prize in economics. Tversky and Kahneman identified several common **heuristics**, which are mental shortcuts or rules of thumb, that people typically use during inductive reasoning and decision making. Heuristic thinking often occurs at the unconscious level; we are not aware of taking these mental shortcuts as we go through our day. Indeed, using heuristic processing allows us to focus our attention on other things, which is an important part of its usefulness. The processing capacity of the conscious mind is limited. You can think about only so many things at once, and it is difficult to do too many things at the same time, such as driving in heavy traffic and using a cellular phone (which is why it is now illegal in some places). Heuristics are valuable because they require minimal cognitive resources.

heuristics In problem solving, shortcuts used to minimize the amount of thinking that must be done when moving from step to step in a solution space.

In terms of decision making, heuristic thinking can be adaptive because it allows for quick decisions rather than weighing all of the evidence each time. One of our colleagues always buys the second-cheapest item, no matter what he is buying, which he believes saves money and avoids purchasing the absolute worst products. Other people follow the heuristic of buying only brand names. These quick rules of thumb are often enough to provide reasonably good decisions. However, Tversky and Kahneman demonstrated that heuristics can also result in specific biases, which may lead to errors or faulty decisions. Consider the heuristic that "expensive equals high quality," which many people believe to be true. Although laboratory studies show that one type of soap is basically as good as any other, attaching a fancy French name and a large price tag has convinced a great number of consumers to buy really

When are heuristics adaptive?

expensive soap. Let's examine some of the most common heuristics that people use during inductive reasoning and decision making.

AVAILABILITY HEURISTIC Is it more common for the letter *R* to be the first letter in a word, or the third letter? If you were asked this question, how would you go about determining an answer? If you are like most people, you might try to think of as many words as you can with *R* as a first letter (for example, *right* or *read*) or *R* as a third letter (for example, *care* or *sir*) and come to the conclusion that *R* is more often the first letter in a word. However, *R* is much more likely to be the third letter of a word.

Why do people have difficulty with this type of reasoning? This error in reasoning occurs because it is much easier to think of words beginning with *R* than words with *R* as the third letter. When we try to determine an answer, the most readily available instances are those with *R* as a first letter. Making a decision based on the most available answer—the one that comes most easily to mind—is known as the **availability heuristic**. Although this heuristic can be useful because information that is important comes most easily to mind, it can also create a bias in reasoning and result in the wrong conclusion.

availability heuristic Making a decision based on the answer that most easily comes to mind.

representativeness heuristic A rule for categorization based on how similar the person or object is to our prototypes for that category.

Using the availability heuristic in real-world circumstances can lead to significant consequences. As mentioned at the beginning of the chapter, when there is a major airplane disaster, such as the 9/11 hijackings, people who are planning to travel may be biased to drive rather than fly. Even though statistics indicate that the likelihood of dying in a car accident far exceeds that of dying in a plane crash, the salient and publicized examples of airline disasters make them more noticeable when people are reasoning about the relative safety of these two modes of travel. In the time period following 9/11 in particular, the availability heuristic created a bias in reasoning that led more people to drive on their vacations and trips. This ultimately resulted in an increase in the overall number of traffic fatalities.

REPRESENTATIVENESS HEURISTIC Helena is an intelligent, ambitious woman who describes herself as scientific-minded. She enjoys working on mathematical puzzles and talking with other people. She is an avid reader and gardener. Do you think that Helena is more likely a cognitive psychologist or a postal worker? If you are like most people, you would guess that Helena is a psychologist because these characteristics seem to be more representative of psychologists than of postal workers. Basing the selection on the extent to which Helena reflects what we believe about psychologists is an example of the **representativeness heuristic**. The problem with following this heuristic is that it can lead to faulty reasoning if you fail to take other information into account. For instance, overall there are many more postal workers in the world than cognitive psychologists. The frequency of an event's occurring is known as a *base rate;* in this case the base rate for postal workers is higher than that for cognitive psychologists. This base rate information should result in the guess that Helena is a postal worker, since there are thousands of postal workers and only a few hundred cognitive psychologists. In other words, it is something like 200 times more likely that any given person is a postal worker, and surely many postal workers enjoy reading, math, and gardening. Although the traits listed for Helena may be more representative of cognitive psychologists overall, they also are likely to apply to a large number of postal workers. People tend to pay insufficient attention to base rates when they are reasoning, instead focusing on whether or not the information presented is representative of one conclusion or another.

But what if postal workers don't like math problems or gardening? In that case you might wonder whether the base rate matters. Consider Ashley, whom you know to be an extremely reserved, sensitive, and literary person. Is it more likely that (a) Ashley is an engineering major, or (b) Ashley is majoring in engineering and minoring in English? When answering this question, people become distracted by information that is inconsistent with their stereotypes about engineers. Given that she is reserved and literary, it makes sense that if Ashley was an engineering major, she would also likely take courses in English. However, the probability of a *conjunction*—two events happening together—must always be less than the probability of either event. Thus, engineering majors who minor in English are a subset of engineering majors. The representativeness heuristic can be an important factor in racial stereotypes, such as when people constantly see black faces as crime suspects in the media, which alters their beliefs both about black people and about criminals. We will discuss stereotypes in greater detail in Chapter 15.

confirmation bias A tendency to search for and believe evidence that fits our existing views.

framing The effect of presentation on how information is perceived.

CONFIRMATION BIAS The tendency to focus selectively on some information and not use other information underlies the heuristic called **confirmation bias**. Have you ever tried to make a point by citing a single specific instance? For instance, a smoker, when trying to justify her choice to smoke, might point to an aunt who smoked every day and lived a healthy life until she died in her 90s. A politician arguing for welfare reform might try to gain support by providing an example of a specific individual who managed to cheat the current system. In general, people tend to seek out information that is consistent with whatever hypothesis, view, or theory they hold. At the same time, they tend to discount or ignore information that is inconsistent with their beliefs.

Confirmation bias is a natural tendency that can be difficult to guard against. Although scientists are trained to overcome inductive reasoning biases, a strong belief in the validity of a hypothesis can have an influence on their conclusions. For example, most scientists design experiments to confirm their hypotheses. Only rarely does a scientist conduct an experiment that is specifically designed to disconfirm his or her favoured hypothesis. More often than not, scientists who have their own favourite alternative hypotheses conduct the studies that disconfirm others' hypotheses. An important component of critical thinking is to try to avoid confirming prior beliefs.

FRAMING EFFECTS Every weekend it seems that one of the local merchants is having a gigantic sale on jewelry, with bold advertisements proclaiming "80 percent off everything." Although this seems like a good deal, it turns out that the regular prices are substantially higher than any other store's. Thus, the sale simply lowers the price so it is more competitive. Nonetheless, people love to think they are getting a bargain. The way information is presented can alter how people perceive it, which is known as **framing**. Framing a decision to emphasize the losses or the gains when introducing alternatives can significantly alter the choice made. Consider the following problem: Imagine that Canada is preparing for the outbreak of a disease that is expected to kill 600 people. Two alternative programs are proposed to combat the disease. The scientific estimates of the consequences of the two programs are as follows:

If Program A is chosen, 200 of these people will be saved.

If Program B is chosen, there is a one-third probability that 600 people will be saved and a two-thirds probability that nobody will be saved.

Which would you choose? When asked this question, 72 percent of respondents chose program A (Kahneman & Tversky, 1984). Weighing the probabilities would result in 200 people saved for both programs, but people showed a clear preference for a sure gain as outlined in Program A, as opposed to a chance of a larger gain with the additional chance of no gain as outlined in Program B. Now consider these two alternatives:

> If Program A is chosen, 400 people will die.
>
> If Program B is chosen, there is a one-third probability nobody will die and a two-thirds probability that 600 people will die.

When asked this question, 78 percent of respondents chose program B. In this case, most people felt that the certain death of 400 people, outlined in Program A, was a worse alternative than the likely, but uncertain, death of 600 people outlined in Program B. The idea of a sure loss was less appealing than an uncertain, but possibly greater, loss. However, notice that the probabilities and outcomes are identical to those presented earlier, with the exception that the gains are emphasized in the first set of alternatives and the losses are emphasized in the second set of alternatives. Framing a decision so that the gains or losses are emphasized can have a clear impact on the ultimate choice. Research on the framing effect indicates that when people make choices, they may weigh losses and gains differently. Indeed, people are generally much more concerned with costs than with benefits, which is known as *loss aversion.* A schematic illustration of this concept is shown in Figure 8.8.

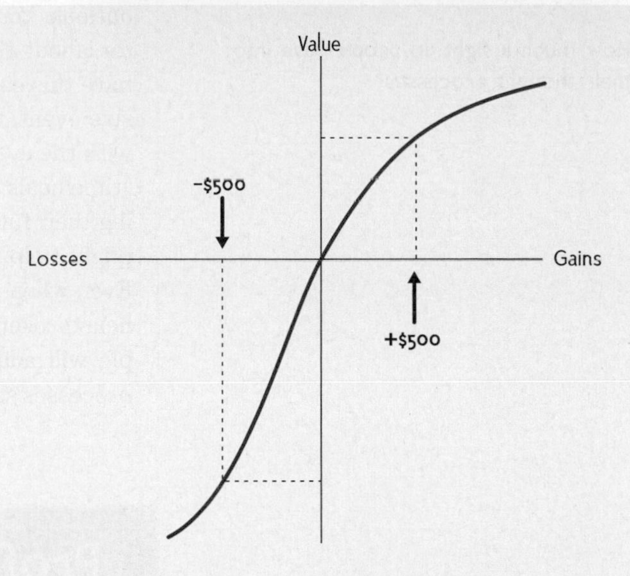

8.8 Loss Aversion The critical aspect of this figure is that with decision "value" plotted on the vertical axis, potential losses have a bigger impact on perceived value. That is, the potential loss of $500 has a greater effect on our decision making than a potential gain of $500.

This tendency to view costs and benefits differently is exploited by companies that offer people the opportunity to test products on a trial basis before purchasing. These companies are banking on the probability that once their product is being used, the "cost" of giving it up at the end of the trial period will be perceived to be greater than the actual monetary cost of purchasing it. In this manner, the company has successfully changed the reference point from which the decision to purchase the product has been made: A cost that originally might have been viewed only in terms of dollars has now been juxtaposed with a cost viewed in terms of losing access to the product. If the latter is perceived as the greater cost, people will end up making a purchase they might not have made if they had considered only the cost in dollars.

AFFECTIVE HEURISTICS People often decide to do things that will make them happy, while they avoid doing things they will later regret. Expectation for how a decision will affect emotional state in the future is a powerful force in decision making. Unfortunately, people are not particularly good at *affective forecasting*—predicting how they will feel in the future. During the 2004 U.S. presidential election, many Americans claimed they would move to Canada if George W. Bush was reelected. As the Harvard psychologist Daniel Gilbert noted in the *New York Times,* "[M]y unofficial tally suggests that the number of disgruntled Democrats who actually emigrated northward is roughly zero, plus or minus none" (2005, p. 23). Gilbert and his colleague Timothy Wilson have found compelling evidence that people overestimate the extent to which negative events will affect them in the future. People mispredict how bad they will feel following a romantic breakup, losing a job, being diagnosed with a serious medical illness, or their favourite politician's losing an election (Gilbert, Pinel, Wilson, Blumberg, & Wheatley, 1998; Wilson & Gilbert, 2003).

According to Gilbert and Wilson, following a negative event people engage in a number of strategies that help them feel better, such as rationalizing why it happened and minimizing the event's importance. Essentially, they figure out a way that the event makes sense, which reduces its negative consequences. Humans have an amazing capacity for manufacturing happiness, and so even if a negative event causes them temporary anguish, most people adapt and return to their typical positive

How much insight do people have into their thought processes?

outlook. You will learn much more about the various biases that help people feel better about themselves in Chapter 15. Interestingly, people seem unaware that they have this capacity, and so when asked to predict how they will feel following an aversive event, they overestimate their pain and underestimate how well they will cope with the event (Gilbert, Morewedge, Risen, & Wilson, 2004). Even though the phrase "time heals all wounds" is familiar, people neglect to consider this idea when forecasting their future negative reactions. Knowing that we readily cope with most setbacks might help people keep their perspective when they face failure or other setbacks. Even when it seems as if you will never recover from failing an important test, being heartbroken, or receiving a devastating medical diagnosis, the fact is that most people will adapt and find new meaning in their lives. How emotions affect thought processes such as decision making is discussed in greater detail in Chapter 10.

Profiles in Psychological Science

The Tragedy of Winning Olympic Silver

The dream of any Olympic athlete is to a win a gold medal. As the American tennis player Mardy Fish prepared for his final match at the 2004 Athens Olympics, he fantasized about standing on the podium listening to "The Star-Spangled Banner" while a gold medal was placed around his neck. Instead, he stood a step down on the podium with a silver medal and tears of sadness rather than joy, as he listened to the national anthem of Chile. Following the match he spoke to reporters about his disappointment: "I think the most disappointing thing about it was that I just really wanted to hear the national anthem . . . I think that the Olympics is the biggest thing, and a gold medal is the biggest prize in sports. And I felt like if, you know, if I heard the national anthem, and I felt like some tears would come there as well. Sure enough, they came, but it was a song that I had never heard of" (as cited on www.usoc.org/11604_25477.htm). Fish's Olympic performance was amazing. He was an unseeded player who made it to the medal stand. Yet the experience was bittersweet as he reflected on what he might have done differently to win the gold.

Perhaps ironically, Fish might have been happier if he had won a bronze medal instead of silver. According to research by the social psychologists Victoria Medvec, Scott Madey, and Thomas Gilovich, the subjective outcome of winning a silver medal can be more negative than that of winning a bronze. To test their hypothesis, Medvec and her colleagues (1995) examined videotapes of Olympic athletes at the moment they found out about the finish. For instance, they watched tapes of the instant that Jackie Joyner-Kersee completed her last long jump to win the bronze medal. They also examined video footage of the awards ceremonies where the athletes received their various medals, as when Matt Biondi received the silver for the 50-meter freestyle and the Lithuanian men's basketball team received bronze medals (in tie-dye uniforms to honour the Grateful Dead, who provided sponsorship for the team). Students watched the tape and rated the emotions expressed in the faces on a 10-point scale "from agony to ecstasy." The researchers found consistent evidence that bronze medal winners were happier than those who won silver, both at the moment of finding out how they did and also later on the stand.

What explains these reactions? The difference in medalists' emotional responses can be explained by counterfactual thinking, which is the act of imagining an outcome that didn't happen. For instance, "What would the Canadian economy be like if the North American Free Trade Agreement had not been passed in 1994?" is a counterfactual in that it considers something happening that did not. People use counterfactual reasoning to gauge how a decision may feel, basing the actual decision on which imagined outcome produced the greatest elation and least regret. For example, when game-show contestants have to decide whether to keep the winnings they have already earned and quit, or risk losing all in the hopes of gaining an even bigger prize, many will choose to "quit while they're ahead," for fear of experiencing the inevitable regret that would result if they failed to win the bigger prize and ended up with nothing.

Let's use the idea of counterfactuals to explain the thoughts of bronze and silver medalists. For the bronze medalists, the most salient alternative outcome was not winning any medal, so compared with that situation, they are pretty happy. By contrast, the silver medalists are focused on having just missed the gold medal, not on having just beaten the bronze-medal winner (Figure 8.9). Indeed, when Medvec and colleagues had students listen to statements the athletes made during subsequent interviews, it was clear that silver medalists made comments such as "I almost," whereas bronze medalists made comments such as "At least I," indicating a completely different reference for interpreting the value of their medal.

You have probably engaged in similar counterfactual reasoning in your life, for instance, when you are happy with a B in a course in which you are just getting by but are frustrated by a B in a course in which you have higher expectations. Similarly, people report that they would feel more upset if they missed a flight by five minutes than if they missed one by an hour, even though in both cases they missed the flight. This is because in both cases we focus on how we might have done something different to get an A or make the flight. This counterfactual thinking leads to the paradox of greater disappointment for those who come closest to success. ■

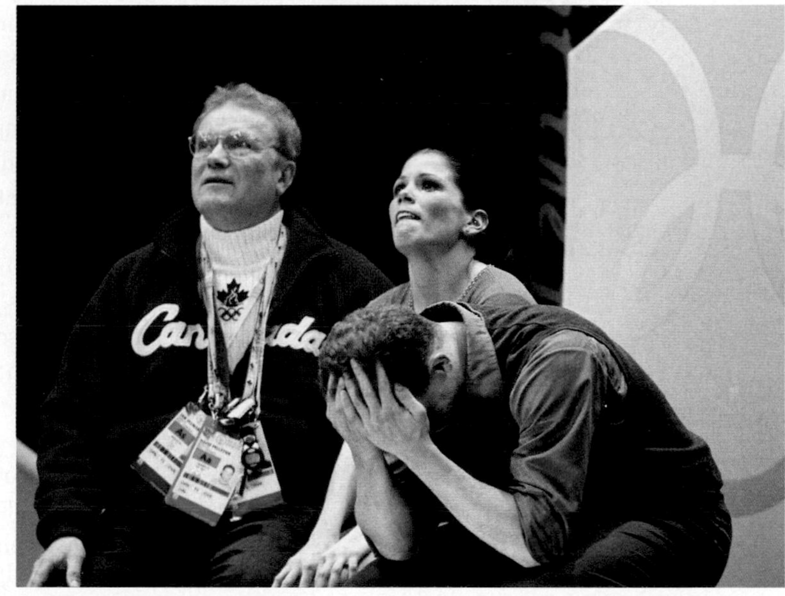

8.9 David Pelletier and Jamie Sale Even though this pair initially won the silver medal at the 2002 Olympics, they were disappointed with the outcome.

Problem Solving Achieves Goals

We have considered how reasoning leads to beliefs and conclusions and how this knowledge can aid decisions. Much of the time our thoughts are focused on our goals and how we might achieve them. What do you have to do to get an A in this course? How do you get into your car when you've locked the keys inside? How can you make enough money to spend your spring break somewhere nice? Problem solving explores how people use knowledge to achieve goals. For our purposes, a person has a problem when faced with a goal with no simple and direct means for attaining that goal. The task in problem solving is to determine how to move from the current initial state to the goal state, which often requires overcoming obstacles and devising strategies. In this section we examine some of the best ways to solve problems. We will see that how we think about problems can help or hinder our ability to find solutions. Sometime we think the wrong thing, and sometimes perhaps we think too much.

ORGANIZATION OF SUBGOALS One approach to the study of problem solving is to identify the steps people take to find a solution to a problem. The idea is that people can solve most problems in a number of different ways. In turn, each solution

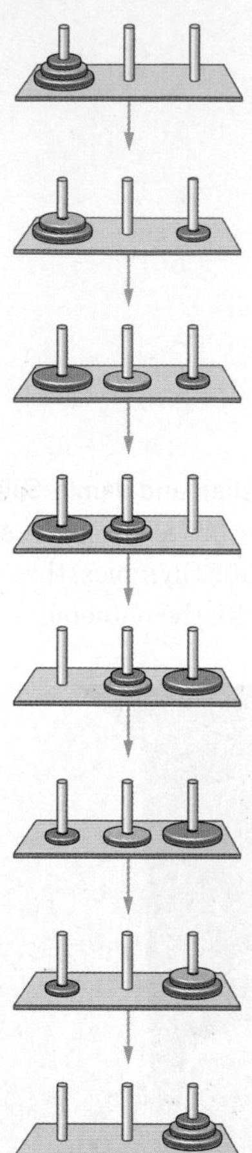

8.10 The Tower of Hanoi Problem To solve this problem you need to break the task down into subgoals.

insight The realization of a solution to a problem.

can be broken down into a series of stages or steps. Researchers examine how people go from one step to the next, the typical errors people make when tricky or unintuitive steps must be negotiated, and how people converge on more efficient—or, in some cases, less efficient—solutions.

For example, in the classic "Tower of Hanoi" problem, participants are given a board that has a row of three pegs on it, with the peg on one end having three discs stacked on it in a descending order of size, such that the largest disc is on the bottom (Figure 8.10). The task is to move the ordered stack of discs to the peg on the other end, with the following constraints: Only one disc can be moved at a time, and a larger disc can never be placed on top of a smaller disc. Go ahead and try it. The easiest way to simulate this problem is to use a quarter, nickel, and penny to represent the different-sized disks. Solving the "Tower of Hanoi" problem requires you to break the task down into *subgoals*. For instance, your first subgoal is to move the largest ring to the farthest peg. You can do this by first moving the smallest ring to the farthest peg, the middle ring to the middle peg, and then the smallest ring to the middle peg on top of the middle ring. You can now move the large ring to the farthest peg. Your next subgoal is to get the middle ring on top of the large ring, which you can do by moving the small ring to the first peg. And then you are pretty much done.

Using subgoals is important for many problems. Imagine that Sarah, a high school senior, has decided she would like to become a doctor. In order to achieve this goal, she needs first to attain the more immediate subgoal of being admitted to a good college. In order to get into a good college, she needs to earn good grades in high school. This additional subgoal would require developing good study skills and paying attention in class. Breaking down a problem into subgoals is an important component of problem solving, although identifying the appropriate steps or subgoals and their order can be challenging for complex problems for which there is no obvious next step.

SUDDEN INSIGHT Problems are not often identified as problems until they seem unsolvable and the problem solver is, more or less, "stuck." It is when we spot the keys in the ignition when we encounter the locked door that we know we have a problem. As we stand helplessly trying to figure out a solution, sometimes one will just "pop" into our heads. How does this happen? **Insight** is the stereotypical mental lightbulb going on in one's head, a metaphor that is used to capture the phenomenon of suddenly realizing the solution to a problem. In 1925, the Gestalt psychologist Wolfgang Köhler conducted one of psychology's most famous examples of research on insight; perhaps ironically, he did not use humans as his research subjects. Köhler was convinced that some animals could behave in an intelligent fashion, and so he set out to study whether chimpanzees could solve problems. Köhler would place a banana out of reach outside the chimp's cage, while providing several sticks that, if used properly, could be used to move the banana within grabbing distance. In one situation there were two sticks, but neither was long enough to reach the banana. One chimpanzee, who sat looking at the sticks for some time, suddenly grabbed the sticks and joined them together. This longer stick did the trick and the chimp was able to capture the banana. Köhler argued that, after pondering the problem poised by the distant banana, the chimp had eventually had the insight to join the sticks in order to make a tool long enough to reach the banana. Once the chimp solved the problem, it was able to transfer this solution to other similar problems and solve them quickly.

In another classic study of insight, Norman Maier (1931) brought participants, one at a time, into a room that had two strings hanging from the ceiling and a table in the corner. On the table were several random objects, including a pair of pliers. The task given to each participant was to tie the two strings together. However, it was impossible to grab both strings at once—if a participant was holding one string, the other was too far away to grab as well. The solution to the problem was to tie the pliers onto one of the strings, so that the string could be used as a pendulum; one could then hold the other string and then grab the pendulum string as it swung by. Although a few participants eventually figured out this solution on their own, most people were stumped by the problem. After letting people ponder the problem for ten minutes, Maier casually crossed the room and brushed up against the string, causing it to swing back and forth. Maier reported that once the participants saw the brushed string swinging, most immediately solved the problem, as if they had had new insight. However, none of the subjects reported that Maier gave them the solution. They all believed that they had come up with it on their own. As Gladwell notes in *Blink,* Maier's hint was so subtle that it registered only at the unconscious level. As we will see a bit later, unconscious processes might be quite good at helping us solve certain types of problems.

restructuring A new way of thinking about a problem that aids its solution.

The study by Maier also provides an example of how insight can be achieved when a problem initially seems unsolvable. In this case, the possible functions of the string and pliers needed to be reconsidered to solve the problem. How we view or represent a problem can have a significant impact on how easily we solve it. For example, Maier's "pendulum" experiment is based on the expectation that most people will fail to see the pliers as a pendulum weight; that is, participants initially think about the problem in ways that fail to consider the solution based on creating a pendulum. If our current view isn't working, are there other, less obvious ways we could structure the problem? Although terms such as "thinking out of the box" and "think different" have become recent clichés in our culture, the ideas they embody have been around for a long time and continue to be of great value.

CHANGING REPRESENTATIONS TO OVERCOME OBSTACLES

There are some common ways in which people become stuck in solving a problem, as well as common strategies that they can use to help overcome obstacles in problem solving. Insight into a problem can arise when we suddenly *restructure* a problem in a novel way, discovering a solution that was not available under the old problem structure. **Restructuring** the representation of a problem can lead to the sudden "aha!" moment that is characteristic of insight. In one now-famous study, Scheerer (1963) gave participants a sheet of paper that had a square of nine dots on it (Figure 8.11). The task was to connect all the dots using only four straight lines, without lifting the pencil off the page. As shown in Figure 8.11, the solution is to structure the problem such that it is permissible to draw the lines beyond the box formed by the dots. The difficulty most people have when first encountering this task is that their representation of the problem fails to consider solutions that do not keep all lines within the square. Although keeping the lines within the box formed by the dots was not a requirement of the problem, most people have a tendency to regard the problem this way. Solving the problem requires restructuring their representation of it. Thinking differently about the constraints of the problem and eliminating some assumed constraints allows for a solution.

8.11 Scheerer's Nine-Dot Problem The problem (a) is solved by realizing that the lines can extend beyond the boundary formed by the dots (b).

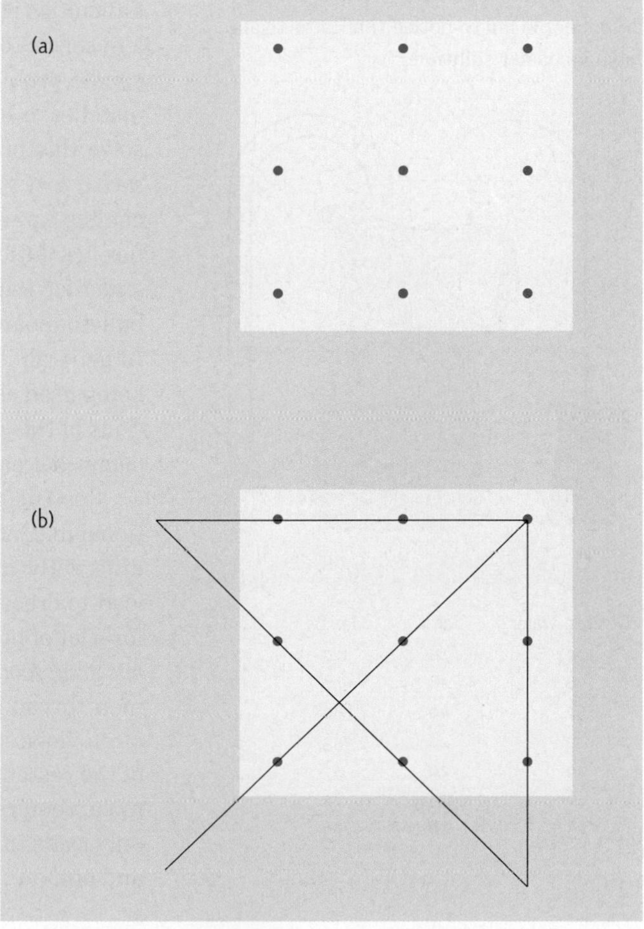

What role do prior expectations play in problem solving?

When we consider a solution to a problem, we commonly think back to how we solved similar problems in the past. Thus, we tend to persist with a previous strategy, known as a **mental set**, which sometimes makes it difficult to find the best solution. Abraham Luchins demonstrated a classic example of a mental set in 1942 (Figure 8.12). He gave participants three jars of different sizes and indicated that they had to use these jars to measure out specified amounts of water. For example, participants might be told that Jar A holds 21 cups, Jar B holds 127 cups, and Jar C holds 3 cups. Using these three jars, participants are asked to measure exactly 100 cups of water. The solution to this problems is to fill up Jar B with 127 cups, and then use Jar A to remove 21 cups of water, and use Jar C to remove 3 cups of water twice, which would leave 100 cups remaining in Jar B. The structure to the solution is B − A − 2(C). Participants are then given many similar problems in which the jar sizes and goal measurements differ, but in each the same formula applies. Participants are then given another problem, Jar A holds 23 cups of water, Jar B holds 49 cups of water, and Jar C holds 3 cups of water. Participants are asked to measure 20 cups of water. Luchins (1942) found that even though the simplest solution to the task would be to fill Jar A and remove 3 cups with Jar C, participants usually came up with the much more complicated solution using all three jars. Having developed a mental set of using three jars in combination to solve this type of problem, it was difficult for them to settle on the simpler solution using only two jars. Surprisingly, when given a problem that had a simple solution and for which the original formula did not work (see the last line of Figure 8.12), many participants failed to solve the problem.

Our mental representations about the typical functions of objects can also create difficulties in problem solving. Overcoming **functional fixedness** requires the problem solver to reinterpret the potential use or function of an object. For example, the candle problem, introduced by Karl Duncker (1945), gave people a candle, a box of matches, a bulletin board, and a box of tacks (Figure 8.13). Imagine how you would solve this problem: *Using only these objects, attach the candle to the bulletin board in such a way that the candle can be lit and burn properly.* Most people have difficulty in coming up with an adequate solution. However, if we reinterpret the function of the box, a solution emerges. Viewing the box not as a container for the matches but as a stand for the candle allows for an easy solution: The side of the box is tacked to the bulletin board so that it creates a stand; the candle is then placed on the box and lit. In general, it is difficult for subjects to view the box as a possible stand when it is being used as a container for the matches. When participants are shown representations of this problem with an empty box and the matches on the table next to it, it is somewhat easier to solve.

Restructuring the problem, breaking out of a mental set, and overcoming functional fixedness are all techniques that require the problem solver to view a problem differently and restructure the initial mental representations. Restructuring can lead to creative insights into how to solve difficult problems. It also forms the basis for a lot of humour. *Have you heard about the new restaurant that opened on the moon? It has great food, but no atmosphere!* Whether or not you found this joke amusing, a common thread in humour is that expectations based on common representations are often violated. The premise of this joke is that *atmosphere,* when interpreted in light of the restaurant schema, means one thing, but the fact that the restaurant is on the moon requires another interpretation. The ability to restructure representations not only leads to insight into problem solutions, but is also a key component of creativity and humour.

mental set A problem solving strategy that has worked in the past.

functional fixedness A tendency in problem solving to think of objects only as they are most commonly used.

8.12 Luchins's Mental Set Try to solve the water-jar problem. As described in the text, the solution for the early trials is B − A − 2(C). People fail to notice that later trials have an easier solution.

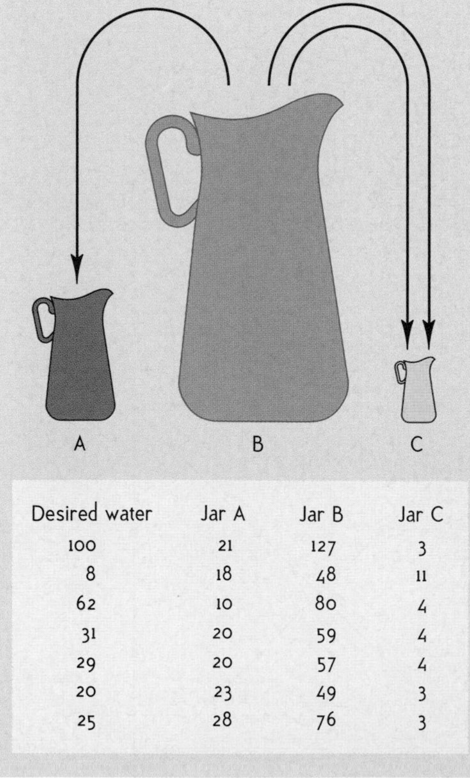

Desired water	Jar A	Jar B	Jar C
100	21	127	3
8	18	48	11
62	10	80	4
31	20	59	4
29	20	57	4
20	23	49	3
25	28	76	3

CONSCIOUS STRATEGIES Although restructuring mental representations is a valuable means to develop insight into solving a problem, it is often difficult to enact this as a conscious strategy when we are stuck. However, we can systematically apply strategies that may help lead to a solution. Two common heuristic strategies to overcome obstacles in problem solving are *working backward* and *finding an appropriate analogy.*

For complex problems, the appropriate steps to the solution are not always clear. One means of generating different strategies about how to solve a problem is to work backward from the goal state to the initial state. Consider the "water lily" problem (Sternberg & Davidson, 1983):

> Water lilies double in area every 24 hours. On the first day of summer there is only one water lily on the lake. It takes 60 days for the lake to be completely covered in water lilies. How many days does it take for half of the lake to be covered in water lilies?

One way to solve this problem is to figure that on day 1 there is one water lily, so on day 2 there are two water lilies, on day 3 there will be four water lilies, and so on, until you discover how many water lilies there are on day 60 and see which day had half that many. But if you work backward, you will realize that if the water lilies double every 24 hours and if on day 60 the lake is covered in water lilies, then on day 59 half of the lake must have been covered in water lilies. The natural tendency when problem solving is to focus on how to move from the initial state toward the goal state. However, at times it helps to consider working backward from the goal state to the initial state when trying to find the best solution.

Another common strategy to overcoming a problem-solving obstacle is to *find an appropriate analogy* (Reeves & Weisberg, 1994). For example, imagine that a doctor is dealing with a patient who has a tumour. The doctor has a laser that, when used at a high intensity, could destroy the tumour. However, an intensity high enough to destroy the tumour would also destroy the surrounding living tissue. How can the doctor safely use the laser without damaging the living tissue through which the rays must travel? The vast majority of people have a hard time when this type of problem is presented by itself (Duncker, 1945). However, when they are given an example of an analogous situation, their ability to solve this problem improves substantially.

Now imagine that the doctor reads a story about a general who wanted to capture a fortress. A large number of soldiers were needed to capture the fortress, but all the roads to the fortress were planted with mines. Small groups of soldiers could safely travel the roads, but the mines would be detonated by a larger group. In order to move all the soldiers he needed to the fortress, the general divided his army into smaller groups and had each group take a different road to the fortress where they would converge and attack together.

From reading this story about a problem with analogous constraints, the doctor might get the idea to aim several lasers at the tumour from different angles. Each laser by itself can be weak enough so that it does not destroy the living tissue in its path, but the combined intensity of all the lasers converging at the tumour could be strong enough to destroy it. When we are given analogous problems, the ability to find a solution is enhanced. According to some researchers, those participants who are given two or more analogous descriptions develop a schema for the solution, which helps them solve similar problems in the future (Gick & Holyoak, 1983). However, analogous solutions work only if people recognize that the problem they face is similar to those they have solved in the past (Keane, 1987; Reeves & Weisberg, 1994).

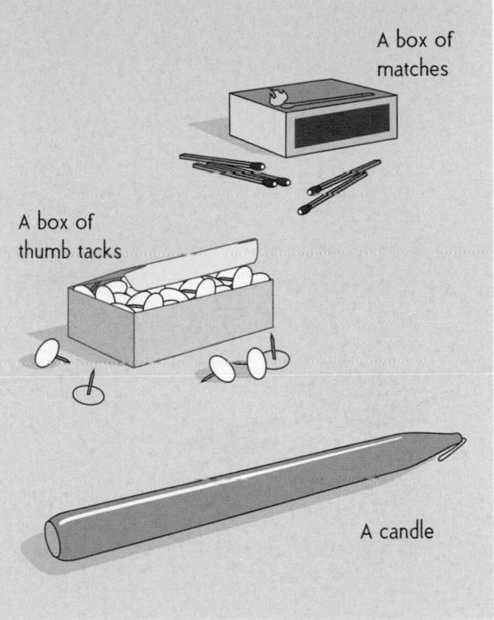

8.13 Overcoming Functional Fixedness
People are given the task of attaching a candle to the wall using only these materials. The solution requires restructuring our concept of the matchbox (see the following page for the solution).

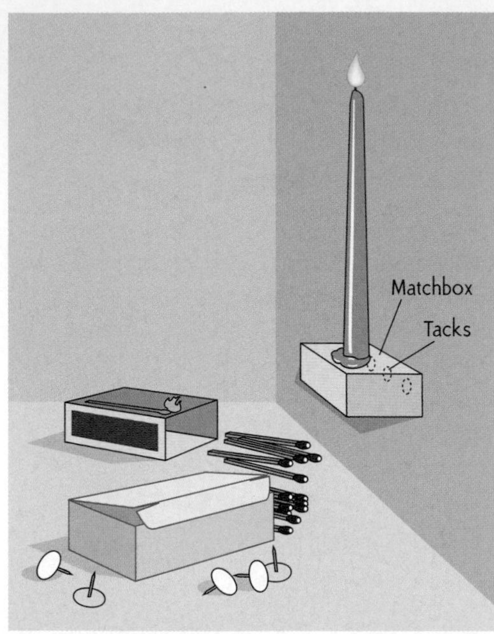

How might conscious thought interfere with problem solving?

We can also benefit from finding an appropriate analogy when trying to discover how to achieve life goals. For example, recent immigrants to Canada often have difficulty navigating the customs and expectations of the new culture and country. Other immigrants with a similar background who have succeeded in their new home can provide examples of paths that lead to success. We look to role models to provide examples of achieving certain goals because we believe that we may achieve a similar outcome if we follow an analogous path.

UNCONSCIOUS PROCESSING Common sense tells us that consciously thinking about a problem or deliberating about our options is the best strategy for solving problems and making decisions. We have noted, however, that insight and intuition seem to benefit from unconscious processing. In *Blink,* Gladwell makes the bold claim that fast decisions are often superior to those that follow deliberative thought. Clearly this is not always true. Which number is larger, the square root of 6724 or nine squared? You probably can't provide an immediate answer—you have to solve this problem consciously by doing the calculations. But does not consciously thinking sometimes produce superior outcomes to thinking? A recent study by Ap Dijksterhuis (2004) at the University of Amsterdam required people to evaluate complex information regarding real-world choices, for example, selecting an apartment. In each case they had to choose between various alternatives that had many negative and positive features, such as a nice landlord, cheap rent, bad location, and so on, but objectively, one apartment was a better choice. Participants were assigned to one of three conditions, which required them either to make an immediate choice (no thought), think for three minutes and then choose (conscious thought), or work for three minutes on a difficult, distracting task and then choose (unconscious thought). Across three separate studies, those in the unconscious thought condition made the best decisions. According to Dijksterhuis, unconscious processing is especially valuable for complex, difficult decisions in which it is difficult to weigh the pros and cons consciously. A similar phenomenon has also been reported in the creativity literature. Anecdotal reports suggest that allowing an idea to incubate over time helps in problem solving. Perhaps this is why for really important decisions people are often told to "sleep on it."

Can conscious thought undermine good decision making? Tim Wilson and Jonathan Schooler (1991) had research participants rate jam. When research participants simply tasted the jam, their ratings were very similar to ratings made by experts. However, when the participants had to explain their rating for each product, their ratings differed substantially from those of the experts. Having to talk about their reasons apparently altered their perception of the jam. Jonathan Schooler introduced the concept of *verbal overshadowing* to describe the impairment of performance that occurs when people try to explain verbally perceptual experiences that are not easy to describe (Schooler & Engslter-Schooler, 1990; Schooler, 2002). For example, participants who had to provide a verbal description of a person they saw in a simulated bank robbery were less able to pick the person out of a lineup than participants who did not provide a description. In another study, people who were asked to describe the taste of wines were less able to pick out the wine later than those who simply tasted the wine without describing it (Melcher & Schooler, 1996). The gist of these studies is that we really aren't good at using verbal labels to describe perceptual experiences, and so when we are forced to do so it impairs our processing of that perceptual information.

Consider one more example of how thinking too much can cause negative outcomes. Imagine looking at two lights, one red, the other green. Your job is to predict, on each trial, which light will come on. If you are correct you will receive some small reward. As you are performing this task, the lights flash in a random sequence, but overall the red light flashes 70 percent of the time. People are pretty quick to notice that the red light comes on more often. So to receive the most reward, what is the best strategy to follow? After doing this task a number of times, most nonhuman animals simply choose the most probable response 100 percent of the time; thereafter they are rewarded on 70 percent of the trials. This makes great sense in terms of adaptiveness. Humans do something much different. They appear to try to figure out patterns in the way the lights are presented, and they choose the red light about 70 percent of the time. That is, overall their choices match the frequency of how often red flashes, but because the lights are randomly presented, on any given trial they are often wrong. Indeed, in the 70 percent scenario, it turns out that they receive rewards on only 58 percent of the trials. Why don't humans follow the optimal strategy, which even rats can figure out? According to George Wolford and colleagues (2000), people search for patterns that might not even exist. These researchers attribute this human tendency to the workings of the left hemisphere *interpreter,* which as you recall from Chapter 4 tries to make sense out of the world. To test this idea, Wolford and colleagues had two split-brain patients perform a version of this task. They found that the patients' right hemispheres tended to respond in the optimal way that animals did (choosing the same thing 100 percent of the time), whereas the patients' left hemispheres matched that frequency but continued to guess both outcomes. The tendency of the left hemisphere interpreter to seek patterns and try to find relations between events may be adaptive in some contexts, but it can produce less-than-optimal outcomes when a pattern doesn't really exist. The greater reliance on verbal processes on the left hemisphere may also contribute to the impairments caused by verbal overshadowing (Schooler, 2002). Although exactly why thinking too much can impair memory and judgment is unclear, it appears that some things are best left unsaid, and that sometimes it is better to "sleep on it" before making important decisions. Unconscious processes may have a greater influence over our decisions that has previously been recognized.

"This isn't a hasty decision. A lot of daydreaming went into it."

Thinking Critically

Are More Choices Always Better?

In our modern society, people widely believe that the more choices we have, the better. But today we face an ever increasing array of choices for just about everything, from types of cars to career options. Daily life requires deciding among countless options. Instead of ordering a cup of coffee, we now have to decide whether we want a latte, cappuccino, or mocha, as well as whether we want it hot or cold. Ice cream comes in dozens of flavours. We can choose from hundreds of different brands of breakfast cereal, and even numerous brands of bottled water. Although some choice is better than none, is there any downside to having too much choice? According to some scholars, too much choice can be frustrating, unsatisfying, and ultimately debilitating (Schwartz, 2004). Is there a tyranny of choice? What is the relation between choice and well-being? Are more choices always better?

Imagine if you were told which car to drive, which kind of food to eat, or what brand of shoes to wear. Not being able to choose violates our sense of freedom, at least in Western cultures. Indeed, having a sense of control in our lives and feeling free to choose how we act are generally viewed as beneficial to mental health. But when too many options are available, especially when all of them are attractive, people experience conflict and indecision. Consider a study in which shoppers were presented with a variety of gourmet jams (Iyengar & Lepper, 2000). In a grocery store, shoppers encountered a display containing either six or 24 varieties of jam to sample. They were also given a discount coupon to encourage them to make a purchase. Whereas the greater variety of jams attracted more shoppers, it failed to produce more sales; 30 percent of those with the limited choice bought jam compared with only 3 percent who did so with the greater variety. In a subsequent study, these investigators found that people choosing among a limited selection of chocolates were more satisfied with their selection than those who chose from a wider variety.

According to the psychologist Barry Schwartz, having too much choice can make some people miserable. He divides the world into groups he calls "satisficers" and "maximizers." Satisficers live according to a "good enough" philosophy. Schwartz borrowed the term from the late Nobel laureate psychologist Herbert Simon, who described satisficing as choosing an option that sufficiently satisfies needs. It isn't that satisficers lack standards, but they look around until they find something that most closely matches what they want and then buy it. They don't worry about whether there are better or cheaper products available. They like good things, but they don't care if those things are not the absolute best.

By contrast, maximizers always seek to make the best possible choices. They devote serious time and effort to reading labels, checking out consumer magazines, reading Internet reviews, considering alternatives, and comparing prices. They often feel frustrated by the countless options available to them and feel paralyzed by indecision, as they have to select between equally attractive choices. For maximizers, costs are associated with making the wrong choice. Choosing to order the fish special for dinner means that you won't have the pleasure of tasting the chicken special. What if the fish isn't good tonight? What if the chicken is a better choice? Not only do they hesitate in making decisions, but maximizers continue to evaluate options even after they have made their selection. As a result, they are generally more disappointed with their decisions and more likely to experience regret. They continue to analyze and question their choice, often ruminating about its negative features.

The paradox of choice might also be responsible for a cultural shift in the age at which people are settling into jobs and marriage. Historically, people were set in their career paths and often married and having children before age 25. A pattern emerging in industrialized nations is for young adults to delay all of this for many years, as they explore the vast number of options they have for working and seek a mate who matches their ideals (Grossman, 2005). In countries such as Canada, England, France, Germany, Italy, and Japan, the average age of marriage is now approaching 30. One possible explanation for this phenomenon is that young adults are trying to maximize their life choices. They want just the right job and just the right marriage. After all, picking badly in either of these areas can lead to serious consequences. The burden of making the right decision means that many young adults fail to make any decisions. The growth of options has turned more and more young adults into maximizers.

According to Schwartz, the consequence of nearly unlimited choice may explain the increase in clinical depression in modern countries. "If the experience of disappointment is relentless, if virtually every choice you make fails to live up to expectations and aspirations, and if you consistently take personal responsibility for the disappointments, the trivial looms larger and larger, and the conclusion that you can't do anything right becomes devastating" (2004, p. 215).

What can we do about the problem of maximizing? Schwartz suggests that people restrict their options when the decision isn't crucial and to settle for a choice that meets their needs even if it isn't the absolute best. He also encourages people not to dwell on their choices but instead to focus on the positive aspects of the decision they made. Does this sound to you as if Schwartz is advising people to strive for mediocrity? How much effort should you put into making good choices? Do you think you would be happier if you had fewer choices? Do you think having too many options is responsible for the recent increase in depression and other cultural changes, as when people put off careers and marriage? ■

REVIEWING the Principles | How Do We Make Decisions and Solve Problems?

People use deductive and inductive reasoning to draw conclusions and then use decision making to choose among alternatives. Although normative models, such as the expected utility theory, have often been assumed to be accurate because they describe how decisions should occur if humans are performing optimally, it is increasingly clear that a number of factors that influence our choices are not completely rational. The descriptive approach to understanding reasoning, decision making, and problem solving shows that people often use heuristics that can lead to faulty outcomes, that they fail to take into account information such as base rates, and that their decisions tend to confirm that which they already believed. To solve problems, people often need to reconsider the problem and break out of their mental set. Working backward and finding an appropriate analogy are useful conscious strategies. Unconscious processes also play an important role in problem solving and decision making.

How Do We Understand Intelligence?

So far, we have considered the manner in which knowledge is represented in the mind and how we use this knowledge to make decisions and solve problems. One of the most hotly debated topics in psychological science concerns how knowledge and its application in everyday life translate into *intelligence,* and the relative degree to which intelligence is determined by our genes and our environment (Neisser et al., 1996). **Intelligence** is the human ability to use knowledge to solve problems, understand complex ideas, learn quickly, and adapt to environmental challenges. As modern society becomes more complex, with greater amounts of information available and greater demands on people to understand it all, being able to adapt to change and deal with increasing complexity is becoming more and more important (Gottfredson, 2004a; Lubinski, 2004). To succeed, people need to be innovative and "stay ahead of the curve." We are all intelligent beings. But there are individual differences among us in terms of intelligence, just as there are differences in our physical and psychological characteristics (Vernon, 1979; Vernon et al., 2000). So how do we know someone's level of intelligence?

intelligence The human ability to use knowledge, solve problems, understand complex ideas, learn quickly, and adapt to environmental challenges.

The earliest efforts to study intelligence were led by Sir Francis Galton in the late 1800s. Galton believed that intelligence was related to the speed of neural responses and the sensitivity of sensory/perceptual systems—the quicker our responses and the keener our perceptions, the smarter Galton believed us to be. Since that time researchers have taken numerous different approaches to defining and studying intelligence. Today three general approaches to understanding intelligence cross the major levels of analysis. The first focuses on how people perform on standardized achievement tests. This *psychometric* approach examines what people know and the skills they have for solving problems. The *cognitive* approach examines the particular mental abilities that allow people to operate in an intelligent fashion. In this approach researchers examine how information is processed, for instance, the speed at which people can react, the amount of information they can hold in memory, and the extent to which they can stay focused on a task. Finally, the *biological* approach is concerned with how the brain processes information and the extent to which differences in brain activity are affected by genes and the environment.

Intelligence Is Assessed by Psychometric Tests

For much of the last century, the psychometric approach to intelligence has been the most dominant and influential, especially in terms of how intelligence is viewed in everyday life. Tests that focus on *achievement* assess current levels of skill and knowledge. *Aptitude* tests examine whether people will be good at various tasks in the future. Scores on aptitude tests such as those used for law or medical school are widely used to predict how you will perform academically, which is why taking the tests makes most people very anxious. Aptitude tests also predict what kinds of jobs people might be good at, for example, whether they might make good accountants, pilots, or even spies. Performance on these tests can have a huge impact on people's lives. Given these high stakes, what do we know about intelligence testing?

8.14 Alfred Binet Binet launched the psychometric approach to intelligence.

The measurement of intelligence began in earnest just over a century ago. At the encouragement of the French government, Alfred Binet (Figure 8.14) developed the first method of assessing intelligence to identify children in the French school system who needed extra attention and instruction. Binet proposed that intelligence is best understood as a collection of high-level mental processes—in today's terms, attributes such as "verbal," "mathematical," and "analytical" abilities. Accordingly, with the help of his assistant Théodore Simon, Binet developed a test for measuring children's vocabulary, memory, skill with numbers, and other mental abilities—the *Binet-Simon Intelligence Scale.* One assumption underlying the test was that although students might do better on some of the specific components by chance, how they performed on average across the different components would be a good indicator of students' overall level of intelligence. Indeed, Binet found that scores on his tests were consistent with teachers' beliefs about children's abilities and the grades that they received.

A number of different intelligence tests have been developed since Binet's test. In the United States, the psychologist Lewis Terman of Stanford University modified the Binet-Simon test and established normative scores for American children. This test, revised for the fifth time in 2003, remains among the most widely used for children in the United States and is also used in Canada. In 1939 David Wechsler developed a test for use among adults, which is called the Wechsler Adult Intelligence Scale (WAIS), the most current version being the WAIS-IV. The WAIS has two separate parts, each consisting of several tasks. The *verbal* part measures such aspects as comprehension, vocabulary, and general knowledge. The *performance* part involves nonverbal tasks, such as arranging pictures in proper order, assembling parts to make a whole object, and identifying missing features from a picture. There are seperate norms used for the WAIS in Canada, as well as a French version of the test.

INTELLIGENCE QUOTIENT An important concept introduced by Binet was the idea of **mental age**, which is an assessment of a child's intellectual standing compared with that of peers of the same age. Binet noticed that some children seem to think like normal children of a younger or older age. Thus, for instance, an eight-year-old who reads Shakespeare and performs calculus might score as well as the average 16-year-old, thus giving the eight-year-old a mental age of 16. Mental age is determined by comparing a given child's test score with the average score for children of each chronological age. For example, the **intelligence quotient (IQ)**, developed by the German psychologist Wilhelm Stern, is computed by dividing a child's estimated mental age by the child's chronological age and then multiplying this number by 100: For our child prodigy, this would mean an IQ of 200!

mental age An assessment of a child's intellectual standing relative to that of his or her peers; determined by a comparison of the child's test score with the average score for children of each chronological age.

intelligence quotient (IQ) The number computed by dividing a child's estimated mental age by the child's chronological age, and then multiplying this number by 100.

Although the term *IQ* continues to be used, it is no longer conceptualized in the same way, because using a strict ratio has serious flaws. For instance, consider a child who has a mental age of six. The day before his sixth birthday his IQ is 120, and then overnight his IQ drops to 100. Likewise, a 40-year-old professor would have to have an IQ of 230 just to be as smart as a typical undergraduate. Thus, the ratio concept of IQ does not apply equally to all different age groups. Today the average IQ is set at the value of 100, with norms for assessing where people stand relative to this average. Across large groups of people, the distribution of IQ scores forms a bell curve, which is known as a *normal distribution* (Figure 8.15). Most people are close to the average, with fewer and fewer people at the tails of the distribution. Recall from Chapter 2 the statistical concept of standard deviation, which indicates how far people are from the average. The standard deviation for most IQ tests is 15, such as that approximately 68 people fall within 1 standard deviation (they score from 85 to 115) and just over 95 percent of people fall within 2 standard deviations (they score from 70 to 130).

VALIDITY Are intelligence tests valid? That is, do they really measure what they claim to measure? The overall evidence indicates that IQ is a reasonably good predictor of grades at school and performance at work (Gottfredson, 2004b). For example, students who score high on tests tend to do better than those who score less well. In a recent analysis of more than 20,000 people who took the Millers Analogy Test, researchers found that high scores not only predicted graduate students' academic performance, but also their productivity, creativity, and job performance (Kuncel, Hezlett, & Ones, 2004). Similarly, people who have professional careers, such as attorneys, accountants, and physicians, tend to have high IQs. Those with jobs such as a miner, farmer, lumberjack, or barber tend to have lower IQs (Jencks, 1979; Schmidt & Hunter, 2004). Of course, this refers to averages rather than to individuals, but overall the data suggest a modest correlation between IQ and work

8.15 The Distribution of IQ Scores Across the population, IQ is distributed in a bell-curve shape, such that most people score close to the mean. The height of the curve reflects the percentage of people scoring at each level of IQ. The mean IQ is set at 100 with a standard deviation of 15. Just over 68 percent of people have an IQ within one standard deviation of the mean, which means they will score between 85 and 115. Fewer than 3 percent of people have an IQ of more than 130, which is 2 standard deviations of the mean.

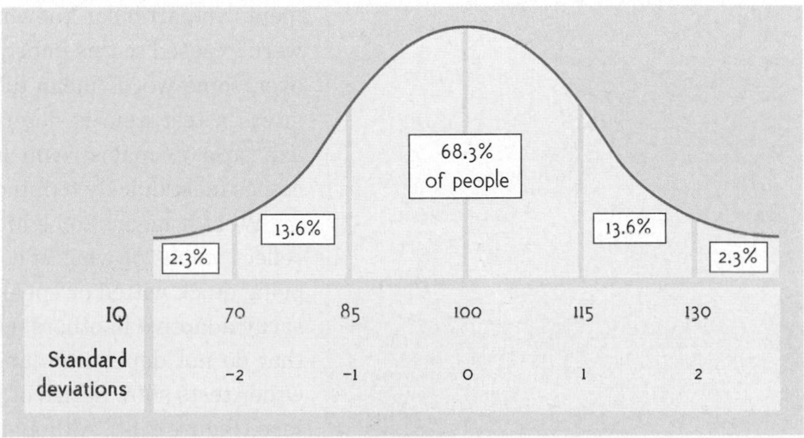

"I don't have to be smart, because someday I'll just hire lots of smart people to work for me."

performance as well as income. The relation between IQ and work performance also holds within professions, at least for those jobs requiring complex skills. Although IQ does not predict who will be a better truck driver, it does predict who will be a better computer programmer (Schmidt & Hunter, 2004).

We should note a number of things regarding these findings. First, IQ scores typically predict only about 25 percent of the variation in work or school performance, which means that other factors also matter (Neisser et al., 1996). Children who come from privileged backgrounds tend to have higher IQs, but it is possible that these students have other advantages that help them succeed, such as family contacts who can help them find good summer internships or access to better schools that can cater to their needs. Moreover, there are large individual differences in motivation, and some people are willing to spend more time working to get ahead. One 20-year follow-up study of nearly 2,000 gifted 13-year-olds (i.e., those with IQs in the top 1 percent of their age group) revealed huge differences in how much people reported working as well as what they said they were willing to work. At age 33, some individuals refused to work more than 40 hours per week, whereas others reported regularly working more than 70 (Lubinski & Benbow, 2000). You can imagine how much more someone working 60 or 70 hours per week can accomplish compared with someone working 25 or 30, and therefore perhaps how such a person might be more likely to make partner in a law firm or earn a higher income (Lubinski, 2004). In other words, although IQ may be important, it is one among several factors that contribute to success in the classroom or the workplace.

Why might students with privileged backgrounds score especially high on IQ tests?

CULTURAL BIAS One important criticism of intelligence tests is that they may unfairly penalize people from some cultures or groups. Doing well on many tests of intelligence requires intimate knowledge of mainstream culture and language. For instance, consider this analogy:

STRING is to GUITAR as REED is to:
(a) TRUMPET
(b) OBOE
(c) VIOLIN
(d) TROMBONE

Being able to solve this analogy requires specific knowledge of these instruments, in particular knowing that an oboe uses a reed to make music. But unless you were exposed to this information, it is unlikely you could answer the question. Moreover, some words mean different things to different groups, and how a person answers a test item is determined by its meaning in his or her culture. A person's exposure to mainstream language and culture affects which meaning of a word comes most quickly to mind, if the person knows the meaning at all.

What it means to be intelligent also varies across cultures. Most measures of IQ reflect values of what is considered important in modern Western culture, such as being quick witted or speaking well. But what is adaptive in one society is not necessarily adaptive in others. One approach to dealing with cultural bias is to use items that do not depend on language, such as the performance measures of the WAIS. Other tests show series of patterns and asks the person to identify the missing pattern (Figure 8.16). Although these tests may be fairer than those that rely on understanding language, it is difficult to remove all forms of bias from the testing situation, which even includes the extent to which test takers are motivated to do well. Doing well on IQ tests matters more to some than to others, and therefore tests are biased to favour those who wish to do well.

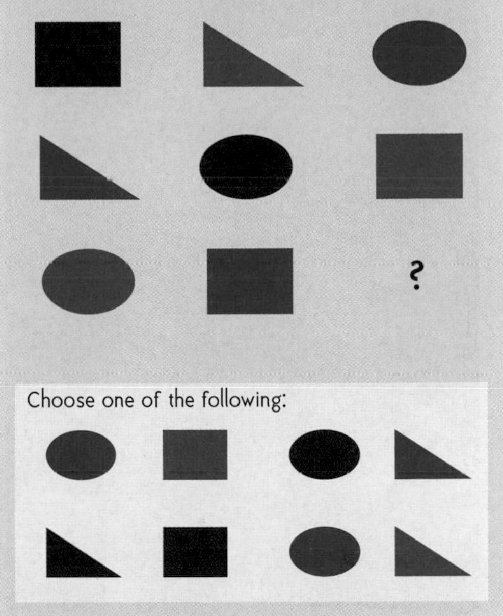

8.16 Removing Bias from Tests Test takers are asked to identify the missing pattern in the sequence. The ability to answer this question does not rely on verbal knowledge and is therefore not culturally biased. The correct answer is the first triangle in the second row.

General Intelligence Involves Multiple Components

Binet's original conception viewed intelligence as a general overall ability. However, we all know people who are especially talented in some areas but have weaknesses in others. Some of us feel more confident, for example, in our ability to write poetry than in our ability to solve difficult calculus problems. The question then is whether intelligence reflects one talent or many. One early line of evidence indicating that intelligence reflects a general capacity used a statistical procedure that analyzes the extent to which different test items are related to each other. By examining the correlations among all items, *factor analysis* creates clusters of items that are most similar to one another; these clusters are referred to as factors. Using this method, Charles Spearman (1904) found that most intelligence test items tended to cluster as one factor, and that people who scored high on one particular type of item also tended to score highly on other types of items; people who are very good at math are also good at writing, problem solving, and other mental challenges. Spearman viewed **general intelligence**, known as **g**, as a factor that contributed to performance on any intellectual task. In a sense, providing a single IQ score reflects the idea that one general factor underlies intelligence. Although most psychological scientists agree that some form of g exists, they also recognize different forms of intelligence, which we now examine.

FLUID VS. CRYSTALLIZED INTELLIGENCE Raymond Cattell (1971) agreed with Spearman but believed that g is made up of two different types of intelligence. **Fluid intelligence** involves information processing, especially in novel or complex circumstances, such as reasoning, drawing analogies, and thinking quickly and flexibly. It is often assessed in nonverbal culture-free intelligence tests, such as completing a series of complex patterns (see Figure 8.16). In contrast, **crystallized intelligence** concerns knowledge we acquire through experience, such as accumulated vocabulary and cultural information, and the ability to use this knowledge to solve problems. Distinguishing between fluid and crystallized is somewhat analogous to distinguishing between working memory (which is more like fluid intelligence) and long-term memory (which is more like crystallized intelligence). As would be expected because both intelligences are components of g, people who score high on one factor also score high on the other, which suggests that a strong crystallized intelligence is likely aided by a strong fluid intelligence.

MULTIPLE INTELLIGENCES Whereas Cattell argued that two different types of intelligence contribute to g, Howard Gardner (1983) proposed his theory of **multiple intelligences**, in which he identified different types of intellectual talents that are independent from one another. For example, he proposed that musical intelligence enables some people to discriminate subtle variations in pitch or timbre and also have a greater than average appreciation of music. Gardner also proposed bodily-kinesthetic intelligence, which is possessed by some individuals, such as athletes and dancers, who are highly attuned to their bodies and able to control their motions with exquisite skill. Gardner also described other intelligences, such as linguistic (excellent verbal skills), mathematical/logical, spatial (thinking in terms of images

general intelligence (g) The idea that one general factor underlies all mental abilities.

fluid intelligence Information processing in novel or complex circumstances.

crystallized intelligence Knowledge acquired through experience and the ability to use that knowledge.

multiple intelligences The idea that people can show different skills in a variety of different domains.

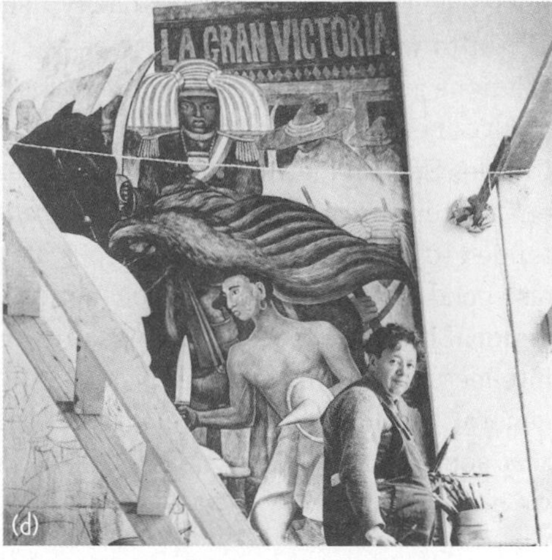

8.17 Multiple Intelligences Howard Gardner suggested that intelligence can take many different forms. Here, four different types of intelligence are represented by (a) athlete Mario Lemieux, (b) musician Yo Yo Ma, (c) scientists David H. Hubel and Torsten N. Wiesel, and (d) artist Diego Rivera.

and pictures), intrapersonal (or self-understanding), and interpersonal (or social understanding) (Figure 8.17).

One important feature of Gardner's theory is that it recognizes that people can be average or even deficient in some domains and outstanding in others. Consider the Juilliard School student Jay Greenberg, who by age 12 had written five full-length symphonies, including *The Storm,* which was commissioned by the New Haven Symphony and composed in just a few hours. Greenberg, a typical boy in all other regards, has been described as perhaps the greatest *prodigy* in the last two hundred years; his composing skills are said to rival those of historical greats such as Mozart and Beethoven. How does he do this? Greenberg told *60 Minutes* that he doesn't know where the music comes from, but he hears it fully written, as if it were being played by an orchestra in his head (*60 Minutes,* November 28, 2004).

We currently know very little about how prodigies come to be. Nor do we know much about another interesting group, known as *savants,* people with minimal intellectual capacities in most domains, but who at a very early age show an exceptional ability in some sort of "intelligent" process, such as math, music, or art. The talents of savants can be quite striking. Oliver Sacks (1995) recounts the story of an artistic savant named Stephen Wiltshire who could reproduce highly accurate drawings of buildings and places years after having only glanced at them (Figure 8.18). Further, his drawings were so remarkable that by the time he was a young teenager, he had a book of his artwork published. However, Stephen has *autism* (see Chapter 13), and it was only with the utmost effort that he was able to acquire language sufficient even for the simplest verbal communication.

According to Gardner, each person has a unique pattern of intelligences so that no one should be viewed as "smarter" than others, just differently talented. As appealing as this idea is, some psychological scientists have questioned whether being able to control body movements or compose music is truly a form of intelligence or whether such abilities should more appropriately be considered specialized talents. If a person is a bit of a klutz or tone deaf, does that make him unintelligent? To date we still have no standardized ways to assess many of Gardner's intelligences, in part because Gardner believes that the tests we do have fail to capture the true essence of intelligence. Instead, Gardner provides exemplars, such as the artist Pablo Picasso, the dancer Martha Graham, the physicist Albert Einstein, and the poet T. S. Eliot. There is no denying that each is especially talented in his or her field, but these individuals were likely talented in many respects, and all might be considered to have been high in general intelligence (Gottfredson, 2004b).

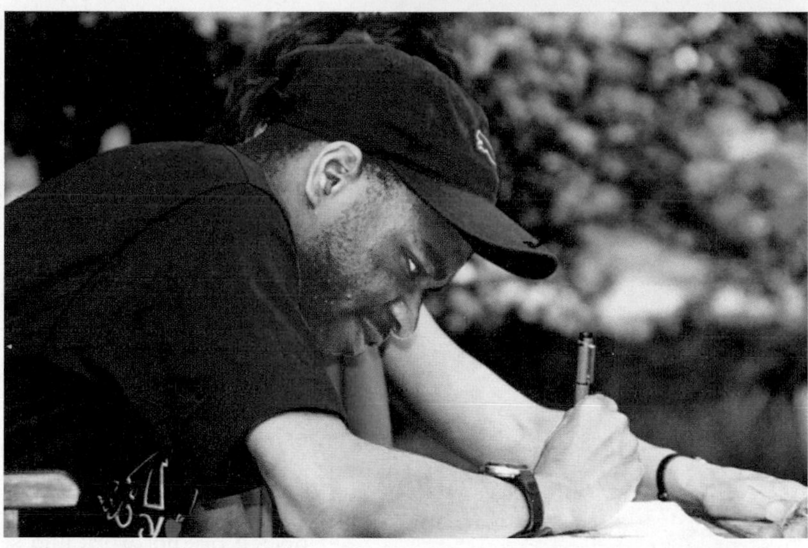

8.18 Stephen Wiltshire Although he has autism, Stephen Wiltshire can produce amazingly accurate drawings after glancing at something.

The Yale psychologist Robert Sternberg (1999) proposed a different type of theory that supports the notion of multiple intelligences. His theory involves three types of intelligence: analytical, creative, and practical. *Analytical intelligence* is similar to that measured by psychometric tests, such as being good in problem solving, analogies, puzzles, and other academic challenges. *Creative intelligence* involves the ability to gain insight and solve novel problems, to think in new and interesting ways. Finally, *practical intelligence* refers to understanding how people deal with everyday tasks, from knowing whether a parking space is large enough for your vehicle, to "reading" people, to becoming an effective leader. Each of these tasks is a challenge in daily life and people who are capable of performing them tend to get ahead. For instance, many successful business leaders and politicians received only marginal grades in college. Bill Gates dropped out of college, yet has developed one of the world's largest companies and now is the richest person alive. The current U.S. president, George W. Bush, was a mediocre student at best. The flamboyant British billionaire Richard Branson did not even attend college; neither did the cosmetic giant Estée Lauder. Each of these individuals clearly has talents that led to phenomenal success in spite of not doing well at traditional academics.

EMOTIONAL INTELLIGENCE Another type of intelligence is *emotional intelligence.* Conceived by the psychological scientists Peter Salovey and John Mayer, and subsequently popularized by the science writer Daniel Goleman, **emotional intelligence (EQ)** is a form of social intelligence that emphasizes the ability to perceive, understand, manage, and use emotions to guide thoughts and actions (Salovey & Mayer, 1990). Emotional intelligence involves such skills as being able to manage frustration, control one's emotions, and get along with other people. People high in EQ recognize emotional experiences in themselves and others and are able to respond to those emotions in a productive fashion. As you will discover in Chapters 9 and 10, emotions can sometimes overwhelm cognition and undermine motivation. When people are upset they sometimes act impulsively and thoughtlessly, for instance, lashing out at others, eating a cupboard full of chocolate, or doing other things they later regret. Being able to regulate mood, resist impulses and temptations, and control behaviour are all important components of EQ. The idea of emotional intelli-

emotional intelligence (EQ) A form of social intelligence that emphasizes the ability to perceive, understand, manage, and use emotions to guide thoughts and actions.

gence has had a large impact in schools and industry, and programs have been designed to increase the emotional intelligence of students and workers. At the same time, some critics have questioned whether EQ really is a type of intelligence or whether it stretches the idea of intelligence too far. What the concept of emotional intelligence highlights is that many human qualities are important, and that independent of whether it is a type of intelligence, emotional intelligence is advantageous for those who have it.

IMPORTANCE OF G Even as intelligence may be defined in different ways, there is likely a general capacity that supports most of them, namely g. Some scholars have viewed g as a statistical anomaly that has no factual basis. Within the last few years, however, a growing number of psychological scientists have concluded that g exists and that it exerts an important influence over life's outcomes (Conway, Kane, & Engle, 2003; Deary, 2001; Garlick, 2002; Gray & Thompson, 2004; Haier, Jung, Yeo, Head, & Alkire, 2005; Plomin & Spinath, 2002). Research has shown that not only does g predict work and school performance, but it also predicts longevity. Indeed, low g is related to early death from a variety of causes, including heart disease, diabetes, stroke, Alzheimer's disease, traffic accidents, drownings, and other accidents (Gottfredson, 2004a; Gottfredson & Deary, 2004). Why might this occur? It may be the case simply because a number of different environmental forces are at work on each of us, such as differential access to health care, dangerous jobs, or other social factors. According to Linda Gottfredson, however, g may directly affect our own health. People who score high on psychometric measures of g may have greater health knowledge, be better able to follow medical advice, understand the link between behaviour and health, and generally be more literate about health issues. As medical knowledge itself rapidly advances and becomes more complex, trying to keep up with and process all of this new information is a daunting challenge. From this perspective, people who are higher in g have an advantage in doing so. This is a provocative idea that warrants further investigation. Should it be true, it has a number of important implications for the medical system and how doctors communicate medical advice.

General intelligence, especially fluid intelligence, seems to be most important for predicting performance in jobs that require fast and creative thinking. Why is this? A number of theorists have proposed that the value of g is that it allows people to adapt quickly to environmental challenges, and the more complex the challenge, the greater the importance of g. Satoshi Kanazawa (2004) suggests that general intelligence is relevant only to evolutionarily novel situations. He notes that for most of human evolution, our ancestors on the savannah experienced few changes in their daily routines. Indeed, the life of a prehistoric hunting-gathering woman was pretty much the same as that of her mother, and of her grandmother, and of her great-grandmother, and so on for thousands of years. Only once in a while did some novel event occur, such as a drought or other natural disaster, and it was at these moments that those high in g had an advantage. According to Kanazawa, g does not matter for recurring adaptive challenges, and so it has little influence over many aspects of daily human life, such as figuring out what to eat, finding mates, recognizing our friends, and raising children. Indeed, people with high g have no advantage in these domains. For instance, there is no evidence that people with high g make better parents (Herrnstein & Murray, 1994). But today we encounter a multitude of phenomena that could scarcely be imagined even two hundred years ago, such as computers, wireless telephones, televisions, airplanes, automobiles, electric appliances, high-rise buildings, international conglomerates, and so on. With these new inventions

Why might people with high IQs live longer?

and institutions, there is great value in being able to adapt to novel challenges and to think creatively. The overall evidence is consistent with the idea that general intelligence is most valuable for tasks that require understanding novel, complex information (Lubinski, 2004). But what aspects of g provide these abilities? We now examine the evidence for the contribution of cognitive and brain processes to intelligence.

Intelligence Is Associated with Cognitive Performance

What cognitive processes are involved in producing intelligence? Being able to take a three-hour exam requires us to stay focused and pay attention, whereas recalling the definition of the word *heraldry* requires us to remember information about the Middle Ages. According to Galton, intelligence is related to the efficiency of how our brains work as well as the possession of keen perceptual skills. He speculated that intelligent people had larger, more efficient brains. Is there any evidence to support these ideas? Some psychological scientists believe that intelligence is supported by low-level cognitive processes, such as speed of mental processing, working memory, and attention. But can we equate these types of brain performance with our level of intelligence?

SPEED OF MENTAL PROCESSING People who are not very intelligent are sometimes described as being "a bit slow." Is that description accurate? Perhaps. On reaction-time tests, people who score high on intelligence tests respond more quickly and consistently than those who score lower on such tests; their responses are slower and more variable (Deary, 2000). One type of reaction-time task might require a person to press a button whenever a stimulus appears on the screen to test what is known as *simple reaction time*. A more difficult task requires a person to choose among various possible responses, again as quickly as possible, depending on the stimulus presented (Saklofske et al., 2000). Scores on intelligence tests are related even more strongly to this *choice reaction time* (Jensen, 1998). It is not simply being fast that explains this finding, because reaction times correlate with intelligence test items that do not require people to respond quickly.

Another line of evidence that supports a relation between general intelligence and speed of mental processing can be found in *inspection time* tasks. Note in Figure 8.19 that a stimulus can be shown in which either side *A* or side *B* is longer. If we present one of these quickly and then quickly cover it up, how much time do you need to inspect the stimulus to be able to identify correctly which side was longer? According to research by the psychologist Ian Deary and his colleagues, those people who need very little time for this task tend to score higher on psychometric tests of intelligence (Deary, 2001). Additional evidence that highly intelligent people's brains work faster is found in studies where the electrical activity of the brain is measured in response to the presentation of some stimulus (recall the psychophysiological methods described in Chapter 2). Brain event-related potentials to various stimuli occur more quickly for people high in intelligence, and the pattern of brain responses is somewhat different than for people low in intelligence (Deary & Caryl, 1997).

The relation between intelligence and mental speed appears to be involved in the greater longevity of people with high IQs. In a recent longitudinal study, those higher in intelligence and those who had faster reaction times at age 56 were much less likely to die in the next 14 years, even after controlling for factors such as smoking, social class, and education. The effect for reaction time was somewhat stronger than for scores on standardized intelligence tests (Deary & Der, 2005). Why people who are mentally quick live longer is uncertain, as is why they tend to be more intelligent.

8.19 Inspection Time Tasks A stimulus (a) is presented and then quickly followed by a mask (b). The participant is asked to decide whether side A or B is longer. Although easy when you have lots of time to examine the stimulus, the mask quickly covers it, making the task more difficult.

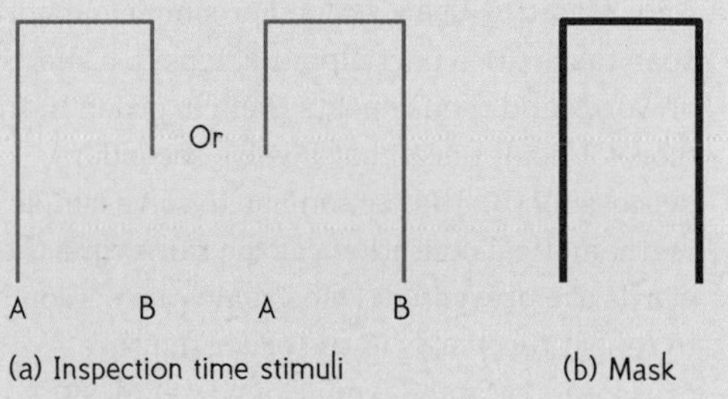

Although the various response-time measures lead to the same conclusion—that intelligence is associated with speed of mental processing—researchers are far from knowing what this actually means. Perhaps being able to process information quickly is just one of the many talents that people who are high in g possess. It may allow them to solve problems or make decisions quickly when doing so is advantageous. As the ultimate fictional example, think about someone such as agent Jack Bauer on the TV show *24*. Bauer is constantly forced to make snap judgments in life-and-death situations. In real life, think about soldiers in combat or emergency-room doctors and their support staff. People's lives hang on their abilities to make good snap judgments.

WORKING MEMORY Over the past decade, a growing number of researchers have noted that general intelligence scores are closely related to working memory (Conway et al., 2003), although the two are not identical (Ackerman, Beier, & Boyle, 2005). Recall from Chapter 7 that working memory is the active processing system that holds information on line so that we can use it for activities such as problem solving, reasoning, and comprehension. In that capacity, working memory might be related to intelligence (Kyllonen & Cristal, 1990; Süß, Oberauer, Wittman, Wilhelm, & Schulze, 2002). Many of the studies examining how working memory and intelligence are related differentiate between simple memory-span tests and those that require some form of secondary processing (Figure 8.20). Simple memory-span tasks require a person to listen to a list of words and then repeat them back in the same order; performance on this task is weakly related to general intelligence (Engle, Tuholski, Laughlin, & Conway, 1999). Working-memory tasks that have dual components, however, show a strong relation between working memory and general intelligence (Gray & Thompson, 2004; Kane, Hambrick, & Conway, 2005; Oberauer, Schulze, Wilhelm, & Süß, 2005). According to Randy Engle and his colleagues, the link between working memory and intelligence is due to the role of attention. In particular, being able to pay attention, especially while being bombarded with competing information or other distractions, allows people to stay on task until successful completion (Engle & Kane, 2004). Given our earlier discussion of how general intelligence is most important in the face of novel, complex tasks, the importance of being able to stay focused and ignore distracting information makes a great deal of sense. This raises the issue of whether brain regions that support working memory are involved in intelligence. We now consider this issue.

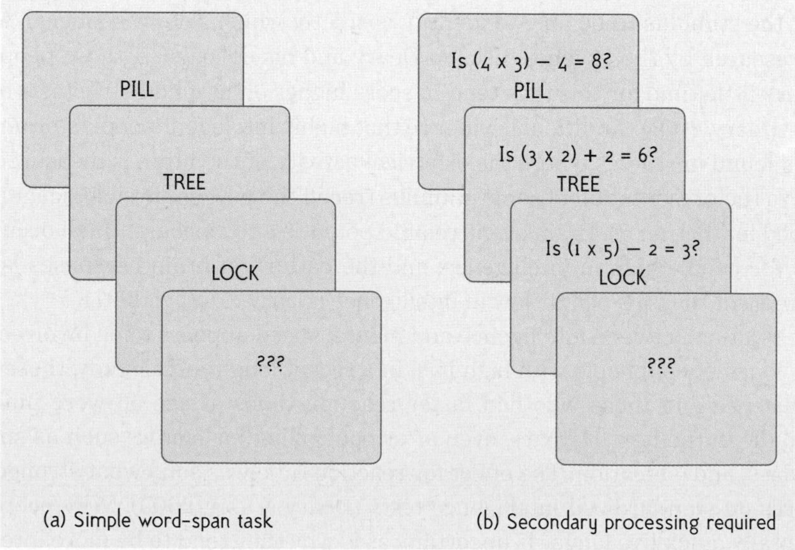

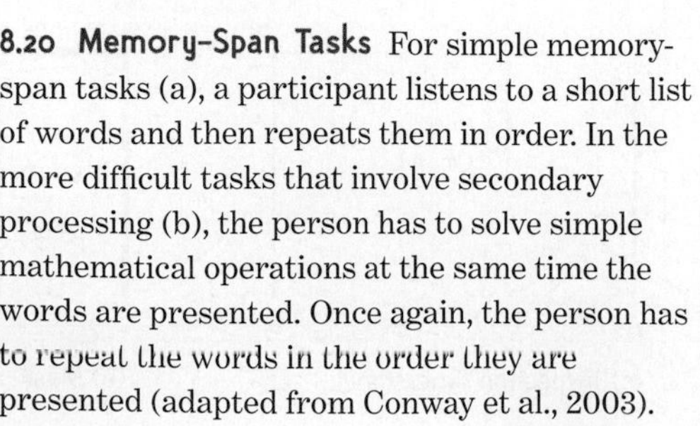
8.20 Memory-Span Tasks For simple memory-span tasks (a), a participant listens to a short list of words and then repeats them in order. In the more difficult tasks that involve secondary processing (b), the person has to solve simple mathematical operations at the same time the words are presented. Once again, the person has to repeat the words in the order they are presented (adapted from Conway et al., 2003).

BRAIN STRUCTURE AND FUNCTION What is the relation between the brain and intelligence? We sometimes call intelligent people "brainy." Maybe it won't surprise you, then, that many studies have documented a clear relationship between head circumference and scores on intelligence tests (Vernon, Wickett, Bazana, Stelmack, & Sternberg, 2000). Head circumference also predicts school performance, although the correlations are quite small (Ivanovic et al., 2004). What researchers are really trying to get at by measuring the circumference of the head is brain size. Studies using magnetic resonance imaging have found a small but significant correlation between brain size and scores on intelligence tests (Haier, Jung, Yeo, Head, & Alkire, 2004). A recent meta-analysis of 37 samples containing more than 1,500 people found that brain volume, as assessed by MRI, explained about 10 percent of the variability between people in general intelligence (McDaniel, 2005). You might wonder if the overall size of the brain matters or whether the sizes of some brain regions (those associated with working memory, planning, reasoning, and problem solving) are related to intelligence. Indeed, studies have found that a larger volume of neuronal cell bodies (gray matter) in the frontal lobes and other brain regions that support attentional control (the anterior cingulate) is related to measures of fluid intelligence (Frangou, Chitins, & Williams, 2004; Haier et al., 2005; Wilke, Sohn, Byars, & Holland, 2003), but not to crystallized intelligence (Gong et al., 2005). This is consistent with evidence that injury to the frontal lobes causes impairments in fluid intelligence, whereas crystallized intelligence remains intact (Duncan, Burgess, & Emslie, 1995).

Consistent with the anatomical and injury findings, studies using functional brain imaging have established that the frontal lobes are particularly important for fluid general intelligence. John Duncan and colleagues (2000) examined activity in people's brains as they performed three different cognitive tasks. Each of these tasks had two forms or levels, based on prior research: a high correlation with g and a low correlation with g. They found that regardless of the type of task, the high g version produced more frontal lobe activity. Jeremy Gray and colleagues (2003) used a working-memory task that on some trials required participants to ignore a distracting stimulus. As you might expect in light of the work on attention described above, the activity in frontal brain regions when the participants had to ignore the distracting information was strongly associated with scores on a measure of fluid intelligence. Considered together, the evidence is substantial that the frontal lobes play an important role in the components of intelligence involving reasoning and problem solving, namely fluid intelligence. Whether these or other brain regions support other components of intelligence, such as practical intelligence, is not yet known.

How might environmental factors affect brain development?

An obvious question is whether differences in brain structure and function are determined by genes or by the environment. Consider head circumference, which we noted was correlated with IQ. Correlation exists between parents and their children in the size of their heads. Yet, childhood nutritional status is highly related to head size, with malnourished children having smaller than expected head circumference and brain growth. Moreover, all of the findings linking brain size and structure to intelligence are correlational, and so it is possible that more education and frequent exposure to intellectual challenges lead to selective increases in some brain regions, just as we observed in Chapter 4 for London taxi drivers with larger hippocampuses. That is, perhaps more intelligent individuals seek out mental challenges, which in turn increase the volume of their frontal lobes (Gray & Thompson, 2004). One recent theory suggests that general intelligence reflects greater plasticity of the brain, such that some people more readily form neural connections during development. According to this view, general intelligence reflects the ability of the brain to adapt quickly to new environments (Garlick, 2002).

How much of our brain organization and cognitive capacity is due to genes and how much to environmental challenges? In the next section we consider the roles of genes and the environment in determining cognitive processing and intelligence.

Genes and Environment Both Influence Intelligence

One of the most contentious battles in psychological science has been over the role of genes in intelligence. This is especially true for the question of whether biological processes can explain IQ differences among groups. This is the quintessential nature-nurture debate: To what extent are individual differences in intelligence due to the effects of genes or to the environment? Throughout this book we have emphasized that both nature and nurture are important; this is especially so for intelligence. A key lesson in Chapter 3 was that gene expression is strongly influenced by external factors, and so to the extent that intelligence has a genetic basis, the environment will influence it. Thus, even if intelligence has a genetic component, how intelligence becomes expressed will be affected by various situational circumstances. The size of someone's vocabulary is considerably heritable, for example, but she has learned every word in it (Neisser et al., 1996). Moreover, which words she learns is also affected by the culture in which she is raised as well the amount of schooling that she receives. With both nature and nurture being crucial to intelligence, it becomes pointless to try to demonstrate which is more important. Instead, psychological scientists recognize that both are important, so they try to identify how genetic and environmental factors contribute to intelligence (Wahlsten, 1997).

BEHAVIOURAL GENETICS Recall from Chapter 3 that the field of behavioural genetics uses twin and adoption studies to estimate the extent to which traits are heritable, that is, the portion of variance in a trait that can be attributed to genes. Numerous studies have been conducted on the behavioural genetics of intelligence, and the conclusion is clear—genes matter. In studies of many thousands of twins, siblings, and adoptees, the evidence suggests that at least 50 percent of the variance in intelligence is because of genetic differences (Plomin & Spinath, 2004; Figure 8.21). Please remember that this doesn't mean that you received half of your intelligence from your genes! It means that genes account for half of the variation in intelligence among people.

Interestingly, the genetic contribution to intelligence becomes stronger over time, such that genes matter the least in childhood, when perhaps parents and teachers exert great influence, and matter the most in adulthood. Thus, adopted children raised in the same household show some similarity in intelligence as children, but as adults they are no more alike than strangers raised in different homes (McGue, Bouchard, Iacono, & Lykken, 1993). Likewise, correlations in IQ for dizygotic (nonidentical) twins drop off over time, whereas the correlations in IQ for identical twins remain strong throughout life (Figure 8.22). Why might gene effects grow stronger over time? One reason is that as children grow older, they are more in control of their environments, and it is possible that the environments people select as they get older are more in keeping with their genes, as when a very bright child spends time at the library rather than playing outside. This highlights the important role that people have in selecting, changing, and creating their own environments.

The quest now is to find the various genes that contribute to intelligence and to understand the conditions under which those genes are expressed. So, for instance, is there a genetic component to brain development? One research team examined monozygotic and dizygotic twins and found that the frontal lobes appeared to be under strong genetic influence, in that monozygotic twins were more similar in frontal lobe

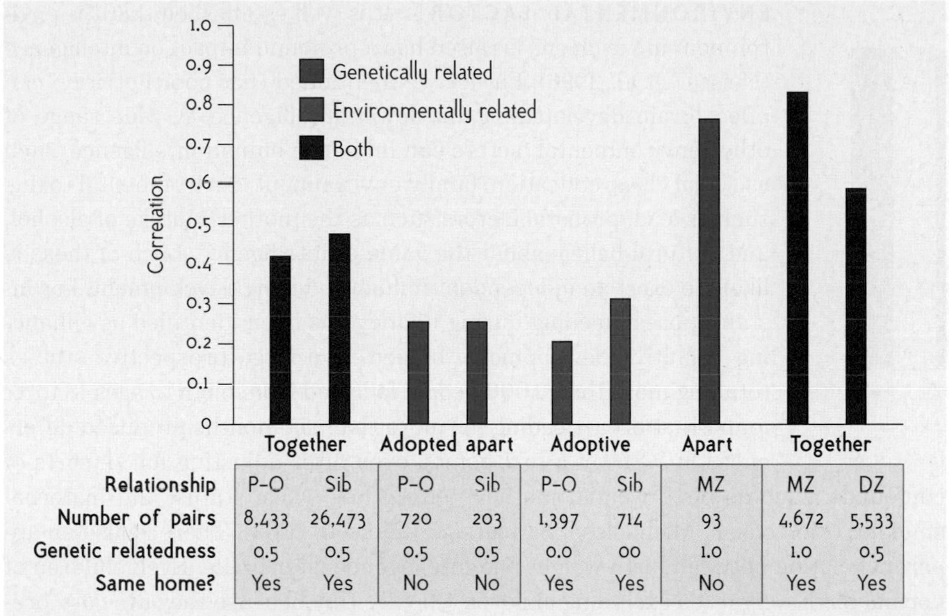

Relationship	P-O	Sib	P-O	Sib	P-O	Sib	MZ	MZ	DZ
Number of pairs	8,433	26,473	720	203	1,397	714	93	4,672	5,533
Genetic relatedness	0.5	0.5	0.5	0.5	0.0	00	1.0	1.0	0.5
Same home?	Yes	Yes	No	No	Yes	Yes	No	Yes	Yes

8.21 Genes and Intelligence Shown are average IQ correlations for family, adoption, and twin study designs. What is clear is that the greater the degree of genetic relation, the greater the correlation in intelligence. The highest correlations are found among monozygotic twins, and this is so whether they are raised in the same household or not. What can also be observed is that siblings raised together show more similarity than siblings raised apart. A similar pattern occurs for the correlation between parent and child; they are more similar when the parent raises the child than when the child is raised by someone else. (P-O = parent, offspring; Sib = siblings; DZ = dizygotic twins; MZ = monozygotic twins)

brain volume than dizygotic twins were, and that for both types of twins there was a relation between the volume of the frontal lobes and intelligence (Thompson et al., 2001). Another approach is to seek specific genes that might contribute to intelligence, for example, by obtaining profiles of gene expression. Although attempts are under way to identify candidate genes that contribute to intelligence, so far researchers have been unsuccessful. One problem with this approach is that perhaps thousands of genes contribute to intelligence, and each gene might have only a small effect. In spite of these difficulties, researchers are actively pursuing efforts to identify these genes using large samples, and it is considered a promising area for future research (Plomin & Spinath, 2004). Finally, another method involves manipulating genes through knock-out or knock-in studies. Recall the Doogie Howser mice from Chapter 7 that became fast learners following genetic manipulation. Researchers have made great advances in understanding genes and their expression over the past few years. It is likely that we will make considerable progress moving beyond the demonstration of a genetic component to intelligence to understanding which genes work together to support cognition and intelligence.

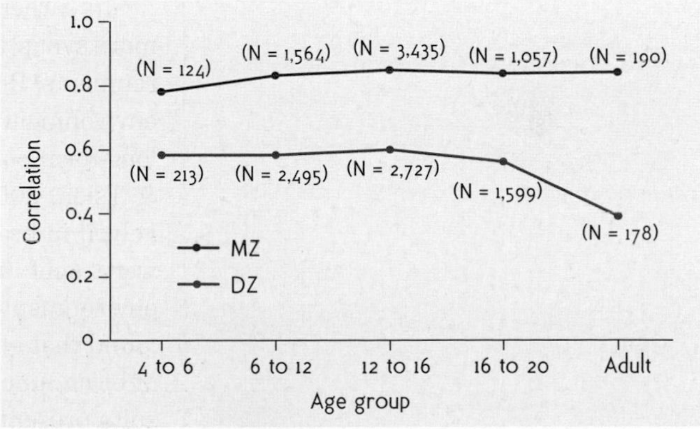

8.22 The Effects of Genetic Influence over Time The correlations between monozygotic twins for intelligence remains stable over time, whereas correlations decrease for dizygotic twins. In other words, nonidentical twins become less similar to each other over time.

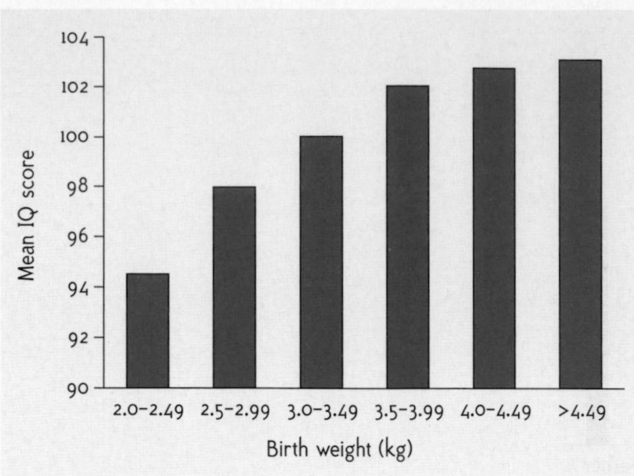

8.23 Birth Weight and Intelligence Among children of normal birth weight, mean IQ scores increase with weight.

ENVIRONMENTAL FACTORS It is well established that the environment in which one is raised has a profound impact on intelligence (Neisser et al., 1996). Earlier we mentioned that poor nutrition can affect brain development and lower intelligence. A wide range of other environmental factors can influence human intelligence, such as social class, education, family environment, environmental toxins such as lead, prenatal factors such as the mother's intake of alcohol, and cultural beliefs about the value of intelligence. Each of these is likely to exert an independent influence during development. For instance, breastfeeding during infancy has been identified as enhancing cognitive development. Indeed, two large prospective studies totalling more than 3,000 people followed from birth to ages 18 to 27 found that breastfeeding for more than six months produced differences in IQ from 5 to 7 points, even after adjusting for all sorts of confounds, such as birth weight, mother's education, social status, and maternal smoking (Mortensen, Michaelsen, Sanders, & Reinisch, 2002). There is also an apparent relation between birth weight and intellegence later in life. Even children of normal birth weight (so excluding children who are tiny because they are very premature) show a small positive association between birth weight and intelligence (Figure 8.23), although this effect is smaller than that of social class (Shenkin, Starr, & Deary, 2004).

The intellectual opportunities a child receives also affect intelligence. Stephen Ceci (1999) notes that schooling, for instance, makes an important contribution to intelligence. The longer that children remain in school, the higher their IQs will be. Students who start school early because of their birthdate have higher test scores than their same-age peers who start school later. It is clear that schooling not only builds knowledge but also teaches critical thinking skills such as being able to stay focused, think abstractly, and learn strategies for solving problems (Neisser et al., 1996). Considered together, the evidence is considerable that environmental factors contribute to intelligence. One sign of this is that IQ scores have risen quite dramatically during the last century of intelligence testing; this rise has been called the *Flynn effect* after James R. Flynn, the researcher who first described it (Flynn, 1981, 1987). Since genes cannot have changed much during this period, the increase must be due to environmental factors. Some possible explanations for the Flynn effect include better nutrition and health care, the refinement of education methods, longer school years, and smaller families with more intensive parenting, as well as exposure to technology and computers. The mean IQ score has not changed, however, because the various tests of intelligence have been re-standardized on a number of occasions over time so that the mean remains at 100.

In earlier chapters we noted that rats raised in enriched environments showed more synaptic connections and larger neurons than those raised in impoverished environments. Research from a number of different laboratories has shown that enriched environments enhance learning and memory as well, for example, by sustaining long-term potentiation (Lambert, Fernandez, & Frick, 2005; Tang, Wang, Feng, Kyin, & Tsien, 2001). The implication is that the environment influences how genes involved in brain development are expressed. In one study, genetically identical mice were split into groups that were exposed to different levels of an "enriched" environment—a cage with toys, tunnels, boxes, and the like. These researchers found that enrichment was associated with activation of genes involved in a number of brain functions, including forming new synapses (Rampon et al., 2000). These results present a clear affirmation that our environment can affect properties associated with intelligence by influencing—at least in part—the expression of our genes. Although it would be unethical to assign children randomly to enriched and impoverished environments, it is clear that humans as well as mice gain clear advantages from living in stimulating environments.

GROUP DIFFERENCES The most controversial aspect of intelligence testing over the last century has been the idea that genetics can explain overall differences in intelligence scores between different racial groups. In particular, one ongoing debate concerns differences in how American blacks and whites score on measures of intelligence. Multiple studies over the past 30 years have found that whites score about 10 to 15 points higher on average than blacks on most measures of intelligence. At issue is what causes the difference, not whether the difference exists. Given the importance of intelligence to educational and career attainment, claims that some groups are superior to others require close scrutiny because of the values of our society. This hotly debated topic will also challenge our critical thinking skills.

The first issue to be considered is whether "race" is a biologically meaningful concept. Many psychologists believe that it is not. Indeed, a recent special issue of the flagship journal of the American Psychological Association, *American Psychologist* (January 2005), was devoted to the subject. Various authors noted that the vast majority of genes are identical among people (perhaps as many as 99.9 percent). Although we differ in physical attributes (hair colour, skin colour, and so on), we do so to varying degrees depending on our entire ancestry, which raises troublesome issues of classification. Is Tiger Woods black or Asian? He has ancestors from both groups. Most methods of classifying race depend on self-report, in which people group themselves into categories. The rise in interracial families is making it harder for some people to select only one category. In regard to this discussion, it is clear that some genetically based biological differences do exist among people who identify themselves as black or white, such as skin colour. What is not clear is whether these biological differences have any relevance for the mental capacities that underlie intelligence.

Why do minority group members score lower on IQ tests in different cultures around the world?

Whether race is real or not, it is not scientifically appropriate to conclude that genes cause differences among groups if there are any environmental differences among the groups. Richard Lewontin (1976) provided an excellent example of the difficulties of contrasting groups that differ in their circumstances. Consider seeds that are planted in two separate containers (Figure 8.24). In one, the soil is rich, and the seeds receive regular watering, nutrients, and sunlight. The other planter has poor soil, restricted water, and intermittent sunlight. Within both planters, differences

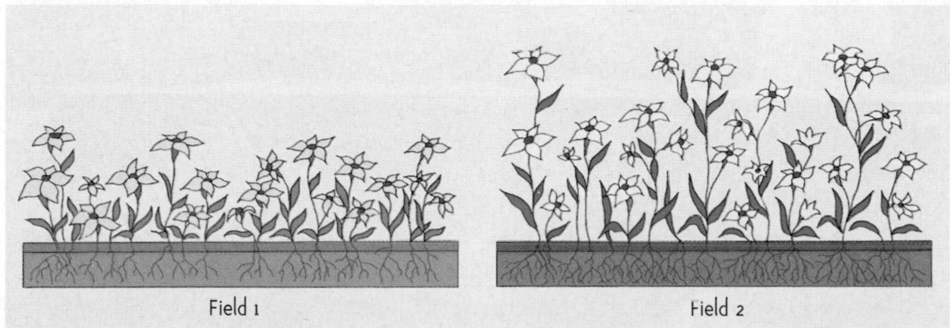

8.24 Variability in Fields Crops grown from similar seeds will vary in both fields, but on average, the crops will grow better when the soil is rich, as is the case in Field 2.

between the plants in their growth can be attributed to some genetic difference in the seeds. After all, the environment is identical, so only genes can explain the differences. But obviously the flowers in the separate containers will differ because of their different environments. The enriched environment helps the seeds reach their potential, whereas the impoverished environment will stunt growth.

There are many differences between black and white Americans. On average, blacks make less money and are more likely to live in poverty. On average, they also have fewer years of education and lower-quality health care, and they are more likely to face prejudice and discrimination. Around the world, minority groups that are the target of discrimination have lower intelligence scores on average, such as the Maori in New Zealand, the Burakumin in Japan, and the Dalits, or "untouchables," in India. John Ogbu (1994) argues that the treatment of minority-group members can make them pessimistic about their chances of success within their cultures. The potential result is that they are less likely to believe that hard work will pay off for them, which may lower their motivational levels in terms of performance. Although each of these explanations is plausible, at this time there is no clear-cut basis for understanding the differences in test scores between black and whites (Neisser et al., 1996). This does not mean, however, that we should not work to make adjustments in the way we measure intelligence to try to remove as much bias as possible against different groups of peoples.

Psychological Science *in Action*

Introducing the New SAT

On Saturday, March 12, 2005, many anxious students who were applying to study at American universities sat down to take a test that will play a vital role in determining which schools will admit them. The test takers were especially nervous because this was the first time that the new SAT was being administered (Figure 8.25). Doing well on the SAT is important, and many high school students spend considerable time and money preparing. Numerous SAT tutoring services are available for those

8.25 The New SAT American high school students face considerable pressure to perform well on the SAT, a standard component of the college application process.

who can afford them. This is not the first change for the SAT. Originally designed to measure innate intelligence and called the Scholastic Aptitude Test, it later became the Standardized Assessment Test. Now SAT doesn't stand for anything. The initials have no meaning—it is just the SAT, temporarily the *new* SAT. How did the old SAT become the new SAT?

In 2001, the psychologist Richard Atkinson launched a vigorous debate on the use of the SAT in college admission decisions. As president of the entire University of California system, Atkinson was in a powerful position to influence the makers of the SAT. Atkinson noted in a speech to the American Council on Education that his concerns about the SAT were sparked when he visited his 12-year-old granddaughter's private school and found the students practicing specific drills for the SAT. The 12-year-olds, he said, "spend hours each month—directly and indirectly—preparing for the SAT, studying long lists of verbal analogies such as 'untruthful is to mendaciousness' as 'circumspect is to caution.' " Atkinson believed not only that the SAT was distorting how children were being taught in schools, but also that it was not doing a good job of predicting how students would do in college, which raised doubts about its validity. As a good psychological scientist, he conducted research to examine how the SAT predicted student performance. His research found that high school grades and scores on tests of class material were better predictors than the SAT of first-year college grades.

In addition to his concern that the SAT was not a good predictor of college performance, Atkinson also believed that the current system of administering the tests may be unfair to those unable to pay for SAT preparation courses. Indeed, black and Hispanic students tend to score behind other groups on the SAT, and this might be attributable partly to their limited access to preparation programs. According to Stanford's Claude Steele, another problem affecting minority student performance on standardized exams is *stereotype threat,* anxiety that results from fears of doing badly and confirming societal stereotypes of racial inferiority in intelligence. According to Steele (1997), minority students who worry that they might confirm negative stereotypes become anxious and preoccupied, and this interferes with their ability to perform well on tests. For instance, when equally intelligent (as assessed by SAT scores) black and white Stanford students took a difficult test of verbal skills and were told that the test was unrelated to intelligence, black and white students performed at the same level. However, when the students were told the test was a measure of intelligence, black students did quite poorly (Figure 8.26; Steele & Aronson, 1995). Similarly, women do more poorly when taking an exam on which they believe men typically outscore women, but they do as well as men on the same test if they do not hold such a belief (Spencer, Steele, & Quinn, 1999). In an especially intriguing example of stereotype threat, researchers found that Asian American women did well on a math test when the "Asians are good at math" stereotype was primed and poorly when the "women are bad at math" stereotype was primed (Shih, Pittinsky, & Ambady, 1999). Stereotype threat is a problem because it contaminates the scores of minority test takers. The possible inaccuracy of these scores undermines the goal of the SAT, which is to provide a standardized assessment of all students taking the test. Recent research demonstrates that teaching test takers about stereotype threat may eliminate its negative effects (Johns, Schmader, & Martens, 2005). This provides another example of how research in psychological science contributes practical solutions to real-world problems.

8.26 Stereotype Threat When the test was described as diagnostic, stereotype threat led the black students to perform poorly.

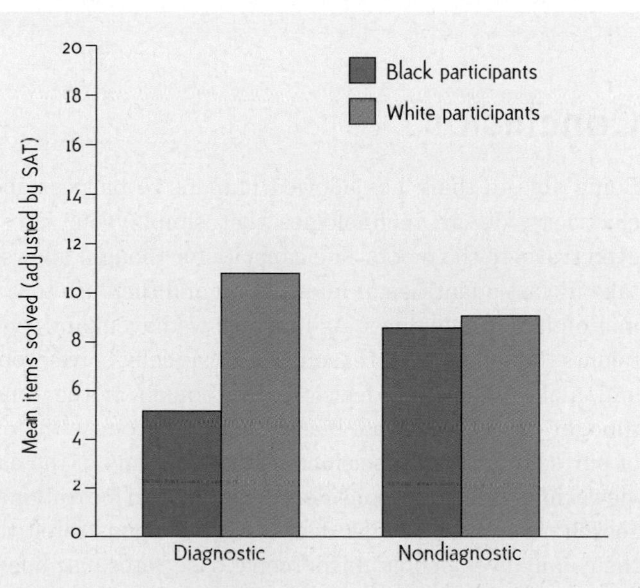

In response to various concerns about the SAT, many American colleges and universities have placed more emphasis on school performance during the admissions process. In addition, the new SAT now includes an assessment of classroom learning and deemphasizes innate aptitude. According to the makers of the SAT, the reasoning component is now designed to assess critical thinking skills that students will need in college. It also emphasizes achievement more than aptitude. Through careful research documenting problems with using the SAT as the major determinant of college admission, Richard Atkinson was able to use his psychological training to instigate sweeping changes both in the admissions process and in how relevant learning skills are assessed in students. ■

REVIEWING the Principles | How Do We Understand Intelligence?

Intelligence refers to human ability to reason, solve problems, think quickly and efficiently, and adapt to environmental challenges. The psychometric approach reveals multiple components to intelligence, but also a central dimension that has been called general intelligence, or g. Fluid intelligence is related to cognitive measures, such as working memory, as well as other measures of speed of mental processing. Crystallized intelligence reflects knowledge that we acquire. Other forms of intelligence have been identified, although there is debate regarding whether they reflect the same kind of intelligence as that which is more traditionally defined. Both genes and the environment influence intelligence, and researchers are now seeking to understand how environmental factors lead to differential gene expression. Finally, genes cannot be used to account for group differences in intelligence when there are environmental differences.

Conclusion

Being able to think has allowed humans to harness and use electricity, design technologies that simplify our lives, and even travel to the moon. The capacity for thought allows us to take in vast quantities of information and then use that information in our daily lives. As humans, we have many cognitive talents that allow us to reason, think critically, solve problems, and make important decisions. But much of the time our thought processes reveal biases that compromise the quality of our solutions and decisions. Although many of the earliest views in psychology considered humans to be rational, research over the past few decades has demonstrated that in their daily lives people misperceive base rates, use anecdotal information, engage in counterfactuals, and fail to make accurate predictions about how future events will affect them. Much of the time people do not think deeply, but rather rely on heuristics, many of which are unconscious thought processes. Although heuristics are often adaptive and efficient, they can sometimes lead to unfortunate outcomes. People can also get caught in mental ruts that keep them from solving problems. We need to use knowledge effectively to solve problems and achieve goals, but doing so is not always easy.

Some people use information more effectively than others. Understanding the nature of intelligence has long been of interest to psychological scientists. We have vigorously debated most features of it, such as whether there is one monolithic way to be smart or whether there are multiple types of intelligence. The reality is likely in the middle ground. An underlying dimension—general intelligence, or g—predicts performance above and beyond our specific abilities, such as being good at algebra or poetry. Research in cognitive psychology and cognitive neuroscience has revealed new insights about general intelligence and its relation to working memory and attention. Throughout the chapter we have seen that living in the modern world places more and more emphasis on adapting to novelty, and therefore modern society places a premium on g.

How do we make people more intelligent? The answer appears to be that we are doing it already. By emphasizing early childhood and prenatal care and nutrition, by building good schools that stimulate and challenge, and by developing new technologies, society has made possible considerable gains in intelligence. An important goal, of course, is to ensure that each person, irrespective of race, ethnicity, or social group, can benefit from these circumstances, a goal that remains unfulfilled. Even very smart people fail to act rationally sometimes. That appears to be just as much a feature of human cognition as the talents that allow us to travel to the moon.

Summarizing the Principles of Thinking and Intelligence

How Does the Mind Represent Information?

1. **Mental images are analogical representations:** Thoughts can take the form of visual images. The primary visual cortex is activated proportionately to the size of an image in the mind's eye. Symbolic knowledge affects the ways in which we use visual imagery.

2. **Concepts are symbolic representations:** Concepts are mental representations of categories of subtypes and may be formed by defining attributes, prototypes, or exemplars. The boundaries of a category may be fuzzy.

3. **Schemas organize useful information about the environment:** We develop schemas based on our real-life experiences. Scripts are schemas that allow inferences about the sequence of events.

How Do We Make Decisions and Solve Problems?

4. **People use deductive and inductive reasoning:** Deductive reasoning follows from a general statement to specific applications. Syllogisms are formal structures of deduction. Inductive reasoning draws a general conclusion from specific instances.

5. **Decision making often involves heuristics:** Expected utility models assume normative, logical processes. Descriptive models highlight reasoning shortcomings—specifically the use of mental shortcuts (i.e., heuristics) that easily come to mind. We also select information to confirm our conclusions, avoid loss or regret, and to be consistent with the framing of a problem.

6. **Problem solving achieves goals:** Problem solving involves reaching a goal that is usually broken down into subgoals. Insights come suddenly by seeing elements of the problem in new ways. Restructuring aids solutions; mental sets and functional fixedness inhibit solutions. Unconscious processing (i.e., incubating) aids problem solving, and verbal overshadowing can confuse.

How Do We Understand Intelligence?

7. **Intelligence is assessed by psychometric tests:** The Binet-Simon was the first test of mental ability and lead to the concept of IQ as a ratio of mental age and chronological age. This was later normed to a distribution with a mean of 100 and standard deviation of 15; therefore average ability is from 85 to 115. The question of the validity of these tests persists, with cultural bias as a significant criticism.

8. **General intelligence involves multiple components:** Using the factor analysis of test performance, Charles Spearman concluded that a general intelligence component existed, known as g. Fluid intelligence involves problem-solving strategies whereas crystallized intelligence is accumulated knowledge. Howard Gardner has proposed a theory of multiple intelligences that include linguistic, mathematical/logical, spatial, bodily-kinesthetic, intrapersonal, and interpersonal abilities. Additionally, Robert Sternberg has proposed that there are three types of intelligence: analytical, creative, and practical. Finally, emotional intelligence is the ability to understand and use emotions.

9. **Intelligence is associated with cognitive performance:** Speed of mental processing is part of intelligence (i.e., choice reaction times and inspection times). The relationship of working memory to intelligence seems to be with

regard to attention. Various components of the brain (i.e., size and activity of the frontal lobes) have been found to relate to qualities of intelligence.

10. **Genes and environment both influence intelligence:** Behavioural genetics has revealed the substantial influence of genetics in setting the limits of the expression of intelligence. Environmental factors, including breastfeeding, intellectual opportunities, nutrition, schooling, and parenting, seem to establish where within the genetic limits IQ falls. Assumed racial differences in intelligence based on genetics are invalid.

Applying the Principles

1. Think about what the brain looks like. Now rotate the image so that you are only examining the brain's left hemisphere. Your ability to do this demonstrates what type of mental representation? What is one region of your brain that is involved in processing this representation?

    ____ a) analogous representation; primary visual cortex
    ____ b) symbolic representation; temporal lobe
    ____ c) iconographic representation; right hemisphere of the cortex
    ____ d) virtual representation; frontal cortex

2. You want to go to a movie this weekend but you can't decide which one to see. If you ask a variety of friends if they have seen and liked a particular film and base your decision on their answers, you would be using ________. If you know the film stars an actor you like and therefore conclude that it will be a good movie, you are using ________.

    ____ a) inductive reasoning; deductive reasoning
    ____ b) the scientific method; the representative heuristic
    ____ c) loss aversion; confirmation bias
    ____ d) expected utility; the availability heuristic

3. Dave spent the weekend away from campus with his grandparents. He enjoyed discussing with his grandmother, who is a retired educator, what he has been studying in his psychology class. He was impressed by how much she knows. Furthermore, he found it interesting how quickly she understood the complexities of his new cell phone. His grandmother's general knowledge exemplifies her ________ intelligence, whereas her ability to figure out the cell phone exemplifies her ________ intelligence.

    ____ a) verbal; spatial
    ____ b) crystallized; fluid
    ____ c) fluid; spatial
    ____ d) g; multiple

ANSWERS: *1. a 2. a 3. b*

Key Terms

analogical representation, p. 298
availability heuristic, p. 309
cognition, p. 298
concept, p. 301
confirmation bias, p. 310
crystallized intelligence, p. 325
deductive reasoning, p. 306
defining attribute model, p. 301
emotional intelligence (EQ), p. 327
fluid intelligence, p. 325
framing, p. 310
functional fixedness, p. 316
general intelligence (g), p. 325
heuristics, p. 308
inductive reasoning, p. 307
insight, p. 314
intelligence, p. 321
intelligence quotient (IQ), p. 323
mental age, p. 323
mental set, p. 316
multiple intelligences, p. 325
prototype model, p. 302
representativeness heuristic, p. 309
restructuring, p. 315
symbolic representation, p. 298

… Readings

…telligence: A very short introduction. …rsity Press.

… The power of thinking without … Little, Brown.

… Image and brain. Cambridge, MA: …

Me… …oss, B. H., & Markman, A. B. (2005). *Cog… psychology* (4th ed.). New York: Wiley.

Murphy, G. L. (2002). *The big book of concepts.* Cambridge, MA: MIT Press.

Neisser, U. (1976). *Cognition and reality: Principles and implications of cognitive psychology.* New York: Freeman.

Pinker, S. (1997). *How the mind works.* New York: Norton.

Schwartz, B. (2004). *The paradox of choice: Why more is less.* New York: Ecco.

Further Readings: A Canadian Presence in Psychology

Darou, W. G. (1992). Native Canadians and intelligence testing. *Canadian Journal of Counseling, 26,* 96–99.

Harvey, C. B., Manshu, Z., Bio, K. C., & Jue, Z. F. (1986). Spatial conceptions in Chinese and Canadian children. *Journal of Genetic Psychology, 147,* 457–464.

Paulhus, D. L., Lysy, D. C., & Yik, M. S. M. (1998). Self-report measures of intelligence: Are they useful as proxy IQ tests? *Journal of Personality, 66,* 525–554.

Pelissier, M. C., & O'Connor, K. P. (2002). Deductive and inductive reasoning in obsessive-compulsive disorder. *British Journal of Clinical Psychology, 41 (Pt. 1),* 15–27.

Saklofske, D. H., Hildebrand, D. K., & Gorsuch, R. L. (2000). Replication of the factor structure of the Wechsler Adult Intelligence Scale—Third edition with a Canadian sample. *Psychological Assessment, 12,* 436–439.

Vernon P. A., Wickett, J. C., Bazana, P. G., Stemack, R. M., & Sternberg, R. J. (2000). The neuropsychology and psychophysiology of human intelligence. In R. J. Sternberg (Ed.), *Handbook of intelligence* (pp. 245–264). Cambridge: Cambridge University Press.

Vernon, P. E. (1979). *Intelligence: Heredity and environment.* San Francisco, W .H. Freeman.

Wahlsten, D. (1997). The malleability of intelligence is not constrained by heritability. In B. Devlin, S. E. Fienberg, D. P. Resnick, & K. Roeder (Eds.), *Intelligence, genes and success*. New York: Copernicus.

Overcoming the Challenges of Life

Setting goals helps to motivate, or energize, our behaviour. Many psychological scientists assert that we are more persistent when pursuing especially challenging goals than when we pursue less difficult ones. At age 18, Terry Fox was diagnosed with osteogenic sarcoma (bone cancer), and had to have his right leg amputated 15 centimeters (6 inches) above the knee. After that, Terry set a goal to run from one end of Canada to the other in an effort to raise money for cancer research. He began his legendary journey on April 12, 1980, and traversed more than 5,300 kilometers (3,300 miles) from St. John's, Newfoundland to Thunder Bay, Ontario, ultimately having to end his trek early when his cancer returned. Accomplishing such a feat has inspired thousands of others, and to date the Terry Fox Foundation has raised over $400 million for cancer research.

CHAPTER

9

Motivation

Oprah Winfrey is one of the most successful women in North American history (Figure 9.1). Starting as a junior reporter in Nashville, Tennessee, Oprah rose through the ranks to become one of the most influential and important people in the entertainment industry. In 2004, *Time* magazine named her one of the 100 most influential people of the twentieth century. Her television show has changed daytime television; her power is so great that a single positive reference on her show can make any book a bestseller. Her many other successes include acting and producing, becoming one of the few women to own a thriving production company, and she is also a restauranteur, opening a popular restaurant in Chicago. Raised in poverty and bounced from house to house, she is now a wealthy humanitarian whose Angel Network has provided financial support to thousands.

Outlining the Principles

How Does Motivation Activate, Direct, and Sustain Behaviour?

- Needs, Drives, and Arousal Motivate Behaviour
- Pleasure Can Motivate Adaptive Behaviours
- Some Behaviours Are Motivated for Their Own Sake

Why Are Human Beings Social?

- Humans Have a Fundamental Need to Belong
- People Seek Others When They Are Anxious
- People Are Motivated to Detect and Reject Cheaters

How Do People Achieve Personal Goals?

- Good Goals Enhance Confidence
- Self-Regulation Requires Self-Awareness and Delay of Gratification
- The Frontal Lobes Are Important for Self-Regulation

What Determines How We Eat?

- Time and Taste Play Roles
- Culture Determines What We Eat
- Multiple Neural Processes Control Eating

What Is Addiction?

- Addiction Has Psychological and Physical Aspects
- People Use—and Abuse—Many Psychoactive Drugs
- Alcohol Is the Most Widely Abused Drug

What Factors Motivate Sexual Behaviour?

- Biological Factors Influence Sexual Behaviour
- Cultural Scripts and Rules Shape Sexual Interactions
- Mating Strategies Differ between the Sexes

9.1 Oprah Winfrey

Oprah Winfrey is the embodiment of the modern dream. What has enabled this talented woman to achieve such success? It is clear from biographies of Oprah that she has been driven to succeed since childhood. She is a self-proclaimed workaholic, once stating about work: "This is all I do. I do this and I do it till I drop. I work, and on weekends I go as many places as I can to speak." Oprah's aspirations and goals are lofty, but she has attained many of them and then sought to achieve even more. By all accounts, she is as motivated as is humanly possible.

Yet Oprah faces ongoing personal battles in her everyday life. Like many of us, she has struggled throughout her life with her body weight. In 1988 Oprah slimmed down from nearly 220 pounds to 145 pounds, and exclaimed that doing so was the "greatest single achievement" of her life. Unfortunately, as is the case for most dieters, Oprah regained much of the weight. Over the past decade she has endured a constant struggle to control her eating and to maintain a body weight she finds acceptable. A self-described food addict, Oprah has found dieting success only when her eating and exercise habits are severely regimented and her actions supervised by personal trainers and chefs.

This raises a large question for many of us. How can we be so successful in some aspects of life and yet struggle so mightily in others? The simplest answer is that many different forces compel our behaviour. We struggle even more when these forces conflict. Listen to what Oprah had to say in a recent interview on this very topic. When asked during an interview, "What keeps you motivated?" Oprah responded, "To work or to work out? To work—it is a mission for me, this show, more than just a talk show. I want people to see things on our show that make them think differently about their lives. That makes them better. For working out, my motivation is I don't want a fat butt!" Oprah's eating problems can be viewed as a basic conflict between her desire to eat good-tasting food and her desire to see herself as attractive. But why is this so difficult, especially for someone who has such an amazing capacity for success in other aspects of her life?

Oprah Winfrey is hardly alone in her quest to lose weight; indeed, more and more people in many Western countries are becoming overweight. The struggle to lose weight can sometimes seem a battle between body and mind—the mind trying to resist the body's urges to eat. As biological machines shaped by evolutionary forces, our bodies require certain things to

1890
Instincts as Motives Greatly influenced by Darwin's ideas, William James proposes that human beings are endowed with behavioural instincts that require no experience or education, only environmental triggers to activate them.

1908
Yerkes-Dodson Law Robert Yerkes and John Dodson are the first to observe that there is an inverted U-shaped function between arousal and performance, with the best performance on learning tasks occurring at moderate levels of arousal.

1920s
Homeostasis In *The Wisdom of the Body*, Harvard physiologist Walter B. Cannon describes the process by which bodily systems, such as temperature and digestion, are regulated to maintain a steady state.

1940s
Drives Satisfy Needs Clark Hull proposes that states of biological deprivation, or drives, stimulate behaviours to satisfy needs. Drive theory becomes one of the most influential theories in the history of psychological science.

keep going, and to get those things the body encourages us to engage in specific behaviours. Many of these behaviours are things we enjoy, such as eating and having sex. As you will see, common brain systems make these behaviours rewarding. But modern living has provided new challenges that our bodies were not designed to handle, such as easily available food and substances such as drugs and alcohol that activate our basic reward circuitry. Each day we feel a tug between engaging in these behaviours that bring us pleasure and living up to the demands of society. William James noted that on a cold morning it takes an enormous amount of willpower to haul oneself out of a cozy bed. Throughout this chapter, we will consider how people struggle for self-control every day, how often we need to choose one action over another, forgoing immediate satisfaction to achieve our longer-term goals. Understanding motivation will provide insights into many aspects of our daily lives. ■

Motivation (from Latin, "to move") is the area of psychological science that studies the factors that energize, or stimulate, behaviour. Specifically, it is concerned with how behaviour is initiated, directed, and sustained. This concern leads to the study of physical factors such as the need for sleep and food, as well as the psychological factors that inspire people to set goals and try to achieve them. Many times people attempt to override their biological needs, as when they try to go without sleep or food, and often they find that doing so is difficult. How much can people prevail over their bodies in attempting to reach personal goals? As a student, you probably are highly motivated to succeed, and therefore you spend many of your waking hours reviewing notes, reading textbooks, and writing papers, perhaps resisting the temptation to meet with friends or have some fun. Although doing well in this course may not aid your immediate survival, it does help you achieve long-term aspirations.

motivation Factors that energize, direct, or sustain behaviour.

This chapter explores different conceptions of the basis of motivation and discusses some of the most commonly motivated behaviours. These behaviours relate to immediate biological survival (such as responding to hunger and thirst), as well as to long-term social needs (such as affiliation and achievement). Issues of motivation are spread throughout the many levels of analysis of psychological science. For instance, the concepts of reward and reinforcement discussed in Chapter 6 and the physiological mechanisms discussed in Chapter 3 are central to theories of motivation. As you read through this chapter, keep in mind two important questions: What motivates me in my daily life? How can I manage, direct, and sustain those motivating forces so that I can live a meaningful life?

1948
Kinsey Reports Alfred Kinsey conducts landmark research on human sexuality, publishing *Sexual Behaviour in the Human Male* in 1948 and *Sexual Behaviour in the Human Female* in 1953.

1954
Dual-Centre Theory Eliot Stellar proposes that motivated behaviour is regulated by excitatory and inhibitory centres in the hypothalamus. The hypothalamus is later found to be crucial for many motivated behaviours.

1959
Affiliation and Anxiety Stanley Schachter examines how anxiety affects the human desire to affiliate with others. He concludes that the need for social comparison is an important motivator of affiliation.

1960
Plans and Goals George Miller, Eugene Gallanter, and Karl Pribram develop a cognitive model of self-regulation, in which people are motivated by discrepancies between their situation and an ideal state represented as a goal.

1960s
Delay of Gratification Walter Mischel and his students conduct research showing that children who are best able to delay gratification are later able to perform at a superior level in many domains.

need State of biological or social deficiencies within the body.

need hierarchy Maslow's arrangement of needs, in which basic survival needs are lowest and personal growth needs are highest in terms of ultimate priority.

self-actualization A state that is achieved when one's personal dreams and aspirations have been attained.

drive Psychological state that motivates an organism to satisfy its needs.

arousal Term to describe psychological activation, such as increased brain activity, autonomic responses, sweating, or muscle tension.

homeostasis The tendency for bodily functions to maintain equilibrium.

How Does Motivation Activate, Direct, and Sustain Behaviour?

Theories of motivation seek to answer questions such as: Where do needs come from? How are goals established? What determines which needs are most important? How are motives converted into action? Most general theories of motivation emphasize four essential qualities of motivational states. First, motivational states are *energizing* in that they activate or arouse behaviours—they cause animals to do something. For instance, the desire for fitness might motivate you to get out of bed and go for a run on a cold morning. Second, motivational states are *directive* in that they guide behaviours toward satisfying specific goals or needs. Hunger motivates eating; thirst motivates drinking; pride (or fear or many other things) motivates studying hard for exams. Third, motivational states help people to *persist* in their behaviour until goals are achieved or needs are satisfied. Hunger gnaws at you until you find something to eat; persistence drives you to practice foul shots until you succeed. Fourth, most theories agree that motives differ in *strength,* depending on both internal and external factors. In this next section we will look at a wider range of forces that motivate our behaviours.

Needs, Drives, and Arousal Motivate Behaviour

What do we really need to do to stay alive? For one, we have to satisfy our biological needs. A **need** is a state of deficiency, such as a lack of air or food. You *need* air and food to survive. But satisfying our basic needs is not enough to live a fully satisfying life. We also have social needs, such as the need for achievement and the need to be with others. People *need* other people, as we shall see. Needs, however they are defined, lead to goal-directed behaviours. Failure to satisfy the need leads to psychosocial or physical impairment.

Abraham Maslow proposed an influential need theory of motivation in the 1940s. Maslow believed that humans are driven by many needs, which he arranged into a **need hierarchy** (Figure 9.2), in which survival needs (such as hunger and thirst) were lowest and personal growth needs were highest in terms of ultimate priority. Maslow believed that satisfaction of lower needs in the hierarchy allowed humans to function at a higher level. People must have their biological needs met, feel safe and secure, feel loved, and have a good opinion of themselves in order to experience personal growth.

Which needs are most important?

Maslow's theory is an example of *humanistic psychology,* in which people are viewed as striving toward personal fulfillment. Humanists focus on the *person* in motivation—it is John Smith who desires food, not John Smith's stomach. According to the humanist perspective, human beings are unique among animals because we continually try to improve ourselves. A state of **self-actualization** occurs when someone achieves his or her personal dreams and aspirations. The self-actualized person has lived up to his or

1960s
Optimal Arousal Daniel Berlyne proposes that people seek to maintain an optimal level of arousal.

1966
Human Sexual Response William Masters and Virginia Johnson record physical changes that occur during human sexual activity and find four basic phases of physical responding. Their work launches the therapeutic treatment of sexual disorders.

1974
Undermining Intrinsic Motivation Mark Lepper and his colleagues demonstrate that giving external rewards for behaviours that are enjoyable for their own sake undermines the natural motives to engage in those behaviours.

1987
Dopamine and Addiction Roy Wise and Michael Bozarth propose that dopamine is the primary neurotransmitter involved in reward and addition: Activation of dopamine receptors is associated with feelings of pleasure.

1990s
Neural Basis of Self-Regulation Neuroscience research reveals that various regions of the prefrontal cortex interact with the limbic system as people develop and use strategies and plans.

her potential and therefore is truly happy. Maslow writes, "A musician must make music, an artist must paint, a poet must write, if he is ultimately to be at peace with himself. What a man *can* be, he *must* be" (Maslow, 1968, p. 46).

Maslow's need hierarchy has long been embraced in education and business, but it is generally lacking in empirical support. Independent of whether one needs to be self-actualized in order to be happy, the ranking of needs is not so simple as Maslow suggests. For instance, some people starve themselves to death to demonstrate the importance of their personal beliefs, whereas others who have satisfied physiological and security needs are loners who do not seek out the companionship of others. In addition to many exceptions such as these, the concept of self-actualization is difficult to define and measure precisely. Maslow's hierarchy, therefore, is more useful at the descriptive level than at the empirical level. It does, however, make the important point that some needs are more compelling than others and are therefore stronger motivators. The greater the need, the greater the motivation to satisfy it.

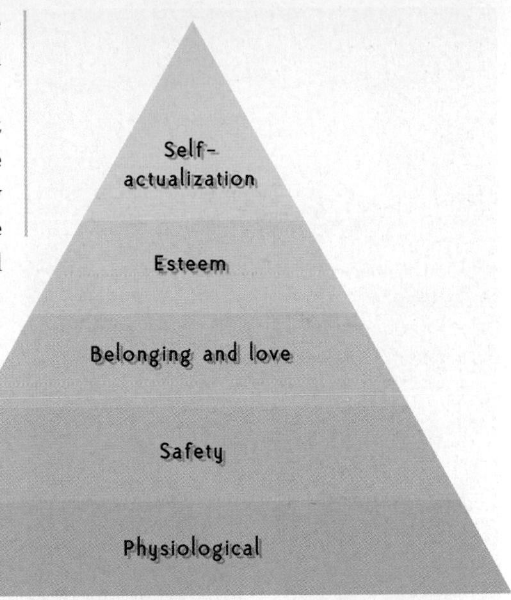

9.2 Need Hierarchy Maslow's hierarchy of needs indicates that basic needs, such as food and water, are satisfied before higher needs, such as self-actualization.

DRIVES How do we satisfy our needs? **Drives** are psychological states activated to satisfy needs. For example, people can be described by their level of sex drive, or characterized (like Oprah Winfrey) as being driven to succeed. Needs create *arousal,* which motivates behaviours that will satisfy these needs (Figure 9.3). **Arousal** is a generic term used to describe physiological activation (such as increased brain activity) or increased autonomic responses (such as quickened heart rate, increased sweating, or muscle tension). Try holding your breath as long as you can. If you continue for more than a minute or two, you probably will begin to feel a strong sense of urgency, even anxiety. This state of arousal is drive.

For biological states, basic drives help animals maintain a steady state, or *equilibrium.* Walter B. Cannon, a brilliant young physiologist at Harvard during the 1920s, coined the term **homeostasis** to describe the tendency for body functions to maintain equilibrium. For example, in the *negative feedback model,* people respond to deviations from equilibrium. Consider a home heating and cooling system controlled by a thermostat set to some optimal level, which we will call a set-point (see Figure 9.4). A *set-point*

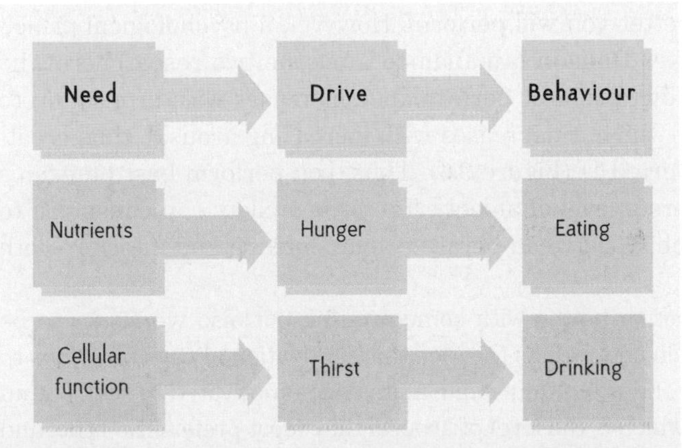

9.3 Needs and Drives Needs create drives that motivate specific behaviours.

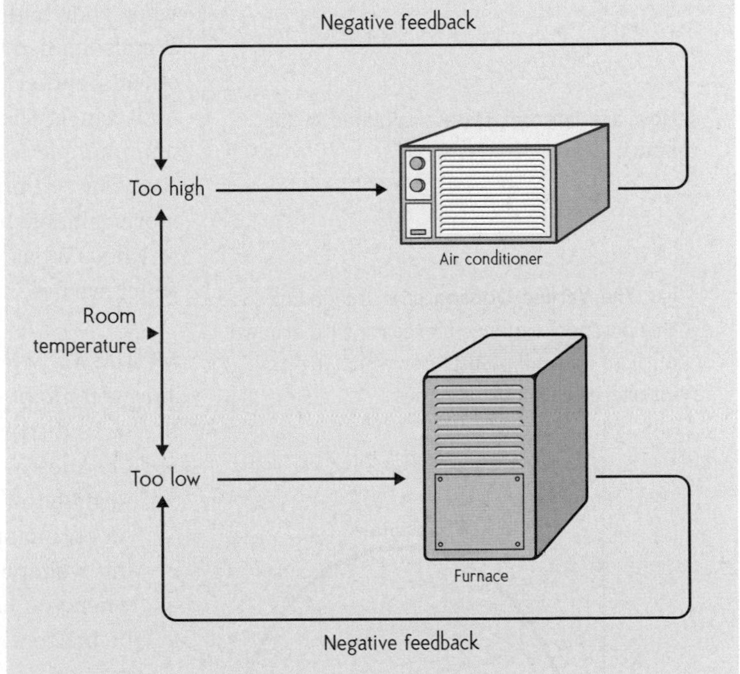

9.4 A Negative Feedback Model of Homeostasis When the temperature rises above the set-point (room temperature), a discrepancy is detected and the air conditioning is activated. The air is cooled until the temperature returns to the set-point. Conversely, when the temperature falls below the set-point, the furnace is activated. When the temperature reaches the set-point, the furnace quits working.

9.5 Clark Hull Hull was one of the pioneers of behavioural models of drive reduction. His drive theory of motivation is among the most influential theories in the history of psychological science.

How are internal states regulated in the body?

9.6 The Yerkes-Dodson Law According to this law, performance increases with arousal until an optimal point, after which arousal interferes with performance.

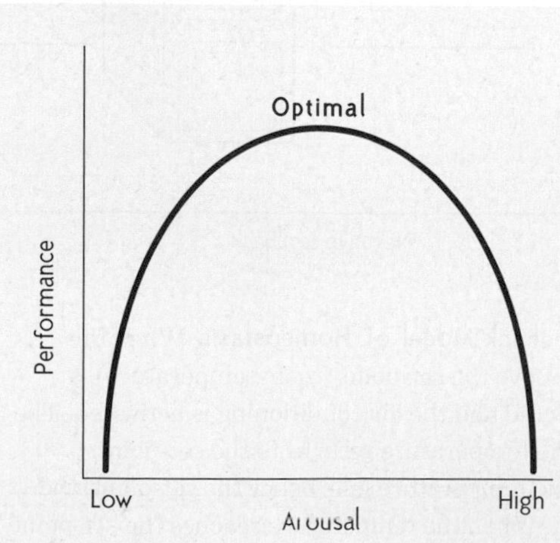

is a hypothetical state that indicates homeostasis. If the actual temperature is different from the set-point, the furnace or air conditioner operates to adjust the temperature. Similarly, the human body regulates a set-point of around 37° Celsius. When people are too warm or too cold, brain mechanisms, particularly the hypothalamus, initiate responses such as sweating (to cool the body down) or shivering (to warm the body up). At the same time, behaviours such as putting on or taking off clothes are also motivated. Negative feedback models are useful for describing a number of basic biological processes, such as eating, fluid regulation, and sleep.

Building on Cannon's work, Clark Hull (Figure 9.5) proposed that when an animal is deprived of some need (such as water, sleep, or sex), a specific drive increases in proportion to the amount of deprivation. The drive state creates arousal, which activates behaviours until performing one of them reduces the drive. According to this model, the initial behaviours in which the animal engages are arbitrary, but any behaviour that satisfies a need is reinforced and therefore is more likely to occur. Over time, if a behaviour consistently reduces a drive, it becomes a *habit*; the likelihood that a behaviour will occur is due to both drive and habit. For example, you might find that watching television makes you forget your troubles, which then reinforces further television viewing. Over time, you might develop the habit of watching television, perhaps especially when you are stressed.

Hull's theory had a tremendous impact on psychological research. One criticism, however, was that Hull's drive theory could not explain why people choose to engage in behaviours that do not appear to satisfy biological needs. Why, for instance, do people stay up all night studying for an exam? Why do people have a second piece of pumpkin pie at Thanksgiving dinner even though they weren't hungry when they had the first one? Although drive states push us to reduce arousal, we are also pulled toward certain things in our environments. **Incentives** are external objects, rather than internal drives, that motivate behaviours. Getting an A on your midterm is the incentive for studying hard; the comforting, sweet taste of the pumpkin pie is incentive for eating two pieces; having money to buy a car or help pay your tuition is incentive for working during summer vacation. Some things are more valuable to us than others, and we are more highly motivated to obtain things of value. What we value, in turn, is determined to a large extent by the culture in which we live.

AROUSAL AND PERFORMANCE If drives create arousal and motivate behaviour, you might expect that the more aroused you are, the more motivated you will be and therefore the better you will perform. However, a psychological principle known as the **Yerkes-Dodson law** (named after the two researchers who formulated it in 1908) dictates that performance increases with arousal up to an optimal point, after which it decreases with increasing arousal, thus creating a shape like an inverted U (Figure 9.6). Thus, you perform best on exams when you have a moderate level of anxiety. Too little anxiety can cause you to be inattentive; too much anxiety can paralyze your thinking and interfere with memory.

Not only do we function better with some arousal, but also we prefer to be somewhat aroused, which contradicts the idea that motivation always functions to reduce levels of tension and arousal. Instead, animals are motivated to seek out an *optimal level of arousal,* which is the level of arousal they most prefer. Too little, and we are bored; too much, and we are overwhelmed. Arousal theories help to explain why people choose different activities—we choose them because they arouse us and absorb our attention. People like to dance, listen to music, read exciting books, and watch horror or adventure movies. Arousal, then, not only motivates us to satisfy basic needs, but an optimal amount of arousal is also desirable on its own.

incentives External stimuli that motivate behaviours (as opposed to internal drives).

Yerkes-Dodson law A psychological principle that dictates that behavioural efficiency increases with arousal up to an optimum point, after which it decreases with increasing arousal.

Pleasure Can Motivate Adaptive Behaviours

Motivation can be viewed as a capability that initiates, directs, and sustains behaviours that promote survival and reproduction. Thus, animals are strongly motivated to repeat adaptive behaviours and to avoid maladaptive behaviours. Motivational states therefore arouse behaviours that solve adaptive problems. Animals need to take in oxygen, water, and nutrients, and therefore breathing, drinking, and eating are motivated behaviours. Engaging in these behaviours, especially when we are in a state of deprivation, often brings great pleasure. Here we discuss how pleasure motivates behaviour.

Sigmund Freud believed that drives are satisfied according to the *pleasure principle,* which tells organisms to seek pleasure and avoid pain. This idea is central to many theories of motivation. Originating with the ancient Greeks, *hedonism,* as a motivational principle, refers to the human experience of pleasantness and unpleasantness. Darwin's contemporary Herbert Spencer wrote, "every animal persists in each act which gives pleasure—and desists from each act which gives pain" (Spencer, 1872). We do things that feel good, and if something feels good, we do it again. A prime example of hedonism and adaptiveness is sex. Sex is essential for the survival of the species, and few people need to be persuaded of the pleasurable aspects of sexual behaviour, whether or not it leads to reproduction. The scientific study of sexual behaviour is discussed later in this chapter.

What role does pleasure play in motivating behaviour?

Recall that one limit of biological drive theories (such as Clark Hull's) is that animals engage in behaviours that do not necessarily satisfy biological needs. This commonly occurs when the behaviour is pleasurable, such as eating two slices of pumpkin pie at Thanksgiving. For instance, one classic study examined food preferences in rats (Sheffield & Roby, 1950). One of the substances that produced a strong response was saccharine, which is sweet and therefore pleasurable but does not satisfy any biological need. Often animals even choose pleasure over biological drive reduction. In the classic Olds and Milner 1954 study at McGill University, electrical stimulation applied to certain brain regions was found to be highly rewarding (see Chapter 6). Animals who could self-administer electrical stimulation by pressing a bar did so up to 2,000 times an hour. Male rats would press the bar even when they were food deprived, had access to a sexually receptive female, or had to cross an electrified grid to do so.

From an evolutionary perspective, behaviours associated with pleasure are often those that promote the animal's survival and reproduction, whereas behaviours associated with pain interfere with survival and reproduction. A good example of this is that animals prefer to eat sweets. Infants given sweet solutions seem to find them pleasurable, as revealed by their facial expressions (Steiner, 1977; Figure 9.7). Sweetness usually indicates that food is safe to eat. By contrast, most poisons and toxins taste bitter, so it is not surprising that animals avoid bitter tastes. Recall

9.7 Early Motivation Even newborns prefer sweet tastes to bitter: 1. resting face, 2. reaction to distilled water, 3. sweet stimuli, 4. sour stimuli, 5. bitter stimuli.

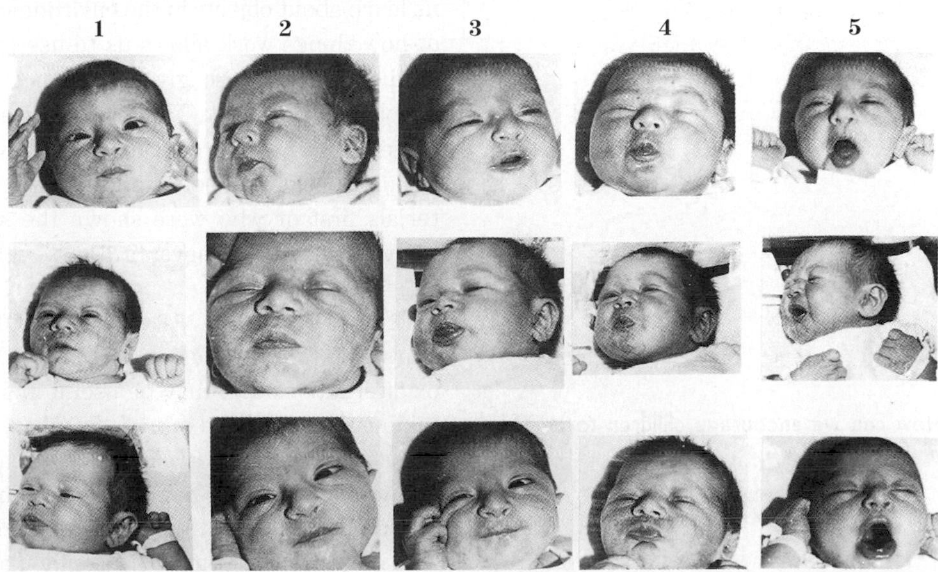

from Chapter 6 that the experience of reward is primarily due to activation of dopamine neurons (Fibiger & Phillips, 1988). Adaptive behaviours also use the dopamine system. Sexual behaviour, for example, even the anticipation of sexual behaviour, leads to increased dopamine activity (Pfaus & Phillips, 1991). Drugs that block dopamine interfere with the rewarding properties of food, water, and various drugs, such as cocaine or amphetamine, that are usually experienced as rewarding. Thus, dopamine activation may help guide adaptive behaviours by rewarding those that promote survival or reproduction. Note in this context that using cocaine or similar drugs is not likely to enhance survival. Rather, evolution did not anticipate that humans would one day create drugs that could activate natural reward systems.

Some Behaviours Are Motivated for Their Own Sake

So far we have discussed the pleasure associated with drive reduction and sustaining biological needs. However, many of the activities we find most satisfying, such as reading a good novel, do not appear to fulfill any obvious purpose other than providing enjoyment. **Extrinsic motivation** emphasizes the external goals toward which an activity is directed, such as drive reduction or reward—for example, working to earn a paycheck at the end of the week. **Intrinsic motivation** refers to the value or pleasure that is associated with an activity but that has no apparent biological goal or purpose—for example, listening to music. Intrinsically motivated behaviours are performed for their own sake (Enzle & Ross, 1978; Vallerand, 1997).

extrinsic motivation Motivation to perform an activity because of the external goals toward which that activity is directed.

intrinsic motivation Motivation to perform an activity because of the value or pleasure associated with that activity, rather than for an apparent biological goal or purpose.

creativity The capacity to generate or recognize ideas, alternatives, or possibilities that may be useful in solving problems, communicating with others, or entertaining ourselves and others.

CURIOSITY AND PLAY One of the hallmarks of childhood is curiosity, a mental state that leads to intrinsically motivated behaviour. Children like to seek out new situations and games. When children encounter new objects, they become fascinated and tinker with them until the object is fully inspected and all its features are displayed. After the child has thoroughly examined the object, interest diminishes and the child moves on to something new. Playful exploration is characteristic of all mammals and especially primates. For instance, the psychologist Harry Harlow and his colleagues showed that monkeys have a strong exploratory drive and will persevere in efforts to solve relatively complex puzzles in the absence of any apparent motivation (Harlow, Harlow, & Meyer, 1950).

What adaptive value might play serve? One obvious function of play is that it lets us learn about objects in the environment; this clearly has survival value. Knowledge of how things work allows us to use those objects for tasks outside of play. In one study children were given a chance to play with materials involved in a later problem-solving task. The task required the children to join two sticks to retrieve a piece of chalk. Children who were allowed to play with the sticks and clamps for ten minutes performed better at the task than children who were not exposed to the materials first or who were shown the solution to the task but not allowed to play (Sylva, Bruner, & Genova, 1976).

How can we encourage children to be creative?

CREATIVITY AND PROBLEM SOLVING Many intrinsically motivated behaviours allow people to express creativity. **Creativity** is the tendency to generate ideas or alternatives that may be useful in solving problems, communicating, and entertaining ourselves and others (Franken, 1998). Creativity involves constructing novel images, synthesizing two or more disparate ideas or concepts, and applying existing knowledge to solving new problems. The human mind also tries to consolidate information into coherent stories. People are motivated to make sense of the world. Just like split-brain patients who confabulate to make sense of conflicting information arriving in their separate hemispheres (see Chapter 4), people actively develop accounts of how the world works and how causal events are connected. Some theorists have pointed out that creativity is an integral component of solving adaptive problems (see Chapter 8). Many creative pursuits do not themselves represent adaptive solutions, but they are modern uses of mechanisms that did evolve for such purposes. For instance, we tend to prefer art that has elements that capture our attention. The capacity to notice and attend to features in the environment, such as food or predators, has obvious adaptive value (Pinker, 1997).

"What do you think . . . should we get started on that motivation research or not?"

REWARDING INTRINSIC MOTIVES Recall that one of the basic principles of learning theory is that rewarded behaviours increase in frequency. You would expect, therefore, that rewarding intrinsically motivated behaviours would reinforce them. Surprisingly, consistent evidence suggests that extrinsic rewards can undermine intrinsic motivation and decrease the likelihood that people will perform the rewarded behaviour. In a classic study, Mark Lepper and his colleagues allowed children to draw with coloured marking pens, an activity that most children find intrinsically motivating (Lepper, Greene, & Nisbett, 1973). One group of children was given an extrinsic motive for drawing by being led to expect a "good player award." Another group of children was unexpectedly rewarded following the task, and a third group was neither rewarded nor given the expectation of reward. During a subsequent free-play period, children who had been given the expectation of extrinsic reward spent much less time playing with the pens than did the group of children who were never rewarded or the children who received an unexpected reward.

CONTROL THEORY AND SELF-PERCEPTION Why do extrinsic rewards sometimes reduce intrinsic value? The social psychologists Edward Deci and Richard Ryan (1987) argue that feelings of *personal control* and competence make people feel good about themselves and inspire them to do their most creative work. Doing something to gain external rewards does not satisfy our need for autonomy. It undermines our feeling that we are choosing to do something for ourselves.

Another explanation is based on Daryl Bem's (1967) self-perception theory, which states that people are seldom aware of their specific motives and draw inferences about their motivation on the basis of what seems to make the most sense. For example, somebody gives you a big glass of water and you drink the whole thing, exclaiming, "Wow, I must have been thirsty." You believe you were thirsty because you drank the whole glass, even though you were not aware of any physical sensations of thirst. When people are unable to come up with an obvious external explanation for their behaviour (such as being rewarded or satisfying a biological drive), they conclude that they simply engage in the behaviour because they like it. Rewarding people for engaging in an intrinsic activity, however, gives them an alternative explanation for why they are engaging in it. It is not because the behaviour is fun; it is because of the reward. Therefore, in the absence of the reward, they have no reason to engage in the behaviour.

Because good grades reward performance in school, you might conclude that we should do away with them in order to preserve young children's curiosity and love of learning. However, Deci and Ryan argue that if the extrinsic reward provides information about how much personal control people have over learning, it will not undermine intrinsic motivation. From this perspective, getting good grades in school provides information that you are doing well, and therefore it does not undermine intrinsic motivation. If anything, receiving good grades further reinforces intrinsic motivation: Now you know schoolwork is something you like to do and are good at doing. Moreover, extrinsic rewards undermine only behaviour that is intrinsically rewarding. Thus, the challenge to educators is to find ways to increase the intrinsic value of schoolwork. Failing this, extrinsic rewards can be used to make an otherwise boring task seem worth pursuing.

Psychological Science *in Action*

Motivation on the Job

What type of career do you hope to have? Do you view a job as basically a way to pay the bills and do the things you really like to do? Or do you feel you have a calling, a greater purpose that will bring meaning and enrichment to yourself as well as others? Does your personal satisfaction matter more than how much money you will make or do you view making a large salary as justifying long hours and personal sacrifice? How will you balance family and career?

For most of us, work occupies a great deal of our time. The amount of time people spend working varies greatly across different countries; people in Canada, the United States, and Australia spend on average around 350 more hours—nine full weeks—working per year than their counterparts in Europe. North American workers make more money on average, but does it make them happier? Given work's importance in our daily lives, how can psychological science contribute to our understanding of work?

One of the central challenges for employers is keeping their employees motivated to work hard and be productive. Not only do companies want motivated workers, but those who are most highly motivated are more likely to enjoy their jobs and feel that their work is meaningful. Many companies hire psychologists to assist them in finding, training, retaining, and motivating employees. These psychologists are trained in industrial and organizational (called I-O for short) psychology, a branch of psychology whose mission is to apply findings from psychological science to work settings in order to enhance human well-being and performance. Key challenges in I-O psychology include helping employers deal with employees fairly, helping to design jobs so that workers find them interesting and satisfying, and helping workers to be more productive.

One of the first tasks of I-O psychologists is to help employers hire workers who have the necessary skills, abilities, and other personal characteristics needed to perform the job well. Many employers use various screening measures, such as personality tests, interviews, and even tests for illegal substances, to try to make sure they make good hires. But I-O psychologists are also concerned that the hiring process treats people from diverse backgrounds fairly. For instance, I-O psychologists use tests that do not favour people from certain cultures or backgrounds. In terms of hiring, many employers make the mistake of placing too much emphasis on the interview. How a person performs in an interview is not a particularly good predictor of whether he or she will be effective in the job. Factors such as likeability, physical attractiveness, and assertiveness can obscure potential weaknesses during interviews. Accordingly, I-O psychologists have developed more structured ways to evaluate employees, such as by trying to identify specific personality traits that match the requirements of the job or testing specific knowledge or experience. Once people are hired, I-O psychologists help develop training programs to ensure that workers have the skills they need to succeed at their jobs, and they design programs to encourage workers to be productive. Recall from Chapter 6 that principles from operant conditioning are used to reinforce effective job performance. Finally, I-O psychologists help employers develop formal methods to determine appropriate promotions and raises or to evaluate whether a worker ultimately must be fired.

An important finding in I-O psychology is that satisfied workers are the best workers, and that people will work hardest when their jobs are meaningful and when they feel that they have some control over what they do. For instance, giving workers a say in decision making makes them more willing to follow the decision. Programs that encourage employees to suggest improvements, that give them part ownership, or that involve them in important decisions are associated with greater employee satisfaction. Similarly, companies that are family friendly, perhaps offering flexible scheduling or on-site day care, often have the most loyal employees. By showing companies how to create superior working conditions, I-O psychologists help bring satisfaction to workers, which in the end is also good for the corporate bottom line. ■

REVIEWING the Principles | How Does Motivation Activate, Direct, and Sustain Behaviour?

Motivational states activate, direct, and sustain behaviours that help satisfy needs or achieve goals. Drive states develop from need deficiencies and motivate behaviours that reduce the need, such as hunger motivating eating or thirst motivating drinking. Many biological needs operate to keep the body in homeostasis, a state of equilibrium. According to Maslow, needs can be arranged in a hierarchy, in which physiological needs take precedence over personal growth needs. A general motivational principle is to perform actions that bring pleasure and avoid those associated with pain. Providing external rewards for behaviours that are intrinsically motivated may undermine the desire to perform them.

Why Are Human Beings Social?

Why are people motivated to be social? Do we really need each other? How much would you suffer if you lived by yourself on some remote island? Consider the following. Suppose you are offered a sizeable amount of money to remain alone in a room. As a motivation, the longer you stay, the more money you make. The room has no windows and only a limited number of amenities, such as a bed, chair, table, and toilet. You will not have contact with other people, and you will not have access to newspapers, books, television, or the Internet. How long could you last? Which motive is stronger? To be social? Or to earn the most money you can? Stanley Schachter (1959) found five men who were willing to try living under such isolated conditions. Most of the men found it very difficult. Indeed, the first experienced an uncontrollable desire to leave after only 20 minutes. Three of the volunteers lasted two uneasy days but then wanted out. Even the fifth man, who lasted eight days, started to feel nervous while alone. The evidence from this and other studies indicates that people feel anxious when socially isolated. Although you may often wish to find a nice, quiet place where you can get away from other people, human nature requires us to have frequent and close contact with others. Psychological scientists have provided substantial documentation that the nature of social relationships has far-reaching implications for emotion, cognition, and mental well-being (Reis, Collins, & Berscheid, 2000).

Humans Have a Fundamental Need to Belong

need to belong theory The need for interpersonal attachments is a fundamental motive that has evolved for adaptive purposes.

Roy Baumeister and Mark Leary (1995) formulated the **need to belong theory,** which states that the need for interpersonal attachments is a fundamental motive that has evolved for adaptive purposes. Over the course of human evolution, those who lived with others were more likely to survive and pass along their genes. Children who stayed with adults (and who resisted being left alone) were more likely to survive until their reproductive years because the group would protect and nurture them (see Chapter 11). Similarly, adults who were capable of developing long-term, committed relationships were more likely to reproduce, and also more likely to have offspring who survived to reproduce. Effective groups shared food, provided mates, and helped care for offspring (including orphans). Some survival tasks (such as hunting large mammals or looking out for predatory enemies) were best accomplished by group cooperation. It therefore makes great sense that, over the millennia, humans have committed to living in groups.

The need to belong theory explains the ease and frequency with which many people form social bonds. Did you ever make friends at summer camp very quickly? University students often make lifelong friendships within days of arriving on campus. Societies differ in their types of groups, but all societies have some form of group membership (Brewer & Caporael, 1990). Not belonging to a group increases a person's risk for a number of adverse consequences, such as illnesses and premature death (Cacioppo, Berntson, Sheridan, & McClintock, 2000). In their theory, Baumeister and Leary argue that the need to belong is a basic motive that drives behaviour and influences cognition and emotion; ill effects follow when it is not satisfied. Just as a lack of food causes hunger, a lack of social contact causes emptiness and despair. In the movie *Cast Away,* Tom Hanks's character has such a strong need for companionship that he begins carrying on a friendship with a volleyball he calls Wilson. As noted by the film reviewer Susan Stark (2000), this volleyball convinces us that "human company, as much as shelter, water, food and fire, is essential to life as most of us understand it."

The need to belong theory is supported by evidence that people feel anxious when excluded from their social groups. According to *social exclusion theory,* anxiety warns individuals that they may be facing rejection from their group (Baumeister & Tice, 1990). People are socially excluded for reasons of immorality, incompetence, or unattractiveness. Breaking group norms and rules, which is the essence of immorality, threatens group structure; incompetence drains group resources; being unattractive or having a stigmatizing condition may suggest inferior genes. People who feel unworthy, incompetent, or unattractive experience anxiety, which may lead them to behave in ways that enhance their value to the group, such as contributing additional effort. People who are rejected show increased social sensitivity, as evidenced by increased memory for social events (Gardner, Pickett, & Brewer, 2000). This sensitivity means that those who fear rejection are especially aware of others' reactions to them. When people are unable to alter what is causing their anxiety, such as their appearance, they may avoid contact with group members or seek less evaluative groups. Thus, one important aspect of motivation is that we avoid behaviours that might lead us to be rejected by others. For instance, our bodies may encourage us to indulge our appetite for fattening foods, but our culture warns us that people may reject us because of our being overweight and so we attempt to limit the amount of such foods we eat. Clearly, our anticipation of how others will react to us is a powerful force in compelling us to control our behaviour.

People Seek Others When They Are Anxious

In the study described earlier, Schachter found that isolation caused anxiety. He reasoned that the opposite might also be true, that anxiety could motivate the desire for company. To test this, he manipulated anxiety levels and then measured how much the participants preferred to be around others (Schachter, 1959). The participants in these studies, all women, thought they were taking part in a routine psychological study. They were greeted at the lab by a serious-looking man named Dr. Zilstein, who had a cold look about him and spoke with a vaguely European accent. He told them that he was from the neurology and psychiatric school and that the study involved measuring "the physiological effects of electric shock." Zilstein told the participants that he would hook them up to some electrical equipment and would administer some electric current to their skin. Those in the low-anxiety condition were told that the shocks would be painless, no more than a tickle. Those in the high-anxiety condition were told, "These shocks will hurt, they will be painful. As you can guess, if we're to learn anything that will really help humanity, it is necessary that our shocks be intense. These shocks will be quite painful, but, of course, they will do no permanent damage." As you might imagine, those who heard this speech were quite fearful and anxious.

When do others make us anxious or calm?

Zilstein then said he needed time to set up his equipment, so there would be a 10-minute period before the shocks would begin. At this point the participants were offered a choice: They could spend the waiting time alone or with others. This was the critical dependent measure, and after the choice was made, the experiment was over. Schachter found that increased anxiety did indeed lead to increased affiliative motivations. As can be seen in Figure 9.8, those in the high-anxiety condition were much more likely to want to wait with other people. Hence, it appears that misery loves company. But does misery love just any company? A further study revealed that high-anxiety participants wanted to wait only with other high-anxiety participants, not with people who supposedly were just waiting to see their research supervisors. So misery loves miserable company, not just any company.

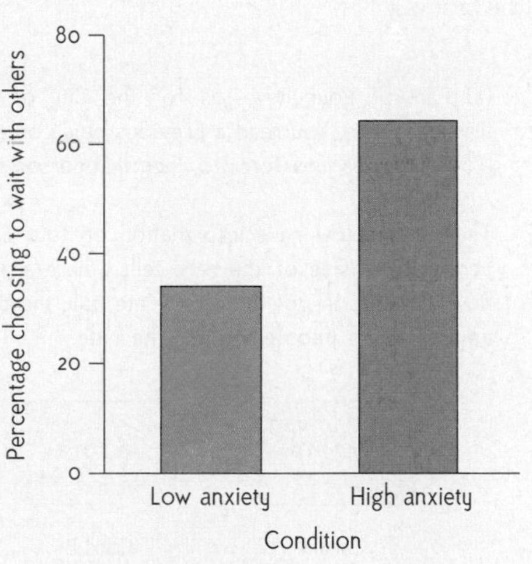

9.8 Anxiety Increases Affiliation People who were told they would receive either a painful or a harmless shock were given the choice to wait alone or with others. People preferred to spend time with others when they were expecting painful shocks.

Why do anxious people prefer to be around other anxious people? According to Schachter, others provide information that helps people to evaluate whether they are acting appropriately. According to Leon Festinger's *social comparison theory* (1954), people are motivated to have accurate information about themselves and others. People compare themselves with those around them in order to test and validate personal beliefs and emotional responses, especially when the situation is ambiguous and they can compare themselves with people who are relatively similar to them. For instance, if you receive a 75 on an exam, you might not know whether it is a good score until you compare it with the class average. The comparisons we make among ourselves can inspire feelings of competitiveness and the desire to outperform others.

People Are Motivated to Detect and Reject Cheaters

We noted earlier that people feel anxious when they face possible group exclusion, as when they act immorally. However, members of groups face a clear and powerful dilemma: whether to primarily give or take resources. On the one hand, the group survives only if members contribute. On the other hand, individuals who get more than they give will have a tremendous advantage. A **social dilemma** occurs when we

social dilemma When there is a motivational conflict both to cooperate and to be selfish.

feel a motivational conflict between cooperating and being selfish; typically, selfishness maximizes short-term interests, whereas cooperation maximizes long-term interests. Psychologists generally find that people act for individual short-term interests in social dilemmas, especially if they believe that others are doing so as well. But much of this research occurs when strangers are used as research participants. If this happened within existing groups, more and more group members would start being selfish, which would eventually threaten the stability of the group. Therefore, group members are vigilant to detect those who cheat, and they punish these individuals accordingly.

People employ a number of methods for detecting and dealing with cheaters. Leda Cosmides and John Tooby (2000) have provided compelling evidence that people have specialized *cheater detectors.* They have found that people are especially good at solving logical problems that represent cheating in a social exchange. For instance, in the standard *Wason selection task* (Figure 9.9), typically only one in four people chooses the correct response in this deductive reasoning task. In the first problem, most people choose the cards that say *Toronto* and *subway* even though the subway card provides no useful information. Even people with training in logical thinking often make this error. In contrast, when the task involves a social contract, as in the second problem, nearly three out of four people answer correctly, and no training is required. Indeed, even people with serious mental impairments, such as schizophrenia, are able to solve this problem.

The finding that people are better at solving the social version of the Wason task implies that humans have an inference mechanism for detecting cheating specific to the social domain. A variety of other evidence supports the general idea that people are sensitive to cheaters. For instance, in one very large study of couples, a partner who suspected the other of infidelity was accurate more than 90 percent of the time (Blumstein & Schwartz, 1983).

What happens when cheaters are detected? People who violate norms of trust, reciprocity, honesty, and morality are stigmatized and socially excluded by group members (Neuberg, Smith, & Asher, 2000). Indeed, the deepest and most horrific circle of hell in Dante's *Inferno* was reserved for those who had betrayed their kin or countries. Traitors are often punished severely and seldom given a second chance. Of course, the stigmatization attached to behaviours such as cheating, stealing, or betrayal refers to ingroup behaviours—stealing from or cheating outgroups can lead to great advantages for the group doing the stealing and cheating. The important point is that people are rejected for cheating group members. The threat of being rejected is sufficient to keep most people in line, allowing them to fulfill their need to belong.

9.9 The Wason Selection Task In this task, only cards A and D provide useful information about whether the person violated the rule. Most people find the second problem (b) to be much easier than the first (a).

(a) Part of your new job for the City of Scarborough is to study the demographics of transportation. You read a previous study on the habits of Scarborough residents that says, "If a person goes into Toronto, then that person takes the subway."

The cards below have information on four Scarborough residents. Each card represents one person. One side of the card tells where a person went, and the other side of the card tells how that person got there. Indicate only the card(s) you definitely need to turn over **to see if any of these people violate the rule.**

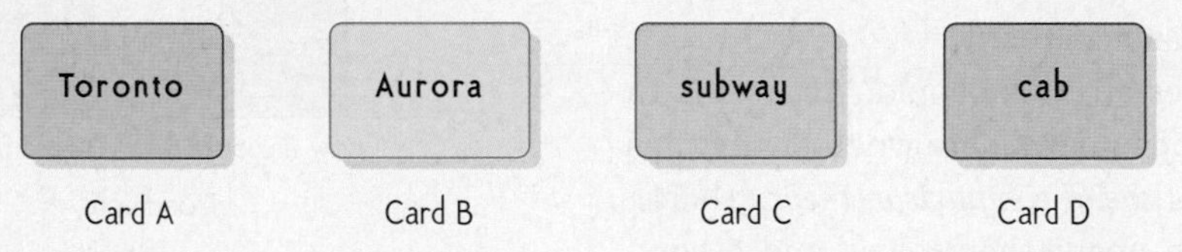

(b) Part of your new job as bar manager requires you check whether the bartender is properly checking proof of age. The rule says, "If a person drinks beer, then that person must be 21 or over."

The cards below have information on four bar patrons. Each card represents one person. One side of the card tells what the person is drinking, and the other side of the card tells the person's age. Indicate only the card(s) you definitely need to turn over **to see if any of these people violate the rule.**

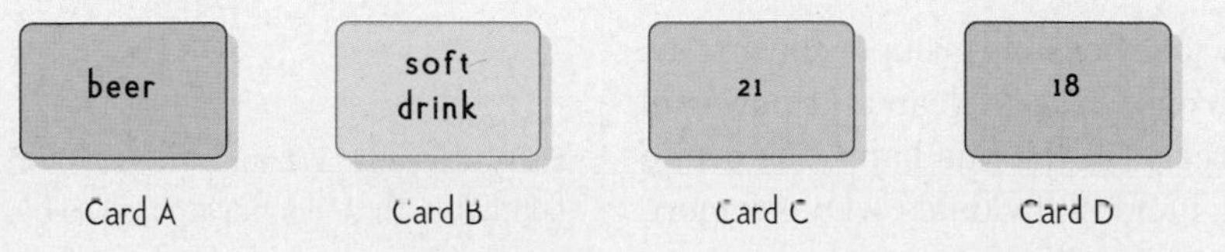

REVIEWING
the Principles | Why Are Human Beings Social?

The need to belong exerts a powerful force on behaviour. People are motivated to form relations with others; those who fail to do so suffer alienation and loneliness. Because people are motivated to belong to groups, those who act in ways that place them at risk for group exclusion become anxious. This anxiety motivates them to control their behaviours. People reject those who lie, cheat, steal, or violate group standards. This behaviour may occur because individuals who lived in well-functioning groups had an evolutionary advantage over those who lived in discordant groups.

How Do People Achieve Personal Goals?

A question we have asked from the beginning of this chapter is how people accomplish meaningful things. For instance, consider your future aspirations: What would you like to be doing in ten years? What things about yourself would you change? Here we are referring to psychosocial needs, such as for power, self-esteem, and achievement. The study of psychosocial motives began with Henry Murray, a personality psychologist at Harvard University. In the 1930s, Murray proposed 27 basic psychosocial needs, such as the need for power, autonomy, achievement, sex, and play. The study of psychosocial needs has provided a number of important insights into what motivates human behaviour. Chief among these is that people are especially motivated to achieve personal goals. **Self-regulation** of behaviour is the process by which people initiate, adjust, or stop actions in order to attain personal goals. A **goal** is a desired outcome and is usually associated with some specific object (such as tasty food) or some future behavioural intention (such as getting into medical school).

Good Goals Enhance Confidence

Sometimes when you are up late studying for an exam, you might wonder why you work as hard as you do. Many people's long-term goals motivate their hard work. According to an influential theory developed by the organizational psychologists Edwin Locke and Gary Latham (1990), goals direct attention and effort, encourage long-term persistence, and help people develop strategies. The types of goals people set influence their efforts at self-regulation. Locke and Latham suggest that *challenging, difficult,* and *specific* goals are the most productive. Challenging goals arouse the greatest effort, persistence, and concentration, whereas goals that are too easy or too hard can undermine motivation and often lead to poor outcomes. Goals that are divisible into specific, concrete steps also foster success. A person interested in running the Ottawa Race Weekend must first gain the stamina to run one mile. On achieving the goal of running a mile, the person can subsequently work up to the 42-kilometer marathon. Focusing on concrete, short-term goals facilitates achieving long-term goals.

FEELINGS OF SELF-EFFICACY Albert Bandura (see Chapter 6) has argued that people's personal expectations for success play an important role in motivation. For instance, if you believe that your efforts at studying will lead to a good grade on an exam, you will be motivated to study. **Self-efficacy** is the expectancy that your efforts

self-regulation The process by which people initiate, adjust, or stop actions in order to promote the attainment of personal goals or plans.

goal A desired outcome associated with some specific object of desire or some future behavioural intention.

self-efficacy The expectancy that one's efforts will lead to success.

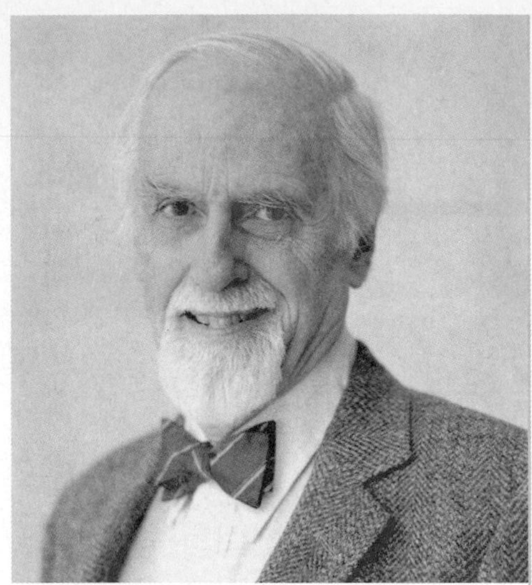

9.10 David McClelland McClelland studied the achievement motive in many cultures.

will lead to success; this belief helps to mobilize your energies. If you have low self-efficacy and do not believe that your efforts will pay off, you may be too discouraged to even try to study. People with high self-efficacy tend to set lofty goals and obtain favourable outcomes. However, sometimes people whose self-views are inflated set goals that they cannot possibly achieve. Goals that are challenging but not overwhelming usually are most conducive to success.

ACHIEVEMENT MOTIVATION People differ in the extent to which they pursue challenging goals. The **achievement motive** is the desire to do well relative to standards of excellence. One of Henry Murray's students, David McClelland (Figure 9.10), spent most of his career studying people who were high in achievement motivation. In research spanning nearly 50 years, McClelland and his students found that the desire to achieve helped people to succeed. Compared to those low in achievement need, high-need students sit closer to the front of classrooms, score higher on exams, and obtain better grades in courses relevant to their career goals (McClelland, 1987). Interestingly, high-achievement-need students tend to be more realistic in their career aspirations than students low in achievement need. They set personal goals that are challenging but attainable, whereas those low in achievement need tend to set goals that are either extremely easy or almost impossible. Parents of students who are high in achievement need tend to set very high goals for their children, and they encourage persistence. These parents discourage their children from complaining or making excuses for poor performance, and they encourage them to try novel solutions. The one potential downside for those high in achievement motivation is that they like to achieve personal goals on their own, and therefore in situations in which they have to delegate, such as managerial positions, they sometimes try to do too much themselves. Indeed, one research team found that need for achievement was inversely related to effectiveness among political leaders (Spangler & House, 1991).

Self-Regulation Requires Self-Awareness and Delay of Gratification

Throughout this chapter, we've noted the basic motivational conflict between satisfying our personal desires and living up to societal standards of conduct. For the individual, doing whatever feels good at the time is highly motivating. If you enjoy loud music, you might want to play it all the time. But at the group level, people need to be discouraged from actions that negatively affect other group members, and so you control your urge to crank up the volume when others are sleeping. We also often need to control our immediate desires in order to achieve our longer-term goals, such as continuing to study for an important exam rather than going out with friends.

How do people actually control their behaviour to achieve their goals? According to the cognitive psychologists George Miller, Eugene Galanter, and Karl Pribram (1960), people possess mental representations of their goal states and compare how they are doing to those goals. The cognitive mechanism that directs behaviour is based on negative feedback, like a home heating system, and is similar to the homeostatic models described earlier. Miller and his colleagues called this mechanism a *TOTE unit,* which stands for test-operate-test-exit (Figure 9.11). In this model, a comparison is made between the person's current state and the goal state. A discrepancy between the two states motivates behaviour to minimize the discrepancy, and this process continues until the goal is achieved, at which point the person exits the process. The **TOTE model** requires people to keep track of their goals and to be self-aware in order to monitor their progress toward achieving those goals. For example,

achievement motive The desire to do well relative to the standards of excellence.

TOTE model A model of self-regulation in which people evaluate progress in achieving goals.

for a C student to become a B student you have to know what scores you need to get a B and pay attention to your performance on exams and assignments. If your scores are below the B criteria, you will have to work harder or more efficiently to achieve your goal.

The social psychologists Charles Carver and Michael Scheier (1981, 1998) developed an influential self-regulatory theory that builds on the TOTE model by emphasizing the influence of self-awareness. When self-awareness is high, people act in accordance with personal standards; when self-awareness is low, their inhibitions disappear and they lose touch with those standards, a mental state known as **deindividuation.** In one study, Edward Diener and his colleagues found that students were less likely to cheat on a test if they were seated in front of a mirror—which presumably increased self-awareness (Diener, 1979). Similarly, Diener found that children who wore costumes that made them anonymous stole more candy at Halloween than children who were not anonymous. This is why many gambling casinos are noisy and devoid of mirrors. The casinos try to eliminate all self-focusing cues so that people become deindividuated and overcome their inhibitions to gamble away their money.

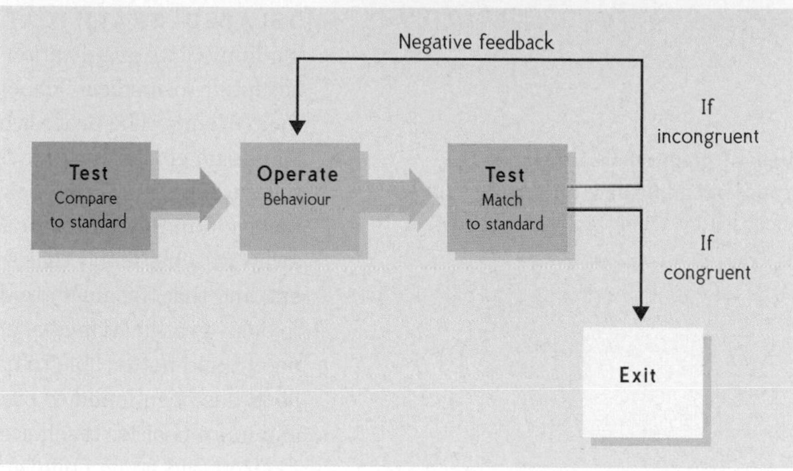

9.11 The TOTE Model A TOTE unit is a homeostatic model of self-regulation. Negative feedback from discrepancies between current state and ideal standing motivate efforts to reduce the discrepancies.

deindividuation A phenomenon of low self-awareness in which people lose their individuality and fail to attend to personal standards.

According to the Carver and Scheier model, self-awareness of discrepancies between ideal and current states leads to either negative or positive affect, which then guides subsequent behaviour. When people perform above some standard, they experience positive affect and therefore stop evaluating themselves on that dimension. If you are performing better than your goal, you do not adjust your behaviour to move back toward the goal (as would seem to be suggested by the TOTE model).

When people are performing below their ideal standards, they experience negative affect, such as feelings of sadness, frustration, and anxiety. Negative affect is a signal that you are not satisfying personal goals. A key point of the Carver and Scheier model is that future attempts at self-regulation depend on the likelihood of being able to narrow the gap between where you are and where you want to be. If the perceived probability of successfully reducing the discrepancy is high, then the negative affect motivates attempts to reduce the discrepancy. If, however, the perceived probability of success is low, then people may give up trying.

SELF-DESTRUCTIVE BEHAVIOUR According to Carver and Scheier, one way to reduce negative affect is to avoid self-awareness through *escapism.* Some people drink alcohol or take drugs to help them forget their troubles; others go for a run, read a book, or watch a movie. The selective appeal of escapist entertainment is that it distracts people from reflecting on their failures to live up to their goals, and therefore it helps them to avoid feeling bad about themselves. While some escapist activities are relatively harmless distractions, indeed often positive, such as going for a run, reading a book, or watching a movie, others, such as drinking excessive alcohol or taking drugs, move beyond distraction to become means of actually forgetting one's troubles and avoiding self-awareness for a much longer time than the hour spent jogging at the track. Unfortunately, evidence suggests that escaping from self-awareness is associated with a variety of self-destructive behaviours, such as binge eating, unsafe sex, and suicide. According to the social psychologist Roy Baumeister (1991), these problems occur because people do things when they have low self-awareness that they would never do if self-aware.

delay of gratification When people transcend immediate temptations to successfully achieve long-term goals.

DELAYED GRATIFICATION One common challenge in self-regulation is postponing immediate gratification in the pursuit of long-term goals. Students who want to be accepted to medical school often have to stay in and study rather than go out with their friends. The process of transcending immediate temptations in order to achieve long-term goals is known as **delay of gratification.** In a series of now-classic studies, the developmental psychologist Walter Mischel gave children the choice of waiting to receive a preferred toy or food item or having a less preferred toy or food item right away. Mischel found that some children are better at delaying gratification than others, and that the ability to do so is predictive of success in life. Children who were able to delay gratification at age four were rated ten years later as being more socially competent and better able to handle frustration. The ability to delay gratification in childhood has been found to predict higher standardized test scores, better school grades, and more positive teacher evaluations (Mischel, Shoda, & Rodriguez, 1989).

Why is self-regulation difficult and why does it fail so often?

How did some children manage to delay gratification? One strategy was simply ignoring the tempting item rather than looking at it. Older children, who were better at delaying gratification, tried covering their eyes or looking away, whereas very young children tended to look directly at the item they were trying to resist, making the delay especially difficult. A related strategy was self-distraction, through singing, playing games, or pretending to sleep. The most successful strategy involved what Mischel and his colleague Janet Metcalfe refer to as turning *hot cognitions* into *cold cognitions.* This sophisticated strategy involves mentally transforming the desired object into something undesired, such as viewing a tempting pretzel as a brown log or viewing marshmallows as clouds (Figure 9.12). Hot cognitions focus on the rewarding, pleasurable aspects of objects, whereas cold cognitions focus on conceptual or symbolic meanings. Metcalfe and Mischel (1999) proposed that this hot/cold distinction is based on how the information is processed in the brain, with the hot system being amygdala based and the cold system hippocampus based. According to this theory, the amygdala processes the reward features of biologically significant stimuli, whereas the hippocampus processes plans, strategies, and goals and is therefore responsible for self-control.

9.12 Delaying Gratification Children who are able to turn hot cognitions into cold cognitions have an easier time delaying gratification.

SELF-REGULATORY STRENGTH Humans have the capacity to delay gratification, control appetites and impulses, and persevere in order to attain goals, but many people have difficulties with self-control from time to time. A theory developed by Roy Baumeister and Todd Heatherton (1996) suggests that self-regulation may best be conceptualized as an individual strength. This means that self-regulatory strength is a limited resource that is renewable over time and can be increased with practice. It is only by exercising regular self-discipline that people become able to exert self-control when faced with temptations and persevere when goal attainment becomes difficult. Self-regulatory strength also can be depleted by situational demands. The dieter confronted by a party host offering an array of tasty treats may find him- or herself unable to resist. Moreover, people who put a great deal of effort into controlling one aspect of their lives often have problems with self-control in other areas. This may be why New Year's resolutions fail: People vow to give up smoking, drinking, eating high-calorie foods, and watching television; and they set for themselves the impossible task of doing all of this at the same time. People have only so much self-regulatory strength, and to spread it too thin is to invite failure in multiple self-regulatory domains (Muraven & Baumeister, 2000).

The Frontal Lobes Are Important for Self-Regulation

Self-regulation requires us to control behaviours when they are not appropriate. This requires two things. First, it requires us to be aware of societal standards that tell us what is appropriate. This includes understanding how others will react to our actions, such as what will happen if we play our music very loud at 3:00 A.M. Second, it also involves the ability to inhibit doing those actions, such as overriding the impulse to turn up the music. The frontal lobes of our brains are important for both of these abilities. As you read in Chapter 4, people with damage to their frontal lobes often engage in inappropriate behaviour. They are impulsive and have little self-control. If our dopamine system revs up the engine, our frontal lobes act as the brakes.

The frontal lobes are also involved in formulating goals, plans, and strategies. Studies using brain imaging have found significant activation of the prefrontal region during difficult tasks but not during simple recall tasks (Schacter & Buckner, 1998; Spence & Frith, 1999). The difficult tasks required effort, which links motivation to the prefrontal cortex. Damage to the prefrontal cortex is associated with difficulties in making plans and performing behaviours necessary to achieve goals. Patients with prefrontal damage often have intact intelligence and comprehension but lack the ability to put their knowledge to use. They might be able to tell you how to achieve a goal, but they seem incapable of following their own suggestions and are often unwilling even to try. It is possible that motivational deficits observed after brain injury occur because of damage to dopamine circuits that link the prefrontal cortex to the limbic system.

Scientists believe that people with prefrontal damage have difficulty focusing on a task because they are unable to ignore other stimuli in the environment (Knight & Grabowecky, 1995). For instance, if you sustained damage to your prefrontal region, you might be unable to study because you could not tune out background noise or other distractions. The prefrontal cortex also plays a critical role in working memory (see Chapter 7), which allows people to compare current performance with past standards and future goals. Working memory is important for the temporal organization of memory, which is often required for self-regulatory tasks. Consider the task of cooking a meal, which requires doing things in a certain order. People with frontal-lobe dysfunction often have difficulty following plans that require temporal ordering of behaviours. They might be able to tell you all the ingredients for a recipe but be incapable of following the steps of the recipe in the necessary order.

ADAPTIVENESS OF SELF-REGULATION Not only are the frontal lobes involved in inhibiting behaviour, but they are also critical for living with other people. Living in social groups requires cooperation and the suppression of personal desires for the good of the group. Hunters and gatherers who returned to the group to share their bounty, rather than eating it immediately after the kill, would not only have increased the likelihood that their relatives received sufficient nutrition to survive and reproduce, but also would demonstrate their value as group members. Living in groups requires people to control their emotional reactions and violent urges as well as their appetites. Those who are unable to control their emotions or behaviour are usually excluded from social groups and are therefore not likely to pass along their genes.

That we can consciously override biological urges is an important point to make about human nature. Some people object to evolutionary approaches to understanding behaviour because of the assumption that genes are the sole determinant of behaviour. This assumption implies that certain behaviours are inevitable and uncontrollable. Research on self-regulation has established that people (and perhaps some nonhuman animals) have the capacity to override biology, at least within reason. You can choose to postpone eating, going to the bathroom, or breathing, albeit for brief periods of time—but you can do it. Socialization is an important mechanism that teaches group members the appropriate times and places to satisfy or inhibit various biological drive states.

REVIEWING the Principles | How Do People Achieve Personal Goals?

People are motivated to set and achieve goals and to control their behaviour in pursuit of those goals. The best goals are challenging, concrete, and divisible into steps. The process by which people achieve goals is known as self-regulation, which depends on self-awareness of goals and behaviours. Self-regulatory strength allows people to delay immediate gratification in the service of long-term goals. The capacity for self-regulation depends on having intact frontal lobes. People who suffer damage to the frontal lobes often have trouble staying on task, making decisions, and inhibiting behaviour. The ability to inhibit desires helps people live harmoniously with others.

What Determines How We Eat?

One of life's greatest pleasures is eating, and we do a lot of it: Most North Americans consume between 75,000 and 85,000 meals during their lives—more than 40 tons of food! Ingesting food is something everyone needs to do to survive, but eating is much more than simply survival. Special occasions often involve elaborate feasts, and much of the social world revolves around eating. With eating we can observe complex interactions between biology and cultural influences on cognition. Common sense dictates that most eating is controlled by hunger and *satiety*. People eat when they feel hungry and stop eating when they are full. However, some people eat a lot even when they are not hungry, whereas others avoid eating even though they are not full (Pinel et al., 2000). What determines human eating behaviour?

Time and Taste Play Roles

Eating is greatly affected by learning. Have you ever noticed that most people eat lunch at approximately the same time? This makes little sense, for people differ greatly in metabolic rate, the amount they ate for breakfast, and the amount of fat they have stored for long-term energy needs. Yet it occurs because people have been classically conditioned to associate eating with regular mealtimes. We eat not because we have deficient energy stores, but because it is time to eat. The clock indicating mealtime is much like Pavlov's bell, in that it leads to a number of anticipatory responses that motivate eating behaviour and prepare the body for digestion. For instance, an increase in insulin promotes glucose utilization and increases short-term hunger signals. The sight and smell of tasty foods can have the same effect, and even the mere thought of fresh-baked bread, a favourite pizza, or a decadent dessert may initiate bodily reactions that induce hunger.

One of the main factors that determines feeding is flavour; good-tasting food motivates eating. When it comes to flavour, variety really is the spice of life. Animals presented with a variety of foods tend to eat a great deal more than animals presented with only one food type. For instance, rats who normally maintain a steady body weight when eating one type of food eat huge amounts and become obese when presented with a variety of high-calorie foods, such as chocolate bars, crackers, and potato chips (Sclafani & Springer, 1976; Figure 9.13). Humans show the same effect, eating much more when a variety of foods is available than when only one or two types of food are available.

One reason that rats and people eat more when presented with a variety of foods is because we quickly grow tired of any one flavour. *Sensory-specific satiety* is a phenomenon in which animals will become full relatively quickly if they just have one type of food to eat, but will eat more if presented with a different type of food. This may explain the behaviour of people at times like Thanksgiving, when even though they couldn't imagine eating another bite of turkey they apparently can find room for a piece of pumpkin pie, or two, to finish the meal. From an evolutionary perspective, animals that ate a diversity of food types were more likely to satisfy nutritional requirements and survive than those that relied on a small number of foods. Sensory-specific satiety encourages consumption of varied foods, which increases the likelihood that nutritional needs will be met. It also encourages animals to eat a great deal, a behaviour that may have been adaptive when the food supply was scarce or unpredictable.

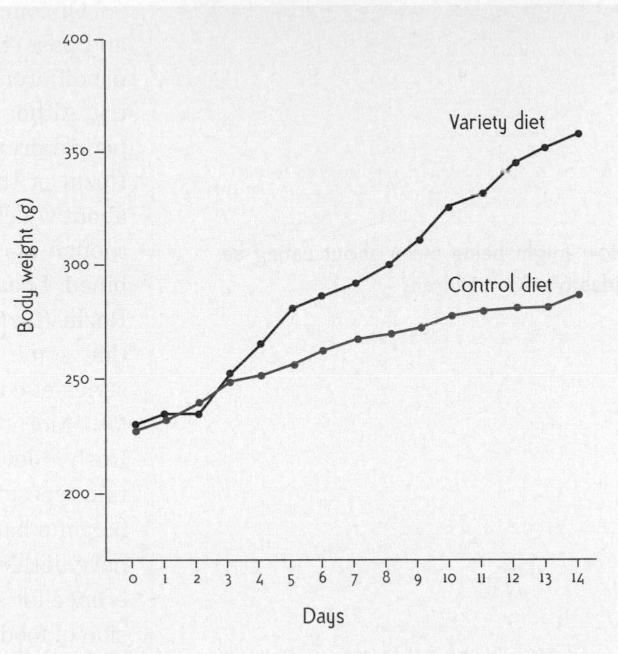

9.13 Tastes and Eating Behaviour Rats eat a great deal more when given a variety of food and will become obese if given access to food such as marshmallows, salami, and cookies.

Culture Determines What We Eat

Would you eat bat? Bat is a delicacy in the Seychelles Islands—it tastes something like chicken. What we will eat has little to do with logic and everything to do with what our minds believe is food. Some of the most nutritious foods are not eaten in North America because they are viewed as disgusting. For instance, fried termites, a favourite in Zaire, have more protein than beef; spiders are considered tasty and nutritious treats in many countries. At the same time, people from other cultures might be nauseated by some of our favourite foods, such as hamburgers and ice cream. Even when people are starving to death, they will refuse to eat perfectly nutritious substances: People died in Naples in 1770 because they were suspicious of the potatoes sent to relieve their famine, whereas many of those who died in the potato famine in Ireland refused to eat corn that was sent over from America. Three million died in Bengal in 1943 in spite of being supplied with wheat, which they rejected because it was not familiar as food. What factors determine what is regarded as food?

What people will eat is determined by a combination of personal experience and cultural beliefs. Infants have an inborn preference for sweets, but they can learn to like just about anything. A general rule that determines preferences is familiarity: Animals like what is familiar and tend to avoid that which is unfamiliar (Galef & Whiskin, 2000). This makes great sense, because unfamiliar foods may be dangerous or poisonous, and therefore avoiding them is adaptive for survival. Getting children to like new foods often involves exposing them to small amounts at a time, until they get used to the taste. The avoidance of unfamiliar foods is an example of *neophobia,* the fear of novel things. Infants learn to try foods by observing their parents and siblings—if Mom eats something, it must be okay. Children are much more likely to eat a new food offered by their mothers than the same food offered by a friendly stranger. This also makes great sense from an evolutionary point of view, in that the rule "eat what my relatives are eating" would typically have been a good one.

How might being picky about eating be adaptive for children?

Of course, what the mother prefers to eat is determined by her own upbringing and past experiences, and therefore families tend to like specific types of food. Ethnic differences in food preference often continue when a family moves to a new country. Although we may enjoy the occasional meal of some novel ethnic food, most people prefer the food of their own culture for their regular diet. Moreover, as Paul Rozin of the University of Pennsylvania has pointed out, there are cultural rules about which food is appropriate in different contexts. He gives the example that even though most Americans like chocolate and french fries, few people like them combined. Local norms of not only what to eat but also how to prepare it, part of what Rozin (1996) calls *cuisine,* reinforce many food preferences. Oprah Winfrey noted that some of her problems with food might have developed because of Southern styles of cooking: "I come from a culture of people where you cook a green in pork fat." Moreover, religious and cultural values often tell people which foods to avoid: Kosher Jews eat beef but not pork, and Hindus eat pork but not beef. Taboos on certain types of food may have been adaptive because those foods were likely to contain harmful bacteria. However, many food taboos and preferences are idiosyncratic and have nothing to do with harm avoidance. They simply reflect an evolved group preference for specific foods prepared and eaten in certain ways. The cultural transmission of food preferences is a powerful factor in choosing what foods to eat.

Multiple Neural Processes Control Eating

The hypothalamus is the brain structure with the greatest influence on eating. Although it does not act alone to elicit eating behaviour, the hypothalamus integrates the various inhibitory and excitatory feeding messages and organizes behaviours involved in eating. In the first half of the twentieth century, research revealed that damage to the hypothalamus leads to dramatic changes in eating behaviour and body weight, depending on which specific area is damaged.

One of the first observations occurred in 1939, when researchers discovered that patients with tumours of the hypothalamus became obese. To examine whether obesity could be induced in normal-weight animals, researchers selectively damaged specific hypothalamic regions through lesioning. Experimentally lesioning the middle or *ventromedial* region (*VMH*) of the hypothalamus causes rats to eat great quantities of food, referred to as *hyperphagia.* Rats with VMH lesions grow extremely obese. In contrast, lesioning the outer or *lateral* area of the hypothalamus (*LH*) is associated with a condition known as *aphagia,* which is diminished eating behaviour that leads to weight loss and eventual death, unless the animal is force fed.

Brain structures other than the hypothalamus are also involved in eating behaviour. For instance, taste cues such as sweetness or saltiness are processed in part of the prefrontal cortex. This suggests that this region processes information about the potential reward value of food. Damage to either the limbic system or the right frontal lobes sometimes produces *gourmand syndrome,* in which people become obsessed with fine food and food preparation. One 48-year-old stroke patient who previously had shown little interest in food became preoccupied with food and eventually quit his job as a political correspondent to become a food critic (Regard & Landis, 1997). In spite of their fascination with good-tasting food, those who have gourmand syndrome do not necessarily become overweight; their obsession is not with eating but with the quality and variety of the food itself and how it is prepared.

INTERNAL SENSATIONS Earlier you learned how negative-feedback processes contribute to homeostasis, and for many years it was believed that eating was a classic homeostatic system. From this perspective, some sort of detector would notice deviations from the set-point and would signal an animal either to begin eating or to stop eating. But where do hunger signals come from? The search for energy-depletion detectors has led scientists from the stomach to the bloodstream to the brain.

The initial assumption, based on common sense, was that hunger derived from receptors in the stomach. Many people describe the subjective experience of hunger as emanating from their stomachs, and even our common descriptions of not being hungry refer to "feeling full." The first empirical test of the idea that hunger comes from the stomach was conducted by Walter Cannon and his research assistant Arthur Washburn. One day while working in the lab, Cannon noticed that Washburn's stomach was making rumbling sounds. Cannon theorized that perhaps it was doing so because Washburn was hungry, and so he quickly rigged up an experiment to test his hypothesis. Rather than let Washburn eat, Cannon had him swallow an empty balloon and then inflated the balloon so that he could register changes in the stomach walls caused by contractions. At the same time, he asked Washburn to report each time he felt a pang of hunger. Cannon found that Washburn reported feeling hungry at the end of each stomach contraction and therefore proposed that hunger was the subjective feeling associated with stomach contractions. Although these findings were intuitively appealing, it was quickly apparent that stomach contractions are not the only factors that determine hunger. Eating a small amount of food stops stomach contractions but is more likely to lead to additional eating than to terminate eating. Moreover, people who have had their stomachs removed continue to report being hungry. Research over the past few decades has established that stomach contractions and distensions are relatively minor determinants of hunger and eating.

In the 1940s and 1950s, a number of researchers postulated the existence of receptors in the bloodstream that monitor levels of vital nutrients. One of the best-known of these theories is the *glucostatic theory,* which proposes that specialized glucose receptors, called *glucostats,* monitor the extent to which glucose is taken up into cells so it can be used for energy. Because glucose is the primary fuel for metabolism and is especially crucial for neuronal activity, it makes sense for animals to be sensitive to deficiencies in glucose. Evidence supports the existence of glucoreceptors, and various experiments have demonstrated that injections of glucose in the bloodstream can postpone eating in hungry animals. Similarly, the *lipostatic theory* proposes a set-point for body fat in which deviations from set-point initiate compensatory behaviours to return to homeostasis. For instance, when an animal loses body fat, hunger signals motivate eating and a return to set-point.

The recently discovered hormone *leptin* is involved in fat regulation. Leptin is released from fat cells at a rate approximately equal to the amount of fat being stored in those cells. Leptin travels to the hypothalamus, where it acts to inhibit eating behaviour. The message from leptin is slow acting, so it takes considerable time after eating before leptin levels change in the body. Thus, leptin is more important for long-term body-fat regulation than for short-term eating control. Animals that lack the gene necessary to produce leptin become extremely obese, and injecting leptin into these animals leads to a rapid loss of body fat. The extent to which leptin contributes to human obesity is unclear. As you will learn in the next chapter, many factors contribute to obesity. You will also learn why most of us who diet, like Oprah, have difficulty in losing weight and keeping it off.

REVIEWING the Principles | What Determines How We Eat?

The consumption of food is necessary for survival. A number of overlapping, redundant physiological systems motivate eating behaviour to satisfy nutritional requirements. The motivation for variety helps animals achieve a nutritious diet, whereas fear of new foods and acquired taste aversions help animals avoid potential harm. People learn from those around them to like certain flavours. There is great variety in the types of substances that people consume as food.

What Is Addiction?

What motivates people to risk death or injury to achieve momentary pleasure? Throughout history, people have discovered that ingesting all sorts of substances can alter their mental state in desirable ways. Societal problems stemming from drug abuse are well known. Considering only the commonly abused substances, such as illegal drugs, alcohol, and prescription drugs, it is likely that most people know and care about someone who is an addict. If we include nicotine and caffeine, it is likely that addicts are in the majority in most countries in the world. The questions psychological science asks about addiction include, Why do people use drugs? Why do some people become addicted to drugs? Why do individuals with drug addiction continue to abuse drugs when doing so causes such personal turmoil and suffering? To gain a full understanding of addiction, it is necessary to address these questions on multiple levels, from the biological to the individual to the social.

Addiction Has Psychological and Physical Aspects

physical dependence Synonymous with addiction, the physiological state in which failing to ingest a specific substance leads to bodily symptoms of withdrawal.

psychological dependence Habitual substance use, despite consequences, and a compulsive need to continue using the drug.

The term **physical dependence** is synonymous with *addiction,* a physiological state in which failing to ingest a substance leads to symptoms of *withdrawal,* a state characterized by anxiety, tension, and cravings. Physical dependence is associated with *tolerance,* so that a person needs to consume more of the drug to achieve the same subjective effect. **Psychological dependence** refers to habitual and compulsive substance use despite the consequences. People can be psychologically dependent without showing tolerance or withdrawal. People can also show psychological dependence on other behaviours, such as gambling or shopping. Here we focus on addiction to substances.

As mentioned in Chapter 6, drug addiction has both negative and positive reinforcement aspects. Addicts habitually abuse drugs to escape withdrawal, thereby negatively reinforcing drug use. People also take drugs to forget their problems or to deal with stress, and this too involves negative reinforcement. Drug use is also positively reinforced. People take drugs to feel good, which occurs because drugs activate dopamine receptors. Recall that dopamine is the most important neurotransmitter for reinforcement and dependence (Koob, 1999; McGregor & Roberts, 1993).

How do people become addicted? Many theories related to the initiation of drug or alcohol use among children or adolescents focus on the social level of analysis. They explore social-learning processes that emphasize the role of parents, peers, and mass media. For example, the Joe Camel advertising campaign of the early 1990s in the United States heralded a tenfold increase in market share among American adolescents for Camel cigarettes; at one point more young children could identify Joe Camel than Mickey Mouse (Mizerski, 1995). Social-learning theories also emphasize self-identification with high-risk groups (e.g., "stoners" or "druggies") as central to the initiation of drug and alcohol use. Teenagers want to fit in somewhere, even with groups that society perceives as deviant. As you will recall from Chapter 6, children imitate the behaviour of role models, especially those they admire or with whom they identify. For children whose parents smoke, the modeling of the behaviour is continuous through early childhood and elementary school. It is not surprising then that multiple studies show that when parents smoke, it leads to positive attitudes about smoking and the early adoption of smoking behaviour in their children (Rowe, Chassin, Presson, & Sherman, 1996).

At the individual level, some adolescents are especially likely to experiment with illegal drugs and to abuse alcohol. Children who are high in sensation seeking (a personality trait that involves attraction to novelty and risk taking) are more likely to associate with deviant peer groups and to use alcohol, tobacco, and drugs (Wills, DuHamel, & Vaccaro, 1995). These children also tend to have poor relationships with their parents, which, in turn promotes greater deviant-peer-group association, and so on. So does the family environment determine alcohol and drug use? Some theorists, operating at the biological level of analysis, turn this around and suggest that an inherited predisposition to sensation seeking may predict behaviours that increase the possibility of substance abuse, such as affiliating with drug users. Indeed, some evidence points to genetic components to addiction, especially for alcoholism. But it is important to point out that little direct evidence points to a *single* "alcoholism" or "addiction" gene. Rather, what might be inherited is a cluster of characteristics, including certain personality traits, a reduced concern about personal harm or injury, a nervous system that is chronically low in arousal, or a predisposition to find chemical substances pleasurable. In turn, these factors make some people more likely to explore drugs and to like them.

Why don't all drug users become addicts?

One of the most fascinating aspects of addiction is that only about 5 to 10 percent of those who use drugs other than tobacco or alcohol become addicted. Indeed, the vast majority of Canadians have tried alcohol, and many have experimented with illicit drugs. Yet most either use alcohol or drugs in moderation (especially alcohol) or try them for a while and then give them up completely. In a major longitudinal study, Jonathan Shedler and Jack Block (1990) found that those who had experimented with drugs and alcohol as adolescents were better adjusted in adulthood than those who never tried them at all. Both complete abstainers and heavy drug users had adjustment problems compared with the experimenters. However, this does not mean that everyone should rush out to try drugs and alcohol, or that parents should encourage such behaviour. After all, you have no way of knowing if you will become addicted or whether you are prepared to handle drugs' effects on your behaviour.

In this section we examine some of the most commonly used drugs. One of the most common drugs—nicotine in cigarettes—is discussed in Chapter 10, along with a discussion of how addictions can be treated. Here we are concerned with the motivation to use drugs that alter mental state.

People Use—and Abuse—Many Psychoactive Drugs

Drugs are one of the most mixed blessings of our lives. Taking them when we have a severe injury or even a moderate headache can provide soothing relief. The right drugs, given to people who are depressed, can lift their moods and help them lead better lives. Other drugs given to children with attention deficits or hyperactivity disorders can help them settle down and be better classroom learners. But many of these same drugs can be used for recreational purposes—to alter physical sensations, levels of consciousness, thoughts, moods, and behaviours in ways that users believe to be desirable—and this recreational use can sometimes lead to severe negative consequences.

Psychoactive drugs are mind-altering substances that change the brain's neurochemistry by activating neurotransmitter receptors. The effects of a particular drug depend on which receptors are activated. Common psychoactive drugs include stimulants, depressants, narcotics, and hallucinogenics. In this section we consider drugs that often have legitimate medical uses, but are frequently abused outside of treatment. Later we will examine the most widely used legal drug that is seldom employed medically, namely alcohol.

STIMULANTS *Stimulants* are drugs that increase behavioural and mental activity and include substances such as cocaine, nicotine, caffeine, and amphetamines. These drugs activate the sympathetic nervous system (increasing heart rate and blood pressure), improve mood, cause people to become restless, and disrupt sleep. They are also highly reinforcing. Stimulants generally work by interfering with the normal reuptake of dopamine by the releasing neuron, which allows dopamine to remain in the synapse and thus prolongs its effects, although sometimes stimulants also increase the release of dopamine. Drugs that block the action of dopamine reduce the reinforcing properties of stimulants.

Cocaine is a stimulant that is often inhaled as a powder or smoked in the form of *crack cocaine.* However it is used, cocaine is derived from the leaves of the coca bush, which grows primarily in South America. After snorting or smoking cocaine, users experience a wave of confidence and good feeling. They are alert, energetic, sociable, and feel little need of sleep. These effects are especially intense for users of crack cocaine. A pharmacist from Georgia, John Pemberton, was so impressed with cocaine that in 1886 he added it to soda water for easy ingestion, thus creating Coca-Cola (Figure 9.14). In 1906, the U.S. government outlawed cocaine, so it was removed from the drink. To this day, however, coca leaves from which the cocaine has been removed are used in the making of Coca-Cola. In contrast to the positive effects of cocaine, habitual use of large quantities can lead to paranoia, as well as psychotic and even violent behaviour (Ottieger, Tressell, Inciardi, & Rosales, 1992).

9.14 Everyday Stimulants
An early ad for Coca-Cola.

Amphetamines are stimulants synthesized using simple lab methods. Amphetamines go by street names such as *speed, meth, ice,* and *crystal.* Amphetamines have a long history of use for weight loss and staying awake. However, amphetamines have a number of negative side effects—such as insomnia, anxiety, and heart problems—and people quickly become addicted to them. They are seldom used for legitimate medical purposes.

MDMA One drug that has recently become very popular is *MDMA,* or *ecstasy,* which produces an energizing effect similar to that of stimulants, but also causes slight hallucinations. The drug first became popular among young adults in nightclubs or at all-night parties known as raves. Compared with amphetamines, MDMA is associated with less dopamine release and more serotonin release. The latter aspect may explain its hallucinogenic properties. As you might recall from Chapter 3, LSD's hallucinogenic effects are due to activation of serotonin neurons. Although many users believe it to be relatively safe, researchers have documented a number of impairments from long-term ecstasy use, especially memory problems and a diminishment of the ability to perform complex tasks. Because ecstasy also depletes serotonin, users often feel depressed when the rewarding properties of the drug wear off.

MARIHUANA The most widely used illegal drug in North America is marihuana, which is the dried leaves and flower buds of the hemp plant. The active psychoactive ingredient in marihuana is the chemical THC (tetrahydrocannabinol), which alters the working mind in a number of complex ways, but in particular leads to a relaxed mental state, uplifting or contented mood, and some perceptual and cognitive distortions (Fried et al., 2005). Users report that it makes perceptual experiences more vivid, with some extolling its effects especially on taste perception. Most first-time users of marihuana do not experience the "high" obtained by more experienced users. Part of this may be that novice smokers use inefficient techniques and may have trouble inhaling, but it is also appears that users need to learn how to *appreciate* the drug's effects. This makes marihuana different from most other drugs, which have stronger effects on first-time users, with subsequent usage leading to tolerance.

Although the brain mechanisms that marihuana affects remain somewhat mysterious, investigators have recently discovered a class of receptors that are activated by naturally occurring THC-like substances. Activation of these cannabinoid receptors appears to adjust and enhance mental activity, and perhaps alter the perception of pain. There is a large concentration of these receptors in the hippocampus, which may explain in part why marihuana impairs memory (Ilan, Smith, & Gevins, 2004).

OPIATES Heroin, morphine, and codeine belong to the family of drugs known as opiates. These drugs provide enormous reward value by increasing dopamine activation in the nucleus accumbens and binding with opiate receptors, producing feelings of relaxation, analgesia, and euphoria. Prior to the turn of the twentieth century, heroin was widely available without prescription and was marketed by the Bayer Aspirin Company. Heroin provides a rush of intense pleasure that most addicts describe as similar to orgasm, which then evolves into a pleasant, relaxed stupor. The dual activation of opiate receptors and dopamine receptors may explain why heroin and morphine are so highly addictive.

Profiles in Psychological Science

Soldiers Kick the Narcotics Habit

American soldiers who fought in Vietnam often took drugs to cope with the hellish conditions of war. By the late 1960s, estimates suggested that drug abuse among American soldiers, including the use of narcotics such as heroin and opium, had reached epidemic proportions. The widespread use of drugs was not surprising—the late 1960s were a time of youthful drug experimentation, soldiers in Vietnam had easy access to a variety of drugs, and drugs helped the soldiers cope temporarily with fear, depression, homesickness, boredom, and the repressiveness of army regulations. The soldiers were also attracted to the euphoric and rewarding qualities of the drugs; they enjoyed getting "high." Although the military commanders were aware of drug use among soldiers, they mostly ignored it, viewing it as soldiers blowing off steam.

Beginning in 1971, the military began mandatory drug testing of soldiers in order to identify and detoxify drug users before they returned to the United States. More than 1 in 20 soldiers tested positive for narcotics, though they knew in advance that they would be tested. With speculation that many of these soldiers were unable to go without drugs for even a short time, concern grew that a flood of addicted soldiers returning from Vietnam would swamp treatment facilities back home. The White House asked a team of behavioural scientists to study a group of returning soldiers to assess the extent of the addiction problem.

Led by the behavioural epidemiologist Lee Robins, the research team examined a random sample of 898 soldiers who were leaving Vietnam in September 1971. Robins and her colleagues found extremely high levels of drug use among them (Robins, Helzer, & Davis, 1975). Over 90 percent reported drinking alcohol, nearly three-quarters smoked marihuana, and nearly half used narcotics such as heroin, morphine, and opium. About half of the soldiers who used narcotics either had symptoms of addiction or reported that they believed they would be unable to give up their drug habits. Robins's findings suggested that approximately 1 soldier in 5 returning from Vietnam was an addict. Given the prevailing view that addiction was a biological disorder with a low rate of recovery, these results indicated that tens of thousands of heroin addicts would soon be inundating American towns and cities. But this didn't happen.

Robins and her colleagues examined drug use among the soldiers after they returned to the United States. Of those who were apparently addicted to narcotics in Vietnam, only half sought out drugs when they returned to the States, and fewer still maintained their narcotic addictions. Approximately 95 percent of the addicts no longer used drugs within months of their return, which is an astonishing quit rate considering that the success rate of the best treatments is typically only 20 to 30 percent. A long-term follow-up study conducted in the early 1990s confirmed that only a handful of those who were addicts in Vietnam remained addicts.

Why did coming home help the addicts recover? One possible explanation comes from classical conditioning—all of the cues for drug use were connected to Vietnam and its associated military environment. When the soldiers returned home, their environment contained none of those cues, and therefore the soldiers did not experience conditioned withdrawal effects. But it is likely, too, that their motives for taking the drugs were also absent on their return. No longer needing the drugs to escape the horrors of combat, they focused on other needs and goals that became more important, such as family obligations and careers. An important lesson from this case study is that we cannot ignore the environment when we try to understand addiction. Knowing the physical actions of drugs in the brain may give us insights into the biology of addiction, but it fails to account for how these biological impulses can be overcome by other motivations. ■

Alcohol Is the Most Widely Abused Drug

North Americans have a love-hate relationship with alcohol. On the one hand, moderate drinking is an accepted aspect of normal social interaction and may even be good for health. On the other, alcohol is a major contributor to many of our societal problems, such as spousal abuse and other forms of violence. Moreover, one-quarter of suicide victims and one-third of homicide victims have blood-alcohol levels that meet legal criteria for impairment. A recent study found that approximately one-third of college students reported having had sex during a drinking binge, and the heaviest drinkers were likely to have had sex with a new or casual partner (Leigh & Shafer, 1993). This places heavy college drinkers at great risk for exposure to AIDS.

Most individuals who drink do so in moderation, although 60 percent of Canadians report alcohol use, with the heaviest drinkers found to be males, outnumbering females 5 to 1 (Kelner, 1997). Some people who drink alcohol meet criteria for **alcoholism,** defined as abnormal alcohol seeking with some loss of control over drinking, accompanied by physiological effects of tolerance and withdrawal. Some 90 percent of high school seniors have tried alcohol, and in one recent survey almost one-third of seniors reported having had a bout of heavy drinking (five or more drinks) within the preceding two weeks.

alcoholism Abnormal alcohol seeking characterized by loss of control over drinking and accompanied by physiological effects of tolerance and withdrawal.

One reason people consume alcohol is the anticipated effects that alcohol will have on their emotions and behaviour. Both light and heavy drinkers believe that alcohol reduces anxiety, and so many people regularly have a drink or two after a difficult day. Unfortunately, the available evidence does not support this belief; although moderate doses of alcohol are associated with temporarily increased positive mood, larger doses are associated with a worsening of mood. Also, although alcohol can interfere with cognitive processing of threat cues, such that anxiety-provoking events are less troubling when people are intoxicated, this occurs only if people drink *before* the anxiety-inducing event. According to this research, drinking after a hard day can actually increase people's focus on and obsession with their problems (Sayette, 1993).

Alan Marlatt, a leading researcher on alcohol and drug abuse, has commented that people view alcohol as the "magic elixir," capable of increasing social skills, sexual pleasure, confidence, and power (Marlatt, 1999). Expectations about the effects of alcohol are learned very early in life; children see that people who drink have a lot of fun and that drinking is an important aspect of many celebrations (Figure 9.15). Teenagers may view drinkers as sociable and grown up, two things that they desperately want to be. Thus, through observation, children learn to expect that consuming alcohol will have positive effects. It has been shown that children who have very positive expectations about alcohol are more likely to start drinking and to be heavy drinkers than children who do not share those expectations.

According to the social psychologists Jay Hull and Charles Bond, expectations have profound effects on behaviour. To study the true effects of alcohol, researchers give individuals either tonic water or alcohol and then tell them that they are getting tonic water or alcohol, though not necessarily in correlation to the actual contents of the drink (Figure 9.16). This balanced-placebo design allows for a comparison of those who think they are drinking tonic water but are actually drinking alcohol with those who think they are drinking alcohol but are actually drinking tonic water. Researchers thus can separate drug effects from personal beliefs. This study has demonstrated that alcohol truly does impair motor processes, information processing, and mood, independent of whether people think they have consumed it. In contrast, the belief that one has consumed alcohol leads to disinhibition of a variety of social behaviours, such as sexual arousal and aggression, whether or not the person actually has consumed alcohol (Hull & Bond, 1986). Thus, some of people's behaviours when drunk are accounted for by learned beliefs about intoxication rather than by the pharmacological properties of alcohol. Sometimes the pharmacology and learned

9.15 Alcohol and Behaviour Many alcohol and beer companies promote the idea that good times will follow from drinking.

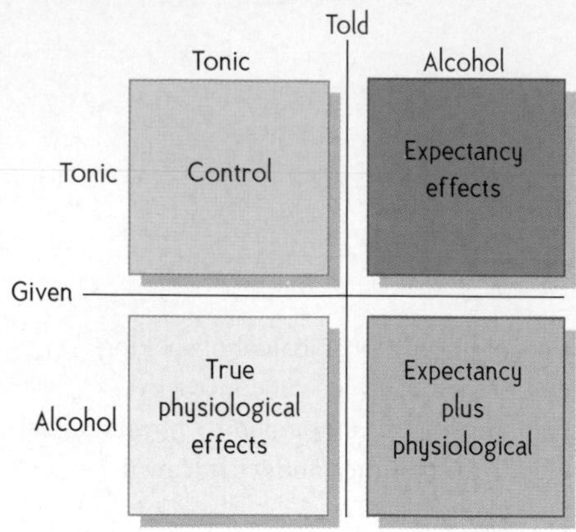

9.16 True Effects of Alcohol The consumption of alcohol typically includes both physiological and expectancy effects. The balanced-placebo design allows researchers to examine these influences on behaviour separately.

expectancies work in opposite ways. For instance, alcohol tends to increase sexual arousal, but it interferes with sexual performance.

Alcohol involves a number of different neurotransmitter systems and activates receptors for GABA, opiates, and dopamine (see Chapter 3). A variety of evidence suggests that the reinforcing aspects of alcohol are due to its activation of dopamine receptors, as is the case with other addictive drugs. Alcohol also interferes with the neurochemical processes involved in memory, which explains why memory loss can follow excessive alcohol intake. Heavy long-term alcohol intake can cause extensive brain damage. *Korsakoff's syndrome* is an alcohol-related disorder characterized by severe memory loss and intellectual deterioration.

REVIEWING the Principles | What Is Addiction?

People have long ingested drugs that alter the way they think, feel, and act. Drug use and abuse has enormous societal costs, from impaired driving, to illness, to crime. Although moderate use of drugs such as alcohol and caffeine may have few consequences, excessive use can lead to physical or psychological dependence. The term *addiction* refers to physical dependence, in which not using the drug produces withdrawal and more of the drug is needed to obtain its effects. Psychoactive drugs affect a variety of neurotransmitter systems, but personal beliefs about a drug also affect the user's experience and behaviour.

What Factors Motivate Sexual Behaviour?

Sexual desire has long been recognized as one of humanity's most durable and powerful motivators because most human beings have a significant desire for sex. However, sex drives vary substantially across individuals and circumstances. This variation in sexual practices can be explained both in terms of individual differences and the fact that society has a dominating influence over how and when individuals engage in sexual activity.

For much of the history of psychological science, the study of sex was something of a taboo. The idea that women were motivated to have sex was almost unthinkable; many theorists believed that women were even incapable of enjoying the sexual act. The pioneering work of Alfred Kinsey and his colleagues in the 1940s provided—for the times—shocking evidence that women were in many ways similar to men in their sexual attitudes and behaviours. In Kinsey's surveys of thousands of Americans, he found that more than half of both men and women reported premarital sexual behaviour, that masturbation was common in both sexes, that women enjoyed orgasms, and that homosexuality was much more common than most people expected. More than 50 years later, we know a great deal more about sexual behaviour, although the topic still makes some people uncomfortable. Public discussion of sexual activity is rare, except perhaps on talk radio and daytime televison shows. However, more than 85 percent of parents agree that sex education should be provided in schools, as do 92 percent of youth (Society of Obstetricians and Gynaecologists of Canada). In this section we describe what psychological science has learned about the motivation for sex.

Biological Factors Influence Sexual Behaviour

Kinsey's research demonstrated how little most people knew about human sexual behaviour. Despite Kinsey's eye-opening contributions, ignorance about the physiology of sex persisted into the 1960s, when William Masters and Virginia Johnson began to study sexual behaviour in the laboratory. It needs to be acknowledged that their sample was a little bit unusual, in that only people who were willing to be filmed while having intercourse or masturbating served as research participants. Nonetheless, their work provided considerable insights into the physiological aspects of human sexual behaviour. The most enduring contribution of this research was the identification of the **sexual response cycle**, consisting of a predictable pattern of physical responses that occur in four stages. The *excitement phase* occurs when people contemplate sexual activity or begin to engage in behaviours such as kissing and touching in a sensual manner. During this stage, blood flows to the genitals and they become aroused. The male penis begins to become erect. The female clitoris becomes swollen, the vagina expands and secretes fluids, and the nipples enlarge.

sexual response cycle A pattern of physiological responses during sexual activity.

As excitement continues, people reach the *plateau phase,* during which pulse rate, breathing, and blood pressure increase, as do the various signs of arousal. For many people, this is the frenzied phase of sexual activity, in which inhibitions are lifted and passion takes control. The plateau phase culminates in the *orgasm phase,* which consists of involuntary muscle contractions throughout the body, dramatic increases in breathing and heart rate, rhythmic contractions of the vagina for females, and ejaculation of semen for males. Although orgasm nearly always occurs for healthy males, it is more variable for females. When it does occur, however, the pleasurable sensations are nearly identical to those experienced by males. Following orgasm there is dramatic release of sexual tension and a slow return to a normal state of arousal. In this *resolution phase,* the male enters a refractory period during which he is temporarily unable to maintain an erection or have orgasm. Females do not have such a refractory period and may experience multiple orgasms with short resolution phases between each one. This sexual response cycle involves both biological and psychological processes. Here we focus on the biological.

HORMONES As you recall from Chapter 3, hormones are involved in the production and termination of sexual behaviours. In nonhuman animals, hormones profoundly influence sexual activity. In many species, females are sexually receptive only when fertile, and estrogen is believed to control reproductive behaviours. As noted in Chapter 3, however, estrogen appears to play only a small role in human female sexuality. What role do hormones play for people?

Hormones influence human sexual behaviour in two ways. First, they influence physical development of the brain and body. During the developmental phase of *puberty,* increases in hormones throughout the body stimulate bodily changes, such as the development of *secondary sexual characteristics,* including pubic hair, body hair, and, for females, breasts. In North America and Europe, puberty typically begins at age 11 or 12, and it typically begins earlier for females.

The second way hormones influence sexual behaviour is through motivation; they activate reproductive behaviour. Recall that sex hormones are released from the gonads (testes and ovaries) and that androgens are more commonly released in males and estrogens are more commonly released in females. As noted in Chapter 3, androgens appear to be much more important for reproductive behaviour than estrogens, at least for humans. In both men and women, the androgen testosterone is involved in sexual functioning. Males need a certain amount of testosterone to be able to engage in sex, although they do not perform better if they have more testosterone. It also supports aggression in both sexual and other enounters (Bain, 1987). For women, it appears that the more testosterone they have, the more likely they are to have sexual thoughts and desires (Meston & Frohlich, 2000). Adolescent females with high testosterone levels are more likely to engage in sexual intercourse (Halpern, Udry, & Suchindran, 1997). Another hormone that is important for both men and women is *oxytocin,* which is released during sexual arousal and orgasm. Some researchers believe that oxytocin may promote feelings of love between partners; it also seems to be involved in social behaviour more generally. Given the important role of the hypothalamus in controlling the release of hormones into the bloodstream, it should come as no surprise that the hypothalamus is the brain region considered most important for stimulating sexual behaviour. Studies have shown that damaging the hypothalamus in rats interrupts sexual behaviour, although the area that is most affected differs slightly for males and females.

NEUROTRANSMITTERS A variety of other neurotransmitters can affect various aspects of the sexual response. For instance, dopamine receptors in the limbic system are involved in the physical experience of pleasure. Serotonin has also been implicated in sexual behaviour. The most common pharmacological treatments for depression enhance serotonin function, but they also cause serious reductions in sexual interest, especially for women. Why this happens is currently unclear, although small doses of drugs that enhance serotonin are useful for treating premature ejaculation.

One chemical that acts as a neurotransmitter in the brain and is also critical for sexual behaviour is *nitric oxide.* Sexual stimulation leads to nitric oxide production, which then promotes blood flow to the penis and clitoris and subsequently plays an important role in the arousal of sexual organs, especially penile erections. When this system fails, males are unable to maintain an erection. Various drugs that enhance this system, such as Viagra, have been developed to treat erectile disorders. Although it is unknown whether drugs such as Viagra can be used to treat women's sexual disorders, they do appear to enhance the sexual experience for healthy women (Caruso, Intelisano, Farina, Di Mari, & Agnello, 2003).

VARIATIONS ACROSS THE MENSTRUAL CYCLE Women differ from men in how the hypothalamus controls the release of sex hormones. Whereas hormones for males are released at the same rate over time, the release of hormones for women varies according to a cycle that repeats itself approximately every 28 days—the menstrual cycle. Although research has found only minimal evidence that women's sexual behaviour varies across the menstrual cycle, recent evidence indicates that women may process social information differently depending on whether they are in a fertile phase of the cycle or not. For instance, when researchers used a computer program to alter masculinity and femininity in male faces, they found that women preferred the more masculine faces during ovulation compared with other phases of the menstrual cycle (Pentok-Voak et al., 1999). Macrae and colleagues (2002) demonstrated that during ovulation women have an enhanced ability to identify faces as being male and are faster to judge behaviours as masculine, such as males being more athletic, decisive, and vulgar. In another recent study, women who watched videotapes of men competing to be lunch dates rated self-assured men as more desirable potential sex partners, but only if the woman were ovulating (Gangestad, Simpson, Cousins, Garver-Agar, & Christensen, 2004). These studies add to a growing literature suggesting that women evaluate men differently across the menstrual cycle.

PHEROMONES One interesting line of research examines the influence of pheromones, chemicals secreted from various glands throughout the body, on the initiation of sex. Recall from Chapter 5 that pheromones are detected through the olfactory system. Evidence from a variety of species suggests that pheromones can activate sexual behaviours. What about in humans (Pfaus et al., 2003)? At this point some evidence suggests that pheromones might influence behaviour, at least indirectly. In one study, female participants who overnight wore necklaces that contained male underarm sweat spent more time in social interactions with males the next day than did women who wore necklaces containing inert substances (Cowley & Brooksbank, 1991). In another study, men who mixed synthetic pheromone in their aftershave reported higher levels of sexual intercourse and other sexual behaviours, but no increase in masturbation. The authors concluded that the greater number of sexual encounters resulted because the pheromones increased the sexual attractiveness of the men to women (Cutler, Friedmann, & McCoy, 1998).

NEURAL CORRELATES OF VIEWING EROTICA Some recent imaging studies indicate that viewing erotica activates brain reward regions, such as various limbic structures. This effect was greatest for the men who had higher blood levels of testosterone (Stoleru et al., 1999). As noted in Chapter 4, Hamann and colleagues (2004) found that when men and women viewed sexually arousing stimuli, such as film clips of sexual activity or pictures of opposite-sex nudes, men showed more activation of the amygdala, which the authors suggest contributes to the arousing nature of the stimulus. Interestingly, men are typically more likely than women to report visual erotic stimulation as pleasurable (Herz & Cahill, 1997). However, this could be because much of erotica is produced for men. Research has shown that women prefer erotica produced specifically for women, which tends to emphasize more of the emotional factors of the sexual interaction. A recent study found that although both sexes prefer erotica that is aimed at their own sex, men were aroused by both types of movies (Janssen, Carpenter, & Graham, 2003).

Cultural Scripts and Rules Shape Sexual Interactions

In the movies, sexual relationships often start when one attractive young person meets another by chance, some exciting time is spent together, attraction develops, and sexual behaviour ensues—often within a day or two. In real life, however, the course of action is often quite different. For one thing, most people rely on social networks to meet their sex partners. It is also apparent that people do not generally fall into bed together as fast as they do in the movies; most people know someone a long time before having sex. But the depiction of sexual behaviour in movies and other media does shape beliefs and expectancies about what sexual behaviours are appropriate and when they are appropriate. *Sexual scripts* are cognitive beliefs about how a sexual episode should be enacted (see Chapter 8). For instance, the sexual script indicates who should make the first move, whether one should resist, the sequence of sexual acts, and even how one should act afterward. In North America, the sexual script involves initial flirtation through nonverbal actions, the male initiating physical contact, the female controlling whether sexual activity takes place, and refusals typically being verbal and direct (Berscheid & Regan, 2005).

DOUBLE STANDARDS At the social level, the double standard is a well-known pattern of cultural influence. It stipulates that certain activities (such as premarital or casual sex) are morally and socially acceptable for men but not for women. The sexual revolution of the late twentieth century entailed a large change in sexual behaviours, most of which must be attributed to changing cultural pressures and expectations. Although sexual customs and norms vary to some degree across cultures, all known cultures have some form of sexual morality, which indicates the importance to society of providing some regulation of sexual behaviour. Cultures may seek to restrain and control sex for a variety of reasons, including maintaining control over the birth rate, establishing paternity, and reducing conflicts.

Perhaps because of the mixed messages that women receive, they tend to be higher than men in erotophobia, which is the disposition to respond negatively to sexual cues. Those high in erotophobia report parental strictness about sex, conservative attitudes, and the avoidance of masturbation (Fisher, Byrne, White, & Kelley, 1988). Paradoxically, those high in erotophobia may be at greater risk for unwanted pregnancies and sexually transmitted diseases, because they are less likely to carry condoms or take other precautions.

LOSING VIRGINITY The age at which people first lose their virginity has varied considerably across cultures and history. In recent decades in North America, this age has become younger. Miller and Benson (1999) concluded from multiple studies that about half of today's young women and two-thirds of the men have had intercourse by their eighteenth birthdays. A substantial minority remain virgins longer than that, however—one out of five is still a virgin at age 20. In Canada, the average age of initial sexual experiene for both males and females is 16; 28 percent of teens aged 15–17 reported having intercourse at least once. This number jumps to 80 percent by age 20–24 (Society of Obstetricians and Gynaecologists of Canada). Even so, the average can conceal wide variations. Interestingly, many people who start having sex at an early age express regret later on. Losing virginity is a major event in most lives, although the actual experience can be anticlimactic and disappointing. Sprecher, Barbee, and Schwartz (1995) found that men reported both more anxiety and more pleasure during the first experience than did women, and the men were also more likely to have an orgasm. Women were more likely to feel guilty. For both genders, alcohol use was correlated with lower enjoyment of the first sexual experience, whereas having the first sex in the context of a committed relationship was associated with having a better experience.

How much of human sexual behaviour is determined by biology?

SEX DIFFERENCES IN SEXUAL MOTIVES A noticeable and consistent finding in nearly all measures of sexual desire is that men, on average, have a higher level of sexual motivation than women, although there are many individual exceptions. Research studies have found that in general men masturbate more frequently than women, want sex earlier in the relationship, think and fantasize about sex more often, spend more time and money (and other resources) in the effort to obtain sex, desire more different sexual activities, initiate sex more and refuse sex less, and rate their own sex drives as stronger than women's (Baumeister, Catanese, & Vohs, 2001). In one study, researchers asked college-age men and women how many sex partners they would ideally like to have in their lives, if they were unconstrained by fears about disease, social pressures, and the like (Miller & Fishkin, 1997). The majority of women wanted one or two, whereas the average answer by the men was several dozen. A large study of more than 16,000 people from 10 major regions around the world found that the greater male motive for sexual activity and sexual variety occurs in all cultures (Schmitt et al., 2003).

sexual strategies theory Evolutionary theory that suggests men and women look for different qualities in their relationship partners due to the gender-specific adaptive problems they've faced throughout human history.

The relative influence of nature and culture on sexual motivation may vary with gender. Roy Baumeister (2000) used the term *erotic plasticity* to refer to the degree to which the sex drive can be shaped by social, cultural, and situational factors. Evidence suggests that women have higher erotic plasticity than men. A woman's sexuality may evolve and change throughout her adult life, whereas men's desires remain relatively constant (except for the gradual decline with age). Women's sexual desires and behaviours depend significantly on social factors such as education and religion, whereas men's sexuality shows minimal relationship to such influences.

Mating Strategies Differ between the Sexes

All humans desire to enter into romantic relationships, and for the most part they want the same thing in their mates. Both men and women seek attractive mates who are kind, honest, and good-natured. In addition to liking the same set of features in a potential mate, both sexes avoid certain characteristics, such as showing insensitivity, displaying bad manners, speaking too loudly or in a shrill voice, and bragging about sexual conquests (Cunningham, Barbee, & Druen, 1996). In spite of these similarities, males and females have been found to differ in the relative emphasis they place on physical appearance and social status, with men being more concerned with appearance and women being more concerned with status. How does psychological science explain these differences?

According to the **sexual strategies theory** of the evolutionary psychologist David Buss (Figure 9.17), these differences are due to the different adaptive problems that have faced men and women throughout human history (Buss & Schmitt, 1993). From this perspective, women differ from men in how they maximize the passing along of their genes to future generations. Women's basic strategy is intensive care of a relatively small number of infants. Their commitment is to nurture rather than to simply maximize production. Once a woman is pregnant, additional matings are of no use, and once she has a small child, an additional pregnancy can put her current offspring at risk. Thus, biological mechanisms—for example, nursing typically prevents ovulation—ensure spacing between children. On purely reproductive grounds, men have no such interludes. For them, all matings may have a reproductive payoff, since they bear few of the personal costs of pregnancy.

9.17 David Buss Buss uses evolutionary theory to examine mating strategies. His research shows that there is great cross-cultural consistency in what men and women desire in their mates, and that mating strategies are adaptive.

Because having offspring is a much more intensive commitment for women, they will likely be more cautious about having sex. Indeed, there is evidence that women are much less willing than men to have sex with someone they don't know well. In one study, an attractive stranger approached people of the opposite sex and said, "I have been noticing you around campus. I find you attractive. Would you go to bed with me tonight?" Although not one woman said yes, three-quarters of the men agreed to the request (Figure 9.18). Indeed, the men were less likely to agree to go out on a date with the attractive woman than they were to agree to have sex with her (Clark & Hatfield, 1989). In another study, people were asked to state how long a couple should be together before it is acceptable to engage in sexual intercourse, given mutual desire. Women tend to think that couples should be together for a month or so before sex is appropriate, whereas males believe that even relatively short periods of acquaintanceship make having sex acceptable (Buss & Schmitt, 1993).

9.18 Sexual Motives and Response When men and women were propositioned by an attractive stranger, men were much more willing than women to agree to have sex or to go home with the stranger. Note that not a single woman agreed to have sex.

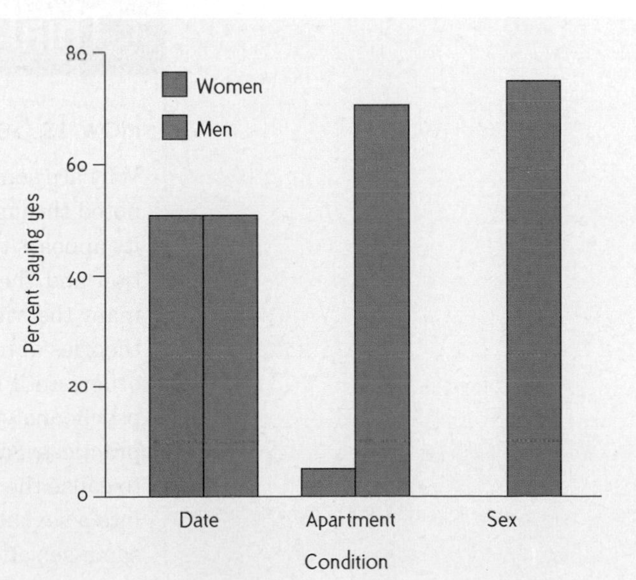

Although men and women value most of the same things in a relationship partner, men and women diverge in terms of which features they emphasize. Strictly in terms of maximizing reproduction, men primarily need to be concerned with whether a woman looks like she could bear healthy children. Conversely, women are better able to maximize reproduction by seeking mates who are likely to provide resources and assistance to help nurture offspring. In one study, Buss asked 92 married couples about which characteristics they valued in their spouse. Women generally preferred men who were considerate, honest, dependable, kind, understanding, fond of children, well liked by others, good earners, ambitious, career oriented, from a good family, and fairly tall. Men tended to value good looks, cooking skills, and sexual faithfulness. In general, men rate physical attractiveness as much more important than do women,

"I'm rich, you're thin. Together, we're perfect."

who demonstrate a preference for education and good earning capacity—in other words, status. Across some 37 cultures studied, females valued a good financial prospect to a greater extent than did men (Buss, 1989). In addition, women in all 37 cultures tended to marry older men, who are often more settled and financially stable. The relative male preference for attractiveness and the female preference for status can be observed in personal ads—studies show that males tend to seek out good looks and females tend to seek out maturity and financial stability. It is important to note, however, that both men and women value physical attractiveness highly. It is the relative emphases that conform to evolutionary predictions.

The evolutionary account of human mating is controversial. Some researchers believe that behaviours shaped by evolution have little impact on contemporary relationships. We must consider two important factors. First, the modern era is only a tiny fraction of human evolutionary history, and the modern mind resides in a Stone Age brain, solving adaptive problems that have faced our species for thousands of years. Thus, remnants of behaviours that were adaptive in prehistoric times may still linger even if they are not adaptive in contemporary society. What is more important, however, is that natural selection bestows not only biological urges, but also a strong sensitivity to cultural and group norms, so that instinctive behaviours are constrained by social context. The frontal lobes work to inhibit people from breaking social rules, which are determined to a great extent by culture. The current social context differs greatly from that of millions of years ago; human mating strategies are indeed influenced by these contemporary norms. For example, from a biological view, it might seem advantageous for humans to reproduce as soon as they are able. But many contemporary cultures discourage sexual behaviour until people are older and better able to care for their offspring. The critical point is that human behaviour emerges to solve adaptive problems, and to some degree the modern era introduces new adaptive challenges based on societal standards of conduct. These standards shape the context in which both men and women view sexual behaviour as desirable and appropriate.

Thinking Critically

How Is Sexual Orientation Determined?

Why are some people homosexual? Homosexual behaviour in various forms has been noted throughout recorded history. From an evolutionary perspective, homosexuality appears to make little sense; exclusive homosexuality would not lead to reproduction and therefore would not survive in the gene pool. Although there have been many theories of homosexuality, none to date has received conclusive support. Early theories of homosexuality regarded it as deviant, abnormal, and a psychological disorder; until 1973 it was officially viewed as a mental illness by psychiatrists. Classic psychoanalytic theories of homosexuality emphasized the importance of parenting practices. Families with a domineering mother and a submissive father were thought to cause the children to identify with the opposite-sex parent, which would translate into a sexual attraction toward the gender opposite of their identification—that is, a same-sex attraction. However, the overwhelming majority of studies have found little evidence that how parents treat their children has anything to do with sexual orientation. Likewise, no other environmental factor has been found to account for homosexuality. So does biology determine sexual orientation?

Remember that when we ask whether something is biological, we are really talking about the relative contribution of biological factors compared with that of environmental factors. Every behaviour is the result of biological processes, such as gene expression, that are themselves influenced by events in the environment. A number of different approaches examine the extent to which biological factors contribute to sexual orientation. One such approach explores how much hormones affect the development of sexual orientation. For instance, early theorists speculated that lesbians had higher levels of testosterone and gay males had higher levels of estrogen, but those were wrong; the levels of circulating hormones do not differ. Rather, the best available evidence suggests that exposure to hormones, especially androgens, in the prenatal environment might play some role in sexual orientation (Mustanski, Chivers, & Bailey, 2002). For example, because of a mother's medical condition, some females are exposed to much higher levels of androgens than normal during prenatal development, and these women often have masculine characteristics, both physical and psychological, and are more likely to report being lesbians later in adulthood. An intriguing finding is that gay males are more likely to have older male siblings than are heterosexual males. One explanation for this is that the mother's body responds in some way to being pregnant with a boy and this response alters the level of hormones in the prenatal environment when the mother becomes pregnant with another boy (Blanchard & Ellis, 2001). But most males with older brothers are not homosexual, so why would this response affect only some males?

A second approach to understanding the biological contribution to sexual orientation is through behavioural genetics. The idea that gene expression might be involved in sexual orientation is supported by a recent study that used fruit flies. Researchers found that altering expression of a single "master" gene reversed the sexual orientation of both male and female flies (Demir & Dickson, 2005). But what about in humans? In 1993, the biologist Dean Hamer reported that he had found a link between a marker on the X chromosome and sexual orientation in males, which the media quickly dubbed "the gay gene." Is homosexuality inherited? Twin studies provide some support for the idea of a genetic component to homosexuality, particularly for males. In a review of previous studies, Mustanski and colleagues (2002) report the heritability of homosexuality as being greater for males than for females, but there was a significant genetic component for both. What is inherited? At this time it is unclear how human sexual orientation might be encoded in the genes.

Some research suggests that the hypothalamus may be related to sexual orientation. In postmortem examinations, neuroscientist Simon LeVay (1991) found that an area of the hypothalamus that typically differs between men and women was only half as large in homosexual men as heterosexual men. In fact, the size of this area in homosexual men was comparable to its size in heterosexual women. Likewise, in a recent brain-imaging study, heterosexual males showed greater activation of the hypothalamus when they sniffed a female pheromone substance than they did when they sniffed a male pheromone substance, whereas females showed greater activation when they sniffed a male rather than female pheromone substance (Savic, Berglund, & Lirdström, 2005). Male homosexuals showed a pattern more similar to women than to heterosexual men, with greater activation of the hypothalamus in response to the male pheromone substance. Although intriguing, both of these studies can be criticized on the grounds that correlation does not equal causation. That is, a size difference or activation difference in any one part of the brain cannot establish whether this area determines sexual orientation, whether being heterosexual or homosexual results in changes to brain structure or function, or whether a third variable is responsible for all these effects. For instance, some researchers believe that the size of the hypothalamus is determined by prenatal exposure to androgens. Thus, although the findings of these two studies are suggestive, we currently have insufficient evidence to establish a causal connection between brain regions and sexual orientation. Considered together, the evidence is consistent that biological processes do play some role in sexual orientation. The question now is how and when biology contributes and to what degree. It seems likely to many psychological scientists that multiple processes may affect one's sexual orientation in more than one subtle way.

How might biology interact with aspects of the environment to determine sexual orientation? Daryl Bem (1996) proposed that feeling different from opposite- or same-sex peers predicts later sexual orientation. A notable finding in the literature is that homosexuals report preferring the leisure activities of their opposite-sex peers, which Bem believes may be due to biological differences in temperament. As you will learn, children tend to play in same-sex groups, so that boys and girls are often segregated. In terms of sexual orientation, Bem believes that people are attracted to that which is different, so most girls are attracted to boys and vice versa. However, to some children, what seems different is peers of their own sex. According to Bem, that which is different creates arousal, and this arousal is the essence of sexual attraction. Because the opposite sex is different, it becomes attractive. Or, in Bem's words, "the exotic becomes erotic." For homosexuals, it is the same sex that is exotic, and that therefore becomes erotic. The interesting part of Bem's model is that it is based on the idea of initial biological differences, but then proposes that the social environment shapes what is sexually attractive. Although the theory is supported by a variety of anecdotal evidence, there is so far no concrete proof for it. But how can researchers ever really test such theories in humans?

What evidence is needed to conclude whether sexual orientation is determined or chosen? Are you pleased or troubled by evidence suggesting that homosexuality is biologically determined? How might your sexual orientation affect how you interpret these data and how you feel about them? ■

REVIEWING the Principles | What Factors Motivate Sexual Behaviour?

Pioneering research by Alfred Kinsey launched the study of human sexual behaviour and shattered many myths regarding the sexual lives of men and women. Masters and Johnson found that there were four stages to the sexual response cycle: excitement, plateau, orgasm, and resolution. Sexual behaviour involves both physiological and psychological factors. Hormones influence the development of secondary sex characteristics as well as motivate sexual behaviour. However, behaviours are constrained by sexual scripts, which dictate appropriate actions for men and women. Men and women look for different qualities in potential partners, which may be due in part to the different adaptive problems faced by the sexes over the course of human evolution. On average and across nearly all measures, men have a higher level of sexual motivation than women.

Conclusion

The study of motivation deals with explaining the occurrence of diverse behaviours, from composing poetry to working late to finish an important project to ingesting mind-altering chemicals. Motivation is concerned not only with immediate survival, but also with our efforts to achieve our loftiest goals. Many of our motivations arise to satisfy the basic biological needs of our bodies, such as drinking to quench our thirst or eating when we are hungry. But other needs are also important in our lives, such as the need for other people and the need to feel competent. An important principle in motivation is that over the course of human evolution, those things that brought pleasure were often adaptive, in that doing them increased survival and reproduction; sexual behaviour is perhaps the most obvious example. However, evolution did not anticipate that people would be able to sit down one day and eat an entire box of cookies or order pizza delivery through the Internet, or that we would discover drugs that would artificially activate brain reward circuits that had been adaptive for millions of years. Now our efforts often require overriding the urge to seek pleasure, such as restricting our consumption of tasty foods or resisting the urge to use drugs. Maintaining a reasonable body weight is an example of a long-term goal, and long-term goals are another important part of motivation. Humans differ from other animals because we have the capacity to set personal goals and try to achieve them. All of this requires the ability to have some control over our biological urges as well as the ability to make choices, such as choosing to study rather than sleep or go out with friends. Each day brings new challenges in motivation as we try to satisfy our immediate needs, keeping in mind that we want to live meaningful and satisfying lives. The study of motivation helps us to make sense of why people desire to do certain things and why sometimes they need to inhibit those desires.

Summarizing the Principles of Motivation

How Does Motivation Activate, Direct, and Sustain Behaviour?

1. **Needs, drives, and arousal motivate behaviour:** Satisfying a need (a state of deficiency) motivates, with survival needs being the most motivating. According to the humanist perspective, we are compelled upwards in our needs hierarchy to achieve self-actualization. Drives are activated to satisfy needs and maintain homeostasis. Arousal affects performance, depending on the nature of the task and level of arousal.
2. **Pleasure can motivate adaptive behaviours:** The pleasure principle is central to motivation and pleasurable behaviours will be selected over satisfying biological needs. Satisfying pleasure needs may facilitate survival and reproduction. Dopamine activation may guide survival behaviours.
3. **Some behaviours are motivated for their own sake:** Extrinsically motivated behaviours emphasize external goals; intrinsically motivated behaviours are performed for their own sake, not biological needs. Curiosity, play, creativity, and problem solving are intrinsically motivated and are reduced by external reward.

Why Are Human Beings Social?

4. **Humans have a fundamental need to belong:** The formation of social bonds is adaptive because it aids survival and reproductive success. We experience anxiety at the prospect of rejection.
5. **People seek others when they are anxious:** Isolation causes anxiety and when anxious, we choose to be with others. Social comparison allows us to validate our self-perceptions.
6. **People are motivated to detect and reject cheaters:** A social dilemma arises when we experience a conflict between cooperating and being selfish. People act selfishly if they think others are as well, which threatens group stability. Therefore we are vigilant in detecting cheaters in the group and will stigmatize and reject them.

How Do People Achieve Personal Goals?

7. **Good goals enhance confidence:** Challenging, difficult, and specific goals are most productive. A sense of self-efficacy leads to success. Being motivated to achieve is a powerful influence on behaviour.
8. **Self-regulation requires self-awareness and delay of gratification:** People use cognitive strategies to evaluate progress towards a goal. They act in accordance with their own goals when self-aware; deindividuation reduces that action. Avoiding self-awareness can lead to self-destructive behaviour. Delay of gratification and self-regulatory strength are necessary to achieve goals.
9. **The frontal lobes are important for self-regulation:** The frontal lobes are involved in more complex tasks and therefore is involved in self-regulation. This is an adaptive function.

What Determines How We Eat?

10. **Time and taste play roles:** Eating schedules are conditioned. Sensory-specific satiety increases the likelihood of meeting nutritional needs during scarcity because it encourages consumption of a variety of foods but it can also account for overeating.

11. **Culture determines what we eat:** Specific food choices and how they are eaten are determined and vary by culture.

12. **Multiple neural processes control eating:** The hypothalamus influences eating behaviour. Taste cues are processed in the prefrontal cortex. Receptors in the body also detect glucose and leptin levels as signals to eat.

What Is Addiction?

13. **Addiction has psychological and physical aspects:** Physical addiction involves the body's responses and tolerance. Psychological dependence involves habit and compulsion despite consequences.

14. **People use—and abuse—many psychoactive drugs:** Stimulants increase behavioural and mental activity. MDMA (ecstacy) produces energizing and hallucinogenic effects. THC (the active ingredient in marihuana) alters perception. Opiates provide high reward value by increasing dopamine activation.

15. **Alcohol is the most widely abused drug:** Though believed to reduce anxiety, alcohol consumption can increase anxiety and negative mood. A drinker's expectation can have significant effects on behaviour.

What Factors Motivate Sexual Behaviour?

16. **Biological factors influence sexual behaviour:** The sexual response cycle includes excitement, plateau, orgasm, and resolution. Hormones determine physical development and they motivate sexual behaviour. Neurotransmitters play a role in motivating sexual behaviour. Women's sexual behaviour varies across the menstrual cycle. Pheromones might play a role in sexual activity. Viewing erotica activates the amygdala.

17. **Cultural scripts and rules shape sexual interactions:** The cognitive sexual scripts inform sexual activity and are influenced by culture. A double standard of sexual behaviours exists for men verses women. Loss of virginity is culturally significant. Erotic plasticity is shaped by culture.

18. **Mating strategies differ between the sexes:** Female reproductive strategies revolve around intensive care of a few infants. Male strategies are less restrictive. Cross-cultural evidence support the different behaviours of males and females regarding reproductive imperatives.

Applying the Principles

1. Madison's third-grade teacher consistently tells the class that they are the best and smartest class ever. Madison's self-perception is that she is smart and she works hard on her school tasks. Mason's third-grade teacher gives students who do good work a sticker on their paper, but after a couple of months she notices the overall quality of Mason's work has leveled off to average. What difference in motivation may account for the difference in these two children's performance?

    ____ a) Madison is intrinsically motivated; Mason is externally motivated
    ____ b) Madison is responding to internally regulated drives; Mason to conditioning
    ____ c) Madison needs approval; Mason does not need approval
    ____ d) Madison needs no motivation; Mason needs increasing motivation

2. You have just received the results of the chemistry test you took last week. When class is over, you ask your classmates how they did. Why is this information useful in judging your performance?

    ____ a) It really isn't useful since true academic accomplishments are internally driven.
    ____ b) We are only able to determine satisfaction if it is shared by others.
    ____ c) Social comparison offers us accurate information about ourselves.
    ____ d) We are motivated to figure out who is truthful about their performance relative to our own.

3. Trang has just returned from the grocery store and is tempted to eat the chocolate that she bought to make brownies later in the day. Which of the following strategies might best enable her to resist?

    ____ a) putting a note on the chocolate that says "step away from the chocolate!"
    ____ b) thinking of herself as a good person
    ____ c) thinking of the chocolate as an unappetizing lump of brown clay
    ____ d) repeatedly saying out loud "I won't eat chocolate."

4. Ann says that she is addicted to chocolate. Her statement is most likely

_____ a) true because she has a strong psychological need for chocolate.
_____ b) false because chocolate is not a substance that leads to strong physical addiction.
_____ c) true because chocolate has been found to create permanent changes in levels of neurotransmitters.
_____ d) false because although she has a strong psychological desire for chocolate, she gives no indication of physical dependence.

5. The saying "women give sex to get love and men give love to get sex" exemplifies the idea

_____ a) that we are attracted to what is different.
_____ b) that men have greater sexual motivation.
_____ c) that women and men have different reproductive scripts.
_____ d) that men are only capable of love if it leads to sexual arousal.

ANSWERS: *1.a 2.c 3.c 4.d 5.c*

Key Terms

achievement motive, p. 358
alcoholism, p. 371
arousal, p. 347
creativity, p. 350
deindividuation, p. 359
delay of gratification, p. 360
drive, p. 347
extrinsic motivation, p. 350
goal, p. 357
homeostasis, p. 347
incentives, p. 348
intrinsic motivation, p. 350
motivation, p. 345
need, p. 346
need hierarchy, p. 346
need to belong theory, p. 354
physical dependence, p. 366
psychological dependence, p. 366
self-actualization, p. 346
self-efficacy, p. 357
self-regulation, p. 357
sexual response cycle, p. 373
sexual strategies theory, p. 377
social dilemma, p. 355
TOTE model, p. 358
Yerkes-Dodson law, p. 348

Further Readings

Berscheid, E., & Regan, P. (2005). *The psychology of interpersonal relationships.* New York: Prentice-Hall.

Capaldi, E. D. (1996). *Why we eat what we eat.* Washington, DC: American Psychological Association.

Franken, R. E. (1998). *Human motivation.* Pacific Grove, CA: Brooks-Cole.

Ilgen, D., & Pulakos, E. D. (1999). *The changing nature of performance.* San Francisco: Jossey-Bass.

Further Readings: A Canadian Presence in Psychology

Bain, J. (1987). Hormones and sexual aggression in the male. *Integrative Psychiatry, 5,* 82–89.

Enzle, M. E. & Ross, J. M. (1978). Increasing and decreasing intrinsic interest with contingent rewards: A test of cognitive evaluation theory. *Journal of Experimental Social Psychology, 14,* 588–597.

Fibiger, H. C., & Phillips, A. G. (1988). Mesocorticolimbic dopamine systems and reward. *Annals of the New York Academy of Sciences, 537,* 206–215.

Fried, P. A., Watkinson, B., & Gray, R. (2005). Neurocognitive consequences of marihuana—a comparison with pre-drug performance. *Neurotoxicology & Teratology, 27,* 231–239.

Galef, B. G., Jr., & Whiskin, E. E. (2000). Social influences on the amount eaten by Norway rats. *Appetite,* 34, 327–332.

Koob, G. F. (1999). Drug reward and addiction. In M. J. Zigmond, F. E. Bloom, S. C. Landis, J. L. Roberts, & L. R. Squire (Eds.), *Fundamentals of neuroscience* (pp. 1261–1279). San Diego, CA: Academic Press.

McGregor, A., & Roberts, D. C. S. (1993). Dopaminergic antagonism with the nucleus accumbens or the amygdala produces differential effects on intravenous cocaine self-administration under fixed and progressive ratio schedules of reinforcement. *Brain Research, 624,* 245–252.

Olds, J., & Milner, P. (1954). Positive reinforcement produced by electrical stimulation of septal areas and other regions of rat brains. *Journal of Comparative and Physiological Psychology, 47,* 419–427.

Pfaus, J. G., Kippin, T. E., & Coria-Avila, G. (2003). What can animal models tell us about human sexual response? *Annual Review of Sex Research, 14,* 1–63.

Pfaus, J. G., & Phillips, A. G. (1991). Role of dopamine in anticipatory and consummatory aspects of sexual behavior in the male rat. *Behavioral Neuroscience, 105,* 727–743

Pinel, J. P. J., Assanand, S. & Lehman D. R. (2000). Hunger, eating, and ill health. *American Psychologist, 55,* 1105–1116.

Vallerand, R. J. (1997). Toward a hierarchical model of intrinsic and extrinsic motivation. In M. P. Zanna (Ed.), *Advances in experimental social psychology* (vol. 29, pp. 271–360). San Diego, CA: Academic Press.

Coping with Stress

Stress is a basic component of our daily lives, and a small amount of stress helps us to determine the boundaries of our emotional well-being. Extreme duress over an extended period of time, however, can lead to serious health problems. Combat soldiers often must use a combination of coping strategies to deal with the stress of their situation.

CHAPTER

10

Emotions and Health

Elliot was in his early thirties when he began to suffer from severe headaches. He was happily married, a good father, and doing well professionally. His headaches increased until he could no longer concentrate, so he went to see his doctor. Sadly, it turned out that Elliot had a tumour the size of a small orange growing behind his eyes. As the tumour grew, it forced his frontal lobes upward into the top of his skull. A group of skilled surgeons removed the benign tumour, but in doing so they could not help but remove some of the surrounding frontal lobe tissue. Although the surgery appeared at first to be a great success—Elliot's physical recovery was quick and he continued to be an intelligent man with a superb memory—Elliot changed in a way that baffled his friends and family. He no longer experienced emotion.

Outlining the Principles

How Are Emotions Adaptive?

- Facial Expressions Communicate Emotion
- Emotions Serve Cognitive Functions
- Emotions Strengthen Interpersonal Relations

How Do People Experience Emotions?

- Emotions Have a Subjective Component
- Emotions Have a Physiological Component
- Emotions Have a Cognitive Component
- People Regulate Their Moods

What Is the Neurophysiological Basis of Emotion?

- Emotions Are Associated with Autonomic Activity
- The Amygdala and the Prefrontal Cortex Are Involved in Emotion
- Emotion Systems Are Lateralized in the Brain

How Do People Cope with Stress?

- There Are Sex Differences in Response to Stressors
- The General Adaptation Syndrome Is a Bodily Response to Stress
- People Encounter Stress in Daily Life
- Stress Affects Health
- Coping Is a Process

What Are Some of the Behaviours That Affect Physical Health?

- Obesity Results from a Genetic Predisposition and Overeating
- Restrictive Dieting Can Be Problematic
- Smoking Often Begins in Childhood
- People Do Not Get Enough Exercise

On the surface, Elliot seemed a reasonable, intelligent, and charming man, and yet his life fell apart. He lost his job; entered a doomed business venture with a sleazy character against the advice of his family and lost all his savings; divorced, remarried, and then quickly divorced again; and so on. He was incapable of making even trivial decisions, and he failed to learn from his mistakes. Though he had previously been a caring and compassionate man, he was now detached from his problems, reacting to them as if they had happened to someone else about whom he didn't care that much.

The neurologist Antonio Damasio was asked to examine Elliot to ascertain whether his emotional problems were real and whether they might have been due to the surgery. Damasio noted that Elliot displayed few emotional responses: "I never saw a tinge of emotion in my many hours of conversation with him: no sadness, no impatience, no frustration with my incessant and repetitious questioning" (Damasio, 1994, p. 45). Damasio's research team showed Elliot a series of disturbing pictures, such as severely injured bodies, and found that none of these elicited an emotional reaction from him. Elliot was not oblivious to his loss of emotion. He was able to report that he knew the pictures were disturbing, and that before the surgery he would have had an emotional response, but now he had none.

Imagine what your life would be like if you did not have any feelings. Like the android character from *Star Trek: The Next Generation,* Commander Data, you would be a rational being capable of sensation, perception, thought, memory, decision making, and so on, but you would not have any emotions. What sort of life would that be? For us human beings, every action and thought is coloured by emotional reactions. As you read in the last chapter, emotions are a primary source of motivation, as we seek objects and activities that make us feel good and avoid doing or saying things that make us feel bad. Emotions permeate human life as we fall in love, achieve success, and enjoy friendships, but they also underlie painful episodes that many of us would just as soon forget.

Sometimes our emotions can overwhelm us, as when we become stressed by demands on our time. When people are stressed, they often act in ways that are bad for their health, such as eating too much or smoking cigarettes. People who feel time pressured to get things done, such as completing school or work projects, fail to get proper rest and exercise (Shepard, 1997). We will learn that stress also directly affects physical health through its impact on various bodily systems, making people vulnerable to infection and disease. Even apparently minor daily hassles and frustrations can cause serious physical and mental health problems, especially if the frustration builds up over time. Learning to

1872

Emotions Are Adaptive Charles Darwin publishes *Expression of Emotion in Man and Animals*, in which he argues that for many species, emotions are adaptive and are hard wired through the processes of natural selection.

1884

Reactions Cause Emotions William James and Carl Lange, working separately, assert that physical reactions can cause emotions.

1927

Emotions Cause Bodily Reactions Walter Cannon, in contrast to James, argues that people experience emotional states that subsequently lead to physiological responses.

1930s

General Adaptation Syndrome The endocrinologist Hans Selye proposes that stressors lead to bodily reactions and that prolonged stress could lead to deficiencies in the immune system. Selye's work launches the field of health psychology.

cope with negative emotional states is important for our health and well-being. People can use a number of constructive strategies to reduce the stress in their lives, from seeking the support of others to getting sufficient sleep and taking care of their bodies through exercise and good nutrition.

This chapter explores how emotions influence the human experience, including where they come from and how they are experienced, as well as how people cope with stress and handle negative events in their lives. We also consider the most important behaviours that influence physical health and describe ways to promote healthy living. We will see that daily habits—such as diet, exercise, and smoking—play an important role in the most common causes of death in our modern society, and that emotional states often encourage people to develop bad habits. Understanding how emotions influence cognition and behaviour helps us to gain insight into what makes us mentally and physically healthy. ■

Emotions are a fundamental part of the human experience (Jenkins et al., 1998). They warn of danger, create bonds between people, and bring joy to life. However, they can also cause problems. People who feel overly anxious may be too afraid to meet new people or even to leave the house. People indulge in unhealthy behaviours trying to cope with stress and other negative feelings. For thousands of years, people have reflected on why we have emotions and what they do for us. Many scholars have viewed cognition and emotion to be separate, with emotion occasionally overwhelming reason and causing people to act in impulsive or inappropriate ways. However, an important lesson from Chapter 8 is that everyday cognition is far from cold and rational. Decisions and judgments are affected by how people feel. We now know that emotion and cognition are completely intertwined. There has been a growing emphasis in psychological science in understanding how emotion influences our daily lives. Important research is being conducted across all levels of analysis, and the biological revolution is producing exciting new findings about how neural and cognitive processes are involved in the experience of emotion. Case studies of people such as Elliot provide ample evidence about the role of various brain regions in producing and regulating emotional responses, as well as how people use emotional information.

Almost everyone has an intuitive sense of what is meant by the term *emotion*, but it has proven a difficult concept to define precisely. For psychological scientists, **emotion** (or *affect*) refers to feelings that involve subjective evaluation, physiological processes, and cognitive beliefs. Emotions are immediate responses to environmental events, such as being cut off in traffic or getting a nice gift. It is useful to distinguish emotion from mood, since the two are often used equivalently in everyday language. **Moods** are diffuse and long-lasting emotional states that influence rather

emotion Feelings that involve subjective evaluation, physiological processes, and cognitive beliefs.

mood A diffuse and long-lasting emotional state that influences rather than interrupts thought and behaviour.

1937
Neural Basis of Emotions James Papez hypothesizes that emotional expression is mediated by several neural systems that form a circuit (known as the "Papez loop").

1962
Two-Factor Theory Stanley Schachter and Jerome Singer propose that emotions are the result of physical arousal and the attribution, or cognitive explanation, of the source of the arousal.

1963
Facial Feedback Hypothesis On the basis of Darwin's writings, Silvan Tomkins proposes that facial expressions trigger the experience of emotions.

1970s
Universal Emotional Expressions Paul Ekman and his colleagues find that people from diverse cultures recognize similar facial expressions. Carroll Izard demonstrates that infants also display basic emotions in their facial expressions.

1977
Health Psychology An interdisciplinary meeting of scientists at Yale University produces a vibrant new field of research that is focused on how behavioural and psychological factors affect physical health.

than interrupt thought and behaviour. Many times people who are in good or bad moods have no idea why they feel the way they do. According to some views, mood reflects people's perceptions of whether they have the personal resources necessary to meet environmental demands (Morris, 1992). As people begin to feel overwhelmed by the demands placed on them, their moods become negative and they experience stress. **Stress** is defined as a pattern of behavioural and physiological responses to events that match or exceed an organism's abilities (Hayley et al., 2005).

stress A pattern of behavioural and physiological responses to events that match or exceed an organism's abilities.

health psychology The field of psychological science concerned with the events that affect physical well-being.

This chapter considers first the functions served by emotion and then how emotion is experienced and regulated. We also examine the implications of emotional processes for daily living. How people deal, or *cope,* with stress has implications for physical and mental health, and these coping strategies are discussed later in the chapter. In the final section we also consider how many behaviours can affect physical health. **Health psychology** is the field of psychological science concerned with the events that affect physical well-being. It is concerned with how people remain healthy, when they become ill, and what they can do to regain their health. Most of the large health problems facing contemporary society are the direct result of our lifestyles, as we indulge in things that are bad for us and fail to do what is good for us. In modern society, the major causes of death are heart attacks, strokes, cancer, and so on (Frasure-Smith & Lesperance, 2003). Poor nutrition, smoking, and a lack of exercise contribute to virtually all of them. These daily behaviours are influenced by our emotional states, as when people tend to overeat or smoke cigarettes when they are under stress. Let's now examine what psychological scientists have discovered about human emotional experiences and how they affect our physical and mental health.

How Are Emotions Adaptive?

Our theme that the mind helps to solve adaptive problems is well illustrated by research on emotions. Negative and positive experiences guide behaviour that increases the probability of surviving and reproducing. Emotions are adaptive because they prepare and guide behaviours, such as running away when you encounter dangerous animals. Emotions provide information about the importance of stimuli to personal goals and then prepare people for actions aimed at helping achieve those goals (Frijda, 1994).

What functions do emotions serve?

Because humans are social animals, we should not be surprised that many emotions involve interpersonal dynamics. People feel hurt when teased, angry when insulted, happy when loved, proud when complimented, and so on. Moreover, people interpret facial expressions of emotion to predict the behaviour of other people. Facial expressions provide many clues about whether our behaviour is pleasing to others or whether it is likely to cause them to reject, attack, or cheat us. Thus, both emotions and emotional expressions provide adaptive information.

1980s

Coping Styles Susan Folkman and Richard Lazarus propose that the way people cope with stress determines its impact. They posit a distinction between coping strategies that focus on emotion and those that focus on problem solving.

1990

The Genetics of Obesity Claude Bouchard and colleagues conduct the first experiment demonstrating that genes play an important role in body weight. This finding confirmed research conducted in the 1980s that genes contribute to obesity.

1990s

Interpersonal Functions of Emotions Researchers begin to focus attention on emotions that serve interpersonal functions, such as shame, guilt, and embarrassment, with the idea that emotions evolved to solve adaptive problems.

1990s

The Emotional Brain Researchers such as Joseph LeDoux develop models of how the brain processes emotions. LeDoux's work demonstrates that the amygdala is especially important for the experience and perception of emotion.

2000s

Affective Neuroscience Brain-imaging techniques allow researchers to study the emotional brain in action. Emotions rely on the activity of interrelated brain structures, especially those in the limbic system and prefrontal cortex.

Facial Expressions Communicate Emotion

In 1872, Charles Darwin wrote *Expression of Emotion in Man and Animals,* in which he argued that expressive characteristics were adaptive in all forms of life, from the dog's hard stare and exposed teeth when defending territory to the red-facedness of humans when preparing to fight. Being able to tell when other people or other species are threatening is of obvious survival value.

Emotional expressions are powerful nonverbal communications. Although over 550 words in the English language refer to emotions (Averill, 1980), humans can communicate emotions quite well without verbal language. Consider human infants. Because infants cannot talk, they must communicate their needs largely through nonverbal action and emotional expressions (Figure 10.1). At birth, an infant is capable of expressing joy, interest, disgust, and pain. By two months of age, infants can express anger and sadness. By six months, they can express fear (Izard & Malatesta, 1987). The social importance of emotional expressions can be seen even in ten-month-old infants, who have been found to smile more while their mothers are watching (Jones, Collins, & Hong, 1991). Thus, in the absence of verbal expression, nonverbal displays of emotions signal inner states, moods, and needs. Interestingly, the lower half of the face may be more important than the upper half of the face in communicating emotion. In a classic study conducted in 1927, Dunlap demonstrated that the mouth better conveys emotion than the eyes, especially for positive affect (Figure 10.2). However, the eyes are also extremely important for the communication of emotion. If people are presented with pictures of just eyes or just mouths and asked to identify the emotion expressed, they are more accurate using the eyes (Baron-Cohen, Wheelwright, & Jolliffe, 1997). But if the whole face is presented at once, the mouth appears to be most important in determining how people perceive the emotional expression (Kontsevich & Tyler, 2004).

The display of emotions alters behaviour in observers. For instance, people avoid those who look angry and approach those who look happy or in need of comfort. Even among chimpanzees, a smile from a subdominant to a dominant chimp can ward off a potential attack. Hence, emotions provide information to others as to how people are feeling and, in addition, can prompt them to respond in accordance with others' wants and needs.

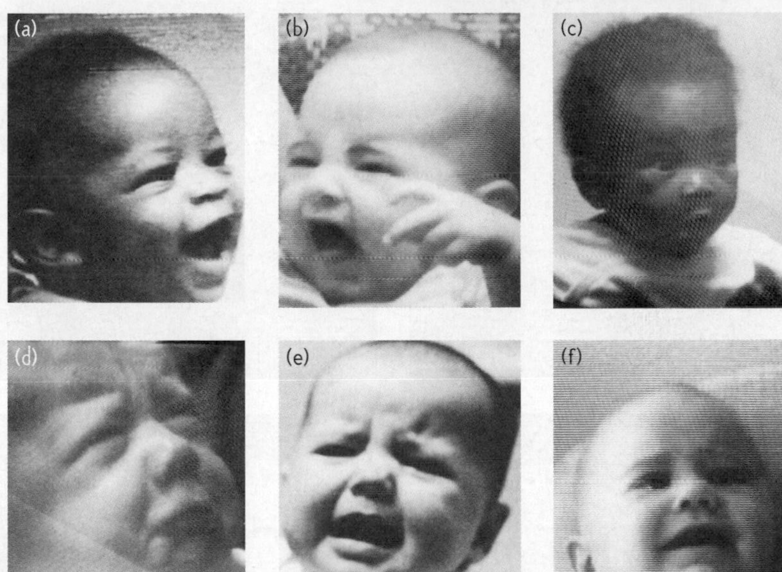

10.1 Early Display of Emotions Infants display emotions that are distinguishable and similar to facial displays among adults such as (a) joy, (b) disgust, (c) surprise, (d) sadness, (e) anger, and (f) fear.

10.2 How Faces Convey Emotion Based on Dunlap's classic study of the effects of facial expression, these photos show that the mouth and not the eyes better conveys emotion. The first two photos show the original face expressing pleasure (a) and sadness (b). In the next two photos (c, d) the lower halves of the faces have been interchanged.

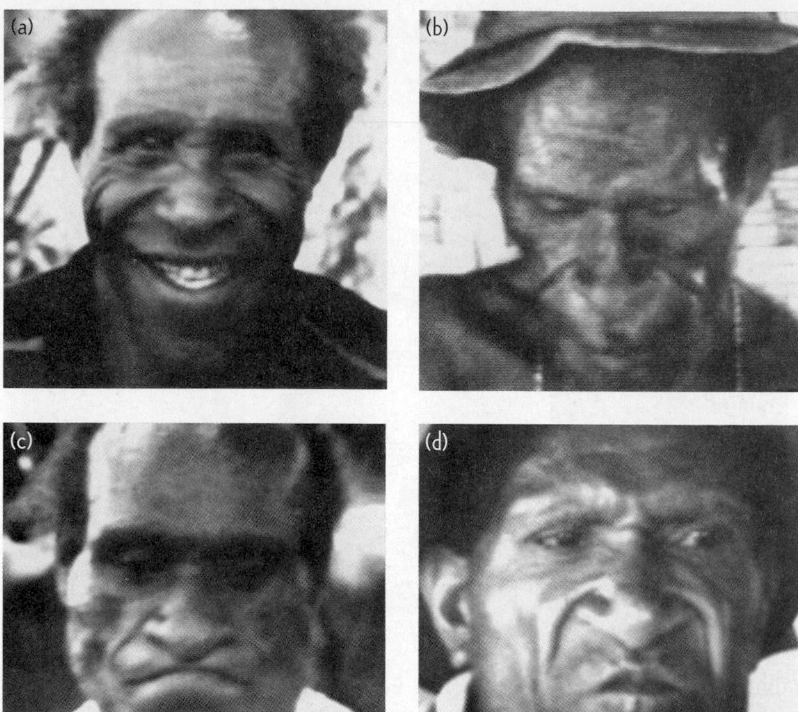

10.3 Facial Expressions across Cultures People across cultures largely agree on the meaning of different facial expressions. These data indicate that recognition of facial expressions may be universal and therefore biologically based. Here the man is expressing (a) happiness, (b) sadness, (c) anger, and (d) disgust.

display rules Cultural rules that govern how and when emotions are exhibited.

FACIAL EXPRESSIONS ACROSS CULTURES Darwin argued that the face innately communicates emotion to others and that these communications are understandable by all people, regardless of culture. To test this hypothesis, Paul Ekman and colleagues (1969) went to Argentina, Brazil, Chile, Japan, and the United States and asked people to identify the emotional responses displayed in photographs of posed emotional expressions. They found that people from all of these countries recognized the expressions as anger, fear, disgust, happiness, sadness, and surprise. However, it could be argued that all of the people tested in these countries have extensive exposure to each other's cultures and that learning, not biology, could be responsible for the cross-cultural agreement. So the researchers traveled to a remote area in New Guinea that had little exposure to outside cultures and where the people received only minimal formal education. The New Guinea natives were able to identify the emotions seen in the photos fairly well, although agreement was not quite so high as in other cultures. The researchers also asked participants in New Guinea to display certain facial expressions and found that these were identified by evaluators from other countries at better than chance level (Figure 10.3; Ekman & Friesen, 1971). Subsequent research finds general support for cross-cultural congruence in identification of some facial expressions, most strongly for happiness and least strongly for fear and disgust (Elfenbein & Ambady, 2002). Some scholars believe that the cross-cultural consistency results may be biased by cultural differences in the use of emotion words and by the way people are asked to identify emotions (Russell, 1994). Overall, however, the evidence is sufficiently consistent to indicate that some facial expressions are universal, and therefore likely to be biologically based.

DISPLAY RULES AND GENDER Although basic emotions seem to be expressed similarly across cultures, the situations in which they are displayed differ substantially. **Display rules** govern how and when emotions are exhibited. These rules are learned via socialization and dictate which emotions are suitable to a given situation. Differences in display rules help to explain cultural stereotypes, such as of Americans as loud and obnoxious, of the British as cold and bland, and of the French as refined and snobbish. This also may explain why the identification of facial expressions is much better within cultures than between cultures (Elfenbein & Ambady, 2002).

There are gender differences in display rules that guide emotional expression, particularly for smiling and crying. It is generally believed that women more readily, frequently, easily, and intensely display emotions (Plant, Hyde, Keltner, & Devine, 2000), and the current evidence suggests that they do, except for perhaps emotions related to dominance (LaFrance & Banaji, 1992). There are evolutionary reasons to think that men and women may vary in their emotional expressiveness: The emotions most closely associated with women are those related to caregiving, nurturance, and interpersonal relationships, whereas emotions associated with men are related to competitiveness, dominance, and defensiveness. But just because women are more likely to display emotion does not mean that they actually experience emotions more intensely. Although the evidence generally indicates that women report more intense emotions, this might reflect societal norms about how women are supposed to feel (Grossman & Wood, 1993). Moreover, in modern Western society, women tend to be better than men at articulating their emotions (Feldman Barrett, Lane, Sechrest, & Schwartz, 2000)—perhaps due to their upbringing—which might account for their more intense descriptions.

Emotions Serve Cognitive Functions

Psychological scientists have for many years studied cognitive processes without giving consideration to emotional processes. Studies on decision making, memory, and so on were conducted as if people were evaluating the information from a purely rational perspective. Yet our immediate affective responses arise quickly and automatically, colouring our perceptions at the very instance we notice an object. Robert Zajonc points out, "We do not just see 'a house': We see a *handsome* house, an *ugly* house, or a *pretentious* house" (1980, p. 154). These instantaneous evaluations subsequently guide decision making, memory, and behaviour.

Moreover, people's moods can alter ongoing mental processes. When people are in good moods they tend to use heuristic thinking (see Chapter 8), which allows them to make decisions more quickly and efficiently. Positive moods also facilitate creative, elaborate responses to challenging problems and motivate persistence (Isen, 1993). During the pursuit of goals, positive feelings signal that satisfactory progress is being made, thereby encouraging additional effort. One recent theory proposes that increased dopamine levels mediate the effects of positive affect on cognitive tasks (Ashby, Isen, & Turken, 1999). According to this view, positive affect leads to higher levels of dopamine production, which subsequently leads to heightened activation of dopamine receptors in other brain areas, which appears to be crucial for the advantageous cognitive effects of positive affect.

DECISION MAKING Would you rather go rock climbing in the Alps or attend a performance of a small dance troupe in Paris? Anticipated emotional states are an important source of information that guide decision making (see Chapter 8). In the face of complex, multifaceted situations, emotions are heuristic guides, providing feedback for making quick decisions (Slovic, Finucane, Peters, & MacGregor, 2002). Moreover, emotion appears to have a direct effect that does not depend on cognitive processes. For instance, people might decide to cancel air travel shortly after hearing about a plane crash, even if the news did not change their outward belief about the likelihood of their own plane's crashing. Events that are recent or particularly vivid have an especially strong influence on behaviour. Thus, risk judgments are strongly influenced by current feelings, and when cognitions and emotions are in conflict, emotions typically have more impact on decisions (Loewenstein, Weber, Hsee, & Welch, 2001).

How do emotions affect our decisions?

The *affect-as-information* theory posits that people use their current emotional state to make judgments and appraisals, even if they do not know the source of their moods (Schwarz & Clore, 1983). For instance, the researchers asked people to rate their overall life satisfaction, a question that involves consideration of a lifetime of situations, expectations, personal goals, and accomplishments, as well as a multitude of other factors. Schwarz and Clore note that people do not labour through all these elements to arrive at an answer but instead seem to rely on current mood state. People who are in a good mood rate their lives as satisfactory, whereas people in bad moods give lower overall ratings. People's evaluations of plays, lectures, politicians, and even strangers are influenced by their moods, which themselves are influenced by day of the week, weather, health, and the like. Interestingly, if people are made aware of the source of their mood (as when the researcher calls attention to the fact that their good mood might be caused by the bright sunshine), their feelings no longer influence judgment.

SOMATIC MARKERS The neuroscientist Antonio Damasio has suggested that reasoning and decision making are guided by the emotional evaluation of an action's consequences. In his influential book *Descartes' Error* (1994), Damasio sets forth the *somatic marker theory,* which posits that most self-regulatory actions and decisions are affected by the bodily reactions, called **somatic markers,** that arise from contemplating their outcomes. For Damasio, the term, *gut feeling* almost can be taken literally. When you contemplate an action you experience an emotional reaction, based in part on your expectation of the action's outcome, which itself is determined by your past history of performing that action or similar actions. To the extent that driving fast has led to speeding tickets, you will be motivated to slow down. Damasio has found that people with damage to the frontal lobes tend not to use past outcomes to regulate future behaviour. For instance, in studies using a gambling task, patients continued to follow a risky strategy that had proven faulty in previous trials. The absence of somatic markers might have been one of the problems that Elliot, the patient described at the beginning of the chapter, had with making decisions.

somatic markers Bodily reactions that arise from the emotional evaluation of an action's consequences.

In terms of adaptiveness, emotional reactions help us select responses that are likely to promote survival and reproduction. Thus, the anticipation that an event, action, or object will produce a pleasurable emotional state motivates us to approach it, whereas anticipation of negative emotions motivates us to avoid other situations. Hence, somatic markers may guide organisms to engage in adaptive behaviours.

EMOTIONS CAPTURE ATTENTION If emotions truly are adaptive, then people should be especially sensitive to emotional information. Indeed, research has demonstrated that emotional information captures attention. For example, research using the *emotional Stroop task* shows that cognitive processes are biased toward emotional stimuli (Williams, Mathews, & MacLeod, 1996). In these studies, participants are asked to name the colour of the ink that a word is printed in, which is difficult because it requires overriding a habitual desire to speak the word itself. Typically, words that are emotionally arousing (such as "anger") are more difficult to override than are neutral words (such as "pencil"), suggesting that there is an attentional bias for encoding affective stimuli.

Emotion also appears to lessen a common phenomenon known as attentional blink. In research on this phenomenon, people are given the task of remembering target words from a number of words that are printed on cards and presented rapidly. For instance, they would be quickly shown 15 words, two of which were green. Participants would have to remember the two green words. If the target words are presented closely together, people often have difficulty remembering the second one. This *attentional blink* occurs because attention was focused on the first word, and there is a temporary impairment in processing subsequent words. However, if the second word contains emotional information, as with the word *rape,* then people are much better at remembering it. The emotional content of the word captures attention and reduces the attentional blink (Anderson & Phelps, 2001).

EMOTIONS AID MEMORY People have improved memory for emotion producing events or stimuli. Think back to your childhood. What memories come to mind most rapidly? Research has found that important, clear personal memories are typically those that are highly emotional. The link between emotionality and memory was tested directly in an experiment using the *remember/know* procedure, in which participants are asked about their recognition of an item from a previous trial. Participants state whether they have a feeling that the item is familiar, which is a *know* judgment, or whether their recollection of the item is accompanied by sensory, semantic, or emotional detail, which is a *remember* judgment. This study found that highly negative photographs were more likely to be identified as "remember" items than were neutral or positive photos (Ochsner, 2000).

Considerable research demonstrates that increased emotional arousal enhances memory across a variety of tasks for many species. For instance, creating stress or administering drugs that produce arousal leads to enhanced memory formation. Recall from Chapter 7 that the neurotransmitter norepinephrine enhances memory for emotional events. Drugs that block norepinephrine receptors—such as certain beta blockers that are normally used to control blood pressure—impair memory for emotional words but have little effect on memory for neutral words (Strange & Dolan, 2004). Interestingly, these beta blockers also impair the ability to recognize emotional expressions (Harmer, Perrett, Cowen, & Goodwin, 2001), perhaps because blocking norepinephrine interferes with emotional processing. This finding is important because it suggests that administering drugs that block norepinephrine might help prevent the development of posttraumatic stress disorder, which as you will recall from Chapter 7 involves reoccurring unwanted memories of traumatic events. Indeed, in one study those who were given beta blockers within six hours of a traumatic event had fewer symptoms of PTSD one month later than those who were given placebo (Pitman et al., 2002).

Emotions Strengthen Interpersonal Relations

For most of the past century, psychologists paid little attention to interpersonal emotions. Guilt, shame, and the like were associated with Freudian thinking and therefore not studied in mainstream psychological science. However, recent theories have reconsidered interpersonal emotions in terms of the evolutionary need of humans to belong to social groups. Given that survival was enhanced for those who lived in groups, those who were expelled would have been less likely to survive and pass along their genes. According to this view, people were rejected primarily because they drained group resources or threatened group stability. Thus, as you will recall from Chapter 9, those who tried to cheat others, steal mates, or freeload were rejected. Accordingly, people feel anxious when engaging in behaviours that could lead them to be expelled from groups. Hence, anxiety serves as an alarm function that motivates people to behave according to group norms (Baumeister & Tice, 1990). This new approach views interpersonal emotions as evolved mechanisms that facilitate interpersonal interaction, such as helping to appease and repair interpersonal transgressions.

GUILT STRENGTHENS SOCIAL BONDS **Guilt** is a negative emotional state associated with anxiety, tension, and agitation. The experience of guilt, including its initiation, maintenance, and avoidance, rarely makes sense outside of the context of interpersonal interaction. For instance, the prototypical guilt experience occurs when someone feels responsibility for another person's negative affective state. Thus, when people believe that something they did either directly or indirectly caused another person harm, they experience feelings of anxiety, tension, and remorse, which

guilt A negative emotional state associated with an internal experience of anxiety, tension, and agitation, in which a person feels responsible for causing an adverse state.

can be labeled guilt. Guilt can occasionally arise even when individuals do not feel personally responsible for others' negative situations (for example, survivor guilt).

A recent theoretical model of guilt outlines its benefits to close relationships. Roy Baumeister and colleagues (1994) contend that guilt protects and strengthens interpersonal relationships through three mechanisms. First, feelings of guilt keep people from doing things that would harm their relationships, such as cheating on their partners, while encouraging behaviours that strengthen relationships, such as phoning their mothers on Sundays. Second, displays of guilt demonstrate that people care about their relationship partners, thereby affirming social bonds. Third, guilt is an influence tactic that can be used to manipulate the behaviour of others. For instance, you might try to make your boss feel guilty so that you don't have to work overtime.

SOCIALIZATION IS CRUCIAL FOR INTERPERSONAL EMOTIONS Evidence indicates that socialization is more important than biology for the specific manner in which children experience guilt. One longitudinal study examined the impact of socialization on the development of a variety of negative emotions, including guilt, in monozygotic and dizygotic twins at 14, 20, and 24 months (Zahn-Waxler & Robinson, 1995). The study found that all the negative emotions showed considerable genetic influence (as evidenced by higher concordance rates for identical twins), but guilt was unique in being highly influenced by the social environment. With age, the influence on guilt of a shared environment became stronger, whereas the evidence for genetic influences disappeared. These findings support the hypothesis that socialization is the predominant influence on moral emotions such as guilt. Perhaps surprisingly, parental warmth is associated with greater guilt in children, suggesting that feelings of guilt arise in healthy and happy relationships. Thus, as children become citizens in a social world they develop the capacity to empathize, and they subsequently experience feelings of guilt when they transgress against others.

EMBARRASSMENT AND BLUSHING Embarrassment is a naturally occurring state that usually occurs following social events such as violations of cultural norms, loss of physical poise, teasing, and self-image threats (Miller, 1996). Some theories of embarrassment suggest that it rectifies interpersonal awkwardness and restores social bonds after a transgression. Embarrassment represents submission to and affiliation with the social group and a recognition of the unintentional social error. Research supports these propositions in showing that individuals who look embarrassed after a transgression elicit more sympathy, forgiveness, amusement, and laughter from onlookers (Cupach & Metts, 1990). Hence, like guilt, embarrassment may serve to reaffirm close relationships after a transgression.

Why do some people become easily embarrassed?

Mark Twain once said, "Man is the only animal that blushes. Or needs to." Darwin, in his 1872 book, called blushing the "most peculiar and the most human of all expressions," thereby separating it from emotional responses he deemed necessary for survival. Recent theory and research suggests that blushing occurs when people believe that others view them negatively, and that blushing communicates a realization of interpersonal errors. This nonverbal apology is an appeasement that elicits forgiveness in others, which repairs and maintains relationships (Keltner & Anderson, 2000).

JEALOUSY In his book *The Dangerous Passion: Why Jealousy Is as Necessary as Love and Sex,* David Buss (2000) argues that jealousy serves adaptive functions. Although no one enjoys feeling jealous and threatened, Buss contends that jealousy is an indispensable component of long-term relationships because it keeps mates together by sparking passion and commitment. Buss theorizes that when faced with the possibility of a sexual rival, a person feels and displays jealousy as a sign of commitment to the relationship. This hypothesis is supported by research showing that people who are more invested in the relationship often purposely provoke jealousy as a test of their partners' commitment. Buss also proposes that jealousy revives sexual passion in the threatened partner. Of course, jealousy has negative sides. Jealousy is one of the most common reasons for spousal/partner abuse and homicide, and when unfounded it can ruin a relationship by exposing a lack of trust.

REVIEWING the Principles | How Are Emotions Adaptive?

Emotions are adaptive because they bring about a state of behavioural readiness. The evolutionary basis for emotions is supported by research on the cross-cultural recognition of emotional displays. Facial expressions communicate meaning to others and enhance emotional states. Emotions aid in memory processes by garnering increased attention and deeper encoding of emotionally relevant events. Positive and negative emotions serve as guides for action. Emotions also repair and maintain close interpersonal relationships.

How Do People Experience Emotions?

Emotions are difficult to define because they defy language. Imagine trying to describe the concept of emotions to an alien from another planet. What would you say? You might say that emotions make you "feel," but the alien doesn't understand the word "feel." You could demonstrate the behaviours that accompany certain emotions, such as smiling, frowning, or crying, but the alien doesn't understand how water spouting from the eyes is associated with this thing we call "emotions." In the end, it would be akin to trying to describe colour to someone who was born blind.

What would life be like without emotions?

Psychologists generally agree that emotions consist of three components. One is the feeling state that accompanies emotions—the *subjective experience* that psychologists and laypeople alike refer to when they ask "How are you feeling?" Psychologists also consider *physical changes*—such as increased heart rate, skin temperature, or brain activation—to be an integral part of what makes an emotion. A third component, *cognitive appraisal,* involves people's beliefs and understandings about why they feel the way they feel.

Emotions Have a Subjective Component

Emotions are *phenomenological,* meaning that we experience them subjectively. You know when you're experiencing an emotion because you *feel* it. The intensity of emotional reactions varies; some people report many distinct emotions every day, whereas others report only infrequent and minor emotional reactions. People who are either over- or underemotional tend to have psychological problems (Anisman & Merali, 2003). Among the former are people with *mood disorders* such as depression or panic attacks. People with mood disorders experience such strong emotions that they can become immobilized.

At the other extreme are those who suffer from **alexithymia,** a disorder in which people do not experience the subjective component of emotions. Elliot, considered

alexithymia A disorder involving a lack of the subjective experience of emotion.

at the beginning of this chapter, suffered from alexithymia. The explanation for the disorder is that the physiological messages associated with emotions do not reach the brain centres that interpret emotion. Damage to certain brain regions, especially the prefrontal cortex, is associated with a loss of the subjective component of mood.

primary emotions Evolutionarily adaptive emotions that humans share across cultures; they are associated with specific biological and physical states.

secondary emotions Blends of primary emotions, including states such as remorse, guilt, submission, and anticipation.

circumplex model An approach to understanding emotion in which two basic factors of emotion are spatially arranged in a circle, formed around the intersections of the core dimensions of affect.

DISTINGUISHING AMONG TYPES OF EMOTIONS How many emotions does a person experience, and how do they relate to each other? Many emotion theorists distinguish between primary and secondary emotions, an approach conceptually similar to viewing colour as consisting of primary and secondary hues. Basic or **primary emotions** are evolutionarily adaptive, shared across cultures, and associated with specific biological and physical states. They include anger, fear, sadness, disgust, and happiness, as well as possibly surprise and contempt. **Secondary emotions** are blends of primary emotions; they include remorse, guilt, submission, and anticipation.

One approach to understanding the experience of emotion is the **circumplex model,** in which emotions are arranged in a circle around the intersections of two core dimensions of affect (Russell, 1980). James Russell and Lisa Feldman Barrett (1999) developed one such model that posits that emotions can be mapped according to their *valence,* or degree of pleasantness or unpleasantness, and their *activation,* which is the level of arousal or mobilization of energy (Figure 10.4). Thus, "excited" is an affective state that includes pleasure and arousal, whereas "depressed" describes a state of low arousal and negative affect. There has been some debate about naming the dimensions, but circumplex models have proven useful for providing a basic taxonomy of mood states.

The psychological scientists David Watson, Lee Anna Clark, and Auke Tellegen make a distinction between *positive activation* (pleasant affect) and *negative activation* (unpleasant affect), which can be plotted on a circumplex. They also propose that negative and positive affect are independent, such that people can experience both simultaneously. For example, consider the bittersweet feeling of being both happy and sad, such as the way you might feel when reminiscing about a good friend or family mem-

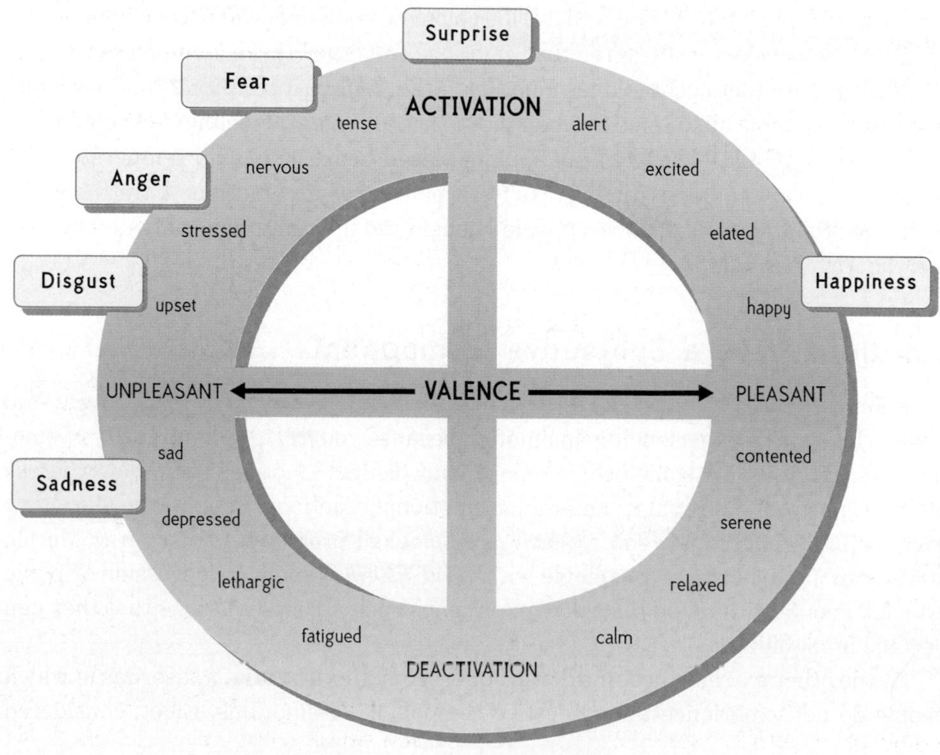

10.4 James Russell and Lisa Feldman Barrett's Model of Emotion The figure depicts a circumplex map of the structure of emotions.

ber who has just died. In one study, research participants who were surveyed just after seeing the movie *Life Is Beautiful,* moving out of their dormitories, or graduating from college felt happy and sad at the same time (Larsen, McGraw, & Cacioppo, 2001). Neurochemical evidence suggests that positive activation states are associated with an increase in dopamine and that negative activation states are associated with an increase in norepinephrine, which supports the idea that positive and negative affect are independent. Furthermore, Watson and colleagues argue that the distinction between positive and negative activation is adaptive. For instance, they link affect to motivational states of approach and avoidance—the motivation to seek out food, sex, and companionship is typically associated with pleasure, whereas the motivation to avoid dangerous animals is associated with pain (Watson, Wiese, Vaidya, & Tellegen, 1999).

Emotions Have a Physiological Component

A clumsy mistake causes most people to feel embarrassed and brings a rush of blood to the face that warms the cheeks. Indeed, emotions are associated with physical changes. But which causes which? Common sense suggests that the emotion leads to the physical change. But in 1884, William James argued that it was just the opposite. In a proposal similar to one made by Descartes in the 1600s, James asserted that it is how people interpret the physical changes in a situation that leads to the feeling of an emotion. In James's words, "we feel sorry because we cry, angry because we strike, afraid because we tremble, and not that we cry, strike, or tremble because we are sorry, angry, or fearful." James believed that physical changes occur in distinct patterns that translate directly into a specific emotion. Around the same time, a similar theory was independently proposed by a Danish psychologist named Carl Lange. Thus, the idea that felt emotion is the result of perceiving specific patterns of bodily responses is called the **James-Lange theory of emotion.**

James-Lange theory of emotion A theory that suggests that the experience of emotion is elicited by a physiological response to a particular stimulus or situation.

facial feedback hypothesis The idea that facial expressions trigger the experience of emotion.

One implication of the James-Lange theory is that if you mold the facial muscles to mimic an emotional state, you activate the associated emotion. According to the **facial feedback hypothesis,** first proposed by Silvan Tomkins in 1963, facial expressions trigger the experience of emotions, not the other way around. James Laird tested this idea in 1974 by having people hold a pencil between their teeth or with their mouths in a way that produced a smile or a frown (see Figure 10.5). When participants rated cartoons, those in a posed smile found the cartoons funniest. Further support comes from the results of studies by Paul Ekman and colleagues (1983), who asked professional actors to portray feelings of anger, distress, fear, disgust, joy, and surprise. Physiological changes recorded during the actors' portrayals were in fact different for various emotions. Heart rate changed little with surprise, joy, and disgust, but it increased with distress, fear, and anger. Anger was also associated with higher skin temperature, whereas the other emotions resulted in little change in skin temperature. Thus, these results give some support to James's theory that specific patterns of physical changes are the basis for emotional states. However, subsequent evidence suggests that emotional reactions are not specific enough to fully explain the subjective experience of emotions. We experience more emotions than there are distinct bodily reactions.

The counterintuitive James-Lange theory quickly attracted criticism. In 1927, Harvard University's Walter Cannon (Figure 10.6) noted that although the human mind is quick to experience emotions, the body is much slower,

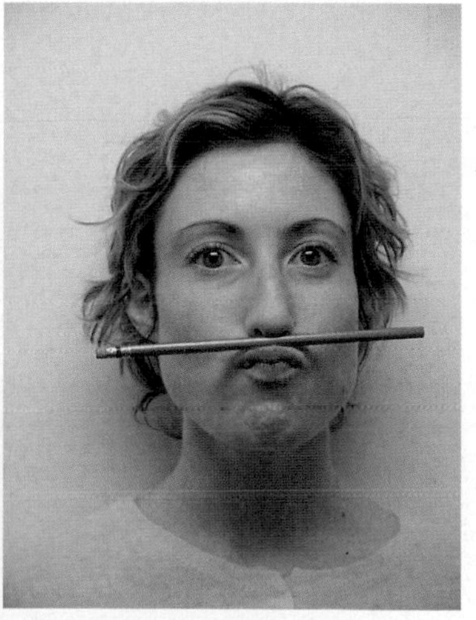

10.5 Facial Feedback Hypothesis Alteration of facial expression can lead to changes in the subjective experience of emotions. The person on the left (pen in mouth) is more likely to report feeling happy than the person on the right (pencil on lip).

10.6 **Walter Cannon** A physiologist at Harvard University, Cannon formulated the ideas of regulatory homeostasis, the role of the body in emotion, and the flight-or-fight autonomic response.

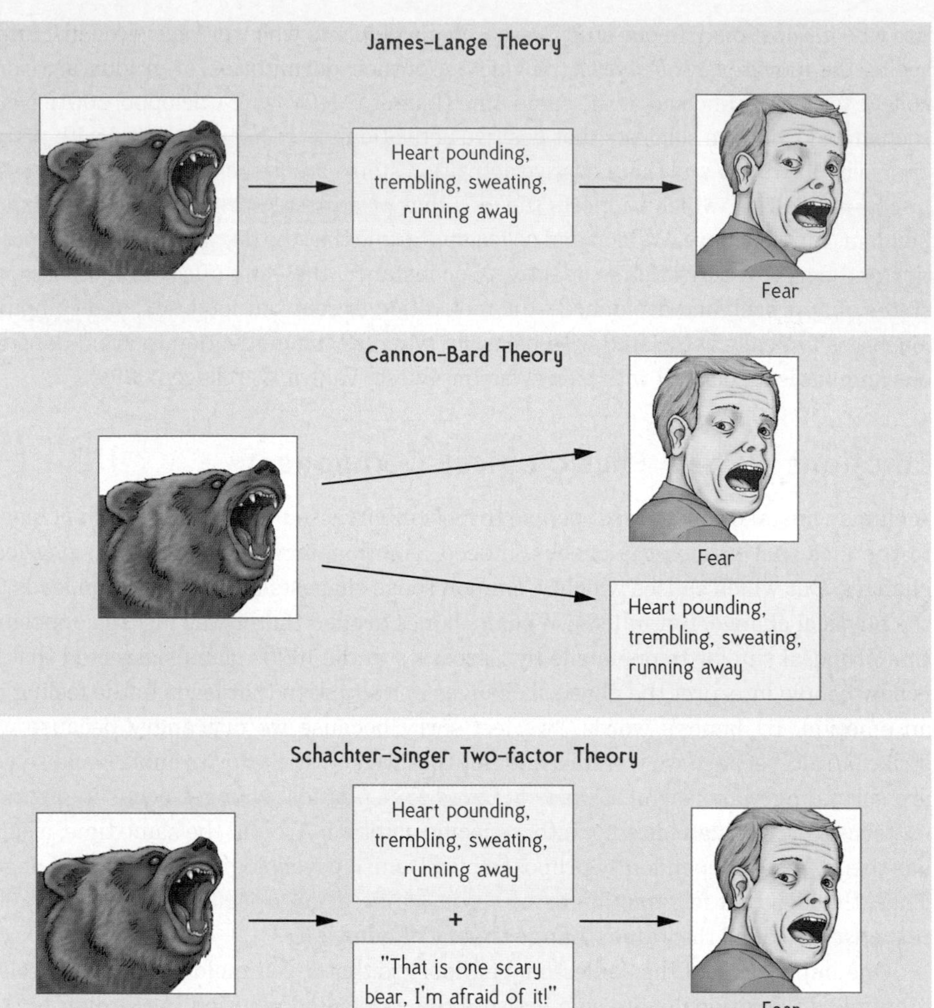

10.7 **Three Major Theories of Emotion** The theories differ not only in their relative emphasis on physiology and cognition, but also in terms of when emotional state is determined.

taking at least a second or two to respond. Cannon also noted that many emotions produced similar visceral responses, making it too difficult for people to determine quickly which emotion they were experiencing. For instance, anger, excitement, and sexual interest all produce similar changes in heart rate and blood pressure. Cannon, along with Philip Bard, proposed instead that the mind and body operate independently in experiencing emotions. According to the **Cannon-Bard theory of emotion,** the information from an emotion-producing stimulus is processed in subcortical structures, causing the experience of two separate things at roughly the same time: an emotion and a physical reaction. When you see a grizzly bear, you simultaneously feel afraid, begin to sweat, experience a pounding heart, and run (Figure 10.7). Everything happens together. As you will see later, recent evidence from brain research provides support for the idea that there are separate systems for the processing of emotional information. Therefore, scientists tend to support the Cannon-Bard theory rather than the James-Lange theory.

Cannon-Bard theory of emotion A theory that asserts that emotion-producing stimuli from the environment elicit both an emotional and a physical reaction.

two-factor theory of emotion A theory that proposes that a situation evokes both a physiological response, such as arousal, and a cognitive interpretation.

Emotions Have a Cognitive Component

Stanley Schachter (Figure 10.8) developed the hypothesis that emotions are the interaction of physiological arousal and cognitive appraisals. His **two-factor theory of emotion** proposed that a situation evokes both a physiological response, such as arousal, and a cognitive interpretation, or *emotion label.* When people experience arousal they initiate a search for its source. Although the search for a cognitive explanation is often quick and straightforward, since we generally recognize what event led to our emotional state, sometimes we are incorrect in our conclusions.

Schachter and his student Jerome Singer (1962) devised an ingenious experiment to test the two-factor theory. First, participants were injected with either a stimulant or a placebo. The stimulant was adrenaline, which produced symptoms such as sweaty palms, increased heart rate, and the shakes. Then either participants were told that the drug they took would make them feel aroused, or they were not given any information. Finally, each participant was left to wait with a confederate of the experimenter. In the euphoric condition, participants were exposed to a confederate who played with a hula hoop and made paper airplanes. In the angry condition, they were exposed to a confederate who asked them very intimate, personal questions, such as "With how many men (other than your father) has your mother had extramarital relationships?" [Making the question even more insulting were the three choices: (a) four or fewer, (b) five to nine, or (c) ten or more.] Note that those participants who received the adrenaline but were told what to expect would have an easy explanation for their arousal. However, those participants who received adrenaline but were not given information would be aroused but would not know why. Thus, they would look to the environment to explain or label their mood. Participants in the uninformed condition reported being happy with the euphoric confederate but less happy with the angry confederate (Figure 10.9).

10.8 Stanley Schachter Schachter emphasized the role of cognitive beliefs in the experience of emotion.

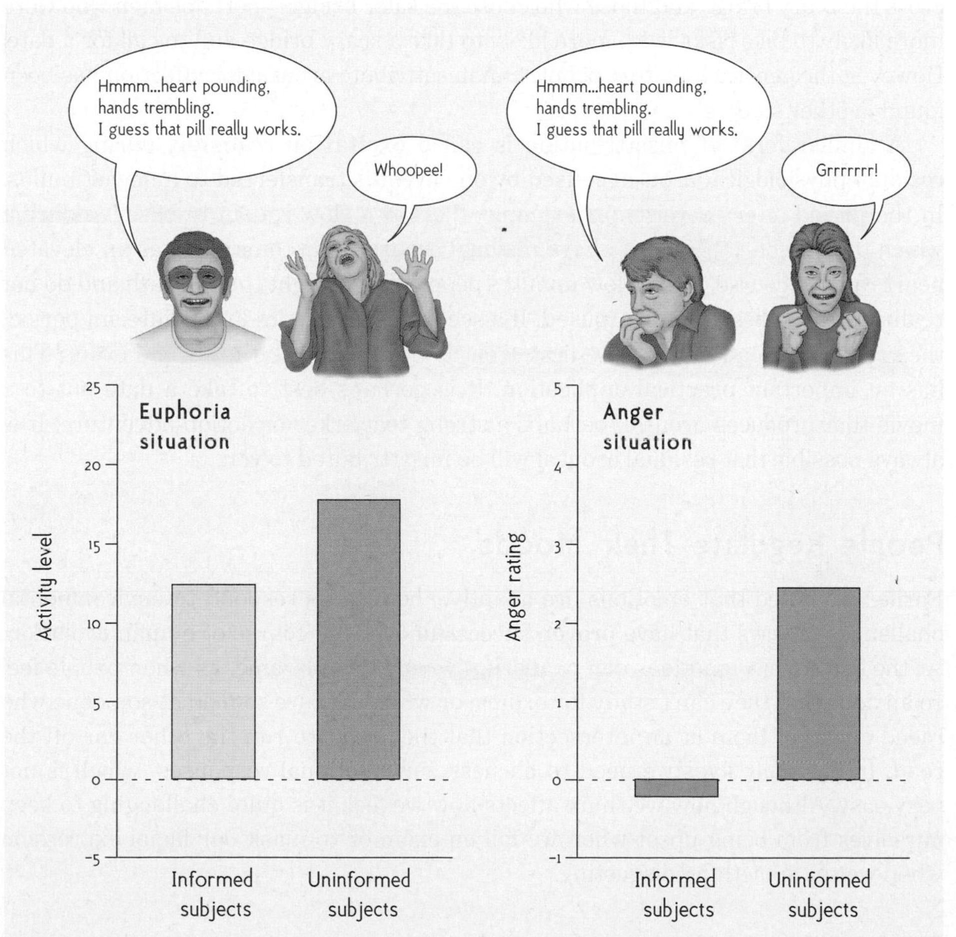

10.9 Testing the Two-Factor Theory The results of Schachter and Singer's experiment to test their theory of emotion showed that participants' subjective experience of emotion was a combination of the situation they were in, the physiological arousal of the stimulant pill, and whether they knew the purported effects of the pill.

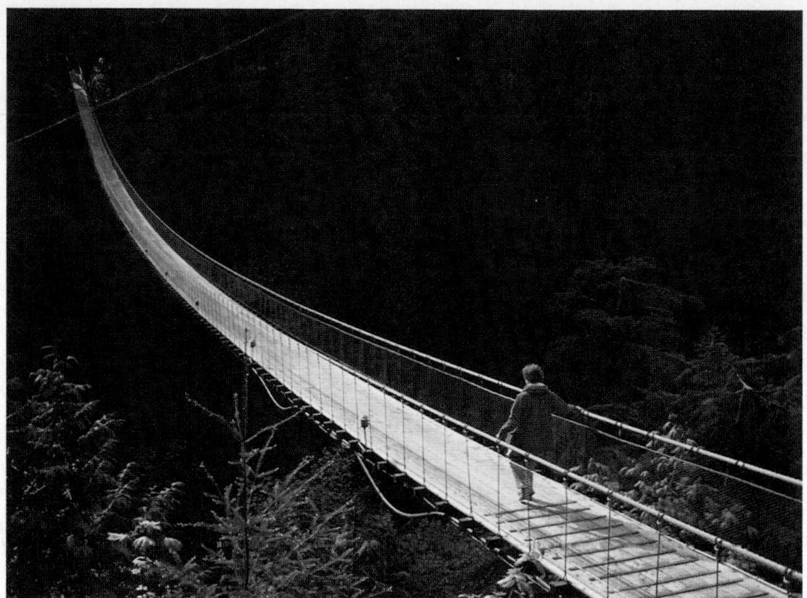

10.10 Excitation Transfer Men who walked across a narrow and scary bridge (like the Capilano Bridge) displayed more attraction to a female experimenter than those who walked across a stable bridge.

PEOPLE CAN MISATTRIBUTE THE SOURCE OF EMOTIONAL STATES One interesting implication of the Schachter theory is that emotion can be mistakenly attributed to something that did not actually cause the arousal. *Misattribution of arousal* is a term used when an emotion label is derived from the wrong source. In one of the most amazing demonstrations of this phenomenon, researchers tried to see whether people could fall in love through misattribution (Dutton & Aron, 1974). Male participants were asked to meet an interviewer on one of two bridges over the Capilano River in British Columbia. One was a narrow suspension bridge with a low rail that swayed 230 feet above raging, rocky rapids (Figure 10.10); the other was a sturdy modern bridge just above the river. An attractive female research assistant approached the men and interviewed them in the middle of the bridge. She gave them her phone number and offered to explain the results of the study to them at a later date. According to the two-factor theory of emotion, the scary bridge would produce arousal that could possibly be misattributed to the interviewer. Indeed, this is what happened: Men who were interviewed on the scary bridge were more likely to call the interviewer and ask her for a date. It has to be noted that there could have been initial differences among the men who chose to cross the scary bridge and those who chose the safer bridge—perhaps men who were more likely to take risks were more likely to take a scary bridge and to call for a date. However, the general idea that people can misattribute arousal for affection has been found in other studies.

excitation transfer A form of misattribution in which residual physiological arousal caused by one event is transferred to a new stimulus.

A similar form of misattribution is called **excitation transfer,** during which residual physiological arousal caused by one event is transferred to a new stimulus. In the period after exercise, for example, there is a slow return to baseline during which the person continues to have residual arousal symptoms, such as an elevated heart rate. Of course, after a few minutes people have caught their breath and do not realize their bodies are still aroused. If a second event occurs in this interim period, the residual excitation from the first event is transferred to the second event. This has an important practical application. It is perhaps best to take a date out to a movie that produces arousal, perhaps a strong tearjerker or action adventure. It is always possible that residual arousal will be misattributed to you.

People Regulate Their Moods

Why is it hard to control emotions?

Earlier we noted that emotions are adaptive, helping us respond to environmental challenges in ways that have proven successful over the course of human evolution. At the same time, emotions can be disruptive and troublesome, as when people feel so anxious that they can't study for exams, or when they are so mad at someone who raced ahead of them at an intersection that they want to run the other car off the road. In our daily lives we need to harness our emotional responses, which is not very easy. Although how we think affects how we feel, it is quite challenging to keep ourselves from being upset when we fail an exam or to mask our facial expression when we see something disgusting.

Successfully regulating our emotional states depends on several strategies, some of which we can use to prevent or prepare for bad events and some of which we can use to deal with those events after they occur. James Gross (1999) outlined a number of strategies people use to regulate their emotions. For instance, we can try to put ourselves in certain situations and avoid others; if you want to feel romantic before proposing to your girlfriend, you are better off going to a quiet, intimate bistro than to a fast-food diner; if you don't want to feel jealous of your sister's athletic skill, you could choose not to attend her soccer, basketball, and softball games. People can also focus their attention on certain aspects of a situation that will help manage their emotional state. If you are scared of flying, you can distract yourself from your anxiety by helping the woman next to you entertain her restless toddler. People can also try to directly alter their emotional reactions to events by trying to change, or reappraise, the event in more neutral terms. If you get scared while watching a movie, you can remind yourself that it is all staged and that no one is actually being hurt. Recent studies have found that having people engage in reappraisal changes the activity of brain regions involved in the experience of emotion (Ochsner, Burge, Gross, & Gabrieli, 2002).

"Next question: I believe that life is a constant striving for balance, requiring frequent tradeoffs between morality and necessity, within a cyclic pattern of joy and sadness, forging a trail of bittersweet memories until one slips, inevitably, into the jaws of death. Agree or disagree?"

The way in which we think about an event, how we *cognitively frame it,* can contribute to the intensity of our emotional response as well as influence what label we place on it (Smith & Ellsworth, 1985). As an example, if the roof of your house is blown off during a storm and you attribute the situation to external forces, you will likely feel sad. However, if you see the outcome as having been under your personal control because you failed to replace what you knew to be a faulty roof, you will instead feel guilty and angry. How people appraise and frame events plays an important role in how they cope with the daily stress in their lives, which we consider later in this chapter.

HUMOUR Humour is a simple and effective method of regulating negative emotions, one that has numerous mental and physical health benefits. Most obviously, humour increases positive affect. When we find something humourous, we smile, laugh, and enter a state of pleasurable, relaxed excitation. Research shows that laughter stimulates endocrine secretion; an improved immune system; and the release of hormones, catecholamines, and endorphins. When people laugh, they experience a rise in circulation, blood pressure, skin temperature, and heart rate, along with a reduction in perception of pain. All these responses are similar to those resulting from physical exercise, and they are considered beneficial to both short-term and long-term health.

People sometimes laugh in situations that do not seem very funny, such as at funerals or wakes. According to one theory, laughing in these situations helps people distance themselves from their negative emotions and strengthens their connections to other people. Dacher Keltner and George Bonanno (1997) interviewed 40 people who had recently lost a spouse. They found that genuine laughter during the interview was associated with positive mental health and fewer negative feelings, such as grief.

SUPPRESSION AND RUMINATION People make two common mistakes when trying to regulate mood. The first is *thought suppression,* in which they attempt not to respond or feel the emotion at all. Research by Daniel Wegner and his colleagues has demonstrated that suppressing any thought is extremely difficult and often leads to a *rebound effect*, in which people think more about something after suppression than before. Thus, for example, those who try not to think of a white bear end up obsessed with thoughts of white bears (Wegner, Shortt, Blake, & Page, 1990). **Rumination,** the second mistake, involves thinking about, elaborating, and focusing on undesired thoughts or feelings, which prolongs the mood. Moreover, rumination impedes successful mood-regulation strategies, such as focused problem solving or distraction (Lyubomirsky & Nolen-Hoeksema, 1995).

rumination Thinking about, elaborating, and focusing on undesired thoughts or feelings, which prolongs, rather than alleviates, a negative mood.

Overall, distraction is the best way to avoid the problems of suppression or rumination, since it absorbs attention and temporarily helps people to stop thinking about their problems. But some distractions backfire, such as thinking about other problems or engaging in maladaptive behaviours. Watching a movie that captures your attention helps you to escape your problems, but watching a movie that reminds you of your troubled situation may lead you to wallow in mental anguish.

REVIEWING the Principles | How Do People Experience Emotions?

Emotions comprise a subjective experience, physiological changes, and cognitive interpretation. The three main theories of emotion differ in their relative emphasis on these components. The James-Lange theory states that specific patterns of physical changes give rise to the perception of associated emotions. The Cannon-Bard theory proposes two separate pathways, physical changes and emotion, occur at the same time. Schachter's two-factor theory emphasizes the combination of generalized physiological arousal and cognitive appraisals in determining specific emotions. People also use a number of strategies to alter their moods. The best methods for regulating negative affect include humour and distraction.

What Is the Neurophysiological Basis of Emotion?

There has been a recent increase in research that crosses levels of analysis to study emotion. Here we focus on the biological level. As was apparent in the three major theories discussed in the preceding section, the body plays an important role in emotional experience. People who have spinal-cord injuries report feeling less intense emotions than prior to injury, which occurs because the messages have difficulty reaching the brain. The closer the damage is to the brain, the greater the loss of sensation, and consequently, the greater the decrease in emotional intensity. As you will recall from Chapter 3, various hormones, neurotransmitters, and drugs affect mood states. For instance, drugs that increase the activity of serotonin receptors lead to a reduction in depressed mood; drugs that activate dopamine receptors lead to feelings of euphoria.

During the biological revolution at the end of the twentieth century, researchers found that specific emotional states are associated with unique patterns of brain activity, although many of the same brain structures are involved in multiple emotional experiences. For instance, imaging studies have shown that disgust, sadness, and happiness all activate the thalamus and prefrontal cortex. However, differential activation in surrounding structures can be distinguished among these three emotions. For example, happiness and sadness cause increased activity in the hypothalamus, whereas disgust does not (Lane, Reiman, Ahern, & Schwartz, 1997; Figure 10.11). Similarly, processing of faces showing fear and disgust may lead to differential patterns of brain activity. Fear activates the amygdala, whereas disgust activates another brain region that has previously been related to gustatory reactions to unpleasant tastes and smells (Phillips et al., 1997). We understand much about the neurophysiology of emotion, but we have a great deal still to learn. In this section we will examine several key issues regarding the link between physiology and emotion.

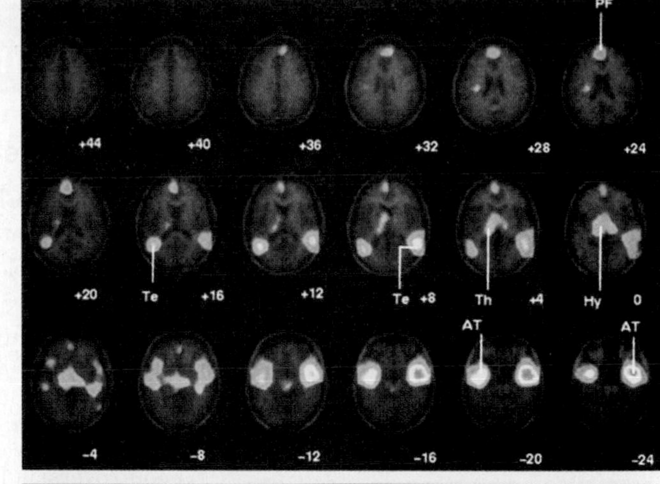

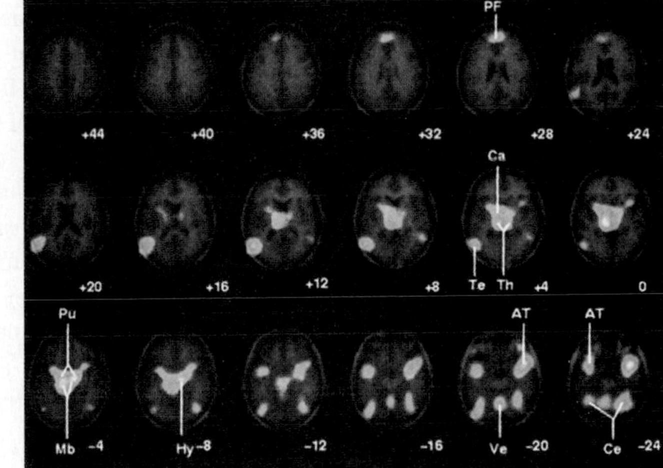

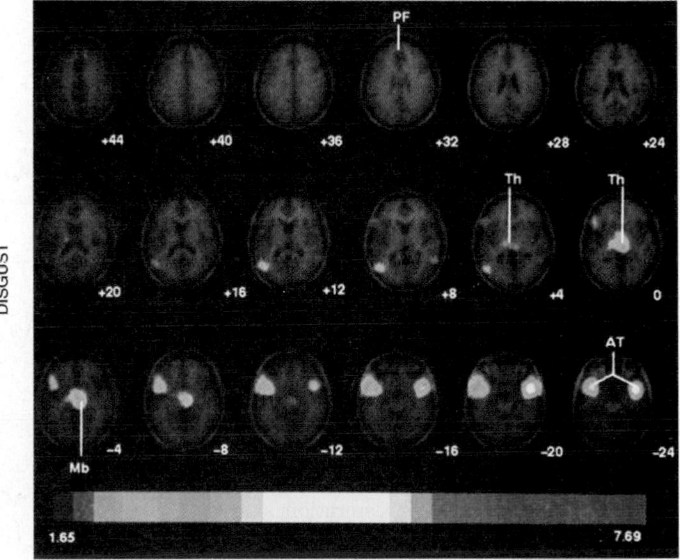

10.11 Brain Activation and Emotions Significant increases in regional brain activity occur during happiness, sadness, and disgust. Note that these emotions produce different patterns of brain activation.

Emotions Are Associated with Autonomic Activity

Emotions tend to overlap in their pattern of autonomic nervous system activity, although there are some differences between emotional states. For instance, when a person is aroused, whether because of anger or sexual attraction, the face becomes flushed, but the pupils contract during anger and dilate during sexual arousal. As mentioned, actors who portrayed particular emotions showed differential heart-rate and blood-pressure patterns. These findings have been replicated with the Minangkabau people of West Sumatra (Levenson, Ekman, Heider, & Friesen, 1992), in spite of dramatic differences in culture, religion, lifestyles, and display rules. The Minangkabau people showed patterns of autonomic arousal similar to those of stage actors: Heart rate increases with distress, fear, and anger, the last of which was also accompanied by higher skin temperature. However, the overlap in activity among emotions is so great that in most cases it is difficult to distinguish emotions based solely on autonomic responses (Cacioppo, Klein, Berntson, & Hatfield, 1993).

Robert Zajonc and colleagues described another link between physiological states and emotion. Zajonc hypothesized that facial expressions, through facial musculature, control the directional flow of air into the body, warming or cooling the hypothalamus. This in turn affects the release of neurotransmitters that influence emotions. According to Zajonc, cooling the brain produces positive emotions, whereas warming the brain produces negative emotions. To test his theory, participants allowed air to be blown into their nasal passages. In support of his predictions, Zajonc found that people reported more negative emotions when the air was warm and more pleasant emotions when the air was cool (Zajonc, Murphy, & Inglehart, 1989).

Thinking Critically

Can Lies Be Detected from Physiological Responses?

We noted in Chapter 9 that people are motivated to detect liars and cheaters. But how do you catch a liar? For several decades, researchers have hoped that the methods of psychological science can be used to read bodily responses that indicate deception. Increasingly, applicants for certain types of jobs, such as those that involve classified documents, are asked to take a polygraph test, known informally as a lie-detector test. A *polygraph* is an electronic instrument that assesses the body's physiological response to questions. It records numerous aspects of arousal such as breathing rate, heart rate, and so on (Figure 10.12). The use of polygraphs is highly controversial. In 1987, the Supreme Court of Canada decreed that polygraph results would no longer be admissible as evidence in Canadian courts, and few people understand what a lie detector can and cannot do. Let's examine some of the critical issues involved in using polygraphy to detect lies.

The goal of polygraphy is to determine a person's level of emotionality, as indicated by autonomic arousal, when confronted with certain information. For instance, criminals might be asked about specific illegal activities, whereas job applicants might be asked about drug use or previous job performance. Psychologists have long known that lying is stressful; therefore autonomic arousal should be higher when people are lying than when they are telling the truth. But can factors other than lying cause arousal? Of course, some people become nervous simply because the whole procedure is new and scary, or they are upset that someone thinks they are guilty when they are innocent. Even being attracted to the person giving the polygraph exam could produce arousal.

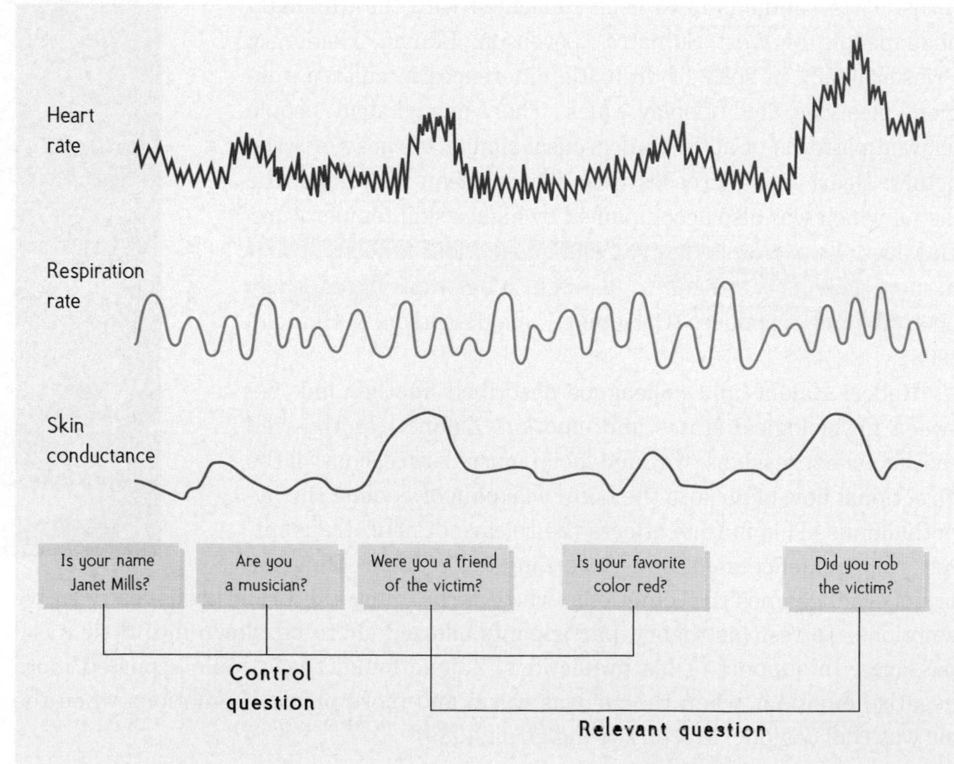

10.12 Lie Detection A polygraph measures autonomic systems such as respiration, skin conductance from sweating, and heart rate. Differences in autonomic reactions to critical questions, compared to control questions, indicate arousal, which in turn may indicate nervousness as a result of lying. However, the arousal might be due to general nervousness and may falsely indicate that the person is lying.

No absolute measure of autonomic arousal can indicate the presence or absence of a lie, because each person's level of autonomic arousal is different. So simply being highly aroused does not indicate guilt. Instead, to assess physiological arousal, polygraphers employ a control-question technique in which they ask a variety of questions, some of which are relevant to the critical information and some of which are not. Control questions, such as "What is your home address?" and "Have you ever travelled outside of Canada?" are selected with the assumption that they should not produce a strong emotional response. Critical questions, such as "Did you steal the money?" or "Do you use drugs?" are those of specific interest to the investigators. The difference between the physiological responses to the control and critical questions is the measure used to determine whether the person is lying. Sometimes the questions contain information that only a guilty party would know, such as how a person was killed or where the body was found. Thus, simply having guilty knowledge should produce arousal and therefore be detected by the polygraph.

How valid is the polygraph for detecting lies? How successful is it? Two researchers examined 50 criminal investigations in which a polygraph was administered to two suspects and later one of them confessed to the crime (Kleinmuntz & Szucko, 1984). When comparing the polygraph results with the suspects' true guilt or innocence, the researchers found that the polygraph correctly identified the guilty person 76 percent of the time but also wrongly accused someone of lying 37 percent of the time. Thus, polygraphy is far from infallible. Indeed, an expert panel convened by the United States National Academy of Sciences in 2002 found no scientific evidence to justify the regular use of polygraphs. The committee noted that although thousands of employees of the FBI, CIA, and other governmental agencies have been given polygraphs, not one person has been discovered to be a spy—including Americans such as Aldrich Ames and Robert Hansen, who passed polygraphs but subsequently were convicted of selling secrets to the Russians. Indeed, part of the reason that they were able to spy for so long was that they always passed the polygraph. It turns out that people can cheat the test in a number of ways. One method is to tense up on the control questions and try to relax on the critical questions, thus lessening the difference between one's physiological responses. Ways of increasing arousal include biting the tongue, pressing fingernails into the palms, or tightening the sphincter muscle for several seconds to increase blood pressure. Studies of college students who were taught these techniques found that they avoided detection by two expert polygraphers 50 percent of the time.

As you might expect, researchers are seeking new strategies to uncover deception. For instance, psychological scientists using event-related brain potentials have found that a brain-wave pattern typically associated with rare, meaningful stimuli, such as when a female name appears on a long list of male names, may indicate deception (Rosenfield, 2001). In a recent study using brain imaging, deceptive responses were associated with increased activation of specific frontal lobe regions (Langleben et al., 2002). However, the deception task in this study was relatively trivial, and the participants did not report feeling anxious about being deceptive. Whether the activation of the frontal lobe regions indicates genuine deception or simply reflects general inhibitory responses during the experimental task is unclear.

An important problem with polygraph tests is that most people who fail them are actually telling the truth. Polygraphy produces a high rate of false-positives (recall the discussion of signal detection in Chapter 5). Out of the many thousands of people who are required to take the test as part of their employment, many will be regarded as liars although they actually are not. As noted earlier, people who are especially anxious about the procedure or who are nervous in general may fail the test. Consider a person who doesn't use drugs but has a family member with a serious addiction. When asked about drug use, the person might start thinking of the relative and become upset, subsequently showing autonomic arousal. The basic problem is that the polygraph can't tell whether the response is due to lying, sexual fantasizing, or anything else arousing.

Given all this, why do you think that so many people remain convinced that the polygraph provides accurate information? Why do employers use it when it is unlikely to really tell them anything useful about their employees? What if a system could detect 99 percent of liars correctly, but falsely identified 5 percent of those telling the truth? Is that acceptable? Do you think the polygraph might be useful in scaring criminals into confessing to their crimes? Why or why not? ■

The Amygdala and the Prefrontal Cortex Are Involved in Emotion

In 1937, James Papez proposed that many different subcortical brain regions were involved in emotion. Fifteen years later, Paul MacLean expanded this list and called it the *limbic system*. (Recall from Chapter 4 that the word "limbic" refers to brain structures that border the cerebral cortex.) We now know that many brain structures outside the limbic system are involved in emotion and that many limbic structures are not central to emotion per se. For instance, the hippocampus is important mostly for memory and the hypothalamus mostly for motivation. Thus, the term *limbic system* is mainly used in a rough, descriptive way rather than in terms of directly linking brain areas to specific emotional functions (Figure 10.13). For understanding emotion, the two most important brain regions are the amygdala and the prefrontal cortex (Livingston & Hornykiewicz, 1978).

THE AMYGDALA The amygdala processes the emotional significance of stimuli and generates immediate emotional and behavioural reactions. According to Joseph LeDoux (Figure 10.14), affective processing in the amygdala is a circuit that has developed over the course of evolution to protect animals from danger. LeDoux (1996) has established the amygdala as the brain structure most important for emotional learning, such as the development of classically conditioned fear responses. Removal of the amygdala in animals produces a disorder known as *Kluver-Bucy syndrome*, after the scientists who first identified it in 1939. Animals with Kluver-Bucy syndrome engage in unusual behaviours, such as hypersexuality and putting objects into their mouths, and they are fearless.

Humans with damage to the amygdala do not develop the more severe symptoms associated with Kluver-Bucy syndrome, but they do experience a variety of deficits in processing and responding to emotional cues. More important, they show impairments in fear conditioning. They show fear when confronted with dangerous objects, but they do not develop conditioned fear of objects associated with dangerous objects (see Chapter 6). For instance, if you give people an electric shock each time they see a picture of a blue square, they will normally develop a conditioned response, evidenced by greater physiological arousal, when they see the blue square. But people with damage to the amygdala do not show classical conditioning of these fear associations. Consider patient S.P., who had damage to the amygdala (Anderson & Phelps, 2000). S.P. first showed signs of neurological impairment around age three and was later diagnosed with epilepsy. At age

10.13 The Limbic System The limbic system is important for assessing and responding to emotionally relevant stimuli. The two most important structures are the amygdala and prefrontal cortex.

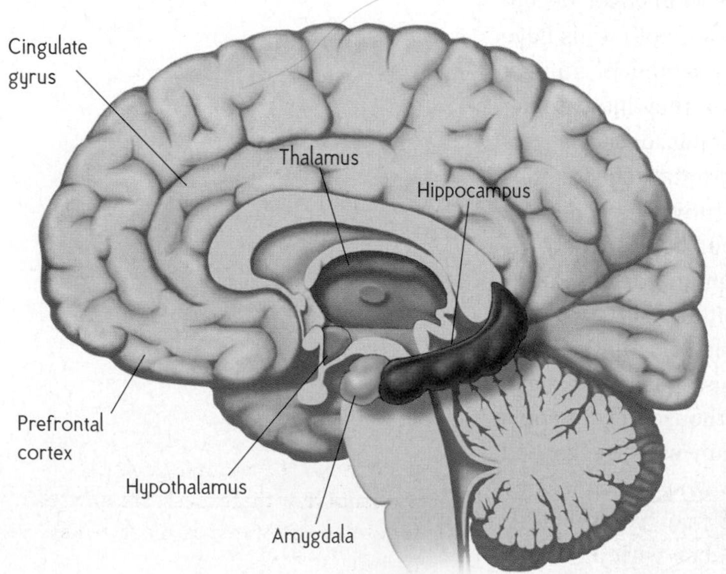

48, she had her right amygdala removed to reduce the frequency of the seizures. The surgery was reasonably successful, and S.P. retained most of her intellectual faculties. She has a normal IQ, has taken college courses, and performs well on standardized tasks of visual attention. However, she does not show fear conditioning. Strangely, S.P. can tell you that the blue square is associated with shock, but her body shows no physiological evidence of having acquired the fear response.

Information reaches the amygdala along two separate pathways. The first path is a quick and dirty system that processes sensory information nearly instantaneously. Sensory information travels quickly through the thalamus to the amygdala for priority processing. The second pathway is somewhat slower, but it leads to evaluations that are more deliberate and thorough. Sensory material travels from the thalamus to the sensory cortex, where the information is scrutinized in greater depth before it is passed along to the amygdala. Contemporary thinking is that the fast system prepares the animal to respond should the slower pathway confirm the threat.

10.14 Joseph LeDoux LeDoux has conducted important work showing that the amygdala is central to emotional learning.

Recall from the earlier discussion that emotional events are especially likely to be stored in memory. Indeed, brain-imaging studies demonstrate that increased activity in the amygdala during an emotional event is associated with improved long-term memory for that event (Cahill et al., 2001; Hamann, Ely, Grafton, & Kilts, 1999). Researchers currently believe that the amygdala modifies how the hippocampus consolidates memory (Phelps, 2004). This idea is supported by research that shows a correlation between activity in the amygdala and the hippocampus for emotional stimuli but not for neutral stimuli (Dolcos, LaBar, & Cabeza, 2004).

Another role of the amygdala in emotion processing is its involvement in perceiving social stimuli, such as deciphering the affective meaning of facial expressions. For instance, fMRI studies demonstrate that the amygdala is especially sensitive to the intensity of fearful faces (Dolan, 2000; Figure 10.15). This effect occurs even if the participants are unaware that they have seen a face at all (Whalen et al., 1998).

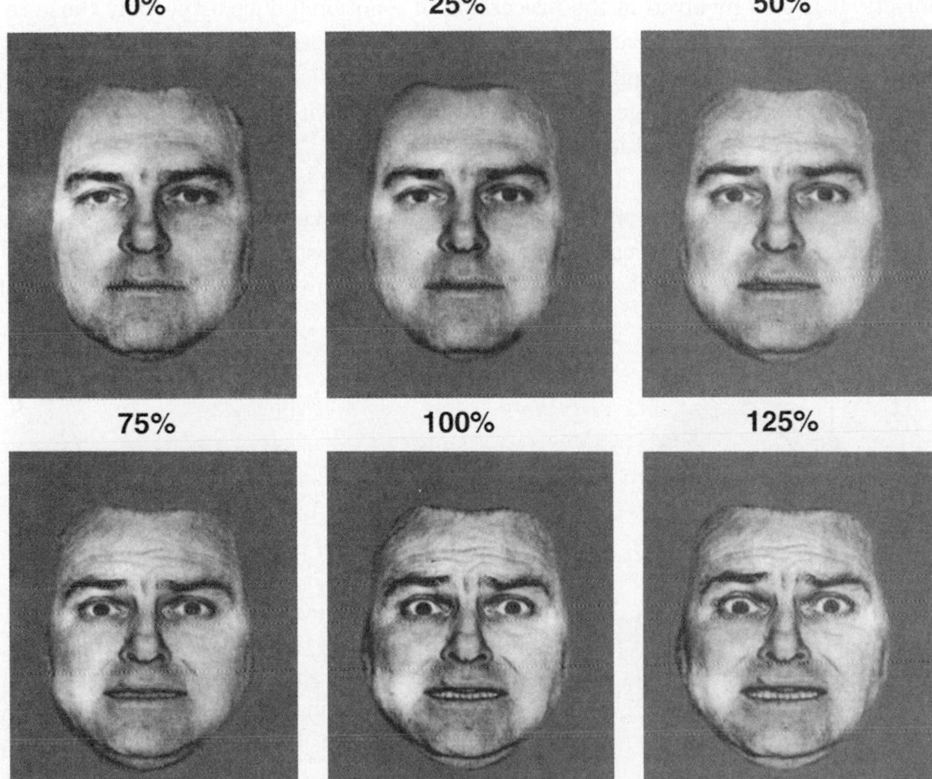

10.15 Activation of the Amygdala As participants view these faces, the amygdala becomes more activated as the face displays more fear.

Why does the amygdala respond most strongly to facial expressions of fear?

Perhaps surprisingly, the amygdala reacts more when people observe faces displaying fear than those displaying anger. On the surface, this doesn't make a great deal of sense, because a person looking at you with an angry expression may be more likely to be dangerous. According to some researchers, the greater activity of the amygdala when people look at a frightened face is due to the ambiguity of the situation (Whalen et al., 2001)—that is, not being sure about what the person fears. This is not to say that the amygdala reacts only to expressions of fear; rather, it responds to a variety of emotional expressions, even happiness, but generally the effect is greatest for fear. One recent study showed that the amygdala is activated even by neutral facial expressions, but only in people who are chronically anxious (Somerville, Kim, Johnstone, Alexander, & Whalen, 2004).

Given that the amygdala is involved in processing the emotional content of facial expressions, it is not surprising that damage to it leads to social impairments. Those with damage to the amygdala often have difficulty evaluating the intensity of fearful faces even though they do not show impairments in judging the intensity of other facial expressions, such as happiness. One interesting study suggests that those with damage to the amygdala fail to use information contained within facial expressions to make accurate interpersonal judgments (Adolphs, Tranel, & Damasio, 1998). For instance, those with damage to the amygdala have difficulty using photographs to assess people's trustworthiness, a task that most people can do without any trouble (Figure 10.16). People with amygdala damage can tell a smile from a frown, but they do not seem to use this information in making social judgments (Adolphs, Sears, & Piven, 2001). They also tend to be unusually friendly with people they don't know, which may mean they lack the normal mechanisms for caution around strangers and the feeling that some people should be avoided.

THE PREFRONTAL CORTEX Recall from the motivation chapter that the prefrontal cortex is involved in assessing the potential reward value of situations and objects. It is also involved in the processing of emotional cues, especially those related to interpersonal interactions. People with damage to this region often act inappropriately and are generally insensitive to the emotional expressions of others. Moreover, prefrontal damage is sometimes associated with excessive aggression and violence, suggesting difficulties with emotional control.

Antonio Damasio has found that patients, such as Elliot, with damage to the middle of the prefrontal region often fail to use somatic markers. When these regions are damaged, people still can recall information, but it has lost most of its affective meaning. They might be able to describe their current problems or talk about the death of

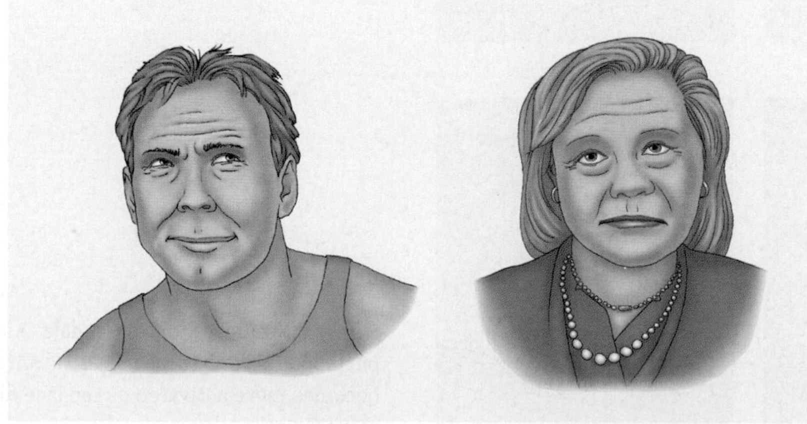

10.16 Amygdala Damage and Judgment People with damage to the amygdala can tell a smiling face from a frowning face, but they have difficulty judging whether the person is trustworthy. Which of these people would you trust?

a loved one, but they do so without experiencing any of the emotional pain that normally accompanies such thoughts.

Emotion Systems Are Lateralized in the Brain

The psychological scientist Richard Davidson (Figure 10.17) has shown that unequal activation of the left and right hemispheres is associated with specific emotional states, a pattern known as **cerebral asymmetry.** In a series of studies, Davidson and his colleagues found that greater activation of the right prefrontal cortex is associated with negative affect, whereas greater activation of the left hemisphere is associated with positive affect. A study of responses to film clips found that people who were left-hemisphere dominant showed the most positive response to pleasant scenes, whereas those who were right-hemisphere dominant showed the most negative response to unpleasant scenes. Another study found that those who reported the strongest negative emotion showed the greatest activation of the right amygdala in response to unpleasant pictures (Davidson, 2000a; Figure 10.18).

Cerebral asymmetry is associated with general motivation (Davidson, 2000b). For instance, greater activation of the left hemisphere is associated with increased confidence and effort to achieve goals. A greater activation of the right hemisphere is associated with lack of motivation, a symptom of clinical depression. Imaging studies have shown diminished left-hemisphere activation among depressed patients, and depression is more common among those who have brain injuries to the left hemisphere. Remarkably, as early as three months of age, children of depressed mothers also show asymmetrical brain activation (Field, Fox, Pickens, & Nawrocki, 1995). Although the functional significance of cerebral asymmetry is not known, recent work by Davidson and his colleagues suggests it may have to do with the regulation of emotional states. For instance, anti-anxiety drugs decrease fearful behaviour and increase left frontal activation in infant monkeys, suggesting that negative affective states such as anxiety may suppress the left frontal lobe and thus reduce motivation. The lateralization of amygdala activity and anxiety is also evident in rodents (Adamec & Shallow, 2000).

Research has demonstrated that the right hemisphere is more involved than the left in the interpretation and comprehension of emotional material. Neuroimaging studies have found that although emotional stimulation activates both hemispheres, activation

10.17 Richard Davidson Davidson has proposed that affective style is influenced by cerebral asymmetry, with the right hemisphere associated with negative moods and the left hemisphere associated with positive moods.

cerebral asymmetry An emotional pattern associated with unequal activation of the left and right frontal lobes.

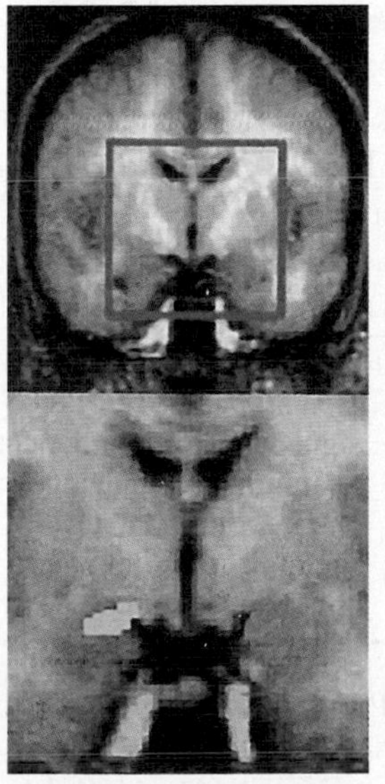

(a)

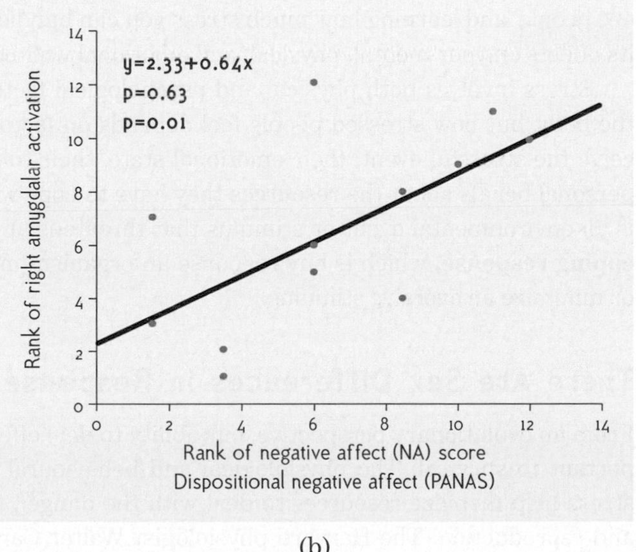

(b)

10.18 Cerebral Asymmetry (a) The amygdala region is magnified in the bottom image and shows the greater activation of the right amygdala (on the left side of the image) during viewing of unpleasant pictures. (b) This activation corresponds with greater negative affect.

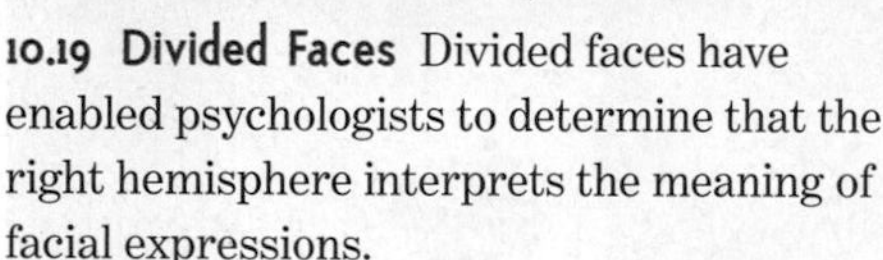

10.19 Divided Faces Divided faces have enabled psychologists to determine that the right hemisphere interprets the meaning of facial expressions.

is much greater on the right side. The right hemisphere is also more accurate at detecting the emotional tone of speech (such as whether the voice sounds sad or happy), whereas the left hemisphere is more accurate at decoding semantic content. As an example, consider that when we look at facial expressions, the left half of the face is projected to the right hemisphere and the right half of the face is projected to the left hemisphere. Researchers who create divided faces, such as those shown in Figure 10.19, have found that the emotion "seen" by the right hemisphere heavily influences the interpretation of an emotional expression. Which face looks sadder to you? Which face looks happier? As researchers continue to conduct research at the biological level of analysis, it is likely that we will learn a great deal more about how the brain produces emotion.

REVIEWING the Principles | What Is the Neurophysiological Basis of Emotion?

Emotions lead to specific autonomic reactions, but there is considerable overlap among emotions. A number of brain structures, especially the amygdala and the prefrontal cortex, are involved in emotion and social judgment. Research indicates that left-hemisphere activity is related to positive emotions and right-hemisphere activity to negative emotions.

How Do People Cope with Stress?

Stress is a common component of everyday emotional life, and it consists of a number of emotional responses. Although the term *stress* has negative connotations, a moderate level of stress is beneficial. Different levels of stress are optimal for different people, and learning how much stress you can handle is essential for recognizing its effects on your mental, physical, and emotional well-being.

Stress involves both physical and psychological factors. It has direct effects on the body, but how stressed people feel depends on factors such as how people perceive the stressful event, their emotional state, their tolerance for stress, and their personal beliefs about the resources they have to cope with the stressor. A **stressor** is an environmental event or stimulus that threatens an organism and that elicits a **coping response,** which is any response an organism makes to avoid, escape from, or minimize an aversive stimulus.

stressor An environmental event or stimulus that threatens an organism.

coping response Any response an organism makes to avoid, escape from, or minimize an aversive stimulus.

fight-or-flight response The physiological preparedness of animals to deal with danger.

tend-and-befriend response The argument that females are more likely to protect and care for their offspring and form social alliances than flee or fight in response to threat.

There Are Sex Differences in Response to Stressors

From an evolutionary perspective, our ability to deal effectively with stressors is important to survival. The physiological and behavioural responses that accompany stress help mobilize resources to deal with the danger, thereby facilitating survival and reproduction. The Harvard physiologist Walter Cannon coined the term **fight-or-flight response** to describe the physiological preparation of animals to deal with any attack. This physical reaction includes increased heart rate, contraction of the spleen, redistribution of the blood supply from skin and viscera to muscles and brain, deepening of respiration, dilation of the pupils, inhibition of gastric secretions, and an increase in glucose released from the liver. This response to a stressor occurs within seconds or minutes and allows an organism to direct all energy to dealing with the threat at hand while postponing less critical autonomic activities.

The generality of the fight-or-flight response has been questioned by Shelley Taylor (Figure 10.20) and her colleagues, who point out that the vast majority of both human and nonhuman animal research has been conducted using males, which has distorted the scientific understanding of responses to stress. Most rat studies use male rats to avoid female hormonal cycles such as estrus. Although the reasons are unclear, a similar sex inequality exists for past human studies, with women representing fewer than 1 in 5 participants in laboratory stress studies.

10.20 Shelley Taylor Taylor has conducted important research in health psychology. She and her colleagues have revolutionized thinking about how animals respond to threat.

Taylor and her colleagues argue that females respond to stress by protecting and caring for their offspring, as well as by forming alliances with social groups to reduce individual risk. They coined the phrase **tend-and-befriend response** to describe this pattern. Tend-and-befriend responses make great sense from an evolutionary point of view. Females typically bore a greater responsibility for the care of offspring, and responses that protected offspring as well as the self were maximally adaptive. When a threat appears, quieting the offspring and hiding may be more effective means of avoiding harm than trying to flee while pregnant or with a clinging infant. Furthermore, those females who selectively affiliate with others, especially other females, might acquire additional protection and support. Thus, Taylor's group proposes that women respond adaptively to stress by becoming more nurturing and by seeking out friendships.

How do women and men differ in response to stress?

A wide variety of evidence supports the tend-and-befriend hypothesis. At the biological level, females have low levels of testosterone, the male hormone that is implicated in physically aggressive behaviours, such as those used to fight predators or rivals (Bain, 1997). Rather, in response to stress, females tend to show greater release of the hormone oxytocin. Oxytocin plays an important role in maternal behaviour, as well as in relaxation and feelings of love (Panksepp, 1992). Thus, a large release of oxytocin during stress may calm females as well as promote nurturing behaviour toward offspring. At the behavioural level, mothers who are stressed by their jobs tend to pay more attention to their children, whereas stressed fathers tend to withdraw from their families. Indeed, children report receiving the most love and nurturance from their mothers on the days that their mothers report the highest levels of stress (Repetti, 1997, cited in Taylor et al., 2000). The tend-and-befriend stress response is an excellent example of how thinking about psychological mechanisms in terms of their evolutionary significance leads us to question long-standing assumptions about how the mind works. Females who responded to stress by nurturing and protecting their young and by forming alliances with other females apparently had a selective advantage over those who tried to fight or flee.

10.21 Hans Selye Selye popularized the term *stress* and demonstrated that stress could affect physical health. Although Hungarian, he spent most of his career in Montreal.

The General Adaptation Syndrome Is a Bodily Response to Stress

In the early 1930s in Montreal, Hans Selye (Figure 10.21) began studying the physiological effects of sex hormones by injecting rats with samples from other animals. When he examined the rats, he found enlarged adrenal glands, decreased levels of lymphocytes in the blood, and stomach ulcers. Selye surmised that the foreign hormones must have been the cause of these changes, so he conducted further tests, using different types of chemicals and even physically restraining the animals. He

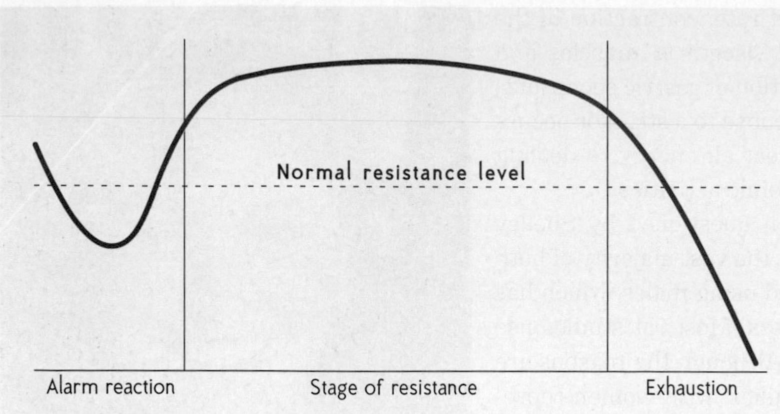

10.22 The General Adaptation Syndrome Selye's general adaptation syndrome describes the three stages of physiological response to stress. Initially, resistance is reduced as the body prepares to fight or flee. However, the body eventually adapts and resistance increases. Eventually, the system becomes exhausted and resistance plummets.

general adaptation syndrome A consistent pattern of responses to stress that consists of three stages: alarm, resistance, and exhaustion.

found slight variations in some physiological effects, but each manipulation produced this pattern of bloated adrenal glands, damaged lymphatic structures, and stomach ulcers. He concluded that these three responses were the hallmarks of *nonspecific stress response.* But these changes reduce the organism's potential ability to resist additional stressors. Selye borrowed the term *stress* from engineering, in which it is used to describe a force applied against resistance.

The **general adaptation syndrome,** a consistent pattern of responses that Selye identified, consists of three stages: the *alarm stage,* the *resistance stage,* and the *exhaustion stage* (Figure 10.22). It occurs in addition to specific physiological responses to particular stressors. The first stage is the *alarm stage,* an emergency reaction that prepares the body to fight or flee. In this stage, physiological responses are aimed at boosting physical abilities while reducing activities that make the organism vulnerable to infection after injury. This is the stage in which the body might be exposed to infection and disease, so the immune system kicks in and the body begins to fight back. During the *resistance stage,* the defenses are prepared for a longer, sustained attack against the stressor; immunity to disease continues to increase somewhat as the body maximizes its defenses. However, the body eventually reaches the *exhaustion stage,* during which a variety of physiological and immune systems fail. Bodily organs that were already weak before the stress are the first to fail. The general principles of this theory have been confirmed by scientific research (Selye, 1985).

Scientists now generally agree that the presence of a stressor leads to activation of the hypothalamic-pituitary-adrenal (HPA) system. During a stress response, the hypothalamus (H) secretes a hormone that triggers the pituitary (P) to release another hormone into the bloodstream, which in turn causes cortisol, still another hormone, to be released from the adrenal glands. Cortisol produces many of the bodily reactions to stress, such as breaking down protein and converting it to glucose, which helps meet immediate energy needs. Cortisol also helps the body deal with inflammation, which would occur if there were physical injuries. When cortisol levels are sufficiently high, a feedback loop is triggered that shuts down the HPA system (Anisman & Merali, 2003). Because it takes 15 minutes to an hour to produce a feedback loop, cortisol continues to affect the body even after the stressor has ended. In addition to cortisol, the adrenal glands release norepinephrine and epinephrine, which activate the sympathetic nervous system, increasing blood pressure, heart rate, and other sympathetic responses (see Chapter 3). Thus, stress has a powerful effect on bodily systems and behaviour.

People Encounter Stress in Daily Life

Stress occurs when aspects of the environment overwhelm people. That is, people feel stressed when too much is expected of them, or when events seem scary or worrisome. Stress occurs when we perceive a discrepancy—whether real or not—between the demands of the situation and the resources of our biological, psychological, and social systems. Psychologists typically think of stressors as falling into two categories: major life stressors and daily hassles. *Major life stressors* are changes or disruptions that strain central areas of people's lives. Positive events can be as stressful as, or even more stressful than, negative events. For instance, some parents report that having their first child is one of the most joyful—but also one of the most taxing—experiences of their lives. Major life stressors can be choices made by individuals, not just things that happen to them. Nonetheless, research has shown that unpredictable and uncontrollable catastrophic events (such as an earthquakes) are especially stressful. In general, life changes are stressful, whether it is moving to a new college or job, getting married, being fired, losing a parent, or winning a major award. The greater the number of changes, the greater the stress, and the more likely stress is to have an impact on physiological state.

Daily hassles are small, day-to-day irritations and annoyances, such as driving in traffic, dealing with unfair bosses or teachers, or having to wait in line. Daily hassles are stressful, and their effects can be comparable to those of major life changes. These low-level irritations are ubiquitous and, most important, pose a threat to coping responses by slowly wearing down personal resources. Studies that ask people to keep diaries of their daily activities consistently find that the more intense and frequent the hassles, the poorer the physical and mental health of the participant. People appear to habituate to daily problems, but some types, such as interpersonal difficulties, seem to have a cumulative effect on health. Living in poverty or working in a crowded city or in an area with noise or environmental pollution can also have detrimental effects on emotional well-being.

Why is chronic stress harmful?

Stress Affects Health

One of Selye's central points was that the prolonged action of stress hormones, such as cortisol, has a negative impact on health. Indeed, although stress hormones are essential to normal health, over the long term they are associated with such problems as increased blood pressure, cardiac disease, diabetes, declining sexual interest, and dwarfism. People who have stressful jobs—air-traffic controllers, combat soldiers, firefighters—have numerous health problems, presumably in part due to the effects of chronic stress. Robert Sapolsky (1994) chronicles the multiple health problems attributable to chronic stress, especially psychosocial stress. He notes that chronic stress can even lead to memory impairments because cortisol damages neurons in the hippocampus. Recall from Chapter 7 that the hippocampus is involved in the consolidation of memory.

Overwhelming evidence suggests that stress is associated with the initiation and progression of a wide variety of diseases, from cancer to AIDS to cardiac disease. Not only does stress lead to specific physiological responses that affect health, but many people cope with stress by engaging in damaging behaviours. For instance, the number-one reason that problem drinkers give for abusing alcohol is to cope with negative stress in their lives. When people are stressed, they smoke cigarettes, eat junk food, use drugs, and so on. Most of the major health problems in Western society are attributable to unhealthful behaviours, many of which occur when people are stressed. We discuss some of these later in the chapter.

HEART DISEASE Coronary heart disease is the leading cause of death for adults in contemporary society. According to a World Health Organization report in 2002, more than 30 million people around the globe have heart attacks each year, of which more than 12 million are fatal. Heart disease is determined by many factors, including genetics, but two extremely important determinants are health behaviours and stress. Later in this chapter we describe three of the major risk factors for heart disease: a lack of exercise, obesity, and smoking. Here we are concerned with how stress affects the heart.

The heart pumps nearly 2,000 gallons of blood each day, on average beating more than 100,000 times. A vast network of blood vessels carries oxygen and nutrients throughout the body. As people age, the arteries leading from the heart become narrow because of the buildup of fatty deposits, known as plaque; this narrowing makes it more difficult for the heart to pump blood. The buildup of plaque leads to coronary heart disease. When pieces of plaque break off from the wall of the blood vessel, blood clots form around the plaque and interrupt blood flow. If a clot blocks a blood vessel feeding the heart, it causes a heart attack; if the vessel feeds the brain, the blockage causes a stroke.

Stress and negative emotions increase the risk of coronary heart disease (Sirois & Burg, 2003). Being stressed or depressed can cause heart problems in two ways. First, people often cope through behavioural strategies that are bad for health, such as overeating or smoking. But stress also produces direct physiological effects on the heart. Chronic stress leads to overstimulation of the sympathetic nervous system, causing higher blood pressure, the constriction of blood vessels, elevated levels of cortisol, increased release of fatty acids into the bloodstream, and greater buildup of plaque on arteries; all of these contribute to heart disease. Over time, then, being stressed causes wear and tear on the heart, making it more likely to fail.

What aspects of personality are related to heart disease?

Have you ever noticed that some people always seem to be stressed out, whereas others are more laid back? Given what we know about stress and the heart, those stressed-out people should be more likely to have heart disease. This assumption was first tested in 1960, when the Western Collaborative Group began what was to be an eight-and-a-half-year study of the effects of personality on coronary heart disease. Physicians recruited 3,500 men from northern California who were free of heart-disease symptoms and screened them annually for blood pressure, heart rate, cholesterol, and overall health practices. Personal details such as level of education, medical and family history, income, and personality traits were also assessed. The study results indicated that, adjusting for the presence of established risk factors (such as high blood pressure or high cholesterol), a pattern of personality traits predicted heart disease. This pattern, called the **Type A behaviour pattern,** describes competitive, achievement-oriented, aggressive, hostile, time-pressed (constantly feeling hurried, restless, unable to relax), impatient, confrontational people. Men who exhibited these traits were much more likely to develop coronary heart disease than were those who exhibited the **Type B behaviour pattern,** which describes a relaxed, noncompetitive, easygoing, accommodating person. In fact, this study found that Type A personality was as strong a predictor of heart disease as was smoking or high cholesterol or blood pressure (Rosenman et al., 1975).

Type A behaviour pattern A pattern of behaviour characterized by competitiveness, achievement orientation, aggressiveness, hostility, restlessness, inability to relax, and impatience with others.

Type B behaviour pattern A pattern of behaviour characterized by relaxed, noncompetitive, easygoing, and accommodating behaviour.

Since the initial reports, numerous investigators have sought to understand the relation between personality and heart disease. Research over the past three decades has found that only certain components of the Type A behaviour pattern are related to heart disease. Researchers at Duke University found that perhaps the most toxic factor was hostility (Williams, 1987). Hot-tempered people, who are frequently angry, cynical, and combative, are much more likely to die at an early age from heart disease (Eaker, Sullivan, Kelly-Hayes, D'Agostino, & Benjamin, 2004). Indeed, having a high level of hostility while in college predicts greater risk for heart disease later in life (Siegler et al., 2003). Evidence is also considerable that other negative emotional states, especially depression, also predict heart disease (Miller, Freedland, Carney, Stetler, & Banks, 2003). Of course, it could be that having a heart condition makes people feel depressed, but feelings of depression also predict worsening of heart disease, which creates a vicious cycle. In contrast, people who are optimistic about life tend to be at lower risk for

heart disease (Maruta, Colligan, Malinchoc, & Offord, 2002). The take-home message from this research is that people who are frequently angry, depressed, or pessimistic are much more likely to die from coronary problems. Learning to manage stress and anger has been shown to improve outcomes for those who have heart disease (Sirois & Burg, 2003).

immune system The body's mechanism for dealing with invading microorganisms, such as allergens, bacteria, and viruses.

lymphocytes Specialized white blood cells known as B cells, T cells, and natural killer cells that make up the immune system.

THE IMMUNE SYSTEM Stress alters the functions of the **immune system,** which is the physical mechanism that helps the body deal with infection. When foreign substances—such as viruses, bacteria, or allergens—enter the body the immune system launches into action to destroy the invaders. Stress interferes with this natural process. As Esther Sternberg (2000) has noted, "Stress can make you sick because the hormones and nerve pathways activated by stress change the way the immune system responds, making it less able to fight invaders" (p. 131). The field of *psychoneuroimmunology* studies the response of the body's immune system to psychological variables. Over the last 30 years, more than 300 studies have demonstrated that short-term stress boosts the immune system, whereas chronic stress weakens it, leaving the body less able to deal with infection (Segerstrom & Miller, 2004).

The immune system is made up of three types of specialized white blood cells known as **lymphocytes**: B cells, T cells, and natural killer cells. *B cells* produce *antibodies,* protein molecules that attach themselves to foreign agents and mark them for destruction. Some types of B cells are able to remember specific invaders, making for easier identification in the future; this is why you have lifelong immunity to some diseases once you've been exposed to them naturally or through inoculation. The *T cells* are involved in attacking the intruders directly and also with increasing the strength of the immune response. These so-called helper cells are incapacitated by infection with HIV, which eventually leads to the immune disorder AIDS. *Natural killer cells* are especially potent in killing viruses and also help attack tumours. The detrimental effects of both immediate and long-term stress on physical health are due, in part, to decreased lymphocyte production, which renders the body less capable of warding off foreign substances.

In a particularly clear demonstration that stress affects health, Sheldon Cohen and his colleagues (1991) paid healthy volunteers to be exposed to the common cold virus. Those who had reported the highest levels of stress in a questionnaire developed worse cold symptoms and higher viral counts than those who reported being less stressed (Figure 10.23). In another study, participants kept a daily diary for up to 12 weeks in which they recorded their moods and events in their lives. The participants also rated these events as desirable or undesirable. Each day, the participants took an antigen to challenge their immune systems and provided a saliva sample so that researchers could examine their antibody responses. The study revealed that the more desirable events a participant reported, the greater the antibody production. Similarly, the more undesirable events reported, the weaker the antibody production. The effect of a desirable event on antibodies lasted for two days (Stone, Neale, Cox, & Napoli, 1994). These and subsequent findings provide substantial evidence that perceived stress influences the immune system. Chronic stress, especially when associated with changes in social roles or identity (such as becoming a refugee, losing a job, or getting divorced), has the greatest impact on the immune system (Segerstrom & Miller, 2004).

10.23 Stress and the Immune System People who reported the greatest levels of stress were the most likely to catch colds.

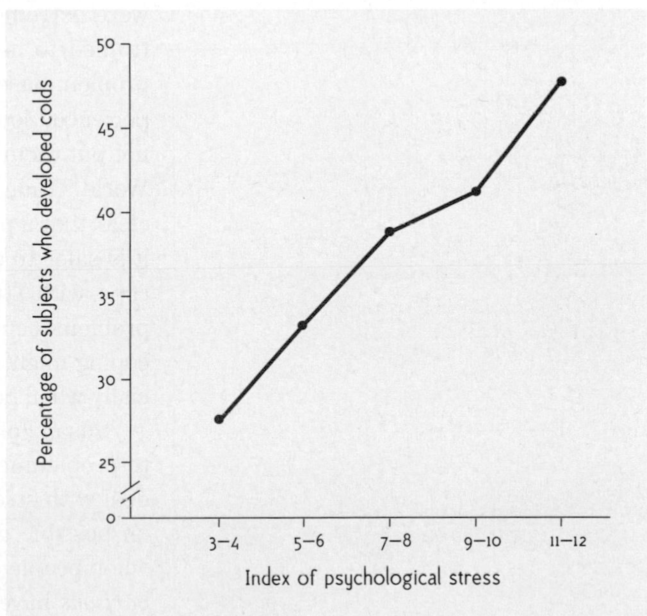

Coping Is a Process

Cognitive appraisals affect people's perception of and reactions to potential stressors. Coping that occurs before the onset of a future stressor is called *anticipatory coping,* as when parents rehearse what they will say when telling their children about plans to divorce. The psychologist Richard Lazarus (1993) conceptualized a two-part appraisal process. People use **primary appraisals** to decide whether the stimulus is stressful, benign, or irrelevant. If the stimulus is deemed stressful, people use **secondary appraisals** to evaluate response options and choose coping behaviours.

primary appraisal Part of the coping process that involves making decisions about whether a stimulus is stressful, benign, or irrelevant.

secondary appraisal Part of the coping process during which people evaluate their options and choose coping behaviours.

emotion-focused coping A type of coping in which people try to prevent having an emotional response to a stressor.

problem-focused coping A type of coping in which people take direct steps to confront or minimize a stressor.

positive reappraisal A cognitive process in which people focus on possible good things in their current situation.

TYPES OF COPING A taxonomy of coping strategies developed by Folkman and Lazarus suggests that people use two general categories of coping style. **Emotion-focused coping** involves trying to prevent having an emotional response to the stressor. It includes strategies such as avoidance, minimizing the problem, trying to distance oneself from the outcomes of the problem, or doing things such as eating or drinking. People adopt these often passive strategies to numb the pain. They do nothing to solve the problem or prevent it from reoccurring in the future. By contrast, **problem-focused coping** involves taking direct steps to solve the problem. People generate alternative solutions, weigh them in terms of costs and benefits, and choose between them. People adopt problem-focused behaviours when they perceive the stressor as controllable and experience only a moderate level of stress. Conversely, emotion-focused behaviours enable people to continue functioning in the face of an uncontrollable stressor or a high level of stress.

Most people report using both problem- and emotion-focused coping. Usually, emotion-based strategies are effective only in the short run. For example, if someone in a bad mood is giving you a hard time, just ignoring him or her can be the best option. Yet ignoring your partner's drinking problem won't make it go away, and eventually you'll need a better coping strategy. However, for problem-focused coping strategies to work, people need to be able to do something about the situation. A fascinating study tested the best way to cope with an extremely threatening situation (Strentz & Auerbach, 1988). In this study, 57 airline workers were taken hostage by five "terrorists," who were actually FBI agents. Even though the participants volunteered to be hostages with full knowledge, the situation was very realistic and they were extremely stressed over the four days of the study. Half of the people were trained to use emotion-based coping whereas the other half were trained to use problem-based coping. Which strategy worked better? The emotion-based group experienced less stress because there was nothing the hostages could do that would not put them in greater danger. However, this study was done in 1988. Following the World Trade Center and Pentagon tragedies on September 11, 2001, it is no longer clear that a passive response would help people cope. The previous assumption that it is safer to cooperate with hijackers may no longer be true. Thus, the best way to cope with stress depends on personal resources and on the situation. Although problem-focused coping is usually more effective over the long run, emotion-focused coping might be a useful strategy in some circumstances over the short term, especially when people have little control over the situation.

Susan Folkman and Judith Moskowitz (2000) have demonstrated that in addition to problem-focused coping, two strategies can help people use positive thoughts to deal with stress. **Positive reappraisal** is a cognitive process in which people focus on possible good things in their current situation, the proverbial silver lining, as when people compare themselves to those who are worse off. These *downward comparisons* have been shown to help people coping with serious illnesses. *Creation of* *positive events* refers to a strategy of infusing ordinary events with positive meaning. Doing such things as taking note of a beautiful sunset, trying to find humour in a situation, or simply taking satisfaction in a recent compliment allows people to focus on the positive aspects of their lives, which helps them deal with their negative stress. For instance, caregivers of those with AIDS are under enormous stress. Folkman and Moskowitz have found that positive appraisal, creation of positive events, and problem-focused coping—which gave the caregivers some sense of control—were instrumental in successful coping.

INDIVIDUAL DIFFERENCES IN COPING People differ widely in the degree to which they perceive life events as being stressful. Some people could be termed "stress-resistant individuals" for their ability to adapt to life changes by viewing the events constructively. An idea that captures this personality characteristic is **hardiness,** developed by Suzanne Kobasa (1979), which has three components: *commitment, challenge,* and *control.* People who are high in hardiness are committed to their daily activities, view threats as challenges or opportunities for growth, and see themselves as being in control of their lives. People who are low in hardiness are typically alienated, view events as under external control, and fear or resist change. Numerous studies have found that people high in hardiness report fewer negative responses to stressful events. In a laboratory experiment in which participants were given difficult cognitive tasks, people high in hardiness exhibited higher blood pressure during the task, a physiological indicator of active coping. Moreover, a questionnaire completed immediately after the task revealed that participants high in hardiness boosted the number of positive thoughts about themselves in response to the stressor.

hardiness A personality trait that enables people to perceive stressors as controllable challenges.

social support A network of other people who can provide help, encouragement, and advice.

SOCIAL SUPPORT One of the most important factors for whether people effectively cope with stress is the level of social support they receive. **Social support** refers to having other people who can provide help, encouragement, and advice. Having social support is an essential component of positive mental and physical health. Ill people who are socially isolated are much more likely to die than those who are well connected to others (House, Landis, & Umberson, 1988), in part because isolation itself is associated with numerous health problems. Conversely, a recent review of more than 80 studies found strong evidence linking social support to fewer health problems (Uchino, Cacioppo, & Kiecolt-Glaser, 1996). For example, Janice Kiecolt-Glaser and Ronald Glaser (1988) found that people with troubled marriages and people going through divorce or bereavement all had compromised immune systems. In a study that categorized newlyweds on the basis of observed interactions, couples who fought more and showed more hostility toward one another exhibited decreased natural-killer-cell activity in the 24 hours after the interaction (Kiecolt-Glaser, Malarkey, Chee, & Newton, 1993).

Social support is important for people of all ages. Studies of children who seem *resilient,* meaning they have good outcomes in spite of being raised in deprived or chaotic situations, show the presence of parental or family support to be especially important (Masten, 2001). Because the elderly are particularly prone to social isolation, the effects of social support on health are particularly strong for this group. In keeping with the tend-and-befriend pattern mentioned earlier, women are much more affiliative under stress than men. They are more likely to seek out social support, to receive social support, and to be satisfied with the social support they receive. Interestingly, when stressed, women are especially likely to seek out other women. Thus, women are more connected to their social networks than are men.

Social support helps people cope with stress in two basic ways. First, people with social support experience less stress overall. Consider single parents who have to deal with job and family demands in isolation. The lack of a partner places more demands on them, which increases the likelihood that they will feel stressed. Second, social support aids coping because other people lessen the negative effects of the stress that occurs. The **buffering hypothesis** (Cohen & Wills, 1985) proposes that others can provide direct support in helping people cope with stressful events. Receiving emotional support is more important than just having others who provide information. Emotional support includes expressions of caring and willingness to listen to another person's problems. Social support can also take more tangible forms, such as providing material help or assisting with daily chores. But to be effective, social support needs to imply that people *care* about the recipient of the support.

buffering hypothesis The idea that other people can provide direct support in helping individuals cope with stressful events.

REVIEWING the Principles | How Do People Cope with Stress?

Stress occurs when people feel overwhelmed by the challenges they face, as when major change happens in their lives. Hans Selye proposed the general adaptation theory to conceptualize the stages of physiological coping. The stress of daily life includes both major life changes and daily hassles. However, cognitive appraisals, such as determining the relevance of the stressor or adopting a problem- versus emotion-focused approach, can alleviate stress or minimize its harmful effects. Perhaps the most important way people cope with stress is through their interactions with others who provide social support.

What Are Some of the Behaviours That Affect Physical Health?

We just looked at how stress affects our bodies, especially our hearts and immune system. We also learned some of the ways people cope with stress. We now look at the effect stress has on our everyday behaviour. Before the twentieth century, most people died from infections and diseases that were transmitted from person to person. But the last century saw a dramatic shift in the leading causes of mortality. People now die from heart disease, cancer, accidents, diabetes, liver disease, and so on, all of which are at least partial outcomes of lifestyle. Our daily habits, such as smoking, poor eating, alcohol use, and lack of exercise, contribute to nearly every major cause of death in developed nations (Smith, Orleans, & Jenkins, 2004). Stress plays an important role in motivating each of these health-threatening behaviours. A study of more than 12,000 people from Minnesota found that high stress was associated with greater intake of fat, less frequent exercise, and greater smoking (Ng & Jeffrey, 2003). The field of *health psychology* was launched more than 25 years ago as psychologists, physicians, and other health professionals came to appreciate the extent to which lifestyle factors played important roles in physical health.

Psychological scientists who seek to understand health behaviours typically view them from a *biopsychosocial* perspective (Suls & Rothman, 2004) that emphasizes biological processes (e.g., the activation of reward circuits by neurotransmitters and the physiology of weight regulation); individual factors (e.g., level of stress, coping skills, personality traits); and social influences (e.g., friends, family, societal values, and environmental factors). Embodying one of our major themes for this book, research that crosses these levels of analyses helps scientists to identify strategies that prevent disease by helping people lead healthier lives. In this section we consider daily behaviours that have important health consequences.

Obesity Results from a Genetic Predisposition and Overeating

Obesity is a major health problem with both physical and psychological consequences. One factor that contributes to obesity is overeating, which some people are especially likely to do when stressed (Heatherton & Baumeister, 1991). Although there is no precise definition of obesity, people are considered obese if they are approximately 20 percent over ideal body weight, as indicated by various mortality studies (Birmingham et al., 1999). One of the most widely used measures of obesity in research is **body mass index (BMI)**, which computes a ratio of body weight to height. Figure 10.24 shows how to calculate BMI and how to interpret the value obtained. According to standard BMI cutoffs, approximately 16 percent of Canadians currently are medically obese. Researchers have spent the last 60 years trying to understand the causes of obesity. Although we have learned a great deal, obesity is clearly a multifaceted problem that is influenced both by genes and by the environment.

body mass index (BMI) A ratio of body weight to height used to measure obesity.

GENETIC INFLUENCE A trip to the local mall will immediately reveal one obvious fact about body weight: Obesity tends to run in families. Indeed, various family and adoption studies indicate that approximately half of the variability in body weight can be considered the result of genetics. One of the best and largest studies, carried out during the 1980s in Denmark, found that the BMI of adopted children was strongly related to the BMI of the biological parents and *not at all* to the BMI of the adoptive parents (Sorensen, Holst, Stunkard, & Skovgaard, 1992). Studies of identical and fraternal twins provide even stronger evidence of genetic control of body weight, with the heritability estimates (see Chapter 3) ranging from 60 to 80 percent. Moreover, the agreement for body weights of identical twins does not differ between twins raised together and twins raised apart, which suggests that environment has far less effect on body weight than genetics has.

If genes primarily determine body weight, why has the percentage of Canadians who are obese more than doubled over the past few decades? Albert Stunkard, a leading researcher of human obesity, points out that genetics determine whether a person *can* become obese, but the environment determines whether that person will actually *become* obese (Stunkard, 1996). Consider an important study, conducted at Laval University by the geneticist Claude Bouchard, in which identical twins were overfed by approximately 1,000 calories each day for 100 days. Most of the twins gained some weight, but there was great variability between pairs in how much was gained (ranging from 4.3 to 13.3 kg) (Bouchard et al., 1990). However, there was a striking degree of similarity within the twin pairs in terms of how much weight they gained and in which parts of the body they stored the fat. Some of the twin pairs were especially likely to put on weight. Thus, genetics determine sensitivity to environmental influences. Genes

10.24 A Nomogram for Determining Body Mass Index You can find your own BMI by drawing a straight line between your height and weight. A BMI under 20 suggests that you are underweight, between 20 and 25 is average, between 25 and 30 is overweight, and 30 and higher is considered obese. According to the latest guidelines a healthy BMI for an adult is between 18.5 and 25. Beyond or below this optimal range, the more you are at risk for health problems.

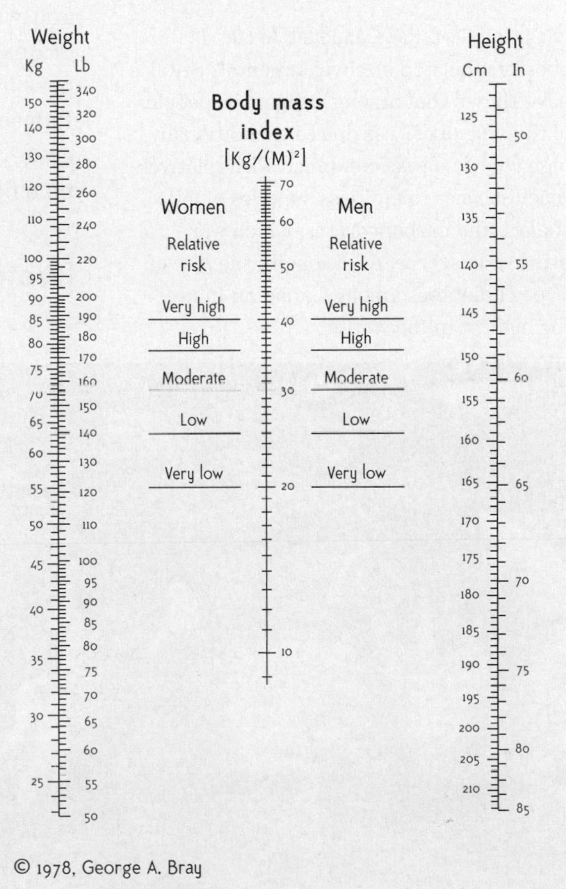

predispose some people to obesity in environments that promote overfeeding, such as contemporary Western societies. Many genes are involved in obesity, as you might expect in such a complex condition. Indeed, more than 300 different genetic markers or genes have been identified as playing some role in human obesity (Snyder et al., 2004).

Why is obesity stigmatized?

THE STIGMA OF OBESITY Obesity is associated with a significant number of medical problems, including heart disease, high blood pressure, and gastric ailments (Figure 10.25). It can also give rise to a variety of psychological problems, primarily because of the extreme stigma associated with being overweight. In most Western cultures, obese individuals are viewed as less attractive, less socially adept, less intelligent, and less productive than their normal-weight peers (Dejong & Kleck, 1986). Moreover, perceiving oneself as overweight is linked to lower self-esteem, depression, and anxiety.

Not all cultures stigmatize obesity (Hebl & Heatherton, 1998). In some developing countries, such as many African nations, being obese is a sign of being upper class. Obesity may be desirable in developing countries because it helps prevent some infectious diseases, reduces the likelihood of starvation, and is associated with having more successful births. Obesity in developing countries may also serve as a status symbol, an indication that one can afford to eat luxuriously. In Pacific island countries such as Tonga and Fiji, being obese is a source of personal pride, and dieting is therefore not common. In most Western cultures, where food is generally abundant, being overweight is associated with lower socioeconomic status, especially for women. The upper classes in Western cultures have a clear preference for very thin body types. This preference can be verified by looking at high-fashion magazines. The typical woman presented by the fashion industry is 180 centimetres tall and weighs approximately 50 kilograms, which is 18 centimetres taller and 14 kilograms lighter than the average woman in Canada. Such an extreme standard of thinness reinforces body-weight ideals that are difficult for most people to obtain. Indeed, women report holding ideals for body weight that are not only lower than average weight, but also lower than what men find attractive *and* what women think men find attractive, as may be seen in Figure 10.26 (Fallon & Rozin, 1985).

10.25 Inuit of the Canadian Arctic Is obesity inherited or environmental? Studies have shown that among indigenous peoples of Canada, obesity is due to a traditionally high caloric intake combined with relatively recent access to limitless supplies of fats, sugars, and carbohydrates, which were previously scarce. Additionally, the rate of Type II diabetes among Canadian Inuit is the highest in the world.

Restrictive Dieting Can Be Problematic

Dieting is a notoriously ineffective means of achieving permanent weight loss. Most individuals who lose weight through dieting eventually regain the weight; very often they gain back *more* than they lost. The primary reason that most diets fail has to do with the body's natural defense against weight loss. Body weight is regulated around a *set-point* that is determined primarily by genetic influence. Consider two examples. In 1966, several Vermont prisoners were given an interesting challenge: to try to increase their body weight by 25 percent (Sims et al., 1968). For six months these prisoners consumed more than 7,000 calories per day, nearly double their usual intake. Assuming each prisoner was eating about 3,500 extra calories a day (the equivalent of seven large cheeseburgers), simple math suggests each should have gained approximately 80 kilograms over the six months.

In reality, few gained more than 20 kilograms, and most lost the weight when they went back to normal eating. Those who did not lose the weight had family histories of obesity, supporting the view that genetics predispose obesity when people are exposed to overfeeding.

At the other end of the spectrum, during World War II, more than 100 American men volunteered to take part in a scientific study as an alternative to military service (Keys, Brozek, Henschel, Mickelsen, & Taylor, 1950). The researchers were interested in the short- and long-term effects of semistarvation. Over six months, the men lost an average of 25 percent of their body weight, but most found it very hard to do and some had great difficulty losing more than 5 kilograms. The men underwent dramatic changes in emotions, motivation, and attitudes toward food. They became anxious, depressed, and listless; they lost interest in sex and other activities; and they became obsessed with eating. Interestingly, many of these outcomes are quite similar to those experienced by people with eating disorders.

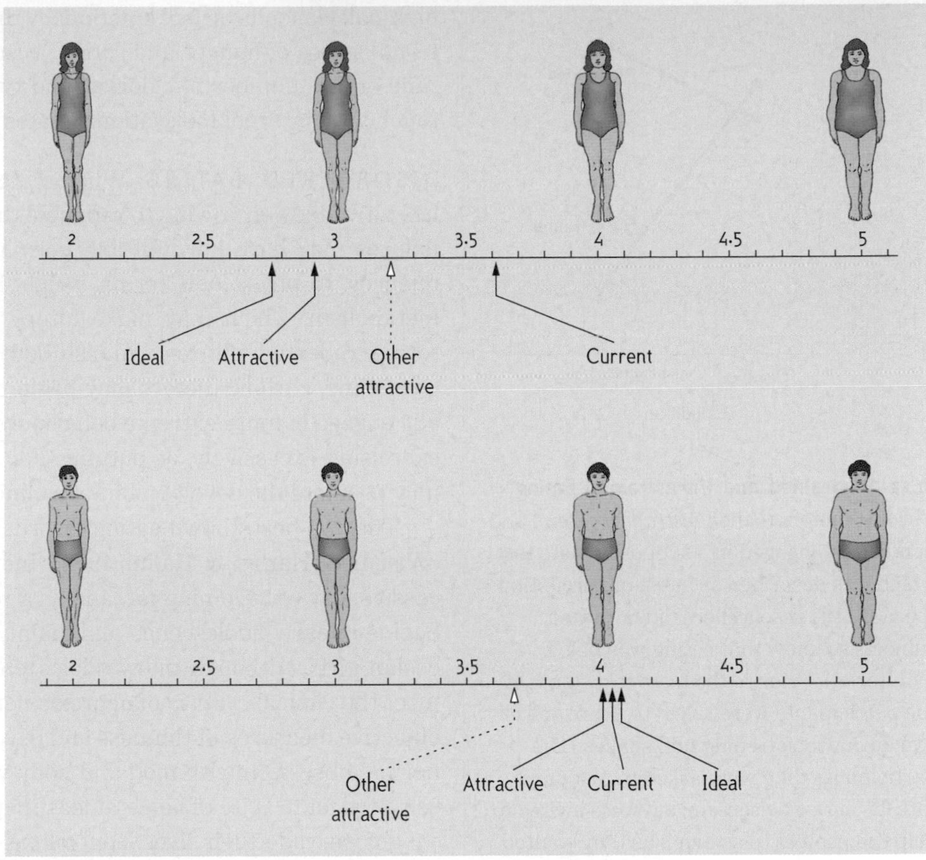

10.26 Rating Attractiveness Men and women were asked to choose which figures represent how they currently view their bodies, what their ideal body shape is, and what they believe is attractive to the opposite sex. Women chose a body-weight ideal that is thinner than what they believe men want and also thinner than what men actually say they want in women. Men chose an ideal that is closer to their current self-rating and close to what they believe women prefer, but heavier than what women say they want in men.

Although it is possible to alter body weight, the body responds to weight loss by slowing down the metabolism and using less energy. Thus, after the body has been deprived of food, it takes less food to maintain a given body weight. Likewise, weight gain occurs much faster in previously starved animals than would be expected by caloric intake alone. In addition, repeated alterations between caloric deprivation and overfeeding have been shown to have cumulative metabolic effects, so that weight loss and metabolic functioning are slowed more each time the animal is placed on caloric deprivation, and weight gain occurs more rapidly with each resumption of feeding. Such patterns might explain why "yo-yo dieters" tend to become heavier over time.

RESTRAINED EATING A second reason that diets tend to fail is occasional, or not so occasional, bouts of overeating. University of Toronto psychologists Janet Polivy and Peter Herman (1985) have demonstrated that chronic dieters, whom they call *restrained eaters,* are prone to excessive eating in certain situations. For instance, if restrained eaters believe they have eaten high-calorie foods, they abandon their diets (see Figure 10.27). The mind-set of restrained eaters is "I've blown my diet, so I might as well just keep eating." Many restrained eaters diet all week only to lose control on the weekends when they are faced with increased food temptations at the same time that they are in a less-structured environment. Being under stress also leads restrained eaters to break their diets (Heatherton, Herman, & Polivy, 1991).

Binge eating by restrained eaters depends on their *perception* of whether their diet is broken or not. Dieters can eat 1,000-calorie Caesar salads and believe that their diets are fine, but if they eat 200-calorie chocolate bars they feel the diet is ruined and they become disinhibited. The problem for restrained eaters is that they rely on cognitive control of food intake, which is likely to break down when they eat

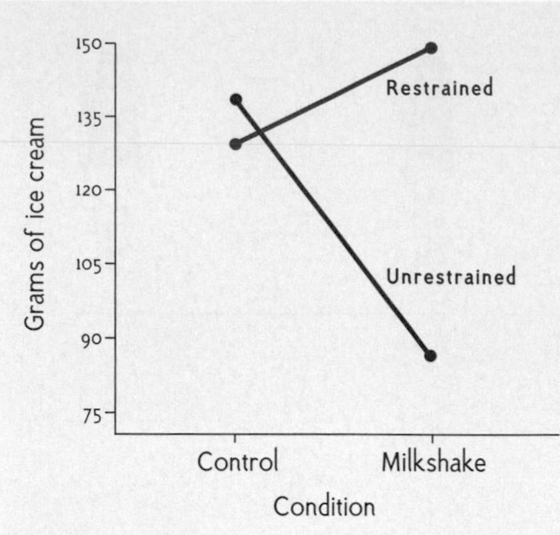

10.27 Restrained and Unrestrained Eating Chronic dieters (called *restrained eaters*) and nondieters engaged in a supposed taste test, which was described as a test of perception (because the researchers did not want subjects to know that eating was being monitored). Prior to the taste test, some of the participants were asked to drink one or two obviously fattening milkshakes. The participants then were asked to taste and rate flavours of ice cream and were invited to help themselves to as much as they wanted. Nondieters (called unrestrained eaters) ate sensibly: Those who had drunk one or two milkshakes ate much less ice cream than those who had not. Restrained eaters, however, did just the opposite, eating much more ice cream if they had had the milkshakes. This has been called the "what the hell effect."

anorexia nervosa An eating disorder characterized by an excessive fear of becoming fat and thus a refusal to eat.

bulimia nervosa An eating disorder characterized by dieting, binge eating, and purging.

high-calorie foods or feel emotionally distressed. Rather than eating according to internal states of hunger and satiety, restrained eaters eat according to rules, such as time of day, number of calories, and type of food. Getting restrained eaters back in touch with internal motivational states is one goal of sensible approaches to dieting.

DISORDERED EATING When dieters fail to lose weight, they often blame their lack of willpower, vowing to redouble their efforts on the next diet. Repeated dietary failures may have harmful and permanent physiological and psychological consequences. In physiologic terms, weight-loss and weight-gain cycles alter the dieter's metabolism, which may make future weight loss more difficult. In psychological terms, repeated failures diminish body-image satisfaction and damage self-esteem. Over time, chronic dieters may begin to feel helpless and depressed. Some eventually engage in more extreme behaviours to lose weight, such as taking drugs, fasting, exercising excessively, or purging. For some vulnerable individuals, chronic dieting may promote the development of a clinical eating disorder.

The two best-known eating disorders are anorexia nervosa and bulimia nervosa (Wiseman, Harris, & Halmi, 1998). Individuals with **anorexia nervosa** have an excessive fear of becoming fat, and as a result they refuse to eat. Anorexia most often begins in early adolescence and mainly affects upper-middle- and upper-class Caucasian girls. Although many adolescent girls strive to be thin, fewer than 1 in 100 meet the clinical criteria of anorexia nervosa (Table 10.1). These criteria include both objective measures of thinness and psychological characteristics that indicate an abnormal obsession with food and body weight. Those who have anorexia view themselves as fat in spite of being at least 15 to 25 percent underweight. Issues of food and weight pervade their lives, controlling not only how they view themselves but also how they view the world. Initially, the results of their self-imposed starvation may draw favourable comments from others, although as the anorexic approaches her emaciated ideal, family and friends usually become quite concerned. In many cases, medical attention is required to prevent death from starvation. Anorexia is very difficult to treat, since patients maintain the belief that they are overweight, or not as thin as they would like to be, even when they are severely emaciated. This dangerous disorder causes a number of serious health problems, especially a loss of bone density, and about 15 to 20 percent of those with anorexia eventually die from the disorder (American Psychiatric Association, 2000).

Individuals with **bulimia nervosa** alternate between dieting and binge eating. Bulimia often develops during late adolescence. Like anorexia, bulimia is most common among upper-middle- and upper-class Caucasian women, but it is more common among minorities and men than anorexia. Approximately 1 to 2 percent of women in high school and college meet the definitional criteria for bulimia nervosa. These women tend to be of average weight or slightly overweight. They regularly binge-eat; feel that their eating is out of control; have excessive worries about body weight issues; and engage in one or more compensatory behaviours, such as self-induced vomiting, excessive exercise, or the abuse of laxatives. Unlike the starvation of anorexia, binge eating behaviour tends to occur secretly. Although bulimia is associated with serious health problems, such as dental and cardiac disorders, it is seldom fatal (Keel & Mitchell, 1997). A variant of bulimia is *binge-eating disorder,* wherein individuals engage in binge eating but do not purge. Many of those with binge-eating disorder are obese.

Smoking Often Begins in Childhood

In spite of overwhelming evidence that smoking cigarettes leads to premature death, millions around the globe continue to light up (Carmody, 1993). Although smoking

TABLE 10.1 Diagnostic Criteria for Anorexia Nervosa and Bulimia Nervosa

CRITERIA FOR ANOREXIA NERVOSA	CRITERIA FOR BULIMIA NERVOSA
A. Refusal to maintain body weight at or above a minimum normal for age and height (e.g., weight loss leading to maintenance of body weight less than 85 percent of that expected; or failure to make expected weight gain during period of growth, leading to body weight less than 85 percent of that expected). B. Intense fear of gaining weight or becoming fat, even though underweight. C. Disturbance in the way in which one's body weight or shape is experienced, undue influence of body weight or shape on self-evaluation, or denial of the seriousness of the current low body weight. D. In postmenarcheal females, amenorrhea, i.e., the absence of at least three consecutive menstrual cycles. (A woman is considered to have amenorrhea if her periods occur only following hormone, e.g., estrogen, administration.)	A. Recurrent episodes of binge eating. An episode of binge eating is characterized by both of the following: (1) Eating, in a discrete period of time (e.g., within any 2-hour period), an amount of food that is definitely larger than most people would eat during a similar period of time and under similar circumstances. (2) A sense of lack of control over eating during the episode (e.g., a feeling that one cannot stop eating or control what or how much one is eating). B. Recurrent inappropriate compensatory behaviour in order to prevent weight gain, such as self-induced vomiting; misuse of laxatives, diuretics, enemas, or other medications; fasting; or excessive exercise. C. The binge eating and inappropriate compensatory behaviours both occur, on average, at least twice a week for 3 months. D. Self-evaluation is unduly influenced by body shape and weight. E. The disturbance does not occur exclusively during episodes of anorexia nervosa.

SOURCE: *Diagnostic and Statistical Manual of the American Psychiatric Association,* 1994.

has declined over the past 40 years, Statistics Canada reports that about 22 percent of Canadians were currently smokers in 2002. Most smokers begin in childhood or early adolescence. Twenty-one percent of Canadian youth between the ages of 12 and 19 smoked with a reported increase of only 5 percent for adult smokers (Statistics Canada, 2002).

Smoking causes a number of health problems, such as heart disease, respiratory ailments, and a variety of cancers, and is blamed for over 45,000 deaths per year in Canada, decreasing the life of the typical smoker by more than 15 years (Physicians for a Smoke-Free Canada). Cigarette smoke also causes health problems for non-smoking bystanders, which has led to bans on smoking in numerous public and private places. It is not uncommon to see smokers huddled outside of office buildings or hospitals, braving both the elements and hostile looks from nonsmokers to satisfy their cravings for nicotine. Smokers also endure scoldings from physicians and loved ones who are concerned for their health and welfare, and in addition to spending money on cigarettes, they pay significantly more for life insurance. Why in the world does anyone smoke?

STARTING SMOKING It is hard to imagine any good reason to start smoking. First attempts at smoking often involve a great deal of coughing, watering eyes, a terrible taste in the mouth, and feelings of nausea. So why do kids persist? Most researchers point to powerful social influences as the leading cause of adolescent smoking (Chassin, Presson, & Sherman, 1990). Research has demonstrated that adolescents are more likely to smoke if their parents or friends are smokers (Biglan & Lichtenstein, 1984; Hansen et al., 1987; Tewolde et al., 2006). They often smoke their first cigarette in the company of other smokers, or at least with the encouragement of their peer group. Moreover, many adolescent smokers appear to show a false-consensus effect, in that they overestimate the number of adolescent and adult smokers (Sherman, Presson, Chassin, Corty, & Olshavsky, 1983). Thus, adolescents who incorrectly believe that smoking is common may take it up to fit in with the crowd.

The real reason dinosaurs became extinct

Others have pointed to the potential meaning of "being a smoker" as having a powerful influence. For instance, research has shown that smokers are viewed as having a number of positive qualities, such as being tough, sociable, and good with members of the opposite sex. Children take up smoking partially to look "tough, cool, and independent of authority" (Leatherdale et al., 2006; Leventhal & Cleary, 1980, p. 384). Thus, smoking may be one way in which adolescents enhance their self-image as well as their public image (Chassin et al., 1990). Of course, it is hard to look tough while gasping and retching, so while most adolescents try one or two cigarettes, the majority do not become regular smokers. By grade 12, 70 percent of adolescents have some experience with tobacco products (Mowery, Brick, & Farrelly, 2000), but fewer than half of those who try cigarettes become established smokers.

Although many of the same factors that lead children to try cigarettes are involved in the establishment of regular smoking, additional factors are important for determining long-term, regular tobacco use. Over time, casual smokers become addicted. It is now widely acknowledged that the drug nicotine in tobacco is of primary importance in motivating and maintaining smoking behaviour (Fagerström & Schneider, 1989; USDHHS, 2004). Once the smoker becomes "hooked" on nicotine, going without cigarettes will lead to unpleasant withdrawal symptoms, including heightened anxiety and distress (Russell, 1990). Some people appear to be especially susceptible to nicotine addiction, which may be due to genetics. Genes that control dopamine pathways are especially influential in determining whether individuals become addicted to nicotine (Sabol et al., 1999). Nicotine may lead to greater activation of dopamine neurons, which would be experienced as more rewarding. Positive reactions to the first smoking experience are associated with a greater likelihood of progression to regular smoking (Blitstein, Robinson, Murray, Klesges, & Zbikowski, 2003; Pomerleau, Pomerleau, & Namenek, 1998).

Profiles in Psychological Science 

Success of a Fast-Food Dieter

One of the most unlikely recent celebrities is Jared Fogle, the formerly obese man who lost weight by walking every day and eating Subway sandwiches. Fogle shed a whopping 110 kilograms and has maintained his weight loss for several years (Figure

10.28). His dramatic success stands in stark contrast to what might be expected from the more pessimistic research literature. A review of published weight loss studies by the United States National Institutes of Health found that nearly two-thirds of those who lose weight regain that weight within one year and that almost no one manages to maintain weight loss for five years (NIH Technology Assessment Conference Panel, 1993). Similarly, the research literature indicates that treatments of addictions are no more than 20 to 30 percent successful over the long term. So how do we explain it when people like Jared Fogle are able to defy the odds and lose weight, stop smoking, or give up other addictions?

10.28 Jared Fogle Fogle lost 110 kilograms from his obese frame, and has stayed slim against the odds.

Do you know someone who has quit smoking or lost a substantial amount of weight and kept it off? Chances are you do. Stanley Schachter (1982) interviewed 83 members of the Columbia University Psychology Department and 78 shopkeepers from Amagansett, New York. Schachter found that more than half of those who had tried to lose weight or quite smoking had done so successfully on their own, without professional treatment. What was impressive was that the average length of abstinence was close to seven years for smoking and that people had maintained weight loss for more than 10 years. How do we reconcile the clinical treatment literature with Schachter's finding? Perhaps people first try to resolve weight and addiction problems on their own, seeking therapy only after repeated failures. Success like Jared's is not counted in any published research because Jared did it on his own.

Even if we can be somewhat optimistic about people successfully changing a problem behaviour, what factors predict whether a person will make the effort? Harald Klingemann (1991) studied the autobiographical accounts of ex-alcoholics and ex-heroin addicts to determine what factors led them to give up their addiction. Approximately three-quarters of the subjects reported extreme levels of stress associated with their addiction (e.g., family tensions, health problems, and feelings of helplessness). Many noted the occurrence of a critical event, such as losing a job or having a car accident. This is the proverbial "hitting rock bottom," what addicts describe as happening just before they realize that they have to do something about their behaviour.

This sequence of hitting bottom and then making a change highlights the important role of thought processes in effecting behavioural changes. Psychologist David Premack provides the example of a man who quit smoking because of something that happened as he was picking up his children at the city library: "A thunderstorm greeted him as he arrived there; and at the same time a search of his pockets disclosed a familiar problem: he was out of cigarettes. Glancing back at the library, he caught a glimpse of his children stepping out in the rain, but he continued around the corner, certain that he could find a parking space, rush in, buy the cigarettes and be back before the children got seriously wet" (Premack, 1970). For the smoker, it was a shocking vision of himself "as a father who would actually leave the kids in the rain while he ran after cigarettes." The man quit smoking on the spot.

Similar critical incidents have been found in other studies of how and why some people are able to stop addictive behaviours. Consider the story of a young pregnant woman who recalls drinking a beer to cure a hangover: "I felt the baby quiver and I poured the rest of the beer out, and I said, 'God, forgive me. I'll never drink another drop.' And from that day to this I haven't" (Tuchfeld, 1981). Or the man who quit smoking after visiting an elderly friend at a cancer ward of a hospital: "While I was sitting a young woman who was approximately my age entered the lobby—she was walking with a walker. Although she was several yards from me I could hear every wheeze of her laboured breathing. When she finally approached her friends, I heard her gasp, 'The doctors think I'm doing really well today.' Right then and there I decided to quit smoking. I guess I realized that it could happen to me" (Heatherton & Nichols, 1994). In each of these cases the sudden realization of the extreme negative consequences associated with the behaviour motivated the person to stop the behaviour.

No particularly dramatic event motivated Jared Fogle; noticing a sign for low-fat sandwiches inspired him to try a new diet. In some unknown way, reading the sign led him to turn his life around. His success has inspired thousands of others to hope they might be able to lose weight. A Web site devoted to Jared contains many success stories. One woman who lost 35 kilograms writes, "I was completely astonished when I learned of Jared's success. Seeing how you could completely change your life in a positive way was such a motivator. At first it was hard to believe that Jared could lose 110 kilograms by eating two subs a day and walking, but it can definitely happen, because it happened to me as well. He was definitely an inspiration." Psychological scientists do not entirely understand these transformative experiences, but it is clear that many people can and do make major life changes to live healthier lives. ■

People Do Not Get Enough Exercise

The evidence is overwhelming that regular exercise profoundly affects physical and mental health. Modern society allows people to exert little physical energy, in comparison with what was required for most of previous human history. People drive to work, take the elevator, spend hours watching remote-controlled television, use various labour-saving devices, and complain about not having time to exercise. Once people are out of shape, it is very difficult for them to start regular exercise. According to the Public Health Agency of Canada, more than 57 percent of Canadians do not exercise regularly, if at all. This is unfortunate: lifelong exercise yields both bodies and minds that deteriorate less radically with age.

In general, the more people exercise, the better their physical health. Aerobic exercise, that which momentarily increases breathing and heart rates, is especially good for cardiovascular health, lowering blood pressure and strengthening the heart and lungs (Lesniak & Dubbert, 2001). Other recent evidence indicates that exercise might help prevent certain cancers and improve the immune system. Because exercise also helps control appetite and metabolism, as well as burn calories, it is an essential element of any weight-control program.

If exercise has so many benefits, why do so few people exercise regularly?

Exercise is important for the mind as well as the body. Research using animals demonstrates that exercise increases blood flow to the brain, the number of synaptic connections, and even the development of new neurons, all of which can lead to better learning and performance. Becoming fit has also been shown to increase human brain functioning. Older adults who had six months of cardiovascular training performed better on a measure of cognitive function and showed increased activation of brain regions involved in the ability to attend to an object and to ignore distractors (Colcombe et al., 2004).

Exercise is also good for mental health, as it has been found to reduce stress and improve mood. Indeed, as little as 10 minutes of exercise can improve feelings of vigor and enhance mood, although 30 minutes daily is associated with the most positive mental state (Hansen, Stevens, & Coast, 2001). The evidence is compelling that exercise can contribute to positive outcomes for the clinical treatment of depression (Craft & Perna, 2004) as well as of addiction and alcoholism (Read & Brown, 2003). How exercise exerts these positive effects is not currently known. It may provide a way for people simply to feel good, or it may build self-confidence and help people cope with stress. It might also affect neurotransmitter systems involved in reward, motivation, and emotion.

Psychological Science *in Action*

Taking Care of Mind and Body

Over the last quarter century, psychological science has learned a great deal about the complex relations between stress, behaviour, and health. Fifty years ago people did not know how bad smoking was for the lungs, or that saturated fats and other dietary factors contribute to cardiovascular disease, or that being under stress could damage our bodies. We now know that people need to cope with stress, regulate their emotions, and control their daily habits to live healthy lives. Let's look at some of the behaviours that health psychologists have determined will improve physical and mental health.

- **Eat natural foods** Various food fads go in and out of fashion, but the essential rules never change. You should eat a varied diet that emphasizes natural foods (e.g., whole grains, fruits, and vegetables) and avoid processed and fast foods. If you choose, various types of animal products can also be part of a natural, healthy diet. Try to avoid foods with trans fatty acids, artificial types of fat that prolong store shelf life. Some studies have found natural oils, such as olive oil, to have benefits for the heart.
- **Watch portion size** Eat a varied diet in moderation—and eat only when you are hungry. Many prepared foods are sold in large portions, which encourage overeating. Many people eat for flavour or other reasons rather than to satisfy hunger. Over time, the extra calories from large portions contribute to obesity. Eating small snacks between meals can prevent people from becoming overly hungry and eating too much at their next meal.
- **Drink alcohol in moderation, if at all** Some research indicates that two or so glasses of wine per day, or similar quantities of other alcohol-containing drinks, can produce cardiovascular benefits. However, excessive alcohol consumption can cause serious health problems, including alcoholism, liver problems, some cancers, heart disease, and immune-system deficiencies—and the list goes on.
- **Keep active** Find a way to engage in moderate physical activity at least four times a week for at least 30 minutes. Ignore the mantra "no pain, no gain." Research shows not only that it is inaccurate, but that pain prevents people from exercising over the long run. Starting with moderate exercise that doesn't leave you breathless, and building up over time, is a safe way to ensure fitness. In addition, find opportunities throughout the day to be active, such as taking the stairs or walking.
- **Don't smoke** This may seem obvious, yet many college students begin smoking each year. Smoking has no known benefits and eventually produces undesirable physical effects for all smokers, such as a hacking cough, unpleasant odour, and bad breath. The major way that smoking helps most people relax is that it reduces withdrawal symptoms in those who are addicted to nicotine. Going without nicotine makes people anxious.
- **Practice safe sex** Numerous sexually transmitted diseases affect millions, including college students. A growing proportion of new HIV cases occurs among those under age 25 who are infected through heterosexual activity. In spite of the devastating consequences of many sexually transmitted diseases, many young adults engage in risky sexual practices, such as not using condoms.

This is especially likely to occur when alcohol or other drugs are involved. If you think that a sexual encounter is possible, be prepared and take appropriate precautions.

- **Learn to relax** As we have seen, daily hassles and stress can cause numerous health problems, as well as conditions such as insomnia that can interfere with your ability to function in daily life. There are many relaxation exercises that can help people soothe their bodies and minds. Seek help from trained counselors who can teach you these methods, such as using biofeedback to measure your physiological activity so that you can learn to control it. You might also try an activity such as yoga.
- **Learn to cope** Negative events are a part of life. You need to learn strategies for dealing with them, assessing them realistically so that you see what might be positive about them as well as accept the difficulties they pose. You can learn specific strategies for dealing with stressors, such as seeking advice or assistance, actively attempting new solutions, distracting yourself with more pleasant thoughts or activities, trying to reinterpret the situation in humorous ways, and so on. Different strategies are effective for different people, so you need to find out what works for you. The important thing is not to allow stress to consume your life. Remember that exercise helps reduce stress and is an excellent daily strategy for keeping it in check.
- **Build a strong support network** Having supportive friends and family can help people deal with much of life's stress, from daily frustrations to serious catastrophes. Avoid people who encourage you to act in unhealthy ways or who are threatened by your efforts to be healthy. Instead, find people who share your values and understand what you want from life, who can listen and who can provide advice, assistance, or simply encouragement. ■

REVIEWING the Principles | What Are Some of the Behaviours That Affect Physical Health?

How people behave in their daily lives has a profound effect on their physical and mental health. With the advent of medical technologies to deal with infectious disease, many causes of mortality in Western culture are related to lifestyle factors, such as diet, lack of physical activity, substance use and abuse, and coping poorly with stress. People engage in many unhealthy behaviours when they are stressed, such as eating fatty foods or smoking. People who eat too much can become obese, especially if they have a genetic predisposition to put on weight, which poses a number of serious health risks. Smoking is especially bad for health and is one of the leading causes of death around the world. Unfortunately, smoking typically starts in adolescence, a time when people give little thought to long-term consequences. One of the best things that people can do for health is exercise. Regular physical activity helps people to cope with stress, enhances their emotional experiences, and builds a strong and healthy body. More recent evidence suggests that exercise can also improve mental functioning through its effects on the brain.

Conclusion

For many years scientists likened the brain to an information-processing device such as a computer. Humans are hardly like this. They react to environmental input with physiological reactions and cognitions, which together produce mental states that are perceived to be positive or negative, or shades in between. This response is known as emotion, and it is a primary force that motivates adaptive behaviours and discourages maladaptive behaviours. Thinking about emotions and stress from an evolutionary viewpoint provides novel perspectives for understanding them. For instance, emotions such as shame and guilt are often viewed as unnecessary emotional baggage. But guilt, embarrassment, and the like help to maintain and affirm social bonds and thereby serve adaptive functions. Similarly, the assumption that fight or flight was a uniform response to stress does not make complete sense from an evolutionary perspective. Scholars have recently shown that females of many species behave in a completely different manner—tending and befriending—that makes more adaptive sense. For both sexes, stress can have a number of negative consequences, beginning with high levels of stress hormones that can alter the brain to causing high levels of anxiety. Both of these effects can cause heart disease and interfere with the immune system. Stress also causes people to engage in the very behaviours that contribute to these health problems, such a eating large quantities of fattening foods, smoking cigarettes, and avoiding exercise. Learning to cope with stress, solve personal problems, develop strong social networks, and modify lifestyle can help people live healthier and more meaningful lives.

Summarizing the Principles of Emotions and Health

How Are Emotions Adaptive?

1. **Facial expressions communicate emotion:** The perception of the nonverbal communication of basic emotions in facial expression is cross culturally similar. However, culture governs the rules for the display of emotions.
2. **Emotions serve cognitive functions:** Emotions serve as heuristic guides in decision making. Emotions command attention and act as somatic markers in making choices. Emotions also aid memory.
3. **Emotions strengthen interpersonal relations:** In relationships, guilt can prevent harmful behaviour, demonstrate caring, and can be used to manipulate. Interpersonal emotions develop out of socialization. Blushing is uniquely human and communicates apology. Jealousy signals commitment.

How Do People Experience Emotions?

4. **Emotions have a subjective component:** Primary emotions are adaptive; secondary emotions are a blend of primary ones. The circumplex model posits that emotions vary by valence and activation. There are neurochemical responses associated with emotions.
5. **Emotions have a physiological component:** The James-Lange theory of emotion states that physical changes are interpreted as specific emotions. According to the facial feedback hypothesis, facial expressions trigger the experience of emotions. The Cannon-Bard theory of emotion states that emotion-producing stimuli simultaneously activates physical and emotional reactions.
6. **Emotions have a cognitive component:** Stanley Schachter's two-factor theory of emotion proposes that a situation evokes a physiological response that is cognitively interpreted with an emotional label. Therefore, people may misattribute arousal states, and transfer their excitement to a new event.
7. **People regulate their moods:** People cognitively appraise and frame events in order to cope effectively. Humour is effective for regulating negative mood. Rebound may occur from emotional suppression. Rumination may prolong negative mood.

What Is the Neurophysiological Basis of Emotion?

8. **Emotions are associated with autonomic activity:** Emotions cause changes in heart rate, blood pressure, and skin temperature. Facial expressions, through facial musculature, may control the warming or cooling of the brain.
9. **The amygdala and the prefrontal cortex are involved in emotion:** The amygdala processes stimuli and generates an immediate emotional response. It deciphers the emotional content of facial expressions. The prefrontal cortex processes emotional cues related to interpersonal experiences.
10. **Emotion systems are lateralized in the brain:** Cerebral asymmetry exists with regard to emotions (i.e., the left brain processes positive emotions and the right brain processes negative ones).

How Do People Cope with Stress?

11. **The general adaptation syndrome is a bodily response to stress:** Hans Selye's general adaptation syndrome model consists of the alarm stage, resistance stage, and exhaustion stage—the three-stage response to stress. Stress has physiological components.
12. **People encounter stress in daily life:** Major life stressors are disruptions that strain people's lives. The more minor, but ubiquitous, daily hassles can have a cumulative effect on health.
13. **Stress affects health:** Stress and negativity increase the likelihood of heart disease. The Type A behaviour pattern of personality traits is a strong predictor of heart disease. Stress compromises the immune system.
14. **Coping is a process:** People engage in primary appraisals of stressors before the secondary appraisal of coping strategies. Emotion-focused coping changes the emotional responses to stress; problem-focused coping seeks to solve the problem. Positive reappraisal helps people deal with stress. Hardiness is a personality characteristic that is helpful in stress resistance. Social support is a stress buffer.

What Are Some of the Behaviours That Affect Physical Health?

15. **Obesity results from a genetic predisposition and overeating:** Genetics sets body mass index range but the environment determines the final values. Culture determines whether obesity is stigmatized.
16. **Restrictive dieting can be problematic:** Set-point values regulate weight gain after restrictive dieting. Eating behaviour that breaks a diet often leads to excessive eating. Disordered eating can lead to anorexia nervosa and bulimia nervosa.
17. **Smoking often begins in childhood:** Social influences are the leading cause of adolescent smoking. "Being a smoker" is associated with some positive qualities and these perceptions can influence smoking onset behaviour. Genetics influence nicotine addiction.
18. **People do not get enough exercise:** In general, the more people exercise, the better their physical and mental health. Older adults who exercise regularly retain mental functioning longer.

Applying the Principles

1. While traveling to different, remote parts of the world, you notice that people display the same basic facial expressions as you do when feeling happiness, sadness, fear, and anger. This also happens to be true of infants from different cultures. Which of the following explains this phenomenon?

    ____ a) emotions are taught to everyone at birth
    ____ b) emotions are innate and adaptive because they signal information to others
    ____ c) there are only four basic emotions; all the others emotions are combinations
    ____ d) cultures have little influence on emotional display

2. You just found out that your essay on civil engagement on campus has been accepted for publication in your college paper! You are smiling and happy. Determining what accounts for the "happy" feeling involves which two components of emotional experience?

    ____ a) physical and cognitive
    ____ b) primary and secondary
    ____ c) emotion and appraisal
    ____ d) arousal and physiology

3. As you are walking home, you hear the sound of gunshots. You are startled and flee in fear. What neurological responses have occurred?

    ____ a) an autonomically governed startle response; a fast-path amygdala response that generated a fear response; a slow-path response that confirmed the fear, thus fleeing was warranted
    ____ b) the thalamus generated an autonomic startle response; a fast-path autonomic response caused you to flee
    ____ c) a sympathetic response generated by the amygdala created fear; a slow parasympathetic response evaluated the fear response and commanded your body to flee
    ____ d) a sympathetic response governed the startle response; the thalamus generated fear; the amygdala evaluated the fear and activated the flee response

4. You enter the classroom a bit late and find everyone is already seated, with a pencil and test form on each desk. You realize that you forgot there is an exam today! After settling in your seat you calm down, gather your thoughts, and do your best to answer the exam questions. This response is consistent with which sequence of the general adaptation syndrome stages?

    ____ a) alarm, avoidance, exhaustion
    ____ b) resistance, exhaustion, termination
    ____ c) alarm, resistance
    ____ d) avoidance, termination

5. Jenna has just joined her junior high school's committee to develop a program to keep kids from smoking, and she asks her older sister Arlene for help. Based on what Arlene has learned from her psychology class, which of the following is likely to be the most effective recommendation?

   ____ a) to tell kids how expensive it is to smoke
   ____ b) to deliver a message showing that smokers are not cool because they are addicts
   ____ c) to explain the health consequences of smoking and the physiological aspects of addiction
   ____ d) to have the anti-smoking message delivered regularly by classroom teachers

ANSWERS: 1. b 2. a 3. a 4. c 5. b

Key Terms

alexithymia, p. 395
anorexia nervosa, p. 422
body mass index (BMI), p. 419
buffering hypothesis, p. 418
bulimia nervosa, p. 422
Cannon-Bard theory of emotion, p. 398
cerebral asymmetry, p. 409
circumplex model, p. 396
coping response, p. 410
display rules, p. 390
emotion, p. 387
emotion-focused coping, p. 416
excitation transfer, p. 400
facial feedback hypothesis, p. 397
fight-or-flight response, p. 410
general adaptation syndrome, p. 412
guilt, p. 393
hardiness, p. 417
health psychology, p. 388
immune system, p. 415
James-Lange theory of emotion, p. 397
lymphocytes, p. 415
mood, p. 387
positive reappraisal, p. 416
primary appraisal, p. 416
primary emotions, p. 396
problem-focused coping, p. 416
rumination, p. 402
secondary appraisal, p. 416
secondary emotions, p. 396
social support, p. 417
somatic markers, p. 392
stress, p. 388
stressor, p. 410
tend-and-befriend response, p. 411
two-factor theory of emotion, p. 398
Type A behaviour pattern, p. 414
Type B behaviour pattern, p. 414

Further Readings

Damasio, A. R. (1999). *The feeling of what happens.* New York: Harcourt Brace.

Ekman, P., & Davidson, R. J. (1994). *The nature of emotion: Fundamental questions.* New York: Oxford University Press.

Glaser, R., & Kiecolt-Glaser, J. K. (1994). *Handbook of human stress and immunity.* San Diego: Academic Press.

Lane, R. D., Nadel, L., & Ahern, G. (1999). *Cognitive neuroscience of emotion.* New York: Oxford University Press.

LeDoux, J. E. (1996). *The emotional brain: The mysterious underpinnings of emotional life.* New York: Simon & Schuster.

Panksepp, J. (1998). *Affective neuroscience: The foundations of human and animal emotions.* New York: Oxford University Press.

Sapolsky, R. M. (1995). *Why zebras don't get ulcers.* New York: Freeman.

Sternberg, E. M. (2000). *The balance within.* New York: Freeman.

Further Readings: A Canadian Presence in Psychology

Adamec, R., & Shallow, T. (2000). Effects of baseline anxiety on response to kindling of the right medial amygdala. *Physiology & Behavior, 70*, 67–80.

Anisman, H., & Merali, Z. (2003). Cytokines, stress and depressive illness: Brain-immune interactions. *Annals of Medicine, 35*, 2–11.

Birmingham, C., Muller, J. L., Palepu, A., Spinelli, J. P., & Anis, A. H. (1999). The cost of obesity in Canada. *Canadian Medical Association Journal, 160*, 483–488.

Frasure-Smith, N., & Lesperance, F. (2003). Depression and other psychological risks following myocardial infarction. *Archives of General Psychiatry, 60*, 627–636.

Livingston, K. E., & Hornykiewicz, O. (1978). *Limbic mechanisms: The continuing evolution of the limbic system concept.* New York: Plenum Press.

Selye, H. (1985). The nature of stress. *Basal Facts, 7*, 3–11.

Tewolde, S., Ferguson, B. S., & Benson, J. (2006). Risky behaviour in youth: An analysis of the factors influencing youth smoking decisions in Canada. *Substance Use and Misuse, 41*, 467–487.

Mysteries of Human Development

Developmental psychology is concerned with changes in physiology, cognition, and social behaviour over the course of a person's life span. Social development refers to how we became social beings: how we learn to interact with others. As the excited (and messy) crowd shown here at the famous Tomatina Festival in Bunyol, Spain, demonstrates, many different factors can influence and dictate "appropriate" behaviour.

CHAPTER

11

Human Development

In 1970, a young girl walked into a welfare office in Los Angeles, California, with her mother, who came seeking help after escaping an abusive husband. When a social worker saw the girl, who was later known as "Genie," she reported the situation to her supervisor, thinking the young girl suffered from autism. On closer examination they discovered that the girl was 13 years old, although at 4 feet 6 inches and 59 pounds she appeared much younger. Her physical state revealed that she had been severely neglected and abused. She could not hop, skip, climb, or do anything requiring the full extension of her limbs (Curtiss, 1977; Rymer, 1993). She was admitted to the hospital and then taken from her parents and placed in foster care.

Genie's early life had been a nightmare. She had been locked in a room for more than ten years, deprived of normal human contact (Figure 11.1). Her father, who was mentally ill, had kept her in a tiny, dark bedroom, tied to a chair, and caged in a crib at night. She had been beaten for making any noise

Outlining the Principles

What Shapes a Child?

- Development Starts in the Womb
- Brain Development Promotes Learning
- Attachment Promotes Survival
- Parental Style Can Affect Children's Well-Being

How Do Children Learn about Their Worlds?

- Perception Introduces the World
- Piaget Emphasized Stages of Development
- Infants Have Innate Knowledge
- Memory Improves over Childhood
- Humans Learn from Interacting with Others
- Language Develops in an Orderly Fashion

How Do Children and Adolescents Develop Their Identities?

- Gender Roles Are Determined by Cultural Norms
- Friends Influence Adolescent Identity and Behaviour
- Identity Includes Moral Values
- People Define Themselves in Terms of Race and Ethnicity

What Brings Meaning to Adulthood?

- Adults Are Affected by Life Transitions
- Aging Can Be Successful
- Cognition Changes during Aging

11.1 Genie A photo of Genie in 1971.

and poorly fed. She had no one to talk to and nothing to listen to or even look at in her barren room. She was raised with essentially no stimulation from the external world. When she was rescued, Genie had a strange gait, almost like a rabbit's, and held her hands out in front of her as if they were paws. She understood just a few words and could form only brief sentences, such as "Stop it" and "No more." Genie also showed few signs of emotion or interest in connecting with those around her.

For the next four years, scientists from Children's Hospital in Los Angeles cared for Genie as they sought to understand the consequences of her social isolation. For decades psychological scientists had struggled to understand the extent to which development was affected by nature and nurture, how much of who we are as humans is hard wired in our genes or is the result of early experience. What is human nature when it is stripped of society and culture? No ethical experiment could be conducted to answer these questions. Could Genie provide insight into what makes us human? Would Genie be able to learn? Could she develop normal social skills that would allow her to become a full member of society? Would a warm, nurturing environment help her recover from her tragic past? If the ability to acquire language is inborn, could she quickly learn how to speak so that she could describe what had happened to her?

In the first few years that Genie was cared for by the scientists, she made some progress in forming social relationships and acquiring minimal language. But as we will explore in greater detail later in the chapter, Genie was able to develop only the most rudimentary language skills. She learned many words, but she was not able to put them together properly in sentences. Sadly, when Genie turned 18 her mother was able to regain custody of her and immediately cut off all contact with the scientific community. Genie lived with her mother for only a short time before she was removed once again and sent to various foster homes. In at least one home she was again abused. She now lives in a small group home for adults who cannot look after themselves. ■

developmental psychology The study of changes in physiology, cognition, and social behaviour over the life span.

The case of Genie is tragic. We will never know what role her social and physical abuse played in the enduring difficulties that she had in interacting with others and acquiring language after age 13. Yet her case provides some tantalizing demonstrations of the role of experience in shaping brain development and psychological capacities. **Developmental psychology** is concerned with changes in physiology, cognition, and social behaviour over the life span. For instance, to study how humans develop

1920s
Cognitive Development View Swiss psychologist Jean Piaget proposes four distinct stages of how "thinking" develops in the brain. This theory guides research for nearly half a century.

1920s
Cultural Context and Cognition Russian psychologist Lev Vygotsky develops the first major theory of how cultural and social context influences cognitive and language development.

1950s
Motherless Monkeys Harry Harlow's experiments with infant rhesus monkeys and their artificial surrogate mothers demonstrate the importance of contact comfort in social development.

1957
The Psycholinguistic Revolution Noam Chomsky transforms linguistics by positing that language must be governed by a universal grammar, a set of rules and principles built into the brain.

cognitive skills, scientists have focused on age-related changes in psychological capacities such as perception, language, and thinking. We will see that the mind develops in ways that are adaptive, as new, useful skills appear at appropriate times, even in the absence of specific training.

This chapter also examines how we develop as social beings. Social development refers to the maturation of skills or abilities that enable us to interact with others. In all cultures and across all time periods, every person's life is touched by the lives of others. Infants innately form bonds with others, an adaptive trait that provides protection and facilitates survival. As children grow, they learn how to communicate with and behave appropriately around others, and how to establish and maintain relationships. In this chapter, we consider how socialization affects human characteristics such as morality, gender, and identity. As we explore these specific aspects of development, we will also examine some of the most fundamental questions about humanity: How do we develop into members of society? How do we develop into who we are and come to value what we value? How do genes interact with early experiences to produce unique individuals? How do we grow and adapt within our own cultures? How do we change across the life span? These are the processes that shape us into who we become as human beings.

What Shapes a Child?

How does culture influence development?

How does the human body grow and change to allow us to perform life's essential functions? How do our physical and social environments influence who we are as we develop over the course of childhood? For the most part, human development follows a predictable progression. Humans all do certain things, such as physically grow and mature at about the same period in the life span. During the prenatal period, the body develops in a fixed sequence, ultimately turning into a male or female infant based on genetic instructions. No newborns talk immediately, nor do any babies walk before they can sit up. Virtually all human babies make eye contact quickly after birth; display their first social smile at around six weeks; learn to roll over, sit up, crawl, stand, walk, and talk in that order (occasionally a step may be skipped or reversed) within a predictable range of ages (Figure 11.2). The consistency of this pattern suggests that our genes set the pace and order of development. However, the environment also influences what happens throughout the development process. For example, children raised in different cultures often achieve developmental milestones at a different pace. For instance, healthy children in Uganda tend to walk by 10 months, whereas those in France often do not walk before 15 months (Berger, 2004). These differences are due in part to different patterns of infant care across cultures, such as whether the infant is carried around during the day, whether the infant sleeps with the parents, or even whether the infant sleeps on her back or on

1960s
Infant Perception and Cognition Pioneering researchers such as Robert Fantz, Paul Eimas, Jacques Mehler, and Tom Bever demonstrate that many perceptual and cognitive abilities are hardwired.

1960s
Attachment John Bowlby proposes that infants are born with proximity-seeking behaviours that act to increase survival odds by keeping the child close to the mother.

1967
Critical Period Hypothesis Eric Lenneberg proposes that there is a biological time period during which environmental input is necessary for a cognitive skill to develop. The term *sensitive periods* is now used for humans.

1968
Stages of Identity Erik Erikson proposes a theory of identity development that considers development in eight stages, from infancy to old age.

1969
Moral Reasoning Lawrence Kohlberg uses dilemmas to examine moral reasoning. His stage model is later criticized for not applying equally across genders and cultures.

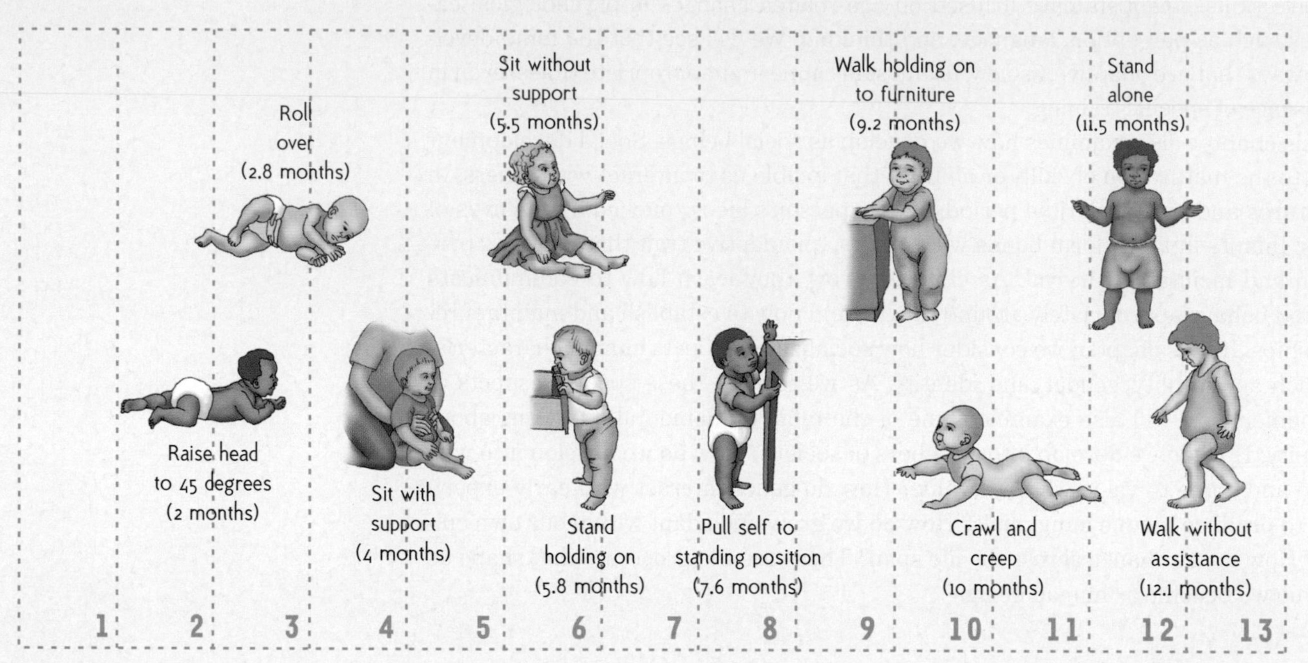

11.2 The Progression of Human Physical Development A baby learns to walk without formal teaching. Learning to walk progresses along a fixed, time-ordered sequence characteristic of all humans.

her stomach. During the past two decades, parents in North America have been advised to place their infants to sleep on their backs, in order to minimize the risk of suffocation. Because it is relatively difficult for infants to turn over, those who sleep on their backs crawl much later than those who sleep on their stomachs. In some other countries, infants are seldom left on the ground to crawl because of hygiene issues. These parental practices influence how motor skills develop. An important lesson from Chapter 3 is that the environment determines which of the genes you possess become expressed and makes you the person you are. The whole nature-nurture dichotomy is now recognized as false—people are the products of both nature and nurture. Genes and experience depend on and affect one another. Without an environment to trigger gene expression, a human does not develop into a fully functioning member of human society. Here we examine how various physical and environmental forces work together to shape early human development.

Development Starts in the Womb

From the moment of conception through birth approximately nine months later, remarkable developments occur. For the first two months, the developing human is known as an *embryo*. During this stage the internal organs, such as heart, lungs, liver

1978
Theory of Mind David Premack and Guy Woodruff coin the term *theory of mind,* a concept that later becomes central to theories of social and cognitive development.

1980s
Gender Roles and Schema A variety of researchers demonstrate that many assumed differences between males and females are the result of gender stereotypes that are transmitted though socialization.

1980s
Innate Cognition Cognitive developmental psychologists such as Rochel Gelman, Dan Osherson, Susan Carey, Renee Baillargeon, and Elizabeth Spelke propose that from birth, children develop skills independently of one another rather than as part of a progression.

1990s–2000s
Intentional Stance Researchers such as Michael Tomasello demonstrate that young infants recognize that people's actions reflect intentions.

2000s
Socioemotional Selective Theory Laura Carstensen demonstrates that many older adults are surprisingly satisfied, in part because they focus more on positive stimuli than do younger people.

and kidneys, the sex organs, and the nervous system begin to form. After two months, the growing human is called a *fetus* and at this time undergoes a great deal of physical growth as the whole body takes its infant form. The final trimester puts the finishing touches on the human, which if healthy is now capable of survival outside the womb.

PHYSICAL DEVELOPMENT Genes govern much of the prenatal development of the human nervous system. Most of the brain's nerve cells develop in a specific sequence in the first seven months of gestation (Rakic, 2000). The forebrain, midbrain, and hindbrain areas begin to form by week 4. The cells that will form the cortex are visible by week 7; the thalamus and hypothalamus are visible by week 10; the basal ganglia and the left and right hemispheres by week 12. By the seventh month, the fetus has a working nervous system. By birth, the brain is complex: it has cortical layers, neuronal connectivity, and myelination. Yet it doesn't stop there—the brain continues to develop throughout childhood into adulthood.

Hormones that circulate in the womb influence the developing fetus. For instance, if the mother's thyroid does not produce sufficient levels of thyroid hormones, the fetus is at risk for lower IQ and diminished intellectual development. The mother's emotional state can also affect the developing fetus. Although pregnancy is often trying, the fetuses of mothers who are unusually anxious and upset may be exposed to high levels of stress hormones, which may interfere with normal development and produce low birth weight and other negative cognitive and physical outcomes that can persist throughout life (Wadhwa, Sandman, & Garite, 2001). Another possible influence of maternal hormones on the developing fetus is that of androgens, such as testosterone, on sexual orientation (as you recall from Chapter 9). Simon Baron-Cohen and his colleagues (2004) have found that levels of testosterone in the womb are correlated with a range of behaviours later in childhood, from making eye contact to developing a vocabulary. They speculate that maternal hormones may play an important role in autism, a communication and cognitive disorder that you will learn more about in Chapter 13.

TERATOGENS Environmental influences may have adverse effects on the developing fetus. For instance, drugs, alcohol, or illness can all impair physical and cognitive development. These are examples of **teratogens**, which are agents (bacteria, viruses, chemicals, and drugs) that can cause abnormal development in the womb. The extent to which a teratogen causes damage depends on when the fetus is exposed to it, as well as the length and amount of exposure. For instance, exposure to a teratogen at about four weeks of age can interfere with the proper development of the brain. A tragic example of birth defects caused by teratogens occurred during the 1950s when women were prescribed the drug thalidomide to ease the symptoms of pregnancy. Thalidomide caused a variety of birth defects, especially limb deformities, the precise nature of which depended on when the mother took the drug.

teratogens Environmental agents that harm the embryo or fetus.

One of the most common teratogens is alcohol. Excessive consumption of alcohol during pregnancy can lead to *fetal alcohol syndrome,* the symptoms of which consist of low birth weight, face and head abnormalities, slight mental retardation, and behavioural and cognitive problems (Janzen et al., 1995; Loney et al., 1994). The Government of Canada reports that there are about 300,000 people with fetal alcohol syndrome in Canada. Although fetal alcohol syndrome is much more likely among infants of women who drink heavily during pregnancy, especially if they binge drink, some evidence suggests that even small amounts of alcohol can be problematic. For that reason, many health workers recommend abstinence from alcohol during

Why do teratogens have different effects depending on the stage of pregnancy?

pregnancy or for those women who are trying to become pregnant. Smoking cigarettes during pregnancy can lead to low birth weight, as well as spontaneous abortion or birth defects. Physical effects of exposure to certain teratogens may be obvious at birth, but language or reasoning disorders may not become apparent until the child is older (Fried, 1989).

Brain Development Promotes Learning

Newborns come into the world able to see, smell, hear, taste, and respond to touch. Although these skills are not fully developed at birth, the newborn is capable of processing some sensory stimuli. For instance, videotapes of two-hour-old infants revealed that they prefer sweet tastes to all other tastes (Rosenstein & Oster, 1988). Young infants also have a reasonably acute sense of smell, at least for those smells associated with feeding. In a number of different studies, infants turned their heads toward a pad containing their own mother's milk but not toward pads containing milk from other breast-feeding mothers (Winberg & Porter, 1998). The sense of hearing is also quite good at birth. Infants are startled by loud sounds and often will turn their bodies toward the source of the sound. The sense of hearing in newborns is much better than their sense of vision, which is quite limited. The range of their visual acuity is 8 to 12 inches, about the distance between the infant's face and the mother's face during breastfeeding. This limited visual range may be adaptive, since it encourages the infant to focus on what is most important, the mother's breast and face, and promotes the beginnings of social interaction for the child. As you will learn in the next section, the perceptual skills of newborns increase tremendously over the first few months of life.

Although newborn infants could not survive on their own, they are not completely helpless. Newborns have a variety of basic reflexes that aid survival. Perhaps you have observed the grasping reflex when a baby held your finger. Some believe this is a survival reflex that has persisted from our primate ancestors, since young apes grasp their mothers. Another innate reflex is the *rooting reflex,* the automatic turning and sucking that infants engage in when a nipple or similar object is near their mouths. Some theorists believe that these reflexes pave the way for learning more complicated behaviour patterns such as feeding oneself or walking. So the brain is sufficiently developed at birth to support basic reflexes, but it appears that further brain development is necessary for cognitive development to occur.

Early brain growth has two important aspects: (1) specific areas within the brain mature and become functional; and (2) regions of the brain learn to communicate with each other through synaptic connections. One important way that brain circuits mature is through myelination, which begins on the spinal cord during the first trimester and on the brain's neurons during the second trimester. Recall from Chapter 3 that myelination is the brain's way of insulating its "wires": Fibres are wrapped with a fatty sheath, much like the plastic coating around electrical wire, to increase the speed with which they are able to transmit signals. Myelination occurs in different brain regions at different stages of development and is believed to reflect the maturation of the fibres (Figure 11.3). Hearing and balance areas are fully myelinated at birth; areas involved in abstract thinking do not become fully myelinated until after age 20.

The myelinated axons form synapses with other neurons. The infant brain grows far more of these connections than it will ever use. Then something remarkable happens—the brain adopts a very strict "use it or lose it" policy. The connections that are frequently used are preserved; those that are not decay and disappear. This process is called **synaptic pruning**, and it occurs in different areas of the brain at different times. Once connections are established, the brain sets about making them

synaptic pruning A process whereby the synaptic connections in the brain that are frequently used are preserved, and those that are not are lost.

more permanent, for example, through increasing myelination. Some developmental psychologists believe that not until certain brain connections are made do infants develop specific cognitive skills. This idea that learning is constrained by, or at least closely tied to, brain development is a new and exciting approach that is energized by the biological revolution in psychological science.

In addition to growth that is determined by genetic instruction, the brain is also a highly "plastic" organ: Part of its hard wiring includes the ability to adapt to different environments (Kolb & Wishaw, 1998b). Though most neurons are already formed at birth, the brain's physical development continues through the growth of neurons and the new connections they make. The brain grows from about 350 grams to about 1,250 grams (about 80 percent of the adult size) by age 4. This size increase is due to myelination and to new synaptic connections among neurons. The plasticity of the developing brain has led researchers to some remarkable findings about the influence of early environment on the brain's physical structure. Recall from Chapter 4 that rats raised in enriched environments show evidence of greater brain development. Early childhood nutrition also affects myelination and other aspects of brain development. Children who are malnourished not only have less myelination, they might also lack the energy to interact with objects and people in their environments. This lack of stimulation also undermines brain development. Thus, although genes provide instructions for the maturing brain, how the brain changes during infancy and early childhood is also very much affected by the environment. Both nature and nurture matter in how the physical brain grows.

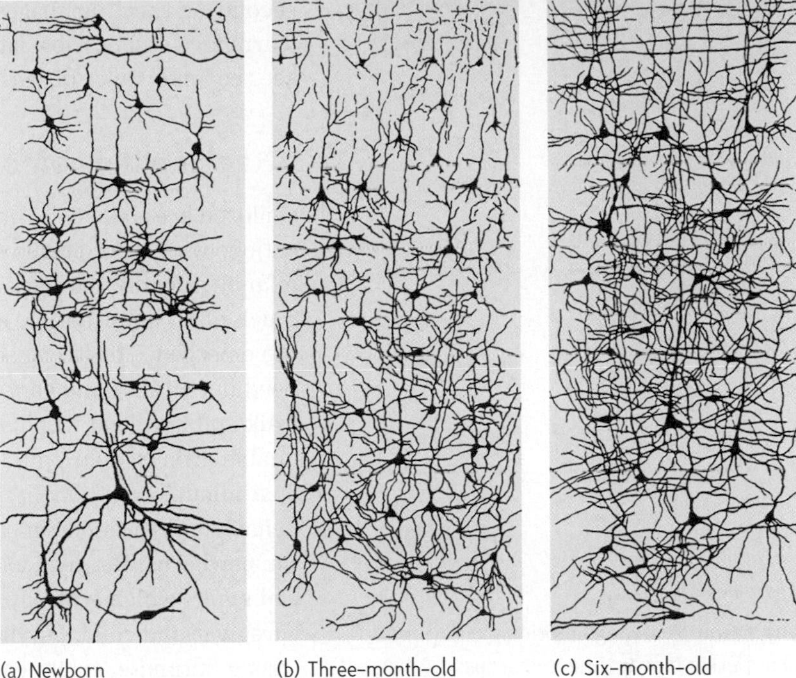

11.3 **Myelination** These drawings of neurons in the visual cortex represent the development of the human brain of (a) a newborn, (b) a three-month-old, and (c) a six-month-old.

CRITICAL LEARNING PERIODS A **critical period** is a development stage during which young animals are able to acquire specific skills and knowledge. If these skills and knowledge are not acquired during the critical period, they cannot be acquired at a later point in development. Psychological scientists believe that the key to learning is creating connections among certain neurons, and that critical learning periods may exist because brain development goes through periods during which certain connections are most easily made, assuming the right stimulus is provided.

critical period Time in which certain experiences must occur for normal brain development, such as exposure to visual information during infancy for the normal development of the visual pathways of the brain.

Consider Genie, who was unable to acquire full language, perhaps because she was not exposed to it early in life. The idea that there are biologically determined time periods when a child must be exposed to language in order to achieve normal brain development was termed the *critical period hypothesis* by Eric Lenneberg (1967). The theory states that environmental input is important, but biology determines when an organism needs to receive particular input in order to make use of it. In the case of language, Genie taught us that the critical period (generally thought to be before age 12 for language) is not so rigid as originally thought; she was able to learn some aspects of language at a later age. Yet her language was especially limited, which suggests that this time period was important. The specific points in development at which some skills are most easily learned are now referred to as "sensitive periods." In general, humans appear to have sensitive periods when the brain is primed to acquire certain skills or knowledge, such as the ability to learn second or third languages. These periods do not seem critical, however, because learning is still possible later, though it may not be as efficient or successful. We will see the importance of sensitive periods throughout our discussion of development.

Attachment Promotes Survival

Children are shaped not only by biological development, but also by their early interactions with other people—especially their caregivers. Social development begins in infancy, roughly the period between birth and 18 to 24 months. Caregivers shape much of an infant's early experience, from what the child eats to where it sleeps to the emergence of strong social connections. Like all young primates, human infants need nurturance and care from adults to survive. Unlike horses and deer, which can walk and find food within hours after birth, humans are born profoundly immature, unable even to hold up their own heads or roll over. But they are far from passive. Just minutes after birth, the cries of infants cause psychological, physiological, and behavioural reactions in caregivers that compel the offering of food and comfort to the newborns. Between four and six weeks of age most infants display their first social smile, which typically enhances powerful feelings of love and bonding between caregiver and child. Psychological scientists refer to this bond as an **attachment**, a strong, intimate, emotional connection between people that persists over time and across circumstances. Infant attachment leads to heightened feelings of safety and security. Even young infants have highly interactive social relationships; for example, infants are profoundly affected by the facial expressions of their caregivers within 10 weeks after birth and may become very upset when their mothers fail to display emotional reactions.

attachment A strong emotional connection that persists over time and across circumstances.

According to John Bowlby (Figure 11.4), who popularized its importance, attachment serves to motivate infants and caregivers to stay in close contact. Bowlby argued that infants have an innate repertoire of attachment behaviours that motivate adult attention. For instance, they put out their arms to be lifted, they smile when they see their caregivers, and they cry when they feel abandoned. Adults also seem to have innate predispositions to respond to infants, as when we pick up a crying child and rock him gently. Adults also automatically respond to infants in a way that infants can understand. Watch adults interacting with infants and notice that they make exaggerated facial expressions and pitch their voices higher than normal. Bowlby argues that these behaviours motivate both infants and caregivers to stay in proximity. Preferring to remain close to caregivers, acting distressed when they leave, and rejoicing when they return are all behaviours aimed at securing caregivers' attention and, ultimately, their protection. Thus, attachment is adaptive. Infants who exhibit attachment behaviours have a higher chance of survival and consequently are more likely to pass along their genes to future generations.

How does attachment aid survival?

11.4 John Bowlby Bowlby's studies of the bonds between children and their caregivers set the groundwork for decades of systematic research in the area of attachment.

ATTACHMENT ACROSS SPECIES The idea that attachment is important for survival applies not only to humans but also to many other species. For instance, infant birds communicate hunger through crying, thereby triggering action on the part of caregivers. Some birds seem to have a critical period in which they develop a strong attachment to whichever adult is nearby, even if that adult is not a member of their species! This pattern, first noticed in the nineteenth century, occurs for birds such as chickens, geese, and ducks, which are able to walk right after hatching and are therefore at risk for straying from their mother's care. Within about 18 hours after hatching, these birds will attach themselves, usually to their mothers, and then follow the object of their attachment. The noted ethologist Konrad Lorenz called such behaviour *imprinting* and noted that goslings that became imprinted on him did not go back to their biological mothers when later given access to them (Figure 11.5). However, such birds do preferentially imprint on a female of their species if one is available.

11.5 Attachment Konrad Lorenz shown walking with goslings that had imprinted themselves on him. These little geese followed Lorenz as if he was their mother.

During the late 1950s, Harry Harlow provided one of the most striking examples of nonhuman attachment. At that time, psychologists generally believed that infants needed their mothers because the mother was the primary source of food. For Freudians, the mother was the source of libidinal pleasures (see Chapter 12). From the behaviourist perspective, the mother was valued as the result of secondary reinforcement, given her role as the provider of food. But an explanation based on either Freudian thinking or learning theory was unsatisfactory to Harlow, who recognized that, in addition to food, infants needed comfort and security.

In a now-famous series of experiments, Harlow placed infant rhesus monkeys in a cage with two different "mothers." One surrogate mother was made of bare wire and could give milk through an attached bottle. The second surrogate mother was made of soft terrycloth and had a monkeylike head (rudimentary eyes, nose, mouth, and ears attached to a flat circle) but could not give milk. The monkeys' responses were unmistakable: They clung to the cloth mother most of the day and went to it for comfort in times of threat. The monkeys approached the wire mother only when they were hungry (Figure 11.6). Harlow tested the monkeys' attachment to the mother in various ways, such as introducing a strange object into the cage. He repeatedly found that the infants were calmer, braver, and overall better adjusted when near the cloth mother. Hence, the mother-as-food theory of mother-child attachment was debunked. Harlow's findings established the importance of contact comfort—allowing an infant to cling to and hold something soft—in social development.

11.6 Harlow's Monkeys and Their "Mothers" The infant rhesus monkey is clinging to its cloth mother, which it uses for contact comfort and security (left). The monkey still clings to the cloth mother while it takes milk from the wire mother (right).

11.7 Mary Ainsworth Ainsworth developed the Strange Situation Test, a 20-minute evaluation to observe the attachment between children and their caregivers.

ATTACHMENT STYLE Attachment behaviours begin during the first months of life and have been observed in children around the world. If Bowlby was correct that attachment encourages proximity between infant and caregiver, then we might expect attachment responses to increase when children naturally become mobile. Indeed, at around 8 to 12 months, just as children are starting to crawl, they typically display separation anxiety, in which they become very distressed when they can't see or are separated from their attachment figure. This pattern occurs in all human cultures.

To study attachment, Mary Ainsworth (Figure 11.7) developed the Strange Situation Test. The test involves observing the child, caregiver, and a friendly but unfamiliar adult in a series of eight semistructured episodes in a laboratory playroom. The crux of the procedure is a standard sequence of separations and reunions between the child and each of the two adults. Over the course of the eight episodes, the child experiences increasing distress and a greater need for caregiver proximity. The extent to which children cope with the distress and need and the strategies they use to do so are considered to indicate the quality of attachment. During the episodes, the child is observed through a one-way mirror, and actions such as crying, playing, level of activity, and attention to the mother and stranger are recorded. Using the Strange Situation Test, Ainsworth originally identified three types of child attachment:

Secure attachment applies to the majority of children (approximately 65 percent). A secure child is happy to play alone and is friendly to the stranger while the attachment figure is present. When the attachment figure leaves, the child is distressed, whines or cries, and shows signs of looking for the attachment figure. When the attachment figure returns, the child is happy and quickly comforted (Figure 11.8), often wanting to be held or hugged. The child then returns to playing (Pederson & Moran, 1996; Deoliveira et al., 2005).

Avoidant attachment applies to approximately 20 to 25 percent of children. Avoidant children do not appear distressed or upset by the attachment figure's departure. If upset, many of them can be comforted by the stranger. When the attachment figure returns, the child does not want a reunion but rather ignores or snubs the attachment figure. If the child approaches the attachment figure, it is often in a tentative manner.

Anxious-ambivalent attachment applies to approximately 10 to 15 percent of a given sample. A child with an anxious-ambivalent style is anxious throughout the test. The child clings to the attachment figure after entering the room; when the attachment figure leaves, the child becomes inconsolably upset. When the attachment figure returns, the child will both elicit and reject caring contact; for instance, the child may want to be held but then fight to be released. The child may cling to the attachment figure even while trying to hit the person. Ambivalence marks their relationship.

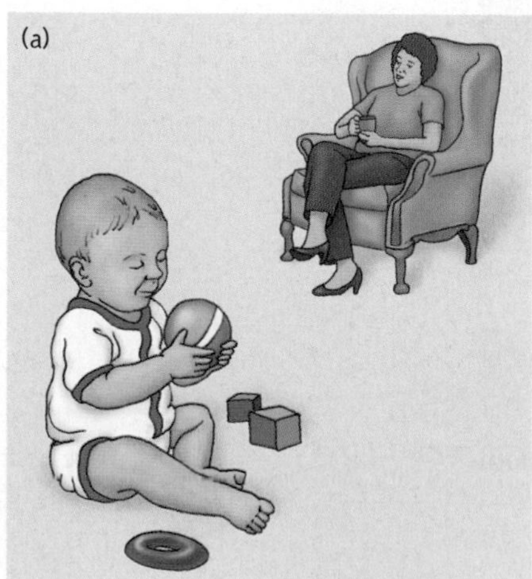
(a)

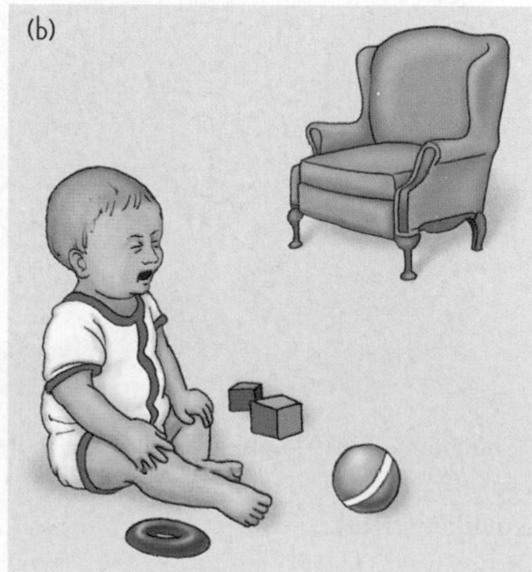
(b)

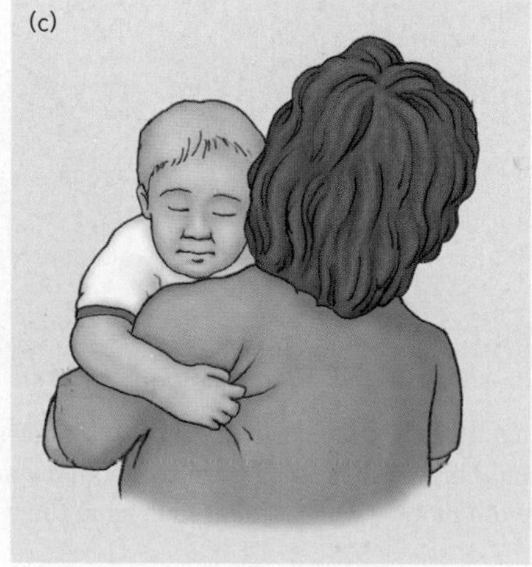
(c)

11.8 Scenes from the Strange Situation Test In (a), the caregiver engages in solitary activity while the child plays. In (b), the child cries and is distressed because the caregiver has left. In (c), the child touches and clings to the caregiver after her return.

Other researchers have identified variants of these attachment styles. For instance, some children show inconsistent or contradictory behaviours, such as smiling when seeing the caregiver but then displaying fear or avoidance. These children have been described as having disorganized attachment (Main & Solomon, 1986).

Researchers examining the role of the child's personality or temperament in determining attachment style have found that children with behavioural problems—such as those who rarely smile, who are disruptive, or who are generally fussy—are more likely to be insecurely attached (that is, anxious-ambivalent or avoidant). The caregiver's personality also contributes to the child's attachment style. Caregivers who are emotionally or behaviourally inconsistent tend to have children with an anxious-ambivalent attachment style, whereas those who are rejecting tend to have children with an avoidant attachment style.

CHEMISTRY OF ATTACHMENT Scientists working at the biological level of analysis have recently discovered that the hormone oxytocin is related to social behaviours, including infant-caregiver attachment (Carter, 2003). Oxytocin plays a role in maternal tendencies, feelings of social acceptance and bonding, and sexual gratification. In terms of the mother and infant, oxytocin affects both of them, promoting behaviours that ensure the survival of the young. For instance, in both animal and human studies, infant suckling triggers the release of oxytocin, which in turn leads to biological processes that move milk into the milk ducts so the infant can nurse. Oxytocin also facilitates infant attachment to the mother. Research using rat pups that have been separated from their mothers and later reunited with them has found that pups who formed an association between a specific odour and maternal reunion show a preference for that smell. However, the odour is not preferred among pups that have been given an oxytocin inhibitor, indicating that oxytocin is important for these attachment associations.

Parental Style Can Affect Children's Well-Being

Thinking about the biological and social influences on behaviour (nature-nurture), how important are caregivers in shaping the developing child? Do parents affect the well-being of their children? Most parents believe that their actions have profound effects on how their children turn out. Indeed, you can imagine how your parents would react if they were told that nothing they did had much influence on you. What might you be like if you had been raised by different people? Clearly parenting matters at the extremes, such as when children are abused. It is likely that Genie would have turned out quite differently had her parents not abused her. Here we will consider the evidence for how caregivers influence social development. Later in the chapter, we will examine the role of parents versus that of peers. We will see that some scholars believe that the importance of parents has been overemphasized.

One good demonstration of the significance of the child-parent interaction is the New York Longitudinal Study, begun in 1956 by Stella Chess and Alexander Thomas. The study ran for six years, assessing 141 children from 85 middle- to upper-middle-class families. Chess and Thomas focused on children's biologically based temperament as the most important aspect of the parent-child interaction (see Chess & Thomas, 1984). Temperament can be characterized as a person's typical mood, activity level, and emotional reactivity. Thus, fussiness on its own is not necessarily a signal of a child's temperament type; instead, the frequency of the fussiness as well as the intensity of the fussiness and how easily it can be controlled are better indicators (the concept of temperament is explored more fully in Chapter 12).

How important are parents to a child's developmental and behavioural outcomes?

"They got extinct because they didn't listen to their mommies."

Chess and Thomas found that the "fit" between the child's temperament and the parents' behaviours is most important in determining social development. For instance, consider difficult children, who tend to have negative moods and a hard time adapting to new situations. Most parents find it frustrating to raise such children. Parents who openly demonstrate their frustration or insist on exposing the child to conflict often unwittingly encourage negative behavioural outcomes. For example, if the child is extremely uneasy about entering a new setting, pushing the child can lead to behavioural problems. If the child is very distractible, forcing him or her to concentrate for long periods of time may lead to emotional upset. In the study, parents of difficult children who responded in a calm, firm, patient, and consistent style tended to have the most positive outcomes. These parents did not engage in a lot of self-blame for their children's negative behaviours, and they managed to cope with their own feelings of frustration and disappointment. Chess and Thomas also noted that overprotectiveness can encourage a child's anxiety in response to a new situation, thereby escalating the child's distress. Ultimately, then, the best style of parenting takes into account the parents' own personalities, the child's temperament, and the situation; it is a dynamic style that emphasizes flexibility.

Other research has shown that parents have multiple influences on their children, such as on their attitudes, values, and religious beliefs. Children learn about the world in part from the attitudes expressed by parents, such as prejudice toward certain groups of people. We will see later that parents who are especially nurturing tend to raise children who experience more social emotions, such as guilt, perhaps because an empathetic attitude toward others is encouraged. Parents also determine the neighborhoods in which children live, the schools that they attend, and the extracurricular activities that provide exercise and stimulation—all of which are likely to influence the child in subtle and not-so-subtle ways. Thus, parents who avoid television, regularly go on family hikes, and participate in a variety of sports may have children who are more physically fit than those who plunk their children down in front of the television for hours per day. Of course nature might contribute to this as well. Athletes possess genes that promote fitness, which they pass along to their children. The child might respond positively to being active because she has a genetic makeup to like exercise and to benefit from it. Again, we see the interaction between genes and environmental influences.

Parents play an important role in shaping the way their children view themselves as members of society. Now that we've looked at some of the larger biological and social forces that shape young children, let's next look at how they develop an understanding of the world around them.

Thinking Critically

Does Divorce Harm Children?

The rise in divorce rates since the 1960s has prompted concern about divorce's effects on children. More than one-third of all marriages in Canada end in divorce. The overall picture that emerges from research suggests that divorce is associated with numerous problems for children (Sareen et al., 2005). Although some cope very well with the divorce of their parents—especially children who are intelligent, socially mature, and responsible—the experience can be quite difficult for many (Hetherington, Bridges, & Insabella, 1998). Compared with those whose parents stay together, children whose parents divorce tend to do less well in school, have more conduct disorders and psychological problems, and have poor social relations and low self-esteem (Amato, 2001). Children of divorce are also more likely to become divorced themselves, suggesting that the negative effects of divorce continue after childhood—although another possible explanation for the perpetuation of divorce in families is that genes lead to temperaments that predispose divorce, and that parents pass those genes along to their children. Either way, divorce is associated with negative outcomes.

Let's begin by thinking about possible explanations for why divorce is associated with negative outcomes. Divorce may damage a child's relationship with one or both parents, such as when the parent with custody moves far away from or restricts contact with the noncustodial parent. The child then loses a potentially important source of emotional support and guidance. But it could also be that children who live in households filled with conflict have psychological problems whether or not their parents stay together. Perhaps people who get divorced differ in important ways from those who stay married, such as being more irritable or depressed or having difficulty coping with conflict, and these personal factors may interfere with their ability to be effective parents. Maybe all of these things contribute to negative outcomes for the children of divorce. So what does research show about why divorce is harmful to children?

The past 30 years have seen a significant decline in two-parent families. It is possible that being raised by only one parent is simply not as good for children. This is important because approximately 10 percent of all children born in Canada have unwed mothers. If having one parent is not as good as having two, then children who grow up in single-parent homes or who have a parent die should experience similar outcomes to those whose parents divorce. But the evidence indicates that children of divorce have more problems than children who lose a parent through death, although both groups fare more poorly than those who grow up with both parents (Amato & Keith, 1991). Children who are raised by single mothers, though, appear to share many of the problems of children of divorce (Clarke-Stewart, Vandell, McCartney, Owen, & Booth, 2000). One study found that girls raised without fathers were much more likely to initiate early sexual activity and were much more likely to become pregnant at a young age (Ellis et al., 2003). The overall evidence indicates that the absence of a biological father is associated with a number of negative outcomes (Dunn, 2004). Interestingly, however, having a stepfather does not resolve these problems—if anything, living with a stepparent may be associated with an increase in psychological problems (Amato & Keith, 1991). Thus, two parents are not always better than one.

Why might children of divorce and children raised by single parents be at a disadvantage compared with children from two-parent homes? An alarming number of single mothers live in poverty, and many rely on assistance to provide for their children's basic needs. For custodial mothers and their children, divorce is often associated with a significant decline in financial resources, and such economic decline predicts at least some of the negative outcomes of divorce on children. Mothers threatened with poverty need to devote considerable time and energy to obtaining money, which may leave them feeling so overburdened, stressed, and hassled that they have little time or energy for parenting. When the father stays involved in the family after the divorce, both economically and emotionally, the effect of divorce is reduced considerably (Pett, Wampold, Turner, & Vaughan-Cole, 1999).

The economic consequences of divorce may make some think it is better to stay married for the sake of the children. Is that a good idea? Just because people do not get divorced does not mean that they are happily married. Frequent arguments, fighting, and even physical abuse characterize some long-term marriages. The available evidence suggests that living in a high-conflict family is associated with even greater negative outcomes than parental divorce (Amato & Keith, 1991). If it is not divorce but living in conflict that causes problems, then children should show the negative outcomes associated with divorce well before the parents separate. Longitudinal studies do offer evidence of greater behavioural and emotional problems among children whose parents later divorce compared with those whose parents stay together (Clarke-Stewart et al., 2000). This seems to indicate that living in homes with conflict inflicts psychological harm on children well before divorce. And children exposed to parents who physically abuse one another may be especially likely to develop emotional and psychological problems.

Do you think that parents should stay together for the sake of their children? Why is being a single parent associated with negative outcomes? Why are some children more affected by divorce than others? ■

REVIEWING the Principles | What Shapes a Child?

The human genome contains instructions for building a functioning human being, but from the earliest moment of development environmental factors influence how that human is formed. Throughout the prenatal period, various environmental agents, from the mother's hormones to substances she consumes, can alter the formation of the physical body and cognitive capacities. Once a child is born, learning is constrained by development of the brain and body. Except in cases of abuse, children crawl and walk when their bodies develop the appropriate musculature and when the brain matures sufficiently to coordinate motor actions. Almost from the moment of birth humans are social creatures, forming bonds of attachments with caregivers. The quality of this attachment, as well as the child's attitudes, values, and beliefs, are shaped by interactions with their caregivers. Parents have an important influence over many aspects of a child's life.

How Do Children Learn about Their Worlds?

Children need to learn a great deal in the first few years of life. Despite the fact that you were once an infant and have since learned how to walk, talk, read, reason, and do countless other things, you most likely have very little knowledge about how you acquired those skills. For years, psychologists have been trying to explain how children learn about their worlds. Do children learn through their interactions with the environment? Or are they born with innate cognitive abilities? Throughout psychology's history, scientists have carried on a vigorous debate regarding the contributions of nature and nurture. Although nearly everyone agrees that both are important, current research focuses on how genes and experience interact to produce the people that we become. In this section, working at the individual level of analysis, we consider the development of cognitive skills of young children.

Perception Introduces the World

To learn, children need to obtain information from the world. The principal way they do this is through their senses. Recall that newborns have all their senses at birth, even if some are poorly developed. But how can we tell what a baby perceives? Psychological scientists have devised clever experiments for gauging what infants know about objects in their environments. These experiments are based on two simple observations about infant behaviour: First, infants tend to look more at stimuli that interest them, and second, they will look longer at novel (new) stimuli than at familiar stimuli. Thus, the *preferential looking technique* is used in many perceptual tests. In these tests, infants are shown two things. If they look longer at one, researchers know that infants can distinguish between the two things. Thus, experiments can be designed to measure whether or not an infant treats two stimuli as different and if so which stimulus an infant "chooses."

How does preferential looking demonstrate infant skills?

Other experiments are based on the **orienting reflex**, which is the tendency of humans, even from birth, to pay more attention to novel stimuli than to stimuli to which they have become *habituated,* or grown accustomed (Fantz, 1966). This means that if you time the number of seconds an infant looks at things, you will find that the infant looks away more quickly from something familiar than from something unfamiliar or puzzling. Using *habituation* allows researchers to create a response preference for one stimulus over another. An infant is shown a picture or an object until she is familiar enough with it that the amount of time she looks at it declines (The infant is bored by or adapts to the stimulus). Once the infant is habituated to the stimulus, researchers can measure whether she reacts to a change in the stimulus: If a new stimulus is now shown, does the infant look longer at the novel stimulus? If so, it indicates that she notices a difference between the two stimuli. If the amount of time the infant looks at the new stimulus is the same as the amount of time she looks at the old stimulus, it is assumed that she does not distinguish between the two. Such tests are used to gauge everything from infants' perceptual abilities—how and when they can perceive colour, depth, and movement, for instance—to their understanding of words, faces, numbers, and laws of physics. So what do we know about infants' early perceptual abilities? What can they learn about their worlds?

orienting reflex The tendency for humans to pay more attention to novel stimuli.

VISION The ability to distinguish differences among shapes, patterns, and colours develops early in infancy (Montreal Symposium, 1994). The preferential looking technique is used to determine an infant's visual acuity—how well an infant can see. Have you ever wondered why mobiles and other playthings for infants are covered with bold black-and-white patterns rather than the colourful images we associate with children's toys and books? Such patterns are used because developmental psychologists have discovered that infants respond more to them than to other stimuli (Figure 11.9). In the early 1960s, Robert Fantz and other developmental psychologists showed infants patterns of black-and-white stripes as well as patches of gray and observed the infants' reactions. The mother was asked to hold the infant in front of a display of the two images (Figure 11.10). Then the experimenter, not knowing which image was on which side, would observe through a peephole to see where the infant preferred to look. It was discovered that infants look at bold stripes more readily than at gray images. The smaller the stripes get—that is, the less contrast between the images—the more difficult it becomes for infants to distinguish them from the gray patches. When infants look at both images equally, it is assumed that they cannot tell the difference between the two. Research has indicated that though infants have poor visual acuity for distant objects when they are first born, it increases rapidly over the first six

11.9 Vision in Infancy Robert Fantz was the first to determine that infants prefer bold patterns. A more thorough understanding of infants' visual abilities has led parents to buy mobiles and toys that use some of Fantz's testing patterns.

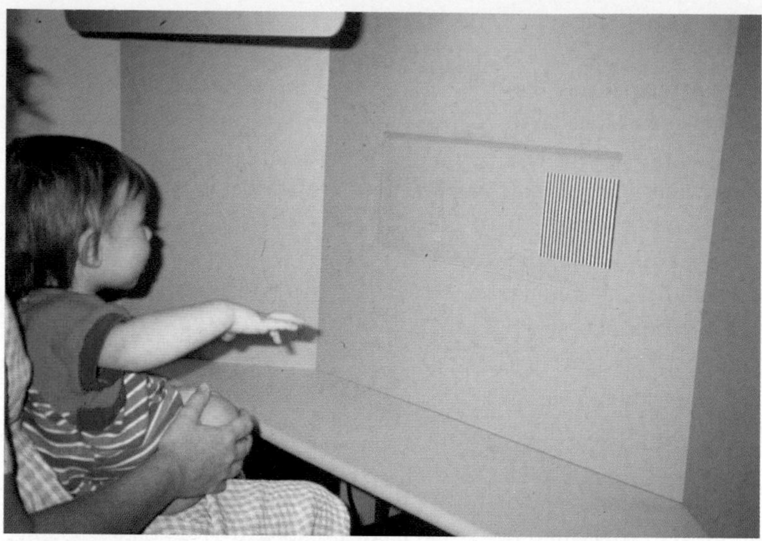

11.10 Testing Visual Acuity in Infants An infant's visual acuity being tested by the preferential looking procedure. The mother holds the infant in front of the display. From the other side of the display, an experimenter looks through a peephole (barely visible between the two stimuli) and notes whether the infant is looking to the right or to the left.

months (Teller, Morse, Borton, & Regal, et al., 1974). Adult levels of acuity are not reached until about one year. The increase in visual acuity is probably due to the development of the infant's visual cortex as well as the development of cones in the retina. Recall from Chapter 5 that the cones are important for detecting detail. Thus, as the cortex and cones develop over the first six months, the infant's vision becomes more capable of perceiving visual detail.

To assess when depth perception emerges in infancy, the perceptual psychologist Robert Fox and his colleagues (1980) showed infants stereograms. Recall that stereograms work because we see one view of an image with one eye and another view with the other, then convert this information into depth perception. If infants cannot use the disparity information to perceive depth, they will see only a random collection of dots. To determine whether infants can see stereograms, Fox devised an experiment in which a baby wearing special viewing glasses looks at a screen while seated on the parent's lap. If the baby has binocular disparity, he should see a three-dimensional rectangle moving back and forth, and he will presumably follow the movement of the rectangle with his eyes. If he does not have binocular disparity, he will see only dots and will not be able to follow the rectangle's movement (Figure 11.11). Fox's results indicate that the ability to perceive depth develops between three and a half and six months of age.

AUDITORY PERCEPTION IN INFANTS Newborns can hear and appear capable of locating a sound's general source. When infants are presented with rattle sounds in their right or left ears, they will turn in the direction of the sound, indicating that they have perceived it (Kelly, 1992). Detailed analysis of the tone levels that infants can hear indicates that by six months babies have nearly adult levels of auditory function (DeCasper & Spence, 1986).

Infants also seem to have some memory for sounds. Using habituation techniques, researchers have determined that infants can recognize sounds they have heard before. By measuring an infant's rate of sucking on a rubber nipple, researchers are able to determine if the infant is aroused in response to a specific sound. In an experiment by Anthony DeCasper and William Fifer (1980), two-day-old infants wore earphones and were given a nipple linked to recordings of their mother's voice and a stranger's voice. If they paused longer between their sucking bursts, the mother's voice played. If they paused for a shorter time, the stranger's voice played. Even at this young age, the newborns altered their sucking patterns to hear their mother's voices more often. In a similar vein, these researchers had one group of pregnant women read *The Cat in the Hat* aloud. A second group read the book aloud but replaced the words "cat" and "hat" with "dog" and "fog." In subsequent tests, the women's newborns changed their sucking patterns to hear the version they had heard in the womb. These experiments, along with others, provide compelling evidence that infants learn their mothers' voice in the womb. These developing sensory capacities allow infants to observe and evaluate objects and events around them. They then use the information gained from perception to try to make sense of how the world works (Eggermont, 2001).

11.11 Fox's Experiment for Binocular Disparity If the infant can use disparity information to see depth, he sees a rectangle moving back and forth in front of the screen and will respond by visually tracking it.

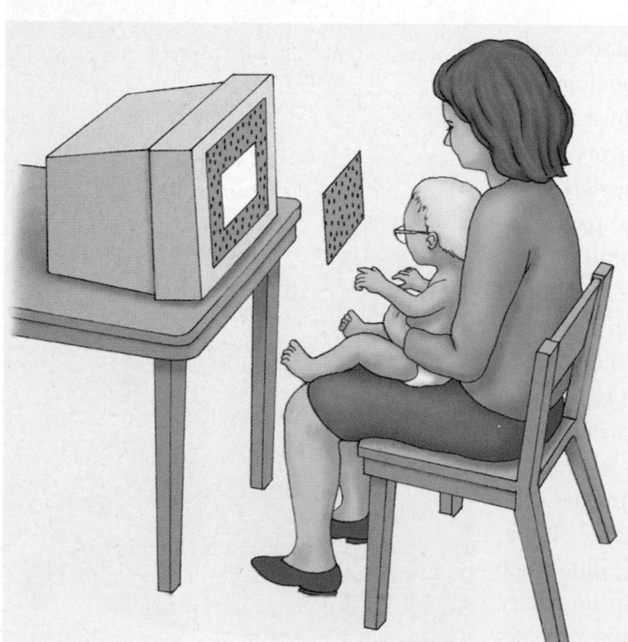

Piaget Emphasized Stages of Development

Obviously, infants differ from adults. They can't read this book, for example. But are infants merely inexperienced humans who have not yet learned the skills they will develop over time? Or do their minds work in completely different ways from those of adults? Through careful observations of young children, Jean Piaget (Figure 11.12) devised an influential theory of how "thinking" develops in the brain. One critical aspect of Piaget's research is that he paid as much attention to how children made errors as to how they succeeded on tasks. These mistakes, although illogical by adult standards, provided insights into how the young mind makes sense of the world. By systematically analyzing children's thinking, Piaget developed the theory that children go through *stages* of cognitive development. These stages reflect different ways of thinking about the world. From this perspective, it is not that infants know less than adults, but rather that their view of how the world works is based on an entirely different set of assumptions than those held by adults. We will see later that others have challenged the idea of stages of development.

Piaget proposed four distinct stages of development (Figure 11.13), each characterized by a different way of thinking, which he referred to as *schemas.* According to Piaget, **schemas** are conceptual models of how the world works that children form at each stage of development. Piaget believed that each stage builds on the previous one through learning by assimilation and accommodation. **Assimilation** is the process through which a new experience is placed into an existing schema; **accommodation** is the process through which a schema is adapted or expanded to incorporate the new experience.

11.12 Jean Piaget Piaget introduced the idea that cognitive development occurs in stages.

SENSORIMOTOR STAGE (BIRTH TO TWO YEARS) The first stage Piaget identified was the **sensorimotor stage**. Piaget's theory was that from birth until about age two, children acquire information about the world only through their senses—they react reflexively to objects. Thus, infants understand objects only when they reflexively react to those objects' sensory input—such as when they suck on a nipple, grasp a finger, or recognize a face. As they begin to control their movements, they develop their first schemas—conceptual models consisting of mental representations of the kinds of actions that can be performed on certain kinds of objects. For instance, consider the sucking reflex. This begins as a reaction to the sensory input from the nipple: The infant simply responds reflexively by sucking. Soon the child realizes it can suck other things—a bottle, a finger, a toy, a blanket. Piaget explained that sucking other objects is an example of assimilation to the schema of sucking. But sucking a toy or a blanket does not result in the same experience as the reflexive sucking of a nipple. This leads to accommodation with respect to the sucking schema—the child must adjust its understanding of sucking. Piaget believed that all of the sensorimotor schemas eventually merge into an exploratory schema. In other words, infants learn that they can act on objects—manipulate them in order to understand them—rather than simply react to them.

One of the cognitive concepts in this stage that Piaget explored was **object permanence**, the understanding that an object continues to exist even when it is hidden from view. Piaget noted that not until nine months of age will most infants search for objects that they have seen being hidden. Even at nine months, when infants will begin to snatch a blanket away to find a hidden toy, their search skills still have limits. For instance, suppose a child is given several trials in which an experimenter hides a toy under a blanket while the child watches, and the child then finds the toy. If the toy is then hidden under a different blanket, in full view of the child, the child will still look

schemas Hypothetical cognitive structures that help us perceive, organize, process, and use information.

assimilation The process by which a new experience is placed into an existing schema.

accommodation The process by which a schema is adapted or expanded to incorporate a new experience that does not easily fit into an existing schema.

sensorimotor stage The first stage in Piaget's theory of cognitive development, during which infants acquire information about the world through their senses and respond reflexively.

object permanence The understanding that an object continues to exist even when it cannot be seen.

Stage	Characterization
Sensorimotor (birth–2 years) 1	Differentiates self from objects Recognizes self as agent of action and begins to act intentionally; for example, pulls a string to set a mobile in motion or shakes a rattle to make a noise Achieves object permanence: realizes that things continue to exist even when no longer present to the senses
Preoperational (2–7 years) 2	Learns to use language and to represent objects by images and words Thinking is still egocentric: has difficulty taking the viewpoint of others Classifies objects by a single feature; for example, groups together all the red blocks regardless of shape or all the square blocks regardless of color
Concrete operational (7–11 years) 3	Can think logically about objects and events Achieves conservation of number (age 6), mass (age 7), and weight (age 9) Classifies objects according to several features and can order them in series along a single dimension, such as size
Formal operational (11 years and up) 4	Can think logically about abstract propositions and test hypotheses systematically Becomes concerned with the hypothetical, the future, and ideological problems

11.13 Piaget's Stages of Cognitive Development

for the toy in the first hiding place. Full comprehension of object permanence was, for Piaget, one of the key accomplishments of the sensorimotor period.

PREOPERATIONAL STAGE (TWO TO SEVEN YEARS) In the **preoperational stage,** children can think about objects that are not in their immediate view and have developed various conceptual models of how the world works. During this stage, they begin to think symbolically—for example, taking a stick and pretending it is a gun. Piaget believed that what they cannot do yet is think "operationally"—they cannot imagine the logical outcome of certain actions on objects. They base their reasoning on immediate appearance, rather than logic. For instance, children at this stage have no understanding of the *law of conservation* of quantity: that the quantity of a substance remains unchanged, even if its appearance changes. For instance, if you pour a short, fat glass of water into a tall, thin glass, we all know that the amount of water has not changed. However, if you do this and ask a child in the preoperational stage which glass contains more, she will pick the tall, thin glass because the water is at a higher level. Children make this error even when they have seen someone pour the same amount of water into each glass—or even when they do it themselves (Figure 11.14).

CONCRETE OPERATIONAL STAGE (7 TO 12 YEARS) At about seven years of age, children enter the **concrete operational stage**, where they remain until adolescence. Piaget believed that humans do not develop logic until they begin to think about and understand *operations.* In other words, they can figure out the world by thinking about how events are related. A classic operation is an action that can be undone: A light can be turned on and off; a stick can be moved across the table and then moved back. He suggested that the ability to understand that an action is reversible enables children to begin to understand concepts such as conservation. Although this is the beginning of logic, Piaget believed that children at this stage reason only about concrete things—objects they can act on in the world. They do not yet have the ability to reason abstractly, or hypothetically, about what might be possible.

FORMAL OPERATIONAL STAGE (12 YEARS TO ADULTHOOD) The **formal operational stage** is Piaget's final stage of cognitive development. Formal operations involve abstract thinking, characterized by the ability to form a hypothesis about something and test it through deductive logic. For instance, at this stage teens can systematically begin to test a theory or solve a problem. Piaget used the example of giving students four flasks of colourless liquid and one flask of coloured liquid. He told them that by combining two of the colourless liquids, they could obtain the coloured liquid. Adolescents were able to try different combinations in a systematic fashion and obtain the correct result. Younger children, who just randomly combined liquids,

(a)

(b)

(c)

11.14 The Preoperational Stage and the Law of Conservation In the preoperational stage, children cannot yet understand conservation tests. In this example, a six-year-old understands that the two short glasses contain the same amount of water (a). She carefully pours the water from one of the short glasses into a taller glass (b). Yet when asked "Which has more?" she points to the taller glass (c).

could not. If you think back to your school curricula, you will recall that subjects that require abstract reasoning, such as algebra or the scientific method, are not usually taught until around eighth grade. Here we see the influence of Piaget's research on education, providing an excellent example of psychological science in action.

Piaget revolutionized the understanding of cognitive development. His theories have dominated thinking for many years. However, many challenges have arisen to Piaget's views of how cognition develops. A key criticism disputes the idea that every person goes through the stages of development in the same order. Piaget believed that as children progress through each stage, they all use the same kind of logic to solve problems. This leaves little room for differing cognitive strategies or skills among individuals—or cultures. Some developmental psychologists have revised many of Piaget's theories while preserving his basic ideas. These theorists believe that different areas in the brain are responsible for different skills, and that the development of different skills therefore does not have to follow strict stages (Bidell & Fischer, 1995; Case, 1992; Fischer, 1980).

Infants Have Innate Knowledge

Earlier we asked if children are simply immature adults or whether their thinking styles are qualitatively different from the thinking styles of those who are older. Piaget suggested that children are not capable of understanding much of the world around them until they go through the various stages of cognitive development. However, recent research indicates that children understand much more and at much earlier ages than was previously believed.

Tests using the preferential looking technique have revealed that infants as young as three months of age are able to remember an object, even when it is no longer in plain sight, which seemingly contradicts Piaget's ideas about object permanence. Similarly, infants' reactions to novel stimuli indicate that they have cognitive skills quite early in life. For example, if you show three-month-old infants events that are impossible or unexpected, as if by magic (that is, that do not make rational sense), they will stare at the magic result longer than if you show them something that does make sense. For instance, suppose you show an infant an apple and then lower a screen. When you raise the screen again, the infant sees one of two situations: a possible event (the apple is still there) or an impossible event (a carrot has magically replaced the apple). If an infant looks longer at the impossible event, the experimenter deduces that the infant can tell the difference between the two scenes (Baillargeon, 1995). By responding differently to possible and impossible events, infants demonstrate some understanding that an object continues to exist when it is out of sight, even though they may not reach for a hidden object in Piaget's object-permanence test. Here we see that Piaget's task probably required too much of infants, especially in terms of having to reach for the object. Baillargeon's more implicit test revealed that object permanence occurs earlier than Piaget believed. In this section we consider some of the cognitive skills that appear to develop early in life.

preoperational stage The second stage in Piaget's theory of cognitive development, during which children think symbolically about objects, but reason is based on appearance rather than logic.

concrete operational stage The third stage in Piaget's theory of cognitive development, during which children begin to think about and understand operations in ways that are reversible.

formal operational stage The final stage in Piaget's theory of cognitive development; it involves the ability to think abstractly and to formulate and test hypotheses through deductive logic.

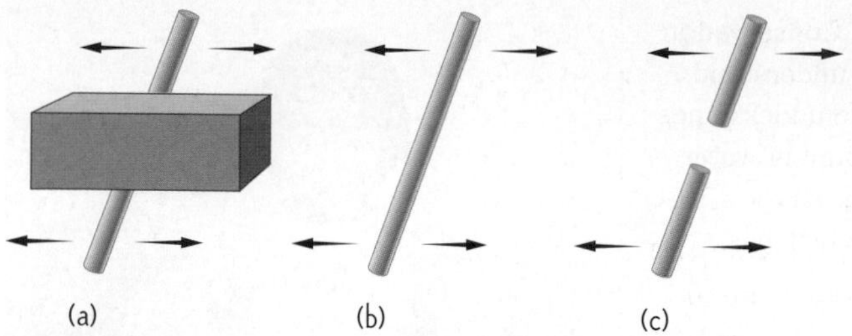

11.15 The Perceptual Effect of Occlusions in Early Infancy Four-month-olds were shown a rod that moved back and forth behind an occluding block as shown in (a). After being habituated to this stimuli they were shown two events, one in which a solid rod (b) moved behind the occluding block, another in which two separate rods (c) moved back and forth behind the block. The infants spent much more time looking at the unexpected event (c).

UNDERSTANDING THE LAWS OF NATURE: PHYSICS The developmental psychologist Elizabeth Spelke and her colleagues have conducted numerous studies indicating that infants have a primitive understanding of some of the basic laws of physics. Consider one such example. Humans are born with the ability to perceive movement: A newborn will follow moving stimuli with his or her eyes and head and prefer to look at a moving stimulus over one that is stationary. Experiments by Spelke and Philip Kellman indicate that as infants get older, they use movement information to determine if an object is continuous (Kellman, Spelke, & Short, 1986). In one experiment, the researchers showed four-month-old infants a rod moving back and forth behind a block. Once habituated, the infants were shown two scenes—one in which the block was removed and there was a single rod, and another in which the block was removed and there were two small rods (Figure 11.15). The infants looked longer at the scene with two smaller rods. This indicated that they expected the rod moving behind the block to be one continuous object rather than two smaller ones. If the experiment is conducted with the rod remaining stationary, the infants do not look longer at the two small rods. So infants appear to use movement to infer that objects that move together are continuous, whereas two stationary objects may or may not be continuous. Understanding the relation between movement and the physical properties of the rod requires cognitive skills. It requires the ability to see the rod as an object separate from the block, and to surmise that since the two ends are moving together, they must be part of the same whole rod, even though it is partially hidden from view.

How can we test infants' innate knowledge?

In a series of studies that use impossible events, Renee Baillargeon has found that children begin to understand what is necessary to support an object in space very early. Working with infants of different ages, the experimenters determined how well they understand whether a box will be stable when it is released on a platform versus when it is released off the platform (Figure 11.16). They found that by three months of age, infants expect the box to be stable if it is released on the platform, but not if it is released off the platform. Thus, children seem to have an intuitive sense of the laws that govern the physical world.

11.16 Understanding the Laws of Nature In Baillargeon's test, infants looked longer during the impossible event, in which the box was placed in midair without any contact support but did not fall.

Possible event

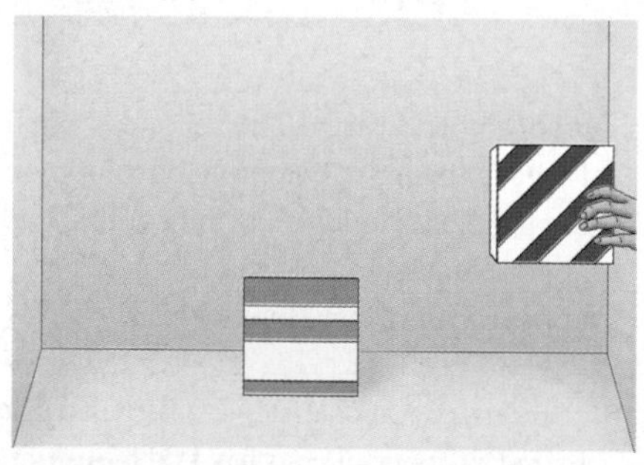

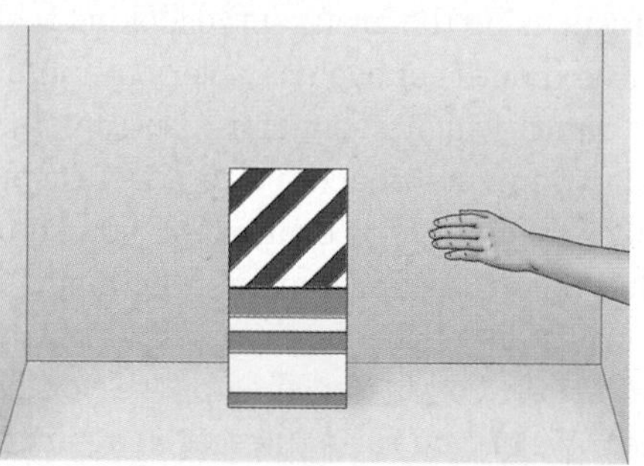

Impossible event

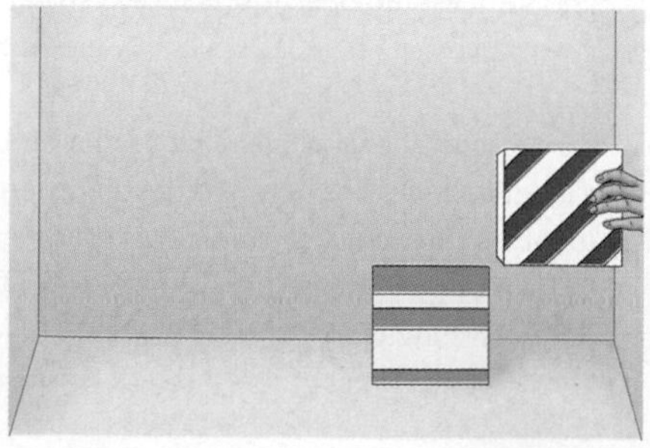
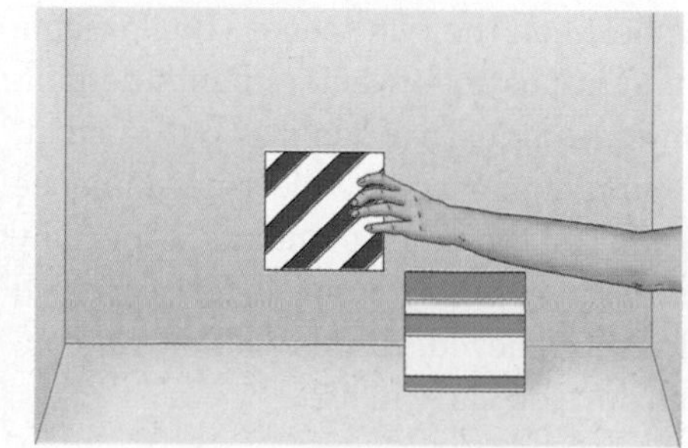
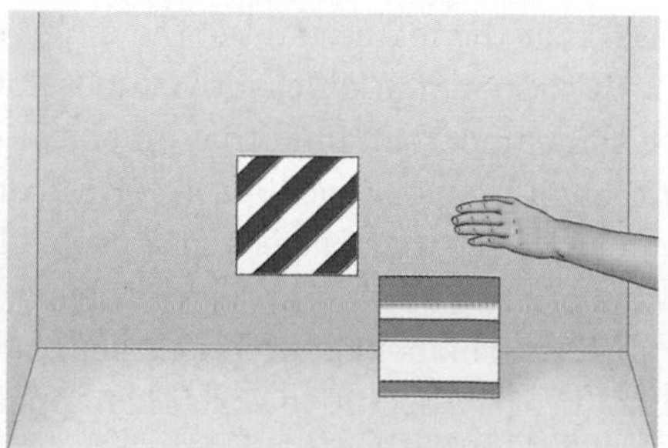

UNDERSTANDING THE LAWS OF NATURE: MATHEMATICS Research over the past 25 years has shown that we are born with certain numerical abilities, such as the ability to understand quantity. Piaget believed that young children do not understand numbers, and therefore must learn counting and other number-related skills through memorization. His experiments consisted of, for example, showing children four to five years old two rows of marbles (Figure 11.17). He found that if the marbles were spread out, then the child would usually say that the longer row had more marbles. Piaget concluded that children had no inherent sense of number—rather, their understanding of "more than" had to do with their perception of the length of the rows. This view was challenged by Jacques Mehler and Tom Bever (1967), who argued that children under three years old could understand the concept of more than or less than. To demonstrate their point, they repeated Piaget's experiment, but they used M&Ms candy (Figure 11.18). They showed the children the two rows of four M&Ms each and asked if the rows were the same. When the children said yes, the researchers then transformed the rows. For instance, they would add two candies to the second row, but they would compress that row so it was shorter than the row with fewer candies. Then they would tell the children to pick the row they wanted to eat. More than 80 percent picked the row with more M&Ms, even though it was the visually shorter row. Thus children do appear to understand the concepts of more than or less than, when properly motivated.

The growing evidence that infants have innate knowledge raises a serious challenge to Piaget's theory. Clearly, Piaget got many things right and his contributions to understanding cognitive development are enormous. The accumulation of scientific findings, however, indicates that Piaget underestimated the mental capacities of young children.

11.17 Piaget's Marble Test

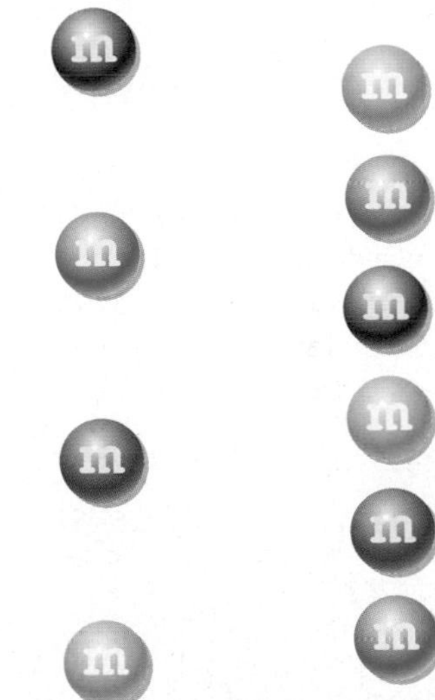

11.18 The M&M Version of Piaget's Marble Test Children who did not succeed on Piaget's marble test were able to choose the row that contained more items when those items were M&Ms and the test question was phrased as "Which row would you like to eat?"

Memory Improves over Childhood

The development of memory also helps children learn about the world around them, as they use new information to build on what they already know. Once again, clever experiments have revealed that infants do possess some types of memory from an early age—though that memory is quite rudimentary. Carolyn Rovee-Collier and her colleagues devised two such experiments. In one, infants learn to move a mobile hanging over the crib by kicking. During the task, a ribbon attaches their ankle to the mobile so the infants learn that the mobile moves when they kick (Figure 11.19). The rate at which an infant kicks when the mobile is not attached serves as the baseline. When the infant is tested later, the ribbon is attached to the ankle but not to the mobile (so the kicks do not move the mobile). If the baby recognizes the mobile, it will presumably kick faster than the baseline rate to try to make the mobile move. Infants ranging in age from 2 to 18 months have been trained for two days on the mobile and then tested after different lengths of time have passed. Findings indicate that infants remember longer as they get older. By 18 months, they can remember the event for several weeks (Rovee-Collier, 1999).

INFANTILE AMNESIA For years it was believed that infants did not remember anything, primarily because adults can seldom remember events that occurred before they were three or four years old. Freud referred to this inability to remember events from early childhood as **infantile amnesia**. Psychologists have offered a variety of explanations for why adults cannot recall their earliest memories (Eacott, 1999). Some believe that memories are retained only when we have the ability to create autobiographical memory based on personal experience—for instance, recalling that "a cat scratched me" versus the fact that "cats scratch." Other psychologists

infantile amnesia The inability to remember events from early childhood.

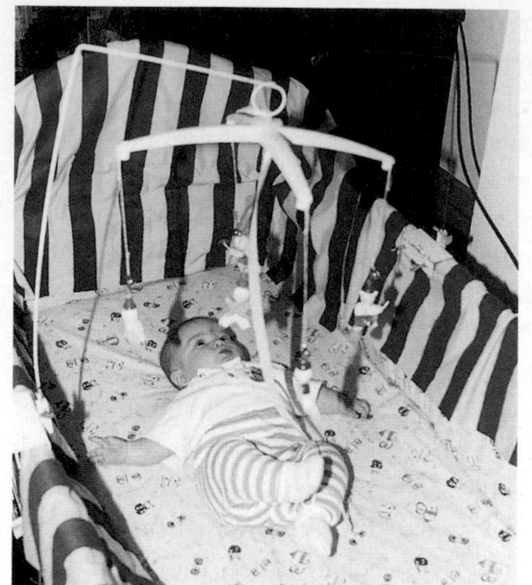

11.19 The Memory-Retention Test A three-month-old was trained in the mobile task and later given a retention test. During training (left), the infant's kicks move the mobile by means of the ankle ribbon attached to the mobile hook. During baseline and all retention tests (right), the ankle ribbon and the mobile are connected to different hooks so that kicks cannot move the mobile.

suggest that childhood memory increases with the acquisition of language because the ability to use words and concepts aids in memory retention.

INACCURATE MEMORY Young children often have difficulty knowing where they learned something, which is called **source amnesia**. Even when tested immediately after being presented with information, young children (age three) forget the source of the information faster than do older children (age five) (Gopnik & Graf, 1988). Evidence from investigations of source amnesia suggests that many of your earliest memories may come from watching home movies or hearing stories from your parents—not from an actual memory you have of the events.

Children are also known to confabulate—make things up. The fact that children have underdeveloped frontal lobes may explain why they are more likely than adults to engage in this behaviour. Confabulation happens most when children are asked about personal experiences rather than general knowledge. In one study, preschool children were interviewed repeatedly and asked to think hard to remember if they had ever gotten their fingers caught in a mousetrap. They were asked to visualize the event, including thinking about the scene (who was with them, what they were wearing, etc.). After the children spent ten weeks thinking about the event, a new adult interviewed them and asked them if they had ever gotten their fingers caught in a mousetrap. Sixty percent of the children provided false narratives, telling a story about this happening to them and behaving as if it really had. Many of the children developed very elaborate stories with numerous details about why their fingers had been caught in a mousetrap and how it had felt (Bruck & Ceci, 1993).

Humans Learn from Interacting with Others

To protect ourselves and to succeed in life, we need to be aware of the intentions of other people. For example, defensive driving is based on the idea that you can predict the potential erratic actions of others. When people are deciding whether to go on a date or begin a physical relationship, they often want to know what the other person has in mind. According to contemporary thinking among developmental psychologists, the early social interactions between infant and caregiver are essential for understanding other people and being able to communicate with them through language. Knowing that other people have mental states and using that information to infer what another person is feeling or thinking is termed **theory of mind**.

source amnesia A type of amnesia that occurs when a person remembers an event but cannot remember where they encountered the information.

theory of mind The term used to describe the ability to explain and predict other people's behaviour as a result of recognizing their mental state.

Piaget proposed that very young children are not good at understanding how others think. He characterized those in the preoperational stage as *egocentric:* They are not able to see another person's point of view, only their own. For instance, children at this stage might stand in front of a television and not understand that they are blocking the screen for others, because they can see the picture just fine. Evidence from the past two decades, however, has shown that young children are less egocentric than Piaget believed. Beginning in infancy, young children come to understand that other people's actions are performed for a reason—that the actions are intentional (Gergely & Csibra, 2003; Somerville & Woodward, 2005). Recognizing the intentionality of actions reflects the growing capacity for theory of mind. This is what allows us to understand, predict, and attempt to influence other people's behaviour (Baldwin & Baird, 2001). In one recent study, infants watched as an adult handed them a toy. On some trials, the child did not receive the toy because the adult was unwilling to give it (e.g., she teased the child with the toy or played with it herself) or because she was unable to give it (e.g., she was distracted by a ringing telephone or accidentally dropped the toy). Children older than nine months showed greater signs of impatience (such as reaching for the toy) when the adult was unwilling than when she was unable (Behne, Carpenter, Call, & Tomasello, 2005). Consider another study in which 15-month-old children imitated the actions of an adult trying to pull apart two halves of a dumbbell. On some trials, the children watched the adult's hand slip off the end of the dumbbell before she was successful in separating it. However, the children imitated the goal of separating the dumbbells, rather than simply imitating the adult's actual movements, which included slipping (Meltzoff, 1995). These studies and others provide strong evidence that children begin to read intentions in the first year of life, and that they become very good at doing so by the end of their second year.

How does theory of mind enable social interaction?

Understanding intentions helps children predict and understand other people's actions, but they do not develop the ability to take another person's perspective until they are around four or five years of age. For instance, children younger than four have difficulty understanding that others can have beliefs that are not true, or beliefs that differ from their own. Consider the *false belief test,* in which, in order to predict a person's actions, children are required to understand that people can sometimes act based on false information. In the classic example (Figure 11.20), Sally places a marble in a basket and then leaves the room. Ann comes in, removes the marble from the basket, and places it in a box. A child who has watched both actions take place is asked to guess where Sally will look for the marble when she comes back in the room. In order to do this correctly, the child must develop a theory: Sally put the marble in the basket; Sally doesn't know Ann moved the marble, so Sally will still look in the basket. Normally, developing children are able to solve this by age four.

Children's development of theory of mind and success at the false belief test appear to coincide with the frontal lobes becoming relatively mature. The importance of the frontal lobes is also supported by research with adults. In brain-imaging studies, prefrontal brain regions become active when people are asked to think about others' mental states. People with damage to this region have difficulty attributing mental states to others in stories (Stone, Baron-Cohen, & Knight, 1998). Futhermore, if the capacity for theory of

11.20 An Example of a Child's Theory of Mind When a child acquires theory of mind, she is able to understand that different individuals have different perspectives and knowledge based on their individual experiences.

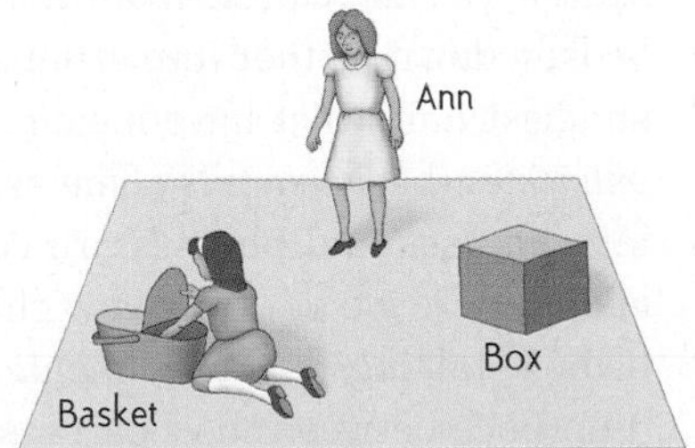
Ann
Box
Basket
Sally puts her marble in the basket.

Sally goes away.

Ann moves the marble.

?
"Where will Sally look for her marble?"

mind is linked to brain development, then children in different cultures ought to solve the false belief test at about the same age. Indeed, a recent study comparing children from Canada, India, Peru, Samoa, and Thailand found that children crossed the false-belief milestone at around age five in each culture (Callaghan et al., 2005).

Finally, theory of mind, like math or physics, may develop independent of other brain functions, such as the ability to reason or general intelligence. Evidence for this comes from the study of children with the communications disorder autism. They are not able to solve the Sally/Ann problem. In contrast, children with Down syndrome can solve this problem, indicating that it is not a general intelligence or reasoning mechanism that governs theory of mind, since Down children are impaired in these areas. You will learn more about autism in Chapter 13.

Language Develops in an Orderly Fashion

The ability to communicate through language allows humans to learn a great deal more than can other animals. Through language we are able to communicate across cultures, sharing the new inventions and technologies that have shaped modern civilization. How does this remarkable ability develop?

As the brain develops, so does the ability to speak and form sentences. For the most part, the stages of language development are remarkably uniform across individuals, although the rate at which language develops does vary. We saw in the case of Genie that the ability to speak can be disrupted by social isolation and lack of exposure to language. According to Michael Tomasello (1999), the early social interactions between infant and caregiver are essential to understanding other people and being able to communicate with them through language. Research has demonstrated that infants and caregivers attend to objects in their environment together; this joint attention facilitates learning to speak (Baldwin, 1991). This capacity to share our mental actions with others and the motivation to do so makes humans unique in the animal kingdom. Language allows us to live in complex societies because it teaches us the history, rules, and values of our culture. We now examine how humans learn language.

FROM ZERO TO 60,000 Language is a system of using sounds and symbols to communicate. It can be viewed as a hierarchical structure (Figure 11.21) in which sentences can be broken down into smaller units consisting of phrases, which can be broken down further into groups of words, each of which consists of *morphemes* (the smallest units that have meaning, including suffixes and prefixes), which consist of *phonemes* (basic sounds). The system of rules that govern how words are combined into phrases and phrases are combined to make sentences is the *syntax* of a language. Here we describe how children learn the syntax of language. Babies may not be born talking, but experiments have revealed that they are born understanding the difference between their parents' language and another language. Babies can distinguish between different *phonemes,* which are the minimal speech sounds that signal a difference in meaning between words. For example, "cat" and "pat" mean different things in spite of /c/ and /p/ not carrying any meaning in themselves. The psychologists Peter Eimas and Peter Jusczyk devised a version of the habituation technique that allows experimenters to study when very young infants prefer to listen to one thing over another or note a difference between sounds. They inserted the switch for a tape recorder into a rubber nipple. When infants sucked on the nipple, the sucking turned on the recorder, which played a syllable such as "ba" each time the sucking occurred. Once the babies tired of the syllable, their sucking rate decreased. Once the infants were habituated, the researchers changed the syllable, to "pa," for

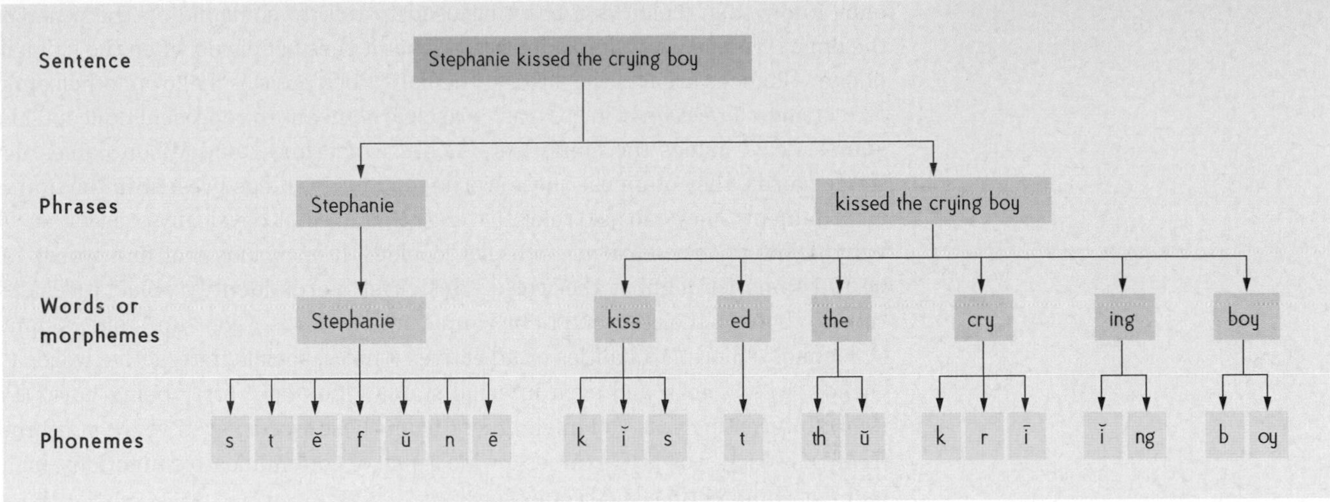

11.21 The Units of Language

instance. The sucking rate increased when the babies heard the new sound (Eimas, Siqueland, Jusczyk, & Vigorito, 1971). Researchers used this same technique to show that French babies as young as four days old sucked more in response to French than to Russian, indicating that they had learned the sound of their native language in the womb or in the first days outside the womb (Mehler et al., 1988). Without hearing or special training techniques, however, speech development and language acquisition are seriously affected (Wong & Shah, 1979).

English has over 40 phonemes, and each phoneme has a distinct sound. For example, the distinction between /r/ and /l/ is important in English: "read" means something quite different than "lead." However, Japanese does not distinguish those two sounds and instead makes other distinctions that English does not make. After several months of exposure to their own language, infants lose the ability to distinguish between sounds that do not matter in their language (Kuhl, 2004; Werker, Gilbert, Humphrey, & Tees, 1981).

Why do children lose the ability to discriminate sounds of foreign languages?

Given that babies can hear the difference between sounds immediately after birth and continue to learn the sounds of their own languages, how does this affect the development of spoken words? A normal adult knows about 60,000 words. Humans appear to go from babbling as babies to a full adult vocabulary without working very hard at it. Speech production follows a distinct path. During the first months of life, newborns' actions generate all of their sounds: crying, fussing, eating, and breathing. The sounds are cries, gurgles, or grunts. Cooing and laughing appear at three to five months. From five to seven months babies begin babbling, using consonants and vowels. From seven to eight months they babble in syllables (ba-ba-ba, dee-dee-dee). By the first year, the syllables are mixed (ba-dee, dah-dee), and those babbles begin to take on the sounds and rhythms of the infant's native language.

Babbling may be an infant's way of testing the system—checking that the basic parts exist and how they get tossed together. The onset of language is marked by the first words that a baby utters and appears to understand. Words consist of *morphemes,* which are, as mentioned, the smallest units of speech that have meaning. Sometimes a word is a morpheme, such as *cat,* but prefixes and suffixes are also morphemes, such as *-ed,* which changes a verb to past tense (see Figure 11.21).

First words tend to appear around age one. Babies most often first utter two types of words. One type is known as *performatives,* wordlike sounds that are learned in a context and that a baby may not be using to represent a meaning. For instance, a baby says, "Hello" or something that sounds like it when holding a phone. Does the baby know that "hello" is a greeting used to talk to someone on the other end of the line? Or is the baby simply imitating what it sees people do when they pick up the phone? Chances are it is the latter if the baby does not say "hello" in other appropriate settings. *True words,* in contrast, are clearly meant to represent concepts. For instance, "cat" names the family pet, "book" a bedtime story. When babies develop single words, they often use those words to identify things present in the immediate environment. They can also refer to things that are not physically present, as when a baby points to the cookie jar and says "cookie." It is curious that first words tend to be the same throughout the world. Most first words identify objects ("cat," "sky," "nose," "book"), the rest tend to be simple action words ("go," "up," "sit"), quantifiers ("all gone!" "more!"), qualities or adjectives ("hot"), socially interactive words ("bye," "hello," "yes," "no"), and even internal states ("boo-boo" after being hurt) (Pinker, 1984). Interestingly, as you might guess from these words, even very young children freely use early words to express a wide range of communicative functions, including naming, commenting, and requesting.

By about 18 months children begin to put words together, and their vocabularies start to grow rapidly. Rudimentary sentences of roughly two words or more emerge. The fascinating thing is that these minisentences, though missing words and grammatical markings, actually have a logic or "syntax." Typically the order of the words indicates what has happened or should happen: "Throw ball. All gone" roughly translates as "I threw the ball and now it's gone." In fact, the psychologist Roger Brown, often referred to as the father of child language for his pioneering research, called these utterances **telegraphic speech** because children speak as if they are sending a telegram—just bare-bones words without any frills but put together according to conventional rules.

telegraphic speech The tendency for children to speak using rudimentary sentences that are missing words and grammatical markings but follow a logical syntax.

As children begin to use language in more sophisticated ways, they make interesting errors in word formation and syntax. Sometimes children start to make mistakes at ages three to five with words they used correctly at age two or three. Psycholinguists believe this occurs because when children learn a new grammar rule, they tend to overapply it. For example, they learn that adding *-ed* to a verb in English makes something past tense. They then add *-ed* to every verb, including irregular verbs that do not follow that rule. For instance, they will say "runned" or "holded" even though they may have said "ran" or "held" at an earlier age. A similar pattern of errors is found with plurals. The rule to add *s* leads to errors such as "mouses" and "mans" when the rule is overapplied. Children do this despite having used the correct irregular plural at a younger age.

Although overgeneralizations are quite rare (Marcus, 1996; Marcus et al., 1992) in children's speech, they reflect an important aspect of language acquisition. Children are not simply repeating what they have heard others say, because they most likely have not heard anyone say "runned." Instead, these errors occur because children are able to use language in a generative way. Simply put, children have discovered a general rule of grammar, but they have not yet discovered that there are exceptions to the rule. Children make more errors with words that are used less frequently (such as "drank" and "knew") because they have heard the irregular form less often. Adults tend to do the same thing; we are more likely to make errors on the past tenses of words such as "trod," "strove," or "slew" (saying "treaded," "strived," or "slayed") because those are words we do not use often (Pinker, 1994).

UNIVERSAL GRAMMAR Much of the research by linguists and psycholinguists is aimed at breaking down language production and comprehension into detailed steps to understand how language is assembled, produced, and understood. The linguist Noam

Chomsky (Figure 11.22) transformed the field when he argued that language must be governed by "universal grammar," or innate knowledge of a set of universal and specifically linguistic elements and relations that form the heart of all human languages.

Is language hardwired?

Until Chomsky came on the scene in the late 1950s, linguists had focused on analyzing language and identifying basic components of grammar. In his early work, Chomsky argued that the way people combine these elements to form sentences and convey meaning is only the *surface structure* of a language. He introduced the concept of *deep structure:* the implicit meanings of sentences. For instance, "The fat cat chased the rat" implies that there is a cat, it is fat, and it chased the rat. "The rat was chased by the fat cat" implies the same idea even though on the surface it is a different sentence. Chomsky believed that we automatically and unconsciously transform surface structure to deep structure. Research has shown that the underlying meaning of a sentence is what we remember, not its surface structure. For example, in studies in which subjects are shown sentences with similar meanings but different syntax, like the examples above, they can remember the point of the sentences, but not which version they saw (Sachs, 1967). For Chomsky, humans are born with a "language acquisition device" that contains universal grammar, but exposure to one's native language narrows down which grammatical rules are learned. All languages contain similar elements, such as nouns and verbs, but how they are arranged varies considerably across languages. Chomsky's ideas are supported by research that shows young infants can understand simple grammatical rules. Gary Marcus and colleagues (1999) familiarized seven-month-old infants with sentences of an ABA design, such as "ga ti ga" or "li na li." Subsequently, they found that infants showed a preference for sentences that followed a new pattern (ABB), such as "wo fe fe," indicating that they noticed the inconsistency. This pattern suggests that young infants have a built-in readiness to acquire grammar.

There is little doubt that environment greatly influences a child's acquisition of language, such as fine-tuning which grammar is used. Indeed, the fact that you speak English rather than Swahili is determined entirely by your environment. The Russian psychologist Lev Vygotsky developed the first major theory that emphasized the role of social and cultural context in cognition and language development. According to Vygotsky, humans are unique because they use symbols and psychological tools, such as speech, writing, maps, art, and so on, through which they create culture. Culture, in turn, dictates what people need to learn and the sorts of skills they need to develop. For example, some cultures might value science and rational thinking, whereas others might emphasize supernatural and mystical forces. Some cultures emphasize keeping social distance, whereas others encourage people to be in close proximity. These cultural values shape how people think about and relate to the world around them. Vygotsky distinguished between elementary mental functions, such as innate sensory experiences, and higher mental functions, such as language, perception, abstraction, and memory. As children develop, their elementary capacities are gradually transformed, primarily through the influence of culture.

11.22 Noam Chomsky The father of modern linguistic theory.

Central to Vygotsky's theories is the idea that social context influences language development, which in turn influences cognitive development. Children start by directing their speech toward specific communications with others, such as asking for food or toys. As children develop, they begin directing speech toward themselves, as when they give themselves directions or talk to themselves while they play. Eventually, children internalize their words into inner speech, which comprises verbal thoughts that direct behaviour and cognition. From this perspective, your thoughts are based on the language that you acquire through your culture, and this ongoing inner speech reflects higher-order cognitive processes.

SENSITIVE PERIODS As we noted earlier in the chapter, there are critical learning periods for certain cognitive skills. But for humans, these periods are not rigid or inflexible but rather are sensitive periods in which skills such as language are easiest to learn. Cases such as those of Genie and other children raised in deprived environments indicate that humans are best able to acquire language prior to puberty. The extent to which individuals can become multilingual after adolescence also supports a sensitive learning period. People who are truly bilingual or multilingual have generally acquired multiple languages before the age of 12. Adults who move to a new country rarely become as fluent as their children (Zatorre, 1989).

Interaction across cultures can shape language. *Creole* describes a language that evolves over time from the mixing of existing languages. Creole languages develop, for example, when several different cultures colonize a new place, as when the French established themselves in southern Louisiana and acquired slaves who were not native French-speakers. Creoles, which follow formal grammatical rules, develop out of more rudimentary communications. Populations that speak several different languages develop a new means for communicating, often mixing words from each group's native language into what is called a *pidgin,* which is an informal creole that lacks consistent grammatical rules. The linguist Derek Bickerton has found that the pidgin is then fully developed by the children of the colonists into its own language—a creole language—as the children hear the pidgin spoken by their parents and impose rules on it. Bickerton argues that this is evidence for built-in, universal grammar: The brain takes a nonconforming language, applies rules to it, and changes it.

Bickerton also found that creole languages formed in different parts of the world, with different combinations of languages, are more similar to each other in grammatical structure than to long-lived languages. Although somewhat controversial, he sees this as further evidence for built-in grammar, arguing that older languages have evolved away from the grammatical constructions most natural for the brain to impose, whereas creole languages have not been around long enough for this to happen. Thus creole languages may be similar because the brain imposes the same basic rules on a new language (Bickerton, 1998).

ACQUIRING LANGUAGE ON THE HANDS What would happen if we stripped the human brain of sound and speech? If sound perception and production are the key neurological determinants of early language acquisition, then babies exposed to signed languages should acquire these languages in fundamentally different ways. If, however, what makes human language special is its highly systematic patterns and the human brain's sensitivity to them, then signing and speaking babies should acquire their respective languages in highly similar ways.

To test this hypothesis, Laura Ann Petitto and her students at McGill University videotaped deaf babies of deaf parents in households using two entirely different signed languages—American Sign Language (ASL) and the signed language of Quebec, langue des signes quebecoise (LSQ). They found that deaf babies exposed to signed languages from birth acquire these languages on an identical maturational time table as hearing babies acquire spoken languages (Petitto, 2000). This research teaches us that speech cannot alone be driving all of human language acquisition. It provides an exciting glimpse into language and the brain by showing us that humans must possess a biologically endowed sensitivity to aspects of the patterns of language—a sensitivity that launches a baby into the course of acquiring language.

ANIMAL COMMUNICATION We know that animals have ways of communicating with each other. But no other animal uses language the way humans do. Scientists have tried for years to teach language to chimpanzees, one of our nearest living relatives (along with bonobos). Chimps lack the vocal abilities to speak aloud, so studies have used sign language or visual cues to determine whether they understand words or concepts such as causation. Although chimpanzees can learn some words and have some sense of causation, other research challenges the idea that this means they have innate language abilities. Consider the work of psychologists Herbert Terrace, Laura Ann Petitto, and Tom Bever, who set out to test Noam Chomsky's assertion that language was a uniquely human trait by attempting to teach a chimpanzee language. Hopeful of success, they named the chimp Neam Chimpsky (nicknamed Nim) (Figure 11.23). But after years of teaching Nim American Sign Language, the team had to admit that Chomsky might be right.

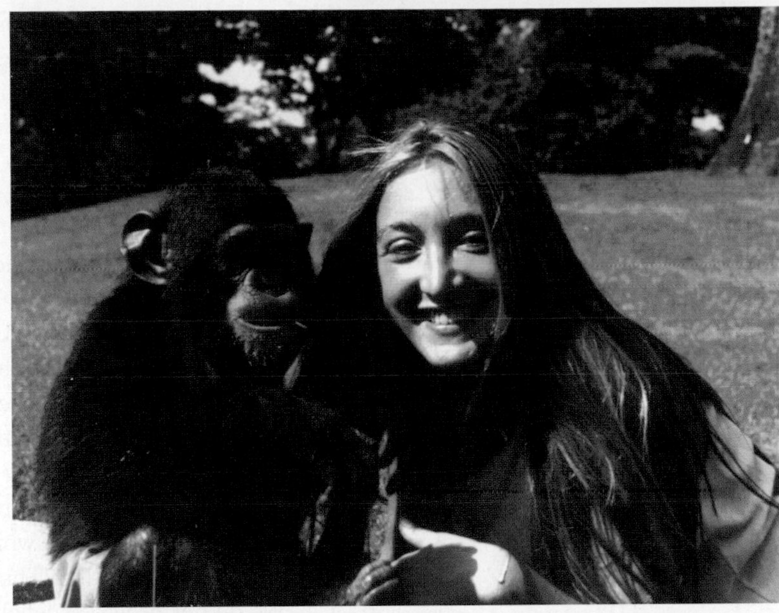

11.23 Laura Ann Petitto with Nim Chimpsky

Nim, like all other language-trained chimps, consistently failed to master key components of human language syntax. While Nim was quite adept at communicating with a small set of basic signs ("eat," "play," "more"), he never acquired the ability to generate creative, rule-governed sentences; he was like a broken record, talking about the same thing over and over again in the same old way. Crucially, unlike even the young child who names, comments, requests, and more with his or her first words, all the chimps used ASL almost exclusively to get something, to request. What the chimps seemed to appreciate was the power of language to obtain outcomes. In the end, the chimps could use bits and pieces of language only to get something from their caretakers (food, more food) rather than to truly express meanings, thoughts, and ideas by generating language (Petitto & Seidenberg, 1979).

How can we tell if animals use language?

Psychological Science *in Action*

Learning to Read

Reading, like speaking, is nearly effortless for most adults. When we look at letters, we see words and automatically derive meaning from them, even if they are misspelled. Try to look at the word *chocolate* without knowing what it means. But learning to read is challenging for most children, who struggle to figure out which symbols are letters, which letters clump together to make words, and which words go together to make meaningful sentences. So what are the best ways to teach children to read?

There are two major schools of thought regarding how to teach reading. Traditional methods use *phonics,* which teaches an association between letters and their phonemes (or sounds). Children learn to spell out words by how they sound. Because of the complexity of the English language, in which the sounds of letters can vary across words, some educators have advocated *whole-language* approaches, in which the emphasis is on meaning and understanding how words are connected in sentences. The whole-language approach has dominated in many Canadian schools for the past two decades, perhaps in part because the philosophy that guides its instruction emphasizes student interest and enjoyment of reading. But does it do a better job than phonics in helping children learn to read? Across Canada, those who advocate whole-language instruction have vigorously debated those who argue that such methods are inferior to the traditional method of phonics. Let's look at the two methods in more detail.

Some form of the phonics method has been popular for teaching reading for over 200 years. In phonics, children learn the relation between letters of the alphabet and the sounds they make, including when they are combined. Thus a *t* is changed when it becomes *th* or *tr* or *st*. Children learn a small number of simple words that teach the sounds of letters across most words of the English language. Because of the irregularities in English, children first learn the general rules and then learn to recognize exceptions to those general rules. The emphasis with this approach is on memorizing the mappings between letters and their sounds, rather than on building vocabulary or processing the meaning of words.

The general idea behind whole-language instruction is that reading should follow the form of language, which requires deriving meaning from listening to a continuous stream of sound. Thus, children should learn to read the way they learn to talk. We do not process speech by breaking down the stream into phonemes; instead, we understand speech as a series of connected words that have meaning in the context of the entire sentence. According to this philosophy, people naturally and unconsciously learn to read by first learning individual words and then stringing them together. Proponents argue that breaking down words into sounds is unnatural, frustrating, and boring—therefore, using phonics is to be avoided.

Psychological science has made tremendous contributions to understanding not only how children learn to read, but also which educational practices work best for most children. A group of psychologists who are experts in reading reviewed the vast literature on how children learn to read and published a comprehensive report for the benefit of educators. From this research they conclude that some methods for teaching children to read are essential. Teachers who use these methods will be most effective in helping their students to become skilled readers (Rayner et al., 2001).

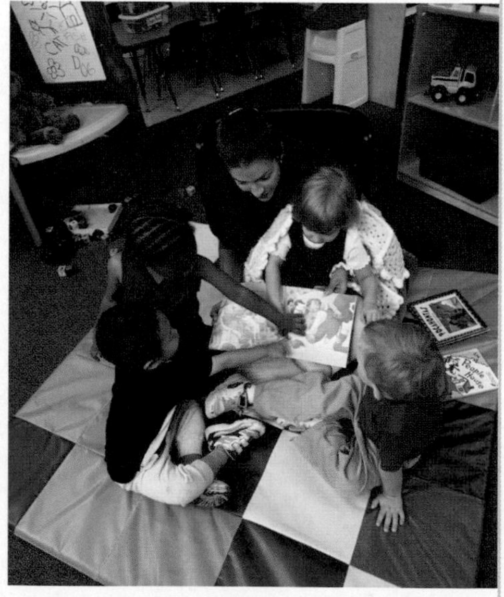

11.24 Phonics versus Whole-Language Approaches Although the whole-language method of teaching children to read has been popular in the last 20 years, researchers have found that phonics is a more effective way to create proficient readers.

So, which is better? Classroom and laboratory research over the past three decades has consistently found evidence that phonics instruction is superior to whole-language methods in creating proficient readers (Rayneer, Foorman, Perfetti, Pesetsky, & Seidenberg, 2001). This is especially true for children who are at risk of becoming poor readers, such as those whose parents do not read to them on a regular basis (Senechal & LeFevre, 2002). At the same time, the whole-language approach has clear benefits, with its emphasis on making reading fun and encouraging students to be creative and thoughtful. However, although whole-language approaches can motivate students to want to read, the students need the basic skills to do so, which are best provided through phonics. After all, reading is not fun when children struggle to identify words.

The phonics-versus-whole-language debate is an excellent example of how scientific research can contribute to figuring out important societal issues. The whole-language movement reflected a progressive philosophy that tried to foster a love of learning. Yet school districts that used this approach observed declines in national assessment tests, and a growing number of children developed reading problems. Parents responded by buying home reading programs that emphasized phonics and forming advocacy groups to change how reading is taught. The science of reading is clear—phonics works best (Figure 11.24). By sharing research with those who set education policy, psychological scientists are helping to make children more capable readers. ■

REVIEWING **the Principles** | How Do Children Learn about Their Worlds?

Children acquire information through perception. Tests based on the fact that young infants look longer at novel stimuli than at familiar stimuli indicate that they are capable of learning very early. Piaget emphasized the consistent stages of cognitive development displayed by most young children. His influential theory suggested that each stage of learning builds on previous stages. However, recent evidence suggests that infants understand a great deal more about the physical properties of objects than was previously believed. For instance, infants can use the laws of physics and even perform simple arithmetic. An important part of learning occurs through social interaction, as young children develop the ability for theory of mind. This capacity allows people to survive in human society. A developing memory system helps children build a store of useful knowledge. The human capacity for language is innate, as there appear to be built-in methods of acquiring words and forming them into sentences. Although the development of language occurs in an orderly fashion, the specific language a child develops is influenced by environmental and cultural factors, such as exposure to the language spoken by caregivers and the creoles and pidgins that are used when groups who speak different languages interact. There are also sensitive periods during which language must be acquired in order to develop normally. These processes develop together to enable young children to learn and survive as they become members of society.

How Do Children and Adolescents Develop Their Identities?

As we go through childhood, we learn more about our worlds. We also begin to develop a sense of identity, of who we are. Identify formation is an important part of **social development**, which, as we described earlier, refers to the maturation of skills and abilities that enable us to interact with others. Many factors, including gender, play a role in shaping our sense of personal identify. What is amazing, however, is how unique each of us is even though we share so many genetic similarities and have so many common experiences. We each possess different habits, customs, personality characteristics, beliefs, skills, and even distinctive quirks. How people define themselves is influenced by the culture in which they are raised as well as the beliefs they hold about personal characteristics, such as race, sex, and age. The quest for identity is an important challenge during development. As children grow older and enter into **adolescence**, they seek to understand how they fit into the world and to imagine what kind of person they will become later in life. In this section we examine how a sense of identity develops.

social development The maturation of skills or abilities that enable people to live in a world with other people.

adolescence The transitional period between childhood and adulthood.

Gender Roles Are Determined by Cultural Norms

How does gender differ from sex?

When people are asked to describe themselves, they often indicate whether they are female or male. Imagine for a second what you would be like if you were the same person but of the opposite sex—for many of us this is hard to do. Being a female or a

"Do you know its sexual identity?"

male is central to how most people define themselves. But how different are men and women? Although the physical differences are obvious, such as in height and weight, how do women and men differ in psychological terms? Psychological scientists use the term *sex* to refer to biological differences between males and females, whereas the term **gender** is used for differences between males and females that result from socialization. According to evolutionary theory, sex differences ought to reflect different adaptive problems men and women face, and this is generally supported by research. But since men and women face similar adaptive problems, they are similar on most dimensions. Many of the differences between males and females have as much to do with socialization as with genetics.

Whether you think of yourself as a male or a female is part of your **gender identity**, which subsequently shapes how you behave. Children as young as one or two years old can tell you whether they are boys or girls. Once boys and girls discover that they are boys or girls, they seek out activities that are culturally sex appropriate. In North American culture, if not everywhere, most parents and teachers discourage girls from playing too roughly and boys from crying. The separation of boys and girls into different play groups is also a powerful socializing force.

Which behaviours boys and girls display is determined in part by cultural rules about sex-appropriate behaviour. **Gender roles** are behaviours that differ between men and women because of cultural influences, expectancies, or learning. For instance, you may believe that it is more appropriate for men than for women to initiate sexual encounters or to repair cars for a living. Although many people now object to such beliefs, their existence is easy to observe. As mentioned in Chapter 8, when young children are asked to draw a picture of a scientist, most draw a man (Chambers, 1983). Children develop their expectations about gender through observing their parents, peers, and teachers, as well as through media (Lytton & Romney, 1991). Cultural norms and media have a strong influence on the development of gender roles. Most news anchors are male, most firefighters are male, most nurses in films and on television are female, and so on. These portrayals help to develop and reinforce gender stereotypes, commonly held beliefs about men and women. We can find many examples of stereotyping in children's toys (Figure 11.25).

Gender schemas are cognitive structures that influence how people perceive the behaviours of females and males. They play a powerful role in establishing gender identity. A gender schema acts as a lens through which people see the world. People raised in environments that make clear distinctions between men and women become *gender typed;* that is, they expect large differences between males and females. The psychologist Sandra Bem measures degrees of gender typing by administering a questionnaire that asks people the extent to which traits that are typical of men and women describe them. People who rate themselves high on stereotypically masculine traits (e.g., competitiveness) and low on stereotypically feminine traits (e.g., tenderness) are labeled *masculine;* those who rate themselves high on feminine and low on masculine traits are labeled *feminine;* and those who rate themselves equally on both types of traits are *androgynous.* Bem believes that being androgynous is advantageous because it allows for more flexibility in social behaviour. Conversely, her studies show that being sex typed can be cognitively efficient during tasks that benefit from encoding by gender, such as predicting whether a mechanic is likely to be male or female (Bem, 1975; Bem, Martyna, & Watson, 1976).

gender A term that refers to the culturally constructed differences between males and females.

gender identity Personal beliefs about whether one is male or female.

gender roles The characteristics associated with men and women because of cultural influence or learning.

gender schemas Cognitive structures that influence how people perceive the behaviours of men and women.

Situational factors also contribute to gender identity. According to a theory proposed by social psychologists Kay Deaux and Brenda Major (1987), the interaction between people and a specific situation *creates* gender-related behaviours. Deaux and Major accept the idea that people have beliefs and expectations about how men and women differ, but they emphasize that people of both sexes act differently depending on the situation. For instance, men may talk about football and cars around their male friends, but they are likely to change the topic when women are around.

A study of young women talking to either their boyfriends or their casual male friends on the telephone illustrates how situation alters gender-related responses. When the women talked to their boyfriends, their voices changed to a higher pitch and became softer and more relaxed, relative to the way they talked to their male friends. The way they spoke to their boyfriends was also more babylike, feminine, and absentminded, as rated by objective judges. When asked, the women said they knew they had taken on a different manner of speaking to their boyfriends relative to how they spoke to their male friends. The women reported doing this in order to communicate affection for their boyfriends (Montepare & Vega, 1988). Likewise, experiments have shown that heterosexual young men behave in a more masculine fashion when they think they are talking to a homosexual man versus when they talk to a heterosexual man (Kite & Deaux, 1986).

Although males and females are similar in many respects, there are sex differences based on biology and gender differences influenced by culture (Marcia, 1980). Research in psychological science has identified some differences in how the brains of men and women differ, as when men show greater responsiveness to pictures of sexual imagery (see Chapter 4). However, research has not yet determined whether this is due to some inherent property of the male brain or to the way boys are treated during development. In terms of how we define ourselves as males and females, both nature and nurture play prominent roles (Kimura, 1999).

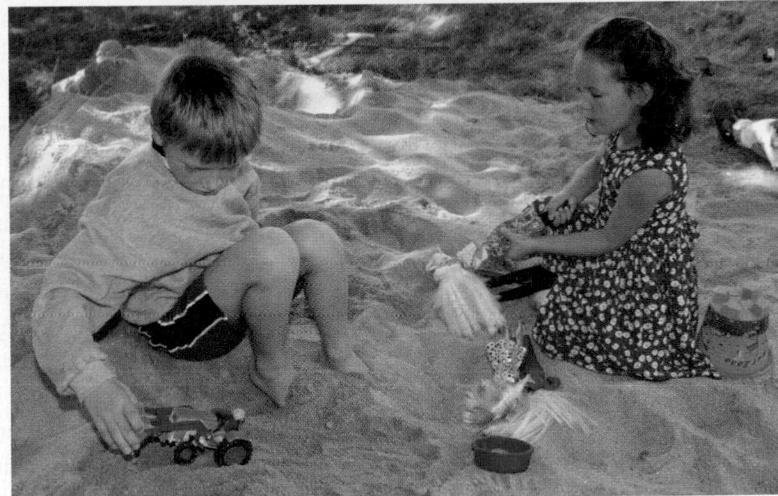

11.25 The Development of Gender Roles Many children's toys may promote gender stereotyping.

Profiles in Psychological Science

How Nature Made Him

How much of gender identity is determine by nature and nurture? On April 27, 1966, seven-month-old twin boys named Bruce and Brian Reimer were brought to St. Boniface Hospital in Winnipeg, for routine circumcision. Unfortunately, during the first operation, Bruce's penis was badly damaged. Its condition deteriorated over the next several days, and within the week it shriveled up and disappeared. (It should be noted that after what happened to Bruce, no attempt was made to circumcise Brian.) The accident and the events that followed changed not only the lives of the Reimer family but also psychologists' beliefs about the concept of gender.

As recounted in John Colapinto's 2000 book *As Nature Made Him: The Boy Who Was Raised as a Girl,* Bruce's parents, Janet and Ron, decided to investigate whether he should undergo sexual reassignment and be raised as a girl. The process had never been attempted on a child born with normal genitalia; previous sexual reassignment cases had involved hermaphrodite children, who were born with ambiguous genitalia. They contacted the world-renowned (and controversial) sexologist John Money at Johns Hopkins University, who convinced the Reimers that sexual reassignment was the best course of action for Bruce's psychological well-being. Bruce was castrated when he was 22 months old, renamed Brenda, and raised as a girl (Figure 11.26). Throughout the 1970s and 1980s, media accounts and psychology textbooks recounted the story as demonstrating that gender was the result of socialization rather than biology. Whether one was a boy or a girl was the result of how one was treated, not of genes or biology. Unfortunately, Colapinto's recent analysis indicates that Brenda's sexual reassignment was a failure from the start.

Brenda's life can be described as tumultuous at best, hellish at worst. Although her parents let her hair grow long, dressed her in feminine colours and clothing styles, and encouraged her to play with other girls, by all accounts Brenda was not comfortable or happy being a girl. She was teased incessantly for her rough and aggressive ways. It got worse when Brenda was given hormones at age 11 to initiate development of secondary sexual characteristics. The development of breasts was a source of intense embarrassment and horror for Brenda that did not fade with time.

As the years passed, Janet and Ron were finally forced to consider that Brenda was not—nor would ever truly be—a girl. After 15 years of family and peer problems and intense psychological troubles, Brenda was told the truth about what had happened. A flood of emotions welled up within her, but the most overwhelming feeling was relief. In her own words, Brenda recalled that day: "I was relieved. Suddenly it all made sense why I felt the way I did. I wasn't some sort of weirdo. I wasn't crazy."

Brenda immediately decided to return to being male. She stopped hormone therapy. She changed her name to David, which she chose because of the biblical story of David and Goliath. Although he had serious adjustment problems for a few years, David eventually sorted them out. New surgical techniques allowed physicians to provide David with a functional artificial penis that could be used for sexual intercourse. At the age of 23, he met a woman with three children. Jane and David fell in love. For many years, David lived a happy family life (Figure 11.27), but a series of financial setbacks and the death of his twin brother led him to become despondent. David killed himself in May 2004 at age 38. Whether the stress of being a boy raised as a girl contributed to problems with identity that troubled him throughout his adult life is unclear. What we learned from David's life is that one's gender identity is not shaped solely by whether one is treated as a boy or as a girl. ■

11.26 Brian and Bruce Reimer A photo of Brian (at left) and Bruce (Brenda) Reimer taken when they were children. The fact that they were identical twins made the situation ideal for studying the effects of culture on gender identity.

11.27 David (Bruce) Reimer

Friends Influence Adolescent Identity and Behaviour

Developmental psychologists increasingly recognize the importance of peers in shaping identity. Children, regardless of their culture, tend to spend a large amount of their time interacting with other children, usually playing. Play takes numerous forms and helps children learn and practice skills that will be valuable in adulthood.

In developmental terms, attention to peers begins at the end of the first year of life, when infants begin to imitate other children and to smile and make vocalizations and other specific social signals to their peers (Brownell & Brown, 1992). Children of all ages learn how to behave from their friends, in part because they receive social rewards or punishments for behaving appropriately or inappropriately. Thus, early friends are both playmates and teachers.

How children compare themselves with peers also influences the development of identity as they come to know their personal strengths and weaknesses. Adolescence is a period of struggle to answer such essential questions as Who am I? and Where am I going? It is thought that adolescents often question who they are because of three changes: (1) changing physical appearance, which leads to examining self-image; (2) more sophisticated cognitive abilities, which prompt increased introspection; and (3) heightened pressure to prepare for the future and, in particular, to make career choices. Establishing a personal identity means breaking away from childhood beliefs by questioning and challenging parental and societal ideas (Erikson, 1968). Adolescents explore alternative belief systems and wonder what they would be like if they were raised in other cultures or in other times. The crisis is the search for an identity that "fits." As part of the search for identity, they may form friendships with other adolescents with similar values and worldviews. Interestingly, despite wide differences in geographic regions, peer groups tend to be described by a fairly small set of stereotypic names: jocks, brains, loners, druggies, nerds, and other not-so-flattering designations.

PARENTS AND PEERS In the last decade, the impact of parents versus peers on young people has become a controversial topic in developmental psychology. People often describe individuals as "coming from a good home" or as having "fallen in with the wrong crowd"; these clichés reflect the importance that is placed on both parents and peers in influencing an individual. In 1995, Judith Rich Harris suggested that beyond choosing where to live and what schools to send their children to, parents contribute surprisingly little to a child's social development in adolescence. Harris reviewed a number of psychological studies and concluded that parents have "no important long-term effects on the development of their child's personality" (p. 458). As mentioned in Chapter 3 and discussed in greater detail in Chapter 12, once you control for the genes they share, siblings raised in the same house are no more alike than strangers plucked off the street. Likewise, twin studies indicate that growing up with the same parents has only a small influence on personality. Harris posited that a child's peers are the most important influence when it comes to socialization. Harris's work is based largely on her *group socialization theory,* which proposes that children learn two sets of behaviours, one for inside the home and one for outside. The behaviours and responses children learn inside the home, such as those typically taught by the parents early in life, are not useful in outside social contexts. According to Harris, only those behaviours learned outside the home have long-term effects on personality and adult behavioural outcomes. Although Harris's theoretical stance has received a great deal of criticism, it has stimulated a fresh look at the social lives of children and the importance of peers.

Are parents or peers more important for social development?

In contrast to Harris's theory, a substantial amount of research has confirmed the influence of parents throughout the individual's life. Significantly, researchers have emphasized that neither the peer group nor the family can be assigned the primary role in a child's social development. Instead, the two contexts play complementary roles. B. Bradford Brown and his colleagues (1993) argue that the influence of parents is actually multifaceted. Parents not only contribute to specific individual behaviours but also affect social development indirectly by influencing the choices the child makes about what kind of crowd to join. In observations of 695 young people from childhood through adolescence, Robert and Beverly Cairns (1994) found that parents and teachers played a major role in realigning social groups so that they were consistent with family norms.

Identity Includes Moral Values

Do you view yourself as a moral person? Morality is an important component of social development because it involves choices people make that affect the lives of others, such as whether to take actions that may harm another or that may break implicit or explicit social contracts. Thus, moral development concerns the way in which people learn to decide between behaviours with competing social outcomes (Walker, 1989; Wark & Krebs, 1996). Typically, morality has been divided into moral reasoning, which rests on cognitive processes, and moral emotions. Of course, cognition and affect are intertwined: Research has shown that if people lack adequate cognitive abilities, their moral emotions may not translate into moral behaviours; similarly, moral reasoning is enhanced by moral emotions.

Psychologists who study the cognitive processes of moral behaviour have focused largely on Lawrence Kohlberg's stage theory. Kohlberg tested moral reasoning skills by asking people to respond to hypothetical situations in which a main character is faced with a moral dilemma, such as having to steal a drug to save his dying wife. Kohlberg was most concerned with the reasons children provided for their answers. He devised a theory of moral judgment that involved three levels of moral reasoning. Children in the *preconventional level* classify answers in terms of self-interest or pleasurable outcomes. For example, a child at this level may say, "He should steal the drug if he really likes his wife." At the *conventional* level responses conform to rules of law and order or focus on others' disapproval, such as "He shouldn't take the drug because it is wrong to steal, so everyone will think he is a bad person." At the highest, *postconventional* level, responses centre around complex reasoning about abstract principles and values, such as "Sometimes people have to break the law if the law is unjust."

Moral reasoning theories have been faulted for emphasizing the cognitive aspects of morality. Some theorists contend that moral reasoning, as such, fails to predict moral behaviour. These psychological scientists believe that moral actions, such as helping others in need, are influenced more by emotions than by cognitive processes (Carpendale, 2000).

MORAL EMOTIONS Research on the emotional components of moral behaviour has focused largely on empathy, sympathy, guilt, and shame. These are called moral emotions because they relate to moral behaviours (as opposed to the cognitive processes, which predict moral reasoning). Along with embarrassment, they are considered "self-conscious" emotions, because they require comprehension of the self as a causal agent and an evaluation of one's own responses. Moral emotions form early in life, though they emerge later than primary emotions (such as happiness or anger), which is why they are also called secondary emotions.

Psychologists differentiate between empathy and sympathy. Empathy is an emotional state that arises from understanding another's emotional state in a manner similar to what the other person is feeling or would be expected to feel in a given situation. In contrast, sympathy arises from feelings of concern, pity, or sorrow for another (Eisenberg, 2000). When someone feels empathy, he feels *with* the other person, whereas when someone feels sympathy, she feels *for* the other person.

Recent research has asked whether parents' behaviours influence their children's level of moral emotions or prosocial behaviour (see also Chapter 10). Parents of sympathetic children tend to be high in sympathy themselves, to allow their children to express negative emotions in a manner that does not harm others, to not express hostility in the home, to help their children cope with negative emotions, and to promote an understanding of and focus on others. In contrast, parents whose children were high in shame tended to show frequent anger, to be lax in discipline, and to not respond positively to appropriate behaviour in the child (Ferguson & Stegge, 1995).

The interaction between the child and the parents is important in the development of moral emotions. Research has shown the value of reasoning with the child about his or her behaviour, which involves inductive reasoning, as exemplified by the statements, "You made Chris cry. It's not nice to hit." Inductive reasoning on the part of the parents promotes sympathetic attitudes, feelings of guilt, and an awareness of others' feelings. One study reported that when mothers of four-year-olds commented on emotions and intentions ("you didn't mean to hurt him") and made evaluative statements ("you were good to do that for her"), their children were more likely to show appropriate forms of guilt and regret after misbehaviour (Laible & Thompson, 2000). Attachment styles, too, have been linked to the presence of moral emotions. A secure attachment style between mother and child has been shown to promote appropriate guilt, empathy, and sympathy.

PHYSIOLOGICAL BASIS OF MORALITY Some evidence from the biological level of analysis indicates that moral emotions are based in physiological mechanisms that help people make decisions. Recall from Chapter 10 that Damasio's somatic-marker hypothesis states that people have a visceral response to real or imagined outcomes and that this response aids decision making. Damasio found that patients with damage to the prefrontal cortex fail to become emotionally involved in decision making because their somatic markers are not engaged. Damasio and colleagues (Anderson, Bechara, Damasio, Tranel, & Damasio, 1999) recently studied two people who had experienced prefrontal damage during infancy. Both of these individuals showed serious deficiencies in moral and social reasoning. When given Kohlberg's moral dilemma task, both patients scored at the preconventional level. These patients also neglected social and emotional factors in their life decisions. Both failed to express empathy, remorse, or guilt for wrongdoing, and neither had particularly good parenting skills. One engaged in petty thievery, was verbally and physically threatening (once to the point of physical assault), and frequently lied for no obvious reason. Thus, the frontal lobes appear to support the capacity for morality. However, it is also clear that children develop a sense of morality through socialization, especially from their parents. The development of a personal code of ethics plays a prominent role in how people define themselves.

People Define Themselves in Terms of Race and Ethnicity

There are approximately six billion people on earth. Given that these people fall into a variety of groups, how individuals develop a sense of their own and others' racial or ethnic identities is an important component of social development.

Researchers have studied how children acquire the racial and ethnic categories prevalent in their community, identify their own race or ethnic group, and form stable attitudes toward their own and others' groups (Spencer & Markstrom-Adams, 1990).

Several studies have demonstrated that by the time children are four years old, they are already beginning to categorize others ethnically and racially. In one study, children demonstrated that they can sort dolls and pictures into racial categories (Bigler & Liben, 1993). These results, however, do not reveal how this ability contributes to the way a child develops ethnic stereotypes or self-concepts. Understanding how children come to form categories about race and ethnicity has been a controversial topic. The majority opinion to date is that shortly after children acquire the ability to make categorical distinctions about ethnicity and race, they become aware of their own ethnicity and form judgments about it. Their attitudes toward their own and other people's ethnicity depend on both the attitudes of their adult caregivers and their perceptions of the power and wealth of their own group in relation to other groups.

The process of identity formation in a country like Canada, where people of so many ethnicities live together, is particularly complicated. Due to prejudice and discrimination and the accompanying barriers to economic opportunities, children of ethnic minorities often face challenges with regard to the development of their ethnic identity. Children entering middle childhood have acquired an awareness of their ethnic identity to the extent that they know the label and attributes that the dominant culture applies to their ethnic group. Many researchers believe that during middle childhood and adolescence, children in ethnic minority groups often engage in additional processes aimed at ethnic identity formation (Phinney, 1990).

In this section we have examined how identity develops from childhood into adolescence. But development does not stop when people become adults. Psychologists now are interested in how people continue to develop across the life span. In the next section we will examine how cognition and social behaviour change as people mature through adulthood.

REVIEWING the Principles | How Do Children and Adolescents Develop Their Identities?

How people define themselves is influenced by cultural beliefs about factors such as race, sex, and age. From an early age, children classify themselves on the basis of their biological sex, but how they come to understand the meaning of being a boy or a girl is largely determined through the socialization of gender roles, in which children adopt behaviours that are viewed as appropriate for their sex. Adolescents struggle for identity by questioning social values and personal goals, as they try to figure out who they are and what they want to become. Moral development helps adolescents define what they value and what is important to them. Finally, race and ethnicity shape our beliefs and attitudes regarding how we fit into the culture in which we are raised.

What Brings Meaning to Adulthood?

For many years developmental psychologists focused on childhood and adolescence, as if by age twenty most important aspects of development had occurred and little was left to consider. In recent decades, researchers working in a wide range of fields have demonstrated that important changes in physiology, cognition, and social behaviour continue throughout adulthood into old age. Therefore many contemporary psychological scientists consider development from a *life-span perspective,* trying to understand how mental activity and social relations change over the entire course of life. It's worth noting here that the research shows that we should not automatically equate growing old with despair. In fact, there are many positive things that happen as we grow older. Although aging is associated with significant cognitive and physical decline, it remains an important part of life that can be very meaningful. As people live longer, understanding old age becomes especially important.

One of the first researchers to take this approach was Erik Erikson (Figure 11.28), who proposed a theory of development that emphasizes age-related psychological processes and their effects on social functioning across the life span. Erikson thought of development as composed of eight stages, ranging from an infant's first year to old age. Further, he conceptualized each stage as having a developmental "crisis" to be confronted, although the term refers more to a development challenge than to some form of emergency. Each challenge provides an opportunity for psychological progression, but if progress is not made, further psychosocial development is impaired (see Table 11.1). Three of Erikson's stages take place during adulthood and for our purposes we are going to focus on those stages exclusively.

11.28 Erik Erikson Erikson proposed a stage theory of identity development. This picture was taken in 1988, with his wife Joan.

TABLE 11.1 Erikson's Stages of Identity

STAGE	AGE	IDENTITY CHALLENGE	SUCCESSFUL RESOLUTION OF CHALLENGE
Infancy	0–2	Trust vs. mistrust	Children learn that the world is safe and that people are loving and reliable.
Toddler	2–3	Autonomy vs. shame and doubt	Children are encouraged to explore the environment and gain feelings of independence and positive self-esteem.
Preschool	4–6	Initiative vs. guilt	Children develop a sense of purpose by taking on responsibilities, but also develop the capacity to feel guilty for misdeeds.
Childhood	7–12	Industry vs. inferiority	By working successfully with others, children learn to feel competent.
Adolescence	13–19	Ego identity vs. role confusion	Through exploration of different social roles, adolescents develop a sense of identity.
Young adulthood	20s	Intimacy vs. isolation	Young adults gain the ability to commit to long-term relationships.
Middle adulthood	30s to 50s	Generativity vs. stagnation	Adults gain a sense that they are contributing to the future and caring for future generations.
Old age	60s and beyond	Integrity vs. despair	Older adults feel a sense of satisfaction that they have lived a good life and developed wisdom.

Erikson's sixth stage involves close relationships and the challenge of *intimacy versus isolation,* which means the formation and maintenance of committed friendships and romantic relationships. Essentially, it involves finding someone with whom to share your life. The seventh stage takes place during middle life. Termed *generativity versus stagnation,* this stage involves producing or giving back to society, and usually includes parenthood or engaging in activities, such as volunteering, that bring additional meaning to life. Erikson's last stage, *integrity versus despair,* takes place in old age. In this stage, older adults reflect on their lives and respond either positively to having had a worthwhile life or with regret and sadness at what has passed. Erikson's theory highlights the importance of considering how people care about different things as they grow older. In this section we examine the physical, cognitive, and social changes that occur during adulthood and how they affect people's quality of life.

Adults Are Affected by Life Transitions

For many young people, university is a magical time of life. Meeting new friends, learning new ideas, having a good time, all occurring as adolescence emerges into adulthood. People in their twenties and thirties undergo incredible changes as they pursue career goals and make long-term commitments in relationships, such as getting married and raising children. All this corresponds to the Eriksonian idea that challenges face us as we mature through adulthood. In essence, the major challenges of adulthood reflect the need to find meaning in our lives. If life involved only getting up in the morning, working, eating, and sleeping, what would be the point? Most people seek more than that from life. They want jobs that fulfill them and they want family members to share their lives. Let's examine how the quest to satisfy these needs influences the course of human lives.

CAREER Children are commonly asked what they want to be when they grow up. Although some university students are committed to achieving specific career goals, as when someone has wanted to be a physician since age ten, many others struggle to find careers that they hope will be satisfying. You will work around 100,000 hours in your life. Imagine spending all that time on something you don't like. Career plays an important role in most people's lives. A good job not only provides material rewards, such as money to buy things you like, but also can bring a sense of accomplishment and purpose. The right career helps you feel that you contribute to society and that other people recognize you for doing so. It brings meaning to your life and helps you grow as a person and become better able to fulfill your life's goals. So how do you find the right job?

Although selecting a career path is rarely easy, one of the most important principles is to be honest with yourself. Although you don't necessarily need a specific plan upfront, you have to assess your strengths and weaknesses and try to develop some general ideas of what kind of career you can pursue effectively and what kind is likely to be fulfilling to you personally. Career counselors often advise people to follow their passions rather than the wishes of parents or family members. It is not uncommon for people who have been toiling away in corporate or professional jobs to have midlife crises in which they realize their jobs fail to bring them any meaningful satisfaction. Many times this happens because people focus on money rather than how the right job would allow them to be the type of parent, spouse, or community member they want to be. From an Eriksonian perspective, the motive for generativity inspires us to want to give something of ourselves to society. In the end, the right career allows you to pursue your interests and gain a sense of accomplishment as you balance family and other life goals.

MARRIAGE In adulthood, people devote a great deal of effort to achieving and maintaining satisfying relationships. Indeed, the vast majority of people around the world marry at some point in their lives or form some type of permanent bond with a relationship partner. The search for the right partner is an important feature in contemporary Western society, although in other cultures parents arrange marriages and decide whom their son or daughter will marry. Although it is hard for many Westerners to imagine, arranged marriages tend to be stable and the people within them quite satisfied.

One clear advantage to marriage is that married people tend to be healthier and to live longer (Berkman, 1984; Gottman, 1998). For example, unmarried heart patients are much more likely to die within five years of being diagnosed with heart disease (Williams et al., 1992). Men who are married survive much longer with prostate cancer than those who are unmarried, divorced, separated, or widowed (Krongrad, Lai, Burke, Goodkin, & Lai, 1996). A number of theories try to explain why married people live longer, but most of them focus on how marital partners can support each other, as, for example, helping each other deal with stress or assisting each other in meeting life's demands. Married people may also influence the behaviour of their partners by encouraging them to be healthy, such as by eating properly and getting exercise.

Given the likelihood that most people will marry, what are the chances that their marriages will be satisfying and stable? According to national surveys, at any given time the vast majority of married people report satisfaction with their marriages. The people who are most satisfied are those who have sufficient economic resources, share decision making, and together hold the view that marriage should be a lifelong commitment (Amato, Johnson, Booth, & Rogers, 2003). Yet in contemporary Canadian society, approximately one-third of marriages end in divorce. Of those couples who stay together, a sizeable portion experience marital problems. It has been estimated that the chance of ending up in a happy and intact marriage is about 1 in 3 (Glenn, 1998). In Chapter 15 we will discuss ways to have a more satisfying relationship.

HAVING CHILDREN According to Erikson, one of the major ways that people satisfy the challenge of generativity is through having children. The arrival of a first child is a profound event for most couples. It changes their lives in almost every respect, from sleep being interrupted nightly to not being able to go out with friends at a moment's notice. First-time parents are also often anxious about how to respond to the child's cries and can be frustrated in trying to figure out why the child is distressed. But new parents also experience great joys—seeing a baby's first social smile, watching the first few tentative steps, and hearing a child proclaim "mama" or "dada" provide powerful reinforcement for becoming parents. Parents often become immersed in their children's lives, making sure the children have playmates, exposing them to new experiences, and trying to make sure they are happy and healthy. Being a parent is central to how many adults come to define themselves.

Does having children bring meaning or happiness?

Although many people are eager to have children, research shows that children can be a strain on marriages, especially when time and money are tight. A consistent finding is that couples with children, especially adolescents, report less marital satisfaction than those who are childless (Belsky, 1990). Although satisfaction might be lower, the marriages of those with children appear to be more stable. For many people, having children is not about happiness, but about how they define their lives, reflecting the value they place on family. They look forward to coaching their kids in sports, encouraging them in school, and watching them grow up, have careers, get married, and provide grandchildren. In other words, along with career and marriage, children bring meaning to life.

Aging Can Be Successful

How do older adults seek out meaning in their lives? According to Erikson, older adults look back and evaluate what they have done with their lives. To the extent that they consider their time well spent, older adults are satisfied and able to live their final years gracefully. The crisis of this stage can be triggered by events that highlight the mortal nature of human life, such as the death of a spouse or close friend, or by changing social and occupational roles, such as retirement. Resolving such challenges allows people to come to terms with the reality of death. Although people of all ages are concerned with the meaning of life, it often becomes a preoccupation for the elderly. Here we examine the psychological processes that characterize the final years of life.

With the "graying" of the population in Western societies, much greater research attention has been given to the lives of people over age 60. By 2015, more than 1 in 5 North Americans will be over age 65. People are living much longer these days, with the number of people over age 85 growing dramatically. Indeed, it is becoming commonplace for people to live beyond 100 years of age. Not only are there more elderly people, but within North America they are increasingly ethnically diverse, well educated, and physically fit.

Stereotypes regarding aging differ greatly across cultures. Just think of the mixed messages that are reflected in Canadian popular culture. For example, films and television often depict older First Nation Canadians as wise and knowing, whereas Anglo-Canadian grandparents are often shown as doddering and senile. Yet the elderly contribute a great deal to modern society. Many older adults live with vitality. Consider Fred Hale, who died in November 2004, twelve days shy of his one hundred fourteenth birthday. He passed a driver's examination at age 104, and his grandson reported that he still found slower drivers annoying when he was 108. When he was 109, he traveled to Japan to visit a grandson. Our view of the elderly is likely to change a great deal as the baby boomer generation ages. Consider music stars, such as Bruce Springsteen, Tina Turner, and the Rolling Stones, who remain popular and vibrant well into their fifties, sixties, and beyond, certainly in defiance of most stereotypes of old people.

However, we cannot ignore the inevitable physical changes that accompany aging, even trivial changes such as graying hair and wrinkled skin. The brain also changes with age, with the frontal lobes shrinking proportionally more than other brain regions. Although scientists once believed that cognitive problems such as confusion and memory loss were an inevitable normal part of aging, they now recognize that most older adults remain alert as they grow older, although they do everything a bit more slowly. Older adults who experience a dramatic loss in mental ability often suffer from a brain condition called *dementia,* which is a progressive deterioration of thinking, memory, and behaviour. Dementia has many causes, including excessive alcohol intake and HIV, but for older adults the major causes are Alzheimer's disease and small strokes that affect blood supply in the brain. After age 70, the risk of dementia increases with each year of life. Approximately 3 to 5 percent of people will develop Alzheimer's disease by age 70 to 75, and then the proportion of people with the disease doubles every five years. The disease is named after Dr. Alois Alzheimer, who in 1906 observed changes in the brain tissue of a woman who had died of an unknown mental illness. Upon inspection, Alzheimer noticed abnormal clumps and tangled bundles of fibre in the brain. The presence of such *plaques* and *tangles* is now considered telltale signs of Alzheimer's disease. Recall from Chapter 3 that acetylcholine neurons are damaged in Alzheimer's disease patients, which may explain the impairments in thought and memory (Feldman et al., 2006). The cause of Alzheimer's is not currently known although there is evidence that genetics contributes to its development. One gene involved in cholesterol functioning has been shown to be predictive of Alzheimer's, although how this gene is related to the disease is not currently known (Corder et al., 1993). The initial symptoms of Alzheimer's disease are typically minor memory impairments, but the disease eventually progresses to more serious difficulties, such as forgetting daily routines. Eventually, the person loses all mental capacities, including memory and language (Figure 11.29).

Despite the physical, social, and emotional challenges of aging, most older adults are surprisingly healthy and happy. Indeed, some individuals thrive in old age, especially those with adequate financial resources and good health (Crosnoe & Elder, 2002). Although it is true that the body and mind do slowly deteriorate, starting at about age 50, the majority of older adults consider themselves to be in good health. Moreover, older adults report being just as satisfied with life as, if not more satisfied than, younger adults (Mroczek & Kolarz, 1998). Indeed, except for dementia, older adults have fewer mental health problems than younger adults, including depression.

According to psychologist Laura Carstensen's *socioemotional selectivity theory,* as people grow older they perceive time to be limited and therefore adjust their

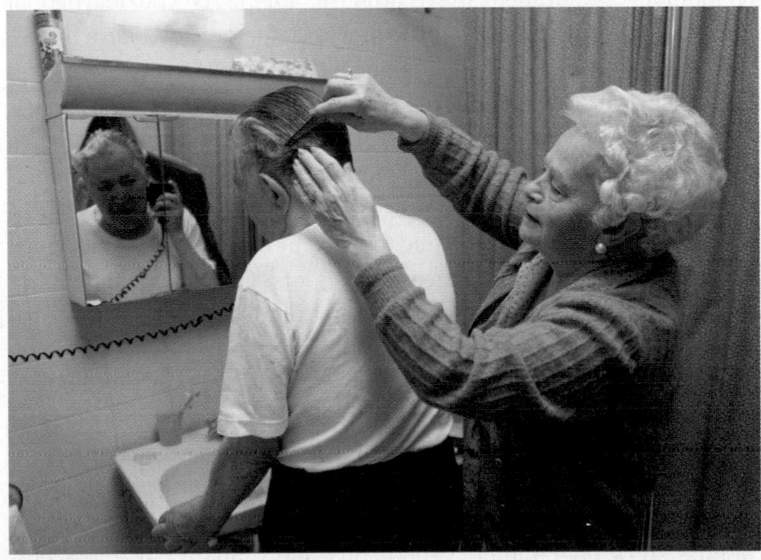

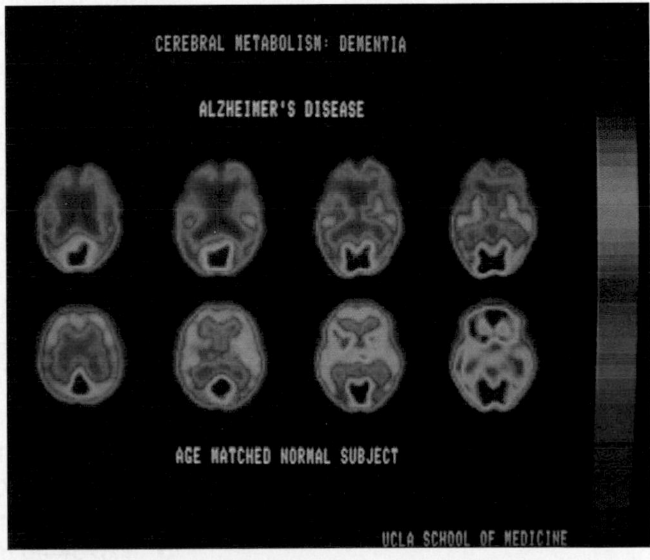

11.29 Alzheimer's Disease A wife helps her husband, who is an Alzheimer's patient (left). PET scans comparing the brain of an Alzheimer's patient with a healthy brain (right).

Why do older people's motives for novelty change?

priorities to emphasize emotionally meaningful events and experiences (Carstensen, 1995; Fung & Carstensen, 2004). According to this theory, people focus more on emotional and rewarding goals as they near the end of life. For instance, they may choose to spend more time with a smaller group of close friends and avoid new people whom they might find annoying. They may spend an increasing amount of time reflecting on their lives and sharing memories with family members and friends. As they reminisce about their lives, older adults report more positive than negative emotions (Pasupathi & Carstensen, 2003). In essence, older adults want to savour their final years by putting their time and effort into having meaningful and rewarding experiences.

Cognition Changes during Aging

Cognitive abilities eventually decline with age, but it is difficult to pinpoint exactly what causes the decline. One of the most consistent and identifiable changes is a slowing of mental-processing speed. Experiments that test the time it takes to process a sensory input and react with a motor response show an increase in response time as early as an individual's mid-twenties (Era, Jokela, & Heikkinen, 1986). This increase becomes more rapid as the individual ages.

Some sensory-perceptual changes occur with age and may account for some of the observed decline. For instance, as we age our sensitivity to visual contrast decreases. Thus activities such as climbing stairs or driving at night may become more difficult and more dangerous. Sensitivity to sound also decreases with age, especially the ability to tune out background noise. This change may make older people seem confused or forgetful when they simply are not able to hear adequately. Here we examine two mental processes that change with age.

MEMORY Older people have difficulty with memory tasks that require the ability to juggle multiple pieces of information at the same time. Tasks in which attention is divided, such as driving while listening to the radio, also prove difficult. Some scientists believe these deficits reflect a decreased ability to store multiple pieces of information in working memory simultaneously (Salthouse, 1992). As you learned in Chapter 7, the frontal lobes play an important role in working memory. There is consistent evidence that the frontal lobes typically shrink as people grow older, and many of the cognitive skills that rely on the frontal lobes, such as working memory, show impairment with advancing age.

Generally speaking, long-term memory is less affected by aging than is short-term memory, although certain aspects of long-term memory do appear to suffer in advanced age. Older people often need more time to learn new information, but once they learn it, they use it as efficiently as younger subjects. It is also clear that elderly subjects are better at recognition than at retrieval tasks. For example, they have no trouble recognizing words that have been shown to them if they are asked, "Did you see the word *cat?*" But if they are asked what word they saw, or whether they saw an animal name, they do not do as well. Consistent with the socioemotional selection theory, older people also show better memory for positive than for negative information (Kennedy, Mather, & Carstensen, 2004), perhaps because older adults selectively ignore negative events (Mather & Carstensen, 2003).

Why do older adults show poor memory? Is it possible that they use less efficient strategies for encoding information to be remembered? In an intriguing study, Jessica Logan and colleagues (2002) examined memory processes in younger (in their twenties) and older adults (in their seventies and eighties). The older adults performed worse than young adults. In terms of brain processes, they had less activation in left-hemisphere brain areas known to support memory and greater activation in right-hemisphere areas that do not aid memory. Next, these researchers sought to determine whether the memory deficit could be reduced if they gave the older subjects a strategy to improve memorization. Recall from Chapter 7 that the more deeply an item is encoded, the better it is remembered. Accordingly, in a second study the researchers asked older participants to classify words as concrete or abstract, a strategy that leads to deeper encoding. Undertaking this classification produced better memory and greater activation of the left frontal regions. These findings suggest that one reason for the decline in memory observed with aging is that older adults tend not to use strategies that facilitate memory, which raises the possibility that cognitive training might be useful for postponing age-related memory deficits.

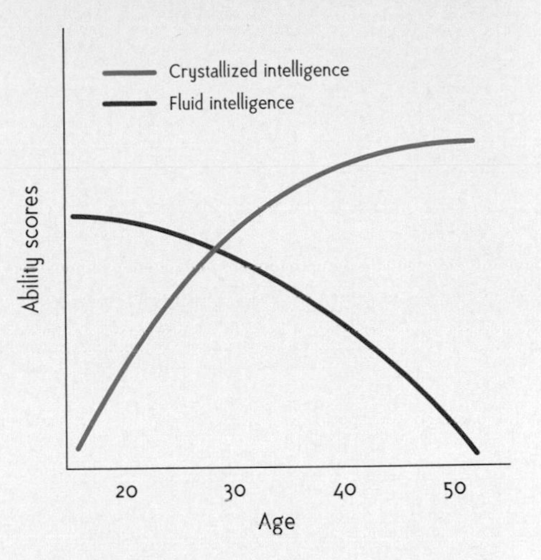

11.30 Fluid and Crystallized Intelligence throughout the Life Span Unlike fluid intelligence, which shows a steady decline with age, crystallized intelligence increases through at least the 40s and 50s and then levels off.

Does intelligence deteriorate with aging?

INTELLIGENCE Research has consistently indicated that intelligence, as measured on standard psychometric tests, declines with advanced age. As we age, do we really lose IQ points? Or do older people just have a shorter attention span or lack the motivation to complete such tests? Some researchers have made a distinction between different forms of intelligence: fluid intelligence and crystallized intelligence (Horn & Hofer, 1992). As you will recall from Chapter 8, *fluid intelligence* is the ability to process new general information that requires no specific knowledge. Many standardized tests measure this kind of intelligence: the ability to recognize an analogy or to arrange blocks to match a picture. This type of intelligence is associated with speed of mental processing. Fluid intelligence tends to peak in early adulthood and decline steadily as we age. *Crystallized intelligence* refers to more specific knowledge that must be learned or memorized, such as vocabulary, or to knowledge of specialized information or reasoning strategies. This type of intelligence usually increases throughout life and breaks down only when declines in other cognitive abilities prevent new information from being processed (Figure 11.30).

These two different types of intelligence also seem to reflect the two different cultural views of aging. The decline in fluid intelligence is in keeping with the stereotype of the doddering old grandparent, whereas the continued increase in crystallized intelligence represents the respected elder to whom young people turn for guidance. So which is more accurate? The Seattle Longitudinal Study tracked adults from age 25 to 81 over seven years, administering tests of cognitive abilities such as verbal and mathematical skills (Schaie, 1990). They found that intellectual decline does not actually occur until the sixties or seventies. Further, people who were healthy and remained mentally active demonstrated less decline. Thus, although memory and the speed of processing may decline, the continued ability to learn new information may mitigate those losses in terms of daily functioning. Further, we can use alternative cognitive strategies to compensate. There has been a great deal of excitement lately about the possibility that active social engagement may help older adults maintain their cognitive abilities. For instance, older adults who have more extensive social contacts and remain active through leisure pursuits may be less susceptible to dementia and Alzheimer's disease (Fratiglioni, Paillard-Borg, & Winblad, 2004). With people now living much longer, it is likely that researchers will place even more emphasis on understanding how people can maintain their cognitive capacities to get the most out of their final years.

REVIEWING

the Principles | What Brings Meaning to Adulthood?

Interest has been growing in studying humans across the entire life span, from conception to old age. Adulthood requires people to meet certain challenges, such as establishing a career, getting married, and raising a family. An overriding theme that emerges from studying life transitions is that people seek meaning in their lives and that they do so increasingly as they age. Although older adults are often characterized as feeble and senile, they are for the most part healthy, alert, and vital. Indeed, one of the biggest surprises in recent years is the finding that older people are often more satisfied than younger adults. In spite of declines in memory and speed of mental processing, older adults generally maintain their intelligence into very old age, especially if they engage in social and mental activities that help keep their mental skills in shape.

Conclusion

Born nearly helpless, infants blossom into inquisitive children, who later become questioning adolescents and stressed-out adults, and finally emerge by and large as satisfied older individuals. This is the journey of life for most of us. Throughout this journey, the interacting forces of nature and nurture work to shape our development as we emerge and grow as unique persons. As we mature, so do our mental processes. For centuries, people thought that children were born with blank minds. We now know that this is far from true. We now recognize that young children think about the world in completely different ways from adults, but now we also know that they are capable of incredible feats of cognition at very early ages, such as understanding the physical laws of the universe. As the brain develops, and especially as the frontal lobes become organized, we become social creatures, capable of sharing attention and communicating through language. Our capacity for theory of mind truly gives us a unique place in the animal kingdom as the most social of all species. Arising through development are the things that define us as people—our values, morals, and beliefs, as well as our hopes and dreams. Shaped by our interactions with parents and peers, who themselves are embedded within culture and society, we develop a sense of identity that recognizes our unique place in the world. This identity guides how we spend our lives. It shapes our career choices, helps us figure out whom to marry (if anyone), and guides us as we raise our children. And at some point we get old and our bodies and minds start to deteriorate. But there is good news: Older adults typically are much happier than we realized, which is a good thing because we are going to be spending increasingly long periods of our lives in what has traditionally been called old age. The cycle of birth, life, and death is remarkable. Even as psychological scientists explore the nature of physical and psychological development, we cannot help but marvel both at the vast amount of knowledge we acquire as children, and at how much we continue to learn throughout our lives.

Summarizing the Principles of Human Development

What Shapes a Child?

1. **Development starts in the womb:** Many factors in the prenatal environment (i.e., nutrition and hormones) can affect development. The most adverse consequences result from teratogens (i.e., drugs, alcohol, or viruses), exposure to which can result in death, deformity, or mental disorders.
2. **Brain development promotes learning:** Brain development proceeds in an interaction between maturation and experience. The plasticity of the brain allows changes in the development of connections and the synaptic pruning of unused neural connections. The timing of requisite experiences for brain development is particularly important in the early years.

3. **Attachment promotes survival:** The emotional bond that develops between a child and the caregiver aids the survival of the child. Attachment styles define the dynamic relationship between caregivers and infants, and are generally categorized as secure, avoidant, anxious-ambivalent, and disorganized. At the biological level, the hormone oxytocin facilitates attachment.

4. **Parental style can affect children's well-being:** Children's temperaments interact with their parents' behaviours and can result in a good "fit" or less optimal outcomes. Furthermore, parents influence attitudes, values, and beliefs and thus shape their childrens' views of themselves and others.

How Do Children Learn about Their Worlds?

5. **Perception introduces the world:** Experiments using the preferential looking technique and habituation have revealed the considerable perceptual ability of infants. Vision and hearing develop rapidly as neural circuitry develops.

6. **Piaget emphasized stages of development:** Jean Piaget proposed that through interaction with the environment, children develop mental schemas and proceed through stages of cognitive development. Initially, infants experience their world in a sensorimotor manner and develop object permanence; next they enter the preoperational stage, during which the appearance of objects dominates thinking; the third stage is marked by the logic of concrete operations; and finally, the formal operational stage is characterized by abstract, complex thinking.

7. **Infants have innate knowledge:** Experiments using the habituation paradigm have revealed that infants understand some of the basic laws of physics and mathematics.

8. **Memory improves over childhood:** Infantile memory is limited by a lack of language ability and autobiographical reference. Source amnesia is common in children. Confabulation, common in young children, may result from underdevelopment of the frontal lobes.

9. **Humans learn from interacting with others:** Being able to infer the mental states of others is known as theory of mind. Through socialization, children move from very egocentric thinking to being able to take another's perspective.

10. **Language develops in an orderly fashion:** Infants can discriminate phonemes. Language proceeds from making different sounds to words to telegraphic speech to sentences. Noam Chomsky asserts that language is governed by "universal grammar." According to Lev Vygotsky, social interaction is the force that develops language. Exposure to language in the sensitive period is required for its development. Interaction across cultures can shape language. Sign language develops like spoken language.

How Do Children and Adolescents Develop Their Identities?

11. **Gender roles are determined by cultural norms:** Gender identity develops in children and shapes their behaviours (i.e., gender roles). Gender schemas are the cognitive constructs of gender. People tend to rate themselves as masculine, feminine, or androgynous in stereotypic thinking. Situations affect gender-related response. Biology and culture shape gender differences.

12. **Friends influence adolescent identity and behaviour:** Social comparisons shape children's identities. Parents and peers shape identity.

13. **Identity includes moral values:** Moral reasoning develops through a series of levels from self-interest to logical, rule-bound reasoning to abstract principles and values. Parents influence the understanding and expression of moral emotions (i.e., guilt, sympathy). There is evidence of a physiological and thus developmental component of morality.

14. **People define themselves in terms of race and ethnicity:** By age four, children begin to categorize themselves and others with regard to race or ethnicity. Ethnic identity is complicated by social prejudice.

What Brings Meaning to Adulthood?

15. **Adults are affected by life transitions:** Erikson considered development from a life-span perspective, and theorized that there are important social issues to be resolved at each stage of life. For adults, development focuses on generativity: being productive with regard to career and family. Marriage is a central issue, though about half of contemporary marriages fail.

16. **Aging can be successful:** With the "graying" of the population in western societies, more research has been done on aging. There are inevitable physical changes, but dementia is not normal and has many causes, including Alzheimer's disease. Most older adults are healthy and remain productive, and often become socioemotionally selective about their relationships and activities.

17. **Cognition changes during aging:** Short-term memory is affected by aging, thus complex memory and divided-attention tasks are affected. Crystallized intelligence increases; fluid intelligence declines in old age as speed in processing declines. Being mentally active and socially engaged preserves functioning.

Applying the Principles

1. While out jogging one day in a park near campus, you stop and rest on a bench at the playground. A group of mothers are seated on the benches, chatting with one another. You notice that most of the toddlers are happily playing and occasionally run to their moms for help or comfort. One girl seems totally independent and keeps some distance from her mother. Another girl is at one moment tugging on her mother to be picked up, and then almost immediately wants to be put down. What you are observing in these children is

    ____ a) the dual roles of temperament and attachment style in development.
    ____ b) the role of biologically based temperament alone in determining social interaction.
    ____ c) the importance of play groups for developing attachment.
    ____ d) different attachment styles.

2. Baby Jennie is sitting on the floor when an orange rolls toward her from the kitchen. She picks it up and puts it in her mouth. One of her tiny teeth pierces the orange rind, causing her to grimace from the bitter taste. She tosses the orange under the couch, out of sight. When a second orange rolls toward her, she picks it up, finds the other one under the couch, and takes both to her mother in the kitchen. Jennie has demonstrated a variety of cognitive functions. These are ________ for mouthing the object; ________ for locating it under the couch; and ________ for not mouthing the second one.

    ____ a) assimilation; object permanence; accommodation
    ____ b) sensorimotor; concrete operations; preoperations
    ____ c) egocentrism; orienting; conservation
    ____ d) source amnesia; theory of mind; egocentrism

3. When you were in high school, it was clear which kids were jocks, brainiacs, Goths, hippies, etc. Whatever group you belonged to, you had some certainty that despite various individual differences, there were great similarities in interests and attitudes among those in your clique. These similarities are important for

    ____ a) being sure that people called your group the right name.
    ____ b) defining who you were in comparison to others.
    ____ c) creating an identity that was distinctly different from anyone in your group.
    ____ d) having a distinct group identity different from your personal identity.

4. Like many university students, Annie often listens to music while studying. When she visits her grandparents, they complain when she does this, and even comment that the loud music makes it impossible for them to think. From what you know about changes in adulthood, how do you explain this difference?

    ____ a) older adults spend more time thinking about unpleasant things and the upbeat style of the music interferes
    ____ b) older adults have more difficulty with divided attention
    ____ c) older adults have more complex thoughts because of their increased crystallized intelligence
    ____ d) younger adults have better memory

ANSWERS: *1.d 2.a 3.b 4.b*

Key Terms

accommodation, p. 449
adolescence, p. 463
assimilation, p. 449
attachment, p. 440
concrete operational stage, p. 450
critical period, p. 439
developmental psychology, p. 434
formal operational stage, p. 450
gender, p. 464
gender identity, p. 464
gender roles, p. 464
gender schemas, p. 464
infantile amnesia, p. 453
object permanence, p. 449
orienting reflex, p. 447
preoperational stage, p. 450
schemas, p. 449
sensorimotor stage, p. 449
social development, p. 463
source amnesia, p. 454
synaptic pruning, p. 438
telegraphic speech, p. 458
teratogens, p. 437
theory of mind, p. 454

Further Readings

Birren, J. E., & Schaie, K. W. (Eds.). (2001). *Handbook of the psychology of aging* (5th ed.). San Diego, CA: Academic Press.

Bloom, P. (2004). *Descartes' baby: How the science of child development explains what makes us human.* New York: Basic Books.

Colapinto, J. (2000). *As nature made him: The boy who was raised as a girl.* New York: HarperCollins.

Goldberg, S., Muir, R., & Kerr, J. (1997). *Attachment theory: Social, developmental, and clinical perspectives.* Hillsdale, NJ: Analytic Press.

Harris, J. R. (1998). *The nurture assumption: Why children turn out the way they do.* New York: Free Press.

Maccoby, E. E. (1998). *The two sexes: Growing up apart, coming together.* Cambridge, MA: Belknap Press/Harvard University Press.

Pinker, S. (1994). *The language instinct.* New York: Morrow.

Tomasello, M. (2001). *The cultural origins of human cognition.* Cambridge, MA: Harvard University Press.

Further Readings: A Canadian Presence in Psychology

Callaghan, T., Rochat, P., Lillard, A., Claux, M. L., Odden, H., Itakura, S., Tapanya, S., & Singh, S. (2005). Synchrony in the onset of mental-state reasoning: Evidence from five cultures. *Psychological Science, 16*, 378–384.

Carpendale, J. I. M. (2000). Kohlberg & Piaget on stages and moral reasoning. *Developmental Review, 20*, 181–205.

Deoliveira, C. A., Moran, G., & Pederson, D. R. (2005). Understanding the link between maternal adult attachment classifications and thoughts and feelings about emotions. *Attachment and Human Development, 7*, 153–170.

Eggermont, J. J. (2001). Between sound and perception: Reviewing the search for a neural code. *Hearing Research, 157*, 1–42.

Feldman, H. H., Gauthier, S., Chetkow, H., Conn, D. K., Freedman, M., & Chris, M. (2006). Canadian guidelines for the development of anti-dementia therapies: A conceptual summary. *Canadian Journal of Neurological Sciences, 33*, 6–26.

Fried, P. A. (1989). Cigarettes and marijuana: Are there measurable long-term neurobehavioral teratogenic effects? *Neurotoxicology, 10*, 577–583.

Janzen, L. A., Nanson, J. L., & Block, G. W. (1995). Neuropsychological evaluation of preschoolers with fetal alcohol syndrome. *Neurotoxicology & Teratology, 17*, 273–279.

Kelly, J. B. (1992). Behavioral development of the auditory orientation response. In R. Romand (Ed.), *Development of auditory and vestibular systems II.* Amsterdam: Elsevier Press.

Kimura, D. (1999). *Sex and cognition.* Cambridge, MA: MIT Press.

Kolb, B, & Wishaw, I. Q. (1998). Brain plasticity and behavior. *Annual Review of Psychology, 49*, 43–64.

Loney, E. A., Green, K. L., & Nanson, J. L. (1994). A health promotion perspective on the House of Commons' report "Foetal alcohol syndrome: a preventable tragedy." *Canadian Journal of Public Health, 85*, 248–251.

Lytton, H., & Romney, D. M. (1991). Parents' sex-related differential socialization of boys and girls: A meta analysis. *Psychological Bulletin, 109*, 267–296.

Marcia, J. E. (1980). Identity in adolescence. In I. Adelson (Ed.), *Handbook of adolescence*. New York: Wiley.

Montreal Symposium (1994). Development and plasticity of the visual system. Proceedings of a symposium, Montreal, Quebec, Canada. *Canadian Journal of Physiology and Pharmacology, 73*, 1294–1405.

Pederson, D. R., & Moran, G. (1996). Expressions of the attachment relationship outside of the strange situation. *Child Development, 67*, 915–927.

Sareen, J., Fleisher, W., Cox, B. J., Hassard, S., & Stein, M. B. (2005). Childhood adversity and perceived need for mental health care: Findings from a Canadian community sample. *Journal of Nervous and Mental Disease, 193*, 396–404.

Senechal, M., & LeFevre, J. A. (2002). Parental involvement in the development of children's reading skill: A five year longitudinal study. *Child Development, 73*, 445–460.

Walker, L. J. (1989). A longitudinal study of moral reasoning. *Child Development, 60*, 157–166.

Wark, G. R., & Krebs, D. L. (1996). Gender and dilemma differences in real-life moral judgment. *Developmental Psychology, 32*, 220–230.

Wong, D., & Shah, C. P. (1979). Identification of impaired hearing in early childhood. *Canadian Medical Association Journal, 121*, 529–532, 535–536.

Identifying Influences on Personality

On their MTV reality show, *The Osbournes,* Kelly and Jack Osbourne often seemed to be imitating the behaviour of their notoriously wild father, rock musician Ozzy Osbourne. Cases like this one interest psychological scientists, who examine to what extent personality is inherited and to what extent it is subject to cultural and social influences. Recent evidence supports the idea that personality is rooted in genetics, although how personality traits are expressed also depends upon a person's environment.

CHAPTER 12

Personality

Tom Hanks is among the most likeable and talented actors of our time, winning Academy Awards for his performances in *Philadelphia* and *Forrest Gump*. Given everything that you know about Hanks, what kind of a person do you think he is? Do you think he is as nice off screen as he is on screen? Where does he get the confidence to act in movies that will be seen and judged by millions? The word *personality* comes from the Latin *persona,* a mask. In ancient Greece and Rome, actors would speak through masks, with each different mask representing a complete separate personality. So, who is the real Tom Hanks? We are constantly trying to figure other people out, so that we can understand why they behave the way they do and so that we can predict their future behaviour. Many students report that one reason they take psychology courses is that they want to know what makes other people tick. We love to analyze people to figure

Outlining the Principles

How Have Scientists Studied Personality?

- Psychodynamic Theories Emphasize Unconscious and Dynamic Processes
- Humanistic Approaches Emphasize Integrated Personal Experience
- Type and Trait Approaches Describe Behavioural Dispositions
- Personality Reflects Learning and Cognitive Processes

How Is Personality Assessed, and What Does It Predict?

- Personality Refers to Both Unique and Common Characteristics
- We Can Use Objective and Projective Methods to Assess Personality
- Observers Show Accuracy in Trait Judgments
- People Sometimes Are Inconsistent
- Behaviour Is Influenced by the Interaction of Personality and Situations

What Is the Biological Basis of Personality?

- Personality Is Rooted in Genetics
- Temperaments Are Evident in Infancy
- Personality Is Linked to Specific Neurophysiological Mechanisms
- Personality Is Adaptive

Can Personality Change?

- Traits Remain Stable over Time
- Characteristic Adaptations Change across Time and Circumstances
- Brain Injury and Pharmacological Interventions Affect Personality

out their personalities. One question that we will ask in this chapter is, "What do we need to know to know a person well?"

Would it surprise you to learn that Tom Hanks was painfully shy when he was growing up? He attributes this shyness to the frequent changes he experienced during childhood. By age 10, Hanks had had three different mothers, had attended five different schools, and had lived in 10 different houses, all of which hindered his chances for making lasting friendships. Lacking the social skills to make friends, Hanks found that he could get attention by being the class clown. He would take on different personas and make people laugh, which was excellent training for someone destined to become an actor.

Hanks soon found that he had a natural skill for changing his personality to suit any situation. In 1997 he told the *Guardian* newspaper that he had learned to adjust his behaviour to what was expected: "If I was supposed to be repressed, I could repress myself; if I was supposed to be outgoing, I could be outgoing; if I was supposed to be raucous, I could be raucous. I think I still do that sometimes, I'm still good at it." So who is the real Tom Hanks?

One of the challenges of figuring people out is that they sometimes act differently, often depending on the situation. How much can a person's behaviour tell us about his or her personality? What else do we need to know to really understand someone? Does knowing that Hanks moved around a lot as a child help us to understand his unique nature? Each person is unique. What determines this uniqueness? For instance, what factors during childhood affect personality? Have you ever wondered what makes you the person you are?

Does knowing that Hanks was shy as a child tell us anything about what he is like as an adult? An important question for researchers who study personality is the extent to which personality changes throughout a person's life. People who are shy spend a great deal of time worrying about what others think of them. Although most people feel shy on some occasions, some people feel shy all the time, so much so that it interferes with making friends or achieving their goals. If Hanks is still shy, how can he stand in front of thousands of people to give an acceptance speech at the Academy Awards? Did becoming involved in acting as a teenager allow Hanks to overcome his shyness? Can people just decide not to be shy? These are the sorts of questions that psychologists grapple with as they seek to understand what makes people unique. This chapter is concerned with the scientific understanding of human personality. ■

200 C.E.
Humoural Theory of Temperament The physician Galen formalizes Hippocrates' idea that personality is rooted in the body, arguing that it is determined by the relative amounts of different humours, or fluids.

1890s–1930s
Freudian Psychodynamics Sigmund Freud proposes that unconscious processes that originate in early childhood experiences determine much of human personality.

1920s
Neo-Freudians Some of Freud's followers, such as Carl Jung and Karen Horney, reject his emphasis on sex, although they share his belief in the importance of unconscious processes.

1930s
Study of Lives Henry Murray uses projective measures to examine unconscious needs and champions the scientific study of whole persons.

Understanding personality may be among the oldest quests in psychology. Since antiquity, an incredibly wide array of grand theories have been proposed to explain basic differences between individuals. **Personality** refers to an individual's characteristics, emotional responses, thoughts, and behaviours that are relatively stable over time and across circumstances. Personality psychologists study the basic processes that influence the development of personality on a number of different levels of analysis, such as the influence of culture, learning, biology, and cognitive factors. At the same time, those who study personality are most interested in understanding *whole persons*. That is, they try to understand what makes each person unique. People differ greatly in many ways, as you have no doubt noticed. Some are hostile, some are loving, and others are withdrawn. Each of these characteristics is a **personality trait**, a dispositional tendency to act in a certain way over time and across circumstances. What sort of person are you? More to the point, Why are you who you are?

personality Characteristics, emotional responses, thoughts, and behaviours that are relatively stable over time and across circumstances.

personality trait A characteristic; a dispositional tendency to act in a certain way over time and across circumstances.

How do we really know a person?

How Have Scientists Studied Personality?

Dan McAdams (Figure 12.1), a leading personality researcher, has posed the interesting question, "What must we know to know a person well?" The specific ways that psychologists try to answer this question vary greatly, often depending on their overall theoretical approach. Some psychological scientists emphasize the biological and genetic factors that predispose behaviours. Others might emphasize culture, patterns of reinforcement, or mental and unconscious processes. To really understand people is to understand everything about them, from their biological makeups, to their early childhood experiences, to the way they think, to the cultures in which they were raised. All of these factors work together to shape a person in a unique way. Thus, personality psychologists approach the study of personality on many levels.

Gordon Allport (Figure 12.2), who published the first major textbook on personality, in 1937, gave perhaps the best working definition of personality: "the dynamic organization within the individual of those psychophysical systems that determine his characteristic behavior and thought" (1961, p. 28). This definition includes many of the concepts most important to a contemporary understanding of personality. The notion of *organization* indicates that personality is not just a list of traits but a coherent whole. Moreover, this organized whole is *dynamic*, in that it is goal seeking, sensitive to context, and adaptive to the environment. By emphasizing *psychophysical systems*, Allport highlights the psychological nature of personality while clearly recognizing that personality arises from basic biological processes. Finally, Allport's definition stresses that personality *causes* people to think, behave, and feel in relatively

12.1 Dan McAdams McAdams believes that to know people well, we need to know everything about them, including the personal narratives of their whole lives.

1937
Allport Defines Personality Gordon Allport publishes the first major textbook in personality psychology, helping to shape personality as the scientific study of the individual.

1950s
Humanism and Phenomenology Dissatisfied with deterministic views of personality, a number of psychologists emphasize personal experience and belief systems, focusing on the potential for individual growth.

1954
Locus of Control Julian Rotter helps establish the social-cognitive perspective on personality by showing that people's beliefs and expectancies play an important role in how they behave.

1960s
Biological Basis of Personality Hans Eysenck demonstrates physiological differences between introverts and extraverts, setting the foundation for contemporary research on the biological basis of personality.

1968
Trait Consistency Is Overrated Walter Mischel charges that personality traits are not predictive of behaviour across different situations, which launches the person-situation debate.

12.2 Gordon Allport Allport published the first major textbook of personality psychology, which defined the field. He also championed the study of individuals and established traits as a central concept in personality research.

How much is personality influenced by unconscious processes?

psychodynamic theory Freudian theory that unconscious forces, such as wishes and motives, influence behaviour.

consistent ways over time. Psychological researchers use diverse approaches to explore different aspects of personality.

Psychodynamic Theories Emphasize Unconscious and Dynamic Processes

Sigmund Freud, an Austrian physician whose theories dominated psychological thinking for many decades, developed one of the most influential theories of human personality. Focusing on the individual level of analysis, Freud developed many of his ideas about personality by observing people he was treating for various psychological disturbances, such as patients who experienced paralysis without any apparent physical cause. The central premise of Freud's **psychodynamic theory** of personality is that unconscious forces, such as wishes and motives, influence behaviour. Freud referred to these psychic forces as *instincts* (although he used the term in a way slightly different from its contemporary use), defining them as mental representations arising out of biological or physical need. For instance, Freud proposed that people have a *life instinct* that is satisfied by following the *pleasure principle*, which directs people to seek pleasure and avoid pain. The energy that drives the pleasure principle is called *libido*. Although nowadays the term has a strong sexual connotation, Freud used it to refer more generally to the energy that promotes pleasure seeking. Instincts can be viewed as wishes or desires to satisfy libidinal urges for pleasure. These psychological forces can be in conflict, which was what Freud viewed as the essential cause of mental illness.

A TOPOGRAPHICAL MODEL OF MIND Freud believed that most of the conflict between various psychological forces occurred below the level of conscious awareness. In his *topographical model* (Figure 12.3), Freud proposed that the structure of the mind, the topography as it were, was divided into three different zones of mental awareness. At the *conscious* level, people are aware of their thoughts. The *preconscious* consists of content that is not currently in awareness but could be brought to awareness; it is roughly analogous to long-term memory. The *unconscious* contains material that the mind cannot easily retrieve. According to Freud, the unconscious mind contains wishes, desires, and motives that are associated with conflict, anxiety, or pain and are therefore not accessible to protect the person from distress. Sometimes, however, this information leaks into consciousness, such as occurs during a *Freudian slip*, in which a person accidentally reveals a hidden motive, such as the person who introduces her- or himself to someone attractive by saying "Excuse me, I don't think we've been properly seduced."

DEVELOPMENT OF SEXUAL INSTINCTS An important component of Freudian thinking was the idea that early childhood experiences had a major impact on the

1980s
The Big Five Building on earlier work by Warren Norman, researchers such as Lewis Goldberg, Paul Costa, Robert McCrae, and Oliver John argue that the universe of personality traits can be reduced to five basic factors.

1982
Inhibition and Approach Jeffrey Gray at Oxford University proposes that approach and inhibition motives, residing in distinct brain structures and processes, are the neural underpinnings of introversion and extraversion.

1990s
Behavioural Genetics Beginning with studies of twins, researchers demonstrate that personality traits are determined more by inheritance than by family environment.

1990s
Personology Led by Dan McAdams, researchers focus on how people's personal stories form coherent life narratives.

1990s
The Return of Temperaments Converging evidence indicates that basic temperaments, which can be observed in infancy and throughout childhood, are biologically based and have consequential impact on personality and behaviour.

development of personality. Freud believed that children went through developmental stages that corresponded to their pursuit of satisfaction of libidinal urges. At each **psychosexual stage**, libido is focused on one of the *erogenous zones*: the mouth, anus, and genitals. The *oral stage* lasts from birth to approximately 18 months, during which time pleasure is sought through the mouth. Hungry infants experience relief when they breastfeed and come to associate pleasure with sucking. When children are two to three years old, toilet training leads them to focus on the anus; therefore learning to control the bowels is the key focus of the *anal phase*. From ages three to five, children enter the *phallic stage*, during which libidinal energies are directed toward the genitals. Children often discover the pleasure of rubbing their genitals during this time, although they have no sexual intent per se.

One of the most controversial Freudian theories applies to children in the phallic stage. According to Freud, children desire an exclusive relationship with the opposite-sex parent. Because the same-sex parent is therefore a rival, children develop hostility toward the same-sex parent, which for boys is known as the *Oedipus complex*, after the Greek legend in which Oedipus unknowingly kills his father and marries his mother. Freud believed that children develop unconscious wishes to kill the one parent in order to claim the other, and that they resolve this conflict through identification with the same-sex parent, taking on many of his or her values and ideals. This theory was mostly applicable to boys. Freud's theory for girls was more complex and even less convincing.

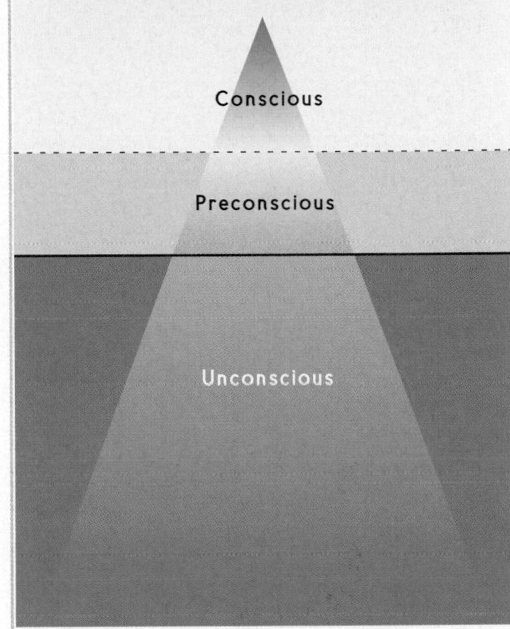

12.3 Levels of Consciousness Freud differentiated levels of consciousness into three zones. He believed that much of human behaviour was influenced by unconscious processes.

Following the phallic stage, children enter a brief *latency stage*, in which libidinal urges are suppressed or channeled into schoolwork or building friendships. Finally, in the *genital stage*, adolescents and adults work to attain mature attitudes about sexuality and adulthood. Libidinal urges are centred on the capacity to reproduce and contribute to society.

According to Freud, progression through these psychosexual stages has a profound impact on personality. For example, some people become *fixated* at a stage during which they have received excessive parental restriction or indulgence. Those fixated at the oral stage develop *oral personalities*; they continue to seek out pleasure via the mouth, such as by smoking, and they are excessively needy. Those fixated at the anal phase may have *anal-retentive* personalities, meaning they are stubborn and overly regulating. The latter may be due to overly strict toilet training, or overly rule-based rearing more generally.

STRUCTURAL MODEL OF PERSONALITY Freud proposed an integrated model of how the mind is organized, which consists of three theoretical structures that vary across the levels of consciousness. At the most basic level, and completely submerged in the unconscious, is the **id**, which operates according to the pleasure principle, acting on impulses and desires. The innate forces driving the id are sex and aggression. Acting as a brake on the id is the **superego**, which is the internalization of societal and parental standards of conduct. Developing during the phallic phase, the superego is a rigid structure of morality, or human conscience. Mediating between superego and id is the **ego**, which tries to satisfy the wishes of the id while being responsive to the dictates of the superego. The ego operates according to the *reality principle*, which involves rational thought and problem solving.

Conflicts between the id and superego lead to anxiety, which the ego copes with by employing a variety of **defense mechanisms**, unconscious mental strategies that the mind uses to protect itself from distress. For instance, people often *rationalize* their behaviour by blaming situational factors over which they have little control, as when you explain that you didn't call your parents because you were too busy studying for

psychosexual stages According to Freud, the developmental stages that correspond to the pursuit of satisfaction of libidinal urges.

id In psychodynamic theory, the component of personality that is completely submerged in the unconscious and operates according to the pleasure principle.

superego In psychodynamic theory, the internalization of societal and parental standards of conduct.

ego In psychodynamic theory, the component of personality that tries to satisfy the wishes of the id while being responsive to the dictates of the superego.

defense mechanisms Unconscious mental strategies the mind uses to protect itself from conflict and distress.

an exam. Much of the theoretical work on defense mechanisms can be credited to Sigmund's daughter, Anna Freud. Several common defense mechanisms are listed in Table 12.1. Research in psychology over the past 40 years has provided considerable support for the existence of many of these defense mechanisms (Baumeister, Dale, & Sommers, 1998), although contemporary researchers believe that these mechanisms protect self-esteem rather than relieve unconscious conflict over libidinal desires. For instance, *reaction formation* occurs when people ward off an uncomfortable thought about the self by embracing its opposite. In one study of homophobic men, those who expressed the most negative views of homosexuality showed greater physiological arousal when showed video depictions of homosexual sex than did men who were more accepting of homosexuality (Adams, Wright, & Lohr, 1996).

PSYCHODYNAMIC THEORY SINCE FREUD Although Freud is most closely identified with psychodynamic theory, a number of scholars, while rejecting certain aspects of Freudian thinking, embraced the notion of unconscious conflict. These *neo-Freudians* include Carl Jung, Alfred Adler, and Karen Horney, each of whom modified Freud through the development of their own psychodynamic theories. For instance, Adler and Horney strongly criticized Freud's view of women, believing many of his theories to be misogynistic. Many neo-Freudians rejected Freud's emphasis on sexual forces and instead focused on social interactions, especially the emotional attachments that children develop to their parents. This focus is embodied in *object relations* theory, in which the object of attachment is another person, such as a parent or spouse. In addition, some neo-Freudians, including Horney and Erik Erikson, emphasized the influence of culture, which Freud saw monolithically as "civilization."

TABLE 12.1 | Common Defense Mechanisms

MECHANISM	DEFINITION	EXAMPLE
Denial	Refusing to acknowledge source of anxiety	Ill person ignores medical advice
Repression	Excluding source of anxiety from awareness	Person fails to remember an unpleasant event
Projection	Attributing unacceptable qualities of the self to someone else	Competitive person describes others as supercompetitive
Reaction formation	Warding off an uncomfortable thought by overemphasizing its opposite	Person with unacknowledged homosexual desires makes homophobic remarks
Rationalization	Concocting a seemingly logical reason or excuse for behaviour that might otherwise be shameful	Person cheats on taxes because "everyone does it"
Displacement	Shifting the attention of emotion from one object to another	Person yells at children after a bad day at work
Sublimation	Channeling socially unacceptable impulses into constructive, even admirable, behaviour	Sadist becomes a surgeon or dentist

Psychological scientists have largely abandoned psychodynamic theories because of the lack of scientific evidence for Freud's central premises, since many of the ideas have not been amenable to empirical examination. However, Freud has to be understood in the context of the time in which he was working and the methods he had at his disposal. He was an astute observer of behaviour and an amazingly creative theorist. His observations and ideas continue to have an important impact on personality psychology and have framed much of the research in personality over the last century (Cleghorn, 1984; Westen, 1998).

Humanistic Approaches Emphasize Integrated Personal Experience

By the early 1950s, most psychological theories of personality were heavily deterministic; that is, personality and its associated behavioural characteristics were considered to be caused by forces beyond the person's control. As we have seen, Freud believed that personality was determined by unconscious conflicts. In a different vein, behaviourists such as B. F. Skinner argued that patterns of reinforcement determined response tendencies, which were the basis of personality. Against this backdrop emerged a view of personality that emphasized the uniqueness of the human condition. **Humanistic approaches** to personality emphasize personal experience and belief systems and propose that people seek to fulfill their human potential. At its core, humanism emphasizes subjective human experience, or *phenomenology*, and views each person as inherently good. Humanistic approaches to personality encourage people to fulfill their individual potential for personal growth through greater self-understanding; this process is referred to as *self-actualization*. Maslow's theory of motivation, discussed in Chapter 9, is an example of a humanistic approach to personality. Maslow believed that the desire to become self-actualized was the ultimate and most important human motive.

humanistic approaches Approaches to studying personality that emphasize personal experience and belief systems, and propose that people seek personal growth to fulfill their human potential.

The most prominent humanistic psychologist was Carl Rogers (Figure 12.4), whose *person-centred* approach to personality emphasized people's personal understandings or phenomenology. As a therapeutic technique, Rogers focused on the need for the therapist to create a warm, supportive, and accepting environment and to deal with clients' problems and concerns as clients understood them.

12.4 Carl Rogers Rogers emphasized people's subjective understandings of their whole lives.

Rogers's theory highlights the importance of how parents show affection for their children and how parental treatment affects personality development. Rogers speculated that although most parents provide love and support, it is often conditional. That is, parents love their children as long as the children do what the parent wants them to do. Parents who do not approve of their children's behaviour often behave in ways that indicate they may withhold their love from the child. As a result, children quickly abandon their true feelings, dreams, and desires and accept only those parts of themselves that elicit parental love and support. Thus, people lose touch with their true selves in their pursuit of positive regard from others. By contrast, Rogers encouraged parents to raise their children with *unconditional positive regard*, in which children are accepted, loved, and prized no matter how they behave. Parents are to express disapproval with bad behaviour, but in a context that ensures that children feel loved, as epitomized in the saying "Love the sinner, hate the sin." According to Rogers, a child raised with unconditional positive regard will develop a healthy sense of self-esteem and will be able to become, in Rogers's term, a *fully functioning person*.

By its nature, humanistic psychology has not been overly concerned with the scientific study of personality, since it emphasizes subjective personal experience. Recently, however, psychologists have begun to use the methods of science to study the positive aspects of humanity. The *positive psychology movement* was launched by the clinical psychologist Martin Seligman when he was president of the American Psychological Association (Seligman & Csikszentmihalyi, 2000). Seligman and others have encouraged the scientific study of qualities such as faith, values, creativity, courage, and hope. For instance, Ed Diener (2000) has conducted more than three decades of research on *subjective well-being*, a general term for how much happiness and satisfaction people have in their lives. He has found that well-being varies across cultures such that the wealthiest countries often have higher levels of satisfaction, a finding that fits in well with Maslow's proposal that people need to satisfy basic needs (such as food, shelter, and safety) before they can focus on self-esteem needs. More recently, Michele Tugade and Barbara Frederickson (2004) found that people who are resilient, who are most able to bounce back from negative events, experience positive emotions even when under stress. According to the *broaden-and-build theory*, positive emotions prompt people to consider novel and creative solutions to their problems, which may in the end help resilient people to cope effectively with setbacks or negative life experiences (Frederickson, 2001).

Type and Trait Approaches Describe Behavioural Dispositions

How do personality types differ from personality traits?

Psychodynamic and humanistic approaches focus on explaining the mental processes that shape personality. The same underlying processes are thought to occur within each person, but people differ because they experience different conflicts, situations, parental treatment, and so forth. Other approaches to personality focus more on description than explanation, which is actually similar to the way most people intuitively view personality. For example, if asked to describe a friend, you would probably not delve into unconscious conflicts; rather, you would describe the person as a certain type, such as an *introvert* or an *extravert*. **Personality types** are discrete categories into which we place people. Subsequently, we fill in gaps in our knowledge with our beliefs about what behaviours and dispositions are associated with these types. Our tendency to assume that personality characteristics go together, and therefore to make predictions about people on the basis of minimal evidence, is referred to as *implicit personality theory*. For example, we think that introverts don't like to go to parties, that they like to read books, and that they are sensitive (Rousseau et al., 1989).

In addition to typologies, many personality psychologists are concerned with *traits*, defined earlier as behavioural dispositions that endure over time and across situations. Traits are on a continuum, with most people toward the middle and relatively few at the extreme ends. Thus, people range from being very unfriendly to very friendly, but most people are moderately friendly. The **trait approach** to personality provides a method for assessing the extent to which individuals differ in personality dispositions, such as sociability, cheerfulness, and aggressiveness (Funder, 2001).

personality types Discrete categories based on global personality characteristics.

trait approach An approach to studying personality that focuses on the extent to which individuals differ in personality dispositions.

How many traits are there? In the earliest stages of his career, Gordon Allport, along with his colleague Henry Odbert of Dartmouth College, went through the dictionary to count the number of words that could be used as personality traits. They counted nearly 18,000. Even weeding out synonyms and archaic words left 4,500 apparent traits.

During the 1950s, Raymond Cattell set out to ascertain the basic elements of personality. Cattell believed that by using statistical procedures he could take the scientific study of personality to a higher level and perhaps uncover the basic structure of personality. Cattell had participants fill out personality questionnaires containing many trait items, which he reduced from the larger set produced by Allport and Odbert. Cattell then performed *factor analysis*, grouping items based on their similarities. For instance, all the terms that referred to friendliness ("nice," "pleasant," "cooperative," and so on) were grouped together. Through this procedure Cattell came to believe that there were 16 basic dimensions of personality, one of which was intelligence. The others were given rather unusual names to avoid confusion with everyday language; most personality psychologists no longer use these terms.

EYSENCK'S HIERARCHICAL MODEL Reducing the number of basic traits even further was the British psychologist Hans Eysenck, who in the 1960s proposed a hierarchical model of personality. As you can see in Figure 12.5, the basic structure begins at the *specific response level*, which consists of observed behaviours. For instance, a person buys an item because it is on sale. A person might then repeat the behaviour on different occasions, which is the *habitual response level*. Some people find it hard to pass up sale items, whether they need them or not. If people are observed on many occasions to behave in same way, they are characterized as possessing a *trait*. Traits such as impulsiveness and sociability can then be viewed as components of *superordinate traits*, of which Eysenck proposed there were three: *introversion-extraversion, emotional stability,* and *psychoticism.*

The dimension of *introversion-extraversion*, terms originally coined by psychoanalyst Carl Jung, refers to the extent that people are shy, reserved, and quiet versus sociable, outgoing, and bold. As you will see later in this chapter, Eysenck believed that this dimension reflects differences in biological functioning. *Emotional stability* refers to the extent that people's moods and emotions change; people who are *neurotic* experience frequent and dramatic mood swings, especially toward negative emotions, relative to those who are more stable. People who are high in neuroticism report often feeling anxious, moody, and depressed, and they also tend to hold very low opinions of themselves. Finally, *psychoticism* describes a mix of impulse control, empathy, and aggression, with those who are high in psychoticism being more aggressive, impulsive, and self-centred than those low in psychoticism. The term "psychoticism" was perhaps not a wise choice, since it implies a level of psychopathology that Eysenck

12.5 Eysenck's Hierarchical Model of Personality Extraversion is a superordinate trait made up of sociability, dominance, assertiveness, activity, and liveliness at the trait level. Each trait is made up of habitual and specific responses.

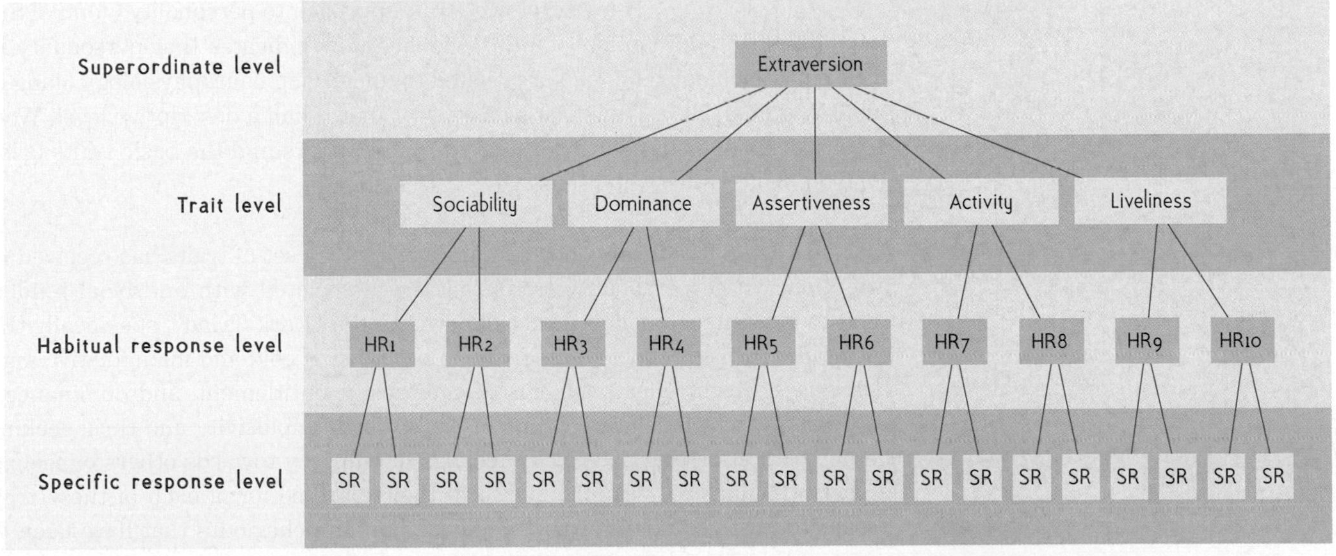

did not intend. More recent conceptions of this superordinate trait call it *constraint*, with people ranging from restrained to disinhibited (Watson & Clark, 1997).

five-factor theory The idea that personality can be described using five traits: openness to experience, conscientiousness, extraversion, agreeableness, and neuroticism.

THE BIG FIVE In the last 20 years, many personality psychologists have agreed that there are five basic personality traits, which is referred to as the **five-factor theory** (McCrae & Costa, 1999). The *Big Five*, as they are known, consist of *extraversion, neuroticism* (similar to Eysenck's model), *conscientiousness, agreeableness,* and *openness to experience.* A good way to remember the five is that they can be arranged to spell the acronym OCEAN. Each of these five factors is a higher-order trait comprising lower-order traits that are related to each other (Jackson et al., 1996a, b). For instance, as can be seen in Table 12.2, conscientiousness is determined by how careful and organized someone is, while agreeableness refers to the extent that a person is or is not trusting and helpful. Those high in openness to experience are imaginative and independent, whereas those low in this basic trait are down to earth and conformist.

TABLE 12.2 The Big Five

OPENNESS TO EXPERIENCE	Imaginative vs. down to earth Variety vs. routine Independent vs. conforming
CONSCIENTIOUSNESS	Organized vs. disorganized Careful vs. careless Self-disciplined vs. weak-willed
EXTRAVERSION	Social vs. retiring Fun-loving vs. sober Affectionate vs. reserved
AGREEABLENESS	Softhearted vs. ruthless Trusting vs. suspicious Helpful vs. uncooperative
NEUROTICISM	Worried vs. calm Insecure vs. secure Self-pitying vs. self-satisfied

Considerable evidence supports the five-factor theory (John, 1990). The Big Five emerge across cultures, among adults and children, even when vastly different questionnaires are used to assess traits, and the same five traits appear whether people rate themselves or are rated by others. Furthermore, people's responses to the Big Five traits have been shown to predict a wide variety of different behaviours (Paunonen & Ashton, 2001), as well as their satisfaction with their jobs, marriages, and life more generally (Heller, Watson, & Ilies, 2004). Nonetheless, some researchers have questioned whether the five-factor theory really helps us understand personality, since the trait terms are descriptive rather than explanatory, and because reducing all of human personality to five descriptions inevitably ignores human subtleties. For instance, evidence shows that many of these traits apply to animals as well as humans, as discussed later. However, one value of the five-factor theory is that it is an organizational structure for the vast number of traits that can be used to describe personality; by providing a common descriptive framework, the theory has helped integrate and invigorate the trait approach to personality (John & Srivastava, 1999). In addition, substantial evidence now indicates that personality traits have genetic components and are related to underlying neurophysiology (Jang et al., 1996; Wiggins, 1996). Thus, traits do exist at more than a descriptive level. Working across levels of analysis provides new ways to understand the basic traits (Clark & Watson, 1999), as we will discuss later in this chapter.

THE DARK TRIAD Beyond the Big Five, another set of traits has received much attention. Three personality traits that are associated with antisocial tendencies tend to cluster together in what is known as the "Dark Triad", specifically, *Machiavellianism*, which reflects a tendency to behave in a cold and manipulative manner, *narcissism*, which reflects feelings of superiority, entitlement, and dominance, and *psychopathy*, which reflects a combination of high impulsivity and thrill-seeking together with a striking absence of feelings of empathy towards others or feelings of anxiety (Paulhus & Williams, 2002). In their extreme form, each of these traits is associated with some of the most heinous human behaviours that have been documented. For example, convicted murderers who scored at clinical levels of psychopathy committed approximately 14 times as many instrumental murders (in which someone is killed for the purpose of pursuing a premeditated instrumental goal such as obtaining money, sex, or revenge) as they did *reactive murders* (in which someone is killed in an impulsive exchange, usually centred around a fight that emerged spontaneously). In contrast, convicted murderers who did not score at clinical levels of psychopathy committed approximately as many instrumental as they did reactive murders (Woodworth & Porter, 2002). In sum, clinical levels of the Dark Triad traits, in particular psychopathy, lead to behaviours in which the welfare of others is largely ignored if it stands in the way of a desired goal.

However, research reveals that even within normal populations (i.e., populations in which the range of scores on the Dark Triad measures are below clinical levels), the traits of Machiavellianism, narcissism, and psychopathy predict a number of deviant behaviours. For example, computer programs that detect cheating on exams reveal that students' scores on the three measures of the Dark Triad each correlate with the likelihood that they copied answers from a neighbour on an exam (Nathanson, Paulhus, & Williams, 2006b). Likewise, people who score high on the Dark Triad are more likely than others to report engaging in various other kinds of misconduct such as drug abuse, bullying, criminal activity, and driving dangerously (Nathanson, Paulhus, & Williams, 2006a). The Dark Triad reliably predicts which people are most likely to engage in antisocial and criminal behaviours.

Profiles in Psychological Science

Paul Bernardo

Paul Bernardo led a double life. Most of the world saw him as a charming, handsome, boyish, former Boy Scout who had a promising career at a major accounting firm (Figure 12.6). Behind this façade, however, lurked the all-consuming rage and hatred of a serial killer who had committed some of the most ruthless and heinous crimes in Canadian history. Bernardo himself bragged about this double identity in many of the rap songs that he wrote, where he took on the name "Deadly Innocence," and sang about how his innocent appearance concealed a vicious sex predator and killer. Paul Bernardo is one of the most horrifying examples of how the Dark Triad can manifest in its most extreme and disturbing forms.

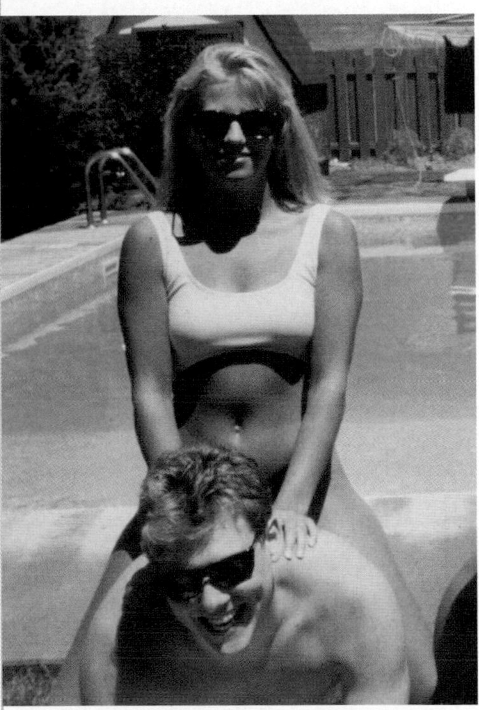

12.6 Paul Bernardo Seemingly normal on the outside, Bernardo committed such heinous acts as raping the 15-year-old sister of his wife (and even convinced his wife to participate), which later led to the young girl's death.

As described in Nick Pron's book, *Lethal Marriage*, Bernardo comes across as a classic case of a Machiavellian. He had his desires, twisted as they often were, and he would act in whatever ways necessary to charm or coerce others (often his wife Karla Homolka) into acting in ways that allowed him to satisfy those desires. He was able to succeed at his dastardly crimes often because he was so effective at manipulating people to do whatever he wanted. He spared no expense at his efforts to win over and cajole others. Even though he was bankrupt, he still always offered to treat others to expensive dinners, wore expensive suits, rented a house in an expensive neighbourhood, and learned how to charm the border guards as he made his daily trips smuggling cigarettes across the border. One of his friends had nicknamed him as the "king of first impressions." Even today in jail, his prison guards say Bernardo still seeks to charm everyone around him.

There are few people with narcissistic tendencies as pronounced as Bernardo. He insisted that his wife and victims refer to him as "the king," and that his wife leave him daily messages on his pillow describing how terrific he was and how much she loved him. He had trouble tearing himself away from a mirror, and would refer to himself as a "specimen of a man, a superior organism." Despite having gone bankrupt, he continued to spend enormous sums of money, because in his view, he always deserved the very best, regardless of whether he could afford it.

But the most dangerous aspect of Bernardo's personality is his psychopathy. Psychopathy is far more pronounced among prison inmates than the normal population, and Bernardo's psychopathy score falls among the top 5 percent of inmates. He shows virtually no empathy towards anyone, and has not shown the least bit of remorse about any of the horrible crimes that he committed. His only apprehensions about his regular violent beatings of his wife, Homolka, appear to be that they might lead her to leave him. He has never showed any concern for her well-being, let alone any of his many victims. He appears to be unable to care about the feelings of anyone around him.

Fortunately, there are not many individuals who have the same dangerous combination of the Dark Triad traits to the degree that Bernardo has. However, if you look at the personalities of many of the most dangerous criminals in the world, the same pattern of high levels of Machiavellianism, narcissism, and psychopathy is often apparent. ■

Personality Reflects Learning and Cognitive Processes

In contrast to those who saw personality as a result of internal processes, behaviourists such as B. F. Skinner viewed personality as little more than learned responses to patterns of reinforcement. However, growing dissatisfaction with strict models of learning theory led researchers to incorporate cognition into the understanding of personality. For instance, the early cognitive theorist George Kelly emphasized the importance of people's understandings, or *personal constructs*, of their circumstances. According to Kelly, personal constructs develop through people's experiences and represent their interpretations and explanations for events in their social worlds. Building further on the cognitive approach, Julian Rotter introduced in the 1950s the idea that behaviour is a function of people's *expectancies* for reinforcement, as well as the *value* they ascribed to the reinforcer. Thus, if a person is deciding whether to study for an exam or go to a party, he or she will weigh the likelihood that studying will lead to a good grade, as well as how much that grade matters, against the likelihood that the party will be fun and the extent to which the person values having fun. Rotter proposed that people differ in the belief that their efforts will lead to positive outcomes, which he referred to as *locus of control*. People with an internal locus of control believe that they bring about their own rewards, whereas those with an external locus of control believe that rewards, and therefore their personal fate, are the result of forces beyond their control. These generalized beliefs have been shown to have a powerful effect on psychological adjustment (Lefcourt, 1966, 1992).

The incorporation of cognition into learning theories led to the development of *cognitive-social theories* of personality, which emphasize how personal beliefs, expectancies, and interpretations of social situations shape behaviour and personality. For instance, Albert Bandura (1977) accepts many of the tenets of learning theory but argues that humans possess mental capacities, such as beliefs, thoughts, and expectations, that interact with the environment to influence behaviour. For Bandura, the extent to which people believe they can achieve specific outcomes, called *self-efficacy*, is an important determinant of behaviour. Moreover, as discussed in Chapter 6, Bandura has proposed that people develop expectancies in part through *observational learning*, such as by noticing whether others are rewarded or punished for acting in certain ways.

One of the most influential cognitive-social theorists is Walter Mischel, who approached the study of personality at the social level of analysis. As you will read later in this chapter, Mischel sparked controversy by proposing that personality traits often fail

How do personal beliefs affect behaviour?

to predict behaviour across different circumstances. According to Mischel's *Cognitive-Affective Personality System* (*CAPS*) (Mischel & Shoda, 1995), people's responses in a given situation are influenced by how they encode or perceive the situation, their affective (emotional) response to the situation, the skills and competencies they have to deal with challenges, and their anticipation of the outcomes that their behaviour will produce.

Imagine, for instance, a person who walks into a party with an expectation of making a good impression, having done so many times in the past. This person will act very differently from a person whose past experiences of awkwardness, discomfort, and shyness lead to the expectation of rejection. Consider the personality style of *defensive pessimism*, studied by Julie Norem and Nancy Cantor. Defensive pessimists expect to fail and therefore enter test situations with dread. By contrast, optimists enter test situations with great expectations. You have probably encountered both of these types at your exams, with the pessimists predicting their imminent academic demise. And yet, both pessimists and optimists tend to perform similarly on exams (Norem, 1989). These two personality styles reflect different motivational strategies, with pessimists expecting the worst so they can be relieved when they succeed, and optimists focusing on positive outcomes.

The CAPS model also emphasizes *self-regulatory capacities*, in which people set personal goals, evaluate their progress, and adjust their ongoing behaviour in pursuit of those goals. In addition, the values that people attach to goals, such as the importance of obtaining good grades, is part of the dynamic system. The pursuit of personal goals is an important component of cognitive-social theories of personality. Indeed, many personality psychologists believe that personal motives and strivings, such as those for achievement, power, or intimacy, are an essential aspect of personality (Snyder & Cantor, 1998). Personality, then, represents behaviour that emerges from how people interpret their social worlds, and from the beliefs that they have about how they will affect and be affected by their social situations.

Thinking Critically

Do Animals Have Personality?

If you've ever owned a pet, you probably had the intuitive feeling that your pet had a personality of its own. For most of the history of psychological science, your intuition would have been regarded with skepticism, the assumption being that you were projecting your own sense of personality onto your pet. But if we are to take the principles of evolution seriously, there is continuity across species, raising the possibility that animals might display consistent behaviours across circumstances that reflect an underlying personality (Gosling, 2001). Developing animal models of personality may aid our understanding of general processes in personality perception. How do psychological scientists determine the personality of animals?

Sam Gosling studied the behaviour of a group of 34 spotted hyenas at a field station of the University of California at Berkeley (Gosling, 1998). He created a personality scale consisting of 44 traits, applicable to both humans and hyenas. Four observers who knew hyenas well used the scale independently to rate the animals. It turned out that the agreement among the raters was as high as is typically found in personality studies of humans, suggesting that the raters could assess the hyenas reliably. Using factor analysis, Gosling found that the traits clustered into five factors, albeit not exactly the same five factors found for humans. Although there were rough similarities between humans and hyenas in traits related to agreeableness, neuroticism, and openness to experience, hyenas showed no evidence of a conscientiousness factor, and extraversion seemed to exist mainly in the form of assertiveness, which makes sense given that hyenas are a species that form dominance hierarchies.

What about other animals? Gosling and John (1999) summarized the findings of 19 studies that assessed multiple personality traits in modestly large samples of nonhuman animals, ranging from household pets to monkeys and other primates, to pigs and donkeys, and even to aquatic animals. They found evidence that traits similar to extraversion, neuroticism, and agreeableness could be seen in most species. Extraversion reflected different levels of energy, approach, and sociability. Neuroticism indicated differences in emotional reactivity, fearfulness, and excitability, whereas agreeableness reflected differences in aggression, hostility, and affinity for mates. What about openness and conscientiousness? In about half of the species, the animals did display individual differences in curiosity and play, and Gosling and John thought that openness in animals might be similar to behaviours one observes in young children. Conscientiousness, a core human trait, was found in a narrow form among only chimpanzees, with some showing more unpredictability and disorganized behaviour. The finding that only chimpanzees showed any signs of conscientiousness is perhaps not surprising, since they are humans' closest relatives.

12.7 Do Dogs Have Personality? Which is friendlier, a rottweiler or a golden retriever? How might you test your hypothesis?

As a critical thinker, you might be wondering whether these judgments of personality in animals are accurate. Do they reflect true variations in personality or are they just stereotypes of certain species or breeds (Figure 12.7)? To test whether differences in personality traits among animals exist and can be measured, Gosling and his colleagues (2003) examined personality judgments for domestic dogs and compared those to judgments made for humans. How do you tell if personality ratings are accurate for animals? After all, they cannot answer for themselves and tell whether raters are correct. Gosling and his colleagues selected three measures of accuracy: (1) the extent to which ratings by the same judges were consistent for similar behaviours over time; (2) whether there was consensus, or agreement, among independent judges viewing the same behaviour; and (3) the extent to which the behaviour corresponds to the definition of the trait, as when a dog is rated aggressive for baring its teeth in an angry fashion. To test these measures of accuracy, dog owners rated their own animals and provided the name of a friend who knew the dog well enough to rate it. Independent judges also assessed the behaviour of the dog while it played in a park and performed a wide array of tasks. The dogs were assessed on four personality factors that roughly correspond to the Big Five without conscientiousness. In general, the findings indicated that the ratings of dogs were consistent over time, that the judges generally were in agreement, and that the trait behaviours displayed by the dogs corresponded to ratings of those traits made by independent judges. To rule out the possibility that breed stereotypes based solely on appearance influenced the ratings, another group of judges rated only pictures of the dogs. The personality ratings remained accurate even after statistically controlling for the ratings made of the pictures. This suggests that dogs can be rated with impressive accuracy. According to Gosling, knowing that animal personality can be assessed may permit studies of personality that could not be conducted with humans, such as breeding studies, altering the early environment, or manipulating neurotransmitter levels. Research might also address issues of animal welfare, such as matching pets to suitable owners.

Why do you think dogs of the same breed might have different personalities? What factors do you think are most important for raising a dog to be affectionate and calm around children? Do you believe that the observed differences in personality in dogs reflect the same sense of uniqueness that we assign to individual humans? Why or why not? ■

REVIEWING the Principles | How Have Scientists Studied Personality?

According to the psychodynamic approach, unconscious motives and conflicts that are experienced throughout life, but especially in childhood, shape personality. Humanists believe that each person is unique and capable of fulfilling great potential. Trait theorists describe the behaviour of people based on trait dispositions. Cognitive-social theorists focus on how cognitive interpretations and beliefs affect people's perceptions of their social environments. These varied approaches are not necessarily in opposition to each other. They share the common goal of trying to understand the ways in which people are similar to and different from one another.

How Is Personality Assessed, and What Does It Predict?

Personality researchers have yet to agree on the best method for assessing various aspects of personality. To understand people you need to examine them in some way. But unlike behaviour or biological responses, personality cannot be directly assessed or observed. Psychological scientists measure personality by asking people to report on themselves, by asking people's friends or relatives to describe them, or by watching how people behave. Each of these methods has strengths and limitations. This section considers how psychological scientists assess personality, and how these different methods of assessment influence how we understand individuals.

Personality Refers to Both Unique and Common Characteristics

As differentiated by Allport, there are two approaches to studying personality: idiographic and nomothetic. **Idiographic approaches** are *person-centred* in that they focus on individual lives and how various characteristics are integrated into unique persons. If your classmates were to write down 10 personality traits that described them, although there would be some overlap, each person would probably have a unique list of traits. People like to be distinctive, and therefore they tend to choose traits that are particularly descriptive of them as compared to other people. These *central traits* are especially important for how individuals define themselves. In contrast, *secondary traits* are those that people consider less personally descriptive or not applicable at all. In general, central traits are more predictive of behaviour than are secondary traits.

idiographic approaches Person-centred approaches to studying personality that focus on individual lives and how various characteristics are integrated into unique persons.

Researchers who use idiographic approaches often examine case studies of individuals, perhaps through interviews or biographical information. For example, many scholars have tried to account for Adolf Hitler's behaviour in Nazi Germany by asking questions about his early childhood experiences, his physical stature, and his personal motivations. Henry Murray pioneered this approach at Harvard University. The study of people's lives emphasizes the idea that personality unfolds over the life course, as people react to their circumstances and come to define themselves.

There has recently been a resurgence of interest in the narrative approach to understanding human lives (McAdams, 1999, 2001). According to Dan McAdams,

nomothetic approaches Approaches to studying personality that focus on characteristics that are common to all people, although there is individual variation.

projective measures Personality tests that examine unconscious processes by having people interpret ambiguous stimuli.

TAT (Thematic Apperception Test) A projective measure of personality where a person is shown an ambiguous picture and asked to tell a story about the picture.

humans weave a *life story* that integrates self-knowledge into a coherent whole. Life stories help bring meaning to life and help people make sense of the world. Like all good stories, the ones people tell about their lives contain characters, settings, acts, plot twists, and themes. The life story is a reconstructive and imaginative process in which people link together personal motives, goals, and beliefs with events and people, and the circumstances in which they find themselves. People define themselves by creating *personal myths* that bind together past events and future possibilities. To study personality, then, we need to pay attention to the stories that people tell about their lives. The method of *psychobiography* uses personal life stories to develop and test theories about human personality.

Nomothetic approaches differ from idiographic approaches in that they focus on characteristics that are common among all people, but on which people vary—in other words, traits. For instance, every human can be rated along a continuum from very disagreeable to very agreeable. Researchers who follow this tradition tend to compare people by using common trait measures, such as questionnaires or other objective data. For example, they might give participants a list of 100 personality traits and ask them to rate themselves on each, on a scale from 1 to 10. From the nomothetic perspective, individuals are unique because of their unique combinations of common traits.

We Can Use Objective and Projective Methods to Assess Personality

Researchers use numerous methods to assess personality, ranging from self-reports to clinical interviews to observer reports. Aside from the manner by which the data are collected, how researchers choose to measure personality depends to a great extent on their theoretical orientations. For instance, trait researchers use personality descriptions to predict specific behaviours, whereas humanistic psychologists use more holistic approaches. At the broadest level, assessment procedures can be grouped into *projective* and *objective* measures.

How might you confirm the validity of projective measures?

PROJECTIVE MEASURES According to psychodynamic theory, personality is influenced by conflicts that people aren't aware of. So how can they tell you about something they don't know? **Projective measures** are tools that attempt to delve into the realm of the unconscious by presenting people with ambiguous stimuli and asking them to describe the stimulus items or tell stories about them. The general idea is that people will *project* their mental contents onto the ambiguous items, thereby revealing otherwise hidden aspects of personality, such as motives, wishes, and unconscious conflicts. Many of these procedures are used to assess psychopathology. One of the best-known projective measures is the *Rorschach inkblot test*, in which people look at an apparently meaningless inkblot and describe what it might be. How a person describes the inkblot is supposed to reveal unconscious conflicts and other problems. Unfortunately, many normal adults and children are found to be psychologically disturbed when they take the Rorschach, and it does a poor job of diagnosing specific psychological disorders (Wood, Garb, Lilienfeld, & Nezworski, 2002).

One classic projective measure used by personality psychologists is the **Thematic Apperception Test**, or **TAT**, developed by Henry Murray and Christiana Morgan, who used the TAT to study achievement motivation. In the TAT, a person is shown an ambiguous picture (Figure 12.8) and asked to tell a story about it. The story is then scored based on the motivational schemes that emerge, which are assumed to reflect the storyteller's personal motives. Many projective measures have been criticized for being too subjective and poorly validated. The TAT, however, has been shown to be useful for measuring motivational states, especially those related to achievement, power, and affiliation, and therefore continues to be used in contemporary research (McClelland, Koestner, & Weinberger, 1989). Indeed, evidence exists that the TAT, if used properly, reliably predicts behaviour (Bornstein, 1999), including culture-based differences (Barbopoulos et al., 2002).

12.8 Thematic Apperception Test The type of picture used in a TAT.

OBJECTIVE MEASURES **Objective measures** of personality are straightforward assessments, usually made by self-report questionnaires or observer ratings. For instance, the *NEO Personality Inventory* consists of 240 items designed to assess the Big Five personality traits (Costa & McCrae, 1992). Although called objective, the tests require people to make subjective judgments. Self-reports can be affected both by desires to avoid looking bad and by biases in self-perception that arise through self-relevant motives.

It can be difficult to compare self-reported objective measures directly, because individuals do not have objective standards against which to rate themselves. Just because two individuals report a "5" on a 7-point shyness scale does not mean they are equally shy. After all, the term can mean different things to different people.

One technique that assesses personal meanings of traits is the *California Q-sort*, which requires people to sort 100 statements printed on cards into nine piles according to what extent the statement is descriptive of them. These piles represent categories ranging from "not at all descriptive" to "extremely descriptive." A person is allowed to place only so many cards in each pile, usually with fewer cards allowed for the extreme ends of the scale. Because most of the cards must be piled in the moderately descriptive categories, the Q-sort has a built-in procedure for getting at central dispositions. The Q-sort, like most objective measures, can also be used by observers, such as parents, teachers, therapists, and friends.

objective measures Relatively unbiased assessments of personality, usually based on information gathered through self-report questionnaires or observer ratings.

Objective tests make no pretense of uncovering hidden conflicts or secret information. They measure only what the raters believe or observe. Personality researchers use them to compare people's responses and assess the extent to which the answers predict behaviour. In addition to large personality inventories, personal researchers often use self-report questionnaires that target specific traits, such as how much excitement you seek out of life, a trait known as *sensation seeking*.

Observers Show Accuracy in Trait Judgments

Do you think other people know you pretty well? Imagine that you often feel a bit shy in new situations, as many people do. Would others know that shyness is part of your personality? Some shy people, such as Tom Hanks, force themselves to act in an outgoing manner in order to mask their inner feelings. Their friends might have no indication that they feel shy. Others react to the fear of social situations by remaining quiet and aloof. Observers might believe them to be cold, arrogant, and unfriendly. People are judged by others throughout their lives. How accurate are those judgments? That is, how well do the personality judgments people make predict the behaviour of others?

An important study by the personality psychologist David Funder (1995) found a surprising degree of accuracy for trait judgments, at least under certain circumstances. For instance, it turns out that your close acquaintances may be more accurate at predicting your behaviour than you are. In one study, ratings of assertiveness made by friends predicted assertive behaviour in the lab better than did the person's own ratings (Kolar, Funder, & Colvin, 1996). The same was true for a variety of other traits. This may occur because friends observe how you behave in situations, whereas you may be preoccupied with evaluating other people. In other words, you may pay more attention to others than to yourself, and therefore you may fail to notice how you actually behave. Another possibility is that your subjective perception may diverge from your objective behaviour. In either case, the study implies that there is some disconnect between how people view themselves and how they behave.

How stable is personality across situations?

People Sometimes Are Inconsistent

Imagine again that you are a shy person. Are you shy in all situations? Probably not. Shy people tend to be most uncomfortable in new situations in which they are being evaluated; they tend not to be shy around family and close friends. Walter Mischel (Figure 12.9) dropped a bombshell on the field of personality in 1968 by proposing that behaviours are determined as much by situations as by personality traits, a theory referred to as **situationism**. For evidence, working at the social level of analysis, he referred to studies in which people who are dishonest in one situation are completely honest in another. For instance, the student who is not totally honest with a professor in explaining why a paper is late is probably no more likely to steal from classmates or cheat on her taxes than is the student who admits he overslept. Mischel's critique of personality traits affected the field for more than a decade and caused considerable rifts between social psychologists, who tended to emphasize situational forces, and personality psychologists, who focused on individual dispositions. After all, the most basic definition of personality is that it is relatively stable across situations and circumstances. If Mischel was correct and there was relatively little stability, then the whole concept of personality seemed empty.

situationism The theory that behaviour is determined as much by situations as by personality traits.

As you might expect, there was a rather vigorous response to Mischel's critique. The basic argument made by personality researchers in the *person-situation debate* is that the extent to which traits predict behaviour depends on the *centrality* of the trait, the *aggregation* of behaviours over time, and the type of trait being evaluated. People tend to be more consistent for central traits than for secondary traits, since the former are most relevant to them. In addition, if behaviours are averaged across many situations, then personality traits are more predictive of behaviour. Shy people may not be shy all the time, but on average they are shy more often than not. Moreover, people who report being shy in college continue to report being shy many years later, which indicates that there is stability to the trait of shyness. Finally, some people may be more consistent than others. Consider the trait of *self-monitoring,* in which some people are highly sensitive to cues of situational appropriateness. Those who are high in self-monitoring alter their behaviour to match the situation, so that they exhibit low levels of consistency. By contrast, those low in self-monitoring are less able to alter their self-presentations to match situational demands, and they tend to be much more consistent across situations.

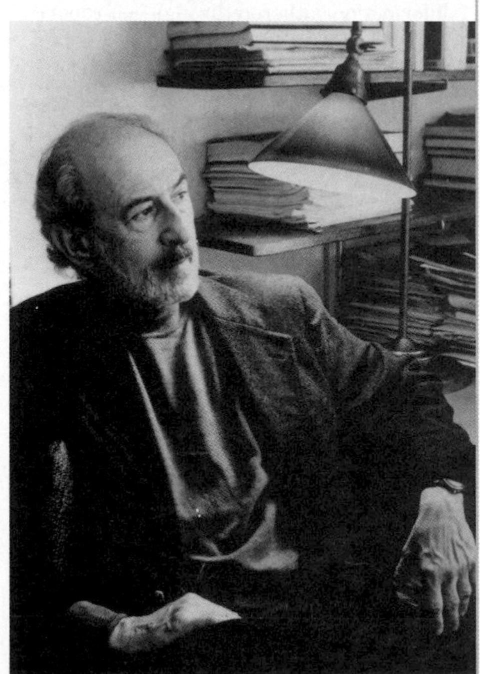

12.9 Walter Mischel Mischel proposed that situations are important in predicting behaviour.

Behaviour Is Influenced by the Interaction of Personality and Situations

Considerable evidence now exists that personality traits are predictive of behaviour. For instance, people who are high in neuroticism tend to be more depressed and have more illness, are more likely to have a midlife crisis, and so on. Indeed, being highly neurotic is the single best predictor of marital dissatisfaction and divorce

(Karney & Bradbury, 1995). Likewise, those high in sensation seeking are more likely to smoke, use drugs, have sex, be impulsive, watch erotic movies, begin conversations, and engage in physically risky activities such as mountain climbing. The past three decades have found strong and compelling evidence that personality dispositions are meaningful constructs that predict people's behaviour over time and across many circumstances. Yet people are also highly sensitive to social context, and most conform to situational norms. Few people would break the law right in front of a police officer or drive on the wrong side of the road just because they felt like it. The situation dictates behaviour irrespective of personality.

How might situations affect the display of personality traits?

Situational influences can be subtle. Consider your own behaviour. Do you behave in the exact same way around your family, friends, and teachers? Although you remain the same person, you may reveal different aspects of your personality during your interactions with different people. Your goals for social interaction change, as do the potential consequences of your actions. For example, your family may be more tolerant of your bad moods than are your friends. Thus, you may feel freer to express your bad moods around your family.

Situations themselves differ in the extent to which they constrain the expression of personality (Kenrick & Funder, 1991). Consider two people, one highly extraverted, aggressive, and boisterous, and the other shy, thoughtful, and restrained. At a funeral, it would be hard to tell them apart just from their behaviour. But at a party there would be obvious differences between them. Personality psychologists differentiate between *strong* situations (elevators, religious services, job interviews) and *weak* situations (parks, bars, your house); the former tend to mask differences in personality because of the power of the social environment. Most trait theorists are **interactionists**, in that they believe that behaviour is jointly determined by situations and underlying dispositions. In other words, you need to know something about both the person and the environment to predict behaviour.

interactionists Theorists who believe that behaviour is jointly determined by underlying dispositions and situations.

However, it is important to point out that people also have an impact on their environments. First, people choose their situations. Introverts tend to avoid parties or other situations in which they would feel anxious, whereas extraverts seek out social opportunities. Moreover, once they are in the situation, their behaviour affects those around them. Some extraverts may draw people out and encourage them to have fun, whereas others might act aggressively and turn people off. Some introverts might create an intimate atmosphere and encourage people to open up and reveal personal concerns, whereas others might make people uncomfortable and anxious as both actors try to fill awkward gaps in the conversation. A reciprocal interaction occurs between the person and the environment so that they simultaneously influence each other. The important point is that personality is a dynamic system that reflects both underlying dispositions and the activation of goals and affective responses in given situations.

Psychological Science *in Action*

Finding the Right Career for Your Personality

Are scientists introspective, salespeople gregarious, and judges sober and responsible? Research by psychologists has demonstrated that some occupations seem better suited to certain personality types. What type of career suits you? Thousands of careers are possible today—some of them might bring you joy and fulfillment, whereas others may lead you to feel trapped or headed down a dead-end road. How can you find the career that is best for you? How can you determine what occupation best fits your unique abilities, interests, and personality?

Occupational psychologists help people find the right careers. Among their primary tools are psychological tests that identify personality factors indicating suitedness to specific careers. One of the most widely used personality measures for occupational counseling is the Myers-Briggs Type Indicator (MBTI). The scale of the MBTI is based on Carl Jung's theories about how people approach the world, how they perceive information, and how they make decisions. The MBTI assesses four dimensions: (1) whether people focus attention on the outside world of people and objects or on the inner world of ideas; (2) whether people prefer to rely on their objective senses or on intuition; (3) whether people make decisions based on logic or on feelings; and (4) whether their style of dealing with the world is planned and organized or flexible and spontaneous. On the basis of how a person is classified on each of these four dimensions, he or she is placed into one of sixteen possible personality types. More than three million people complete the MBTI each year, much of the time for corporations that use it for team building and management development (Gardner & Martinko, 1996). Although there are questions about the psychometric properties of the MBTI, it at least allows people to think about how they typically function in the world, and how this might make some jobs more satisfying than others. For example, people who base their decisions on intuition and feelings may be happier in business than in the more objective world of science.

Another tool widely used to assess vocational compatibility relies on the extent to which people in certain occupations tend to share similar preferences. The *Strong Interest Inventory* measures an individual's interest in a broad range of occupations, leisure activities, and school subjects; those measurements are then compared with the measurements of people who are successfully employed in a variety of occupations. Thus, for example, you might find that people who like to engage in the same activities as you have fulfilling careers as artists or musicians but are miserable as accountants or investment bankers. Of course, career counselors need to take into account other personal factors because, for example, the relation between interests and occupation varies across cultures depending on how occupations are perceived (Fouad & Walker, 2005).

Among the most influential occupational psychologists was John Holland, who pioneered the scientific study of how careers match personality types. As a personnel clerk in the army during World War II, Holland noted that certain types of people had similar occupational histories. He examined this more formally while working on his doctorate in psychology at the University of Minnesota, which he obtained in 1952. During the next 20 years he conducted groundbreaking research demonstrating that certain personality types are best matched to particular work environments. His theoretical work has made a number of important contributions to occupational psychology, such as the core idea that occupational choice expresses one's personality (Gottfredson, 1999). After 40 years of research, Holland and his colleagues have identified six occupational personality types, each of which has distinctive key personal attributes and activity preferences (Table 12.3). These occupational types are associated with varying job satisfaction across different work settings (Holland, 1997). Holland's careful research has provided effective assessment tools that have helped millions of people around the world make successful career choices. ■

TABLE 12.3 | Holland's Occupational Types

OCCUPATION TYPE	PERSONALITY CHARACTERISTICS	PREFERRED ACTIVITIES	CAREER EXAMPLES
Realistic	Practical, stable, reliable, athletic, mechanical, honest, frank, humble	Repair things, enjoy physical activity, work outdoors, work with animals	Engineer, veterinarian, electrician, carpenter, farmer, mechanic, firefighter, geologist
Investigative	Curious, analytical, intellectual, reserved, independent, rational	Use computers, play chess, explore ideas, work alone, solve problems	Scientist, medical technician, systems analyst, software programmer, physician
Artistic	Creative, expressive, innovative, imaginative, impulsive, emotional	Attend music and dance events, write poetry, draw, take photographs	Artist, writer, musician, interior designer, reporter, actor, architect, photographer
Social	Helpful, kind, generous, friendly, responsible, concerned for others	Teach, volunteer, organize social events, engage in group activities, talking, entertaining	Teacher, social worker, nurse, clergy, speech therapist, aerobics instructor
Enterprising	Ambitious, confident, energetic, persuasive, enthusiastic, leader	Public speaking, managing others, promoting ideas, debating	Lawyer, sales and marketing, executive, public relations, manager, coach, stockbroker
Conventional	Efficient, organized, conscientious, precise, methodical, careful	Collect stamps or coins, build models, perform civic duties, organize things	Accountant, banker, secretary, financial analyst, statistician, computer operator

REVIEWING the Principles | How Is Personality Assessed, and What Does It Predict?

Personality is assessed through either projective or objective measures, depending on the goals of the researcher. Sometimes personality psychologists examine individuals, for example, by examining personal myths that people use to explain their lives. Other times they examine many people using a common measure to assess whether individual differences in those measures predict behaviour. People are relatively good at assessing personality traits in others, and some evidence suggests that observers might even be better at predicting people's behaviour than are the people themselves. This occurs when observers are sensitive to environmental cues that might shape behaviour. Indeed, traits interact with environments, such that situations sometimes constrain behaviour, and personality sometimes influences situations.

What Is the Biological Basis of Personality?

Where does personality come from? Why are you the person that you are? Freud emphasized early childhood experiences in his theory of psychosexual stages. Certainly most people assume that the way parents treat their children will have a substantial

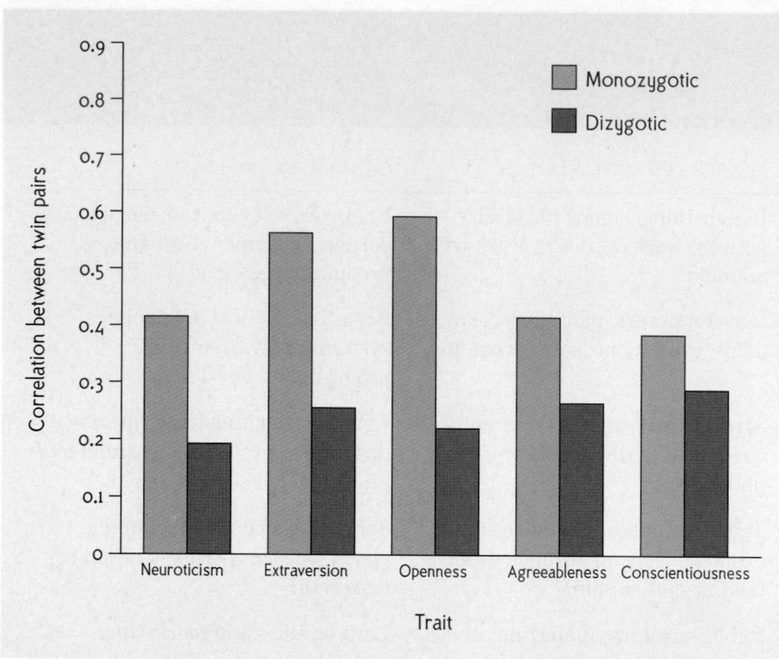

12.10 Correlation in Twins Researchers examined correlations between 123 pairs of identical twins (monozygotic) and 127 pairs of fraternal twins (dizygotic) in Vancouver. For each Big Five trait, the correlations for monozygotic twins were higher than for dizygotic twins, resulting in heritability estimates of 41 percent, 53 percent, 61 percent, 41 percent, and 44 percent of the variance in neuroticism, extraversion, openness, agreeableness, and conscientiousness.

impact on personality, for example, Rogers's belief that unconditional positive regard leads to positive mental health. But what role does biology play? Are you born with certain predispositions? How do the workings of your body and brain affect the sort of person you are?

Over the past few decades, evidence has emerged that biological factors—such as genes, brain structures, and neurochemistry—play an important role in determining personality. This is not to say that these factors are insensitive to experience. Recall from Chapter 3 that every cell in the body contains the genome, or master recipe, that provides detailed instructions for physical processes. Gene expression—whether the gene is turned off or on—underlies all psychological activity, such as how a person thinks or acts. Ultimately, genes have their effects only if they are expressed. The environment determines when or if this happens. In terms of personality, a person's genetic makeup may predispose certain traits or characteristics, but whether these genes are expressed depends on the unique circumstances that each child faces during development. For instance, we learned in Chapter 3 that children with a certain gene variation were more likely to become violent criminals as adults if they were abused during childhood. An important theme throughout this book is that nature and nurture work together to produce unique persons; this is particularly true for personality. In this section we consider the role of biological processes in personality.

Personality Is Rooted in Genetics

The evidence is overwhelming from researchers working at the biological level of analysis that nearly all personality traits have a genetic component (Plomin & Caspi, 1999). One of the earliest studies to document the heritability of personality was conducted by James Loehlin and Robert Nichols (1976). They looked at similarities in personality between more than 800 pairs of twins. Across a wide variety of traits they found that monozygotic twins were much more similar than dizygotic twins. Recall from Chapter 3 that this pattern reflects the actions of genes, since monozygotic twins share the same genes whereas dizygotic twins do not. Numerous twin studies have subsequently found that genetic influence accounts for approximately half of the variance (40–60 percent) between individuals in personality traits, including the Big Five (Figure 12.10), as well as specific attitudes that reflect personality traits, for example, attitudes towards the death penalty, abortion on demand, and enjoying roller coaster rides (Olson, Vernon, Harris & Jang, 2001). These genetic patterns can be found whether the twins rate themselves or whether friends, family, or trained observers rate them (Vernon et al., 1997; Jang et al., 2003).

Of course, you might wonder whether identical twins are treated more similarly than other siblings and whether that explains the similarities in personality. The best evidence refuting this idea was obtained by Thomas Bouchard and his colleagues from the University of Minnesota. Their studies of twins raised apart (such as the twins described in Chapter 3) reveal that twins raised apart are often as similar, or even more similar, than twins raised together. Why might twins raised apart be more similar than twins raised together? One possibility is that parents strive to bring out individual strengths in each twin so that each feels unique and special.

Thus, parenting style may foster differences rather than similarities. If this is true, one might expect that the correlations between personality traits would be stronger for older twins than for younger twins, since the effects of parenting would diminish over time. And indeed, identical twins become more alike as they grow older. By contrast, siblings and dizygotic twins do not.

"Oh, he's cute, all right, but he's got the temperament of a car alarm."

ADOPTION STUDIES Further evidence for the genetic basis of personality can be found in adoption studies, although such studies usually yield lower estimates of gene influence than do twin studies. Siblings who are adopted (and not biologically related) and raised in the same household are no more alike in personality than any two strangers randomly plucked off the street (Plomin & Caspi, 1999). Moreover, the personalities of adopted children bear no significant relationship to those of the adoptive parents who raised them. These findings suggest that parenting style may have relatively little impact on personality. Is that true? The current evidence suggests that parental style has much less impact than has long been assumed. For instance, studies typically find only small correlations in personality between biological siblings or between children and their biological parents. Although small, however, these correlations are larger than for adopted children, suggesting that the similarities have a genetic component. Why are children raised together in the same household (who are not monozygotic twins) so different? One explanation is that the lives of siblings diverge as they establish friendships outside the home. Thus, siblings' personalities slowly grow apart as their initial differences become magnified through their interactions with the world.

Although the small correlations in personality between siblings might imply that parenting style has little effect, there are reasons to believe that caretakers matter a great deal. David Lykken (2000), a leading behavioural genetics researcher, has argued that children who are raised with inadequate parenting, such as those raised in poverty by single parents who are not able to spend time parenting, are not socialized properly and are therefore much more likely to become delinquent or display antisocial behaviour. Thus, a minimum level of parenting is crucial, but the particular style of parenting does not appear to have a major impact on shaping personality.

ARE THERE SPECIFIC GENES FOR PERSONALITY? Research has revealed genetic components for particular behaviours such as television viewing and getting divorced, and even for specific attitudes, such as feelings about capital punishment and appreciation of jazz music. This does not mean, of course, that a gene lurking in your DNA determines the amount or type of television that you watch. Rather, genes predispose certain personality traits that are associated with behavioural tendencies. In most cases researchers are referring to the influence of multiple genes that each interact independently with the environment to produce general dispositions, such as a preference for quiet and passive activities over active outdoor pursuits.

Yet growing evidence indicates that genes can be linked with some specificity to personality traits. For instance, a gene that regulates one particular dopamine receptor has been associated with novelty seeking in a number of studies (Cloninger, Adolfsson, & Svrakic, 1996; Ekelund, Lichtermann, Jaervelin, & Peltonen, 1999). The theory is that people with one form of this gene are deficient in dopamine and seek out novel experiences to increase the release of dopamine. Research on the traits of neuroticism and agreeableness implicates a particular gene that regulates serotonin, although the effect is not large (Jang et al., 2001).

Within the last ten years, numerous studies have tested various candidate genes to assess their relation to personality. Although specific genes do seem to be involved in personality, it is likely that the measured effects will continue to be small. This is not surprising, as there are thousands of genes that could affect personality, and each one is likely to contribute to various personality traits in distinctive ways. For instance, some genes seem to work in opposite ways, such a making people more or less neurotic, and therefore they may cancel each other out (Savitz & Ramesar, 2004). David Lykken and his colleagues (1992) noted that it may be the chance aggregation of genes that produces unique persons. They provide the interesting analogy of a poker hand, in which you receive the ten and king of hearts from your mother and the jack, queen, and ace of hearts from your father. Although neither parent alone has a meaningful hand, together they pass you a royal flush. Thus, your personality reflects the resulting hand you were dealt from both of your parents. Moreover, each person lives in different circumstances that may cause the selective expression of certain genes (Jang et al., 1998). Given the complexity of most people's personalities, the complexity of their underlying physiology is hardly surprising. As we gain a greater understanding of the human genome in general, it is likely that we will continue to make progress in identifying how specific genes interact with the environment to produce various aspects of personality.

Temperaments Are Evident in Infancy

temperaments Biologically based tendencies to feel or act in certain ways.

Genes work by affecting biological processes. Thus, to the extent that genes influence personality, there ought to be corresponding biological differences, referred to as **temperaments**, in personality. Temperaments are considered much broader than traits and are described as general tendencies to feel or act in certain ways. Most of the research on temperaments focuses on infants, in that personality differences that exist very early in life indicate the actions of biological mechanisms. Although life experiences may alter personality traits, temperaments represent the basic innate biological structure of personality.

Arnold Buss and Robert Plomin (1984) argue that essentially three personality traits can be considered temperaments. *Activity level* refers to the overall amount of energy and behaviour a person exhibits. For example, some children zoom around the house at great speeds whereas others are less vigorous and still others slow-paced. *Emotionality* describes the intensity of emotional reactions, or how easily and frequently people become aroused or upset. Children who cry often or become easily frightened and adults who are quickly angered are likely to be high in emotionality. Finally, *sociability* refers to the general tendency to affiliate with others. Those high in this temperament prefer to be with others rather than to be alone. According to Buss and Plomin, these three temperamental styles are the main personality factors influenced by genes. Indeed, evidence from twin studies, adoption studies, and family studies indicates a powerful effect of heredity on these core temperaments.

"I could cry when I think of the years I wasted accumulating money, only to learn that my cheerful disposition is genetic."

SHYNESS AND INHIBITION Throughout this chapter we have used the example of shyness, which refers to feelings of discomfort and inhibition during interpersonal situations (Henderson & Zimbardo, 1998). People who are shy are excessively self-focused and spend a great deal of time worrying about what others think of them. The extent to which people are shy has been linked to early differences in temperament. Research conducted at Harvard University by Jerome Kagan, Nancy Snidman, and their colleagues has shown that children as young as six weeks of age can be identified as likely to be shy

(Kagan & Snidman, 1991). Approximately 15 to 20 percent of newborns react to new situations or strange objects by becoming startled and distressed, crying, and vigorously moving their arms and legs. Kagan refers to these children as *inhibited*, which he views as being biologically determined. Showing signs of inhibition at two months of age predicts later parental reports that the children are shy at four years of age, and such children are likely to be shy well into their teenage years. Indeed, measures of brainstem reactivity at ages 10 to 12 correspond to ratings made of these children when they were four months old (Woodward et al., 2001), with inhibited children showing greater reactivity. Similar genetic findings for inhibited behaviour have been found in a variety of species, including dogs and monkeys. The biological evidence suggests that the amygdala is involved in shyness. For instance, in one study people who were socially phobic—extremely shy—had much greater amygdala activation when shown unfamiliar faces than did control participants (Birbaumer et al., 1998).

Although shyness has a biological component, it clearly has a social component as well: Approximately one-quarter of behaviourally inhibited children are not shy later in childhood. This development typically occurs when parents create supportive and calm environments in which children are able to deal with stress and novelty at their own pace. At the same time, these parents do not completely shelter their children from stress, so that the children learn to deal with their negative feelings in novel situations. Moreover, at the social level of analysis, shyness is related to being highly feminine in sex-role orientation and varies across cultures (Chen et al., 1998); it is quite common in Japan and less common in Israel, once again highlighting the interplay between nature and nurture.

LONG-TERM IMPLICATIONS OF TEMPERAMENTS To what extent do infant temperaments predict adult personality? If you were a fussy child, are you now a fussy adult? Recent research has documented compelling evidence that early childhood temperaments have a pervasive and powerful influence over behaviour and personality structure throughout development (Caspi, 2000). In a particularly impressive study that we introduced in Chapter 3, a team of researchers in Dunedin, New Zealand, investigated the health, development, and personality of over 1,000 children born during a one-year period. These individuals were examined approximately every two years, with an impressive 97 percent remaining in the study through their twenty-first birthdays (Silva & Stanton, 1996). Children were classified at three years of age into temperamental types based on examiners' ratings. The classification of children at age three predicted personality structure and a variety of behaviours in early adulthood. For instance, those judged undercontrolled at age three were later more likely to have alcohol problems, to be criminals or unemployed, to attempt suicide, to be antisocial and anxious, and to have less social support (see Figure 12.11). Inhibited children were much more likely to become depressed. Early childhood temperaments do appear to be good predictors of later behaviours (Kerr et al., 1997).

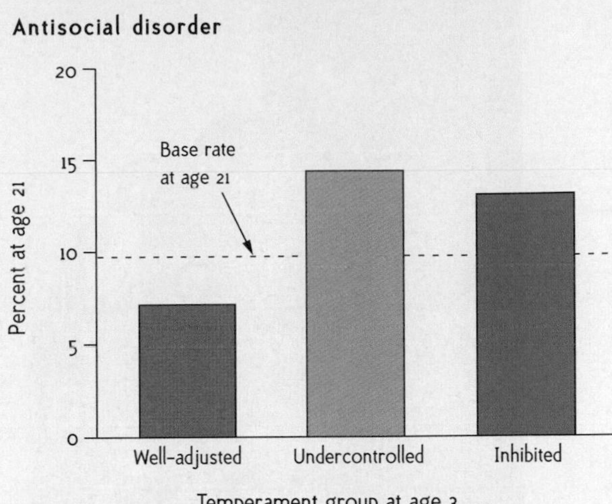

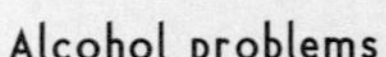

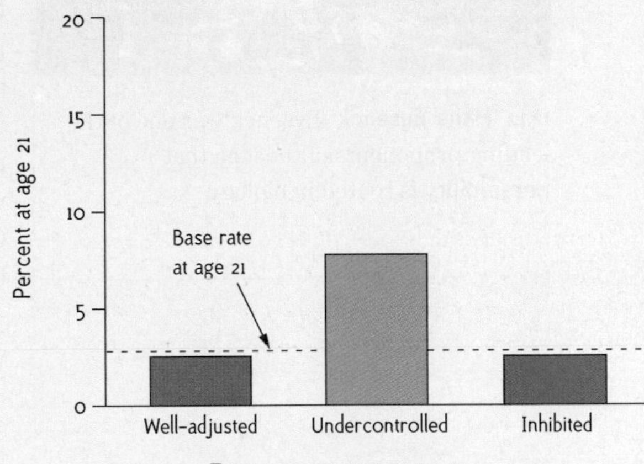

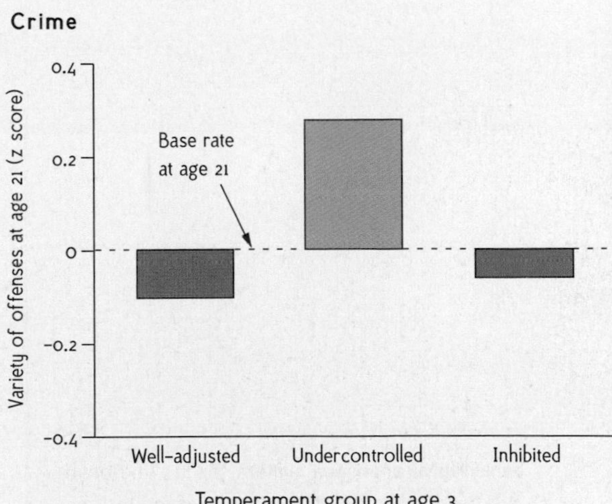

12.11 Predicting Personality Temperament group at age 3 predicts antisocial personality disorder, alcohol problems, and criminal activity at age 21.

Personality Is Linked to Specific Neurophysiological Mechanisms

Genes act to produce temperaments, which affect how children respond to and shape their environments, which in turn interact with temperament to shape personality. But how do these genetic predispositions produce personality? That is, what neurophysiological mechanisms are linked to personality? Some theories focus on the underlying biological processes that produce the thoughts, emotions, and

12.12 Hans Eysenck Eysenck was one of the leading proponents of the idea that personality is rooted in biology.

behaviours that make up personality. From this perspective, personality differences may reflect differences in the relative activation of different biological systems. Most research on the neurobiological underpinnings of personality has focused on the dimension of extraversion/introversion, and it is there that we focus as well.

AROUSAL AND EXTRAVERSION/INTROVERSION The intellectual founder of the modern biological approach to personality was Hans Eysenck (Figure 12.12). Eysenck believed that underlying differences in cortical arousal produce the observed behavioural differences between extraverts and introverts. Cortical arousal, or alertness, is regulated by the *ascending reticular activating system* (*ARAS*), and Eysenck proposed that this system differs between extraverts and introverts. Eysenck noted that extraverts seem constantly to be trying to seek out additional arousal—for example, by attending parties or seeking out new people. By contrast, introverts seem to avoid arousal by preferring solitary, quiet activities, such as reading. According to earlier psychological theories, each person prefers to operate, and operates best, at some *optimal level of arousal* (see Chapter 9). Eysenck proposed that the resting levels of the ARAS are higher for introverts than for extraverts, which places them above their optimal levels. In contrast, extraverts are typically below their optimal levels, which means that they are chronically underaroused. To operate efficiently, extraverts have to find arousal. Thus, they impulsively seek out new situations and new emotional experiences. Introverts don't want any additional arousal. They prefer quiet solitude with few incoming stimuli. So if you are an introvert, a noisy environment will distract you, whereas if you are an extravert, quiet places are boring. Consistent with Eysenck's theory, research has demonstrated that extraverts prefer to perform, and actually do perform better, in noisy settings (Geen, 1984).

If introverts are chronically more aroused than extraverts, they ought to be more sensitive to stimuli at all levels of intensity. This appears to be generally true. For instance, introverts are more sensitive to pain, and they salivate more than extraverts do if you place a drop of lemon juice on their tongues. However, evidence for baseline differences in arousal has been more difficult to produce. Thus, it appears that what differentiates introverts from extraverts is level of *arousability*, or reactivity to stimuli, with introverts being more arousable.

The psychologist Marvin Zuckerman describes the arousal-based trait of *sensation seeking* as similar to extraversion, but with an impulsive element that more closely matches Eysenck's psychoticism superordinate trait. According to Zuckerman, sensation seekers have a neurochemical deficiency that motivates them to seek arousal through adventures and new experiences. Moreover, those who are high in sensation seeking are easily bored and escape boredom through the use of drugs and alcohol.

NEUROPHYSIOLOGY OF EXTRAVERSION/INTROVERSION Since Eysenck's initial work on the biological underpinnings of personality, a number of theorists have offered refinements that reflect more recent understandings of functional neuroanatomy. Although a variety of theories have been proposed, they share common features, including a basic differentiation between approach and avoidance learning. Jeffrey Gray of Oxford University incorporated this distinction in his approach/inhibition model of learning and personality. Gray proposed that personality was rooted in motivational functions that had evolved to help organisms respond efficiently to reinforcement and punishment. In Gray's model, the **behavioural approach system (BAS)** consists of the brain structures that lead organisms to approach stimuli in pursuit of rewards. This is the *Go* system. There is also a *Stop* system, known as the **behavioural inhibition system (BIS)**, which is sensitive to punishment and therefore inhibits behaviour that might lead to danger or pain (Figure 12.13). According to

behavioural approach system (BAS) The brain system involved in the pursuit of incentives or rewards.

behavioural inhibition system (BIS) The brain system that is sensitive to punishment and therefore inhibits behaviour that might lead to danger or pain.

Gray, extraverts have a stronger BAS than BIS, so they are more influenced by rewards than by punishments. Indeed, extraverts tend to act impulsively in the face of strong rewards, even following punishment (Patterson & Newman, 1993). By contrast, introverts have a more active BIS. Their chronic anxiety often leads them to avoid social situations in which they anticipate possible negative outcomes.

The BIS is associated with activity in the frontal lobes, which are known to help inhibit inappropriate social behaviour. Those with injury to the frontal lobes, especially the prefrontal cortex, exhibit social incompetence, disinhibition, impaired social judgment, and a lack of sensitivity to social cues. One imaging study found that introversion was associated with greater activation of the frontal lobes (Johnson et al., 1999), which supports Gray's model of BIS. The amygdala is another brain region involved in both social sensitivity and processing of cues related to possible punishment. Some researchers believe that personality traits such as fearfulness, anxiousness, and shyness are associated with excessive activation of the amygdala (Zuckerman, 1991). People who are usually anxious show a heightened amygdala response when observing pictures of neutral facial expressions, perhaps because they are especially sensitive to any signs of social rejection; faces looking at us with neutral expressions may trigger evaluation concerns (Somerville, Kim, Johnstone, Alexander, & Whalen, 2004).

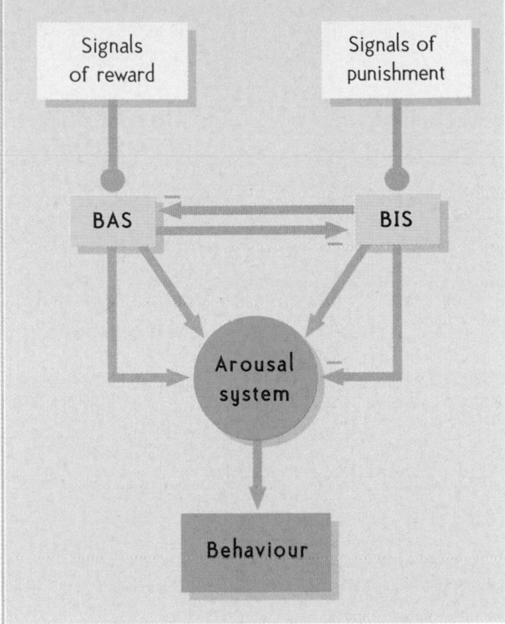

12.13 Behavioural Approach System (BAS)/ Behavioural Inhibition System (BIS) Signals of potential reward and punishment are processed by the BAS and the BIS. This information affects behavioural responses, with the BAS activating behaviour and the BIS inhibiting behaviour.

Studies of the neurochemistry of the behavioural activation system implicate heightened activation of dopamine circuits for extraverts compared with introverts. Dopamine is involved in reward (as you will recall from Chapter 6), and extraversion appears to be associated with greater activation of dopamine receptors in the nucleus accumbens. Links between various brain regions that are rich in dopamine receptors process information about the reward value of objects and seem to be involved in extraversion (Depue & Collins, 1999). A second line of evidence for the role of dopamine in extraversion is that they are both associated with positive affect (Ashby, Isen, & Turken, 1999). Extraverts report high levels of energy, desire, and self-confidence (Figure 12.14). Indeed, the experience of positive affect may be the fundamental feature of extraversion (Lucas, Diener, Grob, Suh, & Shao, 2000). Finally, the link between dopamine and extraversion is also supported by the finding that a gene involved in dopamine reception is an indicator for novelty seeking, which itself is related to the greater willingness of extraverts to approach novel stimuli.

We still have much to learn about the biological bases of personality. As recently as three decades ago personality psychologists largely ignored the question of biology. Only with recent advances in technology have researchers started to explore the genetic and neurophysiological correlates of personality. We can anticipate many exciting new discoveries relevant to personality and temperament as the biological revolution spreads throughout psychological science (Canli, 2004). Consider a study in which only people possessing one particular form of a serotonin gene showed greater amygdala activity when looking at pictures of people showing emotional expressions than did those with the other form of the gene (Hariri et al., 2002). Interestingly, the first form of the gene has been associated with greater overall fearfulness and negativity. Using brain imaging and gene data together may become a powerful new way to assess the biological basis of personality.

12.14 Jim Carrey Carrey appears to have all the hallmarks of an extraverted personality.

Personality Is Adaptive

Natural selection has shaped the human genome over the course of evolution. Adaptive characteristics were likely to spread through the gene pool and have been passed along to increasing numbers in future generations. Thus, in terms of personality, we might expect that traits that were useful for survival and reproduction would have been favoured. It is easy to imagine how being competitive might lead a person to obtain great rewards, or how being cooperative might increase a person's value to the group. But traits also provide important information about desirable and undesirable qualities in mates, such as whether a person is conscientious and agreeable or neurotic. David Buss (1999a) has argued that the Big Five personality traits emerged as foundational traits because they each provide important information regarding mate selection.

But if traits are adaptive, why are there such great individual differences among people? Natural selection ought to make people more similar rather than more different. After all, if a trait increased survival and reproduction, that trait should predominate in future generations. One possibility is that individual differences reflect characteristics that were of trivial importance over the course of evolution, perhaps resulting from random processes (Tooby & Cosmides, 1990). However, Buss and Greiling (1999) have proposed that individual differences may reflect the inheritance of alternative strategies that become activated depending on situational context. For instance, consider a situation in which most people are honest and cooperative and therefore routinely trust others to be honest and cooperative as well. A dishonest person in such a system could do well by exploiting the basic trust of others. Of course, if too many people did this, then the system would change or collapse. However, the important point is that evolution has bestowed multiple strategies that are differentially adaptive depending on the environment, and that a built-in calibration system is sensitive to environmental demands. According to Buss and Greiling, the *early experiential calibration* system locks a person into the strategy chosen, to the exclusion of others that might have been pursued under different circumstances.

Another possible explanation for individual differences is that human groups whose members possess diverse skills have a selective advantage over other groups (Caporael, 2001). Members of successful groups and their relatives would all benefit from being members of a strong group, since they would be more likely to survive and reproduce. Consider the trait of novelty seeking. Having group members who seek out and explore new territory might lead to the discovery of new resources, such as an abundant food supply. At the same time, novelty seekers expose themselves to greater risks, and the group would suffer if all its members followed this strategy. Therefore it is to the advantage of the group to have cautious members in addition to bold members; these individuals may enhance the group in other ways, such as being more thoughtful or providing social support.

REVIEWING the Principles | What Is the Biological Basis of Personality?

Evidence from behavioural genetics has demonstrated that personality has a substantial genetic component explaining approximately half of the variance in most traits. Temperaments, the biological bases of traits, are evident in early childhood and have long-term implications for adult behaviour. Temperamental styles reflect underlying differences in biological processes, suggesting that people are different in part because they have different underlying neurophysiology. Such differences may reflect adaptive advantages over the course of human evolution.

Can Personality Change?

The Jesuit maxim "Give me a child until he is seven, and I will show you the man" is the thesis of Michael Apted's well-known documentary series of films (the *Up* series), which follows the development of a group of British school children through interviews at ages 7, 14, 21, 28, 35, and 42. A striking aspect of these films is the apparent stability of personality over time. The child interested in the stars and science becomes a professor of physics; the boy who finds his childhood troubling and confusing develops an apparent schizo-affective personality; the reserved, well-mannered, upper-class girl at age 7 grows into the reserved, well-mannered woman in her pastoral retreat at age 35; a second-grader successfully predicts not only his future career, but also the schools that he will eventually attend. Are people really so stable? Is personality at age 70 preordained at age 7?

Why should personality be stable?

From a functional perspective, it makes sense that personality should remain relatively stable over time. People need to predict the behaviour of those they care about or rely on. When people choose friends or relationship partners, they do so with the general expectation that their interactions with these people will be somewhat predictable across time and situations. Yet people sometimes expect others to be able to change. Marital partners often hope that unfaithful spouses will change their wandering ways, and the penal system releases prisoners with the expectation that they will adopt less deviant roles in society. Are these expectations realistic?

Earlier you learned that childhood temperaments predict behavioural outcomes in early adulthood. But what about change during adulthood? If you are shy now, are you doomed to be shy forever? Indeed, the foundation of clinical psychology is the belief that people can and do change important aspects of their lives. People exert considerable energy trying to change—they attend self-help groups, read self-help books, buy time with therapists, and struggle with themselves and others. But how much do people really change over the course of their lives?

Traits Remain Stable over Time

How we define the essential features of personality has tremendous implications for whether it is fixed or changeable. Continuity over time and across situations is inherent in the definition of "trait," and accordingly it is not surprising that most research finds personality traits to be remarkably stable over the adult life span (Heatherton & Weinberger, 1994). For instance, over many years the relative rankings of individuals on each of the Big Five personality traits remain stable (McCrae & Costa, 1990). People who are very extraverted tend to stay very extraverted; people who are very introverted tend to remain that way as well. A recent meta-analysis of 150 studies consisting of nearly 50,000 participants who had been followed for at least one year found strong evidence for stability in personality (Roberts & Friend-DelVecchio, 2000). The rank orderings of individuals on any personality trait were quite stable over long periods of time across all age ranges. However, stability was lowest for young children and highest for those over age 50 (Figure 12.15). This suggests that personality does change somewhat in childhood, but that it becomes more stable by middle age. Such findings tend to support the contention of William James, who stated in 1890, "for most of us, by age 30, the character has set like plaster and will never soften again." According to the meta-analysis, James was right that personality becomes set, but this appears to happen a little later than age 30.

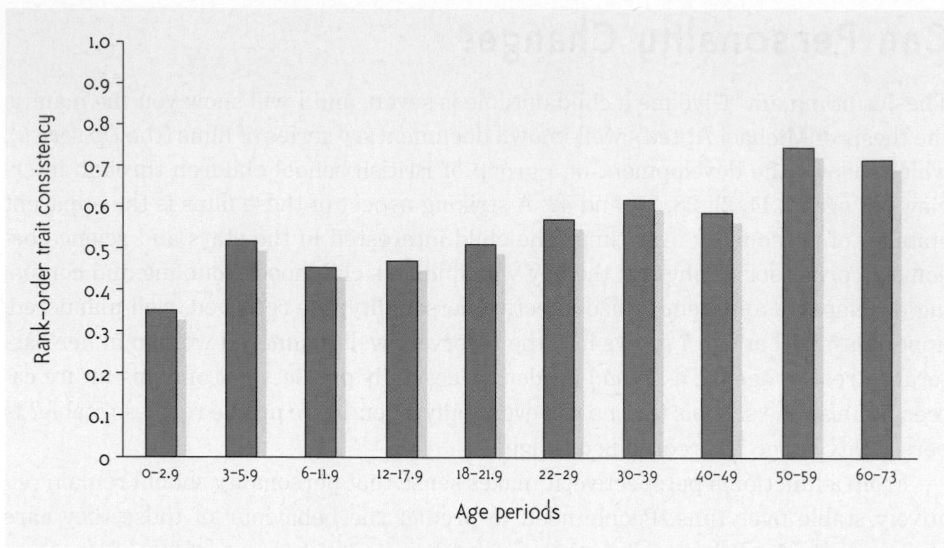

12.15 The Stability of Personality This graph shows consistency across time measures for personality at different ages. Consistency is lowest in childhood and highest after age 50.

Stability in rank ordering means that individuals stay the same compared to others. However, is it possible that all people change in personality as they age, while retaining their relative rankings? For instance, consider the stereotype that people become wiser and more cautious as they get older. Is this true? In general, people do become less neurotic, less extraverted, and less open to new experiences as they get older. People also tend to become more agreeable and much more conscientious with age (Srivastava, John, Gosling, & Potter, 2003). These effects are not large, but they are consistent. What is most amazing is that the pattern holds in different cultures (Figure 12.16; McCrae et al., 2000). This suggests that age-related changes in personality occur independent of environmental influences, and therefore that personality change itself may be based in human physiology. Indeed, some evidence shows that personality change has a genetic component, such that the extent of change is more similar in monozygotic twins than in dizygotic twins (McGue, Bacon, & Lykken, 1993).

The reason personality is so stable, especially among adults, appears to be due to a number of factors. If personality is determined in part by biological mechanisms,

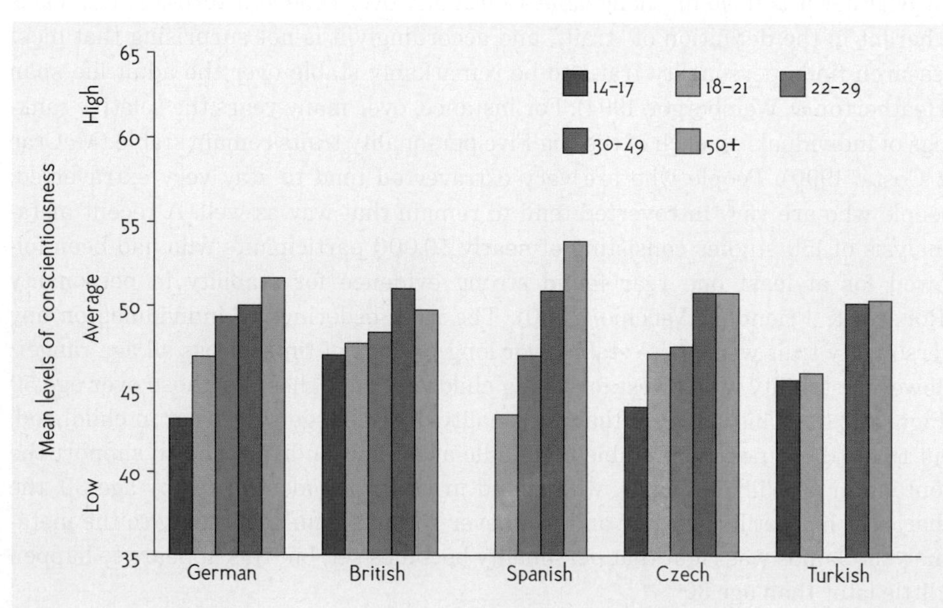

12.16 Mean Levels of Conscientiousness at Different Ages in Five Cultures Note that in all five cultures, conscientiousness increases as people age.

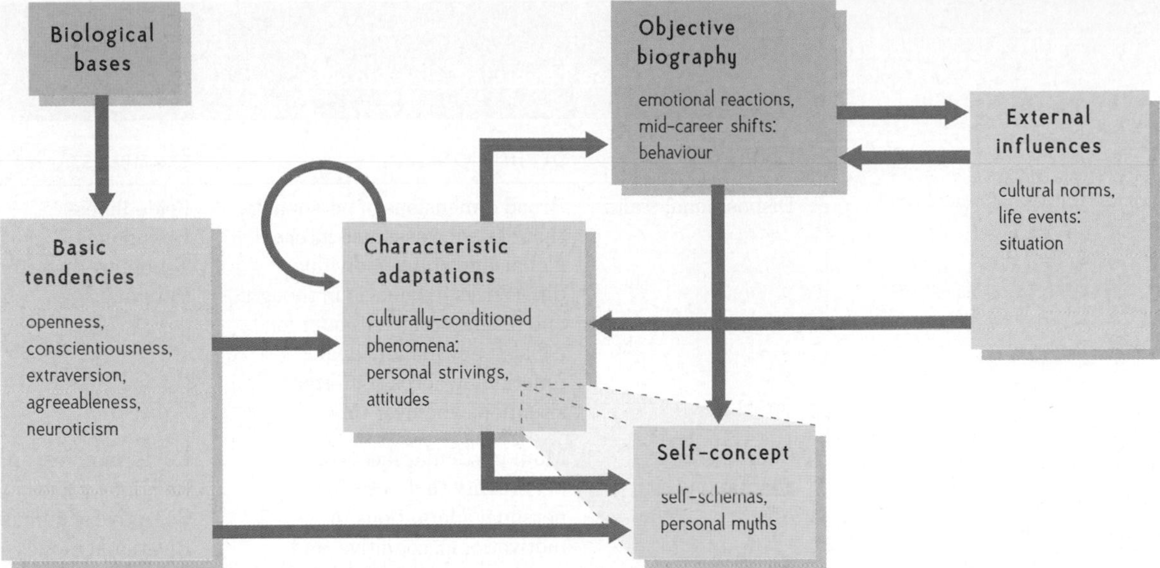

12.17 McCrae and Costa's Model of Personality Basic tendencies are biologically based, but characteristic adaptations are influenced by the situation and how the person behaves. The lines with arrows indicate some of the ways in which the different components interact. The important point is that basic tendencies do not change across situations, but observable behaviour (objective biography) does because it is influenced by personal goals and motives as well as by the situation.

then we can expect changes only to the extent that there are changes in biological makeup. The fact that the brain develops well into early adulthood may explain the greater evidence of personality change before age 30. Perhaps more important, environments tend to be relatively stable, especially after early adulthood. People tend to marry those who are similar in attitudes and personality, and people tend to have successive jobs that have the same level of status. It seems likely that the stability of situations contributes to the stability of personality.

Characteristic Adaptations Change across Time and Circumstances

In their research on potential change in personality, Robert McCrae and Paul Costa (1999) emphasize an important distinction between basic tendencies of personality and characteristic adaptations (Figure 12.17). *Basic tendencies* are dispositional traits that are determined to a great extent by biological processes. As such, they are very stable. *Characteristic adaptations* are the adjustments people make to situational demands, which tend to be consistent because they are based on skills, habits, roles, and so forth. These changes in behaviour do not indicate changes in the underlying core disposition (Figure 12.18). For instance, consider a young adult who is high in extraversion. She may go to a lot of parties, be a thrill seeker, and have multiple sexual partners. As she approaches 90, she is less likely to do these things, but she may still have many friends and enjoy traveling. Although the behaviours are different, they continue to reflect the underlying trait of extraversion.

Dan McAdams (1999; 2001) has proposed that some aspects of personality do change, depending on how you define personality. According to McAdams, personality can be examined at three levels (Table 12.4). At the basic level are *dispositional traits*, such as the Big Five, which by their nature are relatively stable. At the

12.18 Celine Dion How similar has Celine Dion's personality remained over the years?

TABLE 12.4 | Three Levels of Personality

LEVEL	DEFINITION	EXAMPLES
Dispositional traits	Broad dimensions of personality that describe assumed internal, global, and stable individual differences in behaviour, thought, and feeling. Traits account for consistency in individual functioning across different situations and over time.	Friendliness Dominance Tendency toward depression Punctuality
Characteristic adaptations	More particular facets of personality that describe personal adaptations to motivational, cognitive, and developmental challenges and tasks. Characteristic adaptations are usually contextualized in time, place, situation, or social role.	Goals, motives, and life plans Religious values and beliefs Cognitive schemas Relational modes Psychosocial stages Developmental tasks
Life stories	Internalized and evolving narratives of the self that people construct to integrate the past, present, and future and provide life with some sense of unity, purpose, and meaning. Life stories address the problems of identity and integration in personality—problems especially characteristic of modern adulthood.	Earliest memory Reconstruction of childhood Perceptions of future self "Rags to riches" stories Imagery and theme in story

second level are *personal concerns*, which include goals, motives, and social roles; these reflect the tasks and challenges that people face within specific contexts or situations. Unlike dispositional traits, McAdams argues, personal concerns ebb and flow as people mature into new roles (such as career and family) and encounter new challenges. For instance, as people grow older they may express more concern with the well-being of future generations, for example, by making plans to educate their children or by donating to environmental charities. At the third level are *life narratives*, the personal myths and stories that people use to define their lives and identities. McAdams argues that life narratives evolve as people develop coherent stories from the different places, opportunities, and positions they encounter (Baver, McAdams, & Sakaeda, 2005).

Indeed, some people tell stories of sudden, dramatic changes in personality, much like the transformation of Ebenezer Scrooge in Charles Dickens's *A Christmas Carol*. William Miller and Janet C'de Baca (2001) describe these stories as examples of **quantum change**, which is a transformation of personality that is sudden, profound, and enduring. The people telling the stories claimed to remember the episodes vividly, down to the date and time of day when the change occurred. The quantum change experiences tended to be unexpected, often originating from some external source or event. The specific antecedents were amazingly varied and ranged from taking a walk to cleaning a toilet to having an abortion. Those who reported quantum change emphasized that everything about them changed, including

quantum change A transformation of personality that is sudden, profound, and enduring, and that affects a wide range of behaviours.

their temperaments, perceptual styles, goals, and values. After the change, most people reported an increased sense of meaning and happiness, an increased sense of spirituality, and increased overall satisfaction. Of course, without an objective measure of personality before the experience it is hard to know whether the reported changes are accurate, but family members and friends substantiated the stories, which gives them some credence.

Brain Injury and Pharmacological Interventions Affect Personality

To the extent that biological processes determine personality, we might anticipate that physical changes would produce personality changes. Indeed, damage to specific regions of the brain is associated with dramatic changes in personality. For example, damage to the frontal lobes—such as in the case of the railroad worker Phineas Gage, described in Chapter 4, or Elliot, who had brain surgery, described in Chapter 10—has been found to produce a variety of changes in personality, including causing people to become more extraverted, impulsive, socially inappropriate, and moody (Stuss, 1991; Stuss, Gow, & Hetherington, 1992). Patients with temporal-lobe damage also experience personality change, often becoming humourless, obsessive, paranoid, and rule bound. One sign of a possible brain tumour is a sudden and profound change in personality. Indeed, the whole point of psychosurgery, which was popular in the 1950s, was to alter the brain surgically to change abnormal aspects of personality (see Chapter 4). Some diseases that cause damage to the brain, such as Alzheimer's disease, also produce changes in personality that can be dramatic and surprising to family members.

NEUROCHEMISTRY AND PERSONALITY CHANGE In his 1993 book *Listening to Prozac*, Peter Kramer claimed that a drug used to treat depression could lead to changes in personality. His book includes stories of patients who underwent dramatic transformations after taking Prozac, a drug that treats depression by enhancing the activity of the neurotransmitter serotonin through selectively blocking its reuptake into the presynaptic neuron, thereby prolonging its effects in the synapse. In one story, a dour, shy, pessimistic woman suddenly became outgoing and sociable after she took Prozac. Her bubbly new personality, Kramer claimed, is a result of her altered serotonin levels.

Research has demonstrated that serotonin is related to some aspects of personality. For instance, low levels of serotonin are associated with hostility, aggressiveness, and criminality. Low levels of serotonin have also been linked to impulsivity and sensation seeking (Depue & Collins, 1999; Zuckerman, 1995). Thus, we might expect that drugs that enhance the activity of this neurotransmitter would lead to decreased impulsiveness (LeMarquand, 1998). Indeed, evidence has been found that serotonin-enhancing drugs do cause significant changes in personality, such as increased social dominance and decreased hostility (Brody et al., 2000). One research team found that use of drugs that enhanced serotonin led to changes in two of the Big Five personality traits: neuroticism decreased and extraversion increased (Bagby, Levitan, Kennedy, Levitt, & Joffe, 1999). This effect apparently occurs independent of changes in mood or depression (Ekselius & von Knorring, 1999). That is, the personality change does not appear to occur simply because depression is lifted (Bezchlibnyk-Butler et al., 2000).

In perhaps the best empirical test of the effects of serotonin-based drugs, researchers had normal volunteers take a selective serotonin reuptake inhibitor (SSRI) and assessed its effects on social behaviour and personality (Knutson et al., 1998). Animal research had previously shown that SSRIs led to decreased aggression and increased social affiliation among monkeys, and the researchers sought to examine whether it would produce similar effects in nondepressed humans. Following a careful screening for any psychiatric disorders, volunteers received either an SSRI or a placebo for four weeks. The results indicated that administration of the SSRI led to reductions in hostility and increases in cooperative behaviour. Thus, altering brain neurochemistry can change personality even among normal, nondepressed individuals.

REVIEWING the Principles | Can Personality Change?

The evidence that personality is generally stable over the life course is considerable, especially for basic dispositions such as traits. People adapt to novel situations by altering behaviour, but they tend to do so in ways that are consistent with their basic personalities. As shown by their life stories, people believe that they change, and such change narratives might help predict motives and behaviour. Indeed, the story of change is itself a part of personality whether or not change occurs. Although it is unclear whether people can purposefully change, brain injury or disease and alterations in neurochemistry can transform personality. This supports the general idea that biological processes are important to personality.

Conclusion

Studying personality is complex because each person is unique. Researchers have contributed to the accumulated principles underlying personality by examining cognitive processes, both conscious and unconscious; behavioural dispositions; and people's own life narratives. Knowing a person well requires knowing a great deal about him or her. At the same time, it is now apparent—from research that crosses levels of analysis—that the basic blueprint for personality is genetically determined and is manifest through biologically based temperaments. Yet these temperaments do not predetermine personality; interaction between people and their social worlds creates unique individuals, perhaps by influencing brain processes that determine personality. What is clear is that personality is coherent and stable. Although people change jobs, relationships, and living circumstances, the fundamental core of their personalities stays with them throughout their lives.

Summarizing the Principles of Personality

How Have Scientists Studied Personality?

1. **Psychodynamic theories emphasize unconscious and dynamic processes:** Freud believed that personality was in part the result of unconscious conflicts. The personality is comprised of the id, ego, and superego. Stages of psychosexual development occur from birth to adolescence. The ego uses defense mechanisms to reduce the anxiety of the oppositional demands of the id and superego. Neo-Freudians have focused on relationships, especially the emotional attachments that children develop to their parents, rather than sexual forces (libido).

2. **Humanistic approaches emphasize integrated personal experience:** Humanists view personality as the result of human experiences and beliefs. Humans strive to realize

their full potential, and may be hampered in doing so if they do not receive unconditional positive regard from their parents. The positive psychology movement researches subjective well-being.

3. **Type and trait approaches describe behavioural dispositions:** Personality type theories focus more on description than explanation. Trait theories are based on the assumption that personality is a collection of traits that individually vary, and exist in a hierarchy of importance. Eysenck proposed a model of personality that showed that the biologically based traits (extraversion, emotional stability, psychoticism) are the important traits, under which lesser traits are organized. The Big Five theory considers personality to be the result of the composition of the basic traits: openness to new experience, conscientiousness, extraversion, agreeableness, and neuroticism.
4. **Personality reflects learning and cognitive processes:** Through interaction with their environment, people learn patterns of responding that are guided by expectancies and values. The extent to which people believe they can achieve specific outcomes, called self-efficacy, is an important determinant of behaviour. The Cognitive-Affective Personality System (CAPS) emphasizes self-regulation.

How Is Personality Assessed, and What Does It Predict?

5. **Personality refers to both unique and common characteristics:** Idiographic approaches are person-centred; they evaluate personality from the perspective of assessing the unique pattern of characteristics of an individual. Nomothetic approaches focus on characteristics that are common among all people, but on which people vary (i.e., traits).
6. **We can use objective and projective methods to assess personality:** Projective measures (e.g., the Thematic Apperception Test, the Rorschach inkblot test) subjectively evaluate the unconscious issues a person projects onto ambiguous stimuli. Objective measures are straightforward assessments, usually made by self-report questionnaires or observer ratings.
7. **Observers show accuracy in trait judgments:** Your close acquaintances may be more accurate at predicting your behaviour than you are.
8. **People sometimes are inconsistent:** Mischel proposed that situations are more important than traits in predicting behaviour (situationism).
9. **Behaviour is influenced by the interaction of personality and situations:** Situations vary in the extent to which they influence behaviour and interact with personality to determine behaviour.

What Is the Biological Basis of Personality?

10. **Personality is rooted in genetics:** Nearly all personality traits have a genetic component. Twin and adoptions studies have found that 40–60 percent of personality variation is due to genetics. Personality characteristics are polygenetic, and their expression is the result of interaction with the environment.
11. **Temperaments are evident in infancy:** Temperaments, the general tendencies of how we behave, are biologically mediated and observable in infants.
12. **Personality is linked to specific neurophysiological mechanisms:** Cortical arousal is regulated through the ascending reticular activating system (ARAS), and results in characteristics of introversion/extraversion. The behavioural approach system (BAS) and the behavioural inhibition system (BIS) affect variations in arousal and the behavioural responses. Dopamine pathway activation is greater in extraverts.
13. **Personality is adaptive:** Variations in personality and skills of individual members benefit a group, and provide an advantage for group survival.

Can Personality Change?

14. **Traits remain stable over time:** Trait consistency is lowest for young children and highest for those over age 50. Biological and environmental factors are more stable in adulthood.
15. **Characteristic adaptations change across time and circumstances:** Basic tendencies are dispositional traits that are determined to a great extent by biological processes, and are therefore stable. Characteristic adaptations are contextualized by time, place, situation, and social role. Quantum-change events, as reported in life stories, are significant in changing personality.
16. **Brain injury and pharmacological interventions affect personality:** Drugs that affect the action of the neurotransmitter serotonin create changes in personality, particularly with regard to social dominance and reduced hostility.

Applying the Principles

1. After studying the different approaches to personality, Diane decides that multiple theories each partially explain her personality. She feels strongly that her childhood relationship with her mother has had an impact. She also realizes that she is basically introverted, conscientious, and cautious. Finally, she knows that she likes quiet situations and feels awkward around strangers. Which three theories of personality, respectively, correspond to Diane's three interpretations of her personality?

   ____ a) psychodynamic; trait; social-cognitive
   ____ b) humanistic; social-cognitive; behavioural
   ____ c) behavioural; trait; psychodynamic
   ____ d) humanistic; trait; behavioural

2. As part of the application process for becoming a police officer, candidates must have a personality evaluation. There are generally three parts to this evaluation. The first requires the candidates to decide which traits listed on a questionnaire apply to them; this is a(n) ____ measure. In the next part, the candidates report what they "see" when viewing some ink blots; this is a(n) ____ measure. Finally, the candidates undergo an extensive life history interview; this is a(n) ________ approach to personality.

   ____ a) situationism; projective; objective
   ____ b) objective; nomothetic; thematic apperception
   ____ c) ideographic, objective; situationism
   ____ d) objective; projective; ideographic

3. Bonita is a bit reckless, drives fast, and plunges into new situations without much forethought. Her sister Anna is almost the complete opposite. All but which of the following biological factors of personality are likely to vary between the two girls?

   ____ a) temperament
   ____ b) genotype
   ____ c) cortical arousal
   ____ d) natural selection

4. Your roommate Todd is very close to his older brother Jim, who is now married and whose wife is expecting a baby. Todd complains that his brother is not as much fun as he used to be and wonders if Jim has had a personality change. Based on what you have learned about the stability of personality, what would you say?

   ____ a) It is likely that Jim has experienced some dramatic changes now that he is becoming a father; he will be his "old self" again soon.
   ____ b) Jim's characteristic adaptations are consistent but his personal concerns have changed.
   ____ c) Jim's dispositional tendencies are biological and have changed as he has aged.
   ____ d) Jim has experienced a quantum change in personality because of an extraordinary life event.

ANSWERS: 1. a 2. d 3. d 4. b

Key Terms

behavioural approach system (BAS), p. 508
behavioural inhibition system (BIS), p. 508
defense mechanisms, p. 487
ego, p. 487
five-factor theory, p. 492
humanistic approaches, p. 489
id, p. 487
idiographic approaches, p. 497
interactionists, p. 501
nomothetic approaches, p. 498
objective measures, p. 499
personality, p. 485
personality trait, p. 485
personality types, p. 490
projective measures, p. 498
psychodynamic theory, p. 486
psychosexual stages, p. 487
quantum change, p. 514
situationism, p. 500
superego, p. 487
TAT (Thematic Apperception Test), p. 498
temperaments, p. 506
trait approach, p. 490

Further Readings

Funder, D. C. (2004). *The personality puzzle* (3rd ed.) New York: Norton.

Heatherton, T. F., & Weinberger, J. L. (1994). *Can personality change?* Washington, DC: American Psychological Association.

Hogan, R., Johnson, J., & Briggs, S. (1997). *Handbook of personality psychology.* San Diego: Academic Press.

McAdams, D. P. (2005). *The person: An integrated introduction to personality psychology* (4th ed.). Fort Worth: Harcourt.

Pervin, L. A., & John, O. P. (1999). *Handbook of personality: Theory and research.* New York: Guilford Press.

Further Readings: A Canadian Presence in Psychology

Barbopoulos, A., Fisharah, F., Clark, J. M., & el-Khatib, A. (2002). Comparison of Egyptian and Canadian children on a picture apperception test. *Cultural Diversity and Ethnic Minority Psychology, 8*, 395–403.

Chen, X., Hastings, P. D., Rubin, K. H., Chen, H., Cen, G., & Stewart, S. L. (1998). Child-rearing attitudes and behavioural inhibition in Chinese and Canadian toddlers: A cross-cultural study. *Developmental Psychology, 34*, 677–686.

Cleghorn, R. A. (1984). The development of psychiatric research in Canada up to 1984. *Canadian Journal of Psychiatry, 29*, 189–197.

Heller, D., Watson, D., & Ilies, R. (2004). The role of person versus situation in life satisfaction: A critical examination. *Psychological Bulletin, 130*, 574–600.

Jackson, D. N., Ashton, M. E., & Tomes, J. L. (1996a). The six factor model of personality: Facets from the Big Five. *Personality and Individual Differences, 21*, 391–402.

Jackson, D. N., Paunonen, S. V., Fraboni, M., & Goffin, R. D. (1996b). A five factor versus six factor model of personality structure. *Personality and Individual Differences, 20*, 33–45.

Jang, K. L., Hu, S. Livesley, W. J., Angleitner, A., Riemann, R., Ando, J., et al. (2001). Covariance structure of neuroticism and agreeableness: A twin and molecular genetic analysis of the role of the serotonin transporter gene. *Journal of Personality and Social Psychology, 81*, 295–304.

Jang, K. L., Livesley, W. J., & Vernon, P. A. (1996). Heritability of the Big Five personality dimensions and their facets: A twin study. *Journal of Personality, 64*, 577–591.

Jang, K. L., Stein, M. B., Taylor, S., Asmundson, G. J., & Livesley, W. J. (2003). Exposure to traumatic events and experiences: Aetilogical relationships with personality function. *Psychiatric Research, 30*, 61–69.

Lefcourt, H. M. (1966). Internal versus external control of reinforcement: A review. *Psychological Bulletin, 65*, 206–220.

Lefcourt, H. M. (1992). Durability and impact of the locus of control construct. *Psychological Bulletin, 112*, 411–414.

Nathanson, C., Paulhus, D. L., & Williams, K. M. (2006a). Personality and misconduct correlates of body modification and other cultural deviance markers. *Journal of Research in Personality, 40*, 799–802.

Nathanson, C., Paulhus, D. L., & Williams, K. M. (2006b). Predictors of a behavioural measure of scholastic cheating: Personality and competence but not demographics. *Contemporary Educational Psychology, 31*, 97–122.

Olson, J. M., Vernon, P. A., Harris, J. A., & Jang, K. L. (2001). The heritability of attitudes: A study of twins. *Journal of Personality and Social Psychology, 80*, 845–860.

Paulhus, D. L., & Williams, K. M. (2002). The Dark Triad of personality: Narcissism, Machiavellianism, and psychopathy. *Journal of Research in Personality, 36*, 556–563.

Paunonen, S. V., & Ashton, M. C. (2001). Big five factors and facets and the prediction of behaviour. *Journal of Personality and Social Psychology, 81*, 524–539.

Stuss, D. T., Gow, C. A., & Hetherington, C. R. (1992). "No longer Gage": Frontal lobe dysfunction and emotional changes. *Journal of Consulting and Clinical Psychology, 60*, 349–359.

Vernon, P. A., Jang, K. L., Harris, J. A., & McCarthy, J. M. (1997). Environmental predictors of personality differences: A twin and sibling study. *Journal of Personality and Social Psychology, 72*, 177–183.

Wiggins, J. S. (Ed.). (1996). *The five-factor model of personality: Theoretical perspectives*. New York: Guilford Press.

Woodworth, M., & Porter, S. (2002). In cold blood: Characteristics of criminal homicides as a function of psychopathy. *Journal of Abnormal Psychology, 111*, 436–445.

Living with Autism

Jay Jensen, 23, is a high-functioning person with autism who works a regular job doing data entry. Quite focused and indifferent to outside distractions, Jay is able to accomplish many tasks in a much shorter time than his nonautistic colleagues. Awareness of the symptoms of autism has increased so much in recent years that the number of children diagnosed with the disorder has risen dramatically.

CHAPTER

13

Disorders of Mind and Body

Gertrude and Henry met in the 1930s. She found him annoying but eventually agreed to marry him after he threatened suicide. Gertrude became pregnant after three years of marriage, and in the fifth month of pregnancy, Gertrude's doctor predicted that she would have twins. In fact, she had quadruplets: Nora, Iris, Myra, and Hester. (Their names have been changed to protect their privacy.) Henry was frequently unemployed, so they could not afford to support four babies. A local newspaper raised money, and a house was donated to them. A steady stream of visitors wanted to see the babies, and Henry began charging admission. After a while, however, he became worried about the safety of his family and barred all visitors. At one point, he fired a pistol at his wife, mistaking her for an intruder.

Outlining the Principles

How Are Mental Disorders Conceptualized and Classified?

- Mental Disorders Are Classified into Categories
- Mental Disorders Must Be Assessed before Diagnosis
- Mental Disorders Have Many Causes
- The Legal System Has Its Own Definition of Psychopathology

Can Anxiety Be the Root of Seemingly Different Disorders?

- There Are Different Types of Anxiety Disorders
- Anxiety Disorders Have Cognitive, Situational, and Biological Components

Are Mood Disorders Extreme Manifestations of Normal Moods?

- There Are Different Types of Mood Disorders
- Mood Disorders Have Cognitive, Situational, and Biological Components

Is Schizophrenia a Disorder of Mind or Body?

- Schizophrenia Has Positive and Negative Symptoms
- Schizophrenia Is Primarily a Brain Disorder
- Environmental Factors Influence Schizophrenia

Are Personality Disorders Truly Mental Disorders?

- Personality Disorders Are Maladaptive Ways of Relating to the World
- Borderline Personality Disorder Is Associated with Poor Self-Control
- Antisocial Personality Disorder Is Associated with a Lack of Empathy

Should Childhood Disorders Be Considered a Unique Category?

- Autism Is a Lack of Awareness of Others
- Attention-Deficit/Hyperactivity Disorder Is a Disruptive Behaviour Disorder

13.1 The Genain Quadruplets as Adults

Henry and Gertrude both believed that the four girls really represented one person split into four. They consistently preferred the two larger babies, Nora and Myra, to the other two, and Gertrude took a particular dislike to Hester, especially after she caught her masturbating. Gertrude continued to worry excessively about masturbation. After she caught Hester and Iris engaging in mutual masturbation, she had a doctor amputate their clitorises. As the girls got older, Henry insisted on watching them dress and undress, and he even watched them changing their sanitary pads. In high school, Hester became irritable and depressed; her behaviour was at times destructive and bizarre. She "admitted" to a series of sexual activities dating back to elementary school. Rather than fully exploring the causes for this behaviour, her parents concluded that masturbation had made her mentally retarded. She was heavily sedated and kept at home. Although the other girls graduated from high school and even held jobs for a while, they all had psychotic episodes resulting in hospitalization. All four women were eventually diagnosed with schizophrenia. The National Institute of Mental Health in the United States offered free care in return for the opportunity to study them. To preserve their anonymity, the NIMH dubbed them the Genain quadruplets and assigned them first names beginning with the initials of NIMH (Figure 13.1; information from Rosenthal, 1963).

Although today the Genains are recognized as being mentally ill and receive appropriate treatment, the mentally ill have not always been viewed in a humane manner. In the Middle Ages, the Genain quadruplets might have been thought to be possessed by demons and been persecuted. In the 1700s, they would likely have languished in understaffed, overcrowded mental institutions. There would have been little attempt to understand their disorders and even less of an attempt to treat them. In the 1800s, the focus would have been on the environmental factors contributing to their illness, such as the physical abuse they suffered and their father's bizarre and inappropriate behaviour. Although psychological factors probably contributed to the development of the quadruplets' mental disorders, we now understand that biological factors play a critical role in schizophrenia. Advances in research have demonstrated that most mental disorders are ultimately disorders of both mind and body, of both biological and psychological factors. ■

460–377 B.C.E.
Humoral Imbalance Hippocrates explains abnormal behaviour as the result of imbalances of four bodily fluids: blood, phlegm, black bile, and yellow bile.

130–200 C.E.
Psychological Factors Galen postulates that abnormal behaviour could be due to imbalances in bodily fluids but also suggests psychological origins.

1400s
Demonic Possession During the Middle Ages, abnormal behaviour is attributed to demonic possession.

1831
Unacceptable Impulses Johann Christian Heinroth states that mental disorders are the result of unconscious conflict arising from unacceptable impulses.

Over the course of human history, people have struggled with how best to understand **psychopathology**—literally, sickness or disorder of the mind. The prevailing view in the twenty-first century is that psychopathology has its origins in psychological turmoil and biological dysfunction. Philosophers as early as Plato and Aristotle suggested that mental illness might be the result of conflicts between thought and emotion. Sigmund Freud was the first to develop a theory of psychopathology and to propose specific treatment techniques based on his theory.

psychopathology A disorder of the mind.

Understanding the role of both psychological and biological factors in the genesis of mental disorders is now known to be critical. Psychological scientists are working at all levels of analysis on many mental disorders so that we can understand and treat them effectively. The willingness of those who are afflicted with mental illness, such as the Genain sisters, to participate in scientific investigations has allowed researchers to make significant strides in understanding mental disorders. Advances in technology such as structural and functional neuroimaging have given us a window into the workings of the human brain. Our increased understanding of the effects of genes, neurotransmitters, and hormones on brain and behaviour has resulted in remarkable advances in understanding both normal and abnormal human behaviour. At the same time, longitudinal studies have demonstrated that children who are raised in certain types of environments, such as with abusive parents, are more likely than children not raised in such environments to develop some form of psychopathology later in life. In addition, we know that environment can trigger some mental disorders, as when combat leads to posttraumatic stress disorder or the death of a loved one triggers a bout of depression. This chapter examines how we think about mental illness and the circumstances that we believe produce mental disorders.

How Are Mental Disorders Conceptualized and Classified?

How do you know if a person has a mental disorder? It can be quite challenging to decide if a given behaviour is caused by psychopathology, because behaviour, especially unusual behaviour, always needs to be reviewed according to the situation. A person running through the streets, screaming hysterically, sobbing, grabbing people, and hugging them might have some form of mental illness—or could be celebrating following a phone call that told her she'd won a Nobel Prize. Many behaviours that are considered normal in one setting may be considered deviant in other settings. Some tribes in Africa spread feces in their hair as part of rituals. You can predict how this same behaviour would be received in an industrialized society. In some Native American and Far Eastern cultures, having spirits talk to you is

1867
Brain Diseases Wilhelm Griesinger argues that "mental diseases are brain diseases," in that mental disorders result from brain dysfunction.

1883
Classifying Mental Illness Emil Kraepelin suggests a classification system based on a biological conception of abnormal behaviour.

1893
The Unconscious Sigmund Freud and Josef Breuer develop the psychoanalytic approach, emphasizing that mental disorders result from unsatisfied drives and unconscious wishes.

1925
Behaviourism John B. Watson argues that mental illness results from basic principles of conditioning.

1930s
Psychological Testing Researchers develop the Minnesota Multiphasic Personality Inventory (MMPI), which is now widely used to assess various aspects of personality and psychopathology.

considered a great honour. In urban Canada, this would be taken as evidence of auditory hallucinations. However, some criteria are important to consider in determining whether behaviour represents psychopathology: Does the behaviour deviate from cultural norms? Is the behaviour maladaptive? Is the behaviour causing the individual personal distress?

When is a behaviour a sign of a mental disorder?

Because the line between normal and abnormal is hard to draw, there is an increasing trend toward defining psychopathology as thoughts and behaviours that are maladaptive rather than deviant. Excessive hand-washing can be deviant but adaptive—after all, it is the single best way of avoiding contagious disease. The same behaviour, however, can be maladaptive when people cannot stop washing their hands raw. The diagnostic criteria for all of the major disorder categories include the stipulation that the symptoms of the disorder must interfere with at least one aspect of the person's life, such as work, social relations, or self-care. This is a critical component in determining whether a given behaviour or set of behaviours represents a mental disorder or is simply unusual.

Mental Disorders Are Classified into Categories

Despite the problems in conceptualizing mental disorders, there are clear advantages to categorizing them. In order to investigate the etiology (factors that contribute to the development) and possible treatments of mental disorders, we need a way to group these disorders into meaningful categories. Researchers and clinicians have struggled for many years with how best to categorize mental disorders. Emil Kraepelin (Figure 13.2) was one of the first to propose a classification system for mental disorders. He recognized that not all patients with mental illness suffered from the same disorder, and he identified mental disorders based on groups of symptoms that occurred together.

13.2 **Emil Kraepelin** Kraepelin developed the first comprehensive categorization of mental illness.

The idea of categorizing mental disorders in a systematic manner was not officially adopted until the first edition of the ***Diagnostic and Statistical Manual of Mental Disorders (DSM)*** was published in 1952. Although it has undergone several revisions over the years, it remains the standard in the fields of psychology and psychiatry. The earlier versions focused on the presumed causes of mental illness, but beginning with the *DSM-III*, in 1980, there was a return to Kraepelin's approach of classifying psychopathology based on description. In the current edition, the fourth, disorders are described in terms of observable symptoms, and patients must meet specific criteria in order to be given a particular diagnosis. In addition, patients are not given a single label but rather are classified in terms of a set of clinically important factors (see Table 13.1). This **multiaxial system** is based on the growing realization that mental health is affected by a variety of factors. Diagnosing a patient on all five axes provides a more complete picture of the person.

1952
The *DSM* The first edition of the *Diagnostic and Statistical Manual of Mental Disorders* (*DSM*) is published.

1970s
Cognitions and Beliefs The cognitive perspective focuses on the ways in which distorted thought processes can contribute to mental disorders.

1980
Multiaxial Diagnoses The third edition of the *DSM* introduces a new classification system emphasizing description of mental disorders.

1990
Decade of the Brain The National Institutes of Health sponsor programs and publications aimed at introducing the general public to cutting-edge research on the brain.

2003
Genes and Behaviour The Human Genome Project has the potential to contribute greatly to the search for genetic causes of mental illness.

TABLE 13.1 *DSM-IV* Multiaxial Classification System

Axis I	Clinical disorders and other conditions that may be a focus of clinical attention (schizophrenia, mood disorders, anxiety disorders, sexual and gender disorders, sleep disorders, eating disorders).
Axis II	Mental retardation and personality disorders (antisocial personality disorder, paranoid personality disorder, borderline personality disorder).
Axis III	General medical conditions that may be relevant to mental disorders (cancer, epilepsy, obesity, Parkinson's disease, Alzheimer's disease).
Axis IV	Psychosocial and environmental problems that might affect the diagnosis, treatment, and prognosis of mental disorders (unemployment, divorce, legal problems, homelessness, poverty, parental overprotection).
Axis V	Global assessment of functioning (social, psychological, and occupational). Rated on a scale from 1 to 100, with 1 representing danger of hurting self or others and 100 meaning superior functioning in a wide range of areas.

SOURCE: *Diagnostic and Statistical Manual of the American Psychiatric Association*, 1994.

Mental Disorders Must Be Assessed before Diagnosis

Whereas physical illness can often be detected by medical tests, such as blood tests or biopsies, determining whether a person has a mental disorder is not so straightforward. Clinical psychologists often work like detectives, tracking down information from a variety of sources—including self-reports, observations, and interviews—to figure out what is wrong. The process of examining a person's mental functions and psychological health is known as **assessment**. The goal of assessment is to make a *diagnosis* so that appropriate treatment can be provided for the specific disorder. The course and probable outcome, or *prognosis,* of different mental disorders varies, and the correct diagnosis can help the patient and family understand what the future might bring.

The particular method of assessment sometimes depends on how a person comes into contact with mental health workers. People commonly show up in emergency rooms showing confusion, memory problems, or other mental impairments. Patients in this condition are often given a *Mental Status Exam* to provide a snapshot of their psychological functioning. The exam involves evaluating the person for things such as personal grooming, ability to make eye contact, tremors or twitches, mood, speech, thought content, and memory. This subjective evaluation frequently offers insights into whether a person has a mental disorder. For example, a patient who arrives in a disheveled state wearing excess layers of clothing is more likely to suffer from schizophrenia than from an anxiety disorder. The Mental Status Exam is also useful for determining if the mental impairments are due to a psychological condition or to some sort of physical condition, such as stroke or head injury.

Most symptoms of psychological problems develop over fairly long periods of time, and people seeking help for such problems frequently have been encouraged to see a psychologist by family members or personal physicians. The first step in a psychological assessment is for the psychologist to ask the person about current symptoms and any recent experiences that might be causing distress—for example, if the person is feeling depressed, the psychologist will likely ask whether he or she has

Diagnostic and Statistical Manual of Mental Disorders (DSM) A handbook of clinical disorders used for diagnosing psychopathology.

multiaxial system The system used in the *DSM* that provides assessment along five axes describing important mental health factors.

assessment In psychology, examination of a person's mental state in order to diagnose possible mental illness.

recently experienced some sort of loss. This *clinical interview* is the most common method of psychological assessment. The skills of the interviewer matter a great deal in terms of the quantity and value of information that is obtained from the client. Good interviewers express empathy, build rapport quickly, are seen as nonjudgmental and trusting, and are generally supportive of the client's efforts to find out what is wrong and how it might be fixed.

STRUCTURED VS. UNSTRUCTURED INTERVIEWS Since the beginning of modern psychology, most interviews have been *unstructured,* with the topics of discussion varying as the interviewer probes different aspects of the person's problems. The interview is guided by the clinician's past experiences as well as by any beliefs about the client and the type of problems that are most likely. Although this type of interview is very flexible, it is highly idiosyncratic—no two unstructured interviews are likely to elicit identical information. They are also too dependent on the quality of the interviewer. In contrast, *structured interviews* use standardized questions that are asked in the same order each time. How a person answers each question is coded according to a predetermined formula; diagnoses are based on specific patterns of responding. The most commonly used structured interview is known formally as the *Structured Clinical Interview for DSM (SCID),* which makes diagnoses according to *DSM* criteria (Spitzer, Williams, Gibbon, & First, 1992). The SCID begins with general questions, such as "What kind of work do you do?" and then proceeds to a series of questions about symptoms the client might have experienced, including the frequency and the degree of severity. In addition to assessment, the SCID is also valuable for research and treatment because it allows researchers and practitioners to know that results obtained in a research study of one group of patients are likely to apply to other patients diagnosed with the same disorder. Thus, structured interviews facilitate assessment as well as research and treatment.

BEHAVIOURAL ASSESSMENTS A psychological assessor can often gain valuable information simply by observing the client and his or her behaviour. For instance, the avoidance of eye contact during an examination might indicate social anxiety. A client whose eyes dart around nervously may have feelings of paranoia. Behavioural assessments are often useful with children, for instance, observing their interactions with others or seeing whether they can sit still in the classroom (Cjte et al., 2002).

Another assessment method is neuropsychological testing. In this method, the client is asked to perform certain actions, such as copying a picture, drawing a design from memory, sorting cards that show various stimuli into categories based on size, shape, or colour, placing blocks into slots on a board while blindfolded, tapping fingers rapidly, and so forth. Each task requires a particular ability, such as planning, coordination, or memory. By highlighting behaviours that the client performs poorly, the assessment might indicate problems with a particular brain region. For instance, those who have difficulty following a different rule for categorizing objects, such as sorting by shape rather than by colour, are likely to have impairments in the frontal lobes. Subsequent assessment with MRI or PET might be used to examine whether there is specific brain damage, as might be caused by a tumour or an injury.

PSYCHOLOGICAL TESTING Another source of information regarding symptoms of psychopathology is psychological testing. Examples of these types of tests were discussed in Chapter 12 when we considered the assessment of personality. Recall that both objective and projective tests can be used. Indeed, thousands of different psychological tests are available to clinicians. Some are tests for specific mental disorders, such as the Beck Depression Inventory, a relatively short scale that is widely used by researchers to assess symptoms of depression (Beck, Ward, Mendelson, Mock, & Erbaugh, 1961; Beck, Steer, & Brown, 1996). Other psychological tests assess a broad range of mental disorders and general mental health. The most widely used questionnaire for psychological assessment is the *Minnesota Multiphasic Personality Inventory (MMPI).* First developed during the 1930s, the MMPI was updated for language changes in the 1990s. The latest version consists of 567 true/false items that assess emotions, thoughts, and behaviours. The MMPI has ten clinical scales (for example, paranoia, depression, mania, hysteria), and the way a person scores on these scales produces a particular profile that indicates whether he or she may have a particular mental disorder (Bagby et al., 2005).

Recall from Chapter 2 that a common problem with all self-report assessments, such as the MMPI, is that people do not necessarily answer honestly. People can distort the truth or outright lie to make a favourable impression; they can be evasive or defensive to avoid detection of a mental disorder; and they can even untruthfully endorse negative items in order to try to look especially troubled. To counter these response biases, the MMPI contains validity scales that measure the probability that people are being less than truthful. For instance, some people may try to present themselves too positively by endorsing a large number of items like "I always make my bed," "I never tell lies," and "I am always nice to others." Although a person might answer one or two of these in a positive way, few of us are perfect at all times. Thus, a high score on this category of items indicates that a person might be trying to present an overly positive image. Of course, the person might not be doing this consciously, but it does indicate that the answers on the other items need to take into account the positivity bias in responses. Other validity scales examine whether the test taker answers similar questions in the same manner each time, endorses items that are extremely rare, and endorses an especially large number of negative items (known as faking bad).

Although the MMPI has been shown to be a reliable and valid assessment tool for most people in North America, it has been criticized because it may not be appropriate for use in other countries or with groups such as the poor, elderly, or racial minorities. This is because the scores that are considered "normal" on the MMPI are based on studies in which such people were inadequately represented.

Tests such as the MMPI are widely used in psychological assessment, but they are seldom the sole source of information. Indeed, most psychological clinicians would not make a diagnosis until they had consistent results from psychological tests and structured interviews. Like any good detective, the psychological assessor needs to consider all of the relevant evidence before reaching a conclusion, which in this case is a diagnosis of a mental disorder.

Thinking Critically

Can a Person Be Diagnosed with Multiple Identities?

In 1978 Billy Milligan was found innocent of robbery and rape charges on the grounds that he had been diagnosed with multiple personality disorder. Milligan clearly committed the robberies and rapes, but his lawyers successfully argued that he had multiple personalities and that different ones committed the crimes; therefore, Billy could not be held responsible. In his 1981 book *The Minds of Billy Milligan,* Daniel Keyes describes the 24 separate personalities sharing the body of 26-year-old Billy Milligan. One is Arthur, who at age 22 speaks with a British accent and is self-taught in physics and biology. He reads and writes fluent Arabic. Eight-year-old David is the keeper of the pain; anytime something physically painful happens, David experiences it. Christene is a three-year-old dyslexic girl who likes to draw pictures of flowers and butterflies. Regan is 23 and Yugoslavian; speaks with a marked Slavic accent; and reads, writes, and speaks Serbo-Croatian. He is the protector of the family and acknowledges robbing his victims, but denies raping them. Adalana, a 19-year-old lesbian who writes poetry, cooks, and keeps house for the others, later admitted to committing the rapes.

You might wonder how these different personalities share one brain—one person. According to a description by Arthur, there is a bright spotlight around which all the personalities stand or lie sleeping in their beds. Whoever steps in the spot is connected to the external world, possessing Milligan's consciousness. According to Arthur and Regan, the most dominant personalities, Billy himself had been kept asleep for seven years to protect him. After his acquittal, Milligan spent close to a decade in various mental hospitals. In 1988 psychiatrists believed that Milligan's 24 personalities had been merged into one and that he was no longer a danger to society. Since that time he is reported to be living quietly in California. Many people respond to reports such as this with astonishment and incredulity, believing that people such as Billy Milligan must be faking. Let's examine what is known about his condition and how it is diagnosed.

Dissociative identity disorder (DID), commonly called multiple personality disorder, involves the occurrence of two or more distinct identities in the same individual. DID is an example of the *DSM* category of dissociative disorders, which involve disruptions of identity, memory, and conscious awareness. Another example of such a disorder is dissociative amnesia: After experiencing overwhelming trauma, the person forgets that an event happened or loses awareness of substantial blocks of time, for example, suddenly losing memory for personal facts, including who he is and where he lives. In these *fugue* states, people can sometimes move to new cities and even assume new identities. When the person regains his identity, he often does not remember what occurred during the fugue state. The commonality among dissociative disorders is confusion over personal identify and the splitting off of some parts of memory from conscious awareness.

Most of those diagnosed with DID are women who report being severely abused as children. According to the most common theory of DID, children cope with abuse by pretending that it is happening to someone else and entering a trancelike state in which they dissociate their mental states from their physical bodies (Ross et al., 1991). Over time, this dissociated state takes on its own identity. Different identities develop to deal with different traumas. Often the identities have periods of amnesia, and sometimes only one of the identities is aware of the others. Indeed, diagnosis often occurs only when a person has difficulty accounting for large chunks of time in his or her day. The separate identities often differ in substantial ways, such as gender, sexual orientation, age, language spoken, interests, physiological profiles, and even different patterns of brain activation (Reindeers et al., 2003). Even their handwriting can be different (Figure 13.3).

In spite of this evidence, many researchers remain skeptical about whether DID is a genuine mental disorder, if it exists at all (Spanos et al., 1985; Kihlstrom, 2005).

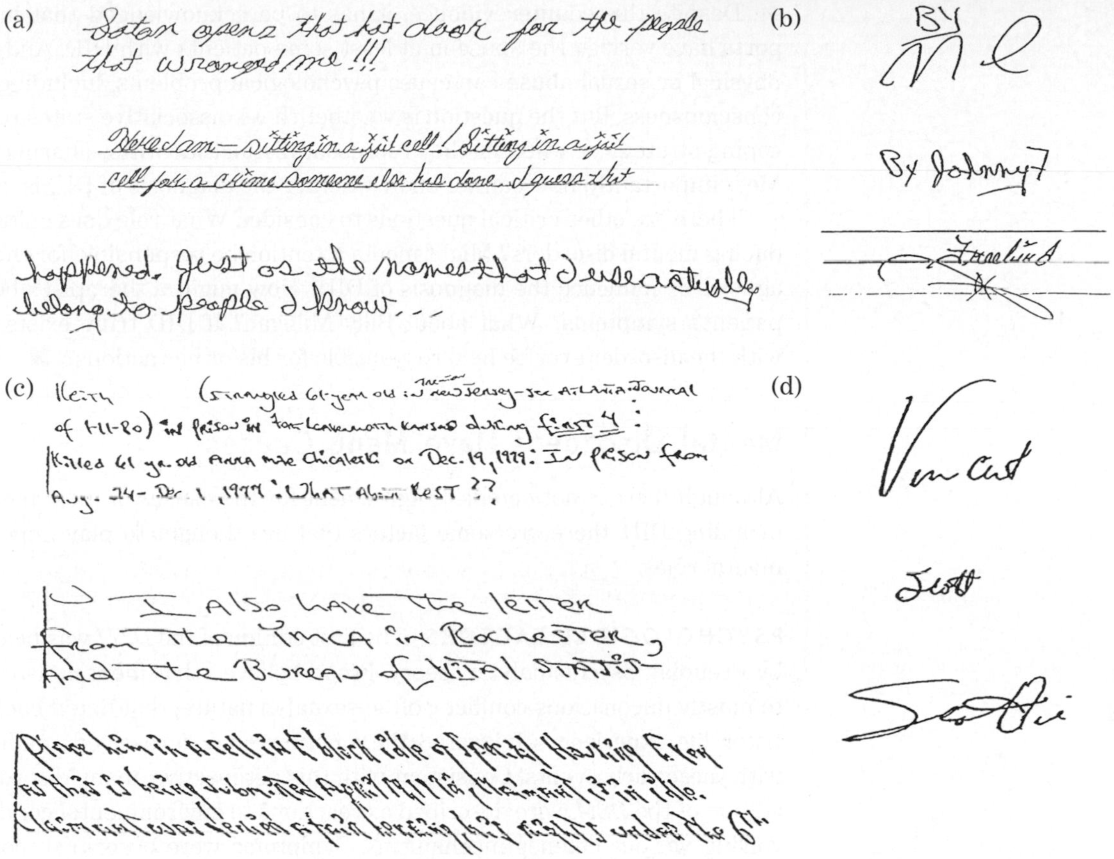

13.3 Handwriting Samples of Four People Diagnosed with Dissociative Identity Disorder Handwriting has been shown to differ in the different identities of people with DID. Researchers studied 12 murderers diagnosed with the disorder. Writing samples from 10 of the participants revealed markedly different handwriting in each of their identities. We see here handwriting samples and signatures from one subject (a and b), as well as handwriting samples from another subject (c), and signatures from a different subject (d). The handwriting is obviously very different depending on which identity (personality or alter) is doing the writing.

Most of the studies on DID have failed to use adequate controls, such as comparing supposed cases to individuals pretending to have DID (Merckelbach, Devilly, & Rassin, 2002). A person could alter physiological indicators or brain-imaging results simply by thinking disturbing thoughts or by changing her mood state. These different patterns of activity do not prove the existence of separate personalities. Moreover, people sometimes seem to have ulterior motives for claiming DID. Diagnoses of DID often occur after people have been accused of committing crimes, which raises the possibility that they are pretending to have multiple identities in order to avoid conviction. This is a problem whenever diagnoses are made in a legal context, as we will see shortly.

Other skeptics point to the sharp rise in reported cases as evidence that the disorder might not be real, or at least is diagnosed far too often. Prior to the 1980s, there were only sporadic reports of this disorder. Famous cases include the *Three Faces of Eve* (Thigpen & Cleckley, 1954) and *Sybil* (Schreiber, 1974). However, in the 1990s the number of cases skyrocketed into the tens of thousands. Moreover, those evincing the disorder went from having one or two identities to having several dozen, or even hundreds. What can explain these changes? One possibility is that patients develop DID after seeing therapists who believe strongly in the disorder. In the 1980s and 1990s, there was a surge of therapists who believed that childhood trauma was frequently repressed and that it needed to be uncovered during treatment. These therapists tended to use hypnosis to discover traumatic events from the patient's childhood. Indeed, in most cases those diagnosed as having DID were unaware of their other identities until they had had many sessions of therapy. During diagnosis, the assessor needs to guard against suggesting symptoms to the person being assessed (Lindsay, 1996).

Despite this counterevidence, it has to be acknowledged that independent reports have verified the abuse in at least some patients with DID. And it is likely that physical or sexual abuse can cause psychological problems, including distortions of consciousness. But the question is whether these dissociative states reflect a learned coping strategy or whether they represent distinct identities sharing a single brain. Most importantly, how can we know whether the diagnosis of DID is valid?

There are other critical questions to consider. What role does culture play in producing mental disorders? Might media attention be responsible for the rise in the diagnosis or influence the diagnosis of DID? How might a therapist's beliefs change a patient's symptoms? What about Billy Milligan? If DID truly exists, can a person with the disorder ever be held responsible for his or her actions? ■

Mental Disorders Have Many Causes

Although there is not complete agreement for the causes of most mental disorders, including DID, there are some factors that are thought to play important developmental roles.

PSYCHOLOGICAL FACTORS The first edition of the *DSM* was heavily influenced by Freudian psychoanalytic theory. Freud believed that mental disorders were due to mostly unconscious conflicts, often sexual in nature, that dated back to childhood. Later life experiences triggered the emotions and unresolved conflicts associated with these early events. Consistent with this perspective, many disorders in the first edition of the *DSM* were described as reactions to environmental conditions or as involving various defence mechanisms. Symptoms were described more in terms of inner causes than of external behaviours.

How does socioeconomic class affect the development of mental disorders?

Psychological factors play an important role in the manifestation and treatment of mental disorders. At the social level of analysis, thoughts and emotions are shaped by the environment and can profoundly influence behaviour, including abnormal behaviour. Not only traumatic events but also less extreme circumstances can have long-lasting effects. The **family systems model** is based on the idea that the behaviour of an individual must be considered within a social context, in particular the family. Problems that arise within an individual are manifestations of problems within the family, such as was the case for the Genain quadruplets, who we described at the beginning of the chapter. Developing a profile of an individual's family interactions can be important not only in understanding possible factors contributing to the disorder, but also in determining whether the family is likely to be helpful or detrimental to the client's progress in therapy. Similarly, the **sociocultural model** views psychopathology as the result of the interaction between individuals and their cultures. Some disorders, such as schizophrenia, are more common among the lower socioeconomic classes, whereas disorders such as anorexia nervosa are more common among the middle and upper classes. From the sociocultural perspective, these differences in prevalence are due to differences in lifestyles, expectations, and opportunities among the classes of society. Note, however, that it is possible that there are cultural biases in the willingness to ascribe disorders to different social classes.

family systems model A diagnostic model that considers symptoms within an individual as indicating problems within the family.

sociocultural model A diagnostic model that views psychopathology as the result of the interaction between individuals and their cultures.

cognitive-behavioural approach A diagnostic model that views psychopathology as the result of learned, maladaptive cognitions.

COGNITIVE-BEHAVIOURAL FACTORS At the level of the individual, the central principle of the **cognitive-behavioural approach** is that abnormal behaviour is learned. Whereas the psychoanalytic approach focuses on unconscious internal factors, the behavioural approach is based on observable variables. As you recall from Chapter 6, in classical conditioning, an unconditioned stimulus produces an unconditioned response. For example, a loud noise produces a startle response. If a neutral stimulus is paired with this unconditioned stimulus, it can eventually by itself produce a similar response. If a child is playing with a fluffy white rat and is frightened by a loud noise, as was the case with Little Albert that was presented in Chapter 6, the white rat alone can later cause fear in the child. In fact, this is how John B. Watson, the founder of behaviourism, demonstrated that many fears are learned rather than innate.

Proponents of strict behaviourism argue that mental disorders are the result of classical and operant conditioning. Originally, behaviour was defined as only overt observable actions, but this view was later challenged. The revised cognitive-behavioural perspective includes the idea that thoughts and beliefs should be considered as another type of behaviour that can be studied empirically. The premise of this approach is that thoughts can become distorted and produce maladaptive behaviours and emotions. In contrast to the psychoanalytical perspective, thought processes are believed to be available to the conscious mind. From this perspective, individuals are aware of, or can be easily made aware of, the thought processes that give rise to maladaptive behaviour.

BIOLOGICAL FACTORS The biological perspective on mental disorders focuses on how physiological factors, such as genetics, contribute to mental illness. Chapter 3 described how comparing the rates of mental illness between identical and fraternal twins and studying individuals who have been adopted have revealed the importance of genetic factors to the development of mental illness. Other biological factors also influence the development and course of mental illness. The fetus is particularly vulnerable, and there is evidence that some mental disorders may arise from prenatal problems such as maternal illness, malnutrition, and exposure to toxins. Environmental toxins and malnutrition during childhood and adolescence can also put the individual at risk for mental illness. All of these biological factors are thought to contribute to mental disorders because of their effects on the central nervous system. Evidence is emerging that neurological dysfunction contributes to the manifestation of many mental disorders, although a causal link has not been proven.

The recent use of imaging to identify brain regions associated with psychopathology has allowed researchers to generate hypotheses about the types of subtle deficits that might be associated with different mental disorders. Structural imaging has revealed neuroanatomical differences, perhaps due to genetics, between those with mental disorders and those without, but functional neuroimaging is currently at the forefront of research into the neurological correlates of mental disorder. PET and fMRI have revealed brain regions that may function differently in those with mental illnesses (Figure 13.4). Another source of insights into neural dysfunction has been research on the role of neurotransmitters in mental disorders. In

13.4 Biological Factors in Mental Disorders Although these men are twins, the one on the right has schizophrenia. Compared to his normal twin, he has larger ventricles, as revealed through MRI. This same pattern has emerged in the study of other twin pairs in which one has schizophrenia and the other does not.

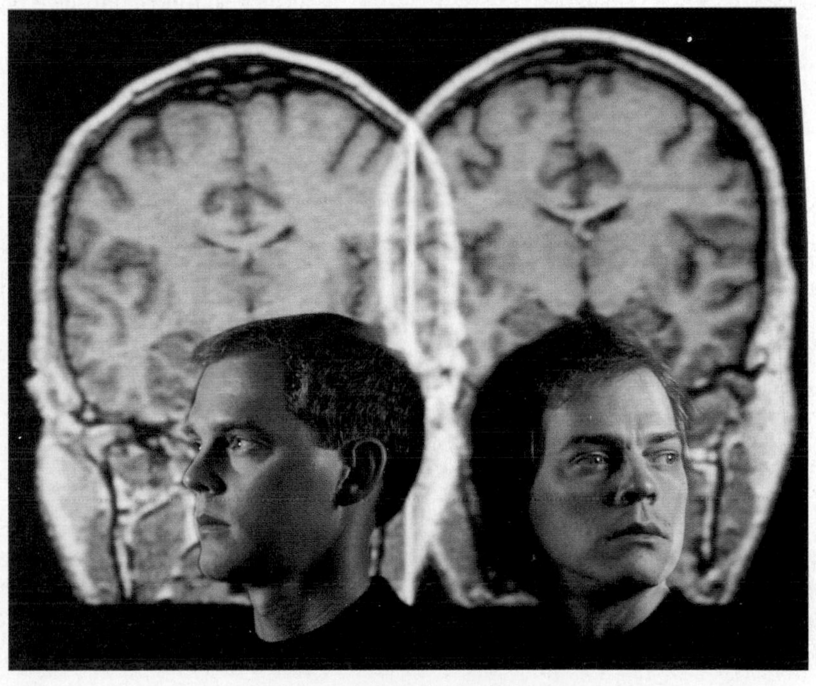

some cases, medications have been developed based on what is known about the neurochemistry of mental illness. In other cases, however, the unexpected effects of medications have led to discoveries about the neurotransmitters involved in mental disorders.

INTEGRATING THE FACTORS INVOLVED IN MENTAL DISORDERS Perhaps the most useful way to think about the causes of mental disorders is as an interaction among multiple factors. The **diathesis-stress model** provides one such way of thinking about the onset of mental disorders. In this model, an individual can have an underlying vulnerability or predisposition (diathesis) to a mental disorder. This can be biological, such as a genetic predisposition to a specific disorder, or environmental, such as childhood trauma. The vulnerability by itself may not be sufficient to trigger mental illness, but the addition of stressful circumstances can tip the scales. If the stress level exceeds an individual's ability to cope, the symptoms of mental illness will manifest. This is the theory that has been proposed for dissociative identity disorder. To this way of thinking, a family history of mental illness suggests vulnerability rather than destiny. The diathesis-stress model encourages researchers and practitioners to cross levels of analysis to understand the nature and nurture dynamic of mental disorders.

diathesis-stress model A diagnostic model that proposes that a disorder may develop when an underlying vulnerability is coupled with a precipitating event.

The Legal System Has Its Own Definition of Psychopathology

In clinical psychology, the *DSM* provides the structure for conceptualizing and categorizing mental disorders. The legal system, however, has a fundamentally different approach. Whereas clinical psychologists are concerned with diagnosing and treating mental illness, the legal profession is focused on the issue of personal responsibility for actions. According to the current legal system, a person is not responsible if, at the time of the crime, a mental disorder or defect led to an inability to appreciate the criminality of the act or to an inability to conform to the

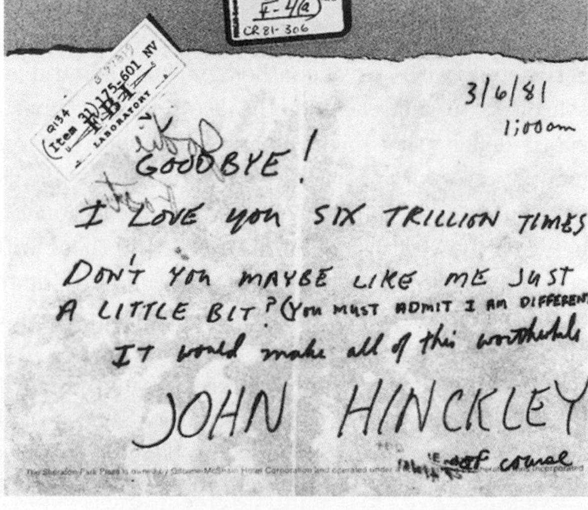

DEFENDANT'S EXHIBIT F-4(a) CR 81-306

FBI LABORATORY Item 31 175-601 NV

3/6/81
1:00 am

GOODBYE!

I LOVE you SIX TRILLION TIMES

DON'T YOU MAYBE LIKE ME JUST A LITTLE BIT? (YOU MUST ADMIT I AM DIFFERENT) IT would make all of this worthwhile

JOHN HINCKLEY

of course

13.5 Hinckley and the Insanity Defence John Hinckley Jr. and a letter he wrote to actress Jodie Foster, with whom he was obsessed.

requirements of the law. This is known as being not guilty by reason of mental disorder, previously referred to as insanity. The notion of *insanity* is actually legal, not psychological.

Contrary to popular perception, the mental disorder defence is not used often, but its use in high-profile trials such as that of John Hinckley Jr. has caused public outcry (Figure 13.5). In an attempt to assassinate U.S. President Ronald Reagan in 1981, Hinckley shot Reagan in the lung and Reagan's press secretary, James Brady, through the head. Both survived, but Brady sustained permanent severe physical and cognitive impairments. There was no doubt that Hinckley planned and committed the crime. The only question at the trial was whether he was legally insane at the time. After many hours of testimony, it became clear that Hinckley had had psychological problems beginning in early childhood. The two sides disagreed, however, about the severity of those problems and whether they met the American criteria for legal insanity. According to the law at the time, the prosecution had to prove that the defendant was sane if the defence claimed that he was not. The jury in the Hinckley trial concluded that there was reasonable doubt about his sanity at the time of the crime and returned a verdict of "not guilty by reason of insanity." As a result of the public outrage following the verdict, the law was revised so that the burden of proof now falls on the defence to prove that a defendant was not sane at the time of a crime. For both the prosecution and the defence, psychologists are often called as expert witnesses regarding possible psychological problems that may have existed at the time a person committed a crime.

In Canada, being found not guilty by reason of insanity was very rare prior to 1992, because it usually resulted in the severe punishment of the person being held indefinitely in a mental institution at the pleasure of the lieutenant governor. Indeed, it was usually the Crown rather than the defence who raised the issue of sanity. The legal system was changed in 1992, switching from the term *insanity* to *mental disorder* and preventing the Crown from alleging that the accused has any disease of the mind. Moreover, those found not criminally responsible because of mental disorder can no longer be held indefinitely, but rather must be assessed regularly by a review board. As might be expected, there has been an increase in the use of the mental disorder defence over the past decade. However, fewer than 1 percent of those charged with a criminal offence are found not criminally responsible due to mental disorder.

Psychological Science *in Action*

Forensic Assessment and Profiling

The field of *forensic psychology* involves the application of psychological science to the criminal justice system. Forensic psychologists, who typically have a Ph.D. in clinical psychology, are called on to perform several different functions, ranging from assisting in child-custody disputes, to examining eyewitness testimony, to advising on jury selection, to improving interrogation techniques, to helping design correctional facilities. Another common duty is determining whether a defendant is mentally competent to stand trial—that is, whether the defendant

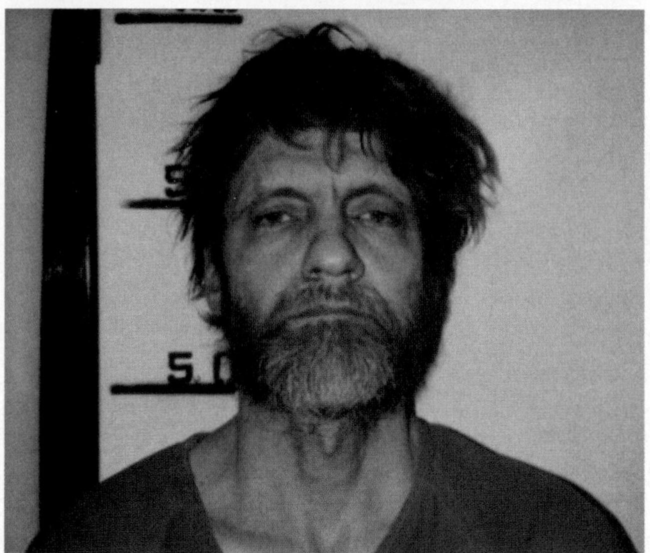

13.6 Theodore Kaczynski Kaczynski, known as the Unabomber, shown here (top) as a young professor and (bottom) after his arrest. Kaczynski pled guilty rather than use the insanity defense.

can understand the legal proceedings and contribute to his or her own defence. The American case of Theodore Kaczynski, referred to in the media as the Unabomber, hinged on his competence to stand trial (Figure 13.6). Kaczynski's lawyers planned to argue that he was insane at the time of the crimes (the planting of bombs at various locations around the United States), but Kaczynski refused to consider this approach. Kaczynski then requested permission to act as his own attorney, but this request was denied. Forensic psychologists examined Kaczynski and still determined that although he met the DSM criteria for paranoid schizophrenia, he was still competent to stand trial. Faced with the choice of having his lawyers present a case for insanity or pleading guilty, Kaczynski pled guilty.

Another challenging task for forensic psychologists is predicting whether individuals are dangerous, such as whether they are likely to become violent in the workplace, be sexual predators, or stalk and kill their spouses. However, determining whether a person is dangerous is fraught with ethical, legal, and clinical difficulties. Although forensic psychologists have worked hard to develop and refine ways of accurately predicting violent behaviour, it remains virtually impossible to do so. In part because of the difficulty of predicting dangerousness, if a forensic assessment indicates that a person poses a danger to someone, the clinician has the obligation to warn that person. This duty to warn is one of the few situations in which clinicians are required to violate patient confidentiality (Anfang & Appelbaum, 1996).

Forensic psychologists also assist law enforcement officers in pursuing criminals, sometimes by developing psychological profiles of criminals. For example, forensic psychologists sometimes examine murder scenes for clues to the relationship between the murderer and victim. Certain types of wounds indicate that the murderer was acting in a rage, suggesting some prior relationship with the victim. Compared with such *expressive* crime scenes, *instrumental* crime scenes are those in which the victim appears simply to have been used so that the murderer could achieve other aims, such as obtaining money or sex. Research indicates that the crime scene often provides basic information regarding the murderer, which helps law enforcement officers identify potential suspects (Salfati, 2000).

How accurate are criminal profilers? Psychologist Richard Kocsis and his colleagues in Australia have conducted thorough studies of criminal profiling (Kocsis, 2004; Kocsis, Hayes, & Irwin, 2002). Kocsis finds that profiling is most successful when it follows scientific principles, such as being objective and using critical thinking. Those who rely on gut instincts or hunches rarely develop successful profiles. In Kocsis's tests, undergraduate science students often outperform seasoned detectives in producing accurate profiles, perhaps because the students rely almost entirely on facts, while the detectives perhaps draw too much on past experience and intuition and pay too little attention to what actually happened. Through careful research like that conducted by Kocsis, forensic psychologists can help establish methods to develop psychological profiles of criminals. ■

REVIEWING the Principles | How Are Mental Disorders Conceptualized and Classified?

Psychopathology takes many forms, and mental disorders are consequently difficult to define and categorize. Although the behavioural manifestations of mental disorders vary widely, those who are diagnosed with these disorders have two things in common: Their behaviour is maladaptive, and it interferes with some important aspect of their lives. Diagnoses of specific mental disorders are based on the checklist system of the *DSM.* Clinical assessments typically consisting of interviews and psychological tests are used to examine a person's mental functions and psychological health. The specific cause of most mental disorders is unknown and may arise from a complex interaction of psychological, biological, and cognitive-behavioural factors. Psychopathology is conceptualized differently in the legal system than in the medical field: the former focuses on responsibility for one's actions, the latter on understanding and treating the disorder.

Can Anxiety Be the Root of Seemingly Different Disorders?

What does the fear of spiders have in common with the need to repeatedly check that the stove is turned off? They are both manifestations of anxiety disorders. Anxiety itself is normal and even useful. It can prepare us for upcoming events and motivate us to learn new ways of coping with life's challenges. For some people, however, anxiety can become debilitating and can interfere with every aspect of life. **Anxiety disorders** are characterized by excessive anxiety in the absence of true danger. It is normal to be anxious in stressful or threatening situations. It is abnormal to feel strong chronic anxiety without cause.

anxiety disorders Disorders characterized by the experience of excessive anxiety in the absence of true danger.

There Are Different Types of Anxiety Disorders

People suffering from anxiety disorders feel tense, anxious, and apprehensive. They are often depressed and irritable because they cannot see any solution to the anxiety they feel. Constant worry can make falling asleep and staying asleep difficult, and attention span and concentration can be impaired. Problem solving and judgment may suffer as well. Chronic anxiety also causes a variety of somatic symptoms, due to the arousal of the autonomic nervous system. Sweating, dry mouth, rapid pulse, shallow breathing, increased blood pressure, and increased muscular tension are all consequences of autonomic arousal. Chronic arousal can also result in hypertension, headaches, and intestinal problems and can even cause illness or tissue damage. Because of their high levels of autonomic arousal, those who suffer from anxiety disorders also exhibit restless and pointless motor behaviours. Exaggerated startle response is typical and behaviours such as toe tapping and excessive fidgeting are common. Recent research has shown that chronic stress can produce atrophy in the hippocampus, a brain structure involved in learning and memory (McEwen, 2000). It is not yet known whether the damage is reversible, or if there are functional implications of hippocampal atrophy. The fact that chronic stress can damage the body and brain, however, indicates the importance of identification and effective treatment of disorders that involve chronic anxiety. Although different anxiety disorders share some emotional, cognitive, somatic, and motor symptoms, the behavioural manifestations of these disorders are quite different (Barlow, 2002).

Are phobias exaggerations of adaptive reactions?

PHOBIC DISORDER A **phobia** is a fear of a specific object or situation. Fear can be adaptive, as it can lead us to avoid potential danger, such as poisonous snakes and rickety bridges. In phobias, however, the fear is exaggerated and out of proportion to the actual danger. Phobias are classified based on the object of the fear. *Specific phobias* (formerly known as simple phobias) involve particular objects and situations, such as snakes (ophidiophobia), enclosed spaces (claustrophobia), and heights (acrophobia). *Social phobia* is a fear of being negatively evaluated by others and includes fear of public speaking and of eating in front of others (Wallace & Alden, 1997). Specific phobias affect about 8 percent of the population versus 6 percent for social phobia (Offord et al., 1996). Women are diagnosed with specific phobias at least twice as often as men, but with social phobia only slightly more often.

GENERALIZED ANXIETY DISORDER Whereas the anxiety in phobic disorders has a specific focus, the anxiety in **generalized anxiety disorder** is diffuse and omnipresent. People with this disorder are constantly anxious and worry incessantly about even minor matters (Sanderson & Barlow, 1990). Because the anxiety is not focused, it can occur in response to almost anything, and so the sufferer is constantly on the alert for problems. This hypervigilance results in distractibility, fatigue, irritability, and sleep problems, as well as headaches, restlessness, and muscle pain. Probably 1 to 2 percent of the Canadian population is affected by this disorder, with women more likely to be diagnosed than men.

PANIC DISORDER It has been estimated that panic disorder affects about 1–2 percent of the population, with women twice as likely to be diagnosed as men. **Panic disorder** sufferers experience attacks of terror that are sudden and overwhelming. The terror can seemingly come out of nowhere or can be cued by external stimuli or internal thought processes. Panic attacks typically last for several minutes. The victim of such an attack begins to sweat and tremble; her heart races, she begins to feel short of breath, and her chest hurts; she can feel dizzy and light-headed with numbness and tingling in her hands and feet. People experiencing panic attacks often feel that they are going crazy or that they are dying, and people who suffer from persistent panic attacks attempt suicide much more frequently than those in the general population (Fawcett, 1992; Korn et al., 1992; Noyes, 1991; Shear & Maser, 1994). A related disorder is **agoraphobia**, a fear of being in situations in which escape is difficult or impossible, such as a crowded shopping mall. Being in those situations causes panic attacks. Those who experience panic attacks during adolescence are especially likely to develop other anxiety disorders, such as agoraphobia, PTSD, and general anxiety disorder, in adulthood (Goodwin et al., 2004).

OBSESSIVE-COMPULSIVE DISORDER **Obsessive-compulsive disorder (OCD)** involves frequent intrusive thoughts and compulsive actions, and affects 2 to 3 percent of the population. It is more common in women than men and generally begins in early adulthood (Robins & Regier, 1991; Weissman et al., 1994; Horwath & Weissman, 2000). Obsessions and compulsions plague sufferers of this disorder. *Obsessions* are recurrent, intrusive, and unwanted thoughts, ideas, or images. They often include fear of contamination, accidents, or one's own aggression. *Compulsions* are particular acts that the OCD patient feels driven to perform over and over again. The most common compul-

phobia An irrational fear of a specific object or situation.

generalized anxiety disorder A diffuse state of constant anxiety not associated with any specific object or event.

panic disorder An anxiety disorder characterized by sudden, overwhelming attacks of terror.

agoraphobia An anxiety disorder marked by fear of being in situations in which escape may be difficult or impossible.

obsessive-compulsive disorder (OCD) An anxiety disorder characterized by frequent intrusive thoughts and compulsive actions.

sive behaviours are cleaning, checking, and counting. For instance, people may constantly check to make sure they locked the door because of their obsession that their home may be invaded, or they may engage in superstitious counting to protect against accidents. Those with OCD anticipate catastrophe and loss of control. However, as opposed to those who suffer from other anxiety disorders, who fear what might happen to them, those who suffer from OCD often fear what they might do or might have done:

> While in reality no one is on the road, I'm intruded with the heinous thought that I *might* have hit someone . . . a human being! God knows where such a fantasy comes from. . . . I try to make reality chase away this fantasy. I reason, "Well, if I hit someone while driving, I would have *felt* it." This brief trip into reality helps the pain dissipate . . . but only for a second. . . . I start ruminating, "Maybe I did hit someone and didn't realize it. . . . Oh my God! I might have killed somebody! I have to go back and check." Checking is the only way to calm the anxiety. (Rapoport, 1990, pp. 22–27)

Anxiety Disorders Have Cognitive, Situational, and Biological Components

Although the behavioural manifestations of anxiety disorders can be quite different, they share some causal factors (Barlow, 2002). When presented with ambiguous or neutral situations, anxious individuals tend to perceive them as threatening, whereas nonanxious individuals assume them to be nonthreatening (Eysenck, Mogg, May, Richards, & Matthews, 1991). The ambiguous sentence "The doctor examined little Emma's growth" is interpreted by anxious individuals as "The doctor looked at little Emma's cancer" and by nonanxious individuals as "The doctor measured little Emma's height." Anxious individuals also tend to focus excessive amounts of attention on perceived threats (Rinck, Reinecke, Ellwart, Heuer, & Becker, 2005). Threatening events are thus more easily recalled than nonthreatening events, increasing their perceived magnitude and frequency.

In addition to cognitive components, situational factors also play a role in the development of anxiety disorders. In Chapter 6 you learned that monkeys will develop a fear of snakes if they observe other monkeys responding to snakes in a fearful way. Similarly, a person could develop a fear of elevators by observing another person's reaction to the closing of the elevator doors. Such a fear might then generalize to other enclosed spaces, resulting in claustrophobia. Biological factors also seem to play an important role in the development of anxiety disorders; recent investigations have resulted in a number of exciting findings. For instance, children who have an inhibited temperamental style during infancy and childhood are more likely to have anxiety disorders later in life (Fox, Henderson, Marshall, Nichols, & Ghera, 2005). Recall from Chapter 12 that these children are usually shy and tend to avoid unfamiliar people and novel objects. These inhibited children are especially at risk for developing social phobia (Biederman et al., 2001). In one recent study, adults who were categorized as inhibited during their second year of life received brain scans while viewing pictures of familiar faces and of novel faces. Compared with adults who were categorized as uninhibited before age two, the inhibited group showed greater activation of the amygdala to the novel faces (Schwartz, Wright, Shin, Kagan, & Rauch, 2003). This suggests that some aspects of childhood temperament are preserved in the adult brain. Next we will consider some of the ways these three factors interact in the development of two anxiety disorders.

"Is the Itsy Bitsy Spider obsessive-compulsive?"

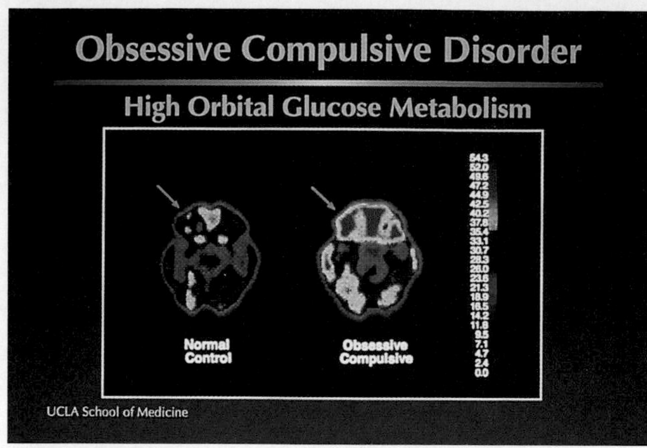

13.7 PET Scans of People with OCD The scans of people with OCD show greater activity in the frontal lobes, basal ganglia, and thalamus than do those of normal controls.

THE CAUSES OF OBSESSIVE-COMPULSIVE DISORDER A paradoxical aspect of OCD is that people are aware that their obsessions and compulsions are irrational and yet they are unable to stop them. One explanation is that the disorder results from operant conditioning. Anxiety is somehow paired to a specific event, and the person engages in behaviour that reduces anxiety. This reduction of anxiety is reinforcing and increases the chance of engaging in that behaviour again. For example, say you are forced to shake hands with someone who has a bad cold, and you have just seen that person wiping his nose with his right hand. Shaking that hand might cause you to be anxious or uncomfortable because you don't want to get sick yourself. As soon as the pleasantries are over, you run to the bathroom and wash your hands. You feel relieved. You have just paired hand washing with a reduction in anxiety, thus increasing the chances of hand washing in the future.

There is good evidence that the etiology of OCD is at least in part genetic (Crowe, 2000; Zai et al., 2004). Recently, it has been demonstrated that the caudate nucleus is dysfunctional in people with OCD (Baxter, 2000). The caudate is part of the basal ganglia, a region that helps to suppress impulses. In patients with OCD, the caudate is smaller and has structural abnormalities (Figure 13.7). In addition, patients with basal ganglia disease often manifest symptoms of OCD. Because this region is involved in impulse suppression, it is thought that dysfunction in this region results in the leak of impulses into consciousness. The prefrontal cortex then becomes overactive in an effort to compensate (Whiteside, Port, & Abramowitz, 2004). It has been demonstrated that there are alterations in brain waves over the prefrontal cortex in OCD patients; severing the connections between the prefrontal cortex and the caudate can often result in dramatic improvement in the symptoms of the disorder.

There is also growing evidence that OCD can sometimes be triggered by environmental factors. In particular, it appears that a streptococcal infection can cause symptoms of OCD in some young children. Originally identified in 1998 by Susan Swedo and her colleagues at the United States National Institute of Mental Health, this syndrome strikes virtually overnight, with children suddenly displaying odd symptoms of OCD, such as engaging in repetitive behaviours, developing irrational fears and obsessions, and having facial tics. Researchers have speculated that an autoimmune response damages the caudate, thereby producing the symptoms of OCD (Snider & Swedo, 2004). Importantly, treatments that enhance the immune system have been found to diminish the symptoms of OCD in children with this syndrome. Why some children are susceptible to this autoimmune response is currently unknown.

It may be that biological and cognitive-behavioural factors interact to produce the symptoms of OCD. A dysfunctional caudate nucleus allows impulses to enter consciousness, and these impulses may give rise to the obsessions of OCD. The prefrontal cortex becomes overactive in an attempt to compensate, and this overactivity can establish associations between obsessions and behaviours that reduce the anxiety arising from the obsessions. These behaviours thus become compulsions through conditioning.

THE LINK BETWEEN PANIC DISORDER AND AGORAPHOBIA People who suffer from agoraphobia avoid going into open spaces or to places where there might be crowds. Specifically, they seem to fear being in places from which escape might be difficult or embarrassing. In extreme cases, sufferers of this type of phobia may feel unable to leave their homes:

> Ms. Watson began to dread going out of the house alone. She feared that while out she would have an attack and would be stranded and helpless. She stopped riding the subway to work out of fear she might be trapped in a car between stops when an attack struck, preferring instead to walk the 20 blocks between her home and work. She also severely curtailed her social and recreational activities—previously frequent and enjoyed—because an attack might occur, necessitating an abrupt and embarrassing flight from the scene. (Spitzer, Skodol, Gibbon, & Williams, 1983)

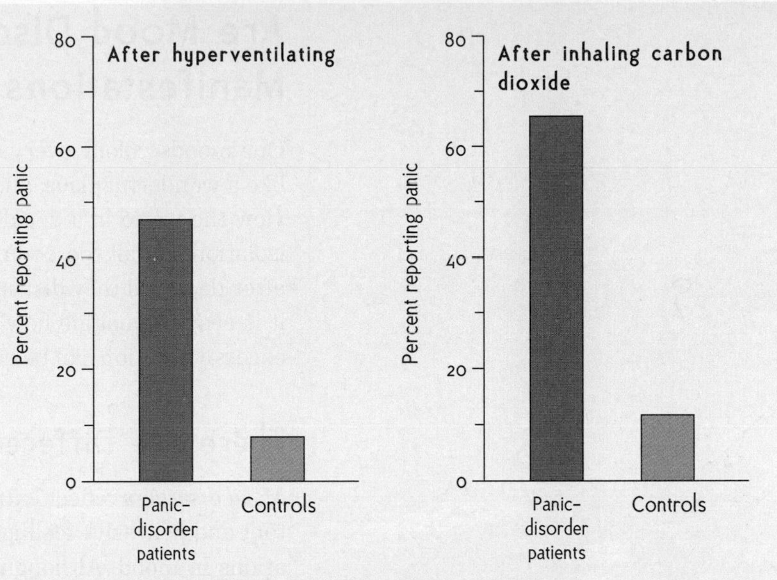

13.8 Panic Disorders and Biological Factors Research participants prone to panic attacks are more likely than normal controls to have a panic attack after hyperventilating or inhaling carbon dioxide.

This sort of description is typical of those with agoraphobia, and such descriptions have led many to explore the relationship between panic attacks and agoraphobia. The symptoms of sympathetic nervous system arousal that occur during a panic attack often lead victims to believe that they are having a heart attack. From a cognitive-behavioural perspective, a tendency to catastrophize (to expect the worst) paired with some kind of trigger stimulus can produce a panic attack. It has been speculated that agoraphobia is the result of untreated panic attacks (Barlow, 1988). People who have had panic attacks begin to fear having them again. This fear results in avoidance of situations in which it might be embarrassing to have an attack and difficult to escape. The fear then develops into a phobia.

Although agoraphobic symptoms are thought to arise from learned associations, panic attacks themselves seem more influenced by biological factors. Research suggests that panic disorder is in part genetic and linked to abnormalities that result in increased arousal of the central nervous system (Crowe, 2000). Studies show that infusions of lactate and inhalation of carbon dioxide, which heighten arousal, can produce panic attacks in those with a family history of panic disorder, but not in those without a family history (Rapee, Brown, Antony, & Barlow, 1992; Figure 13.8). Interestingly, although medication can be quite effective in reducing or eliminating panic attacks, it has no effect on agoraphobia. The fear of open or public places has been learned, and treating the neural basis for the fear has no effect. The conditioned response must be unlearned via cognitive-behavioural therapy.

REVIEWING the Principles | Can Anxiety Be the Root of Seemingly Different Disorders?

On the surface, many of the anxiety disorders seem to be quite different. However, they share many of the same emotional, cognitive, somatic, and motor symptoms. All of the disorders in this category are associated with anxiety that is out of proportion with reality. This anxiety may arise from biological factors, but many of the behavioural and cognitive manifestations of the disorders are strongly influenced by psychological and environmental factors.

Are Mood Disorders Extreme Manifestations of Normal Moods?

Our moods colour every aspect of our lives. When we are happy, the world seems like a wonderful place, and we are filled with boundless energy. When we are sad, we view the world in a decidedly less rosy light. Feelings of sadness, hopelessness, and isolation are all too common. Few of us, however, experience these symptoms day after day until they disrupt our ability to work, learn, and play. In addition, although it is easy to imagine how periods of sadness can interfere with daily life, periods of excessive elation can be equally devastating.

There Are Different Types of Mood Disorders

Mood disorders reflect extreme emotions, with depressive disorders featuring persistent and pervasive feelings of sadness, and bipolar disorders involving radical fluctuations in mood. Although some of their characteristics overlap, the two categories represent fundamentally different disorders.

major depression A disorder characterized by severe negative moods and a lack of interest in normally pleasurable activities.

dysthymia A form of depression that is not severe enough to be diagnosed as major depression.

DEPRESSIVE DISORDERS Within the category of depressive disorders are major depression and dysthymia. In order to be diagnosed with **major depression**, a person must have one of two symptoms: depressed (often irritable) mood or loss of interest in pleasurable activities. In addition, the person must have other symptoms such as appetite and weight changes, sleep disturbances, loss of energy, difficulty concentrating, feelings of self-reproach or guilt, and frequent thoughts of death and suicide. The following case study is of a 56-year-old woman diagnosed with depression:

> She described herself as overwhelmed with feelings of guilt, worthlessness, and hopelessness. She twisted her hands almost continuously and played nervously with her hair. She stated that her family would be better off without her and that she had considered taking her life by hanging herself. She felt that after death she would go to hell, where she would experience eternal torment, but that this would be a just punishment. . . . (Andreasen, 1984, p. 39)

Episodes of depression can last from a few weeks to many years. On average, depressive episodes last approximately six months. In many cases, the episodes are self-limiting and resolve without intervention. Although 90 percent of sufferers will recover from depression, they have a 50 percent chance of experiencing another depressive episode in the future (Keller, Shapiro, Lavori, & Wolfe, 1982).

Unlike major depression, **dysthymia** is of mild to moderate severity. Those diagnosed with dysthymia must have a depressed mood most of the day, more days than not, for at least two years. The depression is not, however, severe enough to merit a diagnosis of major depression. Periods of dysthymia last from 2 to 20 or more years, although the typical duration is about 5 to 10 years. Because the depressed mood is so long lasting, some psychological scientists consider it to be a personality disorder rather than a mood disorder. In fact, earlier editions of the *DSM* included a category of "affective personality" in the personality disorders section. In later editions, this term was dropped and dysthymia was added to the mood disorders category. The distinction between a depressive personality, dysthymic disorder, and major depression is unclear. They may be points along a continuum rather than distinct disorders. In support of this view, it has been found that dysthymia often precedes major depression (Griffiths et al., 2000; Lewinsohn, Allen, Seeley, & Gotlib, 1999; Lewinsohn, Rodhe, Seeley, & Hops, 1991).

Approximately 8–9 percent of Canadians will develop major depression in their lifetimes (Canadian Psychiatric Association, 2001). Indeed, during any given year, between 4–5 percent of Canadians are clinically depressed. These numbers are for major depression; the number of persons affected by less-severe depression is much higher. It has been estimated that 1 in 4 university students experience a bout of depression during their time at school.

BIPOLAR DISORDERS Although we all experience variations in our moods, normal fluctuations from sadness to exuberance seem minuscule in comparison to the extremes experienced by those with bipolar disorder (Figure 13.9). Bipolar disorder was previously known as manic depression, a term that captured the essence of the disorder. Those who are diagnosed with **bipolar disorder** have periods of major depression but also experience episodes of mania. *Manic episodes* are characterized by elevated mood, increased activity, diminished need for sleep, grandiose ideas, racing thoughts, and extreme distractibility. Kay Redfield Jamison, whom we will discuss in greater detail shortly, describes one of her own episodes of mania:

> With vibrissae twinging, antennae perked, eyes fast-forwarding and fly faceted, I took in everything around me. I was on the run. Not just on the run but fast and furious on the run, darting back and forth across the hospital parking lot, trying to use up a boundless, restless, manic energy. I was running fast, but slowly going mad. (Jamison, 1995, p. 3)

During episodes of mania, the heightened levels of activity and euphoria often result in excessive involvement in pleasurable but foolish activities—such as sexual indiscretions, buying sprees, and risky business ventures—that the individual will come to regret once the mania has subsided. Whereas some sufferers of bipolar disorder experience true manic episodes, others may experience less extreme mood elevations. These *hypomanic episodes* are often characterized by heightened creativity and productivity and can be extremely pleasurable and rewarding. People experiencing episodes of major depression and mania are diagnosed with *bipolar I* disorder, whereas people fluctuating between major depression and hypomania are diagnosed with *bipolar II.* A third category of bipolar disorder is **cyclothymia**, in which individuals experience hypomania and mild depression. Bipolar disorder is much less common than depression; the lifetime prevalence is estimated at 1 percent. In addition, whereas depression is more common in women, the prevalence of bipolar disorder is equal in women and men.

bipolar disorder A mood disorder characterized by alternating periods of depression and mania.

cyclothymia A less extreme form of bipolar disorder.

13.9 Margot Kidder In 1996, Margot Kidder was found bewildered, dazed, and frightened, hiding in a stranger's backyard where she was mistakenly identified as a homeless woman. She was taken to a nearby psychiatric ward, where she was quickly diagnosed as manic depressive, or bipolar.

Mood Disorders Have Cognitive, Situational, and Biological Components

Mood disorders can be devastating. The sadness, hopelessness, and inability to concentrate that characterize major depression can result in the loss of jobs, friends, and family relationships. Because of the profound effects of this disorder, and the danger of suicide, a great deal of research has been focused on understanding the causes of and treatment of major depression. Suicide is also a risk with those with bipolar disorder. In addition, errors in judgment during manic episodes can have devastating effects on the lives of those who suffer from the disorder as well as on their families and friends.

DEPRESSION IS CAUSED BY A COMBINATION OF FACTORS Studies of twins, families, and adoptions support the notion that depression has a genetic component. Although there is some variability among studies, concordance rates between identical twins are generally around four times higher than rates between fraternal twins (Gershon, Berrettini, & Goldin, 1989). The existence of a genetic component implies that biological factors are involved in depression, but there is as yet no consensus as to what these factors are. We know that medications that increase the availability of norepinephrine alleviate depression, whereas those that decrease levels of this neurotransmitter can cause depression. More recent medications, such as Prozac, selectively increase serotonin, so there is increased interest in understanding the role of this neurotransmitter in modulation of mood. Studies of brain function have suggested neural structures that may be involved in mood disorders. Damage to the left prefrontal cortex often leads to depression, but this is not true of damage to the right hemisphere. The brain waves of depressed persons show low activity in these same regions in the left hemisphere. Interestingly, this pattern persists in patients who have been depressed but are currently in remission. It therefore may be a kind of biological marker of predisposition to depression.

Biological rhythms have also been implicated in depression. Depressed patients enter REM sleep more quickly and have more of it. There is also evidence for abnormalities in a number of other biological rhythms, such as body temperature. Many people show a cyclical pattern of depression depending on the season. This disorder, known as **seasonal affective disorder (SAD)**, results in periods of depression corresponding to the shorter days of winter in northern latitudes (Figure 13.10). Treatment with an artificial light source seems to alleviate the depression, possibly by resetting biological rhythms (Michalak et al., 2000).

seasonal affective disorder (SAD) A disorder in which periods of depression occur during the times of year with less sunlight.

Although biological factors may play a role in depression, situational factors are also important. A number of studies have implicated life stressors in many cases of depression (Hammen, 2005). One source of stress seems to be particularly relevant for depression is interpersonal loss, such as having someone close die or getting divorced (Paykel, 2003). In one classic study, depression was especially likely in the face of multiple negative events (Brown & Harris, 1978; Figure 13.11). Another study found that depressed patients had more negative life events during the year before the onset of their depression (Dohrenwend, Shrout, Link, Skodol, & Martin, 1986). How an individual reacts to stress, however, can be influenced by interpersonal relationships, which play an extremely important role in depression (Joiner, Coyne, &

How does the social environment affect depression?

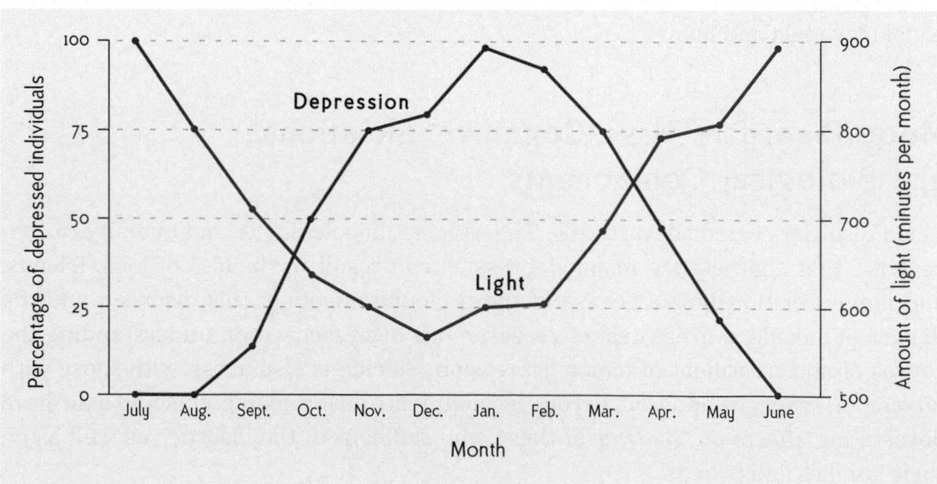

13.10 Depression and Light For some people, there is a direct correspondence between the amount of sunlight and their levels of depression.

Blalock, 1999). Regardless of any other factors, relationships contribute to the development of depression, alter people's experiences when depressed, and ultimately may be damaged by the constant needs of the depressed person. Many people report negative reactions to depressed people, perhaps because of their poor social skills. Over time, people may avoid interactions with those who are depressed, which in turn makes them even more depressed, resulting in a downward spiral (Dykman, Horowitz, Abramson, & Usher, 1991). By contrast, a person who has a close friend or group of friends is less likely to become depressed when faced with stress. This protective factor is not related to the number of friends, but to the quality of the friendships. One good friend is more protective than a large number of casual acquaintances.

Cognitive processes also play a role in depression. Aaron T. Beck proposed that depressed people think about themselves, their situation, and the future in a negative manner, which he refers to as the *cognitive triad* (Beck, 1967, 1976; Beck, Brown, Seer, Eidelson, & Riskind, 1987; Beck, Rush, Shaw, & Emery, 1979). Misfortunes are blamed on personal defects, whereas positive occurrences are seen as the result of luck. Nondepressed people tend to do the opposite. Beck also notes that depressed people make *errors in logic,* such as overgeneralizing based on single events, magnifying the seriousness of bad events, and personalizing, or taking responsibility for, bad events in the world that have little to do with them.

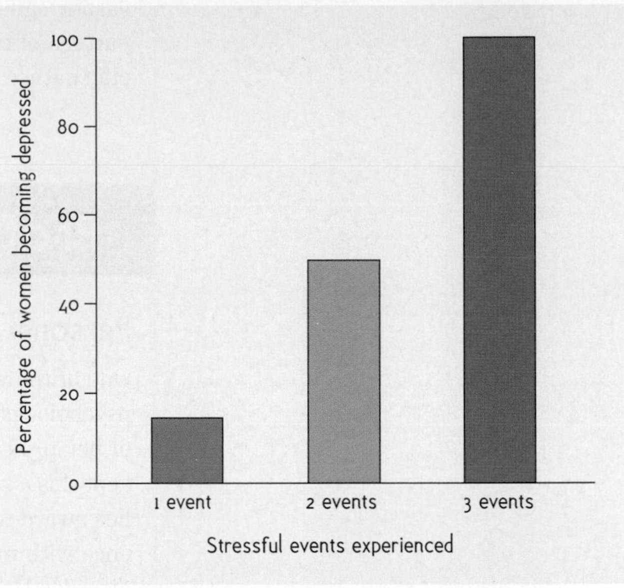

13.11 **Depression and Stress** In this study, the greater the number of stressors, the more likely it was that people felt depressed (Brown & Harris, 1978).

A second cognitive model of depression is the **learned helplessness model** (Seligman, 1974, 1975), in which people see themselves as unable to have any effect on events in their lives. Seligman's model is based on years of research demonstrating that animals placed in aversive situations that they cannot escape eventually become passive and unresponsive, lacking the motivation to try new methods of escape when given the opportunity. People suffering from learned helplessness come to expect that bad things will happen over which they will have little control. The attributions, or explanations, they make for negative events tend to refer to personal factors that are stable and global, rather than to situational factors that are temporary and specific. This attributional pattern leads people to feel hopeless about making positive changes in their lives (Abramson, Metalsky, & Alloy, 1989). The accumulation of evidence suggests that dysfunctional cognitive patterns are a cause rather than a consequence of depression.

learned helplessness model A cognitive model of depression in which people feel unable to control events around them.

BIPOLAR DISORDER IS A BIOLOGICAL DISORDER Although depression may arise from a variety of factors, evidence is strong that bipolar disorder is predominantly a biological disorder. Twin studies reveal that the concordance for bipolar disorder in identical twins is upward of 70 percent, versus only 20 percent in fraternal or dizygotic twins (Nurnberger, Goldin, & Gershon, 1994). In the 1980s, a genetic research study was carried out on the American Amish, an ideal population because they keep good genealogical records and few outsiders marry into the community. In addition, substance abuse is virtually nonexistent among adults, so mental illness is easier to detect. The results revealed that bipolar disorder ran in a limited number of families and that all of those afflicted had a similar genetic defect (Egeland et al., 1987).

Genetic research suggests that the hereditary nature of bipolar disorder is complex and not linked to a single gene (Maziade et al., 2001). Current research is focusing on identifying genes that may be involved. In addition, it appears that in families with bipolar disorder, successive generations tend to have more severe illness and earlier age of onset (McInnis et al., 1993; Petronis & Kennedy, 1995). Focus on this pattern of transmission may help to reveal the genetics of the disorder, but the specific nature of the heritability of bipolar disorder remains to be discovered.

Profiles in Psychological Science

Personal Insight into Psychopathology

One of the world's foremost authorities on bipolar disorder has unique insight among psychological scientists—she has suffered from the disorder since childhood. For much of her professional career, Johns Hopkins University professor Kay Redfield Jamison kept this a secret, but eventually acknowledged her struggles with manic depression in her award-winning 1995 memoir *An Unquiet Mind.* It is possible that Jamison's experience with mania and depression helped guide her research into the nature of bipolar disorder. Her work has helped shape the study of the disorder, and her 1990 textbook on manic-depressive illness, coauthored with Frederick Goodwin, is considered the standard for the field. She has been cited as one of the "Best Doctors in the United States," was chosen by *Time* magazine as a "Hero of Medicine," has received numerous awards for her research and books, and won a MacArthur "genius grant" in 2001.

It is truly amazing that Jamison (Figure 13.12) has been so successful in spite of having bipolar disorder. In *An Unquiet Mind,* she details how she was intensely emotional and occasionally obsessive as a child. When she was 17, she had her first serious bout of what she describes as psychotic, profoundly suicidal depression, which seemed to strike out of the blue. Although she would later suffer even worse depression, at the time the experience was beyond anything she had ever imagined. In a recent essay she describes it this way: "The experience forever changed how I looked at myself, the gods, and the mentally ill; it was clear that the line between the mad and the not-mad had dissolved for me, and along with its dissolution went a packet of dreams and my faith in what the future might bring" (Jamison, 2004).

13.12 Kay Redfield Jamison Jamison was able to overcome her crippling bipolar disorder to succeed as a professor and researcher and an author.

Jamison experienced deepening swings from nearly psychotic exuberance to paralyzing depression throughout her undergraduate years at the University of California at Los Angeles (UCLA). In part because of personal experiences, Jamison decided to pursue clinical psychology in graduate school. She was highly successful, being selected UCLA Graduate Woman of the Year and winning a National Science Foundation Research Fellowship. In 1975, after obtaining her Ph.D., she joined the UCLA Department of Psychiatry, where she directed the Affective Disorders Clinic. Within months after she began this job, her condition deteriorated dramatically. She began hallucinating and feared that she was truly losing her mind. This so terrified her that she sought out a psychiatrist with whom she felt safe, who quickly diagnosed Jamison as having manic-depressive illness and prescribed a drug called lithium. (You will read more about lithium and other treatments for bipolar disorder in Chapter 14.)

One of the unfortunate side effects of lithium is that it blunts positive feelings. Those who have bipolar disorder experience profoundly enjoyable highs during their manic phases. In a talk at the University of Melbourne, Jamison describes it this way: "But my manias, at least in their early and mild forms, were absolutely intoxicating states that gave rise to great personal pleasure, an incomparable flow of thoughts, and a ceaseless energy that allowed the translation of new ideas into papers and projects" (July 26, 2000). Even though patients with bipolar disorder know that lithium helps them, they resent the drug and often refuse to take it. Jamison took her medication only occasionally for the first few years and continued to experience episodes of mania, during which she made bad decisions and engaged in risky behaviours. These episodes were always followed by intense depression that left her unable to get out of bed. During one period of particularly severe depression, Jamison attempted suicide by taking an overdose of lithium. She survived only because she reflexively answered a phone call from a relative, who realized that something was wrong and called Jamison's psychiatrist. Left untreated, approximately 20 percent of those with bipolar disorder commit suicide.

Although lithium has helped Jamison, she also credits the psychological support she receives from her psychiatrist, as well as the general support of her family and friends. Jamison herself has made the point that lithium can rob people of creative energy. In her 1993 book *Touched With Fire,* she asks whether lithium would have dampened the genius of artists and writers who may have had mood disorders, such as Vincent van Gogh, Georgia O'Keeffe, Michelangelo, Ernest Hemingway, and Emily Dickinson. Jamison demonstrates that there is a strong association between manic depression and artistic genius, and she raises the disturbing question of whether eradicating the disease would rob society of much great art. Jamison herself embodies the type of a genius plagued by bipolar disorder. Her early career benefited from the energy and creativity of her manic phases even as her personal life was threatened by devastating depression. Her intelligent, beautiful writings have provided inspiration and hope to her fellow sufferers and have given the rest of us a glimpse into the lives of those suffering from mental illness. ■

REVIEWING the Principles | Are Mood Disorders Extreme Manifestations of Normal Moods?

Although all people experience extreme moods and fluctuations in mood, mood disorders are qualitatively different from normal variations in emotion. Mood disorders disrupt the person's ability to function for significant lengths of time, and they are accompanied by a number of psychological and physiological elements. Depressive and bipolar disorders are included in the same *DSM* category and share some behavioural characteristics, but they are distinctly different disorders. Depressive disorders seem to result from a variety of psychological, cognitive, and biological factors, whereas bipolar disorders seem to arise from predominantly biological factors.

Is Schizophrenia a Disorder of Mind or Body?

schizophrenia A mental disorder characterized by alterations in perceptions, emotions, thoughts, or consciousness.

Schizophrenia literally means "splitting of the mind" and refers to a split between thought and emotion. **Schizophrenia** is a *psychotic disorder,* which means that it is characterized by alterations in thoughts, perceptions, or consciousness. Current estimates are that about 1 percent of the population has schizophrenia (Keith, Regier, & Rae, 1991). This rate remains approximately the same in all countries of the world, although the prognosis tends to be better in developing than developed cultures (Kulhara & Chakrabarti, 2001), perhaps because there is more tolerance for symptoms or greater sympathy for unusual or different people in those countries. During

TABLE 13.2 | *DSM-IV* Subtypes of Schizophrenia

SUBTYPE	CHARACTERISTICS
Paranoid type	Preoccupation with delusion(s) or auditory hallucinations. Little or no disorganized speech, disorganized or catatonic behaviour, or inappropriate or flat affect.
Disorganized type	All the following—disorganized speech, disorganized behaviour, and inappropriate or flat affect—are prominent in behaviour, but catatonic-type criteria are not met. Delusions or hallucinations may be present, but only in fragmentary or noncoherent form.
Catatonic type	At least two of the following: extreme motor immobility; purposeless excessive motor activity; extreme negativism (motionless resistance to all instructions) or mutism (refusing to speak); peculiar or bizarre voluntary movement; echolalia.
Undifferentiated type	Does not fit any of the subtypes above, but meets the symptom criteria for schizophrenia.
Residual type	Has experienced at least one episode of schizophrenia, but currently does not have prominent positive symptoms (delusions, hallucinations, disorganized speech or behaviour). However, continues to show negative symptoms and a milder variation of positive symptoms (odd beliefs, eccentric behaviour).

SOURCE: *Diagnostic and Statistical Manual of the American Psychiatric Association,* 1994.

the 1940s and 1950s, researchers and clinicians broadened the definition of schizophrenia by proposing new subtypes and suggesting that schizophrenia could sometimes masquerade as other disorders. Because of these changes, by 1952, 80 percent of the patients in the New York Psychiatric Institute were diagnosed with schizophrenia, up from 20 percent in the 1930s. With more specific guidelines described in the recent editions of the *DSM,* the rate of schizophrenia diagnoses has stabilized (Manderscheid et al., 1985). Clinicians and researchers now rely on lists of symptoms to diagnose various subtypes of schizophrenia (Table 13.2).

Schizophrenia Has Positive and Negative Symptoms

Schizophrenia is arguably the most devastating of the mental disorders for the victim and for the family. It is characterized by a combination of motor, cognitive, behavioural, and perceptual abnormalities that result in impaired social, personal, and/or vocational functioning. Some researchers have grouped these characteristics into two categories: **Positive symptoms** are excesses, whereas **negative symptoms** are deficits in functioning.

positive symptoms Symptoms of schizophrenia, such as delusions and hallucinations, that are excesses in behaviour.

negative symptoms Symptoms of schizophrenia marked by deficits in functioning such as apathy, lack of emotion, and slowed speech and movement.

delusions False personal beliefs based on incorrect inferences about reality.

POSITIVE SYMPTOMS OF SCHIZOPHRENIA Delusions and hallucinations are the symptoms most commonly associated with schizophrenia. **Delusions** are false personal beliefs based on incorrect inferences about external reality (Table 13.3). Delusional people will persist in their beliefs in spite of evidence to the contrary. An example of a *paranoid delusion* arose in response to the Domino's Pizza advertising campaign "Avoid the Noid." The Noid was depicted as an elflike creature who turned pizzas

TABLE 13.3 Delusions and Associated Beliefs

Persecution	Belief that others are persecuting, spying on, or trying to harm them.
Reference	Belief that objects, events, or other people have particular significance to them.
Grandeur	Belief that they have great power, knowledge, or talent.
Identity	Belief that they are someone else, such as Jesus Christ or the prime minister of Canada.
Guilt	Belief that they have committed a terrible sin.
Control	Belief that their thoughts and behaviours are being controlled by external forces.

cold. In 1989, Kenneth Noid, a 22-year-old Georgia man, concluded that this ad campaign was directed at him, and in retaliation he took two Domino's employees hostage. The hostages escaped and Kenneth Noid surrendered to the police.

Although delusions are characteristic of schizophrenia regardless of the culture, the type of delusion can be influenced by cultural factors (Tateyama et al., 1993). When the delusions of German and Japanese schizophrenia patients were compared, the two groups had similar rates of *delusions of grandeur*. However, there were significant differences between the two groups for other types of delusions. The German patients had delusions involving guilt and sin, particularly as these concepts related to religion. They also suffered from *delusions of persecution,* whereas the Japanese patients had delusions of harassment, such as the belief that they were being slandered by others.

Hallucinations, the other hallmark of schizophrenia, are perceptions with no clear external cause. Hallucinations are frequently auditory, although they can also be visual, olfactory, or somatosensory:

> I was afraid to go outside and when I looked out of the window, it seemed that everyone outside was yelling, "kill her, kill her." ... Things continued to get worse. I imagined that I had a foul body odor and I sometimes took up to six showers a day. I recall going to the grocery store one day, and I imagined that the people in the store were saying "Get saved, Jesus is the answer." (O'Neal, 1984)

Auditory hallucinations are often voices giving a running commentary of what a person is doing. They can be accusatory, telling the person he is evil or inept, or they can command the person to do dangerous things. Sometimes the person hears a cacophony of environmental sounds with voices intermingled. Although the cause of hallucinations remains unclear, recent neuroimaging studies suggest that hallucinations are associated with activation in cortical areas that process external sensory stimuli. For example, auditory hallucinations accompany increased activation in brain areas that are also activated in normal subjects when they engage in inner speech (Stein & Richardson, 1999). This has led to speculation that auditory hallucinations might be caused by a difficulty in distinguishing inner speech from external sounds.

Loosening of associations is another characteristic associated with schizophrenia, in which patients shift between seemingly unrelated topics as they speak, making it difficult or impossible for the listener to follow their train of thought:

hallucinations False sensory perceptions that are experienced without an external source.

loosening of associations A speech pattern among schizophrenic patients in which their thoughts are disorganized or meaningless.

"They're destroying too many cattle and oil just to make soap. If we need soap when you can jump into a pool of water, and then when you go to buy your gasoline, my folks always thought they could get pop, but the best thing to get is motor oil, and money. May as well go there and trade in some pop caps and, uh, tires, and tractors to car garages, so they can pull cars away from wrecks, is what I believed in." (Andreasen, 1984)

In more extreme cases, *clang associations* become apparent—the stringing together of words that rhyme but that have no other apparent link. Speech problems make it very difficult for people with schizophrenia to communicate (Docherty, 2005).

DISORGANIZED AND INAPPROPRIATE BEHAVIOUR Another common symptom of schizophrenia is bizarre behaviour, such as wearing multiple layers of clothes even on hot summer days. Those with schizophrenia may walk along muttering to themselves, alternating between anger and laughter, or they might pace, wringing their hands as if extremely worried. Those who have *catatonic* schizophrenia might mindlessly repeat back words that they hear, a behaviour called *echolalia;* or they might remain immobilized in one position for hours, with a rigid, masklike facial expression and eyes staring off into the distance. It has been speculated that catatonic behaviour may be an extreme fear response—akin to how animals respond to sudden dangers—in which the person is literally "scared stiff" (Moskowitz, 2004).

NEGATIVE SYMPTOMS OF SCHIZOPHRENIA A number of behavioural deficits associated with schizophrenia result in patients' becoming isolated and withdrawn. People with schizophrenia often avoid eye contact and seem apathetic. They do not express emotion even when discussing emotional subjects, and they tend to use slowed speech, reduced speech output, and a monotonous tone of voice. This is characterized by such things as long pauses before answering, failing to respond to a question, or being unable to complete an utterance after initiating it. There is often a similar reduction in overt behaviour; a patient's movements may be slowed and overall amount of movement reduced, with little initiation of behaviour and no interest in social participation. These symptoms, though less dramatic than delusions and hallucinations, can be equally serious. Negative symptoms are more common in men than women (Raesaenen, Pakaslahti, Syvaelahti, Jones, & Isohanni, 2000) and are associated with a poorer prognosis.

Interestingly, although the positive symptoms of schizophrenia can be dramatically reduced or eliminated with antipsychotic medications, the negative symptoms often persist. The fact that negative symptoms are often intractable to medications has led researchers to speculate that positive and negative symptoms may result from different organic causes. Since positive symptoms do respond to antipsychotic medications that act on neurotransmitter systems, these symptoms are thought to be the result of neurotransmitter dysfunction. In contrast, it has been speculated that negative symptoms are associated with neuroanatomical factors, since structural brain deficits are not affected by changes in neurochemistry. Some researchers believe that schizophrenia with negative symptoms is a separate disorder from schizophrenia with positive symptoms (Messias et al., 2004).

Schizophrenia Is Primarily a Brain Disorder

Although schizophrenia runs in families, the etiology of the disorder is complex and not well understood. It has been established that genetics plays a role in the development of schizophrenia (Figure 13.13). If one twin develops schizophrenia, the likelihood of the other's also succumbing is almost 50 percent if the twins are identical, but only 14 percent if the twins are fraternal. If one parent has schizophrenia, the risk of the child's developing the disease is 13 percent. If, however, both parents have schizophrenia, the risk jumps to almost 50 percent (Gottesman, 1991). However, the genetic component of schizophrenia represents a predisposition rather than destiny. If schizophrenia was caused solely by genetic factors, concordance in identical twins would approach 100 percent. Although this is true of the Genains, the identical quadruplets introduced at the beginning of the chapter, there is a marked difference in the severity of the disorder among the four sisters (Mirsky et al., 2000). Because they are identical genetically, the difference in severity of the disorder cannot be accounted for by genetic factors.

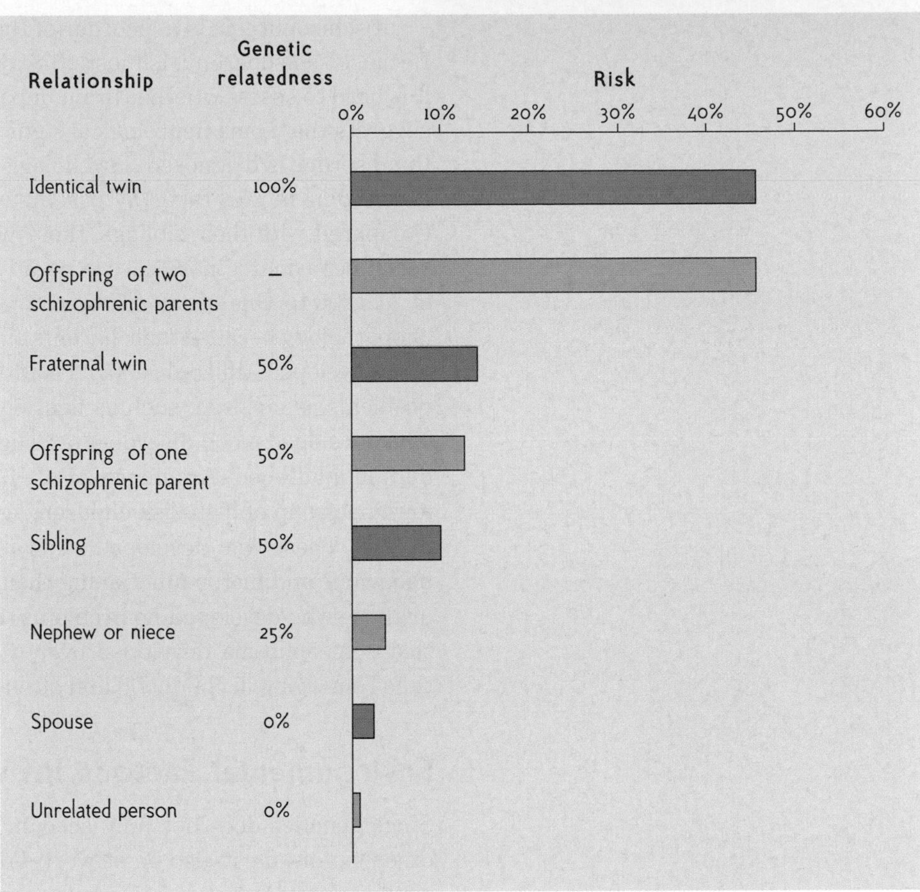

13.13 Genetics and Schizophrenia The more closely someone is genetically related to a person with schizophrenia, the greater is the risk that he or she will develop schizophrenia.

We now know that there is no single gene for schizophrenia, but rather that multiple genes likely contribute in subtle ways to the expression of the disorder. Indeed, each of at least two dozen candidate genes might have a modest influence on the development of schizophrenia (Gottesman & Hanson, 2005).

It is also now well established that schizophrenia is a brain disorder (Walker, Kestler, Bollini, & Hochman, 2004). The advent of techniques that image both the structure and the function of the brain have documented consistent abnormalities among those with schizophrenia. As noted earlier in the chapter, the first evidence of abnormality was that the ventricles, fluid-filled cavities in the brain, were enlarged in patients with schizophrenia, meaning that actual brain tissue is reduced. Moreover, greater reductions in brain tissue are associated with more negative outcomes (Mitelman, Shihabuddin, Brickman, Hazlett, & Buchsbaum, 2004), and longitudinal studies show continued reductions over time (Ho et al., 2003). This reduction of tissue occurs in many regions of the brain, especially the frontal lobes and medial temporal lobes. In functional brain imaging studies, typical results indicate reduced activity in frontal and temporal regions (Barch, Sheline, Csernansky, & Snyder, 2003).

Given that brain abnormalities occur throughout many regions of the brains of people with schizophrenia, some have speculated that schizophrenia is more likely a problem of connection between brain regions than the result of diminished or changed functions of any particular brain region (Walker et al., 2004). One possibility is that schizophrenia results from abnormality in neurotransmitters. Since the 1950s, scientists have believed that dopamine may play an important role in schizophrenia; drugs that block dopamine activity decrease symptoms, whereas drugs that increase the activity of dopamine neurons increase symptoms (Seeman, 2002). However, evidence from the past two decades has implicated a number of other neurotransmitter systems in schizophrenia. More recently, researchers have suggested that schizophrenia might involve abnormalities in the glial cells that make up the myelin sheath (Davis et al., 2003; Moises & Gottesman, 2004). Such abnormalities would impair neurotransmission throughout the brain.

If schizophrenia is a disorder of the brain, when do brain abnormalities emerge? Because schizophrenia is most often diagnosed when people are in their 20s or 30s, it is hard to assess whether brain impairments occurred earlier in life. New evidence suggests that some neurological signs of schizophrenia can be observed long before the disorder is diagnosed. Psychologist Elaine Walker and her colleagues have analyzed home movies taken by parents whose children later developed schizophrenia. Compared with their siblings, those who developed the disorder displayed unusual social behaviours, more severe negative emotions, and motor disturbances (Walker et al., 2004). One study followed a group of children at risk for developing psychopathology because their parents suffered from a mental disorder. Those children who developed schizophrenia as adults were much more likely to display behaviour problems as children, such as fighting or not getting along with others, than those who developed mood disorders or drug abuse problems or did not develop any disorders in adulthood (Amminger et al., 1999). Recently, Walker and her colleagues followed a group of high-risk children, aged 11 to 13, who were videotaped eating lunch in 1972. Those who developed schizophrenia showed greater impairments in social behaviour and motor functioning than those who developed other mental disorders or those who developed no problems (Schiffman et al., 2004). These studies suggest that schizophrenia developed over the life course, with obvious symptoms sometimes emerging in childhood and often in late adolescence.

Environmental Factors Influence Schizophrenia

Why do urban environments have higher rates of schizophrenia?

Since genetics does not fully account for the onset and severity of schizophrenia, other factors must also be at work. Although the rate of schizophrenia tends to be similar across cultures, being born or raised in an urban area approximately doubles the risk of developing schizophrenia later in life (Torrey, 1999). Some have theorized that the increased stress of urban environments can trigger the onset of the disorder. In those at risk for schizophrenia, environmental stress does seem to contribute to its development (Walker et al., 2004). A study of adopted children whose biological mothers were diagnosed with schizophrenia found that if the adoptive families were psychologically healthy, none of the children became psychotic, but if the adoptive families were severely disturbed, 11 percent of the children became psychotic and 41 percent had severe psychological disorders (Tienari et al., 1990, 1994).

Others have speculated that there may be some kind of *schizovirus,* and that living in the close quarters of a big city increases the likelihood of the viruses spreading. In support of the virus hypothesis, some researchers have reported finding in the blood of schizophrenics antibodies that are not found in the blood of those without the disorder (Waltrip et al., 1997). Those diagnosed with schizophrenia are more likely to be born during the late winter and early spring (Mednick, Huttunen, & Machon, 1994; Torrey, Torrey, & Peterson, 1977). Mothers of children born in the late winter and early spring are in their second trimester of pregnancy during flu season. Retrospective studies suggest that the mothers of those with schizophrenia are more likely than other mothers to have contracted influenza during this critical period. Why is the second trimester so important? During this period a great deal of fetal brain development occurs. Trauma or pathogens in the second trimester have the potential of interfering with how brain regions are organized.

Studies of identical twins provide yet another piece of the puzzle. In identical twins, the embryo splits any time until the fifteenth day after conception. If the embryo splits by day 4, the twins are identical genetically and in every physical way. If the embryo splits later, the twins are more likely to be mirror images of each other, with opposite handedness. Early splitters have separate placentas, whereas late-splitting twins usually share the same placenta, which means they share the same blood supply as well as any viruses that pass through that blood supply. Therefore, if prenatal exposure to viruses does play a role in the etiology of schizophrenia, late-splitting twins should be more likely to be concordant for schizophrenia than early splitters. Evidence shows that this is true. For twins who have the same hand preference, the concordance rate for schizophrenia is approximately 10 to 30 percent. In contrast, for twins who have opposite handedness, and therefore likely shared the same placenta, the rate is 60 percent (Davis, Phelps, & Bracha, 1995). This shows not only that there is a strong genetic component to schizophrenia, but that the prenatal environment plays a significant role in the development of the disorder.

REVIEWING the Principles | Is Schizophrenia a Disorder of Mind or Body?

Schizophrenia consists of negative and positive symptoms, and it is at least in part a biological disorder. Twin and adoption studies have highlighted the critical role of genetics in the development of schizophrenia, and recent advances in genetic analysis have begun to yield more insights into the complexities of this disorder. In addition, research suggests the presence of neurochemical and neuroanatomical anomalies in the brains of those afflicted with schizophrenia. Despite all of this evidence, however, the specific ways in which genetics, neurochemistry, and neuroanatomy contribute to this disorder remain unclear. Also, most researchers would agree that environmental factors play a role in the development of schizophrenia, but little is known about how the environment interacts with biological factors.

Are Personality Disorders Truly Mental Disorders?

Personality is each person's unique way of responding to the environment. Although people do change somewhat over time (see Chapter 12), the ways we interact with the world and cope with events tend to be fairly fixed by the end of adolescence. Some people interact with the world in maladaptive and inflexible ways. When this style of interaction is long lasting and causes problems in work and social situations, it becomes a **personality disorder**. All of us at some point have likely exhibited at least some of the symptoms of personality disorders. We might be indecisive, self-absorbed, or emotionally unstable. Why isn't everyone diagnosed with a personality disorder? In general, those with true personality disorders consistently behave in maladaptive ways, whereas most people do so only on occasion. In addition, those with personality disorders show a more extreme level of maladaptive behaviour and consequently experience more personal distress and more problems as a result of their behaviour.

personality disorder A class of mental disorders marked by inflexible and maladaptive ways of interacting with the world.

TABLE 13.4 | Personality Disorders and Associated Characteristics

ODD OR ECCENTRIC BEHAVIOUR	
Paranoid	Tense, guarded, suspicious; holds grudges.
Schizoid	Socially isolated, with restricted emotional expression.
Schizotypal	Peculiarities of thought, appearance, and behaviour that are disconcerting to others; emotionally detached and isolated.
DRAMATIC, EMOTIONAL, OR ERRATIC BEHAVIOUR	
Histrionic	Seductive behaviour; needs immediate gratification and constant reassurance; rapidly changing moods; shallow emotions.
Narcissistic	Self-absorbed; expects special treatment and adulation; envious of attention to others.
Borderline	Cannot stand to be alone; intense, unstable moods and personal relationships; chronic anger; drug and alcohol abuse.
Antisocial	Manipulative, exploitive; dishonest; disloyal; lacking in guilt; habitually breaks social rules; childhood history of such behaviour; often in trouble with the law.
ANXIOUS OR FEARFUL BEHAVIOUR	
Avoidant	Easily hurt and embarrassed; few close friends; sticks to routines to avoid new and possibly stressful experiences.
Dependent	Wants others to make decisions; needs constant advice and reassurance; fears being abandoned.
Obsessive-compulsive	Perfectionistic; overconscientious; indecisive; preoccupied with details; stiff; unable to express affection.

Personality Disorders Are Maladaptive Ways of Relating to the World

Whereas the other disorders discussed in this chapter are classified on Axis I (see Table 13.1), personality disorders are classified on Axis II, along with mental retardation. Personality disorders and mental retardation are grouped together because they usually last throughout the life span with no expectation of significant change.

The personality disorders are generally divided into three groups, as listed in Table 13.4. People with the first group of disorders display odd or eccentric behaviour. *Paranoid, schizoid,* and *schizotypal* personality disorders are included in this group. People with these personality disorders are often reclusive and suspicious. They have difficulty forming personal relationships because of their strange behaviour and aloofness. The second group—*histrionic, narcissistic, borderline,* and *antisocial* personality disorders—is characterized by dramatic, emotional, and erratic behaviours. Borderline and antisocial personality disorders have been the focus of a great deal of research, and they are considered in more detail in the following sections. The third group of personality disorders—*avoidant, dependent,* and *obsessive-compulsive*—involves anxious or fearful behaviours.

Personality disorders remain controversial in modern clinical practice for a variety of reasons. First, personality disorders appear to be extreme or exaggerated versions of normal personality traits. For example, indecisiveness is characteristic of obsessive-compulsive personality disorder, but the *DSM* does not define the degree to which someone must be indecisive in order to be diagnosed as obsessive-compulsive. Second, there is overlap among the traits listed as characteristic of different personality disorders, so that the majority of people who are diagnosed with one personality disorder also meet the criteria for another personality disorder. This suggests that the categories may not be mutually exclusive and that there actually may be fewer types of personality disorders than are listed in the *DSM*.

It may seem that personality disorders do not have so large an impact on daily life as do some of the Axis I disorders discussed in this chapter. However, although people with personality disorders do not hallucinate or have radical mood swings, their ways of interacting with the world can have serious consequences. An in-depth consideration of borderline personality disorder and antisocial personality disorder will illustrate the devastating effect these disorders can have on the individual, family and friends, and society in general.

Borderline Personality Disorder Is Associated with Poor Self-Control

Borderline personality disorder, characterized by identity, affective, and impulse disturbances, was officially recognized as a diagnosis in 1980. The term *borderline* was initially used because these patients were considered to be on the border between normal and psychotic (Knight, 1953). The wide variety of clinical features of this disorder reflects its complexity (see Table 13.5). Approximately 1 to 2 percent of

borderline personality disorder A personality disorder characterized by identity, affective, and impulse disturbances.

TABLE 13.5 Clinical Features of Borderline Personality Disorder

A person having at least five of these characteristics might be considered to have borderline personality disorder.

1. Employment of frantic efforts to avoid real or imagined abandonment.
2. Unstable and intense interpersonal relationships.
3. Persistent and markedly disturbed, distorted, or unstable sense of self (e.g., a feeling that one doesn't exist or that one embodies evil).
4. Impulsiveness in such areas as sex, substance use, crime, and reckless driving.
5. Recurrent suicidal thoughts, gestures, or behaviour.
6. Emotional instability, with periods of extreme depression, irritability, or anxiety.
7. Chronic feelings of emptiness.
8. Inappropriate intense anger or lack of control of anger (e.g., loss of temper, recurrent physical fights).
9. Transient, stress-related paranoid thoughts or severe dissociative symptoms.

SOURCE: *Diagnostic and Statistical Manual of the American Psychiatric Association*, 1994.

adults meet the criteria for borderline personality disorder (Swartz, Blazer, George, & Winfield, 1990; Torgerson, Kringlen, & Cramer, 2001); the disorder is more than twice as common in women than in men.

People with borderline personality disorder do not seem to have a strong sense of self. They cannot tolerate being alone and have an intense fear of abandonment. Because they desperately need an exclusive and dependent relationship with another person, they can be very manipulative in their attempts to control relationships, as shown in the following example:

> A borderline patient periodically rented a motel room and, with a stockpile of pills nearby, would call her therapist's home with an urgent message. He would respond by engaging in long conversations in which he "talked her down." Even as he told her that she could not count on his always being available, he became more wary of going out evenings without detailed instructions about how he could be reached. One night the patient couldn't reach him due to a bad phone connection. She fatally overdosed from what was probably a miscalculated manipulation. (Gunderson, 1984, p. 93)

In addition to problems with identity, borderline individuals also have affective disturbances. Emotional instability is paramount, with episodes of depression, anxiety, anger, and irritability that are often sudden and last from a few hours to a few days. Shifts from one mood to another usually occur with no clear precipitating cause.

Impulsivity is the third hallmark of borderline personality disorder. Impulsive behaviours can include sexual promiscuity, physical fighting, and binge eating and purging. The impulsive behaviour most commonly associated with this disorder, however, is self-mutilation. Cutting and burning of the skin are typical behaviours for those with the disorder, who are also at high risk for suicide.

There is strong evidence from twin studies that borderline personality disorder has a genetic component (Jang et al., 2002; Skodol et al., 2002). Rates of mood disorders in the families of borderline patients also tend to be high, and these patients often show sleep abnormalities characteristic of depression. One possible reason that borderline personality disorder and affective disorders such as depression may be linked is that both appear to involve the neurotransmitter serotonin. Evidence has linked low serotonin levels to the impulsive behaviour seen in borderline personality disorder (Skodol et al., 2002).

A strong relationship exists between borderline personality disorder and trauma or abuse (Lieb, Zanarini, Schmahl, Linehan, & Bohus, 2004). Some studies have reported that 70 to 80 percent of patients with borderline personality disorder have experienced physical or sexual abuse or observed some kind of extreme violence. Consider the following case of a young woman with borderline personality disorder:

> [When Roberta was in fourth grade, her] oldest brother, Sam, turned fifteen and began babysitting his younger siblings two evenings a week while his mother went to Bingo and his father was out with "the boys." On one of these evenings, a few weeks after he'd started babysitting, Sam decided to "help" Roberta with her bath. Overriding her protests with the statement that he was in charge while their parents were gone, he demanded that she stand naked in front of him so that he could "make sure I was clean." Gradually, over the next several months, Sam's demands escalated. He watched her undress. He watched her urinate. By spring he had begun to masturbate as he watched her. (Bernheim, 1997, pp. 257–58)

Other theories implicate early interactions with caretakers. It has been suggested that borderline patients had caretakers who did not accept them or were unreliable or unavailable. The constant rejection and criticism made it difficult for the individual to learn to regulate emotions and understand emotional reactions to events (Linehan, 1987). An alternative theory is that caregivers may have encouraged dependence, and therefore individuals do not adequately develop a sense of self. They become extremely sensitive to the reactions of others, and if they are rejected, they reject themselves.

Do people with antisocial personality disorder "suffer" from their condition?

Antisocial Personality Disorder Is Associated with a Lack of Empathy

In the 1800s, the term "psychopath" was coined to describe people who seem willing to take advantage of and hurt others without any evidence of concern or remorse (Koch, 1891). The *DSM* dropped this pejorative label and adopted the term **antisocial personality disorder**. This disorder is distinguished not so much by particular behaviours as by an approach to life, the most salient feature being a lack of remorse. Those with antisocial personality disorder tend to be hedonistic, seeking immediate gratification of wants and needs without any thought of others. Because they lack empathy, their interpersonal relationships tend to be shallow at best. They are very impulsive and engage in sensation-seeking behaviour, as is illustrated in the case of American Gary Gilmore (Figure 13.14), who was executed for the murder he describes here.

> "I went in and told the guy to give me the money. I told him to lay on the floor and then I shot him. I then walked out and was carrying the cash drawer with me. I took the money and threw the cash drawer in a bush and I tried to push the gun in the bush, too. But as I was pushing it in the bush, it went off and that's how come I was shot in the arm. It seems like things have always gone bad for me. It seems like I've always done dumb things that just caused trouble for me. I remember when I was a boy I would feel like I had to do things like sit on a railroad track until just before the train came and then I would dash off. Or I would put my finger over the end of a BB gun and pull the trigger to see if a BB was really in it. Sometimes I would stick my finger in water and then put my finger in a light socket to see if it would really shock me." (Spitzer et al., 1983, pp. 66–68)

13.14 Gary Gilmore after His Arrest
Gilmore would have been given the diagnosis of antisocial personality disorder under the *DSM-IV*.

CHARACTERISTICS OF ANTISOCIAL PERSONALITY DISORDER Antisocial personality disorder is surprisingly common, with an estimated lifetime prevalence of 4.5 percent in men and 0.8 percent in women (Robins & Regier, 1991). Interestingly, the disorder is most apparent in late adolescence and early adulthood, and it appears to remit around age 40 (Hare, McPherson & Forth, 1988). A diagnosis of antisocial personality disorder cannot be made before age 18, but the person must have displayed antisocial conduct before age 15. This stipulation ensures that only those with a lifetime history of antisocial behaviours can be diagnosed with antisocial personality disorder. They must also meet criteria such as repeatedly performing illegal acts, repeated lying or use of aliases, and reckless disregard for their own safety or the safety of others. Many of these individuals can be quite bright and highly verbal, and as a result many can talk their way out of bad situations. In any event, punishment seems to have very little effect on them (Lykken, 1957, 1995), and they often repeat the problem behaviour a short time later (Hare, 1999).

antisocial personality disorder A personality disorder marked by a lack of empathy and remorse.

It has been estimated that 40 to 75 percent of the prison population meet the criteria for antisocial personality disorder (Hare, 1993; Widiger & Corbitt, 1995), but not everyone with this disorder engages in criminal behaviour. Because of the prevalence of the disorder in the prison population, much of the research on antisocial personality disorder has been conducted in this setting. One researcher, however, came up with an ingenious way of finding research participants outside of prison. She put the following advertisement in a counterculture newspaper:

> Wanted: charming, aggressive, carefree people who are impulsively irresponsible but are good at handling people and at looking after number one. Send name, address, phone, and short biography proving how interesting you are to . . . (Widom, 1978, p. 72)

Seventy-three people responded to the ad, about one-third of whom met the criteria for antisocial personality disorder. These individuals were then interviewed and given a battery of psychological tests. It was found that their characteristics were very similar to those of prisoners diagnosed with antisocial personality disorder, except that the group that responded to the ad had been able to avoid imprisonment.

THE CAUSES OF ANTISOCIAL PERSONALITY DISORDER In 1957, David Lykken reported that psychopaths did not become anxious when subjected to aversive stimuli. He and other investigators have continued this line of work, showing that such individuals do not seem to feel fear or anxiety (Lykken, 1995; Hare, 1965). EEG examinations have demonstrated that criminals who meet the criteria for antisocial personality disorder have slower alpha-wave activity (Raine, 1989). This indicates a lower overall level of arousal and may explain why these people often engage in sensation-seeking behaviour. It also may explain why they do not learn from punishment—because of their low arousal, they do not experience punishment as particularly aversive. This pattern of reduced psychophysiological responding in the face of punishment has been found in adolescents who are at risk for developing psychopathy (Fung et al., 2005). There is also evidence of amygdala abnormalities in those with antisocial tendencies, which include the amygdala being smaller and less responsiveness to negative stimuli (Blair, 2003). Deficits in frontal-lobe functioning have also been found and may account for the lack of forethought and the inability to consider the implications of actions characteristic of antisocial personality disorder (Seguin, 2004).

To what can we attribute these biological correlates of antisocial personality disorder? There is good evidence for both genetics and the environment. Identical twins have a higher concordance of criminal behaviour than do fraternal twins (Lykken, 1995), although this research did not rule out the role of shared environment. A large study of 14,000 adoptions (Mednick, Gabrielli, & Hutchings, 1987) found that adopted male children had a higher rate of crime if their biological fathers had criminal records. In addition, the more entrenched the criminal behaviour of the biological father, the more likely it was that the adopted son had engaged in criminal behaviour. Although genetics may be at the root of antisocial behaviours, such factors as low socioeconomic status, dysfunctional families, and childhood abuse may also play important roles. Indeed, malnutrition at age 3 has been found to predict antisocial behaviour at age 17 (Liu, Raine, Venables, & Mednick, 2004). An enrichment program for children that included a structured nutrition component was associated with less criminal and antisocial behaviour 20 years later (Raine, Mellingen, Liu, Venables, & Mednick, 2003). This raises the possibility that malnutrition or other similar environmental factors might contribute to the development of antisocial personality disorder.

REVIEWING
the Principles | Are Personality Disorders Truly Mental Disorders?

Personality disorders are diagnosed along a different axis than are the other mental disorders discussed thus far. Personality disorders are not considered clinical disorders, yet they can have devastating effects on the individual, the family, and society. Borderline personality disorder is characterized by identity, affective, and impulse disturbances. Those with the disorder often have a history of abuse or rejection by caregivers. Those with antisocial personality disorder are hedonistic and sensation seeking, and they lack empathy for others. Although the outward symptoms of these disorders are not as severe as those of some of the clinical disorders, they are highly resistant to change, and the ways these individuals interact with others cause significant personal and societal problems.

Should Childhood Disorders Be Considered a Unique Category?

When Emil Kraepelin published his classic text on the classification of mental disorders in 1883, no mention was made of childhood disorders. The first edition of the *DSM,* published 70 years later, also considered children to be essentially mini versions of adults, and consequently disorders of childhood were not considered separately from disorders of adulthood. Currently, in response to the belief that cognitive, emotional, and social abilities should be considered in the context of the individual's developmental state, the *DSM* has a category in Axis I called "disorders usually first diagnosed in infancy, childhood, or adolescence." This category includes a wide range of disorders, from those affecting only circumscribed areas of a child's world, such as reading disorders and stuttering, to those affecting every aspect of a child's life, such as autism and attention-deficit/hyperactivity disorder (see Table 13.6). Despite the variety of disorders in this category, all should be considered within the context of knowledge about normal childhood development. Some symptoms of childhood mental disorders are extreme manifestations of normal behaviour or are actually normal behaviours for children at an earlier developmental stage. Bedwetting, for example, is normal for two-year-olds but not for ten-year-olds. Other behaviours, however, deviate significantly from normal development. Two disorders of childhood, autism and attention-deficit/hyperactivity disorder, will be explored to illustrate this issue.

Autism Is a Lack of Awareness of Others

Autism is characterized by deficits in social interaction, impaired communication, and restricted interests (Volkmar, Chawarska, & Klin, 2005). Autism was first described in 1943 by Leo Kanner, who was struck by the profound isolation of some children and coined the term "early infantile autism." Kanner described the syndrome in the following way: "There is from the start an extreme autistic aloneness that, whenever possible, disregards, ignores, shuts out anything that comes to the child from the outside." When people think of autism, they often imagine a beautiful,

autism A developmental disorder involving deficits in social interaction, impaired communication, and restricted interests.

TABLE 13.6 | Childhood Disorders

DISORDER	DESCRIPTION
Attention-Deficit/Hyperactivity	A pattern of hyperactive, inattentive, and impulsive behaviour that causes social or academic impairment.
Autism	Characterized by unresponsiveness; impaired language, social, and cognitive development; and restricted and repetitive behaviour.
Elimination Disorders	The repeated passing of feces or urination in inappropriate places by children from whom continence should be expected.
Learning Disorders	Marked by substantially low performance in reading, mathematics, or written expression with regard to what is expected for age, amount of education, and intelligence.
Mental Retardation	Characterized by below average intellectual functioning (IQ lower than 70) and limited adaptive functioning that begins before age 18.
Rumination Disorder	The repeated regurgitation and re-chewing of partially digested food, not related to nausea or gastrointestinal disorder.
Selective Mutism	Failure to speak in certain social situations, despite the ability to speak in other situations; interferes with social or academic achievement.
Tourette's Disorder	Recurrent motor and vocal tics that cause marked distress or impairment and are not related to a general medical condition.

What factors might be responsible for the increased diagnosis of autism spectrum disorders?

graceful child who seems tragically shut off from the world. Some autistic children are beautiful, and autistics tend to walk on their toes, which might have given rise to the myth of gracefulness. However, autistic children usually look the same as any other children, and they are often clumsier than normal children. Approximately 4 to 6 children out of 1,000 will show signs of autism, with males outnumbering females three to one (Muhle, Trentacoste, & Rapin, 2004). There has been an increase in the number of children diagnosed with autism from 1980s to now. This increase is likely due to greater awareness of the symptoms of autism by parents and physicians as well as a willingness to apply the diagnosis to a wider spectrum of behaviours (Rutter, 2005).

Autism varies in severity, from mild social impairments to severe social and intellectual impairments. *Asperger's syndrome* is high-functioning autism, in which children of normal intelligence have specific deficits in social interaction, such as having impoverished theory of mind (i.e., understanding other's intentions). Recall from Chapter 11 that theory of mind refers to understanding that other people have mental states and possessing the ability to predict their behaviour accordingly. Perhaps the most famous person with Asperger's is Temple Grandin, an accomplished scientist who now designs humane animal facilities (Grandin, 1995). Although extremely intelligent, she has great difficulty understanding the subtle social motives and behaviours of other humans. She finds it easier to relate to animals. Grandin is able to understand the environment from an animal's point of view, which allows her to design facilities that are calming.

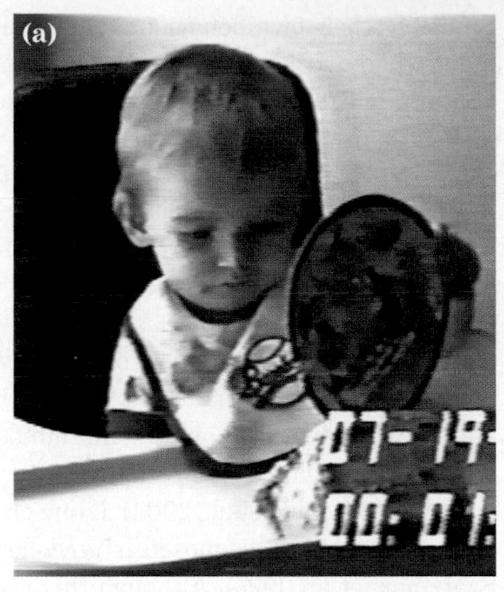

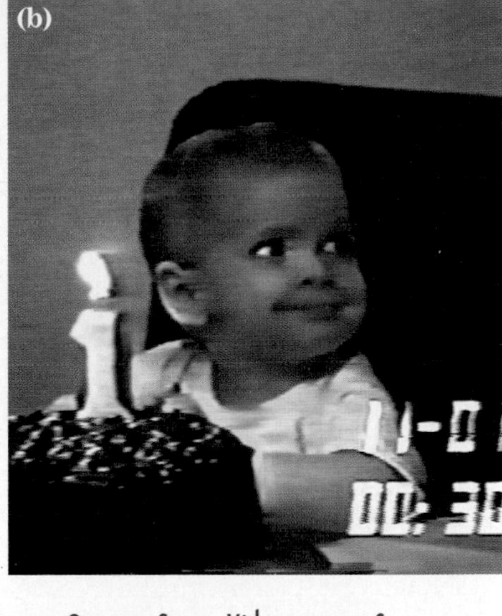

13.15 Scenes from Videotapes of Children's Birthday Parties The child in (a) focused more on objects than on people and was later diagnosed with autism. The child in (b) developed normally.

CORE SYMPTOMS OF AUTISM

Autistic children are seemingly unaware of others. As babies, they do not smile at their caregivers, they do not respond to vocalizations, and they often actively reject physical contact with others. Children with autism do not establish eye contact and do not use gaze to either gain or direct the attention of those around them. One group of researchers had participants view video footage of autistic children's first birthdays to see if autistic characteristics could be detected before the child was diagnosed with autism (Osterling & Dawson, 1994). By considering only the number of times a child looked at another person's face, the participants correctly classified the children as either autistic or normal 77 percent of the time (see Figure 13.15).

Deficits in communication are the second major cluster of behaviours characteristic of autism. Autistic children show severe impairments in both verbal and nonverbal communication. Even if they do vocalize, it is often not with any intent to communicate. Autistics who develop language often exhibit odd speech patterns, such as echolalia and *pronoun reversal.* Echolalia, mentioned earlier as a symptom of catatonia, is the repetition of words or phrases, sometimes including an imitation of the intonation and sometimes using a high-pitched monotone. Pronoun reversal, in which autistic children may replace "I" with "you," may be related to echolalia. Even if a child ceases being echolalic, pronoun confusion often persists. Autistic children who develop functional language also often interpret words in a literal manner, use language inappropriately, and lack verbal spontaneity.

A third category of deficits is restricted activities and interests. Although autistic children appear to be oblivious to the people around them, they are acutely aware of their surroundings. Children with autism tend to focus on small details in their surroundings. Although most children automatically pay attention to the social aspects of a situation, those with autism may focus on seemingly inconsequential details (Klin, Jones, Schultz, & Volkmar, 2003; Figure 13.16). Any changes in the daily routine or in the placement of furniture or toys is very upsetting and can result in extreme agitation and tantrums. Autistic children also do not play in the same way as other children. Their play tends to be stereotyped and obsessive, and the focus is often on the sensory aspects of objects. They may smell and taste objects, or spin and flick them for the visual stimulation.

13.16 Toddler Viewer with Autism A two-year-old with autism focuses on the unimportant details in the scene rather than the social characters.

Similarly, their own behaviour tends to be stereotyped, with ritualistic hand movements, body rocking, and hand flapping. Self-abuse is common, and in some cases children must be forcibly restrained to keep them from hurting themselves.

AUTISM IS A BIOLOGICAL DISORDER Kanner believed autism was an innate disorder that was exacerbated by cold and unresponsive mothers, whom he called "ice box mothers" or "refrigerator mothers." He described the parents of autistic children as insensitive, meticulous, introverted, and highly intellectual. This view is given little credence today, as it is now well established that autism is the result of biological factors, the specific nature of which is still undetermined. Genetic studies of autism have been hampered by the rarity of the disorder and by the fact that autistic persons rarely marry and almost never have children. Despite the limited genetic research, there are indications that the disorder has a strong hereditary component (Muhle et al., 2004). If one child in a family is autistic, the probability of a second also being diagnosed is anywhere from 2 to 9 percent (Jorde et al., 1990; Szatzmari et al., 1993). Although this risk is relatively low, it is significantly higher than the estimated 0.4 percent prevalence of the disorder in the population at large. If two siblings are autistic, the chances of a third sharing the same diagnosis jumps to 35 percent. A number of studies have found concordance rates in twins to be 70 to 90 percent for monozygotic twins and 10 percent for dizygotic twins (Bailey et al., 1995; Steffenburg et al., 1989).

Research into the causes of autism points to prenatal and/or neonatal events that may result in brain dysfunction. The brains of children with autism grow unusually large during the first two years of life, and then growth slows until age five (Courchesne, Redcay, & Kennedy, 2004). Researchers are investigating both genetic and nongenetic factors that might explain this overgrowth-undergrowth pattern. In some cases, mothers of autistic children have experienced significant bleeding during the second trimester of pregnancy, suggesting some kind of trauma during the critical period for neuronal development. Autistic children also have a higher rate of neonatal complications such as apnea, seizures, and delay in breathing. Moreover, autistic children are more likely to have minor physical anomalies, to be the product of a first pregnancy, and to be born to older mothers (Gilberg, 1980).

Two exciting new research developments may help to focus the future of autism research. A deficit in oxytocin, a neuropeptide involved in social behaviour (see Chapter 11), may be related to some of the behavioural manifestations of autism. It has been demonstrated that mice lacking oxytocin behave normally, except that they cannot recognize other mice or their mother's scent. A single dose of oxytocin cured them (Ferguson et al., 2000). In one study, autistic adults who received injections of oxytocin showed a dramatic improvement in their symptoms (Novotny et al., 2000). Injections of oxytocin seem particularly useful for reducing repetitive behaviours, such as repeating oneself, questioning, inappropriate touching, and self-injury (Hollander et al., 2003). A second recent finding suggests that levels of four proteins in the blood are elevated in 97 percent of autistic children and 92 percent of retarded children, but in none of the healthy controls. These elevated protein levels were found in blood samples taken at birth, and all four proteins are involved in brain development (Nelson et al., 2001). It is possible that these proteins may be involved in the pattern of overgrowth and undergrowth of the brain for those with autism. However, both whether these proteins are influenced by genetic or environmental factors and what role they play in the development of autism are matters for future research.

Attention-Deficit/Hyperactivity Disorder Is a Disruptive Behaviour Disorder

attention-deficit/hyperactivity disorder (ADHD) A disorder characterized by restless, inattentive, and impulsive behaviours.

> A hyperactive child's mother might report that he has difficulty remembering not to trail his dirty hand along the clean wall as he runs from the front door to the kitchen. His peers may find that he spontaneously changes the rules while playing Monopoly or soccer. His teacher notes that he asks what he is supposed to do immediately after detailed instructions were presented to the entire class. He may make warbling noises or other strange sounds that inadvertently disturb anyone nearby. He may seem to have more than his share of accidents—knocking over the tower his classmates are erecting, spilling his cranberry juice on the linen tablecloth, or tripping over the television cord while retrieving the family cat—and thereby disconnecting the set in the middle of the Super Bowl game. (Whalen, 1989)

We can laugh at this description, yet most of us have come into contact with such children. Although the symptoms can seem humorous in the retelling, the reality is a different story. Children with **attention-deficit/hyperactivity disorder (ADHD)** are restless, inattentive, and impulsive. They need to have directions repeated and rules explained over and over. Although these children are often friendly and talkative, they can have trouble making and keeping friends because they miss subtle social cues and make unintentional social mistakes. Note that many of these symptoms are exaggerations of the typical behaviour of toddlers. This makes the line between normal and abnormal behaviour hard to draw—as many as 50 percent of mothers of four-year-old boys believe their sons are hyperactive (Varley, 1984).

THE ETIOLOGY OF ADHD It is estimated that 3 to 5 percent of children have ADHD, although estimates from other sources vary widely, with some estimates as high as 20 percent (Gillis, Gilger, Pennington, & DeFries, 1992). The causes of the disorder are unknown; one of the difficulties in pinpointing the etiology is that ADHD is most likely a heterogeneous disorder. In other words, the behavioural profiles of children with ADHD vary, so it is likely that the causes of the disorder vary as well. There is some suggestion that children with ADHD are more likely than other children to grow up in disturbed families. This finding has led researchers to speculate that the disorder may have a psychological component. Factors such as poor parenting and social disadvantage may contribute to the onset of symptoms. Whether psychological factors are the cause or effect of the disorder, however, is not always clear. Most current research is focused on biological factors contributing to the development of ADHD.

ADHD clearly has a genetic component, with a 10 to 35 percent incidence of the disorder in families of those diagnosed with ADHD (Biederman, Faraone, Keenan, Knee, & Tsuang, 1990; Biederman, Faraone, & Lapey, 1992). Concordance is estimated at 55 percent in monozygotic twins and 32 percent in dizygotic twins (Goodman & Stevenson, 1989; Sherman, McGue, & Iacono, 1997). Although it is clear that something is inherited in ADHD, the question is what. In an imaging study, Zametkin and his colleagues (1990) found that adults who had been diagnosed with ADHD in childhood had reduced metabolism in brain regions involved in self-regulation of motor functions and attentional systems (Figure 13.17). These researchers postulated that the connection between the frontal lobes and the limbic system was impaired in ADHD patients. The symptoms of ADHD are similar to those seen in patients with frontal-lobe damage: problems with planning, sustaining concentration, using feedback, and thinking flexibly. Imaging studies

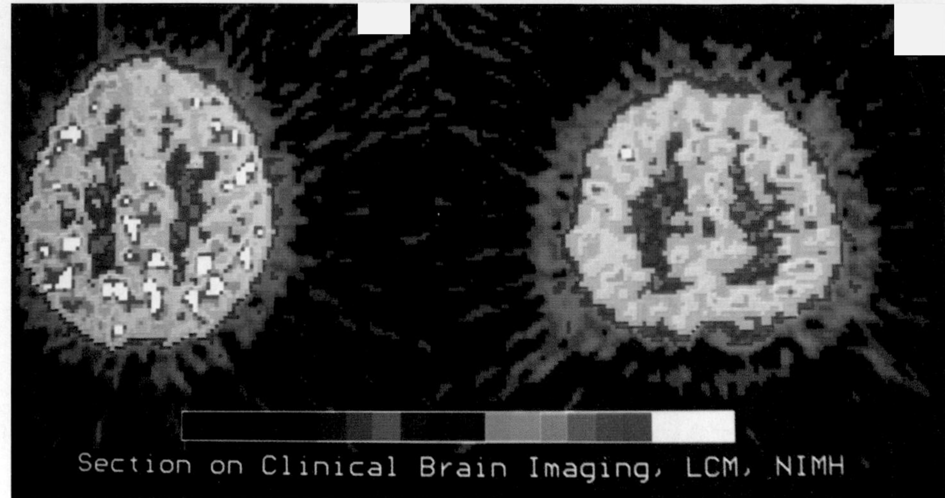

13.17 ADHD and the Brain During PET scanning, the brain of the person on the right, who has a history of ADHD, shows lesser overall activation, especially in frontal and limbic regions.

What are the long-term prospects for children with ADHD?

have found prefrontal abnormalities when adolescents with ADHD perform tasks that require inhibiting a motor response. Greater impairments in performance on the task are associated with more activation of these prefrontal regions (Schulz et al., 2004). Current research continues to focus on underarousal of the frontal lobes (Barkely, 1997; Neidemeyer & Naidu, 1997) but also implicates subcortical structures. In particular, researchers have demonstrated differences in the basal ganglia in the brains of some ADHD patients (Aylward, Reiss, Reader, & Singer, 1996; Castellanos, Giedd, Eckberg, & Marsh, 1998; Fillipek et al., 1997; Hynd, Hern, & Novey, 1993). This region is involved in regulation of motor behaviour and impulse control, and dysfunction in these structures could contribute to the hyperactivity characteristic of ADHD.

ADHD ACROSS THE LIFE SPAN Children are not generally given a diagnosis of ADHD until they enter a structured setting in which they are expected to conform to rules, get along with peers, and sit in a seat for long periods of time. In the past, this happened when children entered school, between the ages of five and seven. Now, with the increasing prevalence of structured day-care settings, the demands on children to conform are occurring much earlier. Although it is not unreasonable to expect a six-year-old to sit quietly and share crayons, it is probably unrealistic to expect the same of a three-year-old. Many preschoolers who are considered inattentive and overactive by their parents and teachers are simply going through a normal developmental stage that will not become the persistent pattern of ADHD. Some characteristics discriminate between children who go on to develop ADHD and those who do not. Infants who are later diagnosed with ADHD have difficulty establishing regular patterns for eating and sleeping (Ross & Ross, 1982). Toddlers exhibit excessive activity and temperamental behaviour quite early in life (Hartsough & Lambert, 1985). They are very curious and engage in vigorous play, and consequently they tend to be quite accident prone. Older children do not generally demonstrate excess motor activity but are instead restless and fidgety (Pelham & Bender, 1982).

"Excuse me, Doc, my attention wandered. What type of deficit disorder did you say I had?"

It was previously thought that children outgrew ADHD by the time they entered adulthood. More recent longitudinal studies, however, dispute this notion (McGough & Barkley, 2004). Between

30 and 80 percent of those with ADHD in childhood continued to show symptoms of this disorder in adulthood (Weiss & Hechtman, 1993). Animal models support this suggestion, and further predict learning problems with a predisposition for seizures (Anisman & McIntyre, 2002). Not surprisingly, therefore, adults with ADHD symptoms may struggle both academically and vocationally. They generally reach a lower-than-expected socioeconomic level and change jobs more frequently (Bellak & Black, 1992; Mannuzza et al., 1991). The impact of ADHD on society is apparently greater than previously believed. However, some ADHD patients do learn to adapt and are successful in their personal and vocational lives.

REVIEWING **the Principles** | Should Childhood Disorders Be Considered a Unique Category?

Until recently, children were considered to be mini versions of adults, and mental disorders in children were classified using adult categories. Currently, it is understood that disorders in children should be considered in the context of normal development. In some cases, mental disorders identified in childhood have a lasting impact on the individual, and the problems apparent early in life continue throughout maturation. This is clearly the case for autism, in which the social and cognitive characteristics of the disorder deviate significantly from normal childhood development and continue to have a major impact throughout the life span. The impact of childhood ADHD on adult functioning is less certain.

Conclusion

Throughout this book, we have emphasized the adaptive nature of the human mind, as it solves problems of survival and performs complex functions such as perceiving and remembering. But sometimes minds fail to work properly, producing maladaptive thoughts, emotions, and behaviours. For instance, people can become too afraid to leave their houses, too depressed to get out of bed, or so manic that they take dangerous risks. Others hear voices or believe that their family members are trying to poison them. Sometimes people are incapable of interacting with others.

In this chapter, we have explored the nature and causes of psychopathology. From the time of Freud, most mental disorders were viewed as arising from mistreatment by caregivers; for instance, autism was considered the result of cold, uncaring mothers. The biological revolution has provided compelling evidence that many mental disorders involve a genetic predisposition, abnormalities in various neurotransmitter systems, or brains that do not function normally. It is now clear that biology plays an important role in psychopathology. At the same time, psychological and social factors are implicated in the development and course of most mental disorders. An obvious example is the distorted thought patterns that afflict those with major depression and other mood disorders.

Various environmental factors also contribute to psychopathology. Parental maltreatment can damage children, as was the case with the Genain quadruplets and with those who report having multiple identities. Prenatal and early childhood factors, such as malnutrition, exposure to toxins and viruses, and poverty, seem to increase the risk that psychopathology will develop. The experience of loss or trauma can produce mood disorders among otherwise mentally healthy adults. Thus, to understand psychopathology thoroughly, we need to recognize the social, psychological, and biological factors that contribute to the expression of mental illness in those who are susceptible. Crossing levels of analysis to understand the causes of mental disorders is especially useful for determining the best ways to treat them, which is the topic of the next chapter.

Summarizing the Principles of the Disorders of Mind and Body

How Are Mental Disorders Conceptualized and Classified?

1. **Mental disorders are classified into categories:** The *Diagnostic and Statistical Manual of Mental Disorders (DSM)* has evolved from Kraeplin's categories of mental illness. It is a multiaxial system for diagnosing groups of symptoms in the context of related factors.

2. **Mental disorders must be assessed before diagnosis:** Assessment is the process of examining a person's mental functions and psychological health in order to make a diagnosis. Assessment is accomplished through interviews, behavioural evaluations, and psychological testing.

3. **Mental disorders have many causes:** Disorders may arise from psychological factors, such as the family system or sociocultural context. They may be the result of learned, maladaptive cognitions. Biological factors also underlie mental illness. The diathesis-stress model looks at mental disorders as an interaction among multiple factors. In this model, stressful circumstances may trigger a disorder in an individual with underlying vulnerabilities.

4. **The legal system has its own definition of psychopathology:** The legal system focuses on insanity as it relates to the issue of personal responsibility for actions.

Can Anxiety Be the Root of Seemingly Different Disorders?

5. **There are different types of anxiety disorders:** Phobias are exaggerated fears of specific stimulus. Generalized anxiety disorder is diffuse and omnipresent. Panic attacks cause sudden overwhelming terror, and may lead to agoraphobia. Obsessive-compulsive disorders involve anxiety-related thoughts and behaviours.

6. **Anxiety disorders have cognitive, situational, and biological components:** The etiology of OCD is at least in part genetic, and also involves brain dysfunction. The irrational thoughts that accompany panic attacks may lead to agoraphobia through cognitive-behavioural connections.

Are Mood Disorders Extreme Manifestations of Normal Moods?

7. **There are different types of mood disorders:** Depressive disorders may be unipolar or bipolar and are more severe than dysthymia. The less extreme form of bipolarity is cyclothymia.

8. **Mood disorders have cognitive, situational, and biological components:** The biological factors of depression include genetics, frontal-lobe functioning, and serotonin modulation, as well as biological rhythms. Negative thinking and poor interpersonal relations also contribute to depression.

Is Schizophrenia a Disorder of Mind or Body?

9. **Schizophrenia has positive and negative symptoms:** Positive symptoms include excesses, such as delusions and hallucinations. Negative symptoms are deficits in functioning, such as social withdrawal and reduced movement.

10. **Schizophrenia is primarily a brain disorder:** The brains of schizophrenics have larger ventricles and less brain mass, with reduced frontal and temporal activation. A variety of neurochemical and neural structural abnormalities exist as well.

11. **Environmental factors influence schizophrenia:** Urban environments may trigger the onset of schizophrenia. Trauma or pathogens encountered by pregnant women may increase the likelihood of the disorder in their children.

Are Personality Disorders Truly Mental Disorders?

12. **Personality disorders are maladaptive ways of relating to the world:** Odd behaviours, extreme emotions, and fearful behaviours are characteristic of personality disorders. It is controversial whether some of these extremes are true psychopathologies.

13. **Borderline personality disorder is associated with poor self-control:** Borderline personality disorder involves disturbances in identity, affect, and impulse control. A strong relationship exists between the disorder and trauma and abuse.

14. **Antisocial personality disorder is associated with a lack of empathy:** Antisocial personality disorder is marked by lack of empathy and remorse, and a tendency to be manipulative. Both genetics and environment seem to be contributing factors.

Should Childhood Disorders Be Considered a Unique Category?

15. **Autism is a lack of awareness of others:** Autism emerges in infancy and is marked by avoidance of eye contact and impairment in verbal and nonverbal communication. Asperger's syndrome is a high-functioning variation of autism. Autism is a biological disorder. The biological factors may include abnormalities in oxytocin, brain growth, and blood proteins.

16. **Attention-deficit/hyperactivity disorder is a disruptive behaviour disorder:** Children with ADHD are restless, inattentive, and impulsive. It is a heterogeneous disorder, so it is likely that the causes of the disorder vary. Environmental factors may include poor parenting and social disadvantages. Genetic factors and brain abnormalities, particularly with regard to activation of the frontal lobes and subcortical basal ganglia, may also play a part. ADHD continues into adulthood, presenting academic and career challenges.

Applying the Principles

1. Dr. Jones, a psychologist, is evaluating a client's negative mood and thoughts to decide if he is clinically depressed. Dr. Jones notes that the client has recently been in a serious car accident and suffered injuries that have left him unable to return to work. The financial stress on the client's family has been significant. Dr. Jones's consideration of the client's physical and mental health as a result of the accident and his concern about family circumstances represent the ________ approach of the *DSM.*

    ____ a) symptom check list
    ____ b) multiaxial
    ____ c) "pigeon hole"
    ____ d) labeling

2. Christina is constantly preoccupied with thoughts that she has left her car unlocked, and repeatedly returns to the parking structure to check. It is likely that she is suffering from ________, which is the result of the brain structure known as the ________ not functioning normally.

    ____ a) generalized anxiety; amygdala
    ____ b) phobia; hippocampus
    ____ c) panic disorder; motor cortex
    ____ d) OCD; caudate nucleus

3. Recently, Sam has started to think more about the nature of depression. First, his Uncle Joe seems depressed since his Aunt Becky passed away. Second, his good friend Mike from high school always seems somewhat melancholy. Third, he's noticed that his good friend Alice seems to go through periods of deep, long-term depression. What do these examples tell us about the nature of depression, as well as many other mental disorders?

    ____ a) There are unique symptoms in each patient.
    ____ b) It is a disorder only if it has been preceded by a triggering event.
    ____ c) Disorders vary along a continuum from mild to severe.
    ____ d) Disorders are only apparent when observed in others, and difficult to recognize in ourselves.

4. Jason and David have both been diagnosed with schizophrenia. Jason has been hearing voices and feeling paranoid. David has become mute and catatonic. Jason's symptoms are ________ and likely due to ________; David's symptoms are ________ and likely due to ________.

    ____ a) negative, neurochemistry; positive, neuroanatomy
    ____ b) positive, environment; negative, neurochemistry

____ c) positive, neurochemistry; negative, neuroanatomy
____ d) negative, environment; negative, neurochemistry

5. Amber makes up a story of being mugged, in order to get the interest and sympathy of her boyfriend. She is emotionally volatile and manipulative. It is likely that Amber suffers from ________ and the prognosis for treatment is ________.

____ a) borderline personality disorder; extremely poor
____ b) antisocial personality disorder; excellent only with medication
____ c) narcissistic personality disorder; excellent
____ d) antisocial personality disorder; extremely poor

6. Shanea works at the campus daycare centre. She has noticed a young boy who does not like to be held, avoids contact with others, does not use language, and gets very upset if the routine is changed. He has been diagnosed with ________, which is likely due to ________.

____ a) extreme shyness; genetics
____ b) autism; attachment and social factors
____ c) Asperger's syndrome; biological factors
____ d) autism; biological factors

ANSWERS: 1.b 2.d 3.c 4.c 5.a 6.d

Key Terms

agoraphobia, p. 536
antisocial personality disorder, p. 555
anxiety disorders, p. 535
assessment, p. 525
attention-deficit/hyperactivity disorder (ADHD), p. 561
autism, p. 557
bipolar disorder, p. 541
borderline personality disorder, p. 553
cognitive-behavioural approach, p. 530
cyclothymia, p. 541
delusions, p. 546
Diagnostic and Statistical Manual of Mental Disorders (*DSM*), p. 524
diathesis-stress model, p. 532
dysthymia, p. 540
family systems model, p. 530
generalized anxiety disorder, p. 536
hallucinations, p. 547
learned helplessness model, p. 543
loosening of associations, p. 547
major depression, p. 540
multiaxial system, p. 524
negative symptoms, p. 546
obsessive-compulsive disorder (OCD), p. 536
panic disorder, p. 536
personality disorder, p. 551
phobia, p. 536
positive symptoms, p. 546
psychopathology, p. 523
schizophrenia, p. 545
seasonal affective disorder (SAD), p. 542
sociocultural model, p. 530

Further Readings

Frith, C., & Johnstone, E. (2003). *Schizophrenia: A very short introduction.* New York: Oxford University Press.

Grandin, T. (1995). *Thinking in pictures.* New York: Doubleday.

Jamison, K. R. (1995). *An unquiet mind.* New York: Vintage Books.

Lezak, M. D. (2004). *Neuropsychological assessment* (4th ed.). New York: Oxford University Press.

Rapoport, J. (1990). *The boy who couldn't stop washing: The experience and treatment of obsessive-compulsive disorder.* New York: Penguin Books.

Sacks, O. (1985). *The man who mistook his wife for a hat.* New York: Harper & Row.

Styron, W. (1990). *Darkness visible: A memoir of madness.* New York: Random House.

Further Readings: A Canadian Presence in Psychology

Anisman, H., & McIntyre, D. C. (2002). Conceptual and cue learning in the Morris water maze in Fast and Slow kindling rats: Attention deficit comorbidity. *Journal of Neuroscience, 22*, 7809–7817.

Cjte, S., Tremblay, R. E., Nagin, D., Zoccolillo, M., & Vitaro, F. (2002). The development of impulsivity, fearfulness, and helpfulness during childhood: patterns of consistency in the trajectories of boys and girls. *Journal of Child Psychology and Psychiatry, 43*, 609–618.

Hare, R. D. (1993). *Without conscience: The disturbing world of the psychopaths among us.* New York: Pocket Books.

Hare, R. D. (1999). Psychopathy as a risk factor for violence. *Psychiatry Quarterly, 70*, 181–197.

Hare, R. D., McPherson, L. M., & Forth, A .E. (1988). Male psychopaths and their criminal careers. *Journal of Consulting and Clinical Psychology, 56*, 710–714.

Jang, K. L., Livesley, W. J., & Vernon, P. A. (2002). The etiology of personality function: The University of British Columbia Twin Project. *Twin Research, 5*, 342–346.

Maziade, M., Roy, M. A., Rouillard, E., Bissonnette, L., Roy, A., et al. (2001). A search for specific and common susceptibility loci for schizophrenia and bipolar disorder: A linkage study in 13 target chromosomes. *Molecular Psychiatry, 6*, 684–693.

Michalak, E. E., Tam, E. M., Majunath, C. V., Levitt, A. J., et al. (2004). Generic and health-related quality of life in patients with seasonal and non-seasonal depression. *Psychiatric Research, 128*, 245–251.

Offord, D. R., Boyle, M. H., Campbell, D., Goering, P., Lin, E., Wong, M., & Racine, Y. A. (1996). One-year prevalence of psychiatric disorder in Ontarians 15 to 64 years of age. *Canadian Journal of Psychiatry, 41*, 559–563.

Ross, C., Miller, S. D., Bjornson, L., & Reagor, P. (1991). Abuse histories in 102 cases of multiple personality disorder. *Canadian Journal of Psychiatry, 36*, 97–101.

Seeman, P. (2002). Atypical antipsychotics: Mechanism of action. *Canadian Journal of Psychiatry, 47*, 27–38.

Shear, M. K., & Maser, J. D. (1994). Standardized assessment for panic disorder research. A conference report. *Archives of General Psychiatry, 51*, 346–354.

Spanos, N. P., Weekes, J. R., Bertrand, L. D. (1985). Multiple personality: A social psychological perspective. *Journal of Abnormal Psychology, 94*, 362–376.

Wallace, S. T., & Alden, L. E. (1997). Social phobia and positive social events: The price of success. *Journal of Abnormal Psychology, 106*, 416–424.

Zai, G., Bezchlibnyk, Y. B., Richter, M. A., Arnold, P., Burroughs, E., Barr, C. L., & Kennedy, J. L. (2004). Myelin oligodendrocyte glycoprotein (MOG) gene is associated with obsessive-compulsive disorder. *American Journal of Medical Genetics, 129*, 64–68.

Using Alternative Therapies

An animal trainer at Dolphin Reef Eilat in Israel holds an autistic child during a therapeutic session involving a bottlenose dolphin. Alternative therapies such as this one are sometimes useful, but their effectiveness must be established through careful research. Various disorders of mind and body require different kinds of treatments. The choice of treatment is based on rigorous scientific observation.

CHAPTER 14

Treating Disorders of Mind and Body

Dennis was a 31-year-old insurance salesman. One day while shopping in a mall with his fiancée, Dennis suddenly felt very sick. His hands began to shake, his vision became blurred, and he felt a great deal of pressure in his chest. He began to gasp and felt weak all over. All of this was accompanied by a feeling of overwhelming terror. Without stopping to tell his fiancée what was happening, he ran from the store and sought refuge in the car. He opened the windows and lay down, and he started to feel better in about ten minutes. Later, Dennis explained to his fiancée what had happened and revealed that he had experienced this sort of attack before. Because of this, he would often avoid places like shopping malls. At the urging of his fiancée, Dennis agreed to see a psychologist. During his first several treatment sessions, Dennis downplayed his problems, clearly concerned that others might think him crazy. He was also reluctant to rely on medications because of the stigma attached. He

Outlining the Principles

How Is Mental Illness Treated?

- Psychotherapy Is Based on Psychological Principles
- Biological Therapies Are Necessary for Some Disorders
- Common Factors Enhance Treatment

What Are the Most Effective Treatments?

- Treatments That Focus on Behaviour and Cognition Are Superior for Anxiety Disorders
- Many Effective Treatments Are Available for Depression
- Lithium Is Most Effective for Bipolar Disorder
- Pharmacological Treatments Are Superior for Schizophrenia

Can Personality Disorders Be Treated?

- Dialectical Behaviour Therapy Is Most Successful for Borderline Personality Disorder
- Antisocial Personality Disorder Is Difficult to Treat

How Should Childhood Disorders Be Treated?

- Children with ADHD Can Benefit from a Variety of Approaches
- Autistic Children Benefit from a Structured Treatment Approach

had read about cognitive-behavioural therapy and was interested in trying this approach to address his problem.

His therapist explained that Dennis was experiencing panic attacks combined with agoraphobia. The therapist believed Dennis's problems were the result of vulnerability to stress combined with thoughts and behaviours that exacerbated anxiety. The first step in treatment was therefore relaxation training, to give Dennis a strategy to use when he became anxious and tense. The next step was to modify his maladaptive thought patterns. Dennis kept a diary for several weeks to identify situations in which distorted thoughts might be producing anxiety, such as meeting with prospective clients. Before meeting with a client, Dennis would become extremely anxious because he felt that it would be catastrophic if he was unable to make the sale. With the help of his therapist, Dennis came to recognize that being turned down by a client was difficult but manageable, and that it was unlikely to have any long-term impact on his career. The final phase of treatment was to address Dennis's avoidance of situations that he associated with panic attacks. Dennis and his therapist constructed a hierarchy of increasingly stressful situations. The first was an easy situation, involving a short visit to a department store on a lightly crowded weekday morning. Dennis's fiancée accompanied him so that he would feel less vulnerable. After completing this task, Dennis moved on to increasingly difficult situations, using relaxation techniques as necessary to control his anxiety. After six months, therapy was discontinued. Dennis's anxiety levels were significantly reduced and he was able to make himself relax when he did become tense. In addition, he had not experienced a panic attack during that period and was no longer avoiding situations that he had previously found stressful (Oltmans, Neale, & Davison, 1999). ■

In the last chapter you learned about various mental disorders. What can be done about them? Psychologists use two essential techniques to treat mental illness, psychological and biological. Research findings show that there is more than one way to treat most mental disorders. The generic name given to formal psychological treatment is **psychotherapy**, although the particular techniques and methods used depend on the theoretical orientation of the practitioner. **Biological therapies** reflect the medical approach to illness and disease. For example, *psychopharmacology,* the use of medications that affect brain or bodily functions, has proven to be very effective in treating many mental disorders. The success of medication in the treatment of mental illness is largely responsible for the era of deinstitutionalization, in which scores of patients were discharged from mental hospitals and treated with drugs as

psychotherapy The generic name given to formal psychological treatment.

biological therapy Treatment based on the medical approach to illness and disease.

460–377 B.C.E.
Trepanning Fossil evidence indicates that holes were drilled into human skulls, possibly to cure mental illness.

1700s
Humane Treatment Phillipe Pinel in France is among the first to release supposedly dangerous and violent mental patients from chains and manacles.

1800s
Reform in the United States Dorothea Dix promotes reform of mental institutions in the United States, leading to the establishment of the state hospital system.

1900
Dreams in Therapy Sigmund Freud advocates dream interpretation as part of psychoanalytic therapy.

outpatients (Lesage et al., 2000). The smaller number of inpatients meant that institutions could provide better treatment for those still under their care.

Both psychotherapy and biological therapies have proven to be beneficial to many of those who suffer from mental illness. The case of Dennis illustrates the effectiveness of cognitive-behavioural therapy, a form of psychotherapy. You will see throughout this chapter that this treatment is extremely effective for many mental disorders and that its effects are long-lasting. A limitation of biological therapies is that although they are often effective in the short term, long-term success requires the person to continue treatment, sometimes indefinitely. Recent focus has been on combining biological therapies with other therapeutic approaches to optimize treatment for each patient. This chapter explores the basic principles of therapy and describes the various treatment approaches to specific mental disorders.

How Is Mental Illness Treated?

As outlined in Chapter 13, a number of theories have been proposed to account for psychopathology. These approaches propose treatment strategies based on assumptions about the causes of mental disorders. However, even if we gain more understanding into the etiology of a mental disorder, this does not always give us further insights into how best to treat it. It is becoming clear that autism, for example, is caused by biological factors. Although this knowledge has helped parents accept that they are not responsible for their children's problems, it has not led to any significant advances in therapies for the disorder. In fact, the best available treatment at this point is based on behavioural, not biological, principles, as will discover later in this chapter.

Psychotherapy Is Based on Psychological Principles

Psychotherapy, regardless of the theoretical perspective of the treatment provider, is generally aimed at changing patterns of thought or behaviour, though the ways in which such changes are effected can differ dramatically. It has been estimated that there are over 400 approaches to treatment (Kazdin, 1994). The discussion that follows highlights the major components of the most common approaches and describes how therapists use these methods to treat specific mental disorders. Today, many practitioners use an eclectic mix of techniques based on what they believe is best for the client's particular condition.

PSYCHODYNAMIC THERAPY FOCUSES ON INSIGHT Among the first to develop psychological treatments for mental illness was Sigmund Freud, who believed

1940s
Treatment of Mental Illness Loses Ground Many overcrowded mental hospitals return to the use of physical restraints, creating barbaric conditions.

1948
Behaviourism as Therapy B. F. Skinner pioneers behaviour modification as a treatment for mental disorders.

1949
Lobotomies as Treatment Egas Moniz receives a Nobel Prize for his work on prefrontal lobotomies as a treatment for mental disorders, a technique later shown to cause further problems.

1950s
Drugs for Depression Antidepressant medications are discovered and revolutionize the treatment of depression, as their benefits outweigh some side effects.

1960s
The Cognitive Approach Aaron T. Beck and Albert Ellis develop cognitive therapy to treat dysfunctional thoughts and beliefs.

14.1 Psychoanalysis Freud sat behind his desk while his patients lay on the couch facing away from him, in order to reduce their inhibitions about revealing their unconscious beliefs and wishes.

that mental disorders were caused by prior experiences, particularly early traumatic experiences. Along with Josef Breuer, he developed the method of psychoanalysis. In early forms of psychoanalysis, the patient typically lay on a couch with the therapist sitting out of view, in order to reduce the patient's inhibitions and allow freer access to unconscious thought processes (Figure 14.1). Treatment was based on uncovering unconscious feelings and drives believed to give rise to maladaptive thoughts and behaviours. Techniques included *free association,* in which the patient says whatever comes to mind, and *dream analysis,* in which the therapist interprets the hidden meaning of dreams. The general goal of psychoanalysis is to increase patients' awareness of these unconscious processes and how they affect daily functioning. With this **insight**, or personal understanding of their own psychological processes, patients are freed from these unconscious influences, and symptoms disappear. Some of Freud's ideas have since been reformulated; these later adaptations are collectively known as *psychodynamic approaches.* Although the couch was replaced with a chair, proponents of the psychodynamic perspective continue to embrace Freud's "talking therapy," in a more conversational format.

What determines the type of treatment?

During the past few decades, the use of psychodynamic therapy has become increasingly controversial. Traditional psychodynamic therapy is expensive and time consuming, sometimes continuing for many years. Some psychological scientists question whether psychodynamic treatments are effective for treating serious forms of psychopathology. A new approach to psychodynamic therapy consists of offering a smaller number of sessions and focusing them more on current relationships than on early childhood experiences. Evidence from a number of studies indicates that this short-term psychodynamic therapy can be useful for treating a variety of psychological disorders, including depression, eating disorders, and substance abuse (Leichsenring, Rabung, & Leibing, 2004).

HUMANISTIC THERAPIES FOCUS ON THE WHOLE PERSON Recall from Chapter 12 that the humanistic approach to personality emphasizes personal experience and belief systems and the phenomenology of individuals. The goal of humanistic therapy is to treat the person as a whole, not as a collection of behaviours or a repository of repressed and unconscious thoughts. One of the best-known humanistic therapies is **client-centred therapy**, developed by Carl Rogers, which encourages people to fulfill their individual potentials for personal growth through greater self-

1970s
Focus on Interaction Interpersonal psychotherapy is used to identify and modify interpersonal problems.

1980s
The Era of Deinstitutionalization With the advent of psychotropic medications, patients are discharged from mental hospitals in large numbers and cared for through community-based treatment.

1987
Therapy on the Border Marsha Linehan introduces dialectical behaviour therapy for the treatment of borderline personality disorder.

1990s
The Failure of Deinstitutionalization? Inadequate support for those released from mental hospitals results in patients not taking their medications, an increase in unemployment, and a surge in the homeless population.

1990s
Prozac and Beyond Researchers pursue new drugs to treat mental illness that have increasing specificity and fewer side effects.

understanding. A key ingredient of client-centred therapy is a safe and comforting setting for clients to access their true feelings. Therapists strive to be empathic, to take the client's perspective, and to accept the client through unconditional positive regard. Rather than directing the client's behaviour or passing judgment on the client's actions or thoughts, the therapist helps the client focus on his or her subjective experience, often by using *reflective listening,* in which the therapist repeats the client's concerns in order to help the person clarify his or her feelings. One current treatment for problem drinkers, known as *motivational interviewing,* uses a client-centred approach over a very short period of time (such as one or two interviews). Motivational interviewing has proven to be a valuable treatment for drug and alcohol abuse, as well as for increasing healthy eating habits and exercise (Burke, Arkowitz, & Menchola, 2003). William Miller (2000) attributes the outstanding success of this brief form of empathic therapy to the warmth expressed by the therapist toward the client. Although relatively few practitioners follow the tenets of humanistic theory strictly, many of the techniques advocated by Rogers currently are used to establish a good therapeutic relationship between practitioner and client (Wystanski, 2000).

BEHAVIOURAL THERAPY FOCUSES ON OBSERVABLE BEHAVIOUR Whereas insight-based therapies consider maladaptive behaviour to be the result of an underlying problem, behavioural therapists see the behaviour itself as the problem and directly target it in therapy. The basic premise is that behaviour is learned and therefore can be unlearned using the principles of classical and operant conditioning (see Chapter 6). **Behaviour modification**, based on operant conditioning, rewards desired behaviours and ignores or punishes unwanted behaviours. In order for desired behaviour to be rewarded, however, the client must exhibit this behaviour. *Social-skills training* is an effective way to elicit desired behaviour. When clients have particular interpersonal difficulties, such as with initiating a conversation, they are taught appropriate ways of responding in specific social situations. The first step is often *modeling,* in which the therapist acts out appropriate behaviour. The client is encouraged to imitate this behaviour, rehearse it in therapy, and later apply the learned behaviour to real-world situations. An approach that integrates insight therapy with behavioural therapy is *interpersonal therapy* (Markowitz & Weissman, 1995), which focuses on relationships that the patient attempts to avoid. Because interpersonal functioning is seen as critical to psychological adjustment, treatment is focused on helping patients express their emotions and explore interpersonal experiences (Blagys & Hilsenroth, 2000).

Many behavioural therapies for phobia include **exposure**, in which the client is repeatedly exposed directly to the anxiety-producing stimulus or situation (Figure 14.2). The theory, based on classical conditioning, is that when clients avoid the feared stimuli or situations, they experience reductions in anxiety that reinforce avoidance behaviour. Repeated exposure to a feared stimulus increases the client's anxiety, but if the client is not permitted to avoid the stimulus, the avoidance response is eventually extinguished. Exposure methods vary in intensity. A gradual form of exposure therapy, known as **systematic desensitization**, uses relaxation techniques to allow the client to imagine anxiety-producing situations while maintaining relaxation. Relaxation is thereby paired with the feared situation via classical conditioning, gradually weakening or replacing the learned anxiety response.

Does treating the behaviour change the underlying causes of a disorder?

insight A goal of some types of therapy; a patient's understanding of his or her own psychological processes.

client-centred therapy An empathic approach to therapy that encourages personal growth through greater self-understanding.

behaviour modification Treatment in which principles of operant conditioning are used to reinforce desired behaviours and ignore or punish unwanted behaviours.

exposure A behavioural therapy technique that involves repeated exposure to an anxiety-producing stimulus or situation.

systematic desensitization An exposure technique that pairs the anxiety-producing stimulus with relaxation techniques.

14.2 **Exposure** A child is encouraged to approach a dog that scares her. This mild form of exposure teaches the girl that the dog is not dangerous, and she overcomes her fear.

COGNITIVE-BEHAVIOURAL THERAPY FOCUSES ON FAULTY COGNITIONS **Cognitive therapy** is based on the theory that distorted thoughts can produce maladaptive behaviours and emotions. Modifying these thought patterns via specific treatment strategies should eliminate the maladaptive behaviours and emotions. A number of approaches to cognitive therapy have been proposed. Aaron T. Beck (Figure 14.3), a leader in cognitive therapy, advocated **cognitive restructuring**, in which clinicians help their patients recognize maladaptive thought patterns and replace them with ways of viewing the world that are more in tune with reality. Albert Ellis, another major thinker in this area, introduced *rational-emotive therapy,* in which therapists act as teachers who explain and demonstrate more adaptive ways of thinking and behaving. In both types of therapies, maladaptive behaviour is assumed to result from individual belief systems and ways of thinking rather than from objective conditions.

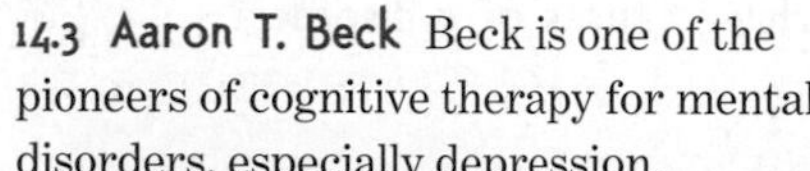

14.3 **Aaron T. Beck** Beck is one of the pioneers of cognitive therapy for mental disorders, especially depression.

Perhaps the most widely used version of cognitive therapy is **cognitive-behavioural therapy (CBT)**, which incorporates techniques from behavioural therapy and cognitive therapy. CBT tries both to correct faulty cognitions and to train clients to engage in new behaviours. For instance, people with social phobia, who fear negative evaluation, might be taught social skills. At the same time, the therapist helps them understand how their appraisals of other peoples' reactions to them might be inaccurate. CBT has proven to be one of the most effective forms of psychotherapy for many types of mental illness, especially anxiety disorders and mood disorders (Deacon & Abromowitz, 2004; Hollon, Thase, & Markowitz, 2002; Dobson & Khatri, 2000; Meichenbaum, 1977, 1993).

GROUP THERAPY BUILDS SOCIAL SUPPORT Group therapy rose in popularity after World War II, when there were more people needing therapy than there were therapists available to treat them. Subsequently therapists realized that group therapy offers advantages that make it preferable to individual therapy in some instances. The most obvious benefit is cost, since group therapy is often significantly less expensive than individual treatment. In addition, the group setting provides an opportunity for members to improve their social skills and to learn from each others' experiences. Group therapies vary widely in the types of patients enrolled in the group, the duration of the treatment, and the theoretical perspective of the therapist running the group. The size of the group also varies, although it is believed by some that eight patients is the ideal number. Many groups are organized around a particular type of problem (such as sexual abuse) or around a particular type of client (such as adolescents). Often, groups continue over long periods of time, with some members leaving and others joining the group at various intervals. Depending on the orientation of the therapist, the group may be highly structured or may be a more loosely organized forum for discussion. Behavioural and cognitive-behavioural groups are often highly structured, with specific goals and techniques designed to modify thought and behaviour patterns of group members. This type of group has been effective for disorders such as bulimia and obsessive-compulsive disorder. In contrast, less structured groups are often more focused on increasing insight and providing social support. The social support that group members can provide each other is one of the most beneficial aspects of group therapy, and attendance at group therapy is often used to augment individual psychotherapy.

cognitive therapy Treatment based on the idea that distorted thoughts produce maladaptive behaviours and emotions.

cognitive restructuring A therapy that strives to help patients recognize maladaptive thought patterns and replace them with ways of viewing the world that are more in tune with reality.

cognitive-behavioural therapy (CBT) A therapy that incorporates techniques from behavioural therapy and cognitive therapy to correct faulty thinking and change maladaptive behaviours.

expressed emotion A pattern of interactions that includes emotional over-involvement, critical comments, and hostility directed toward a patient by family members.

FAMILY THERAPY FOCUSES ON THE FAMILY CONTEXT Although the therapy a patient receives is an important element in treating mental disorders, the patient's family plays an almost equally important role. According to a *systems approach,* an individual is part of a larger context, and any change in individual behaviour will affect the whole system. This is often clearest at the family level. Within the family context, each person plays a particular role and interacts with the other members in specific ways. In the course of therapy, the way the individual thinks, behaves, and interacts with others may change, and this change could have profound effects on family dynamics.

Family members can have a tremendous impact on client outcomes. For instance, the importance of the family to the long-term prognosis of schizophrenia patients has been documented in studies of the attitudes expressed by family members toward the patient. Negative **expressed emotion** includes making critical comments about the patient, being hostile toward him or her, and being emotionally overinvolved. A number of studies have shown that families' levels of negative expressed emotion correspond to the relapse rate for patients with schizophrenia (Hooley & Gotlib, 2000). Schizophrenia patients released from the hospital to families with high levels of negative expressed emotion have high relapse rates, and the relapse rates are highest if the patient has a great deal of contact with the family (more than 35 hours per week) (Figure 14.4). Because family attitudes are often critical to long-term prognoses, some therapists insist that family members be involved in therapy. Indeed, evidence suggests that helping families provide social support leads to better therapy outcomes and reduces relapses.

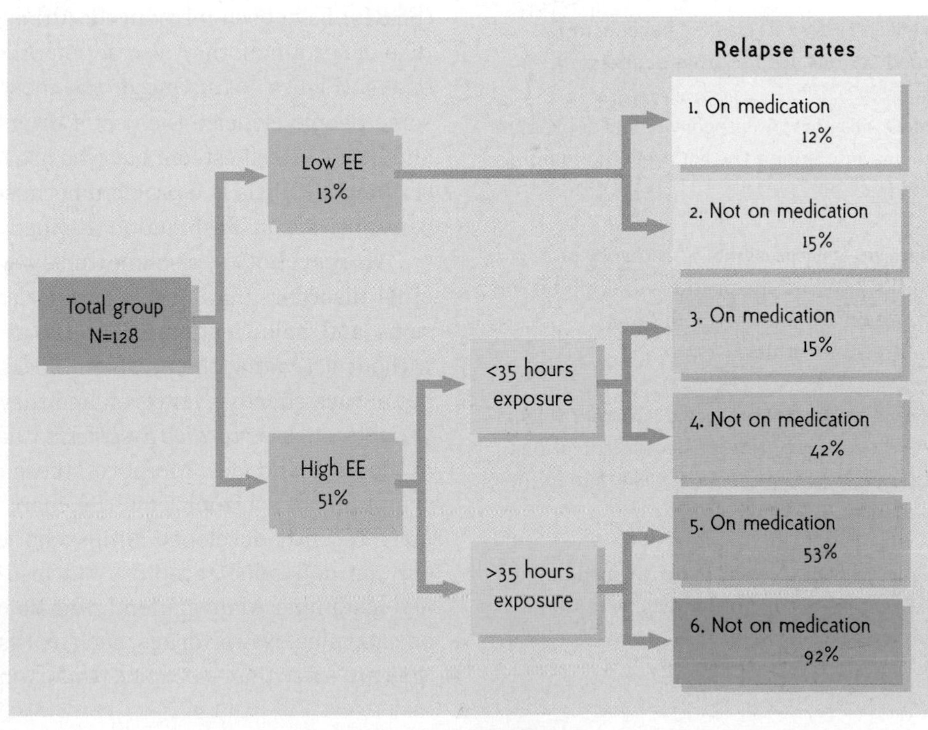

14.4 Relapse Rates After nine months, schizophrenic patients who were most exposed to family members high in negative expressed emotion (EE) were most likely to relapse, particularly if they were not on medication.

Biological Therapies Are Necessary for Some Disorders

Biological therapies are based on the notion that mental illness results from abnormalities in neural and bodily processes, such as imbalances in specific neurotransmitters or malfunctions in certain brain regions. Treatments range from drugs, to electrical stimulation of brain regions, to surgical interventions. In this section we will focus on the most widely used biological treatment. Drugs have proven to be quite effective for treating many mental disorders. Those that act on the brain to affect mental processes are known as **psychotropic medications**. They act by changing brain neurochemistry, for example by inhibiting action potentials or by altering synaptic transmission to increase or decrease the action of particular neurotransmitters.

Although numerous drugs are available, most psychotropic medications fall into three categories. **Anti-anxiety drugs**, commonly called tranquilizers, are used for the short-term treatment of anxiety. Benzodiazepines, such as Xanax and Ativan, increase activity of GABA, an inhibitory neurotransmitter (see Chapter 3). Although these drugs reduce anxiety and promote relaxation, they also induce drowsiness and are highly addictive. Another anti-anxiety medication, buspirone, has fewer side effects and does not appear to have the addictive potential of the benzodiazepines, but it must be taken on a daily basis whether one feels anxious or not.

A second class of psychotropic medications is the **antidepressants**, which as the name implies are used to treat depression. Monoamine oxidase (MAO) inhibitors were the first antidepressants to be discovered. *Monoamine oxidase* is an enzyme that converts serotonin into another chemical form. **MAO inhibitors** therefore result in more serotonin being available in the synapses of the brain. These drugs also raise levels of norepinephrine and dopamine. A second category of antidepressant medications is the **tricyclic antidepressants**, named after their core molecular structure of three rings. These drugs inhibit reuptake of a number of different neurotransmitters, and this inhibition results in more of each neurotransmitter being available in the synapse. More recently, **selective serotonin reuptake inhibitors (SSRIs)** have been introduced. Although these drugs do act by inhibiting the reuptake of serotonin, they also act on other neurotransmitters, to a significantly lesser extent (Figure 14.5). One of the most widely prescribed SSRIs is Prozac, though some people criticize the use of drugs such as Prozac to treat people who are sad and have low self-esteem but who are not clinically depressed. Such widespread prescribing of SSRIs is a problem because all drugs have some side effects, and SSRIs can cause sexual dysfunction (Kennedy et al., 2000).

Antipsychotics, also known as *neuroleptics,* are used to treat schizophrenia and other disorders that involve psychosis. These drugs reduce symptoms such as delusions and hallucinations. Traditional antipsychotics bind to dopamine receptors without activating them, which blocks the effects of dopamine. Antipsychotics are not always effective, however, and they have significant side effects that can be irreversible, such as *tardive dyskinesia,* the involuntary twitching of muscles, especially in the neck and face. Moreover, these drugs are not useful for treating the negative symptoms of schizophrenia (Seeman, 2002) (see Chapter 13). *Clozapine,* one of the more recently developed antipsychotics, is significantly different in that it acts not only on dopamine receptors but also on serotonin, norepinephrine, acetylcholine, and histamine. Many patients who do not respond to the other neuroleptics improve on clozapine. Newer drugs, such as Risperdal and Zyprexa, are widely used because they are safer than clozapine. Unfortunately, they may not be as effective.

psychotropic medications Drugs that affect mental processes.

anti-anxiety drugs A class of psychotropic medications used for the treatment of anxiety.

antidepressants A class of psychotropic medications used to treat depression.

MAO inhibitors A category of antidepressant drugs that inhibit the action of monoamine oxidase.

tricyclic antidepressants A category of antidepressant medications that inhibit the reuptake of a number of different neurotransmitters.

selective serotonin reuptake inhibitors (SSRIs) A category of antidepressant medications that prolong the effects of serotonin in the synapse.

antipsychotics A class of drugs used to treat schizophrenia and other disorders that involve psychosis.

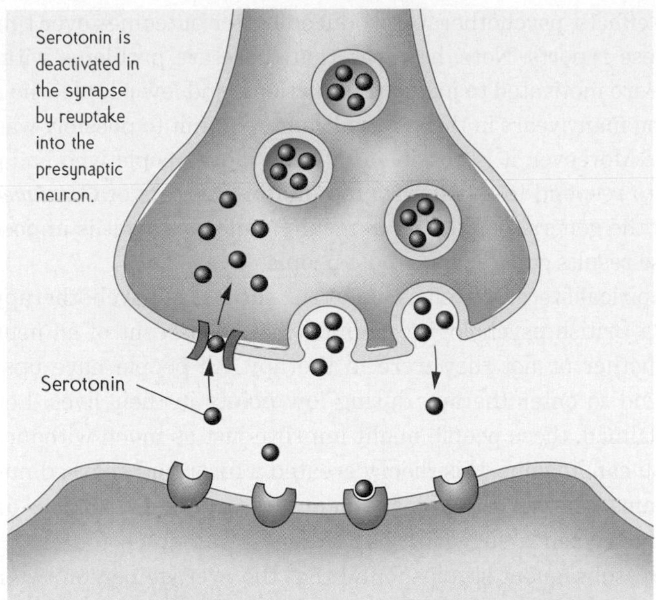
Serotonin is
deactivated in
the synapse
by reuptake
into the
presynaptic
neuron.
Serotonin

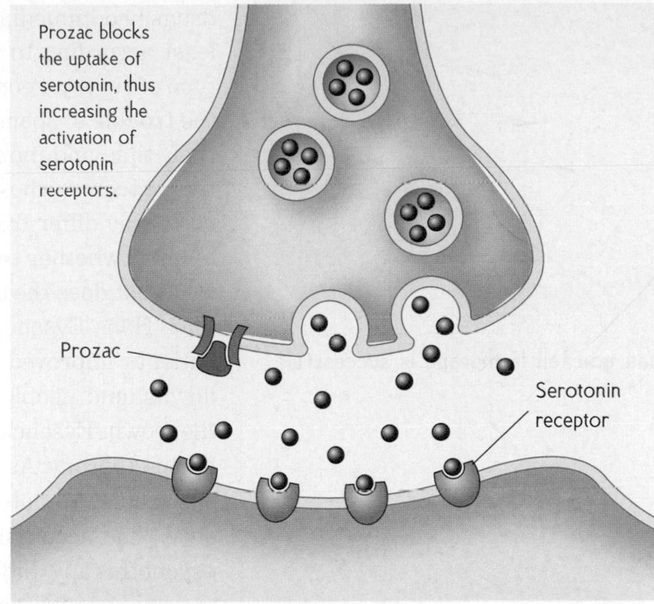
Prozac blocks
the uptake of
serotonin, thus
increasing the
activation of
serotonin
receptors.
Prozac
Serotonin
receptor

14.5 Selective Serotonin Reuptake Inhibitors SSRIs, such as Prozac, work by blocking reuptake into the presynaptic neuron, thereby allowing serotonin to remain in the synapse, where its effects on postsynaptic receptors are prolonged.

Other drugs used to treat mental illness do not fall into traditional categories. *Lithium* is the most effective treatment for bipolar disorder, although the neural mechanisms of how it works are currently unknown. Drugs that work to prevent seizures, called *anticonvulsants,* are also used to regulate moods in bipolar disorder.

Common Factors Enhance Treatment

Although the basic goals of all psychotherapeutic approaches are similar, the paths to these goals vary dramatically. Are some approaches better than others? Which approaches are more effective for specific types of clients or disorders? Perhaps most important, do psychotherapeutic approaches really do any good? These questions have been the focus of research and debate for many years.

EVALUATING THE EFFECTIVENESS OF THERAPY How can we determine if therapy is effective? One way is to ask people who have received therapy if they felt it helped them. In 1994, *Consumer Reports* magazine asked readers to evaluate their mental health treatments. Seven thousand readers responded. Of these, 2,900 sought treatment from mental health professionals, 1,300 joined self-help groups, and 1,000 consulted a family physician. The majority of respondents felt that intervention had helped them, and those who sought help from mental health professionals reported more positive results than those who consulted with a family doctor. No specific type of therapy yielded more positive results than any other, but duration of therapy did make a difference, with longer duration associated with better results. Patients whose choice of therapists or duration of treatment were limited for insurance reasons had the poorest outcome. In a more recent 2004 *Consumer Reports* survey, more than 3,000 readers reported their experiences with treatment for anxiety and mood disorders. Two notable findings emerged. First, many more people received drugs in 2004 (68 percent) compared with 1994 (40 percent), documenting the increasing use of medications to treat psychological disorders. Second, those who received both drugs and psychotherapy reported the best outcomes, followed by those receiving only psychotherapy and then those receiving only drugs. Although drugs had immediate effects, psychotherapy produced better outcomes over time, at least according to these reports. Note, however, that there are problems with this type of survey. People are motivated to justify their actions, and few people who go to the trouble of spending many years in therapy will want to admit to possibly wasting their time and money. Moreover, it is possible that only those people who had positive outcomes chose to respond to the survey. In addition, readers of *Consumer Reports* may differ from the general population in meaningful ways, so it is impossible to know whether these results generalize to most people.

How can you tell if therapy is successful?

What does the empirical literature tell us about the success of psychotherapy? In 1952, Hans Eysenck, a British psychologist, claimed that 75 percent of all neurotic patients improved whether or not they were in therapy. All people have ups and downs, and people tend to enter therapy during low points in their lives. Left on their own, Eysenck claimed, these people might improve just as much without ever having therapy. As you can imagine, this theory created a furor and sparked numerous studies to determine whether therapy for mental disorders did make a difference. The accumulated evidence suggests that Eysenck was much too critical of psychotherapy. Indeed, subsequent studies found that the average person receiving therapy was better off than those suffering from mental disorders who did not receive treatment.

In the 1970s, a number of landmark studies addressed the issue of treatment efficacy. One study followed college students who sought therapy at an outpatient psychiatric facility (Sloane, Staples, Whipple, & Cristol, 1977). Some students received either behavioural therapy or psychodynamic therapy, while the rest were placed on a waiting list for four months. After four months, both treatment groups had improved equally and both more than the control group. After one year, the group receiving behavioural therapy showed more improvement in the specific problem for which they had initially sought therapy, but students in both types of therapies demonstrated the same level of social adjustment. The conclusion from this study was that therapy helps, but the type does not matter. One year after the start of the study, the students were asked what factors they considered most important to the success of their treatment. Regardless of the type of therapy they had undergone, the responses all focused on the same factors: the insight gained into the problem, the relationship with the therapist, the opportunity to vent emotions, and the development of self-confidence. The critical role of the therapist is underscored in a study of 240 patients being seen by 25 therapists at 4 different treatment centres (Luborsky et al., 1986). Surprisingly, it was found that who performed the therapy was more important than the type of therapy. At all of the centres, a few therapists had better success rates than others regardless of the type of treatment or patient. However, as you will see throughout the remainder of this chapter, other research has shown that certain types of treatments are particularly effective for specific types of mental illness (Barlow, 2004). Although the effectiveness of these treatments is not due simply to the qualities of the therapist, *common factors* in therapy might facilitate positive outcomes regardless of the type of therapy used.

A CARING THERAPIST One factor known to affect the outcome of therapy is the relationship between the therapist and patient. This is not limited to mental illness, however—a good relationship with a service provider is important for any aspect of physical or mental health. A good relationship may be important to therapy in part because it can foster an expectation of receiving help (Miller, 2000; Talley, Strupp, Morey, 1990). Most people in the mental health field use the curative power of patient expectation to help their patients achieve success in therapy.

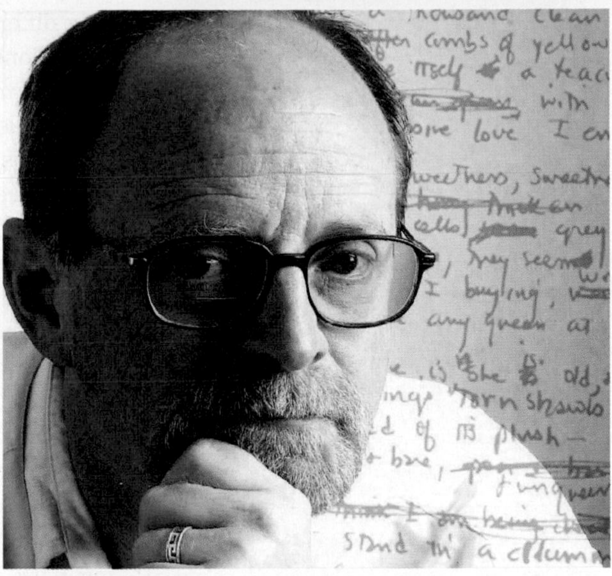

14.6 James Pennebaker Pennebaker has demonstrated that talking about personal traumas and secrets is beneficial to physical and mental health.

CONFESSION IS GOOD FOR THE SPIRIT Aristotle coined the term "catharsis" to describe the way certain messages evoke powerful emotional reactions and subsequent relief. Freud later incorporated this idea into his psychoanalytic approach to the treatment of mental disorders. He believed that uncovering unconscious material and talking about it would bring about catharsis and therefore relief from symptoms. Although other therapeutic approaches do not explicitly rely on this process, the opportunity to talk about one's problems to someone who will listen plays a role in all therapeutic relationships. Just the act of telling someone about your problems can have healing power. James Pennebaker (Figure 14.6) has explored this theory extensively in the laboratory. He finds that when people reveal intimate and highly emotional material, they go into an almost trancelike state. The pitch of their voices goes down and their rate of speech speeds up. In this seemingly hypnotic state, they lose track of time and place. Subsequent research has revealed the far-reaching effects of confession. Talking or writing about emotionally charged events reduces blood pressure, muscle tension, and skin conduction during the disclosure and immediately after (Pennebaker, 1990, 1995). Even writing about emotional topics via e-mail produces positive health outcomes (Sheese, Brown, & Graziano, 2004). In addition to these short-term benefits, there is evidence that writing about emotional events improves immune function, including the immune function of those with HIV (Petrie, Fontanilla, Thomas, Booth, & Pennebaker, 2004). Other research into the uses of such "confessional" therapies has shown that they can lead to better performance in work and school and improved memory and cognition. In addition, the opportunity to talk about troubling events may help clients reinterpret the events in less threatening ways, which is a central component of many cognitive therapies.

Psychological Science *in Action*

Choosing a Mental Health Practitioner

A dizzying array of providers offer treatment for mental disorders, ranging from those with limited training (such as former addicts who provide peer counseling) to those with advanced degrees in psychopathology and its treatment. In addition to mental health specialists, regular health-care providers (e.g., internists, pediatricians), human-service workers (e.g., school counselors), and volunteers (e.g., self-help groups) also provide assistance to those with mental problems.

An important aspect of this picture is that most treatments are based on cumulative research findings in psychological science. That is, no matter who administers the therapy, the specific techniques used have often emerged from psychological laboratories. Several therapies rely on behavioural-conditioning studies conducted by pioneers in psychology, such as John B. Watson and B. F. Skinner. Subsequent studies of cognitive processes provided many insights into the disturbed thinking patterns associated with many mental disorders. Even drug treatments are tested in psychological laboratories. For instance, psychological scientists—some of whom work in the pharmaceutical industry—use methods such as fear learning to examine how different drugs affect fear responses. The major types of specialized mental practitioners include the following:

Clinical psychologists typically have a doctoral degree. The graduate training for a Ph.D. takes four to six years and emphasizes research design and analysis. Many clinical psychologists work in academic or hospital settings where they conduct research in addition to providing treatment. A relatively new training program in clinical psychology leads to the Psy.D. This program tends to emphasize clinical skills over research and is meant for those who intend primarily to provide direct mental health services. Clinical psychologists typically are not able to prescribe medications, although efforts are under way to give them such privileges.

Psychiatrists have a medical degree (M.D.) and three to four additional years of specialized training in residency programs. They often work in hospitals or in private practice. Psychiatrists are the only mental health practitioners legally authorized to prescribe drugs in most states.

Psychiatric social workers most often have a master's degree in social work (M.S.W.), followed by specialized training in mental-health care. In addition to working with patients in psychiatric hospitals, they commonly visit people in their homes and deal with problems that arise from that environment. This might include helping the client receive appropriate resources from social and community agencies.

Psychiatric nurses typically have a bachelor's degree (B.S.N.) in nursing and special training in the care of mentally ill patients. They often work in hospitals or in treatment centres that specialize in serious mental illness.

Counseling psychologists often have a Ph.D. in counseling psychology. They typically deal with problems of adjustment and life stress that do not involve mental illness, such as stress related to school, marital, and occupational problems. Most universities have staff who specialize in problems common to students, such as text anxiety, learning disorders, sleep problems, and family issues.

Paraprofessionals have limited advanced training and usually work under supervision. They assist those with mental health problems in the challenges of daily living. For example, they may work in crisis intervention, pastoral counseling, or community outreach programs, or they may supervise clients of residential treatment centres.

How do people know they need therapy? How do they choose a therapist? Many times concerned family members, friends, or physicians encourage a person to seek help for psychological problems. For example, a doctor might ask a person who complains about being tired all the time whether she or he has been under stress or feeling sad, which might indicate depression. Of course, sometimes people know that they have a psychological problem, such as a fear of flying, and they will seek out a therapist on their own. Most university campuses have counselors who can direct students to appropriate treatment providers.

Choosing the right therapist is difficult but extremely important for ensuring successful treatment. The right therapist is someone with the appropriate training and experience for the specific mental disorder that needs to be resolved. It is critical that the person seeking help believe that the therapist is trustworthy and caring. The initial consultation should make the person feel at ease and hopeful that his or her psychological problem can be resolved. If not, the person should seek another therapist. The question of the ability to prescribe medication should play only a minor role in the choice of therapists, since almost all practitioners have arrangements with physicians who can prescribe medications if necessary. It is more essential to find someone who is both empathetic and experienced in the methods known to be effective in treating specific mental disorders.

The Internet provides a growing number of resources for those trying to deal with or resolve life problems on their own. These self-help programs are meant for minor stresses associated with daily living rather than for more serious mental disorders. ■

REVIEWING

the Principles | How Is Mental Illness Treated?

There are many ways to treat mental disorders. Psychotherapy uses psychological methods that are based on the practitioner's theoretical orientation. Some therapies help people gain insight into why they think, behave, and interact in certain ways. Other psychotherapy methods are more concerned with action than with insight and may try to correct faulty or biased thinking or to teach new behaviours. Psychopharmacology is based on the idea that maladaptive behaviour results from neurological dysfunction, and psychotropic medications are therefore aimed at correcting imbalances of neurotransmitters in the brain. Treatment of mental illness is often effective, perhaps in part because of commonalities among therapies, such as the client-practitioner relationship.

What Are the Most Effective Treatments?

Why is evidence important for choosing psychological treatments?

Earlier we mentioned that certain types of treatments are particularly effective for specific types of mental illness. Because outcomes are influenced by the interaction of client and therapist, it is difficult to make comparisons across disorders and therapists. Nonetheless, the accumulated evidence obtained by psychological scientists indicates that some treatments have empirical support for use with specific disorders and other do not. Moreover, the scientific study of treatment indicates that although some mental disorders are quite easily treated, others are not. For instance, we have highly effective treatments for anxiety disorders, mood disorders, and sexual dysfunction, but few treatments for alcoholism are superior to the natural course of recovery that happens for many people without treatment (Seligman, Walker, & Rosenhan, 2001).

As with all other areas of psychological science, the only way to know whether a treatment is valid is to conduct empirical research. David Barlow, a leading researcher in anxiety disorders, points out that findings from medical studies often lead to dramatic changes in treatment practice (Barlow, 2004). He provides the example of the sharp downturn in the use of hormone replacement therapy following evidence that it caused cardiovascular and neurological problems (see Chapter 3). Similarly, within a year after evidence emerged that arthroscopic knee surgery did not produce better outcomes than sham surgery (in which there was no actual procedure), its use declined dramatically. This reflects the increasing importance of *evidence-based treatments* in medicine. Barlow argues that the same ought to be true of the findings from psychological studies; he calls for the use of the term *psychological treatments* to distinguish evidence-based treatment from more generic "talk" psychotherapy.

Although a number of important issues surround the methods and criteria used to assess clinical research (e.g., Benjamin, 2005; Westen, Novotny, & Thompson-Brenner, 2004), according to Barlow three features characterize psychological treatments. First, they vary according to the particular mental disorder. Just as treatment for asthma differs from that for psoriasis, treatments for panic disorder are likely to differ from those for bulimia nervosa. Treatments should be tailored specifically to the specific psychological symptoms of a client. Second, the techniques used in these treatments have been developed in the laboratory by psychological scientists, especially behavioural, cognitive, and social psychologists. Finally, no overall grand theory guides treatment, but rather it is based on evidence of its effectiveness. In this section we examine the evidence to find out the treatment of choice for major mental disorders.

Treatments That Focus on Behaviour and Cognition Are Superior for Anxiety Disorders

Over the years, treatment approaches to anxiety disorders have met with mixed success. In the era when the classification of mental disorders was based on Freudian psychoanalytic theory, anxiety disorders were thought to be the result of repressed sexual and aggressive impulses. It was the underlying cause, rather than the specific symptoms, that was of interest to the therapist. Ultimately, psychoanalytic theory did not prove useful for treating anxiety disorders. The accumulated evidence suggests that cognitive and behavioural techniques work best to treat most anxiety disorders. The use of anxiety-reducing drugs is also beneficial in some cases, although there are risks of side effects as well as relapse after drug treatment is terminated. For instance, tranquilizers work for generalized anxiety disorder as long as the drug is taken, but they do little to alleviate the source of anxiety. By contrast, the effects of cognitive-behavioural therapy persist long after treatment.

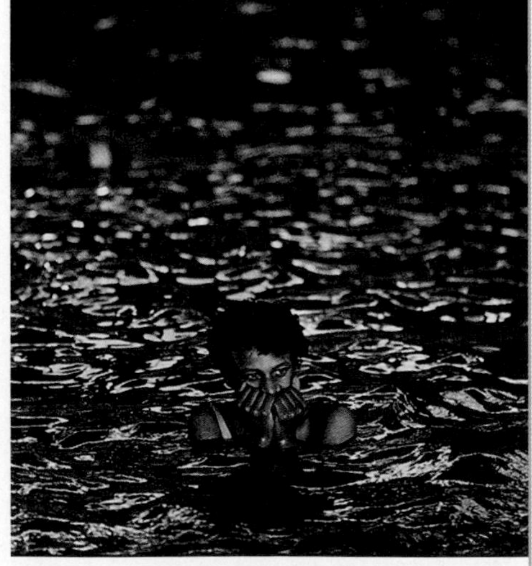

14.7 Treatments for Anxiety Disorders In one type of treatment, called exposure, the subject is forced to directly confront the source of his or her anxiety. This woman is gradually losing her fear of water.

SPECIFIC PHOBIAS Specific phobias are characterized by fear and avoidance of particular stimuli, such as heights, blood, and spiders. Learning theory suggests that these fears are acquired either through experiencing a traumatic personal encounter or by observing similar fear in others. As discussed in the preceding chapter, however, most phobias seem to develop in the absence of any particular precipitating event. Although the development of phobias cannot be completely explained by learning theory, behavioural techniques are the treatment of choice. In systematic desensitization therapy, the client first makes a *fear hierarchy,* a list of situations in which fear is aroused, in ascending order. The next step is relaxation training, in which clients learn to contrast muscular tension with muscular relaxation and to use relaxation techniques. Once the client has learned to relax, exposure therapy is often the next step (Figure 14.7). While the client is relaxed, he is asked to imagine or enact scenarios that become progressively more upsetting (Figure 14.8). New scenarios are not presented until the client is able to maintain relaxation at the previous levels. The theory behind this technique is that the relaxation response competes with and eventually replaces the previously exhibited fear response. The available evidence indicates that it is exposure to the feared object rather than the

Degree of fear	Anxiety hierarchy
5	I'm standing on the balcony of the top floor of an apartment tower.
10	I'm standing on a stepladder in the kitchen to change a lightbulb.
15	I'm walking on a ridge. The edge is hidden by shrubs and treetops.
20	I'm sitting on the slope of a mountain, looking out over the horizon.
25	I'm crossing a bridge 6 feet above a creek. The bridge consists of an 18-inch-wide board with a handrail on one side.
30	I'm riding a ski lift 8 feet above the ground.
35	I'm crossing a shallow, wide creek on an 18-inch-wide board, 3 feet above water level.
40	I'm climbing a ladder outside the house to reach a second-story window.
45	I'm pulling myself up a 30-degree wet, slippery slope on a steel cable.
50	I'm scrambling up a rock, 8 feet high.
55	I'm walking 10 feet on a resilient, 18-inch-wide board, which spans an 8-foot-deep gulch.
60	I'm walking on a wide plateau, 2 feet from the edge of a cliff.
65	I'm skiing an intermediate hill. The snow is packed.
70	I'm walking over a railway trestle.
75	I'm walking on the side of an embankment. The path slopes to the outside.
80	I'm riding a chair lift 15 feet above the ground.
85	I'm walking up a long, steep slope.
90	I'm walking up (or down) a 15-degree slope on a 3-foot-wide trail. On one side of the trail the terrain drops down sharply; on the other side is a steep upward slope.
95	I'm walking on a 3-foot-wide ridge. The slopes on both sides are long and more than 25 degrees steep.
100	I'm walking on a 3-foot-wide ridge. The trail slopes on one side. The drop on either side of the trail is more than 25 degrees.

14.8 Anxiety Hierarchy In systematic desensitization, the client creates a fear hierarchy, in which specific fears are graded from least to most threatening. This client is in therapy to conquer her fear of heights in order to go mountain climbing.

relaxation that extinguishes the phobic response. One way to expose people without putting them in danger is to use *virtual environments,* sometimes called virtual reality. Computers can simulate the environments and objects that are feared, such as by having a person *virtually* stand on the edge of a very tall building or fly in an aircraft (Figure 14.9). Evidence is impressive that exposure to these virtual environments can reduce fear responses (Rothbaum et al., 1999).

Used along with the behavioural methods, some cognitive strategies have also proven useful for the treatment of phobia. If clients are not aware that their fears are irrational, therapy would likely begin by increasing their awareness of the thought processes that maintain the fear of a particular stimulus.

Brain-imaging data indicate that successful treatment with cognitive-behavioural therapy alters how the fear stimulus is processed in the brain. In one study, research participants who were suffering from severe spider phobia received brain scans while looking at pictures of spiders (Paquette et al., 2003). Those whose treatment had been successful showed decreased activation in a frontal brain region involved

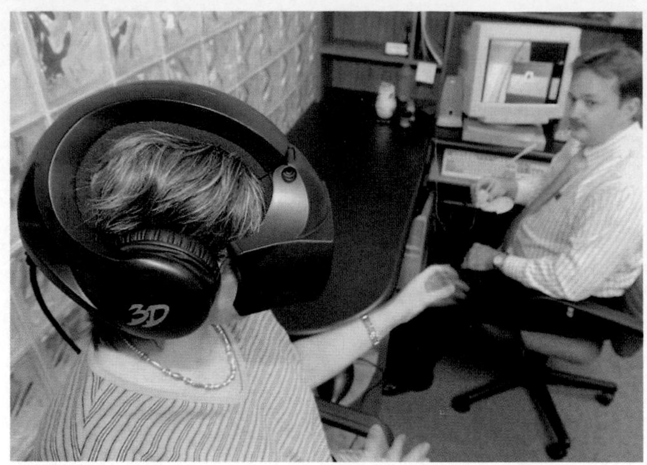

14.9 **Using Virtual Environments to Conquer Fear** Computer-generated images can simulate feared environments, such as heights, flying, or social interactions. Clients can conquer these virtual environments before taking on the fear object in real life.

in emotion regulation. These investigators suggest that psychotherapy effectively rewires the brain.

In terms of pharmacological treatments, tranquilizers can help people handle immediate fear, but as soon as the drug wears off, the phobia returns. Some recent studies have suggested that SSRIs might be useful for social phobia, which is an anxiety disorder characterized by overwhelming anxiety and a high level of self-consciousness in everyday social situations. Indeed, in the most comprehensive study conducted to date, researchers found that both Prozac and cognitive-behavioural therapy were effective in treating social phobia (Davidson et al., 2004). After fourteen weeks, symptoms exhibited by participants in the study did not differ. At the same time, those taking Prozac did have more physical complaints, such as a lack of sexual interest. The available evidence indicates that cognitive and behavioural therapies are the treatments of choice for phobia.

PANIC DISORDER Although many of us at some point experience some of the symptoms of a panic attack, we react to these symptoms in different ways. Some shrug off the symptoms, while others interpret heart palpitations as the beginnings of a heart attack, or hyperventilation as a sign of suffocation. Panic disorder is the result of multiple components, each of which may require a different treatment approach.

This clinical observation is supported by the finding that *imipramine,* a tricyclic antidepressant, prevents panic attacks but does nothing to reduce the anticipatory anxiety that clients have when they fear they might have a panic attack. In order to break the learned association between the physical symptoms and the feeling of impending doom, cognitive-behavioural therapy can be effective, as was illustrated in the opening case study of Dennis.

The most important psychotherapeutic methods for treating panic disorder are based on cognitive therapy. Cognitive restructuring addresses the ways the client reacts to the symptoms of panic attack. First, clients are asked to identify the specific fears they have. Often, people having a panic attack are convinced they are having a heart attack or that they are going to faint. Clients are then asked to estimate how many panic attacks they have had. The therapist helps them to assign percentages to specific fears and then to compare these numbers with the actual number of times these fears have been realized. When people are feeling anxious, they tend to overestimate the probability of danger, and this can contribute to rising feelings of panic. For example, a client might estimate that she fears a heart attack during 90 percent of her panic attacks and fainting during 85 percent of attacks. The therapist can then point out that the actual rate that these occurred was zero. In fact, people do not faint during panic attacks; the physical symptoms of panic attack are the reverse of fainting.

Even if clients recognize the irrationality of their belief systems, however, they often still suffer panic attacks. From a cognitive-behavioural perspective, the panic attacks continue because of a conditioned response to the trigger, whether it is shortness of breath or some other factor. The goal of therapy is to break the connection between the trigger symptom and the resulting panic. This can be done by exposure treatment, which induces feelings of panic, perhaps by asking the client to breathe in and out through a straw to induce hyperventilation or by spinning the client rapidly in a chair. Whatever the method, it is done repeatedly to induce habituation and then extinction. Cognitive-behavioural therapy appears to be as effective or perhaps more effective than medication in the treatment of panic attacks.

A study by David Barlow and his colleagues (2000) found that in the short term, both cognitive-behavioural therapy and imipramine were more effective than placebo for treatment of panic disorder and that they did not differ from one another in results. However, six months after treatment had ended, those who received psychotherapy were less likely to relapse than those who had taken medication. These findings indicate that cognitive behavioural therapy is the treatment of choice for panic disorder. For those who have panic disorder with agoraphobia, the combination of CBT and drugs is better than either treatment alone (Starcevic, Linden, Uhlenhuth, Kolar, & Latas, 2004).

14.10 Howie Mandel Comedian Howie Mandel is a diagnosed sufferer of obsessive-compulsive disorder. Like many people with OCD, Mandel suffers from mysophobia, or the fear of germs. He has admitted that his trademark shaved head helps him with this, as it makes him feel cleaner. Mandel even built a second, sterile house to which he can retreat if he feels he might be contaminated by anyone around him.

OBSESSIVE-COMPULSIVE DISORDER Obsessive-compulsive disorder (OCD) is a combination of recurrent intrusive thoughts (obsessions) and behaviours that the client feels compelled to perform over and over (compulsions) (Figure 14.10). Evidence that OCD is at least in part genetic and appears to be related to Tourette's syndrome, a neurological disorder characterized by motor and vocal tics, convinced many that OCD was a biological disorder that should respond to a biological treatment. However, traditional antianxiety drugs are completely ineffective for OCD. When SSRIs began to be used to treat depression, it was found that they were particularly effective in reducing the obsessional components of some depressions, such as constant feelings of worthlessness. Because of this, SSRIs were tried with clients suffering from OCD and found to be effective (Rapoport, 1989, 1991). The drug of choice for OCD is the potent serotonin reuptake inhibitor *clomipramine*. It is not a true SSRI, since it also blocks reuptake of other neurotransmitters, but its strong enhancement of the effects of serotonin appears to make it effective for OCD.

Researchers can learn a great deal about mental disorders by studying successful treatments. In obsessive-compulsive disorder, the effectiveness of SSRIs implicates the serotonin system. Despite the effectiveness of SSRIs, however, clinical studies have shown that 40 to 60 percent of patients with OCD do not improve when treated with these medications. For some of these patients, combining SSRIs with dopamine receptor blockers has been effective (Denys, Zohar, & Westenberg, 2004). It is interesting to note that the patients who benefit from this combination are those who also suffer from tics or who have a family history of tic disorders, such as Tourette's syndrome. Tic disorders are related to dopamine dysfunction; medications that block dopamine receptors are effective in suppressing tics (Shapiro et al., 1989). Genetic studies have hinted at a link between OCD and Tourette's syndrome, and it has been clinically observed that half to two-thirds of Tourette's patients also have OCD symptoms. Thus, this suggests dopamine plays a role in OCD.

Other research from a different angle implicates the hormone oxytocin in OCD. Researchers have found high levels of oxytocin in the cerebrospinal fluid of people with OCD (Leckman et al., 1994), but only those with no history of tic disorders; patients with tic-related OCD have normal oxytocin levels. Oxytocin receptors are found in a number of brain regions that have been implicated in OCD (Insel, 1992). Levels of oxytocin peak during the third trimester of pregnancy and remain high in breastfeeding women. Since higher levels of oxytocin are related to OCD, we might expect that pregnant and breastfeeding women would be prone to developing obsessive and compulsive symptoms. Anecdotal reports do suggest that women are more likely to develop OCD during these times, and for those who already have OCD, the higher levels of oxytocin during pregnancy and nursing can exacerbate symptoms (Epperson et al., 1995). Taken together, the findings on the roles of dopamine and oxytocin in OCD have an intriguing implication. There may be two distinct types of OCD: one tic related and one non-tic related. These two subtypes may well have

How might drug therapy inform us about the causes of mental disorders?

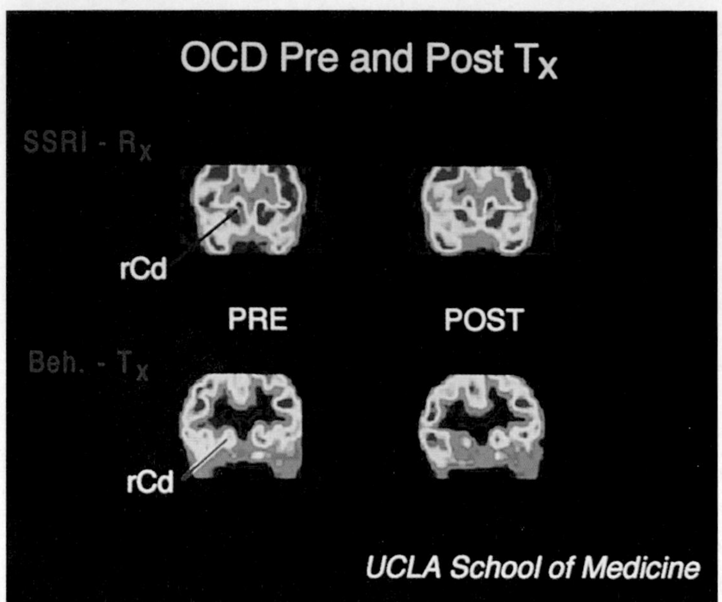

14.11 PET Scans of OCD Patients Arrows point to locations in the basal ganglia that showed similar changes with psychotherapy and drug therapy.

different neural underpinnings, and their differential reactions to medications suggest that this is true. Thus, how people respond to treatment can help us understand the possible causes of a mental disorder.

Cognitive-behavioural therapy is also effective for OCD and is especially valuable for those who do not benefit from or who do not want to rely on medication. Behavioural therapy for OCD differs from therapy for panic disorder in that relaxation training is not typically part of treatment. The two most important components of behavioural therapy for OCD are exposure and response prevention. Clients are directly exposed to the stimuli that trigger compulsive behaviour, but are prevented from engaging in the behaviour. This treatment is based on the theory that a particular stimulus triggers anxiety, and that performing the compulsive behaviour reduces that anxiety (see the hand-washing example in Chapter 13). For example, a client might compulsively wash his hands after touching a doorknob, using a public telephone, or shaking hands with someone. In exposure and response-prevention therapy, the client would be required to touch a doorknob and then would be instructed not to wash his hands afterward. As with exposure therapy for panic disorder, the goal is to break the conditioned link between particular stimuli and compulsive behaviour. Some cognitive therapies are also useful for OCD, such as helping the client learn that most people experience unwanted thoughts and compulsions from time to time. The goal is to help clients recognized that having unwanted negative thoughts is not a catastrophe but rather a normal part of human experience.

In the 1990s, researchers imaged the brains of patients with OCD who were being treated either with Prozac or with cognitive-behavioural therapy. As Figure 14.11 illustrates, patients in both treatment groups showed the same changes in neural activity in the thalamus and basal ganglia (Baxter et al., 1992; Schwartz, Stoessel, Baxter, Martin, & Phelps, 1996). Recall the findings that the spider phobia treatment led to changes in brain activity. These studies provide further evidence that nonbiological therapies change how the brain functions. How does drug treatment compare with cognitive-behavioural therapy for OCD? In one recent study the use of exposure and ritual prevention was found to be superior to the use of clomipramine, the drug of choice for OCD, although both were better than placebo (Figure 14.12; Foa et al., 2005). Cognitive-behavioural therapy may thus be a more effective way of treating OCD than medication, especially over the long term (see also Foa et al., 2005).

Many Effective Treatments Are Available for Depression

Depression, characterized by low mood and loss of interest in pleasurable activities, is one of the most widespread mental disorders among adolescents and adults, and it has become more common over the past few decades (Hollon et al., 2002). Fortunately, a number of effective treatments have been validated through scientific research. There is no "best" way to treat depression; many treatment approaches are available to the depressed patient, and ongoing research is determining which type of therapy works best for individual clients.

PHARMACOLOGICAL TREATMENT In the 1950s, tuberculosis was a major health problem in North America, particularly in urban areas. A common treatment was

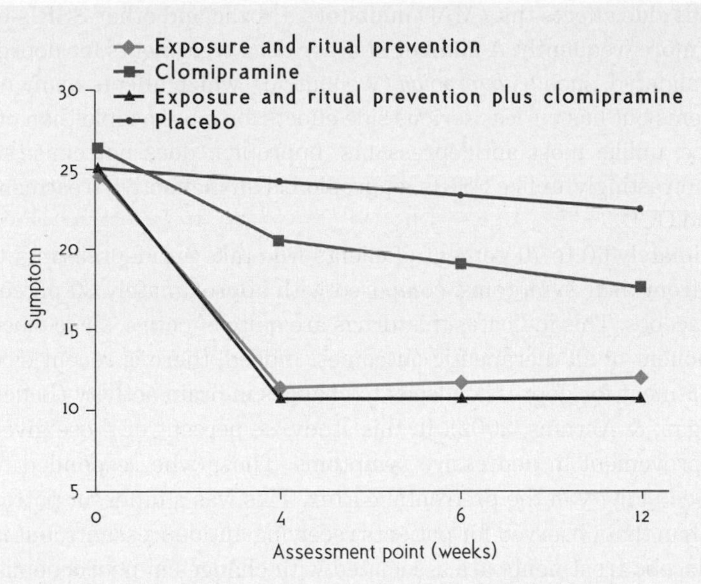

14.12 Treating Obsessive-Compulsive Disorder The most effective treatment for OCD in this study was exposure and ritual prevention. Adding the drug clomipramine did not reduce symptoms any further.

iproniazid, a drug that reduced tubercule bacilli in the sputum of patients. It also stimulated patients' appetites, increased their energy levels, and gave them an overall sense of well-being. In 1957, researchers who had noted the drug's effect on mood reported preliminary success in using it to treat depression. In the following year, nearly half a million depressed patients were given iproniazid, which is an MAO inhibitor. Although they can relieve depression, MAO inhibitors can be toxic because of their effects on a variety of physiological systems. Patients taking these drugs must avoid ingesting any substances containing tyramine, including red wine and aged cheeses, because they can experience severe, sometimes lethal elevations in blood pressure. Interactions with both prescription and over-the-counter medications can also be fatal, so MAO inhibitors are generally reserved for patients who do not respond to other antidepressants.

Another type of antidepressant, *tricyclics,* was also identified in the 1950s. The first tricyclic to be identified—imipramine, a drug developed as an antihistamine—was found to be effective in relieving clinical depression. This drug and others like it act on neurotransmitters as well as on the histamine system. Tricyclics are extremely effective antidepressants. However, as a result of their broad-based action, they have a number of unpleasant side effects. Because of their effect on the histamine system, patients often experience sedation and weight gain. Sweating, constipation, heart palpitations, and dry mouth can result from the effects of tricyclics on acetylcholine.

The discovery of these early antidepressants was largely serendipitous, but subsequently researchers began to search for antidepressants that did not affect multiple physiological and neurological systems and so would not have such troublesome side effects. In the 1980s, researchers discovered fluoxetine hydrochloride, better known as Prozac, which is an SSRI. Because this drug does not affect histamine or cholinergic systems it has none of the side effects associated with the tricyclic antidepressants, although it can occasionally cause insomnia, headache, weight loss, and sexual dysfunction. Because they have

"Before Prozac, she loathed company."

fewer serious side effects than MAO inhibitors, Prozac and other SSRIs began to be prescribed more frequently. A number of other drug treatments for depression have also been validated, such as *bupropion* (Wellbutrin), which affects many neurotransmitter systems but has far less serious side effects for most people than other drugs. For instance, unlike most antidepressants, bupropion does not cause sexual dysfunction. Interestingly, unlike SSRIs, bupropion is an ineffective treatment for panic disorder and OCD.

Approximately 60 to 70 percent of clients who take antidepressants will experience relief from their symptoms, compared with approximately 30 percent who respond to placebos. This indicates that drugs are quite effective, since placebo effects are a component of all therapeutic outcomes. Indeed, there is recent evidence that placebo treatment for depression leads to changes in brain activity (Leuchter, Cook, Witte, Morgan, & Abrams, 2002). In this study, 38 percent of those given placebos showed improvement in depressive symptoms. Those who responded to placebos had increased activity in the prefrontal cortex. This was a different pattern of brain activation than that observed for patients receiving antidepressants, but it does suggest that placebo treatments are associated with changes in neurochemistry, which supports the findings that they do alleviate symptoms for some people. Placebos may work by giving people hope that they will feel better, and these positive expectancies may alter brain activity.

Despite attempts to predict patient response to antidepressants, physicians must often resort to a trial-and-error approach in treating depressed patients. At this time no single drug stands out as being most efficacious. Some evidence suggests that tricyclics might be preferred for the most serious forms of depression, especially for patients who are hospitalized (Anderson, 2000), but SSRIs are generally considered first-line medications because they have the fewest serious side effects (Olfson et al., 2002). The decision of which drug to use often depends on the client's overall medical health and the possible side effects of each medication. Once the depressive episode has ended, should patients continue taking medication? Research has shown that patients who continue taking medication for at least a year have only a 20 percent relapse rate, whereas those who are maintained on a placebo have an 80 percent relapse rate (Frank et al., 1990).

COGNITIVE-BEHAVIOURAL TREATMENT OF DEPRESSION Despite the success of antidepressant medications, not all patients benefit from these drugs. Moreover, others cannot or will not tolerate the side effects associated with drug therapy. Fortunately, the available evidence indicates that cognitive-behavioural therapy is just as effective as biological therapies in treating depression (Hollon et al., 2002). From a cognitive perspective, people who become depressed do so because of automatic, irrational thoughts. According to the cognitive distortion model of Aaron Beck, one of the most influential thinkers in this area, depression is the result of a cognitive triad of negative thoughts about oneself, the situation, and the future. The thought patterns of depressed patients are different from those of people with anxiety disorders. Whereas patients with anxiety disorders worry about the future, depressed people think about how they have failed in the past, how poorly they are dealing with the present situation, and how terrible the future will be.

What factors need to be considered when deciding between drug treatments and psychological treatments?

The goal of the cognitive-behavioural treatment of depression is to help the patient think more adaptively, which in turn should improve mood and behaviour. Although the specific nature of the treatment will be adapted to each individual patient, there are some general principles of this type of therapy for depression. Patients may be asked to recognize and record negative thoughts (Figure 14.13).

Thinking about situations in a negative manner can become automatic, and recognizing these thought patterns can be difficult. Once the patterns are identified and monitored, the clinician can help the patient recognize that there are other ways of viewing the same situation—ways of thinking that are not so dysfunctional.

Although cognitive-behavioural therapy can be effective on its own, research shows that combining it with antidepressant medication is significantly more effective than either one of these approaches alone (McCullough, 2000). In addition, the response and remission rates of the combined-treatment approach were higher than any ever reported for depression (Keller et al., 2000; Kocsis et al., 2003). The issue is not drugs versus psychotherapy, but rather identifying the specific treatments from the available options that provide relief for individual clients. For instance, for clients who are suicidal, in acute distress, or unable to commit to regular attendance with a therapist, drug treatment may be most effective. For many others, especially those who have physical problems, psychotherapy may be the treatment of choice because it is long lasting and does not have the side effects associated with medications. As is the case with other mental disorders, treatment of depression with psychotherapy leads to changes in brain activation similar to those observed in drug treatments (Brody et al., 2001). However, one recent study found that although psychotherapy and drugs involved the same brain regions, activity in those regions was quite different during the two treatments, which suggests that they operate through different mechanisms (Goldapple et al., 2004).

Date	Event	Emotion	Automatic thoughts
April 4	Boss seemed annoyed.	Sad, anxious, worried	Oh, what have I done now? If I keep making him mad, I'm going to get fired.
April 5	Husband didn't want to make love.	Sad	I'm so fat and ugly.
April 7	Boss yelled at another employee.	Anxious	I'm next.
April 9	Husband said he's taking a long business trip next month.	Sad, defeated	He's probably got a mistress somewhere. My marriage is falling apart.
April 10	Neighbor brought over some cookies.	A little happy, mostly sad	She probably thinks I can't cook. I look like such a mess all the time. And my house was a disaster when she came in!

14.13 Recording Thoughts A record of automatic thoughts used in cognitive-behavioural therapy for depression.

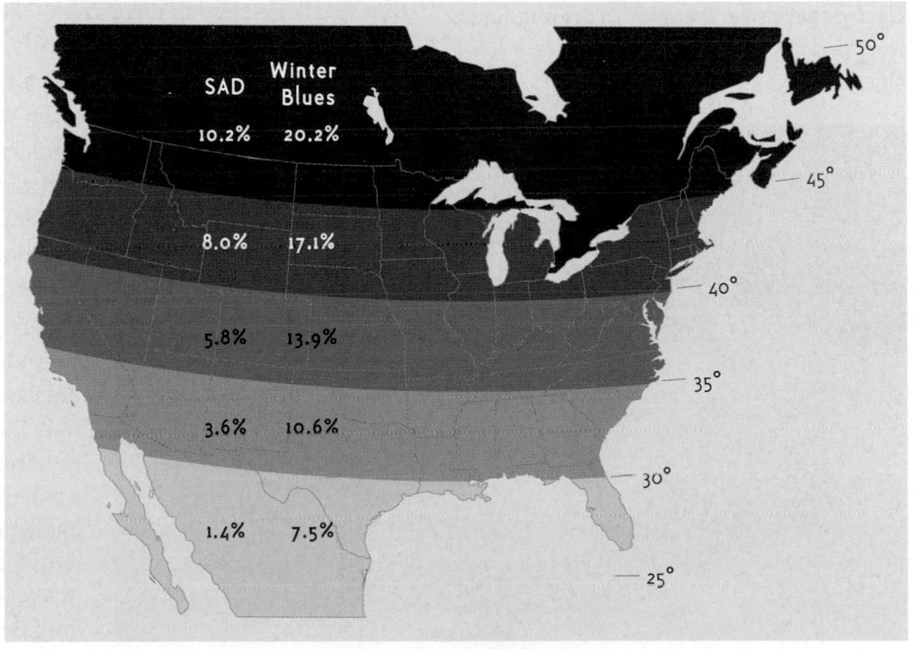

14.14 Seasonal Affective Disorder Incidents of SAD vary by latitude, being much more common in areas that receive less sunlight during the winter.

ALTERNATIVE TREATMENTS In some patients, episodes of depression occur during the winter. The rate of this disorder, known as seasonal affective disorder (SAD), increases with latitude, with rates highest in regions with the fewest hours of daylight during the winter (Figure 14.14). Many of these patients respond favourably to phototherapy, in which patients are exposed to a high-intensity light source for a period of time each day (Figure 14.15). Patients who crave carbohydrates and sleep a great deal in the winter seem to respond better to phototherapy than do SAD patients who do not exhibit these symptoms.

For some depressed patients, regular aerobic exercise can reduce depression and prevent recurrence (Pollock, 2004). It has been speculated that aerobic exercise reduces depression because it releases endorphins, which are chemically related to norepinephrine, a neurotransmitter implicated in depression. Release of endorphins can result in an overall feeling of well-being, which marathoners may recognize as the "runner's high." Aerobic exercise may also serve to regularize bodily rhythms, improve self-esteem, and provide social support if people exercise with others. For depressed patients, however, it may be difficult to find the energy and motivation to begin an exercise regime.

14.15 Phototherapy This woman sits in front of strong lighting for several hours each day to reduce her symptoms of SAD.

In addition to these somewhat benign interventions, more drastic treatments may be warranted for some depressed patients. In the 1930s, **electroconvulsive therapy (ECT)** was developed in Europe and tried on the first human in 1938. ECT involves placing electrodes on the patient's head and administering an electrical current strong enough to produce a seizure (Figure 14.16). Although ECT frequently results in an amelioration of depressed mood, the mechanism by which this occurs is unknown (Fink, 2001; Enns & Reiss, 1992). ECT may affect neurotransmitters or the neuroendocrine system; it has been shown to increase levels of acetylcholine, and drugs that block the action of acetylcholine reverse the beneficial effects of ECT. However it works, ECT is the single most effective treatment for those who are severely depressed and do not respond to conventional treatments (Hollon et al., 2002).

The general public views ECT in a predominately negative manner. The 1975 film *One Flew over the Cuckoo's Nest* did a great deal to expose the abuses in mental health care and graphically depicted ECT as well as the tragic effects of lobotomy. Although care for the mentally ill is still far from perfect, many reforms have been implemented. ECT is now generally done under anesthesia with powerful muscle relaxants. This essentially eliminates the motor convulsions and confines the seizure to the brain. For a number of reasons, ECT might be preferable to other treatments for depression. Antidepressant medication can take weeks to be effective, whereas ECT works quickly. For a suicidal patient, waiting several weeks for relief can literally be deadly. In addition, ECT may be the treatment of choice for depression in pregnant women, since there is no evidence that the seizures harm the developing fetus. Many psychotropic medications, on the other hand, can cause birth defects. Most important, ECT has proven effective in clients for whom other treatments have failed, providing an effective last resort for those who would otherwise continue to suffer.

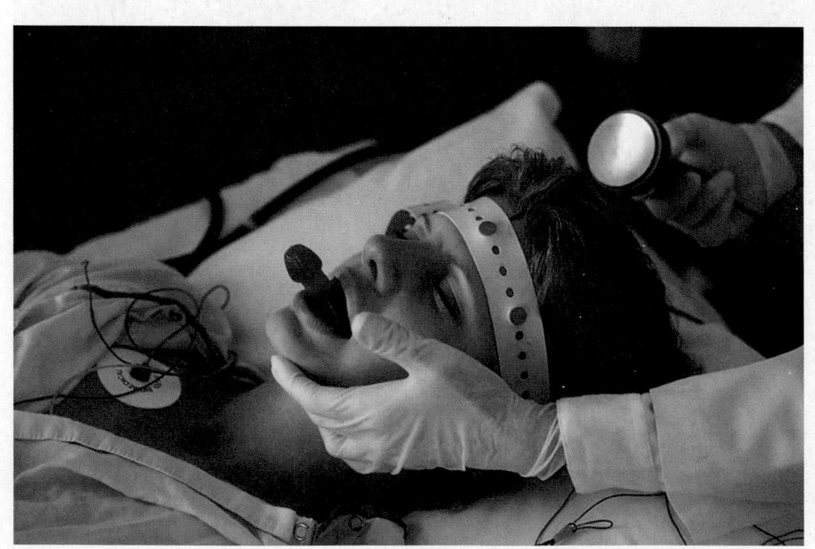

14.16 A Woman Being Prepared for Electroconvulsive Therapy To prevent her from possibly swallowing her tongue, a soft object is placed between her teeth.

ECT does, however, have some serious limitations, including a high relapse rate, often necessitating repeated treatments, and memory impairments (Fink, 2001). In most cases, memory loss is transient and is limited to the day of ECT treatment. Some patients, however, experience substantial memory loss that can be permanent (Donahue, 2000). Some centres perform unilateral ECT over only the hemisphere that is not dominant for language, which seems to reduce any memory disruption (Papadimitrious, Zervas, & Papakostas, 2001). New research suggests that the degree of memory and cognitive impairment resulting from ECT may be related to levels of cortisol, a hormone released in response to physical and psychological stress (see Chapter 10). Depressed patients with higher levels of cortisol show greater impairments in memory and cognitive functions (Neylan et al., 2001). This research may help to identify patients at risk for memory and cognitive losses due to ECT, and these potentially serious side effects can be weighed against the benefits of the treatment.

Recently, research has explored whether **transcranial magnetic stimulation (TMS)** can reduce depressive symptoms. In this procedure, an electromagnetic coil is placed on the scalp and transmits pulses of high-intensity magnetism. The rapid buildup and collapse of the magnetic field induces a momentary electrical current in the brain. The net result is that normal brain activity is disrupted in the brain region beneath the coil. In effect, the coil creates a brief, transient brain lesion. A series of studies has demonstrated that TMS over the left frontal regions results in a significant reduction in depression (Chistyakov et al., 2004; George, Lisanby, & Sackheim, 1999; George et al., 1995; Pascual-Leone, Catala, & Pascual-Leone, 1996). Because TMS does not involve anaesthesia and does not have any major side effects (other than headache), it can be administered outside of hospital settings. It is not likely, however, that TMS will ever completely replace ECT, since the two methods may act via different mechanisms and may therefore be appropriate for different types of patients. TMS seems to be more effective for nonpsychotic depression, whereas ECT seems to be more effective for psychotic depression (George et al., 1999). The long-term value of TMS is that it is effective even for patients who have not responded to treatment with antidepressants (Fitzgerald et al., 2003). This highlights the idea that just as there are multiple causes of depression, so are there multiple effective treatments. The challenge for practitioners is to find the treatment that works best for a particular individual.

electroconvulsive therapy (ECT) A procedure used to treat depression that involves administering a strong electrical current to the patient's brain.

transcranial magnetic stimulation (TMS) A procedure that transmits pulses of high-intensity magnetism to the brain.

lithium A psychotropic medication used to treat bipolar disorder.

Lithium Is Most Effective for Bipolar Disorder

Although major depression and bipolar disorder are both disorders of mood, they are fundamentally different and require different treatments. Bipolar disorder, in which mood cycles between mania and depression, is one of the few mental disorders for which there is a clear optimal treatment (Figure 14.17). Psychotropic medications have been found to be the most effective way to treat this disorder, especially **lithium** (Geddes, Burgess, Hawton, Jamison, & Goodwin, 2004). Only about 20 percent of patients maintained on lithium experience relapses (Keller & Baker, 1991). The mechanisms by which lithium stabilizes mood are not well understood, but the drug seems to modulate neurotransmitter levels, balancing excitatory and inhibitory activities (Jope, 1999). As with other psychotropic drugs, the discovery of lithium for the treatment of bipolar disorder was serendipitous. In 1949, an Australian named John Cade found that the urine of manic patients was toxic to guinea pigs. He postulated that this effect might be due to uric acid. When he gave lithium urate, a salt in uric acid, to the guinea pigs, however, it was not toxic. To his surprise, it protected them against the toxic effects of the manic patients' urine and also sedated them. He next tried lithium salts on himself. When he was assured of their safety, he gave the salts to ten hospitalized manic patients, all of whom recovered rapidly. Lithium does, however, have unpleasant side effects, including thirst, hand tremors, excessive urination, and memory problems. These side effects often diminish after several weeks on the drug.

14.17 Margaret Trudeau The former wife of the late Prime Minister Pierre Trudeau has been diagnosed with bipolar disorder. After her son, Michel, tragically died in a skiing accident in 1998, and Pierre passed away in 2000, she finally sought the appropriate medical treatment. She now publicly advocates that others who live with mental illness get treated, and stresses that there is no shame in coming forward for help.

Although lithium is effective in stabilizing mood, it works better on mania than on depression, and patients are often treated both with lithium and with an antidepressant. In this regard, the available evidence indicates that selective serotonin reuptake inhibitors (SSRIs) are preferable to other antidepressants because they are less likely to trigger episodes of mania (Gijsman, Geddes, Rendell, Nolen, & Goodwin, 2004). Anticonvulsive medications, such as Depakote, also can stabilize mood and may be effective for intense bipolar episodes. As with all psychological disorders, compliance with drug therapy can be a problem for a variety of reasons (Ahrens et al., 1995). Patients may skip doses or stop taking the medications completely in an effort to reduce the side effects of the drugs. Cognitive-behavioural therapy can be an effective way to increase compliance with medication regimes (Miller, Norman, & Keitner, 1989). Patients with bipolar disorder also may stop taking their medications because they miss the "highs" of their hypomanic and manic phases. Recall from Chapter 13 Kay Redfield Jamison's description of the intoxicating pleasure of mania. Psychological therapy can help patients accept their need for medication and understand the impact their disorder has on them and on those around them.

Pharmacological Treatments Are Superior for Schizophrenia

Freud's psychoanalytic theory and treatment were widely touted as the answer to many mental disorders in the early 1900s. Even Freud, however, admitted that his techniques were effective only for what he termed "neuroses" and were unlikely to benefit patients with more severe psychotic disorders such as schizophrenia. Because psychotic patients were difficult to handle and even more difficult to treat, they were generally institutionalized in large mental hospitals. This set up a situation in which the staff and administration of mental hospitals were willing to try any treatment that was inexpensive and had a chance of decreasing the patient population, or that at the very least might make the inmates more manageable.

Brain surgery was considered a viable option in the management of patients with severe mental illness. Although some brain surgeries were performed in the early 1880s, Egas Moniz is generally credited with bringing the practice to the attention of the medical world in the 1930s. His surgical procedure, later known as lobotomy, severed nerve-fibre pathways in the prefrontal cortex. Although Moniz initially reported that the operation was frequently successful, it soon became evident to him that the patients who benefited most from the surgery were those who were anxious or depressed. Schizophrenic patients did not seem to improve following the operation. With the introduction of psychotropic medications in the 1950s, lobotomies were virtually discontinued (Jasper, 1995).

PHARMACOLOGICAL TREATMENT It had been known since the sixteenth century that extracts from dogbane, a toxic herb, could calm highly agitated patients. The critical ingredient was isolated in the 1950s and was named reserpine. When given to schizophrenic patients, it not only had a sedative effect, it also was an effective antipsychotic, reducing the positive symptoms of schizophrenia, such as delusions and hallucinations. Shortly afterward, a synthetic version of reserpine was created that had fewer side effects. This drug, called *chlorpromazine,* acts as a major tranquilizer. It reduces anxiety, sedates without inducing sleep, and decreases the severity and frequency of the positive symptoms of schizophrenia. Later, another antipsychotic, *haloperidol,* was developed that was chemically different and had less of a sedating effect than chlorpromazine.

Haloperidol and chlorpromazine revolutionized the treatment of schizophrenia and became the most frequently used therapy for the disorder. Schizophrenic patients who had been hospitalized for years were able to walk out of mental institutions and live independently. These antipsychotics are not without drawbacks, however. The medications have little or no impact on the negative symptoms of schizophrenia, such as apathy and social withdrawal. In addition, they have significant side effects. Chlorpromazine sedates patients, can cause constipation and weight gain, and causes cardiovascular damage. Although haloperidol does not cause these symptoms, both drugs have significant motor side effects that resemble Parkinson's disease. Immobility of facial muscles, trembling of extremities, muscle spasms, uncontrollable salivation, and a shuffling walk can all occur as a result of antipsychotic medication. **Tardive dyskinesia**, involuntary movements of the lips, tongue, face, legs, or other parts of the body, is another devastating side effect of these medications and is irreversible once it appears (Yassa et al., 1990). Despite these debilitating side effects, chlorpromazine and haloperidol were essentially the only medications available to schizophrenia patients for many years.

In the late 1980s, clozapine was introduced. **Clozapine** was significantly different from the previous antipsychotic medications in a number of ways. First, it acted not only on dopamine receptors, but also on those for serotonin, norepinephrine, acetylcholine, and histamine. Second, it was beneficial in treating the negative symptoms of schizophrenia as well as the positive symptoms. Many patients who had not responded to the previously available neuroleptics improved on clozapine (Figure 14.18). Third, there was no evidence of Parkinsonian symptoms or of tardive dyskinesia in any of the patients taking the drug. Clozapine has fewer side effects than either chlorpromazine or haloperidol, but its side effects are serious. Clozapine is associated with seizures, heart arrhythmias, and substantial weight gain. Of even greater concern, clozapine can cause a fatal reduction in white blood cells. Although the risk of this is low, frequent blood tests are required for patients taking the drug. More recently, other medications similar to clozapine in structure and pharmacology have been introduced. These medications, such as risperidol and olanzapine, appear to be as effective as clozapine but do not reduce white blood cell counts. These new drugs, called second-generation antipsychotics, are now the first line in treatment for schizophrenia (Walker, Kestler, Bollini, & Hochman, 2004). A recent analysis of 11 separate studies containing 2,769 patients found that these second-generation antipsychotics had about one-fifth the risk of producing tardive dyskinesia as first-generation drugs (Correll, Leucht, & Kane, 2004).

tardive dyskinesia A side effect of some antipsychotic medications that produces involuntary movements of the lips, tongue, face, legs, or other parts of the body.

clozapine An antipsychotic medication that acts on multiple neurotransmitter receptors and is beneficial in treating both the negative and positive symptoms of schizophrenia.

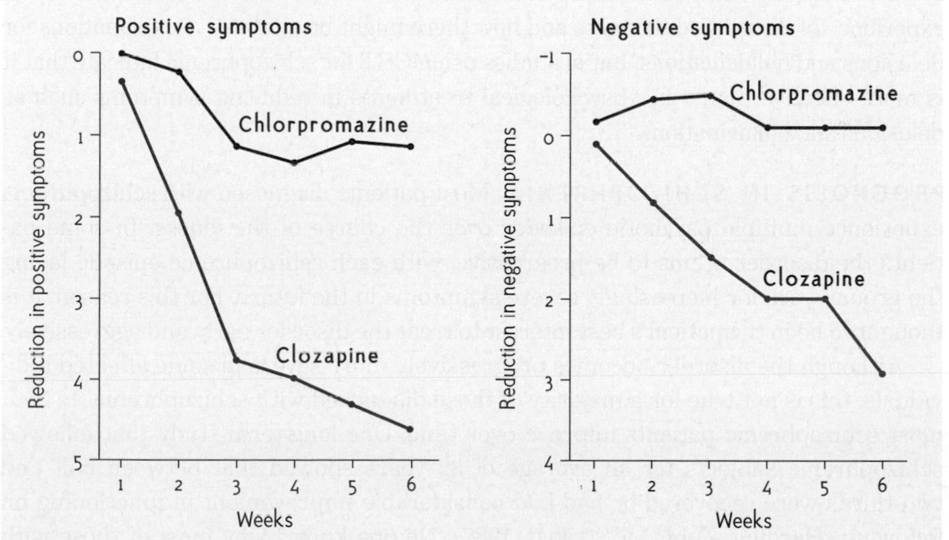

14.18 The Effectiveness of Clozapine Clozapine is more effective than previously available antipsychotic medications in treating both the positive and the negative symptoms of schizophrenia.

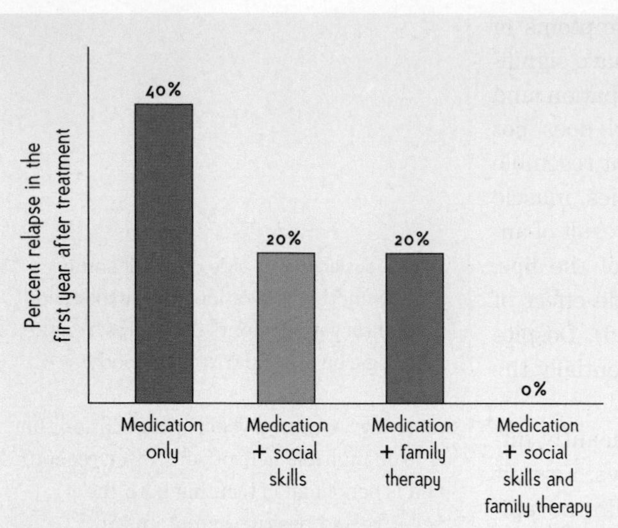

14.19 The Effectiveness of Antipsychotic Medications and Therapy Medication is the most effective way to reduce the rate of relapse for patients with schizophrenia, but adding psychosocial interventions, such as family therapy and social-skills training, improves outcome.

Why might social support enhance drug therapies for schizophrenia?

PSYCHOSOCIAL TREATMENTS It is well established that medication is essential in the treatment of schizophrenia. Without it, patients may deteriorate, experiencing more frequent and more severe psychotic episodes. When antipsychotic drugs became available, other types of therapies for schizophrenia were virtually dismissed. It became clear over time, however, that although medication is effective in reducing delusions and hallucinations, it does not have a substantial impact on patients' social functioning. A study of the clinical outcome of schizophrenia treatments over a 100-year period demonstrated that antipsychotic drugs improved the outcome for schizophrenic patients from just under 40 percent to about 50 percent (Hegarty, Baldessarini, Tohen, Waternaux, & Oepen, 1994). While this is an improvement, it falls far short of a cure.

Antipsychotic medication is most effective when used in combination with other treatment approaches. Social-skills training has been an effective way to address some of the deficits shown by those with schizophrenia. Schizophrenic patients can benefit from intensive training on regulating affect, recognizing social cues, and predicting the effects of their behaviour in social situations (Figure 14.19). With intensive long-term training, schizophrenic patients can generalize the skills learned in therapy to other social environments. Self-care skills can be another area in which schizophrenic patients are deficient. Behavioural interventions can be focused on such areas as grooming and bathing, management of medications, and financial planning. Training in specific cognitive skills, such as modifying thinking patterns and coping with auditory hallucinations, has been less effective.

Recently, psychologist Aaron Beck has proposed that an intensive form of cognitive-behavioural therapy (CBT) is effective for treating schizophrenia (Beck & Rector, 2005). Beck believes that brain dysfunction gives rise to disordered beliefs and behaviours and that schizophrenia may be due in part to limited cognitive resources and an inability to inhibit the intrusion of inappropriate thoughts. From this perspective, delusions and hallucinations reflect biased information processing. In CBT for schizophrenia, a great deal of initial effort is made to get the client to trust the therapist as someone who will be nonjudgmental and who will try to understand the meaning of symptoms from the patient's perspective. Over time the therapy becomes more formal as the therapist seeks to help the client understand how stressful life circumstances contribute to the experience of disordered thoughts and how there might be alternative explanations for delusions and hallucinations. Initial studies using CBT for schizophrenia indicate that it is more effective than other psychological treatments in reducing symptoms such as delusions and hallucinations.

PROGNOSIS IN SCHIZOPHRENIA Most patients diagnosed with schizophrenia experience multiple psychotic episodes over the course of the illness. In some patients the disorder seems to be progressive, with each schizophrenic episode laying the groundwork for increasingly severe symptoms in the future. For this reason, it is thought to be in the patient's best interest to treat the disorder early and aggressively.

Although the disorder becomes progressively more severe in some affected individuals, this is not true for a majority of those diagnosed with schizophrenia. In fact, most schizophrenic patients improve over time. One long-term study that followed schizophrenic subjects for an average of 32 years showed that between half and two-thirds were recovered or had had considerable improvement in functioning on follow-up (Harding, Zubin, & Strauss, 1987). No one knows why most of those with schizophrenia seem to improve as they grow older. It may be that the patients who improve have found the treatment regimen that is most effective for them. Another theory is that the changes in the brain that occur with aging somehow result in fewer psychotic episodes. Dopamine levels may decrease with age and this may be related to the improvement in schizophrenic symptoms.

The prognosis for schizophrenic patients depends on a variety of factors, including age of onset, gender, and culture (Malla et al., 2002). Those diagnosed with schizophrenia later in life tend to have a more favourable prognosis than those who experience their first symptoms during childhood or adolescence (McGlashan, 1988). Women tend to have a better prognosis than men (Hambrecht, Maurer, Hafner, & Sartorius, 1992), perhaps because the onset of schizophrenia in women tends to be later than in men. Interestingly, culture also plays a role in prognosis. Schizophrenia in developing countries is often not so severe as in developed countries (Jablensky, 1989; Leff, Sartorius, Jablensky, Korten, & Ernberg, 1992). More extensive family networks in developing countries may provide more support in caring for schizophrenic patients.

Thinking Critically

Should Drugs Be Used to Treat Adolescent Depression?

Adolescent depression is a serious problem, affecting approximately 1 in 20 North American adolescents at any one time (Angold & Costello, 2001). Untreated depression is associated with drug abuse, dropping out of school, and suicide. In the United States, approximately 5,000 teenagers kill themselves each year, making it the third leading cause of death for that age group (Arias, MacDorman, Strobino, & Guyer, 2003), and the rate is slightly higher in Canada (Leenaars & Lester, 1995). For many years, depression in children and adolescents was ignored or viewed as a typical part of growing up. According to a report of the U.S. Surgeon General, fewer than 30 percent of children who have mental health problems receive any type of treatment (USDHHS, 1999). It is understandable, then, that many in the mental health community reacted favourably to the initial use of SSRIs such as Prozac to treat adolescent depression. Studies had found tricycle antidepressants were ineffective and dangerous for adolescents, but the first studies using SSRIs found them to be effective and safe (e.g., Emslie et al., 1997).

From the time that SSRIs were introduced as treatments for adolescent depression, some mental health researchers raised concerns that the drugs might cause some adolescents to become suicidal (Jureidini et al., 2004). These concerns were based in part on findings that SSRIs cause some adults to feel restless, impulsive, and suicidal. Following a report by one drug company of an increase in suicidal thoughts among adolescents taking its product, the U.S. Food and Drug Administration (FDA) asked all drug companies to analyze their records for similar reports. An analysis of reports on more than 4,400 children and adolescents found that about twice as many of them taking SSRIs reported having suicidal thoughts (4 percent) as those taking a placebo (2 percent), but none of the children or adolescents actually committed suicide. Given evidence of increased thoughts of suicide, the FDA voted in October 2004 to require manufacturers to add to their product labels a warning that antidepressants increase the risk of suicidal thinking and behaviour in depressed children and adolescents, and that physicians prescribing these drugs need to balance risk with clinical need. Physicians in North America were also advised to watch their patients closely, especially in the first few weeks of treatment. Suddenly many parents were wondering whether SSRIs were safe for their children.

Many issues concerning SSRIs and children need to be evaluated critically. First, are SSRIs effective for young people, and if so, are they more effective than other treatments? Second, do these drugs actually cause suicidal feelings, or are depressed adolescents generally likely to feel suicidal whether or not they take medication? Finally, how many adolescents would be suicidal if their depression was left untreated?

An ambitious study supported by the National Institutes of Health in the United States provides clear evidence that the SSRI Prozac is effective in treating adolescent depression. This study, called the Treatment of Adolescents with Depression Study (TADS, 2004), examined 429 patients aged 12 to 17 who had been depressed for an average of 40 weeks before the study began. Participants were randomly assigned to a type of treatment and followed for 12 weeks. The results indicated that 61 percent of participants taking Prozac showed improvement in symptoms, compared with 43 percent receiving cognitive-behavioural therapy, and 35 percent taking a placebo. The group that did best was the group receiving both Prozac and CBT, in which 71 percent improved; this is consistent with other studies showing that the combination of drugs and psychotherapy often produces the strongest results. In terms of suicidality, the results were more mixed. All treatment groups experienced a reduction in thoughts of suicide compared with the baseline. However, participants in the Prozac group were twice as likely to have serious suicidal thoughts or intentions compared with those undergoing other treatments, and of the 7 adolescents who attempted suicide during the study, 6 were taking Prozac. Critics of adolescents receiving drugs point out that these findings are consistent with other studies showing a risk from SSRIs (Antonuccio & Burns, 2004).

We need to keep a few things in mind as we analyze the use of SSRIs for adolescent depression. In the TADS study, suicide attempts were quite uncommon (7 of 429 patients). Indeed, this is consistent with the FDA finding that about 4 percent of adolescents taking Prozac will become suicidal. Moreover, only a small number of adolescents who kill themselves each year are taking antidepressants of any kind. The question is whether the many children who take antidepressants experience more benefits than risks. In addition, the rates of suicide have dropped since the use of SSRIs became widespread (Figure 14.20), which implies that the increased use of drugs such as Prozac may be lowering the overall rate of depression, thereby reducing suicide. The greater the increase in the number of SSRI prescriptions for adolescents within a region, the greater the reduction in teenage suicides (Olfson, Shaffer, Marcus, & Greenberg, 2003). This raises the possibility that not providing SSRIs to adolescents will raise the suicide rate (Brent, 2004).

14.20 Suicide Rates among People 15 to 24 Years of Age The rate of suicide among adolescents and young adults has been in decline for the past decade. This roughly corresponds to the period during which SSRIs have been used to treat depression.

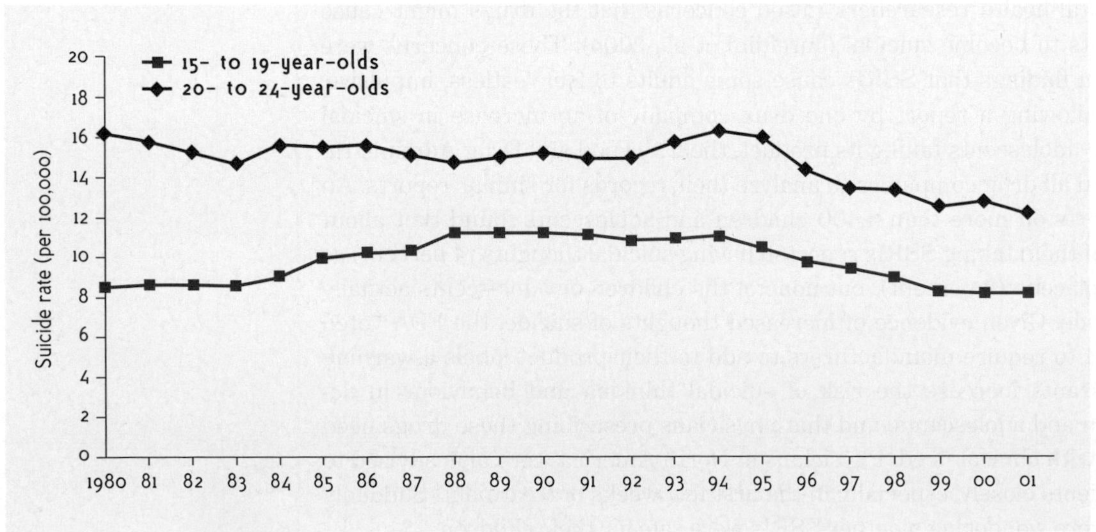

According to some researchers, the relative success of psychotherapy for teenage depression makes it a better treatment choice. Indeed, considerable evidence shows that psychotherapy is effective on its own (Mufson et al., 2004) and also enhances drug treatment. Other psychological treatments, such as interpersonal psychotherapy, are also successful (Hollon et al., 2002). But getting adolescents to comply with psychotherapy can be challenging. In addition, it is unrealistic to expect there to be sufficient resources to provide psychotherapy to all adolescents who need it in the near future. By contrast, it is relatively easy for pediatricians and family physicians to prescribe drugs. Unfortunately, the prescribing of such medications by general practitioners can be problematic because these individuals do not have specific training in treating psychological disorders.

What would you recommend? Do you think the government action is warranted by the data? Do you believe that it is acceptable to provide drugs that benefit many adolescents but place some at risk for suicide? The risk currently looks relatively small. How large a risk is acceptable, if any? ■

REVIEWING the Principles | What Are the Most Effective Treatments?

Psychotherapy and biological therapy are used to treat mental disorders. When the two are equally effective, psychotherapy is preferred because it has fewer side effects and persists beyond treatment. Cognitive and behavioural therapies, especially exposure and response prevention, are particularly useful for anxiety disorders, although drug treatments are also used for panic disorder and OCD. There are a number of treatment strategies for mood disorders, including exercise, antidepressants, cognitive-behavioural therapy, electroconvulsive therapy, and transcranial magnetic stimulation. For mania, however, only lithium appears to be effective. Psychopharmacology is the recommended treatment for schizophrenia, and a variety of drugs are now known to be safe and effective. Pharmacological interventions for schizophrenia are most effective when used in combination with other treatment approaches. Training in social and self-care skills can help patients improve their ability to interact with others and to take care of themselves. Behavioural interventions can also help in improving self-care and in increasing compliance with medication.

Can Personality Disorders Be Treated?

Just as not much is known about the causes of personality disorders, little is known about how best to treat them. There is a growing literature of case studies describing treatment approaches for personality disorders, but few large well-controlled studies. The one thing about personality disorders that most therapists agree on is that they are notoriously difficult to treat. Individuals with personality disorders who are in therapy are usually also being treated for an Axis I disorder, which typically is the problem for which they sought therapy in the first place. People rarely seek therapy for personality disorders because one of the hallmarks of these disorders is that patients tend to see the environment rather than their own behaviour as the cause of their problems. Individuals with personality disorders can be very difficult to engage in therapy since they do not see their behaviour as a problem. We now consider two of the better-researched personality disorders.

Dialectical Behaviour Therapy Is Most Successful for Borderline Personality Disorder

The impulsivity, affective disturbances, and identity disturbances that are characteristic of borderline personality disorder make therapy for those affected very challenging. Traditional psychotherapy approaches have been largely unsuccessful, so there have been efforts to develop approaches specific to borderline personality disorder.

14.21 Marsha Linehan Linehan developed dialectical behaviour therapy, one of the few effective treatments for borderline personality disorder.

The most successful treatment approach to date for borderline personality disorder was developed by Marsha Linehan (Figure 14.21) in the 1980s. **Dialectical behaviour therapy (DBT)** combines elements of the behavioural, cognitive, and psychodynamic approaches (Lieb, Zanarini, Schmahl, Linehan, & Bohus, 2004). Clients are seen in both group and individual sessions, and the responsibilities of both the client and the therapist are made explicit. Therapy proceeds in three stages. In the first stage, the therapist targets the client's most extreme and dysfunctional behaviours. Often these are self-cutting and suicidal threats or attempts. The focus is on replacing these behaviours with others that are more appropriate. Clients are taught problem-solving techniques and more effective and acceptable ways of coping with emotions. In the second stage of the treatment, the therapist helps the client explore past traumatic experiences that may be at the root of emotional problems. In the third stage, the therapist helps the patient develop self-respect and independent problem solving. This is a crucial stage in therapy. Borderline patients are very dependent on others for support and validation, and they must be able to generate these things themselves or they will likely revert to their previous patterns of behaviour.

Because of the depressive and sometimes psychoticlike symptoms experienced by individuals with borderline personality disorder, researchers previously believed that they would develop an Axis I disorder such as schizophrenia or depression. Studies that have followed these individuals over time, however, have demonstrated that this is not the case. Instead, they continue with their symptoms relatively unchanged over time (Plakun, Burkhardt, & Muller, 1985). The only group that does seem to show some long-term improvement is borderline patients of a high socioeconomic level who receive intensive treatment. These individuals show improvement in interpersonal relationships and often achieve full-time employment (Stone, Stone, & Hurt, 1987). In the remainder of borderline patients, however, interpersonal and occupational problems are the norm. Substance abuse is not uncommon, and many patients attempt suicide multiple times.

How do you treat people who thrive on attention or do not believe they have a problem?

Therapeutic approaches specifically targeted at borderline personality disorder, such as DBT, may improve the prognosis for these patients. Studies have demonstrated that clients undergoing DBT are more likely to remain in treatment and less likely to be suicidal than are clients in other types of therapy (Linehan, Armstrong, Suarez, Allmon, & Heard, 1991; Linehan, Heard, & Armstrong, 1993). SSRIs are often prescribed along with DBT to treat feelings of depression. Although DBT was initially developed as an outpatient therapy, it has recently been adapted for use in an inpatient setting (Swenson, Sanderson, Dulie, & Linehan, 2001). Because borderline patients thrive on attention, inpatient settings that are not specifically designed for these patients can inadvertently reinforce dysfunctional behaviour that brings them attention, such as self-injury and suicide attempts. The result is often a worsening of symptoms and long-term hospitalization (Rosenbluth & Silver, 1992). The DBT program begins with a three-month inpatient stay followed by long-term outpatient therapy. It has been found to be very effective in reducing depression, anxiety, and suicidal gestures (Bohus et al., 2004, 2000).

dialectical behaviour therapy (DBT) A treatment for borderline personality disorder that combines elements of behavioural, cognitive, and psychodynamic approaches.

Antisocial Personality Disorder Is Difficult to Treat

Although treating borderline patients may be very difficult, treating those with antisocial personality disorder often seems impossible. These patients lie without thinking twice about it, care little for the feelings of others, and live for the present without consideration of the future. All of these factors make development of a therapeutic relationship and motivation for change remote possibilities at best. Antisocial individuals are often more interested in manipulating their therapists than in changing their behaviour. Therapists working with these patients must be constantly on guard.

THERAPEUTIC APPROACHES FOR ANTISOCIAL PERSONALITY DISORDER

A number of treatment approaches have been tried for antisocial personality disorder. Because antisocial individuals have been reported to have diminished cortical arousal, stimulants have been prescribed to normalize arousal levels. Evidence indicates that these drugs are beneficial in the short term, but they are not a long-term solution. Antianxiety drugs may lower hostility levels somewhat, and lithium has shown some promise in treating the aggressive, impulsive behaviour of violent criminals. Overall, however, psychotropic medications have not proven effective in treating this disorder.

Similarly, traditional psychotherapeutic approaches seem to be of little use in treating antisocial personality disorder. Behavioural and cognitive approaches have met with somewhat more success. Behavioural approaches reinforce appropriate behaviour and ignore or punish inappropriate behaviour in an attempt to replace maladaptive behaviour patterns with more socially appropriate behaviour. This approach seems to work best when the therapist controls reinforcement, the client cannot leave treatment, and the client is part of a group. Individual therapy sessions rarely result in any change in antisocial behaviour. Clearly, the behavioural approach cannot be implemented on an outpatient basis, since the client will receive reinforcement for his behaviour outside of therapy and can leave treatment at any time. For these reasons, therapy for this disorder is most effective in a residential treatment centre or correctional facility.

More recently, cognitive approaches have been tried for antisocial personality disorder. Beck and his colleagues (1990) have conceptualized this disorder as a series of faulty cognitions. The antisocial individual believes that his desire for something justifies any actions he takes to attain it. He believes his actions will not have negative consequences, or if they do that these consequences are not important. He also believes he is always right and that what others think is unimportant. Therapy is therefore focused on making the client aware of these beliefs and challenging their validity. Therapists try to demonstrate that the client's goals can be met more easily by following the rules of society than by trying to get around them, as in the following example (Beck et al., 1990):

> Therapist: How well has the "beat-the-system" approach actually worked out for you over time?
>
> Brett: It works great . . . until someone catches on or starts to catch on. Then you have to scrap that plan and come up with a new one.
>
> Therapist: How difficult was it, you know, to cover up one scheme and come up with a new one?
>
> Brett: Sometimes it was really easy. There are some real pigeons out there.

Therapist:	Was it always easy?
Brett:	Well, no. Sometimes it was a real bitch. . . . Seems like I'm always needing a good plan to beat the system.
Therapist:	Do you think it's ever easier to go with the system instead of trying to beat it in some way?
Brett:	Well, after all that I have been through, I would have to say yes, there have been times that going with the system would have been easier in the long run. . . . But . . . it's such a challenge to beat the system. It feels exciting when I come up with a new plan and think I can make it work.

This dialogue illustrates both the cognitive approach and why these patients are so difficult to work with. Even if they can understand that what they are doing is wrong, they don't care. They live for the thrill of getting away with something.

PROGNOSIS FOR ANTISOCIAL PERSONALITY DISORDER The prognosis that antisocial patients will change their behaviour as a result of therapy is poor. Some of the more recently developed cognitive techniques show promise, but as yet there is no good evidence that the changes they produce are long-lasting or even real. Fortunately for society, however, most antisocial individuals improve after age 40 (Figure 14.22). The reasons for this are not known, but the improvement may be due to a reduction in biological drives. Alternative theories suggest that these individuals may gain insight into their self-defeating behaviours or may just get worn out and be unable to continue their manipulative ways. This improvement, however, is only in the realm of antisocial behaviour. The egocentricity, callousness, and manipulativeness remain unchanged (Harpur & Hare, 1994). In fact, although criminal acts decrease among those with antisocial personality disorder after age 40, over 50 percent of these individuals continue to be arrested after age 40 (Hare, McPherson, & Forth, 1988).

Because of the limited effectiveness of therapy for this disorder, time and effort may be better spent in prevention. Conduct disorder is a childhood condition known to be a precursor to antisocial personality disorder. Some of the environmental and developmental risk factors for conduct disorder have been identified, and focusing on these may reduce the likelihood that a child with conduct disorder will grow up to have antisocial personality disorder.

14.22 Antisocial Personality Disorder Of 521 prisoners studied by Robert Hare and his colleagues, those with antisocial personality disorder (APD) were more likely than other criminals to spend time in jail before age 40. After age 40 they were less likely.

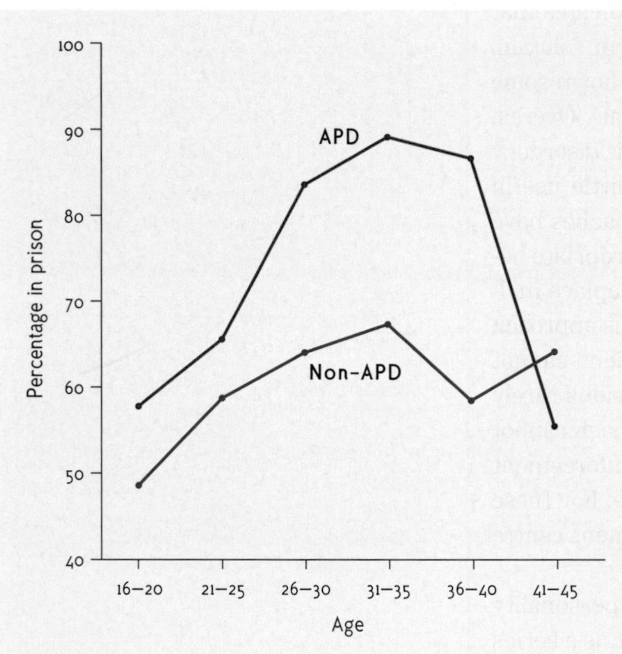

REVIEWING

the Principles | Can Personality Disorders Be Treated?

Personality disorders are characterized by long-standing maladaptive ways of interacting with the world, and they are notoriously difficult to treat. Efforts have been made to develop treatment programs for both borderline and antisocial personality disorders because they can have devastating effects on the individual, the family, and society. Dialectical behaviour therapy is the most successful method of treating borderline personality disorder. At this point, no treatment approach appears to be particularly successful in treating antisocial personality disorder.

How Should Childhood Disorders Be Treated?

Childhood experience and development are critically important to adult mental health. Problems not addressed during childhood will still be there in adulthood and are likely to be more significant and more difficult to treat. Most theories of human development regard children as more malleable than adults and therefore more amenable to treatment. That would suggest that childhood disorders should be the focus of research into etiology, prevention, and treatment. To illustrate the issues involved in treating childhood disorders, we will consider treatment approaches for two of these disorders.

Children with ADHD Can Benefit from a Variety of Approaches

There is some dispute about whether attention-deficit/hyperactivity disorder is a mental disorder or simply a troublesome behaviour pattern that children will eventually outgrow. Some children diagnosed with ADHD do grow out of it. Many more, however, continue to suffer from the disorder throughout adolescence and adulthood. Longitudinal studies show that 70 percent of individuals diagnosed with ADHD during childhood still meet the criteria for the disorder in adolescence. These individuals are more likely to drop out of school and to reach a lower socioeconomic level than expected. They continue to show a pattern of inattention, impulsivity, and hyperactivity, and they are at increased risk for other psychiatric disorders (Wilens, Faraone, & Biederman, 2004). Because of this somewhat bleak long-term prognosis, it is clear that effective treatment early in life is of great importance.

PHARMACOLOGICAL TREATMENT OF ADHD Currently, the most common treatment for ADHD is a central-nervous-system stimulant such as **methylphenidate** (*Ritalin*). The effects of this drug are similar to those of caffeine and amphetamines, but it is more potent than the former and less potent than the latter. Although the actions of methylphenidate are not fully understood, it is thought to affect multiple neurotransmitters, in particular dopamine. Based on the external behaviour of children with ADHD, one might conclude that their brains are overly active, and it may seem surprising that a stimulant would improve their symptoms. In fact, functional brain imaging shows that children with ADHD have underactive brains; their hyperactivity may be a way of raising their arousal levels.

methylphenidate A central-nervous-system stimulant medication used to treat ADHD.

At appropriate doses, central-nervous-system stimulants decrease overactivity and distractibility and increase attention and the ability to concentrate. Children on stimulants are able to work more effectively on a task without interruption and are more academically productive. Even their handwriting seems to improve. They are less disruptive and noisy, and this likely has contributed to the large number of children who take this medication (Figure 14.23). Parents often feel pressured by the school system to medicate children who are an ongoing behaviour problem in the classroom. Parents themselves can pressure physicians to prescribe the drug because it can make home life much more manageable. One study measured the effects of methylphenidate on the behaviour of children during a baseball game (Pelham et al., 1990). Those children with ADHD who were taking the medication would assume

How cautious should we be in prescribing drugs to children with ADHD?

14.23 The Effects of Ritalin The use of methylphenidate (Ritalin) for children with ADHD dramatically reduces negative behaviours while only slightly increasing the amount of positive behaviour.

the ready position in the outfield and could keep track of the game. Children with ADHD who were not taking the drug would often throw or kick their mitts even while the pitch was in progress.

Studies have shown that children taking methylphenidate are happier, more socially adept, and more academically successful. They also interact more positively with their parents, perhaps because they are more likely to comply with requests. The medication has its drawbacks, however. There is evidence that although stimulants are beneficial in the short term, their benefits do not seem to be maintained over the long term. In addition, there is a very real risk of abuse, with numerous cases of children and adolescents buying and selling these stimulants. Side effects include sleep problems, reduced appetite, bodily twitches, and temporary growth suppression. Perhaps most important, some children on medication may see their problem as beyond their control. They may not feel responsible for their behaviour and may not learn coping strategies that they will need later if they discontinue their medication or if it ceases to be effective. Most therapists believe that medication should be supplemented by psychological therapies such as behaviour modification, and some even urge that medication be replaced by other treatment approaches when possible. Nonetheless, the data are clear that treatment progresses much better if drugs are part of the overall treatment plan.

BEHAVIOURAL TREATMENT OF ADHD Behavioural treatment of ADHD aims to reinforce positive behaviours and ignore or punish problem behaviours. The difficulties with this treatment approach are similar to those discussed in the following section on autism. Treatment is very intensive and time consuming. In addition, although it is often not difficult to improve behaviour in a structured setting, the effects of therapy do not necessarily generalize beyond the clinic or classroom. Many therapists advocate combining behavioural approaches with medication. The medication is used to gain control over the behaviour, and then behavioural modification techniques can be taught and the medication slowly phased out. Others argue that medication should be used only if behavioural techniques do not reduce inappropriate behaviours. Although this controversy is ongoing, recent research has shown that medication plus behavioural therapy is more effective than either approach alone.

The U.S. National Institute of Mental Health, in collaboration with teams of investigators, began the Multimodal Treatment of Attention-Deficit Hyperactivity Disorder (MTA) in 1992. The study involved 579 children who were randomly assigned to community care or to one of three treatment groups, each lasting 14 months. The treatment groups were: medical management (usually treatment with a stimulant such as methylphenidate), intensive behavioural treatment, and a combination of the two. Follow-up studies reveal that the children receiving medication and those receiving a combination of medication and behavioural therapy had greater improvement in their ADHD symptoms than did those in the other two groups (Jensen et al., 2001). Children who received both medication and behavioural therapy showed a slight advantage in areas such as social skills, academics, and parent-child relations over those who received only medication.

Autistic Children Benefit From a Structured Treatment Approach

The treatment of autistic children presents unique challenges to mental health professionals. The core symptoms of autism—impaired communication, restricted interests, and deficits in social interaction—make these children particularly difficult to work with. They often exhibit extreme behaviours as well as public self-stimulation. Although these behaviours must be reduced or eliminated before progress can be made in other areas, elimination is difficult because it is hard to find effective reinforcers. Although normal children respond positively to social praise and small prizes, autistic children are often oblivious to these things. In some cases, food is the only effective reinforcement in the initial stages of treatment. Another characteristic of autistic children is an overselectivity of attention. This tendency to focus on specific details while ignoring others interferes with generalizing learned behaviour to other stimuli and situations. A child who learns to set the table with plates may be completely stymied when presented with bowls instead. Generalization of skills must be explicitly taught. For this reason, structured therapies are more effective for these children than unstructured interventions such as play therapy.

BEHAVIOURAL TREATMENT FOR AUTISM One of the best known and perhaps most effective treatments for autistic children was developed by Ivar Lovaas and his colleagues at the University of California at Los Angeles. The program, called applied behavioural analysis, or ABA, is based on principles of operant conditioning. Behaviours that are reinforced should increase in frequency, while behaviours that are not reinforced should be extinguished. This very intensive approach requires a minimum of 40 hours of treatment per week. Preschool-age children with autism were treated by teachers and by their parents, who were given specific training. After more than two years of ABA treatment, the children had gained about 20 IQ points on average and most of them were able to enter a normal kindergarten program (Lovaas, 1987). In contrast, there was no change in IQ in a comparable control group of children who did not receive any treatment. A group of children who received ten hours of treatment per week fared no better than the control group. Initiating treatment at a younger age was also shown to yield better results, as did involving the parents and having at least a portion of the therapy take place in the home. Children with better language skills before entering treatment had better outcomes than those who were mute or echolalic.

Although Lovaas's ABA program has been demonstrated to be effective, it has its drawbacks. The most obvious is the time commitment. The therapy is very intensive and lasts for years. Parents essentially become full-time teachers for their autistic children. The financial and emotional drains on the family can be substantial. If the family has other children, they may feel neglected or jealous because of the amount of time and energy expended on the disabled child.

BIOLOGICAL TREATMENT FOR AUTISM Because there is good evidence that autism is caused by brain dysfunction, there have been many attempts to use this knowledge to treat the disorder. It is easy to find reports of children who benefited from alternative treatment approaches. These case studies are compelling, but when the treatments are assessed in controlled studies, there is little or no evidence that most of them are effective. However, a few biologically based treatment approaches have shown some promise, as we discussed in Chapter 13.

During the 1990s, it was noted that some autistic children had increased levels of peptides in their urine. These peptides were derived from gluten and were thought to have a negative pharmacological effect on learning, attention, brain maturation, and social interaction. Autistic children with elevated peptide levels were put on a special diet, and their progress was monitored for several years (Reichelt, Knivsberg, Lind, & Nodland, 1991). After one year, the level of peptides in their urine was normal. The children demonstrated reductions in odd behaviour and improvements in social, cognitive, and communicative abilities. Four years later, they continued to improve, but their development lagged behind that of their neurologically normal peers. Findings like this are exciting and can provide hope to families with autistic children. However, this treatment was effective only for a subgroup of autistic children, and even within that group it was clearly not a cure.

Another approach to the treatment of autism involves serotonin. SSRIs such as Prozac have been found to reduce compulsions in patients diagnosed with obsessive-compulsive disorder. Because autism involves compulsive and stereotyped behaviour, and because there is some evidence of abnormal serotonin metabolism in autistic children, SSRIs have been tried as a treatment for autism. It has been reported that in some autistic children the drug reduces stereotyped motor behaviour and self-abuse and improves social interactions (McDougle, 1997). There is also evidence that medications that block the action of dopamine have similar effects. Still, at this point, the neurobiology of autism is not well understood, and although attempts to use psychopharmacology to treat the disorder have resulted in some improvements in behaviour, such as by injections of oxytocin, a great deal remains to be learned.

PROGNOSIS FOR CHILDREN WITH AUTISM Despite a few reports of remarkable recovery from autism, the long-term prognosis is considered poor. A recent follow-up study of men in their early twenties revealed that they continued to show the ritualistic behaviour typical of autism. In addition, nearly three-quarters had severe social difficulties and were unable to live and work independently (Howlin, Mawhood, & Rutter, 2000). Several factors affect prognosis for autistic individuals. Although it was once believed that the prognosis was particularly poor for children whose symptoms were apparent before age two (Hoshino et al., 1980), it is possible that only the most severe cases of autism were diagnosed prior to growing public recognition of the disorder. It is now clear that early diagnosis allows for more effective treatments (National Research Council, 2001). Still, severe cases—especially those with notable cognitive deficiencies—are less likely to benefit from treatment. Early language ability is associated with better outcome (Howlin et al., 2000), as is higher IQ. Autistic children have difficulty generalizing from the therapeutic setting to the real world, and this severely limits their social functioning (Handleman, Gill, & Alessandri, 1988). A higher IQ may mean a better ability to generalize learning and therefore a better overall prognosis.

Profiles in Psychological Science

One Boy's Journey out of Autism's Grasp

John O'Neil, a deputy editor at the *New York Times,* recently described what it is like to be the parent of an autistic child (O'Neil, 2004). O'Neil's son James, now eight, began to show signs of being "different" in early childhood. Although he had been an easy baby, as a toddler James seemed to have difficulty looking his parents in the eye, and there wasn't a strong sense of connection. James displayed little interest in objects, even toys, that were given or shown to him, instead repeating behaviours to the point of harming himself, for example, pulling on and off his cowboy boots until his feet were raw. He responded to loud noises by crying. James's behaviour really started to deteriorate when he was two and a half, following the arrival of a baby brother and a move to a new house. His parents simply assumed he was overwhelmed. It took the director of James's new preschool to notice the telltale signs of autism. On her recommendation, a professional assessed James and determined that he did have autism.

The discovery of James's autism follows a familiar pattern. Although most diagnoses of autism are made by age three, the disorder can be detected earlier if parents or pediatricians know what to look for. Many children who will develop autism show abnormal social behaviour in infancy. Other signs include such behaviours as staring at objects for long periods of time and not reaching developmental milestones, such as speaking. Sometimes autism appears to strike out of the blue in an otherwise normally developing child. The child suddenly withdraws from social contact, stops babbling, and may become self-abusive. Parents and pediatricians who notice symptoms of autism in such a child may write them off as quirks or feel that the child is just a little slow to develop. Luckily for James, the preschool teachers noticed his unusual behaviour and recommended a professional evaluation. As we have seen, research has shown that the earlier treatment begins, the better the prognosis.

On finding out that treatments for autism did exist, the O'Neils were relieved. But then they heard the bad news: Treatment is expensive, difficult, and time consuming. As noted earlier, the recommended amount of treatment is over 40 hours per week. During their first visit to a speech therapist, James's mother learned just how much James needed treatment: he had forgotten his name. Although there are different versions of treatment, most are based on Lovaas's original therapy. Known as applied behavioural analysis (ABA), this therapy requires parents and teachers to spend hours helping children with autism learn basic skills, such as saying their names. It uses operant conditioning to reward even small behaviours.

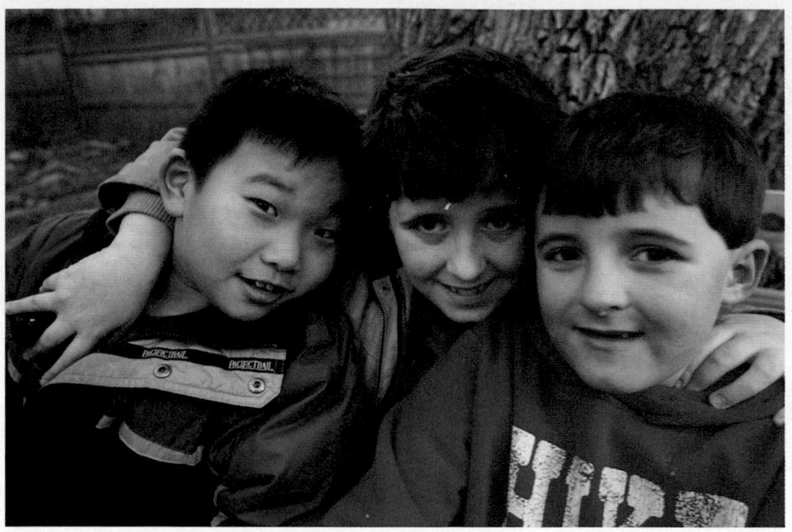

14.24 James at Eight James O'Neil, centre, the eight-year-old autistic son of John O'Neil, with, at left, his friend Larry, also eight, and his brother Miles, six.

As is the case with many school districts, resources were insufficient for James to receive full treatment in school. His mother, Marcia, a physician, gave up her full-time position in order to set up a home-based program for James. James spent up to eight hours every day performing tasks that most children would find extremely boring, such as repeatedly imitating the therapist's placing two blocks next to each other or touching her nose. The same behaviours, repeated over and over, for hours on end: that is the nature of ABA. It teaches each task as a series of simple steps, attempting to actively engage the child's attention in highly structured activities, and provides regular reinforcement of desired behaviour. A day might begin with physical activities to strengthen coordination and build body awareness, followed by a snack break during which appropriate social behaviours are reinforced and language skills are stressed. Each part of the day is designed to work on a child's particular problem areas. Along the way progress is charted to guide subsequent sessions. At one point it became clear that James's language skills had to improve if he was going to be able to attend mainstream school. To accomplish this, James was encouraged to talk by simply being given any treat he asked for, and it worked.

James was able to start school with the assistance of one of his full-time instructors, who attended class with him. Despite some rocky moments, James has made tremendous progress. Not that there still aren't problems, such as continued difficulties with reading comprehension, math, attention, social skills, and James's trying to understand why he has his disorder and why other kids do not. But James's journey is one of triumph. Perhaps his biggest accomplishment, especially for someone with autism, was making friends with a classmate named Larry, who was also eight (Figure 14.24). Why, O'Neil speculates, was Larry attracted to James as a friend? Perhaps they shared a love of potty humour or perhaps they were similarly warm and enthusiastic. One day O'Neil overhead the two friends engaged in silly conversation, telling stupid jokes and gossiping about their "girlfriends." In that moment, O'Neil realized just how many of his dreams for James had been realized. ■

REVIEWING the Principles | How Should Childhood Disorders Be Treated?

Medications such as methylphenidate can be quite effective in treating children with ADHD. However, side effects are associated with this type of medication, as well as ethical concerns. For these reasons, many parents and clinicians have turned to other approaches, such as behavioural therapy. Recent research suggests that the most effective management strategy for ADHD may be a combination of medication and behavioural therapy. In contrast, the most effective treatment for autism currently is structured and intensive behavioural therapy. Biological interventions, such as medication and restricted diet, have shown some promise in subgroups of children with autism but have been largely ineffective.

Conclusion

Although great strides have been made in identifying the etiology of mental disorders, this growing knowledge has not always given rise to new treatment approaches. Treatment of mental disorders is aimed at helping the individual think and behave in more adaptive ways, but the approaches can vary tremendously. For some disorders, several approaches are equally effective. For other disorders, some approaches are more effective than others. New understanding of the ways in which mental disorders affect neural functioning may lead to the discovery of more effective psychotropic medications. History teaches us, however, that medication is not the magic bullet for mental illness. Even with the creation of more effective medications, other interventions remain crucial. A pill may help relieve the symptoms of mental illness, but it cannot help the patient cope with the effects of the disorder, interact with others, or think and behave in more adaptive ways. Psychological, cognitive, and behavioural interventions can address these issues more effectively than any medication. Further research may help identify and implement the treatment approaches or combinations of approaches that are best for each disorder and for each individual.

Summarizing the Principles of Treating Disorders of Mind and Body

How Is Mental Illness Treated?

1. **Psychotherapy is based on psychological principles:** Psychotherapeutic treatments focus on insights. Psychodynamic forms focus on uncovering the unconscious. Humanistic approaches focus on clarifying feelings and motives. Behavioural approaches focus on modifying maladaptive behaviours, and include social skills training and systematic desensitization. Cognitive approaches restructure thinking. Group support is provided with group therapy, and a systems approach is part of family therapy.

2. **Biological therapies are necessary for some disorders:** Psychotropic medications change neurochemistry. Anti-anxiety drugs increase GABA activity. Antidepressants, in their various forms, affect serotonin availability. Antipsychotics reduce positive symptoms.

3. **Common factors enhance treatment:** Therapy helps, although different therapies are useful for different conditions. There are factors common to most successful therapies. Enhancement factors include a caring therapist and the benefit of unburdening one's problems.

What Are the Most Effective Treatments?

4. **Treatments that focus on behaviour and cognition are superior for anxiety disorders:** Behavioural methods are effective in alleviating specific phobias. Cognitive restructuring is effecting in treating panic disorders. Obsessive-compulsive disorder (OCD) responds to medications that block serotonin reuptake and to cognitive-behavioural therapy.

5. **Many effective treatments are available for depression:** Pharmacological treatments include MAO inhibitors, tricyclics, and SSRIs. Cognitive-behavioural treatments are most effective when combined with antidepressants. Alternative therapies include phototherapy for seasonal affective disorder (SAD) sufferers and electroconvulsive therapy and transcranial magnetic stimulation for psychotic depression.

6. **Lithium is most effective for bipolar disorder:** The psychotropic medication lithium has been found to be the most effective in stabilizing mood in bipolar patients but has considerable side effects. Psychological therapy can help support compliance with drug treatment.

7. **Pharmacological treatments are superior for schizophrenia:** Antipsychotic medications are most effective for reducing the positive symptoms of schizophrenia. Tardive dyskinesia and other side effects are common with the older antipsychotic drugs. Clozapine acts specifically on dopamine receptors and reduces both positive and negative symptoms, with fewer side effects. The prognosis for patients depends on a variety of factors, including age of onset, gender, and culture.

Can Personality Disorders Be Treated?

8. **Dialectical behaviour therapy is most successful for borderline personality disorder:** DBT combines elements of behavioural, cognitive, and psychodynamic therapy. Therapy proceeds in three stages. First, the most extreme behaviours are targeted, and replaced with more appropriate behaviours. Next, the therapist explores past traumatic events. Finally, the therapist helps the patient develop self-respect and independence.

9. **Antisocial personality disorder is difficult to treat:** Psychotherapeutic approaches have not proven effective for treating antisocial personality disorder. Behavioural and cognitive approaches have been more effective, but most effectively in a controlled residential treatment environment. Generally, the prognosis is poor. Focusing on prevention by addressing conduct disorder in childhood may be the best strategy.

How Should Childhood Disorders Be Treated?

10. **Children with ADHD can benefit from a variety of approaches:** Methylphenidate, despite its side effects, is an effective pharmacological treatment, and works best as part of an overall treatment plan including psychotherapy, particularly behavioural treatment.

11. **Autistic children benefit from a structured treatment approach:** Behavioural approaches have been effective in improving language and social behaviour. The treatment strategy is very time intensive and extends for years. Dietary restrictions of gluten, which result in elevated peptides, have been effective for some autistic children. Some children also improve with SSRIs. In general, the long-term prognosis is poor.

Applying the Principles

1. Imagine that your roommate is suffering from depression. She is listless, stays in all the time, skips classes, and focuses on sad memories. All but which of the following might be an appropriate approach to treatment?

   ____ a) antipsychotic medications that will stop her irrational depressive thoughts
   ____ b) client-centred therapy that will help clarify her feelings
   ____ c) cognitive-based therapy that will change the ways she thinks about things
   ____ d) psychopharmacology that will change the neurochemical activity in her brain

2. A recent study found that marriage therapy that focused on communication skills was ineffective in improving marriages and avoiding divorce. This finding is consistent with a(n) ________ view of treatment effectiveness. Teaching communication strategies, however, to empower an individual suffering from depression may be effective. This exemplifies that ________.

   ____ a) statistical; treatment must be tailored to the disorder
   ____ b) evidence-based; treatment must be tailored to the disorder
   ____ c) empirical; communication psychotherapies are only effective with individuals
   ____ d) behavioural; communication therapies are only effective for serious psychopathologies

3. Lisa has been caught in an elaborate web of lies about her misfortunes and about having a serious medical condition. These lies have gotten her considerable attention from her roommates, and kept her boyfriend from breaking up with her despite her impulsive and excessive behaviour. What might you expect if you learned Lisa was diagnosed with borderline personality disorder?

   ____ a) She will respond very well to consistent behavioural therapies.
   ____ b) She will need to be institutionalized for effective therapy.
   ____ c) She will likely respond best to a three-step approach directed at behavioural change, insight, and cognitive restructuring.
   ____ d) She simply needs to experience intolerance and rejection by peers on a consistent basis in order to change her patterns of behaviour.

4. When Edgar was in first grade, his teacher complained to his mother that he was "hyperactive" and not able to stay on task. The teacher recommended that Edgar should take medication, like the other children in the class who exhibited similar behavioural patterns. Edgar's mother instead moved him to a school with smaller classes and a more structured environment, and maintained a strict behavioural approach at home. Edgar's behaviour improved and he was able to do well in the classroom. This demonstrates that

   ____ a) medications mimic the effects of behavioural therapy strategies.
   ____ b) ADHD is not a biologically based disorder.
   ____ c) ADHD responds to any treatment because the child needs attention, which he gets with treatment.
   ____ d) biological and psychological treatments may be equally effective when it comes to treating ADHD.

ANSWERS: *1 a 2 b 3 c 4 d*

Key Terms

Further Readings

Dawes, R. M. (1994). *House of cards: Psychology and psychotherapy built on myth.* New York: Maxwell Macmillan International.

Endler, N. (1982). *Holiday of darkness: A psychologist's personal journey out of his depression.* New York: John Wiley & Sons.

Kramer, P. D. (1993). *Listening to Prozac: A psychiatrist explores antidepressant drugs and the remaking of the self.* New York: Viking Penguin.

Linehan, M. M. (1993). *Cognitive-behavioral treatment of borderline personality disorders.* New York: Guilford Publications.

Maurice, C. (1994). *Let me hear your voice: A family's triumph over autism.* New York: Knopf.

Pennebaker, J. W. (1997). *Opening up: The healing power of confiding in others.* New York: William Morrow & Co., Inc.

Rosen, L. E., & Amador, X. F. (1997). *When someone you love is depressed: How to help your loved one without losing yourself.* New York: Free Press.

Further Readings: A Canadian Presence in Psychology

Ahrens, B., Grof, P., Moller, H-J., Muller-Oerlinghaussen, B., & Wolf, T. (1995). Extended survival of patients on long-term lithium treatment. *Canadian Journal of Psychiatry, 40*, 241–246.

Dobson, K. S., & Khatri, N. (2000). Cognitive therapy: Looking backward, looking forward. *Journal of Clinical Psychology, 56*, 907–923.

Enns, N., & Reiss, J. P. (1992). Electroconvulsive therapy. *Canadian Journal of Psychiatry, 37*, 671–678.

Hare, R. D., McPherson, L. M., & Forth, A. E. (1988). Male psychopaths and their criminal careers. *Journal of Consulting and Clinical Psychology, 56*, 710–714.

Harpur, T. J., & Hare, R. D. (1994). Assessment of psychopathology as a function of age. *Journal of Abnormal Psychology, 103*, 604–609.

Jasper, H. H. (1995). A historical perspective: The rise and fall of prefrontal lobotomy. In H. H. Jasper & S. Riggio (Eds.), *Epilepsy and the functional anatomy of the frontal lobe.* New York: Raven Press.

Kennedy, S. H., Eisfeld, B. S., Dickens, S. E., Bacchiochi, J. R., & Bagby, R. M. (2000). Antidepressant-induced sexual dysfunction during treatment with moclobemide, paroxetine, sertraline, and venlafaxine. *Journal of Clinical Psychiatry, 61*, 276–281.

Lehmann, H. H., & Ban, T. A. (1997). The history of the psychopharmacology of schizophrenia. *Canadian Journal of Psychiatry, 42*, 152–162.

Lesage, A. D., Morissette, R., Fortier, L. Reinharz, D., & Contandriopolous, A. (2000). Downsizing psychiatric hospitals: Needs for care and services of current discharged long-stay inpatients. *Canadian Journal of Psychiatry, 445*, 526–531.

Malla, A. K., Norman, R. M., Manchanda, R., Ahmed, M. R., Scholten, D., Harricharan, R., Cortese, L., & Takhar, J. (2002). One year outcome in first episode psychosis: Influence of DUP and other predictors. *Schizophrenia Research, 54*, 231–242.

Meichenbaum, D. (1977). Cognitive behavior modification: An integrative approach. New York: Plenum Press.

Meichenbaum, D. (1993). Changing conceptions of cognitive behavior modification: Retrospect and prospect. *Journal of Consulting and Clinical Psychology, 61*, 202–204.

Seeman, P. (2002). Atypical antipsychotics: Mechanism of action. *Canadian Journal of Psychiatry, 47*, 27–38.

Wystanski, M. (2000). Patient-centred versus client-centred mental health care. *Canadian Journal of Psychiatry, 45*, 670–671.

Yassa, R., Nair, N., Iskandar, H., & Schwartz, G. (1990). Factors in the development of severe forms of tardive dyskinesia. *American Journal of Psychiatry, 147*, 1156–1163.

The Power of Situations

On October 2, 2001, the anniversary of Mahatma Gandhi's birth, children all across India dressed up as the late leader to join in national celebration. As this example shows, situations and cultural norms often dictate human behaviour. As we explore social psychology further, we will learn just how powerful situational forces are in determining people's actions.

CHAPTER

15

Social Psychology

According to the Geneva Convention, prisoners of war are to be treated with respect and dignity. Even amid the horrors of war, we expect the military to behave in a civilized and professional manner. This is why the events at Abu Ghraib prison in Iraq and human rights abuse by Canadian soldiers in Somalia are so shocking to us (Figure 15.1). American soldiers brutalized and humiliated Iraqi detainees, threatening them with dogs, beating them with broom handles and chairs, and forcing them to lie on top of each other naked and simulate oral sex and masturbation. When photographs of these actions appeared on the Internet, outrage and condemnation of the deplorable conduct were immediate. Responding quickly to these furies, U.S. and Canadian government officials stressed that these were isolated incidents carried out by a small group of

Outlining the Principles

How Do We Know Ourselves?

- Our Self-Concept Consists of Self-Knowledge
- Perceived Social Regard Influences Self-Esteem
- We Use Mental Strategies to Maintain Our Views of Self

How Do Attitudes Guide Behaviour?

- We Form Attitudes through Socialization and Experience
- Behaviours Are Consistent with Strong Attitudes
- Discrepancies Lead to Dissonance
- Attitudes Can Be Changed through Persuasion

How Do We Form Our Impressions of Others?

- Nonverbal Actions and Expressions Affect Our Impressions
- We Make Attributions about Others
- Stereotypes Are Based on Automatic Categorization
- Stereotypes Can Lead to Prejudice

How Do Others Influence Us?

- Groups Influence Individual Behaviour
- We Conform to Social Norms
- We Are Compliant
- We Are Obedient to Authority

When Do We Harm or Help Others?

- Aggression Can Be Adaptive
- Aggression Has Social and Cultural Aspects
- Many Factors May Influence Helping Behaviour
- Some Situations Lead to Bystander Apathy

What Determines the Quality of Relationships?

- Situational and Personal Factors Influence Friendships
- Love Is an Important Component of Romantic Relationships
- Making Love Last Is Difficult

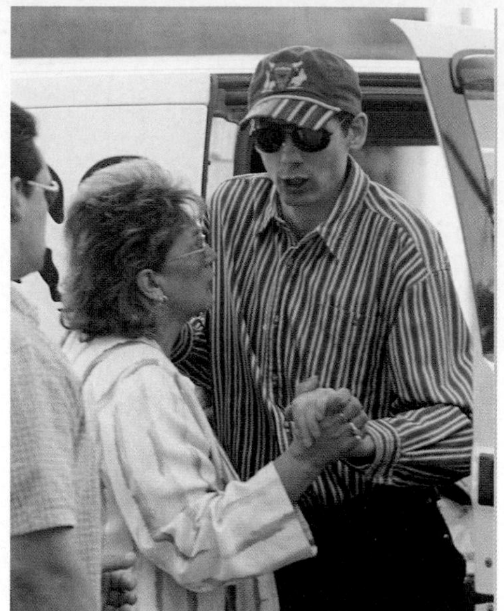

15.1 Clayton Matchee Assisted by family members as he arrived for a military hearing at Court of Queen's Bench in Saskatoon, Saskatchewan, Clayton Matchee faced charges of torturing and beating to death a Somalian teenager in 1993. Sixteen-year-old Shidane Arone was tortured for hours and eventually murdered after being captured in an apparent attempt to steal some Canadian military equipment. Master Corporal Matchee allegedly responded to instructions from his superiors that the soldiers could abuse any prisoners that they caught.

wayward soldiers. The idea that only a few troubled individuals were responsible for torturing and humiliating Iraqi and Somalian prisoners is bizarrely comforting. Somehow we are relieved to know that their deviant behaviour does not reflect on ordinary people. Surely we would not be willing to inflict such humiliation and pain on prisoners, many of whom were just teenagers and young men rounded up for questioning. Or would we?

Would you beat up or humiliate someone simply because you were ordered to do so or were in a situation where others were doing so? Or would you defy authority or resist peer pressure? Although we don't know whether or not they were ordered, how do we explain the actions of the guards at Abu Ghraib? The case of Abu Ghraib challenges many commonsense notions about human nature and forces us to consider questions about the dark side of humanity. People beat, rape, torture, and murder others. What is wrong with these people? What explains the violent side of human nature?

According to social psychologists, nothing is typically wrong with people such as the guards at Abu Ghraib. Rather, they are probably normal people caught up in overwhelming situations that shape their actions. Social psychologists point out a number of situational factors that likely influenced behaviour at Abu Ghraib, such as the absence of any real authority in the prison and therefore a weakening of responsibility. Moreover, social psychologists know that people are typically obedient to authority, especially in times of war. Also, the working conditions promoted aggression. One of the Abu Ghraib units had expected to be in Iraq a short time working in traffic control; instead, they found themselves in an overcrowded prison where they worked long hours, six or seven days a week, in extremely hot temperatures while under frequent mortar attack. Finally, during wartime people are especially likely to view the world as consisting of "us" and "them." Members of the enemy group are viewed as being all the same and are treated in a dehumanized fashion, often portrayed as both evil and inferior. All of these factors contributed to the mistreatment of prisoners at Abu Ghraib.

fundamental attribution error The tendency to overemphasize personal factors and underestimate situational factors in explaining behaviour.

social psychology The branch of psychology concerned with how others influence the way a person thinks, feels, and acts.

One of the most important lessons of social psychology is that people consistently underestimate the power of situations in affecting human behaviour. In a classic study, the social psychologists Philip Zimbardo and Chris Haney had male Stanford undergraduates play the role of prisoners and guards in a mock prison (Figure 15.2; Haney, Banks, & Zimbardo, 1973). The students, who had all been screened and found to be psychologically stable, were randomly assigned to their roles. What happened was unexpected and shocking. Within days, the guards became brutal and sadistic. They constantly harassed the

1898
Social Facilitation Norman Triplett conducts the first social psychology experiment by demonstrating that people ride their bicycles more quickly in the presence of others.

1930s
Attitude Measurement L. L. Thurstone and Rensis Likert develop attitude scales based on the principles of psychophysics. Carl Hovland at Yale University studies how attitudes are changed.

1945
Field Theory Kurt Lewin proposes a field theory based on physics that states behaviour is a function of the person interacting with the environment.

1950s
People as Intuitive Scientists Fritz Heider argues that social psychologists need to understand "commonsense psychology." Heider's ideas form the basis of attribution theory.

prisoners, forcing them to engage in meaningless and tedious tasks and exercises. Although the study was scheduled to last two weeks, it became necessary to stop it after only six days. The Stanford prison study demonstrated the speed at which apparently normal college students could be transformed into the social roles they were playing. Similarly, the guards at Abu Ghraib believed it was their job to "soften up" the prisoners for interrogation, and their actions were strikingly similar to those of the Stanford participants. So is it the person? Or is it the situation?

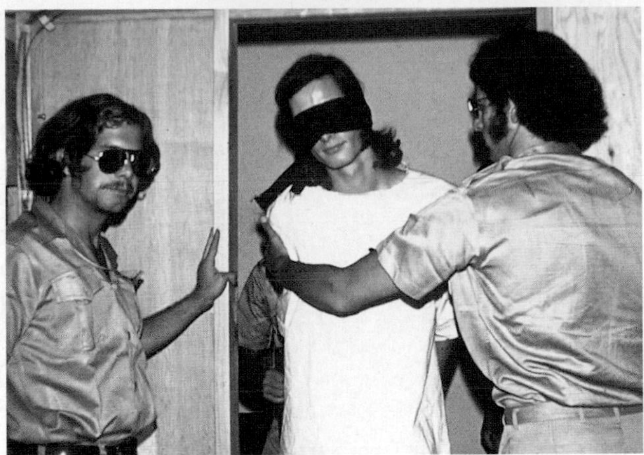

When explaining other people's behaviour, we tend to overemphasize the importance of personality traits and underestimate the importance of situation. This tendency is so pervasive that it has been called the **fundamental attribution error**. As you will learn in this chapter, attributions are explanations for behaviour. We tend to attribute other people's behaviour to their personalities. Even if they are just acting, we believe that their actions reflect their personal dispositions. People tend to think that the *Jeopardy!* host Alex Trebek is smart, that *Star Trek*'s Leonard Nimoy is cold and logical, and that *Friends'* Jennifer Aniston is ditzy and spoiled. We forget that they are actors playing roles and that Alex Trebek is given the answers in advance. Likewise, when people act in a brutal and sadistic way, we assume that they are brutal and sadistic people. We neglect to consider the situation in which they have acted brutally and sadistically and to assess how much power it had in shaping the behaviour. In this chapter you will discover that situations and social rules guide much of our behaviour. You will learn the basic principles of how people interact with each other. You will see that research in social psychology provides profound insights into situations such as Abu Ghraib, revealing to us not that humans are inherently flawed or evil, but rather that social context can be overpowering. Bear in mind that we still must take responsibility for our individual and collective actions, but we also need to learn from situations such as Abu Ghraib and work to create environments that allow individuals and groups to succeed in positive, constructive ways without harming themselves or others. ■

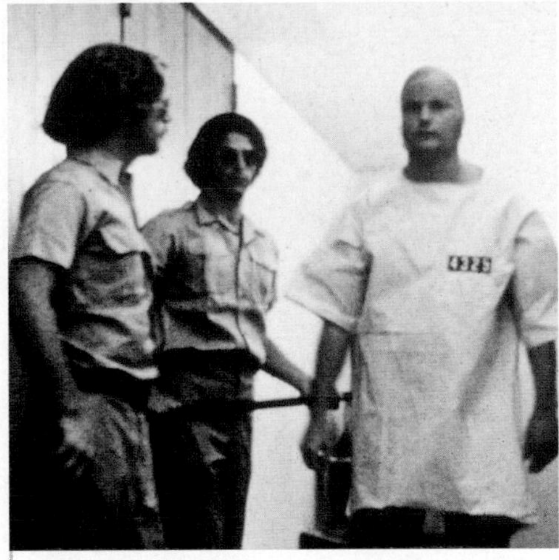

15.2 The Stanford Prison Study The basement of the psychology department at Stanford University was turned into a mock prison, with students randomly assigned to be guards or prisoners. The participants took on their roles with such vigor that the situation became overwhelming for many. The study was terminated early because of concerns for the well-being of the students.

Humans are social animals who live in a highly complex world. Some 6 billion people are right now talking with friends, forming impressions of strangers, arguing with family members, even falling in love with potential mates. Our regular interactions with others—even imagined others—shape who we are and how we understand the world. **Social psychology** is concerned with how people influence other people's

1954
The Nature of Prejudice Gordon Allport provides a comprehensive review of the nature of stereotyping and prejudice. Allport points out that prejudice arises out of normal cognitive processes.

1955
Conformity to Group Norms Solomon Asch demonstrates that people will conform to a group standard. Subsequent research verifies the importance of social influence on human behaviour.

1957
Cognitive Dissonance Leon Festinger argues that people want to maintain consistency between their attitudes and their behaviours, and that discrepancies create tension that needs to be resolved.

1960s
Correspondence Bias and Attributions Edward Jones demonstrates that people overemphasize the importance of personal factors when explaining others' behaviours, a tendency that is the fundamental attribution error.

1965
Obedience to Authority Stanley Milgram shows that normal people will administer painful electric shocks when ordered to do so by an authority figure.

thoughts, feelings, and actions. Because almost every human activity has a social dimension, research in social psychology covers expansive and varied territory: how we perceive ourselves and others, how we function in groups, why we hurt or help people, why we fall in love, why we stigmatize and discriminate against certain people.

As we explore social psychology, we will highlight two important themes that help us understand how humans behave in their social worlds. You have already encountered the first theme, which is that we tend to vastly underestimate the power of situations in shaping both our own and other people's behaviour. As you learned in Chapter 1, cultural rules and norms shape a great deal of our behaviour. Gender roles, for example, dictate appropriate behaviours for women and men, even how we should act in an elevator. But how much do you think situational forces determine your actions? Before reading further, think about a time when you felt compelled to behave in a certain way because of the situation you were in. What forces influenced you to comply with what the group was doing even though you may have felt uncomfortable?

A second important theme of social psychology is that a great deal of mental activity occurs automatically and without conscious awareness or intent, as we have discussed in Chapters 4 and 8 (Bargh & Ferguson, 2000; Gladwell, 2005; Wegner & Bargh, 1998). Within a fraction of a second, we automatically evaluate the people and objects we encounter. Think of yourself walking into a class for the first time or looking over your fellow passengers as you wait to board a plane. We are cognitive "misers" who because of limited mental resources cannot always carefully analyze and scrutinize all our attitudes and behaviours; we sometimes make snap judgments about others based on very little information. Recall from Chapter 8 that these quick judgments can be adaptive, such as helping us avoid dangers. As we noted there, we often are not aware of the extent to which unconscious processes affect our behaviour. We also use a number of mental strategies to protect our self-esteem and to make our attitudes and actions appear in agreement, all of this occurring without our awareness. At this point, you may wonder if we have any control over our thoughts and behaviour. Rest assured: By the time you finish reading about our social lives, you will see that you have plenty of control over your life.

How Do We Know Ourselves?

Sit back for a second and think about your self. What are you visualizing? What questions are you asking? Are you asking yourself, "Who am I?" "What am I?" The notion of self is one of the most commonly held concepts in our society, yet, as this thought exercise shows, it is difficult to define. At a general level, the self involves the mental representation of personal experience and includes thought processes, a physical body, and a conscious experience that one is separate and unique from others. This sense of self is a unitary experience that is continuous over time and space.

1970
Bystander Apathy Bibb Latane and John Darley propose that people are unlikely to help a person in need if there is a diffusion of responsibility among observers.

1972
Stanford Prison Study Researcher Philip Zimbardo "jails" Stanford undergraduates and assigns other students to act as guards. Within a short time, the students assume the social roles of the guards and prisoners.

1977
Social Cognition and the Self-Schema Hazel Markus, heavily influenced by cognitive psychology, conducts research on the self-schema. Her findings contribute to increased interest in social cognition.

1980–1990s
Self-Serving Motivational Biases Social psychologists such as Shelley Taylor, Anthony Greenwald, and Roy Baumeister describe a variety of cognitive biases and illusions that people use to maintain a positive self-image.

2000s
Social Neuroscience Social psychologists and cognitive neuroscientists work together to examine the neural correlates of social cognition and behaviour.

When you wake up in the morning, you don't have to figure out who you are, even if you sometimes do have to figure out where you are, for example, when you are on vacation.

Our Self-Concept Consists of Self-Knowledge

Write down 20 answers to the question "Who am I?" What sorts of things did you list? The information that you provided is part of your **self-concept**, which is everything that you know about yourself. For example, common answers given by university students include gender, age, student status, interpersonal style (shy, friendly), interpersonal characteristics (moody, optimistic), and body image. But how does thinking of one's self as *shy*, or believing one's self to be *optimistic*, or seeing one's self as *overweight* affect how we function and feel from day to day? Many social psychologists view the self-concept as a cognitive knowledge structure that guides our attention to information that is relevant to us and that helps us adjust to the environment. If you think of yourself as shy, you might avoid a raucous party. If you believe yourself to be optimistic, you might easily bounce back from a poor grade in organic chemistry. If you see yourself as overweight, you might choose not to eat dinner at a restaurant with an open buffet.

What effect does being aware of ourselves have on our behaviour?

SELF-AWARENESS Among the first to consider the nature of self, both the psychologist William James and the sociologist George Herbert Mead differentiated between the self as the knower ("I") and the self as the object that is known ("me")—now called the objectified self. In the sense of the knower, the self is the subject doing the thinking, feeling, and acting. The "I" is involved in executive functions, such as choosing, planning, and exerting control. In the sense of the objectified self, the self is the knowledge that you hold about yourself, as when you think about your best and worst qualities. The sense of self as the object of attention is the psychological state known as **self-awareness**. Self-awareness occurs when the "I" thinks about the "me."

self-concept The full store of knowledge that people have about themselves.

self-awareness A state in which the sense of self is the object of attention.

Have you ever noticed that when you are giving a class presentation or having a job interview you become really focused on yourself, perhaps especially noticing any mistakes you make? Social psychologists have long been interested in the consequences of being self-aware. In 1972, psychologists Shelley Duval and Robert Wicklund introduced the theory of *objective self-awareness.* They proposed that self-awareness leads people to act in accordance with their personal values and beliefs. For instance, one study showed that college students are less likely to cheat, when given the opportunity, if they are sitting in front of mirrors. Perhaps seeing our own faces reminds us that we do not personally value cheating. Remember from our earlier discussion (Chapter 9) that discrepancies between personal standards (not cheating) and goals (passing the test) can motivate behaviours (studying) that reduce the discrepancy. According to the social psychologist Tory Higgins's *self-discrepancy theory*, this awareness of differences between personal standards and goals leads to strong emotions (1987). For instance, being self-aware of a discrepancy between how we see ourselves (lazy) and how we would like to see ourselves (hardworking) can lead to disappointment, frustration, and depression; discrepancies between how we see ourselves (lazy) and how we believe we ought to seem to others (hardworking) produce anxiety and guilt. So if you fail an important class because you didn't study enough, you might feel depressed, but if you think about how your parents might feel when you fail an important class because you didn't study enough, you are likely to feel anxious.

We know from studying patients with brain injuries that self-awareness is highly dependent on the normal development of the frontal lobes of the brain. People with damage to the frontal lobes tend to be only minimally self-reflective and seldom report daydreaming or other types of introspection. They also often show a surprising lack of interest in or knowledge about their disorders. Such individuals are not completely unaware of themselves following frontal-lobe damage, but they don't find information about the self personally significant. The neuropsychologist Donald Stuss (1991) reported a highly intelligent patient who had a tumour removed from his frontal lobes. Subsequently, even though his intelligence and knowledge about the world were intact, he had difficulty at work and became extremely unproductive. People with damage to the frontal lobes often have social and motivational impairments that interfere with job performance. Despite 18 months of therapy, the patient continued to do poorly on the job, but he could not realize that he had a problem. When asked to role-play the situation as if he were the boss, he quickly and clearly recognized the problem and made appropriate recommendations—that the worker be put on a disability pension. However, when he was asked to evaluate himself from his own subjective perspective, he disagreed with the recommendation he had just made. This dramatic example—and others like it—shows that frontal-lobe patients often have distortions in how they process information about the self.

self-schema The cognitive aspect of the self-concept, consisting of an integrated set of memories, beliefs, and generalizations about the self.

SELF-SCHEMA Have you ever been at a crowded, noisy party where you could barely hear yourself speak, but when someone across the room mentioned your name, you heard it clearly above the din? As we discovered earlier, the *cocktail party effect* occurs because information about the self is processed deeply, thoroughly, and automatically. According to Hazel Markus (1977), the **self-schema** is the cognitive aspect of the self-concept, consisting of an integrated set of memories, beliefs, and generalizations about the self. Your self-schema also helps filter information, which means you are likely to notice things that are self-relevant, such as your name. Your self-schema consists of those aspects of your behaviour and personality that are important to you. For instance, being a good athlete or a good student may be an important component of your self-schema, whereas having few cavities probably is not. The self-schema helps us perceive, organize, interpret, and use information about the self; it can be viewed as a network of interconnected knowledge about the self (Figure 15.3). Thus, when asked, "Are you ambitious?" we do not have to sort through all occasions in which we acted ambitiously or not to come up with an answer. Rather, self-schemas summarize past information so we can provide an answer automatically.

15.3 Self-Schemas The self-schema consists of interrelated knowledge about the self.

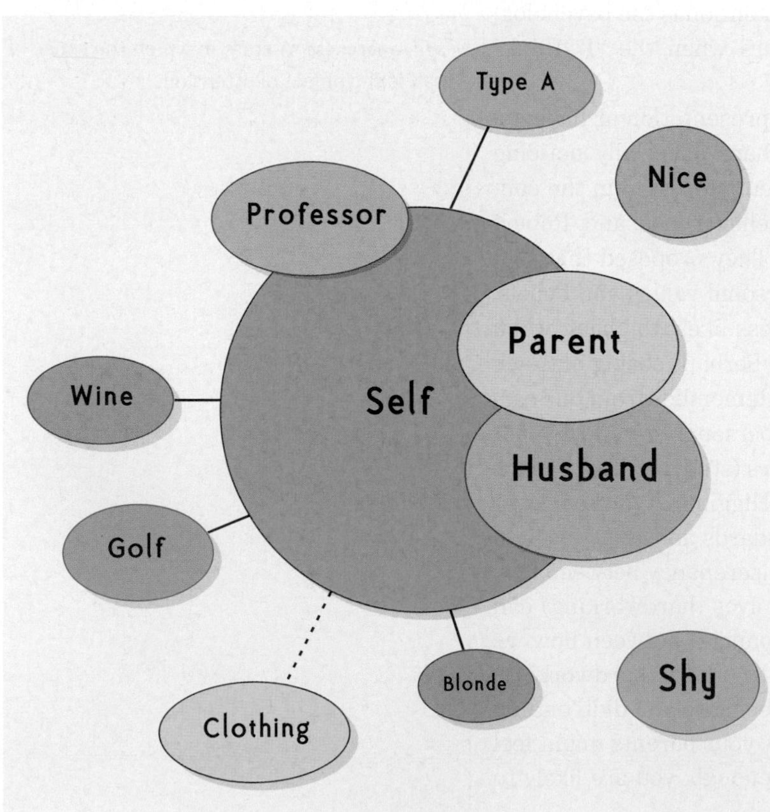

Self-schemas may lead to enhanced memory for information that is processed in a self-referential manner. University of Calgary professor Tim Rogers and colleagues (1977) showed that trait adjectives that were processed with reference to the self (e.g., "Does the word *honest* describe you?") were better recalled than comparable items that were processed only for their general meaning (e.g. "Does the word *honest* mean the same as trustworthy?"). Researchers recently have sought to examine the neural dimensions of self-referential processing (Gillihan & Farah, 2005). A typical finding is activation of the middle of the frontal lobes when people process information about themselves (Kelley et al., 2002). This brain region is more active when you answer questions about your self (e.g., "Are you honest?") than when you answer questions about another person (e.g., "Is Madonna honest?"). The greater the activation of this area during self-referencing, the more likely you are to remember the item later during a surprise memory task (Macrae, Moran, Heatherton, Banfield, & Kelley, 2004). Along with the self-awareness deficit shown in neurological patient studies, this further supports the idea that the frontal lobes are important for processing information about the self.

WORKING SELF-CONCEPT The immediate experience of self, the working self-concept, is limited to the amount of personal information that can be processed cognitively at any given time. Because the working self-concept contains only part of the vast array of self-knowledge, the sense of self varies from situation to situation. For instance, at a party you might think of yourself as fun loving rather than as intelligent, even though both traits are aspects of your self-concept. Thus, your self-descriptions vary as a function of which memories you retrieve, which situation you are in, the people you are with, and your role in that situation. This varying of self-concept is part of the way that situations shape behaviour. Since people act in accordance with their self-concepts, different situations elicit different behaviours because the situation activates different aspects of the self-concept.

(a) Who am I?
I am male.
(b) Who am I?
I am African American.

15.4 Your Self-Concept When answering the question "Who am I?" people are especially likely to mention characteristics that are distinctive.

When people consider who they are, they often emphasize things that make them distinct from others. For instance, think back to your 20 responses to "Who am I?" Which of your answers stressed your similarity to other people or membership in a group? Which stressed your differences from other people, or at least from the people immediately around you? People are especially likely to mention things such as ethnicity, gender, or age if they differ from other people around them at the time (Figure 15.4). Thus, Canadians would be more likely to note their nationality if they were in Boston than if they were in Toronto. Because the working self-concept guides behaviour, this implies that they are also more likely to feel and act like Canadians in Boston than in Toronto. There is, of course, an optimal level of distinctiveness, since most people want to avoid standing out too much; teenagers especially want to stand out and fit in at the same time.

INDEPENDENT AND INTERDEPENDENT SELVES An important way in which people differ in their self-concepts is whether they view themselves as fundamentally separate from or inherently connected to other people. Recall from Chapter 1 that Westerners tend to be "independent" and autonomous, stressing their individuality, whereas Easterners tend to be more "interdependent" with each other, stressing their sense of being part of a collective. Harry Triandis (1989) notes that some cultures (such as Japan, Greece, mid-Africa, Pakistan, and China) emphasize the collective self

How does the basic sense of self vary around the world?

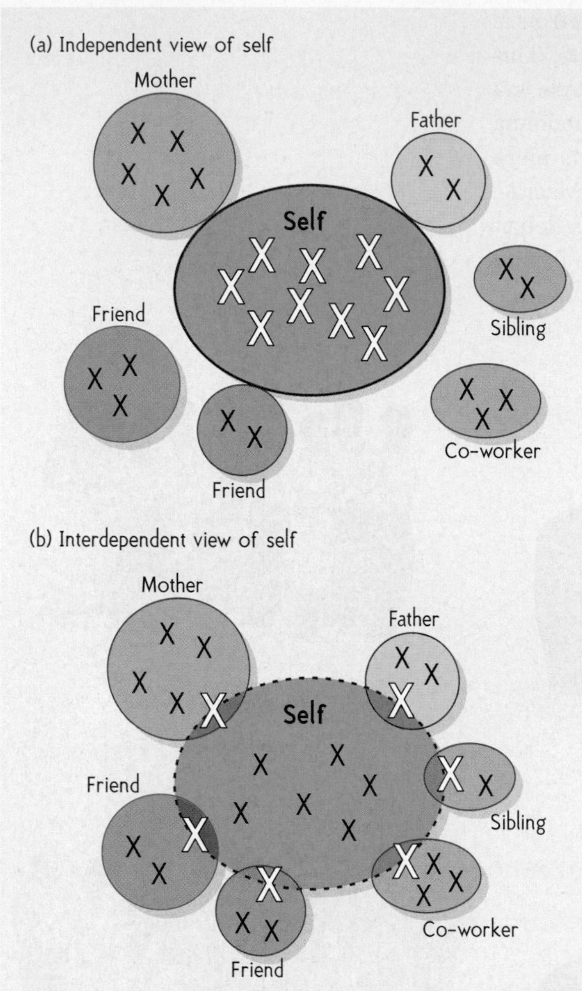

15.5 Self-Construals Self-construals in (a) individualist and (b) collectivist cultures.

more than the personal self. The collective self emphasizes connections to family, social groups, and ethnic groups and conformity to societal norms and group cohesiveness. Individualist cultures (such as in northern and western Europe, Australia, Canada, New Zealand, and the United States) place greater emphasis on the personal self, which stresses personal rights and freedoms, self-expression, and diversity. For example, in Canada people dress differently from one another, cultivate personal interests, and often enjoy standing out from the crowd. In Japan, people tend to dress more similarly and to respect situational norms. When a Canadian family goes to a restaurant, each person orders what he or she prefers. In China, when a family goes to a restaurant, multiple dishes are shared by the entire table.

Hazel Markus and Shinobu Kitayama (1991) noted that people in collectivist cultures tend to have **interdependent self-construals**, in which their self-concepts are determined to a large extent by their social roles and personal relationships (Figure 15.5). As children, they are raised to follow group norms and to be obedient to parents, teachers, and other people in authority. They are expected to find their proper place in society and not to challenge or complain about their status.

By contrast, people in individualist cultures tend to have **independent self-construals**. Children are encouraged by parents and teachers to be self-reliant and to pursue personal success, even at the expense of interpersonal relationships. Their sense of self is based on their feelings of being distinct from others. It is important to note, however, that these are broad and general patterns, and that there is variability in terms of independent/interdependent self-construals within both individualist and collectivist cultures.

Perceived Social Regard Influences Self-Esteem

North American culture has been obsessed with self-esteem for the past two decades. At a basic level, **self-esteem** is the evaluative aspect of the self-concept, referring to whether people perceive themselves to be worthy or unworthy, good or bad. It is people's emotional response as they contemplate and evaluate different characteristics about themselves. Although self-esteem is related to the self-concept, it is possible for people to objectively believe positive things about themselves without really liking themselves very much. Conversely, it is possible for people to like themselves very much, and therefore have high self-esteem, even when objective indicators don't support such positive self-views.

Many theories assume that people's self-esteem is based on how they believe others perceive them, known as *reflected appraisal*. According to this view, people internalize the values and beliefs expressed by important people in their lives. They do this by observing the attitudes and actions of others and adopting these attitudes and behaviours as their own. Consequently, people come to respond to themselves in a manner consistent with how others respond to them. From this perspective, when important figures reject, ignore, demean, or devalue a person, low self-esteem is likely to result.

This social view of self-esteem led some theorists to promote unconditional acceptance of children by their parents, meaning that parents should love their children no matter what the children do. Later theorists have noted, however, that unconditional acceptance needs to occur in the context of relatively strict parenting,

interdependent self-construals Self-concepts determined largely by social roles and personal relationships.

independent self-construals A view of the self as separate from others, emphasizing self-reliance and the pursuit of personal success.

self-esteem The evaluative aspect of the self-concept.

in which parents clearly define limits and enforce them by providing positive reinforcement for behaviours within, and punishment for behaviours outside, those limits (Coopersmith, 1967). Children are accepted and loved no matter what they do, but inappropriate behaviours are corrected through punishment.

SOCIOMETER THEORY A novel and important social account of self-esteem has been proposed by Mark Leary and colleagues (1995). Leary assumes that humans have a fundamental need to belong, a need that is adaptive. For most of human evolution, those who belonged to social groups were more likely to survive and reproduce than those who were excluded and left to survive on their own. According to Leary, self-esteem monitors the likelihood of social exclusion. When people behave in ways that increase the likelihood that they will be rejected, they experience a reduction in self-esteem. Thus, self-esteem is a **sociometer**, an internal monitor of social acceptance or rejection (Figure 15.6). Those with high self-esteem have sociometers that indicate a low probability of rejection, and therefore such individuals do not worry about how they are perceived by others. By contrast, those with low self-esteem have sociometers that indicate the imminent possibility of rejection, and therefore they are highly motivated to manage their public impressions. An abundance of evidence supports the sociometer theory, including the consistent finding that low self-esteem correlates highly with social anxiety (Leary & MacDonald, 2003; Leary, 2004).

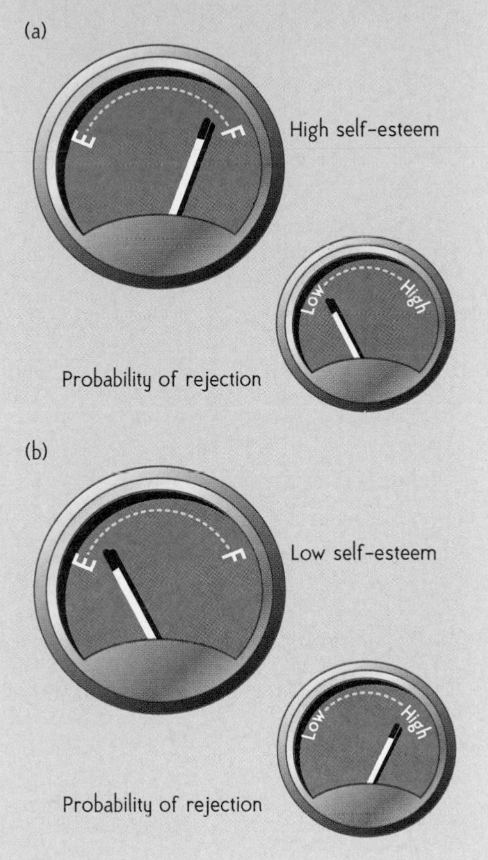

15.6 Sociometers According to sociometer theory, self-esteem is the gauge that measures the extent to which people are (a) being included in or (b) excluded from their social group.

SELF-ESTEEM AND DEATH ANXIETY One provocative theory proposes that self-esteem helps provide meaning for individuals by staving off anxiety over their mortality. According to *terror management theory*, self-esteem protects people from the horror associated with knowing that they will eventually die (Greenberg, Solomon, & Pyszczynski, 1997; Pyszczynski, Greenberg, Solomon, Arndt, & Schimel, 2004). This theory argues that people counter mortality fears by creating a sense of symbolic immortality through contributing to their culture and upholding its values. From this cultural perspective, self-esteem develops from the personal belief that one is living up to criteria that are valued within the culture. Accordingly, exaggerations of personal importance reflect attempts to buffer anxiety about inevitable death. Research has demonstrated that reminding people of their mortality leads them to act in ways that enhance their self-esteem (Goldenberg, McCoy, Pyszczynski, Greenberg, & Solomon, 2000). Can you think of any examples from your own experience? Have the various wars and terrorist incidents over the last few years changed how you have acted and how you feel about yourself?

SELF-ESTEEM AND LIFE OUTCOMES With such emphasis placed on self-esteem within Western culture, you might expect that having high self-esteem is the key to life success. The evidence from psychological science indicates that it may be less important than what is commonly believed. After reviewing several hundred studies, Roy Baumeister and his colleagues (2003, 2005) found that although people with high self-esteem report being much happier with their lives, self-esteem is only weakly related to objective life outcomes. People with high self-esteem who consider themselves smarter, more attractive, and better liked do not necessarily have higher IQs or are considered to be better by others. Many people with high self-esteem are successful in their careers, but so are many people who have low self-esteem. Moreover, to the extent that there is a small relationship between self-esteem and life outcomes, perhaps it is success that causes high self-esteem. Remember that correlation does not prove causation. Some people might have higher self-esteem because they have done well in school or in their careers.

sociometer An internal monitor of social acceptance or rejection.

What might surprise you is that there may be some downsides to having really high self-esteem. It is not uncommon for violent criminals to have very high self-esteem; indeed, they become violent when they feel that others are not treating them with an appropriate level of respect (Baumeister, Smart, & Boden, 1996). School bullies also often have high self-esteem (Baumeister et al., 2003). When people with high self-esteem believe that their abilities have been challenged, they also act in ways, such as being antagonistic or boastful, that cause people to dislike them (Heatherton & Vohs, 2000; Vohs & Heatherton, 2004). So although having high self-esteem does seem to make people happier, it does not necessarily lead to successful social relationships or life success.

We Use Mental Strategies to Maintain Our Views of Self

A consistent finding in social psychology is that most people think of themselves in favourable terms. Indeed, most people automatically show favouritism to anything associated with themselves. For example, people display a consistent preference for their personal belongings over things they don't own (Beggan, 1992), and they even prefer the letters of their own name to other letters (Koole, Dijksterhuis, & van Knippenberg, 2001). This is especially true for their initials (Figure 15.7). Sometimes these positive views of the self seem to be inflated. For instance, 90 percent of adults claim that they are better-than-average drivers, even if they have been hospitalized for injuries caused by car accidents (Guerin, 1994; Svenson, 1981). Similarly, in one College Board Examination Survey in the United States of more than 800,000 college-bound seniors, none rated themselves as below average, and a whopping 25 percent rated themselves in the top 1 percent (Gilovich, 1991). Most people describe themselves as above average in just about every possible way, which is referred to as the *better-than-average effect* (Alicke, Klotz, Breitenbecher, Yurak, & Vredenburg, 1995). People with high self-esteem are especially likely to do so.

According to an influential paper by Shelley Taylor and Jonathan Brown (1988), most people have positive illusions—overly favourable and unrealistic beliefs—in at least three domains. First, people tend to overestimate their own skills, abilities, and competencies, as is the case with the better-than-average effect. Second, most people have an unrealistic perception of their personal control over events. Some fans

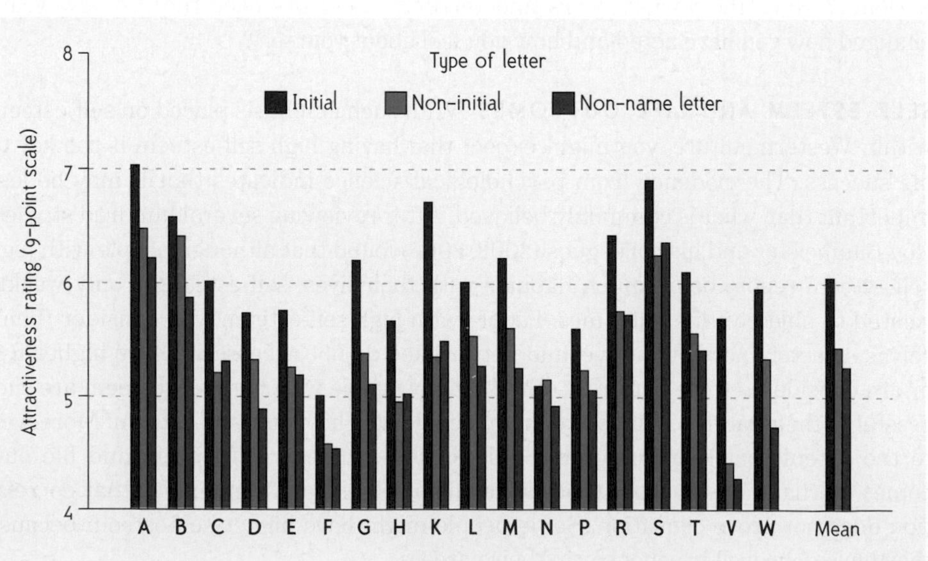

15.7 Favouritism Subjects rated the letters in their name, especially their first and last initials, to be especially attractive.

feel that their favorite sports teams are more likely to win if they are at the game or wear their lucky jersey. Third, most people are unrealistically optimistic about their personal futures, believing that they will probably be successful, marry happily, and live a long life. Although positive illusions can be adaptive when they help people to be optimistic in meeting life's challenges, they can lead to trouble when people overestimate their skills and underestimate their vulnerabilities. This combination might explain the high rate of accidents among novice car drivers.

How do people maintain such positive views of self? Life is filled with failure, rejection, and disappointment. Yet most people feel pretty good about themselves. Over the last 30 years, social psychologists have cataloged (though not necessarily endorsed) a number of automatic and unconscious strategies that people use to maintain their positive sense of self. Among the most common are self-evaluative maintenance, social comparisons, and self-serving biases.

15.8 Avril Lavigne How you feel about Avril Lavigne's successful music career is influenced by whether you know her personally, whether you also desire a music career, and whether you think a successful music career would be attainable for yourself.

SELF-EVALUATIVE MAINTENANCE Abraham Tesser (1988) notes that self-esteem can be affected not only by how people perform, but by how relevant their performances are to their self-concepts and how their performances compare with those of significant people around them. According to the theory of *self-evaluative maintenance*, people can feel threatened when someone close to them outperforms them on a task that is personally relevant. If you had a twin brother who shared your aspiration to be a world-class chef, his brilliant success at cooking would have important implications for how you felt about yourself. To maintain your sense of self-esteem, Tesser argues that you would either distance yourself from the relationship or select a different aspiration. Of course, if your twin brother excels at something you don't find personally relevant, then you might bask in the glow of reflected glory and experience a boost in self-esteem based on your relationship. Similarly, if you had a friend who was a gold medalist in track and field, you would likely experience a boost in self-esteem simply by knowing her. Self-evaluation maintenance causes people to exaggerate or publicize their connections to winners and to minimize or hide their relations to losers.

However, in some situations people will feel good about themselves when they encounter someone who is doing much better than they are on a relevant dimension. For example, Lockwood and Kunda (1997) found that students who planned to be teachers or accountants rated themselves more positively when they read about a high-achieving teacher or accountant than when they read about someone who was successful in a domain that was not relevant to them. At first glance this seems to contradict the findings reported by Tesser. The key difference is that people feel good about themselves by comparing themselves with "superstars" in their chosen fields when they feel that such successful performance is attainable for them. If people can believe that "someday, that will be me," successful others can be very inspirational. In contrast, if the successful performance seems unattainable to people, witnessing superstars tends to make people feel that much worse about themselves (Figure 15.8).

SOCIAL COMPARISONS **Social comparison** occurs when people evaluate their own actions, abilities, and beliefs by contrasting them with other people's. Especially when they have no objective criteria, such as knowing how thin they ought to be or how much money represents a good income, people compare themselves to others to see where they stand. Recall from Chapter 9 that social comparisons are an important way to understand our actions and emotions, such as whether it is reasonable to be afraid in a situation. Obviously, who is chosen as a standard of comparison will have a great influence on self-esteem. In general, people with high self-esteem make *downward comparisons*, contrasting themselves with people who are deficient to them on relevant

social comparison The evaluation of our own actions, abilities, and beliefs by contrasting them with other people's.

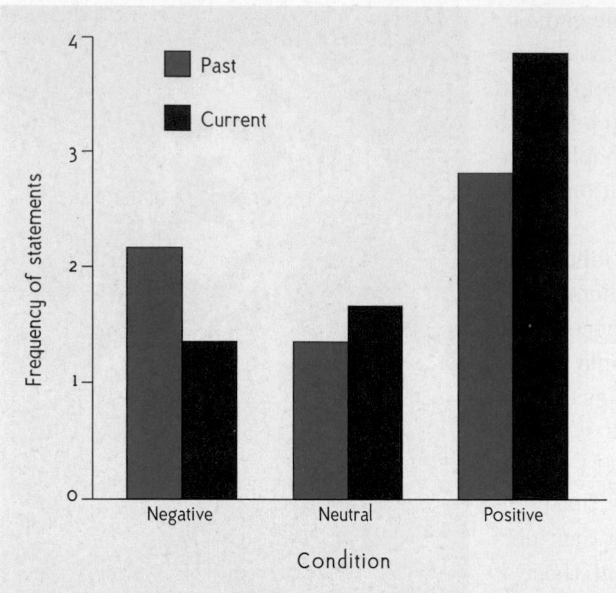

15.9 Making Comparisons College students rated their past selves as having had more negative features than their current selves, which were rated as having more positive features.

dimensions. People with low self-esteem tend to make *upward comparisons* with those who are superior to them. Interestingly, people also use a form of downward comparison when they recall their own past. That is, people often view their current selves as being better than their former selves (Figure 15.9; Wilson & Ross, 2001). For example, many people remember themselves as being awkward and shy as adolescents. Happiness is relative, and people can maintain positive self-feelings by focusing on how much better off they are now than they were in the past. What we can see from these findings is that viewing ourselves as better than others or as better than we used to be makes us feel good about ourselves. But those of us who constantly compare ourselves with people who do better than us may confirm our negative self-feelings.

SELF-SERVING BIASES People with high self-esteem tend to take credit for success but blame failure on outside factors, which is referred to as a **self-serving bias**. For instance, students who do extremely well on exams often explain their performance by referring to personal skills or hard work. Those who do poorly might describe the test as an arbitrary examination of trivial details. People with high self-esteem also assume that criticism is motivated by envy or prejudice. Indeed, members of groups that are prone to discrimination (such as the disabled and ethnic minorities) may have high self-esteem. One theory, proposed by Jennifer Crocker and Brenda Major (1989), suggests that members of these groups maintain positive self-esteem by taking credit for success and blaming negative feedback on prejudice. So if you succeed, you succeed due to personal strengths and despite the odds. If you fail, you fail because of external factors and unfair obstacles. Imagine how this might be useful in maintaining self-esteem for those who regularly face discrimination.

self-serving bias The tendency for people to take personal credit for success but blame failure on external factors.

Thinking Critically

Are Self-Serving Biases Universal?

We have reviewed evidence that people tend to see themselves in especially positive—perhaps even exaggerated—ways, in what is known as the self-serving bias. Over the last thirty years, social psychologists have documented many different ways that people appear to be biased in thinking about themselves compared with how they think about others (Campbell & Sedikides, 1999). When students fail tests, they explain it by blaming their low score on something else, such as the inept teacher who made up the test or their lack of effort, sleep, interest, and so on. We compare ourselves with others who did worse, we diminish the importance of the test, we think about the other things that we are really good at, and we bask in the glow of reflected glory of our family and friends. The overall picture suggests that we are extremely well equipped to protect our positive beliefs about ourselves (Figure 15.10). Some have even argued that these self-serving biases reflect healthy psychological functioning (Taylor & Brown, 1988; Mezulis, Abramson, Hyde, & Hankin, 2004).

Social psychologists have tended to view self-enhancing biases as a universal human trait. Although people who are depressed might fail to show the effect, the assumption is that most healthy, functioning individuals show robust self-enhancement. Over the last few years, however, researchers have increasingly questioned whether the self-serving bias is truly universal across cultures. Steven Heine and his colleagues (1999) argued that self-serving biases may be more common in Western cultures, in which individuality is emphasized, than in Eastern cultures, in which the collective self is emphasized. That is, believing that you are an especially talented individual presupposes that you are better than others. Such an attitude is not acceptable in Eastern cultures in which the group, rather than the individual, is special.

What does the evidence indicate about whether self-serving biases are universal? Consider a study by Endo and Meijer (2004), who simply asked American and Japanese students to list as many success and failure experiences as they could. The Americans showed a bias for listing successes, whereas Japanese students tended to list failure and success equally. They also found that Americans used outside forces to explain failure, but Japanese students used them to explain success. Indeed, Markus and Kitayama (1991) argued that self-criticism rather than self-promotion is the most common social norm in Asian cultures. The overall evidence supports the view that people in individualistic cultures are more concerned with self-enhancement than those in collectivist, particularly Asian, cultures (Heine, 2003). For example, in a recent meta-analysis that involved more than 500 studies, people in Western countries showed a much larger self-serving bias than those in Eastern cultures (Mezulis et al., 2004). Although some Eastern cultures show a modest self-serving bias, especially China and Korea, it is typically much smaller than in most Western countries.

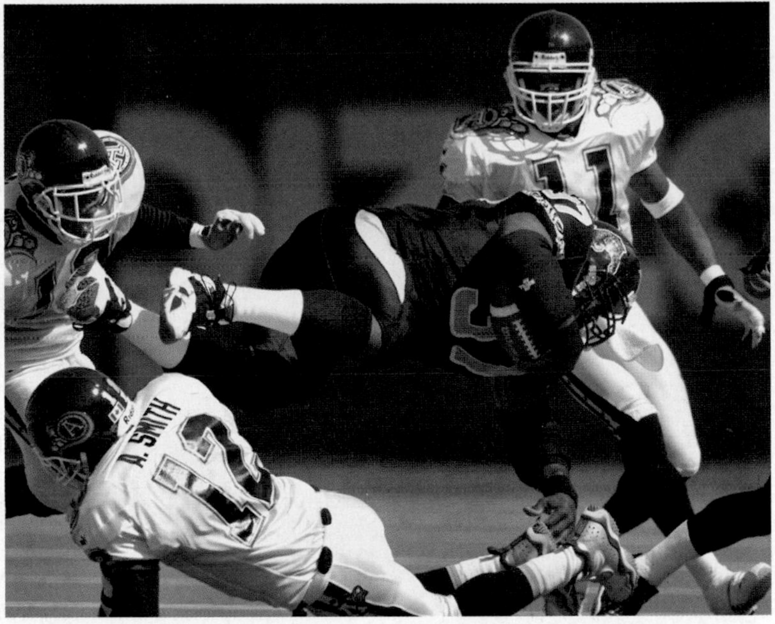

15.10 Self-Serving Bias Athletes often attribute victory to their own skills, and defeat to external factors. When the Toronto Argonauts lost to their bitter rivals, the Hamilton Tiger-Cats, in the 1999 CFL East Division Semi-Finals, they may have attributed their defeat to external factors like poor calls or bad luck. On winning the 92nd Grey Cup in 2004 (triumphing over Hamilton in the East Division Semi-Finals), they most likely attributed their victories to their personal qualities and skills.

One question that you might ask is whether these differences reflect cultural rules about publicly admitting positive self-views. Perhaps people in the East are just more modest in public even though they engage in strategic self-enhancement. However, in studies that have used anonymous reporting, where presumably there is less call for modesty, Easterners continue to show a low level of self-serving bias (Heine, 2003). Thus, it isn't that Easterners are just good at covering up their self-serving tendencies.

It has also been argued that researchers have a Western bias in the domains that they study, and so have failed to assess Eastern research participants in crucial domains. Some findings indicate that those from Asian cultures self-enhance in domains that are particularly important to them (Brown & Kobayashi, 2002), so that if a culture values industriousness people will describe themselves as especially hardworking, but if the culture values kindness, people will emphasize how nice they are. It is possible that people in Eastern cultures emphasize what good group members they are as a method of self-enhancement (Sedikides, Gaertner, & Toguchi, 2003). According to this latter perspective, self-enhancement is universal, but the tactics for achieving it vary across cultures. So when the culture emphasizes personal achievement, people self-enhance as individuals; when the culture emphasizes group achievement, people self-enhance as group members. Perhaps the motivation to be a good person is universal, but what it means to be a good person varies across cultures.

"And now at this point in the meeting I'd like to shift the blame away from me and onto someone else."

The question of whether self-serving biases are universal has led to lively debate among social psychologists, in part because it is connected to important questions about how culture shapes our sense of self (Heine, 2005; Sedikides et al., 2003). Although social psychologists continue to contest the extent to which self-serving biases are universal, how much do you believe people really differ around the world? Do you think it is more important to be respected by others, or to feel good about yourself no matter what others think? Is it possible that people in Eastern cultures feel better about themselves when they demonstrate to others that they are modest and self-effacing whereas Westerners feel better when they can show others that they are successful? What function does self-enhancement serve? How might this function differ across cultures? ■

REVIEWING the Principles | How Do We Know Ourselves?

Everything that we know about ourselves makes up our self-concept. Many times people are especially aware of their objective selves, a tendency that leads them to behave according to their personal standards and beliefs. Information about the self is processed efficiently and quickly through a self-schema, although how people describe themselves can vary depending on the situation. Researchers have found cross-cultural differences in how the self is construed, with collectivist cultures emphasizing the interdependent self and individualistic cultures emphasizing the independent self. Self-esteem is the evaluative aspect of the self-concept. Most people have moderate to high levels of self-esteem, viewing themselves as better than average on a number of dimensions. These and other strategies help people maintain positive self-esteem.

How Do Attitudes Guide Behaviour?

How do you feel about the Abu Ghraib prisoner abuse scandal that you read about at the beginning of this chapter? It is quite likely that, even before you started reading you had opinions about it, just as you have opinions, beliefs, and feelings about yourself, your friends, your favourite television program, or even spinach. These opinions, beliefs, and feelings are called attitudes. The concept of **attitude**, which refers to the evaluation of an object, event, or idea, is central to social psychology. We all have attitudes about all sorts of things in our daily lives, from our deodorant to politics, morals, and religion. Our attitudes range from the trivial to the core beliefs and values that define who we are as human beings. As you will see, some of these attitudes are *explicit*, in that we are aware of them. However, sometimes we are influenced by *implicit* attitudes that we don't even know of or recognize. Some attitudes are complex and involve multiple components. You might enjoy eating ice cream but believe it is bad for your health. You might be disgusted with the Abu Ghraib prisoner abuse but have pity for the U.S. soldiers caught up in the horrific situation. In this section we consider how attitudes affect our daily lives.

attitude The evaluation of objects, events, or ideas.

We Form Attitudes through Socialization and Experience

Direct experience of or exposure to things provides information that shapes attitudes. As we encounter new objects, we explore them and learn about them. In general, people develop negative attitudes about new objects more quickly than they do positive attitudes (Fazio, Eisner, & Shook, 2004).

But how do we come to like something that we could not stand the first time we were exposed to it? We talk about acquiring a taste for foods that we did not like originally, such as coffee or sushi. Typically, the more that people are exposed to an item, the more they tend to like it. Robert Zajonc (1968, 2001), in a classic set of studies, exposed people to unfamiliar items a few times or many times. Greater exposure to the item, and therefore greater familiarity with it, caused people to have more positive attitudes about the item; this is known as the *mere exposure effect*. For example, when people are presented with a normal and a reversed photograph of themselves, they tend to prefer the reversed images, which correspond to what they see when they look in the mirror (Figure 15.11). Friends and family prefer the true photograph, which corresponds to how they perceive the person.

15.11 The Mere Exposure Effect If he is like most people, Stephen Harper will prefer his mirror image (right), with which he is more familiar, to his photographic image (left).

Attitudes can be conditioned. Advertisers often use classical conditioning; seeing an attractive celebrity paired with a brand of toilet-bowl cleaner leads to more positive attitudes about that product. Following classical conditioning, a formerly neutral stimulus, such as toilet-bowl cleaner, leads to the same attitude response as the paired object, such as Brad Pitt. Operant conditioning also shapes attitudes. If each time you study you are rewarded with good grades, you will develop a more positive attitude toward studying.

Attitudes are also shaped through socialization. Caregivers, media, teachers, religious leaders, and politicians guide attitudes about many things. For instance, Hindus often do not eat beef, whereas Jews often do not eat pork. Society socializes many of our attitudes, including which things are edible. Would you eat a worm? Most Westerners would find it disgusting. But in some cultures worms are a delicacy.

Behaviours Are Consistent with Strong Attitudes

When do attitudes predict behaviour?

To the extent that attitudes are adaptive, they should guide behaviour. In general, the stronger, more personally relevant the attitude, the more likely it is to predict behaviour, to be consistent over time, and to be resistant to change. For instance, someone who grew up in a strongly Democratic household, where derogatory comments about Republicans were frequently expressed, is more likely to register as a Democrat and even blindly vote Democratic than someone who grew up in a more politically neutral environment. Moreover, the more specific the attitude, the more predictive it is. For instance, your attitudes toward recycling are more predictive of whether you take your soda cans to a recycling bin than are your general environmental beliefs. Attitudes that are formed through direct experience also tend to predict behaviour better. No matter what kind of parent you think you will be, once you've seen one child through toddlerhood, your attitudes toward different child-rearing techniques will be very strong and will certainly predict how you'll approach the early months and years of your second child.

So far we have been referring mainly to *explicit* attitudes, which are those you know about and can report to other people. When you say that you like bowling, you are stating your explicit attitude toward it. The ease with which memories related to an attitude are retrieved—known as *attitude accessibility*—predicts behaviour consistent with the attitude. Russell Fazio (1995) has shown that attitudes that are easily activated are more stable, predictive of behaviour, and resistant to change. Thus, the more quickly you recall that you like your psychology course, the more likely it is that you will attend lectures and read the textbook.

implicit attitudes Attitudes that influence our feelings and behaviour at an unconscious level.

cognitive dissonance The perceptual incongruity that occurs when there is a contradiction between two attitudes or between an attitude and a behaviour.

Anthony Greenwald and Mahzarin Banaji (1995) noted that many attitudes influence our feelings and behaviour at an unconscious level, and they refer to these as **implicit attitudes**. Implicit attitudes are accessed from memory quickly with little conscious effort or control. Just as implicit memory (see Chapter 7) allows us to ride a bicycle without thinking about it, implicit attitudes shape behaviour without our awareness. For instance, you might purchase a product used by a celebrity when you have no conscious memory of having seen the celebrity use the product. Some evidence suggests that implicit attitudes involve brain regions associated with implicit learning in general (Lieberman, 2000). Also as with implicit memory, researchers assess implicit attitudes through indirect means, such as through behaviour rather than through self-report. Recall from Chapter 1 that the implicit attitudes test (IAT; Greenwald, McGhee, & Schwartz, 1998) is a reaction-time test that can identify implicit attitudes. The IAT measures how quickly we associate concepts or objects with words that are positive or negative. Responding more quickly to the association of female-bad than female-good indicates your implicit attitude about females. Implicit attitudes are also revealed in people's daily behaviours. Those who claim to harbour no racist beliefs still might cross the street to avoid an encounter with a person of colour. Research indicates that people possess dual attitudes about many objects: one is automatic and unconscious; the other is explicit and comes to mind when people contemplate their attitudes (Wilson, Lindsey, & Schooler, 2000). Dual attitudes reflect the complexity of human experience, such as our relationships with siblings. Even if we state explicitly that we can't stand one of our brothers or sisters, we still might show concern and offer aid if he or she is in trouble. The new explicit attitude does not wipe out the implicit attitude that developed over the course of childhood. You will see later that implicit attitudes are important for understanding racial stereotypes.

Discrepancies Lead to Dissonance

Most people expect attitudes to guide behaviour. We expect people to vote for candidates they like and to avoid eating foods they don't like. In 1957, Leon Festinger (Figure 15.12) proposed an elegant theory that was to become one of the most important catalysts of research in experimental social psychology. Festinger was interested in how people resolved situations in which they held conflicting attitudes. He proposed that **cognitive dissonance** occurs when there is a contradiction between two attitudes or between an attitude and a behaviour. People who smoke in spite of knowing that smoking may kill them experience cognitive dissonance. A basic assumption of dissonance theory is that dissonance causes anxiety and tension and therefore motivates people to reduce it and relieve displeasure. Generally, people reduce dissonance by changing their attitudes or behaviours; they sometimes also rationalize or trivialize the discrepancy.

15.12 Leon Festinger Festinger developed the influential theory of cognitive dissonance.

Dissonance theory provides important insights into many perplexing behaviours. Consider the American soldiers who served as prison guards at Abu Ghraib. Their treatment of prisoners likely was dissonant from their views on how people generally should be treated. In such a situation as a poorly run overcrowded wartime prison, it is not uncommon for guards to develop extremely negative attitudes about their prisoners, even viewing them as subhuman. Although this might resolve dissonance for the guards, it encourages mistreatment of the prisoners. Let's examine how dissonance processes affect attitudes and behaviour.

ATTITUDE CHANGE In one of the original dissonance studies, Leon Festinger and Merrill Carlsmith (1959) had students perform an extremely boring task for an hour. The experimenter then paid the student $1 or $20 to lie and tell the next participant that the experiment was really interesting, educational, and worthwhile. Nearly all of the participants subsequently provided the false information. Later, under the guise of a different survey, the same participants were asked how worthwhile and enjoyable the task had actually been. You might think that those paid $20 remembered the task as more enjoyable, but just the opposite happened. Participants who had been paid $1 rated the task much more favourably than those who had been paid $20. According to Festinger, this effect occurred because those in the $1 group had insufficient justification for lying. Therefore, in order to justify why they went along with the lie, they changed their attitude about performing the dull experimental task. The group that was paid $20 had plenty of justification for lying, since $20 was a large amount of money. Thus, they did not experience dissonance and their attitudes about the task were unchanged (Figure 15.13). Thus, one way to get people to change their attitudes is to change their behaviour first, using as few incentives as possible. Recall from Chapter 9 that providing children with rewards for creative drawing with coloured pens actually undermined how much they subsequently used the pens.

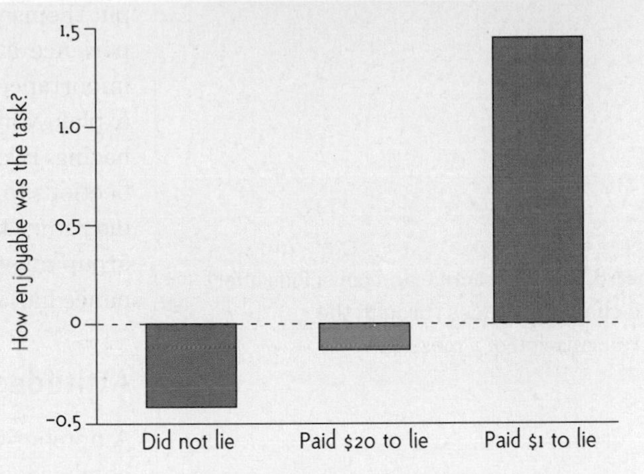

15.13 Cognitive Dissonance Those who were paid only $1 to mislead a fellow student experienced cognitive dissonance, which led them to alter their attitudes about how pleasurable the task had been.

POSTDECISIONAL DISSONANCE Many people find it difficult to choose which university to attend. They narrow down their choice to two or three alternatives, and then they have to decide. According to cognitive dissonance theory, holding positive attitudes about two schools but having to choose just one of them causes dissonance. *Postdecisional dissonance* then motivates the person to focus on the positive characteristics of the school she chose and the negative aspects of the school she did not choose. Similarly, when people choose to purchase a truck rather than a car they suddenly think of a million reasons why owning a truck is better than owning a car. Car owners do just the opposite. This effect occurs automatically and apparently without awareness. Indeed, even patients with long-term memory loss show a postdecisional effect for past choices, although they cannot consciously recall which object they chose (Lieberman, Ochsner, Gilbert, & Schacter, 2001). Thus, dissonance occurs automatically and requires minimal cognitive processing. By focusing on all the positives, people set aside their doubts and feel good about their decision.

JUSTIFYING EFFORT Hazing and other fraternity and sorority initiation rites are a serious problem on many university campuses and on sports teams. For example, some members of the football team at McGill were suspended for the hazing ritual that they performed on a new recruit. Administrators impose rules and penalties to discourage hazing, and yet it continues to occur. What function does hazing serve that makes the groups so resistant to end it? Why don't fraternities, sororities, and sports teams simply select new members and let them in without initiation? It turns out that requiring people to undergo an embarrassing or difficult rite of passage makes membership in the group seem much more valuable and makes the group itself seem cohesive. Eliot Aronson and Judson Mills (1959) required women to undergo an embarrassing test to see if they qualified to take part in a research study. The women were required to read a list of obscene words and sexually explicit passages in front of the male experimenter, whereas a control group read a list of very mild words (such as *prostitute*). The women then listened to an incredibly boring and technical presentation about mating rituals in lower animals. Women who read the embarrassing words reported that the presentation was much more interesting, stimulating, and important than did the women who read the mild list. When people persuasion The active and conscious effort to change attitudes through the transmission of a message.

Why are peripheral cues to persuasion effective?

put themselves through pain, embarrassment, or discomfort to join a group, they experience a great deal of dissonance. They resolve the dissonance by inflating the importance of the group and their commitment to it. This effort justification helps explain why people are willing to subject themselves to humiliating experiences like hazing. Even more tragically, it may also help explain why people who give up connections to families and friends to join cults or follow enigmatic leaders are willing to die rather than leave the group. If they have sacrificed so much to join the group, the group must be extraordinarily important. Here we begin to see why cognitive dissonance has the potential to be an extremely powerful influence on our attitudes.

Attitudes Can Be Changed through Persuasion

A number of forces other than dissonance can conspire to change attitudes. We are bombarded by television advertisements; lectures from parents, teachers, and physicians; public service announcements; politicians appealing for our votes; and so on. **Persuasion** is the active and conscious effort to change attitudes through the transmission of a message. The earliest scientific work on persuasion was conducted by Carl Hovland and his colleagues at Yale University, who emphasized that persuasion was most likely to occur when people paid attention to a message, understood it, and found it convincing; in addition, the message itself had to be memorable so that its impact lasted over time.

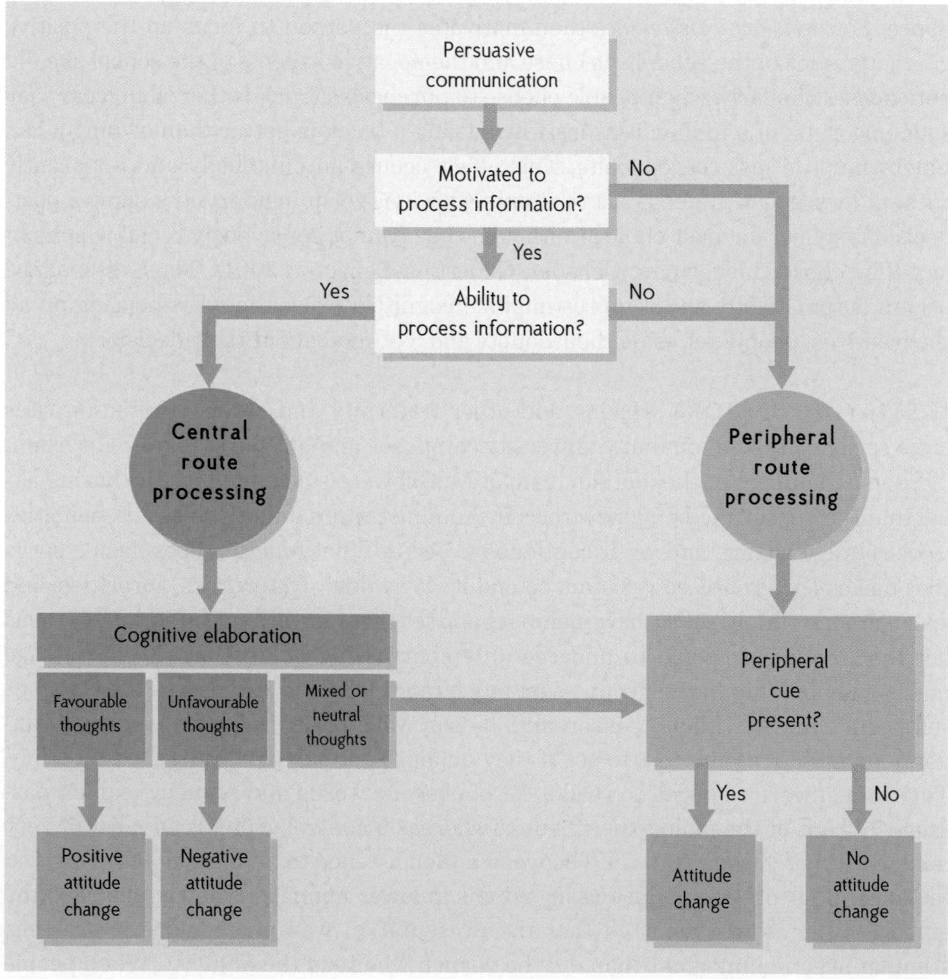

15.14 The Elaboration Likelihood Moduel When people are motivated to consider information carefully, it is processed by the central route, and attitudes are changed accordingly. Otherwise, people process information at a shallow level and are persuaded by peripheral cues.

Researchers have noted that there are two fundamental ways in which persuasion leads to attitude change (Figure 15.14). According to Richard Petty and John Cacioppo's **elaboration likelihood model** (1986), the *central route* to persuasion—in which people pay attention to arguments, consider all the information, and use rational cognitive processes—leads to strong attitudes that last over time and are resistant to change. By contrast, in the *peripheral route* to persuasion, people minimally process the message, as when they are deciding to purchase a new type of bottled water because an advertisement for it featured their favourite soap star (Figure 15.15).

A number of cues influence the extent to which a message is persuasive. These include the *source* (who delivers it), the *content* (what the message says), and the *receiver* (who is processing the message). Sources who are both credible and attractive are the most persuasive. This is why television ads for medicines and medical services often feature unusually attractive people playing the roles of physicians. The message is effective because of peripheral processing. Even better, of course, is when a drug company ad uses a spokesperson who is both attractive *and* an actual doctor! In addition, credibility and persuasiveness are heightened when the receiver perceives similarity between the source and him- or herself.

Of course, the arguments in the message are clearly important for persuasion. Strong arguments that appeal to our emotions are the most persuasive. Advertisers also use the mere exposure effect, repeating the message over and over in the hopes that multiple exposures will lead to increased persuasiveness. As the old saying goes, "If you repeat a lie often enough people will believe it." This is why politicians often make the same statements over and over during campaigns. Those who want to persuade also have to decide whether to deliver a one-sided argument or to consider both sides. One-sided arguments are best if the audience is on the speaker's side or is gullible. When speaking to a more skeptical crowd, speakers who acknowledge both sides but argue that one is superior tend to be more persuasive than those who completely ignore the opposing view.

15.15 Persuasive Advertising This billboard clearly takes the *peripheral route* to persuasion. A person would drive by this quickly, hardly taking notice to anything except for the charming little girl, the beautiful scenery portrayed, and the message "protect my environment." It also comes from a seemingly credible source (Manitoba Hydro), and has powerful content (protection of one's own environment). These characteristics in combination make for extremely persuasive advertising, and lead one to think that Manitoba Hydro is worth using.

elaboration likelihood model A theory of how persuasive messages lead to attitude changes.

How Do Attitudes Guide Behaviour?

Attitudes are evaluations of objects or ideas. They consist of affect, cognition, and behaviour that become linked together. Attitudes are formed through socialization and direct experience and best predict behaviour when they are strong and easily accessible. Discrepancies between attitudes, or between attitudes and behaviour, lead to cognitive dissonance. Dissonance theory can be used to explain a wide range of human behaviour. Attitudes can be changed through persuasion, by either central or peripheral means, depending on whether people think carefully about the issues or not.

How Do We Form Our Impressions of Others?

Social psychologists study attitudes because they are pervasive in our daily lives and influence many of our actions. Not surprisingly, attitudes that we hold about others are especially interesting to social psychologists. In addition to holding attitudes about other people, we also try to understand and predict them, to figure out why they act the way they do. We need to do this both to know how to relate to others and to ascertain whether they will help us or hurt us. Over the course of human evolution, one fact has remained constant: We are social animals who live in groups. Groups provide security from predators and competing groups, as well as mating opportunities and assistance in hunting and gathering food. At the same time, members within a group are essentially competitors for the same food and mates. Hence, mechanisms have evolved for distinguishing members of one's own group from members of other groups, as well as for detecting dangers from within the group, such as deception, coercion, and infidelity. We are constantly required to make social judgments, assessing whether a person is friend or foe, potential mate or potential challenger, honest or dishonest, trustworthy or unreliable, and so on. In this section, we examine how we form impressions of others, as well as how we understand and explain others' actions. We will see that we automatically classify people into social categories, which can have major implications for how we treat people.

Nonverbal Actions and Expressions Affect Our Impressions

What information is contained in nonverbal behaviour?

Over the years, social psychology has confirmed the importance of first impressions on long-term evaluations of people. For instance, if a professor begins a course full of enthusiasm, humour, and energy, students will forgive the odd boring lecture later in the term. However, professors who give a monotonous, dry, dull lecture on the first day are unlikely to be viewed more positively as the course progresses, even if their lecture delivery improves. Many factors influence impression formation, ranging from the observer's expectations and attitudes to what the observed person says, as well as his or her nonverbal gestures and physical appearance.

Living in a social world requires that we understand and predict the behaviour of others. Suppose someone is walking toward you. You make a number of quick judgments, such as whether you know the person, whether the person poses danger, or whether you might like to know the person better. How people initially feel about others is determined mostly by nonverbal behaviours. Facial expressions, gestures, walking style, and fidgeting are all examples of **nonverbal behaviour**, sometimes referred to as body language.

nonverbal behaviour The facial expressions, gestures, mannerisms, and movements by which one communicates with others.

FACIAL EXPRESSIONS The first thing people notice about others is usually the face. Human newborns less than an hour old prefer to look at and will track a picture of a human face over a drawing of a blank outline of a head (Morton & Johnson, 1991). The face communicates a great deal, such as emotional state, interest, and distrust. People use their eyes to indicate anger, to flirt, and to catch the attention of a passing waiter. When people won't meet our eyes, we assume, perhaps incorrectly, that they are embarrassed, ashamed, or lying, whereas we view people who look us in the eye as truthful and friendly. How we perceive this depends on our culture. For some groups, such as some First Nations tribes, making direct eye contact, especially with the elderly, can be considered disrespectful. Social interactions can even be awkward when we can't see someone's eyes. People wearing sunglasses are often described as cold and aloof; police officers sometimes wear sunglasses partly to seem intimidating.

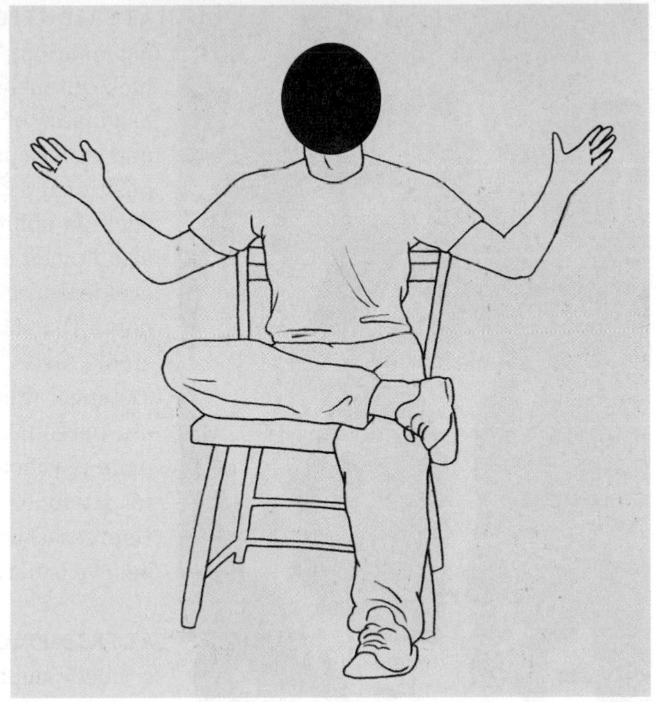

15.16 An Example of a Figural Outline In the study, participants watched a 10-second dynamic clip. Observers correctly guessed the person's sexual orientation at a better-than-chance rate.

BODY LANGUAGE How much can be learned from nonverbal behaviour? Nalini Ambady and Robert Rosenthal have found that people can make accurate judgments based on only a few seconds of observation, what they refer to as *thin slices of behaviour*. For instance, videotapes of judges giving instructions to juries reveal that the nonverbal actions of judges can predict whether juries find defendants guilty or not guilty. Judges, perhaps unconsciously, may indicate their beliefs about guilt or innocence through their facial expressions, tone of voice, and physical gestures. In one study, participants were asked to view 30-second audioless film clips of university teachers lecturing. Based solely on nonverbal behaviours, the participants' ratings corresponded very highly to the ratings given by the instructor's actual students (Ambady & Rosenthal, 1993).

One important nonverbal cue is how people walk, known as gait. Gait provides information about affective state. People with a bounce in their step, who walk along swinging their arms, are seen as happy. By contrast, people who scurry along, taking short steps while stooped over, are perceived as hostile, while those taking long strides with heavy steps are perceived to be angry. In an intriguing study, researchers found that participants accurately judged sexual orientation at a better-than-chance rate after watching a 10-second silent video or a dynamic figural outline of someone walking or gesturing (Figure 15.16; Ambady, Hallahan, & Conner, 1999). These thin slices of behaviour are powerful cues to impression formation.

We Make Attributions about Others

We constantly try to explain other people's motives, traits, and preferences. Why did she say that? Why is he crying? Why does she study so hard? and so on. As we learned earlier, **attributions** are people's causal explanations for why events or actions occur. People are motivated to draw inferences in part by a basic need for order and predictability in their lives. The world can be a dangerous place in which many unexpected things happen. People prefer to think that things happen for a reason, and that therefore they can anticipate future events. For instance, you might believe that you will do well if you study for an exam. Indeed, when events occur that don't seem to make any sense, such as when a person is brutally raped or murdered, people often make attributions about the victim, such as "she deserved it" or "he provoked it." Such attributions are part of what is referred to as the *just world hypothesis* (Figure 15.17). From this perspective, victims must have done something to justify what happened to them. People are applying the just world hypothesis when they say that the Iraqi detainees at the Abu Ghraib prison probably did something that led them to be arrested and therefore were in a sense responsible for and deserving of the abuse they received. Yet some of the prisoners were apparently innocent young men who were in the wrong place at the wrong time. It is much easier for us, however, to believe that they must be guilty of some criminal actions. Such attributions make their mistreatment seem more understandable, more justified.

attributions People's causal explanations for why events or actions occur.

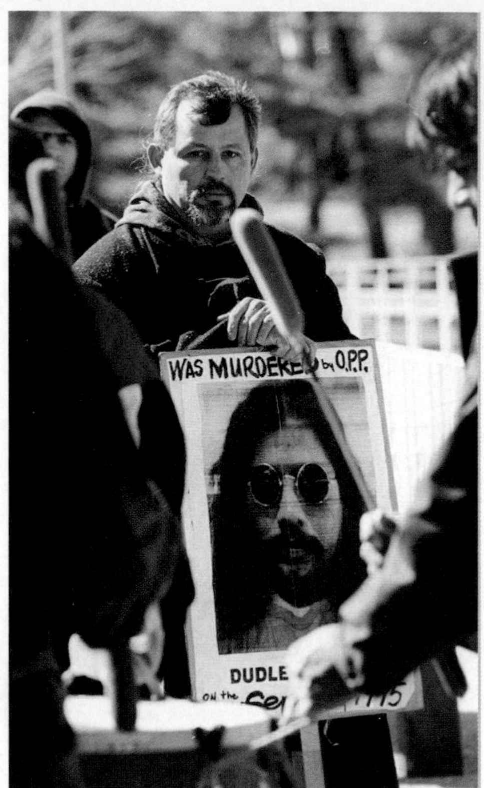

15.17 Dudley George In September 1995, First Nation Dudley George was shot dead by police when he and a group of 30 protesters built barricades outside of military base Camp Ipperwash during a dispute over land ownership rights. Though it is unclear whether or not the protesters were armed, one might try to rationalize George's murder by claiming that he must have done something wrong and got what he deserved, applying the *just world hypothesis* in order to better cope with the situation.

ATTRIBUTIONAL DIMENSIONS In any situation, there are dozens of plausible explanations for specific outcomes. Doing well on a test could be due to brilliance, luck, intensive studying, or the test's being unexpectedly easy. Fritz Heider, the originator of attribution theory, drew the essential distinction between personal and situational attributions. **Personal attributions**, also known as internal or dispositional attributions, are explanations that refer to something within a person, such as abilities, traits, moods, or effort. By contrast, **situational attributions**, also known as external attributions, refer to outside events, such as the weather, accidents, or the actions of other people. Psychologist Bernard Weiner (1974) later noted that attributions can vary on other dimensions, such as whether they are stable over time versus variable, or controllable versus uncontrollable. For instance, weather is situational, variable, and uncontrollable. Good study habits are personal, stable, and controllable. Weiner's theory has also been used to explain psychological states such as depression. People who are not depressed tend to attribute their failures to personal, stable, yet uncontrollable attributes. Depressed people attribute their failures to their own incompetence, which they believe is permanent.

ATTRIBUTIONAL BIAS Theorists such as Fritz Heider and Harold Kelly described people as intuitive scientists, trying to draw inferences about others and make attributions about events. However, unlike the objective scientist, people tend to be systematically biased in their social-information processing. The attributions they make are self-serving and consistent with preexisting beliefs, and they generally fail to take into account that other people are influenced by social circumstances. Recall the fundamental attribution error discussed in the context of Abu Ghraib prison, in which people overemphasize personal factors and underestimate the power of the situation. Edward Jones, who originated the idea of attribution error during the 1960s, actually called it the *correspondence bias*—he wanted to emphasize that people expect others' behaviours to correspond with their own beliefs and personalities. Interestingly, when people make attributions about themselves, they tend to focus on situations rather than on their personal dispositions, which, together with the fundamental attribution error, leads to the *actor-observer discrepancy*. For instance, people explaining their own lateness tend to attribute it to external factors, such as traffic or competing demands. When others are late, people tend to attribute it to personal characteristics, such as laziness or lack of organization. Hence, people are biased toward situational factors when explaining their own behaviour, but toward dispositional factors when explaining the behaviour of others.

How widespread is the correspondence bias?

Is the fundamental attribution error really fundamental? We saw earlier that there are cultural differences in the self-serving attributional bias. Might there also be differences in the attributional styles of Easterners and Westerners? Indeed, recall from Chapter 1 that Easterners tend to be more holistic in how they perceive the world, seeing the forest rather than individual trees. As you might expect, evidence indicates that people from Eastern cultures use much more information when making attributions than do people in Western cultures, and they also are more likely to believe that human behaviour is the outcome of personal and situational factors (Choi, Dalal, Kim-Prieto, & Park, 2003; Miyamoto & Kitayama, 2002). This doesn't mean that Easterners are not susceptible to some of the same attributional biases as Westerners, but only that they are more likely to take situational forces into account, whereas people in the West place overriding emphasis on personal factors.

personal attributions Explanations that refer to internal characteristics, such as abilities, traits, moods, and effort.

situational attributions Explanations that refer to external events, such as the weather, luck, accidents, or the actions of other people.

Stereotypes Are Based on Automatic Categorization

What are Italians like—do they all have fiery tempers? How about Canadians—do they all like hockey? Can white men jump? We hold attitudes and beliefs about groups because they allow us to answer these sorts of questions quickly. These attitudes and beliefs are **stereotypes,** cognitive schemas that organize information about people on the basis of their membership in certain groups. Stereotypes are mental shortcuts that allow for easy, fast processing of social information. Recall from Chapter 8 that heuristic processing allows us to make quick decisions. Stereotyping occurs automatically and, in most cases, outside of our awareness. In and of themselves, stereotypes are neutral and simply reflect efficient cognitive processes. Indeed, some stereotypes are true: Men tend to be more violent than women, and women are more nurturing than men. However, these are true on average; not all men are violent, nor are all women nurturing.

How do stereotypes alter the way we see people?

People construct and use categories to streamline impression formation in order to deal with the limitations inherent in mental processing. Rather than considering each person as unique and unpredictable, we categorize people as belonging to a group, a group about which we hold knowledge in long-term memory. As you walk through the hallways at your school, you might automatically categorize people on the basis of their clothing or hairstyles. Once you have categorized someone, you will have all sorts of beliefs about that person on the basis of your stereotypes of that category.

Stereotypes affect how we view and treat others (Kunda & Spencer, 2003). In Chapter 7, we discussed a study in which participants were asked to judge whether a person is famous or not. Recall that those who had seen the person's name in a prior task were likely to falsely remember the name as that of a famous person, in what is known as the *false fame effect*. However, such misremembering is much more likely to happen for male names than for female names, apparently because of the stereotype that men are more likely to be famous than are women (Banaji & Greenwald, 1995).

stereotypes Cognitive schemas that allow for easy and efficient organization of information about people based on their membership in certain groups.

Once we form stereotypes, we maintain them by a number of processes. As schematic structures, stereotypes guide our attention toward information that confirms the stereotypes and away from disconfirming evidence. People's memories are also biased to match stereotypes. These biases lead to an illusory correlation in which people believe that a relationship exists when it actually does not. The professor who notices a single black student performing poorly but fails to notice other black students who are doing well will confirm a false belief relating race to performance. Similarly, the meaning of a behaviour can be altered so that it is consistent with a stereotype. A white man's success may be attributed to hard work and determination, whereas a black man's success may be attributed to outside factors, such as luck or affirmative action. A lawyer described as aggressive conjures up different images from a construction worker described as aggressive. Moreover, when people encounter someone who does not fit a stereotype, rather than alter the stereotype they put that person in a special category, a process known as subtyping. Thus, a racist who believes that blacks are lazy categorizes Michael Jordan as an exception to the rule rather than as evidence for the invalidity of the stereotype. Forming a subtype that includes successful blacks helps racists maintain the stereotype that most blacks are unsuccessful.

INHIBITING STEREOTYPES Most people do not consider themselves prejudiced, and many are motivated to avoid stereotyping others. Yet according to many researchers in social cognition, categorization and stereotyping occur automatically, without awareness or intent (Bargh & Ferguson, 2000). Patricia Devine (1989) made the important point that people can override the stereotypes they hold and act in a nondiscriminatory fashion. For instance, most people in North America know the negative stereotypes associated with African Americans, and when a nonblack person encounters a black person, the information in the stereotypes becomes cognitively available. According to Devine, people who are low in prejudice override this automatic activation and act in a nondiscriminatory fashion. Clearly, then, simply categorizing people need not entail mistreating them. Unfortunately, some of the automatic consequences of stereotypes alter how we perceive and understand the behaviour of those we stereotype.

You might wonder if automatic stereotyping can be changed. Indeed, numerous studies have shown that they can (Blair, 2002). For instance, Dasgupta and Greenwald (2001) found that presenting positive exemplars of admired black individuals (e.g., Denzel Washington) produced more favourable attitudes toward African Americans on the IAT. In another study, training people to respond in a counter-stereotypical fashion, such as pressing a "no" key when they saw an elderly person paired with an elderly stereotype, led to reduced automatic stereotyping in subsequent tasks (Kawakami, Dovidio, Moll, Hermsen, & Russin, 2000). Telling people that their tests scores indicate that they hold negative stereotypes can motivate people to correct them, and the worse they feel about holding those attitudes, the harder they tried not to be biased (Monteith, 1993). These and other studies show that we can alter automatic stereotyping.

Inhibiting stereotyped thinking in everyday life is difficult and requires self-regulation (Monteith, Ashburn-Nardo, Voils, & Czopp, 2002). Recall from Chapter 9 that the mental functions that allow us to override habitual responses depend on the frontal lobes. In one recent brain-imaging study, showing white participants pictures of black faces produced amygdala activity if the faces were presented briefly. However, if the faces were presented for longer periods of time, the frontal lobes became active and the amygdala response decreased. This happened more for those whose IAT scores indicated negative stereotypes about blacks (Cunningham et al., 2004). In another study, frontal lobe activity when observing black faces was associated with poor performance on a subsequent task of mental function, which implies that controlling stereotypes is mentally taxing (Richeson et al., 2003). These studies and others demonstrate that college students today engage mental processes in order to inhibit their automatic stereotyping.

15.18 Sterotypes and Perception After being primed by black or white faces, white participants had to quickly classify pictures as tools or guns. Participants misidentified tools as guns when primed with black faces, suggesting that stereotypes alter perceptual processes.

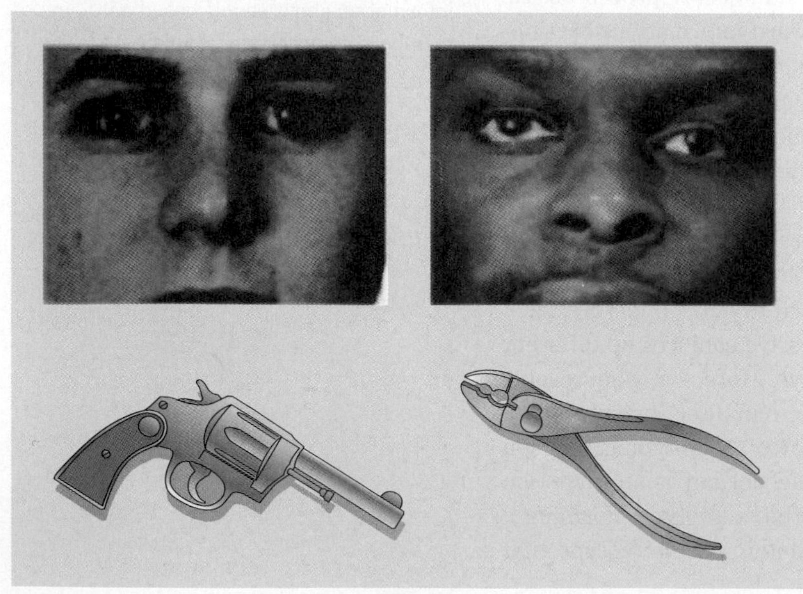

STEREOTYPES AND PERCEPTION We began Chapter 1 by describing the killing of Amadou Diallo by New York City police officers. How can social cognition help us understand this tragedy? In crossing levels of analysis, we find that these implicit social attitudes can influence basic perceptual processes. In two experiments that demonstrated this, Payne (2001) showed pictures of guns or tools to white participants and asked them to classify the objects as quickly as possible (Figure 15.18). Immediately before seeing a picture, the participants were briefly shown a picture of either a white or a black face. Priming by black faces led participants to identify guns more quickly and to mistake tools for guns. A more recent study found that the reverse is true: Priming people with pictures of criminal objects (e.g., guns and knives) leads them to pay greater attention to black faces than to white faces (Eberhardt, Goff, Purdie, & Davies, 2004). Such findings suggest that implicit bias may have been involved in the Diallo shooting. The officers were looking for individuals who fit a specific racial profile. When forced to make a split-second decision, they mistook Diallo's wallet for a weapon.

In a virtual reality simulation of the Diallo shooting, Greenwald, Oakes, and Hoffman (2003) required participants to respond to a simulated criminal holding a gun (by clicking a computer mouse to shoot him), a fellow police officer (by pressing the space bar), or a civilian holding a neutral object (by doing nothing). In some trials, white males holding guns were criminals and black males holding guns were police officers. In the other trials this was reversed—the black simulated persons were criminals. Blacks were more likely to be incorrectly shot in this study, in part because the objects they held were more likely to be identified as weapons. Fortunately, recent evidence suggests that computerized training, in which race is unrelated to the presence of a weapon, can eliminate this racial bias in shooting behaviour among police officers (Plant & Peruche, 2005).

SELF-FULFILLING EFFECTS What effect does being treated as a member of a stereotyped group have on people? Stereotypes that are initially untrue can become true through **self-fulfilling prophecy**, in which people come to behave in ways that confirm their own or others' expectations. One of the most impressive early demonstrations of this process was conducted in the 1960s by Robert Rosenthal with a school principal, Lenore Jacobsen. In this study, elementary-school students took a test that supposedly identified some of them as being especially likely to show large increases in IQ during the school year. Teachers were given a list of the "bloomers" in their classes. At the end of the year, standardized testing revealed that the "bloomers" actually did show a large increase in IQ. As you might have guessed, students were chosen at random rather than through any test, and therefore their increase in IQ was attributed to the extra attention provided by the teachers. In effect, the teachers paid more attention to "bloomers," perhaps giving them extra encouragement. Thus, teacher expectations turned into reality. Of course, negative stereotypes can become self-fulfilling as well. Thus, teachers who expect certain students to fail—for example, minority students—might do things that subtly undermine students' self-confidence or motivation. For instance, offering unwanted help—even with the best of intentions—can send the message that the teacher does not believe the student has what it takes to succeed.

Stereotypes Can Lead to Prejudice

As we have seen, stereotypes are themselves neutral categories. However, negative stereotypes of groups lead to prejudice and discrimination. **Prejudice** refers to the affective or attitudinal responses associated with stereotypes, and it usually involves negative judgments about people on the basis of their group membership. **Discrimination** is the unjustified and inappropriate treatment of people due to prejudice. Prejudice and discrimination are major problems, responsible for much of the conflict and warfare around the world. Within all cultures, some groups of people are treated negatively due to prejudice. Social psychologists have spent the last half century studying the determinants and consequences of prejudice as well as trying to find ways to reduce its pernicious effects.

self-fulfilling prophecy The observation that people may come to behave in ways that confirm their own or others' expectations.

prejudice The affective or attitudinal responses associated with stereotypes.

discrimination The inappropriate and unjustified treatment of people based solely on their group membership.

Why do stereotypes so often lead to prejudice and discrimination? A variety of theories have been proposed over the years, including that only certain types of people are prejudiced, that people treat others as scapegoats to relieve the tensions of daily living, and that people discriminate against others to protect their own self-esteem. An explanation consistent with our theme that the mind is adaptive is that evolution has led to two processes that produce prejudice and discrimination. First, we have a tendency to favour our own groups over other groups. Second, we have a tendency to stigmatize those who pose threats to our groups.

ingroup favouritism The tendency for people to evaluate favourably and privilege members of the ingroup more than members of the outgroup.

INGROUP-OUTGROUP BIAS We are powerfully connected to the groups to which we belong. We cheer them on, we fight for them, and sometimes we are even willing to die for them. Those groups to which we belong are called *ingroups*; those to which we do not belong are called *outgroups*. People treat ingroup members much differently from outgroup members. For instance, in what is called the *outgroup homogeneity effect*, people tend to view outgroup members as less varied than ingroup members. University of Waterloo students may think that McGill students are all alike, but when they think about UW students they can't help but notice the wide diversity of different student types. Of course, the reverse is true as well. Similarly, even as we recognize that people in our home countries differ in substantial ways, we tend to view Arabs, for example, as being very similar, sharing similar values and even attitudes about the West.

The consequence of categorizing people as ingroup or outgroup members is **ingroup favouritism**, in which people are more likely to distribute resources to members of the ingroup than to members of the outgroup. People are more willing to do favours, forgive mistakes or errors, and give tangible resources to ingroup members. The power of group membership is so strong that people will exhibit ingroup favouritism even if the groups are determined by arbitrary processes. British psychologist Henri Tajfel and his American collaborator John Turner (1979) randomly assigned volunteers to one of two groups, using totally meaningless criteria, such as flipping a coin. Participants were then given a task in which they divided up money. Not surprisingly, they gave more money to their ingroup members. But they also tried to prevent the outgroup members from getting any money, even when they were told that the basis of group membership was arbitrary and that giving money to the outgroup would not affect how much money their own group obtained!

15.19 The 1995 Quebec Referendum The importance that people place on identifying with their ingroups is clearly evident in the Quebec separatist movement. Having a distinct identity is valued to such an extent by Quebeckers that many are willing to endure the substantial economic costs of separation.

Why do people value members of their own groups? We can speculate that over the course of human evolution, personal survival has depended on group survival. Especially when there was competition for scarce resources, those who worked together to keep resources within their group and deny resources to outgroup members had a selective advantage over those willing to share. At the psychological level, our group memberships are an important part of our social identity and contribute to our overall sense of self-esteem (Figure 15.19). Interestingly, women show a much greater automatic ingroup bias than men (Rudman & Goodwin, 2004). Why do women like women more than men like men? Rudman and Goodwin speculate that both men and women depend on women for nurturing and that both are threatened by male violence. Moreover, although women can freely express their affection for their female friends, American males appear to be less comfortable doing so, perhaps because it might threaten their sexual identity. Although men generally favour their ingroups, they fail to do so when the category is based on sex.

Psychological Science *in Action*

Cooperation Promotes Harmony between Groups

The first half of this decade has witnessed widespread hostility and violence between various factions in many regions of the world. How might social psychologists use what we have learned in the laboratory to further the cause of world peace? Not surprisingly, we have reason to be pessimistic about whether this is possible. Over the past 50 years, a number of strategies for reducing prejudice and encouraging peace have been attempted, such as bringing people from different groups into contact. Most have failed. It is extraordinarily difficult to change cultural and religious beliefs, and attitudes toward ethnic groups are embedded deeply in both. Around the world, groups clash over disputes that predate the births of most of the combatants, and sometimes people can't even remember the original source of conflict. Yet the global response to the deadly tsunami of late 2004 indicates that people can work together, and in doing so overcome intergroup hostilities if they have a greater purpose, such as dealing with natural catastrophes. What can social psychology tell us about strategies to promote intergroup harmony?

Social psychologists have uncovered a variety of strategies that are effective in producing greater tolerance for and liking of outgroups. The first study to provide evidence about what might work was conducted by Muzafer Sherif and his colleagues in the late 1950s. Sherif arranged for 22 well-adjusted and intelligent white fifth-grade boys from Oklahoma City to attend a summer camp. None of the boys had known each other before that summer. Prior to arriving at camp, the boys were divided into two groups that were essentially the same. During the first week, the boys lived in separate camps, unaware that a similar group of boys was across the lake.

Once the groups were established, Sherif started the experiment by having the groups compete over a four-day period in a number of athletic competitions, such as tug of war, football, and softball. Group pride was extremely strong, with the groups naming themselves the Rattlers and the Eagles. The stakes were high. The winning team would get a trophy, individual medals, and appealing prizes. The losers would receive nothing. Animosity between the groups quickly escalated. The Eagles burned the Rattlers' flag, and the Rattlers retaliated by trashing the Eagles' cabin. Eventually, there were confrontations and physical fights, which had to be broken up by the experimenters. All of the signs of prejudice emerged, including the outgroup homogeneity bias and ingroup favouritism.

Phase 1 of the study was complete. Sherif showed how easy it was to get people to hate each other. Simply divide them into groups and have the groups compete, and prejudice and mistreatment would result. The question in Phase 2 of the study was, Could the hostility be undone?

Sherif first tried what made sense at the time, simply having the groups come into contact with one another. This failed miserably. The hostilities were too strong and skirmishes continued. Sherif reasoned that if competition led to hostility, then cooperation should reduce hostility (Figure 15.20). The experimenters created situations in which members of both groups had to cooperate in order to achieve necessary goals. For instance, the experimenters rigged a truck to break down. Getting the truck moving required all the boys pulling together, ironically on the same rope used earlier in the tug of war. On success a great cheer arose from all, with plenty of backslapping all around. After a series of tasks that required cooperation, the walls

15.20 Cooperation Simon Gagné and Martin Brodeur are bitter rivals throughout the NHL season when they play on competing teams, but at the 2002 Olympics they transcended those rivalries when they played for the Canadian team.

between the two sides broke down and the boys became friends across the groups. From strangers, competition and isolation created enemies. From enemies, cooperation created friends.

Research over the past four decades has indicated that only certain types of contact between hostile groups are likely to reduce prejudice and discrimination. It is clear that having shared superordinate goals—those that require people to cooperate to succeed—reduces hostility between groups. Research has shown that people who work together to achieve a common goal often break down subgroup distinctions as they become one larger group (Dovidio et al., 2004). Athletes on multiethnic teams often have positive attitudes toward other ethnicities. Programs that require cooperation are most effective when they include members who challenge or defy the negative stereotypes of their groups.

The programs that are most successful involve person-to-person interaction. A good example is Eliot Aronson's jigsaw classroom, which he developed with his students in the 1970s. In this program, students work together in mixed-race or mixed-sex groups in which each member of the group is an expert on one aspect of the assignment. For instance, when studying Mexico, one group member might study its geography, another its history, and so on. The various geography experts from each group get together and master the material. They then return to their own groups and teach the material to their team members. Thus, cooperation is twofold: Each group member cooperates not only with members of other groups, but also within the group as well. More than 800 studies of the jigsaw classroom have demonstrated that it does lead to more positive attitudes toward other ethnicities and that students learn the material better and perform at a higher level. According to Aronson, children in jigsaw classrooms grow to like each other more and develop higher self-esteem than do children in traditional classrooms. The lesson is clear: Communal activities working toward superordinate goals can reduce prejudice.

In addition, other strategies have been shown to reduce prejudice. For example, bilingual instruction in schools has been shown to lead to less ingroup favouritism among elementary school children (Wright & Tropp, 2005). Prejudice can also be reduced through explicit efforts to train people about their stereotypic associations. For example, participants who gain practice at associating counter-stereotypic information with women (e.g., strong, dominant), are more likely than a control group to choose to hire women in a subsequent task following a delay (Kawakami, Dovidio, & van Kamp, 2005).

Social psychology has shown us that we can work cooperatively toward reducing intergroup hostility, racism, and prejudice. Now we must take these lessons from the laboratory and the classroom and begin to study and apply them on a global scale. Certainly the outpouring of aid for the tsunami victims shows us that the heart is willing. Now the mind must be so too. ■

REVIEWING the Principles | How Do We Form Our Impressions of Others?

Human social interaction requires people to form impressions of others. People are highly sensitive to nonverbal information and can develop accurate impressions of others on the basis of very thin slices of behaviour. People also are motivated to figure out what causes other people to behave the way they do. People make attributions about others, which are often biased. They tend to attribute other people's behaviour to dispositions rather than to situations, and they use heuristic processing, which biases social judgment. Stereotypes result from the normal cognitive process of categorization. However, negative stereotypes and prejudice lead to discrimination. Humans have a natural tendency to discriminate against those who are threatening, such as outgroup members.

How Do Others Influence Us?

We humans have an overriding motivation to fit in with the group. One way we do this is by presenting ourselves in a positive way, such as being on our best behaviour and trying not to offend others. But people also conform to group norms, obey direct commands by authorities, and are easily influenced by others in their social group. The desire to fit in with the group and avoid being ostracized is so great that under some circumstances people willingly engage in behaviours they would otherwise condemn. As we have noted throughout this chapter, the power of the social situation is much greater than most people believe—perhaps the single most important lesson from social psychology.

What effect do groups have on self-awareness?

Groups Influence Individual Behaviour

The first social psychology experiment was conducted by Norman Triplett in 1897 when he showed that bicyclists pedaled faster when riding with other people than when they were alone. This is due to **social facilitation**, in which the mere presence of others enhances performance. Social facilitation also occurs in other animals, including horses, dogs, rats, birds, fish, and even cockroaches. Robert Zajonc (1965) proposed a model of social facilitation that involves three basic steps (Figure 15.21). First, Zajonc proposed that all animals are genetically predisposed to become aroused by the mere presence of others of their own species, since they are associated with most of life's rewards and punishments. Zajonc then invoked Clark Hull's

social facilitation When the mere presence of others enhances performance.

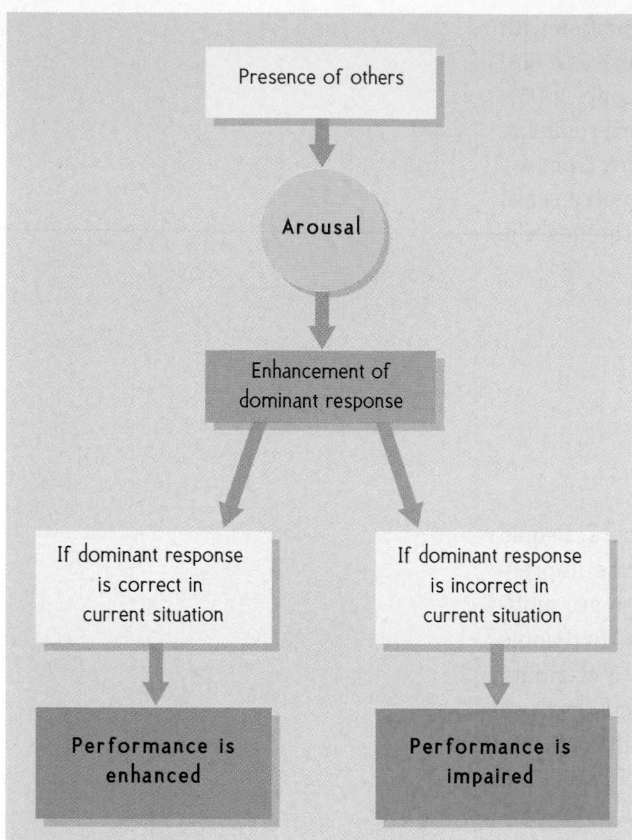

15.21 Social Facilitation The mere presence of other people leads to increased arousal, which in turn favours the dominant response. If this is the correct response, performance is enhanced, but if it is the incorrect response, performance suffers.

well-known learning principle that arousal leads animals to emit a dominant response, that is, the one that is most likely to be performed. In front of food, the dominant response is to eat, for example. Interestingly, Zajonc's model predicts that social facilitation can lead to either enhancement or impairment in performance, depending on whether the dominant response is the correct one. The presence of others improves performance of simple tasks for which the dominant response is well learned, such as adding single digits, but it interferes with performance of complex tasks for which the correct answer requires greater thought, such as differential calculus. As an example, think about the kind of work space that would enhance job performance. If a group of employees is performing a simple task such as database entry, it might be best to have them sit in a fairly open environment, either around a large common table or in cubicles where each employee is aware of the rate at which others are performing. If an employee is performing a more complicated task, such as editing a manuscript, a private office might ensure better performance since, presumably, the editor is the only person working on that particular manuscript, so an awareness of how others are performing is irrelevant to the complex work at hand.

SOCIAL LOAFING In some cases, people work less hard when in a group than when working alone. This **social loafing** occurs when people's efforts are pooled so that no one individual is accountable or feels personally responsible for the group's output. In a classic study, six blindfolded people wearing headphones were told to yell as loudly as they could. Some were informed that they were shouting alone; others were told that they were yelling with other people. Participants did not shout so loudly when they believed others were shouting as well (Latané, Williams, & Harkins, 1979). Of course, making it clear that individual efforts can be monitored wipes out social loafing. Thus, for a group working on a class project, it is important to make each person feel personally responsible for some component of the project if you want her or him to exert maximum effort.

DEINDIVIDUATION People sometimes lose their individuality when they become part of a group. **Deindividuation** occurs when people are not self-aware and therefore are not paying attention to their personal standards. Recall that self-awareness typically causes people to act in accordance with their personal values and beliefs. When self-awareness disappears, so do restraints. Deindividuated people often do things they would not do if they were alone or self-aware. A good example is crowd behaviour. Most of us like to think that we would try to help a person who was threatening suicide. But people in crowds often not only fail to intercede, but also sometimes egg the person on, yelling, "Jump, jump" to someone teetering on a ledge.

People are especially likely to become deindividuated when they are aroused and anonymous, and when there is a diffusion of responsibility. Rioting by fans, looting following disasters, and other mob behaviours are the products of deindividuation. Recall the Stanford Prison Study mentioned at the beginning of this chapter. The study had to be stopped after just six days because the students became so immersed in their roles that many guards acted brutally and many prisoners became listless and apathetic. The situation was sufficiently powerful to radically alter people's behaviour through a process of deindividuation. Not all deindividuated behaviour is violent, of course. Gamblers in crowded casinos, fans doing the wave, and people dancing the

social loafing The tendency for people to work less hard in a group than when working alone.

deindividuation A phenomenon of low self-awareness, in which people lose their individuality and fail to attend to personal standards.

funky chicken while inebriated in nightclubs are also most likely in deindividuated states, and accordingly they act in ways that they would not if they were self-aware.

GROUP DECISION MAKING It has been said that the intelligence of a group can be determined by averaging the IQs of its members and then dividing that average by the number of people in the group. In other words, groups are known for making bad decisions. Social psychologists have shown that being in a group does influence decision making, but in curious ways. For instance, you might think that groups would be especially cautious in making decisions. However, when James Stoner set out to study this assumption in the 1960s, he found that groups often made riskier decisions than individuals did, in what is known as the *risky-shift effect*. An obvious example is a group of children who decide to try something dangerous that none of them would have tried alone. Subsequent research has demonstrated that groups are sometimes riskier than individuals and sometimes more cautious, as groups tend to enhance the initial attitudes of members who already agree. This process is known as *group polarization*. For example, discussion tends to make people on juries believe more strongly in their initial opinions about guilt or innocence. When groups make risky decisions, it is usually because the individuals initially favour a risky course of action, and through mutual persuasion they come to agreement.

Sometimes group members are particularly concerned with maintaining the cohesiveness of the group, and so for the sake of cordiality, groups make bad decisions. Yale social psychologist Irving Janis coined the term *groupthink* in 1972 to describe this extreme form of group polarization. Examples of groupthink include the decision to launch the American space shuttle *Challenger* despite there being clear evidence of a problem with one part and choices made by President Bill Clinton and his advisers following the allegations of his affair with Monica Lewinsky, which ultimately led to his impeachment. Groupthink typically occurs when a group is under intense pressure, is facing external threats, and is biased in a particular direction. The group does not carefully process all the information available to it, dissension is discouraged, and group members assure each other that they are doing the right thing. To prevent groupthink, leaders must refrain from too strongly expressing their opinions at the beginning of a discussion. The group should be encouraged to consider alternative ideas, either by having someone play devil's advocate or by purposefully examining outside opinions. Carefully going through alternatives, and weighing the pros and cons of each, can help people avoid groupthink.

We Conform to Social Norms

Society needs rules. Imagine the problems if you woke up one morning and decided that you would henceforth drive on the wrong side of the road! **Social norms**—expected standards of conduct—influence behaviour in multiple ways, such as indicating which behaviour is appropriate in a given situation. Standing in line is a social norm, and people who violate that norm by cutting in line are treated rudely and often directed to the back of the line. **Conformity**, the altering of one's behaviour or opinions to match those of others or to match social norms, is also a powerful form of social influence.

social norms Expected standards of conduct, which influence behaviour.

conformity The altering of one's opinions or behaviour to match those of others or to match social norms.

How much do people conform to others? Muzafer Sherif was one of the first researchers to demonstrate the power of norms and conformity in social judgment. His studies, conducted in the 1930s, relied on a perceptual phenomenon known as the *autokinetic effect*, in which a stationary point of light appears to move when viewed in a totally dark environment. This effect occurs because people have no frame of reference and therefore cannot correct for small eye movements. Sherif had participants alone in a room estimate how far the light moved. Individual differences were considerable; some saw the light move only an inch or two, whereas others saw it move eight inches or more. In the second part of the study, Sherif put two or more participants in the room and had them call out their estimates. Although there were initial differences, participants very quickly revised their estimates until they agreed. In ambiguous situations, people often compare their reactions with others' reactions in order to judge what is appropriate.

Solomon Asch (1955) speculated that Sherif's effect probably occurred because the autokinetic effect is a subjective visual illusion. If there were objective perceptions, Asch thought, then participants would not conform. To test his hypothesis, Asch assembled participants to take part in a study of visual acuity. He had participants look at a reference line and then decide which of three other lines matched it (Figure 15.22). There were 18 trials, and each man said his answer aloud. This is an easy task, and normally people are able to do it with a high level of accuracy. But in these studies Asch had a naïve participant join a group of six confederates, people pretending to be participants but actually working for the experimenter. The actual participant always went sixth, giving his answer after five of the confederates gave theirs. On 12 of the 18 trials, the confederates deliberately gave the same wrong answer. After hearing five wrong answers, the participant then had to state his answer. Because the answer was obvious, Asch speculated that the participant would give the correct answer, but about one-third of the time, the participant went along with the confederates. More surprisingly, in repeated trials, three out of four people conformed to the incorrect response at least once (Figure 15.23).

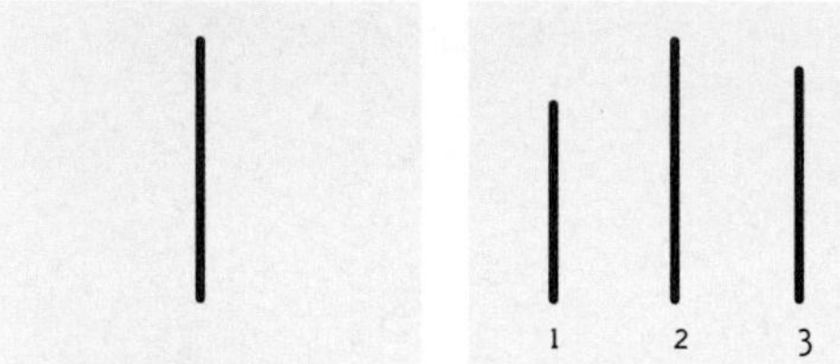

15.22 The Asch Task In the task, participants were shown a standard line (left) and then three comparison lines. Their task was simply to say which of the three lines matched the standard. When confederates gave false answers first, three-quarters of participants conformed by giving the wrong answer.

Research has consistently demonstrated that people tend to conform to social norms. This can be seen outside the laboratory as well: Adolescents conform to peer pressure to smoke; jury members go along with the group rather than state their own opinions; people stand in line to buy tickets. In fact, you might be wondering when people don't conform to social norms. In a series of follow-up studies, Asch and others have identified factors that lessen or decrease the chances of conformity. One factor is group size. If there are only one or two confederates, the participant usually does not conform. But as soon as the confederates number three or more, conformity occurs. Interestingly, there seems to be a limit to conformity. Subsequent research has found that even groups as large as 16 do not lead to greater conformity than groups of 7.

Asch found that lack of unanimity is another factor that diminishes conformity. If even one confederate gave the correct answer, then conformity to the group norm decreased a great deal. Any dissent from majority opinion can diminish the influence of social norms. But dissenters are typically not treated well by groups. In 1951, Stanley Schachter conducted a study in which a group of students debated the fate of a juvenile delinquent, Johnny Rocco. A confederate deviated from the group judgment of how Johnny should be treated. When it became clear that the confederate would not be persuaded by group sentiment, the group began to ostracize him. When group members were subsequently given the opportunity to reduce group size, they consistently rejected the "deviant" confederate. The bottom line is that groups enforce con-

15.23 To Conform or Not to Conform The only true participant is number 6 (in the middle of each photo with glasses). He can't believe his eyes as the other participants, actually confederates of the experiment, give incorrect answers.

formity, and those who fail to go along are rejected. The need to belong, and the anxiety associated with the fear of social exclusion, gives a group powerful influence over its members. Indeed, a recent brain-imaging study that used a modified Asch conformity procedure found activation of the amygdala, perhaps a fear response, in participants whose answers did not conform to the group's incorrect answer (Berns et al., 2005).

We Are Compliant

Many times people influence the behaviour of others simply by asking them to do something. When people do things requested by others, they are exhibiting **compliance**. A number of factors increase compliance. For instance, the Australian psychologist Joseph Forgas (1998) has demonstrated that people are especially likely to comply when they are in a good mood. This may be why people try to "butter up" others when they want something from them. According to the psychologist Robert Cialdini, people are often influenced by others because they fail to pay attention and respond without fully considering their options. For instance, if you give people a reason for a request, they are much more likely to comply, even if the request makes little sense. According to Cialdini, people often comply with requests because they really aren't thinking—they are simply following a standard mental shortcut to avoid conflict.

People can use a number of powerful strategies to influence others to comply. For instance, in the *foot-in-the-door effect*, people are more likely to comply with a large and undesirable request if they have earlier agreed to a small request. In 1962, Jonathan Freedman and Scott Fraser asked homeowners to allow a large, unattractive DRIVE CAREFULLY sign to be placed on their front lawns. As you might imagine, fewer than one in five people agreed to do so. However, another group of homeowners was initially asked to sign a petition to support legislation that would reduce traffic accidents. A few weeks later, these same people were approached about having the large sign placed on their lawns, and more than half of them agreed. Once people are committed to a course of action, they behave in ways consistent with that course.

compliance The tendency to agree to do things requested by others.

obedience The willingness to follow an order given by an authority.

The opposite influence technique is the *door in the face*, in which people are more likely to agree to a small request after they have refused a large request, because the second request seems modest in comparison and people want to seem reasonable. Salespeople often use influence techniques. Another of their favourites is the *low-balling strategy*, which begins when a salesperson offers a product—for example, a car—for a very low price. Once the customer agrees, the salesperson may claim the manager didn't approve the price or that there are additional costs. Whatever the reason, once people are committed to buying the car, they often agree to pay the increased costs.

We Are Obedient to Authority

One of the most famous and most disturbing psychology experiments was conducted in the early 1960s by Stanley Milgram (Figure 15.24), who wanted to understand how apparently normal German citizens would willingly obey orders to injure or kill innocent people during World War II. Milgram was interested in the determinants of **obedience**, why people follow orders given by an authority. Try to imagine yourself as a participant in his experiment. You have agreed to take part in a study of learning. On arriving at the laboratory, you meet another participant, a 50-year-old grandfatherly type (Figure 15.25). The study is described as one in which a teacher will administer electric shocks to a learner during a simple memory task in which word pairs are learned. Your role as the teacher is determined by an apparently random drawing of your name from a hat. On hearing that he may receive electric shocks,

What types of people are obedient to authority?

15.24 Stanley Milgram Milgram, pictured here with his infamous shock generator, demonstrated that average people will obey even hideous orders given by an authority figure.

the learner reveals that he has a heart condition and expresses minor reservations. The experimenter says that although the shocks will be painful, they will not cause permanent tissue damage. You help the experimenter take the learner to a small room and hook him up to the electric shock machine. You then proceed to a nearby room and are seated at a table in front of a large shock generator with switches from 15 to 450 volts. Each voltage level carries a label, ranging from "Slight" to "Danger—Severe Shock," and finally, an ominous "XXX."

Your task is to give the learner a shock each time he makes a mistake, increasing the voltage with each subsequent error. So, whenever the learner errs, you dutifully administer a shock. When you reach 75 volts, the man yelps with pain, which you can hear over an intercom. At 150 volts, he starts screaming, banging on the wall, and demanding that the experiment be stopped. The man is clearly in agony as you apply additional stronger shocks, and at 300 volts he refuses to answer any more questions. After 330 volts, the learner is completely silent. All along you have wanted to leave, and you severely regret ever agreeing to participate in the study. You might have killed the man, for all you know. Indeed, you tried to stop the experiment, but each time you said you were quitting, the experimenter replied, "The experiment requires that you continue"; "It is essential that you go on"; "There is no other choice, you must go on!" So you do.

Does this sound crazy to you? How high do you think you would go in terms of shocking the learner? Do you think you would go up to 450 volts or do you think you would quit as soon as the man started to complain? A group of psychiatrists whom Milgram asked predicted that people would go no higher than 135 volts and that fewer than one in a thousand people would administer the highest level of shock. But that's not what happened. What actually happened would change how people viewed the power of authority.

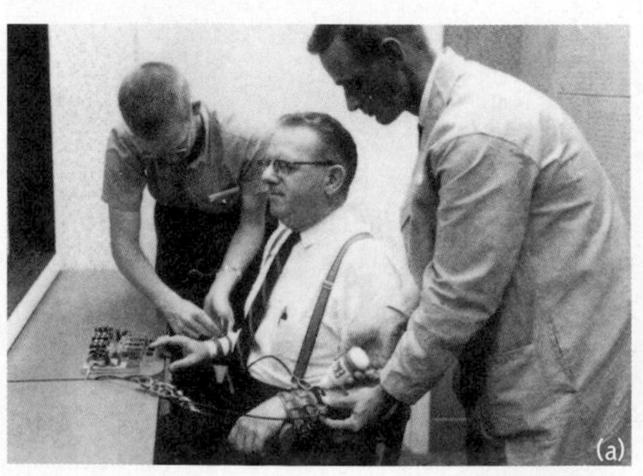

15.25 Testing Obedience (a) The Milgram experiment required participants to "shock" the confederate learner (seated). The research participant (left) helped apply the electrodes that would be used to shock the learner. (b) An obedient participant shocks the learner in the "touch" condition. Fewer than one-third obeyed the experimenter in this condition. (c) After the experiment, all of the participants were introduced to the confederate learner so they could see he was not actually harmed.

Milgram found that although almost all the participants tried to quit, nearly two-thirds completely obeyed the directives of the experimenter (Figure 15.26), believing they were administering 450 volts to a nice old man with a heart condition (actually a confederate). These findings have been replicated by Milgram and others around the world. The conclusion of these studies is that ordinary people can be coerced into obedience by insistent authorities, even when what they're coerced into goes against the way they would usually behave.

You may wonder about the ethics of Milgram's study. It is important to recognize that Milgram himself was surprised by the results. Once he found that people were obedient, he set out to study how to reduce obedience. Milgram did find that some situations produced less compliance. For instance, if participants could actually see or had to touch the learner who was being shocked, obedience decreased. When the authority was more removed from the situation, giving the orders over the telephone, obedience dropped dramatically.

Throughout his study, Milgram was highly concerned with the mental state of his participants. He carefully revealed the true nature of the experiment to the participants in a systematic debriefing, and he made sure that they met the confederate learner and could see that he was not hurt in any way. Milgram also followed his participants over time to ensure that they experienced no long-term negative effects. Actually, many people were glad they had participated, feeling that they had learned something about themselves and about human nature. Most of us assume that only sadistic miscreants would willingly inflict injury on others when ordered to do so. Milgram's research, and that which followed, demonstrated that ordinary people may do horrible things when ordered to do so by an authority. Looking back on the prison guards at Abu Ghraib, it's interesting to consider them as ordinary people. We don't know exactly what they were ordered to do, but a number of social factors, from deindividuation, to social facilitation, to inappropriate norms and conformity, contributed to the situation and likely played prominent roles in the mistreatment of Abu Ghraib prisoners. As we stated earlier, it's important for all of us to be aware of situational influences when we evaluate our own behaviour and that of others, particularly when our core beliefs and values are at risk.

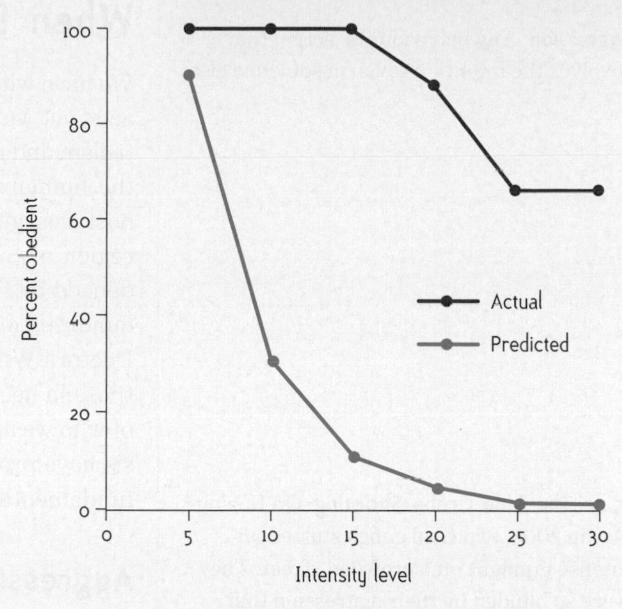

15.26 Predicting the Results Psychiatrists, college students, and middle-class adults predicted that fewer than one-tenth of one percent of participants in the Milgram experiment would be completely obedient and provide the maximum level of shock (each level = 15 volts). They believed that fewer than 10 percent of people would go beyond level 15 (225 volts). In fact, all of the participants were obedient at this level, and 65 percent were obedient to the highest level of shock.

REVIEWING the Principles | How Do Others Influence Us?

For the most part, people follow group norms, are influenced by others' opinions, and are obedient to authority. These effects are powerful, although most people underestimate their importance or do not believe that they themselves are affected by social influence. Yet the evidence is overwhelming that in many situations people will engage in behaviours quite inconsistent with their own standards. This sometimes occurs because people fail to attend to their internal beliefs and values, such as when they are in a state of deindividuation. Those who are aware of the power of social influence often use specific strategies to manipulate the behaviour of others, such as foot-in-the-door, door-in-the-face, and low-balling techniques. Likewise, the more each of us is aware of our own values and beliefs, the more capable we are of upholding our own standards.

aggression Any behaviour or action that involves the intention to harm someone else

When Do We Harm or Help Others?

We have witnessed the extent to which obedience can lead people to commit horrible acts, yet we can also see how the need to belong to a group can lead us to acts of altruism and generosity. The turbulent events of the last few years reveal the extent of the human capacity for harming and helping others. We see terrorists beheading hostages, guards abusing prisoners, and the brutality of tribal warfare in the Darfur region of Sudan. Yet many people are kind, compassionate, and generous, as evidenced by the outpouring of support to the victims of the 2004 tsunami that killed hundreds of thousands and left millions homeless. Similarly, members of the group Doctors Without Borders travel to dangerous regions around the globe to care for those in need. This tension between our aggressive and altruistic sides is at the core of who we are as a species. Psychological scientists working at all the levels of analysis have provided much insight into the roles that nature and nurture play in these fundamental human behaviours.

Aggression Can Be Adaptive

Aggression refers to a variety of different behaviours, but basically it is any behaviour or action that involves the intention to harm someone else (Figure 15.27). Aggression in animals often occurs in the context of fighting over a mate or defending territory from intruders. Sometimes just the threat of aggressive action is sufficient to dissuade invaders; at other times animals have to fight to defend their turf. Physical aggression, although common among young children, is relatively rare among adult humans; their aggressive acts more often involve words or actions designed to threaten, intimidate, or emotionally harm others. Aggression needs to be considered across the levels of analysis, from basic biology to cultural contexts.

15.27 The Jane Creba Shooting On Boxing Day in 2005, two rival gangs started an intense gunfight on a crowded street. They were so blinded by their aggression that they failed to notice an innocent bystander, 15-year-old Jane Creba, who was killed by crossfire while shopping nearby.

BIOLOGICAL FACTORS At the biological levels of analysis, we see that stimulating certain brain regions or altering neurochemicals in animals can lead to substantial changes in behaviour. Stimulating or damaging the septum, amygdala, or hypothalamus regions in the brain leads to corresponding changes in the level of aggression that animals display. For example, stimulating a cat's amygdala with an electric probe causes the animal to attack, whereas damaging the amygdala leads to passive behaviour. In 1937, researchers Heinrich Kluver and Paul Bucy produced a striking behavioural change by removing the amygdalas of normally very aggressive rhesus monkeys. Following the surgery, the monkeys were tame, friendly, and easy to handle. They began to approach and explore normally feared objects, such as snakes. They also showed unusual oral behaviour, putting anything within reach into their mouths, including snakes, matches, nails, dirt, and feces. The behaviour associated with damage to this region is now referred to as Kluver-Bucy syndrome.

In terms of neurochemistry, several lines of evidence suggest that serotonin is especially important in the control of aggressive behaviour. Luigi Valzelli (1973) found increased aggression among mice with low levels of serotonin. Drugs that enhance the activity of serotonin lower aggression, whereas those that interfere with serotonin increase aggressive behaviours (Figure 15.28; Raleigh, McGuire, Brammer, Pollack, & Yuwiler, 1991). Low levels of serotonin in humans have been associated with reports of aggression in adults and hostility and disruptive behaviour in children (Kruesi et al., 1992). In a large sample of men from New Zealand, low levels of serotonin were associated with violence, but not with criminal acts in general (Moffitt et al., 1998). Additionally, postmortem examinations of suicide victims have revealed extremely low levels of serotonin. Obviously, suicide is not the same as aggression, but many psychiatrists believe that suicide and violence toward others are manifestations of the same aggressive tendencies. Indeed, low serotonin levels were found among those who had killed themselves using violent means (such as shooting themselves or jumping from buildings) but not among those using nonviolent means (such as taking a drug overdose) (Asberg, Shalling, Traskman-Bendz, & Wagner, 1987).

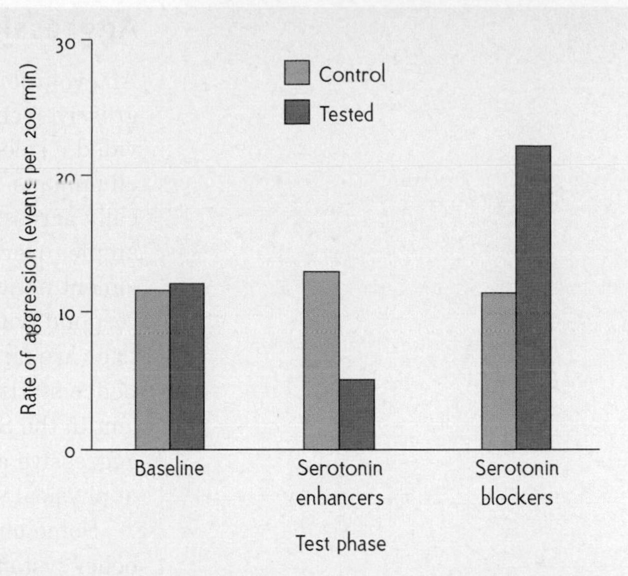

15.28 Serotonin and Aggression Male vervet monkeys were given either serotonin enhancers or serotonin blockers, and a corresponding change occurred in the number of aggressive acts against other members of the group.

Decreased serotonin levels may interfere with good decision making in the face of danger or social threat. For instance, monkeys with the lowest levels of serotonin are likely to pick fights with much larger monkeys. They also are the least socially skilled (Higley et al., 1996). This lack of social competence often leads the other monkeys to attack and kill them. Do these findings have implications for humans? Possibly. In one study, participants given a drug that enhances serotonin activity were found to be less hostile and more cooperative over time, compared with the control group (Knutson et al., 1998).

INDIVIDUAL FACTORS John Dollard and his colleagues proposed the first major psychological model of aggression in the 1930s. According to their **frustration-aggression hypothesis**, the extent to which people feel frustrated predicts the likelihood that they will be aggressive. The more that people's goals are blocked, the greater their frustration and therefore the greater their aggression. Thus, slow traffic is frustrating, and to the extent that it is impeding you from getting somewhere you really want to go or have to go—for instance, a concert you have been looking forward to or your first day in a new job—you may feel especially frustrated. If someone were to cut in front of you, you might feel especially angry and perhaps make some aggressive hand gesture. Indeed, road rage is most likely to occur where traffic is heavy and people feel frustrated.

frustration-aggression hypothesis The extent to which people feel frustrated predicts the likelihood that they will act aggressively.

According to Leonard Berkowitz's *cognitive-neoassociationistic model* (1990), frustration leads to aggression because it elicits negative affect. Thus, any situation that induces negative affect, such as being insulted, afraid, overly hot, or in pain, can trigger physical aggression even if it does not induce frustration. Berkowitz proposed that negative affect leads to aggression because it primes cognitive knowledge associated with aggression. From this perspective, negative events activate thoughts related to escaping or fighting, which prepare people to act in aggressive ways (see Chapter 10). Whether someone behaves aggressively depends on the situational context. If the situation also cues violence—for example, if the person has recently watched a violent movie or been in the presence of weapons—then the person is more likely to act in an aggressive fashion.

We can see the role of frustration and negative affect in the Abu Ghraib prison guards. Frustrated at being assigned to prison detail in Iraq for longer than expected, working in brutal, hot conditions, and being under the constant threat of attack easily primed the guards to be aggressive.

Aggression Has Social and Cultural Aspects

An evolutionary approach to aggression would dictate that similar patterns of aggressive behaviour should exist in all human societies. After all, if aggression provided a selective advantage for human ancestors, then it should have done so for all humans. But an examination of the data shows us that violence varies dramatically across cultures and even within cultures at different points in time. For example, over the course of 300 years, Sweden went from being one of the most violent nations on earth to one of the most peaceable, which certainly did not correspond with a change in the gene pool. Moreover, murder rates in some countries are far higher than in others (Figure 15.29). In the United States, analysis of police statistics reveals that incidents of physical violence are much more prevalent in the South than in the North. Hence, although human nature might have an aggressive aspect, can culture exert a strong influence on the actual commission of physical violence?

Some cultures may be violent because they subscribe to a *culture of honour*, a belief system in which men are primed to protect their reputations through physical aggression (Figure 15.30). Men in the southern United States, for example, were (and perhaps still are) traditionally raised to be ready to fight for their honour and to respond aggressively to personal threats. To determine whether southern males are more likely to be aggressive than northern males, researchers conducted a series of studies at the University of Michigan (Cohen, Nisbett, Bowdle, & Schwarz, 1996). In each study, male participants walking down a narrow hallway had to pass a male confederate who was blocking the hallway at a filing cabinet. As the participant tried to edge past the confederate, the confederate responded angrily and insulted the participant. Compared with participants raised in the North, those raised in the South became more upset and were more likely to feel personally challenged, more physiologically aroused (measured by cortisol and testosterone increases), more cognitively primed for aggression, and more likely to act in an aggressive and dominant manner for the rest of the experiment, for instance, by vigorously shaking the experimenter's hand (Figure 15.31). The culture-of-honour theory of violence supports Bandura's social-learning theory (discussed in Chapter 6) that much aggressive behaviour is learned through social observation of vicarious reward and punishment. People were traditionally admired for defending their honour in the South, and young men are taught that violence is an acceptable way to do so. The implications of this body of evidence suggest that our attitudes toward violence are determined by the cultural norms of our societies.

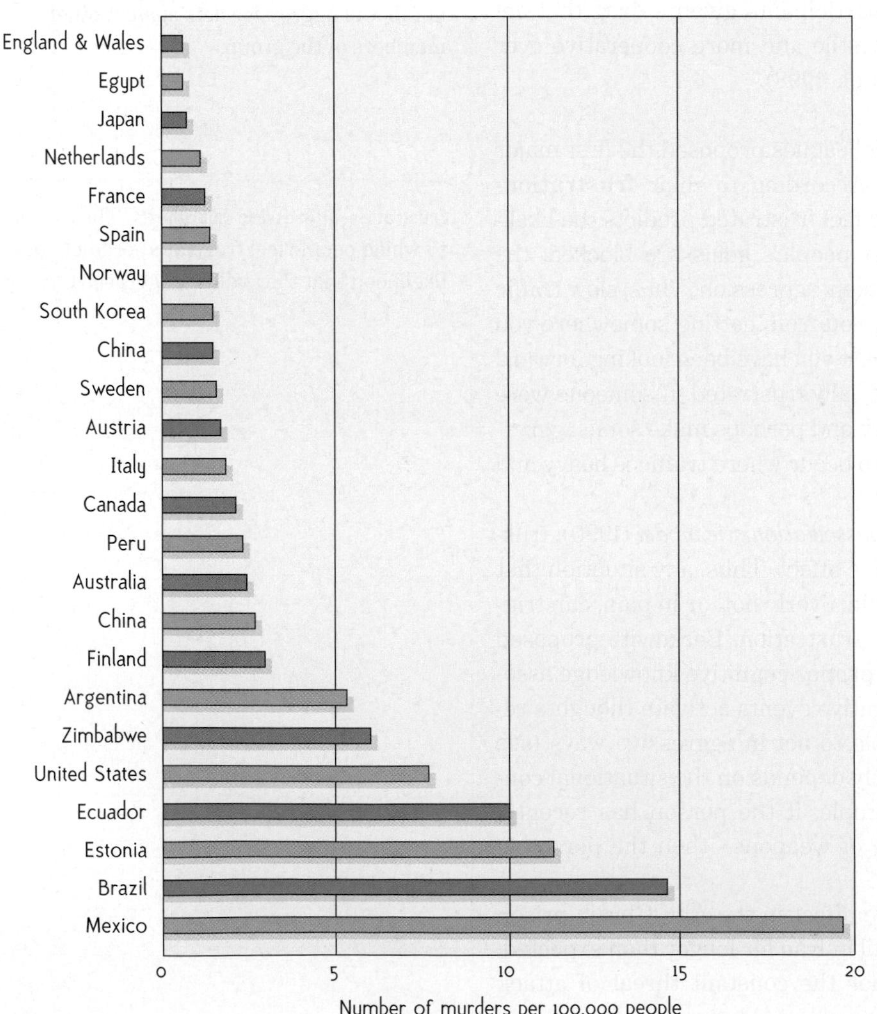

15.29 Aggression across Cultures Murder rates vary dramatically in different countries.

Many Factors May Influence Helping Behaviour

Although people inflict harm on one another in many situations, the opposite is true as well. People often behave in ways that are **prosocial**, meaning that they act in ways that benefit others. Prosocial behaviours include doing favours, offering assistance, paying compliments, subjugating our egocentric desires or needs, resisting the temptation to insult or throttle another person, or simply being pleasant and cooperative. By providing benefits to those around us, prosocial behaviours promote positive interpersonal relationships. Group living, in which people necessarily engage in prosocial behaviours such as sharing and cooperating, may be a central human survival strategy. After all, a group that works well together is a strong group, and belonging to a strong group benefits the individual members.

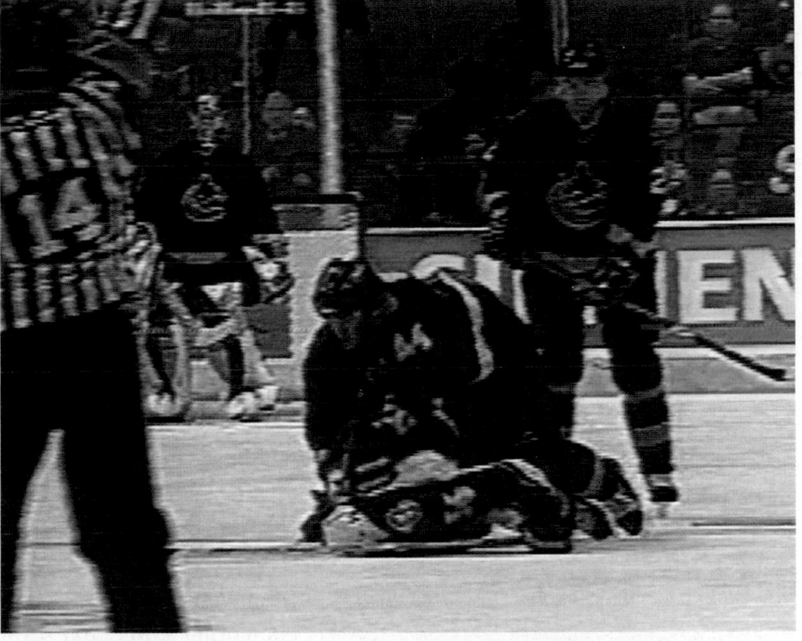

15.30 Culture of Honour On March 8, 2004, Vancouver Canuck Todd Bertuzzi brutally punched Colorado Avalanche Steve Moore from behind in apparent retaliation for an earlier hit on his teammate, Markus Naslund. Bertuzzi inflicted a concussion and three broken vertebrae in Moore's neck. Do you think there is a *culture of honour* in the NHL?

Why are humans prosocial? Theories range from the selfless to the selfish, and from the biological to the philosophical. For instance, Daniel Batson and his colleagues (1988; 1995) argue that prosocial behaviours are motivated by empathy, in which people share the emotions of others. Conversely, Robert Cialdini and his colleagues (1987; Maner et al., 2002) have argued that most prosocial behaviours have selfish motives, such as wanting to manage one's public impression or relieve a negative mood. Others have proposed that people have an inborn disposition to help others. Young infants become distressed when they see other infants crying (Zahn-Waxler & Radke-Yarrow, 1990), and although their early attempts to soothe other children are generally ineffective (for instance, they tend initially to comfort themselves rather than the other children), this empathic response to the suffering of others suggests that prosocial behaviour is hardwired.

Altruism is the providing of help when it is needed, without any apparent reward for doing so. The fact that people help others, and even risk personal safety to do so, may seem contrary to evolutionary principles; after all, those who protect

prosocial Tending to benefit others.

altruism The providing of help when it is needed, without any apparent reward for doing so.

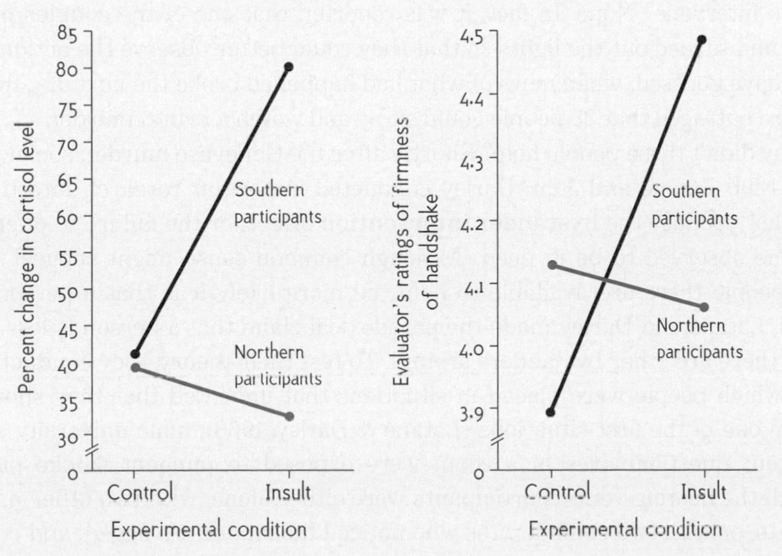

15.31 Cultural Aspects of Aggression When insulted, men from the southern United States had increased cortisol responses and shook hands more vigorously than northern men.

themselves first would appear to have an advantage over those who risk their lives to help others. During the 1960s the geneticist William Hamilton offered an answer to this riddle; he proposed that natural selection occurs at the level of the gene rather than at the level of the individual. Thus, the "fittest" animals are those that pass along the most genes to future generations, through the survival of their offspring. Hamilton introduced the concept of *inclusive fitness* to describe the adaptive benefits of transmitting genes rather than focusing on individual survival. According to this model, people are altruistic toward those with whom they share genes, a process known as **kin selection**. A good example of kin selection occurs among insects such as ants and bees, whose workers feed and protect the egg-laying queen but never reproduce themselves. By protecting the eggs, they maximize the number of their common genes that will survive into future generations (Dugatkin, 2004).

kin selection The tendency to be altruistic toward those who share a genetic bond.

reciprocal helping The tendency to help another because the recipient may return the favour.

bystander intervention effect The failure to offer help by those who observe someone in need.

Of course, sometimes animals help nonrelatives: For example, dolphins and lions will look after orphans. Similarly, a person who jumps into a lake to save a drowning stranger is probably not acting for the sake of genetic transmission. Robert Trivers (1971) proposed the idea of **reciprocal helping** to explain altruism toward nonrelatives. According to Trivers, one animal helps another because the other can return the favour in the future. Consider grooming, in which primates take turns cleaning each other's fur: "You scratch my back, and I'll scratch yours." For reciprocal helping to be adaptive, benefits must outweigh costs, and indeed people are less likely to help when the costs of doing so are high. Reciprocal helping is also much more likely to occur among animals—such as humans—that live in social groups, in which species survival depends on cooperation. This suggests that people will be more likely to help members of their ingroups than those of outgroups, and indeed the data generally supports this. From an evolutionary perspective, then, altruism does confer benefits, either by increasing the transmission of genes or by increasing the likelihood that others in the social group will reciprocate when needed.

Some Situations Lead to Bystander Apathy

15.32 Kitty Genovese Genovese was attacked and killed while 38 of her neighbors listened but did not try to stop the attack or even phone the police.

In 1964, a young woman named Kitty Genovese (Figure 15.32) was walking home from work in a relatively safe area of New York City. An assailant savagely attacked her, eventually killing her during a 30-minute beating. There were 38 witnesses to the crime—all watched it happen from their windows. How many called the police or tried to intervene? None. In fact, it was reported that one of the couples pulled up chairs and turned out the lights so that they could better observe the action. As you might have guessed, when news of what had happened broke the next day, most people were outraged that 38 people could sit by and watch a brutal murder.

Why didn't these people help? Shortly after the Genovese murder, social psychologists Bibb Latané and John Darley conducted important research into the situations that produce the **bystander intervention effect,** or the failure to offer help to someone observed to be in need. Although common sense might suggest that the more people there are available to help the more likely it is that a person will be helped, Latané and Darley made the paradoxical claim that a person is less likely to help if there are other bystanders around. To test their theory, they conducted studies in which people were placed in situations that indicated that they should seek help. In one of the first situations (Latané & Darley, 1968), male university students filling out questionnaires in a room were exposed to pungent smoke puffing in through the heating vents. Participants were either alone, with two other naïve participants, or with two confederates who noticed the smoke, shrugged, and continued filling out the questionnaire. As can be seen in Figure 15.33, when participants were on their own, most went for help. However, with three participants, few initially went for help. With the two calm confederates, only 10 percent of participants went for help in the first six minutes. The other 90 percent "coughed, rubbed their eyes, and opened the window—but they did not report the smoke." Similar results were obtained in subsequent studies in which people were confronted with mock crimes, apparent heart-attack victims in subway cars, and people passed out in public places. The bystander intervention effect, also called bystander apathy, has been shown to occur in a wide variety of contexts—even among divinity students, who while rushing to give a lecture on the Good Samaritan failed to help a person in apparent need of medical attention (Darley & Batson, 1973).

Years of research have indicated four major reasons for the bystander intervention effect. First, a diffusion of responsibility occurs, with people expecting that others who are also around will offer help. Thus, the greater the number of people that witness the need for help, the less likely that someone will actually step forward. Second, people fear making social blunders in an ambiguous situation. All of the laboratory situations had some degree of ambiguity, and people may have worried that they'd look foolish if they sought help that was not needed. There is some evidence that people do actually feel less constrained from seeking help as the need for help becomes clearer. Third, people are less likely to help when they are anonymous and can remain so. If people are not anonymous, they are much more likely to help. Thus, if you need help, it is often wise to point to a specific person and request his or her help.

How might we prevent diffusion of responsibility?

Fourth, a cost-benefit trade-off involves how much harm people risk by helping, or what benefits they would have to forgo if they stopped to help. Imagine you are walking to a potentially dull class on a sunny, warm day, and in front of you somebody falls down, twists an ankle, and needs transportation to the nearest clinic. You probably would be willing to help. Now imagine that you are trying to get to a final exam on time, and you are almost late, and the exam counts for 90 percent of your grade. In this case you probably would be much less likely to offer assistance.

There is evidence of bystander apathy in Abu Ghraib and other prisons. Government agents, such as FBI officers, apparently witnessed some of the mistreatment of prisoners. Although they reported these activities to their superiors, nothing formal was done to try to stop the mistreatment. Another factor that might have contributed to the diffusion of responsibility was the lack of a clear line of authority at Abu Ghraib, as prison guards received orders from various superiors representing various agencies. Thus, even though some people knew about the ongoing mistreatment, apparently little was done to stop it. Finally, one courageous soldier, Joseph Darby, grew disgusted with the abuse, took photographs, and shared them with his superiors. These photographs, subsequently leaked to the press, were concrete evidence of mistreatment that could not be ignored. Without doubt Abu Ghraib demonstrates the extent to which humans harm one another. Yet Darby's actions show humans have the capacity for goodness. Working across levels of analysis, psychological scientists have identified a number of important factors that contribute to how people treat one another. Although humans seem predisposed to act aggressively for personal gain, the world is surprisingly filled with acts of bravery and kindness.

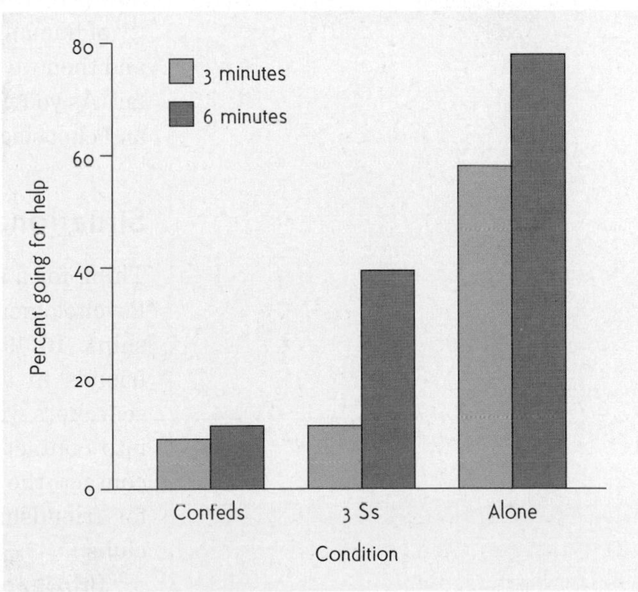

15.33 The Bystander Intervention Effect Student participants waited alone, with two other naïve participants (3Ss), or with two apathetic confederates. When smoke started to fill the room, those who were on their own went for help quickly. Those with the apathetic confederates did not seek assistance.

REVIEWING
the Principles | When Do We Harm or Help Others?

Humans in all cultures engage in aggressive acts. Frustration can lead to aggression, as can other factors that induce negative affect. This may occur because negative emotion primes aggressive thoughts. Cultural norms and beliefs can alter the expression of aggressive behaviour. People are likely to offer help to others in many situations, which is especially true if the person needing help is a relative. However, people are less likely to help when there are personal risks to doing so and when there is a diffusion of responsibility.

What Determines the Quality of Relationships?

Given the fundamental human need to belong (see Chapter 9) and the desire for social contacts, how do people select their friends and relationship partners? You might expect that studying relationships would be a high priority for psychological scientists. But until the last decade or so the topic was given little attention, perhaps due to the difficulty of developing rigorous experiments to test complex and fuzzy concepts such as love, a mysterious state that some think more appropriate for consideration by poets than by scientists. However, researchers have made considerable progress in identifying the factors that lead us to form friendships and close relationships (Berscheid & Regan, 2005). Many of these findings consider the adaptiveness of forming lasting affiliative bonds with others. In Chapter 9 we discussed evidence that humans have a strong need for social contact. We also discussed evidence for how people select mates. This section considers the factors that determine the quality of human relationships. We will first discuss the factors that promote friendships, and then we will look at why people fall in love and why love relationships sometimes fail. As you will see, many of the same principles are involved in choosing our friends and choosing our lovers.

Situational and Personal Factors Influence Friendships

Think for a second about your best friend. How is it that you came to be friends? Psychological scientists have discovered a number of factors that promote friendships. In 1950, Leon Festinger, Stanley Schachter, and Kurt Back examined friends in a college dorm. Because room assignments were random, the researchers were able to examine the effects of proximity, or how often people come into contact, on friendship. They found that the more often students came into contact, the more likely they were to become friends. Indeed, it is commonplace for friendships to form among those who are members of the same groups or clubs.

Proximity might have its effects because of familiarity. People like things that are familiar more than they like what is strange. Humans have a general fear of anything novel, known as neophobia. As discussed earlier in the attitudes section, simply being exposed to something repeatedly leads to increased liking. This mere exposure effect has been demonstrated in hundreds of studies using a variety of objects, including faces, geometric shapes, Chinese characters, and nonsense words (Zajonc, 2001).

BIRDS OF A FEATHER Another factor that increases liking is similarity. Birds of a feather really do flock together. People who are similar in attitudes, values, interests, backgrounds, and personalities tend to like each other. Roommates who are most similar at the beginning of the year are most likely to become good friends. Studies of high-school friendships have found that people tend to be friends with those of the same sex, race, age, and year in school. The most successful romantic couples also tend to be the most physically similar, which is called the matching principle (Bentler & Newcomb, 1978; Caspi & Herbener, 1990).

PERSONAL CHARACTERISTICS People tend especially to like those who have admirable personality characteristics and who are physically attractive, both as friends and as lovers. In a now-classic study conducted in the 1960s, Norman Anderson asked college students to rate 555 trait descriptions by how much they would like others who possessed those traits. Judging from the earlier discussion of who is rejected from social groups, you might expect that people dislike cheaters and others who drain group resources. Indeed, as can be seen in Table 15.1, the least likeable characteristics are dishonesty, insincerity, or lack of personal warmth. Conversely, people especially like those who are kind, dependable, and trustworthy. People like those whom they perceive to be competent much more than those they perceive to be incompetent or unreliable, perhaps because competent people make valuable group members. However, people who seem overly competent or too perfect make others feel uncomfortable or inadequate. In one study, a person who was highly competent but who spilled a cup of coffee on himself was rated more highly than an equally competent person who did not perform this clumsy act (Helmreich, Aronson, & LeFan, 1970). Small mistakes make people more human, and therefore more likeable. Generally, people like those who have personal characteristics valuable to the group.

TABLE 15.1 The Ten Most Positive and Most Negative Personal Characteristics

MOST POSITIVE	MOST NEGATIVE
Sincere	Unkind
Honest	Untrustworthy
Understanding	Malicious
Loyal	Obnoxious
Truthful	Untruthful
Trustworthy	Dishonest
Intelligent	Cruel
Dependable	Mean
Thoughtful	Phony
Considerate	Liar

PHYSICAL ATTRACTIVENESS Most people are also drawn to those who are physically attractive. Attractive people are typically judged to be happier; more intelligent, sociable, and successful; and less socially deviant. Taken together, these findings point to what Karen Dion and her colleagues in 1972 dubbed the "what is beautiful is good" stereotype. People who are physically attractive are less likely to be perceived as criminals, are given lighter sentences when convicted of crimes, are rated as more intelligent by teachers, are paid more for doing the same work, and have greater career opportunities. The preference for physical attractiveness begins early. Children as young as six months prefer to look at attractive faces, and young children prefer attractive over unattractive playmates (Rubenstein, Kalakanis, & Langlois, 1999). Even mothers treat attractive children differently from unattractive children. In one study, researchers examined mothers of over 100 infants feeding and playing with their newborns just after birth (while the mothers were in the hospital) and then again three months later (Langlois, Ritter, Casey, & Sawin, 1995). Mothers of attractive children were much more affectionate and playful than mothers of unattractive children. Those mothers attended to other people more than to their infants. Mothers

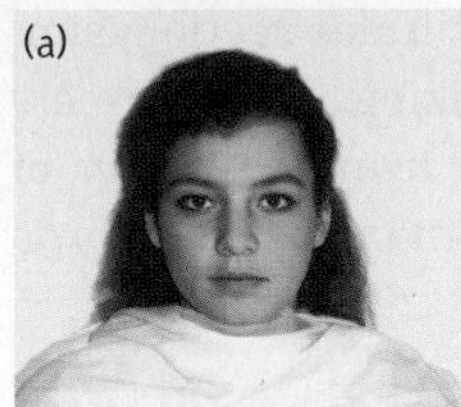

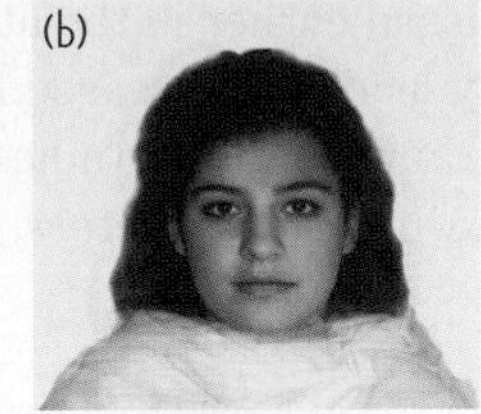

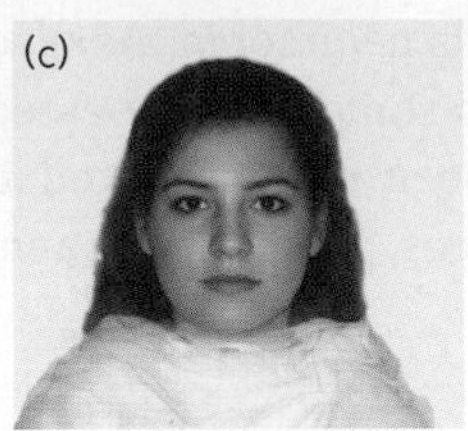

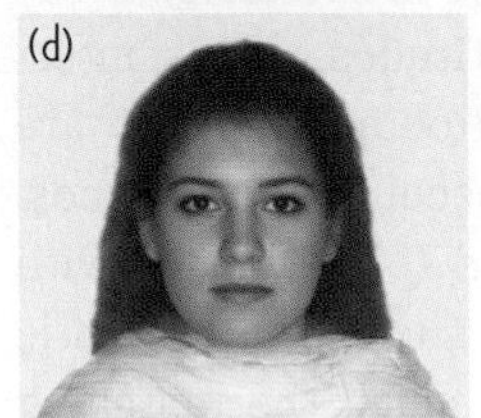

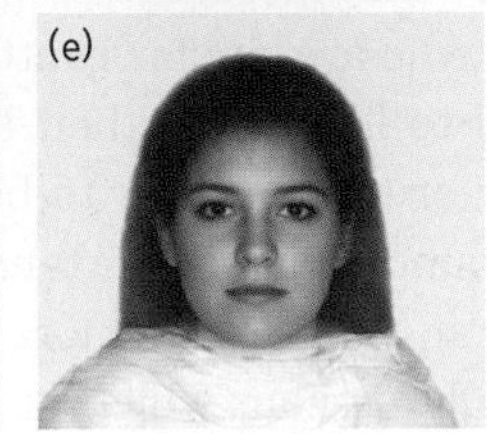

15.34 "Average" Is Attractive The more faces that are averaged together (a–e), the more attractive the outcome. The face in (e), which has the most faces averaged together, is typically rated most attractive.

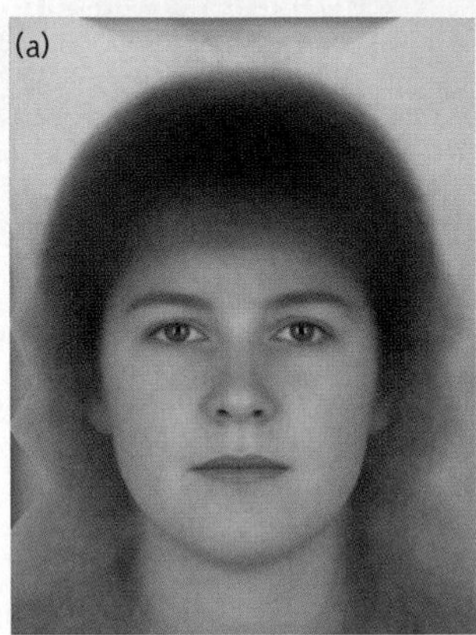

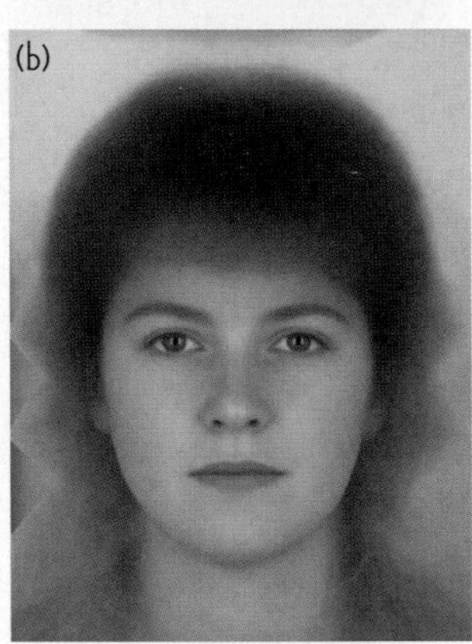

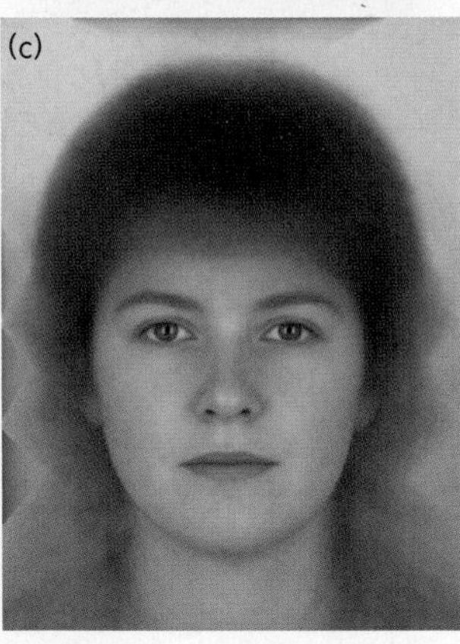

15.35 Which Average Is Preferred? Which of these faces do you prefer? Image (a) represents the averaging of 60 women. Image (b) is the composite of the 15 most attractive faces. Given a choice between image (a) and (b), the vast majority of people prefer image (b). Image (c) exaggerates the subtle differences between images (a) and (b). Given a choice between image (b) and (c), 7 out of 10 participants prefer image (c).

of attractive infants expressed slightly more positive attitudes about their children as well.

Given its importance, what determines physical attractiveness? Although some standards of beauty, such as preferences for body type, appear to change over time and across cultures, how people rate attractiveness is generally consistent across all cultures (Cunningham, Roberts, Barbee, Druen, & Wu, 1995). In terms of facial attractiveness, most people prefer symmetrical faces. This may be adaptive, in that a lack of symmetry could indicate poor health or a genetic defect. In a cleverly designed study of what people find attractive, Langlois and Roggman (1990) used a computer program to combine (or "average") a variety of faces, without regard to their individual attractiveness. They found that as more faces were combined together, participants rated the "average" as more attractive (Figure 15.34). People may view such "average" faces as attractive due to the mere exposure effect, in that such faces may be more familiar than unusual faces. However, other researchers contend that although average faces might be attractive, averaged attractive faces are rated more favourably than averaged unattractive faces (Perrett, May, & Yoshikawa, 1994). As you may notice in Figure 15.35, these female faces may be rated more attractive because they enhance feminine features, resulting in larger eyes, a smaller nose, plumper lips, and a smaller chin.

You might wonder, given this preferential treatment, whether people who are physically attractive actually possess characteristics consistent with the "what is beautiful is good" stereotype. The evidence on this issue is mixed. Among studies of university students, the correlation between objective ratings of attractiveness and other characteristics appears small. In one study that examined physical attractiveness, objectively rated by multiple judges, the researchers did not find any relation between appearance and self-esteem, grades, number of personal relationships, financial resources, or just about anything (Diener, Wolsic, & Fujita, 1995). However, evidence from more diverse settings indicates a robust effect of attractiveness on many outcomes. A meta-analysis of many hundreds of research studies of the effects of attractiveness on interpersonal evaluation and behavioural outcomes found that being attractive confers a number of benefits (Langlois et al., 2000). Not only are attractive people considered especially capable and gifted, but they also enjoy greater school and occupational success, and they are more popular, more socially skilled, and healthier. At the same time, they are similar to less attractive people in intelligence, life satisfaction, and self-esteem. Why does having all the benefits of attractiveness not lead to greater happiness? It is possible that attractive people learn to distrust attention from others, and especially from the opposite sex, because they assume that people like them simply for their looks. Believing that good things happen to them primarily because they are good looking may leave people feeling insecure, in that looks can fade or change with age.

Are attractive people happier?

Love Is an Important Component of Romantic Relationships

As noted earlier, psychological scientists have long neglected the study of love, in part because it seems a mysterious state that defies sensible comprehension. Thanks to the pioneering work of Elaine Hatfield and Ellen Berscheid (Figure 15.36), it is now clear that researchers can use scientific methods to examine this important interpersonal bond. Hatfield and Berscheid drew an important distinction between passionate love and companionate love. *Passionate love* is a state of intense longing and sexual desire, the kind of love that is often stereotypically portrayed in the movies. In passionate love people fall head over heels for one another, feel an overwhelming urge to be together, and are constantly sexually aroused in each other's presence. *Companionate love* is a strong commitment to care for and support a partner that develops slowly over time. It is based on friendship, trust, respect, and intimacy. Although people experience passionate love early in relationships, in most enduring relationships it often evolves into a more companionate love, in which intimacy and commitment dominate (Sternberg, 1986).

One theory of love is based on attachment theory. Recall from Chapter 11 that infants form different levels of attachment with their parents. According to Cindy Hazan and Phillip Shaver (1987), adult relationships, especially romantic relationships, also vary in their attachment style. The type of attachment style you have as an adult is assumed to relate to how your parents treated you as a child (Fraley & Shaver, 2000). Those who believe their parents were warm, supportive, and responsive report having *secure* attachments in their relationships. They find it easy to get close to others and they do not fear being abandoned. Just under 60 percent of adults report having this attachment style (Mickelson, Kessler, & Shaver, 1997). Those who found their parents were cold and distant, about 25 percent of the population, report having *avoidant* attachments. They find it hard to trust or depend on others, and they are wary of those who try to become close to them. Relationship partners make them uncomfortable. Those whose parents treated them in an inconsistent fashion—sometimes warm and sometimes not—have *anxious-ambivalent* attachments. This 11 percent of the population is best described as clingy. They worry that people do not really love them and are bound to leave them. Since people have to recollect how their parents treated them, it is possible that their memories are distorted. At the same time, this theory reinforces the importance of attachments throughout life.

Love is clearly a critical aspect of most people's social lives. Almost all of us enter into romantic relationships of one form or another. As we learned in Chapter 10, people in long-term, committed relationships tend to live longer, healthier lives. But if love is so important to people, why does it sometimes seem so elusive? Many people struggle to find their "one true love," as they date and reject potential partners who fall short of the ideal. As mentioned in Chapter 8, nowadays people marry much later as they spend more and more time looking. Even when people fall in love and form committed relationships, they still encounter challenges to that

15.36 Elaine Hatfield and Ellen Berscheid Hatfield (top) and Berscheid (bottom) are pioneers in the study of human relationships.

relationship. What can social psychology tell us about the factors that make a relationship successful?

Making Love Last Is Difficult

What factors cause marriages to be unhappy?

An unfortunate aspect of contemporary Western marriages is that most of them fail. In North America, approximately 30–50 percent of marriages end in divorce or separation, many within the first few years. In addition, many couples who do not get divorced are unhappy and live together in a constant state of tension or as strangers sharing a house. The social psychologist Rowland Miller notes that "married people are meaner to each other than they are to total strangers" (1997, p. 12). People often take their relationship partners for granted, openly criticize them, and take out their frustrations on them by being cruel or cold. It is perhaps no surprise then that relatively few marriages meet the blissful ideals that newlyweds expect. What factors cause such dissatisfaction in relationships?

As mentioned in Chapter 9, passion typically fades over time. The long-term pattern of sexual activity within relationships shows a rise and then a decline. Typically, for a period of months or even years the two people experience frequent, intense desire for one another and have sex as often as they can arrange it. Past that peak, however, their interest in sex with each other wanes. Frequency of sex declines by about half from the first year of marriage to the second, and it continues to decline, although more gradually, thereafter. Not only does frequency of sex decline, but people typically experience less passion for their partners over time. Unless people develop other forms of satisfaction in their romantic relationships, such as friendship, social support, and intimacy, the loss of passion leads to dissatisfaction and often to the eventual dissolution of the relationship (Berscheid & Regan, 2005).

JEALOUSY AND POSSESSIVENESS Even when people lose some of their sexual desire for one another, however, they generally do not want their mates to have sex with anyone else. Some degree of sexual possessiveness and jealousy is found in all cultures, although with variation as to what makes people jealous and how they express it. Although people may disapprove of sexual infidelity in general, many eventually engage in it. Even so, infidelity is far less frequent than we might think. The best estimates are that one out of four husbands and one out of ten wives have had extramarital sex. Earlier estimates suggesting that half of married men and almost as many women were unfaithful may have reflected the more permissive patterns of the "sexual revolution" of the 1960s and seventies, and people may well be more faithful now than they were a couple of decades ago.

LOVE IS FOSTERED BY IDEALIZATION The Irish playwright, George Bernard Shaw, once commented that "love is a gross exaggeration of the difference between one person and everybody else." Is there any truth in Shaw's cynical remark? Sandra Murray, John Holmes, and Dale Griffin argue that there is. They reason that if people hold unbiased views of their partners they will occasionally observe their partners behaving in rather unlovable ways, such as when they get angry and kick the dog, grow overweight, or lose the family's savings in a foolhardy get-rich-quick scheme. These observations of their partners behaving badly make it challenging for people to reconcile the conflicting thoughts that "I love my partner," and "my partner sometimes does things that drive me crazy!" Attention to a partner's human flaws should make it more difficult to convince oneself that one is in love. One the other hand, if people hold positive illusions about their partners, and see their behaviours in an optimistically biased way, they should encounter fewer conflicting thoughts. So, if instead one can think of their partner's anger as reflecting how "he is in touch with his feelings," his increasing girth as a sign that "he knows how to enjoy life," and his botched investments as indications that "he's trying his best to provide his family with new opportunities," one should identify fewer unlovable behaviours of their partners, and should thus have an easier time keeping the love going.

To investigate this hypothesis, Murray and colleagues (1996) investigated dating and married couples' perceptions of their partners. Consistent with their predictions, they found that those people who idealized their partners the most (i.e., they viewed their partners in the most unrealistically positive terms compared with how they viewed other people, and compared with how their partners viewed themselves) also loved their partners the most. Importantly, those people with the most positively biased views of their partners were more likely to still be in the relation with their partner several months later than were those with more "realistic" views of their partners. A little idealization appears to buffer a relationship against the ugly truths that might threaten it.

DEALING WITH CONFLICT Even in the best relationships, some conflict is inevitable, and couples need to constantly resolve strife. Confronting and discussing important issues is clearly an important aspect of any relationship. The way a couple deals with conflict often determines whether the relationship will last. John Gottman (1994) describes four interpersonal styles that typically lead to marital discord and dissolution. He calls these patterns of interacting the "Four Horsemen of the Apocalypse" to reflect their danger to relationships. These maladaptive strategies include being overly critical, holding the partner in contempt, being defensive, and mentally withdrawing from the relationship. When one partner launches a complaint, the other responds with his or her own complaints and often raises the stakes by recalling all of the other person's failings. People use sarcasm and sometimes insult or demean their partners. Inevitably, any disagreement, no matter how small, escalates into a major fight over the core marital problems, which often centre around a lack of money or sex or both. Couples who are more satisfied tend to express concern for each other, even while disagreeing, try to see each other's point of view, and manage to stay relatively calm. Being able to deliver criticism in a lighthearted, playful fashion is also a strategy for relationship satisfaction (Keltner, Young, Heerey, Demig, & Monarch, 1998).

ATTRIBUTIONAL STYLE AND ACCOMMODATION Unhappy couples also differ from happy couples in attributional style, which refers to how one person explains another's behaviour (Bradbury & Fincham, 1990). Unhappy couples make distress-maintaining attributions in which they view one another in the most negative way possible. In contrast, happy couples make partner-enhancing attributions in which they overlook bad behaviour or respond constructively, a process referred to as accommodation (Rusbult & Van Lange, 1996). Essentially, happy couples attribute good outcomes to one another and bad outcomes to situations, whereas distressed couples do the opposite. If their partners bring them flowers, they wonder what ill deed their partners committed rather than thinking about what sweethearts they married.

Profiles in Psychological Science 

Making Marriage Last

Many people were shocked when the Hollywood power couple Jennifer Aniston and Brad Pitt divorced in 2005. They seemed so perfect for each other. What happened? While wild rumours circulated about possible infidelity, different ideas about wanting children, and possibly some competition about whose career was more successful, one thing is clear: Celebrity marriages seem especially fragile. But despite the relationship implosions that daily feed the tabloids, there are many enduring marriages in Hollywood (Figure 15.37). Consider Danny DeVito and Rhea Perlman, Tom Hanks and Rita Wilson, or, most famously, Joanne Woodward and Paul Newman, who currently have been married for nearly 50 years. As it turns out, marriages outside of Hollywood are fragile as well, with over half of them ending in divorce and only one quarter of them being identified as happy. Although popular culture and the media have focused on negative statistics and why marriages tend to fail—for example, because of infidelity, substance abuse, and money problems—they have ignored the factors responsible for lasting happiness in marriage.

Over the past two decades, a number of psychological scientists have conducted research to examine why marriages succeed or fail. Among the foremost of these researchers is John Gottman, who has studied thousands of married couples to understand what predicts marital outcomes (Gottman, 1998). In his 1994 book, *Why Marriages Succeed or Fail and How You Can Make Yours Last*, Gottman outlines a number of differences between couples who are happy and those who are not. He also dispels a number of commonly held myths, for instance that couples who have the most sex are the happiest. Actually, couples who agree on the *frequency* of sex are happiest. No matter the frequency, if one person thinks it is too often and the other thinks it is not often enough, that spells conflict. Furthermore, many people believe that conflict is a sign of a troubled marriage and that couples who never fight must be the happiest. But this isn't true either. According to research by Gottman and others, fighting, especially when it allows grievances to be aired, is one of the healthiest things a couple can do for their marriage. Conflict is inevitable in any serious relationship, but resolving such conflict in a positive way is the key to marital happiness.

Gottman asserts that the most successful type of couple is what he calls a *validating* couple, with each partner considering the other partner's opinions and emotions valid, even if they disagree. People in such relationships make statements like "I know that it makes you angry when I hang the toilet paper the wrong way, but it was never a big deal to me which way it should hang." Validating couples try to compromise and to demonstrate mutual respect. These couples manage to avoid the "Four Horsemen of the Apocalypse" we mentioned earlier, as their fights typically do not escalate to name-calling, contempt, or one partner fleeing the conflict. Quite simply, they do the hard work of working things out.

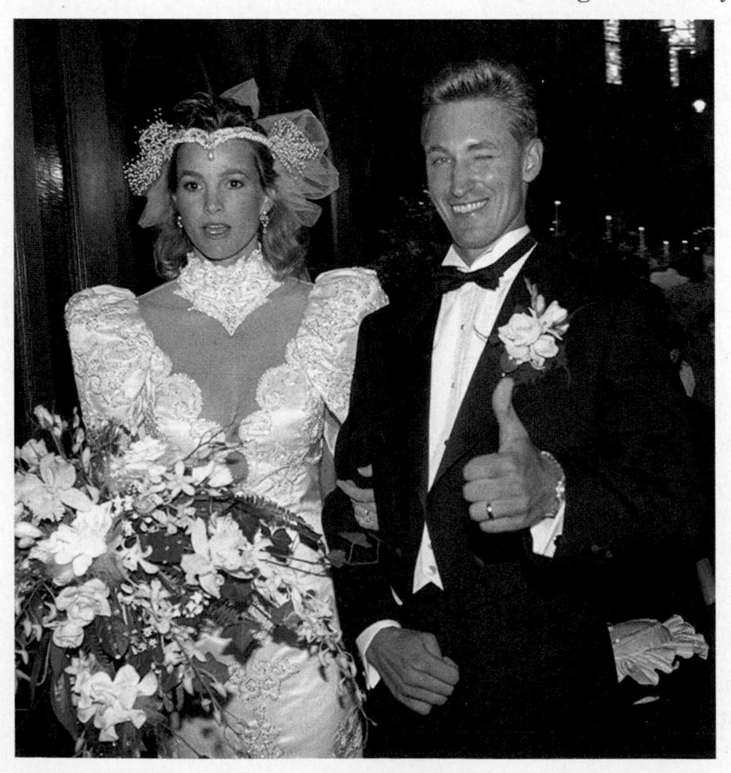

15.37 Wayne Gretzky and Janet Jones Although celebrity marriages often fail, Gretzky and Jones have been married for 20 years.

Based on his research, Gottman believes that as long as there are about five times more positive than negative interactions in a marriage, chances are good that the marriage will be stable. Couples headed for breakup fall below this level, and if there are as many negatives as positives in the marriage, the prognosis is pretty bleak. Thus, the task for any married couple is to seek opportunities for positive feelings within the relationship. Of course, the same principles apply to any long-term committed relationship, including those for gays and lesbians. According to Gottman and others, here are some ways to do that:

- Show interest in your partner. Listen to him or her describe the events of the day. Pay attention while the person is speaking and maintain eye contact. Try to be empathetic; show that you really understand and can feel what your partner is feeling. Such empathy and understanding can't be faked, but saying, "That must have been really annoying" conveys that you understand your partner's feelings.
- Be affectionate. You can show love in very quiet ways, such as simply touching the person once in a while. Reminisce about happy times together. Appreciate the benefits of the relationship. Couples who talk about the joys of their relationship, including comparing their partnership favourably to those of other people, tend to be happier with their relationships.
- Show you care. Many people take their partners for granted. Try to do something spontaneous like buying flowers or calling your partner at an unexpected time just to see how he or she is doing. Such actions let the partner know you think about him or her, even when you're not together. When we are dating people, we flirt with them, give them compliments, and display our best manners. Marriage doesn't mean you don't have to do any of these things. Be nice to your partner and try to make him or her feel that you value your mutual companionship. Praise your partner whenever possible. In turn, he or she will feel free to act in kind, which will help you feel good about yourself.
- Spend quality time together. It is easy for couples to drift apart and develop separate lives, especially in long-distance relationships. Be sure to share intimate details of your day on a regular basis. Find time to explore joint interests, such as hobbies or other activities. This doesn't mean that couples can't pursue independent interests (indeed, they should), but that having some activities and goals in common helps bring a couple closer. Having fun together is an important part of any relationship. Share private jokes, engage in playful teasing, be witty. Enjoy each other.
- Maintain loyalty and fidelity. Outside relationships can be threatening to a marriage or intimate partnership. Believing your partner is emotionally or physically involved with another person can pose harm to even the healthiest relationship. At their core, relationship partners have to trust one another. Anything that threatens that basic sense of trust will harm the relationship.
- Handle conflict. Perhaps most important, learn how to handle conflict. Don't avoid it and pretend you have no serious issues. Rather, calm down, try to control your anger, and avoid name-calling, sarcasm, or excessive criticism. Validate your partner's feelings and beliefs even as you express your own feelings and beliefs. Look for areas of compromise.

Although much of this seems like common sense, many people lose sight of how to express love and commitment in their relationships. It is so easy for couples to get caught up in everything else in their lives, from work to raising children, that it becomes easier to focus on what is wrong in their marriages than on what is right. This is when you know the marriage has taken a wrong turn. To make a marriage stronger people must put considerable effort into recognizing and celebrating all that is good about the relationship. Those affirming experiences make marriages succeed rather than fail. ■

REVIEWING the Principles | What Determines the Quality of Relationships?

People form friendships based on proximity, familiarity, similarity, and personal characteristics, such as personality traits and attractiveness. Love is an important component of romantic relationships. Relationships based solely on sexual passion may fail when that passion starts to wane, which often happens over time. How couples deal with conflict is an important determinant of whether the relationship will endure. Couples who are happy make positive attributions for their partners' behaviours, whereas unhappy couples tend to make negative attributions.

Conclusion

Social psychology teaches us a great deal about human existence. Even as we feel in control of ourselves and believe we can make sense of the actions of others, we often fail to take into account the powerful influence of social situations in shaping behaviour. Throughout this chapter we have learned that situations influence our attitudes, beliefs, and behaviours. Moreover, much of how we think about others and ourselves occurs automatically, with relatively little thought, and is based on minimal information. These judgments are often biased in ways that make people feel good about themselves, frequently at the expense of others. Research that crosses levels of analysis has shown that stereotypes alter the way people perceive and process information about others. Many times brain activity indicates what people themselves do not report, or perhaps even know, such as negative attitudes about ethnic minorities. Using limited mental resources, human minds automatically categorize others, with the outcome that people fail to treat others individually and with respect. It is all too easy for people to develop ingroups and outgroups; all it takes is three people. Yet social psychologists have discovered that cooperative learning systems can benefit both the self and society. The challenge now is to adapt these systems to larger groups in society and to motivate these groups to use them.

Much of the strength of the social situation is that it capitalizes on the human need to belong. People favour their groups, go along with their leaders, and are motivated to avoid conflict to maintain group harmony. However, pressure to conform to the group leaves people susceptible to social influence and causes them to act in ways that conflict with their personal standards. Learning about the power of social situations allows a partial understanding of the long and tragic history of inhumane acts, such as genocide, prisoner abuse, and other organized forms of brutality. The lesson from social psychology is that we need to be aware of the power in a situation in order to understand how it affects our behaviour as well as the behaviour of others. Recognizing social forces provides us with a much greater understanding of how humans behave in their everyday lives—one of the key goals of psychological science.

Summarizing the Principles of Social Psychology

How Do We Know Ourselves?

1. **Our self-concept consists of self-knowledge:** Self-schemas are the cognitive aspects of self knowledge and the working self-concept is what is activated and processed at any given time. Interdependent and independent self-construals are culture-based.

2. **Perceived social regard influences self-esteem:** Self-esteem is influenced by reflected appraisal. In sociometer theory, the basic need to belong influences social anxiety relative to self-esteem. Self-esteem may also be influenced by death anxiety.

3. **We use mental strategies to maintain our views of self:** Positive illusions of self are common. Self-esteem is influenced by our comparisons to others. A self-serving bias helps maintain positive self-esteem and may be culturally influenced.

How Do Attitudes Guide Behaviour?

4. **We form attitudes through socialization and experience:** Attitudes are influenced by familiarity (the mere exposure effect), and can be shaped by conditioning and socialization by others.

5. **Behaviours are consistent with strong attitudes:** Implicit attitudes (those that are automatic and easily activated from memory) influence behaviour and may differ from explicit attitudes (those that we profess).

6. **Discrepancies lead to dissonance:** A mismatch between attitudes or an attitude and a behaviour cause cognitive dissonance, which is often resolved by a change in attitude. A behavioural change is possible but more difficult to accomplish. To justify discrepant behaviour, many people often inflate the positive aspects of the experience.

7. **Attitudes can be changed through persuasion:** Persuasion often works by either focusing on the message (the central route) or the feelings the message generates (the peripheral route).

How Do We Form Our Impressions of Others?

8. **Nonverbal actions and expressions affect our impressions:** Nonverbal behaviour (body language) is quickly interpreted and provides valuable information.

9. **We make attributions about others:** We use personal dispositions and situational factors to explain the behaviour of others. Fundamental attribution error is when the personal attributions are favoured over the situational attributions in explaining behaviour.

10. **Stereotypes are based on automatic categorization:** Stereotypes are cognitive schemas that allow for fast, easy processing of social information, and can lead to bias and illusory correlations. Self-fulfilling prophecies occur when people behave in a way that confirms the biases of stereotypes.

11. **Stereotypes can lead to prejudice:** Prejudice occurs when the associated attitude is negative. Having a negative bias can lead to discriminatory actions. We show a preference for members of our ingroup versus those in the outgroup.

How Do Others Influence Us?

12. **Groups influence individual behaviour:** The presence of others can improve performance (social facilitation) or create laziness (social loafing). Loss of personal identity and self-awareness (deindividuation) can occur in groups. Group decisions can be extreme.

13. **We conform to social norms:** Socially determined influences on behaviour occur through awareness of social norms. Lack of unanimity diminishes conformity.

14. **We are compliant:** A variety of factors influence the likelihood of compliance including what mood we are in and whether we've previously agreed to a lesser request (foot-in-the-door effect).

15. **We are obedient to authority:** People readily behave in ways directed by authorities even to the extent of harming others.

When Do We Harm or Help Others?

16. **Aggression can be adaptive:** Brain structures, neurochemistry, and hormones influence aggression. Biologically based responses can be adaptive. Frustration can lead to aggression.

17. **Aggression has social and cultural aspects:** Aggression is not entirely adaptive and is influenced by the social and cultural experiences we have.

18. **Many factors may influence helping behaviour:** Prosocial behaviours maintain social relations. Altruism toward kin may favour inclusive fitness. Reciprocal helping is more likely in social groups in which survival depends on cooperation.

19. **Some situations lead to bystander apathy:** The presence of others in an emergency may create a diffusion of responsibility causing individual inaction.

What Determines the Quality of Relationships?

20. **Situational and personal factors influence friendships:** People affiliate with those who are similar and possess valued characteristics, including attractiveness.

21. **Love is an important component of romantic relationships:** In romantic relationships, passionate love evolves into companionate love.

22. **Making love last is difficult:** As passion fades, couples must develop other areas of satisfaction. Jealousy arises out of fears of infidelity. How couples deal with conflict influences the stability of the relationship. Generally, happy couples have positive views of their partners and their relationships. Indeed, love appears to be fostered by idealization.

Applying the Principles

1. In the movie *Harold and Kumar Go to White Castle*, Harold feels strongly that jaywalking is wrong but does it anyway. He tells himself it's okay since it is 2 A.M. and there are no cars around. This is an example of

____ a) cognitive dissonance and justification.
____ b) implicit and explicit attitudes.
____ c) persuasion and elaboration likelihood.
____ d) positive self-concept and maintenance.

2. Rex has met Cassie a few times and sees her as a not-very-bright but fun-loving blonde. When she excitedly tells him that she just got an A on her physics exam, he is surprised, and then wonders if perhaps it was an easy exam. What social cognition factor (or factors) account for his attitude?

____ a) self-fulfilling prophecy
____ b) ingroup-outgroup bias
____ c) stereotyping and fundamental attribution error
____ d) stereotyping and discrimination

3. Hazing is part of the tradition of fraternity pledging. Along with several of your frat brothers, you make the pledges strip down, stand out in the snow, and drink a vile concoction. This activity occurs with everyone cheering—and results in the near-death of one pledge. In hindsight, you are shocked by your behaviour. Which of the following is a probable explanation?

____ a) transient deviant personality behaviour
____ b) authoritarian personality attributional style
____ c) conformity and group processes
____ d) compliance behaviours

4. At the end of your English composition class, a student collapses at the front of the room. Although you notice the collapsed student, you rush out without helping because you have an exam to take in your next class. What may explain your behaviour?

____ a) bystander intervention effect and hostility
____ b) time pressure and bystander intervention effect
____ c) anonymity and time pressure
____ d) lack of reciprocity and ambiguity

5. The first day conservative, "preppy" Jenna moved into the dorm, she was shocked and disappointed that her roommate was a liberal, "granola" type. Now at midsemester they have become close friends. What is the most likely explanation for this change?

____ a) proximity and familiarity
____ b) opposites attract
____ c) the effects of crowding
____ d) one changed to be more similar to the other

ANSWERS: *1a 2c 3c 4b 5a*

Key Terms

aggression, p. 646
altruism, p. 649
attitude, p. 624
attributions, p. 631
bystander intervention effect, p. 650
cognitive dissonance, p. 626
compliance, p. 643
conformity, p. 641
deindividuation, p. 640
discrimination, p. 635
elaboration likelihood model, p. 629
frustration-aggression hypothesis, p. 647
fundamental attribution error, p. 613
implicit attitudes, p. 626
independent self-construals, p. 618
ingroup favouritism, p. 636
interdependent self-construals, p. 618
kin selection, p. 650
nonverbal behaviour, p. 630
obedience, p. 643
personal attributions, p. 632
persuasion, p. 628
prejudice, p. 635
prosocial, p. 649
reciprocal helping, p. 650
self-awareness, p. 615
self-concept, p. 615
self-esteem, p. 618
self-fulfilling prophecy, p. 635
self-schema, p. 616
self-serving bias, p. 622
situational attributions, p. 632
social comparison, p. 621
social facilitation, p. 639
social loafing, p. 640
social norms, p. 641
social psychology, p. 613
sociometer, p. 619
stereotypes, p. 633

Further Readings

Berscheid, E., & Regan, P. (2005). *The psychology of interpersonal relationships.* New York: Prentice-Hall.

Cialdini, R. (2001). *Influence: Science and practice.* Boston: Allyn & Bacon.

Franzoi, S. (2005). *Social psychology* (4th edition). Boston: McGraw-Hill.

Gilovitch, T., Keltner, D., & Nisbett, R. E. (2006). *Social psychology.* New York: Norton.

Gottman, J. (1994). *Why marriages succeed or fail.* New York: Simon & Schuster.

Higgins, E. T., & Kruglanski, A. W. (1996). *Social psychology: Handbook of basic principles.* New York: Guilford Press.

Kunda, Z. (2000). *Social cognition.* Cambridge, MA: MIT Press.

Zebrowitz, L. A. (1997). *Reading faces: Window to the soul?* Boulder, CO: Westview Press.

Further Readings: A Canadian Presence in Psychology

Heine, S. J. (2003). An exploration of cultural variation in self-enhancing and self-improving motivations. In V. Murphy-Berman & J. J. Berman (Eds.), *Nebraska symposium on motivation: Vol. 49. Cross-cultural differences in perspectives on the self* (pp. 101–128). Lincoln: University of Nebraska Press.

Heine, S. J. (2005). Where is the evidence for pancultural self-enhancement? A reply to Sedikides, Gaertner, & Toguchi. *Journal of Personality and Social Psychology, 89,* 531–538.

Heine, S. J., Lehman, D. R., Markus, H. R., & Kitayama, S. (1999). Is there a universal need for positive self-regard? *Psychological Review, 106,* 766–794.

Kawakami, K., Dovidio, J. F., & van Kamp, S. (2005). Kicking the habit: Effects of nonstereotypic association training and correction processes on hiring decisions. *Journal of Experimental Social Psychology, 41,* 68–75.

Kawakami, K., Dovidio, J. F., Moll, J., Hermsen, S., & Russin, A. (2000). Just say no (to stereotyping): Effects of training in the negation of stereotypic associations on stereotype activation. *Journal of Personality and Social Psychology, 78,* 871–888.

Kunda, Z., & Spencer, S. J. (2003). When do stereotypes come to mind and when do they color judgment? A goal-based theoretical framework for stereotype activation and application. *Psychological Bulletin, 129,* 522–544.

Lockwood, P., & Kunda, Z. (1997). Superstars and me: Predicting the impact of role models on the self. *Journal of Personality and Social Psychology, 73,* 91–103.

Murray, S. L., Holmes, J. G., & Griffin, D. W. (1996). The benefits of positive illusions: Idealization and the construction of satisfaction in close relationships. *Journal of Personality and Social Psychology, 70,* 79–98.

Wilson, A. E., & Ross, M. (2001). From chump to champ: People's appraisals of their earlier and present selves. *Journal of Personality and Social Psychology, 80,* 572–584.

Wright, S. C., & Tropp, L. R. (2005). Language and intergroup contact: Investigating the impact of bilingual instruction on children's intergroup attitudes. *Group Processes and Intergroup Relations, 8,* 309–328.

The Blending of Cultures

In the modern world, the lines between many similar cultures can blur. In any street corner in Montreal, for example, you can find world-class cheeseburgers. However, some cultures strive to find a solution for the delicate balance between globalization and retaining cultural identity. This monk in Ulan Bator, Mongolia is using a cell phone as he awaits the arrival of the Dalai Lama at the Gandantegcheling monastery in August 2006. The Dalai Lama was visiting Mongolia for the first time in four years to preach words of peace.

CHAPTER

16

Cultural Psychology

It was one of the most widespread and angry demonstrations the world has witnessed. Across the Muslim world in the Middle East, Africa, and Asia, and among Muslim groups in North America, the South Pacific, and Europe, people were furiously, and sometimes violently, expressing their rage. Dozens were killed in the demonstrations, embassies were torched, flags were burned, and many called for the violent death of those against whom they protested. The source of all this outrage: cartoons. In September 2005, a Danish newspaper printed cartoons depicting the Muslim prophet Muhammad, violating a deeply held Muslim belief that the prophet should never be depicted. And to make things even worse, many of the cartoons portrayed the prophet in deliberately offensive ways, such as with a bomb in his turban. Later, citing the importance of freedom of expression, newspapers from around the world reprinted the cartoons, which resulted in even more outrage.

Outlining the Principles

What Is Culture?

- Some Aspects of Culture Are Shared by Other Species and Some Are Unique to Humans
- Humans Have Evolved to Accumulate Cultural Information

How Does Culture Affect the Mind?

- Culture and Mind Are Inextricably Bound
- There Is a Sensitive Period for Learning Culture
- Cultural Differences Become More Pronounced with Age
- The Self-Concept Varies across Cultures

What Happens to People's Sense of Self When They Move to a Different Culture?

- Acculturation Requires Significant and Often Stressful Adjustment
- Psychological Processes Change with Acculturation
- People from Distinct Cultural Backgrounds Often Face Discrimination
- Multicultural People Can Switch between Different Selves

How Does Culture Affect How We Think and Behave?

- Cultures Differ in Analytic and Holistic Thinking
- Cultures Differ in Motivations for Control and Choice
- Cultures Differ in the Bases of Relationships
- Cultures Differ in Group Performance
- Cultures Differ in Moral Reasoning
- Language Influences but Does Not Determine Thought

How Does Culture Influence Mental Health?

- Culture-Bound Disorders Are Largely Limited to Certain Cultural Contexts
- Some Universal Mental Disorders Present Differently in Different Cultures

This issue of whether it was inappropriate to publish caricatures of Muhammad deeply polarized the world, with those on either side of the conflict furious at and puzzled by the reactions of those on the other. The conflict seemed to reflect what political scientist Samuel Huntington had in mind in 1993 when he wrote "The Clash of Civilizations?" in which he argued that the fundamental source of conflict in the new world would not be ideological or economic, but cultural.

So why did Western and Muslim populations see this issue in such fundamentally different terms? The editor of the Danish newspaper justified his decision to publish the cartoons by noting that Christian images are satirized quite frequently—for example, a recent and much publicized Danish painting showed Jesus with an erection. So why not Muhammad? Furthermore, as many Western commentators argued, nothing should be treasured more than freedom of expression, even if it comes at the expense of sometimes offending others. However, these arguments were rejected by many in the Muslim world, with one protestor's sign capturing the feelings of many: "Freedom go to Hell!" ■

16.1 The Müller-Lyer Illusion The line on the left looks longer because the angles at the corners suggest that it is further away than the line on the right. People who are raised in cultures where there are not carpentered corners are not susceptible to this illusion.

In a multicultural world, there is much potential for misunderstanding of the kind described above. Psychological science would appear to be in a good place to facilitate understanding between people. However, a serious limitation of the conclusions that we draw from much psychological research is that we often do not know how well those conclusions generalize to other cultures. People's brains function the same around the world, so we don't have to question how well fundamental neuronal processes generalize. But, in the case of higher order psychological processes, in which social interactions play a key role, it is less clear how these processes will emerge across cultures. Does a finding, say, that American college students tend to rationalize their decisions indicate that people the world over make the same kind of rationalizations? The uncertainty that psychologists have about the generalizability of some of their research findings is due to the fact that many psychological phenomena have been studied mainly in North American and European contexts. Furthermore, attempts to see how well key psychological processes replicate in other cultural contexts indicate that, in many cases, the findings do not generalize across cultures well at all, such as susceptibility to some perceptual illusions (Segall, Campbell, & Herskovits, 1963) (see Figure 16.1), the relation between thinking and speaking (Kim, 2002), preferential learning for nouns over verbs (Tardiff, 1996), preferred decisions in the ultimatum game (Henrich et al., 2005), and some aspects of numerical reasoning (Gordon, 2004).

Fifth century B.C.E.
Exploring Origins The Greek historian Herodotus travels to more than 50 societies to document their distinctive ways of life and how they contrast with classical Greek society.

1904
Folk Psychology Wilhelm Wundt publishes a 10-volume tome entitled "*Volkerpsychologie*" (Folk Psychology); in it he theorizes about how culture is implicated in psychological processes.

1920s
Cultural-Historical Psychology
Alexei Leontiev, Alexander Luria, and Lev Vygotsky found the Russian cultural-historical school of psychology and explore how people interact with their environments through the "tools" of acquired cultural ideas.

1930s–1950s
Culture and Personality Studies Anthropologists such as Ruth Benedict and psychologists such as Henry Murray jointly explore the relations between culture and personality.

The field of cultural psychology emerged because most researchers believe that culture plays a prominent role in our mental lives. It also emerged for a second important reason, which is to see if behaviours and mental processes are the same in all cultures. Cultural psychology is the study of how culture shapes psychological processes. As social psychology emerged out of the fact that humans are ultimately social animals, and thus social environments have a great impact on how we think and feel, so cultural psychology emerged out of the fact that humans are ultimately cultural animals. Like social environments, cultural environments arise from the interactions between people and the information that they share. Cultural environments are social environments writ large, with the norms and shared understandings that develop within them potentially extending to all aspects of an individual's life. Cultural environments are more sweeping and inclusive than social environments. Although you can quite easily leave one social situation and enter a new and different one, most people rarely leave their own cultural environments. Hence, most people are exposed to a single all-encompassing set of norms and shared understandings. Like water to a fish, our own culture remains largely invisible to many of us. We often don't realize that people who grew up in other cultures may have a very different set of norms and shared understandings than we do.

Why is it so wet?

Our exploration of cultural psychology is guided by two key themes. The first addresses how people's sense of self relates to others. Every human is both a distinct, unique individual and a social creature that is fundamentally connected with others. How people accomplish individual goals and how they seek belongingness with others are two sets of behaviours that are emphasized differently across cultures. As we discussed in Chapter 15, people in individualistic cultures are more likely to have independent views of self that accentuate their sense of themselves as distinct, whereas people in collectivistic cultures are more likely to have interdependent views of self that highlight their sense of belongingness with others. This difference in self-concept underlies many of the differences in psychological processes across cultures that we'll discuss in this chapter.

A second theme throughout this chapter is that both universal and culturally specific psychologies exist. People the world over share the same biology, and at some level of analysis, the psychological experiences of people around the world are universally similar. For instance, chili peppers taste spicy to people in Thailand and Canada, in Poland and Brazil. But at other levels of analysis we see pronounced cultural variation in psychological processes, such as the preferences for spicy tastes. Moreover, experiences within a culture can shape psychological experiences—hence people in Thailand, where spicy food is common, may develop a tolerance for strong

1961
The Achieving Society David McClelland extends the work of German sociologist Max Weber by identifying how some motivations to achieve and to acquire wealth are grounded in Protestant religious beliefs.

1960s
Cross-Cultural Research on Perception Research by Marshall Segall and colleagues reveals how susceptibility to some basic perceptual illusions is grounded in particular cultural experiences.

1980
Cultural Values Geert Hofstede conducts a landmark study of values among IBM employees from 40 different countries.

1980s–1990s
Rise of Individualism and Collectivism In a series of papers, Harry Triandis and others build a methodological and theoretical case for the central role of individualism and collectivism in understanding different cultures.

1990
Rise of Cultural Psychology Richard Shweder writes a seminal article that defines the field of cultural psychology.

chili peppers and not experience them as hot at all. A pepper considered mild in Bangkok or Mexico City might bring tears to the eyes of restaurant-goers in Toronto.

Cultural psychologists are interested in the ways that people with different cultural experiences differ in their psychological processes. As you read this chapter, bear in mind some of the key questions that we will be exploring. What is culture? How does culture affect thoughts and behaviours? Are fundamental ways of thinking universal across cultures? Is our sense of self universal across cultures?

How universal are psychological processes?

What Is Culture?

Humans are a cultural species. When you travel to a different country, you quickly detect the various ways that people there live differently than people back home. People create cultural environments wherever they live, and these cultural environments shape the ways that people think and behave. But an important question to consider first is what is culture? This question has been debated by anthropologists, sociologists, and psychologists for decades, and no single answer is accepted in all fields. The definition of culture in Chapter 1 is that it is the beliefs, values, rules, and customs of a group of people who share a language and environment and that these beliefs, values, rules, and customs are transmitted through learning from one generation to another. This definition describes culture in uniquely human terms (i.e., no other species has beliefs, values, rules, or customs, or even much evidence of a common language), and it captures what we refer to as culture throughout this chapter. A broader definition is that culture is any kind of information acquired by individuals through imitative or social learning (Aoki, 1991). According to this broader definition, do you think that any other species can be said to have culture?

Some Aspects of Culture Are Shared by Other Species and Some Are Unique to Humans

Much research has found evidence for cultural learning in other species. One example was discussed in Chapter 6, when the potato-washing behaviour of the monkey Imo spread to the other monkeys in her troupe. The other monkeys learned about potato washing by watching the clever Imo. This is a clear example of imitative learning. And there is solid evidence of cultural learning in various primate species, dolphins, and whales. Indeed, recent research in Scotland indicates that dolphins use distinctive sounds for each other in much the same way humans use names (Figure 16.2) (Janik, Sayigh, & Wells, in press). Plus, cultural learning is not restricted to the most intelligent animal species: even pigeons and guppies have shown limited evidence of learning from others.

1990
Acts of Meaning An architect of the cognitive revolution, Jerome Bruner articulates how the role of meaning is central to an understanding of the functioning of the human mind, and how meaning can best be revealed by studying culture.

1991
Independent and Interdependent Selves Hazel Markus and Shinobu Kitayama provide a theoretical model for integrating individualism and collectivism with the self-concept, thereby laying the foundation for much subsequent empirical cultural psychological research.

1990s
Evolution and Culture In *Culture of Honor* Richard Nisbett and Dov Cohen propose that evolutionary and cultural theory can be used as complementary explanations of human behaviours across time and place.

2000s
Cultural Priming Research by Ying-yi Hong, CY Chiu, Wendi Gardner, and others reveals that priming can activate ways of thinking that tend to be associated with particular cultures.

2000s
Analytic and Holistic Thinking Research by Richard Nisbett and colleagues reveals how cultures vary in their tendencies to think analytically and holistically.

16.2 Cultural Learning in Other Species Dolphins and other animals have been known to communicate with each other, even on a verbal level. Do you think this has implications for the level of imitative and social learning existent among nonhumans?

Humans, then, are not unique in being able to learn cultural information. But, humans do stand out in how much cultural learning they do and how well they do it. Although other species show evidence of cultural learning, none appear to do it well. Yes, the monkeys learned potato-washing from watching Imo, but they weren't especially good at it. It took years for the potato-washing behaviour to spread through the troupe, and quite a few of the monkeys never figured it out. In contrast, many aspects of human culture are learned by virtually every culture member, often after only a single trial. For example, most words are known by everyone within a culture.

How is human culture different from animal culture?

Moreover, human cultural learning is so extensive that, unlike cultural learning in other animals, it pervades all aspects of our lives. The languages we speak, the ways we seek mates, the work we perform, the tools we use, the shelters we live in, the leisure activities we pursue, and the ways we relate to others—all are cultural products. Indeed, it could be said that all human behaviour, or thoughts, have been shaped in some way by cultural learning. In contrast, the cultural learning identified in other species has thus far been limited to specific and often isolated aspects of their behaviours. It is the unique pervasiveness of culture in human lives that allows us to say that humans are the only true cultural species.

Humans Have Evolved to Accumulate Cultural Information

How has cultural learning become so central to our species? Culture must have been adaptive for our human ancestors in order for our species to have evolved to be so dependent on it. What benefits might cultural learning have brought to our ancestors? Humans' social nature appears to have been a key factor in the evolution of the human brain. There is a clear relation between the average group size in which various primate species, including humans, live and the proportion of cerebral cortex in their brains (see Figure 16.3, Dunbar, 1993): the larger the average group size, the larger the cerebral cortex ratio. This suggests that human brains evolved to be as big as they are because the cognitive capacities of a large brain were highly adaptive for social living. Think of all the advantages possessed by primates in large groups. Such groups protect against predators, create significant pools of shared resources, and offer greater opportunities for social learning (Figure 16.4). However, such groups also present complex social dynamics, and individuals need to understand and navigate these dynamics if they are to survive within the group, and particularly ascend

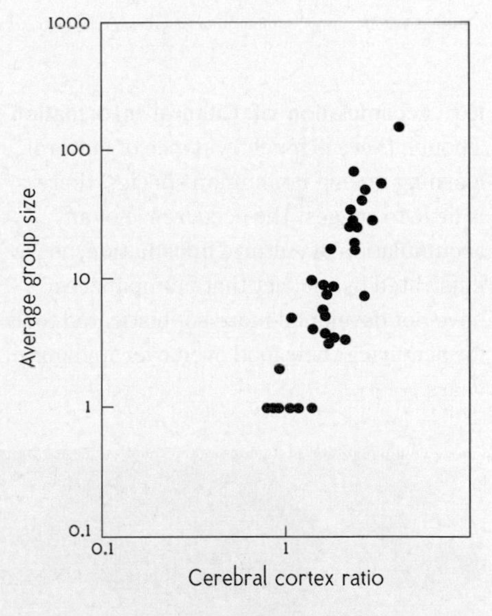

16.3 Cerebral Cortex Ratio The relation between average group size and the ratio of the cerebral cortex to brain volume of various species of primates. Each point represents a species of primates, with humans being the topmost point. The larger the average group that a species lives in, the larger is the proportion of their brain that is made up of the cerebral cortex.

16.4 Social Learning One of the most heavily studied peoples, San bushmen of the Deniui village near Tsumkwe in the Kalahari Desert in eastern Namibia enjoy all the benefits and advantages of living in large groups.

the hierarchy. Understanding and using social dynamics to realize the benefits of social living required the evolution of two capacities that are largely unique to humans.

First, humans evolved sophisticated communication skills that allowed them to convey beliefs, intentions, and complex thoughts—all this, in turn, allowed individuals within a group to coordinate their behaviours (Pinker, 1994). The development of language, the human capabilities for which far transcend the communication skills of other primate species, was a key adaptation for cultural learning. Think of all the advantages that language gave humans. It allowed them to more accurately communicate needs, describe threats or opportunities, understand complex social dynamics, and more precisely describe and perpetuate cultural ideas.

Second, as discussed in Chapter 11, to understand social dynamics one must be able to identify the intentions of fellow group members—that is, one must have a theory of mind. Humans in all cultures develop a theory of mind early in life (Callaghan et al., 2005), whereas evidence for a theory of mind is much weaker, if not largely absent, in other species (there is some conflicting evidence regarding its presence in chimpanzees). Having a theory of mind allows humans to imagine the intentions of others, and this allows cultural learning to occur with a far higher degree of precision than simple observation (Tomasello, 1999). Such high-precision cultural learning allows humans to do something important that no other species can do. Humans can *accumulate* cultural information. That is, after observing a model perform a behaviour, humans are able to reproduce that behaviour accurately, and can then improvise and add to that behaviour, allowing cultural information to accumulate and evolve.

The cumulative nature of human culture means that we can acquire a far richer degree of cultural information than could be learned in any individual's lifespan if each of us had to start from the beginning. So, for example, archaeological records reveal that tens of thousands of years ago humans used very simple tools, such as hammers and knives that were nothing more than rocks with slightly altered edges. Subsequent finds in the archaeological record reveal that these tools slowly became more and more sophisticated over generations and that cultural innovations in tools began to appear at an increasingly rapid pace. Think of how quickly cultural information accumulates today—for example, over the past few decades, the average power of computing chips has been doubling about every eighteen months. This ability to accumulate cultural information, and to create more and more sophisticated, complex, and powerful tools, has been highly adaptive for humans.

16.5 Accumulation of Cultural Information Though there is much evidence of cultural learning among nonhuman species, there is little to suggest the occurrence of any accumulation of cultural information, as illustrated by the fact that chimpanzees have not developed more sophisticated tools for acquiring their food even over millions of years.

In contrast, the simple tools that chimpanzees use, such as a twig for fishing termites out of a mound (Figure 16.5), are not any more complex than those used by chimpanzees thousands of generations ago. Without the ability to easily reproduce a behaviour, chimpanzees and other animals cannot engage in high-precision cultural learning. As a result, each generation of animals can do little more than learn what the previous generation was able to do, and the cultural information does not accumulate. Each generation must start over at square 1.

REVIEWING
the Principles | What Is Culture?

A broad definition of *culture* is that it is any information that is acquired by imitation. Although some other species do show cultural learning, no other species is as dependent on culture as humans. Humans evolved complex communication skills, and they developed the capacity to understand the intentions of others. These capabilities allowed humans to accumulate cultural information, which has many important benefits.

How Does Culture Affect the Mind?

The anthropologist Clifford Geertz (1973) asserted that "we all begin with the natural equipment to live a thousand kinds of life but end in the end having lived only one" (p. 45). This statement nicely captures the tension between the universality and cultural variability of psychological processes. As we discussed in the previous section, humans in all cultures are born with basically the same potentials—we all could learn any language, and adopt any cultural worldview if raised in the respective cultures. Imagine how different your life would be if you were born into a very different cultural environment—for example, a Yanomámo hunting and gathering tribe in the Amazon river basin or a family of nomadic herders on the steppes of Mongolia. At the same time, the psychological experiences of humans are similar around the world. Table 16.1 summarizes some findings from the anthropological literature regarding universal characteristics of humankind.

As Table 16.1 suggests, there is a common foundation to the psychological experiences of all people. However, we also know that by participating in specific cultures, and having experiences particular to those cultures, these universal psychological characteristics are shaped and expressed differently across cultures. For example, marriage is a cultural universal—in all cultures there are formalized and publicly recognized enduring relationships centred around the raising of children (Figure 16.6). But marital practices vary dramatically across cultures, with some cultures practicing monogamy (one husband, one wife), some practicing polygyny (one husband, multiple wives), and some practicing polyandry (one wife, multiple husbands). Imagine the vast difference between married life with one spouse and married life with several. Universal tendencies can thus be expressed in culturally particular ways.

Culture and Mind Are Inextricably Bound

The fact that many universal aspects of the mind are expressed differently across cultures demonstrates that culture and mind are not fully separable. Cultures emerge from the interaction of the minds of the people that live within them, and cultures, in turn shape the kinds of things that those minds think about, and in so doing, influence the ways that those minds process information. For precisely this reason, many cultural psychologists argue that in order to have a rich understanding of the mind it is necessary to have a rich understanding of the culture in which it exists.

TABLE 16.1 | Universal Behaviours, Reactions, and Institutions

A sampling of some of the behaviours and characteristics that anthropologists believe hold for all human cultures, grouped into categories to show general areas of commonality.

SEX, GENDER, AND THE FAMILY		
Copulation normally conducted privately	Rape	Rape proscribed
Live in family (or household)	Sex differences in spatial cognition	Sexual jealousy
Husband usually older than wife	Sexual modesty	Sexual regulation
Males dominate public realm	Division of labour by gender	Males more aggressive
Males more prone to lethal violence	Females do more child care	Marriage
Mother-son incest unthinkable	Incest prevention and avoidance	Preference for own kin
SOCIAL DIFFERENTIATION		
Age statuses	Classification of kin	Ingroup distinguished from outgroup
Division of labour by age	Leaders	
SOCIAL CUSTOMS		
Baby talk	Pretend play	Group living
Dance	Rites of passage	Law (rights and obligations)
Dominance/submission	Tabooed foods	Magic to win love
Feasting	Toys	Practice to improve skills
Gossip	Body adornment	Property
Hygienic care	Death rites	Rituals
Magic to increase and sustain life	Etiquette	Tabooed utterances
Non-bodily decorative art	Food sharing	
EMOTION		
Childhood fear of strangers	Wariness around snakes	Rhythm
Facial expressions of fear, anger disgust, happiness, sadness, and surprise	Envy	Melody
COGNITION		
Aesthetics	Anthropomorphism of animals	Myths
Belief in supernatural, religion	Medicine	Taxonomy
Classification of flora	Language	
Classification of fauna	Narrative	

Source: Compiled by Donald Brown, 1991, appearing in Pinker, 2003.

Consider the cultural practice, common in school systems across North America, of tailoring a child's education to his or her individual needs. In many North American schools, children are promoted or held back depending on their performance in class or on standardized tests. Likewise, many schools use a strategy called "tracking," in which students within a grade are grouped according to their abilities. Such strategies reflect a desire to ensure that education addresses each child's specific and unique needs. In contrast, in school systems across Japan, during the period of mandatory education, *all* students are promoted together to the next grade regardless of whether some are performing at a level above or below their peers (this is sometimes called "social promotion" in North America). Likewise, Japanese students are not grouped according to their abilities. Whether or not one lives in a cultural environment that tailors education to individual needs certainly must influence

16.6 Cultural Differences Norms, practices, and customs can vary widely across cultures, even in remarkably similar situations. Compare the traditional marriage ceremony of South Korea (a) and its counterpart in India (b). Both weddings achieve the same desired result, but yet customarily they contrast strikingly.

one's values. The North American education system highlights the individuality of each student and implicitly suggests that it is more important that children receive an individualized education than it is for them to stay with their peers. In contrast, the Japanese education system suggests that it is more important that children develop a sense of belonging with their peers than it is for them to receive an education tailored to individual needs. Participation in a particular cultural environment (e.g., one that tracks or does not track students through school) can influence the degree to which children attend to their individuality or strive to fit in with their peers. Culture here can be seen to shape the mind.

How are culture and mind related?

At the same time we acknowledge that culture can shape the mind, we have to ask how particular cultures came to be the way they are. How did North American school boards come to emphasize performance-based promotion and Japanese school boards come to emphasize social promotion? Although many factors influence how cultures develop, one is that they develop out of the participation of like-minded individuals. Interactions among individuals create cultural norms, and those norms reflect the underlying values of those individuals. For example, perhaps North American school boards decided to emphasize performance-based promotion because many North Americans believe that it is important to respect the needs of individuals. Likewise, the Japanese Ministry of Education's decision to emphasize social promotion likely emerged from the Japanese people's feeling that it is important to develop a strong sense of belonging with one's peers. That is, culturally shaped ways of thinking influence the education models that countries adopted. In this way, we can see that culture and mind are bound together. The mind is influenced by participating in certain cultural practices, and cultural practices are shaped by the minds of the people who live within them. The underlying thesis of cultural psychology is that culture and mind make each other up (Shweder, 1990) and that we cannot properly study either culture or mind without considering how each affects the other.

There Is a Sensitive Period for Learning Culture

As discussed earlier, humans evolved to engage in cultural learning. Furthermore, in many ways, humans are prewired to acquire cultural information at a young age. Language development provides especially clear evidence for this. Though humans are capable of producing and recognizing approximately 150 phonemes, no language uses more than 70 of them. Interestingly, humans are born with the capacity to recognize all 150 phonemes, but through socialization and being exposed to certain languages, they lose the ability to distinguish between phonemes not in their own languages. For example, as discussed in Chapter 11, the Japanese language does not have separate phonemes for the sounds "la" and "ra," and Japanese who do not learn English as children cannot reliably distinguish between them. Being exposed to a language shapes the ways that we perceive phonemes, and it does so early in life. One study found that native English-speaking children can distinguish between two similar phonemes from the Hindi language when they are 6–8 months old, but that by the time they are 10–12 months old, they have largely lost this ability (Werker & Tees, 1984). Even four-day-old infants show a preference for the rhythms of their own language over other languages (Mehler et al., 1988). Furthermore, our ability to learn new languages diminishes with age. People who acquire a new language after puberty often speak it with an indelible accent of their native tongue.

As with this sensitive period for acquiring languages, there appears to be a sensitive period for acquiring other kinds of cultural information. Yasuko Minoura (1992) studied Japanese-born children who moved to the United States at different ages along with their parents. She investigated the extent to which the children acquired various cognitive, behavioural, and emotional domains of Japanese and American culture. Her results suggest that people begin internalizing cultural information at birth, but that most of it is not permanently retained before the age of 9. That is, Japanese children who moved to the United States before nine reported becoming largely "Americanized" and felt relatively distant from their Japanese heritage. But children who moved after the age of fifteen had greater difficulty acquiring new cultural information, particularly with respect to emotional experiences. Like the accent that people often have when speaking a foreign language, people who learn a new culture later in life appear to preserve an echo of the emotional repertoire of their original culture.

Cultural Differences Become More Pronounced with Age

Humans are born without any culture, and they acquire culture as they are socialized. Although children are particularly sensitive to cultural information from very early, as they age they acquire more and more cultural information, and the universal mind that they are born with comes to be shaped in culturally particular ways. Hence, cultural differences in psychological processes tend to increase as people age and are socialized into their respective cultural worlds.

Why do cultural differences become more pronounced with age?

For example, consider how people develop theories to explain others' behaviours. In Chapter 15 you learned that people of Western descent were more likely to explain other's behaviours by making personal attributions, rather than attending to relevant situational information and making situational attributions—a bias known as the correspondence bias. However, people from a number of other cultural contexts don't show this bias as readily. Westerners tend to view the individual as the source of agency and control, whereas people from various other cultures, including South and East Asians, tend to see behaviour as arising from an individual's interacting with others according to situational demands. Furthermore, this cultural difference in how people make attributions for other people's behaviours increases with age. Joan Miller (1984) asked Americans and Hindu Indians (ages 8, 11, 15, and 19 years old) to describe a situation in which someone had behaved in a prosocial or deviant manner and to then explain why the person had behaved that way. Here is an example of a deviant behaviour as described by an Indian participant:

> This concerns a motorcycle accident. The back wheel burst on the motorcycle. The passenger sitting in the rear jumped. The moment the passenger fell, he struck his head on the pavement. The driver of the motorcycle—who is an attorney—as he was on his way to court for some work, just took the passenger to a local hospital and went on and attended to his court work. I personally feel the motorcycle driver did a wrong thing. The driver left the passenger there without consulting the doctor concerning the seriousness of the injury—the gravity of the situation—whether the passenger should be shifted immediately—and he went on to court. So ultimately the passenger died.

People's explanations were categorized as indicating either that there was something about the person that caused them to act the ways that they did, such as being irresponsible or callous (a personal attribution) or that there was something about the situation that caused the person to act in the ways that they did, such as having important work obligations, or the passenger not looking seriously hurt (a situational attribution). Figure 16.7 reveals that eight-year-olds from the two cultures gave quite similar responses. However, as the American sample got older they tended to make more personal attributions but the same number of situational attributions. In contrast, as the Indian sample aged, they tended to make more situational attributions and their personal attributions did not show much change. American adults showed clear evidence of the correspondence bias, but none of the other American age groups or the Indians showed much evidence of it. Indeed, the Indian adults showed a reverse correspondence bias (they emphasized situational factors more than personal ones). This is one example of how universal potentials come to be shaped by culture and of how those cultural influences grow larger with age (also see Gabrenya et al., 1985; Ji, 2005).

The Self-Concept Varies across Cultures

Being socialized in a particular cultural context affects much of how people think. In particular, it shapes how people come to view themselves. Recall from Chapter 15 that the self is experienced differently across cultures. In particular, in individualistic cultures people are more likely to have independent self-construals in which their identities are grounded in internal aspects of themselves (such as attitudes, personality

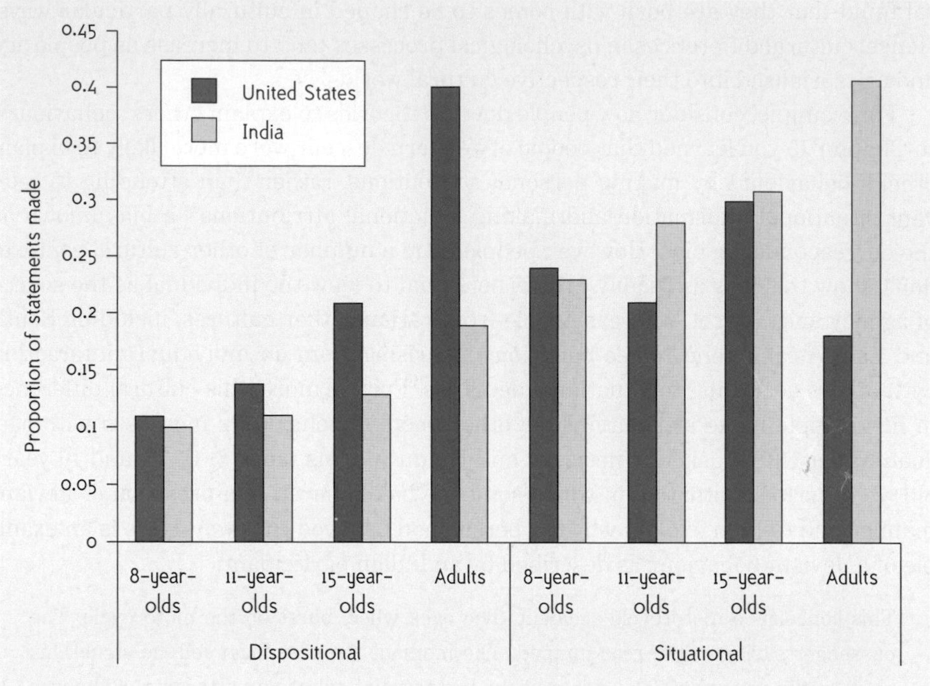

16.7 Dispositional and Situational Attributions Made by Children and Adults in the United States and India Proportions of statements made that explain the behaviour of the target as due to the personal characteristics of the target or the situation that the target was in.

traits, and abilities); they emphasize ways in which they are distinct from others, and they value self-reliance. In contrast, people in collectivistic cultures are more likely to have interdependent self-construals in which their identities are grounded in shared aspects of themselves (such as roles, relationships, and group memberships); they emphasize how they are connected with others, and they value group harmony (Markus & Kitayama, 1991).

Because the self-concept is implicated in how we process information that relates to ourselves, these cultural differences in self-construals are relevant for a wide array of psychological processes. One important domain of research on the self-concept that has been conducted in Western cultures is concerned with the consistency of the self. This research has revealed that people have a powerful motivation to view themselves as consistent. For example, it has been shown that people tend to view themselves and to behave quite similarly across a diverse range of situations (e.g., Funder & Colvin, 1991), and that people strive to make their behaviours consistent with their attitudes (Festinger, 1957; Heider, 1958). Recall that dissonance arises when behaviours and attitudes clash, and that consequently people are motivated to change their attitudes to justify their actions.

How similarly do you view yourself in different situations, say, when you're with your friends or when you're meeting with your professor? Even though you might act differently, do you feel you are being a different person? If your sense of self is primarily identified with attributes that are characteristics within you, such as your traits and attitudes, then it follows that your self-concept should be relatively consistent across situations, as those characteristics should remain constant. But if your self-concept is primarily identified with the roles and relationships that you have with others, then your self-concept should vary depending on what roles and relationships are salient or important in a given situation (Figure 16.8). Indeed, this has been shown to be the case for people from East Asian cultures. A study by Kanagawa, Cross, and Markus (2001) had Japanese and American participants complete a questionnaire in a number of different settings. They were asked to write twenty

16.8 Context and the Sense of Self In *Zelig*, Woody Allen takes on the appearance of those with whom he interacts, providing a dramatic illustration of how we often express different traits and characteristics when in different social contexts. Zelig looks Chinese when next to the Chinese man (a), and takes on African-American features when he stands between the two African-American men (b).

open-ended sentences that described themselves, such as in the exercise described in Chapter 15. Some completed the questionnaire by themselves, some completed it while sitting next to a peer, some completed it in a group, and some completed it in a professor's office. Interestingly, whereas the responses of the American participants were highly similar across the different settings, the Japanese respondents gave different answers depending on who was in the room with them (also see Cousins, 1989; Suh, 2002). The interdependent self-construal appears to be much more contextually variant than the independent self-construal.

In Chapter 15 you learned that when people act in ways that are inconsistent with their inner beliefs they feel anxiety and discomfort and are motivated to reduce it by resolving the inconsistency. This reaction makes sense to the extent that people view their beliefs as determining their behaviours. However, in contrast, if people believe that their behaviours arise to fit social norms, or to meet the expectations of others, as they often appear to be here in East Asian contexts, their experience of dissonance should be different from that experienced in a Western context.

How do motivations for consistency vary across cultures?

Have you ever had to make a difficult decision? How about choosing which university to attend, deciding on a major, figuring out how to deal with a family member's illness? For many people, particularly those from North American and European cultural contexts, difficult decisions are often followed by rationalizations. After making a decision you might convince yourself that you made the right decision by focusing on only the positive aspects of your chosen alternative and considering only the negative aspects of your rejected alternative. This kind of rationalization is known as postdecisional dissonance. Although Westerners show clear evidence of postdecisional dissonance, studies of East Asians reveal that they do not (Heine & Lehman, 1997). However, East Asians do show **postdecisional dissonance** when making choices for others—for example, when ordering food for friends at a restaurant (Hoshino-Browne et al., 2005), suggesting that they are motivated to make their behaviours consistent with others' expectations (also see Kitayama, Snibbe, & Markus, 2004). Likewise, Poles (who also tend to be more interdependent than Americans) strive to act in ways that are consistent with others' actions, whereas Americans are more likely to act consistently with the ways that they have in the past (Cialdini, Wosinka, & Barrett, 1999). In sum, depending on their self-construal, people strive to be consistent in different ways.

postdecisional dissonance The incongruity between one's decision and the undesirable aspects of that decision.

REVIEWING
the Principles | How Does Culture Affect the Mind?

Universal tendencies are expressed in culturally particular ways. Cultural practices shape how people think, and the interaction of people leads to the development of cultural practices. People can acquire cultural information throughout their lives, and cultural differences tend to get larger as people age, but people are especially sensitive to learning new cultures before puberty. Culture shapes the self-concept. People in individualistic cultures develop more independent self-concepts, and people in collectivistic cultures develop more interdependent self-concepts. People with independent self-concepts are motivated to have a consistent self, whereas those with interdependent self-concepts are motivated to act in contextually appropriate ways.

What Happens to People's Sense of Self When They Move to a Different Culture?

Since some humans first walked out of Africa about 60,000 years ago, humans have been a migratory species, and there have likely never been any large-scale societies composed solely of people from one cultural background. Rather, people from different cultural backgrounds have always interacted on a regular basis. Still, the extent to which people are migrating to different countries is higher now than ever before. The city of Toronto, for example, used to be made up largely of people of British ancestry, but today it is as cosmopolitan as they come. Statistics Canada projects that the visible minority population of the city will comprise a majority of the city by 2012 (Figure 16.9). The growing cultural diversity of communities around the world raises an important question. If people think in quite different ways across cultures, what happens to people's sense of self when they move to a new culture?

16.9 Acculturation When the late Pope John Paul II visited Toronto in July 2002, people of all backgrounds came out in droves to greet him. These immigrants share common religious beliefs with the dominant culture that surrounds them, so they are likely to have a close cultural fit and would presumably have an easy time adjusting.

Acculturation Requires Significant and Often Stressful Adjustment

Acculturation is the process by which people migrate to and come to learn a culture that is different from their original (or heritage) culture. Research on acculturation is difficult, and it is often hard to identify basic commonalities in the acculturation process, as individuals have very different experiences depending on where they are coming from, where they are migrating to, their reasons for migrating, their age, and the social network that they have in their new community. For example, the experience of moving to Toronto will be very different for someone coming from a Western culture like Australia, compared to a person coming from a developing, non-Westernized culture like Sri Lanka. Despite the tremendous variability in people's acculturation experiences, all acculturation involves psychological adjustment (Figure 16.10). Furthermore, this adjustment is often extremely stressful.

16.10 Canadian Immigration Thousands of people from all over the world immigrate to Canada every year, and have to work hard to acclimatize to a new culture, while still maintaining their own. These Sikhs are seen celebrating Vaisakhi, an event that marks the baptism of Sikhism in 1699. Though this may look like it was taken in India, this photograph is really from Vancouver, BC.

Have you ever moved to a new culture? If so, think about the kinds of adjustment that you went through. The psychological adjustment associated with acculturation does not occur suddenly, but takes place over many years, and sometimes even over generations. One way to look at this adjustment is to consider people's attitudes to their new (or host) culture. A common pattern of adjustment for acculturating individuals forms a **U-shaped curve** (Figure 16.11) (Lysgaard, 1955). Often, the first few months in the new culture are exciting, as people revel in all of the new experiences in their novel and exotic setting. This is known as the "honeymoon stage." Following this (often from the period of 6 to 18 months in the new culture, although the time varies greatly across individuals), many people develop quite negative views toward their host culture, as the novelty wears off and people find themselves in a foreign place where they often have few close friends from the host culture and lack much of the basic knowledge and many of the language skills necessary to thrive. This stage is called the "crisis" or "culture shock stage."

acculturation The process of adaptation to a culture different from one's own.

U-shaped curve A pattern of acculturation, characterized by three phases: the "honeymoon stage," "crisis" or "culture shock," and "adjustment."

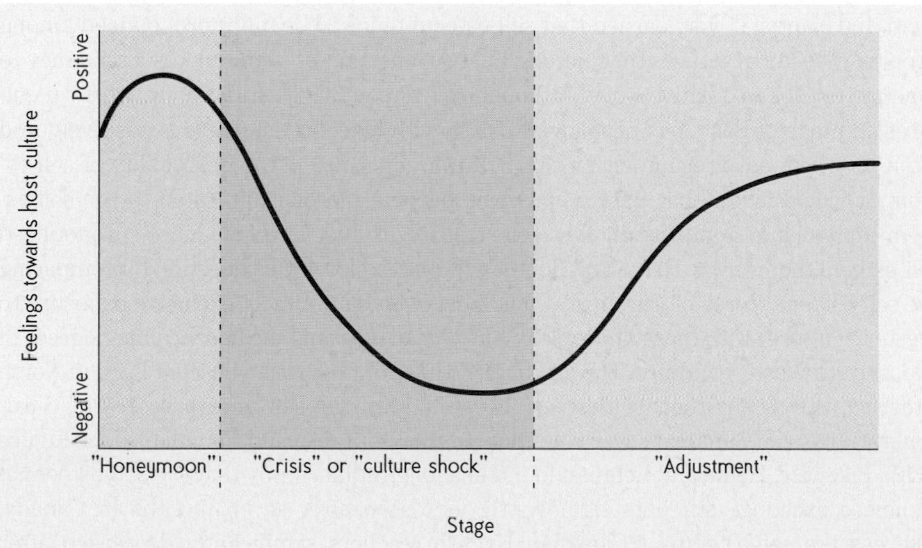

16.11 Pattern of Adjustment for Acculturating Individuals The three stages of adjustment common to acculturating individuals. Most people initially experience very positive feelings, followed by especially negative feelings, and then followed by a gradual increase in their felt positivity toward their host culture.

culture shock Experiencing feelings of anxiety, estrangement, helplessness, and incongruousness with one's surroundings on emigrating to a new culture.

cultural distance The disparity in customs, traditions, beliefs, and general heritage between two different cultures.

cultural fit The degree to which an individual's values and behavioural norms align with those that are common to another culture.

Culture shock is the anxiety, helplessness, irritability, and general homesickness that people experience on moving to a new culture (Church, 1982). Following this period of unhappiness, people typically learn how to exist in the host culture, make friends, improve their language skills, and, over time, slowly develop more and more positive attitudes toward the host culture. This last stage is the "Adjustment" phase, and it tends to extend over a number of years.

Not everyone has the same acculturating experiences. For some, the adjustment is less of an issue. One important factor in predicting acculturative success is **cultural distance**, which is the difference between two cultures in their overall ways of life. Not surprisingly, the smaller the distance, the easier acculturation should be. For example, an immigrant to Quebec from France would have to make fewer adjustments than an immigrant to Quebec from Zaire (Berry & Annis, 1974; Ward & Kennedy, 1995). The greater the differences between an individual's heritage and host cultures, the more significant the adjustment they must make, and the more likely it is that they will experience significant stress. Likewise, some individuals' values and personalities allow them to fit in with their host culture better than other people's do. **Cultural fit** is the degree to which an individual's values and personality are similar to the dominant cultural values in the host culture. Studies find that people with values and personalities that are more similar to those that are common within the host culture have an easier time adjusting (Cross, 1995). And, of course, people's personalities most likely reflect the overall tendencies of their particular cultures. For example, people from English-speaking countries tend to score higher in extraversion than people from some East Asian countries, among them Singapore (McCrae et al., 2005). Consistent with this, Singaporeans who scored high in extraversion adjusted better when they moved to New Zealand than those who scored low; and English-speaking expatriates who scored high in extraversion reported more adjustment problems when they moved to Singapore than those who scored low (Armes & Ward, 1989; Searle & Ward, 1990). The better the fit between an individual's personality and values, and the host culture, the easier will be the adjustment process.

What kinds of people should have an easier time acculturating?

Psychological Processes Change with Acculturation

If cultures shape psychological processes, it follows that as one adjusts to a new cultural environment, one's psychological processes will change. Take the example of self-esteem. In Chapter 15 you learned that self-esteem tends to be more pronounced among people of Western cultures than among East Asians. Self-esteem reflects a tendency to evaluate oneself in positive ways. Although self-esteem is valued in many Western cultures, in much of East Asia people view those with high self-esteem less positively, and often see them as arrogant. Given this cultural difference in the desirability of self-esteem, it would follow that East Asians who become more acculturated to the self-esteem enhancing features of Western culture would develop more pronounced self-esteem themselves. Heine and Lehman (2004) explored this question by comparing the self-esteem scores of European-Canadians (who have almost exclusive exposure to Western culture) with those of people of East Asian descent who had varying degrees of exposure to Western culture. The longer that those of East Asian descent lived in North America, the higher their self-esteem became, until the self-esteem scores of third-generation Asian Canadians were identical to those of European Canadians (see Figure 16.12). Likewise, Heine and Lehman found in a longitudinal study that the self-esteem of Japanese exchange students significantly increased after seven months in Canada, whereas the self-esteem of Canadian English teachers significantly decreased after

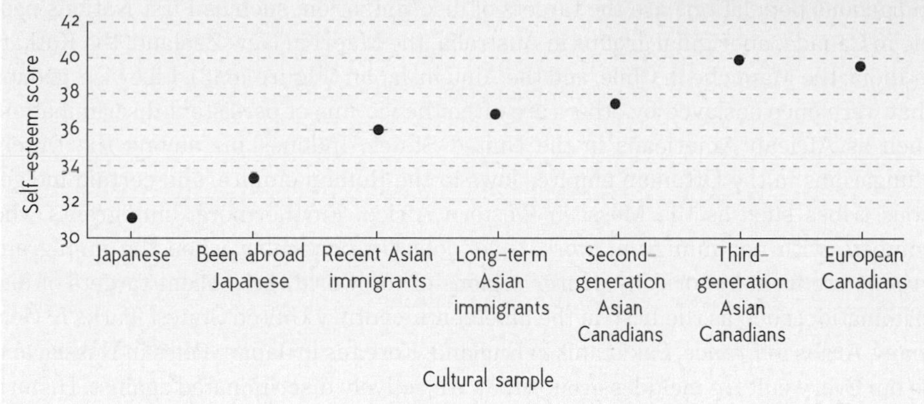

16.12 Cultural Change and Shifts in Self-Esteem The longer people of East Asian descent live in North America, the higher is their self-esteem.

seven months in Japan, although the self-esteem of both groups never reached the level that was common in their host cultures. In a new cultural environment, people over time acquire the associated psychological characteristics of that culture. However, unless they are exposed to the new cultural environment at a young age, the acquisition of new psychological characteristics is usually not complete. These studies show that people come to learn that self-esteem is desirable by participating in Western cultures. In some other cultures, it is not seen as such a good thing.

People from Distinct Cultural Backgrounds Often Face Discrimination

Although immigrants to a new culture do have to make significant psychological adjustments as they learn how to fit in, there is tremendous variation across immigrating populations with regards to the ease with which people adjust to a new cultural environment. For some, the adjustment is gradual and relatively painless. For others it is not. Compounding the difficulties of immigration is the fact that people from different cultures are not always treated with equal respect. As Chapter 15 highlighted, prejudice and discrimination toward many ethnic groups persist around the globe. For many people, moving to a new culture is fraught with active discrimination, systematic disenfranchisement, unjust treatment, mocking and humiliation, violence, and perhaps even threats to their lives. What often makes this even worse is that people in the host country sometimes group people into inaccurate ethnic categories, such as many Canadians viewing immigrants from India and Pakistan or from Korea and China as members of the same group. Moreover, prejudice against minorities is not limited to those who move to a new culture, but also extends to those who have ancestors from a different cultural background.

One consequence of being discriminated against, as discussed in Chapter 8, is stereotype threat. People are vulnerable to stereotype threat whenever there is a negative stereotype about a group to which they belong. Stereotypes represent cultural beliefs—that is, they are shared beliefs among members of a culture. People succumb to stereotype threat whenever they act in a way consistent with a negative stereotype about a group to which they belong and thus are at risk for "proving the stereotype." Stereotyped minorities are particularly vulnerable to falling victim of self-fulfilling prophecies (see Chapter 15) and acting in ways consistent with the negative beliefs that others have about their groups (Steele & Aronson, 1995).

All cultures have negative stereotypes toward certain groups, but the groups that are stereotyped against vary tremendously across cultures. In many countries, indigenous populations are the targets of discrimination, such as First Nations people in Canada, aboriginal groups in Australia, the Maori in New Zealand, the Katkari in India, the Mapuche in Chile, and the Ainu in Japan (Figure 16.13). Likewise, groups that were once enslaved by others are often the victims of persistent discrimination, such as African Americans in the United States, Indonesians among the Dutch, Hungarians in the Ottoman empire, Jews in the Roman empire, and certain indigenous tribes such as the Mossi in Western Africa. Furthermore, immigrants who compete with nonimmigrant workers for jobs (in particular, when the immigrant groups are from historically poorer regions of the world) are salient targets of discrimination, such as the Irish in the nineteenth-century United States, Turks in Germany, Arabs in France, Pakistanis in England, Koreans in Japan, Poles in Russia, and so on. Every culture includes groups that are actively discriminated against. Histori-

16.13 Modern-day Discrimination First Nations people in Canada have been able to make some positive steps toward ending discrimination, as when (a) Assembly of First Nations Chief Phil Fontaine and Canadian Deputy Prime Minister Anne McLellan signed an agreement aimed at reaching a final settlement on compensation for former students at Indian residential schools, on May 30, 2005. Minorities in other places, however, have not been as successful. For example, the indigenous Maori of Waitangi, New Zealand (b), shown here protesting the Waitangi Day celebrations, contend that the European signatories did not honour their side of the 1840 Treaty of Waitangi, in which the Maori yielded land ownership to the settlers. Also, (c) many of the Muslims who live in London and are forced to live in Whitechapel, the city's poorest borough.

cal circumstances, economic competition, and cultural distinctiveness all play an important role in determining which groups become targets of discrimination.

16.14 **Frame-Switching in Action** Stimuli like these were used to actually view frame-switching taking place.

Multicultural People Can Switch between Different Selves

What happens when people with experience in more than one culture switch between different cultural contexts? When people move to a new culture and learn a second language, they typically do not end up speaking a language that is a mixture of their first and second languages—for example, Indonesian speakers who learn English do not speak Englonesian. Rather, they learn to switch between Indonesian and English depending upon the context.

The same kind of **frame-switching** holds for psychological processes that vary between cultures. Frame-switching is changing how you think and behave depending on the context. So, for example, a girl growing up in an inner city African American neighborhood thinks and behaves one way with her neighborhood friends and another when volunteering at a nursing home (Anderson, 1999). Importantly, frame-switching does not occur just for consciously accessible behaviours (such as how one talks), but also for thoughts that lie beyond people's awareness, such as how they make attributions.

frame-switching The shifting of thoughts and behaviours to those appropriate for a given cultural context.

Earlier we discussed how Hindu Indians are more likely to explain behaviours in situational terms than are Americans, who tend to emphasize personal factors. This emphasis on the situation has been identified in other collectivistic cultures, such as Chinese, and it has also been shown to hold for how people explain the behaviours of nonhumans as well, such as fish (Morris & Peng, 1994). Ying-yi Hong and colleagues (Hong, Morris, Chiu, & Benet-Martinez, 2000) reasoned that people who had been exposed to two cultural worlds should frame-switch between making situational or personal attributions depending on which context was made salient to them. They studied Hong Kong Chinese, reasoning that, given the long Western influence over this former British colony, their samples would be familiar with both traditional Chinese ways and Western traditions. Participants were asked to explain the behaviour of a target fish that was swimming in front of other fish (Figure 16.14). Was the target fish being chased by the other fish (i.e., was its behaviour driven by features of the situation), or was the fish leading the other fish (i.e., was its behaviour driven by its own decisions)? Prior to observing the fish, some participants were shown pictures intended to prime traditional Chinese thoughts (e.g., a Chinese dragon, the Great Wall, a rice farmer); some were shown pictures intended to prime American thoughts (e.g., Mickey Mouse, the Statue of Liberty, a cowboy); and some were shown pictures to prime neutral thoughts (e.g., nature scenes). The explanations of the target fish's behaviour were compared across the conditions. As Figure 16.15 shows, those primed to think Chinese thoughts were especially likely to explain the target fish's behaviour in terms of situational attributions, whereas those primed to think American thoughts were especially *un*likely to use situational attributions to explain the target fish's behaviour. The Hong Kong Chinese participants could switch between Chinese and Western ways of explaining the fish's behaviour, and they did so, even

16.15 **Explaining the Behaviour of Fish** Hong Kong participants who were primed with American images were less likely to explain a fish's behaviour in terms of external accounts, whereas those who were primed with Chinese images were especially likely to view the fish as acting because of external forces.

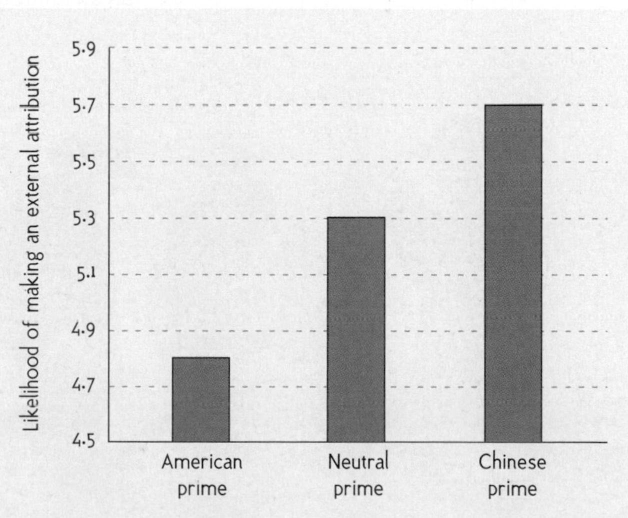

What is the self-concept like for multicultural people?

though they were unaware that Chinese and Westerners tend to make different kinds of explanations for others' behaviours. This tendency to frame-switch has been identified with a variety of other different psychological measures (Ross, Xun, & Wilson, 2002; Wong & Hong, 2005). Multicultural people seem to switch between different self-construals depending on the context.

Profiles in Psychological Science

Movie-Making from a Bicultural Perspective

Ang Lee was born in Taiwan, the first son of a principal of one of the best high schools in the country. Despite his father's desire that he pursue something more professional, Lee became interested in studying film. After graduating from the National Taiwan College of the Arts, Lee moved to the United States to continue his study and received a Master's degree in Film Production from New York University. After several years of struggling to break into the movie business, all the time being supported by his Taiwanese wife, Lee had his first break, with *Pushing Hands* (1992). Since then, Lee has been remarkably successful, winning dozens of international awards for his films.

It is not surprising that Lee has directed some of the most popular Chinese films of all time, such as *The Wedding Banquet* (1993), *Eat Drink Man Woman* (1994), and *Crouching Tiger, Hidden Dragon* (2000). After all, he had a rather typical Chinese upbringing. What is more surprising, is that Lee has had great success in directing movies that would seem to be worlds apart from his Taiwanese background. Lee's *Sense and Sensibility* (1995) is a period piece about nineteenth-century British socialites. *The Ice Storm* (1997) explores the emotional vacuum surrounding two affluent suburban New England families during the 1970's. Lee won an Oscar in 2006 for *Brokeback Mountain,* a film about the hidden love affair of two gay cowboys in Wyoming (Figure 16.16). The critics consistently rave about the depth of the characters that Lee develops in his movies. How could a Taiwanese emigrant come to have such penetrating insight into the minds of nineteenth-century British gentry and gay cowboys?

16.16 Ang Lee Lee's ability to step outside his native culture has allowed him to make a wide variety of award-winning movies. Lee's film *Crouching Tiger, Hidden Dragon*, won the Oscar for Best Foreign Film in 2001, and in 2006 he won Best Achievement in Directing for his film *Brokeback Mountain*, which starred Jake Gyllenhaal (right) and Heath Ledger (middle).

Lee credits some of his success to the fact that he is bicultural. Often referred to as an "outsider's outsider," Lee has been able to approach his subjects in a somewhat naïve way, free of the expectations that govern the perceptions of cultural insiders. *Time* magazine called him "a cosmopolitan chameleon [who] seems at home in any culture while viewing it with an outsider's ironic acuity." Looking in from the outside, Lee feels that he is able to see things differently, and perhaps even more accurately or objectively, than those who are shackled by conventional wisdom. As a distant observer, Lee says, you can notice the subtext faster, because you're not as entangled with the text as cultural insiders.

Many of Lee's films deal with the tension between underlying truths and outward appearances, such as *Hulk* (2003), in which the anger beneath the surface of mild-mannered Bruce Banner boils and occasionally erupts with disastrous consequences. As a Taiwanese native who has established himself in Hollywood, Lee notes that he is often made aware of the two levels of reality that biculturals have to deal with. Chinese thoughts and habits need to be negotiated with American customs. This bicultural predicament leads to a unique perspective that guides Ang Lee's films. ■

REVIEWING the Principles | **What Happens to People's Sense of Self When They Move to a Different Culture?**

Acculturation involves psychological adjustment that can be stressful. Often people who move to a new culture first enjoy it, then go through a period of culture shock, and gradually come to fit in and succeed in the new environment. People who are from more distinct cultures, or who have poor cultural fit, have a harder time during the acculturation process. Over time, people's habitual way of thinking changes with acculturation, as they engage in new cultural practices. People from distinct cultural backgrounds suffer much discrimination and experience stereotype threat. Multicultural people can switch between different selves depending on which one is activated by the context.

How Does Culture Affect How We Think and Behave?

You have a rich set of psychological experiences that contains all of your beliefs, feelings, and thoughts. How similar do you think your thoughts are to the thoughts of people in other cultures? As discussed in Chapter 15, people in individualistic cultures tend to have independent self-construals and emphasize how they are distinct from others, whereas people from collectivistic cultures tend to have more interdependent self-construals and emphasize how they are connected to others. These different self-construals are implicated in many of the ways in which we think. How we understand ourselves influences how we understand others. That is, we often use the theories we have for how our own minds think in trying to understand how other people think. Furthermore, and more profound, how we understand other people appears to influence how we understand other nonsocial aspects of our world as well.

Cultures Differ in Analytic and Holistic Thinking

Which of the following three is least like the other two? A dog, a carrot, a rabbit. Ji, Zhang, and Nisbett (2004) asked this and other similar questions of American and Chinese university students. A clear trend emerged from their responses. Americans tended to group the dog and rabbit together, excluding the carrot, whereas Chinese tended to group the rabbit and carrot together, excluding the dog. The American responses indicate a **taxonomic categorization** strategy. That is, the stimuli are grouped on the basis of perceived similarities of attributes. A dog and a rabbit are both mammals, whereas a carrot is not—that is the reason that the Americans put them together. In contrast, the Chinese responses indicate a **thematic categorization** strategy. That is, the stimuli are grouped on the basis of a relationship among them. Rabbits have a relationship with carrots in that they eat them, whereas

taxonomic categorization A system of grouping stimuli based on perceived similarities of attributes.

thematic categorization A system of grouping stimuli based on perceived relationships among them.

analytic thinking A system of evaluation in which the assessor views objects as independent from context and in terms of individual characteristics, accompanied by an expectation of behaviour in accordance with a fixed set of abstract rules derived from said assessments.

holistic thinking A system of evaluation in which the assessor views objects with regard to context and in terms of the relationships between them, accompanied by an expectation of behaviour in accordance with those relationships.

dogs have little relationship with either rabbits or carrots. These two categorization strategies reflect an underlying difference in the ways that people with independent and interdependent views of self think about their worlds.

One way of thinking is known as **analytic thinking**. Analytic thinking is characterized by a focus on objects and their attributes. Objects are perceived as existing independently from their contexts, and they are understood in terms of their component parts. The attributes that make up objects are used as a basis to categorize them, and a set of fixed abstract rules is used to predict and explain the behaviour of objects. A second way of thinking is known as holistic thinking. **Holistic thinking** is characterized by an orientation to the context as a whole. It is an associative way of thinking in which there is attention to the relations among objects, and among the objects and the surrounding context. Objects are understood in terms of how they relate to their context, and the behaviour of those objects is predicted and explained on the basis of those relationships. In general, analytic thinking is more common among people from individualistic cultures, whereas holistic thinking is more common among people of more collectivistic cultures, in particular those from East Asian cultures, where it has been studied the most.

Psychologists hypothesize that analytic and holistic thinking styles derive from people's interactions with their social worlds. Those with independent views of self tend to think of people as fundamentally distinct from each other, and they consider people's identities to be grounded in inner attributes, such as their personalities, abilities, and attitudes. In contrast, those with interdependent views of self tend to think of people as being fundamentally connected, and they consider people's identities in terms of their relationships, roles, and group memberships. The social experiences that people have growing up, where their attention is directed at either discrete individuals or at relations among individuals, are hypothesized to lead to these broad differences in thinking styles.

For example, look at the two pictures of the caribou (Figure 16.17). Does it look as though it is the same caribou in those two pictures? Japanese and American participants were shown these two pictures as well as a number of other pictures. Unlike you, they never saw the pictures side by side. Participants saw one of the pictures and then they either saw the same picture again, or they saw a picture of the same target animal (the caribou) with a different background. They were asked whether they had seen the

16.17 Identity in Context When viewing these figures separately, East Asians are more likely than Americans not to recognize that they had already seen the caribou in slide (b), after viewing it in slide (a).

same target animal before. Both Japanese and Americans did equally well on this task when they were shown the same picture the second time. However, when they were shown the picture with the new background, Japanese participants were more likely than Americans to say that they hadn't seen the target animal before. This effect generalized across a broad array of different scenes. Apparently, the Japanese participants attend a great deal to the background and they see the focal objects in relation to the background. So when the background changes, the focal object also appears different. American participants, in contrast, focus primarily on the foreground. So they do not readily notice changes in the background (Masuda & Nisbett, 2001). Indeed, research that places East Asians and Americans in eye-trackers (devices that measure where people's gaze is directed) reveals that East Asians look significantly more at the backgrounds of scenes than do Americans (Chua, Boland, & Nisbett, 2005).

Why do Japanese attend more to backgrounds than Americans?

In one study, participants were shown two drawings, such as the flowers shown here (Figure 16.18). The task for participants was to decide whether target flower A or B is more similar to those flowers in Group 1 or in Group 2. What do you think?

There are two ways for participants to solve this problem. One is to apply an abstract rule, and determine an underlying feature that is common to all group members. For example, all the flowers in Group 1 have a curved stem, so flower B, which also has a curved stem, must belong to Group 1. A second way is to use associative reasoning, and to consider which group has the most similar associations with the targets. For example, most of the flowers in Group 1 have a leaf, have curved petals, and have a single circle, which is similar to flower A. This problem is difficult because in this particular case, rule-based reasoning yields the opposite answer as associative reasoning. Rule-based reasoning and associative reasoning are universally found. However, when they are in conflict holistic thinkers are more likely to favour associative reasoning and analytic thinkers are more likely to favour rule-based reasoning. Indeed, Norenzayan, Smith, Kim, and Nisbett (2002) found that American participants tend to search for a rule to solve this problem, and they conclude that flower B belongs to Group 1, whereas East Asian participants tend to look for similar associations and conclude that flower A belongs to Group 1. As we now see, cultural differences in analytic and holistic thinking appear in a wide variety of different tasks.

16.18 To Which Groups Do the Two Target Flowers Belong? This stimulus pits associative reasoning against rule-based reasoning.

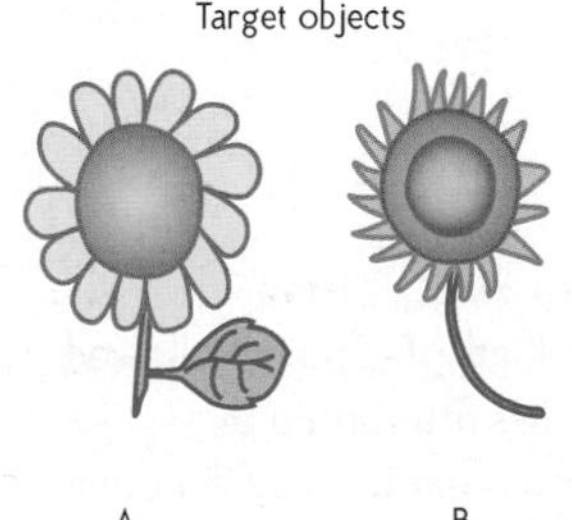

Cultures Differ in Motivations for Control and Choice

Cultures differ in many ways of thinking other than in their reasoning. Take the example of control. Striving for control over life is a universal human concern. However, the ways that people seek control can vary across cultural contexts. Rothbaum, Weisz, and Snyder (1982) proposed two distinct kinds of control: primary control and secondary control. People achieve a sense of **primary control** by striving to shape existing realities to fit their perceptions, goals, or wishes. For example, imagine that you really want to watch the new Tom Cruise movie. You'd be exerting primary control by convincing a friend to go with you and heading off to the Cineplex to see the movie. This is a kind of control that is especially familiar to Westerners. People achieve a sense of **secondary control** when they attempt to align themselves with existing realities, leaving the realities unchanged but exerting control over their psychological impact. For example,

primary control Influencing one's environment to achieve one's goals, desires, or wishes.

secondary control Psychologically aligning oneself with another in order to achieve a sense that one's goals, desires, or wishes are being fulfilled.

16.19 Primary versus Secondary Control In the blockbuster movie *The Matrix* (1999), Neo (Keanu Reeves) is told by Morpheus (Laurence Fishburne) that he is "The One," and that he is destined to end humanity's enslavement by machines. Neo finds that he is limited when he believes he is The One based on Morpheus' words, but is elevated when he decides for himself that the title is actually his. By aligning his views with another's and telling himself his goals are being met, he is only partially successful. However, by taking his own positive steps toward reaching his goals, he is infinitely moreso.

imagine that your friend says no to the Tom Cruise film and states that she really wants to see the new Reese Witherspoon movie. You could exert secondary control by convincing yourself that you would like to see this movie too (Figure 16.19). On the surface, the way we resolve the tensions involved with control seem to relate to cognitive dissonance, which we discussed in Chapter 15. There is a major difference between these two phenomena. Switching from primary to secondary control does not involve resolving conflict, whereas cognitive dissonance always does.

What is the difference between primary and secondary control?

Though everyone experiences both kinds of control, cultures differ in terms of how common each type is. Specifically, primary control is more common and more desired in individualistic cultures, whereas secondary control is more common and more desired in collectivistic cultures. For example, Morling, Kitayama, and Miyamoto (2002) found that primary control experiences were more common than secondary control experiences in the United States and that the reverse was true in Japan. Likewise, Americans viewed primary control experiences as more desirable than did Japanese, and Japanese viewed secondary control experiences as more desirable than did Americans. Americans reported finding secondary control experiences to be rather aversive and would often describe them by saying that they had to change, such as "I had to adjust when my roommate's boyfriend kept coming over." Japanese participants, in contrast, tended to describe their secondary control experiences more positively, such as "My friend wanted to go out for coffee so we went to the coffee shop together."

One way that people exercise primary control is by making choices. By making choices, we try to ensure that our existing realities match our desires. Choice appears to be a greater concern for those from individualistic cultures. In a study by Iyengar and Lepper (1999), fifth-grade children in San Francisco who were of either East Asian or European descent played a computer game (Figure 16.20). The children were randomly assigned to one of three conditions. In the "personal choice" condition, the children were allowed to make a number of choices that were irrelevant to the success of the game (e.g., the name of their spaceship). In the "out-group choice" condition, the children were not allowed to make any choices—they were told that their spaceship had been given the name that was the most popular choice among third-grade students at another school (it turns out that fifth-graders don't have much respect for the opinions of third-graders). In the "in-group choice" condition, the children were not allowed to

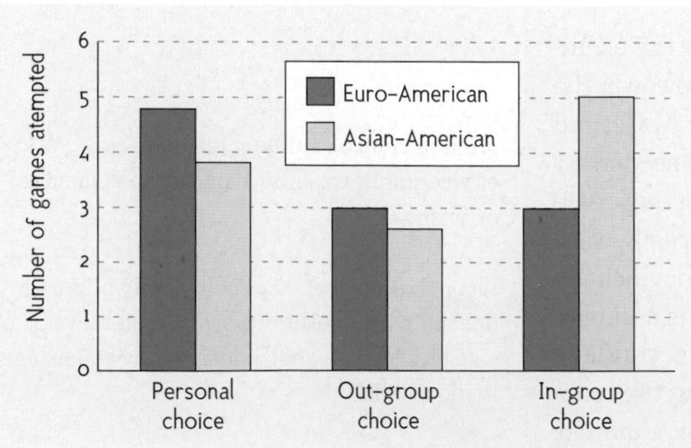

16.20 Motivation to Play a Game Depending on the Kind of Choice Allowed The number of games attempted by children varied across cultures by whether they made some tangential choices to the game themselves, whether the choices were made by an out-group, or whether the choices were made by an in-group.

make any choices, but they were told that their spaceship had been given the name that was most popular with the students in their own class. The experimenters then measured how many games the children attempted during a fixed time period. As Figure 16.20 shows, the European American children played more games when they got to choose the name of their spaceship and played fewer games when members of either an out-group or an in-group made the choice for them. The European American children seemed to react negatively to the idea that others were making choices for them, regardless of who those others were, and lost interest. In contrast, the Asian American students attempted the most games when their own classmates chose the spaceship name for them. However, like the European Americans, they were not very motivated when members of an out-group made the choices for them. Apparently, the Asian American children viewed the in-group's choosing as an opportunity to promote a sense of belongingness with other group members. The European Americans, in contrast, seemed to react as thought they had been stripped of their freedom to choose.

It is useful to remember that most psychology studies are conducted with university students, and that university students tend to differ from other members of the population in certain ways. Most obviously, they tend to be of higher socioeconomic status (SES) than those who do not go to university. What are other ways that people of higher SES might differ from those of lower SES? One way is that they have more choices. Those with lower SES more often have to accept what is given to them rather than being able to choose from a range of alternatives. Snibbe and Markus (2005) demonstrated that upper-middle-class Americans were less satisfied when they had been deprived of making a choice than were working-class Americans. Even within a country, there are clear differences between the ways that people from different SES perceive choices.

Thinking Critically

Is Happiness Necessary for a Good Life?

There is no doubt that people everywhere pursue activities that make them happy. Happiness is a state of mind and an emotion that feels good, and signals to the individual that all is well. Furthermore, an abundance of research reveals that there are clear benefits to being happy, with happiness having been shown to increase longevity and lead to greater career success, at least in North America where this research has been largely conducted (Lyubomirsky, King, & Diener, 2005). Given that happiness has such beneficial consequences, should governments do what they can to maximize their countries' Gross National Happiness?

How one answers this question will likely depend on how one views happiness. Happiness does seem to be a central value among many North Americans. Indeed, the pursuit of happiness has been central enough to American culture that it was described as an "unalienable right" in the Declaration of Independence. Most North Americans report being quite happy (approximately 89 percent of Americans and 85 percent of Canadians), and these numbers have remained quite constant over the past 60 years (Veenhoven, 1993).

But happiness has not always played such a central role in people's lives. In 1843, the British historian Thomas Carlyle wrote that "Happiness our being's end and aim is at bottom, if we will count well, not yet two centuries old in the world." Carlyle is referring to changes during the Enlightenment where the world began to be seen as a more rational place, and that happiness was believed to be achievable through efforts to pursue a good life. Prior to this shift, people's concerns were less about how they could become happy and more about how they could be saved.

Furthermore, there is considerable cultural variation in the extent to which people report being happy. For example, people who are most likely to describe themselves as happy are those from Northern Europe (Iceland, the Netherlands, and Denmark topped a recent survey), followed by those from various English-speaking countries, much of Latin America, and Western Europe (Inglehart & Klingemann, 2000). In contrast, those who are least likely to report being happy are people from former Soviet republics and some impoverished countries in Africa and South Asia. Though the inclusion of poor African and South Asian states suggests that poverty is associated with unhappiness, research reveals that wealth reliably predicts happiness only up to a certain point.

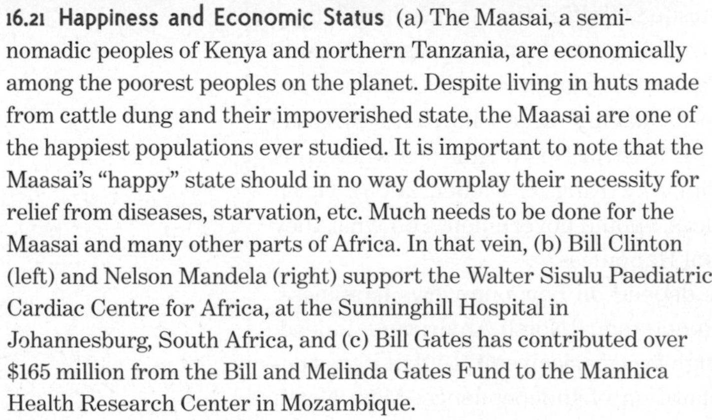

16.21 Happiness and Economic Status (a) The Maasai, a semi-nomadic peoples of Kenya and northern Tanzania, are economically among the poorest peoples on the planet. Despite living in huts made from cattle dung and their impoverished state, the Maasai are one of the happiest populations ever studied. It is important to note that the Maasai's "happy" state should in no way downplay their necessity for relief from diseases, starvation, etc. Much needs to be done for the Maasai and many other parts of Africa. In that vein, (b) Bill Clinton (left) and Nelson Mandela (right) support the Walter Sisulu Paediatric Cardiac Centre for Africa, at the Sunninghill Hospital in Johannesburg, South Africa, and (c) Bill Gates has contributed over $165 million from the Bill and Melinda Gates Fund to the Manhica Health Research Center in Mozambique.

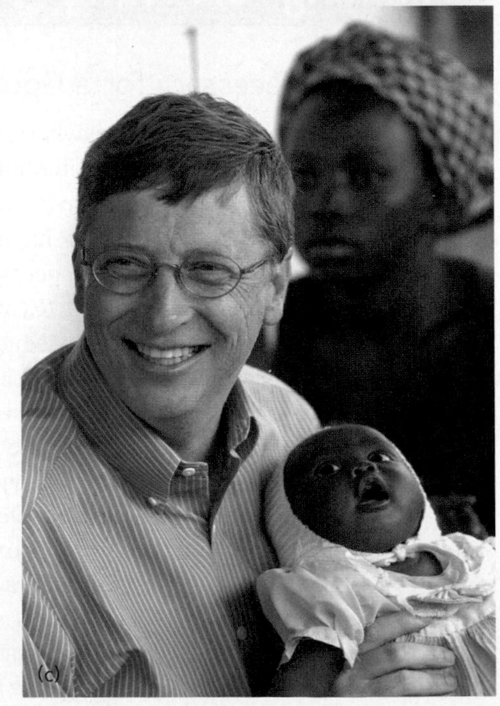

After the point at which average citizens can satisfy their basic needs (indicated by an average GNP at least 40 percent of that of the United States), there is no longer a clear relation between wealth and happiness across nations (Diener & Biswas-Diener, 2002).

Why do people in some cultures report being happier than people in others?

People around the world differ in the extent that they value happiness (Figure 16.21). Among various Latin American nations, such as Puerto Rico, Colombia, and Brazil, people report that it is especially desirable to experience positive emotions. Accordingly, people from these countries report feelings of happiness that are far greater than what one would predict on the basis of their GNP. At the other extreme, people from East Asian countries, such as Japan, Taiwan, and South Korea, are more likely than those from other cultures to report that they do not especially value happiness (Diener, 2001). Furthermore, people from East Asian countries are notably less happy than would be predicted on the basis of GNP. As the Japanese social psychologist Hiroshi Minami has suggested, "It seems that feelings about happiness in life are for some reason diluted among the Japanese. The reason that the word 'happiness' is not used daily is not only because the Japanese masses are not blessed with happiness in daily life but because they have cultivated a habit of hesitation toward happiness" (p. 34). This hesitation toward happiness may have grown out of a concern with balance in East Asian thought, where life is viewed as cyclical, so that happy times are necessarily viewed as leading to unhappy times. With such a view, one cannot systematically work toward increasing happiness without knowing that some day there will be a payback for having had all those good feelings, so many Asians in general work toward maintaining a proper balance in their lives.

What can we make of these cultural differences? Should all people strive to be like those from Latin American nations and maximize their happy feelings, regardless of circumstances? Alternatively, should we view people's approach to happiness as a cultural product, with people from some cultures fully embracing happiness and striving to maximize it and those from other cultures having a more cautious attitude toward it? ■

Cultures Differ in the Bases of Relationships

In some cultures, one of the most important choices that people make is deciding whom they will marry. Indeed, for many people, this choice is their biggest concern, and fretting about it occupies years of their lives. However, reflecting clear cultural differences in the value of choice, and in the nature of marriage, in many cultures people do not choose their marriage partners. Rather, their families do. Table 16.2 shows some of the different kinds of mate selection patterns in preindustrial societies (Broude & Green, 1983). Indeed, in over half of the societies, and especially so for women, parents and kin play a significant role in the decisions. And **arranged marriages** are not a feature of just preindustrial societies: they are still present in China, India, Japan, and Turkey and among orthodox Jews (although the rates of arranged marriages in many cultures have recently been dropping). Furthermore, arranged marriages do not seem to be necessarily less satisfying than **love marriages**. Although in the initial years those in love marriages professed more love for their partners than those in arranged marriages, over time, couples in arranged marriages reported having the most love (see Figure 16.22 from Gupta & Singh, 1982).

arranged marriage A type of marriage in which the bride and groom have been preselected by the bride's and groom's respective families.

love marriage A type of marriage in which the bride and groom have selected each other as marriage partners.

The existence of arranged marriage in some cultures and its absence in others reflects very different attitudes toward what marriage entails. In cultures where extended family systems are strong (i.e., there is much interdependence among parents, grandparents, uncles, aunts, and cousins), there tend to be more arranged marriages than in cultures that have a nuclear family structure (Lee & Stone, 1980).

TABLE 16.2 Mate Selection Practices in Preindustrial Societies	MEN	WOMEN
Parents choose partner; individual cannot object	13%	21%
Parents choose partner; individual can object	17%	23%
Both individual choice and arranged marriages are acceptable	18%	17%
Individuals, parents, kin, and others must agree on an appropriate match	3%	3%
Individual chooses partner autonomously: parental, kin, and/or community approval necessary or highly desirable	19%	29%
Individual chooses partner autonomously: approval by others unnecessary	31%	8%

Based on Broude & Green (1983), pp. 273–274.

Moreover, individualistic societies are more likely to favour love marriages. The notion that each person is a unique individual comes associated with the view that each person has unique needs that can only be fulfilled by another equally unique partner (Dion & Dion, 1993).

The meaning of friendships also varies across cultures. Consider the following poem:

Beware of friends.
Some are snakes under grass;
Some are lions in sheep's clothing;
Some are jealousies behind their façades of praises;
Some are just no good;
Beware of friends.

(Kyei & Schreckenbach, 1975, p. 59)

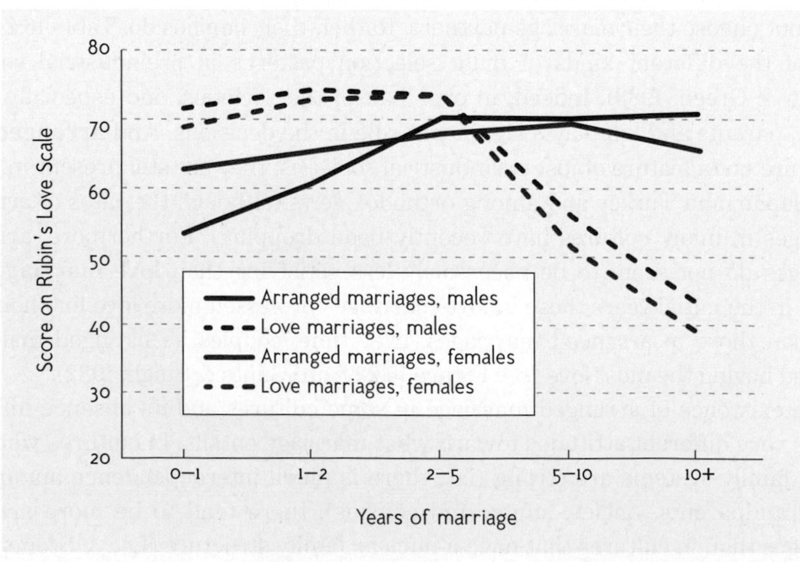

16.22 Arranged and Love Marriages in India Feelings of love tended to start high but subsequently dropped after a few years of marriage among Indians in love marriages. In contrast, the reported love among Indians in arranged marriages started out relatively low but showed a slight increase over time.

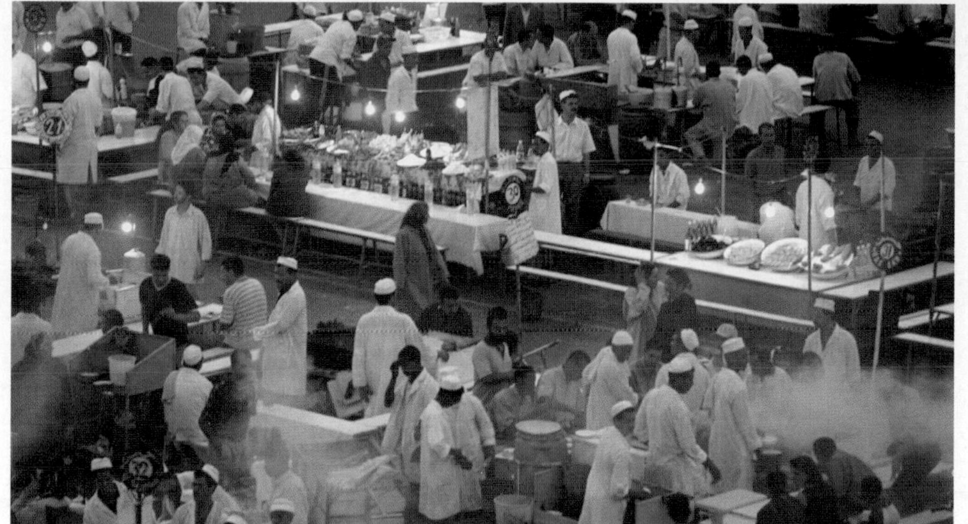

16.23 Relationships within Social Networks Westerners and Easterners tend to view the natural states of relationships with their respective peers differently. While a person from North America may view someone shopping next to him or her as a stranger, someone in the Djem El Fna Square in the centre of Marrakech, Morocco might assume more familiarity with a fellow shopper.

This Ghanaian poem expresses a sentiment that is quite common in West African collectivistic contexts (Adams & Plaut, 2003). Why is it that a cultural context that emphasizes how people are fundamentally connected also stresses that friends should be viewed with suspicion?

Adams argues that relationships are more conditional in individualistic contexts (Figure 16.23). The default state between any two individualistic people is that they have no relationship. A relationship must be developed, and people must make efforts to do so, which they will do only if they perceive benefits from being in that relationship. In contrast, in collectivistic contexts, the default state between any two people in the same in-group is that they have a relationship. Their relationship exists because of their positions within their social networks, regardless of whether that relationship benefits or harms them. As such, in collectivistic contexts close relationships are not automatically assumed to be positive ones. Friendships bring not only good times, but also obligations, and these can be costly. Adams and Plaut (2003) found that Ghanaians are more suspicious of friendships than Americans. Likewise, Ghanaians feel that they have more enemies than do Americans, and they are more likely to view their enemies as coming from their close personal networks than Americans are (Adams, 2005). In sum, the nature of close relationships is shaped by cultural contexts in important, and sometimes surprising, ways.

Why do relatively few Americans report having enemies?

Cultures Differ in Group Performance

People in individualistic and collectivistic cultures differ in how they relate to each other. These differences have a significant impact on how people behave in groups.

Take the example of **social loafing**. Research reveals that Westerners tend to work harder when they are being evaluated as individuals than when their group is being evaluated (Figure 16.24). In a classic study, Ingham and colleagues (1974) asked each member of one group of participants to pull as hard as they could on a rope. On average, each pulled 59 kilograms. Other participants were asked to pull as hard as they could on the rope in groups of three, but each group contained two confederates who stood behind the participant and just pretended to be pulling. When participants thought that other people were pulling on the rope with them, they only pulled 48.5 kilograms. They showed evidence of social loafing. Research conducted in collectivistic cultures, however, reveals that this is not a universal finding. Chinese and Israelis have

social loafing The tendency for people to work less hard in a group than when working alone.

16.24 Social Striving In Japan, Mitsubishi showroom employees work out together before the shop opens. Since Japan is a collectivist culture, chances are that this exercise will promote a feeling of unity and in turn boost productivity.

been shown to work harder in a group than as individuals (a tendency called **social striving**), if they are working in a highly valued group, whereas Americans have been shown to work harder as individuals than when in any kind of group (see Figure 16.25). However, if Chinese and Israelis are working in a less-valued group, they demonstrate the same kind of social-loafing pattern that American participants do (Earley, 1993).

social striving A phenomenon in which people exert more effort to achieve a goal when they are performing in a group than when they are alone.

In Chapter 15, we discussed how participants often show strong conformity motivations and go along with a group, even if their beliefs conflict with those of others in the group. Studies on conformity employing the same kind of line judgment task used by Asch (1956) have been conducted around the world. The results revealed larger conformity effects in collectivistic cultures than in individualistic ones, although the conformity effects are quite pronounced in all cultures. Conformity rates are especially high among collectivistic cultures when people are conforming to those in an in-group (Bond & Smith, 1996). Though conformity is a universal motivation, it is stronger in collectivistic societies where the costs for not conforming tend to be greater.

How do cultures perform differently when they are working in groups?

The kinds of relationships that people have with others also affects their willingness to cooperate. In general, motivations for cooperation with in-group members

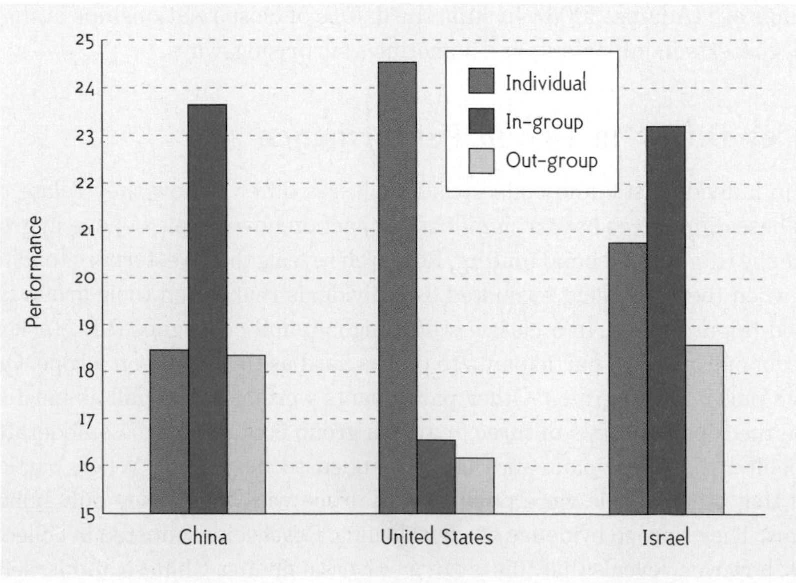

16.25 Social Loafing and Social Striving Whereas performance among Americans is far greater when people are working by themselves compared with when they are working in groups, both Israelis and Chinese perform better when they are working in an in-group compared with when they are working alone. Like the Americans, however, Israelis and Chinese also do not perform so well when they are working in an out-group.

16.26 **Cooperation** The apparatus for the game used by Madsen.

are stronger among people from collectivistic backgrounds than among those from individualistic backgrounds. One study demonstrated this point quite clearly. Madsen (1971) had children (aged 7–10) from small towns in California and Mexico participate in a game that rewarded them for playing cooperatively, but punished them for playing competitively (Figure 16.26). The game worked such that if a child allowed his or her partner to take a turn, the partner would win a marble. Then, if the partner allowed the child to have a turn, the child would receive a marble him- or herself. In contrast, if the children competed with each over who would go first, neither would win any marbles on that turn. The only way to win marbles was to cooperate. In a first trial, the children had to figure out the task for themselves. The American children didn't do too well at this point. Out of a possible 10 marbles that could be won, they won, on average, only 0.25 marbles. In a second trial, the children were explicitly shown a strategy of turn-taking and it was explained to them how that was the way to win marbles. The American children fared better after the training, although they averaged just 4.7 out of a possible 10 marbles. Even when the way to win marbles was made clear to these children, they were often still unable to resist their competitive urges to try to win more marbles than their partners.

In stark contrast, the Mexican children seemed to have an easier time learning how to cooperate to win marbles. In the first 10 trials, they averaged 6.4 out of 10 marbles. After training, they averaged a remarkable 9.5 out of 10 marbles. Among the Mexican children, then, it seems that their motives for cooperation trumped their competitive instincts.

Psychological Science *in Action*

Culture and Advertising

Advertising is a forum in which identified cultural differences in psychology matter a great deal. Prior to the growth of cultural psychology, the dominant view in advertising was that globalization had laid the groundwork for a standardized advertising strategy (e.g., Levitt, 1983). Advertisements that worked in Calgary should work equally well in Calcutta because people's preferences and cognitive styles were universal. However, cross-cultural research in both psychology and marketing has challenged this view of the universally appealing ad.

For example, some research has explored the kinds of content that appears in ads around the world. In collectivistic cultures, such as East Asian ones, magazine ads make appeals that emphasize family integrity (e.g., "A more exhilarating way to provide for your family"), in-group benefits (e.g., "We have a way of bringing people closer together"), trends (e.g., "Forecast for spring: Pastel colors!"), or harmony with group norms (e.g., "Seven out of 10 people are using this product"). In contrast, North American ads frequently emphasize independence (e.g., "She's got a style all her own"), personal benefits (e.g., "How to protect the most personal part of the environment. Your skin"), personal goals (e.g., "Make your way through the crowd"), choice (e.g., "Choose your own view"), and freedom ("Inspiration doesn't keep office hours") (Han & Shavitt, 1994; Kim & Markus, 1999). Likewise, research reveals that people from different cultures respond to these various appeals differently, with collectivistic appeals being more persuasive in East Asia and individualistic appeals being more persuasive in North America.

The presentation of ads also differs across cultures. East Asian ads tend to contain fewer pieces of information than do North American ads. This is apparently because there are more shared understandings among people in collectivistic cultures, making it less necessary to be as explicit when trying to persuade someone (Hall, 1976; Lin, 1993). Logical arguments are prevalent in North American ads, as compared with East Asian ones, reflecting a greater desire for analytic reasoning, whereas East Asian ads will often strive to create a positive impression of the company (Lazer, Murata, & Kosaka, 1985).

Cultural differences in advertising strategies even emerge between cultures that are fairly similar. For example, surveys of consumers' attitudes toward products revealed that French Canadian women were more likely than American women to state that they buy products in order to please others (Woods, Cheron, & Kim, 1985). A comparison of American and British beer ads found that the American ads were more likely to emphasize independence, modernity, and achievement, and to use direct speech, whereas the British ads were more likely to emphasize tradition, eccentricity, humour, and indirect arguments (Caillat & Mueller, 1996). In a comparison of North American and French magazine ads, the North American ads were more likely to provide information cues and logical arguments, whereas the French ads were more likely to make emotional appeals, jokes, and sexual references (Biswas, Olsen, & Carlet, 1992). In sum, most of the advertisements that you see on a daily basis have not emerged from a cultural vacuum; advertisers consistently consider the kinds of appeals that will be most persuasive in your culture. ■

Cultures Differ in Moral Reasoning

One challenging area of study across cultures is moral reasoning. Do people across cultures have different views about right and wrong? This question is especially difficult since our own moral standards were acquired through socialization in our own culture. It's hard to consider the possibility that another set of moral standards might be legitimate, as we are socialized precisely to think in ways consistent with our own cultural values.

The most famous theory of moral reasoning is Lawrence Kohlberg's (1971). Kohlberg proposed that people in all cultures proceed through the same developmental framework of moral reasoning. This framework consists of three hierarchical levels. The first level is the **Preconventional Level**, in which people learn the cultural rules about good and bad, but interpret these rules in terms of the direct consequences of their actions (e.g., I will be punished if I break the rules, and I will get what I want if I obey the rules). The second level is the **Conventional Level**. At this level, people identify with their group and show loyalty toward it by viewing actions as moral to the extent that they facilitate the social order of the group. The third and highest level of moral reasoning is the **Postconventional Level**. Here, moral values exist independently of the authority of social groups. What is viewed as moral is that which is consistent with a set of universal ethical principles that emphasize justice and individual rights.

Kohlberg proposed that all people in all cultures move through these three levels in order, but that some individuals never reach the highest, postconventional, level. Much cross-cultural research has tested Kohlberg's model. One review of studies of moral reasoning in 27 different cultural areas around the world found that all cultures contained adults who reasoned at the preconventional and conventional levels. Hence, at these levels, Kohlberg's model does appear to be universal. However, evidence of postconventional reasoning was not universally found. Although every urban Western sample contained some people who showed postconventional moral reasoning, not a single person from any of the traditional tribal or village folk population that were studied showed evidence of postconventional reasoning (Snarey, 1985). Hence, the highest level of moral reasoning, in Kohlberg's scheme is limited to Western urban populations.

Richard Shweder critiqued Kohlberg's model by arguing that postconventional moral reasoning represents just one of three different codes of ethics that guide moral judgments around the world (Shweder, Much, Mahapatra, & Park, 1997). According to Shweder, the code of ethics in Kohlberg's postconventional level is an **ethic of autonomy**. This ethic views morality in terms of individual freedom and rights violations. An act is seen as immoral under the ethic of autonomy when it directly hurts another person or infringes on another's rights and freedoms as an individual. According to this code, not being free to express one's views, even if they offend others, such as in the cartoons of Muhammad discussed at the beginning of the chapter, is a violation of the ethic of autonomy. A second code of ethics is an **ethic of community**, which emphasizes how individuals have duties that conform with their roles in a community or social hierarchy. According to this code, actions are wrong when individuals fail to perform their duties to others. For example, a man who decides against attending the funeral of his mother would be in violation of this ethic. A third code of ethics is an **ethic of divinity**, which is concerned with sanctity and the perceived "natural order" of things. In this ethic, actions are immoral if they cause impurity or degradation to oneself or others, or if they show any disrespect for God or God's creations. For example, caricaturizing the prophet Muhammad is a blatant violation of this ethic for many Muslims (Figure 16.27).

Research shows that the ethic of autonomy is found in a wide array of cultural contexts around the world (Turiel, 2002). The ethic of autonomy is likely universal, at least at preconventional and conventional levels, as it solves a problem that is universal for humans everywhere: people cannot live in groups if they do not feel remorse about harming others. Humans evolved emotions that signal when one might be harming another, and these moral feelings facilitate group living by reducing conflict. The ethic of autonomy is especially pronounced within Western populations (particularly among liberal, college-educated populations) because there is little competition from the other two ethics in most Western cultures. In contrast, in many other cultures the ethics of community and divinity play a much larger role than they do in the West. For example, Joan Miller and colleagues have shown that Hindu Indians place greater weight on the ethics of community than do Americans. In one study, they asked a group of Indians and

Preconventional Level The first level of moral reasoning from Kohlberg's theory, in which an individual learns cultural rules about right and wrong, and interprets these rules in terms of the direct consequences of his or her actions.

Conventional Level The second level of moral reasoning from Kohlberg's theory, in which an individual identifies with a given group and shows loyalty towards it by viewing his or her actions as right or wrong based on the extent to which those actions facilitate the social order of the group.

Postconventional Level The third level of moral reasoning from Kohlberg's theory, in which an individual views his or her actions as right or wrong not based on the authority of social groups, but on the extent to which those actions are consistent with a set of universal ethical principles that emphasize justice and individual rights.

ethic of autonomy Shweder's code of ethics under which an individual views his or her actions as right or wrong based on whether or not those actions hurt another person or infringe on another person's rights and freedoms.

ethic of community Shweder's code of ethics under which an individual views his or her actions as right or wrong based on the extent to which those actions conform with his or her role in a community or social hierarchy.

ethic of divinity Shweder's code of ethics under which an individual views his or her actions as right or wrong based on whether or not those actions cause impurity or degradation to oneself, or if they show disrespect to a higher being or the higher being's creations.

16.27 **Ethic of Divinity** In response to the caricature of the Muslim Prophet Muhammad (a sin in Islam) in a Danish newspaper, Muslims around the world rioted and burned down buildings such as this Danish mission in Beirut, Lebanon. As Westerners, it is difficult to understand valuing a religious belief over freedom of speech, as most of us are not subscribers to an ethic of divinity as are these Muslim rioters.

Americans to imagine themselves in a dilemma in which the only solution is to either violate an ethic of autonomy in order to fulfill an obligation to a friend or violate an ethic of community to preserve a stranger's rights. A majority of Indians opted to violate the ethic of autonomy in order to preserve the ethic of community, whereas a majority of Americans did the precise opposite (Miller & Bersoff, 1992).

Other research has explored how people respond to violations of the ethic of divinity. In one set of studies, Haidt and colleagues found that when presented with scenarios that involved breaking strongly held taboos (such as not eating the family pet or not engaging in sex with a sibling) people from different cultural contexts reasoned differently. Although participants from all cultures found the behaviours offensive, college-educated populations (especially Americans) tended to focus their moral evaluations on the extent of perceived harm, whereas people of lower socioeconomic status were more likely to base their moral evaluations on how much disgust they felt about the behaviours (Haidt et al., 1993). The ethic of divinity appears more common among those of lower socioeconomic status and in some non-Western cultures. In sum, there appears to be considerable cultural variation in the bases by which people around the world view behaviours to be right or wrong.

How does moral reasoning differ across cultures?

Returning again to the cartoon controversy from the beginning of this chapter, the Muslim protestors and the Western newspaper editors were reasoning about the appropriateness of caricaturizing Muhammad in dramatically different ways. The Muslim protestors appear to be reasoning based on an ethic of divinity, according to which publishing an image of Muhammad is unacceptable. In contrast, the reasoning of the Western newspaper editors appears to be based on the ethic of autonomy, according to which censoring free speech is intolerable. The world was polarized around this issue largely because people were making sense of it according to fundamentally different ethics.

Language Influences but Does Not Determine Thought

One obvious way that cultures differ is in speaking different languages. Does the language that people speak have an impact on their psychology? Does the language that people speak influence how they think?

Whorfian hypothesis The controversial theory that the language one speaks influences the way one thinks.

These questions have been controversial in psychology and linguistics since Benjamin Whorf (1956) first proposed what would later become the **"Whorfian hypothesis."** This hypothesis has been posed in two forms. The strong form of the hypothesis is that language *determines* thought; that is, you can only think through language. There is much evidence that the strong version of the Whorfian hypothesis is not tenable. A weaker version of the hypothesis is that language *influences* thought. Although this remains controversial, much recent research indicates that language does indeed influence thought in a number of different domains.

One domain in which psychologists have tested the Whorfian hypothesis is in colour perception. Colour is determined by the wavelength of light and exists as a continuum that traces through the rainbow. However, colour is expressed in languages in discrete words, such as *blue* and *yellow*. Interestingly, there is great variation in categories of colour words in languages around the world. Some languages, for example, have only three colour words, corresponding to the English terms *black* (or dark), *white* (or light), and *red* (Berlin and Kay, 1969). And some languages divide the colour spectrum into categories that don't map cleanly on to the colour labels used in English. If people have different categories for colours than English-speakers, do they perceive colours differently as well?

For example, Figure 16.28 shows the boundaries of the different colour labels for English speakers, Berinmo speakers (from New Guinea), and Himba speakers (from Namibia). The dots represent the focal colours (e.g., the best example of blue) for the English speaker (as identified by thousands of participants in past research), and the numbers indicate the Berinmo and Himba speakers who selected the particular colour chip as the best example of their respective colour labels (e.g., the best example of Nol). You might notice a couple of things in these three colour grids. First, the Berinmo and Himba colour words do overlap somewhat with English colour terms. For example,

Does language influence thought?

English naming

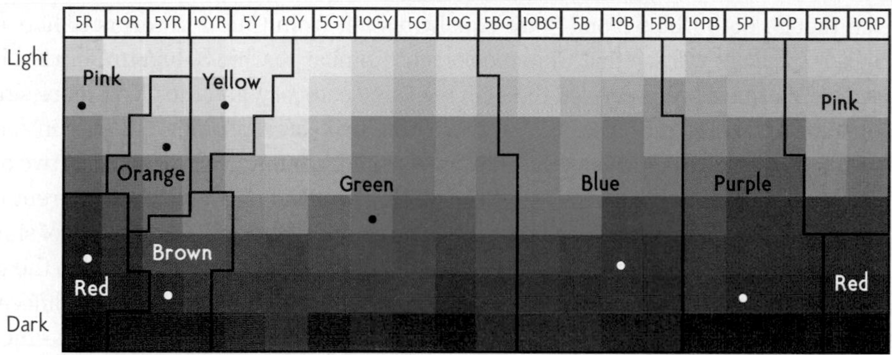

Berinmo naming

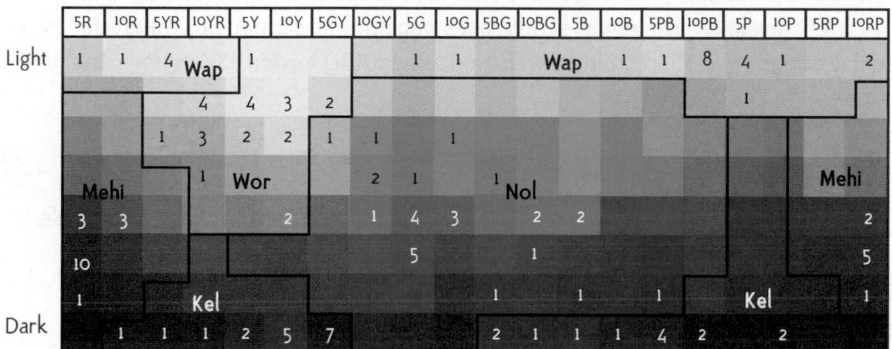

Himba naming

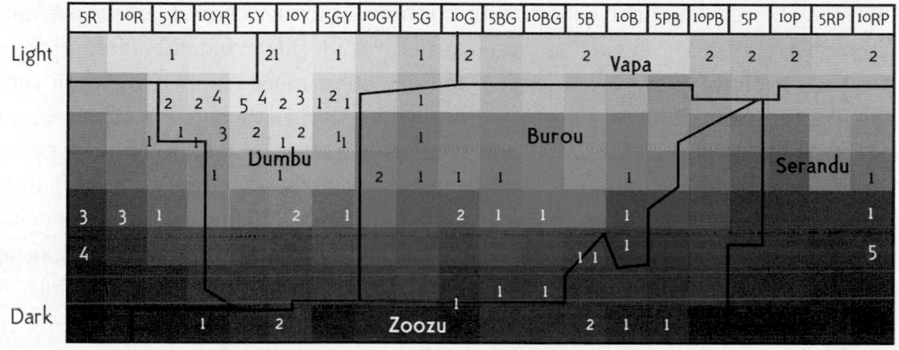

16.28 Colours and Language Colour labels among English, Berinmo, and Himba speakers.

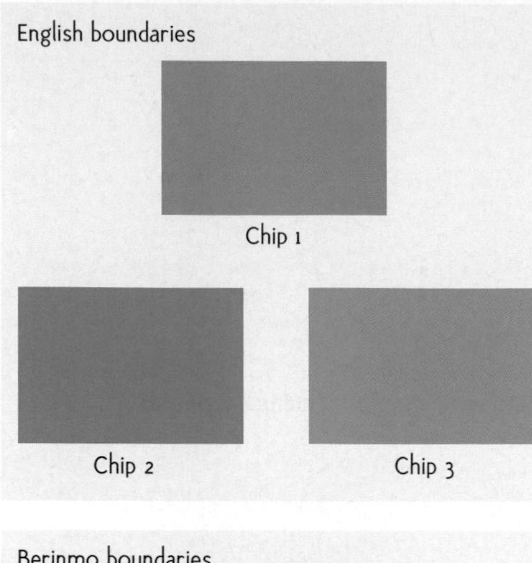

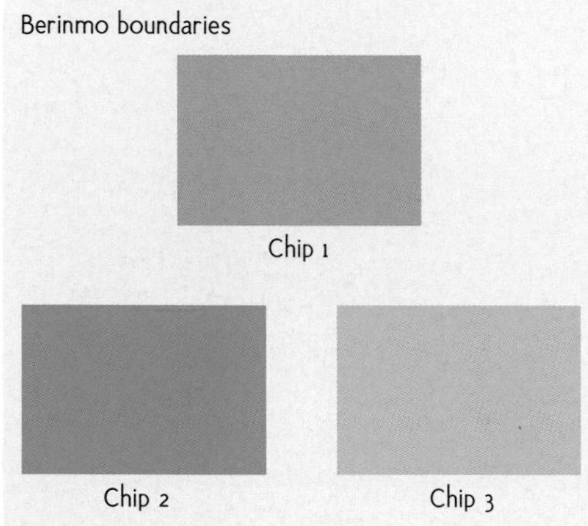

16.29 Perception of Colours Which Chip, 2 or 3, is more similar to the colour of Chip 1?

Red, Mehi, and Serandu cover a fairly similar range of hues across all three languages, and all three groups picked the same colour chip as the best exemplar of this colour. Colour perception is thus not determined by language—if it was, we wouldn't see this kind of similarity (Berlin & Kay, 1969). Second, despite this similarity for some colours, there is also considerable difference in the range for each of the colour terms, and in the focal colours that are selected for some of those colours. The word *Nol* covers the area that coincides with the English terms *green* and *blue*, and the word *Zoozu* covers the area that English speakers would use to categorize both dark colours and light purple.

Roberson and colleagues investigated whether the difference in colour labels affected the perception of colours (Roberson et al., 2000, 2005). They presented participants with series of triads of colour chips, such as those shown in Figure 16.29, and participants were asked which pair of chips was more similar. The chips were selected such that Chip 1 was equally distant from Chips 2 and 3 in terms of hue. However, Chips 1 and 3 fall under the same colour label (*blue* in the English example) and Chip 2 falls under a different colour label (*green* in the English example). When presented with this task, even though both Chips 2 and 3 are equally distant from Chip 1 in terms of hue, most English speakers will say that Chip 1 looks more similar to Chip 3 than to Chip 2. This is because the chips are perceived to fall in the same category, and thus look more similar. Roberson and colleagues also created triads that had corresponding qualities in each of the other two languages. Again, the chips were equidistant in terms of hue, but two of the chips belonged to the same colour category (e.g., Nol) and one fell into a different category (e.g., Wor). When Roberson and colleagues compared people's judgments of similarity, they found that people tended to see two colours as similar if both fell within the same category covered by their own language's colour label, but viewed them as different if they fell within two different colour categories in their language. This effect emerged only when the colours crossed the boundaries of colour terms within participants' own language. Language does appear to influence colour perception, and this challenges some earlier findings that suggested that it does not (Rosch Heider, 1972). Other recent research reveals that language influences people's perceptions of other things as well, such as time, space, and quantities (Boroditsky, 2001; Gordon, 2004; Levinson, 1997).

REVIEWING the Principles | How Does Culture Affect How We Think and Behave?

People from a Western cultural background are more likely to think in analytic terms, whereas people from East Asian cultures often think in holistic terms. People from different cultures experience control differently, with primary control strategies being more common in individualistic cultures, and secondary control strategies being more common in collectivistic cultures. Marriages based on love are more common in cultures with nuclear family structures. People in individualistic cultures are more likely to show social loafing and are often more competitive than those from collectivistic cultures. There are three ethics that guide people's moral reasoning around the world. The ethic of autonomy is most common in individualistic cultures. The ethic of community is common in some collectivist cultures such as India, and the ethic of divinity is common in some other cultural contexts, and especially among those of lower socioeconomic status. The language that people speak appears to influence, but not determine, the ways that they think in some domains.

How Does Culture Influence Mental Health?

People from different cultures differ not only in their normal psychological functioning, but also in the ways that their thinking departs from normal. Recall from Chapter 13 that one major model of the cause of mental disorders is the sociocultural model, which views psychopathology as arising from the interaction between individuals and their cultures. Cultural variations in mental disorders are a real challenge for clinical psychology and psychiatry. If it is difficult to determine what represents a mental disorder within one culture, it is that much more challenging when comparing cultures that differ in their beliefs, values, and motivations.

Culture-Bound Disorders Are Largely Limited to Certain Cultural Contexts

A growing area of research in clinical psychology and in psychiatry is the study of mental disorders that are largely limited to certain cultural contexts. Consider ***dhat* syndrome**, a disorder that is frequently identified in a number of South Asian cultures. *Dhat* syndrome is characterized by a belief among young men that they are leaking semen, which causes them to be morbidly anxious as their culture views semen as a source of vitality. *Dhat* syndrome is often associated with paralyzing guilt and anxiety on the part of its victims, who may become convinced that they have caused a leak through their indulgence in certain disapproved sexual acts, such as masturbation (Obeyesekere, 1985). *Dhat* syndrome is considered a **culture-bound disorder**—a disorder that is greatly influenced by cultural factors and occurs far less frequently, or is manifested in highly divergent ways, in other cultures. *Dhat* syndrome would be a very puzzling disorder if it was identified in the West. Western cultures lack the cultural value system of many South Asian cultures regarding the nature of semen, the kinds of sexual acts that are disapproved, and what happens to people who engage in these disapproved acts. These specific cultural beliefs are necessary for *dhat* syndrome to be meaningful, and for semen maintenance to be a significant concern in people's lives (for further discussion see Kleinman, 1988). In addition to *dhat* syndrome, a variety of other culture-bound disorders have been identified around the world.

***dhat* syndrome** A disorder frequently identified in South Asian cultures characterized by a belief among young men that they are leaking semen, which causes them feelings of anxiety, guilt, and fear, usually as a result of engaging in sexual activities of which their cultures disapprove.

culture-bound disorder A disorder that is influenced by cultural factors and is infrequent or manifests differently in other cultures.

somatic symptoms Side-effects of psychological disorders experienced as physical manifestations.

Some Universal Mental Disorders Present Differently in Different Cultures

Though psychological disorders are found in all parts of the world, the same underlying disorder may present differently in different cultures. Occasionally the presented symptoms are so different that doctors differ about whether the underlying disorder can even be said to be the same (e.g., Kleinman, 1988).

DEPRESSION Though depressive disorder is one of the most common mental disorders, its prevalence varies greatly across cultures. In particular, depression is less commonly identified in China than in Western cultures. Epidemiological surveys have found that the rates of depression in China are approximately one-fifth of those in North America (Kessler et al., 1994). Furthermore, many of the symptoms of Chinese sufferers of depression differ from the symptoms of North American depression patients. Specifically, some symptoms of depression are **somatic**: that is, they are experienced primarily as physical symptoms, such as headaches,

16.30 ***Taijinkyoufushou* (TKS)** This East Asian disorder can take many forms, one of which may manifest in not wanting to be seen laughing by others. A sufferer might cover his or her mouth so as not to be viewed altering his or her face to laugh.

***taijinkyoufushou* (TKS)** A disorder of which victims are morbidly afraid of offending others or causing others discomfort from likely imagined physical flaws.

sleep disorders, and pain. Other symptoms of depression are psychological, such as depressed mood, difficulty concentrating, and feelings of guilt. A number of studies have revealed that Chinese symptoms of depression are more likely to be somatic, whereas North American symptoms of depression are more likely to be psychological (Chang, 1985; Kleinman, 1982). The manifestation of depression can thus vary across cultures.

SOCIAL PHOBIA Humans are a social species, so it is natural that people are often concerned about losing the favour of others. Most likely, we all feel anxious in some kinds of social situations. But people who suffer from social phobia are so concerned about being negatively evaluated by others that it interferes with their lives. In collectivistic cultures, it is extremely important to maintain social harmony and to fit in with others. It thus should not be surprising that people from collectivistic cultural backgrounds (e.g., East Asian) say they experience social phobia symptoms more than Westerners. However, though East Asians tend to score higher than Westerners on self-report measures of social anxiety (Okazaki, 1997), there is far less evidence of people meeting the clinical criteria of social phobia in East Asia than in the West (Hwu, Yeh, & Chang, 1989; Lee et al., 1987). Perhaps this discrepancy is due to symptoms of social anxiety being viewed as more acceptable in collectivistic cultures than in individualistic ones, because of the greater value placed on harmonious social relationships in collectivistic contexts.

At the same time, when social anxieties do become problematic, they may present differently across cultures. One disorder that was identified by a Japanese psychiatrist is ***taijinkyoufushou* (TKS)** (Morita, 1917), which loosely translates as a phobia of confronting others (Figure 16.30). TKS is often viewed as a culture-bound disorder as it very rare for it to be diagnosed outside of East Asia. TKS is similar to social phobia in that it is elicited by social situations. However, the symptoms of TKS are distinct from those of social anxiety. TKS's symptoms include extensive blushing, body odour, sweating, a penetrating gaze, and often these physical symptoms are imagined. A person diagnosed with severe TKS is not only preoccupied with these symptoms, but intensely concerned that they will cause other people to feel uncomfortable. This focus on disturbing others has led TKS to be labeled the "altruistic phobia" (Kasahara, 1986).

ANOREXIA NERVOSA Anorexia nervosa is characterized by a refusal to eat for fear of becoming fat. Its incidence increased during the latter half of the twentieth century. For example, in Denmark, the rates of people diagnosed with anorexia increased by a factor of 4 from the 1970s to the 1980s, and there recently have been instances of children as young as nine showing anorexic symptoms (Pagsberg & Wang, 1994; Rosen, 2003). The dramatic rise in rates of anorexia appears to be largely due to changing cultural norms with regards to ideal body weight. Indeed, studies that have contrasted Miss America pageant contestants and *Playboy* centrefolds across the past few decades have revealed a trend of increasing thinness (Garner et al., 1980). Furthermore, there is less evidence for anorexia in some cultures than there is in North America. These changes across time and the cultural differences in rates of anorexia suggest that culture plays a key role in its manifestation.

At the same time, there is clear evidence for anorexia in all regions of the world (Figure 16.31) (for a review see Keel & Klump, 2003). Indeed, the historical literature reveals many instances of people voluntarily starving themselves in the presence of food. For example, over the past nine centuries in the Italian peninsula about half of

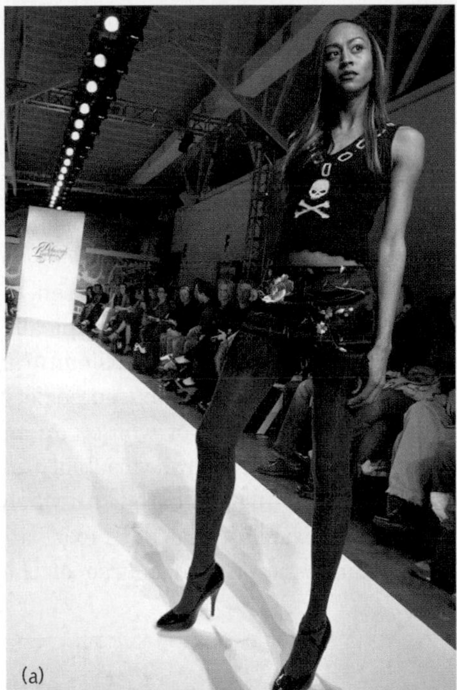

16.31 Shifting Standards The increased frequency of anorexia over the past few decades seems to be because of a shift in the standard of beauty from being curvier to appearing much thinner. This phenomenon, while extremely common in North America for quite some time (a), has actually become an international issue, as public figures from all over the world grow slimmer and slimmer (b).

the people who were canonized as saints refused food and became greatly emaciated because of a belief that their condition reflected divine intervention—a condition that has been termed "holy anorexia" (Bell, 1985). Although the motivation for such self-starvation was not about weight, as it largely is today, some propose that people with temperaments predisposing them toward anorexic symptoms are especially attracted to ascetic lifestyles (Keel & Klump, 2003). To the extent that this is true, it suggests that anorexia is a cultural universal, but that the motivation for self-starvation varies across historical and cultural contexts (e.g., avoiding becoming overweight vs. being a spiritual ascetic).

How similar are mental disorders around the world?

SCHIZOPHRENIA As you learned in Chapter 13, schizophrenia is primarily a brain disorder. And since brains do not differ in any systematic way across cultures, it would seem that schizophrenia should present quite similarly across cultures. Indeed, schizophrenia does present more similarly than any of the other mental disorders discussed thus far. One large-scale study of twelve different cultures found that when a strict definition of schizophrenia was applied, the annual incidence rates of schizophrenia only ranged from 0.7 to 1.4 per 10,000 (Jablensky et al., 1991; WHO, 1973). This reveals rather little cultural variation in worldwide incidences of schizophrenia.

Still, it is important to remember that environmental factors play a significant role in the development of schizophrenia. This role is especially evident when one examines cultural variability in the disorder's presentation. In different cultures, there is considerable variation in which subtypes of schizophrenia are most common. For example, paranoid-type schizophrenia was the most commonly observed subtype in most locations that were examined, but the proportions varied considerably across locations. Seventy-five percent of schizophrenics in the United Kingdom received a diagnosis of paranoid schizophrenia compared with only 15 percent of those in India. Although catatonic schizophrenia was rarely observed in the West (only 1–3 percent of cases in the United States and the United Kingdom), it accounted for over 20 percent of schizophrenia patients in India. Paranoid and catatonic subtypes of schizophrenia are dramatically different manifestations of the disease, and the cultural differences in their proportions reflect the role of culture in the presentation of schizophrenia.

The most striking finding of cultural differences in schizophrenia is that the course of the disease was significantly better for patients in less developed societies than in the more industrially advanced ones (Leff, Sartorius, Jablensky, Korten, & Ernberg, 1992). This is especially noteworthy as rates of schizophrenia are typically higher in the lower social classes than the higher ones. Explanations of this cultural difference have focused on the uncertainty of employment in a wage-based society and on cultural differences in family dynamics that can ameliorate or worsen an individual's condition. In sum, although there is clearly less cultural variability in schizophrenia than in many other mental disorders, reflecting the extent to which schizophrenia is a disorder of the brain, the cultural variability that does exist attests to the role of personal experience in the manifestation and course of the disease.

REVIEWING the Principles | How Does Culture Influence Mental Health?

Cultures shape abnormal thinking as well as normal thinking. Culture-bound disorders are those that appear to be greatly influenced by cultural factors. Some mental disorders, such as depression, social phobia, and anorexia nervosa, show universal features as well as culturally specific manifestations. Though schizophrenia shows the least cultural variation among major psychological disorders, reflecting its biological basis, there is still some variation in its expression across cultures.

Conclusion

Cultural psychology teaches us how growing up in a particular culture shapes the ways that we think, feel, and behave. As a cultural species, humans acquire much of the knowledge necessary for survival by learning it from others. Because children growing up in different cultures have divergent experiences and adapt to different social norms, the ways that people think can emerge quite differently across cultures. Although human psychology is universal in many ways, reflecting the similarity in people's biology and in many of the challenges that they face, a number of important differences have been identified across cultures. Many of these stem from the differences between the experiences that people have in individualistic cultures and those that people have in collectivistic cultures. Cultural differences in psychological processing are not limited to superficial phenomena such as fashion or food preferences, but have been identified in a number of basic and fundamental psychological processes, such as perception, cognition, motivation, and mental disorders. These findings underscore the point that much of how the mind works is a function of our biological potentials being shaped by our experiences.

Summarizing the Principles of Cultural Psychology

What Is Culture?

1. **Some aspects of culture are shared by other species and some are unique to humans:** Some other species have a limited ability to learn by observing others, but cultural learning pervades *all* aspects of human experiences.

2. **Humans have evolved to accumulate cultural information:** Language and a theory of mind allowed humans to develop high-precision cultural learning. This has enabled humans to accumulate cultural information in a way and to a degree that no other species can.

How Does Culture Affect the Mind?

3. **Culture and mind are inextricably bound:** People's ways of thinking are shaped by certain cultural practices. At the same time, cultural practices are shaped by the preferences of people living in those cultures. In this way, culture and mind are fundamentally linked.

4. **There is a sensitive period for learning culture:** Humans are biologically prepared to acquire cultural information at a young age. After puberty, people have a more difficult time adjusting to a new cultural environment.

5. **Cultural differences become more pronounced with age:** Children from different cultures are more similar in their ways of thinking than adults from those cultures. As people age, their accumulated cultural experiences exert a greater influence on their thinking.

6. **The self-concept varies across cultures:** People in individualistic cultures are more likely to develop an independent self-concept, whereas those in collectivistic cultures are more likely to develop an interdependent self-concept. Independent self-concepts remain more similar across situations, whereas interdependent self-concepts are more contextually flexible.

What Happens to People's Sense of Self When They Move to a Different Culture?

7. **Acculturation requires significant and often stressful adjustment:** People are shaped by their cultural experiences, so when people move to a new culture there is much adjustment and often stress. People have an easier time acculturating if they are moving to a similar culture or if their personality is a good match with their host culture.

8. **Psychological processes change with acculturation:** Because culture shapes people's thinking, the longer that people are exposed to a new culture, the more their thinking will change. For example, self-esteem levels depend a great deal on how long people have been exposed to cultural environments in which high self-esteem is common or not.

9. **People from distinct cultural backgrounds often face discrimination:** A frequent source of stress in acculturation is that people from many cultural backgrounds are discriminated against. Such discrimination can lead to stereotype threat, which perpetuates negative stereotypes.

10. **Multicultural people can switch between different selves:** Multicultural people can activate different aspects of their selves in different situations. Some ways of thinking are available to people from different cultures but are activated only when the appropriate situational cue is present.

How Does Culture Affect How We Think and Behave?

11. **Cultures differ in analytic and holistic thinking:** Westerners are likely to think in analytic terms, focusing on objects and their attributes. East Asians are likely to think in holistic terms, focusing on the relations among objects and on the relation between objects and the surrounding context. This difference affects the ways that people think in many different kinds of tasks.

12. **Cultures differ in motivations for control and choice:** People from individualistic cultures are more likely to emphasize primary control, where they strive to change the world to match their desires. People from collectivistic cultures show more evidence of secondary control, where they strive to adjust their desires to existing realities. Likewise, people from individualistic cultures value individual choice more than people from collectivistic cultures.

13. **Cultures differ in the bases of relationships:** Arranged marriages have been common in many cultures, particularly those with strong extended family ties. In individualistic cultures, friendships are typically thought of in positive terms. In collectivistic cultures, friendships are

viewed both in positive terms and in negative terms associated with costly obligations and potential threats.

14. **Cultures differ in group performance:** Social loafing occurs when individuals try less hard in group contexts. Although common in individualistic cultures, social loafing is rarer in collectivistic cultures. People in collectivistic cultures appear to have stronger motivations to cooperate, and weaker motivations to compete as individuals, in some situations.

15. **Cultures differ in moral reasoning:** Postconventional moral reasoning based on justice and individual rights is less common outside of industrialized Western contexts. There appear to be three different moral ethics that people respond to: autonomy, community, and divinity. In non-Western contexts, the ethics of community and divinity are emphasized more than they are in Western contexts.

16. **Language influences but does not determine thought:** Much recent research demonstrates that although language does not determine thought, it does influence it. For example, people perceive colours differently depending on their language's colour labels.

How Does Culture Influence Mental Health?

17. **Culture-bound disorders are largely limited to certain cultural contexts:** Some mental disorders are largely limited to certain cultural contexts. Specific cultural beliefs are necessary for some "culture-bound" disorders to be presented.

18. **Some universal mental disorders present differently in different cultures:** Some mental disorders, such as depression, social phobia, anorexia, and schizophrenia, are found in all cultures. However, these disorders present differently across cultures, although cultural variability is weaker for the more biologically grounded disorders, such as schizophrenia.

Applying the Principles

1. Fritz moved to Germany from Austria and quickly made a number of friends, found a good job, and got engaged to a German woman. Overall, he was very happy with his new life. At the same time, Panadda, from Thailand, also moved to Germany. Panadda felt quite lonely in Germany, made few friends, was unable to find a job, and experienced a great deal of stress in her daily activities. A likely reason for Fritz's and Panadda's different reactions is

    _____ a) cultural fit.
    _____ b) the U-shaped curve.
    _____ c) cultural distance.
    _____ d) frame-switching.

2. Azim read in the newspaper that a young man had been caught trying to have sex with a sheep. Azim was outraged as he saw the man's behaviour as a shameful moral violation. Which of the following is reflected in Azim's moral reasoning?

    _____ a) preconventional level of moral reasoning
    _____ b) the ethic of autonomy
    _____ c) the ethic of community
    _____ d) the ethic of divinity

3. Emi hates to leave her house. She dreads meeting other people as she fears that she will blush excessively, and that this make others feel uncomfortable. It is likely that Emi suffers from

    _____ a) *taijinkyoufushou* (TKS).
    _____ b) social phobia.
    _____ c) depression.
    _____ d) *dhat* syndrome.

4. While hiking, Mira noticed a plant that she had never seen before. She decided to go to the library to try to learn more about it. She observed that the plant was growing on a riverbank in a very sunny location, attracted a certain species of bumblebee, and was downwind from other plants of the same species. Mira's observations reflect what kind of reasoning?

    _____ a) analytic thinking
    _____ b) holistic thinking
    _____ c) taxonomic categorization
    _____ d) rule-based reasoning

5. Imagine that you are asked to create an advertisement to run in Korea for a new beverage called "Cold Snap." Which of the following strategies will likely be most effective?

    _____ a) "Cold Snap" tastes better than Pepsi.
    _____ b) Everyone loves the taste of "Cold Snap."
    _____ c) "Cold Snap" is completely organic and contains 7 essential vitamins.
    _____ d) People with discerning taste prefer "Cold Snap."

ANSWERS: *1. c 2. d 3. b 4. b*

Key Terms

acculturation, p. 679
analytic thinking, p. 686
arranged marriage, p. 691
Conventional Level, p. 697
cultural distance, p. 680
cultural fit, p. 680
culture-bound disorder, p. 701
culture shock, p. 680
dhat syndrome, p. 701
ethic of autonomy, p. 697
ethic of community, p. 697
ethic of divinity, p. 697
frame-switching, p. 683
holistic thinking, p. 686
love marriage, p. 691
Postconventional Level, p. 697
postdecisional dissonance, p. 677
Preconventional Level, p. 697
primary control, p. 687
secondary control, p. 687
social loafing, p. 693
social striving, p. 694
somatic symptoms, p. 701
taijinkyoufushou (TKS), p. 702
taxonomic categorization, p. 685
thematic categorization, p. 685
U-shaped curve, p. 679
Whorfian hypothesis, p. 698

Further Readings

Adams, G. (2005). The cultural grounding of personal relationships: Enemyship in West African worlds. *Journal of Personality and Social Psychology, 88,* 948–968.

Adams, G., & Plaut, V. C. (2003). The cultural grounding of personal relationships: Friendship in North American and West African worlds. *Personal Relationships, 10,* 333–348.

Heine, S. J. (forthcoming). *Cultural psychology.* New York: Norton.

Kitayama, S., & Cohen, D. (forthcoming). *Handbook of cultural psychology.* New York: Guilford.

Kleinman, A. (1988). *Rethinking psychiatry: From cultural category to personal experience.* New York: Free Press.

Nisbett, R. E. (2003). *The geography of thought.* New York: Free Press.

Richerson, P. J., & Boyd, R. (2005). *Not by genes alone.* Chicago: University of Chicago Press.

Shweder, R. (1991). *Thinking through cultures: Expeditions in cultural psychology.* Cambridge, MA: Harvard University Press.

Triandis, H. C. (1995). *Individualism and collectivism.* Boulder, CO: Westview Press.

Further Readings: A Canadian Presence in Psychology

Berry, J. W., & Annis, R. C. (1974). Acculturation stress: The role of ecology, culture, and differentiation. *Journal of Cross-Cultural Psychology, 5,* 382–406.

Dion, K. K., & Dion, K. L. (1993). Individualistic and collectivistic perspectives on gender and the cultural context of love and intimacy. *Journal of Social Issues, 49,* 53–69.

Heine, S. J., & Lehman, D. R. (1997). Culture, dissonance, and self-affirmation. *Personality and Social Psychology Bulletin, 23,* 389–400.

Heine, S. J., & Lehman, D. R. (2004). Move the body, change the self: Acculturative effects on the self-concept. In M. Schaller & C. Crandall (Eds.), *Psychological Foundations of Culture* (pp. 305–331). Mahwah, NJ: Erlbaum.

Hoshino-Browne, E., Zanna, A. S., Spencer, S. J., Zanna, M. P., Kitayama, S., & Lackenbauer, S. (2005). On the cultural guises of cognitive dissonance: The case of Easterners and Westerners. *Journal of Personality and Social Psychology, 89,* 294–310.

Ji, L. (2005). *Culture and thinking across time.* Presentation at the Cultural Psychology Preconference, New Orleans. January 20, 2005.

Ji, L. J., Zhang, Z., & Nisbett, R. E. (2004). Is it culture or is it language? Examination of language effects in cross-cultural research on categorization. *Journal of Personality and Social Psychology, 87,* 57–65.

Norenzayan, A., & Heine, S. J. (2005). Psychological universals: What are they and how can we know? *Psychological Bulletin, 131,* 763–784.

Norenzayan, A., Smith, E. E., Kim, B., & Nisbett, R. E. (2002). Cultural preferences for formal versus intuitive reasoning. *Cognitive Science, 26,* 653–684.

Ross, M., Xun, W. Q. E., & Wilson, A. E. (2002). Language and the bicultural self. *Personality and Social Psychology Bulletin, 28,* 1040–1050.

Werker, J. F., & Tees, R. C. (1984). Cross-language speech perception: Evidence for perceptual reorganization during the first year of life. *Infant Behavior and Development, 7,* 49–63.

GLOSSARY

absentmindedness The inattentive or shallow encoding of events.

absolute threshold The minimum intensity of stimulation that must occur before one can experience a sensation.

accommodation (1) A process by which muscles change the shape of the lens of the eye by flattening it to focus on distant objects or by thickening it to focus on closer objects. (2) The process by which a schema is adapted or expanded to incorporate a new experience that does not easily fit into an existing schema.

acculturation The process of adaptation to a culture different from one's own.

accuracy The extent to which an experimental measure is free from error.

acetylcholine (ACh) The neurotransmitter responsible for motor control at the junction between nerves and muscles; also involved in mental processes such as learning, memory, sleeping, and dreaming.

achievement motive The desire to do well relative to standards of excellence.

acquisition The gradual formation of an association between the conditioned and unconditioned stimuli.

action potential The neural impulse that passes along the axon and subsequently causes the release of chemicals from the terminal buttons.

activation-synthesis hypothesis A theory of dreaming that proposes that neural stimulation from the pons activates mechanisms that normally interpret visual input.

adaptations In evolutionary theory, the physical characteristics, skills, or abilities that increase the chances of reproduction or survival and are therefore likely to be passed along to future generations.

additive colour mixing A way to produce a given spectral pattern in which different wavelengths of lights are mixed. The percept is determined by the interaction of these wavelengths with receptors in the eye and is a psychological process.

adolescence The transitional period between childhood and adulthood.

affiliation The tendency to be in social contact with others.

aggression Any behaviour or action that involves the intention to harm someone else.

agonist Any drug that enhances the actions of a specific neurotransmitter.

agoraphobia An anxiety disorder marked by fear of being in situations in which escape may be difficult or impossible.

alcoholism Abnormal alcohol seeking characterized by loss of control over drinking and accompanied by physiological effects of tolerance and withdrawal.

alexithymia A disorder involving a lack of the subjective experience of emotion.

all-or-none principle The principle whereby a neuron fires with the same potency each time, although frequency can vary; it either fires or not, it cannot partially fire.

altruism The providing of help when it is needed, without any apparent reward for doing so.

amnesia Deficits in long-term memory that result from disease, brain injury, or psychological trauma.

amygdala A brain structure that serves a vital role in our learning to associate things with emotional responses and for processing emotional information.

analogical representation A mental representation that has some of the physical characteristics of an object; it is analogous to the object.

analytic thinking A system of evaluation in which the assessor views objects as independent from context and in terms of individual characteristics, accompanied by an expectation of behaviour in accordance with a fixed set of abstract rules derived from said assessments.

anorexia nervosa An eating disorder characterized by an excessive fear of becoming fat and a refusal to eat.

antagonist Any drug that inhibits the action of a specific neurotransmitter.

anterograde amnesia An inability to form new memories.

anti-anxiety drugs A class of psychotropic medications used for the treatment of anxiety.

antidepressants A class of psychotropic medications used to treat depression.

antipsychotics A class of drugs used to treat schizophrenia and other disorders that involve psychosis.

antisocial personality disorder A personality disorder marked by a lack of empathy and remorse.

anxiety disorders Disorders characterized by the experience of excessive anxiety in the absence of true danger.

arousal Term to describe psychological activation, such as increased brain activity, autonomic responses, sweating, or muscle tension.

arranged marriage A type of marriage in which the bride and groom have been pre-selected by the bride's and groom's respective families.

assessment In psychology, examination of a person's mental state in order to diagnose possible mental illness.

assimilation The process by which a new experience is placed into an existing schema.

attachment A strong emotional connection that persists over time and across circumstances.

attention-deficit/hyperactivity disorder (ADHD) A disorder characterized by restless, inattentive, and impulsive behaviours.

attitude The evaluation of objects, events, or ideas.

attributions People's causal explanations for why events or actions occur.

autism A developmental disorder involving deficits in social interaction, impaired communication, and restricted interests.

autonomic nervous system (ANS) A major component of the peripheral nervous system, which regulates the body's internal environment by stimulating glands and by maintaining internal organs such as the heart, gall bladder, and stomach.

autoreceptors A neuron's own neurotransmitter receptors, which regulate the release of the neurotransmitters.

availability heuristic Making a decision based on the answer that most easily comes to mind.

axon A long narrow outgrowth of a neuron by which information is transmitted to other neurons.

basal ganglia A system of subcorticĬal structures that are important for the initiation of planned movement.

behaviour modification Treatment in which principles of operant conditioning are used to reinforce desired behaviours and ignore or punish unwanted behaviours.

behavioural approach system (BAS) The brain system involved in the pursuit of incentives or rewards.

behavioural inhibition system (BIS) The brain system that is sensitive to punishment and therefore inhibits behaviour that might lead to danger or pain.

behaviourism A psychological approach that emphasizes the role of environmental forces in producing behaviour.

binocular depth cues Cues of depth perception that arise from the fact that people have two eyes.

binocular disparity A cue of depth perception that is caused by the distance between a person's eyes.

biological preparedness The idea that animals are biologically programmed to learn to fear specific objects.

biological therapy Treatment based on the medical approach to illness and disease.

bipolar disorder A mood disorder characterized by alternating periods of depression and mania.

blindsight A condition in which people who are blind have some spared visual capacities in the absence of any visual awareness.

blocking The temporary inability to remember something that is known.

body mass index (BMI) A ratio of body weight to height used to measure obesity.

borderline personality disorder A personality disorder characterized by identity, affective, and impulse disturbances.

bottom-up processing A hierarchical model of pattern recognition in which data are relayed from one processing level to the next, always moving to a higher level of processing.

brainstem A section of the bottom of the brain that houses the most basic programs of survival, such as breathing, swallowing, vomiting, urination, and orgasm.

Broca's area The left frontal region of the brain that is crucial to the production of language.

buffering hypothesis The idea that other people can provide direct support in helping individuals cope with stressful events.

bulimia nervosa An eating disorder characterized by dieting, binge eating, and purging.

bystander intervention effect The failure to offer help by those who observe someone in need.

Cannon-Bard theory of emotion A theory that emotion-producing stimuli from the environment elicit both an emotional and a physical reaction.

case study A research method that involves the intensive examination of one person.

cell body In the neuron, where information from thousands of other neurons is collected and processed.

central nervous system (CNS) The brain and spinal cord.

central tendency A measure that represents the typical behaviour of the group as a whole.

cerebellum A large convoluted protuberance at the back of the brainstem that is essential for coordinated movement and balance.

cerebral asymmetry An emotional pattern associated with unequal activation of the left and right frontal lobes.

cerebral cortex The outer layer of brain tissue that forms the convoluted surface of the brain.

cholecystokinin (CCK) The peptide found in highest concentration in the cerebral cortex; it plays a role in learning and memory, pain transmission, and exploratory behaviour.

chromosomes Structures within the cell body that are made up of genes.

circadian rhythms The regulation of biological cycles into regular patterns.

circumplex model An approach to understanding emotion in which two basic factors of emotion are spatially arranged in a circle, formed around the intersections of the core dimensions of affect.

classical conditioning A type of learned response that occurs when a neutral object comes to elicit a reflexive response when it is associated with a stimulus that already produces that response.

client-centered therapy An empathic approach to therapy that encourages personal growth through greater self-understanding.

clozapine An antipsychotic medication that acts on multiple neurotransmitter receptors and is beneficial in treating both the negative and positive symptoms of schizophrenia.

cochlea (inner ear) A fluid-filled tube that curls into a snail-like shape. The cochlea contains the basilar membrane, which in turn contains auditory receptor cells called hair cells. These transduce the mechanical energy of the sound wave into neural impulses.

cocktail party phenomenon The ability to focus on a single conversation in the midst of a chaotic cocktail party or other similarly noisy situation.

cognition Mental activity such as thinking or representing information.

cognitive-behavioural approach A diagnostic model that views psychopathology as the result of learned, maladaptive cognitions.

cognitive-behavioural therapy (CBT) A therapy that incorporates techniques from behavioural therapy and cognitive therapy to correct faulty thinking and change maladaptive behaviours.

cognitive dissonance The perceptual incongruity that occurs when there is a contradiction between two attitudes or between an attitude and a behaviour.

cognitive map A visual/spatial mental representation of the environment.

cognitive restructuring A therapy that strives to help patients recognize maladaptive thought patterns and replace them with ways of viewing the world that are more in tune with reality.

cognitive therapy Treatment based on the idea that distorted thoughts produce maladaptive behaviours and emotions.

compliance The tendency to agree to do things requested by others.

concept A mental representation that groups or categorizes objects, events, or relations around common themes.

concrete operational stage The third stage in Piaget's theory of cognitive development, during which children begin to think about and understand operations in ways that are reversible.

conditioned response (CR) A response that has been learned.

conditioned stimulus (CS) A stimulus that elicits a response only after learning has taken place.

confabulation The false recollection of episodic memory.

confirmation bias A tendency to search for and believe evidence that fits our existing views.

conformity The altering of one's opinions or behaviour to match those of others or to match social norms.

confound Anything that affects a dependent variable that may unintentionally vary between the different experimental conditions of a study.

consolidation A hypothetical process involving the transfer of contents from immediate memory into long-term memory.

continuous reinforcement A type of learning in which the desired behaviour is reinforced each time it occurs.

conventional level The stage of moral reasoning that reflects conformity to rules of law and order that are learned from others.

coping response Any response an organism makes to avoid, escape from, or minimize an aversive stimulus.

corpus callosum A fibre of axons that transmits information between the two cerebral hemispheres of the brain.

correlation A statistical procedure that provides a numerical value, between +1 and −1, indicating the strength and direction of the relation between two variables.

correlational study A research method that examines how variables are naturally related in the real world, without any attempt by the researcher to alter them.

creativity The capacity to generate or recognize ideas, alternatives, or possibilities that may be useful in solving problems, communicating with others, or entertaining ourselves and others.

critical period The time in which certain experiences must occur for normal brain development, such as exposure to visual information during infancy for normal development of the brain's visual pathways.

critical thinking A systematic way of evaluating information in order to reach reasonable conclusions.

cryptomnesia A type of misappropriation that occurs when people think they have come up with a new idea, yet have only retrieved a stored idea and failed to attribute the idea to its proper source.

crystallized intelligence Knowledge acquired through experience and the ability to use that knowledge.

cultural distance The disparity in customs, traditions, beliefs, and general heritage between two different cultures.

cultural fit The degree to which an individual's values and behavioural norms align with those that are common to another culture.

culture The beliefs, values, rules, and customs that exist within a group of people who share a common language and environment and that are transmitted through learning from one generation to the next.

culture shock Experiencing feelings of anxiety, estrangement, helplessness, and incongruousness with one's surroundings on emigrating to a new culture.

culture-bound disorder A disorder that is influenced by cultural factors and is infrequent or manifests differently in other cultures.

cyclothymia A less extreme form of bipolar disorder.

data Objective observations or measurements.

declarative memory The cognitive information retrieved from explicit memory, knowledge that can be declared.

deductive reasoning A form of reasoning in which logic is used to draw a specific conclusion from given premises.

defense mechanisms Unconscious mental strategies the mind uses to protect itself from conflict and distress.

defining attribute model The idea that a concept is characterized by a list of features that are necessary to determine if an object is a member of a category.

deindividuation A phenomenon of low self-awareness in which people lose their individuality and fail to attend to personal standards.

delay of gratification When people transcend immediate temptations to successfully achieve long-term goals.

delusions False personal beliefs based on incorrect inferences about reality.

dendrites Branchlike extensions of the neuron that detect information from other neurons.

dependent variable In an experiment, the measure that is affected by manipulation of the independent variable.

descriptive study A research method that involves observing and noting the behaviour of people or other animals in order to provide a systematic and objective analysis of behaviour.

developmental psychology The study of changes in physiology, cognition, and social behaviour over the life span.

dhat syndrome A disorder frequently identified in South Asian cultures characterized by a belief among young men that they are leaking semen, which causes them feelings of anxiety, guilt, and fear, usually as a result of engaging in sexual activities of which their cultures disapprove.

Diagnostic and Statistical Manual of Mental Disorders (DSM) A handbook of clinical disorders used for diagnosing psychopathology.

dialectical behaviour therapy (DBT) A treatment for borderline personality disorder that combines elements of behavioural, cognitive, and psychodynamic approaches.

diathesis-stress model A diagnostic model that proposes that a disorder may develop when an underlying vulnerability is coupled with a precipitating event.

difference threshold The minimum amount of change required in order to detect a difference between intensities of stimuli.

discrimination The inappropriate and unjustified treatment of people solely on the basis of their group membership.

display rules Cultural rules that govern how and when emotions are exhibited.

dizygotic twins Twin siblings who result from two separately fertilized eggs (i.e., fraternal twins).

dominant gene A gene that is expressed in the offspring whenever it is present.

dopamine A monoamine neurotransmitter involved in reward, motivation, and motor control.

dreams The product of an altered state of consciousness in which images and fantasies are confused with reality.

drive Psychological state that motivates an organism to satisfy its needs.

dualism The philosophical idea that the mind exists separately from the physical body.

dysthymia A form of depression that is not severe enough to be diagnosed as major depression.

eardrum (tympanic membrane) A thin membrane, which sound waves vibrate, that marks the beginning of the middle ear.

early selection theory A theory that we can choose the stimuli to which we will attend before we process their basic features.

ego In psychodynamic theory, the component of personality that tries to satisfy the wishes of the id while being responsive to the dictates of the superego.

elaboration likelihood model A theory of how persuasive messages lead to attitude changes.

elaborative rehearsal The encoding of information in a more meaningful fashion, such as linking it to knowledge in long-term memory.

electroconvulsive therapy (ECT) A procedure used to treat depression that involves administering a strong electrical current to the patient's brain.

electroencephalography (EEG) A method for measuring the electrical activity of the brain. Electrodes placed on the scalp are able to detect weak electrical signals produced by neural activity.

electrophysiology A method of data collection that measures electrical activity in the brain.

emotion Feelings that involve subjective evaluation, physiological processes, and cognitive beliefs.

emotion-focused coping A type of coping in which people try to prevent having an emotional response to a stressor.

emotional intelligence (EQ) A form of social intelligence that emphasizes the ability to perceive, understand, manage, and use emotions to guide thoughts and actions.

encoding The processing of information so that it can be stored.

encoding specificity principle Any stimulus that is encoded along with an experience can later trigger memory for the experience.

endocrine system A communication system that uses hormones to influence thoughts, behaviours, and actions.

endorphins Peptides involved in natural pain reduction and reward.

enzyme deactivation The process whereby the neurotransmitter is destroyed by an enzyme, thereby terminating its activity.

epinephrine A monoamine, found primarily in the body, which causes a burst of energy after an exciting event.

episodic memory Memory for one's personal past experiences.

ethic of autonomy Shweder's code of ethics under which an individual views his or her actions as right or wrong based on whether or not those actions hurt another person or infringe on another person's rights and freedoms.

ethic of community Shweder's code of ethics under which an individual views his or her actions as right or wrong based on the extent to which those actions conform with his or her role in a community or social hierarchy.

ethic of divinity Shweder's code of ethics under which an individual views his or her actions as right or wrong based on whether or not those actions cause impurity or degradation to oneself, or if they show disrespect to a higher being or the higher being's creations.

evolutionary theory In psychological science, a theory that emphasizes the inherited, adaptive value of behaviour and mental activity throughout the entire history of a species.

excitation transfer A form of misattribution in which residual physiological arousal caused by one event is transferred to a new stimulus.

expected utility theory A model of how humans should make decisions if they were to follow pure reason in their decision making.

experiment A study that tests causal hypotheses by measuring and manipulating variables.

experimenter expectancy effect Actual change in the behaviour of the people or animals being observed that is due to observer bias.

explicit memory The processes involved when people remember specific information.

exposure A behavioural therapy technique that involves repeated exposure to an anxiety-producing stimulus or situation.

expressed emotion A pattern of interactions that includes emotional over-involvement, critical comments, and hostility directed toward a patient by family members.

extinction A process in which the conditioned response is weakened when the conditioned stimulus is repeated without the unconditioned stimulus.

extrinsic motivation Motivation to perform an activity because of the value or pleasure associated with that activity, rather than for an apparent biological goal or purpose.

facial feedback hypothesis The idea that facial expressions trigger the experience of emotion.

family systems model A diagnostic model that considers symptoms within an individual as indicating problems within the family.

fight-or-flight response The physiological preparedness of animals to deal with danger.

filter theory A theory that people have a limited capacity for sensory information and thus screen incoming information, letting in only the most important.

five-factor theory The idea that personality can be described using five traits: openness to experience, conscientiousness, extraversion, agreeableness, and neuroticism.

fixed schedule A schedule in which reinforcement is consistently provided upon each occurrence.

flashbulb memories Vivid memories for the circumstances in which one first learned of a surprising, consequential, and emotionally arousing event.

fluid intelligence Information processing in novel or complex circumstances.

forgetting The inability to retrieve memory from long-term storage.

formal operational stage The final stage in Piaget's theory of cognitive development; it involves the ability to think abstractly and to formulate and test hypotheses through deductive logic.

fovea The centre of the retina where cones are densely packed.

frame-switching The shifting of thoughts and behaviours to those appropriate for a given cultural context.

framing The effect of presentation on how information is perceived.

frontal lobes The region at the front of the cerebral cortex concerned with planning and movement.

frustration-aggression hypothesis The extent to which people feel frustrated predicts the likelihood that they will act aggressively.

functional fixedness A tendency in problem solving to think of objects only as they are most commonly used.

functional magnetic resonance imaging (fMRI) An imaging technique used to examine changes in the activity of the working human brain.

functionalism An approach to psychology concerned with the adaptive purpose, or function, of mind and behaviour.

fundamental attribution error The tendency to overemphasize personal factors and underestimate situational forces in explaining behaviour.

GABA (gamma-aminobutyric acid) The primary inhibitory transmitter in the nervous system.

gender A term that refers to the culturally constructed differences between males and females.

gender identity Personal beliefs about whether one is male or female.

gender roles The characteristics associated with men and women because of cultural influence or learning.

gender schemas Cognitive structures that influence how people perceive the behaviours of men and women.

general adaptation syndrome A consistent pattern of responses to stress that consists of three stages: alarm, resistance, and exhaustion.

general intelligence (g) The idea that one general factor underlies all mental abilities.

generalized anxiety disorder A diffuse state of constant anxiety not associated with any specific object or event.

gene The unit of heredity that determines a particular characteristic in an organism.

genotype The genetic constitution determined at the moment of conception.

Gestalt theory A theory based on the idea that the whole of personal experience is different from simply the sum of its constituent elements.

glutamate The primary excitatory transmitter in the nervous system.

goal A desired outcome associated with some specific object of desire or some future behavioural intention.

gonads The main endocrine glands involved in sexual behaviour: in males, the testes; in females, the ovaries.

gray matter A segment of the spinal cord that is dominated by the cell bodies of neurons.

group polarization A process in which group members conform to the initial attitudes of other members who already agree.

guilt A negative emotional state associated with an internal experience of anxiety, tension, and agitation, in which a person feels responsible for causing an adverse state.

habituation A decrease in behavioural response following repeated exposure to nonthreatening stimuli.

hallucinations False sensory perceptions that are experienced without an external source.

haptic sense The sense of touch.

hardiness A personality trait that enables people to perceive stressors as controllable challenges.

health psychology The field of psychological science concerned with the events that affect physical well-being.

heritability A statistical estimate of the fraction of observed measure of the overall amount of difference among people in a population that is caused by differences in heredity.

heuristics In problem solving, shortcuts used to minimize the amount of thinking that must be done when moving from step to step in a solution space.

hippocampus A brain structure important for the formation of certain types of memory.

holistic thinking A system of evaluation in which the assessor views objects with regard to context and in terms of the relationships between them, accompanied by an expectation of behaviour in accordance with those relationships.

homeostasis The tendency for bodily functions to maintain equilibrium.

hormones Chemical substances, typically released from endocrine glands, that travel through the bloodstream to targeted tissues, which are subsequently influenced by the hormone.

humanistic approaches Approaches to studying personality that emphasize personal experience and belief systems, and propose that people seek personal growth to fulfill their human potential.

hypothalamus A small brain structure that is vital for temperature regulation, emotion, sexual behaviour, and motivation.

hypothesis A specific prediction of what should be observed in the world if a theory is correct.

id In psychodynamic theory, the component of personality that is completely submerged in the unconscious and operates according to the pleasure principle.

idiographic approaches Person-centered approaches to studying personality that focus on individual lives and how various characteristics are integrated into unique persons.

immune system The body's mechanism for dealing with invading microorganisms, such as allergens, bacteria, and viruses.

implicit attitudes Attitudes that influence our feelings and behaviour at an unconscious level.

implicit memory The process by which people show an enhancement of memory, most often through behaviour, without deliberate effort and without any awareness that they are remembering anything.

incentives External stimuli that motivate behaviours (as opposed to internal drives).

independent self-construals A view of the self as separate from others, emphasizing self-reliance and the pursuit of personal success.

independent variable In an experiment, the condition that is manipulated by the experimenter to examine its impact on the dependent variable.

inductive reasoning A form of reasoning in which we develop general rules after observing specific instances.

infantile amnesia The inability to remember events from early childhood.

inferential statistics A set of procedures used to make judgments about whether differences actually exist between sets of numbers.

informed consent A process in which people are given full information about a study, which allows them to make a knowledgeable decision about whether to participate.

ingroup favoritism The tendency for people to evaluate favorably and privilege members of the ingroup more than members of the outgroup.

insight (1) The sudden realization of a solution to a problem. (2) A goal of some types of therapy; a patient's understanding of his or her own psychological processes.

insomnia A sleep disorder characterized by an inability to sleep.

Institutional Review Boards (IRBs) Groups of people responsible for reviewing proposed research to ensure that it meets the accepted standards of science and provides for the physical and emotional well-being of research participants.

intelligence The human ability to use knowledge, solve problems, understand complex ideas, learn quickly, and adapt to environmental challenges.

intelligence quotient (IQ) The number computed by dividing a child's estimated mental age by the child's chronological age, and then multiplying this number by 100.

interactionists Theorists who believe that behaviour is jointly determined by underlying dispositions and situations.

interdependent self-construals Self-concepts that are determined largely by social roles and personal relationships.

interneurons One of the three types of neurons, these neurons communicate only with other neurons, typically within a specific brain region.

interval schedule A schedule in which reinforcement is based on a specific unit of time.

intracranial self-stimulation (ICSS) A procedure in which animals are able to self-administer electrical shock to specific areas of the brain.

intrinsic motivation Motivation to perform an activity because of the value or pleasure associated with that activity, rather than for an apparent biological goal or purpose.

introspection A systematic examination of subjective mental experiences that requires people to inspect and report on the content of their thoughts.

James-Lange theory of emotion A theory that suggests that the experience of emotion is elicited by a physiological response to a particular stimulus or situation.

joint attentional engagement A process whereby caregivers make reference to objects that are part of a child's ongoing actions.

kin selection The tendency to be altruistic toward those who share a genetic bond.

late selection theory A theory that people take in sensory information, process it, and then select which aspects of the stimuli should be attended after processing.

latent content What a dream symbolizes, or the material that is disguised in a dream to protect the dreamer.

latent learning Learning that takes place in the absence of reinforcement.

lateral geniculate nucleus (LGN) A region of the thalamus where visual information first travels, and then relays the information to the visual cortex.

lateral inhibition A visual process in which adjacent photoreceptors tend to inhibit one another.

law of effect Thorndike's general theory of learning, which states that any behaviour that leads to a "satisfying state of affairs" is more likely to occur again, and that those that lead to an "annoying state of affairs" are less likely to recur.

learned helplessness model A cognitive model of depression in which people feel unable to control events around them.

learning An enduring change in behaviour that results from experience.

lithium A psychotropic medication used to treat bipolar disorder.

longitudinal studies A research design that examines the same individuals over time.

long-term memory (LTM) The relatively permanent storage of information.

long-term potentiation (LTP) The strengthening of a synaptic connection so that postsynaptic neurons are more easily activated.

loosening of associations A speech pattern among schizophrenic patients in which their thoughts are disorganized or meaningless.

loss aversion An unequal weighting of costs and benefits such that potential costs weigh more heavily than potential benefits.

love marriage A type of marriage in which the bride and groom have selected each other as marriage partners.

lymphocytes Specialized white blood cells known as, B cells, T cells, and natural killer cells that make up the immune system.

magnetic resonance imaging (MRI) A method of brain imaging that produces high-quality images of the brain.

maintenance rehearsal A type of encoding that involves continually repeating an item.

major depression A disorder characterized by severe negative moods and a lack of interest in normally pleasurable activities.

manifest content The plot of a dream; the way a dream is remembered.

MAO inhibitors A category of antidepressant drugs that inhibit the action of monoamine oxidase.

mean A measure of central tendency that is the arithmetic average of a set of numbers.

median A measure of central tendency that is the value in a set of numbers that falls exactly in the halfway point between the lowest and highest values.

meme The knowledge transferred within a culture.

memory The capacity of the nervous system to acquire and retain usable skills and knowledge, allowing living organisms to benefit from experience.

mental age An assessment of a child's intellectual standing relative to that of his or her peers; determined by a comparison of the child's test score with the average score for children of each chronological age.

mental set A problem solving strategy that has worked in the past.

mesolimbic dopamine system The major brain system involved in reward, it connects the ventral tegmental area (VTA) to the nucleus accumbens.

methylphenidate A central-nervous-system stimulant medication used to treat ADHD.

microsleeps Brief, unintended sleep episodes, ranging from a few seconds to a minute, caused by chronic sleep deprivation.

mind-body problem A fundamental psychological issue that considers whether mind and body are separate and distinct or whether the mind is simply the subjective experience of the physical brain.

mirror neurons Neurons in the premotor cortex that are activated during observation of others performing an action.

modal memory model The three-stage memory system that involves sensory memory, short-term memory, and long-term memory.

mode A measure of central tendency that is the most frequent score or value in a set of numbers.

modeling The imitation of behaviour through observational learning.

monoamines A group of neurotransmitters synthesized from a single amino acid that are involved in a variety of psychological activities.

monocular depth cues Cues of depth perception that are available to each eye alone.

monozygotic twins Twin siblings who result from one zygote splitting in two, and therefore share the same genes (i.e., identical twins).

mood A diffuse and long-lasting emotional state that influences rather than interrupts thought and behaviour.

moral development Concerns the way in which people decide between behaviours with competing social outcomes.

morpheme The smallest unit of speech that has meaning.

motivation Factors that energize, direct, or sustain behaviour.

motor neurons One of the three types of neurons, these efferent neurons direct muscles to contract or relax, thereby producing movement.

multiaxial system The system used in the DSM that provides assessment along five axes describing important mental health factors.

multiple intelligences The idea that people can show different skills in a variety of different domains.

myelin sheath A fatty material, made up of glial cells, that insulates the axon and allows for the rapid movement of electrical impulses along the axon.

natural selection Darwin's theory that those who inherit characteristics that help them adapt to their particular environment have a selective advantage over those who do not.

naturalistic observation A passive descriptive study in which observers do not change or alter ongoing behaviour.

nature-nurture debate The arguments concerning whether psychological characteristics are biologically innate or acquired through education, experience, and culture.

need A state of biological or social deficiency within the body.

need hierarchy Maslow's arrangement of needs, in which basic survival needs are lowest and personal growth needs are highest in terms of priority.

need to belong theory The need for interpersonal attachments is a fundamental motive that has evolved for adaptive purposes.

negative feedback model The body's response to deviations from equilibrium.

negative punishment Punishment that occurs with removal of a stimulus that decreases the probability of a behaviour's recurring.

negative reinforcement The increase in the probability of a behaviour's being repeated through the removal of an aversive stimulus.

negative symptoms Symptoms of schizophrenia marked by deficits in functioning such as apathy, lack of emotion, and slowed speech and movement.

neuron The basic unit of the nervous system that operates through electrical impulses, which communicate with other neurons through chemical signals. Neurons receive, integrate, and transmit information in the nervous system.

neurotransmitter A chemical substance that carries signals from one neuron to another.

nodes of Ranvier Small gaps of exposed axon, between the segments of myelin sheath, where action potentials are transmitted.

nomothetic approaches Approaches to studying personality that focus on characteristics that are common to all people, although there is individual variation.

nonverbal behaviour The facial expressions, gestures, mannerisms, and movements by which one communicates with others.

norepinephrine A monoamine neurotransmitter involved in states of arousal and vigilance.

obedience The willingness to follow an order given by an authority.

object permanence The understanding that an object continues to exist even when it cannot be seen.

objective measures Relatively unbiased assessments of personality, usually based on information gathered through self-report questionnaires or observer ratings.

observational learning Learning that occurs when behaviours are acquired or modified following exposure to others performing the behaviour.

observational technique A research method of careful and systematic assessment and coding of overt behaviour.

observer bias Systematic errors in observation that occur due to an observer's expectations.

obsessive-compulsive disorder (OCD) An anxiety disorder characterized by frequent intrusive thoughts and compulsive actions.

occipital lobes A region of the cerebral cortex at the back of the brain that is important for vision.

olfactory bulb The brain centre for smell, located below the frontal lobes.

operant conditioning A learning process in which the consequences of an action determine the likelihood that it will be performed in the future.

operational definition The quantification of a variable that allows it to be measured.

orienting reflex The tendency for humans to pay more attention to novel stimuli.

ossicles Three tiny bones, the malleus (hammer), incus (anvil), and stapes (stirrup), in the middle ear that transfer the vibrations of the eardrum to the oval window.

outer ear The structure of the ear at which the sound wave arrives.

panic disorder An anxiety disorder characterized by sudden, overwhelming attacks of terror.

parallel play The type of play characteristic of two-year-olds, usually limited to sitting side by side and playing independently.

parasympathetic division of ANS A division of the autonomic nervous system that returns the body to its resting state.

parietal lobes A region of the cerebral cortex lying in front of the occipital lobes and behind the frontal lobes that is important for the sense of touch and the spatial layout of an environment.

Parkinson's disease A neurological disorder that seems to be caused by dopamine depletion, marked by muscular rigidity, tremors, and difficulty initiating voluntary action.

partial reinforcement A type of learning in which behaviour is reinforced intermittently.

partial-reinforcement extinction effect The greater persistence of behaviour under partial reinforcement than under continuous reinforcement.

participant observation A type of descriptive study in which the researcher is actively involved in the situation.

peptides Chains of two or more amino acids found in the brain and the body; they act like classic neurotransmitters or modify the quality of the neurotransmitter with which they are released.

perception The processing, organization, and interpretation of sensory signals that result in an internal representation of the stimulus.

perceptual constancy People correctly perceive objects as constant in their shape, size, colour, and lightness despite raw sensory data that could mislead perception.

peripheral nervous system (PNS) All nerve cells in the body that are not part of the central nervous system. The PNS includes the somatic and autonomic nervous systems.

personal attributions Explanations that refer to internal characteristics, such as abilities, traits, moods, and effort.

personality Characteristics, emotional responses, thoughts, and behaviours that are relatively stable over time and across circumstances.

personality disorder A class of mental disorders marked by inflexible and maladaptive ways of interacting with the world.

personality trait A characteristic; a dispositional tendency to act in a certain way over time and across circumstances.

personality types Discrete categories based on global personality characteristics.

persuasion The active and conscious effort to change attitudes through the transmission of a message.

phenotype Observable physical characteristics that result from both genetic and environmental influences.

pheromones Chemicals released by animals and humans that trigger physiological or behavioural reactions in other members of the same species.

phobia An irrational fear of a specific object or situation.

phonology The study of the set of meaningless sounds and the rules by which we combine them to make words and sentences.

phrenology An early method of assessing personality traits and mental abilities by measuring bumps on the skull.

physical dependence Synonymous with addiction, the physiological state in which failing to ingest a specific substance leads to bodily symptoms of withdrawal.

pituitary gland Located at the base of the hypothalamus, the gland that sends hormonal signals that control the release of hormones from endocrine glands.

place coding A mechanism for encoding high-frequency auditory stimuli in which the frequency of the sound wave is encoded by the location of the hair cells along the basilar membrane.

plasticity A property of the brain that allows it to change as a result of experience, drugs, or injury.

pop-out The phenomenon whereby, when simple stimuli were used, subjects take the same amount of time to find the target, whether there are a few or many distracters.

positive punishment Punishment that occurs with administration of a stimulus that decreases the probability of a behaviour's recurring.

positive reappraisal A cognitive process in which people focus on possible good things in their current situation.

positive reinforcement The increase in the probability of a behaviour's being repeated following the administration of a pleasurable stimulus, referred to as a reward.

positive symptoms Symptoms of schizophrenia, such as delusions and hallucinations, that are excesses in behaviour.

positron emission tomography (PET) A method of brain imaging that assesses metabolic activity by using a radioactive substance injected into the bloodstream.

Postconventional Level The third level of moral reasoning from Kohlberg's theory, in which an individual views his or her actions as right or wrong not based on the authority of social groups, but on the extent to which those actions are consistent with a set of universal ethical principles that emphasize justice and individual rights.

postdecisional dissonance The incongruity between one's decision and the undesirable aspects of that decision.

posttraumatic stress disorder (PTSD) A mental disorder that involves frequent nightmares, intrusive thoughts, and flashbacks related to an earlier trauma.

pragmatics The way people use language to get what they want and to influence their listeners.

Preconventional Level The first level of moral reasoning from Kohlberg's theory, in which an individual learns cultural rules about right and wrong, and interprets these rules in terms of the direct consequences of his or her actions.

prefrontal cortex A region of the frontal lobes, especially prominent in humans, important for attention, working memory, decision making, appropriate social behaviour, and personality.

prejudice The affective or attitudinal responses associated with stereotypes.

preoperational stage The second stage in Piaget's theory of cognitive development during which children think symbolically about objects, but reason is based on appearance rather than logic.

primacy effect In a list, the better memory for items presented first.

primary appraisal Part of the coping process that involves making decisions about whether a stimulus is stressful, benign, or irrelevant.

primary auditory cortex (A1) The region of the temporal lobe concerned with hearing.

primary control Influencing one's environment to achieve one's goals, desires, or wishes.

primary emotions Evolutionarily adaptive emotions that humans share across cultures; they are associated with specific biological and physical states.

primary motor cortex The region of the frontal lobe concerned with movement.

primary reinforcers Reinforcers that are innately reinforcing, such as those that satisfy biological needs.

primary visual cortex (V1) The largest area in the occipital lobe, where the thalamus projects the image.

proactive interference When prior information inhibits the ability to remember new information.

problem-focused coping A type of coping in which people take direct steps to confront or minimize a stressor.

procedural memory A type of implicit memory that involves motor skills and behavioural habits.

projective measures Personality tests that examine unconscious processes by having people interpret ambiguous stimuli.

prosocial Tending to benefit others.

prototype model An approach to object categorization that is based on the premise that within each category, some members are more representative than others.

psychoanalysis A method developed by Sigmund Freud that attempts to bring the contents of the unconscious into conscious awareness so that conflicts can be revealed.

psychodynamic theory Freudian theory that unconscious forces, such as wishes and motives, influence behaviour.

psychological dependence Habitual substance use, despite consequences, and a compulsive need to continue using the drug.

psychological practitioners Those who apply findings from psychological science in order to assist people in their daily lives.

psychological science The study of mind, brain, and behaviour.

psychological scientist One who uses the methods of science to study the interplay between brain, mind, and behaviour and how the social environment affects these processes.

psychomotor stimulants Drugs that activate the sympathetic nervous system and increase behavioural and mental activity.

psychopathology A disorder of the mind.

psychophysiological assessment A research method that examines how changes in bodily functions are associated with behaviour or mental state.

psychosexual stages According to Freud, the developmental stages that correspond to the pursuit of satisfaction of libidinal urges.

psychotherapy The generic name given to formal psychological treatment.

psychotropic medications Drugs that affect mental processes.

punishment A stimulus following a response that decreases the likelihood that the response will be repeated.

qualia The properties of our subjective, phenomenological awareness.

quantum change A transformation of personality that is sudden, profound, enduring, and affects a wide range of behaviours.

random assignment The procedure for placing research participants into the conditions of an experiment in which each participant has an equal chance of being assigned to any level of the independent variable.

ratio schedule A schedule in which reinforcement is based on the number of times the behaviour occurs.

reaction time A quantification of performance behaviour that measures the speed of a response.

reactivity The effect that occurs when the knowledge that one is being observed alters the behaviour being observed.

recency effect In a list, the better memory for words presented later in the list.

receptive field The region of visual space to which neurons in the primary visual cortex are sensitive.

receptors In neurons, specialized protein molecules on the postsynaptic membrane that neurotransmitters bind to after passing across the synaptic cleft.

recessive gene A gene that is expressed only when it is matched with a similar gene from the other parent.

reciprocal helping The tendency to help another because the recipient may return the favour.

reinforcer A stimulus following a response that increases the likelihood that the response will be repeated.

relative risk An important component of the ethical review in which the potential for possible harm to the participant is considered.

reliability The extent to which a measure is stable and consistent over time in similar conditions.

REM sleep The stage of sleep marked by rapid eye movements, dreaming, and paralysis of motor systems.

repetition priming The improvement in identifying or processing a stimulus that has previously been experienced.

replication Repetition of an experiment to confirm the results.

representativeness heuristic A rule for categorization based on how similar the person or object is to our prototypes for that category.

Rescorla-Wagner model A cognitive model of classical conditioning that states that the strength of the CS-US association is determined by the extent to which the unconditioned stimulus is unexpected.

research Scientific process that involves the systematic and careful collection of data.

response performance A research method in which researchers quantify perceptual or cognitive processes in response to a specific stimulus.

resting membrane potential The electrical charge of a neuron when it is not active.

restructuring A new way of thinking about a problem that aids its solution.

reticular formation A large network of neural tissue within the brainstem involved in behavioural arousal and sleep-wake cycles.

retina The thin inner surface of the back of the eyeball. The retina contains the photoreceptors that transduce light into neural signals.

retinotopic organization The systematic ordering of the neuronal pathway from the retina to the occipital lobe; this organization preserves spatial relationships, so that adjacent areas of the retina correspond to adjacent areas in the primary visual cortex.

retrieval The act of recalling or remembering stored information in order to use it.

retroactive interference When new information inhibits the ability to remember old information.

retrograde amnesia The condition in which people lose past memories, such as memories for events, facts, people, or even personal information.

reuptake The process whereby the neurotransmitter is taken back into the presynaptic terminal buttons, thereby stopping its activity.

rumination Thinking about, elaborating, and focusing on undesired thoughts or feelings, which prolongs, rather than alleviates, a negative mood.

schema A hypothetical cognitive structure that helps us perceive, organize, process, and use information.

schizophrenia A mental disorder characterized by alterations in perceptions, emotions, thoughts, or consciousness.

seasonal affective disorder (SAD) A disorder in which periods of depression occur during the times of year with less sunlight.

secondary appraisal Part of the coping process during which people evaluate their options and choose coping behaviours.

secondary control Psychologically aligning oneself with another in order to achieve a sense that one's goals, desires, or wishes are being fulfilled.

secondary emotions Blends of primary emotions, including states such as remorse, guilt, submission, and anticipation.

secondary reinforcers Events or objects that become reinforcers through their repeated pairings with primary reinforcers.

selective serotonin reuptake inhibitors (SSRIs) A category of antidepressant medications that prolong the effects of serotonin in the synapse.

self-actualization A state that is achieved when one's personal dreams and aspirations have been attained.

self-awareness A state in which the sense of self is the object of attention.

self-concept The full store of knowledge that people have about themselves.

self-efficacy The expectancy that one's efforts will lead to success.

self-esteem The evaluative aspect of the self-concept.

self-fulfilling prophecy The observation that people may come to behave in ways that confirm their own or others' expectations.

self-regulation The process by which people initiate, adjust, or stop actions in order to promote the attainment of personal goals or plans.

self-report method A method of data collection in which people are asked to provide information about themselves, such as in questionnaires or surveys.

self-schema The cognitive aspect of the self-concept, consisting of an integrated set of memories, beliefs, and generalizations about the self.

self-serving bias The tendency for people to take personal credit for success but blame failure on external factors.

semantic memory Memory for knowledge about the world.

sensation How sense organs respond to external stimuli and transmit the responses to the brain.

sensitization An increase in behavioural response following exposure to a threatening stimulus.

sensorimotor stage The first stage in Piaget's theory of cognitive development during which infants acquire information about the world through their senses and respond reflexively.

sensory adaptation When an observer's sensitivity to stimuli decreases over time.

sensory memory Memory for sensory information that is stored briefly in its original sensory form.

sensory neurons One of the three types of neurons, these afferent neurons detect information from the physical world and pass that information along to the brain.

serial position effect The ability to recall items from a list depends on order of presentation, with items presented early or late in the list remembered better than those in the middle.

serotonin A monoamine neurotransmitter important for a wide range of psychological activity, including emotional states, impulse control, and dreaming.

sexual response cycle A pattern of physiological responses during sexual activity.

sexual strategies theory Evolutionary theory that suggests men and women look for different qualities in relationship partners due to the gender-specific adaptive problems they've faced throughout human history.

shaping A process of operant conditioning that involves reinforcing behaviours that are increasingly similar to the desired behaviour.

short-term memory (STM) A limited-capacity memory system that holds information in awareness for a brief period of time.

situational attributions Explanations that refer to external events, such as the weather, luck, accidents, or the actions of other people.

situationism The theory that behaviour is determined as much by situations as by personality traits.

social comparison The evaluation of one's own actions, abilities, and beliefs by contrasting them with those of others.

social development The maturation of skills or abilities that enable people to live in a world with other people.

social dilemma A situation in which we feel a motivational conflict between cooperating and being selfish.

social facilitation A situation in which the mere presence of others enhances performance.

social loafing The tendency for people to work less hard in a group than when working alone.

social norms Expected standards of conduct, which influence behaviour.

social psychology The branch of psychology concerned with how others influence the way a person thinks, feels, and acts.

social striving A phenomenon in which people exert more effort to achieve a goal when they are performing in a group than when they are alone.

social support A network of other people who can provide help, encouragement, and advice.

socially desirable responding When people respond to a question in a way that is most socially acceptable or that makes them look good.

social-skills training Treatment designed to teach and reinforce appropriate interpersonal behaviour.

sociocultural model A diagnostic model that views psychopathology as the result of the interaction between individuals and their cultures.

sociometer An internal monitor of social acceptance or rejection.

sodium-potassium pump A mechanism of the neuron that keeps the resting membrane potential at a constant −70 mV, which sets the stage for electrical action.

somatic markers Bodily reactions that arise from the emotional evaluation of an action's consequences.

somatic nervous system A major component of the peripheral nervous system, which transmits sensory signals to the CNS via nerves.

somatic symptoms Side-effects of psychological disorders experienced as physical manifestations.

sound wave The pattern of the changes in air pressure through time that results in the percept of a sound.

source amnesia A type of amnesia that occurs when a person shows memory for an event but cannot remember where he or she encountered the information.

source misattributions Memory distortions that occur when people misremember the time, place, person, or circumstances involved with a memory.

spatial memory Memory for the physical environment and includes such things as location of objects, direction, and cognitive maps.

spinal cord Part of the central nervous system. A rope of neural tissue that runs inside the hollows of the vertebrae from just above the pelvis and into the base of the skull.

split brain A condition in which the corpus callosum is surgically cut and the two hemispheres of the brain do not receive information directly from each other.

spontaneous recovery A process in which a previously extinguished response reemerges following presentation of the conditioned stimulus.

standard deviation A statistical measure of how far away each value is on average from the mean.

stereotypes Cognitive schemas that allow for easy and efficient organization of information about people based on their membership in certain groups.

stimulus discrimination A learned tendency to differentiate between two similar stimuli if one is consistently associated with the unconditioned stimulus and the other is not.

stimulus generalization Occurs when stimuli that are similar but not identical to the conditioned stimulus produce the conditioned response.

storage The retention of encoded representations over time that corresponds to some change in the nervous system that registers the event.

stream of consciousness A phrase coined by William James to describe one's continuous series of ever-changing thoughts.

stress A pattern of behavioural and physiological responses to events that match or exceed an organism's abilities.

stressor An environmental event or stimulus that threatens an organism.

structuralism An approach to psychology based on the idea that conscious experience can be broken down into its basic underlying components or elements.

Substance P A peptide that acts as a neurotransmitter and is involved in pain perception.

subtractive colour mixing A way to produce a given spectral pattern in which the mixture occurs within the stimulus itself and is actually a physical, not psychological, process.

suggestibility The development of biased memories when people are provided with misleading information.

superego In psychodynamic theory, the internalization of societal and parental standards of conduct.

symbolic representation An abstract mental representation that does not correspond to the physical features of an object or idea.

sympathetic division of ANS A division of the autonomic nervous system that prepares the body for action.

synapse The site for chemical communication between neurons.

synaptic cleft The small space between neurons that contains extracellular fluid.

synaptic pruning A process whereby the synaptic connections in the brain that are frequently used are preserved, and those that are not are lost.

systematic desensitization An exposure technique that pairs the anxiety-producing stimulus with relaxation techniques.

***taijinkyoufushou* (TKS)** A disorder of which victims are morbidly afraid of offending others or causing others discomfort from likely imagined physical flaws.

tardive dyskinesia A side effect of some antipsychotic medications that produces involuntary movements of the lips, tongue, face, legs, or other parts of the body.

taste buds Sensory receptors that transduce taste information.

TAT (Thematic Apperception Test) A projective measure of personality where a person is shown an ambiguous picture and asked to tell a story about the picture.

taxonomic categorization A system of grouping stimuli based on perceived similarities of attributes.

telegraphic speech The tendency for children to speak using rudimentary sentences that are missing words and grammatical markings but follow a logical syntax.

temperaments Biologically based tendencies to feel or act in certain ways.

temporal coding A mechanism for encoding the low-frequency auditory stimuli in which the frequency of the sound wave is encoded by the frequency of firing of the hair cells.

temporal lobes The lower region of the cerebral cortex that is important for processing auditory information and also for memory.

tend-and-befriend response The argument that females are more likely to protect and care for their offspring and form social alliances than flee or fight in response to threat.

teratogens Environmental agents that harm the embryo or fetus.

terminal buttons Small nodules at the ends of axons, that release chemical signals from the neuron to an area called the synapse.

thalamus The gateway to the brain that receives almost all incoming sensory information before it reaches the cortex.

thematic categorization A system of grouping stimuli based on perceived relationships among them.

theory A model of interconnected ideas and concepts that explains what is observed and makes predictions about future events.

theory of mind The term used to describe the ability to explain and predict other people's behaviour as a result of recognizing their mental state.

theory of multiple intelligences A theory that attempts to provide practical definitions of intelligence, including musical, verbal, mathematical/logical, spatial, kinesthetic (or body control), intrapersonal (or self-understanding), and interpersonal (or social understanding).

third-variable problem When the experimenter cannot directly manipulate the independent variable and therefore cannot be confident that another, unmeasured variable is not the actual cause of differences in the dependent variable.

tip-of-the-tongue phenomenon When people experience great frustration as they try to recall specific words that are somewhat obscure.

top-down processing A hierarchical model of pattern recognition in which information at higher levels of processing can also influence lower, "earlier" levels in the processing hierarchy.

TOTE model A model of self-regulation in which people evaluate progress in achieving goals.

trait approach An approach to studying personality that focuses on the extent to which individuals differ in personality dispositions.

transcranial magnetic stimulation (TMS) A procedure that transmits pulses of high-intensity magnetism to the brain.

transduction A process by which sensory receptors produce neural impulses when they receive physical or chemical stimulation.

transience The pattern of forgetting over time.

triangular theory of love Proposes that love is made up of differing combinations of passion, intimacy, and commitment.

tricyclic antidepressants A category of antidepressant medications that inhibit the reuptake of a number of different neurotransmitters.

two-factor theory of emotion A theory that a situation evokes both a physiological response, such as arousal, and a cognitive interpretation.

Type A behaviour pattern A pattern of behaviour characterized by competitiveness, achievement orientation, aggressiveness, hostility, restlessness, inability to relax, and impatience with others.

Type B behaviour pattern A pattern of behaviour characterized by relaxed, noncompetitive, easygoing, and accommodative behaviour.

unconditioned response (UR) A response that does not have to be learned, such as a reflex.

unconditioned stimulus (US) A stimulus that elicits a response, such as a reflex, without any prior learning.

unconscious A term that identifies mental processes that operate below the level of conscious awareness.

U-shaped curve A pattern of acculturation, characterized by three phases: the "honeymoon stage," "crisis" or "culture shock," and "adjustment."

validity The extent to which the data collected address the research hypothesis in the way intended.

variability In a set of numbers, how widely dispersed the values are from each other and from the mean.

variable Something in the world that can be measured and that can vary.

variable schedule A schedule in which reinforcement is applied at different rates or at different times.

vicarious learning Learning that occurs when people learn the consequences of an action by observing others being rewarded or punished for performing the action.

visual search task An experiment used to study form perception, in which an observer tries to detect a target stimulus among an array of distracter stimuli.

white matter A segment of the spinal cord that consists mostly of axons and the fatty sheaths that surround them.

Whorfian hypothesis The controversial theory that the language one speaks influences the way one thinks.

working memory An active processing system that keeps different types of information available for current use.

Yerkes-Dodson law A psychological principle that dictates that behavioural efficiency increases with arousal up to an optimum point, after which it decreases with increasing arousal.

REFERENCES

Abramson, L. Y., Metalsky, G., & Alloy, L. (1989). Hopelessness depression: A theory-based subtype of depression. *Psychological Review, 96*, 358–372.

Ackerman, P. L., Beier, M. E., & Boyle, M. O. (2005). Working memory and intelligence: The same or different constructs? *Psychological Bulletin, 131*, 30–60.

Adair, J. G. (1984). The Hawthorne effect: A reconsideration of the methodological artifact. *Journal of Applied Psychology, 69*, 334–345.

Adair, J. G., Paivio, A., & Ritchie, P. (1966). Psychology in Canada. *Annual Review of Psychology, 47*, 341–370.

Adamec, R., & Shallow, T. (2000). Effects of baseline anxiety on response to kindling of the right medial amygdala. *Physiology & Behavior, 70*, 67–80.

Adams, C. E., Fenton, M., Quraishi, S., & David, A. S. (2001). Systematic meta-review of depot antipsychotic drugs for people with schizophrenia. *British Journal of Psychiatry, 179*, 290–299.

Adams, G. (2005). The cultural grounding of personal relationships: Enemyship in West African worlds. *Journal of Personality and Social Psychology, 88,* 948–968.

Adams, G., & Plaut, V. C. (2003). The cultural grounding of personal relationships: Friendship in North American and West African worlds. *Personal Relationships, 10,* 333–348.

Adams, H. E., Wright, L. W., & Lohr, B. A. (1996). Is homophobia associated with homosexual arousal? *Journal of Abnormal Psychology, 105*, 440–445.

Adams, R. B., Jr., Gordon, H. L., Baird, A. A., Ambady, N., & Kleck, R. E. (2003). Effects of gaze on amygdala sensitivity to anger and fear faces. *Science, 300,* 1536–1537.

Adolphs, R. (2003). Cognitive neuroscience of human social behavior. *Nature Reviews Neuroscience, 1,* 165–178.

Adolphs, R., Gosselin, F., Buchanan, T. W., Tranel, D., Schyns, P., & Damasio, A.R. (2005). A mechanism for impaired fear recognition after amygdala damage. *Nature, 433,* 68–72.

Adolphs, R., Sears, L., & Piven, J. (2001). Abnormal processing of social information from faces in autism. *Journal of Cognitive Neuroscience, 13*, 232–240.

Adolphs, R., Tranel, D., & Damasio, A. R. (1998). The human amygdala in social judgement. *Nature, 393*, 470–474.

Ahrens, B., Grof, P., Moller, H-J., Muller-Oerlinghaussen, B., & Wolf, T. (1995). Extended survival of patients on long-term lithium treatment. *Canadian Journal of Psychiatry, 40*, 241–246.

Alicke, M. D., Klotz, M. L., Breitenbecher, D. L., Yurak, T. J., & Vredenburg, D. S. (1995). Personal contact, individuation, and the better-than-average effect. *Journal of Personality and Social Psychology, 68*, 804–825.

Allport, G. W. (1961). *Pattern and growth in personality*. New York: Holt, Rinehart & Winston.

Amato, P. R. (2001). Children of divorce in the 1990s: An update of the Amato and Keith (1991) meta-analysis. *Journal of Family Psychology, 15*, 355–370.

Amato, P. R., Johnson, D., Booth, A., & Rogers, S. J. (2003). Continuity and change in marital quality between 1980 and 2000. *Journal of Marriage and Family, 65*, 1–22.

Amato, P. R., & Keith, B. (1991). Parental divorce and the well-being of children: A meta-analysis. *Psychological Bulletin, 110*, 26–46.

Ambady, N., Hallahan, M., & Conner, B. (1999). Accuracy of judgments of sexual orientation from thin slices of behavior. *Journal of Personality and Social Psychology, 77*, 538–547.

Ambady, N., & Rosenthal, R. (1993). Half a minute: Predicting teacher evaluations from thin slices of nonverbal behavior and physical attractiveness. *Journal of Personality and Social Psychology, 64*, 431–441.

American Psychiatric Association. (1994). *Diagnostic and statistical manual of mental disorders* (4th ed.). Washington, DC: Author.

American Psychiatric Association. (2000). Practice guidelines for the treatment of patients with eating disorders (revised). *American Journal of Psychiatry, 157* (Suppl.), 1–39.

Amminger, G. P., Pape, S., Rock, D., Roberts, S. A., Ott, S. L., Squires-Wheeler, E., et al. (1999). Relationship between childhood behavioral disturbance and later schizophrenia in the New York High-Risk Project. *American Journal of Psychiatry, 156*, 525–530.

Anderson, A. K., Christoff, K., Stappen, I., Panitz, D., Ghahremani, D. G., Glover, G., et al. (2003). Dissociated neural representations of intensity and valence in human olfaction. *Nature Neuroscience, 6*, 196–202.

Anderson, A. K., & Phelps, E. A. (2000). Expression without recognition: Contributions of the human amygdala to emotional communication. *Psychological Science, 11*, 106–111.

Anderson, A. K., & Phelps, E. A. (2001). Lesions of the amygdala impair enhanced perception of emotionally salient events. *Nature, 411*, 305–309.

Anderson, E. (1999). *Code of the street: Decency, violence, and the moral life of the inner city*. New York: Norton.

Anderson, I. M. (2000). Selective serotonin reuptake inhibitors versus tricyclic antidepressants: A meta-analysis of efficacy and tolerability. *Journal of Affective Disorders, 58*, 19–36.

Anderson, S. W., Bechara, A., Damasio, H., Tranel, D., & Damasio, A. R. (1999). Impairment of social and moral behavior related to early damage in human prefrontal cortex. *Nature Neuroscience, 2*, 1032–1037.

Andreasen, N. C. (1984). *The broken brain: The biological revolution in psychiatry*. New York: Harper & Row.

Anfang, S. A., & Appelbaum, P. S. (1996). Twenty years after Tarasoff: Reviewing the duty to protect. *Harvard Review of Psychiatry, 4* (2), 67–76.

Angold, A., & Costello, E. J. (2001). The epidemiology of depression in children and adolescents. In I. M. Goodyer (Ed.), *The depressed child and adolescent* (2nd ed., pp. 143–178). New York: Cambridge University Press.

Anisman, H., & Matheson, K. (2005). Stress, depression and anhedonia: Caveats concerning animal models. *Neuroscience and Biobehavioral Reviews, 29*, 525–546.

Anisman, H., & McIntyre, D. C. (2002). Conceptual and cue learning in the Morris water maze in Fast and Slow kindling rats: Attention deficit comorbidity. *Journal of Neuroscience, 22*, 7809–7817.

Anisman, H., & Merali, Z. (2003). Cytokines, stress and depressive illness: Brain-immune interactions. *Annals of Medicine, 35*, 2–11.

Annau, Z., & Kamin, L. J. (1961). The conditioned emotional response as a function of intensity of the UCS. *Journal of Comparative and Physiological Psychology, 54*, 428–432.

Antonuccio, D., & Burns, D. (2004). Adolescents with depression (Letter to the editor). *Journal of the American Medical Association, 292*, 2577.

Aoki, K. (1991). Some theoretical aspects of the origin of cultural transmission. In S. Osaura & T. Honjo (Eds.), *Evolution of human life: Fossils, molecules, and culture*. New York: Springer-Verlag.

Areni, C. S., & Kim, D. (1993). The influence of background music on shopping behavior: Classical versus top-forty music in a wine store. *Advances in Consumer Research, 20*, 336–340.

Arias, E., MacDorman, M. F., Strobino, D. M., & Guyer, B. (2003). Annual summary of vital statistics—2002. *Pediatrics, 112*, 1215–1230.

Armes, K., & Ward, C. (1989). Cross-cultural transitions and sojourner adjustment in Singapore. *Journal of Social Psychology, 12*, 273–275.

Aronson, E. (2003). *The social animal* (9th ed.). New York: Freeman.

Aronson, E., & Mills, J. (1959). The effects of severity of initiation on liking for a group. *Journal of Abnormal and Social Psychology, 59*, 177–181.

Asberg, M., Shalling, D., Traskman-Bendz, L., & Wagner, A. (1987). Psychobiology of suicide, impulsivity, and related phenomena. In H. Y. Melzer (Ed.), *Psychopharmacology: The third generation of progress* (pp. 655–668). New York: Raven Press.

Asch, S. E. (1955). Opinions and social pressure. *Scientific American, 193*, 31–35 (offprint 450).

Asch, S. (1956). Studies of independence and conformity. A minority of one against a unanimous majority. *Psychological Monographs, 70* (9, Whole No. 416).

Ashby, F. G., Isen, A. M., & Turken, A. U. (1999). A neuropsychological theory of positive affect and its influence on cognition. *Psychological Review, 106*, 529–550.

Averill, J. R. (1980). A constructivist view of emotion. In R. Plutchik & H. Kellerman (Eds.), *Theories of emotion* (pp. 305–339). New York: Academic Press.

Aylward, E. H., Reiss, A. L., Reader, M. J., & Singer, H. S. (1996). Basal ganglia volumes in children with attention-deficit hyperactivity disorder. *Journal of Child Neurology, 11*, 112–115.

Baddeley, A. D. (2001). Is working memory still working? *American Psychologist, 56*, 851–864.

Bagby, R. M., Levitan, R. D., Kennedy, S. H., Levitt, A. J., & Joffe, R. T. (1999). Selective alteration of personality in response to noradrenergic and serotonergic antidepressant medication in a depressed sample: Evidence of non-specificity. *Psychiatry Research, 86*, 211–216.

Bagby, R. M., Marshall, M. B., Basso, M. R., Nicholson, R. A., Bacchiochi, J., & Miller, L. S. (2005). Distinguishing bipolar depression, major depression, and schizophrenia with the MMPI-2 clinical and content scales. *Journal of Personality Assessment, 84*, 89–95.

Bailey, A., Le Couteur, A., Gottesman, I., Bolton, P., Simonoff, E., Yuzda, E., et al. (1995). Autism as a strongly genetic disorder: Evidence from a British twin study. *Psychological Medicine, 25*, 63–78.

Baillargeon, R. (1995). Physical reasoning in infancy. In M. S. Gazzaniga (Ed.), *The cognitive neurosciences* (pp. 181–204). Cambridge, MA: MIT Press.

Bain, J. (1987). Hormones and sexual aggression in the male. *Integrative Psychiatry, 5*, 82–89.

Baldwin, D. A. (1991). Infants' contribution to the achievement of joint reference. *Child Development, 62*, 875–890.

Baldwin, D. A., & Baird, J. A. (2001). Discerning intentions in dynamic human action. *Trends in Cognitive Sciences, 5*, 171–178.

Baltimore, D. (2001). Our genome unveiled. *Nature, 409*, 814–816.

Banaji, M. R., & Greenwald, A. G. (1995). Implicit gender stereotyping in judgments of fame. *Journal of Personality and Social Psychology, 68*, 181–198.

Bandura, A. (1977). *Social learning theory*. Englewood Cliffs, NJ: Prentice-Hall.

Bandura, A., Ross, D., & Ross, S. (1961). Transmission of aggression through imitation of aggressive models. *Journal of Abnormal and Social Psychology, 66*, 3–11.

Bandura, A., Ross, D., & Ross, S. (1963). Vicarious reinforcement and imitative learning. *Journal of Abnormal and Social Psychology, 67*, 601–607.

Barbopoulos, A., Fisharah, F., Clark, J. M., & el-Khatib, A. (2002). Comparison of Egyptian and Canadian children on a picture apperception test. *Cultural Diversity and Ethnic Minority Psychology, 8*, 395–403.

Barch, D. M., Sheline, Y. I., Csernansky, J. G., & Snyder, A. Z. (2003). Working memory and prefrontal cortex dysfunction: Specificity to schizophrenia compared with major depression. *Biological Psychiatry, 53*, 376–384.

Bargh, J. A., Chen, M., & Burrows, L. (1996). Automaticity of social behavior: Direct effects of trait construct and stereotype activation on action. *Journal of Personality and Social Psychology, 71,* 230–244.

Bargh, J. A., & Ferguson, M. J. (2000). Beyond behaviorism: On the automaticity of higher mental processes. *Psychological Bulletin, 126,* 925–945.

Barkley, R. A. (1997). *ADHD and the nature of self-control.* New York: Guilford Press.

Barlow, D. H. (1988). *Anxiety and its disorders: The nature and treatment of anxiety and panic.* New York: Guilford Press.

Barlow, D. H. (2002). *Anxiety and its disorders: The nature and treatment of anxiety and panic* (2nd ed.). New York: Guilford Press.

Barlow, D. H. (2004). Psychological treatments. *American Psychologist, 59,* 869–878.

Barlow, D. H, Gorman, J. M., Shear, M. K., & Woods, S. W. (2000). Cognitive-behavioral therapy, imipramine, or their combination for panic disorder: A randomized controlled trial. *Journal of the American Medical Association, 283,* 2529–2536.

Baron-Cohen, S., Lutchmaya, S., & Knickmeyer, R. (2004). *Prenatal testosterone in mind: Amniotic fluid studies.* Cambridge, MA: MIT Press.

Baron-Cohen, S., Wheelwright, S., & Jolliffe, T. (1997). Is there a "language of the eyes"? Evidence from normal adults and adults with autism or Asperger Syndrome. *Visual Cognition, 4,* 311–332.

Bartoshuk, L. M. (2000). Comparing sensory experiences across individuals: Recent psychophysical advances illuminate genetic variation in taste perception. *Chemical Senses, 25,* 447–460.

Basbaum, A. I., & Fields, H. L. (1984). Endogenous pain control systems: Brainstem spinal pathways and endorphin circuitry. *Annual Review of Neuroscience, 7,* 309–338.

Batson, C. D., Dyck, J. L., Brandt, J. R., Batson, J. G., Powell, A. L., McMaster, M. R., et al. (1988). Five studies testing two new egoistic alternatives to the empathy-altruism hypothesis. *Journal of Personality and Social Psychology, 55,* 52–77.

Batson, C. D., Turk, C. L., Shaw, L.L., & Klein, T. (1995). Information function of empathic emotion: Learning that we value the other's welfare. *Journal of Personality and Social Psychology, 68,* 300–313.

Baumeister, R. F. (1991). *Escaping the self: Alcoholism, spirituality, masochism, and other flights from the burden of selfhood.* New York: Basic Books.

Baumeister, R. F. (2000). Gender differences in erotic plasticity: The female sex drive as socially flexible and responsive. *Psychological Bulletin, 126,* 347–374.

Baumeister, R. F., Campbell, J. D., Krueger, J. I., & Vohs, K. D. (2003). Does high self-esteem cause better performance, interpersonal success, happiness, or healthier lifestyles? *Psychological Science in the Public Interest, 4,* 1–44.

Baumeister, R. F., Campbell, J. D., Krueger, J. I., & Vohs, K. D. (2005). Exploding the self-esteem myth. *Scientific American, 292,* 84–91.

Baumeister, R. F., Catanese, K. R., & Vohs, K. D. (2001). Is there a gender difference in strength of sex drive? Theoretical views, conceptual distinctions, and a review of the relevant literature. *Personality and Social Psychology Review, 5,* 242–273.

Baumeister, R. F., Dale, K., & Sommers, K. L. (1998). Freudian defense mechanisms and empirical findings in modern social psychology: Reaction formation, projection, displacement, undoing, isolation, sublimation, and denial. *Journal of Personality, 66,* 1081–1124.

Baumeister, R. F., & Heatherton, T. F. (1996). Self-regulation failure: An overview. *Psychological Inquiry, 7,* 1–15.

Baumeister, R. F., & Leary, M. R. (1995). The need to belong: Desire for interpersonal attachments as a fundamental human motivation. *Psychological Bulletin, 117,* 497–529.

Baumeister, R. F., Smart, L., & Boden, J. M. (1996). Relation of threatened egotism to violence and aggression: The dark side of high self-esteem. *Psychological Review, 103,* 5–33.

Baumeister, R. F., Stillwell, A. M., & Heatherton, T. F. (1994). Guilt: An interpersonal approach. *Psychological Bulletin, 115,* 243–267.

Baumeister, R. F., & Tice, D. M. (1990). Anxiety and social exclusion. *Journal of Social and Clinical Psychology, 9,* 165–195.

Baumrind, D., Larzelere, R. E., & Cowan, P. A. (2002). Ordinary physical punishment: Is it harmful? Comment on Gershoff (2002). *Psychological Bulletin, 128,* 580–589.

Baver, J. J., McAdams, D. P., & Sakaeda, A. R. (2005). Interpreting the good life: Growth memories in the lives of mature, happy people. *Journal of Personality and Social Psychology, 88,* 203–217.

Baxter, L. R. (2000). Functional imaging of brain systems mediating obsessive-compulsive disorder. In D. S. Charney, E. J. Nestler, & B. S. Bunney (Eds.), *Neurobiology of mental illness* (pp. 534–547). New York: Oxford University Press.

Baxter, L. R., Schwartz, J. M., Bergman, K. S., Szuba, M. P., Guze, B., Mazziota, J. C., et al. (1992). Caudate glucose metabolic rate changes with both drug and behavior therapy for obsessive-compulsive disorder. *Archives of General Psychiatry, 49,* 681–689.

Beck, A. T. (1967). *Depression: Clinical, experimental and theoretical aspects.* New York: Harper & Row.

Beck, A. T. (1976). *Cognitive therapy and the emotional disorders.* New York: International Universities Press.

Beck, A. T., Brown, G., Seer, R. A., Eidelson, J. L., & Riskind, J. H. (1987). Differentiating anxiety and depression: A test of the cognitive content-specificity hypothesis. *Journal of Abnormal Psychology, 96,* 179–183.

Beck, A. T., Freeman, A., & Associates (1990). *Cognitive Therapy of Personality Disorders.* New York: Guilford Press.

Beck, A. T., & Rector, N. A. (2005). Cognitive approaches to schizophrenia: Theory and therapy. *Annual Review of Clinical Psychology, 1,* 577–606.

Beck, A. T., Rush, A. J., Shaw, B., & Emery, G. (1979). *Cognitive therapy of depression.* New York: Guilford Press.

Beck, A. T., Steer, R. A., & Brown, G. K. (1996). *Beck depression inventory manual.* San Antonio, TX: The Psychological Corporation.

Beck, A. T., Ward, C. H., Mendelson, M., Mock, J., & Erbaugh, J. (1961). An inventory for measuring depression. *Archives of General Psychiatry, 4,* 561–571.

Beggan, J. K. (1992). On the social nature of nonsocial perception: The mere ownership effect. *Journal of Personality and Social Psychology, 62,* 229–237.

Beggs, J. M., Brown, T. H., Byrne, J. H., Crow, T., LeDoux, J. E., LeBar, K., et al. (1999). Learning and memory: Basic mechanisms. In M. J. Zigmond, F. E. Bloom, S. C. Landis, J. L.

Roberts, & L. R. Squire (Eds.), *Fundamentals of neuroscience* (pp. 1411–1454). San Diego, CA: Academic Press.

Behne, T., Carpenter, M., Call, J., & Tomasello, M. (2005). Unwilling versus unable: Infants' understanding of intentional action. *Developmental Psychology, 41,* 328–337.

Békésy, G. Von. (1957). The ear. *Scientific American, 197,* 66–78.

Bell, R. M. (1985). *Holy anorexia.* Chicago: University of Chicago Press.

Bellak, L., & Black, R. B. (1992). Attention-deficit hyperactivity disorder in adults. *Clinical Therapeutics, 14,* 138–147.

Belsky, J. (1990). Children and marriage. In F. D. Fincham & T. N. Bradbury (Eds.), *The psychology of marriage: Basic issues and applications* (pp. 172–200). New York: Guilford.

Bem, D. (1967). Self-perception: An alternative explanation of cognitive dissonance phenomena. *Psychological Review, 74,* 183–200.

Bem, D. J. (1996). Exotic becomes erotic: A developmental theory of sexual orientation. *Psychological Review, 103,* 320–335.

Bem, D. J., & Honorton, C. (1994). Does psi exist? Replicable evidence for an anomalous process of information transfer. *Psychological Bulletin, 115,* 4–18.

Bem, S. L. (1975). Sex role adaptability: One consequence of psychological androgyny. *Journal of Personality and Social Psychology, 31,* 634–643.

Bem, S. L., Martyna, W., & Watson, C. (1976). Sex typing and androgyny: Further explorations of the expressive domain. *Journal of Personality and Social Psychology, 34,* 1016–1023.

Benedetti, F., Serretti, A., Colombo, C., Campori, E., Barbini, B., di Bella, D., et al. (1999). Influence of a functional polymorphism within the promoter of the serotonin transporter gene on the effects of total sleep deprivation in bipolar depression. *American Journal of Psychiatry, 156,* 1450–1452.

Benjamin, L. T. (2005). A history of clinical psychology as a profession in America (and a glimpse at its future). *Annual Review of Clinical Psychology, 1,* 1–30.

Bentler, P. M., & Newcomb, M. D. (1978). Longitudinal study of marital success and failure. *Journal of Consulting and Clinical Psychology, 46,* 1053–1070.

Berger, K. S. (2004). *The developing person through the life span* (6th ed.). New York: Worth.

Berkman, L. F. (1984). Assessing the physical health effects of social networks and social support. *Annual Review of Public Health, 5,* 413–432.

Berkowitz, L. (1990). On the formation and regulation of anger and aggression: A cognitive-neoassociationistic analysis. *American Psychologist, 45,* 494–503.

Berlin, B., & Kay, P. (1969). *Basic color terms: Their universality and evolution.* Berkeley and Los Angeles: University of California Press.

Bernasconi, N., Natsume, J., & Bernasconi, A. (2005). Progression in temporal lobe epilepsy: Differential atrophy in mesial temporal structures. *Neurology, 65,* 223–228.

Bernhardt, P. C., Dabbs, J. M., Fielden, J. A., & Lutter, C. D. (1998). Testosterone changes during vicarious experiences of winning and losing among fans at sporting events. *Physiology & Behavior,* 65, 59–62.

Bernheim, K. F. (Ed.). (1997). *The Lanahan cases and readings in abnormal behavior.* Baltimore: Lanahan Publishers.

Berns, G. S., Chappelow, J., Zink, C. F., Pagnoni, G., Martin-Skurski, M. E., & Richards, J. (2005). Neurobiological correlates of social conformity and interdependence during mental rotation. *Biological Psychiatry, 58,* 245–253.

Berry, J. W., & Annis, R. C. (1974). Acculturation stress: The role of ecology, culture, and differentiation. *Journal of Cross-Cultural Psychology, 5,* 382–406.

Berscheid, E., & Regan, P. (2005). *The psychology of interpersonal relationships.* New York: Prentice-Hall.

Bezchlibnyk-Butler, K. Z., Aleksic, I., & Kennedy, S. H. (2000). Citalopram—a review of pharmacological and clinical effects. *Journal of Psychiatry and Neuroscience, 25,* 241–254.

Bickel, W. K., Odum, A. L., & Madden, G. J. (1999). Impulsivity and cigarette smoking: Delay discounting in current, never, and ex-smokers. *Psychopharmacology, 146,* 447–454.

Bickerton, D. (1998). The creation and re-creation of language. In C. B. Crawford & D. L. Krebs (Eds.), *Handbook of evolutionary psychology: Ideas, issues, and applications* (pp. 613–634). Mahwah, NJ: Lawrence Erlbaum Associates, Inc.

Bidell, T. R., & Fischer, K. W. (1995). Between nature and nurture: The role of agency in the epigenesis of intelligence. In R. Sternberg & E. Grigorenko (Eds.), *Intelligence: Heredity and environment.* New York: Cambridge University Press.

Biederman, I. (1987). Recognition-by-components: A theory of human image understanding. *Psychological Review, 94,* 115–147.

Biederman, J., Faraone, S. V., Keenan, K., Knee, D., & Tsuang, M. T. (1990). Family-genetic and psychosocial risk factors, in DSM-III attention deficit disorder. *Journal of the American Academy of Child and Adolescent Psychiatry, 29,* 526–633.

Biederman, J., Faraone, S. V., & Lapey, K. (1992). Comorbidity of diagnosis in attention-deficit hyperactivity disorder. In G. Weiss (Ed.), *Child and adolescent psychiatry clinics of North America: Attention deficit hyperactivity disorder* (pp. 335–360). Philadelphia: Saunders.

Biederman, J., Hirshfeld-Becker, D. R., Rosenbaum, J. F., Herot, C., Friedman, D., Snidman, N., et al. (2001). Further evidence of association between behavioral inhibition and social anxiety in children. *American Journal of Psychiatry, 158,* 1673–1679.

Bierut, L. J., Saccone, N. L., Rice, J. P., Goate, A., Foroud, T., Edenberg, H., et al. (2002). Defining alcohol-related phenotypes in humans: The collaborative study on the genetics of alcoholism. *Genetic Technology in Alcohol Research, 26,* 208–213.

Biglan, A., & Lichtenstein, E. (1984). A behavior-analytic approach to smoking acquisition: Some recent findings. *Journal of Applied Social Psychology, 14,* 207–224.

Bigler, R. S., & Liben, L. S. (1993). A cognitive-development approach to racial stereotyping and reconstructive memory in Euro-American children. *Child Development, 64,* 1507–1518.

Birbaumer, N., Grodd, W., Diedrich, O., Klose, U., Erb, M., Lotze, M., et al. (1998). fMRI reveals amygdala activation to human faces in social phobics. *Neuroreport, 9,* 1223–1226.

Birmingham, C., Muller, J. L., Palepu, A., Spinelli, J. P., & Anis, A. H. (1999). The cost of obesity in Canada. *Canadian Medical Association Journal, 160,* 483–488.

Birren, J. E., & Schaie, K. W. (Eds.). (2001). *Handbook of the psychology of aging* (5th ed.). San Diego, CA: Academic Press.

Bisiach, E., & Luzzatti, C. (1978). Unilateral neglect and representational space. *Cortex, 14,* 129–133.

Biswas, A., Olsen, J. E., & Carlet, V. (1992). A comparison of print advertisements from the United States and France. *Journal of Advertising, 21(4),* 73–81.

Bjork, E. L., & Bjork, R. A. (1996). *Memory.* San Diego, CA: Academic Press.

Blagys, M. D., & Hilsenroth, M. J. (2000). Distinctive feature of short-term psychodynamic-interpersonal psychotherapy: A review of the comparative psychotherapy process literature. *Clinical Psychology: Science and Practice, 7,* 167–188.

Blair, I. V. (2002). The malleability of automatic stereotypes and prejudice. *Personality and Social Psychology Review, 6,* 242–261.

Blair, R. J. (2003). Neurobiological basis of psychopathy. *British Journal of Psychiatry, 182,* 5–7.

Blakemore, C. (1983). *Mechanics of the mind.* Cambridge, UK: Cambridge University Press.

Blakemore, S. J., Wolpert, D. M., & Frith, C. D. (1998). Central cancellation of self-produced tickle sensation. *Nature Neurosciences, 1,* 635–640.

Blakemore, S. J., Wolpert, D. M., & Frith, C. D. (2000). Why can't you tickle yourself? *Neuroreport, 3,* 11–16.

Blanchard, R., & Ellis, L. (2001). Birth weight, sexual orientation, and the sex of preceding siblings. *Journal of Biosocial Science, 33,* 451–467.

Blitstein, J. L., Robinson, L. A., Murray, D. M., Klesges, R. C., & Zbikowski, S. M. (2003). Rapid progression to regular cigarette smoking among nonsmoking adolescents: Interactions with gender and ethnicity. *Preventive Medicine, 36,* 455–463.

Blood, A. J., & Zatorre, R. J. (2001). Intensely pleasurable responses to music correlate with activity in brain regions implicated in reward and emotion. *Proceedings of the National Academy of Science, 98,* 11818–11823.

Bloom, P. (2004). *Descartes' baby: How the science of child development explains what makes us human.* New York: Basic Books.

Blumstein, P., & Schwartz, P. (1983). *American couples.* New York: Simon & Schuster.

Bohus, M., Haaf, B., Simms, T., Limberger, M. F., Schmahl, C., Unckel, C., et al. (2004). Effectiveness of inpatient dialectical behavioral therapy for borderline personality disorder: A controlled trial. *Behavioral Research and Therapy, 42,* 487–499.

Bohus, M., Haaf, B., Stiglmayr, C., Pohl, U., Boehme, R., & Linehan, M. M. (2000). Evaluation of inpatient dialectical-behavioral therapy for borderline personality disorder—A prospective study. *Behaviour Research and Therapy, 38,* 875–887.

Bolles, R. C. (1970). Species-specific defense reactions and avoidance learning. *Psychological Review, 77,* 32–48.

Bond, R., & Smith, P. B. (1996). Culture and conformity: A meta-analysis of studies using Asch's (1952b, 1956) line judgment task. *Psychological Bulletin, 119,* 111–137.

Bornstein, R. F. (1999). Criterion validity of objective and projective dependency tests: A meta-analytic assessment of behavioral prediction. *Psychological Assessment, 11,* 48–57.

Boroditsky, L. (2001). Does language shape thought? English and Mandarin speakers' conceptions of time. *Cognitive Psychology, 43,* 1–22.

Bouchard, C., Tremblay, A., Despres, J. P., Nadeau, A., Lupien, J. P., Theriault, G., et al. (1990). The response to long-term overfeeding in identical twins. *New England Journal of Medicine, 322,* 1477–1482.

Bouchard Jr., T. J., Lykken, D. T., McGue, M., Segal, N. L., & Tellegen, A. (1990). Sources of human psychological differences: The Minnesota study of twins reared apart. *Science, 250,* 223–228.

Bourin, M., Baker, G. B., & Bradwejn, J. (1998). Neurobiology of panic disorder. *Journal of Psychosomatic Research, 44,* 163–180.

Bouton, M. E. (1994). Context, ambiguity, and classical conditioning. *Current Directions in Psychological Science, 3,* 49–53.

Bouton, M. E., Nelson, J. B., & Rosas, J. M. (1999). Stimulus generalization, context change, and forgetting. *Psychological Bulletin, 125,* 171–186.

Bradbury, T. N., & Fincham, F. D. (1990). Attributions in marriage: Review and critique. *Psychological Bulletin, 107,* 3–33.

Bramlett, M. D., & Mosher, W. D. (2002). *Cohabitation, marriage, divorce, and remarriage in the United States* (Vital and Health Statistics, Series 23, No. 22). Hyattsville, MD: National Center for Health Statistics, Centers for Disease Control and Prevention.

Bransford, J. D., & Johnson, M. K. (1972). Contextual prerequisites for understanding: Some investigations of comprehension and recall. *Journal of Verbal Learning and Verbal Behavior, 11,* 717–726. Modified by E. B. Zechmeister & S. E. Nyberg, 1982. *Human memory* (p. 305). Pacific Grove, CA: Brooks Cole.

Braun, A. R., Balkin, T. J., Wesensten, N. J., Gwadry, F., Carson, R. E., Varga, M., et al. (1998). Dissociated pattern of activity in visual cortices and their projections during human rapid eye movement sleep. *Science, 279,* 91–95.

Breland, K., & Breland, M. (1961). The misbehavior of organisms. *American Psychologist, 16,* 681–684.

Brent, D. A. (2004). Antidepressants and pediatric depression—the risk of doing nothing. *New England Journal of Medicine, 351,* 1598–1601.

Brewer, J. B., Zhao, Z., Glover, G. H., & Gabrieli, J. D. E. (1998). Making memories: Brain activity that predicts how well visual experiences will be remembered. *Science, 281,* 1185–1187.

Brewer, M. B., & Caporael, L. R. (1990). Selfish genes vs. selfish people: Sociobiology as origin myth. *Motivation and Emotion, 14,* 237–243.

Brigham, J. C., & Malpass, R. S. (1985). The role of experience and contact in the recognition of faces of own- and other-races persons. *Journal of Social Issues, 41,* 139–155.

Broadbent, D. A. (1958). *Perception and communication.* New York: Pergamon.

Brodal, P. (1998). *The central nervous system.* Oxford, U.K.: Oxford University Press.

Brody, A. L., Saxena, S., Fairbanks, L. A., Alborzian, S., Demaree, H. A., Maidment, K. M., et al. (2000). Personality changes in adult subjects with major depressive disorder or obsessive-compulsive disorder treated with paroxetine. *Journal of Clinical Psychiatry, 61,* 349–355.

Brody, A. L., Saxena, S., Stoessel, P., Gillies, L. A., Fairbanks, L. A., Alborzian, S., et al. (2001). Regional brain metabolic changes in patients with major depression treated with either paroxetine or interpersonal therapy: Preliminary findings. *Archives of General Psychiatry, 58,* 631–640.

Broude, G. J., & Green, S. J. (1983). Cross-cultural codes on husband-wife relationships. *Ethology, 22,* 273–274.

Brown, A. S. (1991). A review of the tip-of-the-tongue phenomenon. *Psychological Bulletin, 109,* 204–223.

Brown, B. B., Mounts, N., Lamborn, S. D., & Steinberg, L. (1993). Parenting practices and peer group affiliations in adolescence. *Child Development, 64,* 467–482.

Brown, D. E. (1991). *Human universals.* Philadelphia: Temple University Press.

Brown, G. W., & Harris, T. O. (1978). *Social origins of depression: A study of psychiatric disorders in women.* New York: The Free Press.

Brown, J. D., & Kobayashi, C. (2002). Self-enhancement in Japan and America. *Asian Journal of Social Psychology, 5,* 145–168.

Brown, J. L., & Pollitt, E. (1996). Malnutrition, poverty and intellectual development. *Scientific American, 274,* 38–43.

Brownell, C. A., & Brown, E. (1992). Peers and play in infants and toddlers. In V. Van Hasselt & M. Hersen (Eds.), *Handbook of social development: A lifespan perspective* (pp. 183–200). New York: Plenum.

Bruck, M. L., & Ceci, S. (1993). Amicus brief for the case of State of New Jersey v. Michaels. Presented by Committee of Concerned Social Scientists. Supreme Court of New Jersey docket no. 36,633. (Reprinted from *Psychology, Public Policy and Law* 1, 1995, 272–322.)

Bruer, J. (1999). *The myth of the first three years: A new understanding of early brain development and lifelong learning.* New York: Free Press.

Buckner, R. L., Kelley, W. M., & Petersen, S. E. (1999). Frontal cortex contributes to human memory formation. *Nature Neuroscience, 2,* 311–314.

Burian, S. E., Hensberry, R., & Liguori, A. (2003). Differential effects of alcohol and alcohol expectancy on risk-taking during simulated driving. *Clinical and Experimental Human Psychopharmacology Clinics, 18,* 175–184.

Burke, B. L., Arkowitz, H., & Menchola, M. (2003). The efficacy of motivational interviewing: A meta-analysis of controlled clinical trials. *Journal of Consulting and Clinical Psychology, 71,* 843–861.

Burman, B., & Margolin, G. (1992). Analysis of the association between marital relationships and health problems: An interactional perspective. *Psychological Bulletin, 112,* 39–63.

Bush, E. C., & Allman, J. M. (2004). The scaling of frontal cortex in primates and carnivore. *Proceedings of the National Academy of Sciences, 101,* 3962–3966.

Bushman, B. J., & Anderson, C. A. (2001). Media violence and the American public: Scientific facts versus media misinformation. *American Psychologist, 56,* 477–489.

Bushman, B. J., & Huesmann, L. R. (2001). Effects of televised violence on aggression. In D. G. Singer & J. L. Singer (Eds.), *Handbook of children and the media* (pp. 223–254). Thousand Oaks, CA: Sage.

Buss, A. H., & Plomin, R. (1984). *Temperament: Early developing personality traits.* Hillsdale, NJ: Erlbaum.

Buss, D. M. (1989). Sex differences in human mate preferences: Evolutionary hypotheses tested in 37 cultures. *Behavioral and Brain Sciences, 12,* 1–49.

Buss, D. M. (1995). Evolutionary psychology: A new paradigm for psychological science. *Psychological Inquiry, 6,* 1–30.

Buss, D. M. (1999a). Human nature and individual differences: The evolution of human personality. In L. A. Pervin & O. P. John (Eds.), *Handbook of personality: Theory and research* (pp. 31–56). New York: Guilford.

Buss, D. M. (2000). *The dangerous passion: Why jealousy is as necessary as love and sex.* New York: Free Press.

Buss, D. M., & Greiling, H. (1999). Adaptive individual differences. *Journal of Personality, 67,* 209–243.

Buss, D. M., & Schmitt, D. P. (1993). Sexual strategies theory: An evolutionary perspective on human mating. *Psychological Review, 100,* 204–232.

Cacioppo, J. T., Berntson, G. G., Sheridan, J. F., & McClintock, M. K. (2000). Multilevel integrative analyses of human behavior: Social neuroscience and the complementing nature of social and biological approaches. *Psychology Bulletin, 126,* 829–843.

Cacioppo, J. T., Klein, D. J., Berntson, G. C., & Hatfield, E. (1993). The psychophysiology of emotion. In M. Lewis & J. M. Haviland (Eds.), *Handbook of emotions* (pp. 119–142). New York: Guilford Press.

Cahill, L. (2003). Sex-related influences on the neurobiology of emotionally influenced memory. *Annals of the New York Academy of Sciences, 985,* 163–167.

Cahill, L., Haier, R. J., Fallon, J., Alkire, M. T., Tang, C., Keator, D., et al. (1996). Amygdala activity at encoding correlated with long-term, free recall of emotional information. *Proceedings of the National Academy of Sciences of the United States of America, 93,* 8016–8021.

Cahill, L., Haier, R. J., White, N. S., Fallon, J., Kilpatrick, L., Lawrence, C., et al. (2001). Sex-related difference in amygdala activity during emotionally influenced memory storage. *Neurobiology of Learning and Memory, 75,* 1–9.

Caillat, Z., & Mueller, B. (1996). The influence of culture on American and British advertising. *Journal of Advertising Research (May/June),* 79–88.

Cain, D. P. (1998). Testing the NMDA, long-term potentiation, and cholingergic hypotheses of spatial learning. *Neuroscience and Biobehavioral Reviews, 22,* 181–193.

Cairns, R. B., & Cairns, B. D. (1994). *Lifelines and risks: Pathways of youth in our times.* Cambridge: Cambridge University Press.

Callaghan, T., Rochat, P., Lillard, A., Claux, M. L., Odden, H., Itakura, S., Tapanya, S. & Singh, S. (2005). Synchrony in the onset of mental-state reasoning: Evidence from five cultures. *Psychological Science, 16,* 378–384.

Campbell, W. K., & Sedikides, C. (1999). Self-threat magnifies the self-serving bias: A meta-analytic integration. *Review of General Psychology, 3,* 23–43.

Canadian Psychiatric Association. (2001). Canadian clinical practice guidelines for the treatment of depressive disorders. *Canadian Journal of Psychiatry.* Suppl. no. 46.

Canadian Psychological Association. (1992a). *Canadian code of ethics for psychologists, 1991.* Old Chelsea, Quebec.

Canadian Psychological Association. (1992b). The history of psychology in Canada. *Canadian Psychology, 33,* 2.

Canli, T. (2004). Functional brain mapping of extraversion and neuroticism: Learning from individual differences in emotion processing. *Journal of Personality, 72,* 1105–1132.

Canli, T., Desmond, J. E., Zhao, Z., Gabrieli, J. D. (2002). Sex differences in the neural basis of emotional memories. *Proceedings of the National Academy of Sciences, 99,* 10789–10794.

Capaldi, E. D. (1996). *Why we eat what we eat.* Washington, DC: American Psychological Association.

Caporael, L. R. (2001). Evolutionary psychology: Toward a unifying theory and a hybrid science. *Annual Review of Psychology, 52,* 607–628.

Carlson, N. R. (2003). *Physiology of behavior* (8th ed.). Needham Heights, MA: Allyn & Bacon.

Carlyle, T. (1843). *Past and present.* New York: W. H. Colyer.

Carmena, J. M., Lebdev, M. A., Crist, R. E., O'Doherty, J. E., Santucci, D. M., Dimitrov, D. F., et al. (2003). Learning to control a brain-machine interface for reaching and grasping by primates. *PLOS Biology, 1,* 193–208.

Carmody, T. P. (1993). Nicotine dependence: Psychological approaches to the prevention of smoking relapse. *Psychology of Addictive Behaviors, 7,* 96–102.

Caro, T. M., & Hauser, M. D. (1992). Is there teaching in nonhuman animals? *Quarterly Journal of Biology, 67,* 151–174.

Carpendale, J. I. M. (2000). Kohlberg & Piaget on stages and moral reasoning. *Developmental Review, 20,* 181–205.

Carstensen, L. L. (1995). Evidence for a life-span theory of socioemotional selectivity. *Current Directions in Psychological Science, 4,* 151–156.

Carter, C. S. (2003). Developmental consequences of oxytocin. *Physiology & Behavior, 74,* 383–397.

Caruso, S., Intelisano, G., Farina, M., Di Mari, L., & Agnello, C. (2003). The function of sildenafil on female sexual pathways: A double-blind, cross-over, placebo-controlled study. *European Journal of Obstetrics & Gynecology and Reproductive Biology, 110,* 201–206.

Carver, C. S., & Scheier, M. F. (1981). *Attention and self-regulation: A control theory approach to human behavior.* New York: Springer-Verlag.

Carver, C. S., & Scheier, M. F. (1998). *On the self-regulation of behavior.* New York: Cambridge University Press.

Case, R. (1992). The role of the frontal lobes in development. *Brain and Cognition, 20* (1), 51–73.

Caspi, A. (2000). The child is father of the man: Personality continuities from childhood to adulthood. *Journal of Personality and Social Psychology, 78,* 158–172.

Caspi, A., & Herbener, E. S. (1990). Continuity and change: Assortative marriage and the consistency of personality in adulthood. *Journal of Personality and Social Psychology, 58,* 250–258.

Caspi, A., McClay, J., Moffit, T. E., Mill, J., Martin, J., Craig, I. W., et al. (2002). Role of genotype in the cycle of violence in maltreated children. *Science, 29,* 851–854.

Castellanos, F. X., Giedd, J. N., Eckberg, P., & Marsh, W. L. (1998). Quantitative morphology of the caudate nucleus in attention deficit hyperactivity disorder. *American Journal of Psychiatry, 151,* 1791–1796.

Cattell, R. B. (1971). *Abilities: Their structure, growth, and action.* Boston: Houghton Mifflin.

Caulfield, T. (2003). Human cloning laws, human dignity and the poverty of the policy making dialogue. *BMC Medical Ethics, 4,* E3.

Ceci, S. J. (1999). Schooling and intelligence. In S. J. Ceci & Wendy M. Williams (Eds.), *The nature-nurture debate: The essential readings* (pp.168–175). Oxford, UK: Blackwell.

Ceci, S. J., & Bruck, M. (1995). *Jeopardy in the courtroom: A scientific analysis of children's testimony.* Washington, DC: American Psychological Association.

Chabris, C. (1999). Prelude or requiem for the "Mozart effect"? *Nature, 400,* 826–827.

Chambers, D. W. (1983). Stereotypic images of the scientist: The Draw-a-Scientist-Test. *Science Education, 67,* 255–265.

Chang, W. C. (1985). A cross-cultural study of depressive symptomatology. *Culture, Medicine, and Psychiatry, 9,* 295–317.

Chapman, P. R., & Underwood, G. (1998). Visual search of driving situations: Danger and experience. *Perception, 27,* 951–964.

Charles, S. T., Mather, M., & Carstensen L. L. (2003). Aging and emotional memory: The forgettable nature of negative images for older adults. *Journal of Experimental Psychology: General, 132,* 310–324.

Chase, W. G., & Simon, H. A. (1973). Perception in chess. *Cognitive Psychology, 4,* 55–81.

Chassin, L., Presson, C. C., & Sherman, S. J. (1990). Social psychological contributions to the understanding and prevention of adolescent cigarette smoking. *Personality and Social Psychology Bulletin, 16,* 133–151.

Cheeseman, J., & Merikle, P. M. (1986). Distinguishing conscious from unconscious perceptual processes. *Canadian Journal of Psychology, 40,* 343– 367.

Cheesman, M. F. (1997). Speech perception by elderly listeners: Basic knowledge and implications for audiology. *Journal of Speech-Language Pathology and Audiology, 21,* 104–119.

Chen, X., Hastings, P. D., Rubin, K. H., Chen, H., Cen, G., & Stewart, S. L. (1998). Child-rearing attitudes and behavioural inhibition in Chinese and Canadian toddlers: A cross-cultural study. *Developmental Psychology, 34,* 677–686.

Cherry, E. C. (1953). Some experiments on the recognition of speech, with one and two ears. *Journal of the Acoustic Society of America, 25,* 975–979.

Cherry, F. (1994). Archivist's corner. *History and Philosophy of Psychology Bulletin, 6(1),* 12.

Chesher, G., & Greeley, J. (1992). Tolerance to the effects of alcohol. *Alcohol, Drugs, and Driving, 8,* 93–106.

Chess, S., & Thomas, A. (1984). *Origins and evolution of behavior disorders: From infancy to early adult life.* Cambridge, MA: Harvard University Press.

Childress, A. R., Mozley, D., McElgin, W., Fitzgerald, J., Reivich, M., & O'Brien, P. C. (1999). Limbic activation during cue-induced cocaine craving. *American Journal of Psychiatry, 156,* 11–18.

Chistyakov, A. V., Kaplan, B., Rubichek, O., Kreinin, I., Koren, D., Feinsod, M., et al. (2004). Antidepressant effects of different schedules of repetitive transcranial magnetic stimulation vs. clomipramine in patients with major depression: Relationship to changes in cortical excitability. *International Journal of Neuropsychopharmacology, 8,* 223–233.

Choi, I., Dalal, R., Kim-Prieto, C., & Park, H. (2003). Culture and judgment of causal relevance. *Journal of Personality and Social Psychology, 84,* 46–59.

Choleris, E., Gustafsson, J. A., Korach, K. S., Muglia, L. J., Pfaff, D. W., & Ogawa, S. (2003). An estrogen-dependent four-gene micronet regulating social recognition: A study with oxytocin and estrogen receptor-alpha and -beta knockout mice. *Proceedings of the National Academy of Sciences, 100,* 6192–6197.

Christianson, S. (1992). Emotional stress and eyewitness memory: A critical review. *Psychological Bulletin, 112,* 284–309.

Chua, H. F., Boland, J. E., & Nisbett, R. E. (2005). Cultural variation in eye movements during scene perception. *Proceedings of the National Academy of Sciences, 102,* 12629–12633.

Church, A. T. (1982). Sojourner adjustment. *Psychological Bulletin, 91,* 540–572.

Cialdini, R. (2001). *Influence: Science and practice.* Boston: Allyn & Bacon.

Cialdini, R. B., Shaller, M., Houlihan, D., Arps, K., Fultz, J., & Beaman, A. L. (1987). Empathy-based helping: Is it selflessly or selfishly motivated? *Journal of Personality and Social Psychology, 52,* 749–758.

Cialdini, R. B., Wosinska, W., & Barrett, D. W. (1999). Compliance with a request in two cultures: The differential influence of social proof and commitment/consistency on collectivists and individualists. *Personality and Social Psychology Bulletin, 25,* 1242–1253.

Cjte, S., Tremblay, R. E., Nagin, D., Zoccolillo, M., & Vitaro, F. (2002). The development of impulsivity, fearfulness, and helpfulness during childhood: Patterns of consistency in the trajectories of boys and girls. *Journal of Child Psychology and Psychiatry, 43,* 609–618.

Clark, A. C., & Watson, D. (1999). Temperament: A new paradigm for trait psychology. In L. A. Pervin & O. P. John (Eds.), *Handbook of personality: Theory and research* (pp. 399–423). New York: Guilford.

Clark, R. D., & Hatfield, E. (1989). Gender differences in receptivity to sexual offers. *Journal of Psychology and Human Sexuality, 2,* 39–55.

Clarke-Stewart, K. A., Vandell, D. L., McCartney, K., Owen, M. T., & Booth, C. (2000). Effects of parental separation and divorce on very young children. *Journal of Family Psychology, 14,* 304–326.

Cleghorn, R. A. (1984). The development of psychiatric research in Canada up to 1984. *Canadian Journal of Psychiatry, 29,* 189–197.

Cloninger, C., Adolfsson, R., & Svrakic, N. (1996). Mapping genes for human personality. *Nature and Genetics, 12,* 3–4.

Cohen, D., Nisbett, R. E., Bowdle, B. F., & Schwarz, N. (1996). Insult, aggression, and the southern culture of honor: An "experimental ethnography." *Journal of Personality and Social Psychology, 70,* 945–960.

Cohen, S., Tyrrell, D. A. J., & Smith, A. P. (1991). Psychological stress and susceptibility to the common cold. *The New England Journal of Medicine, 325,* 606–612.

Cohen, S., & Wills, T. A. (1985). Stress, social support, and the buffering hypothesis. *Psychological Bulletin, 98,* 310–357.

Colapinto, J. (2000). *As nature made him: The boy who was raised as a girl.* New York: HarperCollins.

Colcombe, S. J., Kramer, A. F., Erickson, K. I., Scalf, P., McAuley, E., Cohen, N. J., et al. (2004). Cardiovascular fitness, cortical plasticity, and aging. *Proceedings of the National Academy of Sciences, 101,* 316–321.

Conway, A. R. A., Kane, M. J., & Engle, R. W. (2003). Working memory capacity and its relation to general intelligence. *Trends in Cognitive Sciences, 7,* 547–552.

Conway, M. A., Anderson, S. J., Larsen, S. F., Donnelly, C. M., McDaniel, M. A., McClelland, A. G. R., et al. (1994). The formation of flashbulb memories. *Memory and Cognition, 22,* 326–343.

Conway, M., & Ross, M. (1984). Getting what you want by revising what you had. *Journal of Personality and Social Psychology, 47,* 738–748.

Cook, M., & Mineka, S. (1989). Observational conditioning of fear to fear-relevant versus fear-irrelevant stimuli in rhesus monkeys. *Journal of Abnormal Psychology, 98,* 448–459.

Cooney, J., & Gazzaniga, M.S. (2003). Neurological disorders and the structure of human consciousness. *Trends in Cognitive Sciences, 7,* 161–165.

Cooper, L. A., & Shepard, R. N. (1973). Chronometric studies of the rotation of mental images. In W. G. Chase (Ed.), *Visual information processing* (pp. 75–176). New York: Academic Press.

Coopersmith, S. (1967). *The antecedents of self-esteem.* San Francisco: Freeman.

Corder, E. H., Saunders, A. M., Strittmatter, W. J., Schmechel, D. E., Gaskell, P. C., Small, G. W., et al. (1993). Gene dose of apolipoprotein E type 4 allele and the risk of Alzheimer's disease in late onset families. *Science, 261,* 921–923.

Corkin, S., Amaral, D. G., Gonzalez, R. G., Johnson, K. A., & Hyman, B. T. (1997). H. M.'s medial temporal lobe lesion: Findings from magnetic resonance imaging. *Journal of Neuroscience, 17,* 3964–3979.

Correll, C. U., Leucht, S., & Kane, J. M. (2004). Lower risk for tardive dyskinesia associated with second-generation antipsychotics: A systematic review of 1-year studies. *American Journal of Psychiatry, 161,* 414–425.

Cosmides, L., & Tooby, J. (2000). The cognitive neuroscience of social reasoning. In M. S. Gazzaniga (Ed.), *The new cognitive neurosciences* (pp.1259–1270). Cambridge, MA: MIT Press.

Cosmides, L., & Tooby, J. (2001). Evolutionary psychology: A primer. Retrieved May 24, 2002 from http://www.psych.ucsb.edu/research/cep/primer.html

Costa, P. T., & McCrae, R. R. (1992). *Revised NEO Personality Inventory (NEO-PI-R) and NEO Five-Factor Inventory (NEO-FFI) professional manual.* Odessa, FL: Psychological Assessment Resources.

Courchesne, E., Redcay, E., & Kennedy, D. P. (2004). The autistic brain: Birth through adulthood. *Current Opinion in Neurology, 17,* 489–496.

Cousins, S. D. (1989). Culture and selfhood in Japan and the U.S. *Journal of Personality and Social Psychology, 56,* 124–131.

Cowan, N. (2001). The magical number 4 in short-term memory: A reconsideration of mental storage capacity. *Behavioral and Brain Sciences, 24,* 87–114.

Cowley, J. J., & Brooksbank, L. B. W. (1991). Human exposure to putative pheromones and changes in aspects of social behavior. *Journal of Steroid Biochemistry and Molecular Biology, 39,* 647–659.

Craft, L. L., & Perna, F. M. (2004). The benefits of exercise for the clinically depressed. *Journal of Clinical Psychiatry, 6,* 104–111.

Crocker, J., & Major, B. (1989). Social stigma and self-esteem: The self-protective properties of stigma. *Psychological Review, 96,* 608–630.

Crosnoe, R., & Elder, G. H. (2002). Successful adaptation in the later years: A life-course approach to aging. *Social Psychology Quarterly, 65,* 309–328.

Cross, S. E. (1995). Self-construals, coping, and stress in cross-cultural adaptation. *Journal of Cross-Cultural Psychology, 26,* 673–697.

Crowe, R. R. (2000). Molecular genetics of anxiety disorders. In D. S. Charney, E. J. Nestler, & B. S. Bunney (Eds.), *Neurobiology of mental illness* (pp. 451–462). New York: Oxford University Press.

Crundall, D., Chapman, P., Phelps, N., & Underwood, G. (2003). Eye movements and hazard perception in police pursuit and emergency response driving. *Journal of Experimental Psychology: Applied, 9,* 163–174.

Culler, E. A., Coakley, J. D., Lowy, K., & Gross, N. (1943). A revised frequency-map of the Guinea-pig cochlea. *American Journal of Psychology, 56,* 475–500.

Cunningham, M. R., Barbee, A. P., & Druen, P. B. (1996). Social allergens and the reactions they produce: Escalation of annoyance and disgust in love and work. In R. M. Kowalski (Ed.), *Aversive interpersonal behaviors* (pp. 189–214). New York: Plenum Press.

Cunningham, M. R., Roberts, A. R., Barbee, A. P. Druen, P. B., & Wu, C. (1995). Their ideas of beauty are, on the whole, the same as ours: Consistency and variability in the cross-cultural perception of female physical attractiveness. *Journal of Personality & Social Psychology, 68,* 261–279.

Cunningham, W. A., Johnson, M. K., Raye, C. L., Gatenby, J. C., Gore, J. C., & Banaji, M. R. (2004). Separable neural components in the processing of black and white faces. *Psychological Science, 15,* 806–813.

Cupach, W. R., & Metts, S. (1990). Remedial processes in embarrassing predicaments. In J. Anderson (Ed.), *Communication yearbook* (pp. 323–352). Newbury Park, CA: Sage.

Curtis, C. E., & D'Esposito, M. (2003). Persistent activity in the prefrontal cortex during working memory. *Trends in Cognitive Sciences, 7,* 415–423.

Curtiss, S. (1977). *Genie: a psycholinguistic study of a modern day "wild child."* New York: Academic Press.

Cutler, W. B., Friedmann, E., & McCoy, N. L (1998). Pheromonal influences on sociosexual behavior in men. *Archives of Sexual Behavior, 27,* 1–13.

Dabbs, J. M., & Morris, R. (1990). Testosterone, social class, and antisocial behavior in a sample of 4462 men. *Psychological Science, 1,* 209–211.

Dalton, M. A., Bernhardt, A. M., Gibson, J. J., Sargent, J. D., Beach, M. L., Adachi-Mejia, A. M., et al. (2005). "Honey, have some smokes." Preschoolers use cigarettes and alcohol while role playing as adults. *Archives of Pediatrics & Adolescent Medicine, 159,* 854–859.

Dalton, M. A., Sargent, J. D., Beach, M., Ahrens, M. B., Tickle, J., & Heatherton, T. F. (2003). Effect of viewing smoking in movies on adolescent smoking initiation: A cohort study. *Lancet, 362,* 281–285.

Damasio, A. R. (1994). *Descartes' error.* New York: Avon Books.

Damasio, A. R. (1999). *The feeling of what happens.* New York: Harcourt Brace.

Damasio, H., Grabowski, T., Frank, R., Galaburda, A. M., & Damasio, A. R. (1994). The return of Phineas Gage: Clues about the brain from the skull of a famous patient. *Science, 264,* 1102–1105.

Darley, J. M., & Batson, C. D. (1973). "From Jerusalem to Jericho": A study of situational and dispositional variables in helping behavior. *Journal of Personality and Social Psychology, 27,* 100–108.

Darou, W. G. (1992). Native Canadians and intelligence testing. *Canadian Journal of Counseling, 26,* 96–99.

Darwin, C. (1964). *On the origin of species.* Cambridge, MA: Harvard University Press. (Original work published 1859)

Darwin, F. (Ed.). (1887). *The life and letters of Charles Darwin.* London: Murray.

Dasgupta, A. G., & Greenwald, A. G. (2001). Exposure to admired group members reduces automatic intergroup bias. *Journal of Personality and Social Psychology, 81,* 800–814.

Davidson, J. R., Foa, E. B., Huppert, J. D., Keefe, F. J., Franklin, M. E., Compton, J. S., et al. (2004). Fluoxetine, comprehensive cognitive behavioral therapy, and placebo in generalized social phobia. *Archives of General Psychiatry, 61,* 1005–1013.

Davidson, R. J. (2000a). Affective style, psychopathology, and resilience: Brain mechanisms and plasticity. *American Psychologist, 55,* 1196–1214.

Davidson, R. J. (2000b). The functional neuroanatomy of affective style. In R. D. Lane & L. Nadel (Eds.), *Cognitive neuroscience of emotion* (pp. 371–388). New York: Oxford University Press.

Davis, J. O., Phelps, J. A., & Bracha, H. S. (1995). Prenatal development of monozygotic twins and concordance for schizophrenia. *Schizophrenia Bulletin, 21,* 357–366.

Davis, K. L., Stewart, D. G., Friedman, J. I., Buchsbaum, M., Harvey, P. D., Hof, P. R., et al. (2003). White matter changes in schizophrenia: Evidence for myelin-related dysfunction. *Archives of General Psychiatry, 60,* 443–456.

Davis, M. (1997). Neurobiology of fear responses: The role of the amygdala. *Journal of Neuropsychological and Clinical Neuroscience, 9,* 382–402.

Davison, K., & Pennebaker, J. (1996). Social psychosomatics. In E. T. Higgins & A. W. Kruglanski (Eds.), *Social psychology: Handbook of basic principles* (pp. 102–130). New York: Guilford.

Dawes, R. M. (1994). *House of cards: Psychology and psychotherapy built on myth.* New York: Maxwell Macmillan International.

Deacon, B. J., & Abramowitz, J. S. (2004). Cognitive and behavioral treatments for anxiety disorders: A review of meta-analytic findings. *Journal of Clinical Psychology, 60,* 429–441.

Deacon, T. W. (1997). *The symbolic species: The co-evolution of language and the brain.* New York: Norton.

Deary, I. J. (2000). *Looking down on human intelligence.* New York: Oxford University Press.

Deary, I. J. (2001). *Intelligence: A very short introduction.* New York: Oxford University Press.

Deary, I. J., & Caryl, P. G. (1997). Neuroscience and human intelligence differences. *Trends in Neuroscience, 20,* 365–371.

Deary, I. J., & Der, G. (2005). Reaction time explains IQ's association with death. *Psychological Science, 16,* 64–69.

Deaux, K., & Major, B. (1987). Putting gender into context: An interactive model of gender-related behavior. *Psychological Review, 94,* 369–389.

Debiec, J., LeDoux, J. E., & Nader, K. (2002). Cellular and systems consolidation in the hippocampus. *Neuron, 36,* 527–538.

DeCasper, A. J., & Fifer, W. P. (1980). Of human bonding: Newborns prefer their mothers' voices. *Science, 208,* 1174–1176.

DeCasper, A. J., & Spence, M. J. (1986). Prenatal maternal speech influences newborn's perception of speech sounds. *Infant Behavior and Development, 9,* 133–150.

Deci, E. L., & Ryan, R. M. (1987). The support of autonomy and the control of behavior. *Journal of Personality and Social Psychology, 53,* 1024–1037.

Dejong, W., & Kleck, R. E. (1986). The social psychological effects of overweight. In C. P. Herman, M. P. Zanna, & E. T. Higgins (Eds.), *Physical appearance, stigma and social behavior: The Ontario Symposium* (pp. 65–87). Hillsdale, NJ: Erlbaum.

Demir, E., & Dickson, B. J. (2005). *Fruitless.* Splicing specifies male courtship behavior in *Drosophila*. *Cell, 121,* 785–794.

Dennerstein, L., & Burrows, G. D. (1982). Hormone replacement therapy and sexuality in women. *Clinics in Endocrinology and Metabolism, 11,* 661–679.

Denys, D., Zohar, J., & Westenberg, H. G. (2004). The role of dopamine in obsessive-compulsive disorder: Preclinical and clinical evidence. *Journal of Clinical Psychiatry, 65,* 11–17.

Deoliveira, C. A., Moran, G., & Pederson, D. R. (2005). Understanding the link between maternal adult attachment classifications and thoughts and feelings about emotions. *Attachment and Human Development, 7,* 153–170.

Depue, R. A., & Collins, P. F. (1999). Neurobiology of the structure of personality: Dopamine, facilitation of incentive motivation, and extraversion. *Behavioral and Brain Sciences, 22,* 491–569.

Devine, P. G. (1989). Stereotypes and prejudice: Their automatic and controlled components. *Journal of Personality and Social Psychology, 56,* 5–18.

Diamond, A., & Doar, B. (1989). The performance of human infants on a measure of frontal cortex function, delayed response task. *Developmental Psychobiology, 22,* 271–294.

Dickson, P. R., & Vaccarino, F. J. (1994). GRF-induced feeding: Evidence for protein selectivity and opiate involvement. *Peptides, 15,* 1343–1352.

Diener, E. (1979). Deindividuation, self-awareness, and disinhibition. *Journal of Personality and Social Psychology, 37,* 1160–1171.

Diener, E. (2000). Subjective well-being: The science of happiness and a proposal for a national index. *American Psychologist, 55,* 34–43.

Diener, E. (2001). *Culture and subjective well-being—Why some nations and ethnic groups are happier than others.* Invited address presented at Thirteenth Annual Convention of the American Psychological Society, (June 1–17), Toronto.

Diener, E., & Biswas-Diener, R. (2002). Will money increase subjective well-being? A literature review and guide to needed research. *Social Indicators Research, 57,* 119–169.

Diener, E., Wolsic, B., & Fujita, F. (1995). Physical attractiveness and subjective well-being. *Journal of Personality and Social Psychology, 69,* 120–129.

Dijksterhuis, A. (2004). Think different: The merits of unconscious thought in preference development and decision making. *Journal of Personality and Social Psychology, 87,* 586–598.

Dijksterhuis, A., & van Knippenberg, A. (1998). The relation between perception and behavior, or how to win a game of trivial pursuit, *Journal of Personality & Social Psychology, 74,* 865–877.

Dion, K. K., & Dion, K. L. (1993). Individualistic and collectivistic perspectives on gender and the cultural context of love and intimacy. *Journal of Social Issues, 49,* 53–69.

Dobson, K. S., & Khatri, N. (2000). Cognitive therapy: Looking backward, looking forward. *Journal of Clinical Psychology, 56,* 907–923.

Docherty, N. M. (2005). Cognitive impairments and disordered speech in schizophrenia: Thought disorder, disorganization, and communication failure perspectives. *Journal of Abnormal Psychology, 114,* 269–278.

Dodwell, P. C., & Humphrey, G. K. (1990). A functional theory of the McCollough effect. *Psychological Review, 97,* 78–89.

Dohrenwend, B. P., Shrout, P. E., Link, B. G., Skodol, A. E., & Martin, J. L. (1986). Overview and initial results from a risk factor study of depression and schizophrenia. In J. E. Barrett (Ed.), *Mental disorders in the community: Progress and challenge.* New York: Guilford Press.

Dolan, R. J. (2000). Emotion processing in the human brain revealed through functional neuroimaging. In M. S. Gazzaniga (Ed.), *The new cognitive neurosciences* (pp. 115–131). Cambridge, MA: MIT Press.

Dolcos, F., LaBar, K. S., & Cabeza, R. (2004). Dissociable effects of arousal and valence on prefrontal activity indexing emotional evaluation and subsequent memory: An event-related fMRI study. *Neuroimage, 23,* 64–74.

Dolcos, F., Labar, K. S., & Cabeza, R. (2005). Remembering one year later: Role of the amygdala and the medial temporal lobe system in retrieving emotional memories. *Proceedings of the National Academy of Sciences, 102,* 2626–2631.

Domjan, M. (2004). *The essentials of conditioning and learning.* Belmont, CA: Wadsworth.

Donahue, A. B. (2000). Electroconvulsive therapy and memory loss: A personal journey. *Journal of ECT, 16,* 133–143.

Dovidio, J. F., ten Vergert, M., Stewart, T. L., Gaertner, S. L., Johnson, J. D., Esses, V. M., et al. (2004). Perspective and prejudice: Antecedents and mediating mechanisms. *Personality and Social Psychology Bulletin, 30,* 1537–1549.

Drummond, S. P., Brown, G. G., Gillin, J. C., Stricker, J. L., Wong, E. C., & Buxton, R. B. (2000). Altered brain response to verbal learning following sleep deprivation. *Nature, 403,* 655–657.

Dugatkin, L. A. (2004). *Principles of animal behavior.* New York: Norton.

Dunbar, R. I. M. (1993). The co-evolution of neocortical size, group size and language in humans. *Behavioral and Brain Sciences, 16*, 681–735.

Duncan, J., Burgess, P., & Emslie, H. (1995). Fluid intelligence after frontal lobe lesions. *Neuropsychologia, 33,* 261–268.

Duncan, J., Seitz, R. J., Kolodny, J., Bor, D., Herzog, H., Ahmed, A., et al. (2000). A neural basis for general intelligence. *Science, 289*, 457–460.

Duncker, K. (1945). On problem solving. *Psychological Monographs, 58,* (5, Whole No. 70).

Dunn, J. (2004). Annotation: Children's relationships with their nonresident fathers. *Journal of Child Psychology and Psychiatry, 45*, 659–671.

Dunnett, S. B., & Fibiger, H. C. (1993). Role of forebrain cholinergic systems in learning and memory: Relevance to the cognitive deficits of aging and Alzheimer's dementia. *Progress in Brain Research, 98*, 413–420.

Dutton, D. G., & Aron, A. P. (1974). Some evidence for heightened sexual attraction under conditions of high anxiety. *Journal of Personality and Social Psychology, 30*, 510–517.

Dykman, B. M., Horowitz, L. M., Abramson, L. Y., & Usher, M. (1991). Schematic and situational determinants of depressed and nondepressed students' interpretation of feedback. *Journal of Abnormal Psychology, 100*, 45–55.

Eacott, M. J. (1999). Memory for the events of early childhood. *Current Directions in Psychological Science, 8*, 46–49.

Eaker, E. D., Sullivan, L. M., Kelly-Hayes, M., D'Agostino, R. B., Sr., & Benjamin, E. J. (2004). Anger and hostility predict the development of atrial fibrillation in men in the Framingham Offspring Study. *Circulation, 109*, 1267–1271.

Earley, P. C. (1993). East meets West meets Mideast: Further explorations of collectivistic and individualistic work groups. *Academy of Management Journal, 36*, 319–348.

Eberhardt, J. L., Goff, P. A., Purdie, V. J., & Davies, P. G. (2004). Seeing black: Race, crime, and visual processing. *Journal of Personality and Social Psychology, 87*, 876–893.

Egeland, J. A., Gerhard, D. S., Pauls, D. L., Sussex, J. N., Kidd, K. K., Allen, C. R., et al. (1987). Bipolar affective disorders linked to DNA markers on chromosome 11. *Nature, 325*, 783–787.

Eggermont, J. J. (2001). Between sound and perception: Reviewing the search for a neural code. *Hearing Research, 157*, 1–42.

Eich, J. E., Weingartner, H., Stillman, R. C., & Gillin, J. C. (1975). State-dependent accessibility of retrieval cues in the retention of a categorized list. *Journal of Verbal Learning and Verbal Behavior, 14*, 408–417.

Eichenbaum, H. (2001). The hippocampus and declarative memory: Cognitive mechanisms and neural codes. Behavioural *and Brain Research, 127,* 199–207.

Eichenbaum, H. B., Cahill, L., Gluck, M., Hasselmo, M., Keil, F., Martin, A., et al. (1999). Learning and memory: Systems analysis. In M. J. Zigmond, F. E. Bloom, S. C. Landis, J. L. Roberts, & L. R. Squire (Eds.), *Fundamental neuroscience* (pp. 1455–1486). San Diego, CA: Academic Press.

Eimas, P. D., Siqueland, E. R., Jusczyk, P., & Vigorito, J. (1971). Speech perception in infants. *Science, 171*, 303–306.

Eisenberg, N. (2000). Emotion, regulation, and moral development. *Annual Review of Psychology, 51*, 665–697.

Ekelund, J., Lichtermann, D., Jaervelin, M., & Peltonen, L. (1999). Association between novelty seeking and type 4 dopamine receptor gene in a large Finnish cohort sample. *American Journal of Psychiatry, 156*, 1453–1455.

Ekman, P., & Davidson, R. J. (1994). *The nature of emotion: Fundamental questions.* New York: Oxford University Press.

Ekman, P., & Friesen, W. V. (1971). Constants across cultures in the face and emotion. *Journal of Personality and Social Psychology, 17*, 124–129.

Ekman, P., & Friesen, W. V. (1975). *Unmasking the face: A guide to recognizing emotions from facial clues.* Englewood Cliffs, NJ: Prentice Hall.

Ekman, P., Levenson, R. W., & Friesen, W. V. (1983). Autonomic nervous system activity distinguishes among emotions. *Science, 221*, 1208–1210.

Ekman, P., Sorenson, E. R., & Friesen, W. V. (1969). Pan-cultural elements in facial displays of emotions. *Science, 164*, 86–88.

Ekselius, L., & von Knorring, L. (1999). Changes in personality traits during treatment with sertraline or citalopram. *British Journal of Psychiatry, 174*, 444–448.

Elfenbein, H. A., & Ambady, N. (2002). On the universality of cultural specificity of emotion recognition: A meta-analysis. *Psychological Bulletin, 128*, 203–235.

Ellis, B. J., Bates, J. E., Dodge, K. A., Fergusson, D. M., Horwood, L. J., Pettit, G. S., et al. (2003). Does father absence place daughters at special risk for early sexual activity and teenage pregnancy? *Child Development*, *74*, 801–821.

Emslie, G. J., Rush, A. J., Weinberg, W. A., Kowatch, R. A., Hughes, C. W., Carmody, T., et al. (1997). A double-blind, randomized, placebo-controlled trial of fluoxetine in children and adolescents with depression. *Archives of General Psychiatry*, *54*, 1031–1037.

Endler, N. (1982). *Holiday of darkness: A psychologist's personal journey out of his depression.* New York: John Wiley & Sons.

Endo, Y., & Meijer, Z. (2004). Autobiographical memory of success and failure experiences. In Y. Kashima, Y. Endo, E. S. Kashima, C. Leung, & J. McClure (Eds.), *Progress in Asian social psychology: Vol. 4* (pp. 67–84). Seoul, Korea: Kyoyook-Kwahak-Sa Publishing Company.

Engle, R. W., & Kane, M. J. (2004). Executive attention, working memory capacity, and a two-factor theory of cognitive control. In B. Ross (Ed.), *The psychology of learning and motivation* (pp. 145–199). New York: Academic Press.

Engle, R. W., Tuholski, S. W., Laughlin, J. E., & Conway, A. R. A. (1999). Working memory, short-term memory, and general fluid intelligence: A latent variable approach. *Journal of Experimental Psychology: General, 128,* 309–331.

Enns, J. (2005). *The thinking eye, the seeing brain.* New York: Norton.

Enns, N., & Reiss, J. P. (1992). Electroconvulsive therapy. *Canadian Journal of Psychiatry, 37*, 671–678.

Enzle, M. E., & Ross, J. M. (1978). Increasing and decreasing intrinsic interest with contingent rewards: A test of cognitive evaluation theory. *Journal of Experimental Social Psychology, 14*, 588–597.

Epperson, C. N., McDougle, C. J., Brown, R. M., Leckman, J. F., Goodman, W. K., & Price, L. H. (1995). OCD during pregnancy and the puerperium [Abstract]. *American Psychiatric Association New Research Abstracts, 84*, NR112.

Era, P., Jokela, J., & Heikkinen, E. (1986). Reaction and movement times in men of different ages: A population study. *Perceptual and Motor Skills, 63*, 111–130.

Erikson, E. H. (1968). *Identity: Youth and crisis*. New York: Norton.

Eriksson, P. S., Perfilieva, E., Bjork-Eriksson, T., Alborn, A. M., Nordborg, C., Peterson, D. A., et al. (1998). Neurogenesis in the adult human hippocampus. *Nature Medicine, 4*, 1313–1317.

Erk, S., Spitzer, M., Wunderlich, A. P., Galley, L., & Walter, H. (2002). Cultural objects modulate reward circuitry. *Neuroreport, 13*, 2499–2503.

Eron, L. D. (1987). The development of aggressive behavior from the perspective of a developing behaviorism. *American Psychologist, 42*, 435–442.

Evers, S., & Suhr, B. (2000). Changes of the neurotransmitter serotonin but not of hormones during short time music perception. *European Archives of Psychiatry and Clinical Neuroscience, 250*, 144–147.

Eysenck, M. W., Mogg, K., May, J., Richards, A., & Matthews, A. (1991). Bias in interpretation of ambiguous sentences related to threat in anxiety. *Journal of Abnormal Psychology, 100*, 144–150.

Fagerström, K. O., & Schneider, N. G. (1989). Measuring nicotine dependence: A review of the Fagerström Tolerance Questionnaire. *Journal of Behavioral Medicine, 12*, 159–181.

Fallon, A. E., & Rozin, P. (1985). Sex differences in perceptions of desirable body shape. *Journal of Abnormal Psychology, 94*, 102–105.

Fantz, R. L. (1966). Pattern discrimination and selective attention as determinants of perceptual development from birth. In A. H. Kidd & L. J. Rivoire (Eds.), *Perceptual development in children*. New York: International Universities Press.

Farah, M. J. (1996). Is face recognition "special"? Evidence from neuropsychology. *Behavioural Brain Research, 76*, 181–189.

Farah, M. J., Levine, D. N., & Calvanio, R. (1988). A case study of mental imagery deficit. *Brain and Cognition, 8*, 147–164.

Fawcett, J. (1992). Suicide risk factors in depressive disorders and in panic disorders. *Journal of Clinical Psychiatry, 53*, 9–13.

Fazio, R. H. (1995). Attitudes as object-evaluation associations: Determinants, consequences, and correlates of attitude accessibility. In R. E. Petty & J. A. Krosnick (Eds.), *Attitude strength: Antecedents and consequences* (pp. 247–282). Hillsdale, NJ: Erlbaum.

Fazio, R. H., Eisner, J. R., & Shook, N. J. (2004). Attitude formation through exploration: Valence asymmetries. *Journal of Personality and Social Psychology, 87*, 293–311.

Feldman, H. H., Gauthier, S., Chetkow, H., Conn, D. K., Freedman, M., & Chris, M. (2006). Canadian guidelines for the development of anti-dementia therapies: A conceptual summary. *Canadian Journal of Neurological Sciences, 33*, 6–26.

Feldman Barrett, L., Lane, R. D., Sechrest, L., & Schwartz, G. E. (2000). Sex differences in emotional awareness. *Personality and Social Psychology Bulletin, 26*, 1027–1035.

Ferguson, G. A. (1982). Psychology at McGill. In M. J. Wright & C. R. Myers (Eds.), *History of academic psychology in Canada* (pp. 33–67). Toronto: Hogrefe.

Ferguson, J. N., Young, L. J., Hearn, E. F., Matzuk, M. M., Insel, T. R., & Winslow, J. T. (2000). Social amnesia in mice lacking the oxytocin gene. *Nature Neuroscience, 25*, 284–288.

Ferguson, M. J., & Bargh, J. A. (2004). How social perception can automatically influence behavior. *Trends in Cognitive Science, 8*, 33–39.

Ferguson, T. J., & Stegge, H. (1995). Emotional states and traits in children: The case of guilt and shame. In J. P. Tangney & K. W. Fischer (Eds.), *Self-conscious emotions* (pp. 174–197). New York: Guilford.

Ferrari, P. F., Gallese, V., Rizzolatti, G., & Fogassi, L. (2003). Mirror neurons responding to the observation of ingestive and communicative mouth actions in the monkey ventral premotor cortex. *European Journal of Neuroscience, 17*, 1703–1714.

Festinger, L. (1954). A theory of social comparison processes. *Human Relations, 7*, 117–140.

Festinger, L. (1957). *A theory of cognitive dissonance*. Stanford, CA: Stanford University Press.

Festinger, L. (1987). A personal memory. In N. E. Grunberg, R. E. Nisbett, J. Rodin, & J. E. Singer (Eds.), *A distinctive approach to psychological research: The influence of Stanley Schachter* (pp. 1–9). New York: Erlbaum.

Festinger, L., & Carlsmith, J. M. (1959). Cognitive consequences of forced compliance. *Journal of Abnormal and Social Psychology, 58*, 203–210.

Festinger, L., Schachter, S., & Back, K. W. (1950). *Social pressures in informal groups*. New York: Harper.

Fibiger, H. C., & Phillips, A. G. (1988). Mesocorticolimbic dopamine systems and reward. *Annals of the New York Academy of Sciences, 537*, 206–215.

Field, T., Fox, N. A., Pickens, J., & Nawrocki, T. (1995). Relative right frontal EEG activation in 3- to 6-month-old infants of "depressed" mothers. *Developmental Psychology, 31*, 358–363.

Fillipek, P. A., Semrud-Clikeman, M., Steingard, R. J., Renshaw, P. F., Kennedy, D. N., & Biederman, J. (1997). Volumetric MRI analysis comparing subjects having attention deficit hyperactivity disorder with normal controls. *Neurology, 48*, 589–600.

Fine, I., Wade, A. R., Brewer, A. A., May, M. G., Goodman, D. F., Boynton, G. M., et al. (2003). Long-term deprivation affects visual perception and cortex. *Nature Neuroscience, 6*, 915–916.

Finger, S. (1994). *Origins of neuroscience*. Oxford, UK: Oxford University Press.

Fink, M. (2001). Convulsive therapy: A review of the first 55 years. *Journal of Affective Disorders, 63* (1–3), 1–15.

Fire, A., Xu, S., Montgomery, M. K., Kostas, S. A., Driver, S. E., & Mello, C. C. (1998). Potent and specific genetic interference by double-stranded RNA in Canorhabditis elegans, *Nature, 391*, 806.

Fischer, K. (1980). A theory of cognitive development: The control and construction of hierarchies of skills. *Psychological Review, 87*, 477–531.

Fisher, W. A., Byrne, D., White, L. A., & Kelley, K. (1988). Erotophobia-erotophilia as a dimension of personality. *Journal of Sex Research, 25*, 123–151.

Fitzgerald, P. B., Brown, T. L., Marston, N. A., Daskalakis, Z. J., De Castella, A., & Kulkarni, J. (2003). Transcranial magnetic stimulation in the treatment of depression: A double-blind, placebo-controlled trial. *Archives of General Psychiatry, 60*, 1002–1008.

Flynn, J. R. (1981). The mean IQ of Americans: Massive gains 1932 to 1978. *Psychological Bulletin, 95*, 29–51.

Flynn, J. R. (1987). Massive IQ gains in 14 nations: What IQ tests really measure. *Psychological Bulletin, 101,* 171–191.

Foa, E. B., Liebowitz, M. R., Kozak, M. J., Davies, S., Campeas, R., Franklin, M. E., et al. (2005). Randomized, placebo-controlled trial of exposure and ritual prevention, clomipramine, and their combination in the treatment of obsessive-compulsive disorder. *American Journal of Psychiatry*, *162,* 151–161.

Foley, K. M. (1993). Opioids. *Neurologic Clinics, 11*, 503–522.

Folkman, S., & Moskowitz, J. T. (2000). Positive affect and the other side of coping. *American Psychologist, 55*, 647–654.

Forgas, J. P. (1998). Asking nicely: Mood effects on responding to more or less polite requests. *Personality and Social Psychology Bulletin, 24*, 173–185.

Fossey, D. (1988). *Gorillas in the mist.* New York: Houghton Mifflin.

Fouad, N. A., & Walker, C. M. (2005). Cultural influences on responses to items on the strong interest inventory. *Journal of Vocational Behavior*, *66*, 104–123.

Fox, N. A., Henderson, H. A., Marshall, P. J., Nichols, K. E., & Ghera, M. M. (2005). Behavioral inhibition: Linking biology and behavior within a developmental framework. *Annual Review of Psychology, 56,* 235–262.

Fox, R., Aslin, R. N., Shea, S. L., & Dumais, S. T. (1980). Stereopsis in human infants. *Science, 207*, 323–324.

Fraley, R. C., & Shaver, P. R. (2000). Adult romantic attachment: Theoretical developments, emerging controversies, and unanswered questions. *Review of General Psychology, 4,* 132–154.

Frangou, S., Chitins, X., & Williams, S. C. (2004). Mapping IQ and gray matter density in healthy young people. *Neuroimage, 23*, 800–805.

Frank, E., Kupfer, D., Perel, I., Comes, C., Jarret, D., Mallinger, A., et al. (1990). Three-year outcomes for maintenance therapies in recurrent depression. *Archives of General Psychiatry, 47*, 1093–1099.

Franken, R. E. (1998). *Human motivation*. Pacific Grove, CA: Brooks Cole.

Franzoi, S. (2005). *Social psychology* (4th ed.). Boston: McGraw-Hill.

Frasure-Smith, N., & Lesperance, F. (2003). Depression and other psychological risks following myocardial infarction. *Archives of General Psychiatry, 60*, 627–636.

Fratiglioni, L., Paillard-Borg, S., & Winblad, B. (2004). An active and socially integrated lifestyle in late life might protect against dementia. *Lancet Neurology*, *3*, 343–353.

Fredrickson, B. L. (2001). The role of positive emotions in positive psychology: The broaden-and-build theory of positive emotions. *American Psychologist*, *56*, 218–226.

Freedman, J. L. (1984). Effects of television violence on aggression. *Psychological Bulletin, 96*, 227–246.

Fried, P. A. (1989). Cigarettes and marijuana: Are there measurable long-term neurobehavioral teratogenic effects? *Neurotoxicology, 10*, 577–583.

Fried, P. A., Watkinson, B., & Gray, R. (2005). Neurocognitive consequences of marihuana—a comparison with pre-drug performance. *Neurotoxicology & Teratology, 27*, 231–239.

Frijda, N. H. (1994). Emotions are functional, most of the time. In P. Ekman & R. J. Davidson (Eds.), *The nature of emotion: Fundamental questions. Series in affective science* (pp. 112–122). New York: Oxford University Press.

Frincke, J. L., & Pate, W. E. (2004). Yesterday, today, and tomorrow. Careers in psychology: 2004. What students need to know. Presented at the Annual Convention of the Southeastern Psychology Association, March, Atlanta, GA. Available at http://research.apa.org/

Frith, C., & Johnstone, E. (2003). *Schizophrenia: A very short introduction.* New York: Oxford University Press.

Funder, D. C. (1995). On the accuracy of personality judgment: A realistic approach. *Psychological Review, 102*, 652–670.

Funder, D. C. (2001). Personality. *Annual Review of Psychology, 52*, 197–221.

Funder, D. C. (2004). *The personality puzzle* (3rd ed.). New York: Norton.

Funder, D. C., & Colvin, C. R. (1991). Explorations in behavioral consistency: Properties of persons, situations, and behaviors. *Journal of Personality and Social Psychology, 60*, 773–794.

Fung, H. H., & Carstensen, L. L. (2004). Motivational changes in response to blocked goals and foreshortened time: Testing alternatives to socioemotional selectivity theory. *Psychology and Aging*, *19*, 68–78.

Fung, M. T., Raine, A., Loeber, R., Lynam, D. R., Steinhauer, S. R., Venables, P. D., et al. (2005). Reduced electrodermal activity in psychopathy-prone adolescents. *Journal of Abnormal Psychology, 114*, 187–196.

Gabrenya, W. K., Wang, Y., & Latane, B. (1985). Social loafing on an optimizing task: Cross-cultural differences among Chinese and Americans. *Journal of Cross-Cultural Psychology, 16*, 223–242.

Galef, B. G., Jr., & Whiskin, E. E. (2000). Social influences on the amount eaten by Norway rats. *Appetite, 34*, 327–332.

Gallistel, C. R. (2000). The replacement of general-purpose learning models with adaptively specialized learning modules. In M. S. Gazzaniga (Ed.), *The new cognitive neurosciences* (pp. 1179–1191). Cambridge, MA: MIT Press.

Gallup Organization (1995). *Disciplining children in America: A Gallup poll report.* Princeton, NH: Author.

Gangestad, S. W., Simpson, J. A., Cousins, A. J., Garver-Apgar, C. E., & Christensen, P. N. (2004). Women's preferences for male behavioral displays change across the menstrual cycle. *Psychological Science*, *15*, 203–207.

Garcia, J., & Koelling, R. A. (1966). Relation of cue to consequence in avoidance learning. *Psychonomic Science, 4*, 123–124.

Gardner, H. (1983). *Frames of mind: The theory of multiple intelligences.* New York: Basic Books.

Gardner, W. L., & Martinko, M. J. (1996). Using the Myers-Briggs Type Indicator to study managers: A literature review and research agenda. *Journal of Management*, *22*, 45–83.

Gardner, W. L., Pickett, C. L., & Brewer, M. B. (2000). Social exclusion and selective memory: How the need to belong affects memory for social information. *Personality and Social Psychology Bulletin, 26*, 486–496.

Garlick, D. (2002). Understanding the nature of the general factor of intelligence: The role of individual differences in neural plasticity as an explanatory mechanism. *Psychological Review, 109,* 116–136.

Garner, D. M., Garfinkel, P. E., Schwartz, D., & Thompson, M. (1980). Cultural expectations of thinness in women. *Psychological Reports, 47,* 483–491.

Gazette: Monkeys move matter, mentally. (2004). *Duke Magazine, 90.* Retrieved September 21, 2005, from http://www.dukemagazine.duke.edu/dukemag/issues/010204/depgaz8.html

Gazzaniga, M. S. (1998). *The mind's past.* Berkeley: University of California Press.

Gazzaniga, M. S. (2000). Cerebral specialization and interhemispheric communication: does the corpus callosum enable the human condition? *Brain, 123,* 1293–1326.

Gazzaniga, M. S. (2004). *The cognitive neurosciences* (3rd ed.). Cambridge, MA: MIT Press.

Gazzaniga, M. S., Ivry, R. B., & Mangun, G. R. (2002). *Cognitive neuroscience: The biology of the mind* (2nd ed.). New York: Norton.

Geddes, J. R., Burgess, S., Hawton, K., Jamison, K., & Goodwin, G. M. (2004). Long-term lithium therapy for bipolar disorder: Systematic review and meta-analysis of randomized controlled trials. *American Journal of Psychiatry, 161,* 217–222.

Geen, R. G. (1984). Preferred stimulation levels in introverts and extraverts: Effects on arousal and performance. *Journal of Personality and Social Psychology, 46,* 1303–1312.

Geertz, C. (1973). *The interpretation of cultures.* New York: Basic Books.

George, M. S., Lisanby, S. H., & Sackheim, H. A. (1999). Transcranial Magnetic Stimulation: Applications in neuropsychiatry. *Archives of General Psychiatry, 56,* 300–311.

George, M. S., Wassermann, E. M., Williams, W. A., Callahan, A., Ketter, T. A., Basser, P., et al. (1995). Daily repetitive transcranial magnetic stimulation (rTMS) improves mood in depression. *Neuroreport, 6,* 1853–1856.

Gergely, G., & Csibra, G. (2003). Teleological reasoning in infancy: The naïve theory of rational action. *Trends in Cognitive Sciences, 7,* 287–292.

Gershoff, E. T. (2002). Parental corporal punishment and associated child behaviors and experiences: A meta-analytic and theoretical review. *Psychological Bulletin, 128,* 539–579.

Gershon, E. S., Berrettini, W. H., & Goldin, L. R. (1989). Mood disorders: Genetic aspects. In H. I. Kaplan & B. J. Sadock (Eds.), *Comprehensive textbook of psychiatry* (5th ed). Baltimore: Williams & Wilkins.

Geula, C., & Mesulam, M. (1994). Cholinergic systems and related neuropathological predilection patterns in Alzheimer disease. In R. D. Terry, R. Katzman, & K. Bick (Eds.), *Alzheimer disease* (pp. 263–291). New York: Raven Press.

Gibson, J. J. (1966). *The senses considered as perceptual systems.* Boston: Houghton Mifflin.

Gibson, J. J. (1979). *The ecological approach to visual perception.* Boston: Houghton Mifflin.

Gick, M. L., & Holyoak, K. J. (1983). Schema induction and analogical transfer. *Cognitive Psychology, 15,* 1–38.

Giedd, J. N. (2004). Structural magnetic resonance imaging of the adolescent brain. *Annals of the New York Academy of Sciences, 1021,* 105–109.

Giffiths, J., Ravindran, A. V., Merali, Z., & Anisman, H. (2000). Dysthymia: A review of pharmacological and behavioural factors. *Molecular Psychiatry, 5,* 242–261.

Gigerenzer, G. (2004). Dread risk, September 11, and fatal traffic accidents. *Psychological Science, 15,* 286–287.

Gigerenzer, G., Todd, P. M., & the ABC Research Group (1999). *Simple heuristics that make us smart.* New York: Oxford University Press.

Gijsman, H. J., Geddes, J. R., Rendell, J. M., Nolen, W. A., & Goodwin, G. M. (2004). Antidepressants for bipolar depression: A systematic review of randomized, controlled trials. *American Journal of Psychiatry, 161,* 1537–1547.

Gilberg, C. (1980). Maternal age and infantile autism. *Journal of Autism and Developmental Disorders, 10,* 293–297.

Gilbert, D. T. (2005, January 20). Four more years of happiness. *The New York Times,* p. A23.

Gilbert, D. T., Morewedge, C. K., Risen, J. L., & Wilson, T. D. (2004). Looking forward to looking backward: The misprediction of regret. *Psychological Science, 15,* 346–350.

Gilbert, D. T., Pinel, E. C., Wilson, T. D., Blumberg, S. J., & Wheatley, T. (1998). Immune neglect: A source of durability bias in affective forecasting. *Journal of Personality and Social Psychology, 75,* 617–638.

Gillihan, S. J, & Farah, M. J. (2005). Is self special? A critical review of evidence from experimental psychology and cognitive neuroscience. *Psychological Bulletin, 131,* 76–97.

Gillis, J. J., Gilger, J. W., Pennington, B. F., & DeFries, J. C. (1992). Attention deficit disorder in reading-disabled twins: Evidence for a genetic etiology. *Journal of Abnormal Child Psychology, 20* (3), 303–315.

Gilovich, T. (1991). *How we know what isn't so: The fallibility of human reason in everyday life.* New York: The Free Press.

Gilovitch, T., Keltner, D., & Nisbett, R. E. (2006). *Social psychology.* New York: Norton.

Gilpin, E. A., Choi, W. S., Berry, C., & Pierce, J. P. (1999). How many adolescents start smoking each day in the United States? *Journal of Adolescent Health, 25,* 248–255.

Gladwell, M. (2005). *Blink.* New York: Little, Brown.

Glaser, R., & Kiecolt-Glaser, J. K. (1994). *Handbook of human stress and immunity.* San Diego: Academic Press.

Glenn, N. (1998). The course of marital success and failure in five American 10-year marriage cohorts. *Journal of Marriage and the Family, 60,* 569–576.

Goddard, G. V., McIntyre, D. C., & Leech, C. K. (1969). A permanent change in brain function resulting from daily electrical stimulation. *Experimental Neurology, 25,* 295–330.

Godden, D. R., & Baddeley, A. D. (1975). Context-dependent memory in two natural environments: On land and underwater. *British Journal of Psychology, 66,* 325–331.

Golby, A. J., Gabrieli, J. D. E., Chiao, J. Y., & Eberhardt, J. L. (2001). Differential responses in the fusiform region to same-race and other-race faces. *Nature Neuroscience, 4,* 845–850.

Gold, P. E. (1987). Sweet memories. *American Scientist, 75,* 151–155.

Goldapple, K., Segal, Z., Garson, C., Lau, M., Bieling, P., Kennedy, S., et al. (2004). Modulation of cortical-limbic pathways in major depression: Treatment-specific effects of cognitive behavior therapy. *Archives of General Psychiatry, 61*, 34–41.

Goldberg, S., Muir, R., & Kerr, J. (1997). *Attachment theory: Social, developmental, and clinical perspectives.* Hillsdale, NJ: Analytic Press.

Goldenberg, J. L., McCoy, S. K., Pyszczynski, T., Greenberg, J., & Solomon, S. (2000). The body as a source of self-esteem: The effects of mortality salience on identification with one's body, interest in sex, and appearance monitoring. *Journal of Personality and Social Psychology, 79*, 118–130.

Goldman-Rakic, P. S., Scalaidhe, S., & Chafee, M. (2000). Domain specificity in cognitive systems. In M. S. Gazzaniga (Ed.), *The new cognitive neurosciences* (pp. 733–742). Cambridge, MA: MIT Press.

Gong, Q., Sluming, V., Mayes, A., Keller, S., Barrick, T., Cezayirli, E., et al. (2005). Voxel-based morphometry and stereology provide convergent evidence of the importance of medial prefrontal cortex for fluid intelligence in healthy adults. *Neuroimage, 25*, 1175–1186.

Goodale, M. A., & Milner, A. D. (1992). Separate visual pathways for perception and action. *Trends in Neuroscience, 15*, 22–25.

Goodall, G. (1984). Learning due to the response-shock contingency in signaled punishment. *Quarterly Journal of Experimental Psychology, 36*, 259–279.

Goodman, R., & Stevenson, J. (1989). A twin study of hyperactivity-II. The aetiological role of genes, family relationships, and perinatal adversity. *Journal of Child Psychology and Psychiatry, 30*, 691–709.

Goodwin, D. W., Powell, B., Bremer, D., Hoine, H., & Stern, J. (1969). Alcohol and recall: State dependent effects in man. *Science, 163*, 1358.

Goodwin, R. D., Lieb, R., Hoefler, M., Pfister, H., Bittner, A., Beesdo, K., et al. (2004). Panic attack as a risk factor for severe psychopathology. *American Journal of Psychiatry, 161*, 2207–2214.

Gopnik, A., & Graf, O. (1988). Knowing how you know: Young children's ability to identify and remember the sources of their beliefs. *Child Development, 59*, 1366–1371.

Gordon, P. (2004). Numerical cognition without words: Evidence from Amazonia. *Science, 306*, 496–499.

Gosling, S. D. (1998). Personality dimensions in spotted hyenas (Crocuta crocuta). *Journal of Comparative Psychology, 112*, 107–118.

Gosling, S. D. (2001). From mice to men: What can we learn about personality from animal research? *Psychological Bulletin, 127*, 45–86.

Gosling, S. D., & John, O. P. (1999). Personality dimension in non-human animals: A cross-species review. *Current Directions in Psychological Science, 8*, 69–75.

Gosling, S. D., Kwan, V. S. Y., & John, O. P. (2003). A dog's got personality: A cross-species comparative approach to evaluating personality judgments. *Journal of Personality and Social Psychology, 85*, 1161–1169.

Gosselin, P., & Larocque, C. (2000). Facial morphology and children's categorization of facial expressions of emotions: A comparison between Asian and Caucasian faces. *Journal of Genetic Psychology, 161*, 346–358.

Gottesman, I. I. (1991). *Schizophrenia genesis: The origins of madness.* New York: Freeman.

Gottesman, I. I., & Hanson, D. R. (2005). Human development: Biological and genetic processes. *Annual Review of Psychology*, 56, 263-286

Gottfredson, G. D. (1999). John Holland's contributions to vocational psychology: A review and evaluation. *Journal of Vocational Behavior, 55*, 15–40.

Gottfredson, L. S. (1997). Why g matters: The complexity of everyday life. *Intelligence, 24*, 79–132.

Gottfredson, L. S. (2004a). Intelligence: Is it the epidemiologists' elusive "fundamental cause" of social class inequalities in health? *Journal of Personality and Social Psychology, 86*, 174–199.

Gottfredson, L. S. (2004b). Schools and the g factor. *The Wilson Quarterly, Summer*, 35–45.

Gottfredson, L., & Deary, I. J. (2004). Intelligence predicts health and longevity: But why? *Current Directions in Psychological Science, 13*, 1–4.

Gottman, J. (1994). *Why marriages succeed or fail.* New York: Simon & Schuster.

Gottman, J. M. (1983). How children become friends. *Monographs of the Society for Research in Child Development 48* (3, Serial No. 201).

Gottman, J. M. (1998). Psychology and the study of marital processes. *Annual Review of Psychology, 49*, 169–197.

Gould, E., & Tanapat, P. (1999). Stress and hippocampal neurogenesis. *Biological Psychiatry, 46*, 1472–1479.

Grandin, T. (1995). *Thinking in pictures: And other reports from my life with autism.* New York: Doubleday.

Gray, J. R., Chabris, C. F., & Braver, T. S. (2003). Neural mechanisms of general fluid intelligence. *Nature Neuroscience, 6*, 316–322.

Gray, J. R., & Thompson, P. M. (2004). Neurobiology of intelligence: Science and ethics. *Nature Reviews Neuroscience, 5*, 471– 482.

Green, C. D. (2002). Introduction to "Experimental psychology and the laboratory in Toronto" by Albert H. Abbott (1900). Retrieved from http://psychclassics.yorku.ca/Abbott/intro.htm

Green, L., Myerson, J., Lichtman, D., Rosen, S., & Fry, A. (1996). Temporal discounting in choice between delayed rewards: The role of age and income. *Psychological Science, 11*, 79–94.

Greenberg, J., Solomon, S., & Pyszczynski, T. (1997). Terror management theory of self-esteem and cultural worldviews: Empirical assessments and conceptual refinements. In M. P. Zanna (Ed.), *Advances in Experimental Social Psychology: Vol. 29* (pp. 61–136). New York: Academic Press.

Greenwald, A. G., & Banaji, M. R. (1995). Implicit social cognition: Attitudes, self-esteem, and stereotypes. *Psychological Review, 102*, 4–27.

Greenwald, A. G., McGhee, D., & Schwartz, J. (1998). Measuring individual differences in implicit cognition: The implicit association test. *Journal of Personality and Social Psychology, 74*, 1464–1480.

Greenwald, A. G., Oakes, M. A., & Hoffman, H. (2003). Targets of discrimination: Effects of race on responses to weapons holders. *Journal of Experimental Social Psychology, 39*, 399–405.

Gregory, A. H., Worrall, L., & Sarge, A. (1996). The development of emotional responses to music in young children. *Motivation and Emotion, 20*, 341–348.

Grill-Spector, K., Knouf, N., & Kanwisher, N. (2004). The fusiform face area subserves face perception, not generic within-category identification. *Nature Neurosciences, 7,* 555–562.

Gross, C. G., Rocha Miranda, C. E., & Bender, D. B. (1972). Visual properties of neurons in inferotemporal cortex of the macaque. *Journal of Neurophysiology, 35,* 96–111.

Gross, J. J. (1999). Emotion and emotion regulation. In L. A. Pervin & O. P. John (Eds.), *Handbook of personality: Theory and research* (2nd ed., pp. 525–552). New York: Guilford Press.

Grossman, L. (2005, January 24). Grow up? Not so fast. *Time, 165,* 42–53.

Grossman, M., & Wood, W. (1993). Sex differences in intensity of emotional experience: A social role interpretation. *Journal of Personality and Social Psychology, 65,* 1010–1022.

Gruenewald, P. J., Millar, A. B., Treno, A. J., Yang, Z., Ponicki, W. R., & Roeper, P. (1996). The geography of availability and driving after drinking. *Addiction, 91,* 967–983.

Guerin, B. (1994). What do people think about the risks of driving? Implications for traffic safety interventions. *Journal of Applied Social Psychology, 24,* 994–1021.

Gunderson, J. G. (1984). *Borderline personality disorder.* Washington, DC: American Psychiatric Press.

Gupta, U., & Singh, P. (1982). Exploratory study of love and liking and type of marriages. *Indian Journal of Applied Psychology, 19,* 92–97.

Habib, R., Nybert, L., & Tulving, E. (2003). Hemispheric asymmetries of memory: The HERA model revisited. *Trends in Cognitive Sciences, 7,* 241–245.

Haidt, J., Koller, S. H., & Dias, M. G. (1993). Affect, culture, and morality, or is it wrong to eat your dog? *Journal of Personality and Social Psychology, 65,* 613–628.

Haier, R. J., Jung, R. E., Yeo, R. A., Head, K., & Alkire, M. T. (2004). Structural brain variation and general intelligence. *Neuroimage, 23,* 425–433.

Haier, R. J., Jung, R. E., Yeo, R. A., Head, K., & Alkire, M. T. (2005). The neuroanatomy of general intelligence: Sex matters. *Neuroimage, 25,* 320–327.

Hall, E. T. (1976). *Beyond culture.* New York: Anchor Books.

Hall, W., & Degenhardt, L. (2006). What are the policy implications of the evidence on cannabis and psychosis? *Canadian Journal of Psychiatry, 51,* 566–574.

Halligan, P. W., Marshall, J. C. (1998). Neglect of awareness. *Consciousness and Cognition, 7,* 356–380.

Halpern, C. T., Udry, J. R., & Suchindran, C. (1997). Testosterone predicts initiation of coitus in adolescent females. *Psychosomatic Medicine, 59,* 161–171.

Hamann, S. B., Ely, T. D., Grafton, D. T., & Kilts, C. D. (1999). Amygdala activity related to enhanced memory for pleasant and aversive stimuli. *Nature Neuroscience, 2,* 289–293.

Hamann, S., Herman, R. A., Nolan, C. L., & Wallen, K. (2004). Men and women differ in amygdala response to visual sexual stimuli. *Nature Neuroscience, 7,* 411–416.

Hambrecht, M., Maurer, K., Hafner, H., & Sartorius, N. (1992). Transnational stability of gender differences in schizophrenia: Recent findings on social skills training and family psychoeducation. *Clinical Psychology Review, 11,* 23–44.

Hammen, C. (2005). Stress and depression. *Annual Review of Clinical Psychology, 1,* 293–319.

Han, S., & Shavit, S. (1994). Persuasion and culture: Advertising appeals in individualistic and collectivist societies. *Journal of Experimental Social Psychology, 30,* 326–350.

Hancock, P. J., Bruce, V. V., & Burton, A. M. (2000). Recognition of unfamiliar faces. *Trends in Cognitive Sciences, 4,* 330–337.

Handleman, J. S., Gill, M. J., & Alessandri, M. (1988). Generalization by severely developmentally disabled children: Issues, advances, and future directions. *Behavior Therapist, 11,* 221–223.

Haney, C., Banks, C., & Zimbardo, P. (1973). Interpersonal dynamics in a simulated prison. *International Journal of Criminology and Penology, 1,* 69–97.

Hansen, C. J., Stevens, L. C., & Coast, J. R. (2001). Exercise duration and mood state: How much is enough to feel better? *Health Psychology, 20,* 267–275.

Hansen, W. B., Graham, J. W., Sobel, J. L., Shelton, D. R., Flay, B. R., & Johnson, C. A. (1987). The consistency of peer and parental influences on tobacco, alcohol, and marijuana use among young adolescents. *Journal of Behavioral Medicine, 10,* 559–579.

Harding, C. M., Zubin, J., & Strauss, J. S. (1987). Chronicity in schizophrenia: Fact, partial fact, or artifact? *Hospital and Community Psychiatry, 38,* 477–486.

Hare, R. D. (1965). Temporal gradient of fear arousal in psychopaths. *Journal of Abnormal Psychology, 70,* 442–445.

Hare, R. D. (1993). *Without conscience: The disturbing world of the psychopaths among us.* New York: Pocket Books.

Hare, R. D. (1999). Psychopathy as a risk factor for violence. *Psychiatry Quarterly, 70,* 181–197.

Hare, R. D., McPherson, L. M., & Forth, A. E. (1988). Male psychopaths and their criminal careers. *Journal of Consulting and Clinical Psychology, 56,* 710–714.

Hariri, A. R., Mattay, V. S., Tessitore, A., Kolachana, B., Fera, F., Goldman, D., et al. (2002). Serotonin transporter genetic variation and the response of the human amygdala. *Science, 297,* 400–440.

Harlow, H. F., Harlow, M. K., & Meyer, D. R. (1950). Learning motivated by a manipulation drive. *Journal of Experimental Psychology, 40,* 228–234.

Harmer, C. J., Perrett, D. I., Cowen, P. J., & Goodwin, G. M. (2001). Administration of the beta-adrenoceptor blocker propranolol impairs the processing of facial expressions of sadness. *Psychopharmacology, 154,* 383–389.

Harper, S. Q., Staber, P. D., He, X., Eliason, S. L., Martins, I. H., Mao, Q., et al. (2005). RNA interference improves motor and neuropathological abnormalities in a Huntington's disease mouse model. *Proceedings of the National Academy of Sciences, 102,* 5820–5825.

Harpur, T. J., & Hare, R. D. (1994). Assessment of psychopathy as a function of age. *Journal of Abnormal Psychology, 103,* 604–609.

Harris, J. R. (1995). Where is the child's environment? A group socialization theory of development. *Psychological Review, 102,* 458–489.

Harris, J. R. (1998). *The nurture assumption: Why children turn out the way they do.* New York: Free Press.

Hartsough, C. S., & Lambert, N. M. (1985). Medical factors in hyperactive and normal children: Prenatal, developmental, and health history findings. *American Journal of Orthopsychiatry, 55,* 190–201.

Harvey, C. B., Manshu, Z., Bio, K. C., & Jue, Z. F. (1986). Spatial conceptions in Chinese and Canadian children. *Journal of Genetic Psychology, 147,* 457–464.

Hauk, O., Johnsrude, I., & Pulvermüller, F. (2004). Somatotopic representation of action words in human motor and premotor cortex. *Neuron, 41,* 301–307.

Haydon, P. G. (2001). Glia: Listening and talking to the synapse. *Nature Reviews: Neuroscience, 2,* 185–193.

Hayley, S., Poulter, M., Merali, Z., & Anisman, H. (2005). The pathogenesis of clinical depression: Stressor and cytokine induced alteration of neurplasticity. *Neuroscience, 135,* 659–678.

Hazan, C., & Shaver, P. R. (1987). Romantic love conceptualized as an attachment process. *Journal of Personality and Social Psychology, 52,* 511–524.

Heatherton, T. F., Herman, C. P., & Polivy, J. (1991). Effects of physical threat and ego threat on eating behavior. *Journal of Personality and Social Psychology, 60,* 138–143.

Heatherton, T. F., & Baumeister, R. F. (1991). Binge eating as escape from self-awareness. *Psychological Bulletin, 110,* 86–108.

Heatherton, T. F., & Nichols, P. A. (1994). Personal accounts of successful versus failed attempts at life change. *Personality and Social Psychology Bulletin, 20,* 664–675.

Heatherton, T. F., & Vohs, K.D. (2000). Interpersonal evaluations following threats to self: Role of self-esteem. *Journal of Personality and Social Psychology, 78,* 725–736.

Heatherton, T. F., & Weinberger, J. L. (1994). *Can personality change?* Washington, DC: American Psychological Association.

Hebb, D. O. (1949). *The organization of behavior.* New York: Wiley.

Hebl, M. R., & Heatherton, T. F. (1998). The stigma of obesity in women: The difference is black and white. *Personality and Social Psychology Bulletin, 24,* 417–426.

Hegarty, J., Baldessarini, R., Tohen, M., Waternaux, C., & Oepen, G. (1994). One hundred years of schizophrenia: A meta-analysis of the outcome literature. *American Journal of Psychiatry, 151,* 1409–1416.

Heider, F. (1958). *The psychology of interpersonal relations.* New York: John Wiley and Sons.

Heine, S. J. (2003). An exploration of cultural variation in self-enhancing and self-improving motivations. In V. Murphy-Berman & J. J. Berman (Eds.), *Nebraska symposium on motivation: Vol. 49. Cross-cultural differences in perspectives on the self* (pp. 101–128). Lincoln: University of Nebraska Press.

Heine, S. J. (2005). Where is the evidence for pancultural self-enhancement? A reply to Sedikides, Gaertner, & Toguchi. *Journal of Personality and Social Psychology, 89,* 531–538.

Heine, S. J. (forthcoming). *Cultural psychology.* New York: Norton.

Heine, S. J., & Lehman, D. R. (1997). Culture, dissonance, and self-affirmation. *Personality and Social Psychology Bulletin, 23,* 389–400.

Heine, S. J., & Lehman, D. R. (2004). Move the body, change the self: Acculturative effects on the self-concept. In M. Schaller & C. Crandall (Eds.), *Psychological Foundations of Culture,* (pp. 305–331). Mahwah, NJ: Erlbaum.

Heine, S. J., Lehman, D. R., Markus, H. R., & Kitayama, S. (1999). Is there a universal need for positive self-regard? *Psychological Review, 106,* 766–794.

Heishman, S. J., Taylor, R. C., & Henningfield, J. E. (1994). Nicotine and smoking: A review of effects on human performance. *Experimental and Clinical Psychopharmacology, 2,* 345–395.

Heller, D., Watson, D., & Ilies, R. (2004). The role of person versus situation in life satisfaction: A critical examination. *Psychological Bulletin, 130,* 574–600.

Helmholtz, H. von. (1909–1911). *Treatise on physiological optics* (3rd ed., vols. 2–3, J. P. Southall, Ed. and Trans.). Rochester, NY: Optical Society of America. (Original work published 1866)

Helmreich, R., Aronson, E., & LeFan, J. (1970). To err is humanizing sometimes: Effects of self-esteem, competence, and a pratfall on interpersonal attraction. *Journal of Personality and Social Psychology, 16,* 259–264.

Henderson, L., & Zimbardo, P. G. (1998). Shyness. *Encyclopedia of Mental Health, 3,* 497–509.

Henke, P. G. (1982). The telencephalic limbic system and experimental gastric pathology: A review. *Neuroscience and Behavioral Reviews, 6,* 381–390.

Henrich, J., Boyd, R., Bowles, S., Gintis, H., Fehr, E., Camerer, C., McElreath, R., Gurven, M., Hill, K., Barr, A., Ensminger, J., Tracer, D., Marlow, F., Patton, J., Alvard, M., Gil-White, F., & Henrich, N. (2005). "Economic man" in cross-cultural perspective: Ethnography and experiments from 15 small-scale societies. *Behavioral and Brain Sciences, 28,* 795–855.

Hering, E. (1964). *Outlines of a theory of the light sense* (L. M. Hurvich & D. Jameson, Trans.). Cambridge, MA: Harvard University Press. (Original work published 1878)

Hernstein, R. J., & Murray, C. (1994). *The bell curve: Intelligence and class structure in American life.* New York: Free Press.

Herz, R. S., & Cahill, E. D. (1997). Differential use of sensory information in sexual behavior as a function of gender. *Human Nature, 8,* 275–286.

Hetherington, E. M., Bridges, M., & Insabella, G. M. (1998). What matters? What does not? Five perspectives on the association between marital transitions and children's adjustment. *American Psychologist, 53,* 167–184.

Higgins, E. T. (1987). Self-discrepancy: A theory relating self and affect. *Psychological Review, 94,* 319–340.

Higgins, E. T., & Kruglanski, A. W. (1996). *Social psychology: Handbook of basic principles.* New York: Guilford Press.

Higley, J. D., Mehlman, P. T., Higley, S. B., Fernald, B., Vickers, J., Lindell, et al. (1996). Excessive mortality in young free-ranging male nonhuman primates with low cerebrospinal fluid 5-hydroxyindoleacetic acid concentrations. *Archives of General Psychiatry, 53,* 537–543.

Hill, J. (2002). High-speed police pursuits: Danger, dynamics, and risk reduction. *FBI Law Enforcement Bulletin, 71,* 14–18.

Hingson, R., Heeren, T., Levenson, S., Jamanka, A., & Voas, R. (2002). Age of drinking onset, driving after drinking, and involvement in alcohol related motor vehicle crashes. *Accident Analysis and Prevention, 34,* 85–92.

Ho, B. C., Andreasen, N. C., Nopoulos, P., Arndt, S., Magnotta, V., & Flaum, M. (2004). Progressive structural brain abnormalities and their relationship to clinical outcome: A longitudinal magnetic resonance imaging study early in schizophrenia. *Archives of General Psychiatry, 60*, 585–594.

Hobson, J. A. (1995). *Sleep*. New York: Scientific American Library.

Hobson, J. A. (1999). Sleep and dreaming. In M. J. Zigmond, F. E. Bloom, S. C. Landis, J. L. Roberts, & L. R. Squire (Eds.), *Fundamentals of neuroscience* (pp. 1207–1225). San Diego, CA: Academic Press.

Hobson, J. A., Pace-Schott, E. F., & Stickgold, R. (2000). Consciousness: Its vicissitudes in waking and sleep. In M. S. Gazzaniga (Ed.), *The new cognitive neurosciences* (pp. 1341–1354). Cambridge, MA: MIT Press.

Hoffman, D. D. (1998). *Visual intelligence*. New York: Norton.

Hogan, R., Johnson, J., & Briggs, S. (1997). *Handbook of personality psychology*. San Diego: Academic Press.

Holland, J. L. (1997). *Making vocational choices: A theory of vocational personalities and work environments* (3rd ed.). Odessa, FL: Psychological Assessment Resources.

Holland, P. C. (1977). Conditioned stimulus as a determinant of the form of the Pavlovian conditioned response. *Journal of Experimental Psychology: Animal Behavior Processes, 3*, 77–104.

Hollander, E., Novotny, S., Hanratty, M., Yaffe, R., DeCaria, C. M., Aronowitz, B. R., et al. (2003). Oxytocin infusion reduces repetitive behaviors in adults with autistic and Asperger's disorders. *Neuropsychopharmacology, 28*, 193–198.

Hollis, K. L. (1997). Contemporary research on Pavlovian conditioning: A "new" functional analysis. *American Psychologist, 52*, 956–965.

Hollon, S. D., Thase, M. E., & Markowitz, J. C. (2002). Treatment and prevention of depression. *Psychological Science in the Public Interest, 3*, 39–77.

Hong, Y., Morris, M. W., Chiu, C., & Benet-Martinez, V. (2000). Multicultural minds: A dynamic constructivist approach to culture and cognition. *American Psychologist, 55*, 705–720.

Hooley, J. M., & Gotlib, I. H. (2000). A diathesis-stress conceptualization of expressed emotion and clinical outcome. *Applied and Preventive Psychology, 9*, 135–152.

Horn, J. L., & Hofer, S. M. (1992). Major abilities and development in the adult period. In R. J. Sternberg & C. A. Berg (Eds.), *Intellectual development* (pp. 44–99). New York: Cambridge University Press.

Horwath, E., & Weissman, M. M. (2000). The epidemiology and cross-national presentation of obsessive-compulsive disorder. *The Psychiatric Clinics of North America, 23*, 493–507.

Hoshino, Y., Kumashiro, H., Yashima, Y., Tachibana, R., Watanabe, M., & Furukawa, H. (1980). Early symptoms of autism in children and their diagnostic significance. *Japanese Journal of Child and Adolescent Psychiatry, 21*, 284–299.

Hoshino-Browne, E., Zanna, A. S., Spencer, S. J., Zanna, M. P., Kitayama, S., & Lackenbauer, S. (2005). On the cultural guises of cognitive dissonance: The case of Easterners and Westerners. *Journal of Personality and Social Psychology, 89*, 294–310.

Hothersall, D. (1995). *History of psychology*. New York: McGraw-Hill.

House, J. S., Landis, K. R., & Umberson, D. (1988). Social relationships and health. *Science, 241*, 540–545.

Howlin, P., Mawhood, L., & Rutter, M. (2000). Autism and developmental receptive language disorder—A follow-up comparison in early adult life. II: Social, behavioural, and psychiatric outcomes. *Journal of Child Psychology and Psychiatry and Allied Disciplines, 41*, 561–578.

Hubel, D. H., & Wiesel, T. N. (1962). Receptive fields, binocular interaction, and functional architecture in the cat's visual cortex. *Journal of Physiology (London), 160*, 106–154.

Huesmann, L. R. (1998). The role of social information processing and cognitive schemas in the acquisition and maintenance of habitual aggressive behavior. In R. G. Geen & E. Donnerstein (Eds.), *Human aggression: Theories, research, and implications for policy* (pp. 73–109). New York: Academic Press.

Huff, D. (1993). *How to lie with statistics*. New York: Norton.

Hughes, H. C. (2000). *Sensory exotica*. Cambridge, MA: MIT Press.

Hull, J. G., & Bond, C. F. (1986). Social and behavioral consequences of alcohol consumption and expectancy: A meta-analysis. *Psychological Bulletin, 99*, 347–360.

Hwang, W. S., Roh, S. I., Lee, B. C., Kang, S. K., Kwon, D. K., et al. (2005). Patient-specific embryonic stem cells derived from human SCNT blastocysts. *Science, 308*, 1777–1783.

Hwu, H. G., Yeh, E. K., & Chang, L. Y. (1989). Prevalence of psychiatric disorders in Taiwan defined by the Chinese Diagnostic Interview Schedule. *Acta Psychiatrica Scandinavica, 79*, 136–147.

Hynd, G. W., Hern, K., & Novey, E. S. (1993). Attention deficit-hyperactivity disorder and asymmetry of the caudate nucleus. *Journal of Child Neurology, 8*, 339–347.

Ilan, A. B., Smith, M. E., & Gevins, A. (2004). Effects of marijuana on neurophysiological signals of working and episodic memory. *Psychopharmacology, 176*, 214–222.

Ilgen, D., & Pulakos, E. D. (1999). *The changing nature of performance*. San Francisco: Jossey-Bass.

Ingham, A. G., Levinger, G., Graves, J., & Peckham, V. (1974). The Ringelmann effect: Studies of group size and group performance. *Journal of Experimental Social Psychology, 10*, 371–384.

Inglehart, R., & Klingemann, H. (2000). Genes, culture, democracy, and happiness. In E. Diener & E. Suh (Eds.), *Culture and subjective well-being* (pp. 165–184). Cambridge, MA: MIT Press.

Insel, T. R. (1992). Oxytocin and the neurobiology of attachment. *Behavioral and Brain Sciences, 15*, 515–516.

Insel, T. R., & Young, L. J. (2001). The neurobiology of attachment. *Nature Reviews Neuroscience, 2*, 129–136.

Isen, A. M. (1993). Positive affect and decision making. In M. Lewis & J. M. Haviland (Eds.), *Handbook of emotions* (pp. 261–277). New York: Guilford Press.

Ivanovic, D. M., Leiva, B. P., Perez, H. T., Olivares, M. G., Diaz, N. S., Urrutia, M. S., et al. (2004). Head size and intelligence, learning, nutritional status and brain development. Head, IQ, learning, nutrition and brain. *Neuropsychologia, 42*, 1118–1131.

Iyengar, S. S., & Lepper, M. R. (1999). Rethinking the value of choice: A cultural perspective on intrinsic motivation. *Journal of Personality and Social Psychology, 76*, 349–366.

Iyengar, S. S., & Lepper, M. R. (2000). When choice is demotivating: Can one desire too much of a good thing? *Journal of Personality and Social Psychology, 79*, 995–1006.

Izard, C. E., & Malatesta, C. Z. (1987). Perspectives on emotional development. In J. Osofsky (Ed.), *Handbook of infant development* (pp. 494–554). New York: Wiley.

Jablensky, A. (1989). Epidemiology and cross-cultural aspects of schizophrenia. *Psychiatric Annals, 19*, 516–524.

Jablensky, A., Sartorius, N., Ernberg, G., Anker, M., Korten, A., Cooper, J. E., Day, R., & Bertelsen, A. (1991). *Schizophrenia: Manifestations, incidence and course in different cultures: A World Health Organization ten-country study. Psychological Medicine,* Monograph Supplement, 20, 1–97. Reprinted from Cambridge University Press.

Jackson, D. N., Ashton, M. E., & Tomes, J. L. (1996a). The six factor model of personality: Facets from the Big Five. *Personality and Individual Differences, 21*, 391–402.

Jackson, D. N., Paunonen, S. V., Fraboni, M., & Goffin, R. D. (1996b). A five factor versus six factor model of personality structure. *Personality and Individual Differences, 20*, 33–45.

Jacoby, L. L., Kelley, C., Brown, J., & Jasechko, J. (1989). Becoming famous overnight: Limits on the ability to avoid unconscious influences of the past. *Journal of Personality and Social Psychology, 56*, 326–338.

James, W. (1983). *Principles of psychology.* Cambridge, MA: Harvard University Press. (Original work published 1890)

Jamison, K. R. (1995). *An unquiet mind.* New York: Vintage Books.

Jamison, K. R. (2004). Behavior: Essay by Kay Redfield Jamison. Johns Hopkins Public Health. Retrieved September 21, 2005 from http://www.jhsph.edu/publichealthnews/Mag_Spring04/The_Book/behavior/essay.html

Jang, K. L., Hu, S., Livesly, W. J., Angleitner, A., Riemann, R., Ando, J., et al. (2001). Covariance structure of neuroticism and agreeableness: A twin and molecular genetic analysis of the role of the serotonin transporter gene. *Journal of Personality and Social Psychology, 81*, 295–304.

Jang, K. L., Livesley, W. J., & Vernon, P. A. (1996). Heritability of the Big Five personality dimensions and their facets: A twin study. *Journal of Personality, 64*, 577–591.

Jang, K. L., Livesley, W. J., & Vernon, P. A. (2002). The etiology of personality function: The University of British Columbia Twin Project. *Twin Research, 5*, 342–346.

Jang, K. L., McCrae, R. R., Angleitner, A., Riemann, R. M., & Livesley, W. J. (1998). Heritability of facet-level traits in a cross-cultural twin sample: Support for a hierarchical model of personality. *Journal of Personality and Social Psychology, 74*, 1556–1565.

Jang, K. L., Stein, M. B., Taylor, S., Asmundson, G. J., & Livesley, W. J. (2003). Exposure to traumatic events and experiences: Aetilogical relationships with personality function. *Psychiatric Research, 30*, 61–69.

Janik, V. M., Sayigh, L. S., & Wells, R. S. (in press). Signature whistle shape conveys identity information to bottlenose dolphins. *Proceedings of the National Academy of Sciences.*

Janssen, E., Carpenter, D., & Graham, C. A. (2003). Selecting films for sex research: Gender differences in erotic film preference. *Archives of Sexual Behavior, 32*, 243–251.

Janzen, L. A., Nanson, J. L., & Block, G. W. (1995). Neuropsychological evaluation of preschoolers with fetal alcohol syndrome. *Neurotoxicology & Teratology, 17*, 273–279.

Jasper, H. H. (1995). A historical perspective: The rise and fall of prefrontal lobotomy. In H. H. Jasper & S. Riggio (Eds.), *Epilepsy and the functional anatomy of the frontal lobe.* New York: Raven Press.

Jencks, C. (1979). *Who gets ahead? The determinants of economic success in America.* New York: BasicBooks.

Jenkins, H. H., Barrera, F. J., Ireland, C., & Woodside, B. (1978). Signal-centered action patterns of dogs in appetitive classical conditioning. *Learning and Motivation, 9*, 272–296.

Jenkins, J. M., Oatley, K., & Stein, N. L. (1998). *Human emotions: A reader.* Malden, MA: Blackwell Publishers.

Jensen, A. R. (1998). *The g factor: The science of mental ability.* Westport, CT: Praeger.

Jensen, P. S., Hinshaw, S. P., Swanson, J. M., Greenhill, L. L., Conners, C. K., Arnold, L. E., et al. (2001). Findings from the NIMH Multimodal Treatment Study of ADHD (MTA): Implications and applications for primary care providers. *Journal of Developmental and Behavioral Pediatrics: Special Issue, 22*, 60–73.

Jessell, T. M., & Kelley, D. D. (1991). Pain and analgesia. In E. R. Kandel, J. H. Schwartz, & T. M. Jessell (Eds.), *Principles of neural science* (3rd ed., pp. 385–399). New York: Elsevier.

Ji, L. (2005). *Culture and thinking across time.* Presentation at the Cultural Psychology Preconference, New Orleans. January 20, 2005.

Ji, L. J., Zhang, Z., & Nisbett, R. E. (2004). Is it culture or is it language? Examination of language effects in cross-cultural research on categorization. *Journal of Personality and Social Psychology, 87*, 57–65.

John, O. P. (1990). The "Big Five" factor taxonomy: Dimensions of personality in the natural language and in questionnaires. In L. A. Pervin & O. P. John (Eds.), *Handbook of personality: Theory and research* (pp. 66–100). New York: Guilford Press.

John, O. P., & Srivastava, S. (1999). The Big Five trait taxonomy: History, measurement, and theoretical perspectives. In L. A. Pervin & O. P. John (Eds.), *Handbook of personality: Theory and research* (2nd ed., pp. 102–138). New York: Guilford Press.

Johns, M., Schmader, T., & Martens, A. (2005). Knowing is half the battle. *Psychological Science, 16*, 175–179.

Johnson, D. L., Wiebe, J. S., Gold, S. M., Andreasen, N. C., Hichwa, R. D., Watkins, G. L., et al. (1999). Cerebral blood flow and personality: A positron emission tomography study. *American Journal of Psychiatry, 156*, 252–257.

Joiner, T. E., Coyne, J. C., & Blalock, J. (1999). On the interpersonal nature of depression: Overview and synthesis. In T. E. Joiner & J. C. Coyne (Eds.), *The interactional nature of depression: Advances in interpersonal approaches* (pp. 3–19). Washington, DC: American Psychological Association.

Joint Statement. (2000). *Joint statement on the impact of entertainment violence on children*. Congressional Public Health Summit. July 26, 2000.

Jones, R. A., & Ellis, G. D. (1996). Effect of variation in perceived risk on the secretion of beta-endorphin. *Leisure Sciences, 18*, 277–291.

Jones, S. S., Collins, K., & Hong, H. (1991). An audience effect on smile production in 10-month-old infants. *Psychological Science, 2*, 45–49.

Jope, R. S. (1999). Anti-bipolar therapy: Mechanism of action of lithium. *Molecular Psychiatry, 4*, 117–128.

Jorde, L. B., Mason-Brothers, A., Waldman, R., Ritvo, E. R., Freeman, B. J., Pingree, C., et al. (1990). The UCLA-University of Utah epidemiologic survey of autism: Genealogical analysis of familial aggregation. *American Journal of Medical Genetics, 36*, 85–88.

Jureidini, J. N., Doecke, C. J., Mansfield, P. R., Haby, M., Menkes, D. B., & Tonkin, A. L. (2004). Efficacy and safety of antidepressants for children and adolescents. *British Medical Journal, 328*, 879–883.

Kagan, J., & Snidman, N. (1991). Infant predictors of inhibited and uninhibited profiles. *Psychological Science, 2*, 40–44.

Kahneman, D., & Tversky, A. (1984). Choices, values, and frames. *American Psychologist, 39,* 341–350.

Kanagawa, C., Cross, S. E., & Markus, H. R. (2001). "Who am I?": The cultural psychology of the conceptual self. *Personality and Social Psychology Bulletin, 27*, 90–103.

Kanayama, G., Pope, H. G., Cohane, G., & Hudson, J. L. (2003). Risk factors for anabolic-steroid use among weightlifters: A case-control study. *Drugs and Alcohol Dependent, 71,* 77–86.

Kanazawa, S. (2004). General intelligence as a domain-specific adaptation. *Psychological Review, 111,* 512–523.

Kandel, E. R., Schwartz, J. H., & Jessell, T. M. (1995). *Essentials of neural science and behavior*. Norwalk, CT: Appleton & Lange.

Kandel, E. R., Schwartz, J. H., & Jessell, T. M. (2000). *Principles of neural science* (4th ed.). New York: McGraw-Hill.

Kandel, E. R., Schwartz, J. H., & Jessell, T. M. (2002). *Essentials of neural science and behavior.* New York: Appleton & Lange.

Kane, M. J., Hambrick, D. Z., & Conway, A. R. (2005). Working memory capacity and fluid intelligence are strongly related constructs: Comment on Ackerman, Beier, and Boyle (2005). *Psychological Bulletin, 131*, 66–71.

Kanwisher, N., Tong, F., & Nakayama, K. (1998). The effect of face inversion on the human fusiform face area. *Cognition, 68*, 1–11.

Kapur, S. E., Craik, F. I. M., Tulving, E., Wilson, A. A., Houle, S., & Brown, G. R. (1994). Neuroanatomical correlates of encoding in episodic memory: Levels of processing effects. *Proceedings of the National Academy of Sciences, 91*, 2008–2011.

Karney, B. R., & Bradbury, T. N. (1995). The longitudinal course of marital quality and stability: A review of theory, methods, and research. *Psychological Bulletin, 118*, 3–34.

Kasahara, Y. (1986). Fear of eye-to-eye confrontation among neurotic patients in Japan. In T. Lebra & W. P. Lebra (Eds.), *Japanese Culture and Behavior* (pp. 379–387). Honolulu: University of Hawaii Press.

Kassin, S. (2005). On the psychology of confessions: Does innocence put innocents at risk? *American Psychologist, 60,* 215–228.

Kassin, S. M., & Kiechel, K. L. (1996). The social psychology of false confessions: Compliance, internalization, and confabulation. *Psychological Science, 7*, 125–128.

Kassin, S. M., Goldstein, C. G., & Savitsky, K. (2003). Behavioral confirmation in the interrogation room: On the dangers of presuming guilt. *Law and Human Behavior, 27,* 187–203.

Kawakami, K., Dovidio, J. F., Moll, J., Hermsen, S., & Russin, A. (2000). Just say no (to stereotyping): Effects of training in the negation of stereotypic associations on stereotype activation. *Journal of Personality and Social Psychology, 78*, 871–888.

Kawakami, K., Dovidio, J. F., & van Kamp, S. (2005). Kicking the habit: Effects of nonstereotypic association training and correction processes on hiring decisions. *Journal of Experimental Social Psychology, 41*, 68–75.

Kazdin, A. E. (1994). Methodology, design, and evaluation in psychotherapy research. In A. E. Bergin & S. L. Garfield (Eds.), *International handbook of behavior modification and behavior change* (4th ed., pp. 19–71). New York: Wiley.

Kazdin, A. E., & Benjet, C. (2003). Spanking children: Evidence and issues. *Current Directions in Psychological Science, 12,* 99–103.

Keane, M. (1987). On retrieving analogues when solving problems. *Quarterly Journal of Experimental Psychology, 39A*, 29–41.

Keel, P. K., & Klump, K. L. (2003). Are eating disorders culture-bound syndromes? Implications for conceptualizing their etiology. *Psychological Bulletin, 129*, 747–769.

Keel, P. K., & Mitchell, J. E. (1997). Outcome in bulimia nervosa. *American Journal of Psychiatry, 154*, 313–321.

Keith, S. J., Regier, D. A., & Rae, D. S. (1991). Schizophrenic disorders. In L. N. Robins & D. A. Regier (Eds.)*, Psychiatric disorders in America: The epidemiological catchment areas study*. New York: The Free Press.

Keller, M. B., & Baker, L. A. (1991). Bipolar disorder: Epidemiology, course, diagnosis, and treatment. *Bulletin of the Menninger Clinic, 55*, 172–181.

Keller, M. B., McCullough, J. P., Klein, D. N., Arnow, B., Dunner, D. L., Gelenberg, A. J., et al. (2000). A comparison of nefazodone, a cognitive behavioral analysis system of psychotherapy, and their combination for the treatment of chronic depression. *New England Journal of Medicine, 342*, 1462–1470.

Keller, M. B., Shapiro, R. W., Lavori, P. W., & Wolfe, N. (1982). Recovery in major depressive disorder. *Archives of General Psychiatry, 39*, 905–910.

Kelley, W. M., Miezin, F. M., McDermott, K. B., Buckner, R. L., Raichle, M. E., Cohen, N. J., et al. (1998). Hemispheric specialization in human dorsal frontal cortex and medial temporal lobe for verbal and nonverbal memory encoding. *Neuron, 20*, 927–936.

Kelley, W. T., Macrae, C. N., Wyland, C., Caglar, S., Inati, S., & Heatherton, T. F. (2002). Finding the self? An event-related fMRI study. *Journal of Cognitive Neuroscience, 14,* 785–794.

Kellman, P. J., Spelke, E. S., & Short, K. R. (1986). Infant perception of object unity from translatory motion in depth and vertical translation. *Child Development, 57*, 72–86.

Kelly, J. B. (1992). Behavioral development of the auditory orientation response. In R. Romand (Ed.), *Development of auditory and vestibular systems II*. Amsterdam: Elsevier Press.

Kelner, F. (1997). Alcohol. In P. MacNeil & W. Ikuko (Eds.), *Canada's alcohol and other drugs survey 1994: A discussion of the findings.* Ottawa: Health Canada.

Keltner, D., & Anderson, C. (2000). Saving face for Darwin: The functions and uses of embarrassment. *Current Directions in Psychological Science, 9,* 187–192.

Keltner, D., & Bonanno, G. A. (1997). A study of laughter and dissociation: Distinct correlates of laughter and smiling during bereavement. *Journal of Personality and Social Psychology, 73,* 687–702.

Keltner, D., Young, R. C., Heerey, E. A., Oemig, C., & Monarch, N. D. (1998). Teasing in hierarchical and intimate relations. *Journal of Personality and Social Psychology, 75,* 1231–1247.

Kennedy, S. H., Eisfeld, B. S., Dickens, S. E., Bacchiochi, J. R., Bagby, R. M. (2000). Antidepressant-induced sexual dysfunction during treatment with moclobemide, paroxetine, sertraline, and venlafaxine. *Journal of Clinical Psychiatry, 61,* 276–281.

Kennedy, Q., Mather, M., & Carstensen, L. L. (2004). The role of motivation in the age-related positivity effect in autobiographical memory. *Psychological Science, 15,* 208–214.

Kenrick, D. T., & Funder, D. C. (1991). The person-situation debate: Do personality traits really exist? In V. J. Derlega, B. A. Winstead, & W. H. Jones (Eds.), *Personality: Contemporary theory and research* (pp. 149–174). Chicago, IL: Nelson Hall.

Kerr, M., Tremblay, R. E., Pagni, L., & Vitaro, F. (1997). Boys' behavioural inhibition and the risk of later delinquency. *Archives of General Psychiatry, 54,* 809–816.

Kessler, R. C., McGonagle, K. A., Zhao, S., Nelson, C. B., Hugh, M., Eshleman, S. et al. (1994). Lifetime and 12-month prevalence of DSM-III-R psychiatric disorders in the United States: Results from the National Comorbidity Study. *Archives of General Psychiatry, 51,* 8–19.

Keys, A., Brozek, J., Henschel, A. L., Mickelsen, O., & Taylor, H. L. (1950). *The biology of human starvation.* Minneapolis: University of Minnesota Press.

Kiecolt-Glaser, J. K., & Glaser, R. (1988). Immunological competence. In E. A. Blechman & K. D. Brownell (Eds.), *Handbook of behavioral medicine for women* (Vol. 149, pp. 195–205). Elmsford, NY: Pergamon Press.

Kiecolt-Glaser, J. K., Malarkey, W. B., Chee, M., & Newton, T. (1993). Negative behavior during marital conflict is associated with immunological down-regulation. *Psychosomatic Medicine, 55,* 395–409.

Kihlstrom, J. R. (2005). Dissociative disorders. *Annual Review of Clinical Psychology, 1,* 227–253.

Kim, H. S. (2002). We talk, therefore we think? A cultural analysis of the effect of talking on thinking. *Journal of Personality and Social Psychology, 83,* 828–842.

Kim, H. S., & Markus, H. R. (1999). Deviance or uniqueness, harmony or conformity? A cultural analysis. *Journal of Personality and Social Psychology, 77,* 785–800.

Kimura, D. (1999). *Sex and cognition.* Cambridge, MA: MIT Press.

Kitayama, S., & Cohen, D. (in press). *Handbook of cultural psychology.* New York: Guilford.

Kitayama, S., Snibbe, A. C., & Markus, H. R. (2004). Is there any "free" choice? Self and dissonance in two cultures. *Psychological Science, 15,* 527–533.

Kite, M. E., & Deaux, K. (1986). Attitudes toward homosexuality: Assessment and behavioral consequences. *Basic and Applied Social Psychology, 7,* 137–162.

Klauer, M. H., Musch, J., & Naumer, B. (2000). On belief bias in syllogistic reasoning. *Psychological Review, 107,* 852–884.

Kleinman, A. (1982). Neurasthenia and depression: A study of somatization and culture in China. *Culture, Medicine, and Psychiatry, 6,* 117–190.

Kleinman, A. (1988). *Rethinking psychiatry: From cultural category to personal experience.* New York: Free Press.

Kleinmuntz, B., & Szucko, J. J. (1984). Lie detection in ancient and modern times: A call for contemporary scientific study. *American Psychologist, 39,* 766–776.

Klin, A., Jones, W., Schultz, R., & Volkmar, F. (2003). The enactive mind, or from actions to cognition: Lessons from autism. *Philosophical Transactions of the Royal Society,* Series B, *358,* 345–360.

Klingemann, H. K. (1991). The motivation for change from problem alcohol and heroin use. *British Journal of Addiction, 86,* 727–744.

Knight, R. (1953). Borderline states. *Bulletin of the Menninger Clinic, 17,* 1–12.

Knight, R. T., & Grabowecky, M. (1995). Escape from linear time: Prefrontal cortex and conscious experience. In M. S. Gazzaniga (Ed.), *The cognitive neurosciences* (pp. 1357–1371). Cambridge, MA: MIT Press.

Knutson, B., Fong, G. W., Adams, C. M., Varner, J. L., & Hommer, D. (2001). Dissociation of reward anticipation and outcome with event-related fMRI. *NeuroReport 12,* 3683–3687.

Knutson, B., Wolkowitz, O. M., Cole, S. W., Chan, T., Moore, E. A., Johnson, R. C., et al. (1998). Selective alteration of personality and social behavior by serotonergic intervention. *American Journal of Psychiatry, 155,* 373–379.

Kobasa, S. C. (1979). Personality and resistance to illness. *American Journal of Community Psychology, 7,* 413–423.

Koch, J. L. (1891). *Die psychopathischen Minderwertigkeiten.* Ravensburg, Germany: Maier.

Kocsis, J. H., Rush, A. J., Markowitz, J. C., Borian, F. E., Dunner, D. L., Koran, L. M., et al. (2003). Continuation treatment of chronic depression: A comparison of nefazodone, cognitive behavioral analysis system of psychotherapy, and their combination. *Psychopharmacology Bulletin, 37,* 73–87.

Kocsis, R. N. (2004). Psychological profiling in serial arson offenses: A comparative assessment of skills and accuracy. *Criminal Justice and Behavior, 31,* 341–361.

Kocsis, R. N., Hayes, A. F., & Irwin, H. J. (2002). Investigative experience and accuracy in psychological profiling of a violent crime. *Journal of Interpersonal Violence, 17,* 811–823.

Kohlberg, L. (1971). From is to ought: How to commit the naturalistic fallacy and get away with it in the study of moral development. In L. Mischel (Ed.), *Cognitive development and epistemology* (pp. 151–284). New York: Academic Press.

Kolar, D. W., Funder, D. C., & Colvin, C. R. (1996). Comparing the accuracy of personality judgments by the self and knowledgeable others. *Journal of Personality, 64*, 311–337.

Kolb, B., & Whishaw, I. Q. (1998). *Fundamentals of human neuropsychology.* New York: Freeman.

Kolb, B., & Whishaw, I. Q. (1998b). Brain plasticity and behavior. *Annual Review of Psychology, 49*, 43–64.

Konishi, M. (1993). Listening with two ears. *Scientific American, 268*, 66–73.

Kontsevich, L. L., & Tyler, C. W. (2004). What makes Mona Lisa smile? *Vision Research, 44*, 1493–1498.

Koob, G. F. (1999). Drug reward and addiction. In M. J. Zigmond, F. E. Bloom, S. C. Landis, J. L. Roberts, & L. R. Squire (Eds.), *Fundamentals of neuroscience* (pp. 1261–1279). San Diego, CA: Academic Press.

Koole, S. L., Dijksterhuis, A., & van Knippenberg, A. (2001). What's in a name: Implicit self-esteem and the automatic self. *Journal of Personality and Social Psychology, 80*, 669–685.

Korn, M. L., Kotler, M., Molcho, A., Botsis, A. J., Grosz, D., Chen, C., et al. (1992). Suicide and violence associated with panic attacks. *Biological Psychiatry, 31*, 607–612.

Koshland, D. E. (1989). Drunk driving and statistical mortality. *Science, 244*, 513.

Kosslyn, S., & Koenig, O. (1995). *Wet mind: The new cognitive neuroscience.* New York: Free Press.

Kosslyn, S. M. (1994). *Image and brain.* Cambridge, MA: MIT Press.

Kosslyn, S. M., Thompson, W. L., Kim, I. J., & Alpert, N. M. (1995). Topographical representations of mental images in primary visual cortex. *Nature, 378*, 496–498.

Kovaleski, S. F., & Sheridan, M. B. (2003, January 12). A boy of bright promise and no roots; after transient childhood, sniper suspect latched on to strong father figure. *The Washington Post*, p. C.01.

Kozorovitskiy, Y., & Gould, E. (2004). Dominance hierarchy influences adult neurogenesis in the dentate gyrus. *Journal of Neuroscience, 24*, 6755–6759.

Kramer, P. D. (1993). *Listening to Prozac: A psychiatrist explores antidepressant drugs and the remaking of the self.* New York: Viking Penguin.

Krongrad, A., Lai, H., Burke, M. A., Goodkin, K., & Lai, S. (1996). Marriage and mortality in prostate cancer. *Journal of Urology, 156*, 1696–1700.

Kruesi, M. J., Hibbs, E. D., Zahn, T. P., Keysor, C. S., Hamburger, S. D., Bartko, J. J., et al. (1992). A 2-year prospective follow-up study of children and adolescents with disruptive behavior disorders. Prediction by cerebrospinal fluid 5-hydroxyindoleacetic acid, homovanillic acid, and autonomic measures. *Archives of General Psychiatry, 49*, 429–435.

Krumhansl, C. (2002). Music: A link between cognition and emotion. *Current Directions in Psychological Science, 11*, 45–50.

Kuhl, P. K. (2004). Early language acquisition: Cracking the speech code. *Nature Reviews Neuroscience, 5*, 831– 843.

Kulhara, P., & Chakrabarti, S. (2001). Culture and schizophrenia and other psychotic disorders. *Psychiatric Clinics of North America, 24*, 449–464.

Kuncel, N. R., Hezlett, S. A., & Ones, D. S. (2004). Academic Performance, career potential, creativity, and job performance: Can one construct predict them all? *Journal of Personality and Social Psychology, 86*, 148–161.

Kunda, Z. (2000). *Social cognition.* Cambridge, MA: MIT Press.

Kunda, Z., & Spencer, S. J. (2003). When do stereotypes come to mind and when do they color judgment? A goal-based theoretical framework for stereotype activation and application. *Psychological Bulletin, 129*, 522–544.

Kyei, K. G. & Schreckenbach, H. (1975). *No time to die.* Accra, Ghana: Catholic Press.

Kyllonen, P. C., & Christal, R. E. (1990). Reasoning ability is (little more than) working-memory capacity?! *Intelligence, 14*, 389–433.

LaFrance, M., & Banaji, M. (1992). Toward a reconsideration of the gender-emotion relationship. In M. Clarke (Ed.) *Review of personality and social psychology* (pp. 178–201). Beverly Hills, CA: Sage.

Laible, D. J., & Thompson, R. A. (2000). Attachment and self-organization. In M. D. Lewis & I. Granic (Eds.), *Emotion, development, and self-organization: Dynamic systems approaches to emotional development* (pp. 298–323). New York: Cambridge University Press.

Lambert, J., Fernandez, S., & Frick, K. (2005). Different types of environmental enrichment have discrepant effects on spatial memory and synaptophysin levels in female mice. *Neurobiology of Learning and Memory, 83*, 206–216.

Landau, B. (1994). Where's what and what's where: The language of objects in space. In L. E. Gleitman and B. Landau (Eds.), [special issue] *Lexical acquisition, Lingua, 92*, 259–296.

Lane, R. D., Nadel, L., & Ahern, G. (1999). *Cognitive neuroscience of emotion.* New York: Oxford University Press.

Lane, R. D., Reiman, E. M., Ahern, G. L., & Schwartz, G. E. (1997). Neuroanatomical correlates of happiness, sadness, and disgust. *American Journal of Psychiatry, 154*, 926–933.

Langleben, D. D., Schroeder, L., Maldjian, J. A., Gur, R. C., McDonald, S., Ragland, J. D., et al. (2002). Brain activity during simulated deception: An event-related functional magnetic resonance study. *Neuroimage, 15*, 727–732.

Langlois, J. H., Kalakanis, L., Rubenstein, A. J., Larson, A., Hallam, M., & Smoot, M. (2000). Maxims or myths of beauty? A meta-analytic and theoretical review. *Psychological Bulletin, 126*, 390–423.

Langlois, J. H., Ritter, J. M., Casey, R. J., & Sawin, D. B. (1995). Infant attractiveness predicts maternal behaviors and attitudes. *Developmental Psychology, 31*, 464–472.

Langlois, J. H., & Roggman, L. A. (1990). Attractive faces are only average. *Psychological Science, 1*, 115–121.

Larsen, J. T., McGraw, A. P., & Cacioppo, J. T. (2001). Can people feel happy and sad at the same time? *Journal of Personality and Social Psychology, 81*, 684–696.

Lasagna, L., Mosteller, F., von Felsinger, J., & Beecher, H. (1954). A study of the placebo response. *American Journal of Medicine, 16*, 770–779.

Latané, B., & Darley, J. M. (1968). Group inhibition of bystander intervention in emergencies. *Journal of Personality and Social Psychology, 10*, 215–221.

Latané, B., Williams, K., & Harkins, S. G. (1979). Many hands make light the work: The causes and consequences of social loafing. *Journal of Personality and Social Psychology, 37*, 822–832.

Lazarus, R. S. (1993). From psychological stress to the emotions: A history of changing outlooks. *Annual Review of Psychology, 44*, 1–21.

Lazer, W., Murata, S., & Kosaka, H. (1985). Japanese marketing: Towards a better understanding. *Journal of Marketing, 49*, 2, 69–81.

Leary, M. R. (2004). The function of self-esteem in terror management theory and sociometer theory: Comment on Pyszczynski et al. *Psychological Bulletin, 130*, 478–482.

Leary, M. R., & MacDonald, G. (2003). Individual differences in self-esteem: A review and theoretical integration. In M. R. Leary & J. P. Tangney (Eds.), *Handbook of self and identity* (pp. 401–418). New York: Guilford Press.

Leary, M. R., Tambor, E. S., Terdal, S. K., & Downs, D. L. (1995) Self-esteem as an interpersonal monitor: The sociometer hypothesis. *Journal of Personality and Social Psychology, 68*, 518–530.

Leatherdale, S. T., Sparks, R., & Kirsh, V. A. (2006). Beliefs about tobacco industry (mal)practices and youth smoking behaviour: Insight for future control campaigns (Canada). *Cancer Causes Control, 17*, 705–711.

Leckman, J. F., Elliott, G. R., Bromet, E. J., Campbell, M., Cicchetti, D., Cohen, D. J., et al. (1995). Report card on the National Plan for Research on Child and Adolescent Mental Disorders: The midway point. *Archives of General Psychiatry, 34*, 715–723.

Leckman, J. F., Goodman, W. K., North, W. G., Chappell, P. B., Price, L. H., Pauls, D. L., et al. (1994). The role of central oxytocin in obsessive compulsive disorder and related behavior. *Psychoneuroendocrinology, 19*, 723–749.

LeDoux, J. E. (1996). *The emotional brain: The mysterious underpinnings of emotional life*. New York: Simon & Schuster.

LeDoux, J. E. (2002). *Synaptic self*. New York: Viking.

Lee, C. K., Kwak, Y. S., Rhee, H., Kim, Y. S., Han, J. H., Choi, J. O., & Lee, Y. H. (1987). The nationwide epidemiological study of mental disorders in Korea. *Journal of Korean Medical Science, 2*, 19–34.

Lee, G. R., & Stone, L. H. (1980). Mate-selection systems and criteria: Variation according to family structure. *Journal of Marriage and the Family, 42*, 319–326.

Leenaars, A. A., & Lester, D. (1995). The changing suicide patterns in Canadian adolescents and youth, compared to their American counterparts. *Adolescence, 30*, 539–547.

Lefcourt, H. M. (1966). Internal versus external control of reinforcement: A review. *Psychological Bulletin, 65*, 206–220.

Lefcourt, H. M. (1992). Durability and impact of the locus of control construct. *Psychological Bulletin, 112*, 411–414.

Leff, J., Sartorius, N., Jablensky, A., Korten, A., & Ernberg, G. (1992). The International Pilot Study of Schizophrenia: Five-year follow-up findings. *Psychological Medicine, 22*, 131–145.

Lehmann, H. H., & Ban, T. A. (1997). The history of the psychopharmacology of schizophrenia. *Canadian Journal of Psychiatry, 42*, 152–162.

Leibowitz, H. (1996). The symbiosis between basic and applied research. *American Psychologist, 51*, 366–370.

Leichsenring, F., Rabung, S., & Leibing, E. (2004). The efficacy of short-term psychodynamic psychotherapy in specific psychiatric disorders: A meta-analysis. *Archives of General Psychiatry, 61*, 1208–1216.

Leigh, B. C., & Schafer, J. C. (1993). Heavy drinking occasions and the occurrence of sexual activity. *Psychology of Addictive Behaviors, 7*, 197–200.

LeMarquand, D. G., Pihl, R. O., Young, S. N., Tremblay, R. E., Seguin, J. R., Palmour, R. M., & Benkelfat, C. (1998). Tryptophan depletion, executive functions, and disinhibition in aggressive, adolescent males. *Neuropharmacology, 19*, 333–341.

Lenneberg, E. (1967). *The biological foundations of language*. New York: Wiley.

Lepper, M. R., Greene, D., & Nisbett, R. E. (1973). Undermining children's intrinsic interest with extrinsic reward: A test of the "overjustification" hypothesis. *Journal of Personality and Social Psychology, 28*, 129–137.

Lesage, A. D., Morissette, R., Fortier, L. Reinharz, D., & Contandriopolous, A. (2000). 1. Downsizing psychiatric hospitals: Needs for care and services of current discharged long-stay inpatients. *Canadian Journal of Psychiatry, 445*, 526–531.

Lesniak, K. T., & Dubbert, P. M. (2001). Exercise and hypertension. *Current Opinion in Cardiology, 16*, 356–359.

Leuchter, A. F., Cook, I. A., Witte, E. A., Morgan, M., & Abrams, M. (2002). Changes in brain function of depressed subjects during treatment with placebo. *American Journal of Psychiatry, 159*, 122–129.

LeVay, S. (1991). A difference in hypothalamic structure between heterosexual and homosexual men. *Science, 253*, 1034–1037.

Levenson, R. W., Ekman, P., Heider, K., & Friesen, W. V. (1992). Emotion and autonomic nervous system activity in the Minangkabau of West Sumatra. *Journal of Personality and Social Psychology, 62*, 972–988.

Leventhal, H., & Cleary, P. D. (1980). The smoking problem: A review of research and theory in behavioral risk modification. *Psychological Bulletin, 88*, 370–405.

Levin, J. M., Ross, M. H., Mendelson, J. H., Kaufman, M. J., Lange, N., Maas, L. C., et al. (1998). Reduction in BOLD fMRI response to primary visual stimulation following alcohol ingestion. *Psychiatry Research, 82*, 135–146.

Levinson, S. C. (1997). Language and cognition: The cognitive consequences of spatial description in Guugu Yimithirr. *Journal of Linguistic Anthropology, 7*, 98–131.

Levitin, D. J., & Rogers, S. E. (2005). Absolute pitch: Perception, coding, and controversies. *Trends in Cognitive Science, 9*, 26–33.

Levitt, T. (1983). The globalization of markets. *Harvard Business Review, 61* (May/June), 92–102.

Lewinsohn, P. M., Allen, N. B., Seeley, J. R., & Gotlib, I. H. (1999). First onset versus recurrence of depression: Differential processes of psychosocial risk. *Journal of Abnormal Psychology, 108*, 483–489.

Lewinsohn, P. M., Rodhe, P. D., Seeley, J. R., & Hops, H. (1991). Comorbidity of unipolar depression: I. Major depression with dysthymia. *Journal of Abnormal Psychology, 98*, 107–116.

Lewis, M., & Haviland, J. M. (1993). *Handbook of emotions*. New York: Guilford.

Lewontin, R. C. (1976). Race and intelligence. In N. J. Block & G. Dworkin (Eds.), *The IQ controversy* (pp. 78–92). New York: Pantheon Books.

Lezak, M. D. (2004). *Neuropsychological assessment* (4th ed.). New York: Oxford University Press.

Lieb, K., Zanarini, M. C., Schmahl, C., Linehan, M. M., & Bohus, M. (2004). Borderline personality disorder. *Lancet, 364*, 453–461.

Lieberman, M. D. (2000). Intuition: A social cognitive neuroscience approach. *Psychological Bulletin, 126*, 109–137.

Lieberman, M. D., Ochsner, K. N., Gilbert, D. T., & Schacter, D. L. (2001). Do amnesics exhibit cognitive dissonance reduction? The role of explicit memory and attention in attitude change. *Psychological Science, 121*, 135–140.

Lin, C. A. (1993). Cultural differences in message strategies: A comparison between American and Japanese TV commercials. *Journal of Advertising Research* (July/August), 40–48.

Lindemann, B. (2001). Receptors and transduction in taste. *Nature, 413*, 219–225.

Lindsay, D. S. (1996). Contextualizing and clarifying criticisms of memory work in psychotherapy. In K. Pezdek & W. P. Banks (Eds.), *The recovered memory/false memory debate*. San Diego, CA: Academic Press.

Linehan, M. M. (1987). Dialectical behavior therapy for borderline personality disorder: Theory and method. *Bulletin of the Menninger Clinic, 51*, 261–276.

Linehan, M. M. (1993). *Cognitive-behavioral treatment of borderline personality disorders*. New York: Guilford Publications.

Linehan, M. M., Armstrong, H. E., Suarez, A., Allmon, D., & Heard, H. (1991). Cognitive behavioral treatment of chronically parasuicidal borderline patients. *Archives of General Psychiatry, 48*, 1060–1064.

Linehan, M. M., Heard, H., & Armstrong, H. E. (1993). Naturalistic follow-up of a behavioral treatment for chronically parasuicidal borderline patients. *Archives of General Psychiatry, 50*, 971–974.

Liu, D., Diorio, J., Day, J. C., Francis, D. D., & Meaney, M. J. (2002). Maternal care, hippocampal synaptogenesis, and cognitive development in rats. *Nature Neuroscience, 3*, 799–806.

Liu, J., Raine, A., Venables, P. H., & Mednick, S. A. (2004). Malnutrition at age 3 years and externalizing behavior problems at ages 8, 11, and 17 years. *American Journal of Psychiatry, 161*, 2005–2013.

Livingston, K. E., & Hornykiewicz, O. (1978). *Limbic mechanisms: The continuing evolution of the limbic system concept*. New York: Plenum Press.

Ljungberg, T., Apicella, P., & Schultz, W. (1992). Responses of monkey dopamine neurons during learning of behavioral reactions. *Journal of Neurophysiology, 67*, 145–163.

Lledo, P. M., Gheusi, G., & Vincent, J. D. (2005). Information processing in the mammalian olfactory system. *Physiological Review, 85*, 281–317.

Locke, E. A., & Latham, G. P. (1990). *A theory of goal setting and task performance*. Englewood Cliffs, NJ: Prentice-Hall.

Lockwood, P., & Kunda, Z. (1997). Superstars and me: Predicting the impact of role models on the self. *Journal of Personality and Social Psychology, 73*, 91–103.

Loehlin, J. C., & Nichols, R. C. (1976). *Heredity, environment, and personality: A study of 850 sets of twins*. Austin, TX: University of Texas Press.

Loewenstein, G. F., Weber, E. U., Hsee, C. K., & Welch, N. (2001). Risk as feelings. *Psychological Bulletin, 127*, 267–286.

Loftus, E. (1993). The reality of repressed memories. *American Psychologist, 48*, 518–537.

Loftus, E. F. (2003). Make-believe memories. *American Psychologist, 58*, 867–873.

Loftus, E. F., & Palmer, J. C. (1974). Reconstruction of automobile destruction: An example of the interaction between language and memory. *Journal of Learning and Verbal Behavior, 13*, 585–589.

Logan, J. M., Sanders, A. L., Snyder, A. Z., Morris, J. C., & Buckner, R. L. (2002). Under-recruitment and nonselective recruitment: Dissociable neural mechanisms associated with aging. *Neuron, 33*, 1–20.

Loney, E. A., Green, K. L., & Nanson, J. L. (1994). A health promotion perspective on the House of Commons' report "Foetal alcohol syndrome: A preventable tragedy." *Canadian Journal of Public Health, 85*, 248–251.

Lovaas, O. I. (1987). Behavioral treatment and normal educational and intellectual functioning in young autistic children. *Journal of Consulting & Clinical Psychology, 55*, 3–9.

Lubinski, D. (2004). Introduction to the special section on cognitive abilities: 100 years after Spearman's (1904) "'General Intelligence,' objectively determined and measured." *Journal of Personality and Social Psychology, 86*, 96–111.

Lubinski, D., & Benbow, C. P. (2000). States of excellence, *American Psychologist, 55*, 137–150.

Luborsky, L., Crits-Christoph, P., McLellan, T., Woody, G., Piper, W., Liberman, B., et al. (1986). Do therapists vary much in their success? Findings from four outcome studies. *American Journal of Orthopsychiatry, 56*, 501–512.

Lucas, R. E., Diener, E., Grob, A., Suh, E., & Shao, L. (2000). Cross-cultural evidence for the fundamental features of extraversion. *Journal of Personality and Social Psychology, 79*, 452–468.

Luchins, A. S., (1942). Mechanization in problem solving. *Psychological Monographs, 54*. Whole No. 248.

Lykken, D. T. (1957). A study of anxiety in the sociopathic personality. *Journal of Abnormal Social Psychology, 55*, 6–10.

Lykken, D. T. (1995). *The antisocial personalities*. Hillsdale, NJ: Erlbaum.

Lykken, D. T. (2000). The causes and costs of crime and a controversial cure. *Journal of Personality, 68*, 559–605.

Lykken, D. T., McGue, M., Tellegen, A., & Bouchard, T. J., Jr. (1992). Emergenesis: Genetic traits that may not run in families. *American Psychologist, 47*, 1565–1577.

Lysgaard, S. (1955). Adjustment in a foreign society: Norwegian Fulbright grantees visiting the United States. *International Social Science Bulletin, 7*, 45–51.

Lytton, H., & Romney, D. M. (1991). Parents' sex-related differential socialization of boys and girls: A meta analysis. *Psychological Bulletin, 109*, 267–296.

Lyubomirsky, S., King, L., & Diener, E. (2005). The benefits of frequent positive affect: Does happiness lead to success? *Psychological Bulletin, 131*, 803–855.

Lyubomirsky, S., & Nolen-Hoeksema, S. (1995). Effects of self-focused rumination on negative thinking and interpersonal problem solving. *Journal of Personality and Social Psychology, 69*, 176–190.

Maas, L. C., Lukas, S. E., Kaufman, M. J., Weiss, R. D., Daniels, S. L., Rogers, et al. (1998). Functional magnetic resonance imaging of human brain activation during cue-induced cocaine craving. *American Journal of Psychiatry, 155*, 124–126.

Maccoby, E. E. (1998). *The two sexes: Growing up apart, coming together.* Cambridge, MA: Belknap Press/Harvard University Press.

Macintosh, N. J., & Honig, V. R. (Eds.). (1969) *Fundamental issues in associative learning*. Halifax: Dalhousie University Press.

MacKay, D. G. (1973). Aspects of a theory of comprehension, memory and attention. *Quarterly Journal of Experimental Psychology, 25,* 22–40.

Macrae, C. N., Alnwick, K. A., Milne, A. B., & Schloerscheidt, A. M. (2002). Person perception across the menstrual cycle: Hormonal influences on social-cognitive functioning. *Psychological Science, 13*, 532–536.

Macrae, C. N., Bodenhausen, G. V., & Calvini, G. (1999). Contexts of cryptomnesia: May the source be with you. *Social Cognition, 17,* 273–297.

Macrae, C. N., Moran, J. M., Heatherton, T. F., Banfield, J. F., & Kelley, W. M. (2004). Medial prefrontal activity predicts memory for self. *Cerebral Cortex, 14,* 647–654.

Madden, G. J., Petry, N. M., Badger, G. J., & Bickel, W. K. (1997). Impulsivity and self-control choices in opioid-dependent patients and non-drug-using control participants: Drug and monetary rewards. *Experimental and Clinical Psychopharmacology, 5*, 256–262.

Madsen, M. C. (1971). Developmental and cross-cultural differences in the cooperative and competitive behavior of young children. *Journal of Cross-Cultural Psychology,* 365–371.

Maguire, E. A., Spiers, H. J., Good, C. D., Hartley, T., Frackowiak, R. S., & Burgess, N. (2003). Navigation expertise and the human hippocampus: A structural brain imaging analysis. *Hippocampus, 13,* 250–259.

Maier, N. R. F. (1931). Reasoning in humans II: The solution of a problem and its appearance in consciousness. *Journal of Comparative Psychology, 12*, 181–194.

Main, M., & Solomon, J. (1986). Discovery of a new, insecure-disorganized/disoriented attachment pattern. In T. B. Brazelton & M. Yogman (Eds.), *Affective development in infancy* (pp. 95–124). Norwood, NJ: Ablex.

Malla, A. K., Norman, R. M., Manchanda, R., Ahmed, M. R., Scholten, D., Harricharan, R., Cortese, L., & Takhar, J. (2002). One year outcome in first episode psychosis: Influence of DUP and other predictors. *Schizophrenia Research, 54*, 231–242.

Manderscheid, R. W., Witkin, M. J., Rosenstein, M. J., Milazzo-Sayre, L. J., Bethel, H. E., & MacAskill, R. L. (1985). In C. A. Taube & S. A. Barrett (Eds.), *Mental health, United States, 1985.* Washington DC: National Institute of Mental Health.

Maner, J. K., Luce, C. L., Neuberg, S. L., Cialdini, R. B., Brown, S., & Sagarin, B. J. (2002). The effects of perspective taking on motivations for helping: Still no evidence for altruism. *Personality and Social Psychology Bulletin, 28*, 1601–1610.

Mann, R. F., Smart, R. G., Stoduto, G., Beirness, D., Lamble, R., & Vingilis, E. (2002). The early effects of Ontario's administrative driver's licence suspension law on driver fatalities with a BAC > 80 mg%. *Canadian Journal of Public Health, 93*, 176–180.

Mannuzza, S., Klein, R. G., Bonagura, N., Malloy, P., Giampino, T. L., & Addalli, K. A. (1991). Hyperactive boys almost grown up. Replications of psychiatric status. *Archives of General Psychiatry, 48*, 77–83.

Marcia, J. E. (1980). Identity in adolescence. In I. Adelson (Ed.), *Handbook of adolescence*. New York: Wiley.

Marcus, G. (2004). *The birth of the mind.* New York: Basic Books.

Marcus, G. F. (1996). Why do children say "breaked"? *Current Directions in Psychological Science, 5,* 81–85.

Marcus, G. F., Pinker, S., Ullman, M., Hollander, M., Rosen, T. S., & Xu, F. (1992). Overregularization in language acquisition. *Monographs of the Society of Research and Child Development, 57*, 181.

Marcus, G. F., Vijayan, S., Bandi Rao, S., & Vishton, P. M. (1999). Rule learning by seven-month-old infants. *Science, 283,* 77–80.

Markon, J. (2001, October 8). Elderly Judges Handle 20 Percent of U.S. Caseload. *The Wall Street Journal*, p. A.15.

Markowitz, J. C., & Weissman, M. M. (1995). Interpersonal psychotherapy. In E. E. Beckham & W. R. Leber (Eds.), Handbook of depression (2nd ed., pp. 376–390). New York: Guilford Press.

Markus, H. R. (1977). Self-schemata and processing information about the self. *Journal of Personality and Social Psychology, 35,* 63–78.

Markus, H. R., & Kitayama, S. (1991). Culture and the self: Implications for cognition, emotion, and motivation. *Psychological Review, 98,* 224–253.

Marlatt, G. A. (1999). Alcohol, the magic elixir? In S. Peele & M. Grant (Eds.), *Alcohol and pleasure: A health perspective* (pp. 233–248). Philadelphia: Brunner/Mazel.

Marr, D. (1982). *Vision: A computational investigation into the human representation and processing of visual information.* San Francisco: Freeman.

Marshall, E. (2004). Forgetting and remembering. A star-studded search for memory-enhancing drugs. *Science, 304,* 36–38.

Marshall, L. H., & Magoun, H. W. (1998). *Discoveries in the human brain: Neuroscience prehistory, brain structure, and function.* Totowa, NJ: Humana Press.

Martin, N. G., Eaves, L. J., Heath, A. R., Jardine, R., Feingold, L. M., & Eysenck, H. J. (1986). Transmission of social attitudes. *Proceedings of the National Academy of Science, 83,* 4364–4368.

Martin, P., & Bateson, P. (1993). *Measuring behaviour: An introductory guide.* New York: Cambridge.

Maruta, T., Colligan, R. C., Malinchoc, M., & Offord, K. P. (2002). Optimism-pessimism assessed in the 1960s and self-reported health status 30 years later. *Mayo Clinic Proceedings, 77,* 748–753.

Maslow, A. (1968). *Toward a psychology of being.* New York: Van Nostrand.

Masten, A. S. (2001). Ordinary magic: Resilience processes in development. *American Psychologist, 56,* 227–238.

Masuda, T., & Nisbett, R. E. (2001). Attending holistically vs. analytically: Comparing the context sensitivity of Japanese and Americans. *Journal of Personality and Social Psychology, 81,* 922–934.

Mather, M., & Carstensen, L. L. (2003). Aging and attentional biases for emotional faces. *Psychological Science, 14*, 409–415.

Maurice, C. (1994). *Let me hear your voice: A family's triumph over autism.* New York: Knopf.

Mayer, J. D., Caruso, D., & Salovey, P. (1999). Emotional intelligence meets traditional standards for an intelligence. *Intelligence, 27*, 267–298.

Maziade, M., Roy, M. A., Rouillard, E., Bissonnette, L., Roy, A., et al. (2001). A search for specific and common susceptibility loci for schizophrenia and bipolar disorder: A linkage study in 13 target chromosomes. *Molecular Psychiatry, 6*, 684–693.

Mazur, A., & Booth, A. (1998). Testosterone and dominance in men. *Behavioural and Brain Science, 21*, 353–397.

McAdams, D. P. (1999). Personal narratives and the life story. In L. A. Pervin & O. P. John (Eds.), *Handbook of personality: Theory and research* (2nd ed., pp. 478–500). New York: Guilford Press.

McAdams, D. P. (2001). The psychology of life stories. *Review of General Psychology, 5*, 100–122.

McAdams, D. P. (2005). *The person: An integrated introduction to personality psychology* (4th ed.). Fort Worth: Harcourt.

McCarthy, G., Puce, A., Gore, J. C., & Allison, T. (1997). Face-specific processing in the human fusiform gyrus. *Journal of Cognitive Neuroscience, 9*, 605–610.

McCauley, R. N., & Henrich, J. (2006). Susceptibility to the Muller-Lyer Illusion, theory-neutral observation, and the diachronic penetrability of the visual input system. *Philosophical Psychology, 19*, 1–23.

McClelland, D. C. (1987). *Human motivation.* New York: Cambridge University Press.

McClelland, D. C., Koestner, R., & Weinberger, J. (1989). How do self-attributed and implicit motives differ? *Psychological Review, 96*, 690–702.

McClintock, M. K. (1971). Menstrual synchrony and suppression. *Nature, 229*, 244–245.

McClure, S. M., Li, J., Tomlin, D., Cypert, K. S., Montague, L. M., & Montague, P. R. (2004). Neural correlates of behavioral preference for culturally familiar drinks. *Neuron, 14*, 379–387.

McConnell, S., Roberts, J., Spitzer, L., Zigmond, M., Squire, L., & Bloom, F. (2002). *Fundamental neuroscience* (2nd ed.). San Diego, CA: Academic Press.

McCormick, D. A. (1999). Membrane potential and action potential. In M. J. Zigmond, F. E. Bloom, S. C. Landis, J. L. Roberts, & L. R. Squire (Eds.), *Fundamentals of neuroscience* (pp. 129–154). San Diego, CA: Academic Press.

McCrae, R. R., & Costa, P. T., Jr. (1990). *Personality in adulthood.* New York: Guilford Press.

McCrae, R. R., & Costa, P. T., Jr. (1999). A five-factor theory of personality. In L. A. Pervin & O. P. John (Eds.), *Handbook of personality: Theory and research* (2nd ed., pp. 139–153). New York: Guilford Press.

McCrae, R. R., Costa, P. T., Ostendorf, F., Angleitner, A., Hrebickova, M., Avia, M. D., et al. (2000). Nature over nurture: Temperament, personality, and life span development. *Journal of Personality and Social Psychology, 78*, 173–186.

McCrae, R. R., Terracciano, A., & 79 members of the Personality Profiles of Cultures Project (2005). Personality profiles of cultures: Aggregate personality traits. *Journal of Personality and Social Psychology, 89*, 407–425.

McCrae, R. R., Yik, M. S. M., Trapnell, P. D., Bond, M. H., & Paulhus, D. L. (1998). Interpreting personality profiles across cultures: Bilingual, acculturation, and peer rating studies of Chinese undergraduates. *Journal of Personality and Social Psychology, 74*, 1041–1055.

McCrink, K., & Wynn, K. (2004). Large-number addition and subtraction in infants. *Psychological Science, 15*, 776–781.

McCullough, J. P. (2000). *Treatment for chronic depression: Cognitive Behavioral Analysis System of Psychotherapy (CBASP).* New York: The Guilford Press.

McDaniel, M.A. (2005). Big-brained people are smarter: A meta-analysis of the relationship between in vivo brain volume and intelligence. *Intelligence, 33*, 347–346.

McDougle, C. (1997). Psychopharmacology. In D. Cohen & R. Volkmar (Eds.), *Handbook of autism and pervasive developmental disorders* (2nd ed., pp. 707–729). New York: Wiley.

McEwen, B. S. (2000). The effects of stress on structural and functional plasticity in the hippocampus. In D. S. Charney, E. J. Nestler, & B. S. Bunney (Eds.), *Neurobiology of mental illness* (pp. 475–493). New York: Oxford University Press.

McGaugh, J. L. (2002). Memory consolidation and the amygdale: A systems perspective. *Trends in Neurosciences, 25*, 456–461.

McGlashan, T. H. (1988). A selective review of recent North American long-term follow-up studies of schizophrenia. *Schizophrenia Bulletin, 14*, 515–542.

McGough, J. J., & Barkley, R. A. (2004). Diagnostic controversies in adult attention deficit hyperactivity disorder. *American Journal of Psychiatry, 161*, 1948–1956.

McGregor, A., & Roberts, D. C. S. (1993). Dopaminergic antagonism with the nucleus accumbens or the amygdala produces differential effects on intravenous cocaine self-administration under fixed and progressive ratio schedules of reinforcement. *Brain Research, 624*, 245–252.

McGue, M., Bacon, S., & Lykken, D. T. (1993). Personality stability and change in early adulthood: A behavioral genetic analysis. *Developmental Psychology, 29*, 96–109.

McGue, M., Bouchard, T. J., Jr., Iacono, W. G., & Lykken, D. T. (1993). Behavioral genetics of cognitive ability: A life-span perspective. In R. Plomin & G. E. McClearn (Eds.), *Nature, nurture, and psychology* (pp. 59–76). Washington, DC: American Psychological Association.

McInnis, M. G., McMahon, F. J, Chase, G. A., Simpson, S. G., Ross, C. A., & DePaulo, J. R. (1993). Anticipation in bipolar affective disorder. *American Journal of Human Genetics, 53*, 385–390.

McIntyre, D. C., & Reichert, H. (1971). State-dependent learning in rats induced by kindled convulsions. *Physiology & Behavior, 7*, 15–20.

McNally, R. J. (2003). Recovering memories of trauma: A view from the laboratory. *Current Directions in Psychological Science, 12*, 32–35.

McNally, R. J., Lasko, N. B., Clancy, S. A., Macklin, M. L., Pitman, R. K., & Orr, S. P. (2004). Psychophysiological responding during script-driven imagery in people reporting abduction by space aliens. *Psychological Science, 15,* 493–497.

McQuarrie, E. F., & Mick, D. G. (2003). Visual and verbal rhetorical figures under directed processing versus incidental exposure to advertising. *Journal of Consumer Research, 29,* 579–587.

Meddis, R. (1977). *The sleep instinct.* London: Routledge & Kegan Paul.

Medin, D. L., Ross, B. H., & Markman, A. B. (2005). *Cognitive psychology* (4th ed.). New York: Wiley.

Mednick, S. A., Gabrielli, W. F., & Hutchings, B. (1987). Genetic factors in the etiology of criminal behavior. In S. A. Mednick, T. E. Moffitt, & S. A. Stacks (Eds.), *The causes of crime: New biological approaches* (pp. 267–291). Cambridge, MA: Cambridge University Press.

Mednick, S. A., Huttunen, M. O., & Machon, R. A. (1994). Prenatal influenza infections and adult schizophrenia. *Schizophrenia Bulletin, 20,* 263–267.

Mednick, S., Nakayama, K., & Stickgold, R. (2003). Sleep-dependent learning: A nap is as good as a night. *Nature Neuroscience, 6,* 697–698.

Medvec, V. H., Madey, S. F., & Gilovich, T. (1995). When less is more: Counterfactual thinking and satisfaction among Olympic medalists. *Journal of Personality and Social Psychology, 69,* 603–610.

Mehl, M. R., & Pennebaker, J. W. (2003). The sounds of social life: A psychometric analysis of students' daily social environments and natural conversations. *Journal of Personality and Social Psychology, 84,* 857–870.

Mehler, J., & Bever, T. G. (1967). Cognitive capacity of very young children. *Science, 158,* 141–142.

Mehler, J., Jusczyk, P. W., Lambertz, G., Halsted, N., Bertoncini, J., & Amiel-Tison, C. (1988). A precursor of language acquisition in young infants. *Cognition, 29,* 143–178.

Melcher, J. M., & Schooler, J. W. (1996). The misremembrance of wines past: Verbal and perceptual expertise differentially mediate verbal overshadowing of taste memory. *Journal of Memory and Language, 35,* 231–245.

Meltzoff, A. N. (1995). Understanding the intentions of others: Re-enactment of intended acts by 18-month-old children. *Developmental Psychology 31,* 838–850.

Melzack, R. (1992). Phantom limbs. *Scientific American, 266,* 120–126.

Melzack, R., & Wall, P. D. (1982). *The Challenge of pain.* New York: Basic Books.

Melzack, R., Israel, R., Lacroix, R., & Schultz, G. (1997). Phantom limbs in people with congenital limb deficiency or amputation in early childhood. *Brain, 120,* 1603–1620.

Merckelbach, H., Devilly, G. J., & Rassin, E. (2002). Alters in dissociative identity disorders: Metaphors or genuine entities? *Clinical Psychology Review, 22,* 481–497.

Messias, E., Kirkpatrick, B., Bromet, E., Ross, D., Buchanan, R. W., Carpenter, W. T., Jr., et al. (2004). Summer birth and deficit schizophrenia: A pooled analysis from 6 countries. *Archives of General Psychiatry, 61,* 985–989.

Meston, C. M., & Frohlich, P. F. (2000). The neurobiology of sexual function. *Archives of General Psychiatry, 57,* 1012–1030.

Metcalfe, J., & Mischel, W. (1999). A hot/cool-system analysis of delay of gratification: Dynamics of willpower. *Psychological Review, 106,* 3–19.

Mezulis, A. H., Abramson, L. Y., Hyde, J. S., & Hankin, B. L. (2004). Is there a universal positivity bias in attributions? A meta-analytic review of individual, developmental, and culture differences in the self-serving attributional bias. *Psychological Bulletin, 130,* 711–747.

Michalak, E. E., Tam, E. M., Majunath, C. V., Levitt, A. J., et al. (2004). Generic and health-related quality of life in patients with seasonal and non-seasonal depression. *Psychiatric Research, 128,* 245–251.

Michenbaum, D. (1977). *Cognitive behavior modification: An integrative approach.* New York: Plenum Press.

Michenbaum, D. (1993). Changing conceptions of cognitive behavior modification: Retrospect and prospect. *Journal of Consulting and Clinical Psychology, 61,* 202–204.

Mickelson, K. D., Kessler, R. C., & Shaver, P. R. (1997). Adult attachment in a nationally representative sample. *Journal of Personality and Social Psychology, 73,* 1092–1106.

Miller, B. C., & Benson, B. (1999). Romantic and sexual relationship development during adolescence. In W. Furman, B. B. Brown, & C. Feiring (Eds.), *The development of romantic relationships in adolescence* (pp. 99–121). New York: Cambridge University Press.

Miller, G. A., Galanter, E., & Pribram, K. H. (1960). *Plans and the structure of behavior.* New York: Holt.

Miller, G. E., Freedland, K. E., Carney, R. M., Stetler, C. A., & Banks, W. A. (2003). Cynical hostility, depressive symptoms, and the expression of inflammatory risk markers for coronary heart disease. *Journal of Behavioral Medicine, 26,* 501–515.

Miller, I. W., Norman, W. H., & Keitner, G. I. (1989). Cognitive-behavioral treatment of depressed inpatients: Six- and twelve-month follow-up. *American Journal of Psychiatry, 146,* 1274–1279.

Miller, J. G. (1984). Culture and the development of everyday social explanation. *Journal of Personality and Social Psychology, 46,* 961–978.

Miller, J. G., & Bersoff, D. M. (1992). Culture and moral judgment: How are conflicts between justice and interpersonal responsibilities resolved? *Journal of Personality and Social Psychology, 62,* 541–554.

Miller, L. C., & Fishkin, S. A. (1997). On the dynamics of human bonding and reproductive success: Seeking windows on the adapted-for human-environmental interface. In J. Simpson & D. T. Kenrick (Eds.), *Evolutionary social psychology* (pp. 197–236). Mahwah, NJ: Lawrence Erlbaum.

Miller, M. B., & Wolford, G. L. (1999). Theoretical commentary: The role of criterion shift in false memory. *Psychological Review, 106,* 398–405.

Miller, R. S. (1996). *Embarrassment: Poise and peril in everyday life.* New York: Guilford Press.

Miller, R. S. (1997). We always hurt the ones we love: Aversive interactions in close relationships. In R. M. Kowalski (Ed.), *Aversive interpersonal behaviors* (pp. 11–29). New York: Plenum Press.

Miller, W. R., & C'de Baca, J. (2001). *Quantum changes: When epiphanies and sudden insights transform ordinary lives.* New York: Guilford Press.

Miller, W. T. (2000). Rediscovering fire: Small interventions, large effects. *Psychology of Addictive Behaviors, 14,* 6–18.

Milner, B. (1972). Memory and the temporal regions of the brain. In K. H. Pribram and D. E. Broadbent (Eds.), *Biology of memory* (pp. 215–234). New York: Academic Press.

Milton, J., & Wiseman, R. (2001). Does pi exist? Reply to Storm and Ertel (2001). *Psychological Bulletin, 127,* 434–438.

Minami, H. (1971). *The psychology of the Japanese people.* Tokyo: University of Tokyo Press.

Mineka, S., Davidson, M., Cook, M., & Keir, R. (1984). Observational conditioning of snake fear in rhesus monkeys. *Journal of Abnormal Psychology, 93,* 355–372.

Minoura, Y. (1992). A sensitive period for the incorporation of a cultural meaning system: A study of Japanese children growing up in the United States. *Ethos, 20,* 304–339.

Mirsky, A. F., Bieliauskas, L. A., French, L. M., Van Kammen, D. P., Jonsson, E., & Sedvall, G. (2000). A 39-year followup of the Genain quadruplets. *Schizophrenia Bulletin, 26,* 699–708.

Mischel, W., & Shoda, Y. (1995). A cognitive-affective system theory of personality: Reconceptualizing situations, dispositions, dynamics, and invariance in personality structure. *Psychological Review, 102,* 246–268.

Mischel, W., Shoda, Y., & Rodriguez, M. L. (1989). Delay of gratification in children. *Science, 244,* 933–938.

Mitelman, S. A., Shihabuddin, L., Brickman, A. M., Hazlett, E. A., & Buchsbaum, M. S. (2005). Volume of the cingulate and outcome in schizophrenia. *Schizophrenia Research, 72,* 91–108.

Mittleman, M. A., Lewis, R. A., Maclure, M., Sherwood, J. B., & Muller, J. E. (2001). Triggering myocardial infarction by marijuana. *Circulation, 103,* 2805–2809.

Miyamoto, Y., & Kitayama, S. (2002). Cultural variation in correspondence bias: The critical role of attitude diagnosticity of socially constrained behavior. *Journal of Personality and Social Psychology, 83,* 1239–1248.

Mizerski, R. (1995). The relationship between cartoon trade character recognition and attitude toward product category in young children. *Journal of Marketing, 59,* 58–70.

Mobbs, D., Greicius, M. D., Abdel-Azim, E., Menon, V., & Reiss, A. L. (2003). Humor modulates the mesolimbic reward centers. *Neuron, 40,* 1041–1048.

Moffitt, T. E., Brammer, G. L., Caspi, A., Fawcett, J. P., Raleigh, M., Yuwiler, A., et al. (1998). Whole blood serotonin relates to violence in an epidemiological study. *Biological Psychiatry, 43,* 446–457.

Moises, H. W., & Gottesman, I. I. (2004). Does glial asthenia predispose to schizophrenia? *Archives of General Psychiatry, 61,* 1170.

Monteith, M. J. (1993). Self-regulation of prejudiced responses: Implications for progress in prejudice reduction efforts. *Journal of Personality and Social Psychology, 65,* 469–485.

Monteith, M. J., Ashburn-Nardo, L., Voils, C. I., & Czopp, A. M. (2002). Putting the brakes on prejudice: On the development and operation of cues for control. *Journal of Personality & Social Psychology, 83,* 1029–1050.

Montepare, J. M., & Vega, C. (1988). Women's vocal reactions to intimate and casual male friends. *Personality and Social Psychology Bulletin, 14,* 103–113.

Montreal Symposium (1994). Development and plasticity of the visual system. Proceedings of a symposium, Montreal, Quebec, Canada. *Canadian Journal of Physiology and Pharmacology, 73,* 1294–1405.

Morita, S. (1917). The true nature of shinkeishitsu and its treatment. In *Anthology of theses commemorating the 25th anniversary of Professor Kure's appointment to his chair.* Tokyo: Jikei University.

Morling, B., Kitayama, S., & Miyamoto, Y. (2002). Cultural practices emphasize influence in the United States and adjustment in Japan. *Personality and Social Psychology Bulletin, 28,* 311–323.

Morris, M., & Peng, K. (1994). Culture and cause: American and Chinese attributions for social and physical events. *Journal of Personality and Social Psychology, 67,* 949–971.

Morris, N. M., Udry, J. R., Khan-Dawood, F., & Dawood, M. Y. (1987). Marital sex frequency and midcycle female testosterone. *Archives of Sexual Behavior, 16,* 27–37.

Morris, W. N. (1992). A functional analysis of the role of mood in affective systems. In M. S. Clark (Ed.), *Emotion: Review of personality and social psychology* (Vol. 13, pp. 256–293). Newbury Park, CA: Sage.

Mortensen, E. L., Michaelsen, K. F., Sanders, S. A., & Reinisch, J. M. (2002). The association between duration of breastfeeding and adult intelligence. *Journal of the American Medical Association, 287,* 2365–2371.

Morton, J., & Johnson, M. H. (1991). CONSPEC and CONLERN: A two-process theory of infant face recognition. *Psychological Review, 98,* 164–181.

Moscovitch, M. (1995). Confabulation. In D. L. Schacter (Ed.), *Memory distortions: How minds, brains, and societies reconstruct the past* (pp. 226–251). Cambridge, MA: Harvard University Press.

Moskowitz, A. K. (2004). "Scared stiff": Catatonia as an evolutionary-based fear response. *Psychological Bulletin, 111,* 984–1002.

Moskowitz, H., & Fiorentino, D. A. (2000). *Review of the literature on the effects of low doses of alcohol on driving-related skills.* Washington, DC: National Highway Traffic Safety Administration.

Mowery, P. D., Brick, P. D., & Farrelly, M. (2000). *Pathways to Established Smoking: Results from the 1999 National Youth Tobacco Survey* (Legacy First Look Report 3). Washington, DC: American Legacy Foundation.

Mroczek, D. K., & Kolarz, C. M. (1998). The effect of age on positive and negative affect: A developmental perspective on happiness. *Journal of Personality and Social Psychology, 75,* 1333–1349.

Mufson, L., Dorta, K. P., Wickramaratne, P., Nomura, Y., Olfson, M., & Weissman, M. M. (2004). A randomized effectiveness trial of interpersonal psychotherapy for depressed adolescents. *Archives of General Psychiatry, 61,* 577–584.

Muhle, R., Trentacoste, S. V., & Rapin, I. (2004). The genetics of autism. *Pediatrics, 113,* 472–486.

Munson, M. L., & Sutton, P. D. (2004). *Births, marriages, divorces, and deaths: Provisional data for May 2004* (National Vital Statistics Reports, Vol. 53, No. 8). Hyattsville, MD: National Center for Health Statistics, Centers for Disease Control and Prevention.

Muraven, M., & Baumeister, R. F. (2000). Self-regulation and depletion of limited resources: Does self-control resemble a muscle? *Psychological Bulletin, 126,* 247–259.

Murphy, G. L. (2002). *The big book of concepts.* Cambridge, MA: MIT Press.

Murray, S. L., Holmes, J. G., & Griffin, D. W. (1996). The benefits of positive illusions: Idealization and the construction of satisfaction in close relationships. *Journal of Personality and Social Psychology, 70,* 79–98.

Mustanski, B. S., Chivers, M. L., & Bailey, J. M. (2002). A critical review of recent biological research on human sexual orientation. *Annual Review of Sex Research, 13,* 89–140.

Nathanson, C., Paulhus, D. L., & Williams, K. M. (2006a). Personality and misconduct correlates of body modification and other cultural deviance markers. *Journal of Research in Personality, 40,* 799–802.

Nathanson, C., Paulhus, D. L., & Williams, K. M. (2006b). Predictors of a behavioural measure of scholastic cheating: Personality and competence but not demographics. *Contemporary Educational Psychology, 31,* 97–122.

National Research Council, Committee on Educational Interventions for Children with Autism (2001). *Educating young children with autism.* Washington, DC: National Academy Press.

Neath, I. (1998). *Human memory: An introduction to research, data, and theory.* Pacific Grove, CA: Brooks-Cole.

Neidemeyer, E., & Naidu, S. B. (1997). Attention deficit hyperactivity disorder (ADHD) and frontal-motor cortex disconnection. *Clinical Electroencephalography, 28,* 130–135.

Neisser, U. (1976). *Cognition and reality: Principles and implications of cognitive psychology.* New York: Freeman.

Neisser, U., & Harsch, N. (1993). Phantom flashbulbs: False recollections of hearing the news about Challenger. In E. Winograd & U. Neisser (Eds.), *Affect and accuracy in recall: Studies of "flashbulb" memories* (pp. 9–31). New York: Cambridge University Press.

Neisser, U., Boodoo, G., Bouchard, T. J. Jr., Boykin, A. W., Brody, N., Ceci, S. J., et al. (1996). Intelligence: Knowns and unknowns. *American Psychologist, 51,* 77–101.

Nelson, K. B., Grether, J. K., Croen, L. A., Dambrosia, J. M., Dickens, B. F., Jelliffe, L. L., et al. (2001). Neuropeptides and neurotrophins in neonatal blood of children with autism or mental retardation. *Annals of Neurology, 49,* 597–606.

Neuberg, S. L., Smith, D. M., & Asher, T. (2000). Why people stigmatize: Toward a biocultural framework. In T. F. Heatherton, R. E. Kleck, M. Hebl, & J. G. Hull (Eds.), *The social psychology of stigma* (pp. 31–61). New York: Guilford.

Neylan, T. C., Canick, J. D., Hall, S. E., Reus, V. I., Spolosky, R. M., & Wolkowitz, O. M. (2001). Cortisol levels predict cognitive impairment induced by electroconvulsive therapy. *Biological Psychiatry, 50,* 331–336.

Ng, D. M., & Jeffrey, E. W. (2003). Relationships between perceived stress and health behaviors in a sample of working adults. *Health Psychology, 22,* 638–642.

NIH Technology Assessment Conference Panel. (1993). Methods for voluntary weight loss and control. *Annals of Internal Medicine, 199,* 764–770.

Nisbett, R. E. (2003). *The geography of thought.* New York: Free Press.

Nisbett, R. E., Peng, K., Choi, I., & Norenzayan, A. (2001). Culture and systems of thought: Holistic versus analytic cognition. *Psychological Review, 108,* 291–310.

Nisbett, R. E., & Wilson, T. D. (1977). Telling more than we can know: Verbal reports on mental processes. *Psychological Review, 84,* 231–259.

Norem, J. K. (1989). Cognitive strategies as personality: Effectiveness, specificity, flexibility and change. In D. M. Buss & N. Cantor (Eds.), *Personality psychology: Recent trends and emerging issues* (pp. 45–60). New York: Springer-Verlag.

Norenzayan, A., & Heine, S. J. (2005). Psychological universals: What are they and how can we know? *Psychological Bulletin, 131,* 763–784.

Norenzayan, A., Smith, E. E., Kim, B., & Nisbett, R. E. (2002). Cultural preferences for formal versus intuitive reasoning. *Cognitive Science, 26,* 653–684.

North, A. C., Hargreaves, D. J., & O'Neill, S. A. (2000). The importance of music to adolescents. *British Journal of Educational Psychology, 70,* 255–272.

Norwich, K. H., & Wong, W. (1997). Unification of psychophysical phenomena: The complete form of Fechner's Law. *Perception and Psychophysics, 59,* 929–940.

Novotny, S. L., Hollander, E., Allen, A., Aronowitz, B. R., DeCaria, C., Cartwright, C., et al. (2000). Behavioral response to oxytocin challenge in adult autistic disorders. *Biological Psychiatry, 47,* 523.

Noyes, R. (1991). Suicide and panic disorder: A review. *Journal of Affective Disorders, 22,* 1–11.

Nurnberger, J. J., Goldin, L. R., & Gershon, E. S. (1994). Genetics of psychiatric disorders. In G. Winokur & P. M. Clayton (Eds.), *The medical basis of psychiatry* (pp. 459–492). Philadelphia: W. B. Saunders.

Oberauer, K., Schulze, R., Wilhelm, O., Süss, H. M. (2005). Working memory and intelligence—their correlation and their relation: Comment on Ackerman, Beier, and Boyle (2005). *Psychological Bulletin, 131,* 61–65.

Obeyesekere, G. (1985). Depression, Buddhism and the work of culture in Sri Lanka. In A. Kleinman & B. Good (Eds.), *Culture and depression* (pp. 134–152). Berkeley: University of California Press.

Ochsner, K. N. (2000). Are affective events richly recollected or simply familiar? The experience and process of recognizing feelings past. *Journal of Experimental Psychology: General, 129,* 242–261.

Ochsner, K. N., Bunge, S. A., Gross, J. J., & Gabrieli, J. D. E. (2002). Rethinking feelings: An fMRI study of the cognitive regulation of emotion. *Journal of Cognitive Neuroscienc* e, *14,* 1215–1299.

Offord, D. R., Boyle, M. H., Campbell, D., Goering, P., Lin, E., Wong, M., & Racine, Y. A. (1996). One-year prevalence of psychiatric disorder in Ontarians 15 to 64 years of age. *Canadian Journal of Psychiatry, 41,* 559–563.

Ogbu, J. U. (1994). From cultural differences to differences in cultural frames of reference. In P. M. Greenfield & R. R. Cocking (Eds.), *Cross cultural roots of minority child development* (pp. 365–392). Hillsdale, NJ: Erlbaum.

O'Kane, G. O., Kensinger, E. A., & Corkin, S. (2004). Evidence for semantic learning in profound amnesia: An investigation with the patient H. M. *Hippocampus, 14,* 417–425.

Okazaki, S. (1997). Sources of ethic differences between Asian American and white American college students on measures of depression and social anxiety. *Journal of Abnormal Psychology, 106*, 52–60.

Olanow, C. W., Goetz, C. G., Kordower, J. H., Stoessl, A. J., Sossi, V., Brin, M. F., et al. (2003). A double-blind controlled trial of bilateral fetal nigral transplantation in Parkinson's disease. *Annals of Neurology, 54*, 403–414.

O'Leary, S. G. (1995). Parental discipline mistakes. *Current Directions in Psychological Science, 4*, 11–13.

Olds, J. (1962). Hypothalamic substrates of reward. *Psychological Review, 42*, 554–604.

Olds, J., & Milner, P. (1954). Positive reinforcement produced by electrical stimulation of the septal area and other regions of the rat brain. *Journal of Comparative and Physiological Psychology, 47*, 419–428.

Olfert, E. D., Cross, B. M., & McWilliam, A. A. (Eds.). (1993). *Guide to the care and use of experimental animals* (2nd ed., vols. 1 & 2). Ottawa: Canadian Council on Animal Care.

Olfson, M., Marcus, S. C., Druss, B., Elinson, L., Tanielian, T., & Pincus, H. A. (2002). National trends in the outpatient treatment of depression. *Journal of the American Medical Association, 287*, 203–209.

Olfson, M., Shaffer, D., Marcus, S. C., & Greenberg, T. (2003). Relationship between antidepressant medication treatment and suicide in adolescents. *Archives of General Psychiatry, 60*, 978–982.

Olson, J. M., Vernon, P. A., Harris, J. A., & Jang, K. L. (2001). The heritability of attitudes: A study of twins. *Journal of Personality and Social Psychology, 80*, 845–860.

Oltmans, T. F., Neale, J. M., Davison, G. C. (Eds.). (1999). *Case studies in abnormal psychology*. New York: John Wiley & Sons, Inc.

O'Neal, J. M. (1984). First person account: Finding myself and loving it. *Schizophrenia Bulletin, 10*, 109–110.

O'Neil, J. (2004, December 29). Slow-motion miracle: One boy's journey out of autism's grasp. *The New York Times*, p. B8.

Osterling, J., & Dawson, G. (1994). Early recognition of children with autism: A study of first birthday home videotapes. *Journal of Autism and Developmental Disorders, 24*, 247–257.

Ottieger, A. E., Tressell, P. A., Inciardi, J. A., & Rosales, T. A. (1992). Cocaine use patterns and overdose. *Journal of Psychoactive Drugs, 24*, 399–410.

Overton, D. A. (1978). Basic mechanisms of state dependent learning. *Psychophamacological Bulletin, 14*, 67–68.

Packard, M. G., & Knowlton, B. J. (2002). Learning and memory functions of the basal ganglia. *Annual Review of Neuroscience, 25*, 563–593.

Pagsberg, A. K., & Wang, A. R. (1994). Epidemiology of anorexia and bulimia nervosa in Bornholm County, Denmark, 1970–1989. *Acta Psychiatrica Scandinavia, 90*, 259–265.

Paller, K. A., & Wagner, A. D. (2002). Observing the transformation of experience into memory. *Trends in Cognitive Science, 6*, 93–102.

Panksepp, J. (1992). Oxytocin effects on emotional processes: Separation distress, social bonding, and relationships to psychiatric disorders. *Annals of the New York Academy of Sciences, 652*, 243–252.

Panksepp, J. (1998). *Affective neuroscience: The foundations of human and animal emotions.* New York: Oxford University Press.

Papadimitrious, G. N., Zervas, I. M., & Papakostas, Y. G. (2001). Unilateral ECT for prophylaxis in affective illness. *Journal of ECT, 17*, 229–231.

Paquette, V., Levesque, J., Mensour, B., Leroux, J. M., Beaudoin, G., Bourgouin, P., & Beauregard, M. (2003). "Change the mind and you change the brain": Effects of cognitive-behavioral therapy on the neural correlates of spider phobia. *Neuroimage, 18*, 401–409.

Pascual-Leone, A. Catala, M. D., & Pascual-Leone, P. A. (1996). Lateralized effect of rapid-rate transcranial magnetic stimulation of the prefrontal cortex on mood. *Neurology, 46*, 499–502.

Pasupathi, M., & Carstensen, L. L. (2003). Age and emotional experience during mutual reminiscing. *Psychology and Aging, 18*, 430–442.

Patterson, C. M., & Newman, J. P. (1993). Reflectivity and learning from aversive events: Toward a psychological mechanism for the syndromes of disinhibition. *Psychological Review, 100*, 716–736.

Paulhus, D. L., Lysy, D. C., & Yik, M. S. M. (1998). Self-report measures of intelligence: Are they useful as proxy IQ tests? *Journal of Personality, 66*, 525–554.

Paulhus, D. L., & Williams, K. M. (2002). The Dark Triad of personality: Narcissism, Machiavellianism, and psychopathy. *Journal of Research in Personality, 36*, 556–563.

Paunonen, S. V., & Ashton, M. C. (2001). Big five factors and facets and the prediction of behaviour. *Journal of Personality and Social Psychology, 81*, 524–539.

Paykel, E. S. (2003). Life events and affective disorders. *Acta Psychiatrica Scandinavica, 108*, 61–66.

Payne, B. K. (2001). Prejudice and perception: The role of automatic and controlled processes in misperceiving a weapon. *Journal of Personality and Social Psychology, 81*, 181–192.

Paz-Elizur, T., Krupsky, M., Blumenstein, S., Elinger, D., Schechtman, E., & Livneh, Z. (2003). DNA repair activity for oxidative damage and risk of lung cancer. *Journal of the National Cancer Institute, 95*, 1312–1331.

Pederson, D. R., & Moran, G. (1996). Expressions of the attachment relationship outside of the strange situation. *Child Development, 67*, 915–927.

Pegna, A. J., Khateb, A., Lazeyras, F., & Seghier, M. L. (2005). Discriminating emotional faces without primary visual cortices involves the right amygdala. *Nature Neuroscience, 8*, 24–25.

Pelham, W. E., McBurnett, K., Harper, G. W., Milich, R., Murphy, D. A., Clinton, J., et al. (1990). Methylphenidate and baseball playing in ADHD children: Who's on first? *Journal of Consulting and Clinical Psychology, 58*, 130–133.

Pelham, W., & Bender, M. E. (1982). Peer relationships in hyperactive children: Description and treatment. In K. D. Gadow & I. Bailer (Eds.), *Advances in learning and behavioral disabilities: A research annual*. Greenwich, CT: JAI Press.

Pelissier, M. C., & O'Connor, K. P. (2002). Deductive and inductive reasoning in obsessive-compulsive disorder. *British Journal of Clinical Psychology, 41(Pt. 1)*, 15–27.

Penfield, W., & Jasper, H. (1954). *Epilepsy and the functional anatomy of the human brain.* Boston: Little, Brown.

Pennebaker, J. W. (1990). *Opening up: the healing power of confiding in others.* New York: William Morrow & Co., Inc.

Pennebaker, J. W. (1995). *Emotion, Disclosure, & Health*. Washington, DC: American Psychological Association.

Pennebaker, J. W. (1997). *Opening up: The healing power of confiding in others.* New York: William Morrow & Co., Inc.

Penton-Voak, I. S., Perrett, D. I., Castles, D., Burt, M., Koyabashi, T., & Murray, L. K. (1999). Female preference for male faces changes cyclically. *Nature, 399*, 741–742.

Peretz, I. (1996). Can we lose memory for music? A case of music agnosia in a nonmusician. *Journal of Cognitive Neuroscience, 8*, 481–496.

Peretz, I., & Zatorre, R. J. (2005). Brain organization for music processing. *Annual Review of Psychology, 56*, 89–114.

Perrett, D. I., May, K. A., & Yoshikawa, S. (1994). Facial shape and judgments of female attractiveness. *Nature, 368*, 239–242.

Perrett, D. I., Mistlin, A. J., & Chitty, A. J. (1987). Visual neurons responsive to faces. *Trends in Neurosciences, 10*, 358–364.

Pervin, L. A., & John, O. P. (1999). *Handbook of personality: Theory and research.* New York: Guilford Press.

Petitto, L. A. (1988). "Language" in the pre-linguistic child. In F. Kessel (Ed.)*, Development of Language and Language Researchers: Essays in Honor of Roger Brown* (pp. 187–221). Hillsdale, NJ: Lawrence Erlbaum Associates.

Petitto, L. A. (2000). On the biological foundations of human language. In H. Lane & K. Emmorey (Eds.)*, The signs of language revisited* (pp. 447–471). Mahwah, NJ: Lawrence Erlbaum Associates.

Petitto, L. A., Holowka, S., Sergio, L., & Ostry, D. (2001). Language rhythms in babies' hand movements. *Nature, 413*, 35–36.

Petrie, K. J., Fontanilla, I., Thomas, M. G., Booth, R. J., & Pennebaker, J. W. (2004). Effect of written emotional expression on immune function in patients with human immunodeficiency virus infection: A randomized trial. *Psychosomatic Medicine, 66*, 272–275.

Petronis, A., & Kennedy, J. L. (1995). Unstable genes—Unstable mind? *American Journal of Psychiatry, 152*, 164–172.

Pett, M. A., Wampold, B. E., Turner, C. W., & Vaughan-Cole, B. (1999). Paths of influence of divorce on preschool children's psychosocial adjustment. *Journal of Family Psychology*, *13*, 145–164.

Petty, R. E., & Cacioppo, J. T. (1986). *Communication and persuasion: Central and peripheral routes to attitude change.* New York: Springer-Verlag.

Pezdek, K., & Hodge, D. (1999). Planting false childhood memories in children: The role of event plausibility. *Child Development, 70*, 887–895.

Pfaus, J. G., & Phillips, A. G. (1991). Role of dopamine in anticipatory and consummatory aspects of sexual behavior in the male rat. *Behavioral Neuroscience, 105*, 727–743

Pfaus, J. G., Kippin, T. E., & Coria-Avila, G. (2003). What can animal models tell us about human sexual response? *Annual Review of Sex Research, 14*, 1–63.

Phelps, E. A. (2004). Human emotion and memory: Interactions of the amygdala and hippocampal complex. *Current Opinion in Neurobiology, 14*, 198–202.

Phelps, E. A., O'Connor, K. J., Cunningham, W. A., Funayama, E. S., Gatenby, J. C., Gore, J. C., et al. (2000). Performance on indirect measures of race evaluation predicts amygdala activation. *Journal of Cognitive Neuroscience, 12*, 729–738.

Phillips, M. L., Young, A. W., Senior, C., Brammer, M., Andrew, C., Calder, A. J., et al. (1997). A specific neural substrate for perceiving facial expressions of disgust. *Nature, 398*, 495–498.

Phinney, J. S. (1990). Ethnic identity in adolescents and adults: *Review of research. Psychological Bulletin, 108*, 499–514.

Physicians for a Smoke-Free Canada. Tobacco and the health of Canadians. Retrieved November 28, 2006, from http://www.smoke-free.ca/Health/pscissues_health.htm

Piggott, S., & Milner, B. (1993). Memory for different aspects of complex visual scenes after unilateral temporal- or frontal-lobe resection. *Neuropsychologia, 31*, 1–15.

Pinel, J. P. J. (2005). *Biopsychology* (6th ed.). Needham Heights, MA: Allyn & Bacon.

Pinel, J. P. J., Assanand, S. & Lehman D. R. (2000). Hunger, eating, and ill health. *American Psychologist, 55*, 1105–1116.

Pinker, S. (1984). *Language learnability and language development*. Cambridge, MA: Harvard University Press.

Pinker, S. (1994*). The language instinct*. New York: William Morrow and Company.

Pinker, S. (1997). *How the mind works*. New York: Norton.

Pinker, S. (2003). *The blank slate: The modern denial of human nature.* New York: Viking.

Pitman, R. K., Sanders, K. M., Zusman, R. M., Healy, A. R., Cheema, F., Lasko, N. B., et al. (2002). Pilot study of secondary prevention of posttraumatic stress disorder with propranolol. *Biological Psychiatry, 51*, 189–192.

Plakun, E. M., Burkhardt, P. E., & Muller, A. P. (1985). Fourteen-year follow-up of borderline and schizotypal personality disorders. *Comprehensive Psychiatry, 26*, 448–455.

Plant, E. A., Hyde, J. S., Keltner, D., & Devine, P. G. (2000). The gender stereotyping of emotions. *Psychology of Women Quarterly, 24*, 81–92.

Plant, E. A., & Peruche, M. (2005). The consequences of race for Police Officers' responses to criminal suspects. *Psychological Science, 16*, 180–183.

Plomin, R., & Caspi, A. (1999). Behavioral genetics and personaliity. In L. A. Pervin & O. P. John (Eds.)*, Handbook of personality: Theory and research* (2nd ed., pp. 251–276). New York: Guilford Press.

Plomin, R., & Spinath, F. M. (2002). Genetics and general cognitive ability (g). *Trends in Cognitive Sciences, 6*, 169–176.

Plomin, R., & Spinath, F. M. (2004). Intelligence: Genetics, genes, and genomics. *Journal of Personality and Social Psychology, 86*, 112–129.

Polivy, J., & Herman, C. P. (1985). Dieting and bingeing: A causal analysis. *American Psychologist, 40*, 193–201.

Pollock, K. M. (2004). Exercise in treating depression: Broadening the psychotherapist's role. *Journal of Clinical Psychology*, *57*, 1289–1300.

Pomerleau, O., Pomerleau, C., & Namenek, R. (1998). Early experiences with tobacco among women smokers, ex-smokers, and never-smokers. *Addiction, 93*, 595–599.

Pope, H. G., Jr., Kouri, E. M., & Hudson, J. I. (2000). Effects of supraphysiologic doses of testosterone on mood and aggression in normal men: A randomized controlled trial. *Archives of General Psychiatry, 57,* 133–140.

Posner, M. I., & DiGirolamo, G. J. (2000). Cognitive neuroscience: Origins and promise. *Psychological Bulletin, 126*, 873–889.

Premack, D. (1970). Mechanisms of self-control. In W. A. Hunt (Ed.), *Learning mechanisms in smoking* (pp. 107–123). Chicago: Aldine.

Pylyshyn, Z. (1984). *Computation and cognition*. Cambridge, MA: MIT Press.

Pylyshyn, Z. W. (1999). Is vision continuous with cognition? The case for cognitive impenetrability of visual perception. *Behavioral and Brain Sciences, 22,* 341–365.

Pyszczynski, T., Greenberg, J., Solomon, S., Arndt, J., & Schimel, J. (2004). Why do people need self-esteem? A theoretical and empirical review. *Psychological Bulletin, 130,* 435–468.

Quinones-Vidal, E., Lopez-Garcia, J. J., Penaranda-Ortega, M., & Tortosa-Gil, F. (2004). The nature of social and personality psychology as reflected in JPSP, 1965–2000. *Journal of Personality and Social Psychology, 86,* 435–452.

Raesaenen, S., Pakaslahti, A., Syvaelahti, E., Jones, P. B., & Isohanni, M. (2000). Sex differences in schizophrenia: A review. *Nordic Journal of Psychiatry, 54*, 37–45.

Raine, A. (1989). Evoked potentials and psychopathy. *International Journal of Psychopathology, 8*, 1–16.

Raine, A., Mellingen, K., Liu, J., Venables, P., & Mednick, S. A. (2003). Effects of environmental enrichment at ages 3–5 years on schizotypal personality and antisocial behavior at ages 17 and 23 years. *American Journal of Psychiatry, 160*, 1627–1635.

Rakic, P. (2000). Molecular and cellular mechanisms of neuronal migration: Relevance to cortical epilepsies. *Advances in Neurology, 84*, 1–14.

Raleigh, M. J., McGuire, M. T., Brammer, G. L., Pollack, D. B., & Yuwiler, A. (1991). Serotonergic mechanisms promote dominance in adult male vervet monkeys. *Brain Research, 559*, 181–190.

Ramachandran, V. S. (1993). Behavioral and magnetoencephalographic correlates of plasticity in the adult human brain. *Proceedings of the National Academy of Sciences, 90,* 10413–10420.

Ramachandran, V. S., & Blakeslee, S. (1998). *Phantoms in the brain: Probing the mysteries of the human mind.* New York: William Morrow.

Rampon, C., Jiang, C. H., Dong, H., Tang, Y., Lockhart, D. J., Schultz, P. G., et al. (2000). Effects of environmental enrichment on gene expression in the brain. *Proceedings of the National Academy of Sciences, 97*, 12880–12884.

Rapee, R. M., Brown, T. A., Antony, M. M., & Barlow, D. H. (1992). Response to hyperventilation and inhalation of 5.5% carbon dioxide-enriched air across the DSM III-R anxiety disorders. *Journal of Abnormal Psychology, 101*, 538–552.

Rapoport, J. (1990). *The boy who couldn't stop washing: The experience and treatment of obsessive-compulsive disorder*. New York: Penguin Books.

Rapoport, J. L. (1989). The biology of obsessions and compulsions. *Scientific American, 260*, 83–89.

Rapoport, J. L. (1991). Recent advances in obsessive-compulsive disorder. *Neuropsychopharmacology, 5*, 1–10.

Rauch, S. L., van der Kolk, B. A., Fisler, R. E., & Alpert, N. M. (1996). A symptom provocation study of posttraumatic stress disorder using positron emission tomography and script-driven imagery. *Archives of General Psychiatry, 53*, 380–387.

Rauscher, F. H., Shaw, G. L., & Ky, K. N. (1993). Music and spatial task performance, *Nature, 365,* 611.

Rayner, K., Foorman, B. R., Perfetti, C. A., Pesetsky, D., & Seidenberg, M. S. (2001). How psychological science informs the teaching of reading. *Psychological Science in the Public Interest, 2,* 31–74.

Read, J. P., & Brown, R. A. (2003). The role of exercise in alcoholism treatment and recovery. *Professional Psychology: Research and Practice, 34*, 49–56.

Recarte, M., & Nunes, L. M. (2003). Mental workload while driving: Effects on visual search, discrimination, and decision making. *Journal of Experimental Psychology: Applied, 9,* 119–137.

Reeves, L. M., & Weisberg, R. W. (1994). The role of content and abstract information in analogical transfer. *Psychological Bulletin, 115*, 381–400.

Regard, M., & Landis, T. (1997). "Gourmand syndrome": Eating passion associated with right anterior lesions. *Neurology, 48*, 1185–1190.

Reichelt, K. L., Knivsberg, A. M., Lind, G., & Nodland, M. (1991). Probable etiology and possible treatment of childhood autism. *Brain Dysfunction, 4*, 308–319.

Reinders, A. A., Nijenhuis, E. R., Paans, A. M., Korf, J., Willemsen, A. T., & den Boer, J. A. (2003). One brain, two selves. *Neuroimage, 20*, 2119–2125.

Reis, H. T., Collins, W. A., & Berscheid, E. (2000). The relationship context of human behavior and development. *Psychological Bulletin, 126*, 844–872.

Renfrow, P. J., & Gosling, S. D. (2003). The do re mi's of everyday life: The structure and personality correlates of music preferences. *Journal of Personality and Social Psychology, 84*, 1236–1256.

Repetti, R. L. (1997). *The effects of daily job stress on parent behavior with preadolescents*. Paper presented at the biennial meeting of the Society for Research in Child Development, April, Washington, DC.

Rescorla, R. (1966). Predictability and number of pairings in Pavlovian fear conditioning. *Psychonomic Science, 4*, 383–384.

Rescorla, R. A., & Wagner, A. R. (1972). A theory of Pavlovian conditioning: Variations in the effectiveness of reinforcement and nonreinforcement. In A. H. Black & W. F. Prokosy (Eds.), *Classical conditioning II: Current research and theory* (pp. 64–99). New York: Appleton-Century-Crofts.

Ressler, K. J., Rothbaum, B. O., Tannenbaum, L., Anderson, P., Graap, K., Zimand, E., Hodges L, & Davis M. (2004). Cognitive enhancers as adjuncts to psychotherapy: use of D-cycloserine in phobic individuals to facilitate extinction of fear. *Archives of General Psychiatry, 61,* 1136-1144.

Reynolds, D. V. (1969). Surgery in the rat during electrical analgesia induced by focal brain stimulation. *Science, 164*, 444–445.

Rhodes, G., Byatt, G., Michie, P. T., & Puce, A. (2004). Is the fusiform face area specialized for faces, individuation, or expert individuation? *Journal of Cognitive Neuroscience, 16,* 189–203.

Richerson, P. J., & Boyd, R. (2005). *Not by genes alone.* Chicago: University of Chicago Press.

Richeson, J. A., Baird, A. A., Gordon, H. L., Heatherton, T. F., Wyland, C. L., Trawalter, S., et al. (2003). An fMRI investigation of the impact of interracial contact on executive function. *Nature Neuroscience, 6*, 1323–1328.

Ridley, M. (2003). *Nature versus nurture: Genes, experience, and what makes us human.* New York: HarperCollins.

Rifkin, A., & Rifkin, W. (2004). Adolescents with depression. *Journal of the American Medical Association, 292,* 2577–2578.

Rinck, M., Reinecke, A., Ellwart, T., Heuer, K., & Becker, E. S. (2005). Speeded detection and increased distraction in fear of spiders: Evidence from eye movements. *Journal of Abnormal Psychology, 114,* 235–248.

Rizzolatti, G., & Arbib, M. A. (1998). Language within our grasp. *Trends in Neuroscience, 21,* 188–194.

Rizzolatti, G., & Craighero, L. (2004). The mirror-neuron system. *Annual Reviews in Neuroscience, 27,* 169–192.

Rizzolatti, G., Fadiga, L., Gallese, V., & Fogassi, L. (1996). Premotor cortex and the recognition of motor actions. *Cognitive Brain Research, 3,* 131–141.

Roberson, D., Davidoff, J., Davies, I. & Shapiro, L. (2005) Colour categories in Himba: Evidence for the cultural relativity hypothesis. *Cognitive Psychology, 50,* 378–411.

Roberson, D., Davies, I., & Davidoff, J. (2000). Color categories are not universal: Replications and new evidence from a stone-age culture. *Journal of Experimental Psychology: General, 129,* 369–398.

Roberts, B. W., & Friend-DelVecchio, W. (2000). The rank-order consistency of personality traits from childhood to old age: A quantitative review of longitudinal studies. *Psychological Bulletin, 126,* 3–25.

Roberts, D. F. (2000). Media and youth: Access, exposure, and privatization. *Journal of Adolescent Health 27* (Suppl.), 8–14.

Robins, L. N., Helzer, J. E., & Davis, D. H. (1975). Narcotic use in Southeast Asia and afterward: An interview study of 898 Vietnam returnees. *Archives of General Psychiatry, 32,* 955–961.

Robins, L. N., & Regier, D. A. (1991). *Psychiatric disorders in America: The epidemiological catchment areas study.* New York: The Free Press.

Rock, I. (1984). *Perception.* New York: Scientific American Books.

Roediger, H. L., & McDermott, K. B. (1995). Creating false memories: Remembering words not presented in lists. *Journal of Experimental Psychology: Learning, Memory, and Cognition, 21,* 803–814.

Rogan, M. T., Stäubli, U. V., & LeDoux, J. E. (1997). Fear conditioning induces associative long-term potentiation in the amygdala. *Nature, 390,* 604–607.

Rogers, T. B., Kuiper, N. A., & Kirker, W. S. (1977). Self-reference and the encoding of personal information. *Journal of Personality and Social Psychology, 35,* 677–688.

Rolls, E. T., Burton, M. J., & Mora, F. (1980). Neurophysiological analysis of brain-stimulation reward in the monkey. *Brain Research, 194,* 339–357.

Rolls, E. T., Murzi, E., Yaxley, S., Thorpe, S. J., & Simpson, S. J. (1986). Sensory-specific satiety: Food-specific reduction in responsiveness of ventral forebrain neurons after feeding in the monkey. *Brain Research, 368,* 79–86.

Rosch Heider, E. (1972). Universals in color naming and memory. *Journal of Experimental Psychology, 93,* 10–20.

Rosen, D. S. (2003). Eating disorders in children and young adolescents: Etiology, classification, clinical features, and treatment. *Adolescent Medicine, 14,* 49–59.

Rosen, L. E., & Amador, X. F. (1997). *When someone you love is depressed: How to help your loved one without losing yourself.* New York: Free Press.

Rosenbaum, R. S., Priselac, S., Kohler, S., Black, S. E., Gao, F., Nadel, L., & Moscovitch, M. (2000). Remote spatial memory in the amnesic person with extensive bilateral hippocampal lesions. *Nature Neuroscience, 3,* 1044–1048.

Rosenbluth, M., & Silver, D. (1992). The inpatient treatment of borderline personality disorder. In D. Silver & M. Rosenbluth, *Handbook of borderline disorders.* Madison, CT: International Universities Press.

Rosenfeld, J. P. (2001). Event-related potentials in detection of deception. In M. Kleiner (Ed.), *Handbook of Polygraph* (pp. 265–286). New York: Academic Press.

Rosenman, R. H., Brand, R. J., Jenkins, C. D., Friedman, M., Straus, R., & Wurm, M. (1975). Coronary heart disease in the Western Collaborative Group Study: Final follow-up experience of 8½ years. *Journal of the American Medical Association, 233,* 872–877.

Rosenstein, D., & Oster, H. (1988). Differential facial responses to four basic tastes in newborns. *Child Development, 59,* 1555–1568.

Rosenthal, D., (Ed). (1963). *The Genain quadruplets.* New York: Basic Books.

Rosenzweig, M. R., Bennett, E. L., & Diamond, M. C. (1972). Brain changes in response to experience. *Scientific American, 226,* 22–29.

Ross, C., Miller, S. D., Bjornson, L., & Reagor, P. (1991). Abuse histories in 102 cases of multiple personality disorder. *Canadian Journal of Psychiatry, 36,* 97–101.

Ross, D. M., & Ross, S. A. (1982). *Hyperactivity: Research, theory, and action.* New York: Wiley.

Ross, M., Xun, W. Q. E., & Wilson, A. E. (2002). Language and the bicultural self. *Personality and Social Psychology Bulletin, 28,* 1040–1050.

Rothbaum, B. O., Hodges, L., Alarcon, R., Ready, D., Shahar, F., Graap, K., et al. (1999). Virtual reality exposure therapy for PTSD Vietnam veterans: A case study. *Journal of Traumatic Stress, 12,* 263–271.

Rothbaum, F., Weisz, J. R., & Snyder, S. S. (1982). Changing the world and changing the self: A two-process model of perceived control. *Journal of Personality and Social Psychology, 42,* 5–37.

Rousseau, C., Corin, E., & Renaud, C. (1989). Armed conflict and trauma: A clinical study of Latin-American refugee children. *Canadian Journal of Psychiatry, 34,* 376–385.

Rovee-Collier, C. (1999). The development of infant memory. *Current Directions in Psychological Science, 8,* 80–85.

Rowe, D. C., Chassin, L., Presson, C., & Sherman, S. J. (1996). Parental smoking and the "epidemic" spread of cigarette smoking. *Journal of Applied Social Psychology, 26,* 437–445.

Rozin, P. (1996). Sociocultural influences on human food selection. In E. D. Capaldi (Ed.), *Why we eat what we eat: The psychology of eating* (pp. 233–263). Washington, DC: American Psychological Association.

Rozin, P., & Kalat, J. W. (1971). Specific hungers and poison avoidance as adaptive specializations of learning. *Psychological Review, 78,* 459–486.

Rubenstein, A. J., Kalakanis, L., Langlois, J. H. (1999). Infant preferences for attractive faces: A cognitive explanation. *Developmental Psychology, 35*, 848–855.

Rudman, L.A., & Goodwin, S.A. (2004). Gender differences in automatic in-group bias: Why do women like women more than men like men? *Journal of Personality and Social Psychology, 87*, 494–509.

Rusbult, C. E., & Van Lange, P. A. M. (1996). Interdependence processes. In E. T. Higgins & A. Kruglanski (Eds.), *Social psychology: Handbook of basic principles* (pp. 564–596). New York: Guilford Press.

Russell, J. A. (1980). A circumplex model of affect. *Journal of Personality and Social Psychology, 39*, 1161–1178.

Russell, J. A. (1994). Is there universal recognition of emotion from facial expressions? A review of the cross-cultural studies. *Psychological Bulletin, 115*, 102–141.

Russell, J. A., & L. Feldman Barrett. (1999). Core affect, prototypical emotional episodes, and other things called emotion: Dissecting the elephant. *Journal of Personality and Social Psychology, 76*, 805–819.

Russell, M. A. H. (1990). The nicotine trap: A 40-year sentence for four cigarettes. *British Journal of Addiction, 85*, 293–300.

Rutter, M. (2005). Incidence of autism disorders: Changes over time and their meaning. *Acta Paediatrica, 94*, 2–15.

Ryder, A. G., Yang, J., Zhu, X., Yao, S., Yi, J., Bagby, R. M., & Heine, S. J. (2006). *Culture and depression: Are there differences in Chinese and North American symptom presentation?* Unpublished manuscript. Concordia University.

Rymer, R. (1993). *Genie: A scientific tragedy*. New York: HarperCollins.

Sabol, S. Z., Nelson, M. L., Fisher, C., Gunzerath, L., Brody, C. L., Hu, S., et al. (1999). A genetic association for cigarette smoking behavior. *Health Psychology, 18*, 7–13.

Sachs, J. (1967). Recognition memory for syntactic and semantic aspects of connected discourse. *Perception and Psychophysics, 2*, 437–442.

Sacks, O. (1985). *The man who mistook his wife for a hat*. New York: Summit Books.

Sacks, O. (1995). *An anthropologist on Mars: Seven paradoxical tales*. New York: Knopf.

Saklofske, D. H., Hildebrand, D. K., & Gorsuch, R. L. (2000). Replication of the factor structure of the Wechsler Adult Intelligence Scale—Third edition with a Canadian sample. *Psychological Assessment, 12*, 436–439.

Salfati, C. G. (2000). The nature of expressiveness and instrumentality in homicide and its implications for offender profiling. In P. H. Blackman, V. L. Leggett, B. L. Olson, & J. P. Jarvis (Eds.), *The varieties of homicide and its research: Proceedings of the 1999 meeting of the Homicide Research Working Group*. Washington, DC: Federal Bureau of Investigation.

Salovey, P., & Mayer, J.D. (1990). Emotional intelligence. *Imagination, Cognition, and Personality, 9*, 185–211.

Salthouse, T. (1992). The information-processing perspective on cognitive aging. In R. Sternberg & C. Berg (Eds.), *Intellectual development*. Cambridge, MA: Cambridge University Press.

Sanderson, W. C., & Barlow, D. H. (1990). A description of patients diagnosed with DSM-III-R generalized anxiety disorder. *The Journal of Nervous and Mental Disease, 178*, 588–591.

Sapolsky, R. M. (1994). *Why zebras don't get ulcers*. New York: Freeman.

Sareen, J., Cox, B. J., Afifi, T. O., Clara, I., & Yu, B. N. (2005). Perceived need for mental health treatment in a nationally representative Canadian sample. *Canadian Journal of Psychiatry, 50*, 643–651.

Sareen, J., Fleisher, W., Cox, B. J., Hassard, S., & Stein, M. B. (2005). Childhood adversity and perceived need for mental health care: Findings from a Canadian community sample. *Journal of Nervous and Mental Disease, 193*, 396–404.

Savic, I., Berglund, H., & Lindström, P. (2005). Brain responses to putative pheromones in homosexual men. *Proceedings of the National Academy of Sciences, 102*, 7356–7361.

Savitz, J. B., & Ramesar, R. S. (2004). Genetic variants implicated in personality: A review of the more promising candidates. *American Journal of Medical Genetics, 15*, 20–32.

Sayette, M. A. (1993). An appraisal-disruption model of alcohol's effects on stress responses in social drinkers. *Psychological Bulletin, 114*, 459–476.

Scarr, S., & McCarthy, K. (1983). How people make their own environments: A theory of genotype l environment effects. *Child Development, 54*, 424–435.

Schachter, S. (1959). *The psychology of affiliation*. Stanford, CA: Stanford University Press.

Schachter, S. (1982). Recidivism and self-cure of smoking and obesity. *American Psychologist, 37*, 436–444.

Schachter, S., & Singer, J. (1962). Cognitive, social, and physiological determinants of emotional state. *Psychological Review, 69*, 379–399.

Schacter, D. L. (1996). *Searching for memory: The brain, the mind, and the past*. New York: Basic Books.

Schacter, D. L. (1999). The seven sins of memory: Insights from psychology and cognitive neuroscience. *American Psychologist, 54*, 182–203.

Schacter, D. L. (2001). *The seven sins of memory: How the mind forgets and remembers*. Boston: Houghton Mifflin.

Schacter, D. L., & Buckner, R. L. (1998). On the relations among priming, conscious recollection, and intentional retrieval: Evidence from neuroimaging research. *Neurobiology of Learning and Memory, 70*, 284–303.

Schaie, K. W. (1990). Intellectual development in adulthood. In J. E. Birren & K. W. Schaie (Eds.), *Handbook of the psychology of aging* (3rd ed.). New York: Van Nostrand Reinhold.

Schank, R. C., & Abelson, R. P. (1977). *Scripts, plans, goals, and understanding*. Hillsdale, NJ: Erlbaum.

Scheerer, M. (1963). Problem-solving. *Scientific American, 208*, 118–128.

Schiffman, J., Walker, E., Ekstrom, M., Schulsinger, F., Sorensen, H., & Mednick, S. (2004). Childhood videotaped social and neuromotor precursors of schizophrenia: A prospective investigation. *American Journal of Psychiatry, 161*, 2021–2027.

Schmahmann, J. D, & Sherman, J. C. (1998). The cerebellar cognitive affective syndrome. *Brain, 121*, 561–579.

Schmajuk, N. A., Lamoureux, J. A., & Holland, P. C. (1998). Occasion setting: A neural network approach. *Psychological Review, 105*, 3–32.

Schmidt, F. L., & Hunter, J. (2004). General mental ability in the world of work: Occupational attainment and job performance. *Journal of Personality and Social Psychology, 86*, 162–173.

Schmitt, D. P., Alcalay, L., Allik, J., Angleitner, A., Ault, L., Austers, I., et al. (2003). Universal sex differences in the desire for sexual variety: Tests from 52 nations, 6 continents, and 13 islands. *Journal of Personality & Social Psychology, 85,* 85–104.

Schneider, B. (1997). Psychoacoustics and aging: Implication for everyday listening. *Journal of Speech-Language Pathology and Audiology, 21,* 11–124.

Schoenemann, P. T., Sheehan, M. J., & Glotzer, L. D. (2005). Prefrontal white matter volume is disproportionately larger in humans than in other primates. *Nature Neuroscience, 8,* 242–252.

Schooler, J. W. (2002). Verbalizaton produces a transfer inappropriate processing shift. *Applied Cognitive Psychology, 16,* 989–997.

Schooler, J. W., & Engstler-Schooler, T. Y. (1990). Verbal overshadowing of visual memories: Some things are better left unsaid. *Cognitive Psychology 22,* 36–71.

Schreiber, F. R. (1974). *Sybil: The true story of a woman possessed by sixteen personalities.* London: Penguin.

Schuckit, M. A., Edenberg, H. J., Lakmijn, J., Flury, L., Smith, T. L., Reich, T., et al. (2001). A genome-wide search for genes that relate to a low level of response to alcohol. *Alcoholism: Clinical and Experimental Research, 25,* 323–329.

Schulz, K. P., Fan, J., Tang, C. Y., Newcorn, J. H., Buchsbaum, M. S., Cheung, A. M., et al. (2004). Response inhibition in adolescents diagnosed with attention deficit hyperactivity disorder during childhood: An event-related fMRI study. *American Journal of Psychiatry, 161,* 1650–1657.

Schwartz, B. (2004). *The paradox of choice: Why more is less.* New York: Ecco.

Schwartz, C. E., Wright, C. I., Shin, L. M., Kagan, J., & Rauch, S. L. (2003). Inhibited and uninhibited infants "grown up": Adult amygdalar response to novelty. *Science, 300,* 1952–1953.

Schwartz, J. M., Stoessel, P. W., Baxter, L. R., Martin, K. M., & Phelps, M. E. (1996). Systematic changes in cerebral glucose metabolic rate after successful behavior modification treatment of obsessive-compulsive disorder. *Archives of General Psychiatry, 53,* 109–113.

Schwarz, N., & Clore, G. L. (1983). Mood, misattribution, and judgments of well-being: Informative and directive functions of affective states. *Journal of Personality and Social Psychology, 45,* 513–523.

Sclafani, A., & Springer, D. (1976). Dietary obesity in adult rats: Similarities to hypothalamic and human obesity syndromes. *Physiology and Behavior, 17,* 461–471.

Searle, W., & Ward, C. (1990). The prediction of psychological and socio-cultural adjustment during cross-cultural transitions. *International Journal of Intercultural Relations, 14,* 449–464.

Sedikides, C., Gaertner, L., & Toguchi, Y. (2003). Pancultural self-enhancement. *Journal of Personality and Social Psychology, 84,* 60–79.

Seeman, P. (2002). Atypical antipsychotics: Mechanism of action. *Canadian Journal of Psychiatry, 47,* 27–38.

Segall, M. H., Campbell, D. T., & Herskovits, M. J. (1963). Cultural differences in the perception of geometric illusions. *Science, 193,* 769–771.

Segerstrom, S. C., & Miller, G. E. (2004). Psychological stress and the human immune system: A Meta-analytic study of 30 years of inquiry. *Psychological Bulletin, 130,* 601–630.

Seguin, J. R. (2004). Neurocognitive elements of antisocial behavior: Relevance of an orbitofrontal cortex account. *Brain and Cognition, 55,* 185–197.

Seligman, M. (1970). On the generality of the laws of learning. *Psychological Review, 77,* 406–418.

Seligman, M. E. P. (1974). Depression and learned helplessness. In R. J. Friedman & M. M. Katz (Eds.), *The psychology of depression: Contemporary theory and research.* Washington, DC: V. H. Winston.

Seligman, M. E. P. (1975). *Helplessness: On depression, development, and death.* San Francisco: W. H. Freeman.

Seligman, M. E. P., & Csikszentmihalyi, M. (2000). Positive psychology: An introduction. *American Psychologist, 55,* 5–14.

Seligman, M. E. P., Walker, E. F., & Rosenhan, D. L. (2001). *Abnormal psychology.* New York: Norton.

Selye, H. (1985). The nature of stress. *Basal Facts, 7,* 3–11.

Senechal, M., & LeFevre, J. A. (2002). Parental involvement in the development of children's reading skill: A five year longitudinal study. *Child Development, 73,* 445–460.

Shallice, T., & Warrington, E. (1969). Independent functioning of verbal memory stores. *Quarterly Journal of Experimental Psychology, 22,* 261–273.

Shankar, P., Manjunath, N., & Lieberman, J. (2005). The prospect of silencing disease using RNA interference. *Journal of the American Medical Association, 293,* 1367–1373.

Shapiro, E., Shapiro, A. K., Fulop, G., Hubbard, M., Mandell, J., Nordlie, J., et al. (1989). Controlled study of haloperidol, pimozide, and placebo for the treatment of Gilles de la Tourette's syndrome. *Archives of General Psychiatry, 46,* 722–730.

Shear, M. K., & Maser, J. D. (1994). Standardized assessment for panic disorder research. A conference report. *Archives of General Psychiatry, 51,* 346–354.

Shedler, J., & Block, J. (1990). Adolescent drug use and psychological health: A longitudinal inquiry. *American Psychologist, 45,* 612–630.

Sheese, B. E., Brown, E. L., & Graziano, W. G. (2004). Emotional expression in cyberspace: Searching for moderators of the Pennebaker disclosure effect via e-mail. *Health Psychology, 23,* 457–464.

Sheffield, F. D., & Roby, T. B. (1950). Reward value of a non-nutritive sweet taste. *Journal of Comparative and Physiological Psychology, 43,* 471–481.

Shenkin, S. D., Starr, J. M., & Deary, I. J. (2004). Birth weight and cognitive ability in childhood: A systematic review. *Psychological Bulletin, 130,* 989–1013.

Shephard, R. J. (1997). What is the optimal type of physical activity to enhance health? *British Journal of Sports Medicine, 31,* 277–284.

Sherman, D. K., McGue, M. K., & Iacono, W. G. (1997). Twin concordance for attention deficit hyperactivity disorder: A comparison of teacher's and mother's reports. *American Journal of Psychiatry, 154,* 532–535.

Sherman, S. J., Presson, C., Chassin, L., Corty, E., & Olshavsky, R. (1983). The false consensus effect in estimates of smoking prevalence: Underlying mechanisms. *Personality and Social Psychology Bulletin, 9*, 197–207.

Sherry, D. F., & Schacter, D. L. (1987). The evolution of multiple memory systems. *Psychological Review, 94*, 439–454.

Sherwin, B. B. (1988). A comparative analysis of the role of androgen in human male and female sexual behavior: Behavioral specificity, critical thresholds, and sensitivity. *Psychobiology, 16*, 416–425.

Sherwin, B. B. (1994). Sex hormones and psychological functioning in postmenopausal women. *Experimental Gerontology, 29*, 423–430.

Shettleworth, S. J. (2001). Animal cognition and animal behaviour. *Animal Behaviour, 61*, 277–286.

Shih, M., Pittinsky, T. L., & Ambady, N. (1999). Stereotype susceptibility: Identity salience and shifts in quantitative performance. *Psychological Science, 10*, 80–83.

Shweder, R. (1991). *Thinking through cultures: Expeditions in cultural psychology.* Cambridge, MA: Harvard University Press.

Shweder, R. A. (1990). Cultural psychology: What is it? In J. W. Stigler, R. A. Shweder, & G. Herdt (Eds.), *Cultural psychology: Essays on comparative human development* (pp. 1–43). Cambridge: Cambridge University Press.

Shweder, R. A., Much, N. C., Mahapatra, M., & Park, L. (1997). The "big three" of morality (autonomy, community, divinity) and the "big three" explanations of suffering. In A. M. Brandt & P. Rozin (Eds.), *Morality and health* (pp. 119–169). New York: Routledge.

Siegel, S. (1984). Pavlovian conditioning and heroin overdose: Reports by overdose victims. *Bulletin of the Psychonomic Society, 22*, 428–430.

Siegel, S., Hinson, R. E., Krank, M. D., & McCully, J. (1982). Heroin "overdose" death: Contribution of drug-associated environmental cues. *Science, 216*, 436–437.

Siegler, I. C., Costa, P. T., Brummett, B. H., Helms, M. J., Barefoot, J. C., Williams, R. et al. (2003). Patterns of change in hostility from college to midlife in the UNC Alumni Heart Study predict high-risk status. *Psychosomatic Medicine, 65*, 738–745.

Silva, P. A., & Stanton, W. (1996). *From child to adult: The Dunedin study*. Oxford, UK: Oxford University Press.

Silver, R. L., Cirincione, C., & Steadman, H. J. (1994). Demythologizing inaccurate perceptions of the insanity defense. *Law and Human Behavior, 18*, 63–70.

Simons, D. J., & Levin, D. T. (1998). Failure to detect changes to people during a real-world interaction. *Psychonomic Bulletin and Review, 5*, 644–649.

Sims, H. E. A., Goldman, R. F., Gluck, C. M., Horton, E., Kelleher, P., & Rowe, D. (1968). Experimental obesity in man. *Transactions of the Association of American Physicians, 81*, 153–170.

Singer, T., Seymour, B., O'Doherty, J., Kaube, H., Dolan, R. J., & Frith, C. D. (2004). Empathy for pain involves the affective but not sensory components of pain. *Science, 303*, 1157–1162.

Sirois, B. C., & Burg, M. M. (2003). Negative emotion and coronary heart disease: A review. *Behavior Modification, 27*, 83–102.

Skodol, A. E., Siever, L. J., Livesley, W. J., Gunderson, J. G., Pfohl, B., & Widiger, T. A. (2002). The borderline diagnosis II: Biology, genetics, and clinical course. *Biological Psychiatr y, 51*, 951–963.

Sloane, R. B., Staples, F. R., Whipple, K., & Cristol, A. H. (1977). Patients' attitudes toward behavior therapy and psychotherapy. *American Journal of Psychiatry, 134*, 134–137.

Slovic, P., Finucane, M., Peters, E., & MacGregor, D. (2002). The affect heuristic. In T. Gilovich, D. Griffin, & D. Kahneman (Eds.), *Heuristics and biases: The psychology of intuitive judgment* (pp. 397–420). New York: Cambridge University Press.

Smith, C., & Lapp, L. (1991). Increases in number of REMs and REM density in humans following an intensive learning period. *Sleep, 14*, 325–330.

Smith, C. A., & Ellsworth, P. C. (1985). Patterns of cognitive appraisal in emotion. *Journal of Personality and Social Psychology, 48*, 813–838.

Smith, C. A., & Lapp, L. (1991). Increases in number of REMs and REM density in humans following an intensive learning period. *Sleep, 14*, 325–330.

Smith, S. M., Glenberg, A. M., & Bjork, R. A. (1978). Environmental context and human memory. *Memory and Cognition, 6*, 342–353.

Smith, T. W., Orleans, C. T., & Jenkins, C. D. (2004). Prevention and health promotion: Decades of progress, new challenges, and an emerging agenda. *Health Psychology, 23*, 126–131.

Snarey, J. (1985). The cross-cultural universality of social-moral development: A critical review of Kohlbergian research. *Psychological Bulletin, 97(2)*, 202–232.

Snibbe, A. C., & Markus, H. R. (2005). You can't always get what you want: Social class, agency, and choice. *Journal of Personality and Social Psychology, 88*, 703–720.

Snider, L. A., & Swedo, S. E. (2004). PANDAS: Current status and directions for research. *Molecular Psychiatry, 9*, 900–907.

Snyder, E. E., Walts, B. M., Chagnon, Y. C., Pérusse, L., Weisnagel, S. J., Chagnon, Y. C., et al. (2004). The human obesity gene map: The 2003 update. *Obesity Research, 12*, 369–438.

Snyder, M., & Cantor, N. (1998). Understanding personality and personal behavior: A functionalist strategy. In D. T. Gilbert, S. T. Fiske, & G. Lindzey (Eds.)*, Handbook of social psychology* (pp. 635–679). New York: McGraw-Hill.

Society of Obstetricians and Gynaecologists of Canada (SOGC). Sex facts in Canada 2006. Retrieved December 13, 2006, from http://www.sexuaityandu.ca/media-room/fact-sheets-1.aspx

Solms, M. (2000). Dreaming and REM sleep are controlled by different brain mechanisms. *Behavioral and Brain Sciences, 23*, 793.

Somerville, L. H., Kim, H., Johnstone, T., Alexander, A. L., & Whalen, P. J. (2004). Human amygdala responses during presentation of happy and neutral faces: Correlations with state anxiety. *Biological Psychiatry, 55*, 897–903.

Sommerville, J. A., & Woodward, A. L. (2005). Pulling out the intentional structure of action: The relation between action processing and action production in infancy. *Cognition, 95*, 1–30.

Sorensen, T., Holst, C., Stunkard, A. J., & Skovgaard, L. T. (1992). Correlations of body mass index of adult adoptees and their biological and adoptive relatives. *International Journal of Obesity and Related Metabolic Disorders, 16*, 227–236.

Spangler, W. D., & House, R. J. (1991). Presidential effectiveness and the leadership motive profile. *Journal of Personality and Social Psychology, 60*, 439–455.

Spanos, N. P., Weekes, J. R., & Bertrand, L. D. (1985). Multiple personality: A social psychological perspective. *Journal of Abnormal Psychology, 94*, 362–376.

Spearman, C. (1904). "General intelligence," objectively determined and measured. *American Journal of Psychology, 15*, 201–293.

Spence, S. A., & Frith, C. D. (1999). Towards a functional anatomy of volition. *Journal of Consciousness Studies, 6*, 11–29.

Spencer, H. (1872). *The principles of psychology.* London, England: Williams & Norgate.

Spencer, M. B., & Markstrom-Adams, C. (1990). Identity processes among racial and ethnic minority children in America. *Child Development, 56*, 564–572.

Spencer, S. J., Steele, C. M., & Quinn, D. M. (1999). Stereotype threat and women's math performance. *Journal of Experimental Social Psychology, 35*, 4–28.

Sperling, G. (1960). The information available in brief visual presentations. *Psychological Monographs, 74*, 1–29.

Spetch, M., Wilkie, D. M., & Pinel, J. P. J. (1981). Backward conditioning: A reevaluation of empirical evidence. *Psychological Bulletin, 89*, 163–175.

Spitzer, R. L., Skodol, A. E., Gibbon, M., & Williams, J. B. W. (1983). *Psychopathology, a case book.* New York: McGraw-Hill.

Spitzer, R. L., Williams, J. B., Gibbon, M., & First, M. B. (1992). The Structured Clinical Interview for DSM-III-R (SCID). I: History, rationale, and description. *Archives of General Psychiatry, 49*, 624–629.

Sprecher, S., Barbee, A., & Schwartz, P. (1995). "Was it good for you, too?": Gender differences in first sexual experiences. *Journal of Sex Research, 32*, 3–15.

Squire, L. R., & Kandel, E. R. (1999). *Memory: From mind to molecules.* New York: Scientific American Library.

Squire, L. R., Stark, C. E. L., & Clark, R. E. (2004). The medial temporal lobe. *Annual Review of Neuroscience, 27*, 279–306.

Srivastava, S., John, O. P., Gosling, S. D., & Potter, J. (2003). Development of personality in early and middle adulthood: Set like plaster or persistent change? *Journal of Personality and Social Psychology, 84*, 1041–1053.

Stanovich, K. E. (2003). *How to think straight about psychology* (7th ed.). Boston: Allyn & Bacon.

Starcevic, V., Linden, M., Uhlenhuth, E. H., Kolar, D., & Latas, M. (2004). Treatment of panic disorder with agoraphobia in an anxiety disorders clinic: Factors influencing psychiatrists' treatment choices. *Psychiatry Research, 125*, 41–52.

Stark, S. (2000, December 22). "Cast Away" lets Hanks fend for himself. *The Detroit News*. Retrieved June 27, 2005, from http://www.detnews.com/2000/entertainment/1222/mcastaway/mcastaway.htm

Statistics Canada. Trips by Canadians in Canada, by province and territory. Retrieved December 13, 2006, from http://www40.statcan.ca/101/cst01/arts26a.htm

Steele, C. M. (1997). A threat in the air: How stereotypes shape intellectual identity and performance. *American Psychologist, 52*, 613–629.

Steele, C. M., & Aronson, J. (1995). Stereotype threat and the intellectual test performance of African Americans. *Journal of Personality and Social Psychology, 69*, 797–811.

Steffenburg, S., Gillberg, C., Helgren, L., Anderson, L., Gillberg, L., Jakobsson, G., et al. (1989). A twin study of autism in Denmark, Finland, Iceland, Norway, and Sweden. *Journal of Child Psychological Psychiatry, 30*, 405–416.

Stein, J., & Richardson, A. (1999). Cognitive disorders: A question of misattribution. *Current Biology, 9*, R374-R376.

Steinberg, L., & Scott, E. S. (2003). Less guilty by reason of adolescence: Developmental immaturity, diminished responsibility, and the juvenile death penalty. *American Psychologist, 12*, 1009–1018.

Steiner, J. E. (1977). Facial expressions of the neonate infant indicating the hedonics of food-related chemical stimuli. In J. M. Weiffenbach (Ed.), *Taste and development* (pp. 173–189). Bethesda, MD: U.S. Department of Health, Education, and Welfare.

Stellar, J. R., Kelley, A. E., & Corbett, D. (1983). Effects of peripheral and central dopamine blockade on lateral hypothalamic self-stimulation: Evidence for both reward and motor deficits. *Pharmacology, Biochemistry, and Behavior, 18*, 433–442.

Steriade, M. (1992). Basic mechanisms of sleep generation. *Neurology, 42* (Suppl.), 9–18.

Sternberg, E. M. (2000). *The balance within.* New York: Freeman.

Sternberg, R. J. (1986). A triangular theory of love. *Psychological Review, 93*, 119–135.

Sternberg, R. J. (1999). The theory of successful intelligence. *Review of General Psychology, 3*, 292–316.

Sternberg, R. J., & Davidson, J. E. (1983). Insight in the gifted. *Educational Psychologist, 18*, 51–57.

Stickgold, R., Whidbee, D., Schirmer, B., Patel, V., & Hobson, J. A. (2000). Visual discrimination task improvement: A multi-step process occurring during sleep. *Journal of Cognitive Neuroscience, 12*, 246–254.

Stokstad, E. (2001). New hints into the biological basis of autism. *Science, 294*, 34–37.

Stoleru, S., Gregoire, M. C., Gerard, D., Decety, J., Lafarge, E., Cinotti, L., et al. (1999). Neuroanatomical correlates of visually evoked sexual arousal in human males. *Archives of Sexual Behavior, 28*, 1–21.

Stone, A. A., Neale, J. M., Cox, D. S., & Napoli, A. (1994). Daily events are associated with a secretory immune response to an oral antigen in men. *Health Psychology, 13*, 440–446.

Stone, M. H., Stone, D. K., & Hurt, S. W. (1987). The natural history of borderline patients treated by intensive hospitalization. *Psychiatric Clinics of North America, 10*, 185–206.

Stone, V. E., Baron-Cohen, S., & Knight, R. T. (1998). Frontal lobe contributions to theory of mind. *Journal of Cognitive Neuroscience, 10*, 640–656.

Strange, B. A., & Dolan, R. J. (2004). Beta-adrenergic modulation of emotional memory-evoked human amygdala and hippocampal response: An emotion-induced retrograde amnesia in humans is amygdala- and beta-adrenergic-dependent. *Proceedings of the National Academy of Sciences, 101*, 11454–11458.

Strayer, D. L., Drews, F. A., & Johnston, W. A. (2003). Cell phone-induced failures of visual attention during simulated driving. *Journal of Experimental Psychology: Applied, 9*, 23–32.

Strentz, T., & Auerbach, S. M. (1988). Adjustment to the stress of simulated captivity: Effects of emotion-focused versus problem-focused preparation on hostages differing in locus of control. *Journal of Personality and Social Psychology, 55*, 652–660.

Stunkard, A. J. (1996). Current views on obesity. *American Journal of Medicine, 100*, 230–236.

Stuss, D. T. (1991). Self, awareness, and the frontal lobes: A neuropsychological perspective. In J. Strauss & G. R. Goethals (Eds.), *The self: Interdisciplinary approaches* (pp. 255–278). New York: Springer-Verlag.

Stuss, D. T., Gow, C. A., & Hetherington, C. R. (1992). "No longer Gage": Frontal lobe dysfunction and emotional changes. *Journal of Consulting and Clinical Psychology, 60*, 349–359.

Styron, W. (1990). *Darkness visible: A memoir of madness.* New York: Random House.

Suh, E. M. (2002). Culture, identity consistency, and subjective well being. *Journal of Personality and Social Psychology, 83*, 1378–1391.

Sulin, R. A., & Dooling, D. J. (1974). Intrusion of a thematic idea in retention of prose. *Journal of Experimental Psychology, 103*, 255–262.

Suls, J., & Rothman, A. (2004). Evolution of the biopsychosocial model: Prospects and challenges. *Health Psychology, 23*, 119–125.

Suzuki, D. T., Griffiths, A. J. F., Miller, J. H., & Lewontin, R. C. (1989). *An introduction to genetic analysis* (4th ed.). New York: Freeman Press.

Süß, H. M., Oberauer, K., Wittmann, W. W., Wilhelm, O., & Schulze, R. (2002). Working-memory capacity explains reasoning ability—and a little bit more. *Intelligence, 30*, 261–288.

Svenson, O. (1981). Are we all less risky and more skillful than our fellow drivers? *Acta Psychologica, 47*, 143–148.

Swartz, M. S., Blazer, D., George, L., & Winfield, I. (1990). Estimating the prevalence of borderline personality disorder in the community. *Journal of Personality Disorders, 4*, 257–272.

Swenson, C. R., Sanderson, C., Dulie, R. A., & Linehan, M. M. (2001). The application of dialectical behavior therapy for patients with borderline personality disorder on inpatient units [special issue]. *Psychiatric Quarterly, 72*, 307–324.

Sylva, K., Bruner, J. S., & Genova, P. (1976). The role of play in the problem-solving of children 3–5 years old. In J. S. Bruner, A. Jolly, & K. Sylva (Eds.), *Play: Its role in development and evolution* (pp. 244–257). New York: Penguin.

Szatzmari, P., Jones, M. B., Tuff, L., Bartolucci, G., Fisman, S., & Mahoney, W. (1993). Lack of cognitive impairment in first-degree relatives of children with pervasive developmental disorders. *Journal of the American Academy of Child and Adolescent Psychiatry, 32*, 1264–1273.

(TADS) Treatment for Adolescents with Depression Study Team. (2004). Fluoxetine, cognitive-behavioral therapy, and their combination for adolescents with depression: Treatment for Adolescents with Depression Study (TADS) randomized controlled trial. *Journal of the American Medical Association, 292*, 807–820.

Tajfel, H., & Turner, J. C. (1979). An integrative theory of intergroup conflict. In W. G. Austin & S. Worchel (Eds.), *The social psychology of intergroup relations* (pp. 33–47). Monterey, CA: Brooks/Cole.

Talley, P. R., Strupp, H. H., & Morey, L. C. (1990). Matchmaking in psychotherapy: Patient-therapist dimensions and their impact on outcome. *Journal of Consulting & Clinical Psychology, 58*, 182–188.

Tang, Y. P., Wang, H., Feng, R., Kyin, M., Tsien, J. Z. (2001). Differential effects of enrichment on learning and memory function in NR2B transgenic mice. *Neuropharmacology, 41*, 779–790.

Tardif, T. (1996). Nouns are not always learned before verbs: Evidence from Mandarin speakers' early vocabularies. *Developmental Psychology, 32*, 492–504.

Tateyama, M., Asai, M., Kamisada, M., Hashimoto, M., Bartels, M., & Heimann, H. (1993). Comparison of schizophrenic delusions between Japan and Germany. *Psychopathology, 26*, 151–158.

Taylor, S. E., & Brown, J. D. (1988). Illusion and well-being: A social psychological perspective on mental health. *Psychological Bulletin, 103*, 193–210.

Taylor, S. E., Klein, L. C., Lewis, B. P., Gruenewald, T. L., Gurung, R. A., & Updegraff, J. A. (2000). Biobehavioral responses to stress in females: Tend-and-befriend, not fight-or-flight. *Psychological Review, 107*, 411–429.

Teller, D. Y., Morse, R., Borton, R., & Regal, C. (1974). Visual acuity for vertical and diagonal gratings in human infants. *Vision Research, 14*, 1433–1439.

Tesser, A. (1988). Toward a self-evaluation maintenance model of social behavior. *Advances in Experimental Social Psychology, 21*, 181–227.

Tettamanti, M., Buccino, G., Saccuman, M. C., Welshman, V., Damn, M., Perani, D., et al. (2002). Sentences describing actions activate visuomotor execution and observation systems. 8th Annual HBM Conference.

Tewolde, S., Ferguson, B. S., & Benson, J. (2006). Risky behaviour in youth: An analysis of the factors influencing youth smoking decisions in Canada. *Substance Use and Misuse, 41*, 467–487.

Thigpen, C. H., & Cleckley, H. (1954). A case of multiple personality. *Journal of Abnormal Psychology, 49*, 135–151.

Thompson, P. (1980). Margaret Thatcher: A new illusion. *Perception, 9*, 483–484.

Thompson, P. M., Cannon, T. D., Narr, K. L., van Erp, T., Poutanen, V.P., Huttunen, M., et al. (2001). Genetic influences on brain structure. *Nature Neuroscience, 4*, 1253–1258.

Thompson, W. F., Schellenberg, E. G., & Husain, G. (2001). Arousal, mood, and the Mozart effect. *Psychological Science, 12*, 248–251.

Tickle, J. J., Sargent, J. D., Dalton, M. A., Beach, M. L., & Heatherton, T. F. (2001). Favorite movie stars, their tobacco use in contemporary movies and its association with adolescent smoking. *Tobacco Control, 10*, 16–22.

Tienari, P., Lahti, I., Sorri, A., Naarala, M., Moring, J., Kaleva, M., et al. (1990). Adopted-away offspring of schizophrenics and controls: The Finnish adoptive family study of schizophrenia. In L. Robins & M. Rutter (Eds.), *Straight and devious pathways from childhood to adulthood.* New York: Cambridge University Press.

Tienari, P., Wynne, L. C., Moring, J., Lahti, I., Naarala, M., Sorri, A., et al. (1994). The Finnish adoptive family study of schizophrenia: Implications for family research. *British Journal of Psychiatry Supplement, 23*, 20–26.

Tollesfson, G. D. (1995). Selective serotonin reuptake inhibitors. In A. F. Schatzberg & C. B. Nemeroff (Eds.), *The American psychiatric press textbook of psychopharmacology* (pp. 161–182). Washington, DC: American Psychiatric Press.

Tolman, C. (1994). Archivist's corner: Notes on some sources for the history of psychology in English Canada. *History and Philosophy of Psychology Bulletin, 6(2)*, 10–11.

Tomasello, M. (1999). The *cultural origins of human cognition.* Cambridge, MA: Harvard Press.

Tomasello, M. (2001). *The cultural origins of human cognition.* Cambridge, MA: Harvard University Press.

Tong, F., Nakayama, K., Vaughan, J. T., & Kanwisher, N. (1998). Binocular rivalry and visual awareness in human extrastriate cortex. *Neuron, 21*, 753–759.

Tooby, J., & Cosmides, L. (1990). On the universality of human nature and the uniqueness of the individual: The role of genetics and adaptation. *Journal of Personality, 58*, 17–68.

Torgersen, S., Kringlen, E., & Cramer, V. (2001). The prevalence of personality disorders in a community sample. *Archives of General Psychiatry, 58*, 590–596.

Torrey, E. F. (1999). Epidemiological comparison of schizophrenia and bipolar disorder. *Schizophrenia Research, 39*, 101–06.

Torrey, E. F., Torrey, B. B., & Peterson, M. R. (1977). Seasonality of schizophrenic births in the United States. *Archives of General Psychiatry, 34*, 1065–1070.

Treisman, A. (1988). Features and objects: The Fourteenth Bartlett Memorial Lecture. *Quarterly Journal of Experimental Psychology, 40A*, 201–237.

Treisman, A., & Gelade, G. (1980). A feature-integration theory of attention. *Cognitive Psychology, 12*, 97–136.

Triandis, H. C. (1989). The self and social behavior in differing cultural contexts. *Psychological Review, 96*, 506–520.

Triandis, H. C. (1995). *Individualism and collectivism.* Boulder, CO: Westview Press.

Trivers, R. L. (1971). The evolution of reciprocal altruism. *Quarterly Review of Biology, 46*, 35–57.

Tsien, J. Z. (2000). Building a brainier mouse. *Science, 282,* 62–68.

Tuchfeld, B. S. (1981). Spontaneous remission in alcoholics: Empirical observations and theoretical implications. *Journal of Studies on Alcohol, 42*, 626–641.

Tugade, M. M., & Fredrickson, B. L. (2004). Resilient individuals use positive emotions to bounce back from negative emotional experiences. *Journal of Personality and Social Psychology, 86*, 320–333.

Tulving, E. (1977). Context effects in the storage and retrieval of information in man. *Psychopharmacological Bulletin, 13*, 67–68.

Tulving, E. (1987). Multiple memory systems and consciousness. *Human Neurobiology, 6*, 67–80.

Tulving, E. (2002). Episodic memory: From mind to brain. *Annual Review of Psychology, 53*, 1–25.

Tulving, E. T., Kapur, S., Craik, F. I. M., Moscovitch, M., & Houle, S. (1994). Hemispheric encoding/retrieval asymmetry in episodic memory: Positron emission tomography findings. *Proceedings of the National Academy of Sciences, 91*, 2016–2020.

Turiel, E. (2002). *The culture of morality*. Cambridge: Cambridge University Press.

Uchino, B. N., Cacioppo, J. T., & Kiecolt-Glaser, J. K. (1996). The relationship between social support and physiological processes: A review with emphasis on underlying mechanisms and implications for health. *Psychological Bulletin, 119*, 488–531.

Ungerleider, L. G., & Mishkin, M. (1982). Two cortical visual systems. In D. J. Ingle, R. J. W. Mansfield, & M. S. Goodale (eds.), *The analysis of visual behavior* (pp 549–586). Cambridge, MA: MIT Press.

United States Department of Health and Human Services. (1993). *Alcohol and health.* Washington, DC: U.S. Government Printing Office.

Upton, N. (1994). Mechanisms of action of new antiepileptic drugs: Rational design and serendipitous findings. *Trends in Pharmacological Sciences, 15*, 456–463.

U.S. Department of Health and Human Services. (1999). *Mental health: A report of the Surgeon General.* Washington, DC: Department of Health and Human Services.

U.S. Department of Health and Human Services. (2004). *The health consequences of smoking: A report of the Surgeon General.* Department of Health and Human Services, Public Health Center, Centers for Disease Control and Prevention, National Center for Chronic Disease Prevention and Health Promotion, Office on Smoking and Health.

Vallerand, R. J. (1997). Toward a hierarchical model of intrinsic and extrinsic motivation. In M. P. Zanna (Ed.), *Advances in experimental social psychology* (vol. 29, pp. 271–360). San Diego, CA: Academic Press.

Valzelli, L. (1973). The "isolation syndrome" in mice. *Psychopharmacologia, 31*, 305–320.

Vargha-Khadem, F., Gadian, D. G., Watkinds, K., Connelly, A., Can Paesschen, W., & Mishkin, M. (1997). Differential effects of early hippocampal pathology on episodic and semantic memory. *Science, 277*, 376–380.

Varley, C. K. (1984). Attention deficit disorder (the hyperactivity syndrome): A review of selected issues. *Developmental and Behavioral Pediatrics, 5*, 254–258.

Veenhoven, R. (1993). *Happiness in nations.* Rotterdam, Netherlands: Risbo.

Vernon, P. A., Jang, K. L., Harris, J. A., & McCarthy, J. M. (1997). Environmental predictors of personality differences: A twin and sibling study. *Journal of Personality and Social Psychology, 72*, 177–183.

Vernon, P. A., Wickett, J. C., Bazana, P. G., Stelmack, R. M., & Sternberg, R. J. (2000). The neuropsychology and psychophysiology of human intelligence. In R.J. Sternberg (Ed.). *Handbook of Intelligence* (pp 245–264). Cambridge: Cambridge University Press.

Vernon, P. E. (1979). *Intelligence: Heredity and environment.* San Francisco: W. H. Freeman.

Vohs, K. D., & Heatherton, T. F. (2004). Ego threat elicits different social comparison processes among high and low self-esteem people: Implications for interpersonal perceptions. *Social Cognition, 22*, 168–190.

Volkman, F., Chawarska, K., & Klin, A. (2005). Autism in infancy and early childhood. *Annual Review of Psychology, 56*, 315–336.

von Neumann, J., & Morgenstern, O. (1947). *Theory of games and economic behavior*. Princeton, NJ: Princeton University Press.

Wadhwa, P. D., Sandman, C. A., & Garite, T. J. (2001). The neurobiology of stress in human pregnancy: Implications for prematurity and development of the fetal central nervous system. *Progress Brain Research, 133*, 131-142.

Wager, T. D., & Smith, E. E. (2003). Neuroimaging studies of working memory: A meta-analysis. *Cognitive, Affective, and Behavioral Neuroscience, 3*, 255–274.

Wagner, A. D., Schacter, D. L., Rotte, M., Koutstaal, W., Maril, A., Dale, A. M., et al. (1998). Building memories: Remembering and forgetting of verbal experiences as predicted by brain activity. *Science, 281*, 1188–1191.

Wagner, U., Gals, S., Haider, H., Verleger, R., & Born, J. (2004). Sleep inspires insight. *Nature, 427*, 352–356.

Wahlsten, D. (1997). The malleability of intelligence is not constrained by heritability. In B. Devlin, S. E. Fienberg, D. P. Resnick, & K. Roeder (Eds.), *Intelligence, genes and success*. New York: Copernicus.

Wahlsten, D. (1999). Single gene influences on brain and behaviour. *Annual Review of Psychology, 50*, 599–624.

Walker, E., Kestler, L., Bollini, A., & Hochman, K. M. (2004). Schizophrenia: Etiology and course. *Annual Review of Psychology, 55*, 401–430.

Walker, L. J. (1989). A longitudinal study of moral reasoning. *Child Development, 60*, 157–166.

Wallace, S. T., & Alden, L. E. (1997). Social phobia and positive social events: The price of success. *Journal of Abnormal Psychology, 106*, 416–424.

Waltrip, R. W., Buchanan, R. W., Carpenter, W. T., Kirkpatrick, B., Summerfelt, A., Breier, A., et al. (1997). Borna disease virus antibodies and the deficit syndrome of schizophrenia. *Schizophrenia Research, 23*, 253–257.

Warburton, D. M. (1992). Nicotine as a cognitive enhancer. *Progress in Neuro-Psychopharmacology and Biological Psychiatry, 16*, 181–191.

Ward, C., & Kennedy, A. (1995). Crossing-cultures: The relationship between psychological and sociocultural dimensions of cross-cultural adjustment. In J. Pandey, D. Sinha, & P. S. Bhawuk (Eds.), *Asian contributions to cross-cultural psychology* (pp. 289–306). New Delhi, India: Sage.

Wark, G. R., & Krebs, D. L. (1996). Gender and dilemma differences in real-life moral judgment. *Developmental Psychology, 32*, 220–230.

Watson, D., & Clark, L. A. (1997). Extraversion and its positive emotional core. In R. Hogan, J. Johnson, & S. Briggs (Eds.), *Handbook of personality psychology* (pp. 767–793). San Diego, CA: Academic Press.

Watson, D., Wiese, D., Vaidya, J., & Tellegen, A. (1999). The two general activation systems of affect: Structural findings, evolutionary considerations, and psychobiological evidence. *Journal of Personality and Social Psychology, 76*, 820–838.

Watson, J. B. (1924). *Behaviorism*. New York: Norton.

Wegner, D. M. (2002). *The illusion of conscious will.* Cambridge, MA: MIT Press.

Wegner, D. M., & Bargh, J. A. (1998). Control and automaticity in social life. In D. T. Gilbert, S. T. Fiske, & L. Gardner (Eds.), *Handbook of social psychology* (pp. 446–496). New York: Oxford University Press.

Wegner, D., Shortt, J., Blake, A., & Page, M. (1990). The suppression of exciting thoughts. *Journal of Personality and Social Psychology, 58*, 409–418.

Weiner, B. (1974). *Achievement motivation and attribution theory.* Morristown, NJ: General Learning Press.

Weiss, G., & Hechtman, L. T. (1993). *Hyperactive children grown up*. New York: Guilford Press.

Weissman, M. M., Bland, R. C., Canino, G. J., Greenwald, S., Hwu, H. G., Lee, C. K., et al. (1994). The cross national epidemiology of obsessive compulsive disorder. The Cross National Collaborative Group. *Journal of Clinical Psychiatry, 55*, 5–10.

Wells, G. L., Small, M., Penrod, S., Malpass, R. S., Fulero, S. M., & Brimacombe, C. A. E. (1998). Eyewitness identification procedures: Recommendations for lineups and photospreads. *Law and Human Behavior, 22*, 603–647.

Welsh, D. K., Logothetis, D. E., Meister, M., & Reppert, S. M. (1995). Individual neurons dissociated from rat suprachiasmatic nucleus express independently phased circadian firing rhythms. *Neuron, 14*, 697–706.

Werker, J. F., Gilbert, J. H., Humphrey, K., & Tees, R. C. (1981). Developmental aspects of cross-language speech perception. *Child Development, 52*, 349–355.

Werker, J. F., & Tees, R. C. (1984). Cross-language speech perception: Evidence for perceptual reorganization during the first year of life. *Infant Behavior and Development, 7*, 49–63.

Westen, D. (1998). The scientific legacy of Sigmund Freud: Toward a psychodynamically informed psychological science. *Psychological Bulletin, 124*, 333–371.

Westen, D., Novotny, C. M., & Thompson-Brenner, H. (2004). The empirical status of empirically supported psychotherapies: Assumptions, findings, and reporting in controlled clinical trials. *Psychological Bulletin, 130*, 631–663.

Wexler, N. S., U.S.-Venezuela Collaborative Research Project et al. (2004). Venezuelan kindreds reveal that genetic and environmental factors modulate Huntington's disease age of onset. *Proceedings of the National Academy of Sciences, 101*, 3498–3503.

Whalen, C. K. (1989). Attention deficit and hyperactivity disorders. In T. H. Ollendick & M. Herson (Eds.), *Handbook of child psychopathology* (2nd ed., pp. 131–169). New York: Plenum.

Whalen, P. J., Kagan, J., Cook, R. G., Davis, F. C., Kim, H., Polis, S., et al. (2005). Human amygdala responsivity to masked fearful eye-whites. *Science, 306,* 2061.

Whalen, P. J., Rauch, S. L., Etcoff, N. L., McInerney, N. L., Lee, M. B., & Jenike, M. A. (1998). Masked presentations of emotional facial expressions modulate amygdala activity without explicit knowledge. *Journal of Neuroscience, 18*, 411–418.

Whalen, P. J., Shin, L. M., McInerney, S. C., Fischer, H., Wright, C. I., & Rauch, S. L. (2001). A functional MRI study of human amygdala responses to facial expressions of fear versus anger. *Emotion, 1*, 70–83.

Wheeler, M. E., Petersen, S. E., & Buckner, R. L. (2000). Memory's echo: Vivid remembering reactivates sensory-specific cortex. *Proceedings of the National Academy of Sciences, 97*, 11125–11129.

White, D., & Pitts, M. (1998). Educating young people about drugs: A systematic review. *Addiction, 93*, 1475–1487.

Whiteside, S. P., Port, J. D., & Abramowitz, J. S. (2004). A meta-analysis of functional neuroimaging in obsessive-compulsive disorder. *Psychiatry Research, 15*, 69–79.

Whorf, B. L. (1956). *Language, thought, and reality.* Cambridge, MA: MIT Press.

Widiger, T. A., & Corbitt, E. M. (1995). Are personality disorders well-classified in DSM-IV? In W. J. Livesly (Ed.), *The DSM-IV personality disorders* (pp. 103–126). New York: Guilford Press.

Widom, C. S. (1978). A methodology for studying noninstitutionalized psychopaths. In R. D. Hare & D. A. Schalling (Eds.), *Psychopathic behavior: Approaches to research* (pp. 72ff). Chichester, England: John Wiley.

Wiesel, T. N., & Hubel, D. H. (1963). Single-cell responses in striate cortex of kittens deprived of vision in one eye. *Journal of Neurophysiology, 26,* 1003–1017.

Wiggins, J. S. (Ed.). (1996). *The five-factor model of personality: Theoretical perspectives.* New York: Guilford Press.

Wilens, T. E., Faraone, S. V., & Biederman, J. (2004). Attention-deficit/hyperactivity disorder in adults. *Journal of the American Medical Association, 292,* 619–623.

Wilke, M., Sohn, J.H., Byars, A.W., & Holland, S.K. (2003). Bright spots: correlations of gray matter volume with IQ in a normal pediatric population. *Neuroimage, 20,* 202–215

Williams, J. M. G., Mathews, A., & MacLeod, C. (1996). The emotional Stroop task and psychopathology. *Psychological Bulletin, 120,* 3–24.

Williams, P., Fitzsimons, G. J., & Block, L. G. (2004). When consumers don't recognize "benign" intentions questions as persuasion attempts. *Journal of Consumer Research, 31,* 540–550.

Williams, R. B., Jr. (1987). Refining the type A hypothesis: Emergence of the hostility complex. *American Journal of Cardiology*, 60, 27J–32J.

Williams, R. B., Barefoot, J. C., Califf, R. M., Haney, T. L., Saunders, W. B., et al. (1992). Prognostic importance of social and economic resources among medically treated patients with angiographically documented coronary artery disease. *Journal of the American Medical Association, 267,* 520–524.

Wills, T. A., DuHamel, K., & Vaccaro, D. (1995). Activity and mood temperament as predictors of adolescent substance use: Test of a self-regulation mediational model. *Journal of Personality and Social Psychology, 68,* 901–916.

Wilson, A. E., & Ross, M. (2001). From chump to champ: People's appraisals of their earlier and present selves. *Journal of Personality and Social Psychology, 80,* 572–584.

Wilson, M. A., & McNaughton, B. L. (1994). Reactivation of hippocampal ensemble memories during sleep. *Science, 265,* 676–679.

Wilson, T. D., & Gilbert, D. T. (2003). Affective forecasting. In M. Zanna (Ed.), *Advances in experimental social psychology: Vol. 35* (pp. 345–411). New York: Elsevier.

Wilson, T. D., Lindsey, S., & Schooler, T. Y. (2000). A model of dual attitudes. *Psychological Review, 107,* 101–126.

Wilson, T. D., & Schooler, J. W. (1991). Thinking too much: Introspection can reduce the quality of preferences and decisions. *Journal of Personality and Social Psychology, 60,* 181–192.

Wilson, T. W., & Grim, C. E. (1991). Biohistory of slavery and blood pressure differences in Blacks today. *Hypertension, 17* (Suppl. 1), I122-I128.

Winberg, J., & Porter, R. H. (1998). Olfaction and human neonatal behaviour: Clinical implications. *Acta Paediatrica, 87,* 6–10.

Wise, R. A., & Rompre, P. P. (1989). Brain dopamine and reward. *Annual Review of Psychology, 40,* 191–225.

Wiseman, C. V., Harris, W. A., & Halmi, K. A., (1998). Eating disorders. *Medical Clinics of North America, 82,* 145–159.

Wishaw, I. Q., Drigenberg, H. C., & Comery, T. A. (1992). Rats (rattus norvegicus) modulate eating speed and vigilance to optimize food consumption: Effects of cover, circadian rhythm, food deprivation, and individual differences. *Journal of Comparative Psychology, 106,* 411–419.

Wolford, G. L., Miller, M. B., & Gazzaniga, M. (2000). The left hemisphere's role in hypothesis formation. *Journal of Neuroscience, 20,* 1–4.

Wolpe, J. (1997). Thirty years of behavior therapy. *Behavior Therapy, 28,* 633–635.

Wong, R. Y. M., & Hong, Y. (2005). Dynamic influences of culture on cooperation in the Prisoner's Dilemma. *Psychological Science, 16,* 429–434.

Wong, D., & Shah, C. P. (1979). Identification of impaired hearing in early childhood. *Canadian Medical Association Journal, 121,* 529–532, 535–536.

Wood, J. M., Garb, H. N., Lilienfeld, S. O., & Nezworski, M. T. (2002). Clinical assessment. *Annual Review of Psychology, 53,* 519–543.

Woodruff, G. N., & Hughes, J. (1991). Cholecystokinin antagonists. *Annual Review of Pharmacology & Toxicology, 31,* 469–501.

Woods, S. C., & Stricker, E. M. (1999). Food intake and metabolism. In M. J. Zigmond, F. E. Bloom, S. C. Landis, J. L. Roberts, & L. R. Squire (Eds.), *Fundamentals of neuroscience* (pp. 1091–1109). San Diego, CA: Academic Press.

Woods, W. A., Cheron, E. J., & Kim, D. M. (1985). Strategic implications of differences in consumer purposes for purchasing in three global markets. In E. Kaynak (Ed.), *Global perspectives in marketing* (pp. 155–170). New York: Praeger.

Woodward, S. A., McManis, M. H., Kagan, J., Deldin, P., Snidman, N., Lewis, M., et al. (2001). Infant temperament and the brainstem auditory evoked response in later childhood. *Developmental Psychology, 37,* 533–538.

Woodworth, M., & Porter, S. (2002). In cold blood: Characteristics of criminal homicides as a function of psychopathy. *Journal of Abnormal Psychology, 111,* 436–445.

World Health Organization (WHO). (1973). *The international pilot study of schizophrenia.* Geneva: WHO.

World Health Organization. (2002). *The world health report.* Geneva: WHO.

Wright, M. J. & Myers, C. R. (1982). *History of academic psychology in Canada.* Toronto: Hogrefe.

Wright, S. C., & Tropp, L. R. (2005). Language and intergroup contact: Investigating the impact of bilingual instruction on children's intergroup attitudes. *Group Processes and Intergroup Relations, 8,* 309–328.

Wynn, K. (1992). Addition and subtraction by human infants. *Nature, 358,* 749–750.

Wystanski, M. (2000). Patient-centered versus client-centered mental health care. *Canadian Journal of Psychiatry, 45,* 670–671.

Yassa, R., Nair, N., Iskandar, H., & Schwartz, G. (1990). Factors in the development of severe forms of tardive dyskinesia. *American Journal of Psychiatry, 147,* 1156–1163.

Yates, W. R., Perry, P., & Murray, S. (1992). Aggression and hostility in anabolic steroid users. *Biological Psychiatry, 31,* 1232–1234.

Yuille, J. C., & Cutshall, J. L. (1986). A case study of eyewitness memory of a crime. *Journal of Applied Psychology, 71,* 291–301.

Zahn-Waxler, C., & Radke-Yarrow, M. (1990). The origins of empathic concern. *Motivation and Emotion*, 14, 107–130.

Zahn-Waxler, C., & Robinson, J. (1995). Empathy and guilt: Early origins of feelings of responsibility. In J. P. Tangney & K. W. Fischer (Eds.), *Self-conscious emotions: The psychology of shame, guilt, embarrassment, and pride* (pp. 143–173). New York: Guilford.

Zai, G., Bezchlibnyk, Y. B., Richter, M. A., Arnold, P., Burroughs, E., Barr, C. L., & Kennedy, J. L. (2004). Myelin oligodendrocyte glycoprotein (MOG) gene is associated with obsessive compulsive disorder. *American Journal of Medical Genetics, 129,* 64–68.

Zajonc, R. B. (1965). Social facilitation. *Science 149,* 269–274.

Zajonc, R. B. (1968). Attitudinal effects of mere exposure. *Journal of Personality and Social Psychology Monographs, 9,* 1–27.

Zajonc, R. B. (1980). Feeling and thinking: Preferences need no inferences. *American Psychologist, 35,* 151–175.

Zajonc, R. B. (2001). Mere exposure: A gateway to the subliminal. *Current Directions in Psychological Science 10,* 224–228.

Zajonc, R. B., Murphy, S. T., & Inglehart, M. (1989). Feeling and facial efference: Implications of the vascular theory of emotions. *Psychological Review, 96,* 395–416.

Zametkin, A. J., Nordahl, T. E., Gross, M., King, A. C., Stemple, W. E., Rumsey, J., et al. (1990). Cerebral glucose metabolism in adults with hyperactivity of childhood onset. *New England Journal of Medicine, 323,* 1361–1366.

Zatorre, R. J. (1989). On the representation of multiple languages in the brain: Old problems and new directions. *Brain and Language, 36,* 127–147.

Zatorre, R. J. (2003). Absolute pitch: A model for understanding the influence of genes and development on neural and cognitive function. *Nature Neuroscience, 6,* 692–695.

Zebrowitz, L. A. (1997). *Reading faces: Window to the soul?* Boulder, CO: Westview Press.

Zechmeister, J. S., Zechmeister, E. B., & Shaughnessy, J. J. (2000). *Essentials of research methods in psychology.* Boston: McGraw-Hill.

Zentall, S. S., Sutton, J. E., & Sherburne, L. M. (1996). True imitative learning in pigeons. *Psychological Science, 7,* 343–346.

Zigmond, M. J., Bloom, F. E., Landis, S. C., Roberts, J. L., & Squire, L. R. (2002). *Fundamentals of neuroscience* (2nd ed.). San Diego, CA: Academic Press.

Zihl, J., von Cramon, D., & Mai, N. (1983). Selective disturbance of movement vision after bilateral brain damage. *Brain, 106,* 313–340.

Zuckerman, M. (1995). Good and bad humors: Biochemical bases of personality and its disorders. *Psychological Science, 6,* 325–332.

ACKNOWLEDGMENTS AND CREDITS

Every effort has been made to contact the copyright holders of the material used in *Psychological Science*. Rights holders of any material not credited should contact W. W. Norton & Company, Inc., 500 Fifth Avenue, New York, NY 10110 for a correction to be made in the next reprinting of our work.

Photos

CHAPTER 1: **p. 2:** Aydelotte Rod/Corbis Sygma; **p. 4, 1.1:** AP Photos; **p. 5:** From Phelps, E. A., O'Connor, K. J., Cunningham, W. A., Funayama, E. S., Gatenby, J. C., Gore, J. C., et al. (2000). Performance on indirect measures of race evaluation predicts amygdala activation. *Journal of Cognitive Neuroscience, 12*, 729–738. Images courtesy Elizabeth A. Phelps, Phelps Lab; **p. 9:** The Far Side® by Gary Larson © 1982 FarWorks, Inc. All Rights Reserved. Used with permission; **p. 11, 1.3:** Courtesy Professor Joseph J. Campos, University of California, Berkeley; **p. 17, 1.5:** The Granger Collection, New York; **p. 18, 1.6:** (top) Bettmann/Corbis; **p. 18, 1.7:** (bottom) Bettmann/Corbis; **p. 20, 1.9:** Archives of the History of American Psychology, University of Akron; **p. 21, 1.10:** Archives of the History of American Psychology, University of Akron; **p. 22, 1.11:** Archives of the History of American Psychology. University of Akron; **p. 23, 1.12:** (top) From *American Journal of Psychology*. Copyright 1974 by the Board of Trustees of the University of Illinois. Used with permission of the University of Illinois Press; **p. 23, 1.13:** From *Mind sights*, by Roger N. Shepard. © 1990 by Roger N. Shepard. Henry Holt and Company, LLC; **p. 24, 1.14:** Bettmann/Corbis; **p. 25, 1.15:** Archives of the History of American Psychology. University of Akron; **p. 26, 1.16:** Courtesy George Miller, Princeton University. © 2001, Pryde Brown Photographs, Princeton, NJ; **p. 27, 1.17:** Archives of the History of American Psychology. University of Akron; **p. 29, 1.19:** Reuters/Corbis; **p. 30:** © The New Yorker Collection 1998 Sam Gross from cartoonbank.com. All Rights Reserved; **p. 32, 1.20:** LWA-Dann Tardif/Corbis.

CHAPTER 2: **p. 36:** Laura Dwight/Corbis; **p. 49, 2.3:** Karl Ammann/Corbis; **p. 50, 2.4:** Courtesy Robert Rosenthal, University of California, Riverside; **p. 52:** © The New Yorker Collection 1998 Roz Chast from cartoonbank.com. All Rights Reserved; **p. 54, 2.6:** (left) Reuters/Corbis; (right) The Oregonian/Corbis; **p. 65:** © Sidney Harris from cartoonbank.com. All Rights Reserved; **p. 68, 2.17:** Images.com/Corbis.

CHAPTER 3: **p. 72:** Mario Tama/Getty Images; **p. 77, 3.1:** CNRI/Science Photo Library/Photo Researchers, Inc.; **p. 81, 3.3:** Courtesy Hereditary Disease Foundation; **p. 83, 3.4:** (top row, second from left) Harry B. Clay, Jr., photographer, Bernardsville, NJ; **p. 84, 3.5:** Bob Sacha; **p. 91, 3.9:** James Stevenson, Science Photo Library/Photo Researchers, Inc.; **p. 94, 3.13:** James Stevenson, Science Photo Library/Photo Researchers, Inc; **p. 101, 3.16:** From Widner, H., Rehncrona, S., Snow, B., Brudin, P., Gustavii, B., Bjorklund, A., et al. (1992). Bialateral fetal mesencephalic grafting in two patients with parkinsonism induced by 1-methyl-4-phenyl-1, 2, 3, 6 tetrahydropyridine (MPTP). *The New England Journal of Medicine, 327, (22)*. Copyright © 1992 Massachusetts Medical Society. All rights reserved; **p. 102, 3.17:** From Jon Palfreman and J. William Langston, *The case of the frozen addicts*. Photograph by Russ Lee, © Pantheon Books, 1995; **p. 109:** William Haefeli/NewYorker/cartoonbank.com. All Rights Reserved; **p. 111, 3.21:** Hekimian Julien/Corbis Sygma.

CHAPTER 4: **p. 116:** Ron Sachs/Corbis; **p. 118, 4.1:** With permission of the Curator of the Wilder Penfield Archive, Montreal Neurological Institute; **4.2:** From Penfield, Wilder (1958). *The excitable cortex in conscious man*. Liverpool University Press; **p. 119, 4.3:** From Spurzheim, J. (1825). *Phrenology, or, the doctrine of the mind; and the relations between its manifestations and the body*. London: C. Knight; **p. 120, 4.4:** From the Musée Dupuytren. Courtesy Assistance Publique, Hopitaux de Paris; **p. 128, 4.12:** Warren Anatomical Museum, Francis A. Countway Library of Medicine; **p. 133, 4.15:** Courtesy of the authors; **p. 135, 4.16:** (left) Peter Macdiarmid/Reuters/Corbis; (right) Reuters/Corbis; **p. 136, 4.17:** Reuters/Corbis; **p. 141:** © Sidney Harris; **p. 153:** Robert Mankoff/NewYorker/cartoonbank.com. All Rights Reserved.

CHAPTER 5: **p. 158:** Stone/Getty Images; **p. 160, 5.1:** Bettmann/Corbis; **p. 169, 5.5:** Courtesy History and Special Collections, Louise M. Darling Biomedical Library, University of California, Los Angeles; **p. 170, 5.6:** Shaun Botterill/Allsport/Getty Images; **p. 181, 5.18:** Courtesy of Todd Heatherton, Dartmouth; **p. 183:** Pat Byrnes/NewYorker/cartoonbank.com. All Rights Reserved; **p. 190, 5.28:** From McCarthy, G., Price, A., Allison, T. (1997). Face specific processing in the human fusiform gyrus. *Journal of Cognitive Neuroscience, 9*, 605–610. Images courtesy Dr. Gregory McCarthy, The Duke-UNC Brain Imaging and Analysis Center; **p. 190, 5.29:** Courtesy Peter Thompson, University of York; **p. 191, 5.31:** news.bbc.co.uk; **p. 192, 5.32:** AP Photos; **p. 193, 5.34:** © 2002 Magic Eye Inc., www.magiceye.com; **p. 194, 5.35:** David Muench/Corbis; **p. 195, 5.37:** Courtesy of

Todd Heatherton, Dartmouth; **p. 195, 5.38:** (left) © Exploratorium, www.exploratorium.edu; **p. 198, 5.43:** From *Psychology*, sixth edition, by Henry Gleitman, Alan J. Fridlund, Daniel Reisberg. Copyright © 2004 by W. W. Norton & Company, Inc. Used by permission of W. W. Norton & Company, Inc.; **p. 204, 5.48:** Photo by Frank Both, Courtesy Century Fence Company.

CHAPTER 6: p. 210: Robert Fried/Alamy; **p. 214, 6.1:** Culver Pictures; **p. 215, 6.2:** Bettmann/Corbis; **p. 216:** © The New Yorker Collection 1999 Edward Koren from cartoonbank.com. All Rights Reserved; **p. 220, 6.8:** Getty Images; **p. 221, 6.9:** Benjamin Harris/ University of New Hampshire; **p. 222:** Tom Cheney/NewYorker/cartoonbank.com. All Rights Reserved; **p. 225, 6.11:** Nina Leen/Time Pix/Getty Images; **p. 226, 6.12:** Kirby Lee/AP Photos; **p. 226, 6.13:** Tom Hanson/AP Photos; **p. 230, 6.16:** HO/AP Photos; **p. 234, 6.18:** Albert Bandura, Department of Psychology, Stanford University; **p. 235, 6.19:** Albert Bandura, Department of Psychology, Stanford University; **6.20:** Susan Mineka, Northwestern University; **p. 236, 6.21:** Kevin Winter/Getty Images; **p. 238, 6.22:** © 20th Century Fox Film Corp. All rights reserved. Courtesy Everett Collection; **p. 242, 6.24:** McGill Reporter, PR050904; Biophoto Associates/Photo Researchers, Inc.; **p. 244, 6.27:** Princeton University/AP Photos.

CHAPTER 7: p. 250: Photofest; **p. 258:** The New Yorker Collection 1998 Mick Stevens from cartoonbank.com. All Rights Reserved; **p. 259, 7.6:** Tom Stewart/Corbis; **p. 260:** © Sidney Harris; **p. 262, 7.7:** Courtesy of Barry Shell, Softshell Small Systems Inc.; **p. 271, 7.13:** From Wheeler, M. E., Petersen, S. E., & Buckner, R. L. (2000). Memory's echo: Vivid remembering reactivates sensory-specific cortex. *Proceeding's of the National Academy of Sciences, 97*, 11125–11129; **p. 283, 7.21:** CP Images; **p. 284, 7.22:** AP Photos; **p. 285, 7.23:** Courtesy Elizabeth Loftus, University of Washington; **p. 287, 7.24:** Courtesy Daniel Schacter, Harvard University; **p. 289, 7.25:** Ed Bailey/AP Photos.

CHAPTER 8: p. 294: David Boily/AP Photos; **p. 313, 8.9:** Duomo/Corbis; **p. 319:** © The New Yorker Collection 1998 Bruce Eric Kaplan from cartoonbank.com. All Rights Reserved; **p. 322, 8.14:** The Granger Collection, New York; **p. 324:** © The New Yorker Collection 2002 David Sipress from cartoonbank.com. All Rights Reserved; **p. 326, 8.17: (a)** Duomo/Corbis **(b)** Reuters New Media Inc./Corbis **(c)** AP Photos **(d)** AP Photos; **p. 327, 8.18:** © BBC Picture Archives; **p. 336, 8.25:** Mary Ann Chastain/AP Photos.

CHAPTER 9: p. 342: Bettmann/Corbis; **p. 344, 9.1:** Vince Bucci/Getty Images; **p. 348, 9.5:** Archives of the History of American Psychology, University of Akron; **p. 349, 9.7:** Jacob E. Steiner, PhD, Laboratory of Oral Physiology, The Hebrew University; **p. 351:** © 2005 Bob Zahn from cartoonbank.com. All Rights Reserved; **p. 358, 9.10:** Photo Services, Harvard University; **p. 368, 9.14:** Bettmann/Corbis; **p. 371, 9.15:** Sal Sacco/AP Photos; **p. 377, 9.17:** Courtesy David Buss, University of Texas at Austin; **p. 378:** © The New Yorker Collection 1997 Barbara Smaller from cartoonbank.com. All Rights Reserved.

CHAPTER 10: p. 384: Ahmad Masood/Reuters; **p. 389, 10.1:** From Izard, C. E., Fantauzzo, C. A., Castle, J. M., Hayes, O. M., Rayias, M. F., Putnam, P. H. (1995). The ontogeny and significance of infants' facial expressions in the first nine months of life. *Developmental Psychology, 31*, 997–1013. Photographs courtesy Carroll Izard, University of Delaware; **p. 390, 10.3:** © Ekman 1972–2004; **p. 398, 10.6:** Edgar Fahs Smith Collection; **p. 399, 10.8:** Photo by Bill Apple; **p. 400, 10.10:** Chris Lisle/Corbis; **p. 401:** George Price/NewYorker/cartoonbank .com. All Rights Reserved; **p. 403, 10.11:** From Lane, R. D., Reiman, E. M., Ahern, G. L., Schwatz, G. E., Davidson, R. J. (1997). Neuroanatomical correlates of happiness, sadness, and disgust. *American Journal of Psychiatry, 154*, 926–933. Copyright © 1997 by American Psychiatric Association; **p. 407, 10.14:** Photo by James Prince; **p. 407, 10.15:** From Dolan, R. J. Emotion processing in the human brain revealed through functional neuroimaging. In M. S. Gazzaniga (Ed.), *The new cognitive neurosciences.* Cambridge, MA: MIT Press; **p. 409, 10.17:** Courtesy Richard J. Davidson, University of Wisconsin; **p. 409, 10.18: (a)** American Psychological Association; **p. 411, 10.20:** Courtesy Shelley Taylor, UCLA; **p. 411, 10.21:** Archives of American Psychology, University of Akron; **p. 414:** © The New Yorker Collection 1987 Donald Reilly from cartoonbank.com. All Rights Reserved; **p. 420, 10.25:** blickwinkel/ Alamy; **p. 424:** The Far Side by Gary Larson © 1982 FarWorks, Inc. All Rights Reserved. Used with permission; **p. 425, 10.28:** Ivan Chavez/AP Photos.

CHAPTER 11: p. 432: Jose Fuste Raga/Corbis; **p. 434, 11.1:** Courtesy HarperCollins Publishers Inc.; **p. 440, 11.4:** Robert Marvin, Estate of Mary Ainsworth; **p. 441, 11.5:** Nina Leen/Time & Life Pictures/Getty Images; **p. 411, 11.6:** Harlow Primate Laboratory, University of Wisconsin; **p. 442, 11.7:** Robert Marvin, Estate of Mary Ainsworth; **p. 444:** © The New Yorker Collection 2003 Donald Reilly from cartoonbank.com. All Rights Reserved; **p. 447, 11.9:** Stock Boston, LLC/Spencer Grant; **p. 448, 11.10:** Courtesy Velma Dobson, University of Arizona; **p. 449, 11.12:** Bettmann/Corbis; **p. 451, 11.14:** Michael Newman/PhotoEdit; **p. 454, 11.19:** © Dr. Carolyn Rovee-Collier, Director, Rutgers Early Learning Project; **p. 459, 11.22:** AFP/Getty Images; **p. 461, 11.23:** Courtesy of Laura Ann Petitto; **p. 462, 11.24:** Gaetano/Corbis; **p. 464:** © The New Yorker Collection 2001 Edward Koren from cartoonbank.com. All Rights Reserved; **p. 465, 11.25:** Corbis Sygma; **p. 466, 11.26:** Photo courtesy David Reimer. From Colapinto, John (2000). *As nature made him: The boy who was raised as a girl.* HarperCollins Publishers Inc.; **p. 466, 11.27:** Reuters Newsmedia Inc./Corbis; **p. 471, 11.28:** Sarah Putnam/Index Stock Imagery, Inc.; **p. 472:** © The New Yorker Collection 1994 Sidney Harris from cartoonbank.com. All Rights Reserved; **p. 475, 11.29:** (left) Stephanie Maze/Corbis; (right) Roger Ressmeyer/Corbis.

CHAPTER 12: p. 482: Tom Wagner/Corbis; **p. 485, 12.1:** Dan McAdams, Northwestern University; **p. 486, 12.2:** Bettmann/Corbis; **p. 489, 12.4:** Courtesy Carl Rogers Memorial Library; **p. 493, 12.6:** © Toronto Sun; **p. 496, 12.7:** (top) Royalty-Free/Corbis; (bottom) Ralph A. Clevenger/Corbis; **p. 499, 12.8:** Adapted from Norman, W. T. (1963). Toward an adequate taxonomy of personality attributes: Replicated factor structure in peer nomination personality rating. *Journal of Abnormal and Social Psychology, 66*, 5770; **p. 500, 12.9:** Columbia University; **p. 505:** © The New Yorker Collection 1999 Barbara Smaller from cartoonbank.com. All Rights Reserved; **p. 506:** ©

The New Yorker Collection 1996 J. B. Handelsman from cartoonbank.com. All Rights Reserved; **p. 508, 12.12:** Hulton Archive/Getty Images; **p. 509, 12.14:** Don Emmert/AFP/Getty Images; **p. 513, 12.18:** (left) Jean-Paul Guilloteau/Kipa/Corbis; (right) Getty Images.

CHAPTER 13: **p. 520:** Kat Wade/San Fransisco Chronicle/Corbis; **p. 522, 13.1:** Edna Morlok; **p. 524, 13.2:** The Granger Collection, New York; **p. 529, 13.3:** *Abnormal Psychology* fourth edition, by Seligman, Walker, and Rosenhan. Copyright W. W. Norton & Company, Inc. Used by permission of W. W. Norton & Company, Inc.; **p. 531, 13.4:** Joe McNally; **p. 532, 13.5:** Bettmann/Corbis; **p. 534, 13.6:** (top) Manchester Scott/Corbis Sygma; (bottom) Denver Post/Kent Meireis/Corbis Sygma; **p. 537:** © The New Yorker Collection 2003 Michael Shaw from cartoonbank.com. All Rights Reserved; **p. 538, 13.7:** Courtesy Dr. Lewis Baxter, UAB; **p. 541, 13.9:** Reuters/Corbis; **p. 544, 13.12:** © Richard Wyatt; **p. 555, 13.14:** Peter Aprahamian/Corbis; **p. 559, 13.15:** From Osterling, J., & Dawson, G. (1994). Early recognition of children with autism. A study of first birthday home videotapes. *Journal of Autism and Developmental Disorders, 24,* 247–257. Photographs courtesy Geraldine Dawson; **p. 559, 13.16:** Courtesy of Dr. Ami Klin, (2003) The enactive mind from actions to cognition: Lessons from autism. Philosophical Transactions of the Royal Society; **p. 562, 13.17:** From Zametkin, A. J., Nordhal, T. E., Gross, M., et al. (1990). Cerebral glucose metabolism in adults with hyperactivity of childhood onset. *New England Journal of Medicine, 323(20),* 1361–1366. Images courtesy Alan Zametkin, NIH; (bottom) © The New Yorker Collection 1997 Robert Mankoff from cartoonbank.com. All Rights Reserved.

CHAPTER 14: **p. 568:** Jeffrey L. Rotman/Corbis; **p. 572, 14.1:** Bettmann/Corbis; **p. 573:** © The New Yorker Collection 1989 Danny Shanahan from cartoonbank.com. All Rights Reserved; **p. 574, 14.2:** © 1983 Erika Stone; **p. 574, 14.3:** Courtesy Barbara A. Marinelli; **p. 579, 14.6:** Courtesy James Pennebaker, University of Texas at Austin; **p. 582, 14.7:** Randy Olson; **p. 584, 14.9:** AP Photos; **p. 585, 14.10:** Angela Singer/Everett Collection; **p. 587:** © The New Yorker Collection 1993 Lee Lorenz from cartoonbank.com. All Rights Reserved; **p. 590, 14.15:** © Pascal Goetheluck/Science Photo Library/Photo Researchers, Inc.; **p. 590, 14.16:** © Will McIntyre/Photo Researchers, Inc.; **p. 591, 14.17:** CP Images; **p. 598, 14.21:** Courtesy Marsha Linehan, University of Washington; **p. 606, 14.24:** Richard Perry/The New York Times.

CHAPTER 15: **p. 610:** P. Kumar/AFP/Getty Images; **p. 612, 15.1:** CP Images; **p. 613, 15.2:** © Philip G. Zimbardo, PhD.; **p. 621, 15.8:** Leo La Valle/epa/Corbis; **p. 623:** © The New Yorker Collection 1985 Michael Maslin from cartoonbank.com. All Rights Reserved; **p. 623, 15.10:** Reuters/Corbis; **p. 625, 15.11:** Fred Chartrand/AP Photos; **p. 626, 15.12:** © 1982 Karen Zabuion, Courtesy New School Public Relations Department; **p. 629, 15.15:** Christopher Morris/Corbis; **p. 632, 15.17:** CP Images; **p. 633:** © New Yorker Collection 1996 Tom Cheney from cartoonbank.com. All Rights Reserved; **p. 634, 15.18:** Courtesy Keith Payne, Washington University; **p. 636, 15.19:** Ryan Remiorz/AP Photos; **p. 638, 15.20:** Chris Trotman/Duomo/Corbis; **p. 642, 15.23:** Photos by William Vandivert; **p. 644, 15.24:** Courtesy Alexandra Milgram; **15.25:** Courtesy Alexandra Milgram; **p. 646, 15.27:** Aaron Harris/AP Photos; **p. 649, 15.30:** Rogers Sportsnet/Corbis; **p. 650, 15.32:** New York Times Pictures; **p. 654:** (right) From Langlois, J. H., Ruggerman, L. A., (1990). Attractive faces are only average. *Psychological Science 1,* 115–121. Photographs courtesy Judith Hall Langlois, University of Texas at Austin; **p. 655, 15.36:** (top) Courtesy Elaine Hatfield, University of Hawaii at Manoa; (bottom) Courtesy Ellen Berscheid, University of Minnesota; **p. 658, 15.37:** Dave Buston/AP Photos.

CHAPTER 16: **p. 664:** AFP/Getty; **p. 669, 16.2:** Stephen Frink; **p. 670, 16.4:** Janine Wiedel Photolibrary/Alamy; **16.5:** Anup Shah/Natureplanet; **p. 673, 16.6: (a)** Michel Setboun/Corbis; **(b)** Michel Setboun/Corbis; **p. 678, 16.9:** Reuters/Corbis; **p. 679, 16.10:** Don MacKinnon/Getty Images; **p. 682, 16.13: (a)** Jim Young/Corbis; **(b)** Paul A. Souders/Corbis; **(c)** Sion Touhig/Corbis; **p. 684, 16.16:** Focus Feature/Photofest; **p. 686, 16.17:** Courtesy of the authors; **p. 688, 16.19:** Courtesy The Everett Collection; **p. 690, 16.21: (a)** Wendy Stone/Corbis; **(b)** Louise Gubb/Corbis; **(c)** Naashon Zalk/Corbis; **p. 693, 16.23:** Dung Vo Trung/Corbis; **p. 698, 16.27:** Stephanie Sinclair/Corbis; **p. 702, 16.30:** Christine Schneider/zefa/Corbis; **p. 703, 16.31: (a)** Ted Soqui/Corbis; **(b)** epa/Corbis.

Figures

CHAPTER 1: **p. 19, 1.8:** From Grant, Peter R., *Darwin's Finches: Ecology and evolution of Darwin's finches.* Copyright © 1986 by Princeton University Press. Reprinted by permission of Princeton University Press; **p. 28, 1.18:** The Subfields Doctoral Psychologist.

CHAPTER 3: **p. 92, 3.11:** From Pinel, John P. J. *Biopyschology* 4/e. Published by Allyn and Bacon, Boston, MA. Copyright © 2000 by Pearson Education. Reprinted by permission of the publisher.

CHAPTER 4: **p. 138, 4.19:** From *Cognitive neuroscience: The biology of the mind,* by Michael S. Gazzaniga, Richard Ivry, and George R. Mangun. Copyright © 1998 by W. W. Norton & Company, Inc. Used by permission of W. W. Norton & Company, Inc.; **p. 141, 4.22:** From *Cognitive neuroscience: The biology of the mind,* by Michael S. Gazzaniga, Richard Ivry, and George R. Magun. Copyright © 1998 by W. W. Norton & Company, Inc. Used by permission of W. W. Norton & Company, Inc.; **p. 143, 4.24:** From *Cognitive neuroscience: The biology of the mind,* by Michael S. Gazzaniga, Richard Ivry, and George R. Mangun. Copyright © 1998 by W. W. Norton & Company, Inc. Used by permission of W. W. Norton & Company, Inc.; **p. 151, 4.30:** Aston-Jones, G. and Bloom, F. E. (1981), *The Journal of Neuroscience, 1,* 876–886. Copyright © 1981 by The Society for Neuroscience.

CHAPTER 5: **p. 165, 5.3:** From Zimbardo, P. G. & Gerrig, R. J. *Psychology and life,* 15/e. Published by Allyn and Bacon, Boston, MA. Copyright © 1999 by Pearson Education. Reprinted by permission of the publisher; **p. 175, 5.12:** From *Cognitive neuroscience: The biology of the mind,* by Michael S. Gazzaniga, Richard Ivry, and George R. Mangun. Copyright © 1998 by W. W. Norton & Company, Inc. Used by permission of W. W. Norton & Company, Inc.; **p. 179, 5.16:** From Zimbardo, P. G. & Gerrig, R. J. *Psychology and life,* 15/e. Published by

Allyn and Bacon, Boston, MA. Copyright © 1999 by Pearson Education. Reprinted by permission of the publisher; **p. 179, 5.17:** Bear, *Neuroscience* 2E. Reprinted by permission of Lippincott Williams & Wilkins; **p. 186, 5.21:** *Psychology* by Peter Gray. © 1991, 1994, 1999 by Worth Publishers. Used with permission; **p. 187, 5.24:** From *Psychology*, fourth edition, by Henry Gleitman. Copyright © 1995, 1991, 1986, 1981 by W. W. Norton & Company, Inc.; Used by permission of W. W. Norton & Company, Inc.; **p. 187, 5.25:** Kuhl, J. (1976). Subjective contours. *Scientific American 234*, 44–52. Reprinted by permission of Jerome Kuhl; **p. 188, 5.26:** Osherson, Daniel, adapted from Biedermann, *An invitation to cognitive science* 2nd ed., Vol. 2. Reprinted with the permission of MIT; **p. 197, 5.41:** From *Cognitive neuroscience: The biology of the mind*, by Michael S. Gazzzaniga, Richard Ivry, and George R. Mangun. Copyright © 1998 by W. W. Norton & Company, Inc. Used by permission of W. W. Norton & Company, Inc.; **p. 199, 5.44:** "Turning the tables illusion" from *Mind sights* by Roger N. Shepard, © 1990 by Roger N. Shepard. Reprinted by permission of Henry Holt and Company, LLC; **p. 201, 5.45 and 5.46:** From *Cognitive neuroscience: The biology of the mind*, by Michael S. Gazzaniga, Richard Ivry, and George R. Mangun. Copyright © 1998 by W. W. Norton & Company, Inc. Used by permission of W. W. Norton & Company, Inc.; **p. 202, 5.47:** From *Cognitive neuroscience: The biology of the mind* by Michael S. Gazzaniga, Richard Ivry, and George R. Mangun. Copyright © 1998 by W. W. Norton & Company, Inc. Used by permission of W. W. Norton & Company, Inc.

CHAPTER 6: **p. 227, 6.14:** *Psychology by Peter Gray*. © 1991, 1994, 1999 by Worth Publishers. Used with permission.

CHAPTER 7: **p. 265, 7.8:** Adapted from Craik, F. I. M., & Tulving, E. (1975). Depth of processing and the retention of words in episodic memory. *Journal of Experimental Psychology: General, 104*, 268–294; **p. 267, 7.9:** Adapted from Collins, A. M., & Loftus, E. F. (1975). A spreading-activation theory of semantic processing. *Psychological Review, 82*, 407–428; **p. 268, 7.10:** Adapted from Godden, D. R., & Baddeley, A. D. (1975). Context-dependent memory in two natural environments: On land and underwater. *British Journal of Psychology, 66*, 325–331; **p. 272, 7.14:** From *Cognitive neuroscience: The biology of the mind*, by Michael S. Gazzaniga, Richard Ivry, and George R. Mangun. Copyright © 1998 by W. W. Norton & Company, Inc. Used by permission of W. W. Norton & Company, Inc.; **p. 273, 7.16:** Adapted from Wagner, A. D., Schacter, D. L., Rotte, M., Koutstaal, W., Maril, A., Dale, A. M., et al. (1998). Building memories: remembering and forgetting of verbal experiences as predicted by brain activity. *Science, 281*, 1188–1191.

CHAPTER 8: **p. 299, 8.1:** Eysenck, M. W., figures 7.6 and 7.7, *Cognitive psychology: A student's handbook*, 2nd Edition, Lawrence Erlbaum Associates. Copyright © 1990. Used by permission; **p. 302, 8.5:** Eysenck, M. W., figure 8.3, *Cognitive psychology: A student's handbook*, 2nd Edition, Lawrence Erlbaum Associates. Copyright © 1990. Used by permission; **p. 311, 8.8:** Based on Plous, S., figure 9.1, *Psychology of judgement and decision making*. Copyright © 1993 by The McGraw-Hill Companies; **p. 333, 8.21:** Plomin from *Journal of Personality and Social Psychology 86*. Used by permission; **p. 333, 8.22:** McGue, Behavioral Genetics of Cognitive Ability from Nature, Nurture.

CHAPTER 9: **p. 355, 9.8:** Adapted from Schachter, S. (1959). The psychology of affiliation. Stanford, CA: Stanford University Press; **p. 359, 9.11:** Adapted from Miller, G. A., Galanter, E., & Pribram, K. H. (1960). *Plans and the structure of behavior*. New York: Holt. **p. 377, 9.18:** Adapted from Clark, R. D., & Hatfield, E. (1989). Gender differences in receptivity to sexual offers. *Journal of Psychology and Human Sexuality, 2*, 39–55.

CHAPTER 10: **p. 396, 10.4:** Adapted from Russell, J. A., & Feldman Barrett, L. (1999). Core affect, prototypical emotional episodes, and other things called emotion: Dissecting the elephant. *Journal of Personality and Social Psychology, 76*, 805–819; **p. 398, 10.7:** From *Introduction to psychology* (with Info Trac) 5th edition by Kalat. © 1999. Reprinted with permission of Wadsworth, a division of Thomson Learning; **p. 419, 10.24:** Bray, George A., Nomogram for determining body mass index. Copyright © 1978 George A. Bray. Used by permission; **p. 421, 10.26:** Use of the "Danish Adaptation Register for the Study of Obesity and Thinness: by A. Stunkard, T. Sorensen, and F. Schulsinger, from *The Genetics of Neurological and Psychiatric Disorders*, Raven Press (1980). Copyright © 1980 Dr. Albert Stunkard. Used by permission of Dr. Albert Stunkard; **p. 422, 10.27:** Adapted from Hibscher, J., & Herman, C. P. (1977). Obesity, dieting, and the expression of "obese" characteristics. *Journal of Comparative and Physiological Psychology, 91*, 374–380.

CHAPTER 11: **p. 452, 11.16:** Reprinted from *Cognition*, vol. 47 (2) May 1993, Needham, Amy, and Baillargeon, Renee, Intuitions about support in 4.5 month-old infants, 121–148, Copyright © 1993, with permission from Elsevier Science; **p. 457, 11.21:** Nairne, J. S., figure 9.2 from *Psychology: The adaptive mind*, Wadsworth Group; **p. 477, 11.30:** Sternberg, Robert J., Major abilities and development in the adult period, Intellectual development.

CHAPTER 12: **p. 507, 12.11:** Adapted from Caspi, A. (2000). The child is father of the man: Personality continuities from childhood to adulthood. *Journal of Personality and Social Psychology, 78*, 158–172; **p. 512, 12.15:** (top) Adapted from Roberts, B. W., & Friend-Del Vecchio, W. (2000). The rank-order consistency of personality traits from childhood to old age. A quantitative review of longitudinal studies. *Psychological Bulletin, 126*, 3–25; **p. 512, 12.16:** Adapted from McCrae, R. R., Costa, P. T., Ostendorf, F., Angleitner, A., Hrebickova, M., Avia, M. D., et al. (2000). Nature over nurture: Temperament, personality, and lifespan development. *Journal of Personality and Social Psychology, 78*, 173–186; **p. 513, 12.17:** Adapted from McCrae, R. R., & Costa, P. T., Jr. (1999). A five-factor theory of personality. In L. A. Pervin and O. P. John (Eds.), *Handbook of Personality: Theory and research* (2nd ed., p. 139–153). New York: Guilford Press.

CHAPTER 13: **p. 539, 13.8:** Adapted from Rapee, R. M., Brown, T. A., Anthony, M. M., & Barlow, D. H. (1992). Response to hyperventilation and inhalation of 5.5% carbon dioxide enriched air across the DSM-III-R anxiety disorders. *Journal of Abnormal Psychology, 101*, 538–552; **p. 542, 13.10:** Rosenthal, Sack, Gillin, Lewy, Goodwin, Davenport, Mueller, Newsome, Wehr, Depression and length of daylight in seasonal affective disorder. *Archives of General Psychiatry, 41*, [1984], 72–80; **p. 543, 13.11:** Adapted from Brown &

Harris (1978). *Social origins of depression: A study of psychiatric disorders in women.* New York: The Free Press; **p. 549, 13.13:** Gottesman, *Schizophrenia Genesis: The origins of madness,* W. H. Freeman. Copyright © 1991.

CHAPTER 14: p. 575, 14.4: Vaughn, C. E. (1976). Relapse rates in schizophrenia patients, *British Journal of Psychiatry, 129*; **p. 583, 14.8:** Rudestam, Kjell Erik, *Anxiety Hierarchy, Methods of Self-Change: An ABC Primer*, 1980, Monterey, CA: Brooks/Cole; **p. 587, 14.12:** Reprinted with permission from the *American Journal of Psychiatry*, Copyright 2005. American Psychiatric Association; **p. 589, 14.13:** Nolen-Hoeksema, S., figure 5.16 from *Abnormal Psychology* (1998), The McGraw Hill Companies; **p. 593, 14.18:** Adapted from Kane, J. M., Honingfield, G., Singer, J., & Meltzer, N. (1988). Clozapine for the treatment-resistant schizophrenic. *Archives of General Psychiatry, 45*, 789–796; **p. 594, 14.19:** Adapted from Hogarty, G. E., Anderson, D. M., Reiss, D. J., Kornblith, S. J., Greenwald, D. P., Jaund, D. D., et al. (1986) Family psychoeducation, social skills training, and maintenance chemotherapy in the aftercare treatment of schizophrenia: One-year effects of a controlled study on relapse and expressed emotion. *Archives of General Psychiatry, 43*, 633–642; **p. 600, 14.22:** Adapted from Hare, R. D., McPherson, L. M., & Forth, A. E. (1998). Male psychopaths and their criminal careers. *Journal of Consulting and Clinical Psychology, 56*, 710–714; **p. 602, 14.23:** Adapted from Pelham, W. E., Jr. (1993). Pharmocotherapy for children with attention-deficit hyperactivity disorder. *School Psychology Review, 22*, 199–227.

CHAPTER 15: p. 620, 15.7: Adapted from Johnson, M. M. S. (1986). The initial letter effect: Ego-attachment or mere exposure? Unpublished doctoral dissertation, Ohio State University. Reported in Greenwald, A. G., & Banaji, M. R., (1995). Implicit social cognition: Attitudes, self-esteem, and stereotypes. *Psychological Review, 102*, 4–27; **p. 640, 15.21:** Adapted from Zajonc, R. B. (1965). Social facilitation. *Science, 149*, 269–274; **p. 642, 15.22:** Adapted from Asch, S. E. (1955). Opinions and social pressure. *Scientific American, 193*, 31–35 (offprint 450); **p. 645, 15.26:** Adapted from Milgram, S. (1974). *Obedience to authority: An experimental view.* New York: Harper & Row; **p. 647, 15.28:** Adapted from Raleigh, M. J., McGuire, M. T., Brammer, G. L., Pollack, D. B., & Yuwiler, A. (1991). Serotonergic mechanisms promote dominance in adult male vervet monkeys. *rain Research, 559*, 181–190; **p. 648, 15.29:** Data from United Nations, 1991; **p. 649, 15.31:** Adapted from Cohen, D., Nisbett, R. E., Bowdle, B. F., & Schwartz, N. (1996). Insult, aggression, and the southern culture of honor: An "experimental ethnography." *Journal of Personality and Social Psychology, 70*, 945–960; **p. 651, 15.33:** Adapted from Latané, B., & Darley, J. M. (1968). Group inhibition of bystander intervention in emergencies. *Journal of Personality and Social Psychology, 10*, 215–221; **p. 654, 15.35:** From Perrett, D. I., May, K. A., Yoshikawa, S. (1994). Facial shape and judgments of female attractiveness. Reprinted by permission from *Nature*, 1994, vol. 368, 239–242. Copyright 2002 Macmillan Publishers Ltd. Images courtesy D. I. Perrett, University of St. Andrews.

CHAPTER 16: p. 669, 16.3: Adapted from Dunbar, R.I.M. (1993). The co-evolution of neocortical size, group size and language in humans, *Behavioural and Brain Sciences, 16*, 681–735; **p. 676, 16.7:** Adapted from Miller, J. G., (1984). Culture and the development of everyday social explanation. *Journal of Personality and Social Psychology, 46*, 961–978; **p. 679, 16.11:** Adapted from Lysgaard, S. (1955). Adjustment in a foreign society: Norwegian Fulbright grantees visiting the United States. *International Social Science Bulletin, 7*, 45–51; **p. 683, 16.15:** Adapted from Hong, Y., Morris, M. W., Chiu, C., & Benet-Martinez, V. (2000). Multicultural minds: A dynamic constructivist approach to culture and cognition. *American Psychologist, 55*, 705–720; **p. 687, 16.18:** Adapted from Norenzayan, A., Smith, E. E., & Kim, B., & Nisbett, R. E. (2002). Cultural preferences for formal versus intuitive reasoning. *Cognitive Science, 26*, 653–684; **p. 688, 16.20:** Adapted from Iyengar, S. S., & Lepper, M. R. (1999). Rethinking the value of choice: A cultural perspective on intrinsic motivation. *Journal of Personality and Social Psychology, 76*, 349–366; **p. 692, 16.22:** Adapted from Gupta, U., & Singh, P. (1982). Exploratory study of love and liking and type of marriages. *Indian Journal of Applied Psychology, 19*, 92–97; **p. 694, 16.25:** Adapted from Earley, P. C. (1993). East meets West meets Mideast: Further explorations of collectivistic and individualistic work groups. *Academy of Management Journal, 36*, 319–348; **p. 695, 16.26:** Adapted from Madsen, M. C. (1971). Developmental and cross-cultural differences in the cooperative and competitive behavior of young children. *Journal of Cross-Cultural Psychology*, 365–371; **p. 699, 16.28:** Adapted from Roberson, D., Davies, I., & Davidoff, J. (2000). Color categories are not universal: Replications and new evidence from a stone-age culture. *Journal of Experimental Psychology: General, 129*, 369–398; **p. 700, 16.29:** Adapted from Roberson, D., Davies, I., & Davidoff, J. (2000). Color categories are not universal: Replications and new evidence from a stone-age culture. *Journal of Experimental Psychology: General, 129*, 369–398.

Tables

p. 280, Table 7.1: Adapted from Schacter, D. L. (2001). *The seven sins of memory: How the mind forgets and remembers.* Boston: Houghton Mifflin; **p. 423, Table 10.1:** Adapted from *Diagnostic and Statistical Manual of Mental Disorders DSM-IV*, American Psychiatric Association, 1994; **p. 514, Table 12.4:** Adapted from McAdams. (2000). *The person: An integrated introduction to personality psychology.* Orlando, FL: Harcourt; **p. 525, Table 13.1:** Adapted from *Diagnostic and Statistical Manual of Mental Disorders DSM-IV,* American Psychiatric Association, 1994; **p. 546, Table 13.2:** Adapted from *Diagnostic and Statistical Manual of Mental Disorders DSM-IV*, American Psychiatric Association, 1994; **p. 553, Table 13.5:** Adapted from *Diagnostic and Statistical Manual of Mental Disorders DSM-IV*, American Psychiatric Association, 1994; **p. 672, Table 16.1:** Adapted from Brown, A. S. (1991). A review of the tip-of-the-tongue phenomenon. *Psychological Bulletin, 109*, 204–223; **p. 692, Table 16.2:** Adapted from Broude, G. J., & Green, S. J. (1983). Cross-cultural codes on husband-wife relationships. *Ethology, 22*, 273–274.

NAME INDEX

Page numbers in *italics* refer to figures.

SUBJECT INDEX

Page numbers in *italics* refer to figures.